Signals, Instrumentation, Control, and Machine Learning
An Integrative Introduction

Signals, Instrumentation, Control, and Machine Learning
An Integrative Introduction

Joseph Bentsman
Department of Mechanical Science and Engineering
University of Illinois at Urbana-Champaign, USA

World Scientific
NEW JERSEY • LONDON • SINGAPORE • BEIJING • SHANGHAI • HONG KONG • TAIPEI • CHENNAI • TOKYO

Published by

World Scientific Publishing Co. Pte. Ltd.
5 Toh Tuck Link, Singapore 596224
USA office: 27 Warren Street, Suite 401-402, Hackensack, NJ 07601
UK office: 57 Shelton Street, Covent Garden, London WC2H 9HE

British Library Cataloguing-in-Publication Data
A catalogue record for this book is available from the British Library.

SIGNALS, INSTRUMENTATION, CONTROL, AND MACHINE LEARNING
An Integrative Introduction

ISBN 978-981-125-186-3 (hardcover)
ISBN 978-981-125-231-0 (paperback)
ISBN 978-981-125-187-0 (ebook for institutions)
ISBN 978-981-125-188-7 (ebook for individuals)

For any available supplementary material, please visit
https://www.worldscientific.com/worldscibooks/10.1142/12711#t=suppl

Printed in Singapore

Dedication

To my parents, my children Ben and Michelle, and my wife Osanna — their encouragement and support over a number of years made this book possible.

Preface

This book is written for you, the student, and it aims at making an introductory signal processing, instrumentation, and control course into an adventure in solving a difficult industrial problem by purely dynamics, signal analysis, control, and system integration means, with a glimpse at machine learning In a comedy classic "Throw Momma from the Train" Larry, a writing class instructor suffering from writer's block, and Owen, his student, go through a string of mishaps. In a year they meet again, but now as the authors of two very successful books about their adventures, Larry — of a serious grown-ups' drama and Owen — of a children's comic book, with cute figures popping out!

What eluded me, however, in following Larry's and Owen's footsteps was how to come up with both the drama and the pop-up figures on the course topics — the adventure was missing. My writer's block held full sway when Brian Thomas, my colleague, invited me to join him for a trip to Nucor Steel. Suddenly, the stage for an adventure started being set: there turned out to be a drama — the previous owner of the plant had gone bankrupt and Nucor had bought the plant for pennies on the dollar and fixed many systems, but not the mold oscillator in spite of the long-run effort by several consulting teams. Now it was the university's turn to measure up to what it was cracked up to be.

The adventure began in earnest when Nucor engineer Matthew Smith started flying to Urbana to discuss various distortion scenarios and fixes for them. Initial focus on oscillator beam structure gave way to a suspicion of a nonlinear excitation at the point of hydraulics/structure interaction. However, we still couldn't solve the problem and the drama went on through Nucor again bringing in the consultants, alas, with the same outcome. Then, Nucor, having gotten fed up with the hired guns, miraculously recreated the problem on a much simpler single-beam hardware testbed! Now, chagrined by the physical evidence that something really basic was at play, I undertook two senior design projects, where students started putting together a hardware testbed model by coupling the Euler-Bernoulli beam with the nonlinear hydraulic valve equations, but simulations kept crashing

Then my PhD students jumped into action. Vivek Natarajan changed the beam model to the Timoshenko one and captured the hardware testbed behavior in software through a coupled system of four partial differential equations and two nonlinear ordinary differential equations. The resulting software testbed produced a "digital twin" — made the velocity profile distortion in continuous casting mold oscillators pop up *in silico*! Using both testbeds and a production unit, Vivek showed that it was indeed not the oscillator structure on its own, but rather the interplay of the latter with an almost invisible actuator nonlinearity, that was distorting the motion. Then Vivek solved the problem itself, making the solution pop up too in software and hardware. Bryan Petrus made the simulator pop-ups easy to open, Zhelin Chen developed diagnostics demos, and Oyuna Angatkina made simulator tunable for any system. Meanwhile, thoroughly pleased with the distortion removal, Nucor moved to the "short stroke" — small magnitude high frequency mold oscillator operation.

The task was thus completed, opening up the possibility for taking the class to the adventure of solving the problem that lasted so many years through the magic carpet ride lasting just one semester! However, weaving the magic lecture/text/assignment carpet turned out to be a long multi-year multi-team effort. For the conceptual footing I had to dig back into my "Team Education" wisdom provided by the IIT signal processing, communication, and control masters Jack Deller, Dan Graupe, Dan Costello, and my dear advisor Semyon Meerkov who invited me to the University of Michigan to teach signal analysis. Thanks also go to the authors of textbook classics given in the bibliography.

The “Team Class” — that’s everyone who has taken the course so far – told me what to put in, from frequency definition to Timoshenko beam equations derivation, and corrected tons of errors.

The “Team Department” stood firm over the years behind the evolution of the material, while supplying a stream of outstanding teaching assistants. Here I would like to acknowledge Tad Addy, Richard Buckius, Huseyin Sehitoglu, Placid Ferreira, Tony Jacobi, Dick DeVor, Art Clausing, Chuck Tucker, Craig Dutton, Jim Phillips, Harley Johnson, Ken Christensen, Andrew Alleyne, Liz Hsiao-Wecksler, John Wierschem, and Sunjeev Sinha. Dan Block provided great lab support.

On the faculty side I am grateful to Brian Thomas and his Continuous Casting Center for getting me in close touch with the steel casting industry, Norm Miller for initiating the course and mightily contributing to it, Bill Dunn, Tsu-Chin Tsao, Ping Hsu, Gaurav Bahl, and Vasu Salapaka for contributing to the labs and the course content, Geir Dullerud and Gaurav for co-teaching the course and providing their lecture notes and ideas, Prashant Mehta, Naira Hovakimyan, and Tamer Basar for asking great questions that improved Vivek Natarajan’s PhD thesis, and, finally, Kimani Toussaint, Naira, Harry Dankowicz, and Vasu for forming the winning team in the College Strategic Instructional Innovations Program (SIIP) through which this book took much of its shape.

On the teaching assistants side, Bryan Petrus, Zhelin Chen, and Yonseok Lee converted my hand scribblings into beautiful PowerPoint slides, ported them to MS Word, and tremendously helped with shaping the material into an actual book, Ya Wang and Yongseok helped with appendix on sensors, Ouyna Angatkina – on beam theory, Zhelin Chen – on transforms, and Yongseok – on random signals. Hamza El-Kebir added Savitzky-Golay filter and put together basics of image processing and machine learning in Chapter 10, Colton Willhardt and Donnie Xu put together material on Hilbert-Huang Transform. Donnie also helped with machine learning in casting. Bryan, Zhelin, Hamza, Yongseok, Yang Li, Xinnan Wen, and Andrew Blanco worked out the examples.

The “Team College” — Bruce Litchfield, Luisa Rosu, Cris Migotski, Geoffrey Herman and Laura Hahn — provided SIIP and GATE (Grants for the Advancement of Teaching in Engineering).

The “Team Industry” – Nucor’s Ron O’Malley, Bob Williams, Matthew Smith, Glynn Elliott, Neal Ryan, Dan Stephens, Madeline Rembold, Madison Milligan, and Bryan Petrus gave the substance, the data, and the pictures. Bosch, MTS, National Instruments, Bruel and Kjaer, and Intellectual Ventures Labs permitted inclusion of their instrumentation and system graphics.

The “Team Government” — National Science Foundation helped through CMMI-0900138 and CMMI-1300907 grants. Thanks go to Masayoshi Tomizuka, Mary Tony, Mario Rotea, Suhada Jayasuria, George Chiu, Eduardo Misawa, Jordan Berg, Massimo Ruzzene, and Irina Dolinskaya.

The “Team Publisher”: Roshelle Kronzek, Yolande Koh, Steven Spatt, Chua Hong Koon, Don Dea Teck Hoon, and Tan Siew Lan, have been superb with their great humor, help, and patience.

Many thanks go to “Team Reviewers” — Michael Demetriou, Tsu-Chin Tsao, and Chengyu Cao.

My prime muse is “Team Family and Friends” — my parents, my children Ben and Michelle, my wife Osanna, and my close friend Alex Kurganov. Osanna took lecture notes, Ben introduced me to vocal pitch tuning and computer orchestra, and Alex — to virtual voice assistant, that he brought to life.

This book is an expansion and a massive revision of my book “Introduction to Signal Processing, Instrumentation, and Control: An Integrative Approach,” World Scientific, 2016.

Joseph Bentsman

Contents

Dedication v

Preface vii

Chapter 1 1
Case Study and Course Overview 1
1.1 Introduction 1
1.2 Process Description-Continuous Casting of Steel 1
1.2.1 Production unit description: mold oscillation system at Nucor Decatur 3
1.3 Project Approach and Problem Characterization 5
1.3.1 Project approach and workscope 5
1.3.2 Problem characterization 5
1.3.3 Development of highly simplified experimental mockup — hardware testbed 8
1.4 System Representation 9
1.4.1 Hardware testbed: 9
1.4.2 Hardware testbed schematics, negative feedback, and the loop closure under proportional control (P-control) 10
1.4.3 Hardware testbed: instrumentation and data acquisition 11
1.4.4 Validation of hardware testbed: experimental data exhibits resonance problem 12
1.5 Mathematical System Modeling and Analysis 12
1.5.1 Software testbed 12
1.5.2 Problem investigation: identifying the sources of the distortion 17
1.5.3 Problem definition 20
1.6 Problem Resolution 22
1.6.1 Solving the problem through controller design 22
1.6.2 Controller implementation and testing on the testbeds and the production unit 24
1.7 Problem Solving Capability Buildup 29
1.7.1 Stage I — introduction to signals, basic filtering, loading and system integration, sensing and actuation, instrumentation, data acquisition, system identification, and spectral computation 29
1.7.2 Stage II — introduction to spectral analysis and system identification for system modeling, diagnostics, and decision making; basics of analog and digital filter design and implementation 30
1.7.3 Stage III — introduction to controller design and system integration 31
1.7.4 Stage IV — introduction to nonstationary multisacel signals and machine learning 32
1.7.5 Appendices 32
1.7.6 Homework, project, laboratory, and examination assignments 32
Homework 1: Assignment and Solution Guide 33
References 42

Chapter 2 ... 43
Introduction to Signals ... 43
2.1 Signal Classification and Signal Measures ... 43
2.1.1 Types of signals ... 43
2.1.2 Energy-type signal size measures: integrability and summability ... 52
2.2 Basic Operations on Signals, Signal Representation ... 53
2.2.1 Signal representation ... 53
2.2.2 Basic operations on signals ... 54
2.2.3 Signal representation ... 63
2.3 Delta-function and its Use, Signal Representations, Energy and Power Signals ... 68
2.3.1 Delta-function ... 68
2.3.2 Signal energy, power, and RMS ... 78
Homework 2: Solved Examples and Solution Guide ... 83
2.4 Basic characteristics of a CT sinusoidal signal ... 102
2.4.1 Period and frequency of sinusoid: ... 103
2.4.2 Sums of sinusoids: ... 104
2.4.3 Phase of sinusoid ... 109
2.4.4 Magnitude of sinusoid ... 110
2.5 Real Harmonics, Discrete and Continuous Spectra, Frequency Response, Signal Test Instrumentation ... 110
2.5.1 Fundamental frequency of a sum of sinusoids. Real harmonics ... 110
2.5.2 Discrete real spectra ... 112
2.5.3 Continuous spectra and their application to system analysis: continuous frequency response ... 117
2.5.4 Instrumentation for signal measurement and visualization ... 118
Homework 3: Solved Examples and Solution Guide ... 119

Chapter 3 ... 131
First Look at Signal Processing, Filtering, and Instrumentation ... 131
3.1 Aggregating LTI System Representations for Carrying out Design, Filtering, and Control Tasks ... 131
3.1.1 Integrative portrait of a linear time invariant (LTI) system with a finite zero ... 131
3.1.2 Mechanical systems - a filtering, or input/output viewpoint ... 132
3.1.3 MS2D system force balance ... 133
3.1.4 Implicit input/output operator: state-space representation. State-variable filter implementation ... 135
3.1.5 Impulse response as system representation ... 140
3.1.6 Explicit input/output operator – convolution integral ... 146
3.1.7 Pole-zero map: a look inside the dynamic system behavior ... 150
3.1.8 Frequency response: calculation and use ... 151
3.1.9 Bode plots ... 157
3.1.10 Resonance in the second order systems ... 161
3.1.11 What each representation is good for: ... 164
3.2 Review of Parametric System Modeling, Subsystem Commutativity, Firsts Look at System Integration — Subsystem Loading, Output-Dependent System Properties ... 165
3.2.1 Passive electrical circuits ... 166
3.2.2 System integration problems ... 170
3.3 First Look at Filtering: Basic Instrumentation Techniques — AC versus DC Coupling. Active (Op Amp) Circuits, Loading Mitigation, Logic Components ... 176
3.3.1 Basic filtering: AC versus DC coupling ... 176
3.3.2 Op amp based active circuits ... 177
3.3.3 Other types of signal isolaters ... 182

3.3.4 Two logic components used in signal processing: the 1-bit analog-to-digital converter (ADC) and the 1-bit digital-to-analog converter (DAC) 182
3.3.5 Voltage dividers, potentiometers, and analog-to-digital converter-encoder 183
3.4 Basic Motion Sensing: Strain Gauge, Basic Instrumentation Techniques: Bridge Circuit, Instrumentation Amplifier, and Common Mode Rejection Ratio (CMRR) 185
3.4.1 Strain Gauge 185
3.4.2 Ideal instrumentation amplifier 186
3.4.3 Real instrumentation amplifier 187
3.4.4 Total Harmonic Distortion (THD): a measure of the distortion of a transducer/amplifier response to a pure sine input 190
Homework 4: Solved Examples and Solution Guide 192

Chapter 4 203
Sampling Basics, Harmonic Signals, and Signal Spectrum 203
4.1 Sampling Basics 203
4.1.1 Discrete time sinusoid and digital frequency 203
4.2 Frequency of DT Sinusoid, Principal Range, Aliasing as Folding, Sampling Theorem 208
4.2.1 Frequency of DT sinusoid 208
4.2.2 Digital frequency folding 210
4.3 Sampling Basics: Analog-to-Digital (A/D) and Digital-to-Analog (D/A) Conversion. Harmonic Signals: Review of Complex Numbers, Complex Variables, and Complex Functions 213
4.3.1 Signal conversion 213
4.3.2 Harmonic signals 215
4.3.3 Review of complex numbers, variables, and functions 215
4.4 Complex Harmonic Signals. Discrete Double-Sided Spectrum of CT and DT Periodic Signals. First Look at the Exponential Fourier Series 219
4.4.1 From real harmonics to complex ones. Discrete double-sided spectrum of periodic signals 219
4.4.2 DT complex harmonic signals 222
4.4.3 Adding harmonics after folding and overlap 223
Homework 5: Solved Examples and Solution Guide 227
4.5 Use of Spectrum for Diagnostics and System Identification. Ideal and Real Signal Reconstruction. The Need for Oversampling 238
4.5.1 Use of spectrum for diagnostics 238
4.5.2 Signal reconstruction 242
4.5.3 DT spectrum plots 246
Homework 6: Solved Examples and Solution Guide 250

Chapter 5 269
Function Projection and Fourier Series 269
5.1 Projection of a Signal, Continuous Time and Discrete Time Harmonic Bases, Continuous Time Fourier Series 269
5.1.1 Continuous time harmonic basis 269
5.1.2 Continuous time Exponential Fourier series 273
5.2 Discrete Fourier Series, Power of Periodic CT and DT Signals — Parseval's Relationship 279
5.2.1 Discrete Fourier Series (DFS) 279
5.2.2 Power of periodic signals: Parseval's relationship 282
Homework 7: Solved Examples and Solution Guide 283
5.3 Analytical CT Fourier Series Calculation for Six Basic Waveforms 286

5.3.1 Square wave Exponential Fourier series representation and spectra calculation 286
5.3.2 Exponential Fourier series coefficients calculation for five other basic waveforms 289
5.4 Signal Speed and Bandwidth, Gibbs Phenomenon, Windowing, Filtering as Complex Windowing 293
5.4.1 Link between time and frequency domains: signal speed 293
5.4.2 Convergence of Fourier series: building abruptly changing (infinite speed) signals through smooth basis functions 294
5.4.3 Spectral windowing 296
5.4.4 What are we getting through bandwidth, i.e. means for handling almost abruptly changing (very high speed) signals 297
5.5 Sampling Continuous Signal Spectra: Computation of $H(j\omega)$ from Fourier Series 298
5.6 Computing System Frequency Response from Finite Duration Sampled Data Record: Discrete Fourier Transform (DFT) 303
5.6.1 Sampling CF spectra: computation of $H(j\omega)$ from Fourier series — summary 303
5.6.2 Sampling CF spectra: computation of $H(j\omega)$ from Fourier series using DT data 305
5.6.3 Discrete Fourier Transform 306
5.7 DFS and DFT as Diagnostic Tools for Catching Low Magnitude Harmonics. Fast Fourier Transform and its Use. 307
5.7.2 Fast Fourier Transform (FFT) 313
5.7.3 Developing DFT-based magnifying glass 314
Homework 8: Solved Examples and Solution Guide 316

Chapter 6 335
Linear System Characteristics, Fourier Transform and Introduction to Filters 335
6.1 Linear Systems: Parametric versus Nonparametric Models, Remaining Objectives and Key Relations, Impulse Response of DT System 335
6.1.1 Parametric system models based on physical laws 335
6.1.2 Working with noisy data and poorly parametrizable systems: key objectives 340
6.2 Discrete Time Fourier Transform (DTFT), Fourier Transform: Theory and Practice. Modern PCM and PDM Based Data Acquisition (DAQ) Systems 344
6.2.1 Key relations: frequency response of DT system and its relation to frequency response of CT system 344
6.2.2 Discrete Time Fourier transform (DTFT) from DFS 346
6.2.3 Key time-domain/frequency-domain relations for processing test data 347
6.3 Introduction to Nonparametric System Modeling: Convolution as Both Impulse and Frequency Response Based Filtering of CT and DT Signals 354
6.3.1 Working with field test data: linearity and time-invariance admit standard response-based signal processing techniques for finding nonparametric models 354
6.3.2 Building a noncausal convolution integral 355
6.3.3 CT: impulse response 357
6.3.4 CT and DT system response 361
6.4 Ideal and Real Filters: Continuous and Discrete Time 365
6.4.1 Continuous time ideal filters 366
6.4.2 Continuous time real filters 369
6.4.3 Discrete time filters 374
Project 1: Solved Examples and Solution Guide 377
Exam 1: Solved Examples and Solution Guide 389

Chapter 7 401
CT, DT and Digital Filter Design and Implementation...401
7.1 Introduction to CT Filter Design: Motivation and Ideal Filter Approximation. Classical Filter Transfer Functions. Butterworth Filter: Existence 401
7.1.1 Filter characteristics 401
7.1.2 Low pass filter...404
7.1.3 Higher-order low-pass filters obtained by cascading lower order filters.........405
7.2 Butterworth Filter: Magnitude Response, Existence, Transfer Function, Design and Implementation 407
7.2.1 Butterworth filter magnitude response 407
7.2.2 Butterworth filter: existence...408
7.2.3 Butterworth filter: properties..415
7.2.4 Filter gain and attenuation...417
7.2.5 Butterworth filter – design basis 417
7.2.6 Butterworth filter design example..418
7.2.7 Butterworth filter implementation..419
7.3 CT: High-Pass, Band-Pass and Band-Stop Filters. General Filter Implementation. Classical Low Pass Filters: 5th Order Example 420
7.3.1 High-pass filters 420
7.3.2 Band-pass filter 421
7.3.3 Band-stop filter 422
7.3.4 General active filter implementation...422
7.3.5 Classical low pass CT filter transfer functions and their comparison 425
7.4 CT: Combined Active Filter Implementation. DT Filters: Z-Transform.......................428
7.4.1 Practical filter circuit...428
7.4.2 DT systems and the Z-transform 431
7.4.3 Z-transform 432
7.5 Digital Filter Design and Implementation...435
7.5.1 Digital filter design 436
7.5.2 Bilinear transformation 437
7.5.3 Tustin's rule 438
7.5.4 DT filter implementation...446
Homework 9: Solved Examples and Solution Guide 450

Chapter 8 463
Introduction to Discrete-Continuous Spectral Analysis 463
8.1 Working with Real Signals: Introduction to Discrete-Continuous (Mixed) Spectral analysis. Correlation and Spectral Density for Energy Signals 463
8.1.1 Motivation: a) capturing and tracking periodic signals hidden in noise. b) system model building using random signals...463
8.1.2 Real signals 464
8.1.3 Correlation for energy signals: CT 468
8.2 Correlation and Spectral Density for Power Signals..472
8.2.1 From energy to power 472
8.2.2 Power spectral density..474
8.2.3 Calculating the PSD using pre-limit approximation......................................474
8.2.4 Energy spectral density and correlation are not affected by the time shift 477
8.2.5 Calculating PSD using periodic extension and periodic correlation..............478
8.3 Correlation and Spectral Density of Ideal Mixed Signals..481

8.3.1 Building a power spectral density graph 481
8.3.2 PSD of a mixed ideal signal 485
8.4 Discrete-Continuous (Mixed) Spectral Analysis of a Sampled Signal. Spectral Analysis Applications: Diagnostics, System Identification 491
8.4.1 Spectral analysis of sampled data 491
8.4.2 Detecting CT sinusoids in noise 494
8.5 Applications of PSD and FFT 499
8.5.1 Diagnostics 499
8.5.2 FFT-based detection of hidden CT sinusoids — DFT ⇒ DFS ⇒ FS : summary 503
8.5.3 System identification using PSD: from signal processing to control 504
8.5.4 Filtering action of physical system 504
Project 2: Solved Examples and Solution Guide 509
Exam 2: Solved Examples and Solution Guide 531

Chapter 9 539
Introduction to Control Systems: Basic Control Actions and Basic Controller Design 539
9.1 Control Objectives, System Layout, Open-Loop and Closed-Loop Systems, Block Diagrams, System Performance Assessment: Tracking, Disturbance Rejection, Robustness, Position and Velocity Servomechanisms 539
9.1.1 Control system examples 539
9.1.2 Plant, and closed-loop and open-loop control systems layouts and objectives 543
9.1.3 Block diagrams 545
9.2 Closed-Loop System Topology, Negative and Positive Feedback. Block Diagram Reduction. Closed-Loop System Tracking. Disturbance Rejection, Robustness, Response Shaping (Stabilization), and Output Error 548
9.2.1 General closed-loop system 549
9.2.2 Closed-loop transfer function: unity feedback (perfect sensing) 553
9.3 Basic Control Actions 559
9.3.1 Basic control actions 559
9.3.2 Second-order systems 563
Homework 10: Solved Examples and Solution Guide 567
9.4 Basic Controller Design: Shaping the Second Order System Step Response with P nd PD Controllers. Steady State Error for Unit Step and Unit Ramp Inputs. Closed-Loop Stability and Closed-Loop Poles 576
9.4.1 Closed-loop position and velocity servos with P-controller under unity feedback 576
9.4.2 Response shaping using PD controller 580
9.5 Basic Controller Design: Disturbance Rejection Using Internal Model Principle. Measurement Noise Rejection through Loop Frequency Response Shaping 585
9.5.1 Controller gain over the entire frequency range 585
9.5.2 Closed-loop transfer function for infinite controller gain at zero frequency: I-controller 588
9.5.3 Internal Model Principle (IMP) controller 589
9.5.4 Measurement noise rejection. 2DOF controller topology. RST structure 593
9.6 Feedback-Based Mold Oscillator Problem Solution on the Hardware Testbed. PID Gain Tuning. Integrator Anti-Windup 595
9.6.1 Mold oscillator problem 595
9.6.2 Solution of the mold velocity distortion problem 597
9.6.3 PID gain tuning 602
9.6.4 Anti-windup 608
Project 3: Solved Examples and Solution Guide 611

Exam 3: Solved Examples and Solution Guide ... 632

Chapter 10 ... 641
Introduction to Nonstationary Signal Analysis and Machine Learning ... 641
10.1 Basic Elements of Computer Audio ... 641
10.1.1 Basic audio file formats and storage ... 641
10.1.2 Signal equalization ... 642
10.2 Basic Elements of Machine Vision ... 643
10.2.1 Basic digital image units and formats ... 643
10.2.2 Convolution and linear filters ... 644
10.2.3 Two-dimensional (2D) Fourier analysis in image processing ... 648
10.2.4 Low-pass image filtering Low power harmonics filter for image compression ... 650
10.2.5 High-pass image filtering and its use in image edge detection ... 651
10.3 Beyond Power Spectral Density: Time-Frequency Analysis of Multiresolution and Nonstationary Signals ... 653
10.3.1 Preliminaries: signal norms, Generalized Fourier Series as optimal periodic signal representation ... 653
10.3.2 Introduction to the multiresolution wavelet-based function approximation ... 659
10.3.3 Hilbert-Huang Tarnsform and its application to nonstationary signals ... 667
10.4 Basic Approaches to Machine Learning ... 676
10.4.1 Neural Networks (NN) and Deep Learning (DL) ... 676
10.4.2 Reinforcement Learning (RL) ... 678
10.5 Application of Mechine Learning to System Identification and Control ... 680
10.5.1 The Gradient Descent (GD) function minimization ... 680
10.5.2 Gradient descent based nonlinear system identification ... 681
10.5.3 Machine Learning application to control ... 684
10.6 Application of Machine Learning to Image Processing, Natural Language Recognition, and Material Phase Change Characterization ... 687
10.6.1 Application of ML to image processing using Convolutional Neural Networks (CNN) ... 687
10.6.2 Deep Learning (DL) in natural language recognition ... 688
10.6.3 Deep Learning (DL) in material characterization during the phase change ... 692
Homework 11: Solved Examples and Solution Guide ... 694

Appendix A ... 711
Useful Mathematical Formulas ... 711
A.1 Euler's Identity ... 711
A.2 Trigonometric Identities ... 711
A.3 Indefinite Integrals ... 711
A.4 Integration Properties ... 712
A.5 Definite Integrals Over $(0, \infty)$ ... 712
A.6 Differentiation Formulas ... 713
A.7 Finite Sums ... 713
A.8 Sequences and Series ... 714
A.9 Series Expansions ... 714

Appendix B ... 715
System Classification ... 715
B.1 Systems ... 715
B.2 System Classification ... 716
B.2.1 Continuous time versus discrete time systems ... 716

B.2.2 Static versus dynamic systems 717
B.2.3 Time varying versus time invariant systems 718
B.2.4 Linear versus nonlinear systems 720
B.2.5 Causal versus noncausal systems 721
B.2.6 Unstable versus stable systems 723
B.2.7 Lumped parameter versus distributed parameter systems 724
B.2.8 Stochastic versus deterministic systems 724
B.2.9 Fuzzy versus exact systems 724
B.2.10 Linear operations in LTI systems 725
B.2.11 BIBO stability of LTI Systems 726

Appendix C 727
Basics of Random Signals 727
C.1 Probability basics 727
C.1.1 Probability distribution and density functions 728
C.1.2 Expectation operator 728
C.1.3 The Gaussian probability density function 729
C.1.4 Conditional probability 729
C.2 Stochastic process 730
C.2.1 Autocorrelation and cross-correlation 730
C.2.2 Spectra of continuous-time random processes 731
C.2.3 Spectra of discrete time(DT)random processes 731
C.2.4 White noise 731
C.2.5 Wiener process 732
C.2.6 Series representation of noise 733
C.2.7 Stationary processes and ergodicity-WSS SME process in signal processing 734
C.2.8 Statistical estimatesfor WSS SME random signals 734
C.2.9 The central limit theorem 735
C.2.10 Filtering of random signals 736
C.3 Finite dimensional approximations 736
C.3.1 Markov process 737
C.3.2 Polynomial approximation 738
C.4 Mixed stochastic-deterministic systems 739
C.4.1 CARMA, CARIMA,and Box-Jenkins models 739
C.4.2 State-spacerepresentation 740

Appendix D 741
Fourier Transform 741
D.1 Fourier Transform Definition 741
D.2 Inverse Fourier Transform 741
D.3 Properties of Fourier Transform 742
D.4 Useful Fourier Transform Pairs 745

Appendix E 747
Laplace Transform 747
E.1 Laplace Transform Definition 747
E.2 Convergence 747
E.3 Properties of Laplace Transform 747

E.4 Useful Laplace Transform Pairs 749
E.5 Partial Fraction Expansion 749

Appendix F 753
Basic Types of Sensors and Actuators 753
F.1 Sensors for Temperature 753
F.1.1 Thermocouple 753
F.1.2 Resistance temperature detectors (RTDs) 754
F.1.3 Thermistors 755
F.1.4 Comparative characteristics 756
F.1.5 Thermographic infrared (IR) sensors 757
F.2 Sensors for Displacement 758
F.2.1 Linear variable differential transformer (LVDT) 758
F.2.2 Proximity sensors 758
F.2.3 Strain gauge 759
F.2.4 Extensometer 760
F.2.5 Encoder 761
F.2.6 Magnetostrictive sensor 761
F.2.7 Laser-based sensor - LIDAR 761
F.3 Sensors for Velocity 762
F.3.1 Variable reluctance sensor 762
F.3.2 Hall effect sensor 762
F.3.3 Tachometer generator 763
F.4 Force Sensors 763
F.4.1 Accelerometer 763
F.4.2 Pressure sensor 764
F.4.3 Load cell 764
F.5 PH Sensor 764
F.6 Capacitive Touch Screen Sensor 764
F.7 Gyroscopes 765
F.8 Electromagnetic Actuator 765
F.8.1 DC motor 765
F.8.2 Stepper motor 765
F.8.3 Brushless DC motor 766
F.9 Fluid mechanical actuators 766
F.10 Sensors/actuators for computer audio and nachine vision 766
F.10.1 Microphones and loudspeakers 766
F.10.2 Digital image sensors 766

Appendix G 767
Euler-Bernoulli and Timoshenko Beam Models and Their Use in Software Testbed Development 767
G.1 Introduction 767
G.2 Derivation of the Equations of Motion 767
G.2.1 Equations of motion of the Euler-Bernoulli beam model 767
G.2.2 Equations of motion of the Timoshenko beam model 772
G.3 Modal Analysis of Beam Models 776
G.3.1 Natural frequencies for the Euler-Bernoulli beam model 776
G.3.2 Natural frequencies of the Timoshenko beam model 778

G.4 Forced Beam Response 787
G.5 Application of the Euler-Bernoulli and the Timoshenko Models to the Testbed Modeling: Analytical Description and Numerical Simulation Results 789
G.5.1 Selection of the beam model: Euler-Bernoulli versus Timoshenko 789
G.5.2 Analytical model of Nucor Steel hardware testbed 789
G.5.3 Analytical model of AK Steel (currently Cleveland Cliffs) thick slab software testbed 792
G.5.4 Choice of the beam parameters to match the desired resonance frequency 793
G.5.5 Simulation run-time/approximation error tradeoff 796
G.5.6 Convergence of the numerical solution to the analyitcal one 798
References 800

Appendix H 801
Software Testbed Matlab Programs 801
H.1 Introduction 801
H.2 Beam Dynamics 801
H.2.1 Beam equation 801
H.2.2 Application of the boundary conditions 804
H.3 The Piston Hydraulics 807
H.4 The Controller Simulator 811
H.5 The Software Testbed Program 811

Bibliography 817

Index 819

Chapter 1

Case Study and Course Overview

1.1 Introduction

The goal of the present book is to provide an introduction to signal processing, instrumentation, diagnostics, filtering, control, machine learning, and system integration. To assemble these topics into a coherent package, the book embeds them into a concrete engineering context. This is accomplished by bringing the basic system modeling and interconnection concepts and techniques learned in the earlier courses to the level required for addressing two nontrivial problems of current interest in industrial manufacturing: 1) mold oscillator resonance suppression and 2) mold level fluctuation suppression due to shell bulging, both in continuous steel casting, and demonstrating how the book topics come together to solve the first problem and approach the second one. Material in ***Chapters 1–9*** builds step-by-step the first problem solution. ***Chapter 10*** extends the latter background to the key tools for analyzing and approaching the more challenging second problem.

1.2 Process Description-Continuous Casting of Steel

The continuous casting process layout and its main components are shown in Figures 1.2-1 and 1.2-2.

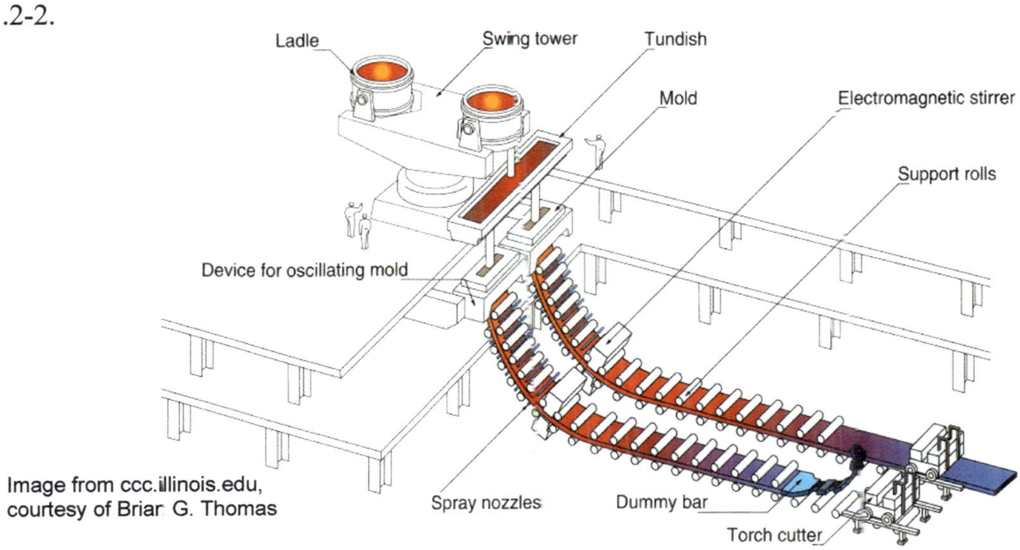

Fig. 1.2-1. Physical layout of the steel continuous casting.

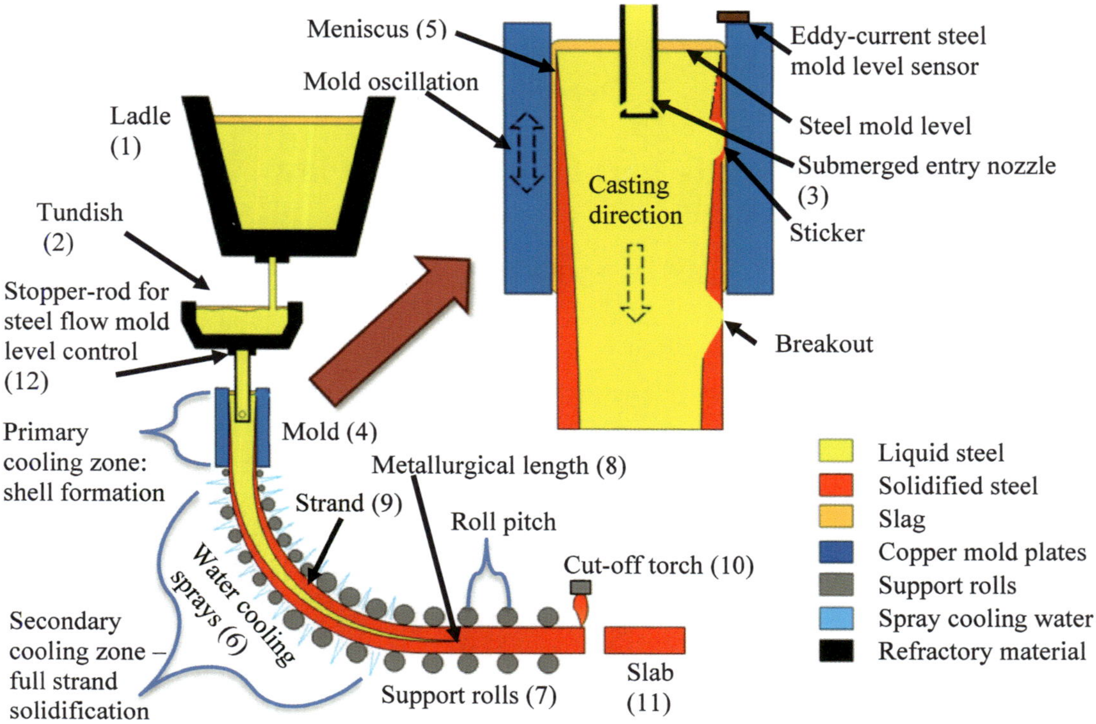

Fig. 1.2-2. Pictorial representation of system components (images adapted from B. G. Thomas, L.C. Hibbeler, and Wikipedia). The numbers in parentheses are used to refer to the corresponding components in the figures below.

The liquid steel is continuously poured through the stopper-rod controlled orifice into submerged nozzle and then through it into the top of the mold, such as that shown in Figure 1.2-3.

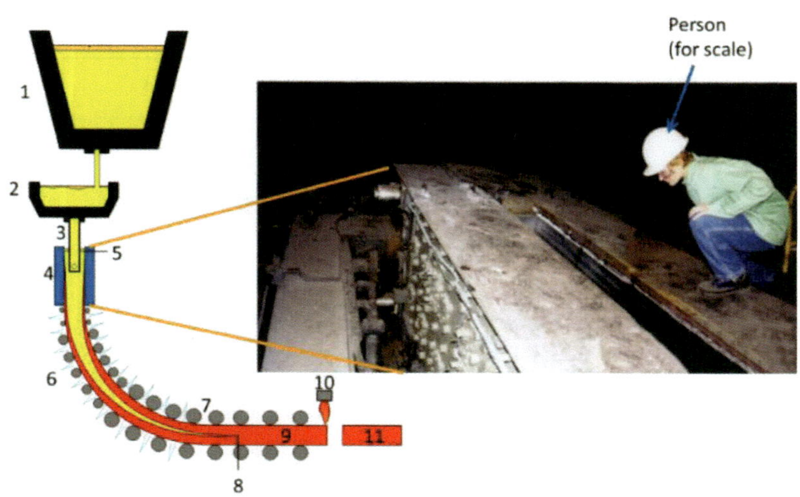

Fig. 1.2-3. Production mold.

The stopper-rod adjusts the steel flow rate to make the ***actual mold level*** match the ***desired mold level*** and eliminate the ***mold level fluctuations*** due to steel shell bulging between the rolls and other disturbances. The stopper-rod is manipulated by the adaptive ***mold level control system*** based on the mold level measurement supplied by the eddy current sensor, closing the loop.

While flowing through the mold, liquid steel freezes against the cooling mold walls, forming a solid shell with liquid steel remaining inside, referred to as ***strand***, shown in Figure 1.2-4. The strand is pulled out at the mold bottom by the rotating motor-driven support rolls at a rate matching the inflow of the liquid steel, resulting in a continuous casting process. ***Vertical sinusoidal oscillation*** of the mold performed by the ***mold oscillation system***, such as that shown in Figure 1.2-5, prevents the sticking of the steel shell to the mold wall. The key process parameter is the ***casting speed*** defined as the rate at which the shell is pulled out of the mold. Another key quantity is ***negative strip time*** — the period during each cycle of oscillation when the mold moves down faster than the shell. Amplitude and frequency of oscillation are dictated by casting speed, negative strip time and oscillation marks.

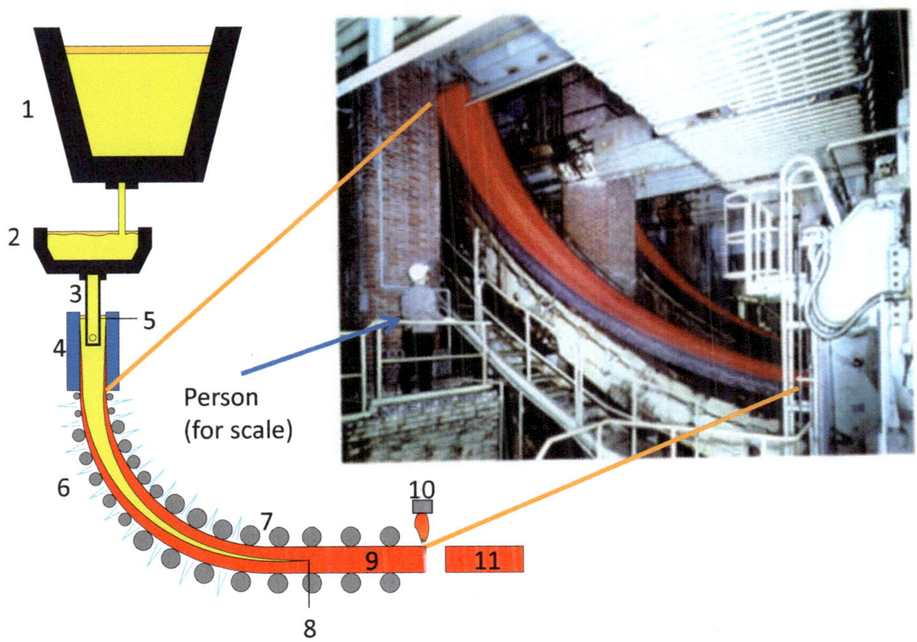

Fig. 1.2-4. Red-hot steel strand.

1.2.1 *Production unit description: mold oscillation system at Nucor Decatur*

The Nucor Steel, Decatur, mold oscillation system consists of subsystem of beams hinged at the center. The mold is attached to the beam subsystem at one end and undergoes sinusoidal displacement induced by an electro-hydraulic actuator with piston attached to the beam subsystem other end. The system exhibits the generic mechanical characteristics — nonlinearity and resonance.

Sideview of mold oscillation system Backview of mold oscillation system

Fig. 12-5. Mold oscillation system.

Why we care about the mold velocity

The negative strip time, or the period during each cycle of oscillation when the mold moves down faster than the shell, is critically important for running the process. In this period, any steel sticking to the mold wall, referred to as "***sticker***", will be stripped off the mold wall and squeezed back onto the steel shell, closing up the holes through which liquid steel could pour out of the shell, an emergency event referred to as "***breakout***". Without proper negative strip time breakout is guaranteed. To maintain consistent negative strip time, the mold velocity profile must have the form of an undistorted sinusoid, especially at high frequencies where the tolerance for error is smaller.

Oscillation marks and general project objective

Mold oscillation leads to the formation of defects on the solidified steel surface called ***oscillation marks***, known to be the sites for the potential crack initiation. A steel surface with representative oscillation marks from a current production run is seen in Figure 1.2-6.

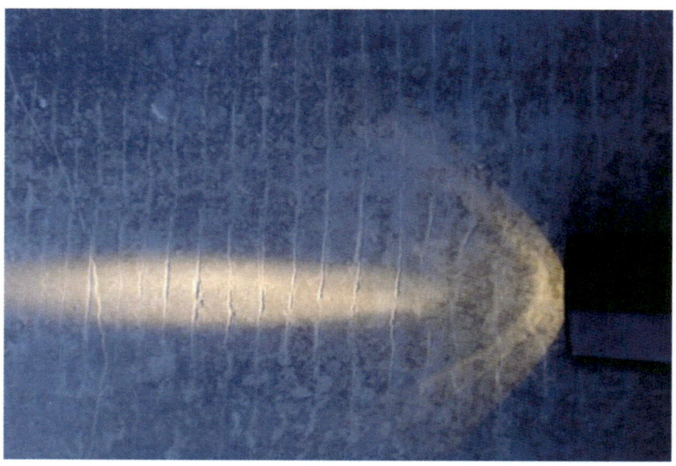

Fig. 1.2-6. Oscillation marks appearing under the original controller for the standard operating parameters: frequency — 3.38 Hz and magnitude — 4.5 mm.

The size of the marks in this figure exceeds the desired limit. These marks can be made shallower by "***shortening the stroke***" — increasing the frequency and reducing the amplitude of oscillation. The latter possibility gives rise to the following project objective.

General Project Objective: ***Provide distortion-free pure sinusoidal mold displacement and velocity profiles in a sufficiently broad frequency range.***

1.3 Project Approach and Problem Characterization

1.3.1 *Project approach and workscope*

To eliminate the mold velocity distortion the task sequence given below has been proposed. The sequence integrates the tools to be studied in the book into a coherent strategy for solving a challenging industrial problem.

(1) Problem characterization. Computational ***diagnostics*** using signal processing technique referred to as ***spectral analysis***.
(2) Production unit prototyping — ***testbed*** development:
 (a) ***hardware testbed***,
 (b) ***software testbed***: i) development of ***analytical*** (mathematical) ***model*** of the hardware testbed (not of the production unit), ii) development of ***numerical code*** that simulates the model.
(3) Selection of instrumentation for sensing, actuation, data acquisition, and real-time computation.
(4) Problem investigation and definition. ***System identification***.
(5) Analytical ***analog controller*** design for distortion elimination.
(6) ***Digital controller*** implementation in testbeds and production unit using real-time digital processor.
(7) Controller ***performance robustness*** assessment and refinement without casting (empty mold) and in production under changes in reference oscillation frequency and production unit dynamics.

1.3.2 *Problem characterization*

1.3.2.1 *Time domain position and velocity profile distortion*

In shortening the stroke — going to higher frequencies while reducing the oscillation amplitude — the velocity profile around 4.4 Hz was found to become highly distorted, as seen below in Figure 1.3-1. In this figure, the violet-hued axis marks the up/down mold displacement in mm, while the pink-hued axis marks the up/down mold velocity in mm/sec. The clearly visible velocity profile distortion has been resulting in failing to provide the required negative strip time and causing violent shaking of the oscillation unit at frequency three times higher than the driving frequency.

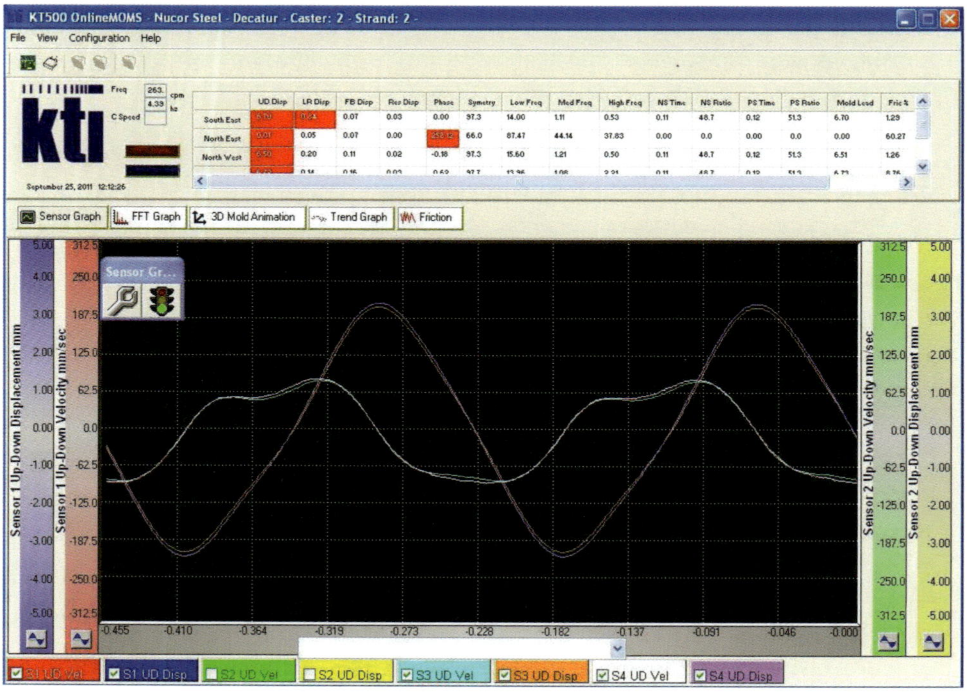

Fig. 1.3-1. Screenshot of mold displacement (magenta) and velocity profiles (reference amplitude 3.5 mm and frequency 4.4 Hz). Position profile is seen to be distorted and velocity profile (time derivative of position profile) — highly distorted.

1.3.2.2 *Frequency domain position and velocity profile distortion*

To plan the next steps, a source of the position and the velocity profile distortion needed to be identified. This was done by mapping a time domain function into frequency domain using spectral analysis technique — the key tool presented in this book. A simple example of this technique is given in Figure 1.3-2: the graph on the left is an observed part (***data record***) of a periodic function of time. The graph on the right shows, in part, the same function drawn in a different way — as a function of frequency.

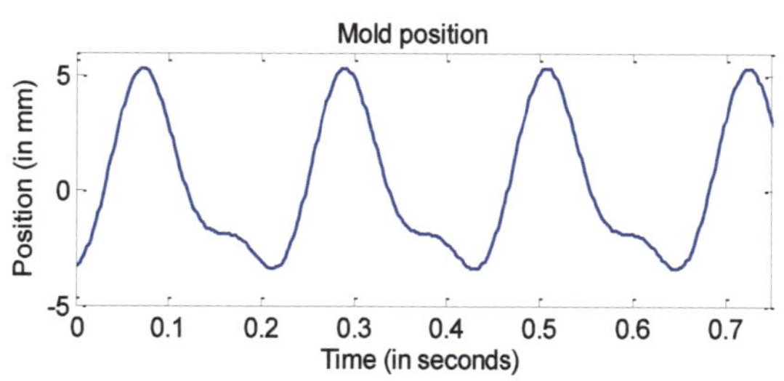

Time domain plot — a direct way of drawing a function

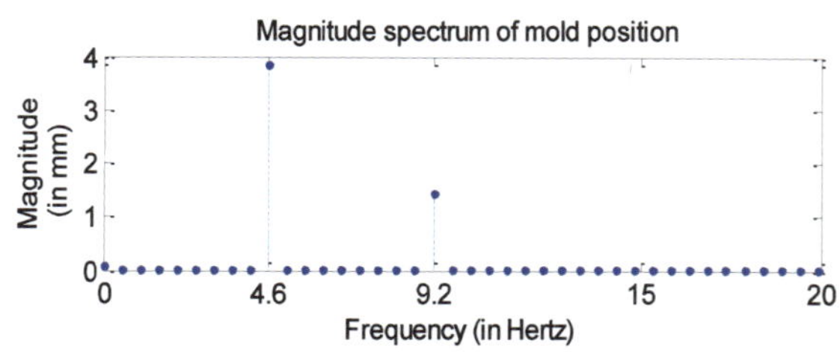

Magnitude spectrum — representing the function on the left as a sum of distinct sine waves

Fig. 1.3-2. Spectral analysis — the frequency domain graph on the right exhibits the frequency domain content of the time domain graph on the left. Relative sine waves position information is also required to complete the description.

The line heights in the "magnitude spectrum" on the right show that the function on the left is a sum of two sine waves — 4 mm and 1.5 mm magnitude ones with respective frequencies 4.6 Hz and 9.2 Hz. A more informative tool of spectral analysis is a waterfall plot given in Figure 1.3-3 below, to be discussed in more detail in Chapters 2 and 7.

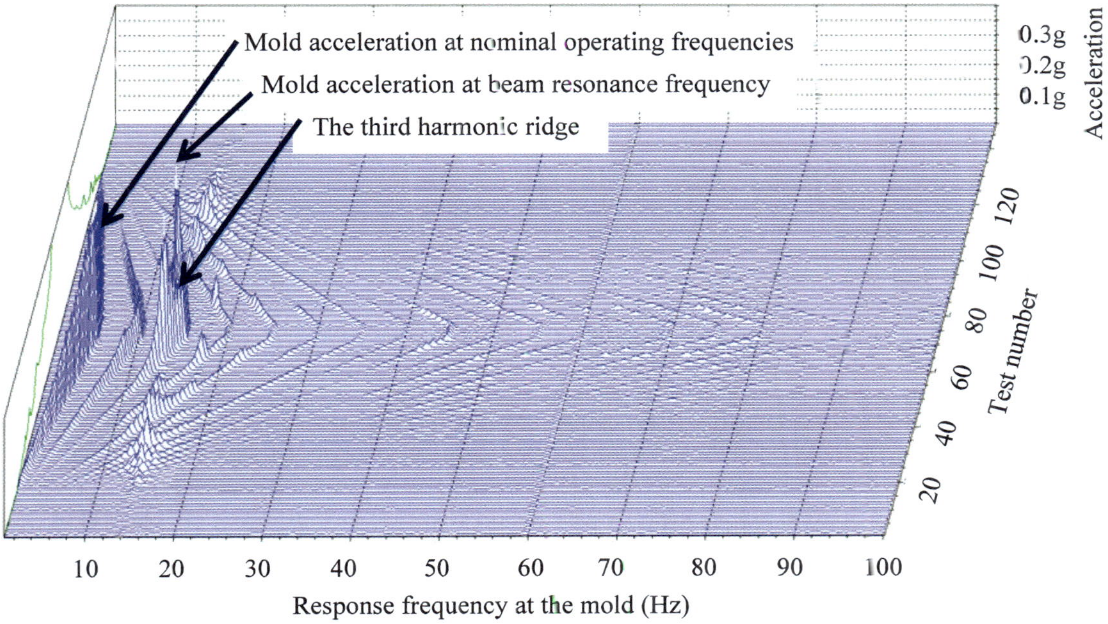

Fig. 1.3-3. Resonance problem: frequency domain waterfall plot based on mold acceleration data shows large force at the primary beam resonance for operating frequency equal to one third of resonance frequency.

Since relation between sinusoidal position, velocity, and acceleration profiles is given, respectively, by $a\sin(\omega t)$, $-a\omega\cos(\omega t)$, and $-a\omega^2\sin(\omega t)$, sinusoidal displacement with even a small magnitude a at primary beam resonance frequency $f = 13.2$ Hz, or $\omega = 2\pi 13.2 = 82.37$ rad/sec, will cause large velocity distortion with magnitude $a\omega = 82.37a$ and very large acceleration (and hence, force) distortion with magnitude $a\omega^2 = 6784.49a$, as seen in the acceleration spike in Figure 1.3-3.

1.3.2.3 *Production unit resonance problem and specific project objective*

Problem — passing through resonance: As seen from system diagnostics, the primary beam main resonance mode was found to start exhibiting excitation when the reference frequency approaches ***one-third*** of the resonance frequency. The mold displacement and the velocity profiles distortion was found to be mainly caused by this resonance. The reason for the onset of beam resonance excitation was not identified. The distortion was not removed, and operation in the desired frequency range was not attained. Thus, the following specific project objective was formulated:

Specific Project Objective: ***i) Model mold oscillation system (Figure 1.3-4), ii) simulate it, iii) identify what causes excitation of the primary beam resonance, and iv) eliminate the distortion.***

Fig. 1.3-4. Position of hydraulic actuator.

1.3.3 *Development of highly simplified experimental mockup — hardware testbed*

Since access to the production unit was limited, an attempt was made to recreate mold oscillation problem on a physical model — the experimental hardware testbed resembling the production unit mold oscillation system. The testbed that was put together consists of a hollow beam that is hinged at the center, supporting a mold-like mass at one end and excited by an electro-hydraulic valve/actuator at the other end, as shown in Figures 1.3-5 and 1.3-6. The testbed actuator was chosen to be similar to that of the production unit, but of a smaller size to match the testbed requirements.

This attempt was successful in that the hardware testbed exhibited similar resonance problem:

- the beam resonance frequency is 9.2 Hz
- the reference input to the piston for tracking is sinusoid at 4.6 Hz
- the mold position profile is highly distorted.

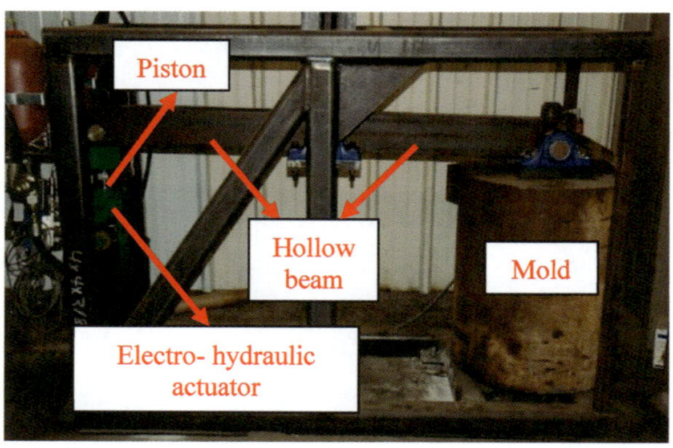

Fig. 1.3-5. Hardware testbed.

1.3.3.1 *Physical variables selection for sensing and control: piston and mold position*

Although the objective was the distortion-free mold ***velocity*** profile, it was decided to focus on removing distortion from the piston and the mold ***position*** profiles observed through sensor signals as shown in Figure 1.3-6, since a distortion-free (pure sinusoidal) displacement guarantees a distortion-free velocity.

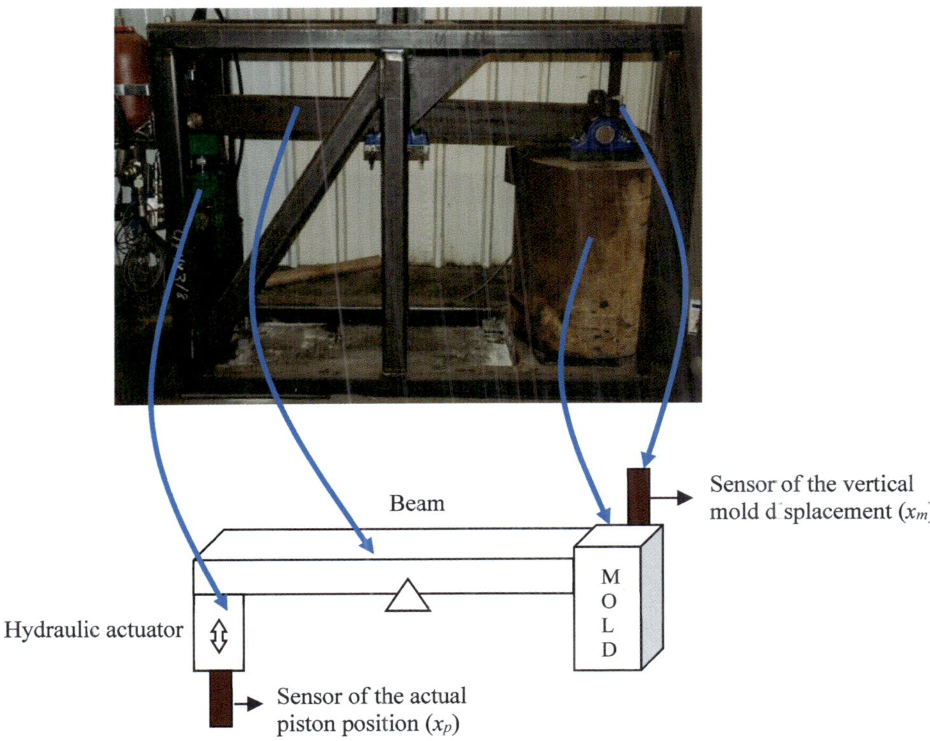

Fig. 1.3-6. Hardware testbed: simplified instrumented layout for sensing and actuating piston and mold position.

1.4 System Representation

1.4.1 *Hardware testbed:*

The basic operation of the servo-beam structure is given in Figure 1.4-1.

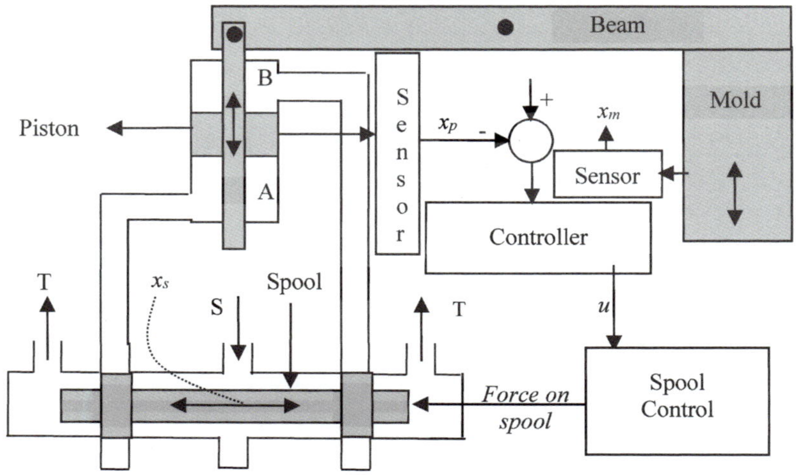

Fig. 1.4-1. Layout and operation of the servo-beam structure.

Important signals in this configuration are:

- x_p — piston position
- x_m — mold position
- x_s — spool position
- r — desired piston position (reference)

How the testbed works:

- Spool moves to the left — S (pressure supply) gets connected to chamber A; T (tank) gets connected to chamber B; piston is pushed up
- Spool moves to the right — S (pressure supply) gets connected to chamber B; T (tank) gets connected to chamber A; piston is pushed down
- Controller uses error between r (desired piston position) and x_p (actual piston position) to generate control signal u that moves spool to induce the desired piston motion
- Typically proportional controller is used. $u = K(r-x_p)$

1.4.2 *Hardware testbed schematics, negative feedback, and the loop closure under proportional control (P-control)*

The corresponding block diagram of the servo-beam structure is given in Figure 1.4-2.

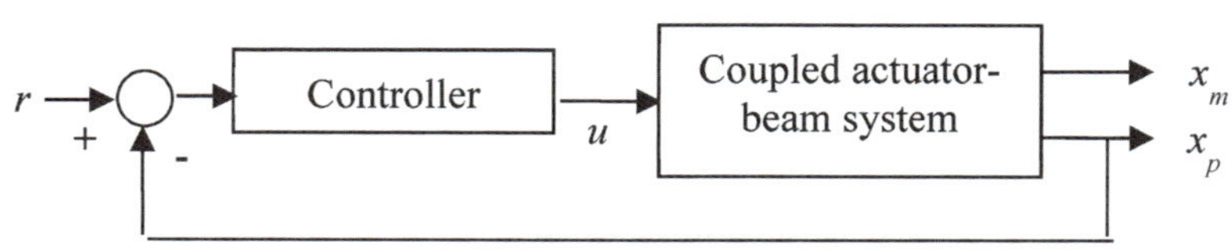

Fig. 1.4-2. Hardware testbed block diagram.

This figure includes a controller block with a ***proportional controller*** $u = K(r-x_p)$, where:

- u is ***control signal***: the force we chose to apply to the spool that in turn moves the piston
- K is ***controller gain***: some positive number
- r is ***reference***: what we want the piston position to be (sinusoidal profile)
- x_p is ***measured system output***: the actual piston position

Control is accomplished in the testbed through measuring the piston position x_p, computing the error between the reference r and the piston position x_p, and passing this error to the controller to generate the control signal u that is then passed to the actuator input. Feeding the actual piston position back to the controller input to compute this error is called ***negative output feedback***, and the resulting configuration where the controller acts on this error and generates control signal u is called the ***closed loop system***. Mold position x_m is measured for monitoring purposes but not used for the closed loop control.

1.4.3 *Hardware testbed: instrumentation and data acquisition*

Hardware testbed actuator position sensor, electro-hydraulic actuator, and actuator controller are shown in Figures 1.4-3–1.4-5. Another identical sensor is used for mold (mass) position measurement.

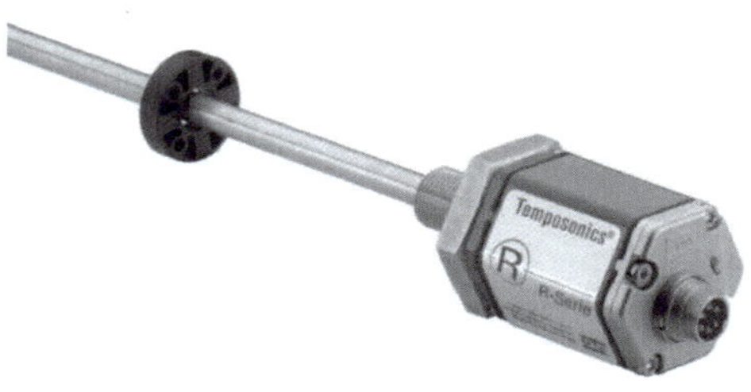

Fig. 1.4-3. Temposonic position sensor; resolution less than 2 microns. MTS Temposonics® R-Series sensor image is provided by MTS Systems Corporation. MTS and Temposonics® are registered trademarks of MTS Systems Corporation.

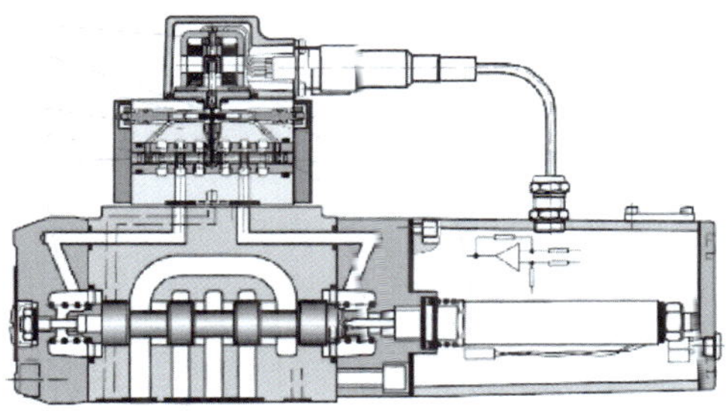

Fig. 1.4-4. 4/3 way directional control valve from Bosch Rexroth; Type 4WRDE. Copyright © Bosch Rexroth.

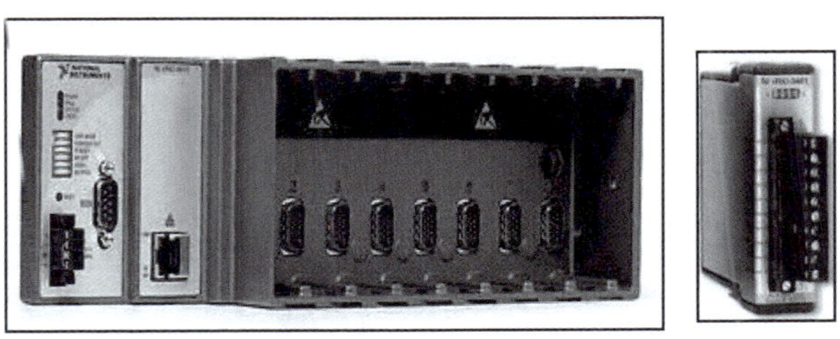

1. Data transfer from the sensor and to the actuator occurs through the I/O (input/output) modules.
2. Control algorithm runs in the FPGA processor and real time CRIO (compact reconfigurable I/O) controller.
3. The software used is LabView from National Instruments.

Fig. 1.4-5. Top view of real time CRIO (compact reconfigurable I/O) controller, FPGA (field programmable gate array) processor, and chassis, and back view of I/O module (right) from National Instruments. Copyright © National Instruments.

1.4.4 *Validation of hardware testbed: experimental data exhibits resonance problem*

The desired position for the piston — reference input — is chosen to be a sine wave of frequency 4.6 Hz (half of 9.2 Hz — the resonance frequency of the beam) and magnitude 3 mm. P-controller is set to have the gain $K = 2$. As seen from Figure 1.4-6, the piston position signal looks almost perfectly sinusoidal, but large distortion of the purely sinusoidal profile is observed at the mold end. This happens when frequency of the piston position reference is near a submultiple of the beam resonance frequency.

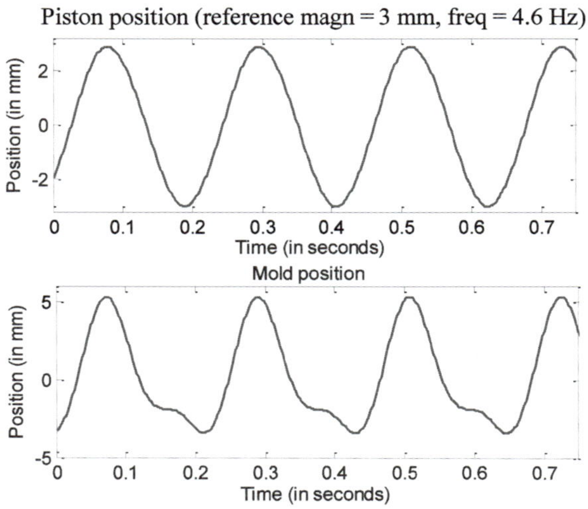

Fig. 1.4-6. Data exhibits resonance problem.

1.5 Mathematical System Modeling and Analysis

To facilitate problem diagnostics and controller development, mathematical — analytical and numerical — models representing the hardware testbed were developed. These models provided an indispensable tool in the problem resolution — the software testbed, development of which is briefly described next, and in more detail — in Appendices G and H.

1.5.1 *Software testbed*

1.5.1.1 *Development of analytical model*

Based on the system characteristics, analytical beam model in the form of four coupled linear partial differential equations (PDEs) using Timoshenko beam theory (Appendix G) and analytical hydraulic system model in the form of a system of nonlinear ordinary differential equations (ODEs) with algebraic turbulent flow constrains, respectively shown in Figures 1.5-1 and 1.5-2, were developed and coupled together.

Beam model

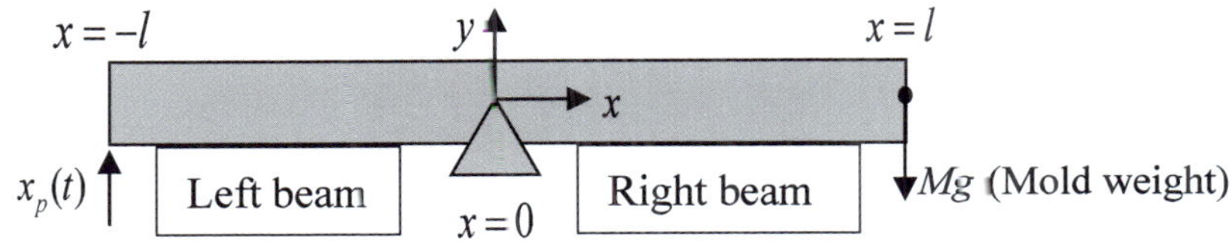

Fig. 1.5-1. Beam model: set of Timoshenko beam (linear) PDEs with boundary conditions.

The overall beam model is composed of the separate models of two beams, the left and the right ones, coupled at the center via boundary conditions. Mold dynamics is incorporated into the rightmost boundary condition. The model equations and their coupling are given below.

(1) 1D dynamic beam transverse displacement PDE:

$$\text{Left beam: } m_b \frac{\partial^2 y_L}{\partial t^2} + \gamma_y \frac{\partial y_L}{\partial t} = \frac{\partial}{\partial x}\left(k' G a_b \left(\frac{\partial y_L}{\partial x} - \psi_L \right) \right) - m_b g, \qquad (1.5\text{-}1)$$

$$\text{Right beam: } m_b \frac{\partial^2 y_R}{\partial t^2} + \gamma_y \frac{\partial y_R}{\partial t} = \frac{\partial}{\partial x}\left(k' G a_b \left(\frac{\partial y_R}{\partial x} - \psi_R \right) \right) - m_b g. \qquad (1.5\text{-}2)$$

(2) 1D dynamic beam bending angle PDE:

$$\text{Left beam: } \frac{I}{a_b} m_b \frac{\partial^2 \psi_L}{\partial t^2} + \gamma_\psi \frac{\partial \psi_L}{\partial t} = \frac{\partial}{\partial x}\left(EI \frac{\partial \psi_L}{\partial x} \right) + k' G a_b \left(\frac{\partial y_L}{\partial x} - \psi_L \right), \qquad (1.5\text{-}3)$$

$$\text{Right beam: } \frac{I}{a_b} m_b \frac{\partial^2 \psi_R}{\partial t^2} + \gamma_\psi \frac{\partial \psi_R}{\partial t} = \frac{\partial}{\partial x}\left(EI \frac{\partial \psi_R}{\partial x} \right) + k' G a_b \left(\frac{\partial y_R}{\partial x} - \psi_R \right). \qquad (1.5\text{-}4)$$

(3) Boundary conditions at $x = -l$:

$$y_L(-l) = x_p(t), \ EI \frac{\partial \psi_L(-l)}{\partial x} = 0. \qquad (1.5\text{-}5)$$

(4) Boundary conditions at $x = 0$:

$$y_L(0) = 0, \ y_R(0) = 0, \ \psi_L(0) = \psi_R(0), \ EI \frac{\partial \psi_R(0)}{\partial x} = EI \frac{\partial \psi_L(0)}{\partial x}. \qquad (1.5\text{-}6)$$

(5) Boundary conditions at $x = l$:

$$EI \frac{\partial \psi_R(l)}{\partial x} = 0, \ k' G a_b \left(\frac{\partial y_R(l)}{\partial x} - \psi_R(l) \right) + Mg + M \frac{\partial^2 y_R(l)}{\partial t^2} + \gamma_m \frac{\partial y_R(l)}{\partial t} = 0. \qquad (1.5\text{-}7)$$

(6) Parameters of testbed:

$E = 200Gpa$ Youngs modulus,

$\rho = 7870\ Kg/m^3$ - density of steel,

$I = 2.2\times10^{-5}m^4$ Moment of inertia of beam,

$\mathrm{M} = 2250\ Kgs$ mold mass,

Beam width $= 5.13'$(hollow with thickness 0.94'),

Beam breadth = 6' (hollow with thickness 0.38'),

Damping coefficients $\gamma_L = \gamma_R = 10,\ \gamma = 10\mathrm{Kg/sec}$,

$G = 82GPa$ Shear Modulus for steel,

$\mathrm{a}_b = 0.0088m^2$ cross section area of beam,

$m_b = 69Kg/m$ Mass per unit length of beam,

$k' = 0.83$ Shear constant,

$l = 34.5'$.

(7) Notation used in the PDEs:

$$\text{right side of the beam}: y_R = \text{ linear displacement, } \psi_R = \text{ angular displacement,}$$
$$\text{left side of the fulcrum}: y_L = \text{ linear displacement, } \psi_L = \text{ angular displacement.}$$

Hydraulic servo model:

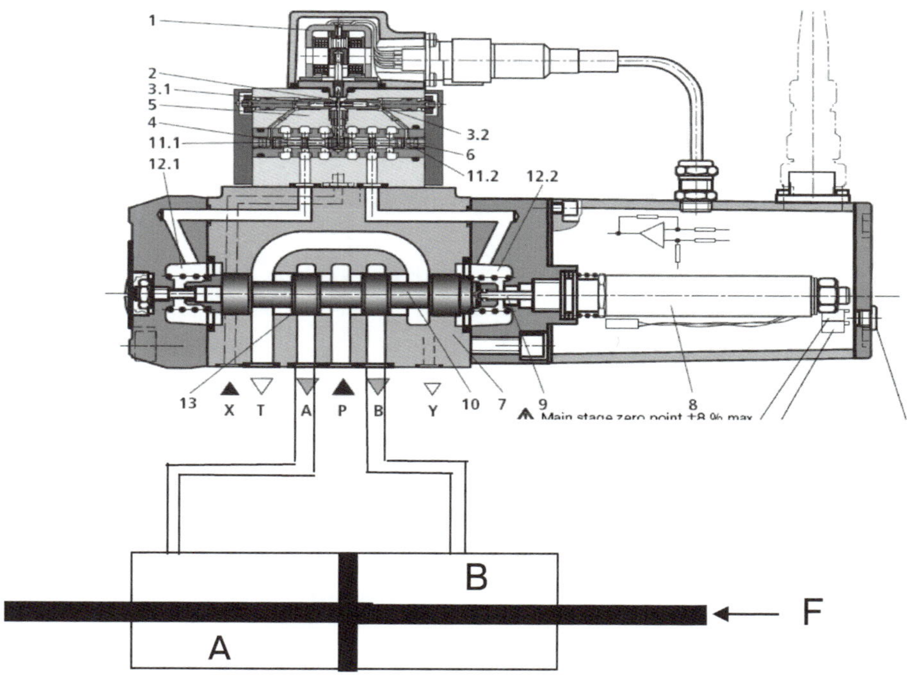

Fig. 1.5-2. Electro-hydraulic servo model: set of nonlinear ODEs and their coupling to beam PDEs. The electro-hydraulic valve is the 4/3 way directional control valve Rexroth, type 4WRDE, from Bosch.

(1) Electronic control equation of spool position is given by

$$\ddot{x}_s + 2\zeta_s \omega_s \dot{x}_s + \omega_s^2 x_s = \omega_s^2 u. \tag{1.5-8}$$

This subsystem has very fast dynamics, and therefore it is not modelled in simulations.

(2) Turbulent flow equations for flow in chambers A and B are:

$$q_A = \begin{cases} c(d-x_s)\sqrt{P_s - P_A}, & x_s < -d, \\ c(d-x_s)\sqrt{P_s - P_A} - c(x_s+d)\sqrt{P_A - P_t}, & -d < x_s < d, \\ -c(x_s+d)\sqrt{P_A - P_t}, & x_s > d, \end{cases}$$
$$q_B = \begin{cases} c(d-x_s)\sqrt{P_B - P_t}, & x_s < -d, \\ c(d-x_s)\sqrt{P_B - P_t} - c(x_s+d)\sqrt{P_s - P_B}, & -d < x_s < d, \\ -c(x_s+d)\sqrt{P_s - P_B}, & x_s > d, \end{cases} \tag{1.5-9}$$

where

x_s is spool position:
- 0 with valve underlap gap of d on both sides,
- positive when the spool moves to the right,

ω_s , ζ_s are spool filter parameters,

q_A is flow rate for champber A – positive when oil flows into A,

q_B is flow rate for chamber B – positive when oil flows out of B,

$\alpha_{\mathrm{A,B,s,t}}$ means Chamber connected to A and B, source, tank.

(3) Pressure equations are:

$$\dot{P}_A = \frac{\beta}{\left(V_A + a_p\left(L + x_p\right)\right)}\left(q_A - a_p\dot{x}_p\right),\ \dot{P}_B = \frac{\beta}{\left(V_B + a_p\left(L - x_p\right)\right)}\left(-q_B + a_p\dot{x}_p\right), \tag{1.5-10}$$

where

β is the bulk modulus of the actuator fluid,

P_A and P_B are pressure values in A and B, respectively,

x_p is piston position – positive if it moves to the right of midpoint of cylinder,

V_A,V_B are, respectively, static volume values of chambers A and B,

$a_p\left(L + x_p\right)$ is dynamic volume of chambers A and B,

L is half the piston stroke length,

a_p is surface area of piston.

(4) Piston dynamics showing the coupling with the beam is given by:

$$m_p \ddot{x}_p + b\dot{x}_p = \left(P_A - P_B\right)a_p - m_p g + k'Ga_b\left(\frac{\partial y_L(-l)}{\partial x} - \psi_L(-l)\right). \tag{1.5-11}$$

1.5.1.2 *Development of numerical model*

The numerical model has been developed using finite-difference approximation of Timoshenko beam PDEs, yielding a system of ODEs coupled to nonlinear piston hydraulics and the control algorithm equations, either given, or to be designed, as shown in Figures 1.5-3 and 1.5-4. The details of the software testbed development and the corresponding programs are described in Appendices G and H.

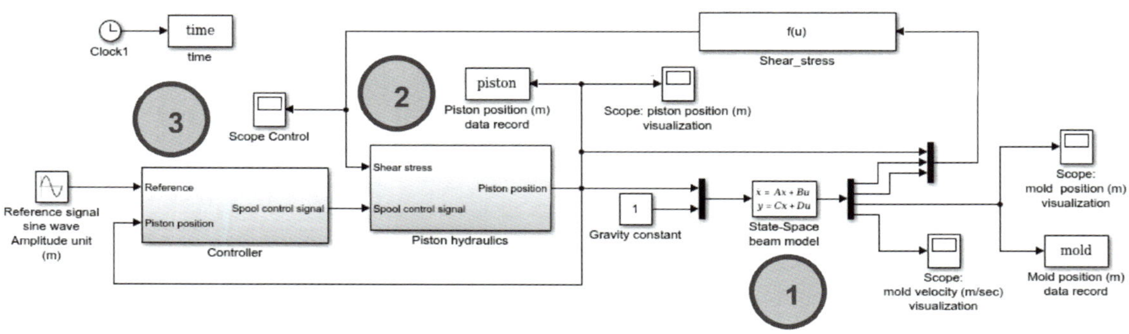

Fig. 1.5-3. MATLAB Simulink diagram of software testbed.

In a finite difference approximation, spatial derivatives in PDE are approximated by difference equations

$$\partial f / \partial x \approx [f(x+\Delta x) - f(x-\Delta x)]/2\Delta x \text{ and } \partial^2 f / \partial x^2 \approx [f(x+\Delta x) - 2f(x) + f(x-\Delta x)]/(\Delta x)^2.$$

The result is a set of coupled time-invariant ODEs, where beam model and controller are implemented in linear state-space form:

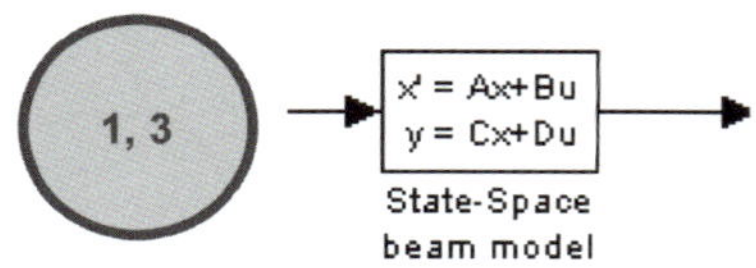

and nonlinear servo model ODE is implemented in Simulink form shown in Figure 1.5-4 below.

1.5.1.3 *Software testbed (numerical code) validation*

Parameters in the beam model were chosen to obtain a resonance frequency at 9.2 Hz that is close to the first production unit resonance of 8.8 Hz (the second ridge from the left on waterfall plot in Figure 1.3-3). Reference input to the piston for tracking was chosen to be a sinusoid at 4.6 Hz, close to the reference frequency 4.4 Hz of the production unit. Proportional controller gain was set to 0.6.

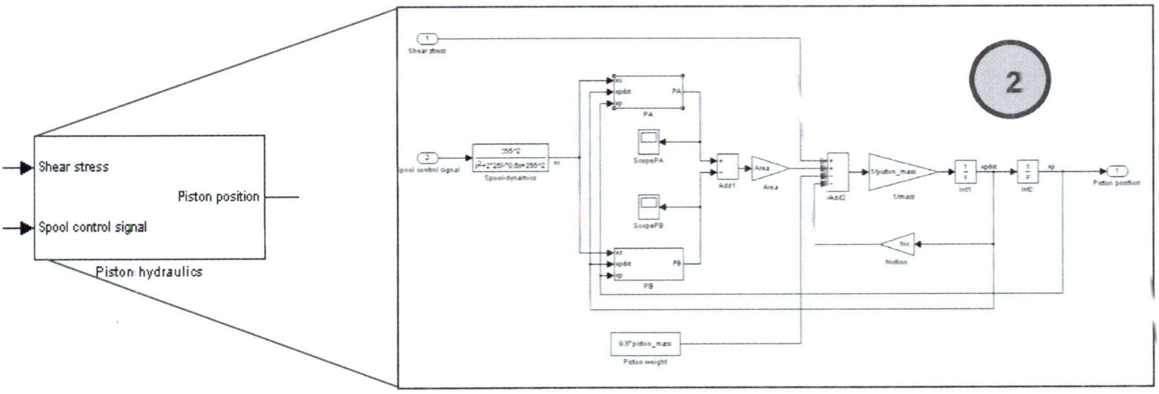

Fig. 1.5-4. MATLAB Simulink diagram of hydraulic actuator.

Mold position exhibited distortions similar to those of the hardware testbed, as shown in Figure 1.5-5. Therefore, model simulator was found to provide a suitable numerical testbed for helping to understand hardware testbed dynamics and *in silico* assessment of controller designs.

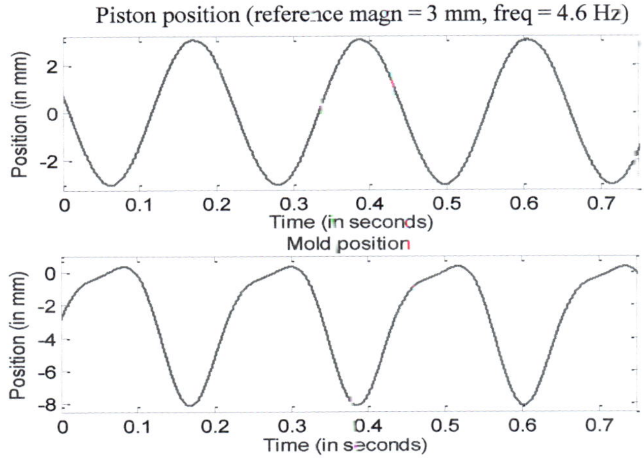

Fig. 1.5-5. Software testbed validation: simulation data exhibits resonance problem.

1.5.2 *Problem investigation: identifying the sources of the distortion*

As we have seen, mold velocity profile in the testbeds exhibits distortions similar to those observed in the mold oscillation system of the actual production unit. Thus, the disturbance sources must be identified based on the similarities between the testbeds and the production mold oscillation system.

Investigation and solution guideline: We conjecture that the causes of distortion in the testbeds and the production unit are identical and proceed with trying to find the distortion source and developing a control technique to eliminate distortions in the mold profile in both testbeds.

Investigation and solution requirements and constraints: Access to the hardware testbed is limited due to the personnel constraints at Nucor, so software testbed would need to be heavily used. Access to the caster is given only after successful demonstration of problem solution on the hardware testbed. Therefore the control technique needs to transition from testbeds to the production unit with minimal hardware experimentation. Due to the different properties (dimensions, materials, etc.) of the hardware testbed and the production unit, this implies that the control scheme should be developed with the performance minimally dependent on the system model.

1.5.2.1 *Hardware testbed performance diagnostics: spectral analysis of piston and mold position signals to determine the source of distortion*

The ***initial distortion source conjecture*** was formulated as follows: nonlinear actuator dynamics produces harmonics of 4.6 Hz distorting the piston position, which in turn distorts the mold position.

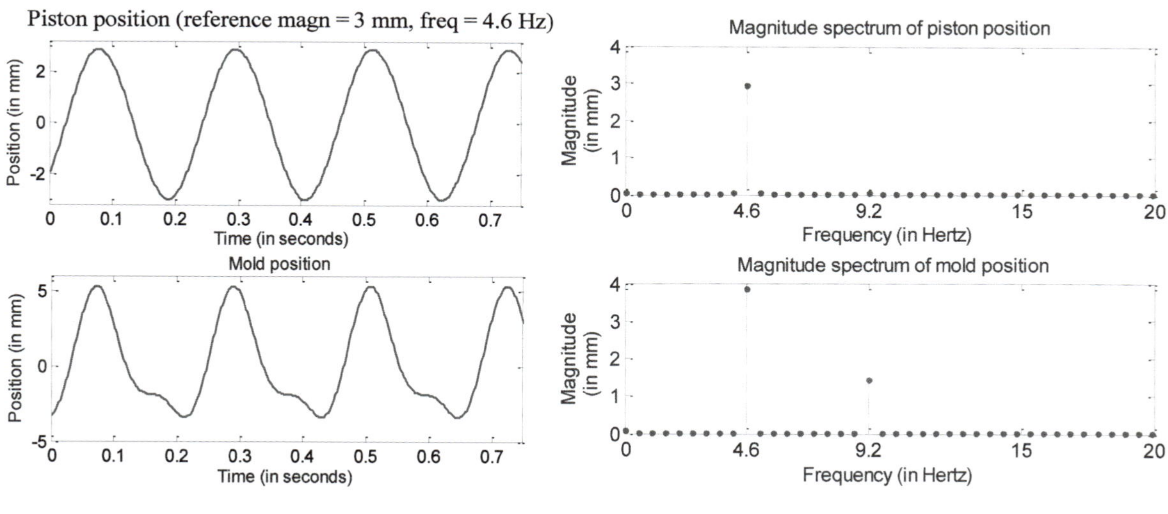

Fig. 1.5-6. Hardware testbed: piston position and mold position discrepancy.

However, with mold position being distorted, *the piston position looks perfect* to the naked eye, i.e. with no visible harmonics both in time and frequency domains, as seen in Figure 1.5-6. Since the actuator is visually observed to perform ideally, *the initial conjecture of the distortion source looks wrong.*

1.5.2.2 *Catching the culprit with a magnifying glass!*

Zooming into the magnitude spectrum of the piston position of Figure 1.5-6 as shown in Figure 1.5-7 reveals a small peak at 9.2 Hz, indicating that the naked eye is easily misled into overlooking both in time and frequency domains a small piston position distortion *caused by nonlinear actuator dynamics.* This peak is small, but critical — at beam resonance frequency it is amplified by about 40 times to cause a large peak at the mold end! Additional experiments show amplification of sinusoidal piston position magnitude of 0.05 mm at 9.2 Hz by the beam by at least a factor of 30. Similar conjecture verification through spectrum based diagnostics was performed using software testbed as well as the production unit data as described below.

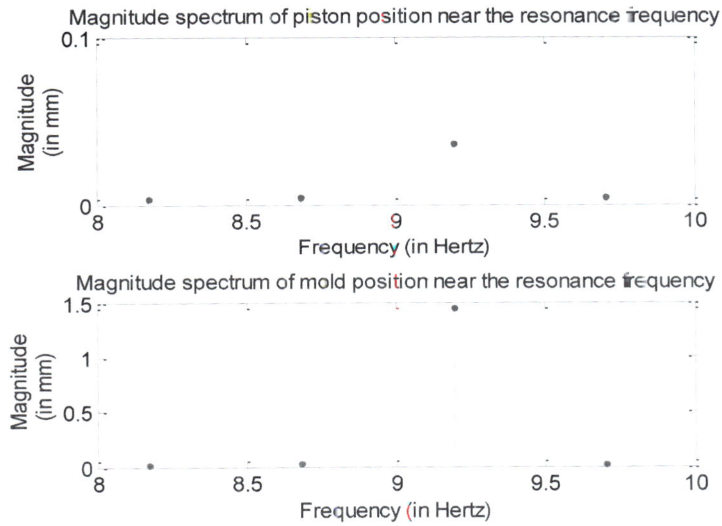

Fig. 1.5-7. Hardware testbed: amplification of mold position disturbance.

> ***Conclusion:*** The small peak in piston position being at resonance frequency excites the beam resonance and creates the large peak in the mold position. *All previous attempts at solving this problem failed because the distortions were thought to originate in the beam. In fact, however, they originate in the piston, and are just amplified by the beam.*

1.5.2.3 *Software testbed: verifying conclusion*

From Figure 1.5-8 below, we can see that above conclusion is also verified through the data obtained from simulation. The piston position looks perfect but has small peak at 9.2 Hz in magnitude spectrum generated by the nonlinear servo model. This small peak is amplified by the beam causing significant distortion in the mold position. The software testbed generates the resonance problem and can be used for the controllers to be developed. As will be shown in Chapter 9, only modeling of this distortion is necessary for the solution developed

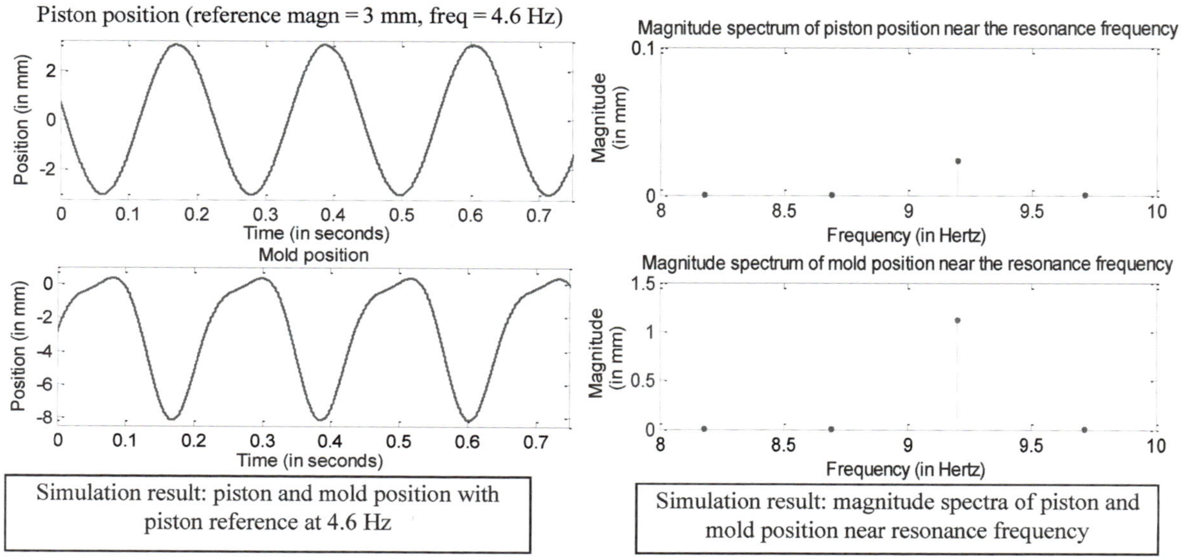

Fig. 1.5-8. Software testbed: piston and mold position; magnitude of distortion of mold position increased at 9.2 Hz.

1.5.2.4 *Summing up diagnostics:*

Answer: The input/output actuator dynamics is slightly nonlinear.

Input signal is the desired piston position, or "reference," — typically a sinusoid with frequency "f_r".

Output signal (actual piston position) is mainly sinusoidal at frequency "f_r"; but in addition contains small magnitude sinusoidal harmonics at frequencies nf_r: $2f_r$, $3f_r$, $4f_r$, …

Consequences: i) at the desired operating frequency 4.6 Hz, the piston position contains a small peak at 9.2 Hz — distorted, but not noticeably so, ii) due to the resonance of the beam, the mold position contains a significantly larger peak at 9.2 Hz, clearly distorted.

1.5.3 *Problem definition*

1.5.3.1 *Production unit: identifying source of distortion*

Through the hardware and the software testbed experiments, and the corresponding spectral analysis it was concluded that the actuator produces integer multiples (*harmonics*) of the reference sinusoid in the *piston position.* These harmonics are amplified by the beam and distort the mold velocity profile in the testbeds. The same phenomenon was conjectured to be the problem in the production unit. Figure 1.5-9 shows the mold velocity profile to be highly distorted with respect to the reference input to the piston — sinusoid of frequency 4.4 Hz.

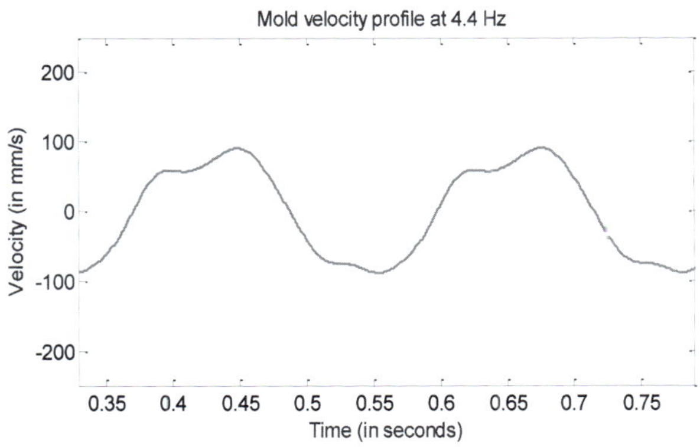

Fig. 1.5-9. Distorted mold velocity profile.

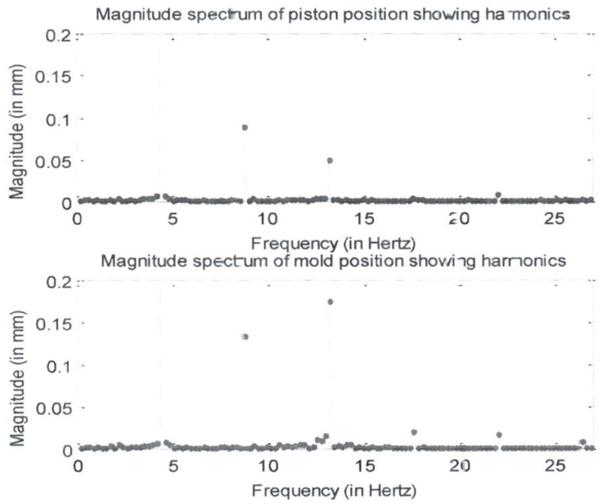

Fig. 1.5-10. Piston position distortion amplified at the mold position by beam resonances.

Zooming in on the piston position spectrum (Figure 1.5-10) reveals small peaks at 8.8 Hz and 13.2 Hz — the second and the third harmonics with respect to the operating frequency 4.4 Hz — in the piston position signal. But the waterfall plot in Figure 1.3-3 shows both small peak frequencies to be beam resonance frequencies! This means that small peaks in piston position create large peaks at the same frequency in the mold position. Thus, the conclusion is verified: the small-magnitude harmonics at the piston output are amplified by the beam into a severe problem at the mold when the *desired operating frequency is a sub-multiple of the resonance frequencies* of the beam.

1.5.3.2 *Testbeds and production unit: multiscale problem formulation*

As we just saw, scaling the magnitude spectrum graph to display the regular magnitude component made the small magnitude piston position components not discernible. To bring out the latter, we "zoomed in", i.e. increased the vertical graph resolution, making the regular magnitude go "off-scale" — outside of the graph.

This inability to clearly visualize signal components on the same graph indicates that they *have* ***wide magnitude scale separation***. Signals like this are called ***multiresolution***, or ***multiscale***. Thus, *the small-scale sinusoids were scaled up by the beam at its resonance frequencies to the regular-scale sinusoids in the mold position.* Hence, instead of trying to remove the latter from the mold position through the direct use of mold position sensor data, *the problem is to remove the small-scale sinusoids in the piston position, which then would eliminate the regular-scale ones in the mold position.* How can this be achieved? Can a hydraulic actuator be sized up to reduce piston position peaks at resonance frequencies? For hardware testbed this might be possible, but not for the production unit.

1.6 Problem Resolution

A different way to approach this problem is by means of ***active control***, i.e. physical action that would reduce the effect of resonance using ***sensing, filtering, and feedback control in the actuator subsystem***. Namely, taking the piston position as an output, we try to meet the following ***filtering and control objective***: design and implement a controller that transforms the closed loop subsystem into a filter that removes the small-magnitude harmonic multiples of the reference signal from the piston output *without affecting the original system performance at the other desired frequencies.* This objective is addressed below in three steps.

1.6.1 *Solving the problem through controller design*

1.6.1.1 *Subsystem simplification*

The first step is simplifying the representation of the existing stable servo system. This is carried out in Figure 1.6-1 through isolating the subsystem of interest and simplifying its block diagram.

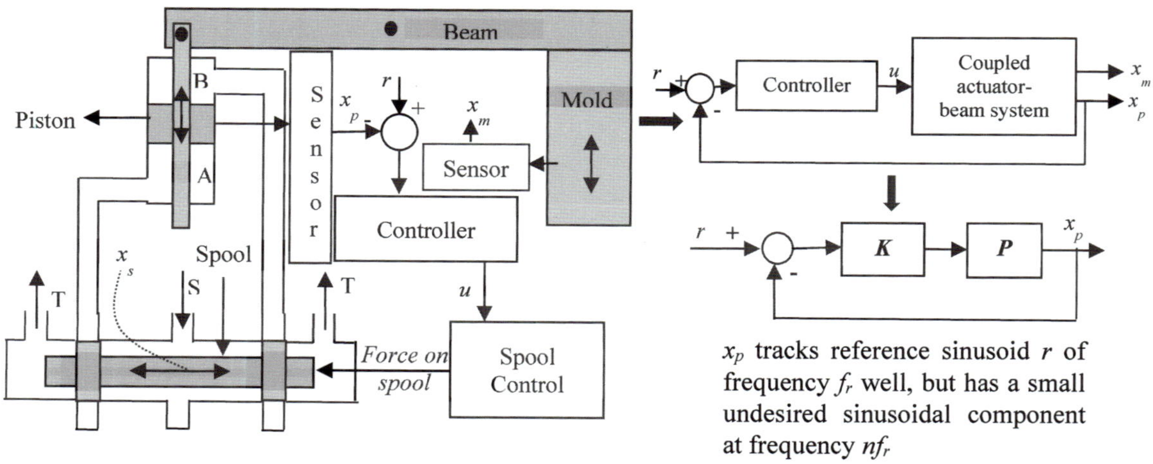

Fig. 1.6-1. Extracting the block diagram of the controlled subsystem from the mold oscillation system.

1.6.1.2 *Loop augmentation: making closed loop act as a filter with the desired properties*

The second step is modifying the controller K in Figure 1.6-1 to ensure the removal of the undesirable harmonic with frequency nf_r from the closed loop servo system (the prefix "sub" is subsequently dropped). This is accomplished through applying to this system the following ***filtering concept*** — transform the entire closed loop system into a filter which has ***"zero"*** at frequency nf_r. Then, the system will *block* any excitation at nf_r from passing to the output x_p and further to x_m.

The zero itself can be generated through a key ***"zero placement" property of feedback:*** the poles of the controller K along with the poles of the plant P are the ***open loop poles***, i.e. the poles prior to the loop closure. However, once the loop is closed these poles become ***the closed loop zeros***, i.e. the zeros of the entire closed loop system.

The choice of the poles for modifying the controller K is based on the ***"internal model principle" (IMP)*** — the requirement that disturbance model be placed in the loop. The latter sets the right poles in the modified controller and, consequently, the right zeros in the closed loop, generating a sinusoid at frequency nf_r that cancels the unwanted sinusoid of the same frequency.

To ensure the desired performance, the controller must also guarantee closed loop stability.

Applying these concepts to system in Figure 1.6-1 yields the controller topology in Figure 1.6-2 with additional feedback loop around filter F that provides the required closed loop properties for the testbeds. The corresponding controller synthesis procedure is described in Chapter 9.

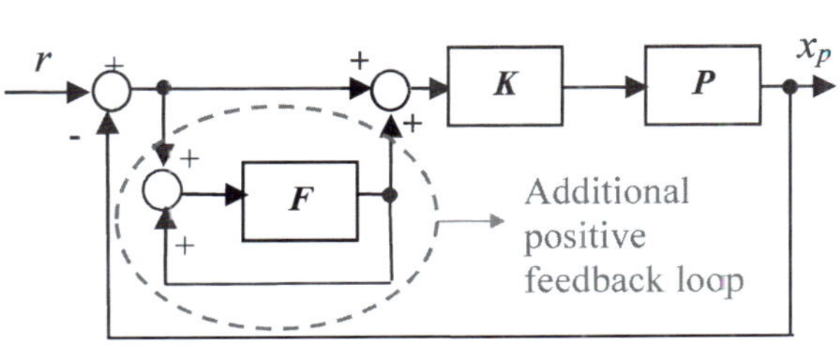

Augmenting a closed loop with an IMP controller given by a positive feedback loop containing filter F as shown ensures that x_p continues tracking r well, while rejecting the undesired sinusoid at frequency nf_r

Fig. 1.6-2. Augmenting a closed loop actuator block diagram with an additional controller.

In the caster, however, stability for the topology of Figure 1.6-2 is degraded. Adding blocks with filter Γ (trivially equal to 1 for the testbeds) as shown in Figure 1.6-3 solves this problem.

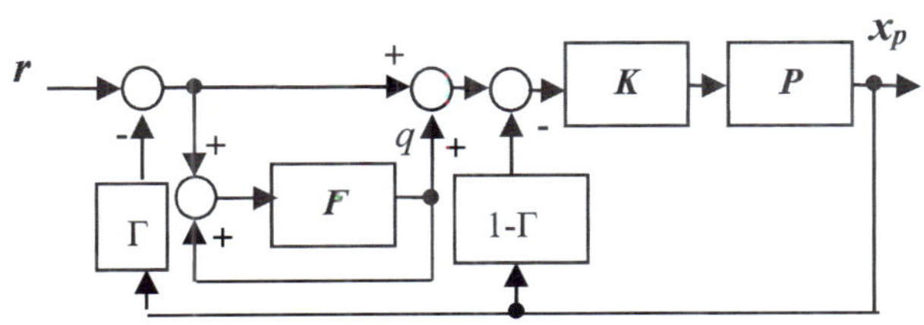

Fig. 1.6-3. Production unit: the control scheme.

Filter F has gain 1 at frequencies that we want to eliminate (8.8 Hz and 13.2 Hz in the example in Figure 1.5-10), making the additional positive feedback loop gain infinite, and thereby ensuring that the unwanted harmonics are removed (cf. Section 9.5.3). Filter Γ gain is negative of the inverse of the plant gain at the frequencies of interest, guaranteeing stability of the scheme. Detailed analysis of the latter topology is given in [1] (references are given at the end of each chapter).

In the latter reference the stability condition is reduced to the bound on the plant gain at the frequency of interest, making the control scheme ***non-plant-model based***. Namely, as long as the plant data at the frequency of interest changes by less than 100%, closed loop stability is guaranteed. This makes the control scheme sufficiently ***robust***, i.e. guarantees closed loop stability under nontrivial plant changes, which indeed take place, as shown in Figure 1.6-8. Robustness, crucial in production, permits going to the third step — controller implementation and assessment.

1.6.2 *Controller implementation and testing on the testbeds and the production unit*

1.6.2.1 *Software testbed: analog controller implementation and performance assessment*

In the third step, controller developed is applied first to the software testbed. The IMP controller takes the form of a positive feedback loop with filter $F = (2\zeta\omega_{res}s)/(s^2 + 2\zeta\omega_{res}s + \omega^2_{res})$, where $\zeta = 0.1$ and $\omega_{res} = 2\pi f_{res} = 2\pi \cdot 9.2$ rad/sec. The poles of F are set to have light damping to move them to the open left half-plane of the s-plane for stability, whereas the poles of the entire controller (Figure 1.6-2) are placed by the positive feedback loop at $\pm j\omega_{res}$. Controller gain is set to $K = 0.6$ and reference sinusoid $r(t)$ is chosen to have magnitude 3 mm and reference frequency $f_r = 4.6$ Hz $= f_{res}/2$. As seen in Figure 1.6-4, the distortions in mold position profile are completely eliminated.

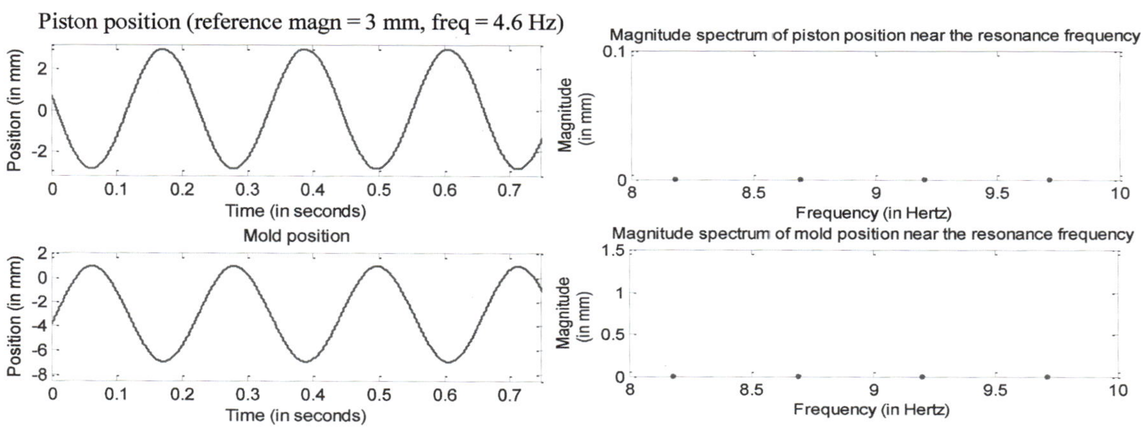

Fig. 1.6-4. Simulation results showing elimination of distortion.

1.6.2.2 *Hardware testbed: digital controller implementation and performance assessment*

Next, the controller developed is discretized with the step size 0.001 sec and its discrete version is implemented in the hardware testbed to remove harmonic at 9.2 Hz from the piston position. The computation is found to require fast ***digital signal processor (DSP)***. Systems combining digital processors with continuous time process dynamics are referred to, depending on the viewpoint, as follows: physical process — ***cyber-physical***, mathematical — ***discrete-continuous***, control systems/signal processing — ***sampled-data***, industrial — ***computer-controlled***. Further, if the computational/signal-processing part is ***clocked*** and ***capable of producing the required computational results within the pre-defined time interval, e.g., sampling interval***, then the system is referred to **as *real-time*, or *online***. When not clocked, the system is said to be ***offline***.

Filter F is chosen to be identical to that used in the software testbed, namely, with $\zeta = 0.1$ and $\omega_{res} = 2\pi \cdot 9.2$ rad/sec. Controller gain has been increased to $K = 2$, and reference sinusoid stayed the same — with magnitude 3 mm and frequency 4.6 Hz. Applying these conditions, the small peak at 9.2 Hz is eliminated from the magnitude spectrum of the piston position signal, reducing the corresponding peak in the mold position from 1.45 mm to 0.45 mm, as shown in Figure 1.6-5, while preserving satisfactory tracking at 4.6 Hz. With the residual peak of 0.45 mm due to hinge friction being insignificant, the results are found to be successful and the decision is made to move testing to the production unit using the testbed control setup, with DSP shown in Figure 1.4-5.

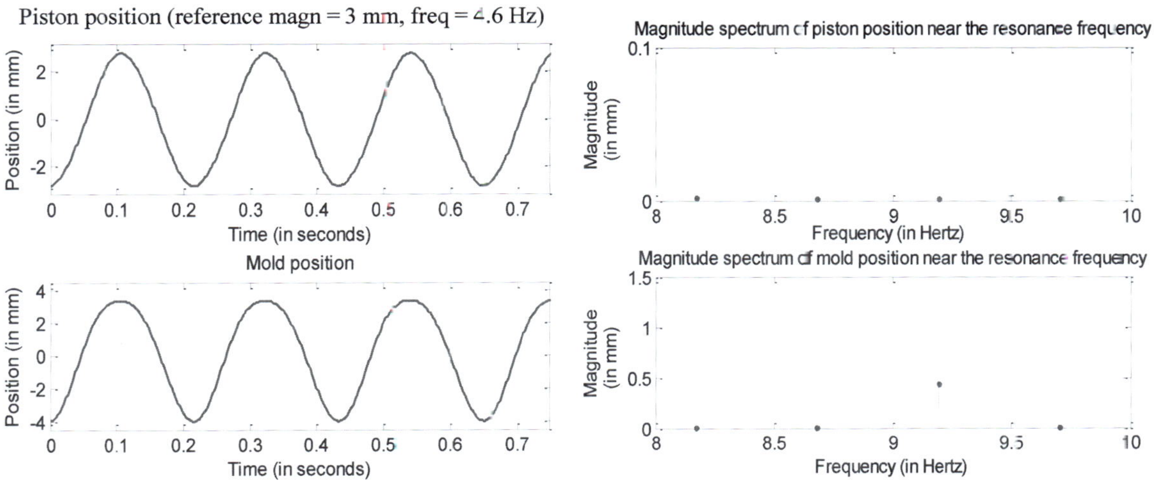

Fig. 1.6-5. Experimental results showing reduction of distortion.

1.6.2.3 *Production unit: digital controller implementation and results of controller testing during dry run and steel casting*

First, the testing is performed on the caster in the dry run, i.e. with no steel being cast, at the reference frequency 4.4 Hz. Controller is implemented in LabVIEW using instrumentation given in Figures 1.5-3 and 1.5-5. Figure 1.6-6 shows that under new controller peaks at 8.8 Hz and 13.2 Hz have been significantly reduced, so mold velocity should be undistorted.

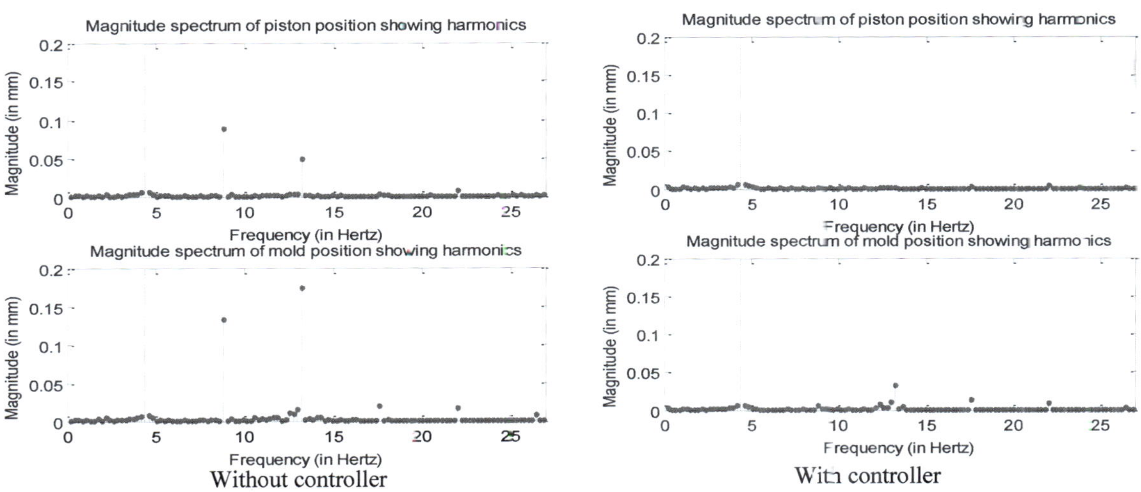

Fig. 1.6-6. Comparison of the results during the dry run (empty mold) — reduction of distortion with controller.

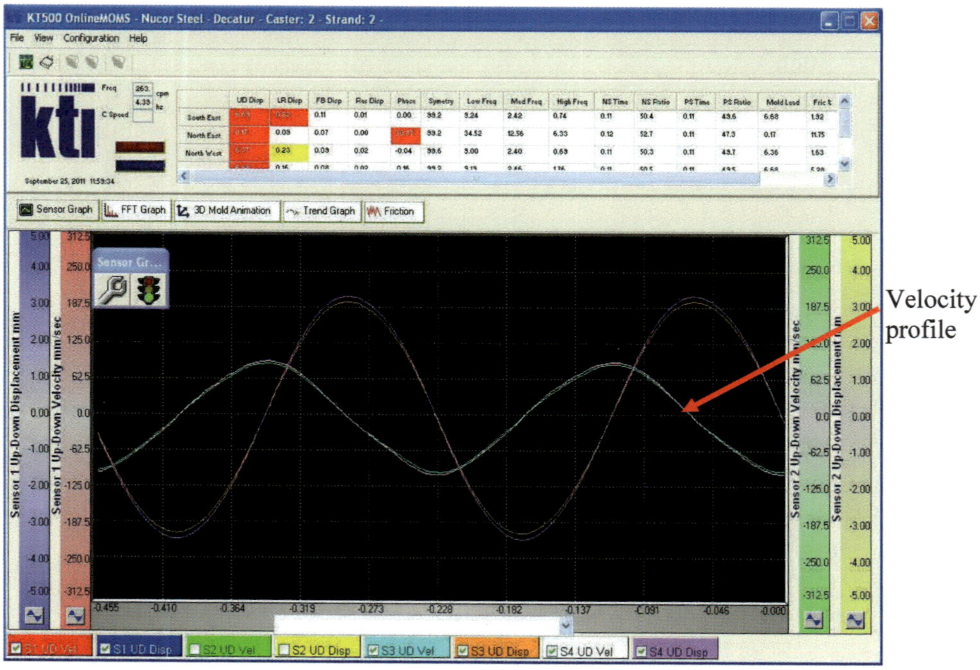

Fig. 1.6-7. Large distortions in the velocity profile during the dry run (no steel) on the caster are eliminated.

Figure 1.6-7 indeed shows that distortion has been largely removed.

The next test was conducted on the caster in production. During casting plant gain slightly changes as shown in Figure 1.6-8, so this gain is typically collected every 2 weeks. Automated collection plant gain code at the frequencies of interest has also been developed. Plant gain changes, however, by less than 100% over a 1-month period, hence the controller is guaranteed to be stable.

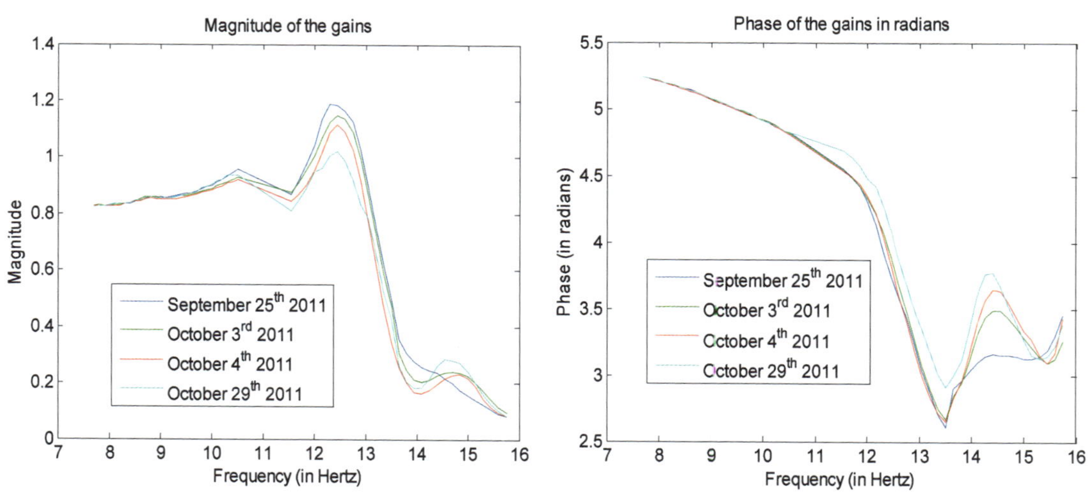

Fig. 1.6-8. Plant gains.

Frequency of oscillation during casting ranges from 3.8 Hz to 5.2 Hz, and the controller coefficients must match that. Such a code has been developed in LabVIEW, as well.

As seen from Figure 1.6-9, the distortion in the velocity profile is almost completely eliminated by the controller, admitting casting speeds of 60 mm/sec and higher.

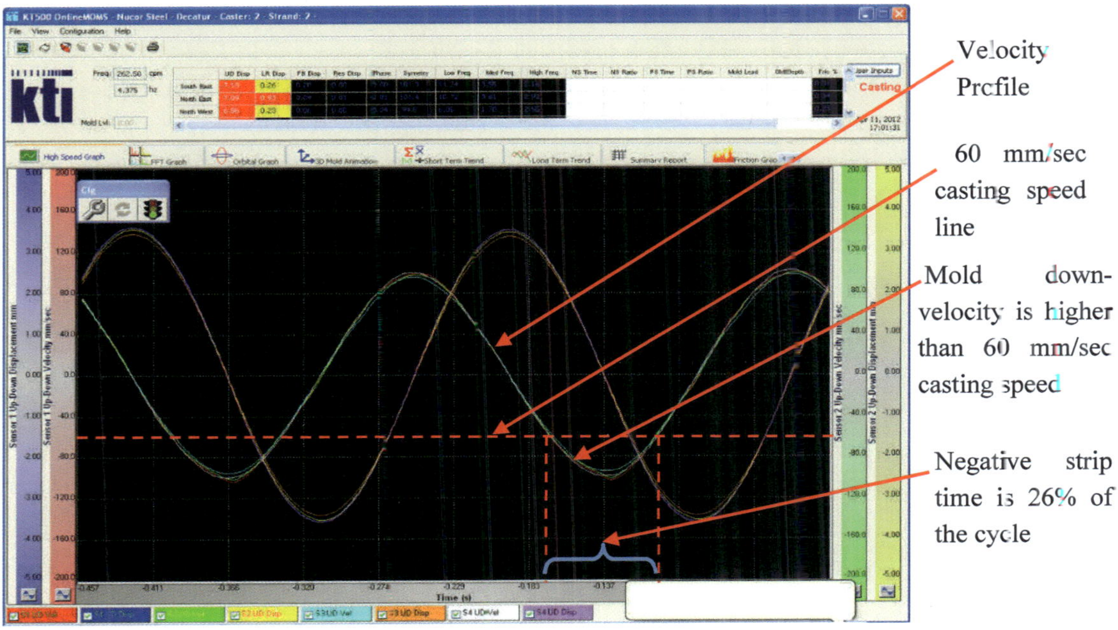

Fig. 1.6-9. Full production: nearly undistorted mold velocity profile at 4.37 Hz provides the desired negative strip time.

Figure 1.6-10 shows that the resulting control signal to the actuator is sufficiently smooth.

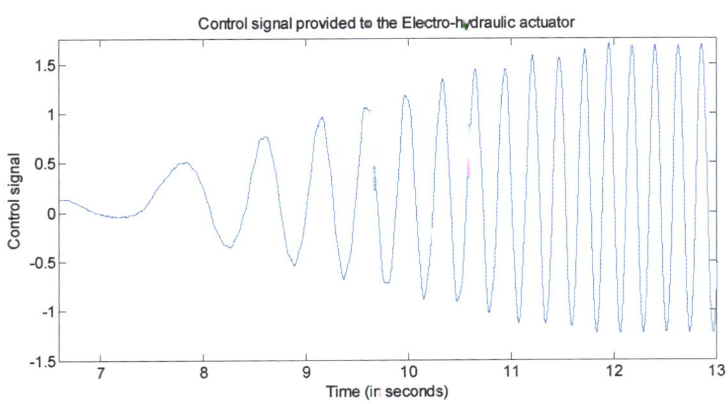

- The control signal shown corresponds to the reference signal frequency increase from 0 Hz to 4.6 Hz.
- The signal is smooth but for the presence of small ripple due to measurement noise.

Fig. 1.6-10. Production unit control signal.

These results demonstrate that the project objective has been attained and the distortion problem has been resolved.

1.6.2.4 *Effect of project results on product quality: cast steel surface defect reduction*

The proposed control structure and technique attained a significant improvement in cast steel surface quality. Indeed, due to the proportional controller inability to avoid beam resonance, the resonance onset prevented the operator from raising frequency of oscillation to reduce oscillation magnitude. The latter, therefore, was kept high minimizing sticker breakout, but resulting in the thin steel slab surface with deep oscillation marks as shown in Figure 1.6-11 (left picture).

The proposed controller has removed this limitation by *keeping the mold velocity, the product of mold displacement and oscillation frequency, but permitting the latter increase and, hence, the corresponding oscillation magnitude reduction.* This yielded shallower oscillation marks, as seen in Figure 1.6-11 (right picture) and, consequently, lesser probability of surface cracks.

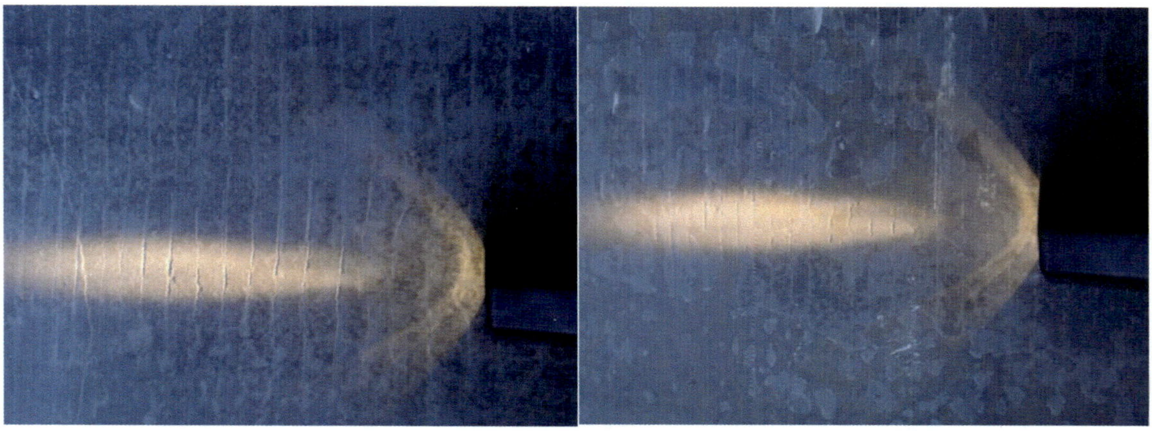

Fig. 1.6-11. Deep oscillation marks characterizing typical previous magnitude 4.5 mm at frequency 3.38 Hz (left), are eliminated by the proposed control system, enabling "short stroke," e.g. magnitude 3.5 mm at frequency 4.33 Hz (right).

1.6.2.5 *Conclusion*

The mold velocity distortion problem at Nucor Steel, Decatur has been solved for a hinged beam-type mold oscillation system through bringing together modeling, instrumentation, signal processing, diagnostics, filtering, control, and system integration. The end result is implementing a new feedback control law based on the disturbance model.

The control law eliminates periodic mold velocity disturbances, caused by excitation of system natural frequencies by actuator nonlinearity. The control law is robust in that it only needs knowledge of the system gain at disturbance frequencies, but otherwise uses no information about a system other than linearity and time-invariance, and it retains stability under perturbations of this gain by 100%. The controller has been implemented in LabVIEW and permanently integrated with the existing hardware into caster operation. According to the 2018 Nucor assessment of the project outcome benefits, the new controller has yielded 50% average cracks reduction across all grades and has enabled operation at previously unattainable casting speeds at or above 60 mm/sec.

1.7 Problem Solving Capability Buildup

The mold oscillator problem described in Sections 1.1–1.6 dramatically brings out the pitfalls of system integration — sizing the components for nominal loads does not always guarantee nominal performance of the integrated system due to the unexpected consequences of component interconnection. It is these consequences, as will be shown in Chapter 5, which have rendered the integrated mold oscillator system not functional in the key part of the intended operating range.

The success in addressing these consequences in the mold oscillator problem has been attained only through incorporation of signal and system thinking and tools, such as system zeros, into the overall system redesign, performance assessment, and refinement. Therefore, without neglecting system poles, typically considered in detail in basic dynamics courses, this book pays close attention at system zeros as being among the main tools in filter and controller design.

The goal of Chapters 1–9 is two-fold: a) develop the signal and system thinking and tools by a buildup of instrumentation, system modeling, signal processing, and control concepts and techniques, and b) integrate all of the latter into a coherent framework through analyzing and solving the mold oscillation resonance problem. Reaching this goal proceeds in three main stages described next, using at every stage analytical and numerical techniques supported by extensive graphical visualization. The latter, depending on the requirements, is generated through Excel, Matlab, and Mathematica. *The goal of Chapter 10* is to further develop the background introduced in the previous chapters to address more complex problems with nonstationary uncertain behavior.

1.7.1 *Stage I — introduction to signals, basic filtering, loading and system integration, sensing and actuation, instrumentation, data acquisition, system identification, and spectral computation*

Stage I, presented in Chapters 2, 3, and 4, introduces basic signals characterization and types, collects distinct linear system representations into an aggregated system portrait with filtering viewpoint, and provides the first look at filtering, system identification and integration, instrumentation, and data acquisition.

Background on linear system resonance frequency calculation is strengthened, and resonance-related problems in integrating hydraulic actuator with the mold oscillator mechanism are introduced.

The focus is on the building blocks highlighted in purple in the diagram in Figure 1.7-1 below, where a digital actuation command signal is converted into analog actuator input, an actuator (hydraulic valve) and the system of interest (plant represented by the mold/beam assembly) are interconnected, a sensor is connected to the plant, measuring mold displacement, an analog signal from the sensor is conditioned for sampling by an analog antialiasing filter, the filtered analog signal is converted to the digital signal through sampling and quantization, and the resulting digital signal is visualized on the display for decision making. The system integration concepts, including loading, ideal (nonloading) sensing, and ideal (nonloadable) actuation are introduced and the first look at spectral computation through continuous and discrete Fourier series is presented.

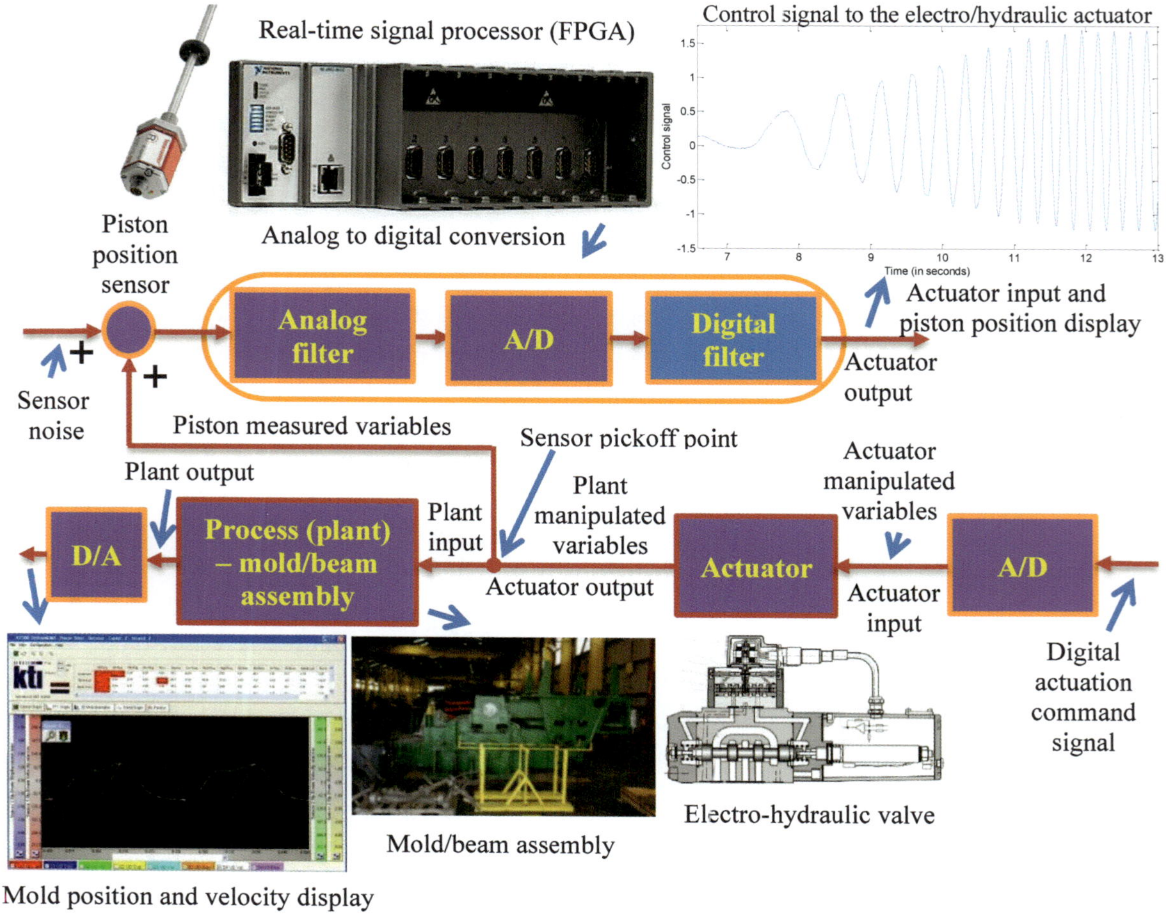

Fig. 1.7-1. System interconnection and data acquisition — going from digital actuator command to digital display of actuator and plant behavior.

1.7.2 *Stage II — introduction to spectral analysis and system identification for system modeling, diagnostics, and decision making; basics of analog and digital filter design and implementation*

At this stage, diagnostics — extraction of critical information from data, as well as system identification — the use of input/output data to build dynamic models of the difficult to model systems, are considered in greater detail. This information is used for bringing out into the open hidden phenomena, such as subtle process changes leading to incipient faults, as in the airplane preventative maintenance (take out of service early — expensive, take out late — dangerous) and system integration pitfalls, such as resonance excitation that gives rise to the mold oscillation system velocity distortion problem. This is done in Chapters 5, 6, 7, and 8.

The blocks considered at this stage are shown in purple in Figure 1.7-2 below. Analog filtering is enhanced through presenting a more general design and implementation technique. Basic digital filter design is presented as well to permit further removal of irrelevant data through digital signal processing. Finally, to enable decision making, the cleaned up digital signal undergoes signal feature extraction in the frequency domain through the spectral analysis techniques — a set of signal processing tools based on mathematics, algorithms, programming, and computation. The techniques presented in the book include continuous time/frequency, discrete-time (DTFT), and discrete (DFT) Fourier transforms, and correlation and power spectral density computations for processing noisy data through discrete-continuous spectra.

An example of the decision-making enabled by diagnostics is the aircraft preventative maintenance: taking the aircraft out of service for preventive maintenance according to an apriori specified schedule makes little sense: if it is too early, then it becomes unnecessarily expensive and disruptive; if, however it is too late, then it puts the flight in danger and reduces the aircraft longevity. Therefore, subtle initial indicators of wear and tear are continually monitored to ensure early detection of the maintenance need.

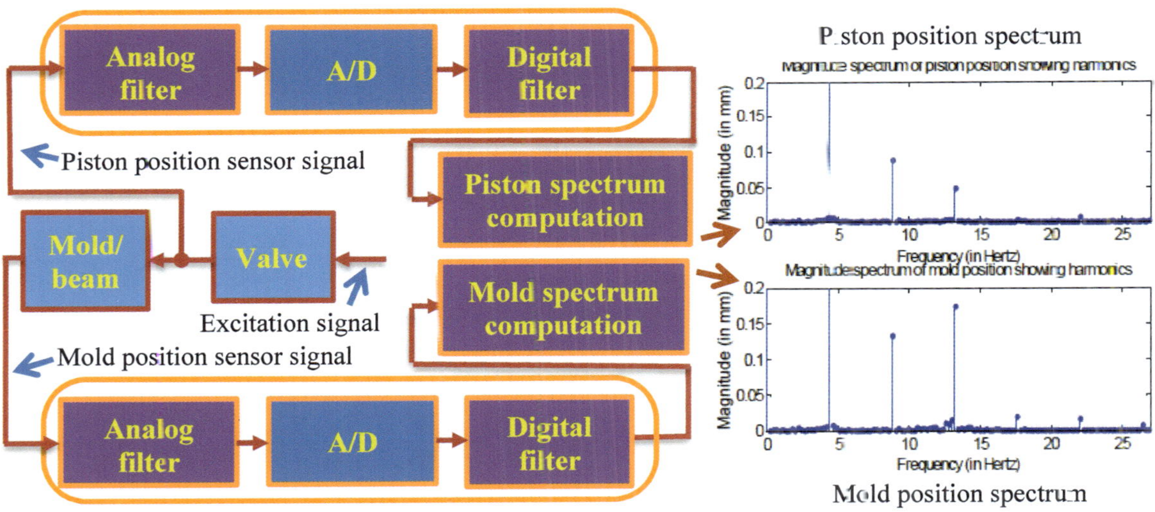

Fig. 1.7-2. System identification and diagnostics — using discrete-continuous spectra for dynamic system modeling and system behavior analysis through signal feature extraction.

1.7.3 *Stage III — introduction to controller design and system integration*

At this stage, the feedback and the closed loop control system concepts and simple design procedures are introduced and shown to solve the mold oscillation problem. This is done in Chapter 9. This stage presents controller design as an element of the overall system design and integration.

The building blocks involved are assembled in the diagram in Figure 1.7-3 below, where the data acquisition blocks of the first stage are followed by the block with control algorithm designed on the basis of application of the model building and analysis techniques learned at the second stage. The plant definition at this stage changes from the mold/beam assembly to the hydraulic actuator itself, with the sensor attached to the piston, rather than the mold. This makes the piston position, rather than the mold position, to be the output variable.

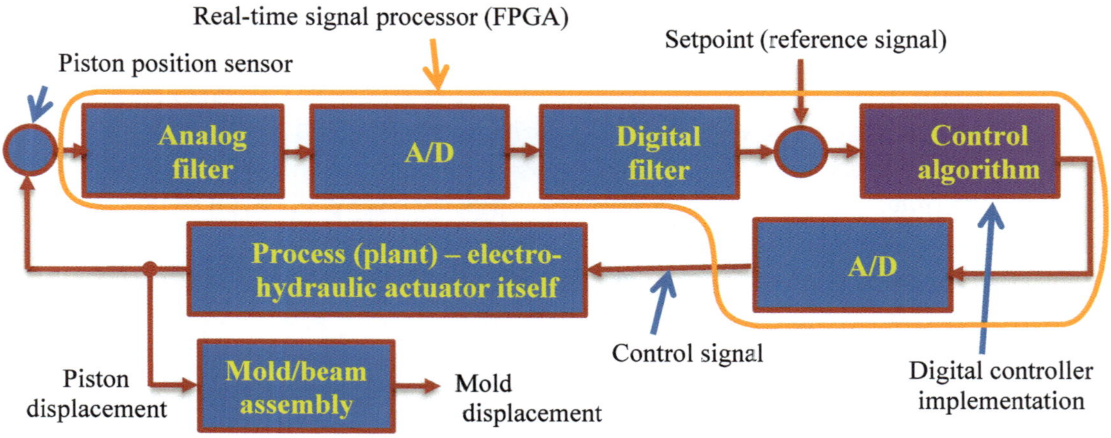

Fig. 1.7-3. Closed loop control and system integration for attainment of project objective.

1.7.4 *Stage IV — introduction to nonstationary multiscale signals and machine learning*

Mold oscillator is a time-invariant system driven by a stationary (with fixed magnitude, frequency, and phase) reference signal. However, the turbulence of the steel flow into the mold and the solidifying steel shell bulging between the rolls, which exerts upward pressure on the liquid steel pool, create irregular spatiotemporal steel mold level fluctuations. The latter give rise to the mold level sensor signal nonstationarity and multiscale structure. Mold level modeling and control then call for time-varying signal analysis, model uncertainty reduction, controller adaptation, and steel quality monitoring through computer vision. These techniques are introduced in Chapter 10.

1.7.5 *Appendices*

Eight appendices provide background and basic reference for Chapters 1–10. Appendix A lists the main trigonometric identities and gives integral and summation tables. Appendix B provides a system classification review. Appendix C presents the basics of random processes and provides a glimpse at constructing white noise type random process (Wiener process). Appendices D and E give the basic Fourier and Laplace transforms theorems and tables. Appendix F presents the basic types of sensors and actuators. Appendix G provides background for Euler-Bernoulli and Timoshenko beam models. Finally, Appendix H lists the Matlab and Simulink codes of all the programs used in the example problems and assignments.

1.7.6 *Homework, project, laboratory, and examination assignments*

The objective of the course based on this book is to merge dynamics, signal processing, instrumentation, and control into a coherent and effective problem-solving package covered in a single semester, with a glimpse at the time-varying signal analysis and machine learning given in the last one or two lectures. This is pursued through gradual blending of reading assignments, virtual-lab-type software testbed simulations, homework assignments, milestone project assignments capping each stage, and examinations, all culminating in completely removing mold oscillator distortion. A complete set of programs used in the book is given in Appendix H and is available by contacting the publisher or the author. Lecture slides are available for instructors.

The key part in material buildup is solving a sequence of homework and milestone project assignments, and examinations. Ten sample homework assignments with solutions and/or solution guides, three sample milestone project assignments, and three examination examples are included. They cumulatively cover Chapters 1–9 and are placed within the book to provide even pace of material exposure. Full versions of typical assignments are available to instructors who adopt the book by contacting the author. Generating assignments that are close to the examples has been found to be an effective teaching tool.

Homework 1: Assignment and Solution Guide

This assignment sets the stage for the course, introducing system model capable of reproducing the resonance excitation problem of the actual thin slab continuous casting mold oscillation system. The utility of subsequent course topics, along with the corresponding assignments, will be demonstrated through solving this problem.

Throughout the course, a number of Matlab and Simulink files will be utilized. One can either use the turn-key files available to instructors though request to the author, or, to get a better grasp of the material and a fuller Matlab experience, re-enter the code of the required file from its listing in Appendix H and try to understand how the program works. Saving the original Matlab files is recommended, since the assignments don't sequentially build on each other.

Assignment submission graph format. To properly document the work, the name of the solution author should be typed, not written by hand, into the Matlab graph header of any graph to be submitted. All axes on all graphs should be properly labeled, with the units indicated, also not by hand, but in the Matlab plotter. This requirement applies to all software generated graphs in all of the subsequent assignments.

Reading. It is recommended to go through Chapter 1, review Appendix A — mathematical notation, trigonometric identities, integrals, finite sums, sequences and series, series expansions, integration by parts, Appendix B — system classification, and brush up on Matlab with tutorials helpful for making plots in Matlab or running m-files: http://coecsl.ece.illinois.edu/me360/matlabtutorial.pdf.

For in-depth understanding it is recommended to take a look at Appendix G — beam theory. It is worth noting that during the initial stage of modeling, combination of hydraulic actuator and Euler-Bernoulli beam models was utilized, but was not numerically well behaved. Replacing the latter with Timoshenko beam model provided a well running testbed software [1], [2], [3], as described in Appendix G.

Problems. For the simulation cases below, plot the input signal (piston position) and output signal (mold position) on separate graphs, but with identical scale for direct quick visual comparison.

1. Briefly describe the mold oscillation testbed and its resonance excitation problem, and explain why the latter takes place.

2. Using Beam_simulation.slx, simulate the response of the coupled beam system (without hydraulic valve) to two input sine waves: (a) one with frequency 4.6 Hz and (b) another with frequency 9.2 Hz. Draw excitation and response graphs and explain why the system acts as a motion amplifier when periodically excited at resonance frequency.
3. Using Testbed_simulation.slx, simulate the system with hydraulic valve and feedback with P-controller only (matching the controller setup to a particular task is described in the detailed procedure below), responding to a 4.6 Hz reference input sine wave. Draw excitation and response graphs and indicate what is the problem with system performance based on the mold displacement and velocity response to periodic input excitation.
4. Simulate the system with hydraulic valve and feedback P-controller with "tracking controller" algorithm from [1], [2], responding to a 4.6 Hz reference input sine wave. Draw excitation and response graphs, examine them, and indicate how well the mold displacement tracks the reference input and how close the velocity response is to a pure sinusoid. Compare system performance (the degree to which the plant response resembles the reference input) with that in task 3 and comment on any performance improvement attained.
Detailed procedure.
Part 1: Explanation of the mold oscillator system, testbed, and distortion problem
In continuous steel casting, molten steel is first poured into a ***mold***, which is usually just four metal plates bolted together in the shape of a box, with no top or bottom. Channels inside the walls pass the cooling water, making the mold *the primary cooling zone*. Due to this cooling, the steel passing through the mold forms an initial solid ***shell***. By the time the steel exits the mold, the outer shell must be thick enough to withstand the ferrostatic pressure (like hydrostatic, but with steel) from the liquid in the middle. One of the major historical problems with designing a continuous casting process was that of ***stickers*** — lumps of steel that occasionally get stuck to the inner sides of the mold. This typically happens only at one place, so the rest of the steel keeps moving, tearing open a hole in the shell and causing a ***breakout*** where liquid steel escapes the shell. This leads to damage to the equipment, work stoppage, and potential danger to anyone nearby when several tons of liquid steel, at over 1500°C, start pouring out of the mold.

The key technology that allowed continuous casting to move forward is called ***mold oscillation***, in which the mold itself is uninterruptedly moved up and down to prevent these breakouts from occurring. The idea is that if the mold is oscillated fast enough, there is some time where the mold is moving downwards faster than the steel, called ***negative strip time*** depicted in Figure Hw1-1.

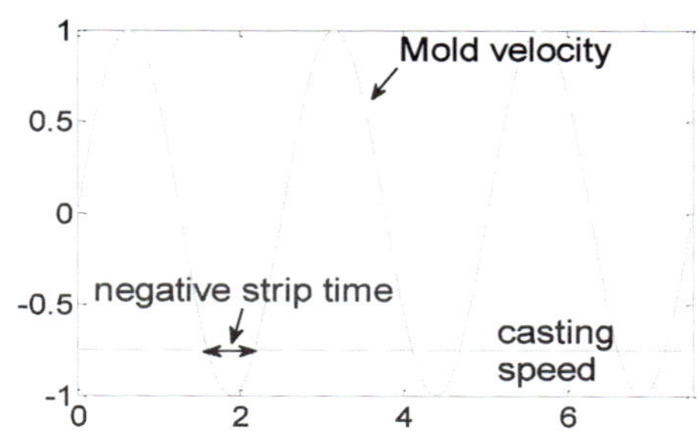

Fig. Hw1-1. Illustration of negative strip time.

During this time, the mold pastes any small stickers that have occurred back onto the steel shell, preventing them from initiating breakouts. At Nucor Steel Decatur, a steel mill in Decatur, Alabama, the mold is attached to a beam assembly, further simply referred to as ***beam***, as shown in Figure Hw1-2. At the other end of the beam is a counterweight that is pushed up and down by a ***hydraulic piston***, inducing mold oscillation, which has an undesirable effect of creating oscillation marks, as shown in Figure 1.2-6.

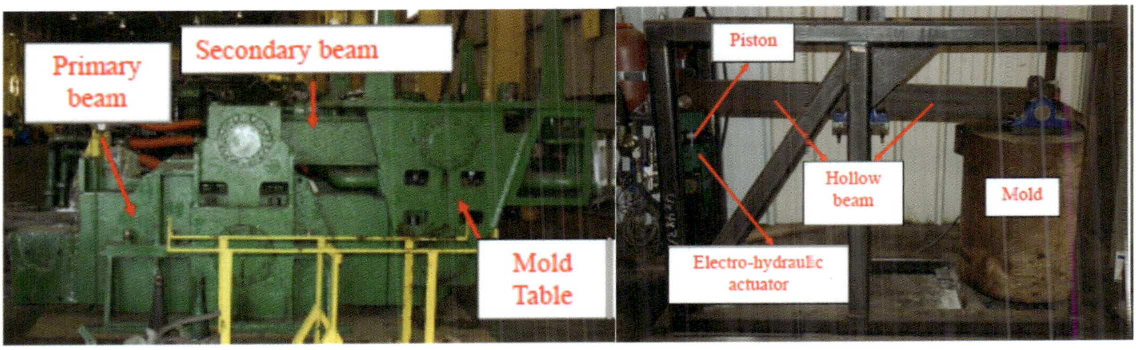

Fig. Hw1-2. Pictures of the actual mold/beam assembly (left) and the hardware testbed (right).

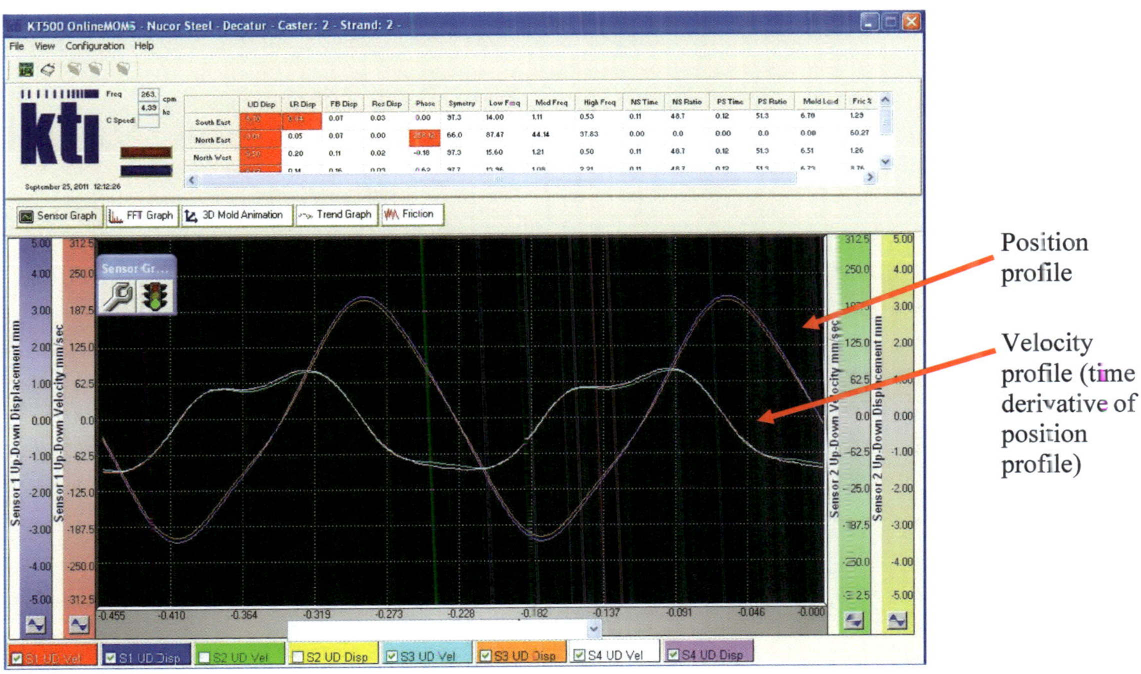

Fig. Hw1-3. Distorted mold velocity profile in a production unit.

Nucor Decatur's metallurgists attempted to increase the operating frequency of their mold oscillator and decrease the amplitude to reduce the height of oscillation marks. However, during the frequency increase they encountered distortions in the velocity profile, as shown in Figure Hw1-3. These distortions meant that metallurgists were no longer providing the negative strip time the process needed — the condition known to virtually guarantee breakouts. Thus, the need appeared for controlling the mold oscillator to attain a clean sinusoidal oscillation in a broader frequency range. This problem was solved using the tools to be introduced in this book, with the main technique presented in [1] and [2].

The engineers at Nucor Decatur built a ***testbed***, a physical mockup of the mold oscillation system on which they could run experiments and test changes to their control system. Like the caster mold oscillator, the testbed is driven by a hydraulic piston attached to one end of the beam, and the goal is to attain a sinusoidal motion of the other side, where a mass is mounted to mimic the effect of the mold. The testbed exhibited similar profile distortions, allowing hands-on problem investigation.

Figure Hw1-4 shows test measurements from an experiment on the testbed. Note that the piston position (top graph) appears to be a clean sinusoid while the mold position (bottom graph) is distorted. The magnitude spectrum (a signal analysis tool to be introduced later in the course) of the mold given by the bottom left graph in Figure Hw1-5 shows that this distorted profile is actually the sum of two sinusoids: one at the frequency the piston seems to be moving (4.6 Hz) and a smaller one at the first resonance frequency of the beam (9.2 Hz).

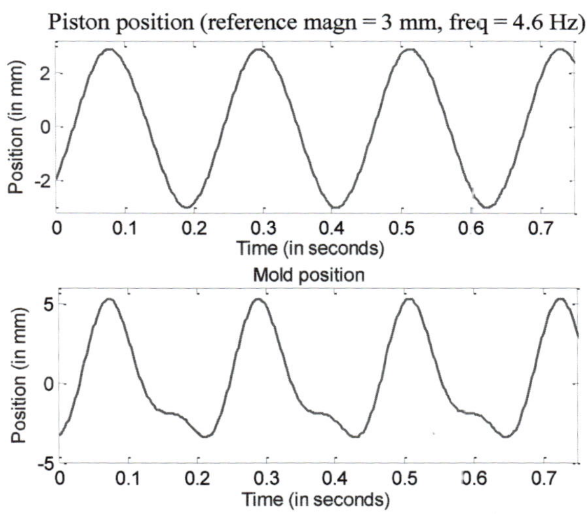

Fig. Hw1-4. Measurements of the piston (top) and the mold (bottom) position during the experiments on the hardware testbed, showing the time history.

A closer look at the magnitude spectra of the piston (top right graph in Figure Hw1-5) shows that the piston position actually does have a small-magnitude sinusoidal component at 9.2 Hz. This component is amplified by the resonance of the beam, causing distortion in the position profile at the mold end.

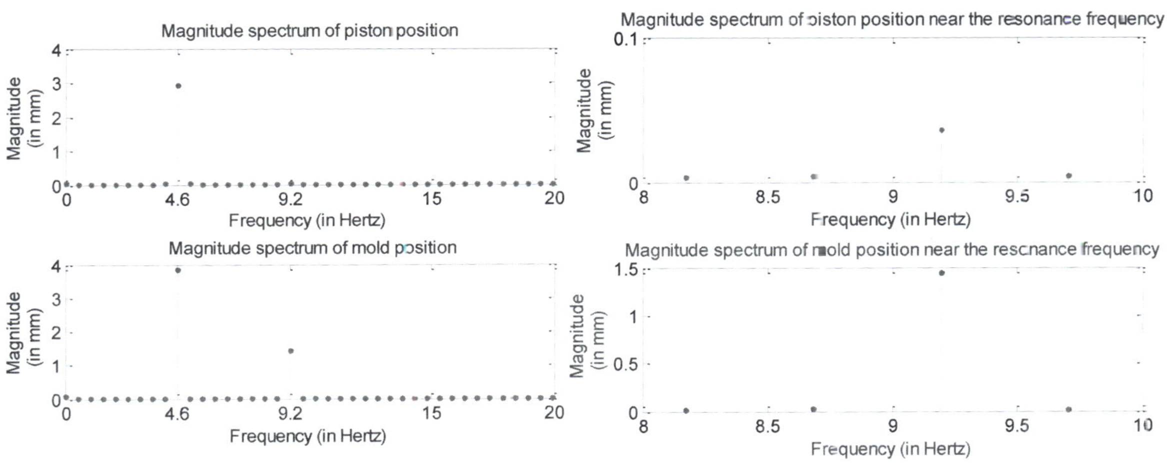

Fig. Hw1-5. Measurements of the piston (top) and the mold (bottom) position during the experiments on the hardware testbed, showing magnitude spectra (left), and a zoomed in portion of magnitude spectra focused on the first resonance frequency of the beam (right).

Mathematical model of this testbed was developed in [1], [2], and finalized in [3], essentially Creating its ***"digital twin"*** — a ***software testbed***, described in detail in Appendices G and H. Figure Hw1-6 shows a schematic drawing of this model. The goal is to get the piston position to follow a ***reference*** signal. The ***controller*** calculates the ***error*** between the ***actual piston position*** and this reference, and uses a ***control algorithm*** to calculate a ***control signal*** to send to the ***piston hydraulics***.

Based on the control signal, the piston hydraulics change the piston position, which in turn moves the ***beam*** and the ***mold*** at the other end. However, the weight of the mold and the dynamics of the beam cause some ***stress*** on the piston hydraulics, referred to as ***loading***, which affects the piston movement.

The original control algorithm that Nucor used is referred to as a ***P*** or ***proportional controller***, to be revisited in more detail Chapter 9. A P-controller just sets the control signal to be directly proportional to the error, but does not eliminate the distortion. To resolve the latter problem, an algorithm which adjusts the control signal to eliminate the distortions and make the output cleanly track the input, referred to as ***tracking controller***, was designed in [1], [2]

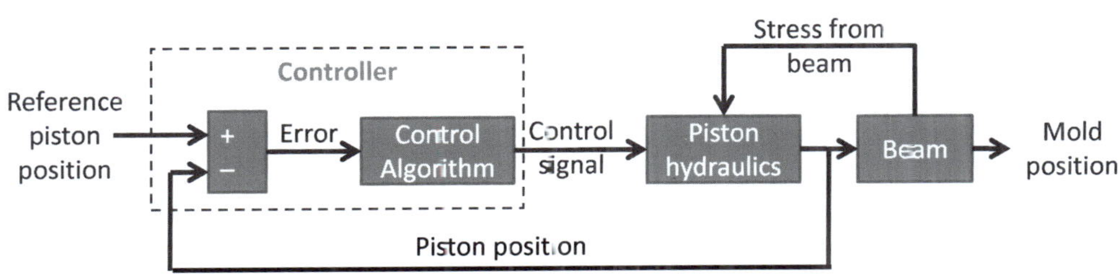

Fig. Hw1-6. Schematic of the Simulink testbed model.

The purpose of this homework is to learn how to i) run this software testbed by introducing a reference signal and examining the response, ii) make some simple changes to the testbed to set various simulation conditions, and iii) obtain the responses of its individual components.

Part 2: Running Simulink models in Matlab and plotting the results

The information below walks one through getting started on the homework. Specific Matlab commands to use will be indicated. These will be written the way they look in the Matlab command window, e.g.

```
>> a = 1 + 1.
```

It should just be possible to copy and paste them (except for the double arrows) into Matlab.

1. Getting the model files

There are three files needed for this homework, all given in Appendix H (MATLAB Programs 1, 2 and 7). The first file, **beam_state_space_form.m** is a Matlab script. The version in the book is a text file. It can be either saved as an .m file from browser or the text could be copied and pasted directly into the Matlab editor. The other two files, **Beam_simulation.slx** and **Testbed_simulation.slx** are the Simulink model files. The first one simulates only the beam, whereas the second one — the entire testbed, including beam, hydraulic piston, and controller.

2. Explanation of the model files

In the Simulink model, to save the output of the simulation to Matlab workspace the blocks referred to as '**to workspace**' are used. These blocks store time and position data and port them into arrays that can be used to make figures in Matlab.

The block name should correspond to the data signal being stored. In provided Simulink models, there are three '**to workspace**' blocks called '**time**,' '**piston**,' and '**mold**.'

If you double click these blocks, you will see a cell 'limit data points to.' You can put the number of data points you want here. For example, if you run the model for 10 sec, and you want to store the data points every 0.002 sec, then limiting data points to 1250 will store the last 0.002*1250 = 2.5 sec of the total 10 sec you run. You can change this number to make different figures. For this HW, the last 2.5 sec would be sufficient.

Next, for Problem 3 and Problem 4, you need to make several changes in the Simulink files. Under a typical default setting, the piston position would appear to be a linear function of time, whereas it should be a sinusoidal one. To fix this problem, double-click the piston block. This allows entering 'sample time' and 'limit data points to.' As indicated in the previous paragraph, set 'limit data points to' to 1250, and 'sample time' to 0.002 sec. This would save the data for the last 2.5 sec, fixing the problem. Now, we can see that the problem arose due to setting the default 'sample time' to 0.00002 sec, which, for 1250 samples, would store only the last 0.025 sec of data. Plotting these data over 1250 samples would yield a linearly looking function of time.

Second, for Problem 2, the unit for the input reference is m, so the magnitude should be 10/1000 for the sinusoid input. However, you will see the output response to be sinusoid but with a shift along the y-axis (i.e. the mold position is not symmetric along x-axis). This is correct, and actually you will see different shift with different input due to the beam model (the reference of the position might be different). But for the mold velocity, since we know that the velocity is 0 while it is static, the reference would be 0, i.e. the x-axis.

3. Setting up the model

Once the files are downloaded or typed in, open up all the files in Matlab. The .m file will open up in the Matlab script editor, and the .slx files will open up in separate Simulink windows. The Simulink models may take a while to open.

The first thing to do is to run the beam_state_space_form.m script. This can either be done by clicking the run button at the top of Matlab editor ( ), or pressing F5 in the editor. If a warning message appears saying something like "**File is not found in the current folder or on the MATLAB path**," click the "**Change Folder**" button to run the file.

This script creates a numerical model of the beam, and sets up some model parameters to match the testbed

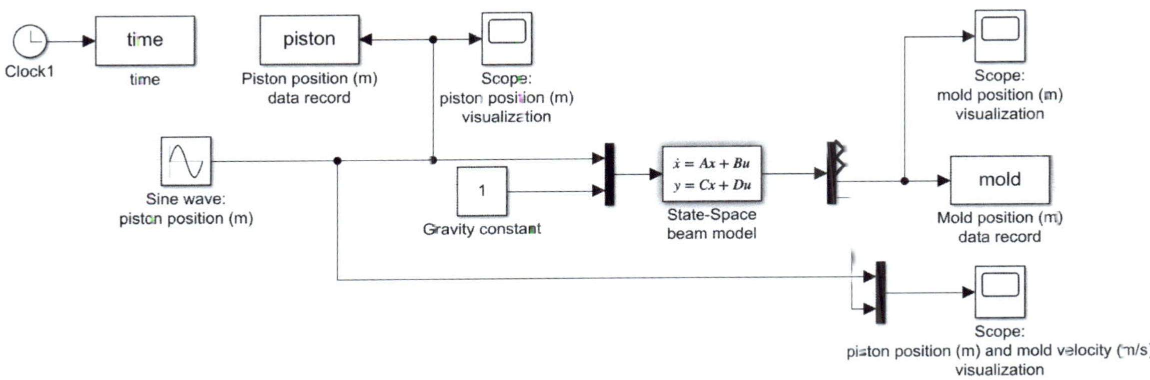

Fig. Hw1-7. Beam_simulation.slx. The displacement units are in m.

Problem 2(a). Now, looking at the Simulink model Beam_simulation.slx, one should see exactly what is shown in Figure Hw1-7. In this model, only the beam itself is simulated. The "State-Space beam model" block is a numerical model of the beam that was created by running beam_state_space_form.m. A sine wave is applied to the input as the piston position, and the simulation time, piston position, and mold position are saved to the Matlab workspace as arrays. Problem 2(a) asks to simulate the beam with an input sinusoid of 4.6 Hz, and the file is already set up to do this.

The simulation is now run by clicking on the run button ( ) at the top of the Simulink window.

A few warnings will come up on the command window, but these can be safely ignored. The simulation will probably take a couple minutes to run. The status can be checked at the bottom of the window (Running in the lower left corner and T=8.852 near the lower right).

When it's done, the homework asks to plot the "excitation" (input, or the piston position for this homework) and the "response" (output, or mold position for this homework) side by side. The Matlab commands to do this are:

```
>> figure
>> subplot(1, 2, 1)
>> plot(time, piston)
```

>> title('Piston position-student's name')
>> subplot(1, 2, 2)
>> plot(time, mold)
>> title('Mold position-student's name')

The homework also asks to use the same scale for each axis. The command to change the axis scale in Matlab is:

>> ylim([ymin ymax])
>> xlim([xmin xmax])

where xmin, xmax, ymin, and ymax should be replaced by the desired limits. Switching between the two subplots is carried out using:

>> subplot(1, 2, #)

where # is either 1 or 2.

>> xlabel('.....')
>> ylabel('.....')

are used to label the axes with proper names and units.

To improve the look of a graph, the property editor can be opened up by clicking on the button at the top of the Figure window ( ). The appearance of most of the plot elements can be changed from there. It can also be changed by commands from the command window or a script. The graph can be either printed directly from Matlab, or saved as a picture in a variety of image file formats.

4. Problem 2(b)

Problem 2(b) asks to rerun the beam simulation using an input sinusoid of 9.2 Hz. The parameters of any block in Simulink can be changed by double-clicking it. To change the frequency of the input sinusoid, double-click the Sine Wave ( ) block. Right now, the "Frequency" setting should be "2*pi*4.6". Change this to "2*pi*9.2".

Run the model again and plot the results. The homework then asks to explain the difference between the results at the two frequencies, remembering that the resonance frequency of the beam is 9.2Hz.

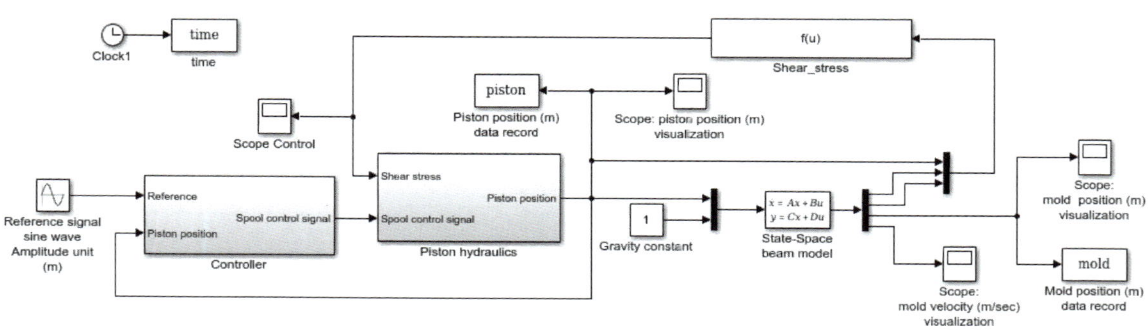

Fig. Hw1-8. Testbed_simulation.slx.

5. Problem 3

Now, looking at the Simulink model Testbed_simulation.slx, one should see exactly what is shown in Figure Hw1-8, which is a model of the full testbed, as sketched in Figure Hw1-6. The model consists of three main parts: a controller, the piston hydraulics, and the beam. Modification of the beam or the hydraulics models is not needed for this homework. However, the change in the "Controller" block is needed for Problems 3 and 4. This block is opened up by double-clicking it.

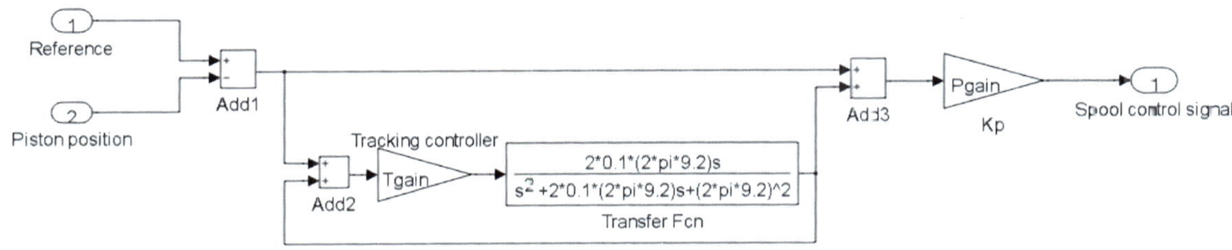

Fig. Hw1-9: Tracking_simulation.slx controller sub-model. Tgain is set to either 1, or 0 (tracking controller on, or off).

The general form here is the same as the "Controller" block of Figure Hw1-6. The algorithm takes as an input the error between the piston position and the reference. The lower loop is "tracking controller" design from [1], [2]. Without the lower loop, the algorithm just multiplies the error by a constant number Kp, called the "proportional gain" or "P-gain", to obtain the control signal.

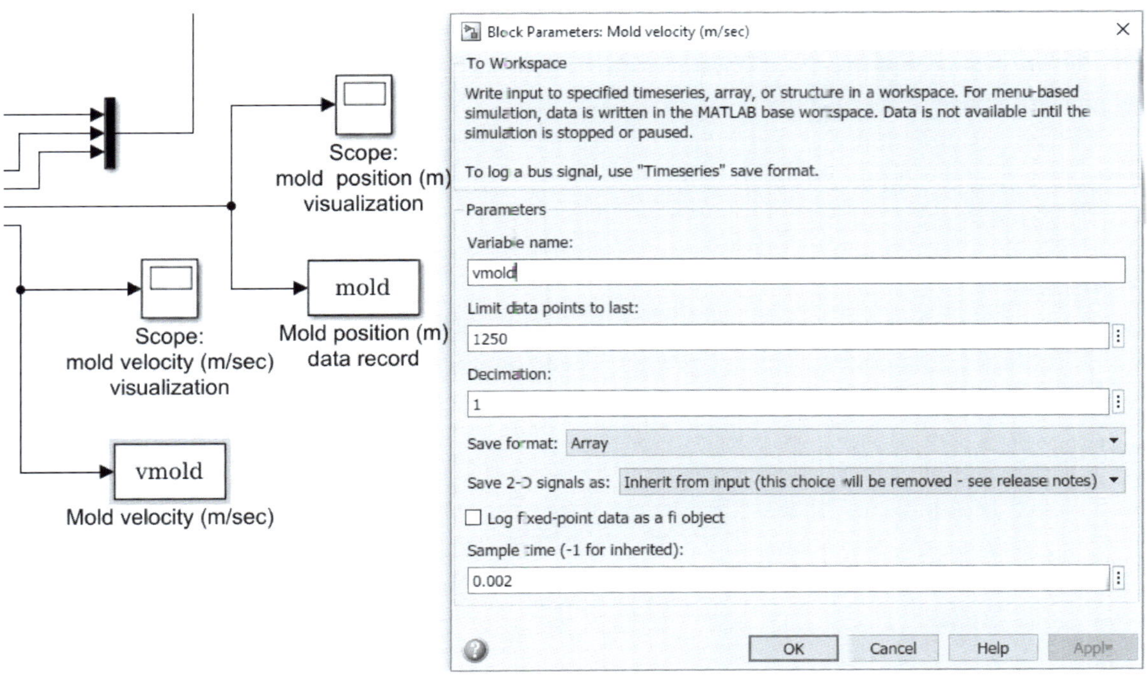

Fig. Hw1-10. Example of adding "To Workspace" block for mold velocity (left) and options for ensuring compatibility with the other outputs (right).

A closer look at the tracking controller in Figure Hw1-9 reveals that it simulates the additional positive feedback loop introduced in Figure 1.6-2, where filter F is given by the 'Transfer Fcn' block, with the only difference being an added switch Tgain, which takes the value of 1 to engage the additional loop or 0 to disengage it.

Problem 3 asks to simulate the controller using only the P-controller. This, then, can be done by disengaging the tracking controller through simply setting 'Tgain' to 0.

Problem 3 also asks to look at the mold velocity, which is not one of the outputs currently included in the file. Although the assignment does not specifically require plotting the mold velocity, it does ask to look at the velocity to answer the question. Notice that one of the Scopes is labeled "ScopeMoldVelocity." The signal going into this block is actually the mold velocity from the beam model. So, that signal needs to be put into the Matlab workspace. The easiest way to do this is to copy and paste any other "To Workspace" block ("Mold position," "Piston position," or "Time"), and then link it to the signal, as shown in Figure Hw1-10 (left). Make sure that it uses the same sampling rate and array size as the other output blocks in order to make the plots. The required parameters are shown in Figure Hw1-10 (right).

Run the simulation as above. Plot the piston and the mold positions side by side and include the graph with the homework. Plot the mold velocity separately to help answer the questions. In particular, remember the idea of negative strip time, discussed above, and that continuous casting requires that there be some guaranteed amount of negative strip time to avoid breakouts.

6. Problem 4

The last problem asks to simulate the system using "tracking controller" algorithm from [1], [2]. This is done by changing the 'Tgain' in the "Controller" back to 1 (Tgain is set to 1 in the beam_state_space_form.m script).

Rerun the simulation, and plot the results. As in Problem 3, although a plot of the mold velocity is not specifically required to hand in, it is needed to answer the questions.

References

1. Natarajan, V. and J. Bentsman (2011). "Robust Rejection of Sinusoids in Stable Nonlinearly Perturbed Unmodelled Linear Systems: Theory and Application to Servo," *Proc. of the 2011 Amer. Contr. Conf.*, pp. 3289–3294.
2. Natarajan, V. (2012). Rejection of periodic disturbances in uncertain nonlinearly perturbed stable infinite dimensional systems, PhD thesis, Dept. of Mech. Science and Engineering, Univ. of Illinois at Urbana-Champaign, 133p.
3. Angatkina, O., V. Natarajan, Z. Chen, M. Ding, and J. Bentsman (2019). "Modeling and Control of Resonance Effects in Steel Casting Mold Oscillators", *Acta Mechanica*, 230, 2087–2104.

Chapter 2

Introduction to Signals

2.1 Signal Classification and Signal Measures

By ***signal*** we will understand a physical event (e.g. movement of a mold) that gives information about the physical phenomenon of interest (e.g., negative strip time and, hence, sticker formation likelihood).

Signal processing consists of two main steps: i) mathematical, i.e. analytical and/or numerical modeling of this event, and ii) the use of the resulting models for the assessment of the properties of interest characterizing this phenomenon. We will refer to mathematical signal models simply as "signals", unless distinction between the actual event and its model is required.

2.1.1 *Types of signals*

Sorting signals out in terms of their basic features considerably helps with their processing. The signals considered in this chapter are classified in terms of the following properties:

A) Continuity in time and magnitude
- i) Analog, or Continuous-Time (CT)
- ii) Sampled-data, or Discrete-Continuous
- iii) Sampled, or Discrete-Time (DT)
- iv) Quantized
- v) Digital

B) Symmetry
- i) Even
- ii) Odd

C) Randomness
- i) Deterministic
- ii) Stochastic
- iii) Mixed

D) Periodicity
- i) Periodic
- ii) Nonperiodic

E) Frequency composition
- i) Single-frequency
- ii) Multifrequency

F) Scale composition
- i) Single-scale
- ii) Multiscale

G) Stationarity
- i) Stationary
- ii) Nonstationary

H) Causality
- i) Causal
- ii) Noncausal

I) Regularity
- i) Regular
- ii) Singular

J) Support
- i) Compact
- ii) Noncompact

K) Energy and power
- i) Energy signal
- ii) Power signal

The corresponding system classification is reviewed in Appendix B.

Analog (CT/CR: continuous-time/continuous-range) signal: An ***analog signal*** $x(t)$ is a function of a continuous independent variable (usually time) that can take any value in a continuous range.

The term ***continuous-time (CT) signal*** instead of analog signal is also used, but continuity refers to time variable only — the signal itself may not necessarily be continuous in time.

Fig. 2.1-1. Temposonic position sensor from MTS Sensors.

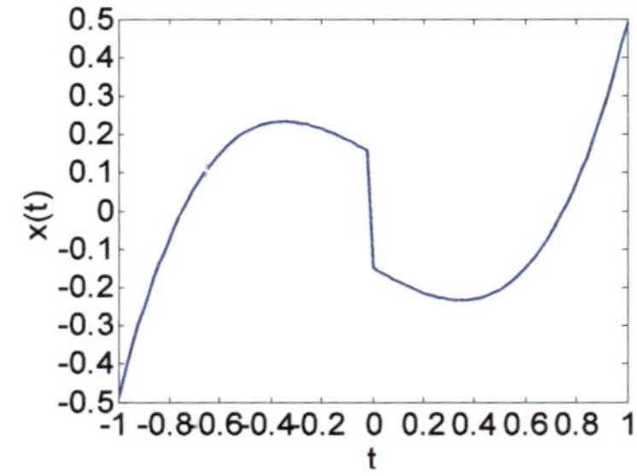

Fig. 2.1-2. Analog signal $x(t)$.

Sampled (DT/CR: discrete-time/continuous-range) signal: A ***discrete-time (DT), or sampled, signal***, $x[n]$, is the signal that is undefined except by the values at a set of discrete time instants. Usually these values are taken from the ***sampled-data signal*** — an analog signal, $x_s(t)$, with all its values omitted (set to 0) except those at discrete time instants nt_s, where t_s is called the ***sampling time interval***. This signal is a special case of a ***discrete-continuous signal*** — the one combining discrete values with continuous ones.

$$x_s(t) = \begin{cases} x(nt_s),\ t = nt_s, \\ n \in [-\infty, \infty]; \\ 0,\ \ t \neq nt_s, \\ n \in [-\infty, \infty]. \end{cases}$$

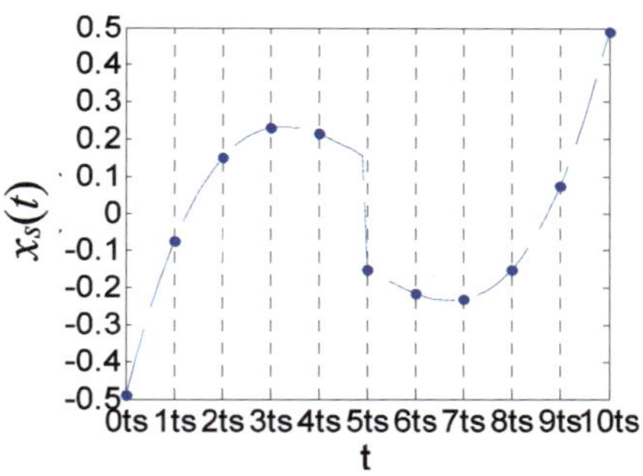

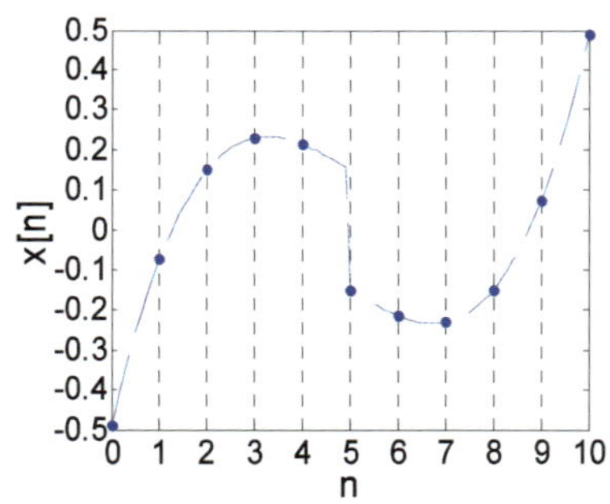

Fig. 2.1-3. Continuous time sampled-data signal $x_s(t)$. Fig. 2.1-4. Discrete time, or sampled signal $x[n]$.

Quantized (CT/DR-discrete-range) signal: A ***quantized signal***, $x_q(t)$, where subscript "q" stands for signal quantization, can take values only in a discrete set at any given time, but is still defined continuously in time.

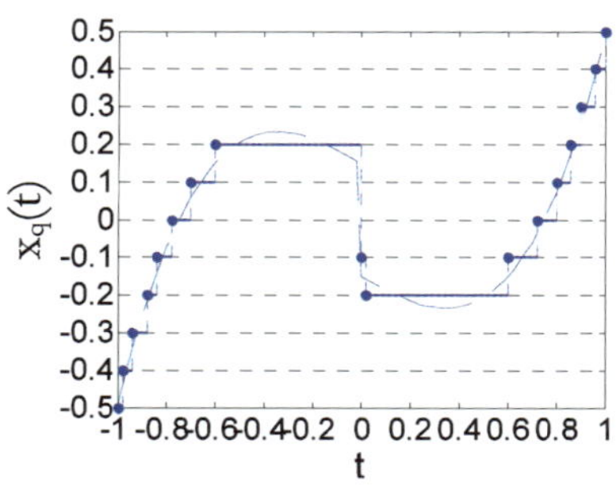

Fig. 2.1-5. Quantized signal $x_q(t)$.

Digital signal (DT/DR): A ***digital signal***, $x_q[n]$, is a DT signal that is quantized. This implies that it is defined on a grid, with vertical lines (representing position in a sequence) labeled through integers and horizontal lines (representing signal value) defined through binary numbers. Thus, digital signal is a sequence of numbers (an array). The difference between two adjacent numbers (magnitude resolution) is determined by the number representation capability of a particular processor.

Fig. 2.1-6. Top view of real time CRIO (compact reconfigurable input/output) processor, FPGA (field programmable gate array), and chassis from NI (National Instruments).

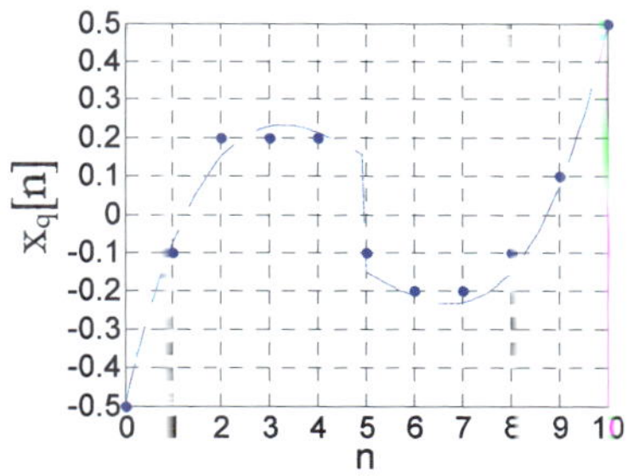

Fig. 2.1-7. Digital signal.

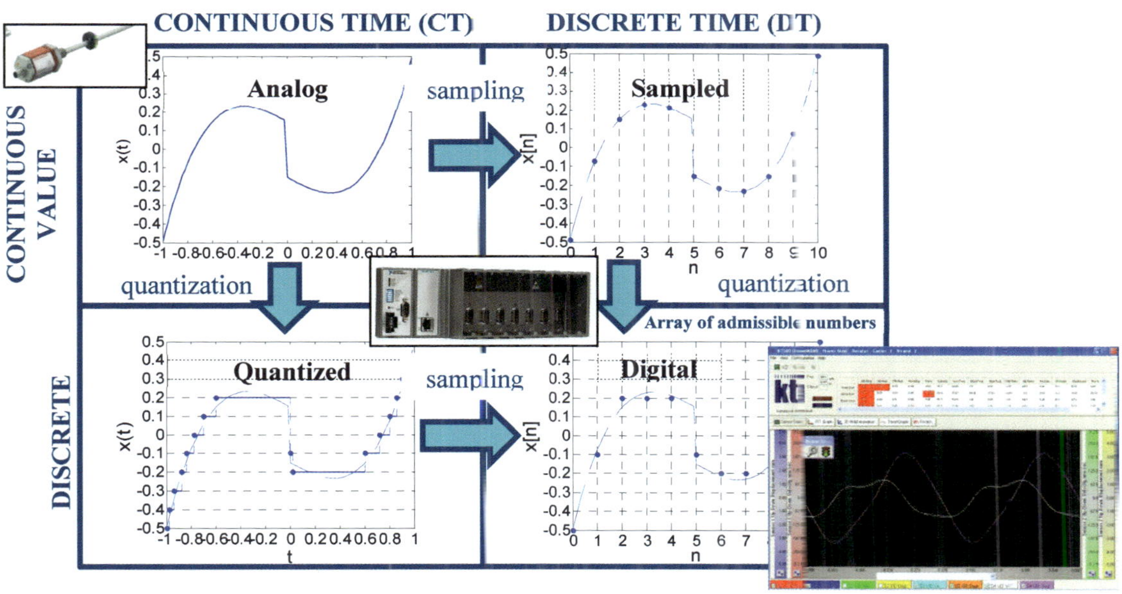

Fig. 2.1-8. Analog to digital signal conversion chart. Subscript "q" is omitted to reduce notational clutter.

Symmetry: Signal is ***even*** iff $f(t)=f(-t)$ (symmetric about vertical axis), e.g., $\cos(t)$ is an even function of t: $\cos(t)=\cos(-t)$. Signal is ***odd*** iff $f(t)=-f(-t)$ (symmetric about origin), e.g., $\sin(t)$ is an odd function of t: $\sin(t)=-\sin(-t)$, *i.e.*, $\sin(-t)=-\sin[-(-t)]=-\sin(t)$.

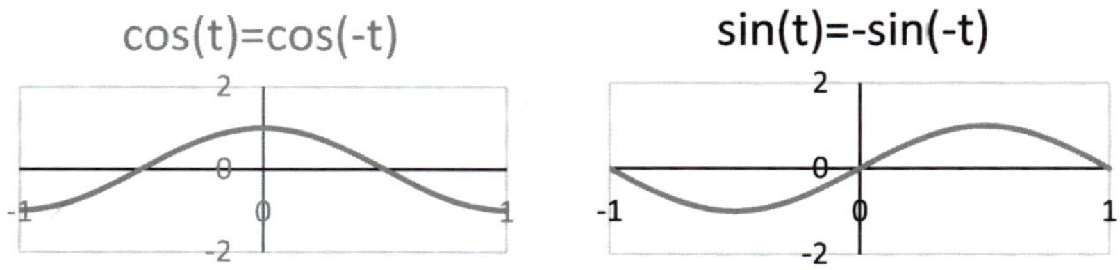

Fig. 2.1-9. Signal symmetry: left graph shows even signal, right graph shows odd signal.

Any function can be decomposed into even and odd parts: $f(t) = f_{ev}(t) + f_{od}(t)$, respectively, where $f_{ev}(t) = (1/2)[f(t) + f(-t)]$ and $f_{od}(t) = (1/2)[f(t) - f(-t)]$.

Example 2.1-1: Find the even and the odd parts of $x(t) = (\sin t + 1)^2$.

Solution: Opening up parentheses yields

$$x(t) = \sin^2 t + 2\sin t + 1. \tag{2.1-1}$$

Finding $x(-t)$ puts the negative sign in front of the odd functions, while not affecting even functions:

$$x(-t) = \sin^2(-t) + 2\sin(-t) + 1 = \sin^2 t - 2\sin t + 1. \tag{2.1-2}$$

Now it remains to cancel the odd parts by adding (2.1-1) and (2.1-2) and the even parts by subtracting (2.1-2) from (2.1-1):

$$x_{ev}(t) = \frac{1}{2}\left[x(t) + x(-t)\right] = \sin^2 t + 1 \text{ and } x_{od}(t) = \frac{1}{2}\left[x(t) - x(-t)\right] = 2\sin t.$$

Periodicity: A signal $x(t)$ is called ***periodic*** if and only if there exists number T, called ***period***, such that $x(t) = x(t + T)$ for all t. Otherwise, signal is called ***nonperiodic***. The examples of periodic signals, referred to as ***pure sinusoids***, are given in Figure 2.1-9, with period $T = 2\pi$ for both $cos(t)$ and $sin(t)$. Periodic signals form the backbone of most of the techniques considered in this book. A detailed discussion of the basic properties of periodic signals is given in Section 2.4, Chapter 4.

Frequency composition: A T-periodic signal $x(t)$ is called ***single-frequency*** if it can be represented by a *single pure sinusoid*; this signal is called ***multifrequency*** if it can be represented only by a *sum of several pure sinusoidal, or single-frequency, signals.* The concept of frequency is considered in detail in Subsection 2.4.1.

Example 2.1-2: Production unit piston position 4.4 Hz reference signal looks like pure sinusoid, but is multifrequency — it has additional pure sinusoids at 8.8 Hz, and 13.2 Hz. The latter arise due to the distortion of the pure sinusoidal 4.4 Hz reference (command) signal by the hydraulic system. Thus, *a hydraulic system acts as a nonlinear signal processor — a multifrequency generator of sinusoidal signals with frequencies which are the integer multiples of the single driving frequency.* It takes in a pure sinusoidal signal with period T and generates a signal that is a sum of several pure sinusoids with periods T/n, n — integer.

Scale composition: We will call a signal ***single-resolution*** or ***single-scale*** if the magnitudes of all the pure sinusoidal components comprising it can be visualized on the same graph. If visualization requires change of resolution, like, for example, zooming in, we will call the signal ***multiresolution***, or ***multiscale***.

Example 2.1-3: Example of a multiscale signal is shown in the upper two graphs in Figure 1.5-6 of Chapter 1. In these graphs, low magnitude of pure sine at 9.2 Hz prevents it from being detectable both in time and frequency domains. The lower two graphs in Figure 1.5-6 show the hardware testbed mold position signal with a regular-scale sinusoid at 9.2 Hz — the result of scaling up a small-scale sinusoid by the beam resonance. *This scale-up converts the multiscale almost undistorted sinusoidal piston position signal into a strongly distorted single-scale mold position signal*, i.e. with all components clearly visible in both domains.

Example 2.1-4: Figure 2.1-10 shows the aggregated hydraulic-actuator-driven mold magnitude response to a pure sine of *fixed frequency X Hz* equally incremented with each test number increase.

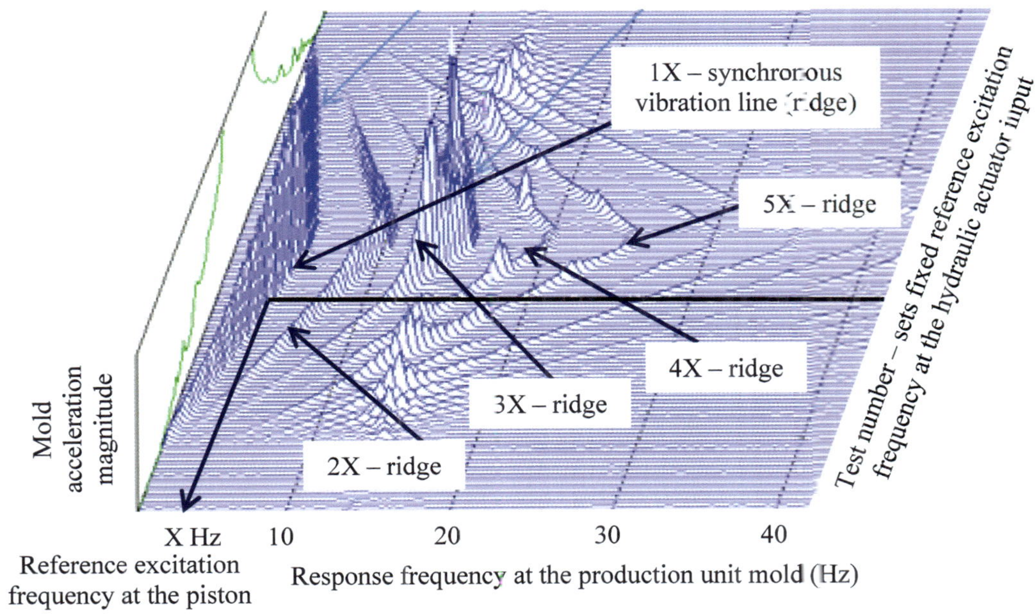

Fig. 2.1-10. Waterfall plot — multifrequency single-scale graph. Piston position small magnitude scale-up is due to i) beam resonance amplification and ii) multiplication of position by frequency square factor to obtain acceleration.

The reference frequency value of each test can be read from the frequency axis using the *1X, or synchronous vibration line*. Visible ridges corresponding to multiples of reference frequency in the multifrequency single-scale mold acceleration indicate nonlinear distortion of the reference signal.

Randomness: Up until now, most signals we have considered have been ***deterministic***, or known for all time, like the mass-spring-damper system position and velocity. However, many processes, like turbulent flow, do not lend themselves to a concise description through some set of equations, and their motion is then described in terms of statistical averages, i.e. is treated, to some degree, as random, or ***stochastic***. Another example is measurement noise in electrical circuits, which is usually modeled as normally-distributed random signal (process). Most real signals consist of deterministic component corrupted by stochastic measurement noise or disturbances, resulting in mixed deterministic/stochastic signal shown in Figure 2.1-13.

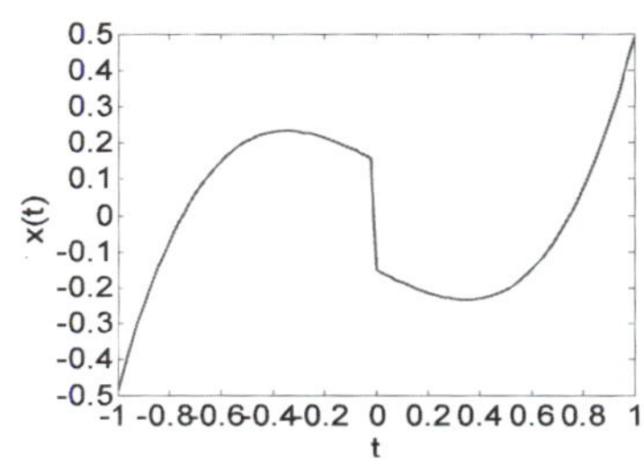

Fig. 2.1-11. Deterministic signal.

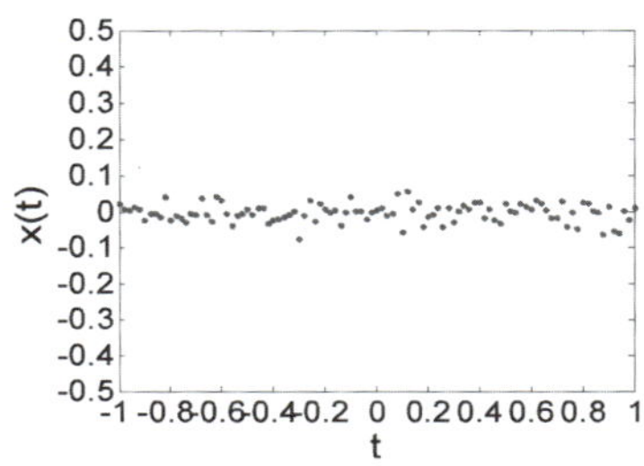

Fig. 2.1-12. Stochastic signal.

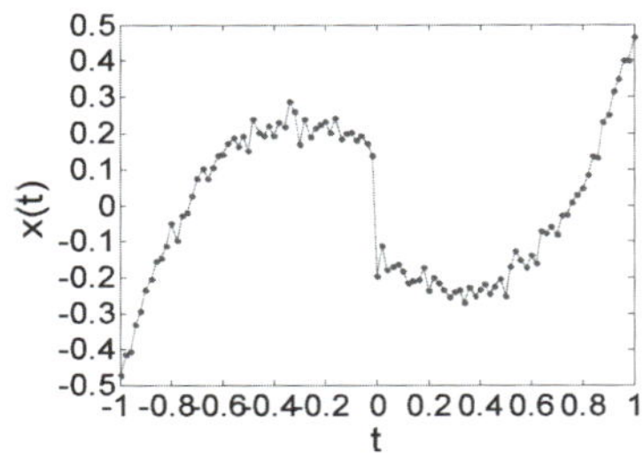

Fig. 2.1-13. Mixed signal.

Stationarity: Systems can be divided into time invariant and time varying ones. For signals, however, such division is meaningless, since this would then separate signals into constants and all others. Instead, signals are divided into stationary and non-stationary ones. A ***stationary signal*** is allowed to be time varying. However, time variation in this signal is assumed to be restricted to such an extent that its statistical (e.g., mean and standard deviation) and/or deterministic (e.g., magnitude and frequency) parameters stay constant with respect to independent variable (time and/or space). Further, ***ergodicity assumption*** is often made: ensemble average and standard deviation of a stochastic signal can be replaced by time average and standard deviation of its single realization (one sufficiently long data record). Such a stationary stochastic signal is shown in Figure 2.1-12. Properly operating rotating machinery produces stationary vibration signals that are close to deterministic ones. Rap music is repetitive enough to be considered stationary stochastic signal.

A ***non-stationary signal*** is the one whose statistical characteristics in stochastic case and basic parameters in deterministic case change over time; e.g. classical music is non-stationary deterministic signal (discounting performance variations), but a jazz improvisation is a non-stationary stochastic one. Non-stationary deterministic and stochastic signals are shown in Figures 2.1-11 and 2.1-13, respectively. To avoid confusion, we will consider signals with decay, even at a constant rate like $\exp(-t)$, to be non-stationary signals.

The concept of stationarity is indispensable in dealing with engineering systems. An operating steady state for these systems typically is not a constant, but rather a stationary regime, like for a mold oscillator, which operates in a repetitive mode at a particular driving frequency and magnitude during a single casting run. Basic random signals background is given in **Appendix C**.

Causality: If a system depends only on what has happened in the past, it is called ***causal***. Analogously, signals that are 0 for $t < 0$, are often called causal. Similarly, signals that are 0 for $t > 0$ are called ***anti-causal***. Signals that are non-zero for at least some $t < 0$ are called ***noncausal***.

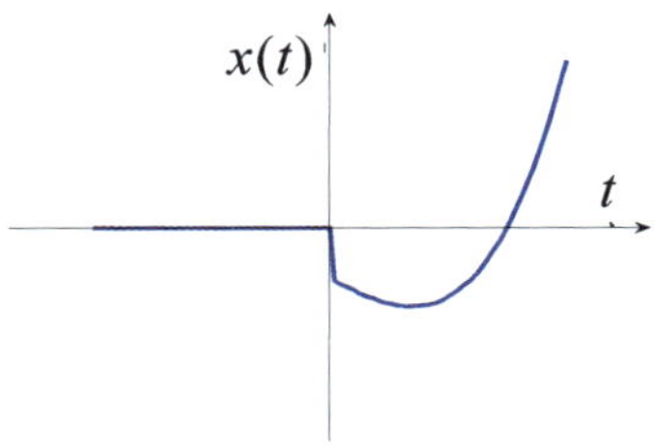

Fig. 2.1-14. Causal signal.

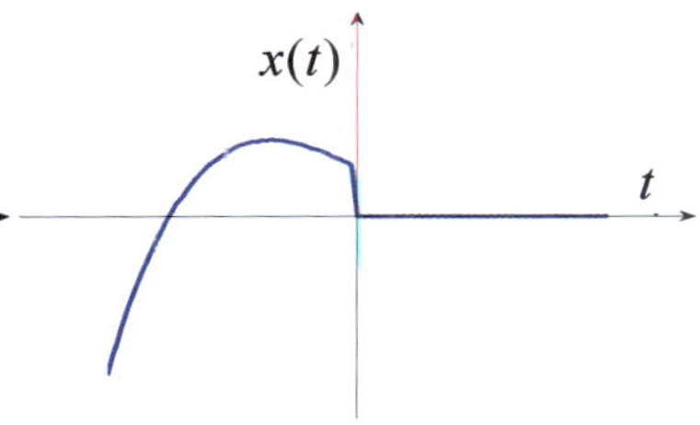

Fig. 2.1-15. Anti-causal signal.

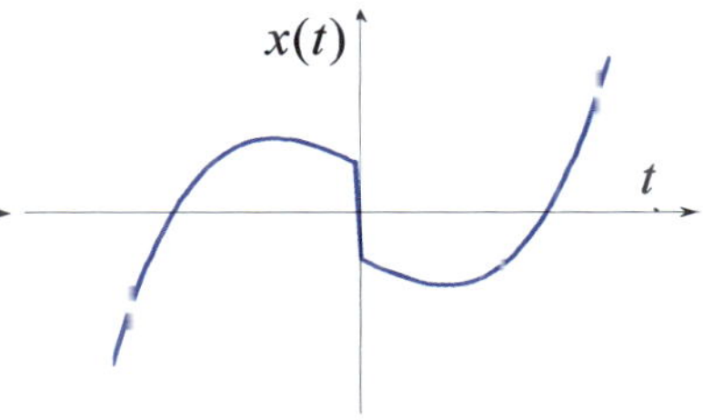

Fig. 2.1-16. Noncausal signal.

Noncausal stationary signal — key mathematical tool: For the majority of this course, we will deal with ***noncausal, periodic signals***. These are ***ideal*** stationary signals, which obviously cannot exist in reality, but which underpin the key mathematical tools for analyzing system behavior.

For example, sinusoid, which by definition is a noncausal signal, represents stationary (identically repeating) signal that, when used as an input to causal system, produces stationary (periodic, i.e. noncausal) output, as shown in Figure 2.1-17.

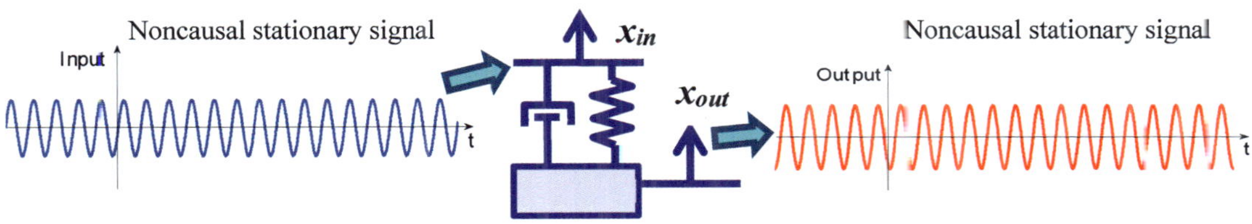

Fig. 2.1-17. Noncausal stationary (sinusoidal) output signal induced by noncausal stationary (sinusoidal) input signal.

Use of causal nonstationary signals to represent noncausal stationary signals. The causal version of a periodic input, when applied to a causal stable linear time-invariant system, will give rise to the output with an initial transient that comes very close to the noncausal (stationary) behavior after some decay time.

Theoretically, the transient completely settles down to the noncausal (stationary) behavior only at $t = \infty$. For real physical systems, however, the transient will usually die out very quickly and the causal nonstationary output, whether deterministic or stochastic, will practically settle down after transient to the (theoretical) noncausal stationary behavior, as shown in Figure 2.1-18, permitting the use of noncausal stationary deterministic and stochastic analytical tools.

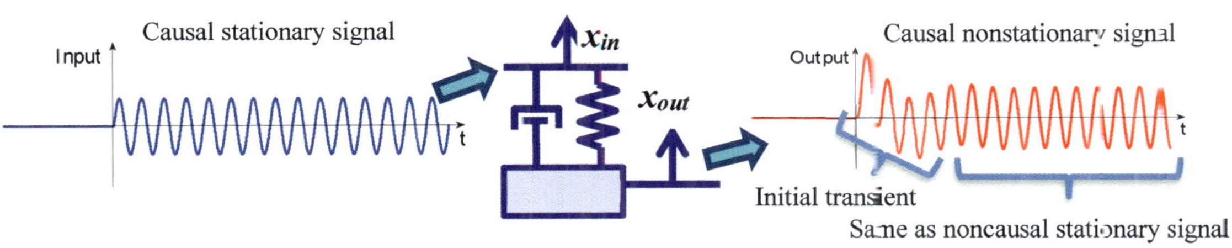

Fig. 2.1-18. Causal nonstationary output signal induced by causal sinusoidal input signal.

Regularity: ***Regular signals*** are those which characterize nice and smooth behavior, say, a light sea breeze. These are usually smooth continuously differentiable functions. ***Singular signals*** are those which characterize behavior with abrupt changes, say meteorite impact. These are usually functions with jumps, discontinuities, zero width and infinite magnitude, not differentiable. In discrete time this distinction blurs, since infinite derivatives correspond to finite differences, therefore we will refer to the discrete time versions of the signals above as the discrete time counterparts.

Basic singular functions and one regular function — sinc function and their discrete-time counterparts are presented in Figures 2.1-19a, b, and 2.1-20, respectively. Integrals and sums of impulse function — step function and ramp function — are given, respectively, in Figures 2.1-21a and 2.1-21b. Singular signals pulse and impulse and interrelations between them are discussed in detail in Section 2.3. Both CT and DT signals presented are extensively used throughout the book.

$$rect(t)=\begin{cases}1, -0.5\le t\le 0.5,\\ 0, \text{ elsewhere.}\end{cases}\qquad tri(t)=\begin{cases}1-|t|, -1\le t\le 1,\\ 0, \text{ elsewhere.}\end{cases}\qquad sinc(t)=\frac{sin(\pi t)}{\pi t}$$

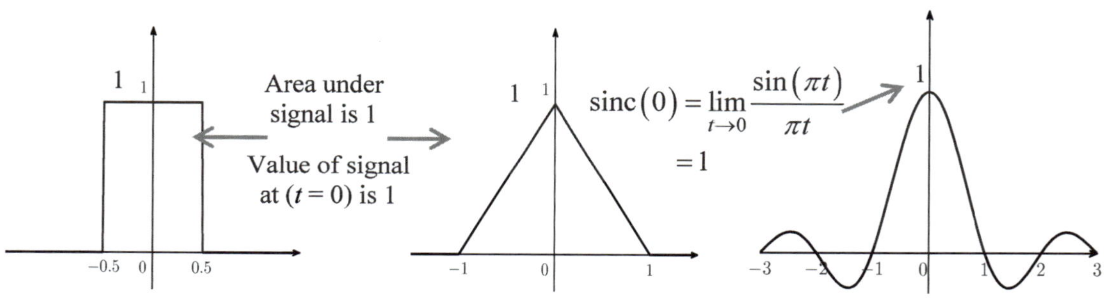

Fig. 2.1-19a. Basic continuous time signals: singular signals rect and tri, and regular signal sinc.

$$\text{pulse: } p_{\Delta}(t)=\begin{cases}\frac{1}{\Delta}, -\frac{\Delta}{2}\le t\le \frac{\Delta}{2},\\ 0, \text{ elsewhere.}\end{cases}\qquad \text{impulse: } \delta(t)\qquad \text{comb}(t)=\sum_{n=-\infty}^{\infty}\delta(t-nt_s)$$

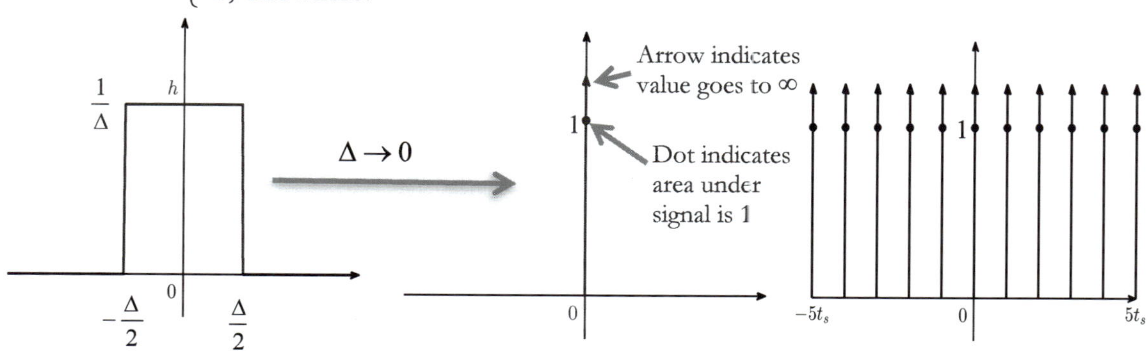

Fig. 2.1-19b. Basic continuous time singular signals: pulse, impulse, and comb.

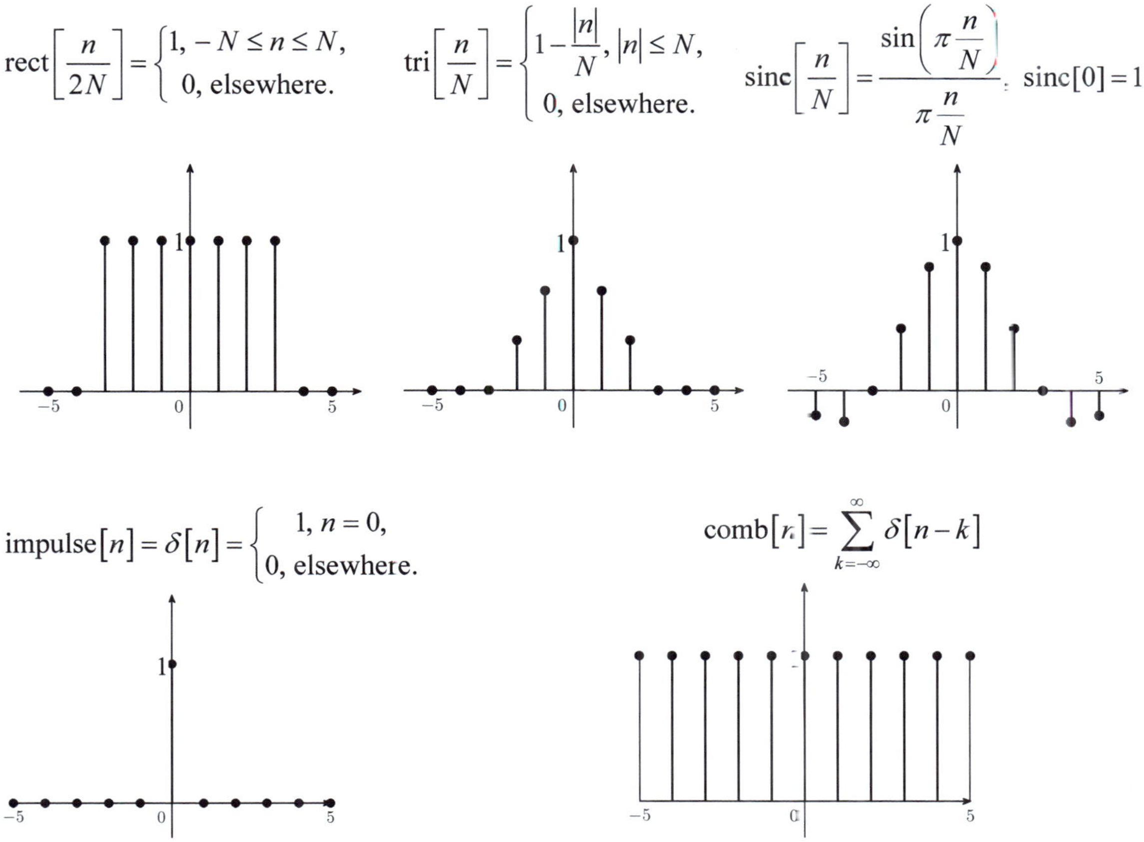

Fig. 2.1-20. Basic singular and regular signal counterparts in discrete time.

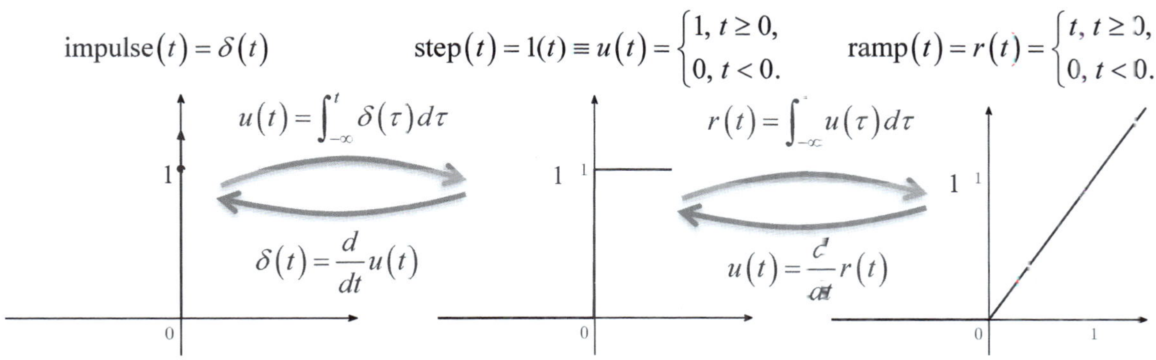

Fig. 2.1-21a. CT ramp and step functions — integrals of impulse function.

$$\text{impulse}[n]=\delta[n]=\begin{cases}1,\ n=0,\\0,\ \text{elsewhere.}\end{cases}\qquad \text{step}[n]=u[n]=\begin{cases}1,\ n\geq 0,\\0,\ n<0.\end{cases}\qquad \text{ramp}[n]=r[n]=\begin{cases}n,\ n\geq 0,\\0,\ n<0.\end{cases}$$

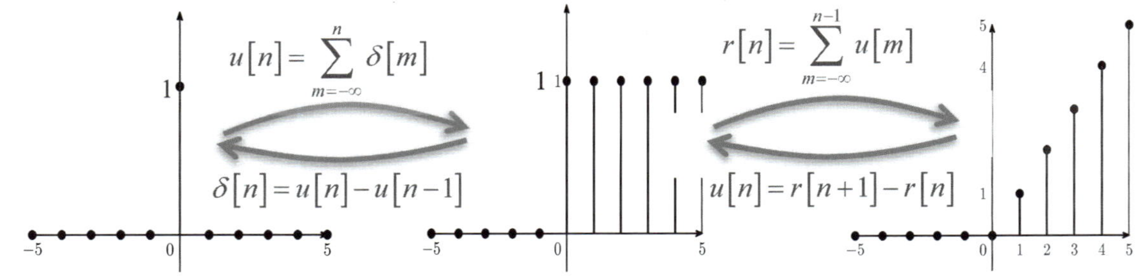

Fig. 2.1-21b. DT ramp and step functions — sums of impulse function.

Function support: If a function takes nonzero values only on a limited range of the values of its argument (on closed and bounded, or compact, set), it is said to have a ***compact support***. Example: *rect*(t). Similarly, a function that takes nonzero values anywhere on $(-\infty, \infty)$ is said to have a ***noncompact support***. Example: *sinc*(t). In this course we are going to use the functions with noncompact support like *cos*(ωt) to analyze the finite data records which are the functions with compact support. This will lead to inconsistencies and numerical errors that we will address.

2.1.2 *Energy-type signal size measures: integrability and summability*

These characterizations are widely employed as conditions for system stability and transform existence. The term "there exists" in the definitions below is used in a sense "is finite".

A continuous-time signal $x(t)$ is, respectively, referred to as ***absolutely integrable*** and ***square-integrable*** if there, respectively, exist $\int_{-\infty}^{\infty}|x(t)|\,dt$ and $\int_{-\infty}^{\infty}|x^2(t)|\,dt$.

A discrete-time signal $x[n]$ is referred to as ***absolutely summable*** and ***square-summable*** if there, respectively, exists $\sum_{k=-\infty}^{\infty}|x[n]|$ and $\sum_{k=-\infty}^{\infty}|x^2[n]|$.

Example 2.1-5: Determine if the following signal is absolutely and square integrable:

$$x(t)=\begin{cases}\dfrac{1}{t},\ t\geq 1,\\0,\ \text{elsewhere.}\end{cases}\tag{2.1-3}$$

Solution: By definition of integrability

$$\int_{-\infty}^{\infty}|x(t)|\,dt=\int_{-\infty}^{\infty}\frac{1}{t}dt=\lim_{T\to\infty}\int_{1}^{T}\frac{1}{t}dt=\lim_{T\to\infty}\left[\ln(t)\right]_1^T=\lim_{T\to\infty}\left(\ln(T)-\ln(1)\right)=\lim_{T\to\infty}\ln(T)$$

does not exist. Therefore, $x(t)$ is not absolutely integrable. However,

$$\int_{-\infty}^{\infty} \left|x^2(t)\right| dt = \int_{-\infty}^{\infty} \frac{1}{t^2} dt = \lim_{T\to\infty} \int_{1}^{T} \frac{1}{t^2} dt = \lim_{T\to\infty} \left[-\frac{1}{t}\right]_1^T = \lim_{T\to\infty} \left(1-\frac{1}{T}\right) = 1.$$

Therefore, $x(t)$ is square-integrable!

Example 2.1-6: Determine if the following signal is absolutely and square summable:

$$x[n] = \begin{cases} \alpha^n, & n \ge 0, \\ 0, & \text{elsewhere.} \end{cases} \tag{2.1-4}$$

Solution: By definition of summability

$$\sum_{n=-\infty}^{\infty} \left|x[n]\right| = \sum_{n=0}^{\infty} |\alpha|^n = \begin{cases} \dfrac{1}{1-|\alpha|}, & |\alpha| < 1, \\ \text{does not exist otherwise.} \end{cases}$$

Therefore, $x[n]$ is absolutely summable if and only if $|\alpha| < 1$. The same condition applies for square-summability, since $\left|\alpha^2\right| < 1$ if and only if $|\alpha| < 1$.

2.2 Basic Operations on Signals, Signal Representation

This section reviews the operations on signals that will provide much of the basis for the signal processing concepts and techniques introduced in subsequent chapters.

2.2.1 *Signal representation*

Real function and graphs: A real function is a rule, typically denoted as a static (no derivatives or integrals) relation $f(\cdot)$, which uniquely matches a real number, say t, with another real number $f(t)$. It can be specified analytically, as a table, or as a graph that for each value on the horizontal axis uniquely assigns the corresponding value on the vertical axis. Often symbol $f(t)$ denotes both a rule (a function, a graph) itself and a value of this function at t. Given two graphs, $f(\alpha)$ and $x(t)$, a graph $y(t)=f[x(t)]$ can be constructed through "point-by-point" matching of t to $x(t)$ and of the latter to $y(t)$, or if $x(t)$ is simple enough, by some rules. When $x(t) = -k(t-t_0)$, k, t_0 – *const.*, the rules take a simple and convenient form of ***folding, scaling, and shifting,*** as described next.

The relation between x and t given by this representation is called ***affine,*** or ***incrementally linear***, or ***linear plus offset***: $x(t) = -k(t-t_0) = -kt + kt_0 \equiv -kt + const.$ Formally, it does not pass any linearity test: indeed, for example, for $t \to ct$ we have $x(t) = -kt + kt_0 \to c(-kt) + kt_0 \ne cx(t)$ $= c(-kt + kt_0) = c(-kt) + ckt_0$, but its nonlinearity is trivial — arising only due to an offset given in this case by a constant t_0. Removing the offset by setting $t_0 = 0$ reveals that the remaining part, $x(t) = -kt$, has linear relation.

2.2.2 Basic operations on signals

2.2.2.1 *Shifting of signals*

Shifting is one of the fundamental operations on signals. It will be later linked to key signal concepts — ***time delay***, ***phase***, and ***signal approximation***, including ***multiresolution*** (wavelet) signal approximations to build the state-of-the-art ***deep learning networks***. Shifting will be also linked to main system characteristics, such as ***time invariance***, also referred to as ***shift invariance***, and ***phase response***, and general system/signal operations, such as ***convolution***, used in ***filtering*** and ***convolutional neural networks (CNNs)*** — a ***machine learning*** tool introduced in Chapter 10.

Example 2.2-1: Construct a graph of a function given by $y(t) \triangleq u(t-t_0)$ where $u(t)$ is a unit step function shown in Figure 2.1-21a. The sign "$\triangleq$" means "defined as" or "a new name for".

Solution: By definition $u(t)=\begin{cases}0, t<0,\\ 1, t\geq 0.\end{cases}$ Therefore, the desired $y(t)=u(t-t_0)$ is obtained simply by replacing t in the definition above by $(t-t_0)$ (often denoted as $t \to t-t_0$, i.e., "t becomes $t-t_0$"):

$$u(t-t_0)=\begin{cases}0, t-t_0<0,\\ 1, t-t_0\geq 0.\end{cases} \Rightarrow u(t-t_0)=\begin{cases}0, t<t_0,\\ 1, t\geq t_0.\end{cases}$$

Expression on the right shows that $y(t)=u(t-t_0)$ shifts the graph of $u(t)$ to the right along t-axis by t_0:

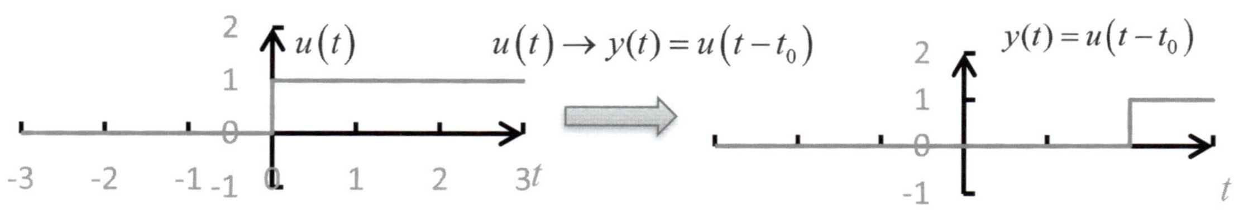

Fig. 2.2-1. Shifted unit step signal.

Looking at the graphs in Figure 2.2-1, we can develop a shifting procedure: thinking of $u(t)$ as a piece of wire bent at a specific location called "zero place", in order to see what is $u(t-t_0)$, we just locate the number t_0 on the horizontal coordinate axis and drag the whole wire along the horizontal axis until this "zero place" reaches t_0. Analogously, thinking of any function $f(t)$, say the one shown in Figure 2.2-2 (left graph) as just another piece of wire, if we want to find $f(t-t_0)$ we first mark the "zero place" $t_0=0$ on the "wire $f(t)$" and then drag this wire along horizontal axis until this "zero place" reaches the point t_0. Shifting of $u(t)$ along with its magnitude scaling will be used for approximating any function by sums of terms of the form $A_i \cdot u(t-t_i)$ as closely as desired.

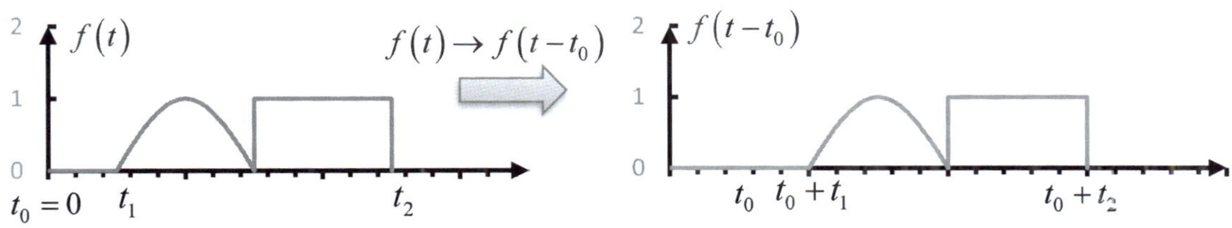

Fig. 2.2-2. Shifted arbitrary signal $f(t\text{-}t_0)$.

2.2.2.2 *Time-scaling of signals*

Signals could be generated at one speed and reproduced at the other. For example, high quality vinyl records could be cut from master tape at half playback speed ("half-speed mastering") to precisely imprint the signal spikiness into the grooves through detailed shapes. When they are then played at normal speed, the sound has a "bite" to it, i.e. crisply reproduced voice and instruments. Time-scaling is also a key operation in building multiresolution (wavelet) networks for deep learning introduced in Chapter 10.

Example 2.2-2: Construct a graph of a function given by $rect(t/\Delta)$, $\Delta > 0$, where $rect(t)$ is a function given in Figure 2.1-19a.

Solution: By definition

$$rect(t) = \begin{cases} 1, & -0.5 \le t \le 0.5, \\ 0, & \text{elsewhere.} \end{cases}$$

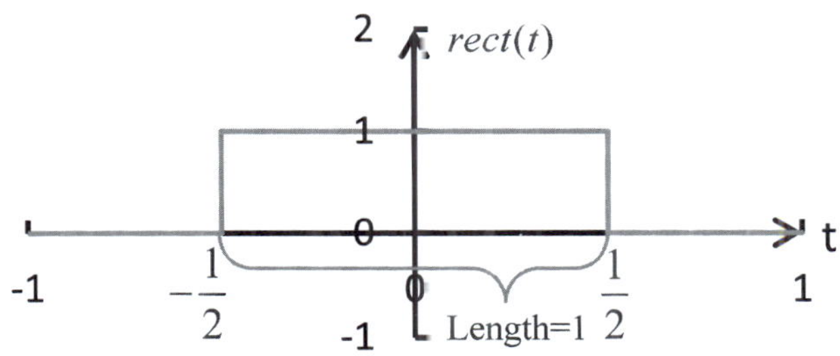

Replacing t by t/Δ: $t \to t/\Delta,\ \Delta > 0$, we obtain

$$rect(t) = \begin{cases} 1, & -0.5 \le t \le 0.5, \\ 0, & \text{elsewhere,} \end{cases} \quad \overset{t \to t/\Delta}{\Rightarrow} \quad rect(t/\Delta) = \begin{cases} 1, & -\frac{1}{2} \le \frac{t}{\Delta} \le \frac{1}{2}, \\ 0, & \text{elsewhere.} \end{cases}$$

Multiplying each term in the inequality on the right by Δ yields $x(t) = rect(t/\Delta)$ as follows:

$$rect(t/\Delta) = \begin{cases} 1, & -\frac{\Delta}{2} \le t \le \frac{\Delta}{2}, \\ 0, & \text{elsewhere.} \end{cases}$$

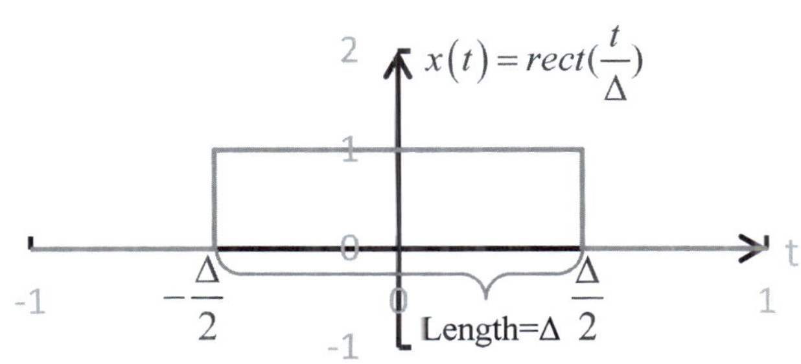

Fig. 2.2-4. Time-rescaled rect signal.

Comparing Figures 2.2-3 and 2.2-4, it is seen that the time-scaling factor Δ sets the graph stretch. Since the scaling factor Δ appears in the example above as the argument divisor, we will refer to it as the ***inverse scaling factor***. The time scaling can be also expressed through the ***direct scaling factor*** $k \triangleq 1/\Delta$ — both forms will come very handy, especially in discussing periodic functions. From Figure 2.2-3 it is seen that if $0 < k < 1$, i.e. $1/k = \Delta > 1$, then rect(kt) "stretches" away from zero by a factor $\Delta = 1/k > 1$. If, however, $k > 1$, i.e. $0 < \Delta < 1$, then rect(kt) "shrinks" towards zero by a factor $\Delta = 1/k > 1$. This scaling of an independent variable should not be confused with the scaling of a function magnitude, which would be $k \bullet \text{rect}(t)$.

Thus, the same signal shape can be reproduced at different time scales through the change of scaling coefficient, k:

$$\text{continuous time: } x(t) \rightarrow x(kt); \quad \text{discrete time: } x[n] \rightarrow x[kn].$$

When $k > 1$, the scaling produces signal ***compression***, or ***speedup***. When $k < 1$, the scaling results in ***stretching***, or ***slowdown***. There is however a significant difference between continuous time and discrete time versions of these operations that is shown next.

Compression: Looking at the signals in Figure 2.2-5, nothing seems to be lost in $x_1(t)$ in comparison to $x(t)$. This, however is not the case in comparing $x_1[n]$ and $x[n]$ in Figure 2.2-6:

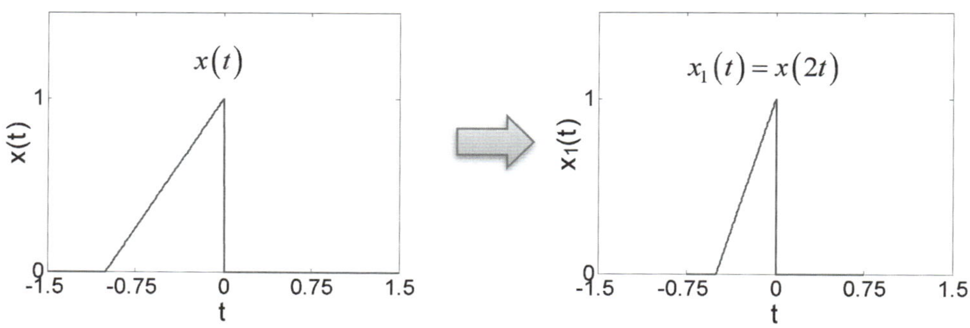

Fig. 2.2-5. Compression of a continuous-time signal.

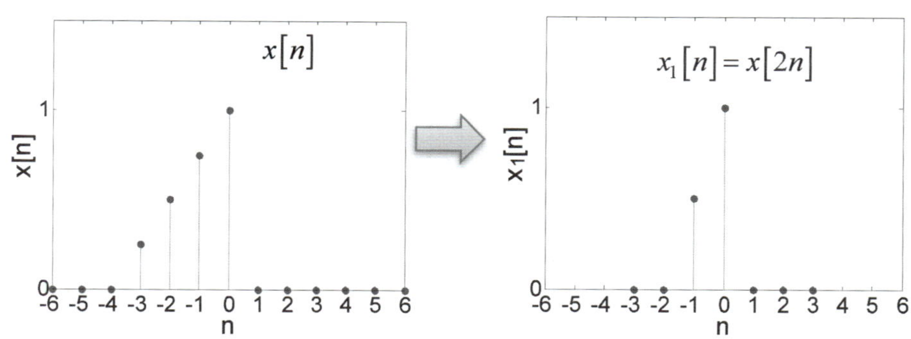

$$x_1[-1] = x[2(-1)] = x[-2],$$
$$x_1[-2] = x[-4].$$

Fig. 2.2-6. Compression of a discrete-time signal.

Indeed, $x_1[n]$ selects every other sample out of $x[n]$, i.e. discards roughly half of the information contained in $x[n]$. Thus, a discrete-time signal compression results in significant loss of information embedded into signal shape. This loss of information is called ***decimation***.

Stretching:

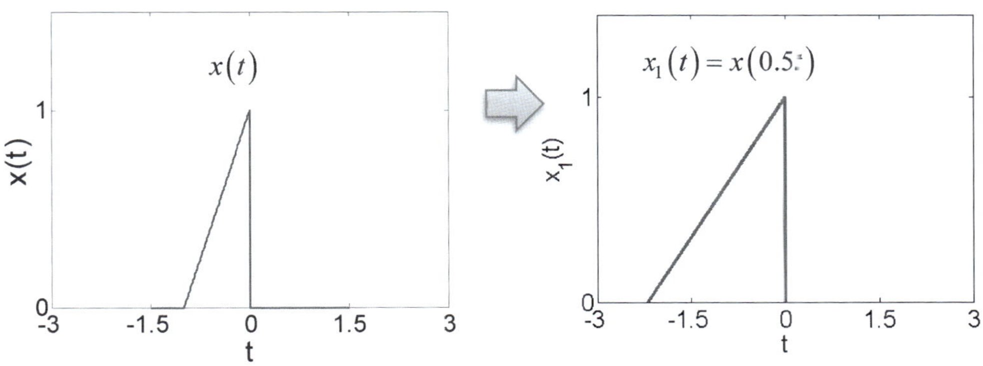

Fig. 2.2-7. Stretching of a continuous-time signal.

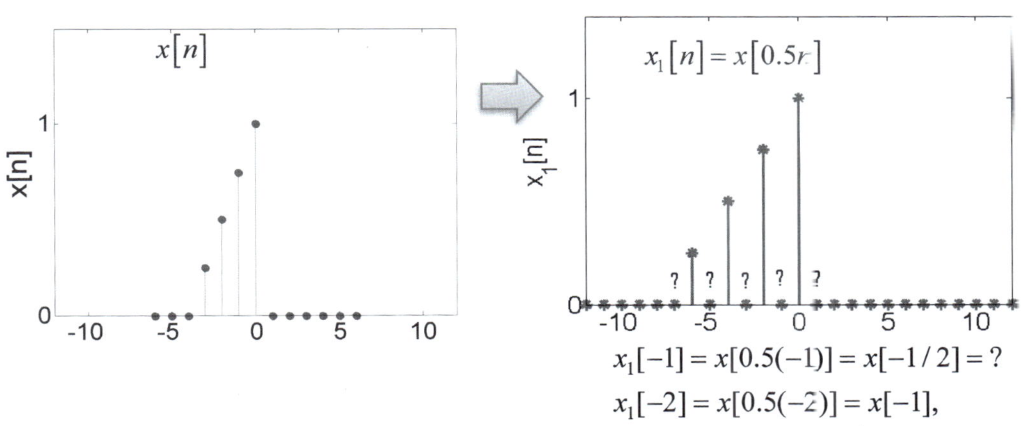

Fig. 2.2-8. Stretching of a discrete-time signal requires filling out the gaps with the "made-up" values.

For $x_1[n]$ we must assign values to the points between the samples of the original signal. This value assignment is called ***interpolation***. The basic interpolation types are:

- ***Zero*** interpolation: use 0 for every undetermined value.
- ***Step*** interpolation: hold the previous value.
- ***Linear*** interpolation: use average of two adjacent values.

Example 2.2-3: Consider $x[n] = \{0.25,\ 0.5,\ 0.75,\ \underset{\downarrow}{1}\}$, where arrow designates the origin. Sketch $y[n] = x[n/2]$ with zero interpolation, step interpolation, and linear interpolation.

Solution: For $y[n]$, we obtain $y[-6] = x[-3] = 0.25$, $y[-4] = x[-2] = 0.5$, $y[-2] = x[-1] = 0.75$, and $y[0] = x[0] = 1$. The values of $y[-7]$, $y[-5]$, $y[-3]$, $y[-1]$, $y[1]$ are undetermined. The sketch of $x[n]$ with the corresponding graphs of $y[n]$ that follow three interpolation rules indicated above are shown in Figure 2.2-9:

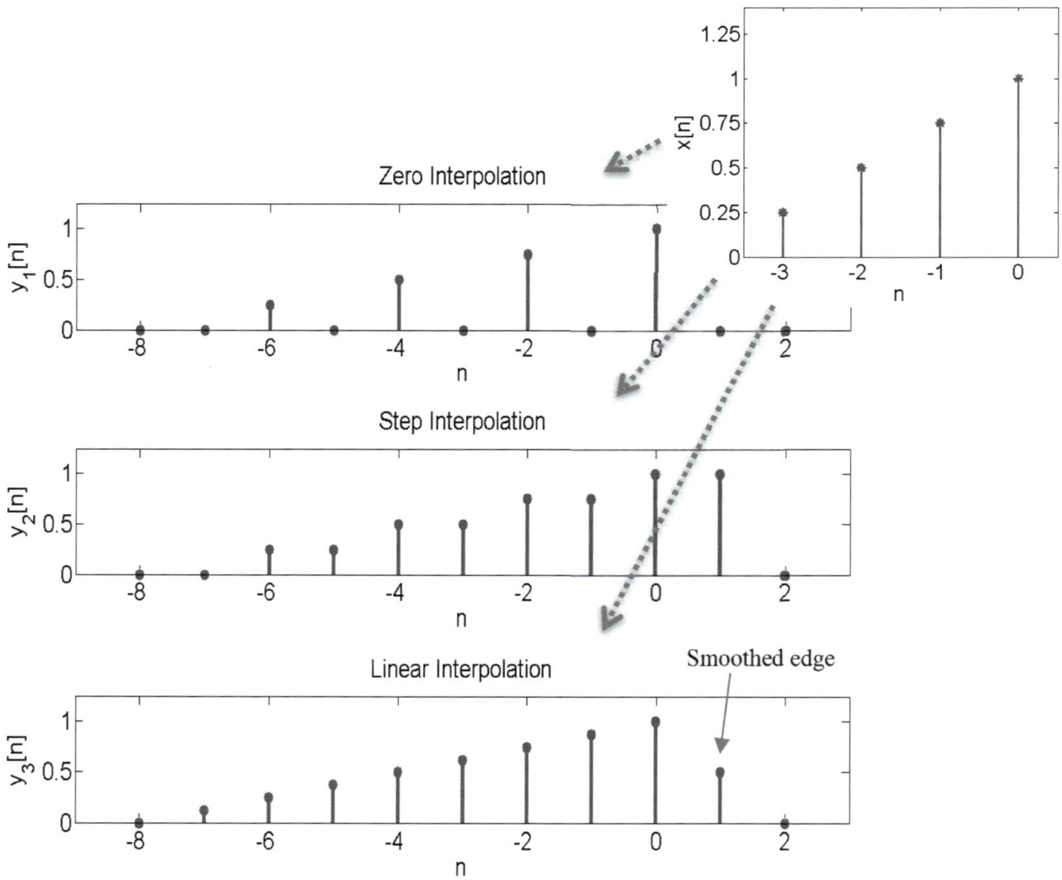

Fig. 2.2-9. $x[n]$ and zero interpolation, step interpolation, and linear interpolation of $y[n]$.

Thus, linear interpolation tends to smooth edges. In audio applications of signal processing, this loss of information and subsequent replacement of the latter with made-up (interpolated) values in analog playback leads, as shown in Figure 2.2-10, to the loss of reproduction fidelity — spikes are smoothed out and "visceral" impact of most instruments is lost.

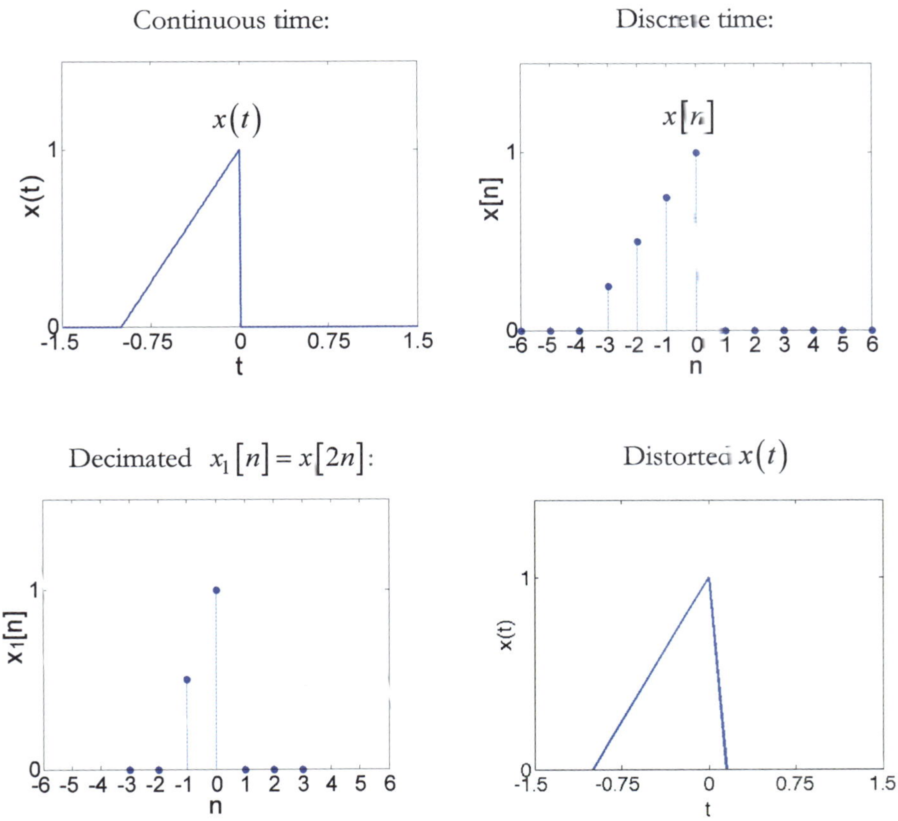

Fig. 2.2-10. Illustration of decimation and the resulting playback smoothing.

Lossy audio compression algorithms, such as mp3, permit trading of compact music data storage against the sound reproduction fidelity and, hence, the emotional richness of the musical piece, with the results pictured in Figure 2.2-11.

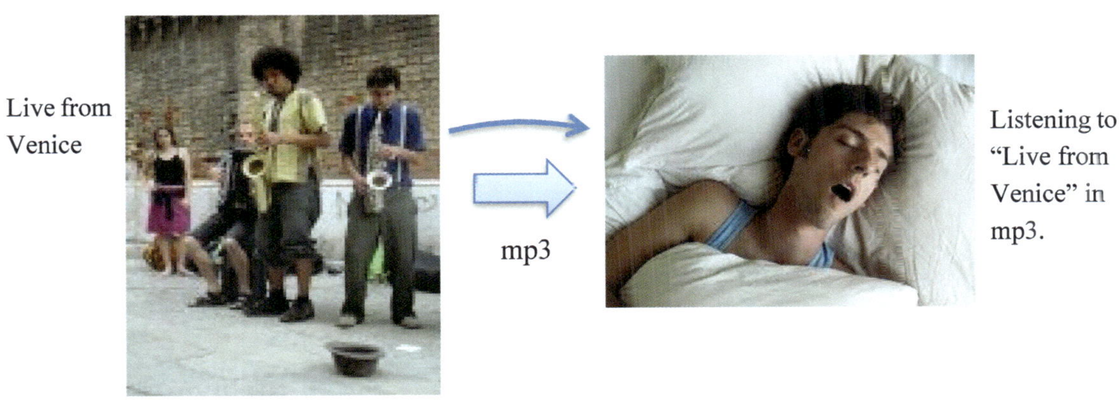

Fig. 2.2-11. Musical impact of recorded live performance played in compressed digital format: gets decimated.

The 1960's saw the peak of fidelity in the mass-marketed musical reproduction in the form of vinyl records and reel-to-reel tapes, after which there was a severe decline in audio reproduction quality. The storage/fidelity trade-off has now receded, permitting the earlier quality to reemerge in full force (after half-century!) in the form of high resolution digital formats to be discussed later.

2.2.2.3 *Folding of signals*

Signal folding — creation of a mirror image of a signal graph with respect to the vertical axis — takes place when function argument changes sign, i.e. under substitution $t \to -t$, as shown next.

Example 2.2-4: Given unit ramp function $r(t)$ defined in Figure 2.1-20, draw $r(-t)$.

Solution: Unit ramp function is defined as $r(t) = \int_{-\infty}^{t} u(\tau)d\tau = \begin{cases} t, t \geq 0, \\ 0, t < 0. \end{cases}$

It shows how area under $u(t)$ is changing as t runs from $-\infty$ to ∞. Argument change $t \to -t$ of $r(t)$ yields

$$r(t) = \begin{cases} t, t \geq 0, \\ 0, t < 0, \end{cases} \Rightarrow r(-t) = \begin{cases} -t, -t \geq 0, \\ 0, -t < 0, \end{cases} = \begin{cases} -t, 0 \geq t, \\ 0, 0 < t, \end{cases} = \begin{cases} -t, t \leq 0, \\ 0, t > 0. \end{cases}$$

For example, for $t = 0$, $r(-t) = -(-0) = 0$ and for $t = -1$, $r(-t) = -(-1) = 1$. The folding of graph of $r(t)$ to graph of $r(-t)$ is shown in Figure 2.2-12:

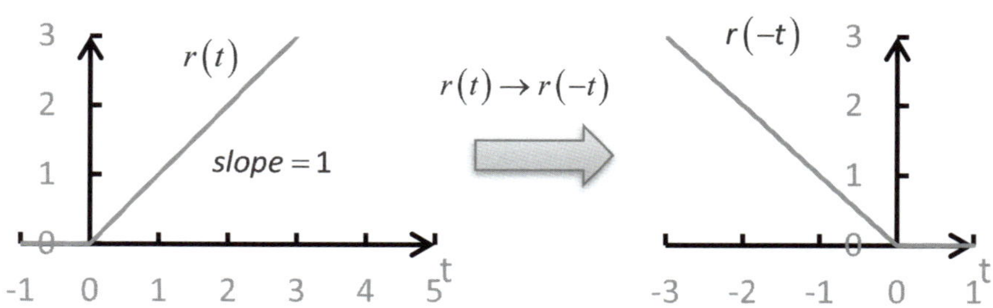

Fig. 2.2-12. Unit ramp signal and its folded version.

Similarly, given any $x(t)$, the graph of $x(-t)$ vs. t will be the graph of $x(t)$ turned around vertical axis.

2.2.2.4 *Combining basic operations on signals*

Given a graph $f(t)$, the sequence of operations for drawing a graph $y(t) = f(-k(t - t_0))$ is graphically shown in Figure 2.2-13 and analytically seen to be as follows:

i) first fold $f(t)$ and draw $y_1(t) = f(-t)$,

ii) then scale $y_1(t)$ and draw $y_2(t) = y_1(kt) = f(-kt)$, and finally

iii) shift $y_2(t)$ and draw $y_3(t) = y_2(t - t_0) = y_1[k(t - t_0)] = f[-k(t - t_0)]$,

which is the desired $y(t)$, as shown in Figure 2.2-13.

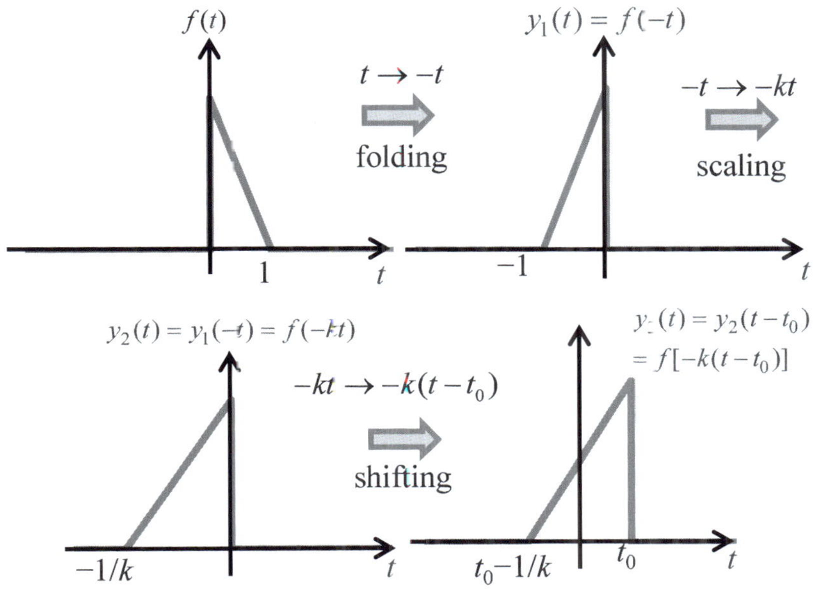

Fig. 2.2-13. Order of operations to draw $y(t)$ given $f(t)$.

<u>Keeping the right order is a must</u>: shifting before scaling would yield the time shift kt_0 instead of t_0; shifting before folding makes the time shifting go to the opposite direction.

The same operations order applies to DT signals, with the scaling step either accompanied by decimation or followed by interpolation.

Example 2.2-5: Given a signal $f(t)$ in Figure.2.2-14, sketch $y(t) = f[-0.5(t-2)]$.

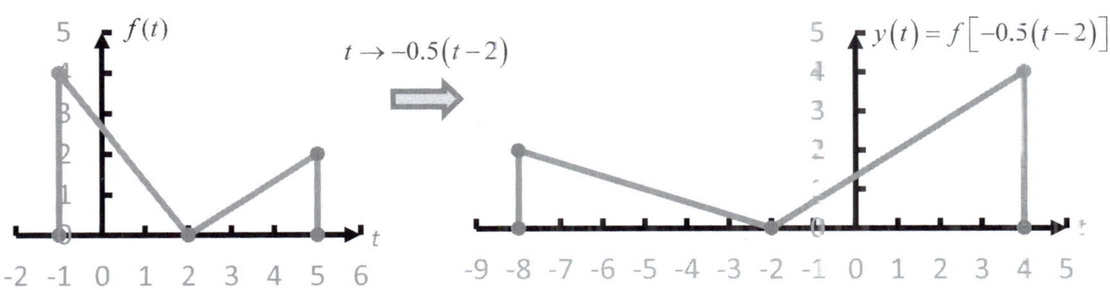

Fig. 2.2-14. Point-by-point mapping of $x(t)$ (left graph) into $y(t)$ (right graph).

Solution: Since the graph $f(\alpha)$ is characterized by only three points: (−1, 4), (2, 0), and (5, 2), let's recalculate the corresponding horizontal coordinates of $y(t)$ through point-by-point matching based on the map $\alpha = x(t) = -0.5(t-2)$. For this purpose we use the inverse of $\alpha = x(t)$, easily seen to be given by $t = x^{-1}(\alpha) = -2\alpha + 2$, as follows:

$$\begin{cases} \alpha_1 = -1 \Rightarrow t_1 = -2\cdot(-1)+2 = 4 \\ \alpha_2 = 2 \Rightarrow t_2 = -2\cdot 2+2 = -2 \\ \alpha_3 = 5 \Rightarrow t_3 = -2\cdot 5+2 = -8 \end{cases} \Rightarrow y: \begin{cases} (4,4) \\ (-2,0) \\ (-8,2) \end{cases}.$$

This yields the corresponding signal $y(t)$ given by the right graph in Figure 2.2-14.

Example 2.2-6: Given a signal $x(t)$ as shown in Figure. 2.2-15a, draw $y(t) = x(-2t-6)$.

Solution: For this example let's use the basic operations. First, we note that in order for shifting as presented earlier to be correct, the shift t_0 must appear with the "−" sign in front of it, i.e. we have to put $x(-2t-6)$ into the form $x[-k(t-t_0)]$. This yields:

$$y(t) = x(-2t-6) = x(-2(t+3)) = x[-2(t-(-3))]$$

where t_0 itself is seen to come out negative.

Now fold $x(t)$ and draw $y_1(t) = x(-t)$: as shown in Figure 2.2-15b.

Then scale $y_1(t)$ and draw $y_2(t) = y_1(2t) = x(-2t)$: as shown in Figure 2.2-15c.

And finally shift $y_2(t)$ and draw $y(t) = y_2(t-(-3)) = y_1[2(t-(-3))] = x[-2(t-(-3))]$ as shown in Figure 2.2-15d.

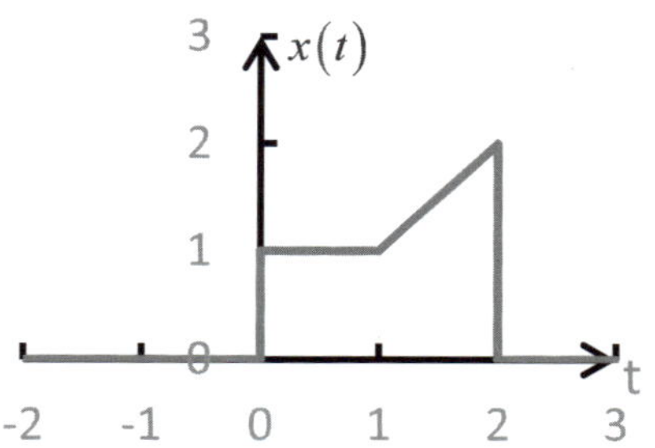

Fig. 2.2-15a. The original signal $x(t)$.

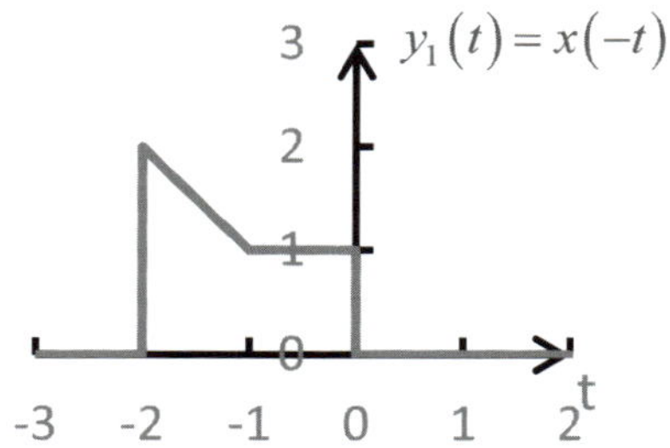

Fig. 2.2-15b. Signal $y_1(t)$ generated by folding signal $x(t)$.

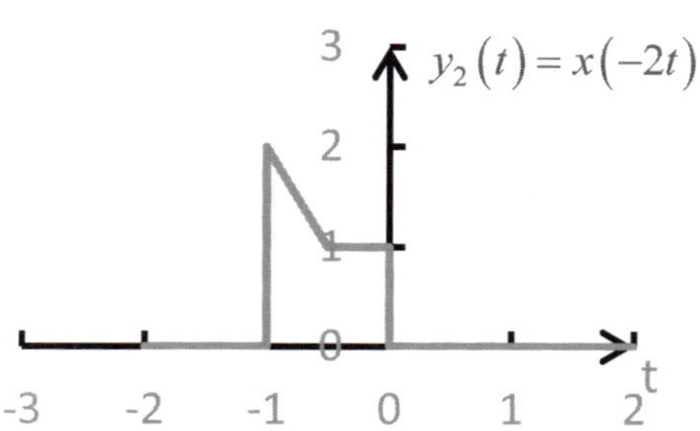

Fig. 2.2-15c. Signal $y_2(t)$ generated by shrinking signal $y_1(t)$.

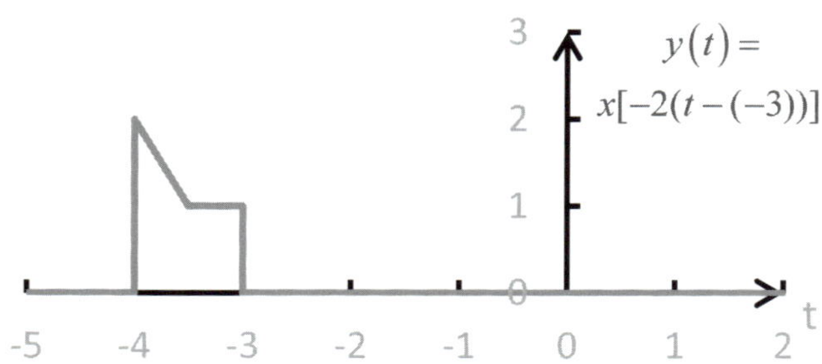

Fig. 2.2-15d. Signal $y(t)$ generated by shifting signal $y_2(t)$.

2.2.3 *Signal representation*

2.2.3.1 *Size preservation of CT signals after sampling*

Let us now sample a CT signal to obtain its DT counterpart and explore the link between them in terms of their corresponding size measures.

Example 2.2-7: Determine if

$$x(t) = \begin{cases} 4e^{t-3}, & t \le 3, \\ 0, & \text{elsewhere}, \end{cases} \tag{2.2-2}$$

is absolutely and square integrable.

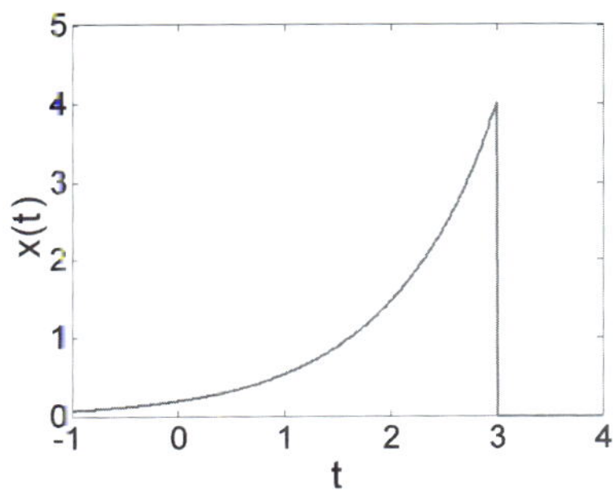

Fig. 2.2-16. Plot of $x(t)$ for Example 2.2-7.

Solution: $x(t)$ is seen to be folded, shifted, and magnitude-scaled version of $e^{-t}, t \ge 0$, plotted for clarity in Figure 2.2-16. By definition:

$$\int_{-\infty}^{\infty}|x(t)|\,dt = \int_{-\infty}^{3}\left|4e^{t-3}\right|dt + \int_{3}^{\infty}0\,dt = \lim_{T\to-\infty}\int_{T}^{3}\left|4e^{t-3}\right|dt \underset{s=t-3,\,t=3\Rightarrow s=0,\,t=T\Rightarrow s=T-3}{=}$$

$$4\lim_{T\to-\infty}\int_{T-3}^{0}e^{s}\,ds = 4\lim_{T\to-\infty}\left[e^{s}\right]_{T-3}^{0} = 4\lim_{T\to-\infty}\left(e^{0}-e^{T-3}\right) = 4\cdot 1 - 4\lim_{T\to-\infty}e^{T-3} = 4.$$

Therefore, $x(t)$ is absolutely integrable. Similarly, $x(t)$ is seen to be square integrable since

$$\int_{-\infty}^{\infty}|x(t)|^2\,dt = \int_{-\infty}^{3}\left|4e^{t-3}\right|^2 dt + \int_{3}^{\infty}0\,dt = \lim_{T\to-\infty}\int_{T}^{3}\left|4^2e^{2(t-3)}\right|dt = 16\lim_{T\to-\infty}\int_{T-3}^{0}e^{2s}\,ds$$

$$= 16\lim_{T\to-\infty}\frac{1}{2}\left[e^{2s}\right]_{T-3}^{0} = 8\lim_{T\to-\infty}\left(e^{2\cdot 0}-e^{2(T-3)}\right) = 8\cdot 1 - 8\lim_{T\to-\infty}e^{2(T-3)} = 8.$$

Example 2.2-8: Obtain DT version $x[n]$ of $x(t)$ in (2.2-2) and determine if it is absolutely and square summable.

Solution: First, sample $x(t)$ with sampling rate t_s to obtain sampled signal $x[n]=x(nt_s)$:

$$x[n]=x(nt_s)=\begin{cases}4e^{(nt_s-3)}, & nt_s\le 3\\ 0, & \text{elsewhere}\end{cases}=\begin{cases}4\left(e^{t_s}\right)^{n-\frac{3}{t_s}}, & n\le\frac{3}{t_s}\\ 0, & \text{elsewhere}\end{cases}=\begin{cases}4\left(e^{-t_s}\right)^{-n+\frac{3}{t_s}}, & n\le\frac{3}{t_s},\\ 0, & \text{elsewhere.}\end{cases} \quad (2.2\text{-}3)$$

Choosing $t_s = 0.36$, arbitrarily at this point, yields

$$x[n]=\begin{cases}4\left(e^{-0.36}\right)^{-n+\frac{3}{0.36}}, & n\le\frac{3}{0.36}\\ 0, & \text{elsewhere}\end{cases}\approx\begin{cases}4(0.698)^{-n+8.33}, & n\le 8.33\\ 0, & \text{elsewhere}\end{cases}$$
$$=\begin{cases}4(0.698)^{-n+8}, & n\le 8\\ 0, & \text{elsewhere}\end{cases}\approx\begin{cases}4\cdot 7^{-n+8}, & n\le 8,\\ 0, & \text{elsewhere.}\end{cases} \quad (2.2\text{-}4)$$

Signal $x[n]$ is seen to be folded, shifted, and magnitude-scaled version of $y[n]=0.7^n, n\ge 0$. Let's first plot $x[n]$ starting from $y[n]$ using a sequence of stretching, folding and shifting:

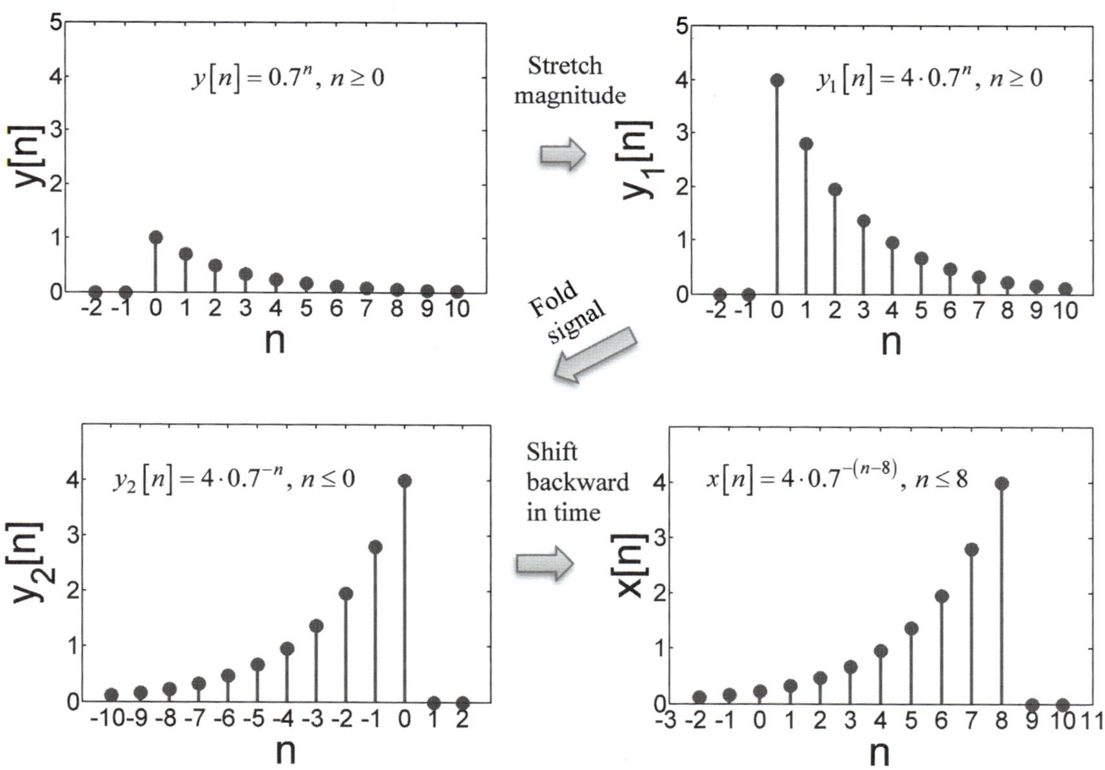

Fig. 2.2-17. Plotting $x[n]$ with a sequence of stretching, folding, and shifting.

To check absolute summability, by definition

$$\sum_{n=-\infty}^{\infty}|x[n]| = \sum_{n=-\infty}^{8}\left|4\cdot 0.7^{-n+8}\right| + \sum_{n=9}^{\infty}0 = \sum_{n=-\infty}^{8}|4|\cdot|0.7|^{-n+8} = 4\sum_{n=-\infty}^{8}0.7^{\overbrace{-n+8}^{m}}.$$

Change of variables $m=-n+8$ yields $n=-\infty \Rightarrow m=\infty$ and $n=8 \Rightarrow m=0$, so that

$$\sum_{n=-\infty}^{\infty}|x[n]| = 4\sum_{n=-\infty}^{8}0.7^{-n+8} = 4\sum_{m=\infty}^{0}0.7^{m} = 4\sum_{m=0}^{\infty}0.7^{m} = 4\frac{1}{1-0.7} \approx 13.333.$$

Thus, $x[n]$ is absolutely summable. Similarly to the previous example, it is also square summable. In general, as seen from Example 2.1-2, the sampled sequence (2.2-3) is absolutely and square-summable if and only if $|e^{-t_s}| = |1/e^{t_s}| < 1$, which is true for any $t_s > 0$.

2.2.3.2 *Assembling arbitrary signal from basic signals*

The experimentally generated data records can be approximated by choosing the appropriate basic signals, adjusting their parameters for example through folding, magnitude and argument scaling, and shifting, and assembling the adjusted signals into arbitrarily close data approximations.

Example 2.2-9: Create an analytical model of a signal $x(t)$ shown in Figure 2.2-18 below:

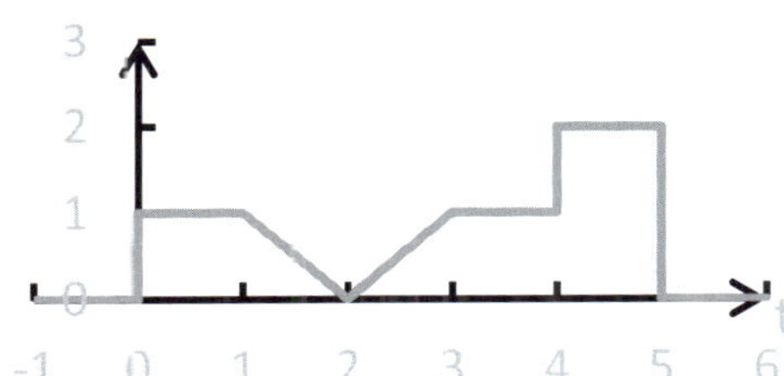

Fig. 2.2-18. Graphically specified signal $x(t)$.

Solution: An analytical signal model is clearly nonunique. Let us choose $r(t)$ and $u(t)$ as the basic functions and use their summation along with shifting and magnitude scaling to develop signal model.

First, we recall that to add two function $x(t)$ and $y(t)$ we fix $t=t_1$ and add $x(t_1)$ and $y(t_1)$, then fix $t=t_2$ and add $x(t_2)$ and $y(t_2)$, and so on until we have enough points to see how $f(t) = x(t) + y(t)$ looks like as a function of t; multiplication of $x(t)$ and $y(t)$ is done analogously. Such operations are called pointwise addition and multiplication.

Next, we proceed as shown in Figure 2.2-19. The graph progression is from top to bottom in each column, from left to right.

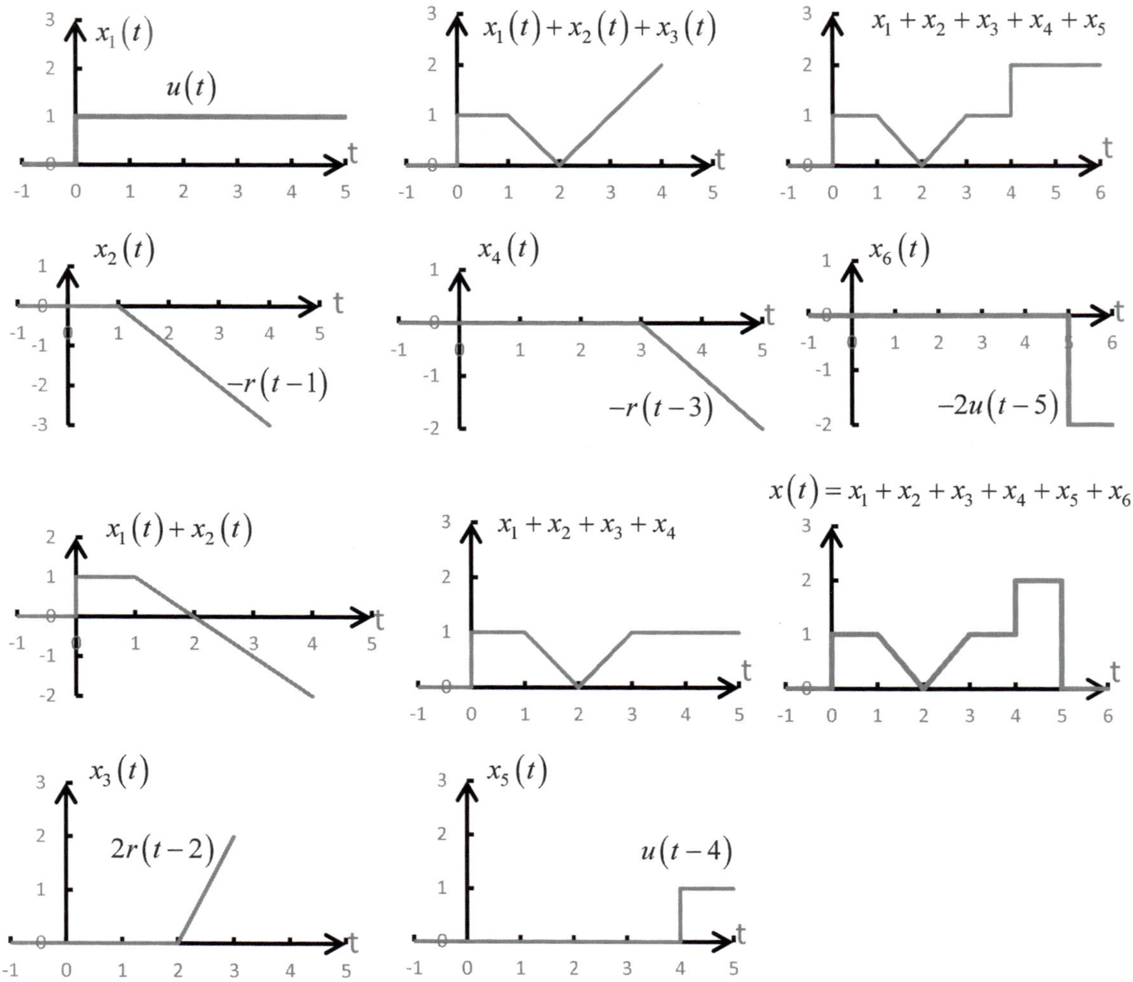

Fig. 2.2-19. Sequence of analytical model building steps for signal in Figure 2.2-18.

Thus, analytical signal model is given by

$$x(t)=u(t)-r(t-1)+2r(t-2)-r(t-3)+u(t-4)-2u(t-5).$$

Example 2.2-10: Create an analytical model of a signal $x(t)$ shown in Figure 2.2-20 below:

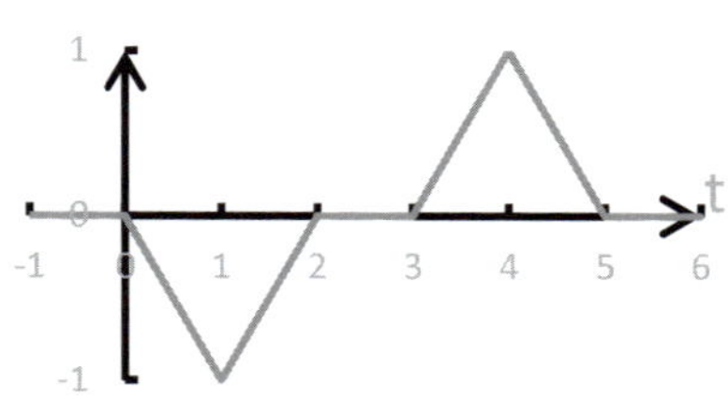

Fig. 2.2-20. Graphically specified signal $x(t)$ for Example 2.2-10.

Solution: Let's use the same functions $r(t)$ and $u(t)$ and operations on them as in the previous example. In this case, we will additionally use the products of the basic functions and proceed as shown in Figure 2.2-21. The graph progression is from top to bottom in each column, from left to right.

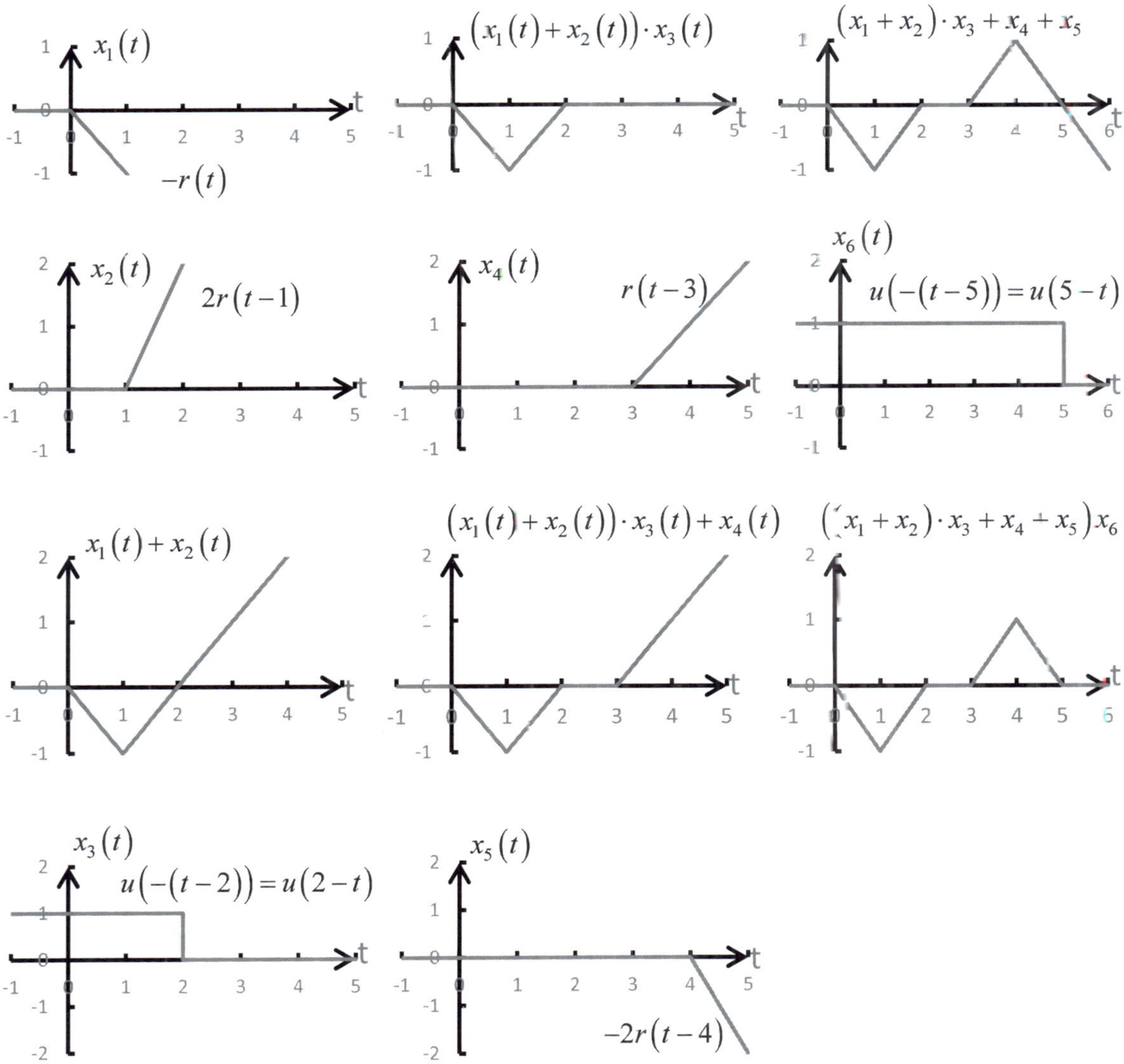

Fig. 2.2-21. Sequence of analytical model building steps for signal in Figure 2.2-20.

Thus, analytical signal model is given by

$$x(t)=\{[-r(t)+2r(t-1)]\cdot u(2-t)+r(t-3)-2r(t-4)\}\cdot u(5-t).$$

2.3 Delta-function and its Use, Signal Representations, Energy and Power Signals

This section provides a glimpse at several analytical and experimental uses of delta function to be discussed in more detail in the subsequent chapters.

2.3.1 *Delta-function*

2.3.1.1 *Delta-function definition*

A **pulse** signal of width Δ is a rectangle signal with width Δ and height $1/\Delta$, i.e. with unit area. We will denote a Δ-width pulse as $p_\Delta(t)$. The CT impulse, denoted by $\delta(t)$ can be thought of as the limit of a pulse as the width goes to 0 (Figure 2.3-1).

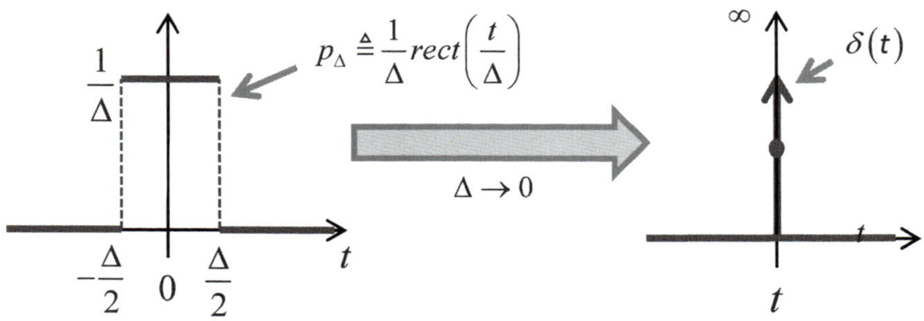

Fig. 2.3-1. Plots of pulse signal (left) and delta function (right).

Setting $\sqrt{1/\Delta}\cdot\text{rect}(t/\Delta)$, with $1/\Delta = 2^j, j-$integer, we obtain $2^{j/2}\text{rect}(2^j t)$ — ***dyadic dilation*** of rect(t), to be used later in Chapter 10 for approximation of signals with many different scales.

The term delta "function" is a misnomer, but we'll continue using it. There is no clear relationship between the independent variable and the value of delta-function: it is zero everywhere except at 0, where it does not exist (jumps to ∞)! Which means it is not a "function", even though the area under it is 1.

In mathematics delta-function is called a "distribution," something that acts on a function. It is best defined through its ***sifting property***: how it acts on a function under the integral.

2.3.1.2 *Delta-function and its properties*

Sifting property: Sifting property can be considered to be a definition of delta-function. To demonstrate sifting property, consider the unit-area pulse $p_\Delta(t)$. If $x(t)$ is continuous and Δ is sufficiently small, $x(t)p_\Delta(t-t_0) \approx x(t_0)p_\Delta(t-t_0)$:

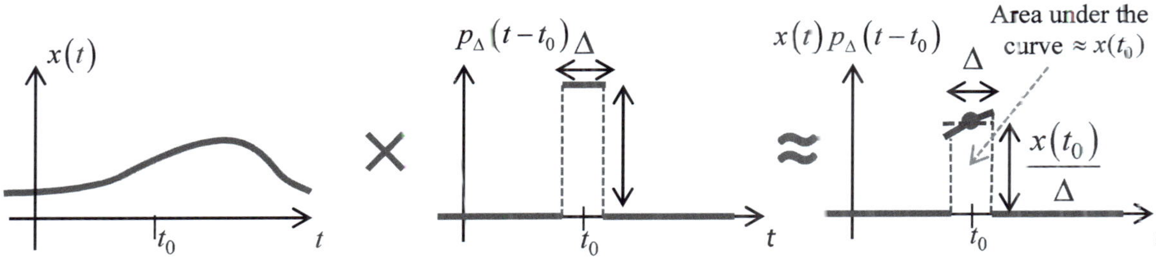

Fig. 2.3-2. Pre-limit (approximate) representation of sifting property.

Then, taking the integral:

$$\int_{-\infty}^{\infty} x(t)p_\Delta(t-t_0)dt = \int_{t_0-\frac{\Delta}{2}}^{t_0+\frac{\Delta}{2}} x(t)\cdot\frac{1}{\Delta}dt \approx \int_{t_0-\frac{\Delta}{2}}^{t_0+\frac{\Delta}{2}} \frac{x(t_0)}{\Delta}dt = \frac{x(t_0)}{\Delta}\Delta = x(t_0).$$

Sifting property

As $\Delta \to 0$: $\int_{-\infty}^{\infty} x(t_0)\delta(t-t_0)dt = x(t_0)\int_{-\infty}^{\infty}\delta(t-t_0)dt = x(t_0)$ or $\int_{-\infty}^{\infty} x(t)\delta(t-t_0)dt = x(t_0)$.

From this, we can draw:

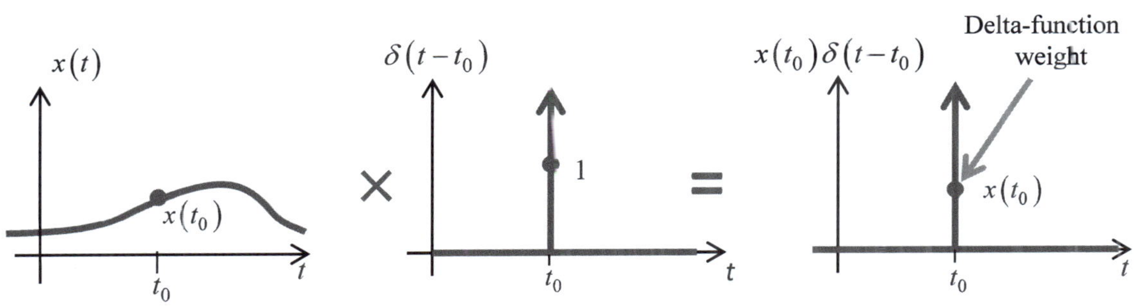

Fig. 2.3-3. Limit (exact) representation of sifting property.

Example 2.3-1: Illustration of delta-function sifting property:

$$\int_{-5}^{5} e^{-t}\delta(t)dt = e^{-0} = 1;\ \int_{-5}^{5} e^{-t}\delta(t-(-1))dt = e^{-(-1)} = e;$$

$$\int_{-5}^{5} e^{-t}\delta(t-10)dt = 0, \quad \text{since } t_0 = 10 \text{ is outside the integration domain.}$$

Also, replacing t by τ and t_0 by t, we can write:

$$\int_{-\infty}^{\infty} x(t)\delta(t-t_0)dt \underset{\substack{t\to\tau \\ t_0\to t}}{=} \int_{-\infty}^{\infty} x(\tau)\delta(\tau-t)d\tau \underset{\substack{\tau-t=s \\ \Rightarrow\tau=t+s}}{=} \int_{-\infty}^{\infty} \underbrace{x(t+s)}_{x_1(s)}\delta(s)ds = x_1(0) = x(t). \quad (2.3\text{-}1)$$

Symmetry property: $\delta(-t)=\delta(t)$. Using this property,

$$x(t)=\int_{-\infty}^{\infty} x(\tau)\delta[-(\tau-t)]d\tau=\int_{-\infty}^{\infty} x(\tau)\delta(t-\tau)d\tau.$$

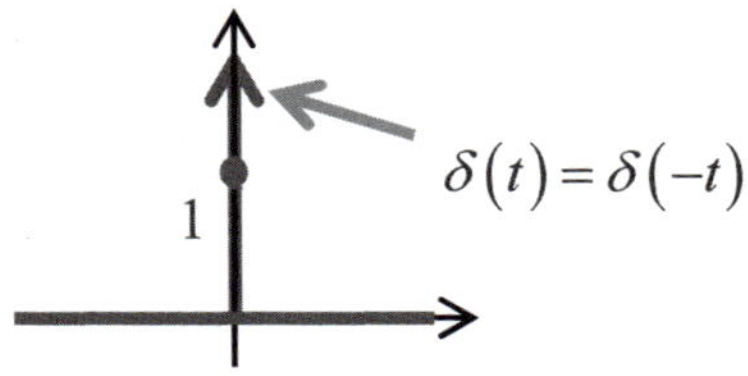

Fig. 2.3-4. Symmetry property.

Time-scaling property: $\delta(kt)=(1/|k|)\delta(t)$. delta-function does not time-scale like a normal function: time rescaling with respect to the original time scale inversely rescales sifting weight when the original time scale is restored. Time-scaling property is derived as follows. Replacing s by τ in the last integral in (2.3-1) and further time-rescaling argument in $\delta(\tau)$ as $\delta(k\tau)$ yields

Symmetry property

$$\int_{-\infty}^{\infty} x(t+\tau)\delta(k\tau)d\tau=\int_{-\infty}^{\infty} x(t+\tau)\delta(|k|\tau)d\tau=\int_{-\infty}^{\infty} x\left(\frac{1}{|k|}(|k|t+|k|\tau)\right)\delta(|k|\tau)\frac{1}{|k|}d(|k|\tau)$$

Change of variables $\lambda=|k|\tau$ Sifting property

$$=\frac{1}{|k|}\int_{-\infty}^{\infty} x\left(\frac{1}{|k|}(|k|t+\lambda)\right)\delta(\lambda)d\lambda=\frac{1}{|k|}x\left(\frac{1}{|k|}|k|t\right)=\frac{1}{|k|}x(t)=\int_{-\infty}^{\infty} x(t+\tau)\left(\frac{1}{|k|}\delta(\tau)\right)d\tau.$$

does not time-scale like a normal function – change of support changes the area under $\delta(t)$:

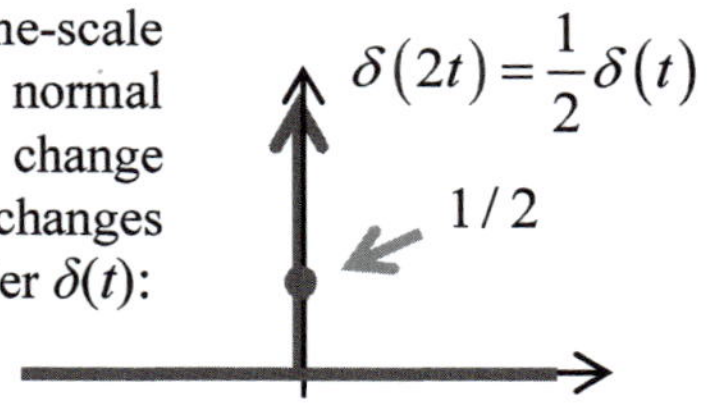

time-rescaling of the delta-function argument inversely rescales the sifting weight:

$$\delta(kt)=\frac{1}{|k|}\delta(t).$$

Fig. 2.3-5. Time-scaling property.

2.3.1.3 *Delta-function use*

Getting useful system representation — impulse response. The delta-function is a key analytical (mathematical) and experimental tool. In its analytical use, it serves as the theoretically most common input signal to characterize a system, especially linear one, through the response to it, referred to as the ***impulse response***:

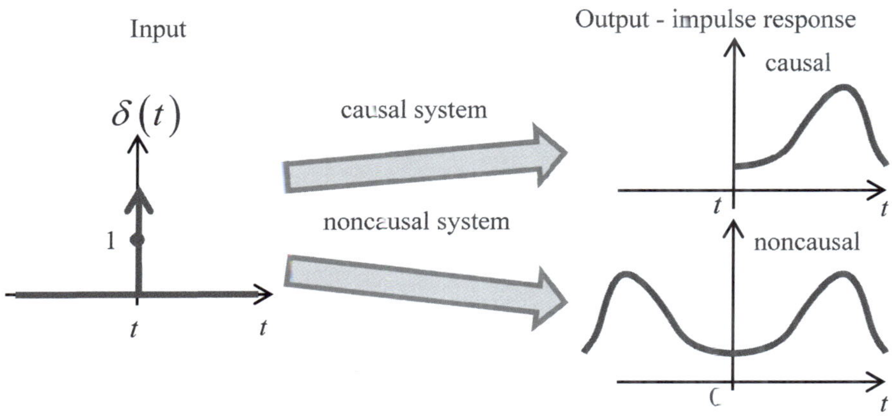

Fig. 2.3-6. Delta function being used as an input signal to characterize systems.

Experimentally obtaining system properties through impact excitation. Experimental use[1]: delta-function-type input can be recreated through the application of the impact hammer to a structure to excite the structure and measure an approximate impulse response. Impact force (input weight) measurement is carried out through the tip sensor, and the structure response (approximate weighted impulse response) measurement can be performed by accelerometer or laser velocity transducer. The record of input/output signal pair is then processed to assess structure condition.

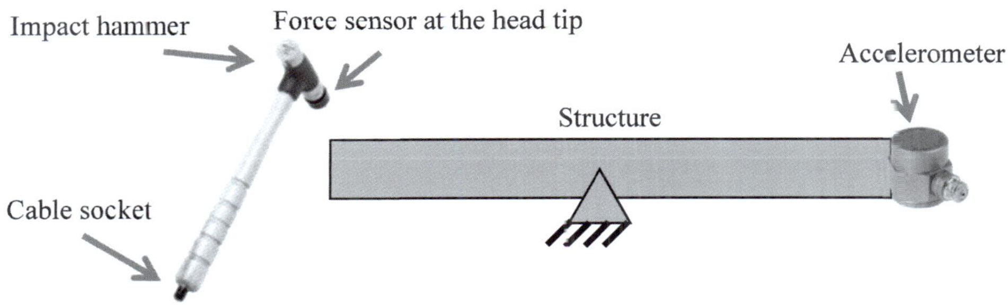

Fig. 2.3-7. Instrumented hammer permits applying to a structure a delta-function-type input that is measurable.

Ideal sampling. Sifting property of delta-function makes it an exact natural sampler with an arbitrarily small sampling interval:

[1]Bruel & Kjaer (http://www.bksv.com/Products/transducers/vibration/impact-hammers.aspx) modal type 8207 sledge hammer with 1 lb. head for medium to very large structures, and general purpose type 4533-B accelerometer with frequency range 0.2 Hz–12.8 kHz. Copyright © Brüel & Kjær.

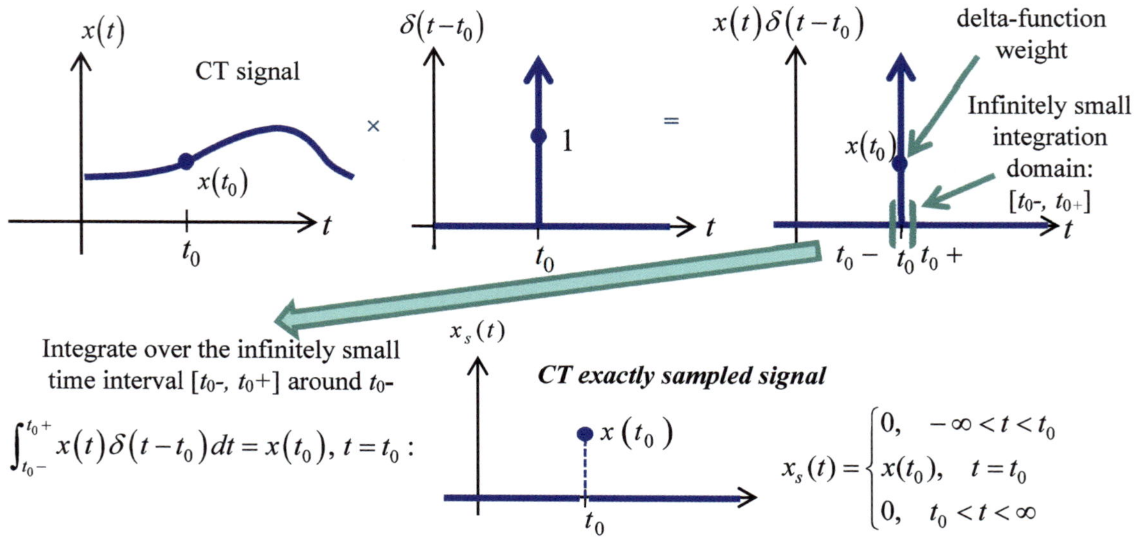

Fig. 2.3-8. Ideal sampling with delta-function: infinitely closely spaced values can be extracted with no error.

Real sampling. In this case, the impulse is replaced by a narrow pulse that does not single out the exact value of function at a given time instant and cannot extract arbitrarily closely spaced values:

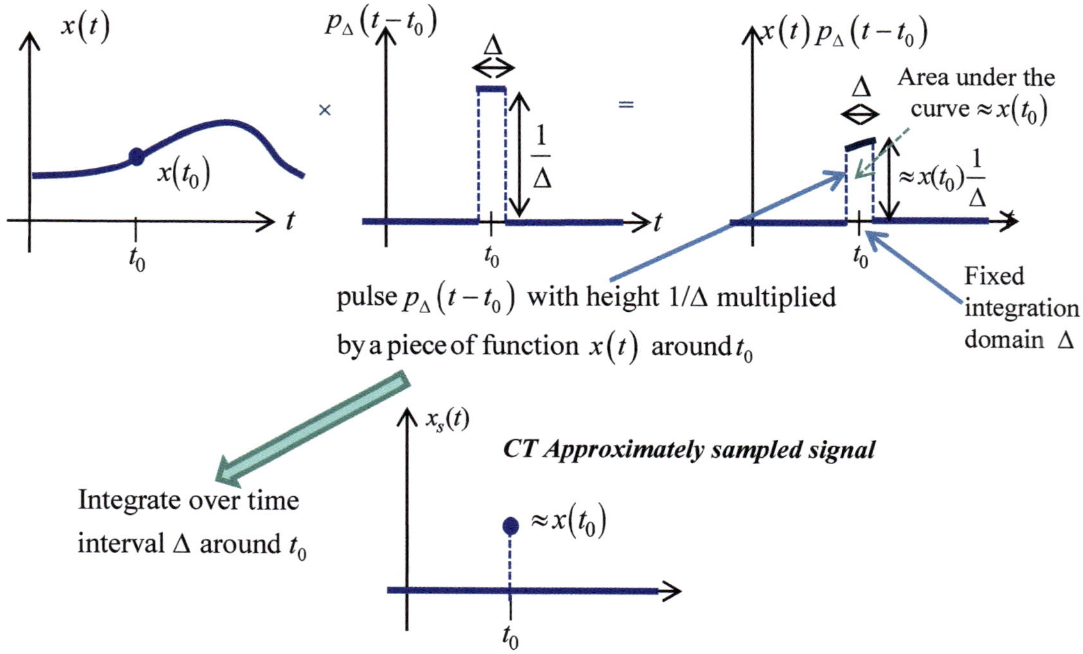

Fig. 2.3-9. Real sampling.

Frequency-based detection of periodic signal hidden in noise through *power spectral density* (PSD): CT. Nonexistence (jump to ∞!) at a single frequency point is a key diagnostic tool in spectral analysis. This property of delta-function, just as important in frequency domain, is briefly sketched below:

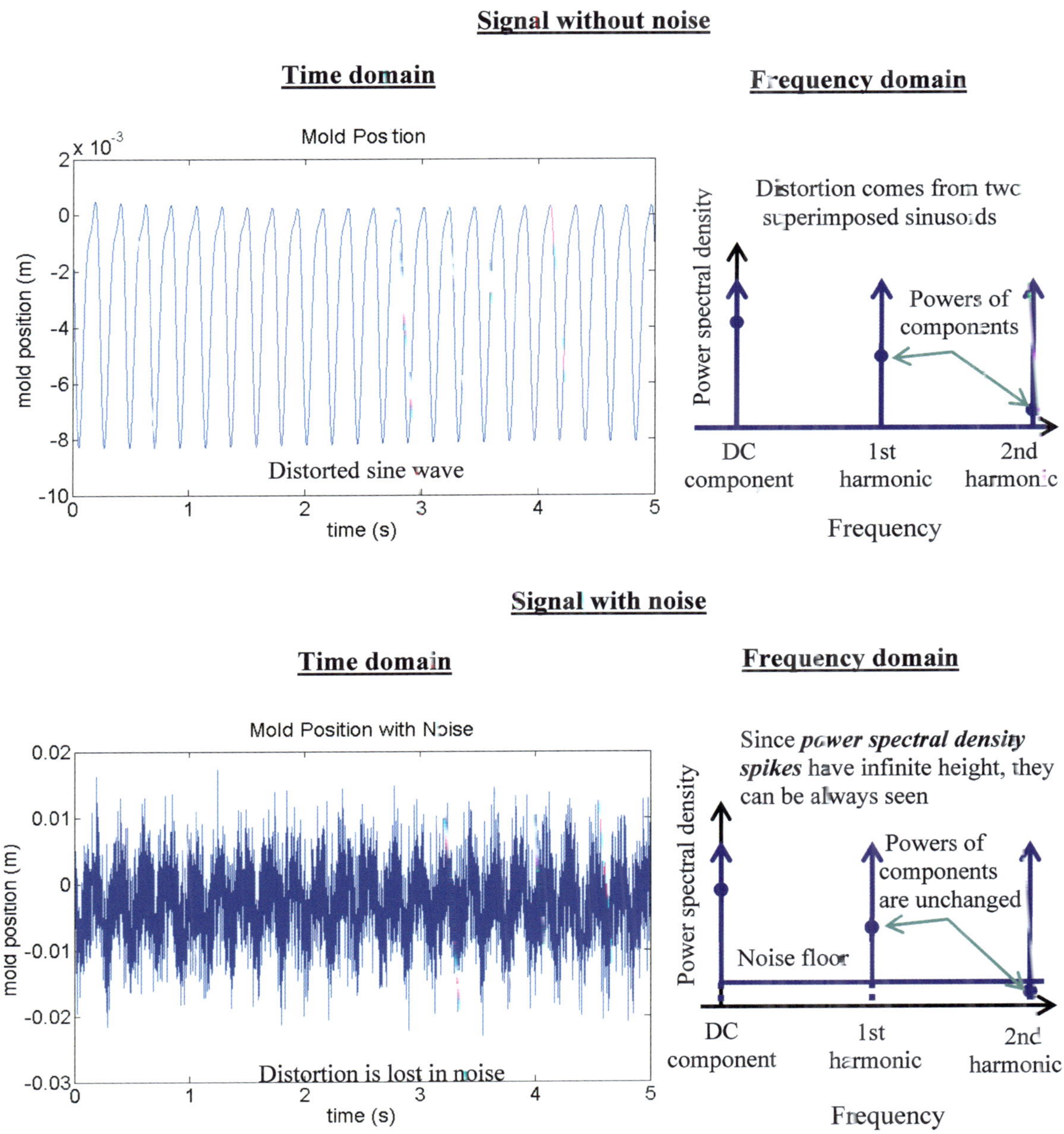

Fig. 2.3-10. Detection of signal hidden in noise: CT.

Frequency-based detection of periodic signal hidden in noise through PSD: DT. Discrete time counterpart of the CT technique depicted above provides widely used computable approximation:

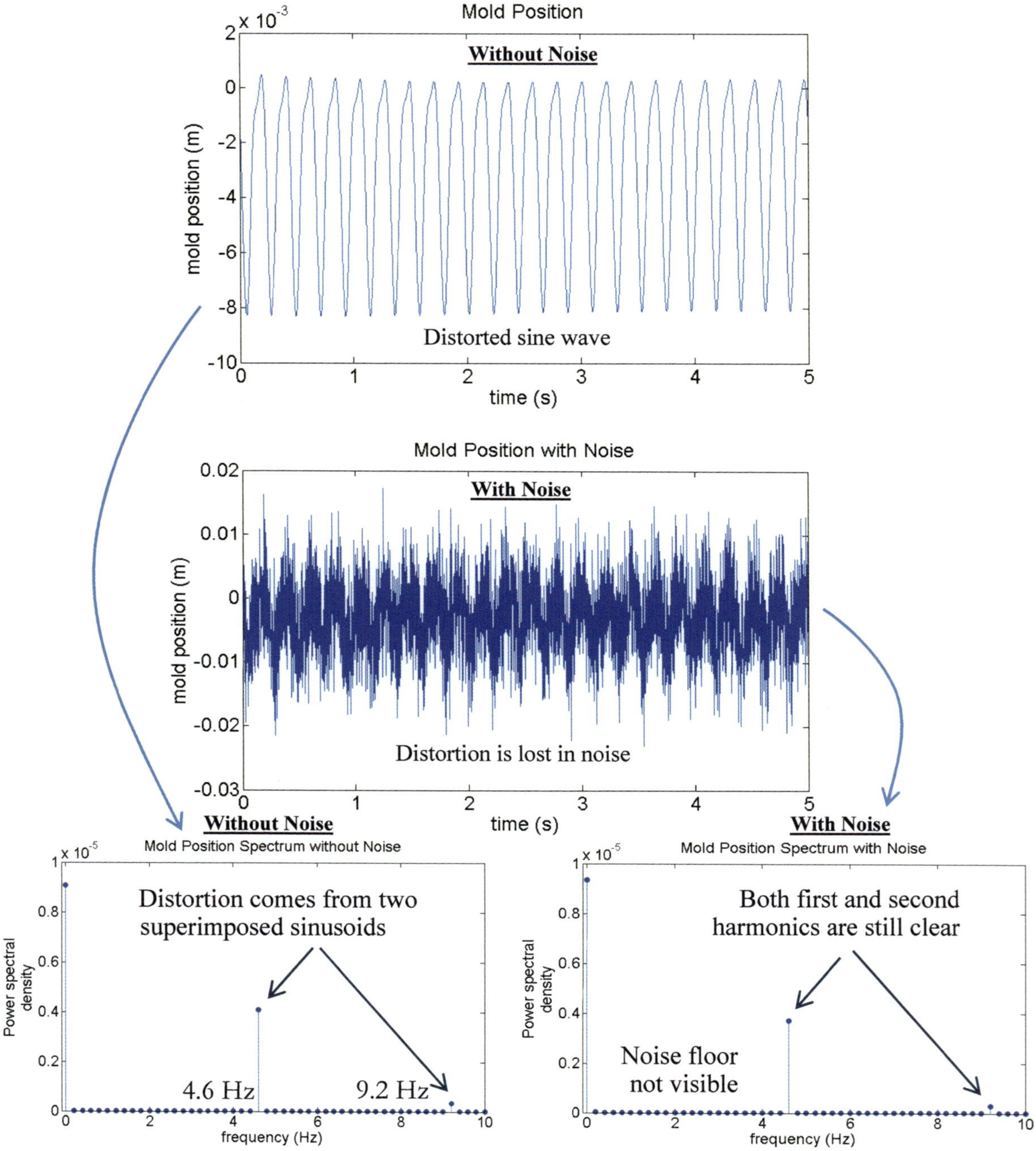

Fig. 2.3-11. Detection of periodic hidden in noise: DT. On the power spectral density graph given in the figure noise floor is not visible. On the magnitude spectrum graph (not given) noise floor would be much more visible.

2.3.1.4 *Sum/integral signal representations: building a pulse train*

To build a pulse train given in Figure 2.3-12, we first represent a single pulse in terms of a rectangle signal as shown in Figure 2.3-13:

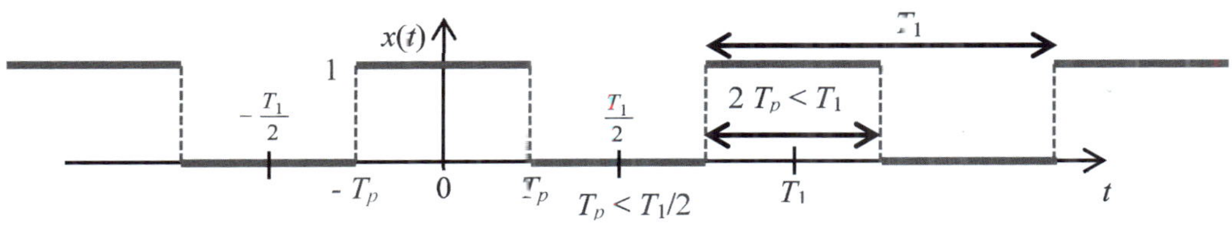

Fig. 2.3-12. Pulse train.

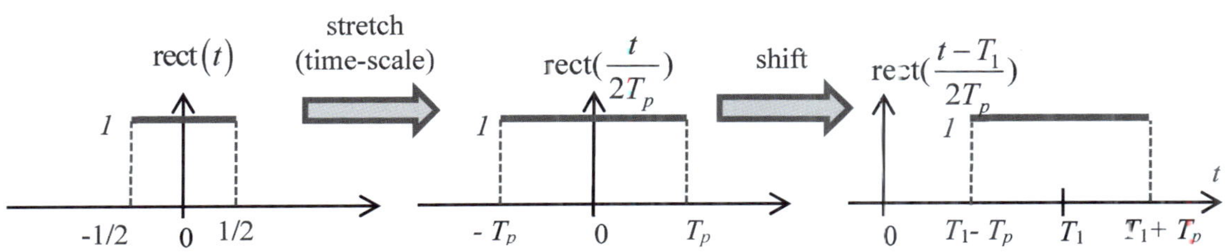

Fig. 2.3-13. Representation of a single pulse in terms of a rectangle signal.

Then, we define periodic extension of a single pulse as follows:

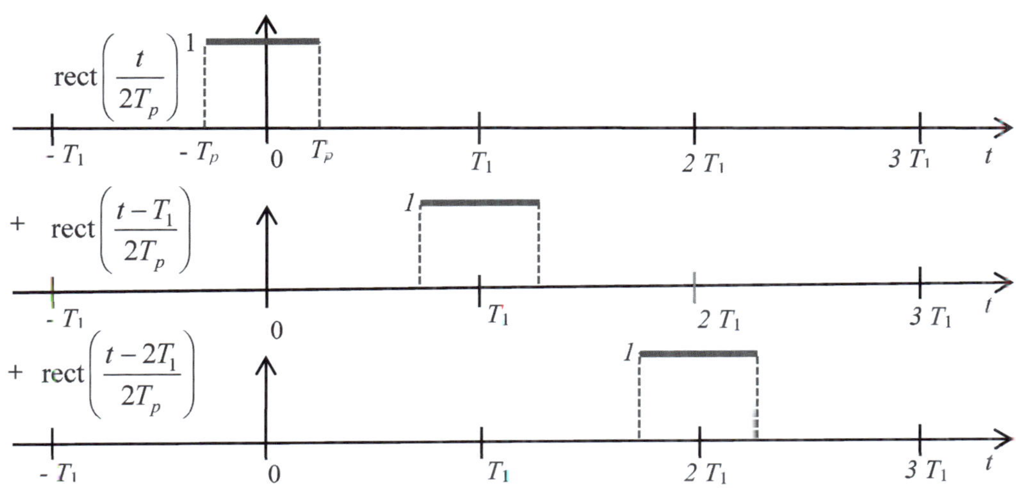

Fig. 2.3-14. Formation of a pulse train — periodic extension of a single pulse.

$$\therefore x_p(t) = \ldots + \text{rect}\left(\frac{t}{2T_p}\right) + \text{rect}\left(\frac{t-T_1}{2T_p}\right) + \text{rect}\left(\frac{t-2T_1}{2T_p}\right) + \ldots = \sum_{n=-\infty}^{\infty} \text{rect}\left(\frac{t-nT_1}{2T_p}\right), T_p < \frac{T_1}{2}, \tag{2.3-2}$$

n – positive or negative integer.

Function $x_p(t)$ represents a periodic extension of a single pulse.

2.3.1.5 *Approximation of a CT signal by pulses*

Let us now augment each pulse in (2.3-2) with individual magnitude scaling capability:

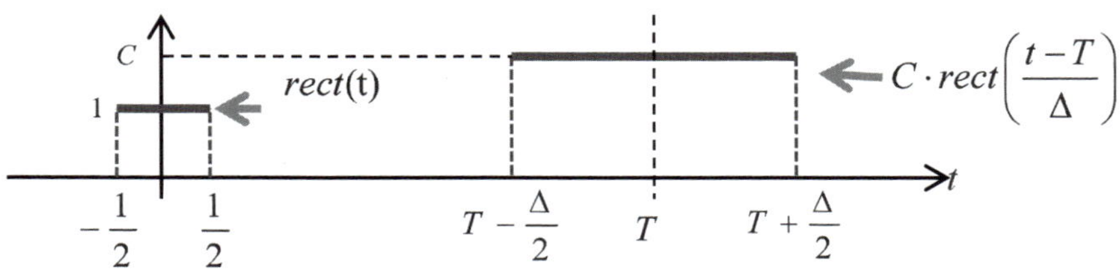

Fig. 2.3-15. Magnitude scaling, stretching, and shifting of rect function.

Using augmented (2.3-2), let's approximate a CT signal by a piecewise constant function.

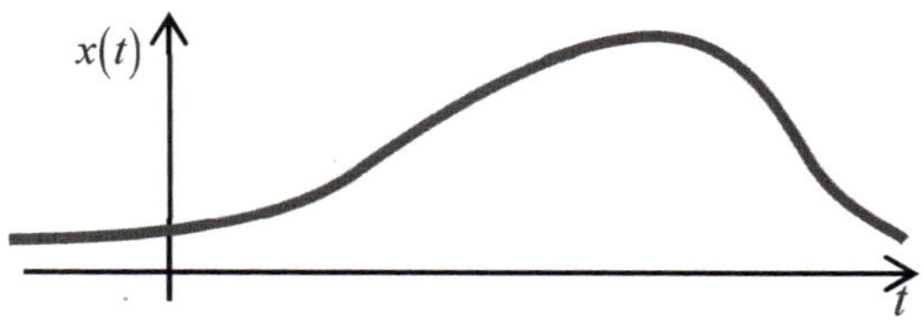

Fig. 2.3-16. CT signal to be approximated.

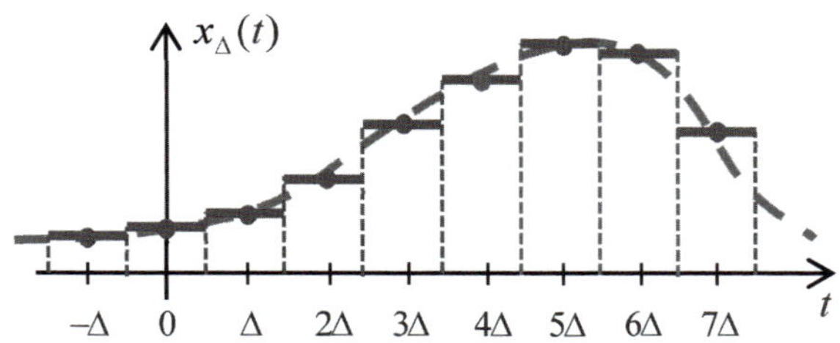

Fig. 2.3-17. CT signal approximated by rectangle functions.

Figure 2.3-16 shows the original signal. Now, we can approximate this function by rectangles with width Δ (Figure 2.3-17) that form scaled pulse train without gaps in terms of the following relation:

$$x(n\Delta) \cdot rect\left(\frac{t-n\Delta}{\Delta}\right) = x(n\Delta) \cdot \frac{1}{\Delta} rect\left(\frac{t-n\Delta}{\Delta}\right) = x(n\Delta) p_\Delta(t-n\Delta)\Delta.$$

2.3.1.6 *Exact representation of a CT signal by impulses*

Writing the above approximation using $p_\Delta(t)$ yields

$$x_\Delta(t) \triangleq \sum_{n=-\infty}^{\infty} x(n\Delta)\cdot \mathrm{rect}\left(\frac{t-n\Delta}{\Delta}\right) = \sum_{n=-\infty}^{\infty} x(n\Delta)p_\Delta(t-n\Delta)\Delta. \tag{2.3-3}$$

Recall now that the impulse $\delta(t)$ is the limit of a rectangle of a pulse with width decreasing to 0: $\delta(t) = \lim_{\Delta\to 0} p_\Delta(t) = \lim_{\Delta\to 0}(1/\Delta)\cdot \mathrm{rect}(t/\Delta)$. As $\Delta \to d\tau$, $n\Delta \to \tau$, and $\Sigma \to \int$, the expression (2.3-3) converges to

$$\int_{-\infty}^{\infty} x(\tau)\delta(t-\tau)d\tau = \int_{-\infty}^{\infty} x(\tau)\delta(\tau-t)d\tau = x(t). \tag{2.3-4}$$

Thus, $x(t)$ is represented by a "wall" (continuum) of infinitely closely spaced weighted impulses, as shown in Figure 2.3-18.

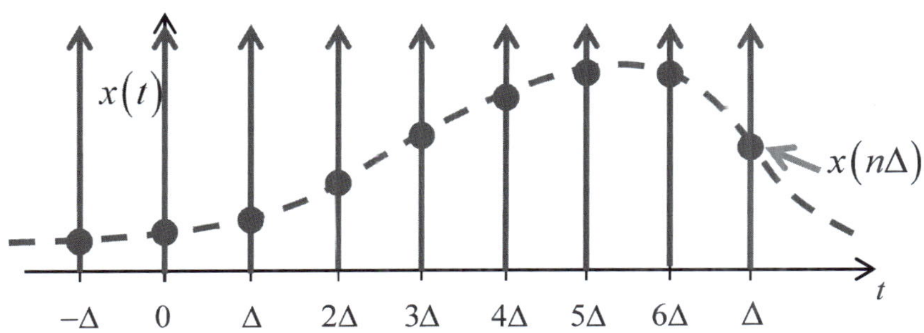

Fig. 2.3-18. CT signal represented by impulses.

2.3.1.7 *Representation of a DT signal by shifted and weighted DT impulses*

The DT counterpart of the CT signal representation (2.3-4) is obtained as follows. Using

$$\delta[n] = \begin{cases} 1, n=0, \\ 0, \text{ otherwise,} \end{cases} \quad \text{yields } \textbf{\textit{Kronecker delta}}\text{: } \delta_{nk} \triangleq \delta[n-k] = \begin{cases} 1, & n=k, \\ 0, & n\neq k, \end{cases}$$

from which

$$x[n]\delta[n-k] = \begin{cases} x[k]\delta[n-k] = x[k], n=k, \\ 0, n\neq k. \end{cases}$$

Therefore, $x[n]$ can be written as the ***convolution sum*** of shifted and weighted DT impulses:

$$\begin{aligned} x[n] &= ... + x[-1]\delta[n-(-1)] + x[0]\delta[n-0] + x[1]\delta[n-1] + x[2]\delta[n-2] + ... \\ &= \sum_{k=-\infty}^{\infty} x[k]\delta[n-k]. \end{aligned}$$

2.3.2 *Signal energy, power, and RMS*

The definitions of signal energy, power and RMS (root mean square) or quadratic mean for continuous time and discrete time are collected in Table 2.3, where for CT $x(t)$ is assumed to be a current flowing through a 1 Ohm resistor.

Table 2.3. Signal measures.

	Continuous time	Discrete time
Energy	$E_x = \int_{-\infty}^{\infty} \lvert x(t)\rvert^2 dt$	$E_x = \sum_{n=-\infty}^{\infty} \lvert x[n]\rvert^2$
Power	**For T-periodic signals,** **i.e.** $x(t+T) = x(t)$ $P_x = \frac{1}{T}\int_T \lvert x(t)\rvert^2 dt$	**For N-periodic signals,** **i.e.** $x[n+N] = x[n]$ $P_x = \frac{1}{N}\sum_{n=0}^{N-1} \lvert x[n]\rvert^2$
RMS	$x_{rms} = \sqrt{P_x} = \sqrt{\frac{1}{T}\int_T \lvert x(t)\rvert^2 dt}$ E.g., AC grid: 120 V rms, 170 V - peak	$\sqrt{P_x} = \sqrt{\frac{1}{N}\sum_{n=0}^{N-1} \lvert x[n]\rvert^2}$
Power	**For non-periodic signals** $P_x = \lim_{T\to\infty}\frac{1}{T}\int_T \lvert x(t)\rvert^2 dt$	**For non-periodic signals** $P_x = \lim_{M\to\infty}\frac{1}{2M+1}\sum_{n=-M}^{M} \lvert x[n]\rvert^2$
RMS	$x_{rms} = \sqrt{P_x} = \sqrt{\lim_{T\to\infty}\frac{1}{T}\int_T \lvert x(t)\rvert^2 dt}$ DC current equivalent through 1 Ohm resistor	$\sqrt{P_x} = \sqrt{\lim_{M\to\infty}\frac{1}{2M+1}\sum_{n=-M}^{M} \lvert x[n]\rvert^2}$

Energy and power are both measures of the "size" of a signal. Energy can be used for comparing signals like $\text{rect}(t)$ and power — for comparing signals like $u(t)$. RMS value of a time-varying signal is its effective action measure in quadratic sense expressed in terms of an identical action measure of a constant signal. For example, the RMS value of a time varying current passing through a 1 Ohm resistor will be the value of a DC current generating the same power when passing through the same resistor.

Example 2.3-2: Calculate energy and power of signals in Figure 2.3-19.

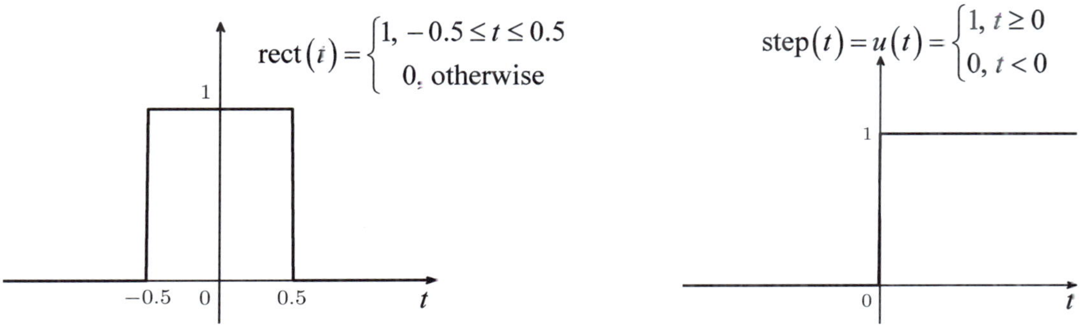

Fig. 2.3-19. Rect and step functions.

Solution: Using definitions given in Table 2.3 yields:

$$E_x=\int_{-\infty}^{\infty}|\text{rect}(t)|^2\,dt=\int_{-0.5}^{0.5}1^2\,dt=1,$$

$$P_x=\lim_{T\to\infty}\frac{1}{T}\int_{-T/2}^{T/2}|\text{rect}(t)|^2\,dt$$

$$=\lim_{T\to\infty}\frac{1}{T}\int_{-0.5}^{0.5}1^2\,dt=\lim_{T\to\infty}\frac{1}{T}=0.$$

Energy is finite, power is zero.

$$E_x=\int_{-\infty}^{\infty}|u(t)|^2\,dt=\int_{0}^{\infty}1^2\,dt=\infty,$$

$$P_x=\lim_{T\to\infty}\frac{1}{T}\int_{-T/2}^{T/2}|u(t)|^2\,dt$$

$$=\lim_{T\to\infty}\frac{1}{T}\int_{0}^{T/2}1^2\,dt=\lim_{T\to\infty}\frac{1}{T}\frac{T}{2}=\frac{1}{2}.$$

Energy is infinite, power is finite.

2.3.2.1 *Energy and power signal classification*

Signals with finite energy are called **energy signals**. Signals with finite power are called **power signals**. Example 2.3-2 shows that rect(t) and step(t) are energy and power signals, respectively.

- Any energy signal has a power of 0.
- Any power signal with non-zero power has infinite energy.

Example 2.3-3: Classify signal $x[n]=\begin{cases}4\cdot 0.7^{-n+8}, & n\le 8\\ 0, & \text{otherwise}\end{cases}$.

Solution: The signal is shown in Figure 2.3-20. By definition

$$E_x=\sum_{n=-\infty}^{\infty}|x[n]|^2=\sum_{n=-\infty}^{8}\left|4\cdot 0.7^{-n+8}\right|^2+\sum_{n=9}^{\infty}0^2=\sum_{n=-\infty}^{8}|4|^2\cdot\left(|0.7|^2\right)^{-n+8}=16\sum_{n=-\infty}^{8}0.49^{-n+8}=16\sum_{m=0}^{\infty}0.49^m$$

$$\approx 31.373.$$

Therefore, $x[n]$ is an energy signal.

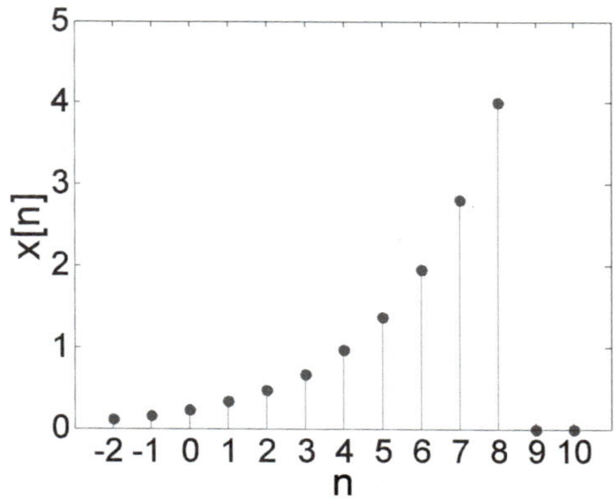

Fig. 2.3-20. Plot of $x[n]$.

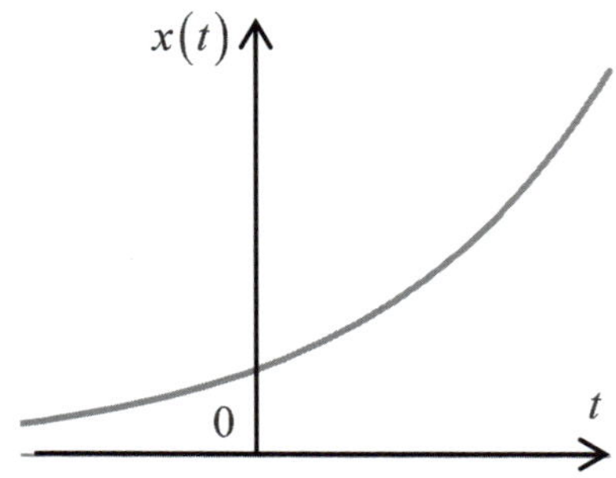

Fig. 2.3-21. Plot of $x(t)$.

Example 2.3-4: Classify signal $x(t) = e^{t/2}$.

Solution: The signal is shown in Figure 2.3-21. By definition

$$E_x = \int_{-\infty}^{\infty} |x(t)|^2 dt = \int_{-\infty}^{\infty} \left(e^{\frac{t}{2}}\right)^2 dt = \int_{-\infty}^{\infty} e^t dt = e^t \Big|_{-\infty}^{\infty} = e^{\infty} - e^{-\infty} = \infty.$$

Therefore, $x(t)$ is not an energy signal.

$$P_x = \lim_{T\to\infty} \frac{1}{T} \int_{-T/2}^{T/2} |x(t)|^2 dt = \lim_{T\to\infty} \frac{1}{T} \int_{-T/2}^{T/2} e^t dt = \lim_{T\to\infty} \frac{1}{T} e^t \Big|_{-T/2}^{T/2} = \lim_{T\to\infty} \frac{1}{T}\left(e^{T/2} - e^{-T/2}\right)$$

$$= \lim_{T\to\infty} \frac{1}{T} e^{T/2} - \lim_{T\to\infty} \frac{1}{T} e^{-T/2} = \lim_{T\to\infty} \frac{1}{T} e^{T/2} - 0 = \lim_{T\to\infty} \frac{1}{T}\left(1 + \frac{T}{2} + \frac{1}{2!}\left(\frac{T}{2}\right)^2 + \dots\right) = \lim_{T\to\infty}\left(\frac{1}{T} + \frac{1}{2} + \frac{1}{8}T + \dots\right) = \infty.$$

Therefore, $x(t)$ is not a power signal either.

Example 2.3-5: Classify signal $x(t)=\cos(t)$.

Solution: Applying the energy and power calculation formulas, we have

$$E_x = \int_{-\infty}^{\infty} (\cos(t))^2 dt = \int_{-\infty}^{\infty} \left(\frac{1+\cos(2t)}{2}\right) dt = \frac{1}{2}\int_{-\infty}^{\infty} (1+\cos(2t)) dt = \frac{1}{2}\left(t + \frac{1}{2}\sin(2t)\right)\Big|_{-\infty}^{\infty} = \infty.$$

Therefore, $x(t)$ is not an energy signal. Now, periodic signal power computation formula yields

$$P_x = \frac{1}{T}\int_{-T/2}^{T/2} |x(t)|^2 dt = \frac{1}{T}\int_{-T/2}^{T/2} (\cos(t))^2 dt = \frac{1}{2T}\left(t + \frac{1}{2}\sin(2t)\right)\Big|_{-T/2}^{T/2}$$

$$= \frac{1}{2T}\left(\frac{T}{2} + \frac{1}{2}\underbrace{\sin(T)}_{0} - \left(-\frac{T}{2} + \frac{1}{2}\underbrace{\sin(-T)}_{0}\right)\right) = \frac{1}{2T}T = \frac{1}{2}.$$

Therefore, $x(t)$ is a power signal.

2.3.2.2 *Making energy signal transform into power signal, and power signal transform into energy signal*

Transitions between power and energy signals serve as one of the key tools in data processing. To illustrate these transitions consider pulse train (2.3-6). $x_p(t)$ is seen to be a power signal with power $P_{x_p}(t) = 2T_p / T_1$. Now, let $T_1\uparrow$ (arrow pointed up indicates that quantity increases) as shown in (2.3-9) and Figure 2.3-22 below.

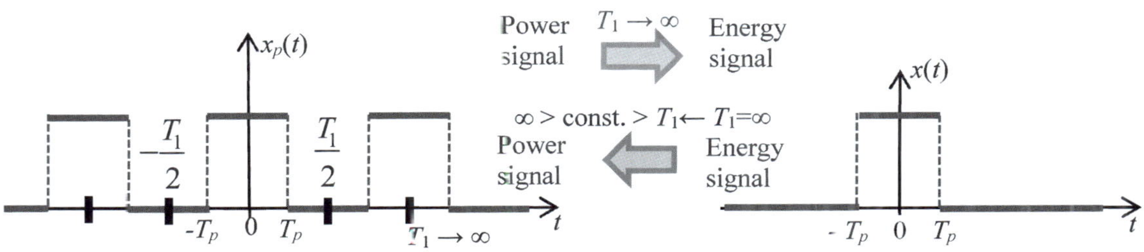

Fig. 2.3-22. Transition between power signal and energy signal.

Then, as T_1 increases, power of $x_p(t)$ decreases, since energy of $x_p(t)$ in each period is fixed at $E_{x_p}(t) = 2T_p, -T_1/2 \le t \le T_1/2$, and independent of period T_1. So, as $T_1 \to \infty$, energy of $x_p(t)$ per time unit goes to zero, while energy of $x_p(t)$ over period stays fixed at $2T_p$, making it in the limit an energy signal:

$$x_p(t) = \sum_{n=-\infty}^{\infty} rect\left(\frac{t-nT_1}{2T_p}\right),\ T_p < \frac{T_1}{2},\ n \in [-\infty, \infty] \underset{T_1 \to \infty}{\to} x(t) = rect\left(\frac{t}{2T_p}\right). \tag{2.3-9}$$

As we can see from Figure 2.3-22, the left signal is power signal and it is periodic, whereas the right signal is energy signal — it is nonperiodic and can be thought of as having "infinite" period, so signal has compact support.

Going in the opposite direction, any nonperiodic signal can be thought of as a periodic signal with infinite period. *Infinite period* is an oxymoron in mathematical sense, meaning that it does not exist, but it is a useful concept to be employed throughout the book.

Now, let $T_1\downarrow$ (arrow pointed down indicates that quantity decreases) from $T_1 = \infty$ to finite decreasing values as shown in (2.3-10) and Figure 2.3-22 above. This results in ***periodic extension*** of energy signal $x(t)$ to signal $x_p(t)$ that is a power signal.

$$x(t) = \mathrm{rect}\left(\frac{t}{2T_p}\right) \overset{\infty}{\underset{}{\overset{T_1\downarrow}{\to}}} x_p(t) = \sum_{n=-\infty}^{\infty} \mathrm{rect}\left(\frac{t-nT_1}{2T_p}\right),\ T_p < \frac{T_1}{2}. \tag{2.3-10}$$

For very large T_1 the power of $x_p(t)$ is close to zero, i.e. to that of an energy signal, but as T_1 decreases, signal power increases.

2.3.2.3 *Non-periodic signal power calculation*

Note that in the power calculation of non-periodic signal we need to calculate the limit of the ratio between two functions that both increase to infinity.

We calculated this using the Taylor series expansion about $T = 0$, which is only valid if the series converges in the range where we took the limit, i.e. for all T up to and including ∞. Taylor series for e^T converges for all T, so taking limit was valid. This is because e^T is analytic (differentiable infinitely many times) for any T. This is not necessarily the case for many functions. Later on, we will encounter cases of non-convergence of series (Gibbs phenomenon).

Counter-example: Taylor series for $1/(1-T)$ about 0 converges only for $|T| < 1$:

$$\frac{1}{1-T} = \sum_{n=0}^{\infty} T^n.$$

Another way to calculate this limit is by using L'Hopital's rule:

$$P_x = \lim_{T\to\infty} \frac{e^{T/2}}{T} = \lim_{T\to\infty} \frac{de^{T/2} / dT}{dT / dT} = \lim_{T\to\infty} \frac{(1/2)e^{T/2}}{1} = \frac{1}{2} \lim_{T\to\infty} e^{T/2} = \infty.$$

Essentially, this result says that e^T grows faster than T.

Homework 2: Solved Examples and Solution Guide

Problem 1: For each signal $x_i(t)$, $i = 1, 2, 3$, in Figure Hw2-1,

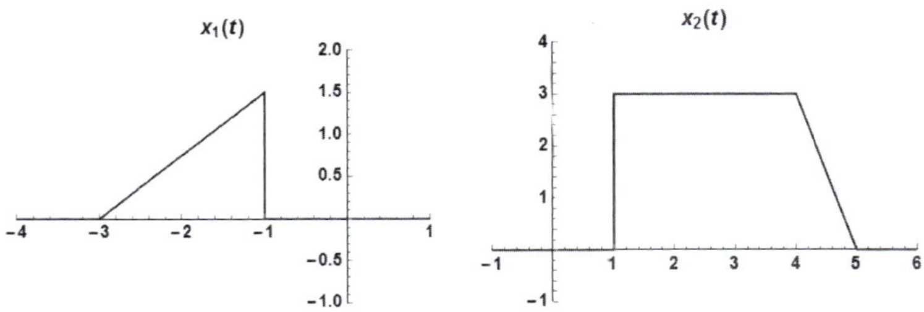

Fig. Hw2-1. Plot of $x_i(t)$, $i = 1, 2, 3$ for Problem 1.

sketch the following: (a) $p_i(t) = x_i[2(t+6)]$, $i = 1, 2, 3$; (b) $h_i(t) = x_i(3-4t)$, $i = 1, 2, 3$; (c) $s_i(t) = x_i(-2t-3)$, $i = 1, 2, 3$.

Solution: (a)

$p_1(t) = x_1\left[2(t+6)\right] = x_1\left[2(t-(-6))\right]$:

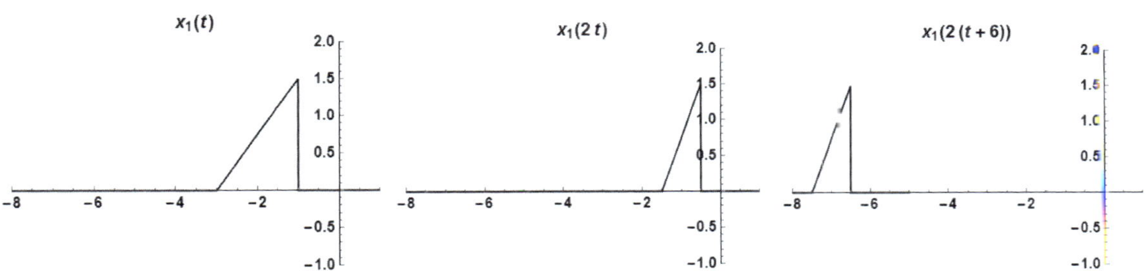

Fig. Hw2-2. Signal $p_1(t)$ for the corresponding $x_1(t)$.

$p_2(t) = x_2\left[2(t-(-6))\right]$:

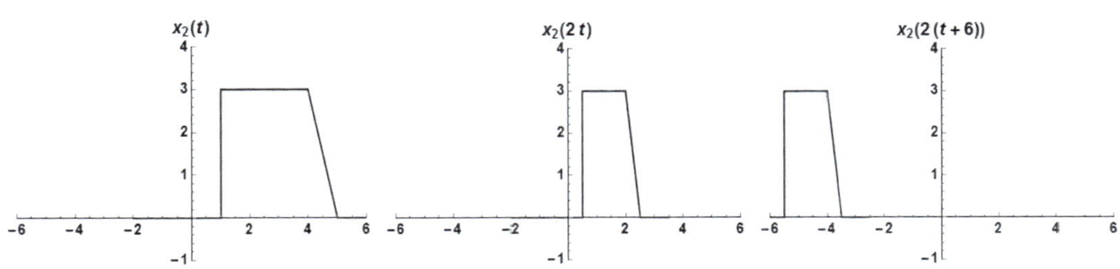

Fig. Hw2-3. Signal $p_2(t)$ for the corresponding $x_2(t)$.

$p_3(t) = x_3\left[2\left(t-(-6)\right)\right]$:

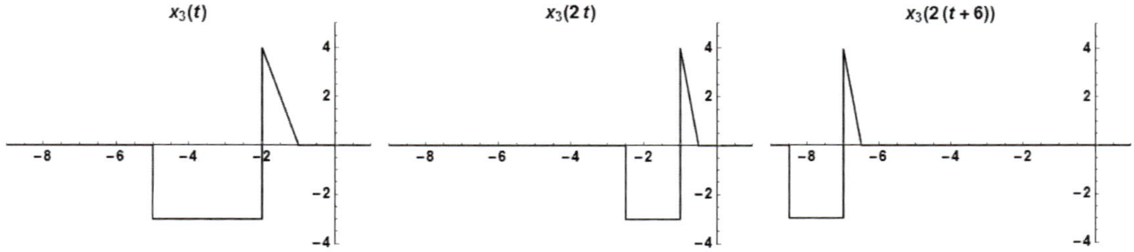

Fig. Hw2-4. Signal $p_3(t)$ for the corresponding $x_3(t)$.

(b) $h_1(t) = x_1(3-4t) = x_1(-4t+3) = x_1\left[-4(t-0.75)\right]$:

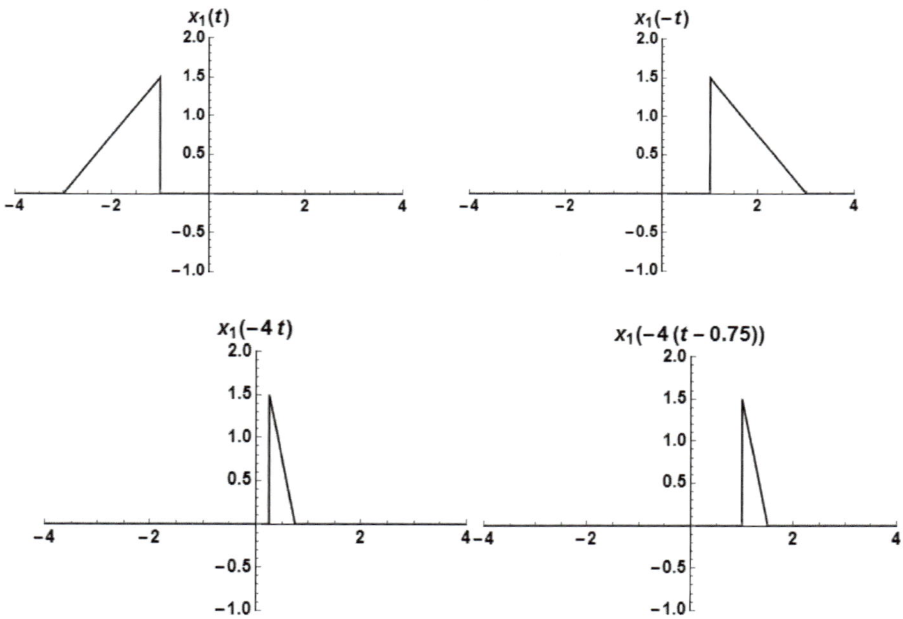

Fig. Hw2-5. Signal $h_1(t)$ for the corresponding $x_1(t)$.

$h_2(t) = x_2\left[-4(t-0.75)\right]$:

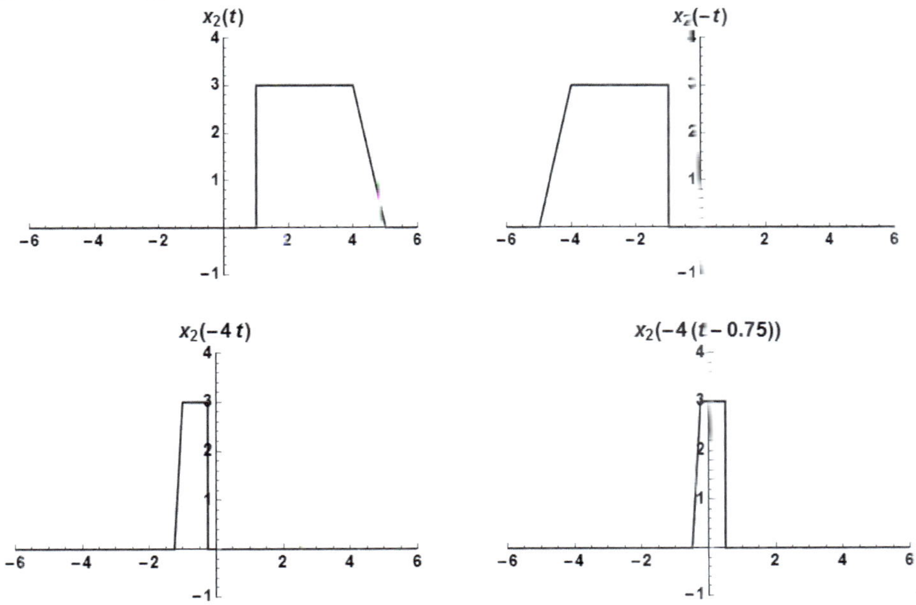

Fig. Hw2-6. Signal $h_2(t)$ for the corresponding $x_2(t)$.

$h_3(t) = x_3\left[-4(t-0.75)\right]$:

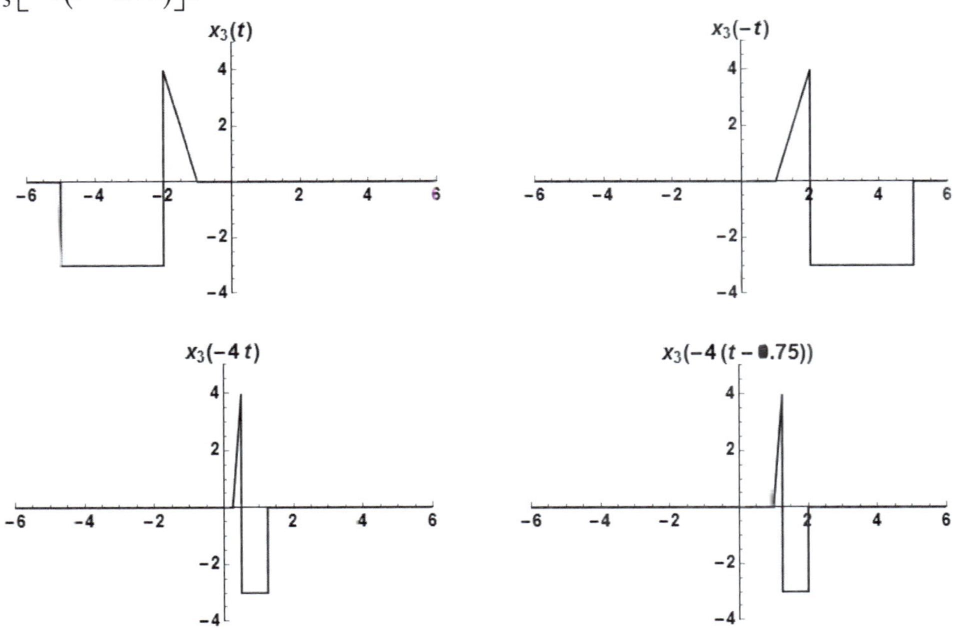

Fig. Hw2-7. Signal $h_3(t)$ for the corresponding $x_3(t)$.

(c) $s_1(t) = x_1(-2t-3) = x_1[-2(t+1.5)] = x_1[-2(t-(-1.5))]$:

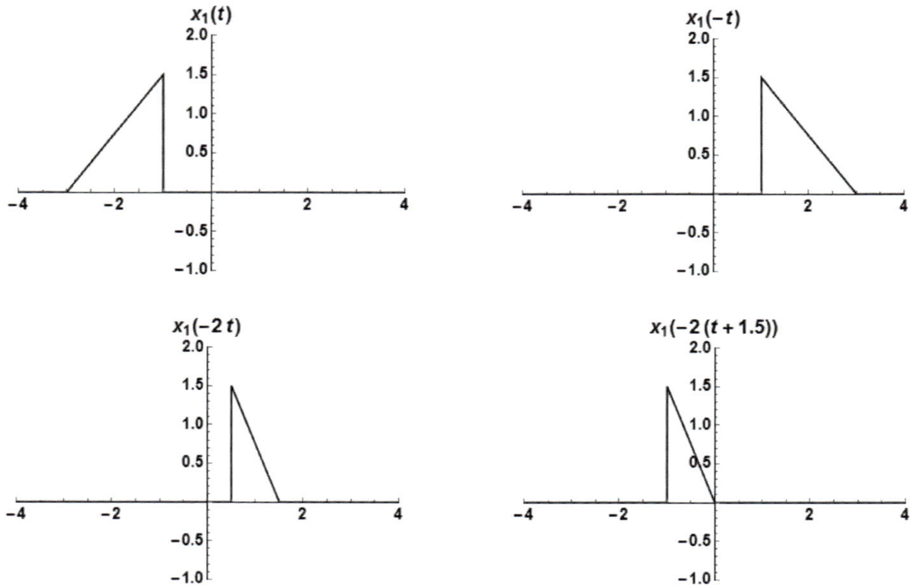

Fig. Hw2-8. Signal $s_1(t)$ for the corresponding $x_1(t)$.

$s_2(t) = x_2[-2(t-(-1.5))]$:

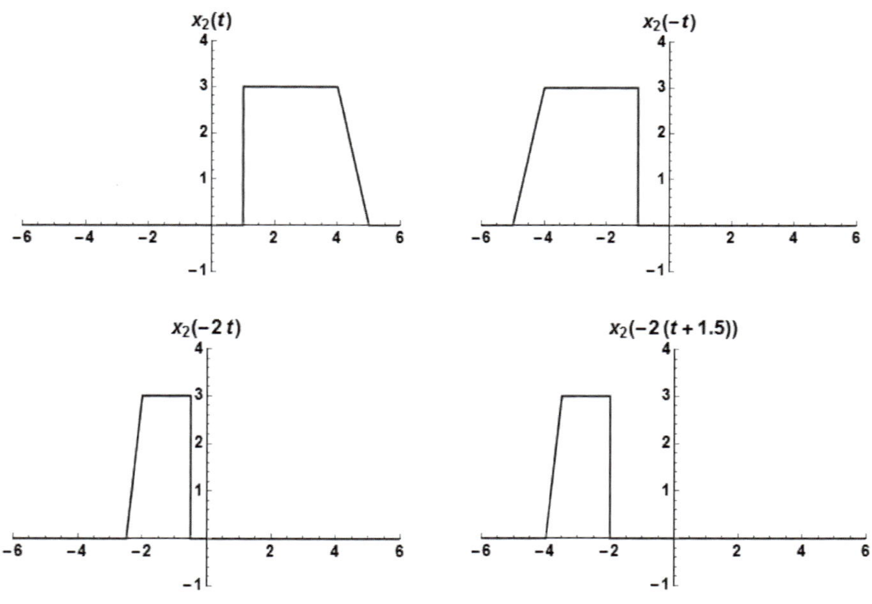

Fig. Hw2-9. Signal $s_2(t)$ for the corresponding $x_2(t)$.

$s_3(t) = x_3\left[-2\left(t-(-1.5)\right)\right]$:

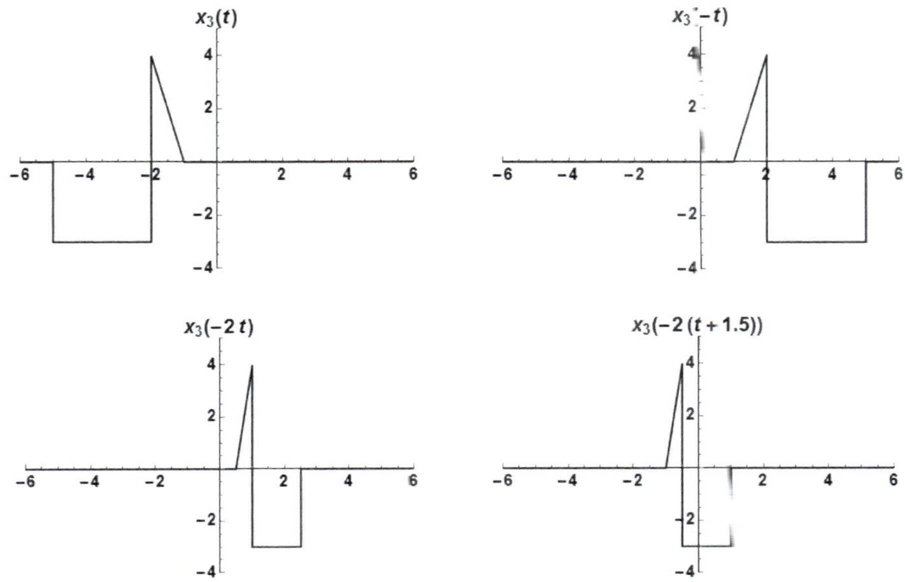

Fig. Hw2-10. Signal $s_3(t)$ for the corresponding $x_3(t)$.

Problem 2: Determine if the following expressions represent power or energy signals, or none of the above, and compute the corresponding signal energy or power: (a) $x(t)=e^{-t}u(-t)$; **(b)** $x(t)=t^2e^{-|2t|}$; **(c)** $x(t)=e^{-3t}\cos(2t)u(t)$.

Solution:

(a) $x(t)=e^{-t}u(-t)$,

$$E_x = \int_{-\infty}^{\infty}\left|x^2(\mathrm{t})\right|dt = \int_{-\infty}^{\infty} e^{-2t}u(-\mathrm{t})dt = \frac{-e^{-2t}}{2}\Big|_{-\infty}^{0} = \infty,$$

$$P_x = \lim_{T\to\infty}\frac{1}{T}\int_{-\frac{T}{2}}^{\frac{T}{2}}\left|x(\mathrm{t})\right|^2 dt = \lim_{T\to\infty}\frac{1}{T}\int_{-\frac{T}{2}}^{0} e^{-2t}dt = \lim_{T\to\infty}\frac{1}{T}\left[\frac{-e^{-2t}}{2}\right]_{-\frac{T}{2}}^{0} = \infty$$

$\therefore$ Since this signal has infinite energy and power, it is neither an energy, nor a power signal.

(b) $x(t)=t^2e^{-|2t|}$,

$$E_x = \int_{-\infty}^{\infty}\left|t^2e^{-|2t|}\right|^2 dt = \int_{-\infty}^{\infty}\left(t^2e^{-|2t|}\right)^2 dt = 2\int_0^{\infty} t^4e^{-4t}dt = -\frac{1}{128}e^{-4t}\left(3+12t+24t^2+32t^3+32t^4\right)\Big|_0^{\infty}$$
$$= -\frac{1}{128}e^{0}(0-3) = \frac{3}{128},$$

$$P_x = \lim_{T\to\infty} \tfrac{1}{T}\int_{-\frac{T}{2}}^{\frac{T}{2}} \left(t^2 e^{-|2t|}\right)^2 dt = \lim_{T\to\infty} \tfrac{2}{T}\int_0^{\frac{T}{2}} t^4 e^{-4t} dt = \lim_{T\to\infty} \tfrac{2}{T}\left[-\tfrac{1}{128}e^{-4t}\left(3+12t+24t^2+32t^3+32t^4\right)\right]_0^{\frac{T}{2}}$$
$$= \lim_{T\to\infty} \tfrac{2}{T}\left[e^{-2T}\left(-\tfrac{T^4}{4}-\tfrac{T^3}{4}-\tfrac{3T^2}{16}-\tfrac{3T}{32}-\tfrac{3}{128}\right)-e^0\left(-0-0-0-0-\tfrac{3}{128}\right)\right]$$
$$= \lim_{T\to\infty} \tfrac{2}{T}\left[e^{-2T}\left(-\tfrac{T^4}{4}-\tfrac{T^3}{4}-\tfrac{3T^2}{16}-\tfrac{3T}{32}-\tfrac{3}{128}\right)+\tfrac{3}{128}\right]=0.$$

$\therefore$ Since this signal has finite energy and zero power, it is an energy signal.

(c) $x(t)=e^{-3t}\cos(2t)u(t)$,

$$E_x = \int_{-\infty}^{\infty}\left|e^{-3t}\cos(2t)\,\mathrm{u(t)}\right|^2 dt = \int_0^{\infty} e^{-6t}\cos^2(2t)dt = \int_0^{\infty} e^{-6t}\,\tfrac{1+\cos(4t)}{2}\,dt$$
$$= \int_0^{\infty} e^{-6t}\,\tfrac{1}{2}dt + \int_0^{\infty} e^{-6t}\,\tfrac{\cos(4t)}{2}dt = \left(-\tfrac{1}{12}e^{-6t}\right)\Big|_0^{\infty} + \left(\tfrac{1}{52}e^{-6t}\left(-3\cos(4t)+2\sin(4t)\right)\right)\Big|_0^{\infty} = \tfrac{1}{12}+\tfrac{3}{52}=\tfrac{11}{78},$$
$$P_x = \lim_{T\to\infty}\tfrac{1}{T}\int_{-\frac{T}{2}}^{\frac{T}{2}}|x(t)|^2 dt = \lim_{T\to\infty}\tfrac{1}{T}\int_{-\frac{T}{2}}^{\frac{T}{2}} e^{-6t}\cos^2(2t)u(t)dt \text{ (since } e^{-6t}\cos^2(2t)\ge 0 \text{ for all } t)$$
$$= \lim_{T\to\infty}\tfrac{1}{T}\int_0^{\frac{T}{2}} e^{-6t}\cos^2(2t)dt = \lim_{T\to\infty}\tfrac{1}{T}\int_0^{\frac{T}{2}} e^{-6t}\,\tfrac{1+\cos(4t)}{2}dt = \lim_{T\to\infty}\tfrac{1}{2T}\left[\int_0^{\frac{T}{2}} e^{-6t}dt - \int_0^{\frac{T}{2}} e^{-6t}\cos(4t)dt\right]$$
$$= \lim_{T\to\infty}\tfrac{1}{2T}\left[\left(\tfrac{1}{12}-\tfrac{1}{12}e^{-3T}\right)+\left(\tfrac{1}{52}e^{-3T}\left(-3\cos(2T)+2\sin(2T)\right)-\tfrac{1}{52}(-3)\right)\right]=0.$$

$\therefore$ Since this signal has finite energy and zero power, it is an energy signal.

Problem 3: Determine if $x(t)=\cos(5\pi t)\sin(10\pi t)$ represents periodic, nonperiodic, or almost periodic signals. Find the signal power and the fundamental period.
Solution:

Recall: $\cos\alpha\sin\beta=\dfrac{1}{2}\left(\sin(\alpha+\beta)+\sin(\beta-\alpha)\right)$

$$\therefore x(t)=\frac{1}{2}\left(\sin\left(15\pi t\right)+\sin\left(5\pi t\right)\right)=\frac{1}{2}\left(\sin\left(2\pi\frac{15}{2}t\right)+\sin\left(2\pi\frac{5}{2}t\right)\right).$$

Signal is periodic: $f_a=15/2; f_b=5/2$; fundamental frequency $f_1=5/2$; common period $T_1=\dfrac{1}{f_0}=\dfrac{2}{5}$.

Signal power

$$P = P_a + P_b = \frac{5}{2}\int_0^{\frac{2}{5}}\left(\frac{1}{2}\sin(15\pi t)\right)^2 dt + \frac{5}{2}\int_0^{\frac{2}{5}}\left(\frac{1}{2}\sin(5\pi t)\right)^2 dt$$

$$= \frac{5}{2}\left(\frac{1}{4}\int_0^{\frac{2}{5}}\frac{1-\cos(30\pi t)}{2}dt + \frac{1}{4}\int_0^{\frac{2}{5}}\frac{1-\cos(10\pi t)}{2}dt\right)$$

$$= \frac{5}{8}\left(\left(\frac{1}{2}t - \frac{1}{60\pi}\sin(30\pi t)\right)\Bigg|_0^{\frac{2}{5}} + \left(\frac{1}{2}t - \frac{1}{20\pi}\sin(10\pi t)\right)\Bigg|_0^{\frac{2}{5}}\right) = \frac{5}{8}\left(\frac{1}{5}+\frac{1}{5}\right) = \frac{1}{4}.$$

Problem 4: Draw the following signals: (a) $x(t) = t^2\, rect(2t)$; **(b)** $x(t) = rect\left[\frac{t-2}{4}\right] tri\left[\frac{t+1}{6}\right]$; **(c)** $x(t) = u(3-|2t|)$.

Solution: (a) $x(t) = t^2\, rect(2t)$:

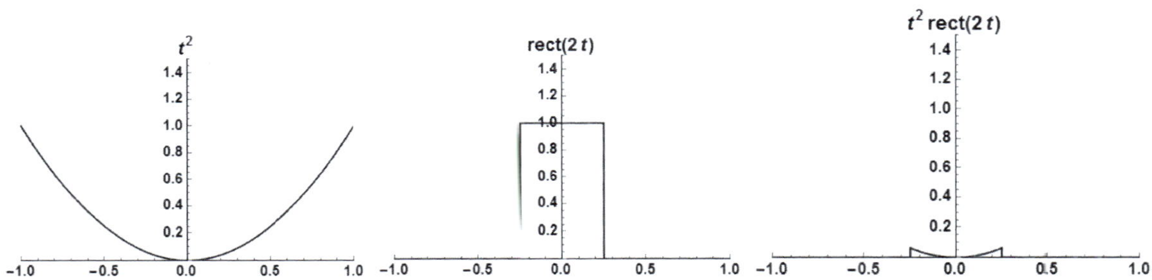

Fig. Hw2-11. Signal $x(t)$ for Problem 4(a).

(b) $x(t) = rect\left[\frac{t-2}{4}\right] tri\left[\frac{t+1}{6}\right]$:

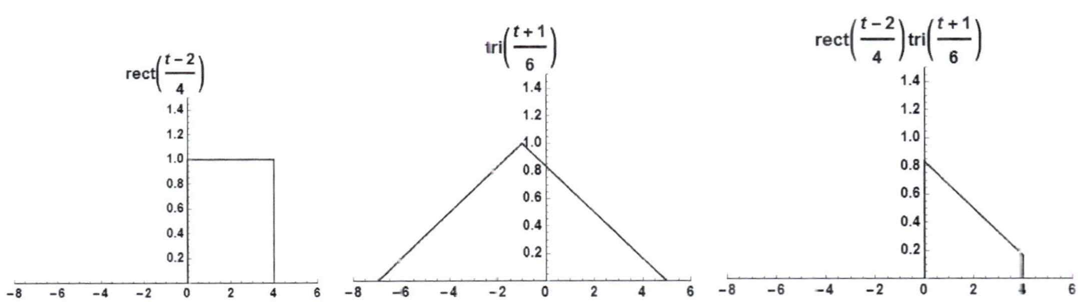

Fig. Hw2-12. Signal $x(t)$ for Problem 4(b).

(c) $x(t) = u(3 - |2t|)$:

denote $u(\tau) = \begin{cases} 1, \tau \geq 0 \\ 0, else \end{cases}$ where $\tau \equiv (3 - |2t|)$. Then

$\tau \geq 0$ for $|2t| \leq 3 \Rightarrow x(t) = 1$ for $|t| \leq 1.5$; $\tau < 0$ for $|t| > 1.5 \Rightarrow x(t) = 0$ for $|t| > 1.5 \therefore x(t) = \begin{cases} 1, |t| \leq 1.5 \\ 0, else \end{cases}$

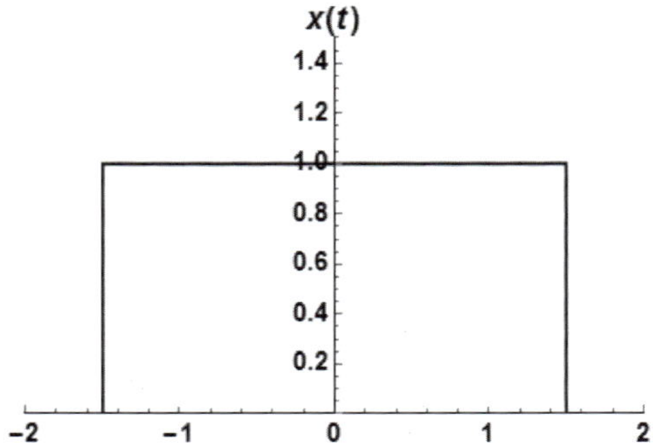

Fig. Hw2-13. Signal $x(t)$ for Problem 4(c).

Problem 5: Find the area under the signal $e^{-t}u(t)$. Use the result and the dependence of the area on the function time scaling to find without integration the area under and the energy of $y(t) = 3e^{-t/2}u(t) - 5e^{-t/4}u(t)$.

Solution:

(a) The area under $y(t) = 3\exp(-t/2)u(t) - 5\exp(-t/4)u(t)$.

If we scale x(t) by α : $x(\alpha t) = \int_0^\infty \exp(-\alpha t)dt, (\alpha > 0)$,

$$x(\alpha t) = \int_0^\infty \exp(-\alpha t)dt = \frac{1}{\alpha}(\text{using integral tables}) \therefore \text{ the area under } x(\alpha t) = \frac{1}{\alpha}.$$

For $y(t) = 3\exp(-t/2)u(t) - 5\exp(-t/4)u(t) = 3x(t/2) - 5x(t/4)$ the areas sum up: the area of $y(t)$ equals the area under $3x(t/2)$ minus the area under $5x(t/4) = |3(2) - 5(4)| = 14$.

(b) The energy of $y(t) = 2\exp(-2t)u(t) - 6\exp(-t)u(t)$.

The energy of $y(t)$ is the area of $(y(t))^2$:

$$\begin{aligned}[y(t)]^2 &= (3\exp(-t/2)u(t) - 5\exp(-t/4)u(t))^2 \\ &= (9\exp(-t)u(t) - 30\exp(-3t/4)u(t) + 25\exp(-t/2)u(t)) \\ &= 9x(t) - 30x(\tfrac{3}{4}t) + 25x(\tfrac{1}{2}t) = 9(1) - 30(\tfrac{4}{3}) + 25(2) = 9 - 40 + 50 = 19.\end{aligned}$$

Problem 6: A linear time-invariant system is said to be *bounded-input-bounded-output (BIBO) stable* (cf. Appendix 1) *if and only if (iff) its impulse response is absolutely integrable*. Determine if systems with the following impulse responses are BIBO stable (a) $x(t)=e^{-t/\tau}u(t)$; **(b)** $x(t)=te^{-t/\tau}u(t)$. **(c)** $x(t)=te^{-t/\tau}\cos(4\pi t)u(t)$, **(d)** $x(t)=\text{sinc}\,(t)$, $-\infty\le t\le\infty$.

Solution:

(a) $\int_{-\infty}^{+\infty}|x(t)|dt=\int_{-\infty}^{+\infty}\left|e^{-t/\tau}u(t)\right|dt=\int_{0}^{+\infty}e^{-t/\tau}dt=(-\tau e^{-t/\tau})|_{0}^{+\infty}=\tau<\infty.$

Since $x(t)=e^{-t/\tau}u(t)$ is absolutely integrable, the corresponding system is BIBO stable.

(b) $\int_{-\infty}^{+\infty}|x(t)|dt=\int_{-\infty}^{+\infty}\left|te^{-t/\tau}u(t)\right|dt=\int_{0}^{+\infty}te^{-t/\tau}dt=(-\tau(te^{-t/\tau}+\tau e^{-t/\tau}))|_{0}^{+\infty}=\tau^2<\infty.$

Since $x(t)=te^{-t/\tau}u(t)$ is absolutely integrable, the corresponding system is BIBO stable.

(c) $\int_{-\infty}^{+\infty}|x(t)|dt=\int_{-\infty}^{+\infty}\left|te^{-t/\tau}\cos(4\pi t)u(t)\right|dt\le\int_{0}^{+\infty}te^{-t/\tau}dt=\tau^2<\infty.$

Since $x(t)=te^{-t/\tau}\cos(4\pi\,\text{t})u(t)$ is absolutely integrable, the corresponding system is BIBO stable.

(d) $\int_{-\infty}^{+\infty}|x(t)|dt=\int_{-\infty}^{+\infty}|\text{sinc}\,(t)|dt=\int_{-\infty}^{+\infty}\left|\frac{\sin\pi t}{\pi t}\right|dt.$

Since the function is even, we have:

$$\int_{-\infty}^{+\infty}\left|\frac{\sin\pi t}{\pi t}\right|dt=\frac{2}{\pi}\int_{0}^{+\infty}\frac{|\sin x|}{|x|}dx=\frac{2}{\pi}\left(\int_{0}^{\pi}\frac{|\sin x|}{|x|}dx+\int_{\pi}^{2\pi}\frac{|\sin x|}{|x|}dx+...\right)=\frac{2}{\pi}\left(\sum_{k=0}^{\infty}\int_{k\pi}^{(k+1)\pi}\frac{|\sin x|}{|x|}dx\right).$$

Since $\frac{|\sin x|}{|x|}>\frac{|\sin x|}{|\max(x)|}=\frac{|\sin x|}{(k+1)\pi}$, $k\pi\le x\le(k+1)\pi$, we have

$$\frac{2}{\pi}\sum_{k=0}^{\infty}\int_{k\pi}^{(k+1)\pi}\frac{|\sin x|}{|x|}dx>\frac{2}{\pi}\sum_{k=0}^{\infty}\int_{k\pi}^{(k+1)\pi}\frac{|\sin x|}{(k+1)\pi}dx=\frac{2}{\pi}\sum_{k=0}^{\infty}\frac{1}{(k+1)\pi}\int_{k\pi}^{(k+1)\pi}|\sin x|dx$$

$$=\frac{2}{\pi^2}\sum_{k=0}^{\infty}\frac{1}{(k+1)}\int_{0}^{\pi}\sin x dx\underset{\substack{k+1=m\\k=0\Rightarrow m=1\\k=\infty\Rightarrow m=\infty}}{=}\frac{2}{\pi^2}\sum_{m=1}^{\infty}\frac{1}{m}\left(-\cos x\Big|_{0}^{\pi}\right)=\frac{4}{\pi^2}\sum_{m=1}^{\infty}\frac{1}{m}\underset{\text{through term-by-term comparison}}{>}$$

$$\frac{4}{\pi^2}\left(1+\frac{1}{2}+\frac{1}{4}+\frac{1}{4}+\frac{1}{8}+\frac{1}{8}+\frac{1}{8}+\frac{1}{8}...\right)=\frac{4}{\pi^2}\left(1+\frac{1}{2}+\frac{1}{2}+\frac{1}{2}+...\right)=\infty\left(\text{harmonic series }\sum_{m=1}^{\infty}\frac{1}{m}\text{ diverges}\right).$$

Thus, $x(t)=\text{sinc}\,(t)$, $-\infty\le t\le\infty$, is not absolutely integrable, and the corresponding system is not BIBO stable.

Problem 7: The response of a stable continuous-time linear time-invariant system is in the general case characterized by an energy signal given by the product of just three types of functions of time: sine, polynomial, and decaying exponential. Compute the energy of the following signals: (a) $x(t)=e^{-t/\tau}u(t)$; (b) $x(t)=te^{-t/\tau}u(t)$. (c) $x(t)=te^{-4t/\tau}\cos(2\pi t)u(t)$.

Solution:

(a) $E_x=\int_{-\infty}^{\infty}\left(e^{-\frac{t}{\tau}}u(t)\right)^2 dt=\int_0^{\infty}e^{-\frac{2t}{\tau}}dt=\frac{\tau}{2}$,

(b) $E_x=\int_{-\infty}^{\infty}\left(te^{-\frac{t}{\tau}}u(t)\right)^2 dt=\int_0^{\infty}t^2e^{-\frac{2t}{\tau}}dt=-\frac{\tau}{2}\int_0^{\infty}t^2de^{-\frac{2t}{\tau}}=-\frac{\tau}{2}\left[t^2e^{-\frac{2t}{\tau}}\Big|_0^{\infty}-2\int_0^{\infty}te^{-\frac{2t}{\tau}}d\right]=-\frac{\tau}{2}\left[0-2\int_0^{\infty}te^{-\frac{2t}{\tau}}d\right]$

$$=\tau\int_0^{\infty}te^{-\frac{2t}{\tau}}d=\tau\left(-\frac{\tau}{2}\int_0^{\infty}tde^{-\frac{2t}{\tau}}\right)=-\tau\cdot\frac{\tau}{2}\left[te^{-\frac{2t}{\tau}}\Big|_0^{\infty}-\int_0^{\infty}e^{-\frac{2t}{\tau}}dt\right]=\frac{\tau^2}{2}\int_0^{\infty}e^{-\frac{2t}{\tau}}dt=\frac{\tau^3}{4},$$

(c) $E_x=\int_{-\infty}^{\infty}\left(te^{-\frac{4t}{\tau}}\cos(2\pi t)u(t)\right)^2 dt=\int_0^{\infty}t^2e^{-\frac{8t}{\tau}}\cos^2(2\pi t)dt=\int_0^{\infty}t^2e^{-\frac{8t}{\tau}}\frac{\cos(4\pi t)+1}{2}dt$

$$=\int_0^{\infty}t^2e^{-\frac{8t}{\tau}}\frac{\cos(4\pi t)}{2}dt+\int_0^{\infty}t^2e^{-\frac{8t}{\tau}}\frac{1}{2}dt=E_{x1}+E_{x2},$$

$$E_{x1}=\int_0^{\infty}t^2e^{-\frac{8t}{\tau}}\frac{\cos(4\pi t)}{2}dt$$

$$=-\frac{\tau}{16}\left[t^2e^{-\frac{8t}{\tau}}\cos(4\pi t)\Big|_0^{\infty}-\int_0^{\infty}e^{-\frac{8t}{\tau}}\left(2t\cos(4\pi t)-4\pi t^2\sin(4\pi t)\right)dt\right]$$

$$=\frac{\tau}{16}\int_0^{\infty}e^{-\frac{8t}{\tau}}\left(2t\cos(4\pi t)-4\pi t^2\sin(4\pi t)\right)dt$$

$$=\frac{\tau}{8}\int_0^{\infty}te^{-\frac{8t}{\tau}}\cos(4\pi t)dt-\frac{\pi\tau}{4}\int_0^{\infty}t^2e^{-\frac{8t}{\tau}}\sin(4\pi t)dt$$

$$=\frac{\tau}{8}\int_0^{\infty}te^{-\frac{8t}{\tau}}\cos(4\pi t)dt+\frac{\pi\tau^2}{32}\left[t^2e^{-\frac{8t}{\tau}}\sin(4\pi t)\Big|_0^{\infty}-\int_0^{\infty}e^{-\frac{8t}{\tau}}\left(4\pi t^2\cos(4\pi t)+2t\sin(4\pi t)\right)dt\right]$$

$$=\frac{\tau}{8}\int_0^{\infty}te^{-\frac{8t}{\tau}}\cos(4\pi t)dt-\frac{\pi\tau^2}{32}\int_0^{\infty}e^{-\frac{8t}{\tau}}\left(4\pi t^2\cos(4\pi t)+2t\sin(4\pi t)\right)dt$$

$$=\frac{\tau}{8}\int_0^{\infty}te^{-\frac{8t}{\tau}}\cos(4\pi t)dt-\frac{\pi^2\tau^2}{8}\int_0^{\infty}e^{-\frac{8t}{\tau}}t^2\cos(4\pi t)dt-\frac{\pi\tau^2}{16}\int_0^{\infty}e^{-\frac{8t}{\tau}}t\sin(4\pi t)dt$$

$$=\frac{\tau}{8}\int_0^{\infty}te^{-\frac{8t}{\tau}}\cos(4\pi t)dt-\frac{\pi^2\tau^2}{8}2E_{x1}-\frac{\pi\tau^2}{16}\int_0^{\infty}e^{-\frac{8t}{\tau}}t\sin(4\pi t)dt.$$

Then we can obtain $E_{x1}=\dfrac{1}{1+\frac{\pi^2\tau^2}{4}}\left(\frac{\tau}{8}\int_0^{\infty}te^{-\frac{8t}{\tau}}\cos(4\pi t)dt-\frac{\pi\tau^2}{16}\int_0^{\infty}e^{-\frac{8t}{\tau}}t\sin(4\pi t)dt\right)$.

Following the similar procedure, using the integration by parts again yields

$$E_{x1}=\frac{4\tau^3-3\tau^5\pi^2}{32(4+\tau^2\pi^2)^3}.$$

The calculation for E_{x2} follows the same procedure given in (b), yielding:

$$E_{x2}=\frac{\tau^3}{512}.$$

Then, the energy of the signal is

$$E_x = E_{x1} + E_{x2} = \frac{4\tau^3 - 3\tau^5\pi^2}{32(4+\tau^2\pi^2)^3} + \frac{\tau^3}{512}.$$

Problem 8: The system rise time is a measure of system agility, defined as the time required to rise from 10% to 90% of the final value in response to a step input. Compute the rise time of the system with an impulse response given by $x(t)=(1-e^{-t}\cos\omega t)u(t)$, where $\omega = 2\pi\cdot 4.6$.

Solution:

(a) The function $x(t)=(1-e^{-t}\cos\omega t)u(t)$ is shown below:

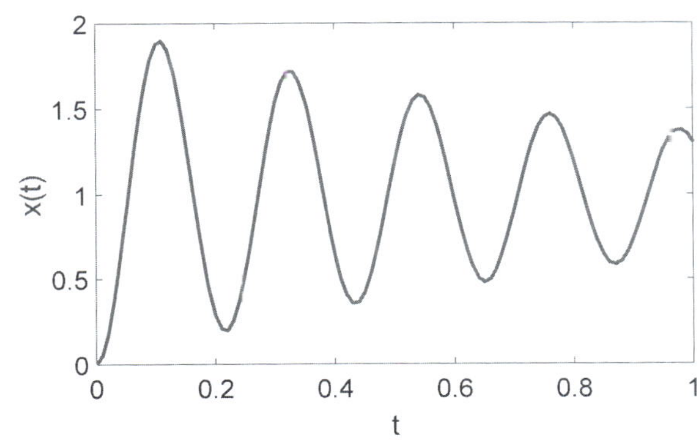

Fig. Hw2-14. Response for Problem 8.

First, we note that

$$\lim_{t\to\infty} x(t) = \lim_{t\to\infty}\left(1-e^{-t}\cos\omega t\right)u(t) = 1 - \lim_{t\to\infty}\left(e^{-t}\cos\omega t\right) = 1-0=1.$$

So 10% of the total rise would be 0.1, and 90% of the total rise would be 0.9.

Now, solving for a specific value $x(t)=x_0$ gives

$$x(t) = 1-e^{-t}\cos\omega t \Rightarrow e^{-t}\cos\omega t = 1-x_0.$$

Now, let's consider the first time that the response reaches 10% and 90% of the final value. Finding the latter two values numerically yields:

$$t_{10\%} = 0.0146s,$$
$$t_{90\%} = 0.0507s.$$

The rise time is then given by $t_{rise} = t_{90\%} - t_{10\%} = 0.0507 - 0.0146 = 0.0361s$.

Problem 9: Let $x[n] = \{1, \hat{5}, 4, 3, 4\}$ **where the hat denotes the position of value at** $n = 0$**. Draw the following signals and find their signal energy: (a)** $y_1[n] = x[n-3]$**; (b)** $y_2[n] = x[n+4]$**; (c)** $y_3[n] = x[-n+3]$**; (d)** $y_4[n] = x[-n-4]$**.**

Solution:

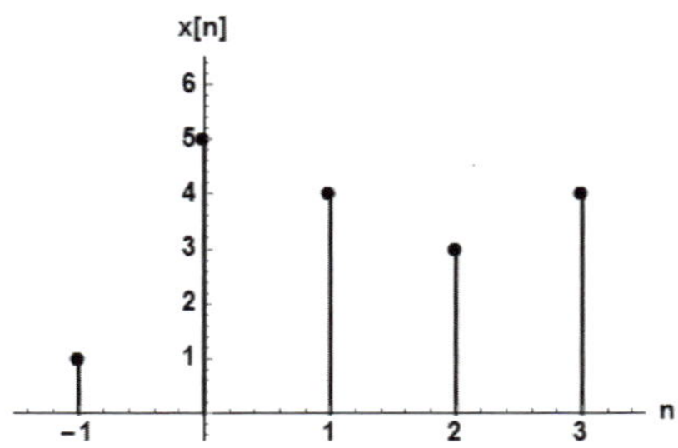

Fig. Hw2-15. Signal $x[n]$ for Problem 9.

(a) $y_1[n] = x[n-3]$:

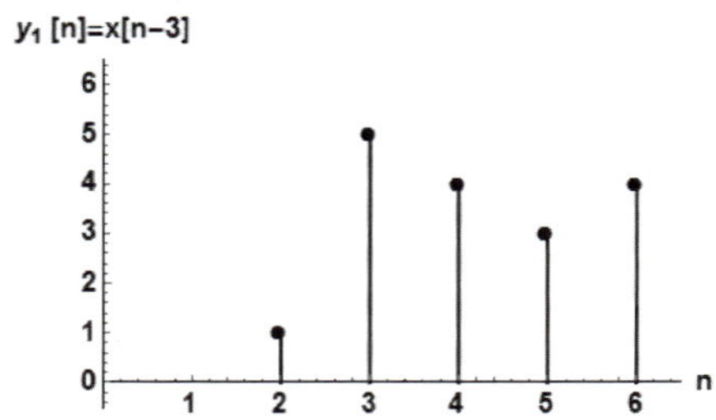

Fig. Hw2-16. Signal $y[n]$ for Problem 9.

(b) $y_2[n] = x[n+4] = x[n-[-4]]$:

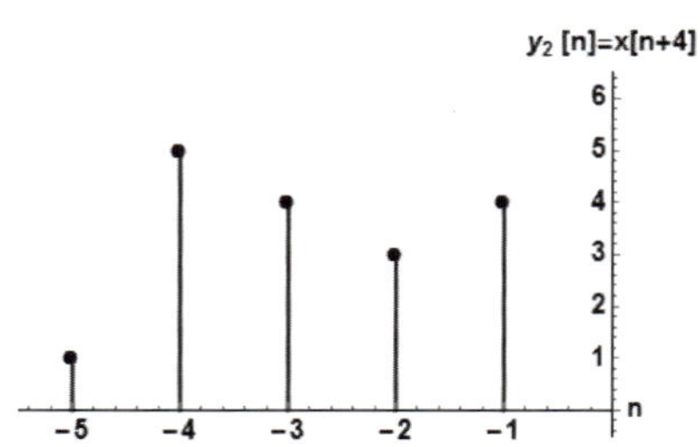

Fig. Hw2-17. Signal $f[n]$ for Problem 9.

(c) $y_3[n]=x[-n+3]=x[-[n-3]]$:

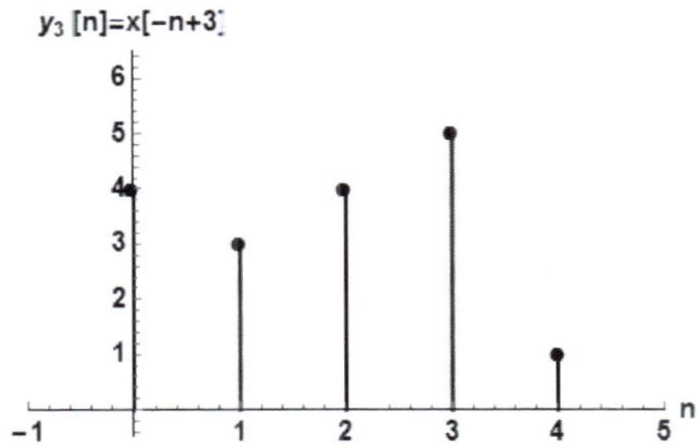

Fig. Hw2-18: Signal $g[n]$ for Problem 9.

(d) $y_4[n]=x[-n-4]=x[-[n+4]]=x[-[n-[-4]]]$:

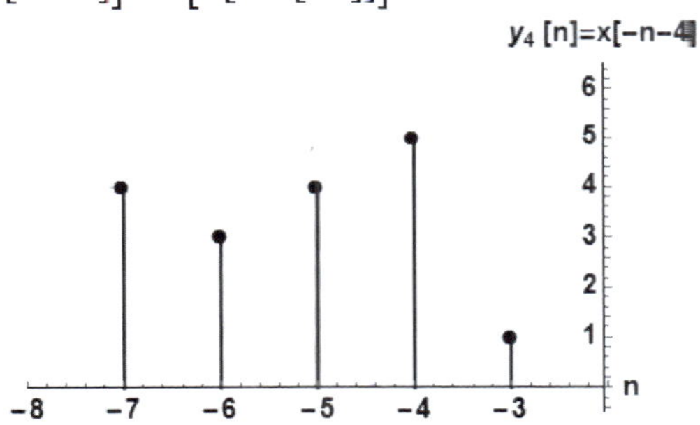

Fig. Hw2-19. Signal $h[n]$ for Problem 9.

The energy of each of them is: $E_x=1^2+5^2+4^2+3^2+4^2=67$.

Problem 10: Let $x[n]=6(0.7)^n\left(u[n+2]-u[n-3]\right)$. **Sketch the following signals:**
(a) $f[n]=x[n+3]$; **(b)** $h[n]=x[-n+3]$.
Solution:

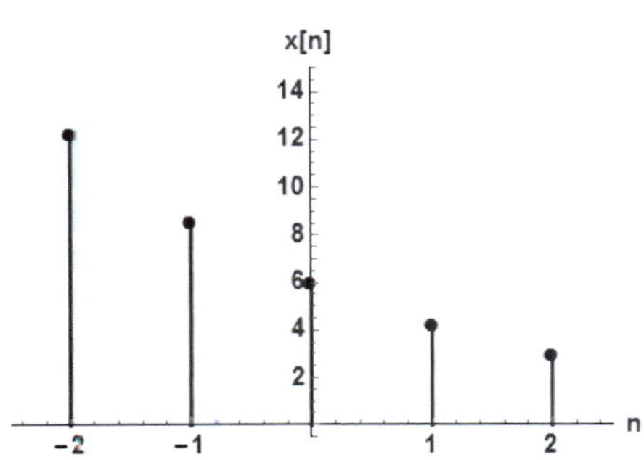

Fig. Hw2-20. Signal $x[n]$ for Problem 10.

(a) $f[n]=x[n+3]=x[n-[-3]]$:

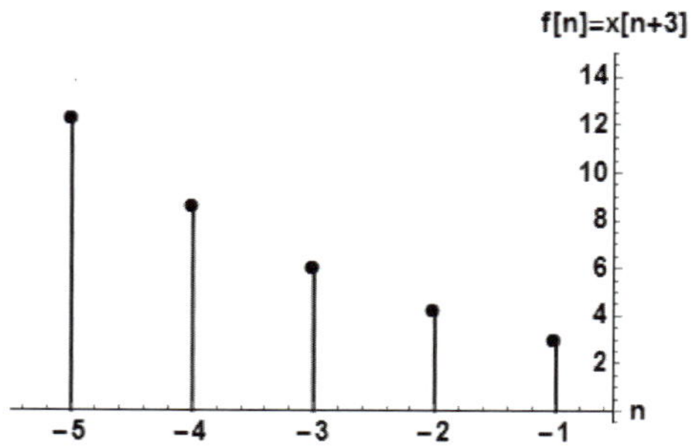

Fig. Hw2-21. Signal $f[n]$ for Problem 10.

(b) $h[n]=x[-n+3]=x[-[n-3]]$:

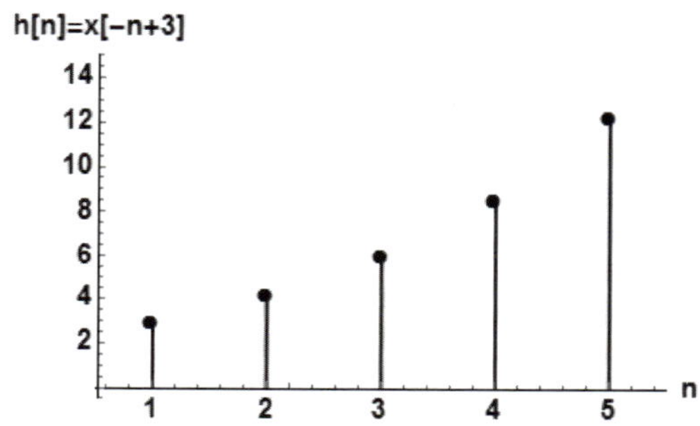

Fig. Hw2-22. Signal $h[n]$ for Problem 10.

Problem 11: Let $x[n]=\{2, 6, -1, \hat{3}, 0, 4, 5, 0, -2, 6\}$ **. Find and sketch the following signals: (a) The decimated signal** $y_1[n]=x[3n]$ **; (b, c, d) The zero-interpolated, step-interpolated, and linearly-interpolated signal** $y_2[n]=x\left[\frac{n}{3}\right]$.

Solution:

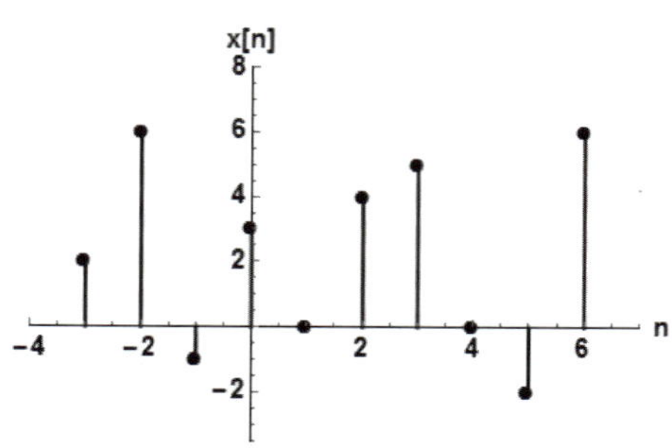

Fig. Hw2-23. Signal $x[n]$ for Problem 11.

(a) The decimated signal $y_1[n] = x[3n]$:

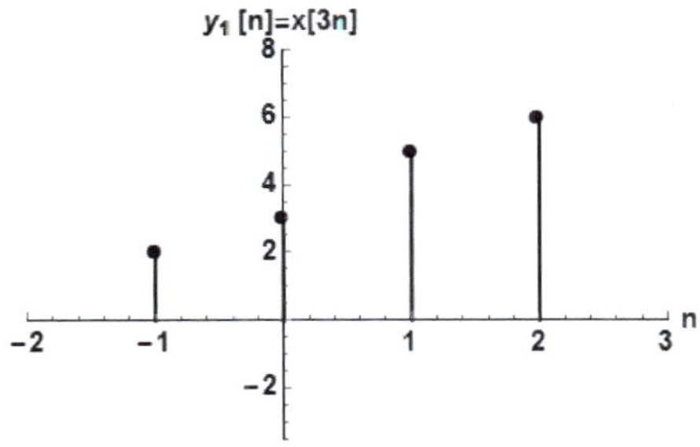

Fig. Hw2-24 Signal *d*[*n*] for Problem 11.

(b) The zero-interpolated signal $y_2[n] = x\left[\frac{n}{3}\right]$:

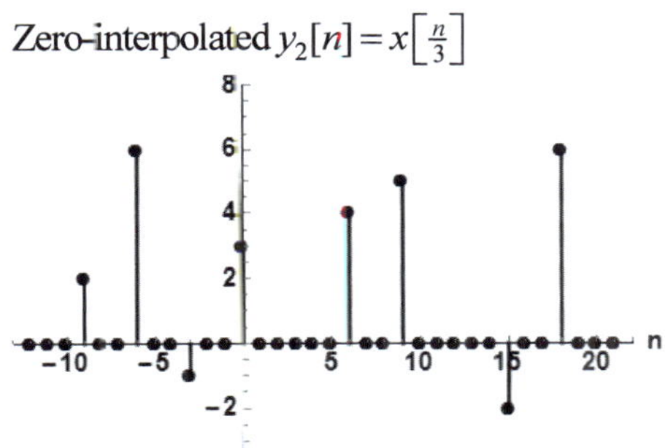

Fig. Hw2-25: Signal *f*[*n*] for Problem 11.

(c) The step-interpolated signal $y_2[n] = x\left[\frac{n}{3}\right]$:

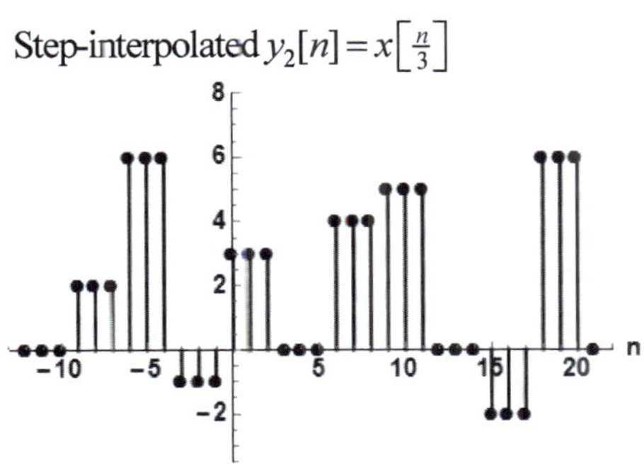

Fig. Hw2-26. Signal *g*[*n*] for Problem 11.

(d) The linearly-interpolated signal $y_2[n] = x\left[\frac{n}{3}\right]$:

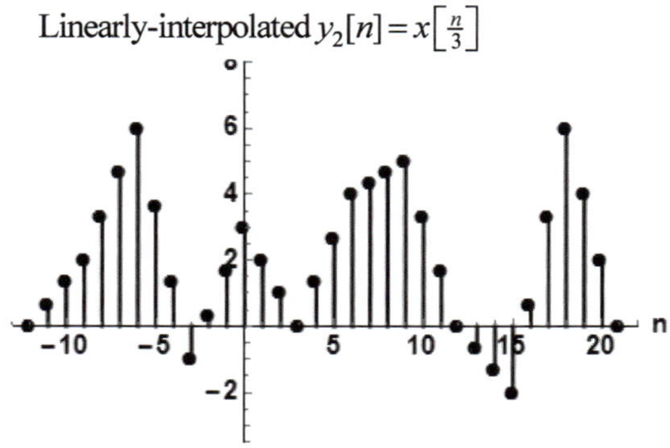

Fig. Hw2-27. Signal $h[n]$ for Problem 11.

Problem 12. One of the properties of a linear system is that "a constant goes through" it. More specifically, let an input signal produce the corresponding output signal. Then, if the input signal is multiplied by a constant and the resulting signal is applied to the system, the resulting output signal is equal to the previous output signal multiplied by the same constant.

A) Check by simulation if the beam of Homework 1, with no mass attached to it, is linear: apply to it a step signal of 1.4 mm magnitude and compute the response, then multiply input signal by 2, apply the new signal, compute and examine the beam response, and comment on beam linearity. B) Repeat task of A) with mass attached to the beam. In this case, check if the system is affine (incrementally linear) due to a constant offset, in which case the beam would provide linear action after the offset removal. C) Do the same for the hydraulic actuator (with P-controller) of Homework 1 not attached to the beam. In A–C tasks above, provide both input and response graphs in an easily comparable format to support conclusions. Follow submission format given in Homework 1.

Problem 12 solution guide.

This problem requires making some changes to the Simulink beam and testbed models from Homework 1. Specifically, the task is to run step-response simulations of:

A. The beam alone without any mass attached
B. The beam with the "mold" mass attached
C. The piston hydraulics alone.

Part A. The best way to do this is to start from the "Beam_simulation.slx" model from Homework 1. Four Simulink model modifications need to be made:

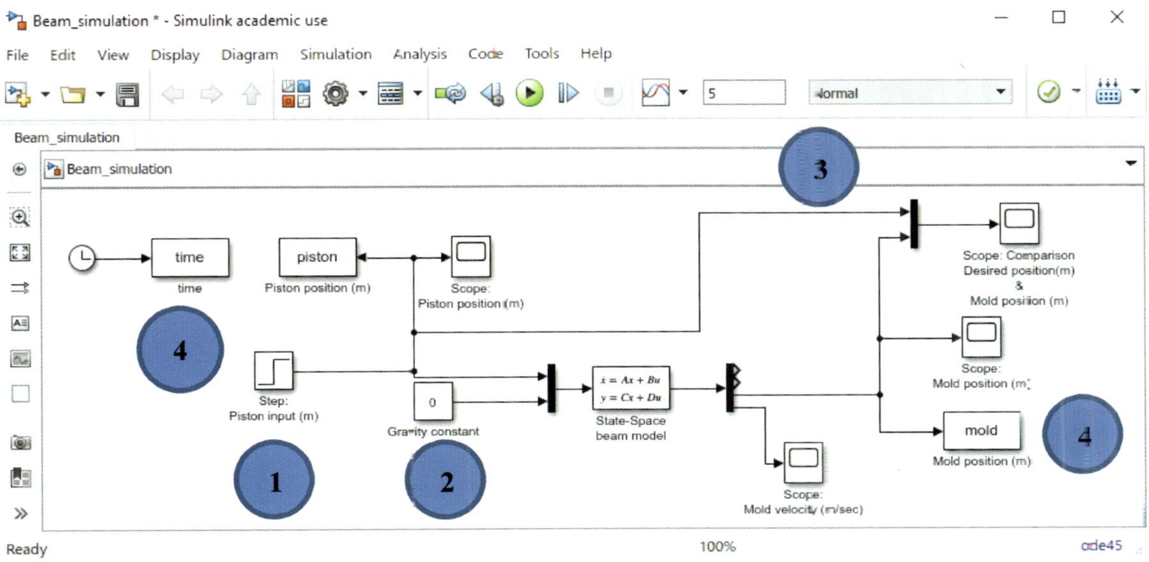

Fig. Hw2-28. Simulink model for beam system.

The first two modifications are required for running the model as specified in the problem tasks.

1. Change the "Sine Wave" input to a "Step" input. Input blocks can be found in the "Sources" sub-menu in the Simulink Library Browser. To open the library, either go to "View\Library Browser" when looking at the model, or type "Simulink" at the command prompt and press enter. The parameters of the Step block also need to be adjusted. Have the step start at time 0 with initial value 0. Since the units of the model are in meters, the final value is needed to be either 0.0014 or 0.0028 for the two cases required to be simulated. See the screenshot in Figure Hw2-29 on the next page for an example.
2. Change the "Gravity constant" to 0. This is not equivalent to having no mass attached to the beam, but close enough for the purposes of this problem. The mass will still have the same inertia, but there will be no gravitational force acting on it.

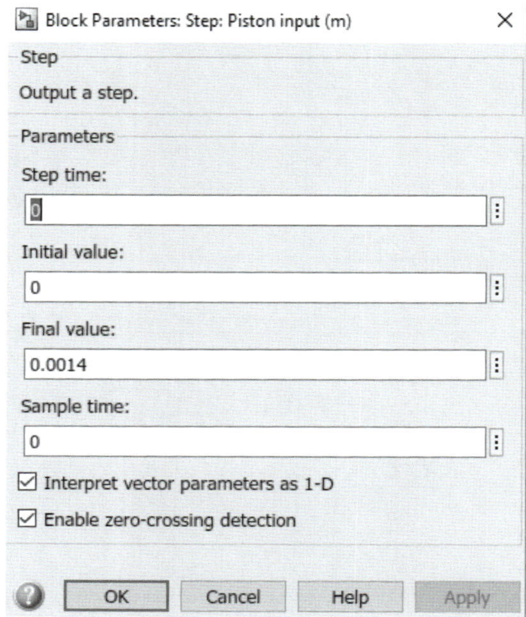

Fig. Hw2-29. Simulink model parameters window.

The next two modifications simply help getting better-looking plots.

3. A sufficient simulation time interval needs to be selected to support conclusions. A 5 seconds interval looks long enough for parts A and B.
4. Obtain the entire system response for the plots. In Homework 1, only the last 2.5 seconds data record was saved since the behavior with the initial transient having died out was of interest. Here, however, the initial transient itself is the focus of the question. So change the "Sample Time" and "Limit data points to last" options to generate a good plot. The screen shot in Figure Hw2-30 on the next page shows an example of collecting 5 seconds of data (the recommended simulation time interval in point 3, above) every 0.002 seconds. Hence, it collects 2500 data points.

The question asks to compare the responses to steps of 1.4 mm and 2.8 mm height. If the system is linear, we expect the output magnitude for the 2.8 mm step to be exactly twice that for the 1.4 mm step. Both outputs should be plotted side-by-side. For direct comparison through graphical visualization, the single unit on the y-axis of the response graph to the 1.4 mm step should exactly correspond to two units on the y-axis of the response graph to the 2.8 mm step. If the system is linear, the two plots should look identical.

Part B. Part B is done in the same manner as Part A, with the exception of setting the "Gravity Constant" block back to 1.

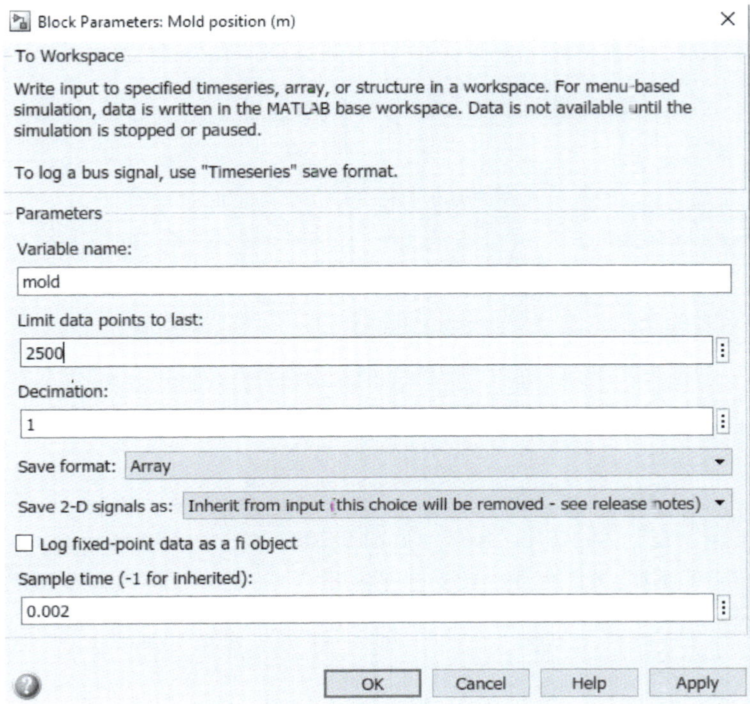

Fig. Hw2-30. Simulink model parameters window.

Part C. Part C asks to simulate the testbed piston hydraulics without the beam attached. This requires starting from the "Testbed_simulation" model and deleting the beam. The end result will look like the screenshot in Figure Hw2-31 on the next page. It doesn't matter that the "Shear stress" input to the piston has no connection.

The simulation interval and "Limit data points to last" options of the output blocks will still need to be adjusted. A shorter time period, for example only 0.5 seconds, will be needed for this simulation.

Since the question asks for "input" (step) and "response" (piston, for Part C) graphs, a "To Workspace" block will need to be added for the input. The easiest way to do this is to copy and paste the "Piston position" block and then change the name. This is shown in the screenshot below.

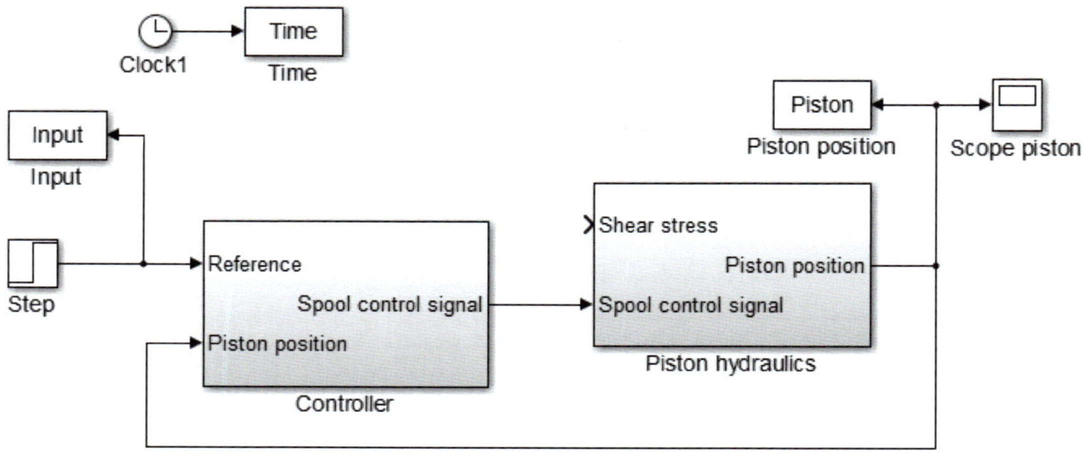

Fig. Hw2-31. Modified Simulink testbed model.

Problem 13. Apply a 3 mm step input to the closed loop testbed without tracking controller (Problem 3, Homework 1). A) Plot the response, indicate its likely composition in terms of elementary functions, and write down its approximate analytical expression. B) Find the system rise time as defined in Problem 8.

2.4 Basic characteristics of a CT sinusoidal signal

Sinusoidal signals form the basis of most of the techniques introduced in this book. ***Basic sinusoid*** is defined as shown in Figure 2.4-1. The "CT" qualifier will be omitted where this causes no confusion. The characteristics of sinusoid are ***period, frequency, phase***, and ***magnitude***, which are really nothing other than the inverse and the direct argument scaling, shifting, and magnitude scaling, as will be shown below. We will also consider ***combinations of sinusoids*** and their simplified graphical representation — ***discrete real spectra***.

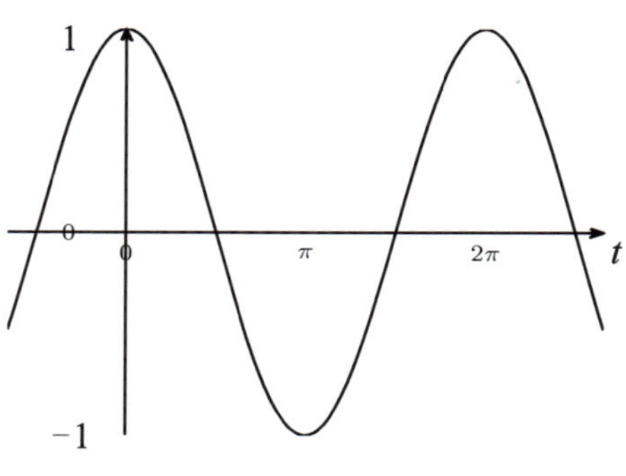

Fig. 2.4-1. Basic sinusoid signal: $x(t) = \cos(t)$.

2.4.1 *Period and frequency of sinusoid*

A periodic signal is the one that repeats itself, so that $x(t+T) = x(t)$ for any time t and some time interval T, called ***period***. The smallest such T is called ***fundamental period***, T_f. A set of values the function takes over its fundamental period will be referred to as a ***cycle.*** The standard argument scaling notation buildup and its effect on the fundamental period of the sinusoid are shown in Figure 2.4-2.

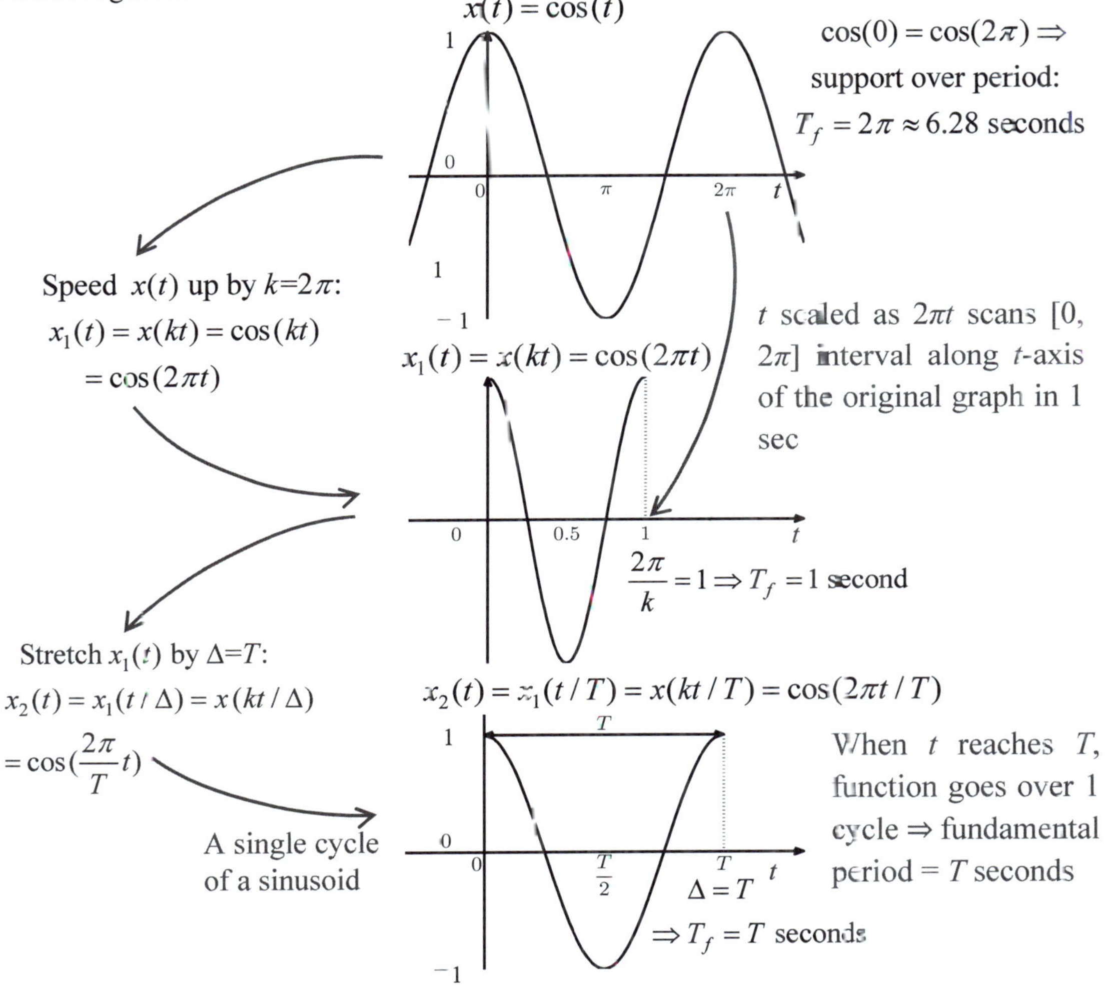

Fig. 2.4-2. Cycle and period of a sinusoid under changes in argument scaling.

Analog frequency of CT sinusoid. The ***frequency*** f in Hz, or ***analog frequency***, is defined as $1/T$, with T in seconds. The ***analog angular frequency*** ω in rad/sec is defined as $2\pi/T = 2\pi f$. "Analog" qualifier will be mostly omitted. Thus,

$$f = \frac{1}{T}\,\text{Hz}\left(\frac{\text{cycle}}{\text{sec}}\right),\ \omega = \frac{2\pi}{T}\,\frac{\text{rad}}{\text{s}},\ \omega = 2\pi f \Rightarrow 1\frac{\text{rad}}{\text{s}} \approx 6.28\,\text{Hz or } 1\,\text{Hz} \approx 0.16\frac{\text{rad}}{\text{s}},\ T = \frac{1}{f} = \frac{2\pi}{\omega}\text{sec}$$

$$\therefore \cos\left(\frac{2\pi}{T}t\right) = \cos(\omega t) = \cos(2\pi f t).$$

To clearly see what frequency in Hz means we note that

$$\cos[2\pi\underbrace{(\frac{1}{T})}_{f}t] = \cos[\underbrace{(\frac{2\pi}{T})}_{\omega}t] \Rightarrow f\,Hz = \frac{1}{T\,\text{sec}} = \frac{1}{T}\frac{1}{1\,\text{sec}} = \frac{1/T}{1\,\text{sec}}.$$

Thus, frequency f in Hz shows how frequently function repeats itself in one second. E.g., T = 0.5 sec ⇒ 1/T = 2 Hz, i.e. 2 Hz indicates two cycles per second, as shown in Figure 2.4-3, while T = 2 sec. ⇒ 1/T = 0.5 Hz, indicating a fraction, 0.5, of one cycle per second.

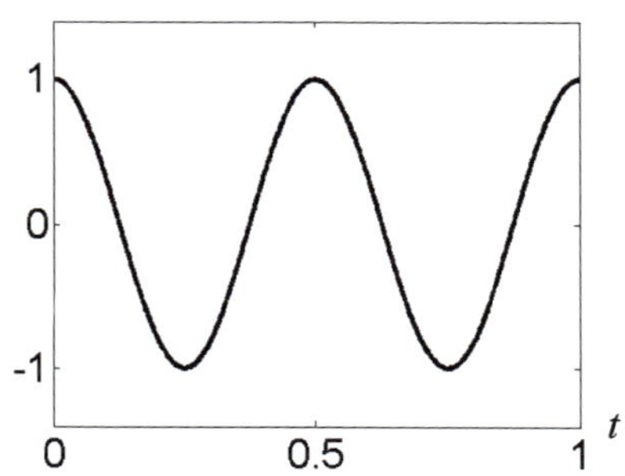

Fig. 2.4-3. Two cycles per second: sinusoidal signal with frequency of 2 Hz.

2.4.2 *Sums of sinusoids*

Periods T_1, T_2, …, T_k, are said to be ***commensurable***, if there exists a number T_{cm}, referred to as a ***common multiple*** of T_1, T_2, …, T_k, such that $T_{cm} = a_1T_1 = a_2T_2 = \ldots = a_kT_k$ where a_1, …, a_k are all integers. The smallest such number is called ***the least common multiple***, T_{lcm}. T_{cm} doesn't always exist. Example: $\cos(t) + \cos(\pi t)$. In this case, the resulting signal has no period and is called ***almost periodic***.

2.4.2.1 *Sums of sinusoids with different frequencies, but commensurable periods*

Given sinusoids $x_1(t)$, $x_2(t)$, …, $x_k(t)$ with commensurable periods T_1, T_2, …, T_k, respectively, their sum is periodic with fundamental period T_f given by the least common multiple T_{lcm} of their periods T_1, T_2, …, T_k.

Example 2.4-1: Find the fundamental period of $x(t) = x_1(t) + x_2(t)$, where $x_1(t)$ and $x_2(t)$ are shown in Figure 2.4-4.

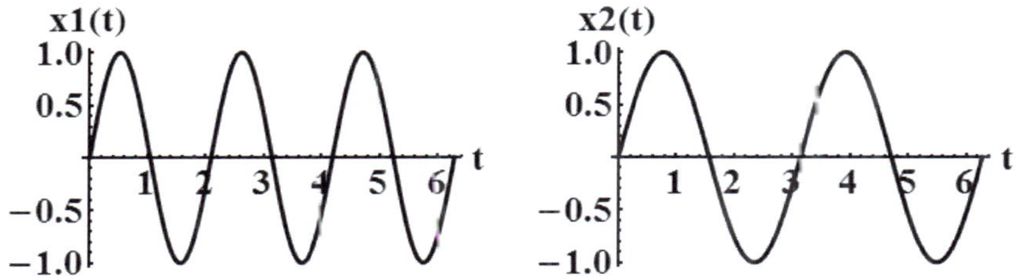

Fig. 2.4-4 Sinusoids for Example 2.4-1.

Solution: It is seen that during the same time interval the left graph goes through three cycles, while the right graph — through two cycles. Therefore, the least common multiple, and hence, the fundamental period, are given by $T_f = T_{lcm} = 3T_1 = 2T_2$.

Example 2.4-2: Find the fundamental period of $x(t) = 2\sin\left(\frac{2}{3}t\right) + 3\cos\left(\frac{1}{2}t\right) + 4\cos\left(\frac{1}{3}t - \frac{1}{5}\pi\right)$ shown in Figure 2.4-5.

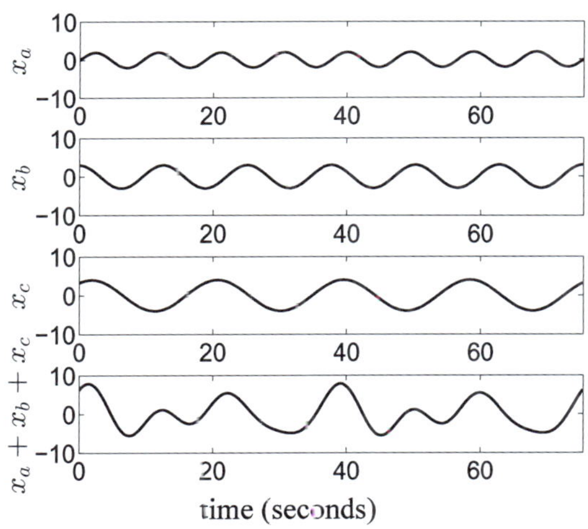

Fig. 2.4-5. Sinusoids and their sum for Example 2.4-2.

Solution: First, we find all periods:

$$\omega_a = \frac{2\pi}{T_a} = \frac{2}{3} \Rightarrow T_a = 3\pi, \omega_b = \frac{2\pi}{T_b} = \frac{1}{2} \Rightarrow T_b = 4\pi, \omega_c = \frac{2\pi}{T_c} = \frac{1}{3} \Rightarrow T_c = 6\pi \,.$$

Then the least common multiple is seen to be $T_f = 4T_a = 3T_b = 2T_c = 12\pi$.

If the periods are given by the integer ratios, the least common multiple is easily seen after decomposing composite numbers into prime factors, as shown next.

Example 2.4-3: Find the fundamental period of $x(t)=\sin(24\pi t)+\cos(40\pi t)$.

Solution: First, we find all the periods and decompose them into prime factors:

$$T_a=\frac{2\pi}{\omega_a}=\frac{2\pi}{24\pi}=\frac{1}{12}=\frac{1}{2\cdot 2\cdot 3},\quad T_b=\frac{2\pi}{\omega_b}=\frac{2\pi}{40\pi}=\frac{1}{20}=\frac{1}{2\cdot 2\cdot 5}.$$

Now, ½ is seen to be a common multiple, but it is not the least common multiple. The latter is seen to be given by ½ ½ = ¼. Thus, the integer multiples for computing the least common multiple are seen to be given by 3 and 5, yielding $T_f=3T_a=5T_b=\dfrac{1}{2\cdot 2}=0.25$ with waveforms shown in Figure 2.4-6.

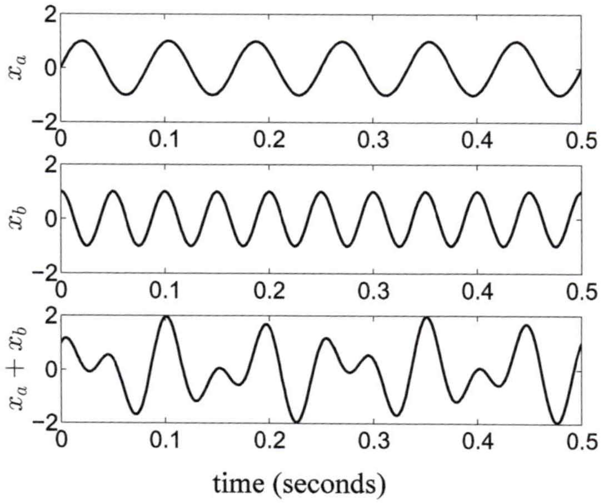

Fig. 2.4-6. Sinusoid and their sum for Example 2.4-3.

2.4.2.2 *Software testbed simulation data*

Let us now take a look at the composition of ***the software testbed simulation data***.

Example 2.4-4: Given the distorted software testbed mass position profile shown in Figure 2.4-7, determine if it is composed of a finite sum of sinusoids at different frequencies and find these sinusoids.

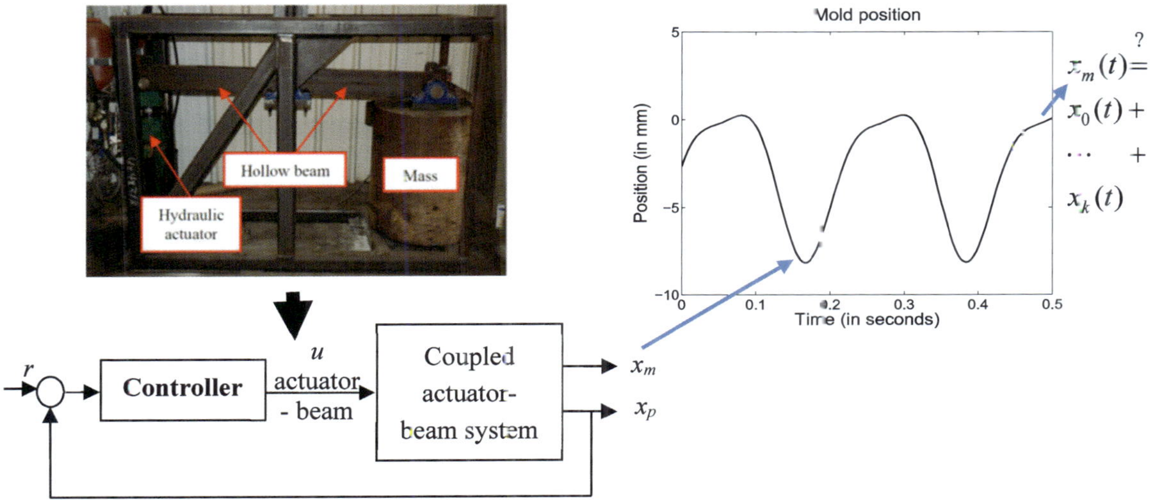

Fig. 2.4-7. Software testbed mass position simulation data.

Solution: We can't yet solve this problem, but we can understand its solution: the distorted mass position profile shown above almost exactly decomposes into a constant (DC component), x_0, sinusoid, x_1, at the reference frequency 4.6 Hz, and sinusoid, x_2, at the 1st beam resonance frequency 9.2 Hz, as shown in Figure 2.4-8:

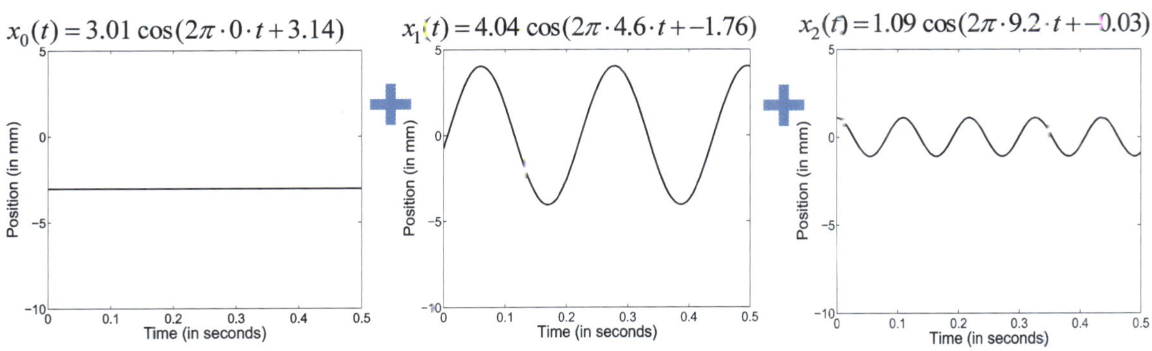

Fig. 2.4-8. Software testbed mass position simulation data decomposition: components from left to right — the DC offset, the reference (desired) frequency, and the 1st beam resonance frequency.

2.4.2.3 *Hardware testbed simulation data*

Let's now set the stage for the search of the mold position profile distortion origin.

Example 2.4-5: In Figure 2.4-9 hardware testbed piston position profile looks like a single undistorted sinusoid, but mass position profile looks distorted and is, in fact, a sum of sinusoids at two frequencies: the reference frequency and the 1st beam resonance frequency.

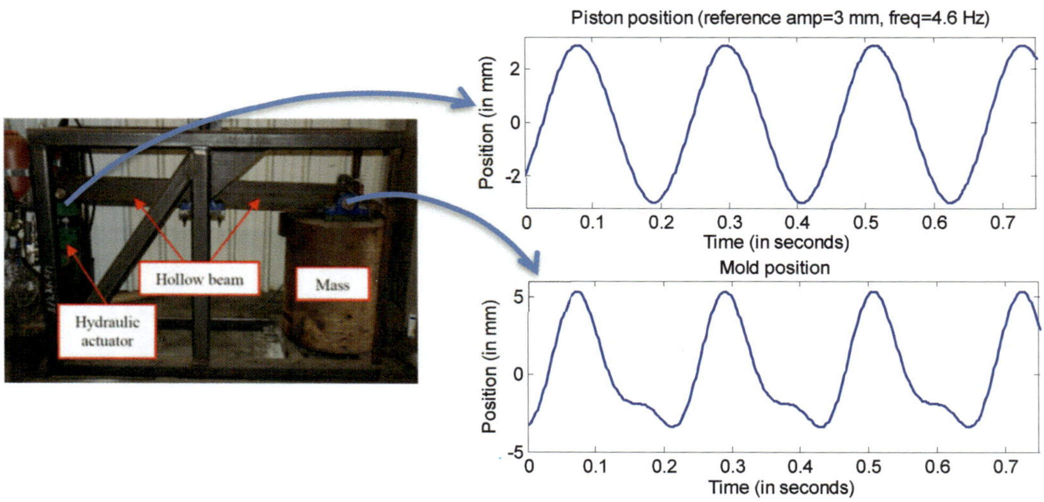

Fig. 2.4-9. Hardware testbed experimental data: positions of piston (looks clean) and mold (distorted).

2.4.2.4 *Production mold experimental data*

The production unit mold exhibits a more complicated distortion than the hardware testbed, as indicated next.

Example 2.4-6: Distorted production mold position and velocity profiles are sums of sinusoids at three frequencies: reference frequency, 1st beam resonance frequency, and 2nd beam resonance frequency. Velocity profile is seen to be a time derivative of position profile.

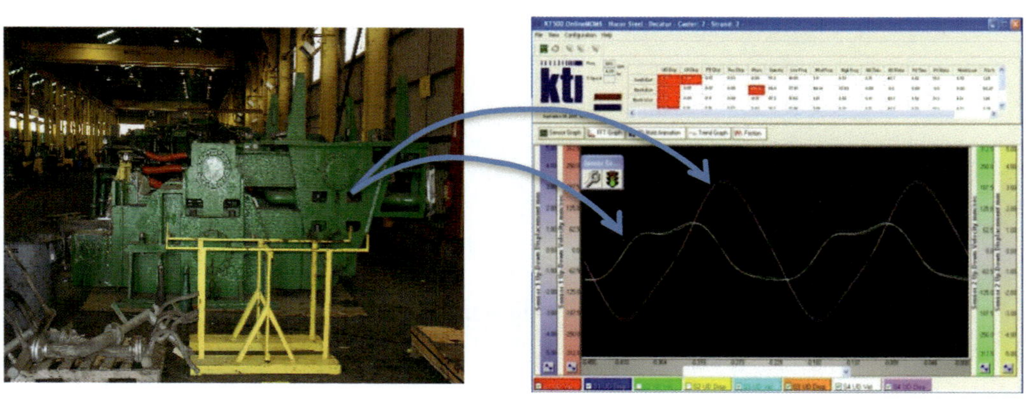

Fig. 2.4-10. Production model experimental data.

2.4.3 *Phase of sinusoid*

Phase, often referred to as ***phase shift*** when used in the context of comparison to a reference sinusoid, provides one of the main signal characterizations that is used in a variety of signal processing and control applications: speed detection, system identification, distortion quantification, musical sound effect processors like phasers and flangers that use phase-shift networks, stability and robustness analysis, and many others. Phase represents a frequency-scaled time shift of the sinusoid with respect to zero taken with negative sign. To see this let us fix ω at $\omega_1 = 2\pi/T_1$ and shift $\cos(\omega_1 t)$ by positive number t_1 as shown in Figure 2.4-11.

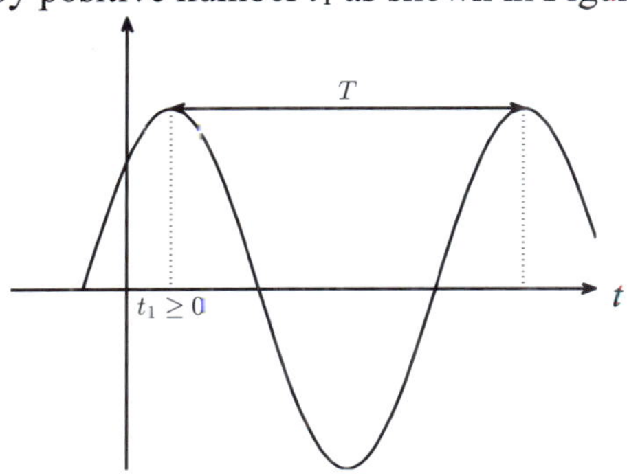

Fig. 2.4-11. Sinusoid with positive time sift.

This yields

$$\cos\left(\omega_1\left(t-t_1\right)\right)=\cos\left(\omega_1 t+(-\omega_1 t_1)\right)=\cos\left(\omega_1 t-\theta_1\right)$$

where

$$\theta_1 = -\omega_1 t_1 \text{ rad} = -\frac{2\pi}{T_1}t_1 \text{ rad} = -\text{ angle fraction } \frac{t_1}{T_1} \text{ of } 2\pi .$$

Thus, ***θ_1 negative ⇒ t_1 positive ⇒ time delay, or time lag***. For the latter reason, negative phase is often referred to as ***phase lag***, and, respectively, positive phase — as ***phase lead***.

Going back to time shift yields

$$\cos\left(\omega_1 t+\theta_1\right)=\cos\left(\omega_1\left(t+\frac{\theta_1}{\omega_1}\right)\right)=\cos\left(\omega_1\left(t-\left(-\frac{\theta_1}{\omega_1}\right)\right)\right)=\cos(\omega_1(t-t_1))$$

where $t_1 = -\dfrac{\theta_1}{\omega_1} = -T_1\dfrac{\theta_1}{2\pi}\text{sec} = -\text{ time fraction } \dfrac{\theta_1}{2\pi} \text{ of } T_1$.

Thus, ***t_1 positive ⇒ θ_1 negative ⇒ phase lag***.

Example 2.4-7: Find the period. Determine the effect of the phase shift in $\cos\left(\frac{\pi}{4}t-\frac{\pi}{8}\right)$ in terms of the time shift with respect to $\cos\left(\frac{\pi}{4}t\right)$.

Solution: The frequency of the waveform is seen to be $\omega_1 = \pi/4$ rad. Therefore, the period is $T_1 = \frac{2\pi}{\omega_1} = \frac{2\pi}{\pi/4} = 8$ sec, and the time shift is $t_1 = -\frac{\theta_1}{\omega_1} = -\frac{-\pi/8}{\pi/4} = \frac{1}{2} > 0$ sec. Thus, the time shift is positive, and the waveform $\cos\left(\frac{\pi}{4}t - \frac{\pi}{8}\right)$ lags the waveform $\cos\left(\frac{\pi}{4}t\right)$ by 0.5 seconds.

2.4.4 *Magnitude of sinusoid*

Magnitude of a sinusoid is the maximal value the sinusoid takes over a period. It can be thought of as a positive scaling coefficient, or a positive gain (amplification) applied to a sinusoid with the maximum value of 1. In the latter context it is used in system identification, resonance characterization, distortion quantification, stability analysis, and many other applications. If we scale and shift the basic sinusoid as $x(t) = A\cos(2\pi f_1 t + \theta)$, the ***magnitude*** of x is given by a non-negative number A as shown in Figure 2.4-12. We will make a distinction between the ***amplitude*** of x that could be positive, A, or negative, $-A$, and the magnitude of x: in

$$-A\cos(2\pi f_1 t + \theta) = A\cos(2\pi f_1 t + \theta \pm \pi)$$

the sign of the signal affects the phase, not the magnitude — the latter always stays non-negative.

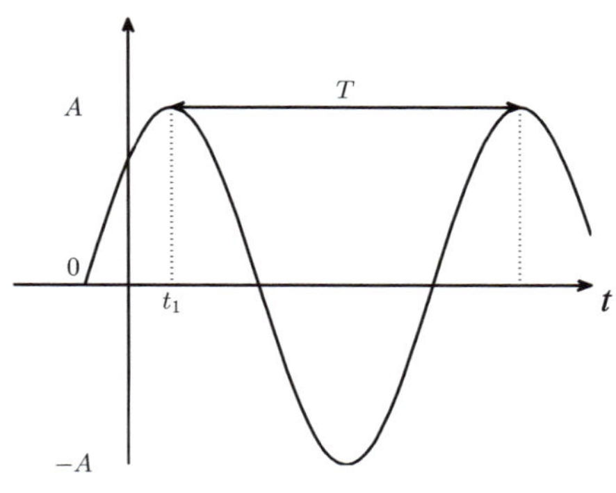

Fig. 2.4-12. Magnitude of a sinusoid.

2.5 Real Harmonics, Discrete and Continuous Spectra, Frequency Response, Signal Test Instrumentation

2.5.1 *Fundamental frequency of a sum of sinusoids. Real harmonics*

Fundamental frequency. As we have seen, if several sinusoids with commensurable periods are added to each other, the result is a periodic signal a period of which is a common multiple of the periods of these sinusoids. The least common multiple then gives the ***fundamental period*** of the resulting signal. This period will be further denoted through the subscript 1 as T_1. The frequency associated with period T_1 is called the ***fundamental frequency***, further denoted by f_1 and ω_1.

The *k'th* real harmonic. Adhering to the notation above, the fundamental period of the sum of the sinusoids is given by $T_{lcm} = T_f = T_1 = a_2T_2 = a_3T_3 \ldots = a_kT_k$.

Then, $1/T_2 = a_2(1/T_1)$, $1/T_3 = a_3(1/T_1)$, …, $1/T_k = a_k(1/T_1)$, i.e., all the sinusoids that add up to the combined signal have frequencies that are ***integer multiples of the fundamental frequency*** of that signal. These sinusoids are called ***harmonics*** of the fundamental frequency. The fundamental frequency is then a ***submultiple*** of each harmonic. Each harmonic is then given by

Always cosine — Always non-negative

$$x_k(t) = A_k \cos(2\pi k f_1 t + \theta_k) = A_k \cos(2\pi f_k t - \theta_k). \tag{2.5-1}$$

Always non-negative — May be positive or negative

The word "integer" is usually assumed and omitted. We will call the sinusoid shown above the ***k'th real harmonic.*** An example of harmonics comprising a signal is shown in Figure 2.5-1.

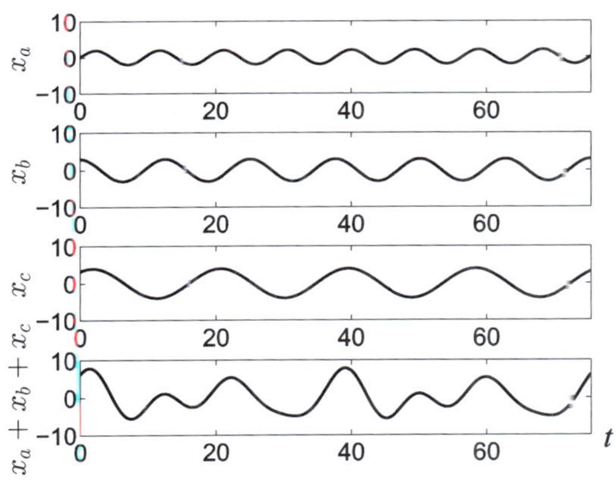

Fig. 2.5-1. Frequencies of harmonics x_a, x_b, and x_c are integer multiples of the frequency of their sum (bottom waveform).

Harmonic numbers for real harmonics. Frequency $f_k = kf_1$, where k is nonnegative integer, will be referred to as ***harmonic frequency***. k will be referred to as the ***harmonic number***:

$$f_k = k f_1$$

Harmonic number

where k can be positive or zero. $k = 0$ (zero'th harmonic) corresponds to the constant (DC) positive or negative signal component embedded into the general harmonic notational (2.5-1) through the following relation:

$$A_0\cos\left(2\pi f_0 t+\theta_0\right)=A_0\cos\left(2\pi k f_1 t+\theta_0\right)_{k=0}=$$
$$A_0\cos\left(2\pi 0 f_1 t+\theta_0\right)=A_0\cos(\theta_0)=\begin{cases}A_0,\theta_0=0,\\ -A_0,\theta_0=\pm\pi.\end{cases} \tag{2.5-2}$$

Sum of real harmonics: graphical time domain representation. Many signal analysis tasks involve visualization of the harmonic contents of a periodic signal. Let us try to develop a convenient visual representation of the latter.

Example 2.5-1: Provide a graphical representation of a signal of Example 2.4-2

$$x(t)=2\cos\left(\frac{2}{3}t-\frac{\pi}{2}\right)+3\cos\left(\frac{1}{2}t\right)+4\cos\left(\frac{1}{3}t-\frac{1}{5}\pi\right)$$

for quick visualization of its contents, including frequencies, magnitudes, and phases.
Solution: Let us first attempt to use direct graphical time domain signal composition representation with explicit frequencies, magnitudes, and phases information, as shown in Figure 2.5-2.

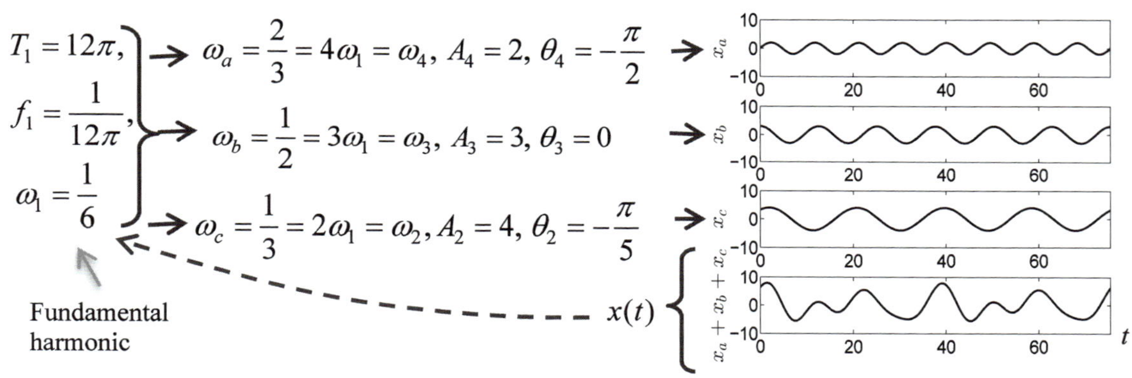

Fig. 2.5-2. Graphical representation of the signal in Example 2.5-1 with explicit frequencies, magnitudes, and phases information.

Obviously, showing signal composition in terms of individual harmonics graphically in time domain is too bulky, and additional need of supplementing the graph with numerical information makes this form unwieldy. For a succinct exposure of harmonics we need a different graphical tool described next.

2.5.2 *Discrete real spectra*

Discrete real spectrum of a sinusoidal signal is not a concept, but rather a graphical representation of this signal after the latter is brought to the form (2.5-1), i.e. to the harmonic form:

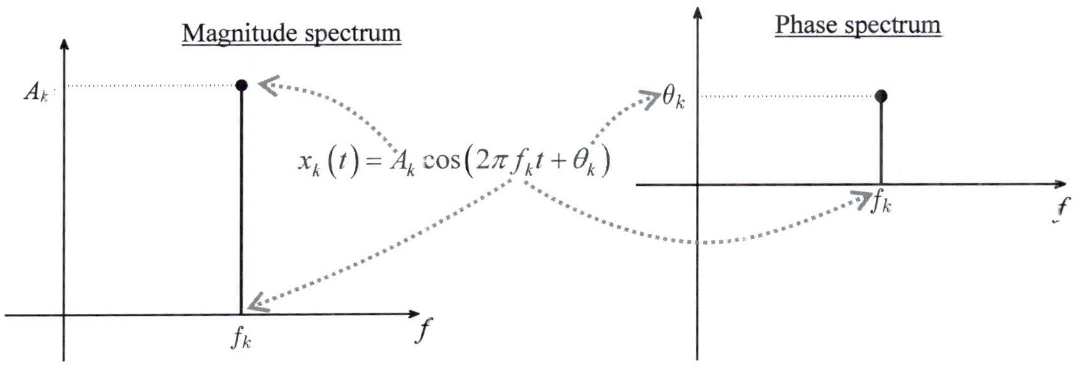

Fig. 2.5-3. Plots of discrete magnitude and phase spectra.

Obviously, this is a much better visualization tool — it provides succinct exposure of harmonics and it can display arbitrary number of harmonic magnitudes and phases along a single respective frequency axis! Why discrete? Because it is defined only for isolated points on frequency axis.

Example 2.5-2: Draw spectra for a signal $x(t) = -5\sin((\pi/4)t - \pi/8)$.

Solution: First, we need to recast this signal into a harmonic form (2.5-1). For this purpose we replace minus sine with cosine through the phase shift:

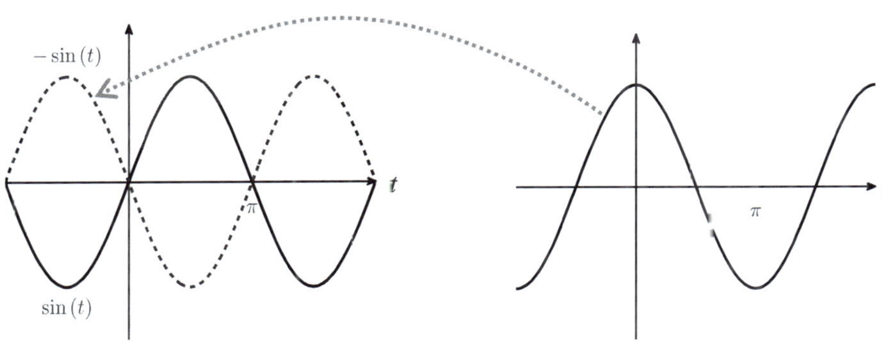

Fig. 2.5-4. Obtaining -sin(α): just shifting cos(α) to the left by $\pi/2$.

From Figure 2.5-4 we see that $-\sin(\alpha) = \cos\left(\alpha - \left(-\frac{\pi}{2}\right)\right) = \cos\left(\alpha + \frac{\pi}{2}\right)$.

$$\therefore x(t) = -5\sin\left(\frac{\pi}{4}t - \frac{\pi}{8}\right) = 5\cos\left(\frac{\pi}{4}t - \frac{\pi}{8} + \frac{\pi}{2}\right) = 5\cos\left(\frac{\pi}{4}t + \frac{3\pi}{8}\right) = 5\cos\left(2\pi(0.125)t + \frac{3\pi}{8}\right).$$

This yields magnitude and phase spectra given in Figure 2.5-5.

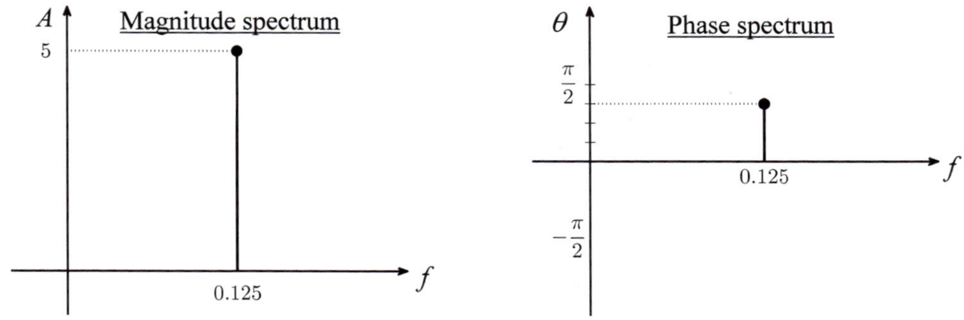

Fig. 2.5-5. Discrete spectra of Example 2.5-2.

Example 2.5-3: Next, we apply frequency domain representation to the simulation data of Figures 2.4-7 and 2.4-8, summarized in Figure 2.5-6. This yields spectral plots shown in Figure 2.5-7.

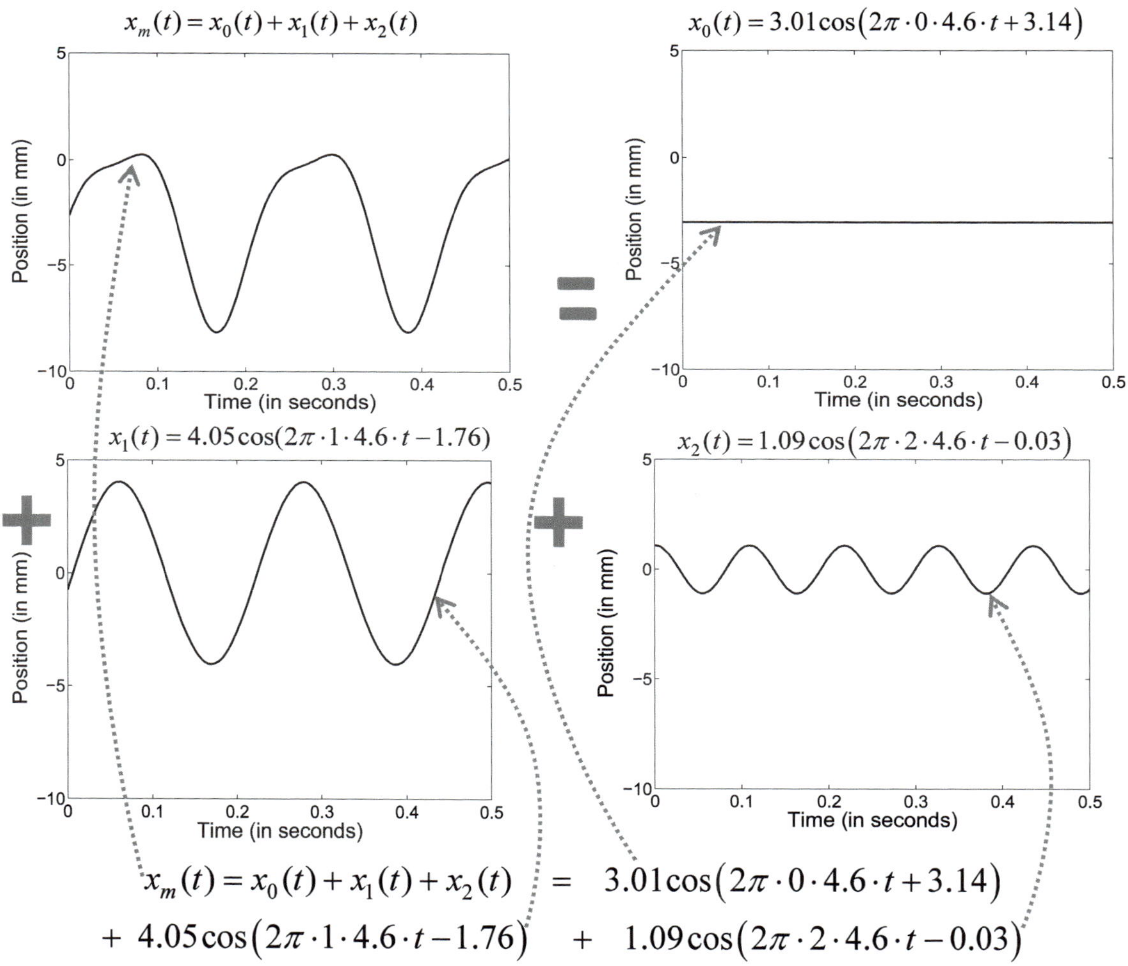

Fig. 2.5-6. Software testbed experimental data (the upper left plot) and its time domain harmonic decomposition.

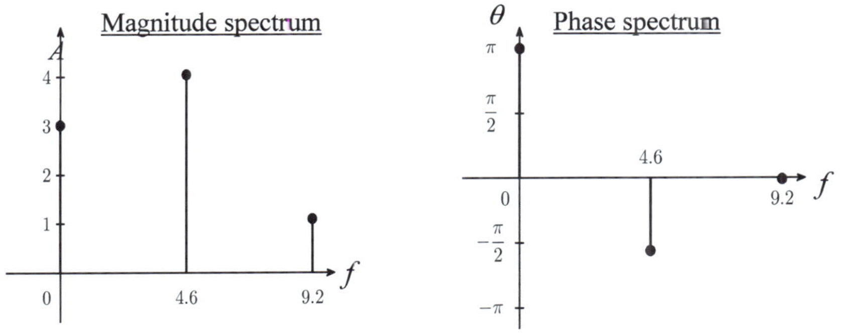

Fig. 2.5-7. Software testbed experimental data spectra.

2.5.2.1 *Application of discrete real spectra to system analysis: discrete frequency response*

Now, let's make use of signal decomposition in Figures 2.5-6 and 2.5-7 in the input/output context. For fixed input signal magnitude and phase, the changes in the output signal magnitude and phase depending on the input signal frequency represent the ***filtering action*** of the system, called, respectively, ***magnitude and phase responses***, and cumulatively referred to as ***frequency response***. Condensing the entire information about the signal composition into magnitude and phase graphs greatly facilitates system analysis and diagnostics through ***discrete frequency response spectra*** as shown in the example below, where the output signal is that of Figures 2.5-6 and 2.5-7.

Example 2.5-4: Examine the magnitude spectra of input and output software testbed signals shown in Figure 2.5-8 and comment on the likely cause of the mold displacement profile distortion.

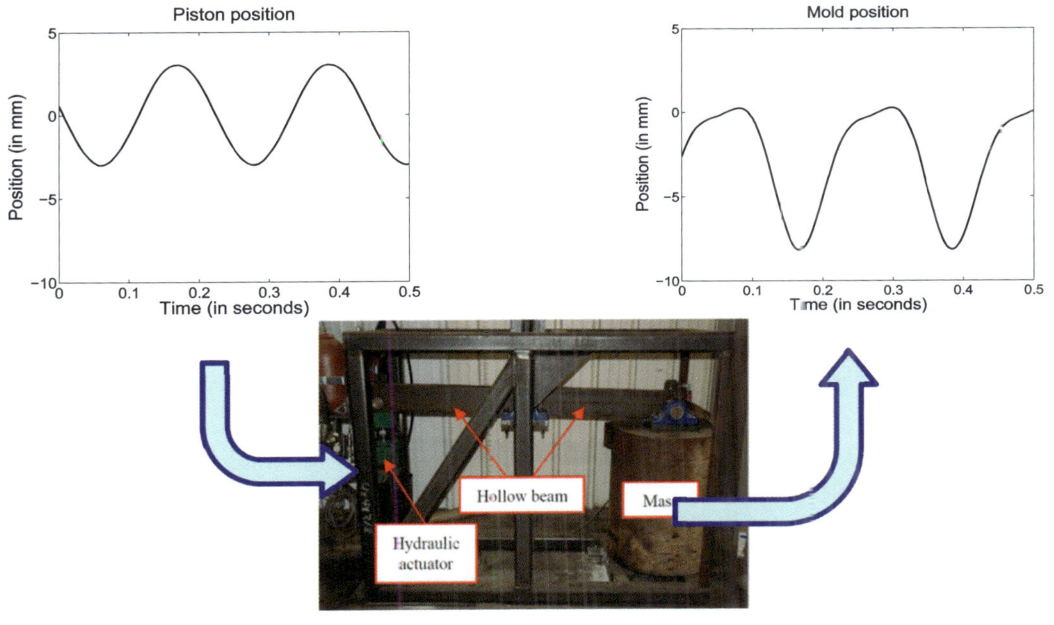

Fig. 2.5-8. Software testbed piston (input) and mold (output) time domain displacement data.

Solution: Composition and magnitude spectra of signals in Figure 2.5-8 are shown in Figure 2.5-9. The signals are seen to be the sums of the sinusoids at the same frequencies.

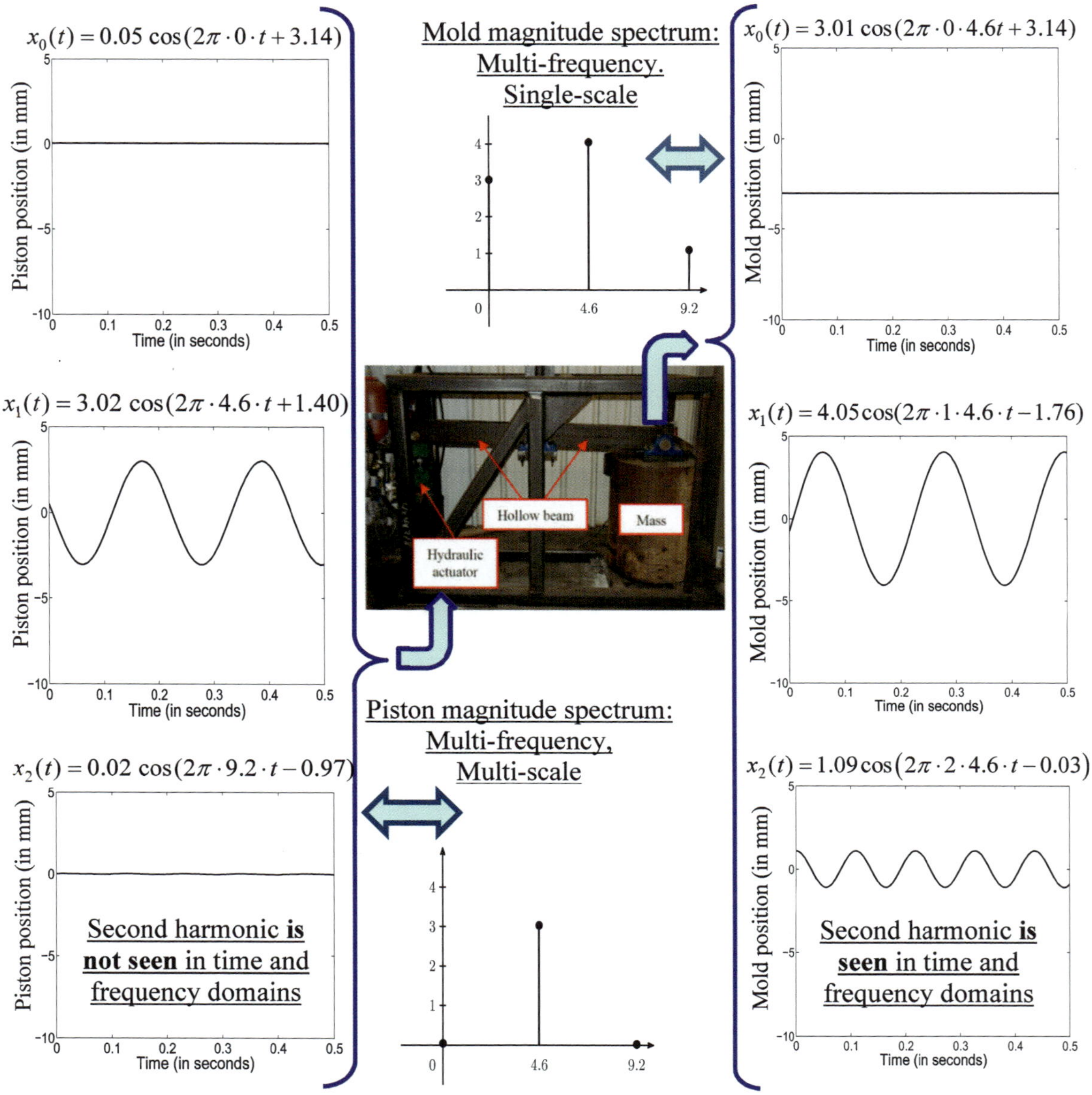

Fig. 2.5-9. Software testbed piston (left) and mold (right) time domain displacement profile compositions and the corresponding magnitude spectra.

With 9.2 Hz being beam resonance frequency, a small input component at 9.2 Hz is likely causing large amplification of the output component at this frequency.

2.5.3 *Continuous spectra and their application to system analysis: continuous frequency response*

Continuous spectra are the key tools for representing and analyzing physical systems as well as designing filters and control systems, both analog and digital. ***Continuous frequency response spectra*** can be thought of as discrete frequency response spectra under continuous frequency change, as shown in Figure 2.5-10. In practice, f_{sw} is varying very slowly to avoid transients.

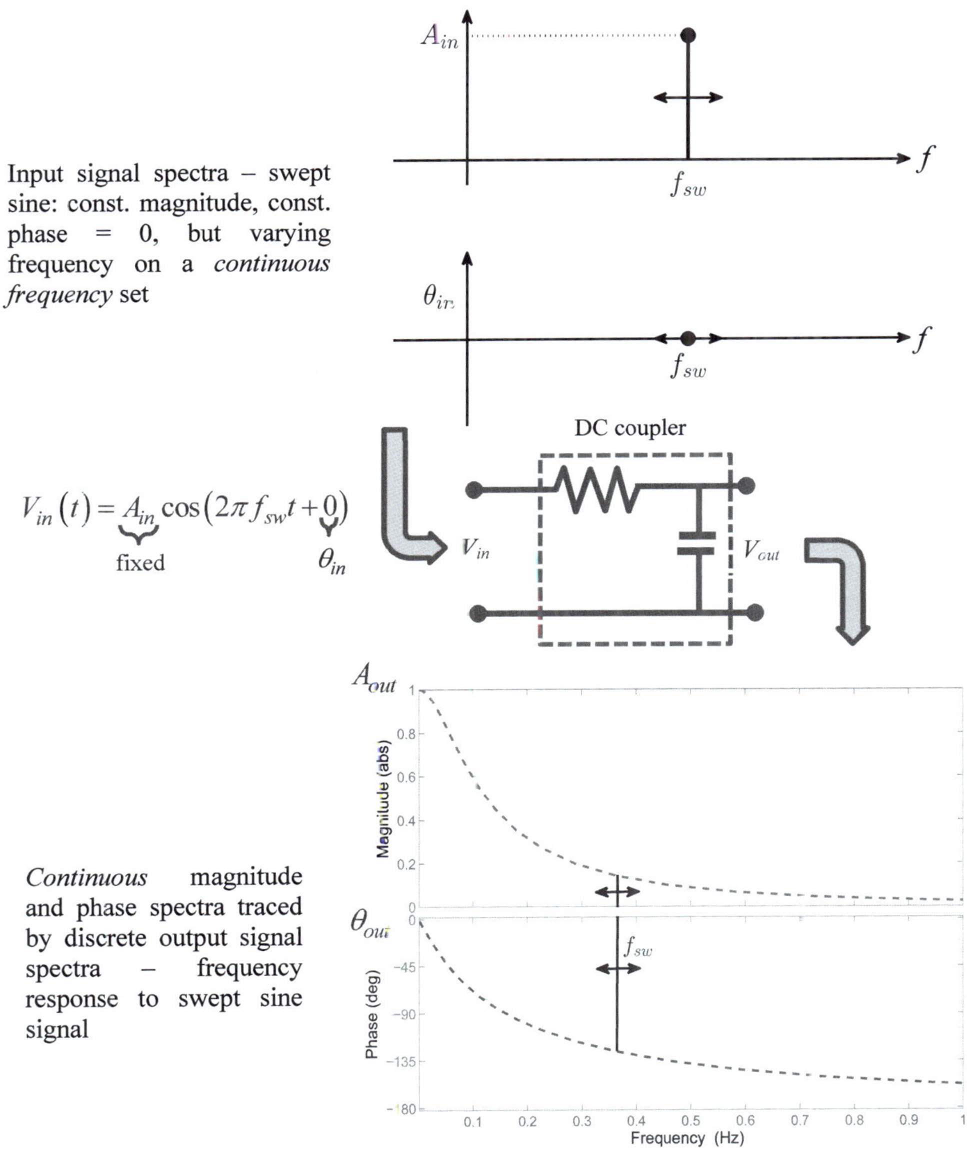

Fig. 2.5-10. Forming continuous spectra through continuously varying frequency of discrete spectra: DC coupler representation in terms of continuous frequency response.

System frequency response can be shaped through component selection, redesign, and/or control, as will be shown in subsequent chapters. An industrial example of frequency response use is given next.

Example 2.5-5: The beam-based mold oscillation production units are not well modelable through either equations or finite elements due to the complex main beams geometry and multi-beam construction. Furthermore, a typically bi-weekly mold replacement and reconditioning necessary due to production mold wear and tear gives rise to noticeable variations in oscillator dynamics resonance properties.

To address these problems, regular frequency response data collection and update has been incorporated into the mold oscillation controller tuning routine in production system at Nucor. A sample of such frequency response update, both magnitude and phase, is shown in Figure 2.5-11.

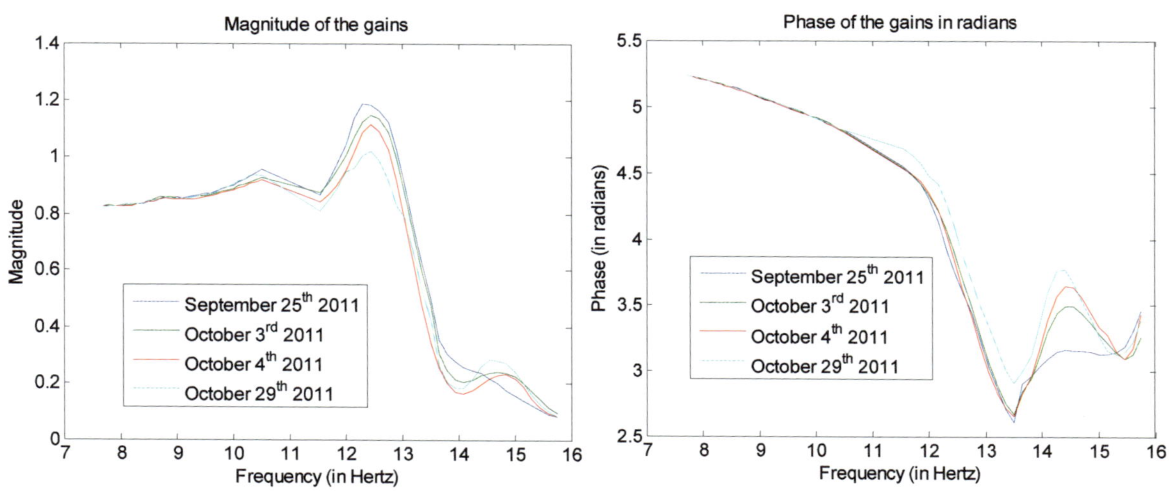

Fig. 2.5-11. Frequency response, both magnitude and phase, data collected over a 1-month period. Frequency of the highest gain (resonance frequency) is used in controller tuning. Gains are seen to change by less than 100%, satisfying condition for stable mold oscillation control system operation.

In practice, continuous frequency response spectra are often obtained by generating sufficiently closely spaced discrete frequency response spectra and interpolating the resulting values.

2.5.4 *Instrumentation for signal measurement and visualization*

Collection of data such as that shown in Figures 2.5-10 and 2.5-11 is accomplished through the use of test instrumentation. The main instruments in a signal processing laboratory include a digital multimeter (DMM), a dynamic signal analyzer (DSA), an oscilloscope, a function generator, and prototyping boards.

They are used to generate excitation signals, take voltage and current measurements of electrical signals, and display the resulting waveforms. DMMs provide high accuracy (5–8 digits) measurements using number display, whereas oscilloscopes can take faster measurements over a much broader frequency range than DMMs, but typically have 3–4 digits accuracy.

Oscilloscopes also have the ability to display complex waveforms, like Lissajous curves formed by plotting $y = b\sin(2\pi ft + \theta)$ against $x = a\sin(2\pi ft)$. For example, if y and x are, respectively, the output and the input of a linear time-invariant system, the Lissajous pattern takes the form of an ellipse with an aspect ratio being a function of a, b, and a phase shift between input and output. An example of using oscilloscopes to observe waveforms can be seen at http://www.youtube.com/watch?v=DkaUsBwe0fo.

A typical schematics for generating and measuring frequency response data of a DC coupler of Figure 2.5-10 is shown in Figure 2.5-12 below.

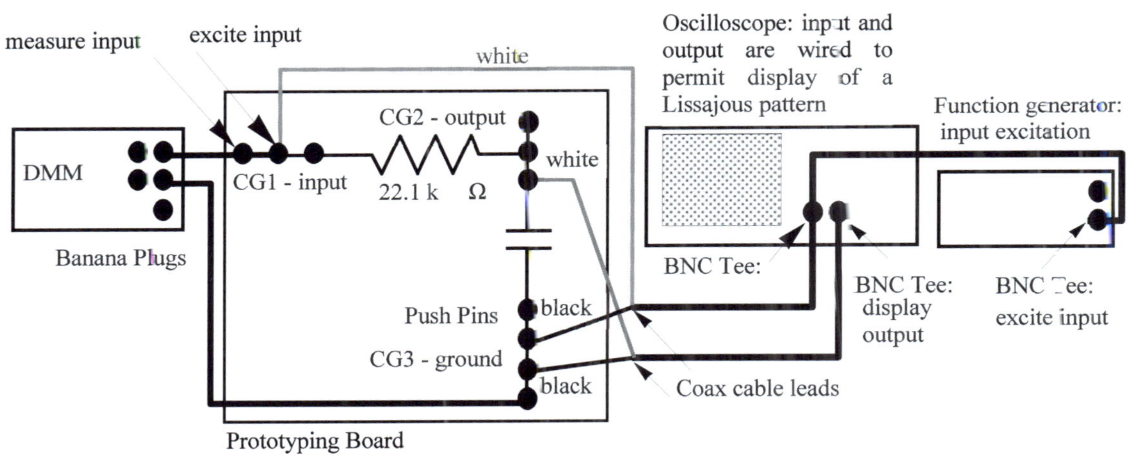

Fig. 2.5-12. Instrumentation and connection schematics wired for measuring DC coupler frequency response. CG stands for a prototyping board connection group, BNC tee is a coaxial cable connector. Credit: UIUC ME360 laboratory handout (W. Dunn).

Homework 3: Solved Examples and Solution Guide

Problem 1: Classify the following signals as periodic, nonperiodic, or almost periodic, and find the signal power if appropriate. If the signal is periodic, find the fundamental period and frequency:

(a) $x(t) = \cos(10\pi t)\sin(10t)$;

(b) $x(t) = \cos(6t) + \sin(15t)$.

Solution:

(a) $x(t) = \cos(10\pi t)\sin(10t) = \frac{1}{2}\sin[\underbrace{(10\pi + 10)}_{\omega_a}t] - \frac{1}{2}\sin[\underbrace{(10\pi - 10)}_{\omega_b}t]$.

$$\left.\begin{aligned} T_a &= \tfrac{2\pi}{\omega_a} = \tfrac{2\pi}{10\pi+10} \\ T_b &= \tfrac{2\pi}{\omega_b} = \tfrac{2\pi}{10\pi-10} \end{aligned}\right\} \Rightarrow \tfrac{T_b}{T_a} \text{ is not rational: there is no common period.}$$

This signal is *nonperiodic.* However, it is the sum of two periodic functions, so it is also almost periodic. Either answer is acceptable.

Signal power: $P_x = \frac{1}{2}\left(\frac{1}{2}\right)^2 + \frac{1}{2}\left(\frac{1}{2}\right)^2 = \frac{1}{4}$.

(b) $x(t) = \cos(6t) + \sin(15t)$;

$$\left.\begin{aligned} \omega_a &= 6 \\ \omega_b &= 15 \end{aligned}\right\} \Rightarrow \frac{\omega_a}{\omega_b} = \frac{3}{5} \Rightarrow \text{ rational fraction. This signal is } \textit{periodic}.$$

Fundamental frequency ω_0 = GCD (6, 15) = 3 rad/sec, common period $T_0 = \dfrac{2\pi}{\omega_0} = \dfrac{2\pi}{3}$ sec .

Signal power: $P_z = \frac{1}{2}(1)^2 + \frac{1}{2}(1)^2 = 1$.

Problem 2: Express the signal in Figure HW3-1, (a) by intervals, (b) as a linear combination of steps and ramps, (c) as a linear combination and/or product of rect and tri functions, (d) plot the first derivative $x'(t)$, (e) find the energy.

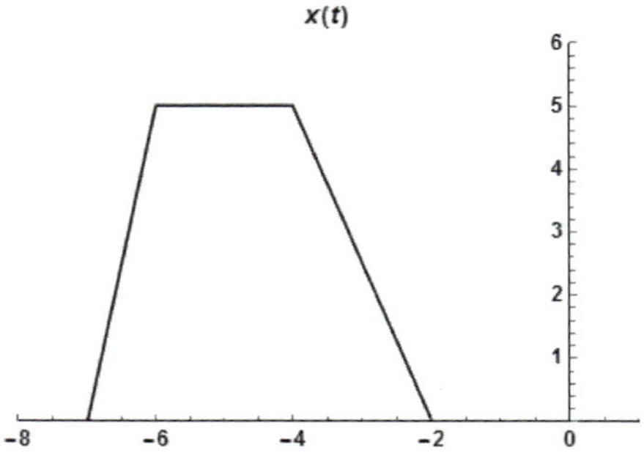

Fig. Hw3-1. Signal $x(t)$ for Problem 2.

Solution:

(a) Express $x(t)$ by intervals:

$$x(t)=\begin{cases}0, \ t<-7,\\ 5(t+7), \ -7\le t\le -6,\\ 5, \ -6<t\le -4,\\ -2.5(t+2), \ -4<t\le -2,\\ 0, \ t>-2.\end{cases}$$

(b) Express *x*(*t*) as a linear combination of steps and/or ramps:

the slope of $r(t)=ramp(t)=1$; $r(t)=ramp(t)=\begin{cases}t, \ t\ge 0,\\ 0, \ else,\end{cases}$

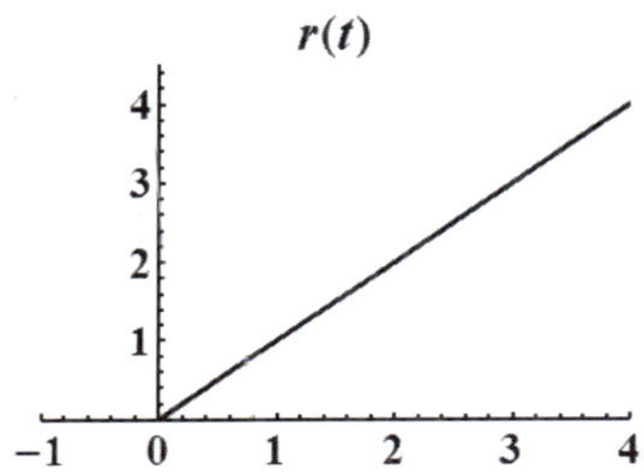

Fig. Hw3-2. Ramp signal *r*(*t*).

$\therefore x(t)=5r(t+7)-5r(t+6)-2.5r(t+4)+2.5r(t+2).$

(c) Express *x*(*t*) as a linear combination and/or product of rect and/or tri functions:

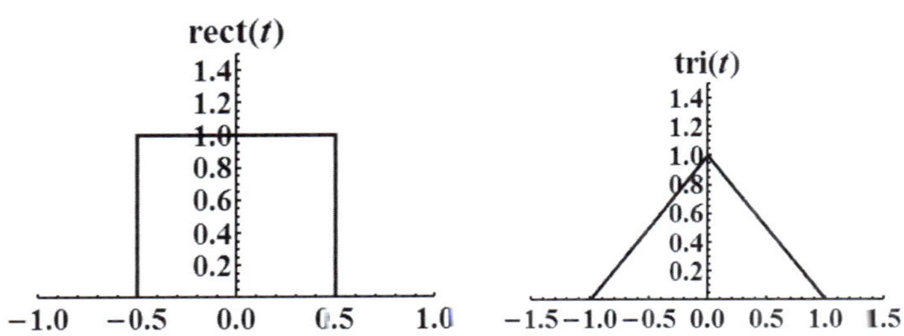

Fig. Hw3-3. Rect and tri signals.

x(*t*) can be seen as: $x(t)=5tri(t+6)rect(t+6.5)+5rect\left(\frac{t+5}{2}\right)+5tri\left(\frac{t+4}{2}\right)rect\left(\frac{t+3}{2}\right).$

(d) Sketch the first derivative $x'(t)$:

$$x'(t)=\begin{cases} 0, \ t<-7, \\ 5, \ -7\le t\le -6, \\ 0, \ -6<t\le -4, \\ -2.5, \ -4<t\le -2, \\ 0, \ t>-2. \end{cases}$$

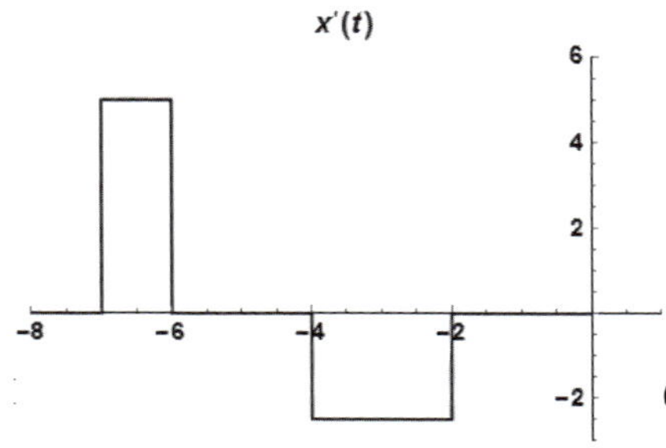

(t).

(e) Find the energy of x(t):

$$E_x=\int_{-\infty}^{\infty}|x(t)|^2\,dt=\int_{-7}^{-6}\left(5(t+7)\right)^2 dt+\int_{-6}^{-4}5^2\,dt+\int_{-4}^{-2}\left(-2.5(t+2)\right)^2 dt=75.$$

Problem 3: Sketch the following signals:
(a) $x(t)=5\delta(2t+3)$, **(b)** $x(t)=comb(0.6t)rect\left(\frac{t+4}{4}\right)$.

Solution:
(a) $x(t)=5\delta(2t+3)=\frac{5}{2}\delta\left(t+\frac{3}{2}\right)$

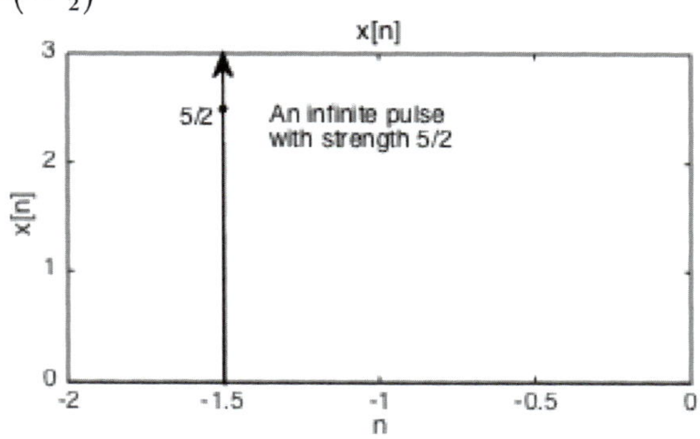

Fig. Hw3-5. Signal $x(t)$ for Problem 3(a).

(b) $comb(t)=\sum_{k=-\infty}^{\infty}\delta(t-k)$, therefore

$$comb(0.6t)=\sum_{k=-\infty}^{\infty}\delta(0.6t-k)=\sum_{k=-\infty}^{\infty}\delta\left[0.6\left(t-\frac{5}{3}k\right)\right]=\frac{5}{3}\sum_{k=-\infty}^{\infty}\delta\left(t-\frac{5}{3}k\right),$$

so that $x(t)=comb(0.6t)rect\left(\frac{t+4}{4}\right)$ is given by

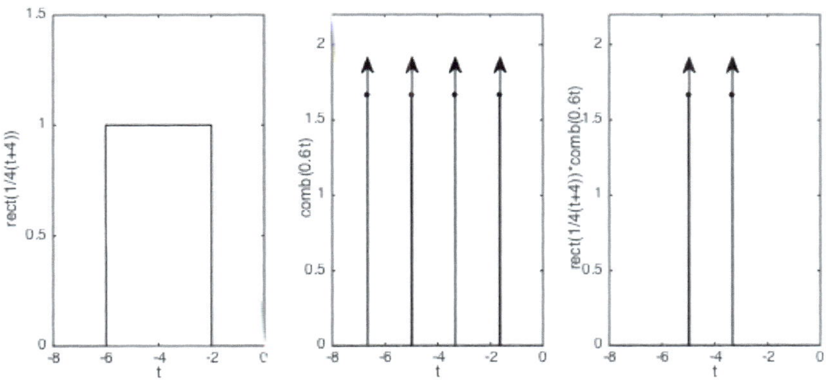

Fig. Hw3-6. Signal $x(t)$ for Problem 3(b).

Problem 4: Evaluate the following integrals:

(a) $\int_{-4}^{4}\left(5+t^3\right)\delta\left(t+5\right)dt$;

(b) $\int_{-7}^{7}\left(5+t^3\right)\delta\left(t+5\right)dt$.

Solution:

(a) $\int_{-4}^{4}\left(5+t^3\right)\delta\left(t+5\right)dt = 0$ (impulse is outside limits of integration),

(b) $\int_{-7}^{7}\left(5+t^3\right)\delta\left(t+5\right)dt = \int_{-7}^{7}\left(5+t^3\right)\delta\left(t-\left(-5\right)\right)dt = 5+\left(-5\right)^3 = -120.$

Problem 5: Draw the signals $x(t), x'(t)$**, and** $x''(t)$ **for the following:** $x(t) = 5tri\left[\frac{1}{3}(t+3)\right]$.

Solution: $x(t) = 5tri\left[\frac{1}{3}(t+3)\right]$ and its derivatives are shown from left to right in Figure Hw3-7:

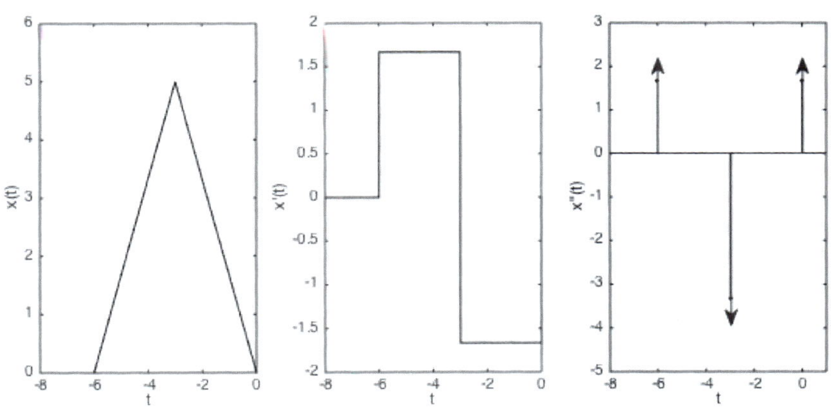

Fig. Hw3-7. The first and the second derivatives of signal $x(t)$.

Problem 6: Find the power and the rms value for a periodic sawtooth waveform given in Figure Hw3-8.

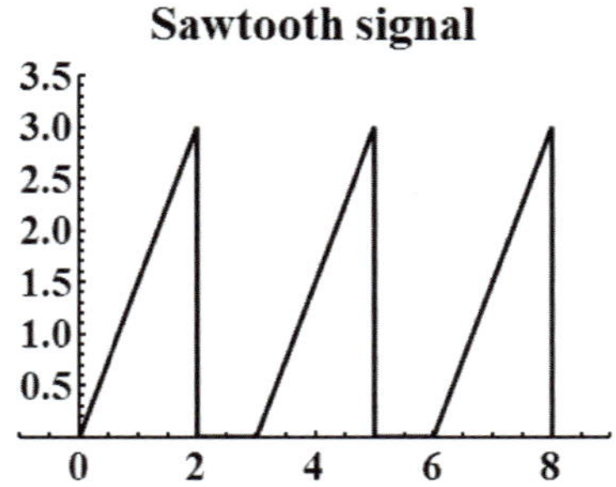

Fig. Hw3-8. Sawtooth signal for Problem 6.

Solution: Signal power:

$$P = \frac{1}{T}\int_T |x(t)|^2\, dt = \frac{1}{3}\int_0^2 \left(\frac{3}{2}t\right)^2 dt = \frac{1}{3}\frac{3}{4}t^3\Big|_0^2 = \frac{1}{3}\left(\frac{3}{4}\cdot 8\right) = 2,\ \text{rms} = \sqrt{P} = \sqrt{2}.$$

Problem 7: Let an energy of signal $x(t)$, with a range $-5 \le t \le 4$, be 20 *J*. Find the range and the energy of $x(\frac{t}{4}+5)$.

Solution:

$$x(\tfrac{t}{4}+5) = x(\tfrac{1}{4}(t-(-20))),\ E_x = 20\,\text{J}.$$

Signal starts at -5 : $\frac{t_s}{4}+5=-5 \Rightarrow t_s=-40$. Signal ends at 4 : $\frac{t_e}{4}+5=4 \Rightarrow t_e=-4$.

$$\therefore range\ : -40 \le t' \le -4.$$

Energy (signal stretched by 4): $E' = 4E = 4(20) = 80$ J.

Problem 8: Consider a periodic signal x(t) with period $T = 8$ and power $P=6$ W. Find the period and the power of $x\left(\frac{t}{4}+5\right)$.

Solution:

$$x\left(\tfrac{t}{4}+5\right) = x\left(\tfrac{1}{4}\left(t-\left(-20\right)\right)\right)\text{: stretched by 4: } T' = 4T = 32.$$

Energy is increased by 4, but so is period, therefore power unchanged by stretching (or shrinking): $P' = P = 6$ W.

Problem 9: Will the product of two periodic signals with commensurable periods T_a and T_b be periodic? Use two real harmonics to determine the answer.

Solution:

The product of two periodic signals with periods T_a and T_b will be periodic if T_a and T_b are commensurate.

Recall: $\cos(\theta)\cos(\psi) = \frac{1}{2}\cos(\theta+\psi) + \frac{1}{2}\cos(\theta-\psi)$ is periodic iff $\theta+\psi$ and $\theta-\psi$ are commensurate.

Problem 10: A signal model is called parsimonious if it is characterized by a small number of parameters compared to the amount of data in the signal it represents. For a signal in Figure Hw3-9, find the values of parameters a and b for each model;
(a) $x(t) = at\exp(-bt)u(t)$ **, (b)** $x(t) = a(\exp(-bt) - \exp(-2bt))$ **, (c)** $x(t) = at/(b+t^2)$ **, and draw the resulting graphs.**

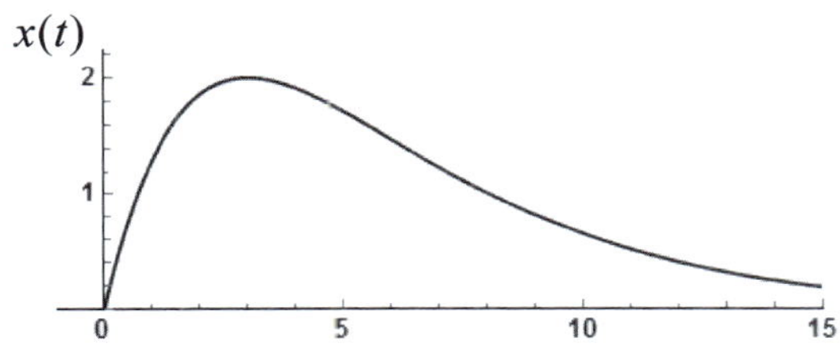

Fig. Hw3-9. Signal for Problem 10.

Solution:

(a) $x(t) = at\exp(-bt)u(t)$:

check end limit. $\rightarrow \lim_{t\to\infty} x(t) = 0$.

To find a and b: $x(t=0) = 0 = a(0)\exp(0)u(0)$ is trivial,

$x(t=3) = 3a\exp(-3b) = 2$ yields $a = (2/3)\exp(3b)$,

$x'(t) = a\exp(-bt) + at(-b)\exp(-bt) = a\exp(-bt)(1-bt)$,

so $x'(3) = 0 = a\exp(-3b)(1-3b)$; $a = 0$ would be trivial $\therefore b = 1/3$, and

$a = (2/3)\exp(1) = 1.8122$.

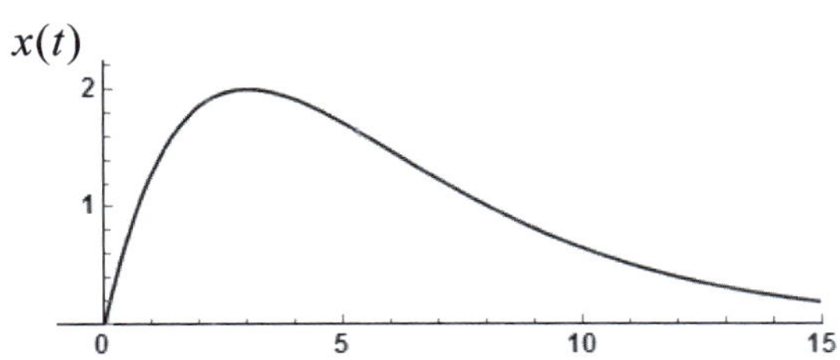

Fig. Hw3-10. Signal for Problem 10(a).

(b) $x(t)=a\left(\exp(-bt)-\exp(-2bt)\right)$: check end limit $\rightarrow \lim_{t\to\infty} x(t)=0$.

To find a and b : $x(t=0)=0=a(e^0-e^0)=0$ is trivial.

$x(t=3)=2=a\left(\exp(-3b)-\exp(-6b)\right)$ yields $a=\dfrac{2}{\exp(-3b)-\exp(-6b)}$,

$$x'(t)=a\left(-be^{-bt}+2be^{-2bt}\right)\Rightarrow x'(t=3)=0=a\left(-be^{-3b}+2be^{-6b}\right).$$

$$-e^{-3b}+2e^{-6b}=0\Rightarrow e^{-3b}=\frac{1}{2}\Rightarrow b=\frac{1}{3}\ln 2=0.231,\ a=\frac{2}{\exp(-3\times 0.231)-\exp(-6\times 0.231)}=8.$$

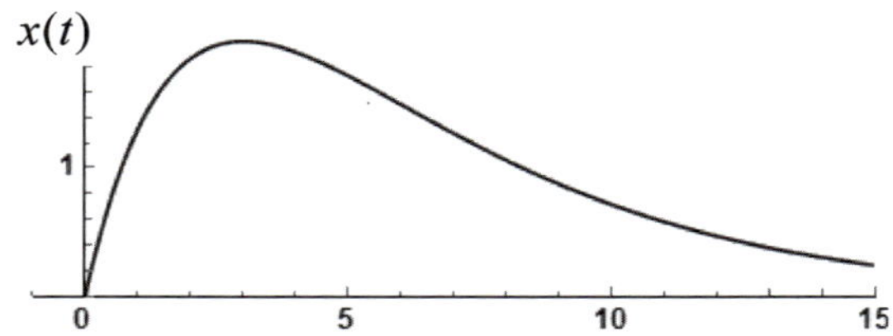

Fig. Hw3-11. Signal for Problem 10(b).

(c) $x(t)=\dfrac{at}{b+t^2}$, check end limit $\rightarrow \lim_{t\to\infty} x(t)=0$.

To find a and b : $x(t=0)=0$ is trivial. $x(t=3)=2=3a/(b+9)\Rightarrow a=2(b+9)/3$.

$$x(t)=\frac{at}{b+t^2}=\frac{f(t)}{g(t)};\ x'(t)=\frac{g(t)f'(t)-f(t)g'(t)}{\left[g(t)\right]^2}=\frac{(b+t^2)a-(at)2t}{(b+t^2)^2}=\frac{(b+t^2)a-2at^2}{(b+t^2)^2}$$

$$x'(t=3)=\frac{a(b+9)-2a\times 9}{(b+9)^2}=0\Rightarrow\frac{3a^2/2-2a\times 9}{(3a/2)^2}=0\ \therefore a=12, b=9.$$

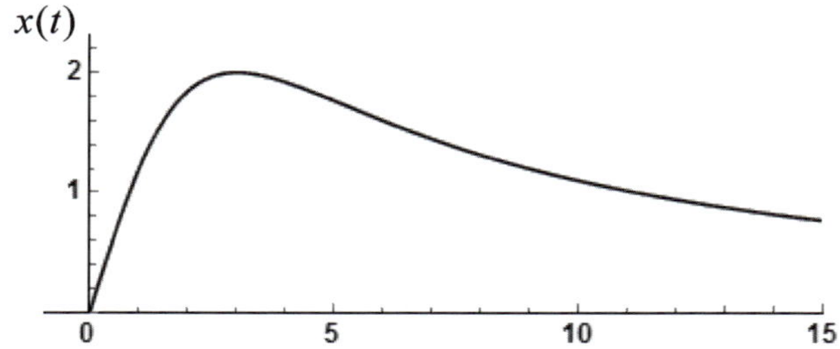

Fig. Hw3-12. Signal for Problem 10(c).

Problem 11: The instantaneous phase and the instantaneous frequency are used to represent time-varying functions. For the sinusoid $x(t)=\cos[\phi(t)]$ at time instant t they are, respectively, given by the values $\phi(t)$ rad and $f(t)=\phi'(t)/2\pi$ Hz where $\phi'(\cdot)$ is the derivative of $\phi(\cdot)$. Consider the signal $y(t)=\cos\left(0.1\pi t^2+\theta\right)$. Show that its instantaneous frequency at time t is $0.1t$.

Solution:

In $y(t)=\cos\left(0.1\pi t^2+\theta\right)$ phase $\phi(t)=0.1\pi t^2+\theta \Rightarrow f(t)=\frac{\phi'(t)}{2\pi}=\frac{\frac{d}{dt}0.1\pi t^2}{2\pi}=\frac{2\cdot 0.1\pi t}{2\pi}=0.1t$.

Problem 12: Signal whose frequency varies linearly with time as $f(t)=f_1+kt,\ t\geq t_1$, where f_1 is the starting frequency at time t_1 and k is the rate of frequency increase, is called swept-frequency, or swept-sine, or chirp signal, or simply chirp. Problem 16 below makes use of this signal for obtaining software testbed frequency response. Consider a sinusoid $x(t)=\cos[\phi(t)]$. Referring to the definitions in Problem 11, (a) find the expression for $\phi(t)$ and $x(t)$ if the instantaneous frequency is constant, f_1 Hz?; (b) Find the expression for $\phi(t)$ and $x(t)$ if the instantaneous frequency varies linearly from 10 Hz to 50 Hz in 5 seconds?

Solution:

(a) What is the expression for $\phi(t)$ and $x(t)$ if the instantaneous frequency is constant, f_1 Hz?

$$\phi'(t)/2\pi=f_1 \text{ Hz} \therefore \phi(t)=2\pi f_1 t+\theta \ \therefore x(t)=\cos\left[2\pi f_1 t+\theta\right].$$

(b) What is the expression for $\phi(t)$ and $x(t)$ if the instantaneous frequency varies linearly from 10 Hz to 50 Hz in 5 seconds?

Assuming 10Hz at t_1=0,

$$\omega_i(t)=2\pi(8t+10),\ \phi(t)=\int\omega_i(t)dt=2\pi(4t^2+10t)+\theta,$$

$$x(t)=\cos\left[\phi(t)\right]=\cos\left[2\pi(4t^2+10t)+\theta\right].$$

Problem 13: Sketch the following signal: (a) $x[n]=tri\left[\left(\frac{1}{8}\right)\text{n}\right]\text{sinc}\left[\left(\frac{1}{6}\right)\text{n}\right]$.

Solution:

(a) $x[n]=tri\left[\left(\frac{1}{8}\right)\text{n}\right]\text{sinc}\left[\left(\frac{1}{6}\right)\text{n}\right]$:

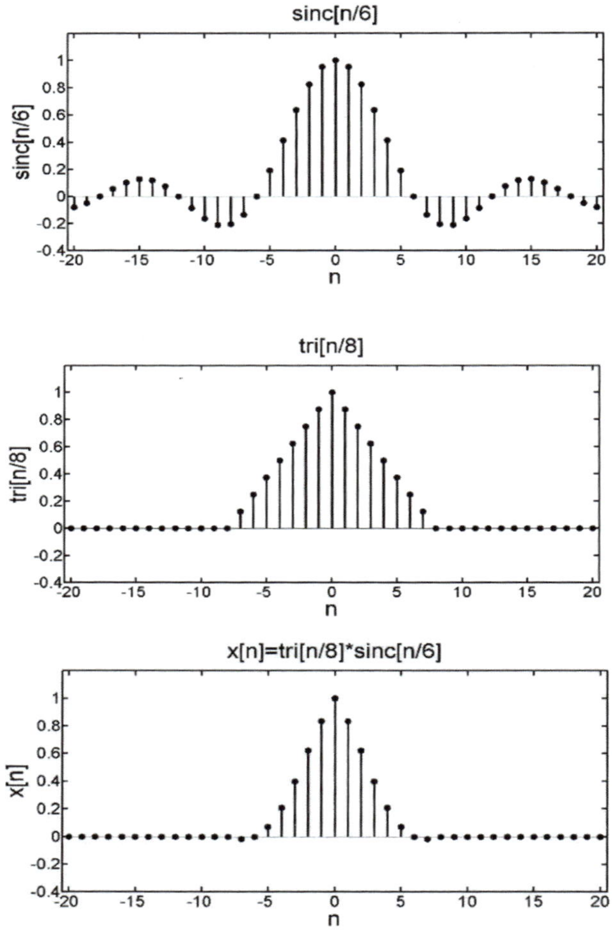

Fig. Hw3-13. Signal tri[(1/8)n], sinc[(1/6)n] and $x[n]$ for Problem 13.

Problem 14: Express the signal shown in Figure HW3-14 in terms of (a) the numeric sequence and mark the index n = 0 by an arrow, (b) impulse functions, (c) steps and ramps; (d) find the signal energy; (e) find the signal power, assuming that the sequence is periodically extended with period N = 8.

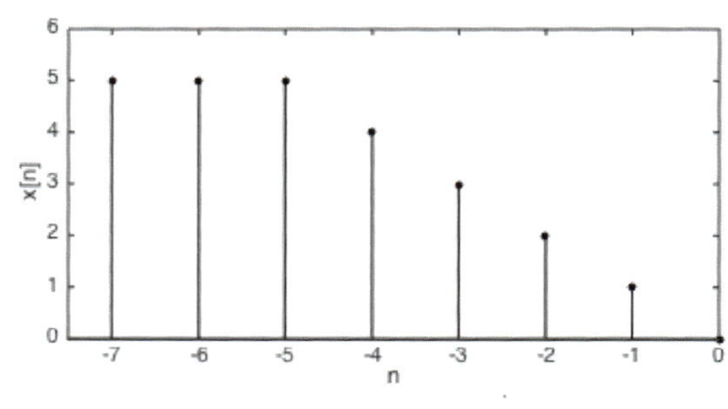

Fig. Hw3-14. Signal $x(t)$ for Problem 14.

Solution: (a) Write out the numeric sequence and mark the index $n = 0$ by an arrow

$$x[n] = \{5,5,5,4,3,2,1,\underset{\uparrow}{0}\}.$$

(b) Write an expression for the signal using impulse functions:

$$x[n] = \delta[n+1] + 2\delta[n+2] + 3\delta[n-3] + 4\delta[n+4] + 5\delta[n+5] + 5\delta[n+6] + 5\delta[n+7].$$

(c) Write an expression for the signal using steps and/or ramps:

$$x[n] = r[-n] - (r[-n-5] + 5u[-n-8]).$$

(d) Find the signal energy:

$$E = \sum_n (x[n])^2 = 0+1+4+9+16+25+25+25 = 105.$$

(e) Find the signal power, assuming that the sequence shown repeats itself:

$$\text{Power (discrete period, } N = 8\text{)}: P = \frac{1}{N}\sum_n (x[n])^2 = \frac{E}{N} = 13.125.$$

Problem 15. <u>System linearity and discrete frequency response.</u> One of the properties of the linear systems is that they admit a sinusoid to "pass through", i.e. a sinusoidal input signal of frequency *f* produces the corresponding sinusoidal output signal at the same frequency *f* and does not produce sinusoidal signals at any other frequency. A) check by simulation if the beam of Homework 1, with no mass attached to it, is linear, incrementally linear (affine), or none (i.e. nonlinear): apply to it a sinusoid of 3 mm magnitude and 3.4 Hz frequency, compute and examine the beam response, and comment on beam linearity. B) repeat the task of a) with mass attached to the beam. C) do the same for the hydraulic actuator (with P-controller) of Homework 1 not attached to the beam. Keep the submission format of Homework 1.

Solution guide: Problem 15 is similar to Problem 12 from Homework 2 in the setup. Here are a few tips:

- The problem specifies a sinusoidal input with amplitude 3 mm and frequency 3.4 Hz. However, the "Amplitude" and "Frequency" parameters in the Simulink model use units of meters and radians per second, respectively. Make sure to do the conversion.
- As in Homework 1, these systems exhibit an initial transient when a sinusoidal input is used. Ignore this transient and plot only the quasi-steady behavior. Run the simulation for 10 seconds, and record only the last 2.5 seconds to plot. (That is, use the data recording options from Homework 1, not Homework 2.)
- To check linearity, find out whether the output is (1) sinusoidal and (2) at the same frequency 3.4 Hz as the input. For this homework, qualitatively supported answer is acceptable. Do your best to show whether or not the output is a 3.4 Hz sinusoid to justify the answer. Subsequent chapters will introduce the tools for giving a more accurate answer.

Problem 16. A) Find and plot, three discrete frequency responses of the closed-loop testbed system without tracking controller (Problem 3, Homework 1) to the reference input given by pure sinusoid with zero phase, magnitude 2 mm, and frequency specified at three values: i) any value between 1 Hz and 3.5 Hz, ii) at 4.6 Hz, iii) any value between 5 Hz and 6 Hz. Draw the response in the format of Figure 2.5-9, showing on the left the graph for the reference input (rather than piston position) and on the right — the graph for the mold displacement. B) Find and plot, as in Figure 2.5-11, the continuous frequency response of the closed-loop system without tracking controller using the swept-sine and comment on the closeness of the simulator frequency response and the production unit frequency response given in Figure 2.5-11. Choose the slow variation rate of the swept-sine frequency to avoid distortion due to nonsteady response.

Solution guide. In Problem 16A, solution is similar to that of Problem 15 in terms of waiting until the transients die out — wait until the phase and the magnitude in the response stop changing and if the response is close enough to pure sinusoid, read off the magnitude and the phase in the response to plot the graph. One would expect such responses to reference inputs with frequencies in the intervals, but not at 4.6 Hz. If the output is not a pure sinusoid (e.g., at 4.6 Hz due to hydraulics nonlinearity exciting beam resonance), one should eyeball and analytically match the best one can the fundamental harmonic in time domain, and try to find at least one other harmonic by subtracting the fundamental one from the output and examining the residual to see if it is sinusoidal enough and what would be its magnitude and frequency. The purpose of this "brute force" calculation of magnitudes and phases of the pure sinusoidal components in a multifrequency signal is a way to experience first-hand that heuristic analysis without the proper analytical tools is probably not a good way to go about the problem.

In Problem 16B the task is to simulate the system with the reference input having sufficiently slowly varying frequency, zero phase, and fixed magnitude, say, 2 mm or 1 mm to get continuous (not discrete) frequency response. Here, some trials would be needed to find a good rate of frequency variation — not slow enough to wait too long, but not fast enough to make sure the response has not transient distortion. This should be done through putting together a Matlab program with swept-sine input which would work well enough to get a reasonable frequency response — as the frequency changes, the values of the output magnitude and phase need to be continually and correctly recorded and plotted to see both responses, or collected and plotted after simulation, with the proper relation to the generating frequency.

Chapter 3

First Look at Signal Processing, Filtering, and Instrumentation

3.1 Aggregating LTI System Representations for Carrying out Design, Filtering, and Control Tasks

3.1.1 *Integrative portrait of a linear time invariant (LTI) system with a finite zero*

A system irrespectively of its physical nature is a filter characterized in the linear range of its behavior by the interrelated nonuplet shown in Figure 3.1-1 below.

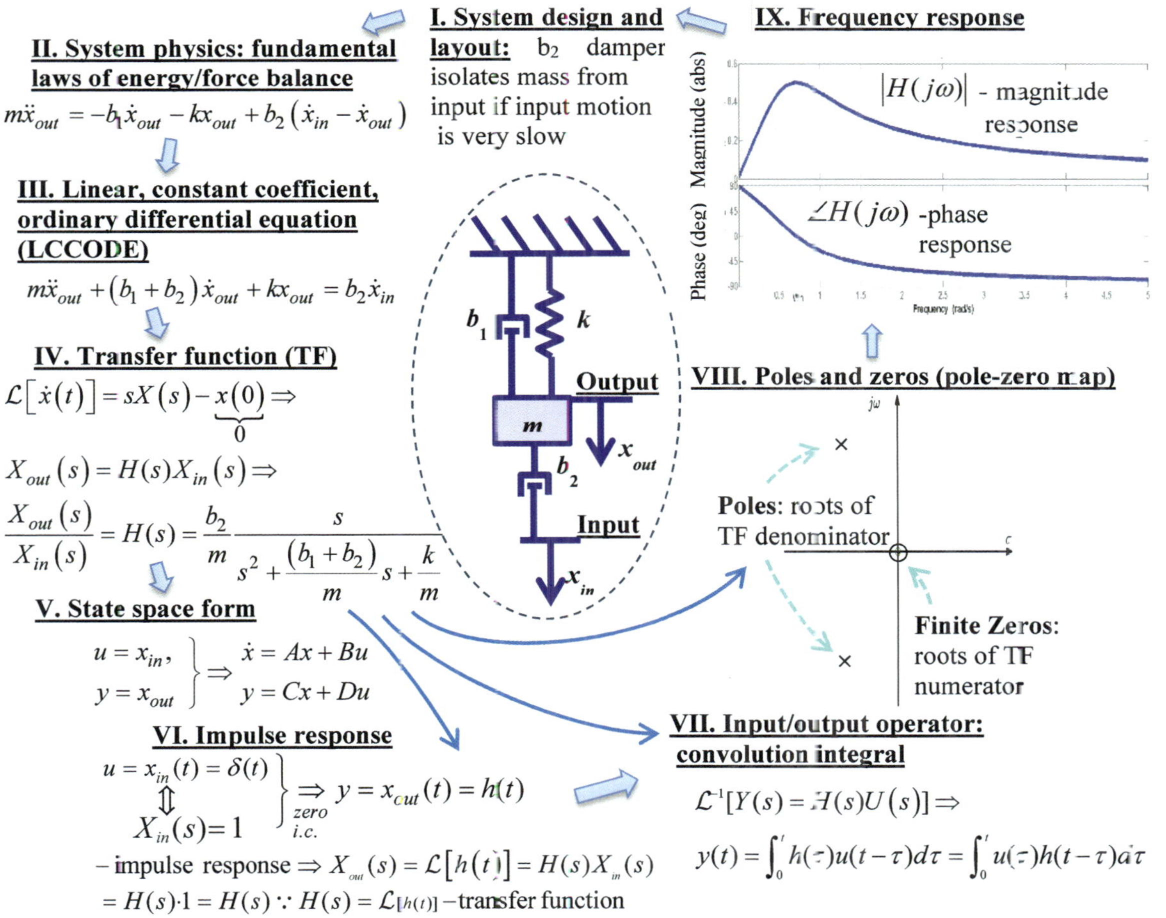

Fig. 3.1-1. Nine LTI filter representations helpful in attaining the desired properties through redesign and/or control

This cumulative portrait enables setting up the design/filtering/control cycle as discussed next.

3.1.2 *Mechanical systems — a filtering, or input/output viewpoint*

In the basic dynamics courses, systems, such as mass-spring-damper (MSD) system, are mainly considered in terms of their response to initial conditions and/or actions representable by basic functions like unit step or sinusoid. In these cases, their response is modelled well by a product of simple functions of time: (exponentials) × (sinusoids) × (polynomials). However, the response to an arbitrary input, considered in the present book, depends on the input complexity, i.e. can be arbitrarily complicated, e.g. stochastic, as in the case of the useful signal corrupted by noise. One of the ways of addressing this complexity is to resort to a ***filtering***, or ***input/output***, system viewpoint depicted in Figure 3.1-2. The key elements of this viewpoint are:

- the ***input*** to the system that can be ***manipulated***, generally denoted by u — in this case position x_{in}; where this causes confusion with a unit step, notation $1(t)$ for the latter is used;
- the ***output*** from the system that can be ***measured***, generally denoted by y — in this case position x_{out};
- the ***operator***, or rule, relating input to output ***under zero initial conditions***.

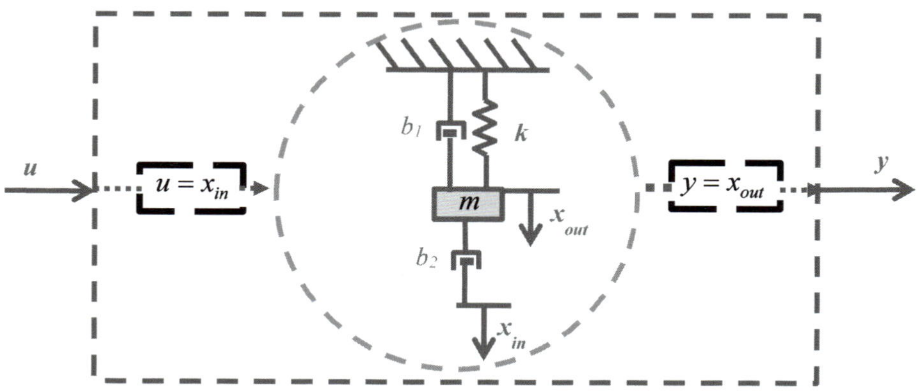

Fig. 3.1-2. Mechanical system as a filter.

In Figure 3.1-2 the operator, or filter, is implemented in terms of the physical components. It processes an incoming signal — displacement of the damper end x_{in} — and produces output, displacement of mass x_{out}. This operator can be thought of as the continuous time (analog) computer that performs calculations and produces the result ***in real time***, i.e. without computational delay.

In the literature $u(t)$ is typically used to denote both general vector or scalar input variable and also unit step function. The same notation is used throughout the book where it causes no confusion. **Changing system filtering properties: system redesign for bringing in finite zeros.** In considering MSD system, attention is typically focused on its transient behavior governed by poles. In filtering and control, however, the focus shifts in large part to frequency response, where a major role is played by system zeros. The latter can be brought in through ***redesign***, i.e. ***passively*** — with no external energy source, or through ***control***, i.e. ***actively*** — using such a source. The mold oscillation distortion problem is solved in Chapter 9 using control. However, to better understand the effects of zeros on system response, we will start with passive approach — attach another damper to the mass in the familiar MSD system and check what happens. We will refer to the resulting system, shown in Figure 3.1-2, as the ***"mass-spring-two-dampers"*** **(MS2D)** system.

3.1.3 *MS2D system force balance*

Unlike MSD system, the system in Figure 3.1-2 includes a damper b_2 with one end driven by the input and another attached to the mass. Therefore, the force balance contains now the force generated by this damper that is proportional to the difference between the derivative of the input and that of the output:

$$m\frac{d^2}{dt^2}x_{out} = -b_1\frac{d}{dt}x_{out} - kx_{out} + b_2\left(\frac{d}{dt}x_{in} - \frac{d}{dt}x_{out}\right). \tag{3.1-1}$$

The derivative at the input implies that the input — displacement x_{in}, if kept constant, does not contribute to the force acting on the mass, and hence to the output — mass displacement x_{out}, only the velocity does.

Implicit input/output operator: linear constant-coefficient ordinary differential equation (LCCODE). Isolating the input term in (3.1-1) yields LCCODE of the form

$$m\frac{d^2}{dt^2}x_{out} + (b_1 + b_2)\frac{d}{dt}x_{out} + kx_{out} = b_2\frac{d}{dt}x_{in}. \tag{3.1-2}$$

This equation contains derivatives of the output, so that in order to compute the output it is necessary to take these derivatives into account in the computational procedure. This makes the input/output relation given by this equation implicit.

Transfer function as input/output system representation. System transfer function (we will also use abbreviation TF), typically denoted as $H(s)$, is the main tool for bringing out LTI system filtering properties. It can be obtained through processing of input/output data record, considered later in the book, or through taking Laplace transform of the LCCODE describing the system, assuming zero initial conditions (at this point it would be helpful to review *Appendix E: Laplace transform*). We will employ the *single-sided Laplace transform that includes the origin*, often denoted as $\mathcal{L}_{-}$. We will subsequently drop the subscript "–" from this notation. Then

$$\mathcal{L}\left[\dot{x}(t)\right] = sX(s) - \underbrace{x(0)}_{0} \Rightarrow X_{out}(s) = H(s)X_{in}(s) \Rightarrow$$

$$\frac{X_{out}(s)}{X_{in}(s)} = H(s) = \frac{b_2}{m}\frac{s}{s^2 + \dfrac{(b_1 + b_2)}{m}s + \dfrac{k}{m}}. \tag{3.1-3}$$

When the input and the output of a system are both scalars, the system is referred to as *single-input/single-output* (*SISO*), and if it is LTI, its transfer function is given by the ratio of numerator and denomination polynomials, referred to as *rational function*:

$$H(s)=\frac{Y(s)}{U(s)}=k\frac{s^m+b_{m-1}s^{m-1}+\ldots+b_0}{s^n+a_{n-1}s^{n-1}+\ldots+a_0}=k\frac{(s-z_1)(s-z_2)\cdots(s-z_m)}{(s-p_1)(s-p_2)\cdots(s-p_n)}, \tag{3.1-4}$$
$$z_i,\ i=1,\ldots,m \text{ - zeros},\quad p_i, i=1,\ldots,n \text{ - poles}.$$

The transfer function is referred to as *proper* or *strictly proper* when $m = n$ or $m < n$, respectively.

The values z_i are indeed seen to be the values of s at which the value of transfer function is zero. In general poles and zeros are complex numbers. Zeros with finite real and imaginary parts are referred to as *finite zeros*. However, strictly proper rational function has still the same number of zeros as poles, counting *zeros at infinity*, i.e. the infinite values of s that make $H(s) = 0$.

When none of the poles is equal to any of the zeros, the numerator and denominator polynomials are called *coprime*. When a pole is equal to a zero, the factors in numerator and denominator become identical and are cancelled. This is referred to as *pole-zero cancellation*. It means that not all internal states are captured by the transfer function, and therefore can be controlled by the input or seen (observed) at the output. The internal states that correspond to cancelled factors are called *hidden modes* and are driven only by the initial conditions. If system has no hidden modes, considering only transfer function is sufficient, since system response to initial conditions can be recreated through input, as will be shown below.

Returning to the MS2D system, it is seen that its transfer function is given by

$z_1 = 0$: one (finite) zero at zero

two complex conjugate poles for underdamped system

$$H(s)=\frac{Y(s)}{U(s)}=\frac{X_{out}(s)}{X_{in}(s)}=\frac{b_2}{m}\frac{s}{s^2+\frac{(b_1+b_2)}{m}s+\frac{k}{m}}=\frac{b_2}{m}\frac{(s-0)}{s^2+\frac{(b_1+b_2)}{m}s+\frac{k}{m}} \tag{3.1-5}$$

where s in the numerator represents the derivative at the input induced by attaching the second damper to the mass. This results in appearance in its transfer function of the finite zero z_1 that has the value $z_1 = 0$ ("zero at zero"). In contrast, in the MSD system the input — force F_{in} acting on a mass — produces a transfer function with no finite zeros:

$$H(s)=\frac{Y(s)}{U(s)}=\frac{X_{out}(s)}{F_{in}(s)}=\frac{1}{m}\frac{1}{s^2+\frac{b\equiv b_1}{m}s+\frac{k}{m}}=\frac{1}{m}\frac{1}{s^2+2\xi\omega_n s+\omega_n^2},\ \omega_n=\sqrt{\frac{k}{m}},\ \xi=\frac{b}{2\sqrt{km}}. \tag{3.1-6}$$

As shown next, the presence of this zero at zero in the transfer function of the MS2D system profoundly changes its properties in comparison with those of the MSD system.

3.1.4 *Implicit input/output operator: state-space representation. State-variable filter implementation*

While focusing on input/output relation brings valuable insights and tools, input/output system description in terms of its internal state gives rise to a state-space representation useful in system behavior are called *state variables*. The number of state variables matches the order of the highest derivative in LCCODE. State-space form is itself LCCODE, but it

— admits nonphysical choice of states to represent the same input-output behavior, and

— in general doesn't admit input derivatives when used for filter implementation.

In the MSD and MS2D systems the output x_{out} is also one of the state variables, but in the general case *the system output is related to the state variables through a "readout map"* matrix.
Let's start with MSD system and then move to MS2D one.
For MSD system:

$$m\ddot{x}_{out} + b\dot{x}_{out} + kx_{out} = F_{in} \tag{3.1-7}$$

In this case input $u(t) = F_{in}$ is the force, and the state variables are given by $x_1 = x_{out}$ and $x_2 = \dot{x}_{out} = \dot{x}_1$, yielding a *state-space* system representation

$$\xrightarrow{u} x=\begin{bmatrix}x_1\\x_2\end{bmatrix}=\begin{bmatrix}x_{out}\\\dot{x}_{out}\end{bmatrix} \Rightarrow \begin{cases}\dot{x}=\begin{bmatrix}\dot{x}_1\\\dot{x}_2\end{bmatrix}=\begin{bmatrix}x_2\\-\frac{b}{m}x_2-\frac{k}{m}x_1+\frac{1}{m}u\end{bmatrix}=\underbrace{\begin{bmatrix}0&1\\-\frac{k}{m}&-\frac{b}{m}\end{bmatrix}}_{A}\underbrace{\begin{bmatrix}x_1\\x_2\end{bmatrix}}_{x}+\underbrace{\begin{bmatrix}0\\\frac{1}{m}\end{bmatrix}}_{B}u\\ y=x_1=\underbrace{[1\ \ 0]}_{C}\underbrace{\begin{bmatrix}x_1\\x_2\end{bmatrix}}_{x}+\underbrace{[0]}_{D}u\end{cases} \xrightarrow{y} \tag{3.1-8}$$

that, with the state x being a vector, can be succinctly written in a vector-matrix form as:

$$\dot{x} = Ax + Bu$$
$$y = Cx + Du$$

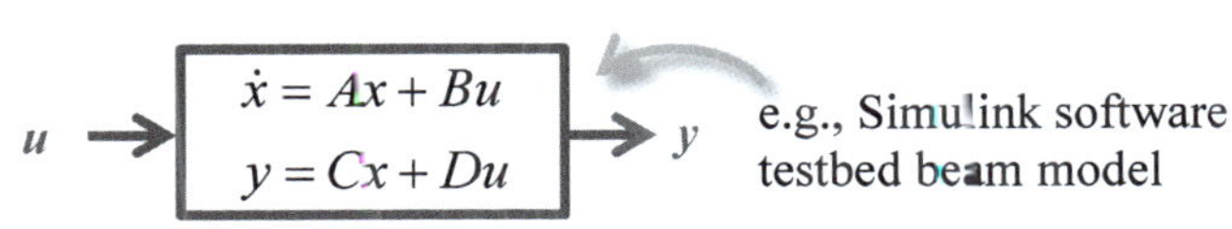

Any linear system can be written in a similar format. In general, input and output are vectors, yielding a *multi-input/multi-output (MIMO)* system, but in this book we will mainly consider systems that have scalar inputs and outputs, commonly referred to, as indicated earlier, as *single-input/single-output (SISO)* systems.

Input/state/output simulation diagram. A Simulink diagram representing equation (3.1-7) is given in Figure 3.1-3 and represents a state-variable implementation of an MSD system filtering action.

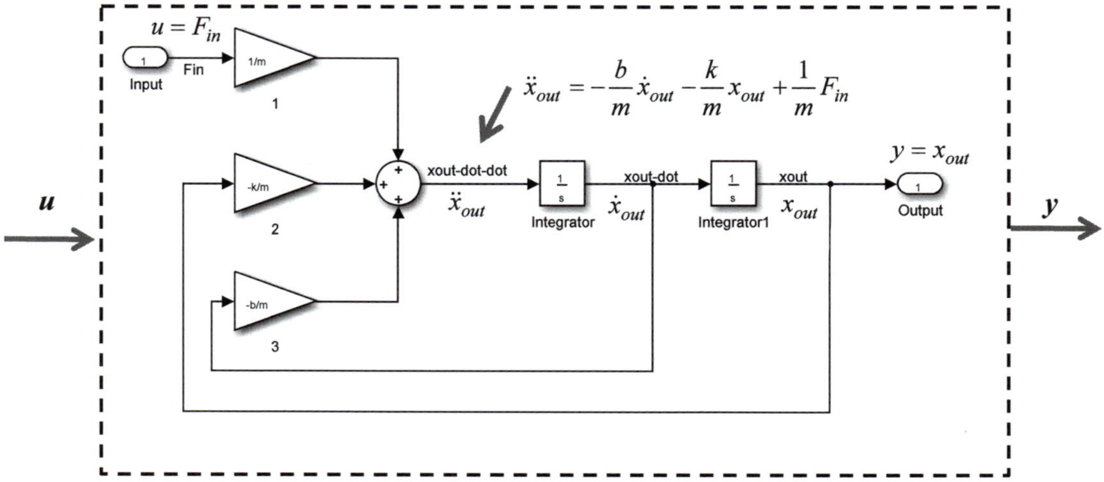

Fig. 3.1-3. Input/state/output simulation diagram for MSD system — state-variable implementation of MSD filter.

This diagram shows system internal structure in terms of its state variables — internal state x_{out} and its time derivative $\dot{x}_{out}$ along with the relation of input and output to the state variables.

3.1.4.1 *Obtaining transfer function from state-space representation*

System transfer function *H*(*s*), a *matrix* for an MIMO system, can be obtained through taking Laplace transform of the state space LCCODE assuming zero initial conditions, $x(0) = 0$, as follows:

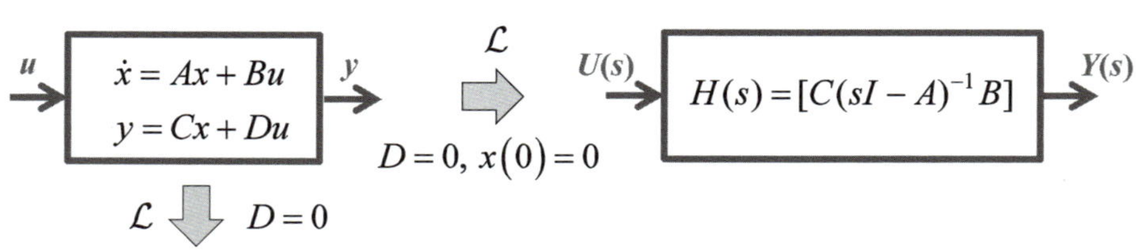

$$\begin{aligned} &sX(s) - x(0) = AX(s) + BU(s) \Rightarrow (sI - A)X(s) = x(0) + BU(s) \Rightarrow \\ &\quad X(s) = (sI - A)^{-1}x(0) + (sI - A)^{-1}BU(s) \Rightarrow \\ &\quad Y(s) = [C(sI - A)^{-1}]x(0) + [C(sI - A)^{-1}B]U(s) \\ &\because x(0) = 0 \Rightarrow Y(s) = [C(sI - A)^{-1}B]U(s) \Rightarrow H(s) = [C(sI - A)^{-1}B]. \end{aligned} \tag{3.1-9}$$

For example, applying (3.1-9) to the state-space MSD system representation (3.1-8) yields

$$Y(s)=[C(sI-A)^{-1}B]U(s)=[1 \quad 0]\left[\begin{bmatrix} s & 0 \\ 0 & s \end{bmatrix}-\begin{bmatrix} 0 & 1 \\ -\frac{k}{m} & -\frac{b}{m} \end{bmatrix}\right]^{-1}\begin{bmatrix} 0 \\ \frac{1}{m} \end{bmatrix}U(s)$$

$$=[1 \quad 0]\begin{bmatrix} \frac{ms+b}{ms^2+bs+k} & \frac{m}{ms^2+bs+k} \\ -\frac{k}{ms^2+bs+k} & \frac{ms}{ms^2+bs+k} \end{bmatrix}\begin{bmatrix} 0 \\ \frac{1}{m} \end{bmatrix}U(s) \tag{3.1-10}$$

$$=\begin{bmatrix} \frac{ms+b}{ms^2+bs+k} & \frac{m}{ms^2+bs+k} \end{bmatrix}\begin{bmatrix} 0 \\ \frac{1}{m} \end{bmatrix}U(s)=\frac{1}{ms^2+bs+k}U(s).$$

3.1.4.2 *Problem with MS2D system — input derivative*

Direct recasting of MS2D equation (3.1-2) into simulation diagram yields filter implementation shown in Figure 3.1-4. At the input, the diagram has the undesirable differentiator block that leads to problems with filtering high frequency noise from the input signal. Indeed, differentiation of the high frequency noisy component of the input will produce large magnitude signals that can saturate b_2/m block or cause large numerical errors.

Therefore, the state-variable filter implementation in general is characterized by the absence of input derivatives. Since MSD system ODE has no input derivatives, recasting it into state-space form is straightforward. However, MS2D system ODE does have input derivative. So it is not clear how to recast it into state space form. The latter task is addressed next.

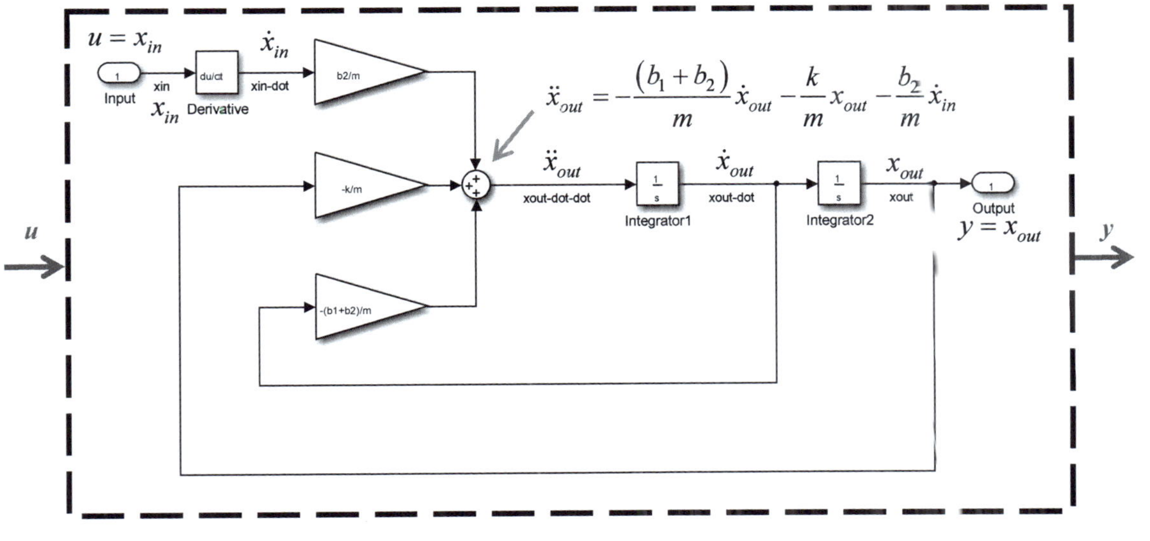

Fig. 3.1-4. Simulink diagram of MS2D equation (3.1-2).

3.1.4.3 *Removing input derivative from simulation diagram — transfer function viewpoint*

To remove the input derivative from the MS2D system Simulink diagram, while retaining the same input/output filtering properties, let's take a look at the MS2D system input/output relation given in terms of its transfer function (3.1-3)

$$X_{out}(s) = H(s)X_{in}(s) = \frac{b_2}{m}\frac{s}{s^2 + \frac{(b_1+b_2)}{m}s + \frac{k}{m}}X_{in}(s). \tag{3.1-11}$$

This relation can be viewed in terms of input derivative:

$$X_{out}(s) = H(s)X_{in}(s) = \frac{b_2}{m}\frac{1}{s^2 + \frac{(b_1+b_2)}{m}s + \frac{k}{m}}\underbrace{sX_{in}(s)}_{\text{input derivative}} \tag{3.1-12}$$

or output derivative of a related MSD-type system:

$$X_{out}(s) = \underbrace{s\tilde{X}_{out}(s)}_{\substack{\text{output}\\ \text{derivative}}} = \underbrace{s\tilde{H}(s)X_{in}(s)}_{\substack{\text{output}\\ \text{derivative}}} = s\,[\overbrace{\frac{b_2}{m}\frac{1}{s^2 + \frac{(b_1+b_2)}{m}s + \frac{k}{m}}}^{\tilde{H}(s)}]\,X_{in}(s). \tag{3.1-13}$$

State variables, however, are not unique — we can shoose them so that the derivative at the input is removed, while preserving the same transfer function. This means that we can represent MS2D system in terms of not one, but two relations:

$$\tilde{X}_{out}(s) = \tilde{H}(s)X_{in}(s), \quad X_{out}(s) = s\tilde{X}_{out}(s). \tag{3.1-14}$$

3.1.4.4 *Implementation of filters with finite zeros in the state-variable form*

Recasting the latter into one differential equation and one derivative relation yields

$$m\ddot{\tilde{x}}_{out} + (b_1+b_2)\dot{\tilde{x}}_{out} + k\tilde{x}_{out} = b_2 x_{in}, \quad x_{out} = \dot{\tilde{x}}_{out}. \tag{3.1-15}$$

This allows us to define state variables as in MSD system: $x_1 = \tilde{x}_{out}$, $x_2 = \dot{\tilde{x}}_{out}$ and system output as $x_{out} = x_2$ yielding a *state-space* system representation

(3.1-16)

$$u \rightarrow \quad x = \begin{bmatrix} x_1 \\ x_2 \end{bmatrix} = \begin{bmatrix} \tilde{x}_{out} \\ \dot{\tilde{x}}_{out} \end{bmatrix} \Rightarrow \begin{cases} \dot{x} = \begin{bmatrix} \dot{x}_1 \\ \dot{x}_2 \end{bmatrix} = \begin{bmatrix} x_2 \\ -\frac{b_1+b_2}{m}x_2 - \frac{k}{m}x_1 + \frac{b_2}{m}u \end{bmatrix} = \underbrace{\begin{bmatrix} 0 & 1 \\ -\frac{k}{m} & -\frac{b_1+b_2}{m} \end{bmatrix}}_{A}\underbrace{\begin{bmatrix} x_1 \\ x_2 \end{bmatrix}}_{x} + \underbrace{\begin{bmatrix} 0 \\ \frac{b_2}{m} \end{bmatrix}}_{B}u \\ y = x_2 = \underbrace{\begin{bmatrix} 0 & 1 \end{bmatrix}}_{C}\underbrace{\begin{bmatrix} x_1 \\ x_2 \end{bmatrix}}_{x} + \underbrace{\begin{bmatrix} 0 \end{bmatrix}}_{D}u \end{cases} \quad \rightarrow y$$

which produces state-variable MS2D filter realization that retains finite zero, while removing the input derivative:

$$Y(s)=[C(sI-A)^{-1}B]U(s)=\begin{bmatrix}0 & 1\end{bmatrix}\left[\begin{bmatrix}s & 0\\ 0 & s\end{bmatrix}-\begin{bmatrix}0 & 1\\ -\dfrac{k}{m} & -\dfrac{b_1+b_2}{m}\end{bmatrix}\right]^{-1}\begin{bmatrix}0\\ \dfrac{b_2}{m}\end{bmatrix}U(s)$$

$$=\begin{bmatrix}0 & 1\end{bmatrix}\begin{bmatrix}\dfrac{ms+(b_1+b_2)}{ms^2+(b_1+b_2)s+k} & \dfrac{m}{ms^2+(b_1+b_2)s+k}\\ -\dfrac{k}{ms^2+(b_1+b_2)s+k} & \dfrac{ms}{ms^2+(b_1+b_2)s+k}\end{bmatrix}\begin{bmatrix}0\\ \dfrac{b_2}{m}\end{bmatrix}U(s)=\frac{b_2 s}{ms^2+(b_1+b_2)s+k}U(s).$$

(3.1-17)

The Simulink diagram corresponding to state space MS2D form (3.1-16) is shown in Figure 3.1-5.

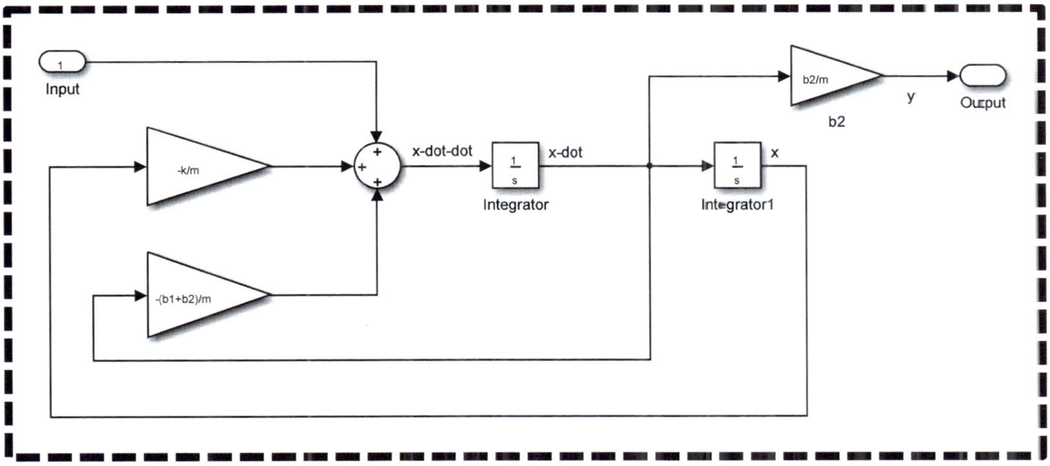

Fig. 3.1-5. Simulink diagram of MS2D equation in the state-variable form with no input derivative.

Similarly redefining system variables permits implementing any filter with finite zeros.

3.1.4.5 *Disturbances and measurement noise*

In the mold oscillator problem we will be dealing with ***input disturbance*** and ***output disturbance***, both known signals, and ***measurement noise*** — an unknown signal. A disturbance is a signal that affects the input and/or the output, and is not under our control:

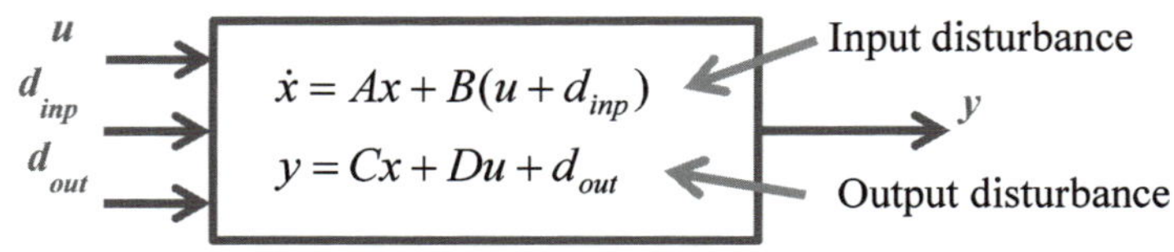

A measurement noise is a signal that affects only the measurement of the output, but not the actual output, meaning that the physical output itself is undistorted by noise, but can't be measured without it.

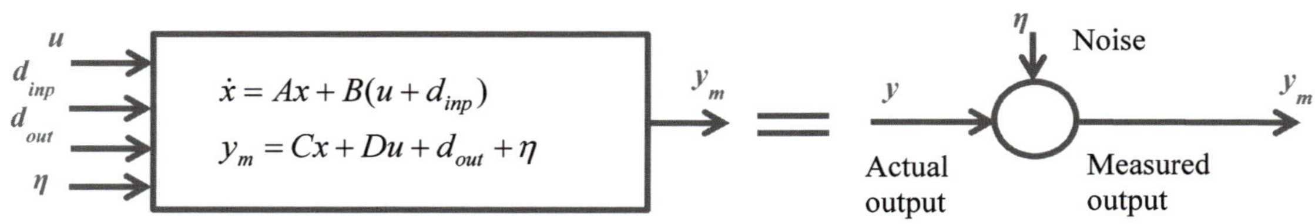

Initial conditions are often viewed as a consequence of impulsive disturbances acting at system startup, as explain below.

3.1.5 *Impulse response as system representation*

Impulse response — brief review. Impulse response, generally denoted as $h(t)$, is used for bringing out the properties of the system components, finding system response to an arbitrary input through the convolution integral, and theoretically and practically determining a system model. To recall a physical meaning of impulse response, let's consider applying a force to a simple mass — say, kicking a football that is at rest just before the impact. Then we have the following setting:

— the time passing between contacting the ball and the ball flying away is too small to measure, so, say, it is 0,

— the impact, however, adds a non-zero, finite amount of momentum over this zero-length time interval with the following implications depicted in Figure 3.1-6 below:

- just before the impact the ball is at rest;
- just after the impact the ball acquires non-zero velocity;
- therefore, the acceleration at the instant of impact, and, hence the impact force are effectively *infinite*;
- but, the position has not changed.

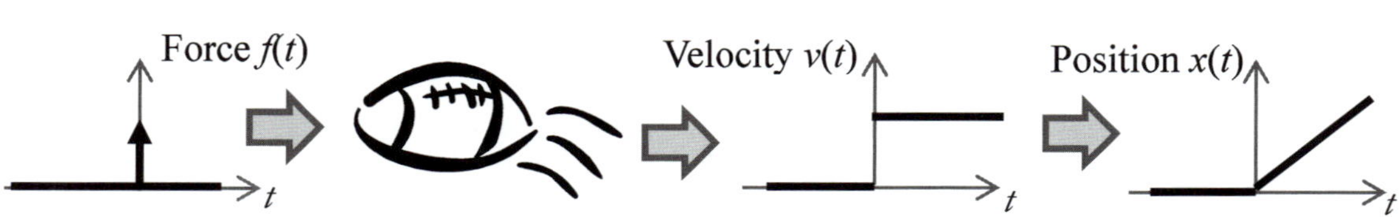

Fig. 3.1-6. Kicking a football.

Thus, the idea of an impulse and impulse response is in that:

— the impulse effectively represents an infinite instantaneous force (acceleration) acting over an effectively zero time interval, but with the finite integral over this time interval, and

— the impulse response represents an instantaneous change in the state of the system: in this case the velocity, but not the position.

Let's model this impulsive impact and the ***impulse response***, denoted as $h(t)$, and taken here as the response of the *unity mass* ball to this impact, based on the latter considerations. First, we model the input — an infinite instantaneous force — by the weighted impulse function:

$$\text{Force}(t) =: f(t) = f_0\delta(t) \underset{\text{Force of impact}}{\rightarrow} \mathcal{L}[f(t)] = f_C.$$

Next, we model the output $h(t)$, ignoring friction and gravity, as taking two forms:

1) the velocity response as the velocity, instantaneously reset from 0 to v_0, in the form of a weighted step function $v(t) = v_0u(t)$, and
2) the position response as the changing position in the form of a weighted ramp function $x(t) = v_0r(t)$ under the constant velocity v_0:

$$v(t) = \dot{x}(t) = \underset{\text{Velocity jump at } t=0}{v_0u(t)} \rightarrow \mathcal{L}[v(t)] = \frac{v_0}{s}\,;\; x(t) = \underset{\text{No position jump at } t=0}{v_0r(t)} \rightarrow \mathcal{L}[x(t)] = \frac{v_0}{s^2}. \quad (3.1\text{-}18)$$

3.1.5.1 *Impulse response and system transfer function*

Next, we model the impact and the response to it through the governing unity mass equation:

$$f = ma = m\dot{v} = m\ddot{x} = \dot{v} = \ddot{x}.$$

Taking Laplace transform (assuming 0 initial conditions, i.e. the ball being at rest) yields two transfer functions — velocity/input and position/input:

$$\mathcal{L}[f] = F(s) = \mathcal{L}[\dot{v}] = sV(s) \rightarrow \frac{V}{F} = \frac{1}{s};\; F(s) = \mathcal{L}[\ddot{x}] = s^2X(s) \rightarrow \frac{X}{F} = \frac{1}{s^2}. \quad (3.1\text{-}19)$$

Thus, the Laplace transform of the system's response to an impulse modeled earlier through empirical considerations matches the input/output relationship of the system, i.e. the system's transfer function obtained through the "first principles" Newton's second law.

Constants in the empirical models are found using conservation of momentum:

$$f\Delta t = m\Delta v \Rightarrow f = m(\Delta v/\Delta t) \underset{\Delta t \to 0}{\Rightarrow} f = m\dot{v} \Rightarrow f_0\delta(t) = d(v_0u(t))/dt = v_0\delta(t) \Rightarrow v_0 = f_0.$$

In the MS2D system mass is affected by the derivative of the impulse, called *doublet*, making, at impact instant, mass velocity infinite and mass position — discontinuous.

In the general case, impulse, impulse response, and system transfer function are characterized by the following relations:

Impulse Input: $x_{in}(t) = \delta(t) \Rightarrow$ zero i.c. $\Updownarrow$ $X_{in}(s) = 1$

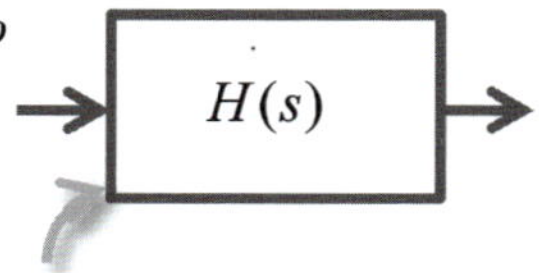

Impulse response: $x_{out}(t) = h(t)$ $\Updownarrow$

$$X_{out}(s) = \mathcal{L}\left[h(t)\right] = H(s)X_{in}(s) = H(s)\cdot 1 = H(s)$$

$\therefore$ transfer function: $H(s) = \mathcal{L}\left[h(t)\right]$.

3.1.5.2 *Impulse response of a system described by a proper 2nd order transfer function*

The effects of system pole/zero structure on system impulse response. The effect of finite zero on the impulse response could be brought out by calculation of the latter through $\mathcal{L}^{-1}\left[H(s)\right]$ — the inverse Laplace transform of a transfer function, using for example, Table E.4 in Appendix E.

1. Assume that TF has the same order of numerator and denominator polynomials, i.e. is proper, and factor out the square term in the numerator. The inverse of a constant will become an impulse:

$$H(s) = \frac{b_2 s^2 + b_1 s + b_0}{s^2 + a_1 s + a_0} = b_2 + \frac{(b_1 - a_1 b_2)s + (b_0 - a_0 b_2)}{s^2 + a_1 s + a_0} = b_2 + \frac{b_1' s + b_0'}{s^2 + a_1 s + a_0}. \quad (3.1\text{-}20)$$

2. Focusing on the second term, factor the denominator in terms of poles. If the TF has real coefficients, it will have either **2 real poles, distinct or repeated,** or **2 complex conjugate poles:**

$$\frac{b_1 s + b_0}{s^2 + a_1 s + a_0} \begin{cases} = \dfrac{b_1 s + b_0}{(s - p_1)(s - p_2)}, \; p_1 \neq p_2, p_1 \text{ and } p_2 \text{ real}, \\[2ex] = \dfrac{b_1 s + b_0}{(s - p_1)^2}, \\[2ex] = \dfrac{b_1 s + b_0}{(s - p_1)(s - p_2)} = \dfrac{b_1 s + b_0}{(s - \sigma - j\omega)(s - \sigma + j\omega)}, \; p_2 = p_1^*. \end{cases} \quad (3.1\text{-}21)$$

Real poles, nonrepeated

1. Using partial fraction expansion, decompose the second order TF into two first order TFs:

$$\frac{b_1 s + b_0}{(s - p_1)(s - p_2)} = \frac{A}{s - p_1} + \frac{B}{s - p_2}. \quad (3.1\text{-}22)$$

2. Take the inverse Laplace transform:

$$H(s) = \frac{b_2 s^2 + b_1 s + b_0}{s^2 + a_1 s + a_0} = b_2 + \frac{A}{s - p_1} + \frac{B}{s - p_2} \Rightarrow h(t) = b_2\delta(t) + Ae^{p_1 t}u(t) + Be^{p_2 t}u(t). \quad (3.1\text{-}23)$$

Real poles, repeated

1. Rewrite the sum as two second order TFs:

$$\frac{b_1 s + b_0}{(s-p_1)^2} = \frac{b_1 s}{(s-p_1)^2} + \frac{b_0}{(s-p_1)^2}. \tag{3.1-24}$$

2. Take the inverse Laplace transform:

$$H(s) = \frac{b_1 s + b_0}{(s-p_1)^2} = \frac{b_0}{(s-p_1)^2} + \frac{b_1 s}{(s-p_1)^2}$$

$$\Rightarrow h(t) = [b_0 t e^{p_1 t} + b_1(\frac{d}{dt} t e^{p_1 t})]u(t) = [b_0 t e^{p_1 t} + b_1 e^{p_1 t}(1 + t p_1)]u(t). \tag{3.1-25}$$

Complex poles:

1. Rewrite the denominator to eliminate complex numbers by completing the squares:

$$\frac{b_1 s + b_0}{(s-\sigma-j\omega)(s-\sigma+j\omega)} = \frac{b_1 s + b_0}{(s-\sigma)^2 - (j\omega)^2} = \frac{b_1 s + b_0}{(s-\sigma)^2 + \omega^2}. \tag{3.1-26}$$

2. Factor the latter into two parts:

$$\frac{b_1 s + b_0}{(s-\sigma)^2 + \omega^2} = \frac{A(s-\sigma)}{(s-\sigma)^2 + \omega^2} + \frac{B\omega}{(s-\sigma)^2 + \omega^2}. \tag{3.1-27}$$

3. Take the inverse Laplace transform:

$$H(s) = \frac{b_2 s^2 + b_1 s + b_0}{s^2 + a_1 s + a_0} = b_2 + A\frac{s-\sigma}{(s-\sigma)^2 + \omega^2} + B\frac{\omega}{(s-\sigma)^2 + \omega^2}$$
$$\Rightarrow h(t) = b_2 \delta(t) + A e^{\sigma t}\cos(\omega t)u(t) + B e^{\sigma t}\sin(\omega t)u(t). \tag{3.1-28}$$

Summary: In general, the impulse response of a second order transfer function decomposes into one of six possible responses or their sums:

(1) Impulses: $H(s) = 1$ ⟹ $h(t) = \delta(t)$;

appear when TF is not strictly proper, i.e. orders of numerator and denominator are the same (Figure 3.1-7).

(2) Exponentials: $H(s) = \dfrac{1}{s - p_1}$ ⟹ $h(t) = e^{p_1 t}u(t)$;

appear when poles of TF are real, nonrepeated (Figure 3.1-8).

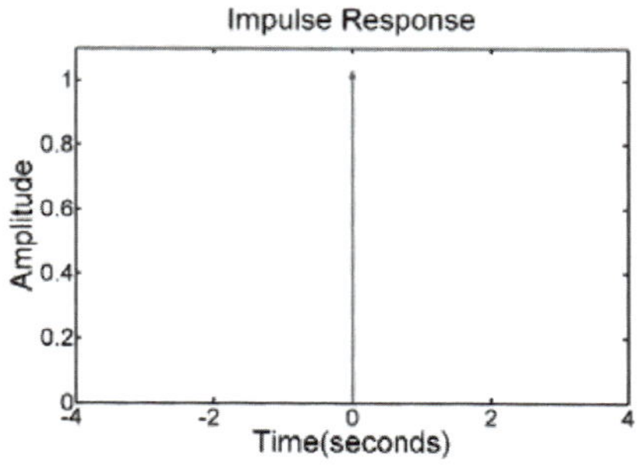

Fig. 3.1-7. Plot of $h(t)$ when H(s) = 1.

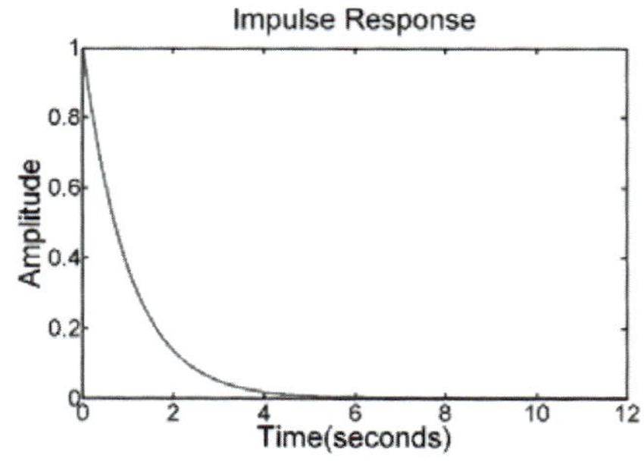

Fig. 3.1-8. Plot of h(t) when H(s) = 1/(s-p_1).

Exponentials × polynomials:

$$H(s)=\frac{1}{(s-p_1)^2} \Longrightarrow h(t)=te^{p_1 t}u(t) \overset{d(\bullet)/dt}{\Longrightarrow} H(s)=\frac{s}{(s-p_1)^2} \Longrightarrow h(t)=(1+tp_1)e^{p_1 t}u(t);$$

appear when poles of TF are real, repeated. If numerator is constant (does not depend on s), then there is no discontinuity at the origin when impulse is applied (Figure 3.1-9). Otherwise, there is a nonzero jump discontinuity at the origin (Figure 3.1-10).

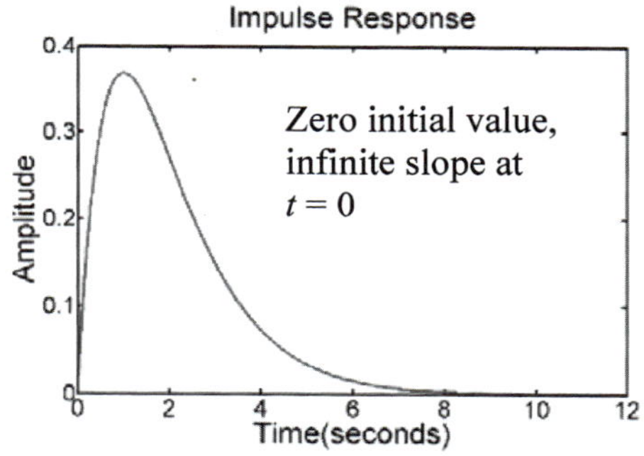

Fig. 3.1-9. Plot of $h(t)$ when $H(s)$ has no s in the numerator and 2 repeated poles.

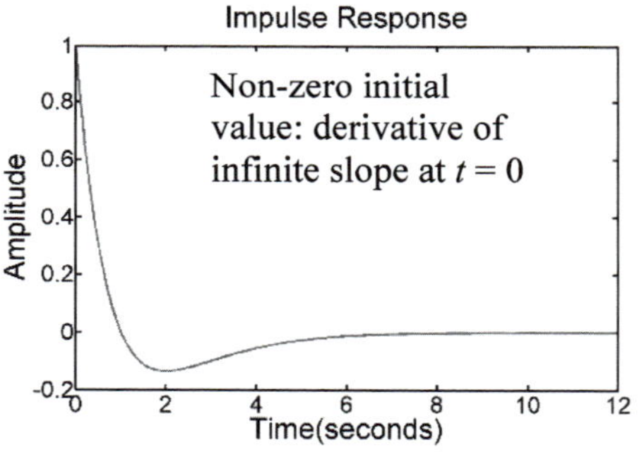

Fig. 3.1-10. Plot of $h(t)$ when $H(s)$ has s in the numerator and 2 repeated poles.

Exponentials × sinusoids:

$$H(s)=\frac{s+\sigma}{(s+\sigma)^2+\omega^2} \Longrightarrow h(t)=e^{-\sigma t}\cos(\omega t)u(t),$$

$$H(s)=\frac{\omega}{(s+\sigma)^2+\omega^2} \Longrightarrow h(t)=e^{-\sigma t}\sin(\omega t)u(t);$$

appear when poles of TF are complex. If numerator is constant (does not depend on s), then there is no discontinuity at the origin when impulse is applied (Figure 3.1-11). Otherwise, there is a nonzero jump discontinuity at the origin (Figure 3.1-12).

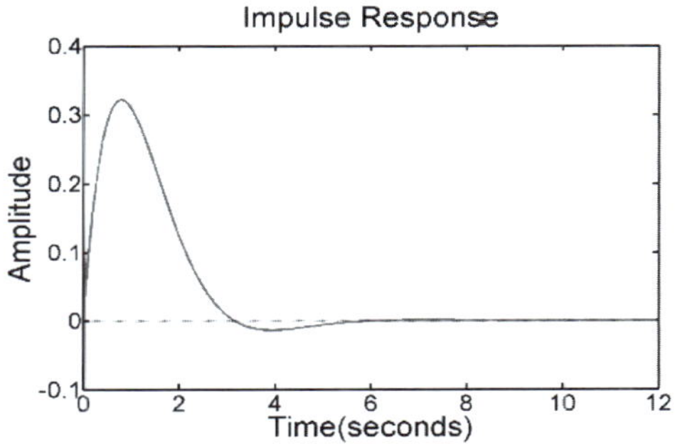

Fig. 3.1-11. Plot of $h(t)$ when $H(s)$ has no s in the numerator.

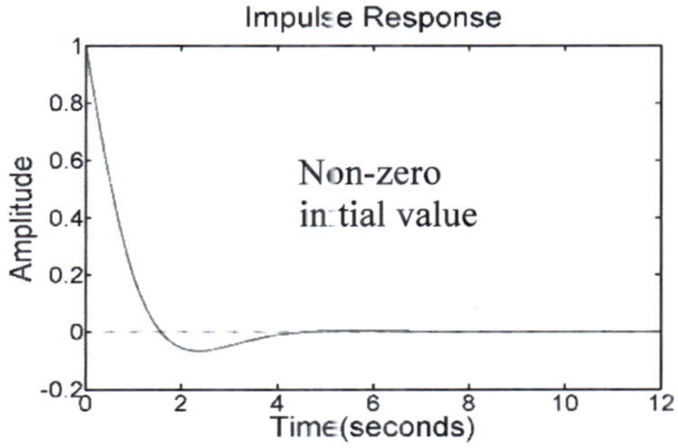

Fig. 3.1-12. Plot of $h(t)$ when $H(s)$ has s in the numerator.

Example 3.1-1: Input derivative effect on impulse response in MS2D system.
Assuming the underdamped case, we rewrite the transfer function (3.1-5) as

$$H(s)=\frac{X_{out}(s)}{X_{in}(s)}=\frac{b_2}{m}\frac{s}{s^2+[(b_1+b_2)/m]s+k/m}=s\frac{K}{s^2+2\zeta\omega_n s+\omega_n^2}=s\tilde{H}(s),\ \zeta<1,$$

where the damped natural frequency ω_d is given by $\omega_d=\omega_n\sqrt{1-\zeta^2}$.

Then, considering only $\tilde{H}(s)$ and completing the squares yields

$$\tilde{H}(s)=\frac{b_2/m}{s^2+2\zeta\omega_n s+\omega_n^2}=\frac{b_2/m}{s^2+2\zeta\omega_n s+\zeta^2\omega_n^2-\zeta^2\omega_n^2+\omega_n^2}$$
$$=\frac{b_2/m}{(s+\zeta\omega_n)^2+\omega_n^2(1-\zeta^2)}=\frac{b_2/m}{\omega_d}\frac{\omega_d}{(s+\zeta\omega_n)^2+\omega_d^2}.$$

Selecting $m=1$ and $b_1=b_2=k=0.5$ yields

$$\tilde{H}(s)=\frac{0.5}{s^2+s+0.5}=\frac{0.5}{s^2+2\cdot\sqrt{0.5}\cdot\sqrt{0.5}\cdot s+(\sqrt{0.5})^2}=\frac{0.5}{(s+0.5)^2+(0.5)^2}$$

and $\tilde{h}(t)=e^{-0.5t}\sin(0.5t)u(t)$.

Now, using Laplace transform differentiation theorem yields

$$H(s)=s\tilde{H}(s)\Rightarrow h(t)=\frac{d}{dt}\tilde{h}(t)=\frac{d}{dt}\left[e^{-0.5t}\sin\left(\frac{1}{2}t\right)u(t)\right]=\frac{1}{\sqrt{2}}e^{-\frac{1}{2}t}\sin\left(\frac{1}{2}t+\frac{\pi}{4}\right)u(t). \quad (3.1\text{-}29)$$

This response is plotted in Figure 3.1-13. Input derivative is seen to affect, along with the magnitude, the phase of $\tilde{h}(t)$, shifting the latter and creating discontinuity at the origin.

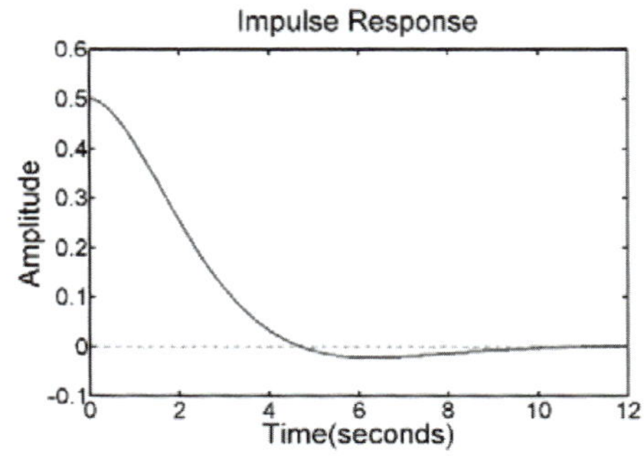

Fig. 3.1-13. Impulse response of the MS2D system.

3.1.6 *Explicit input/output operator — convolution integral*

Using Laplace transform convolution theorem (Section E.3 in Appendix E), the explicit form of the input/output relation for causal system can be obtained by taking the inverse Laplace transform of the input/output relation obtained from LCCODE:

$$\begin{aligned} X_{out}(s) = H(s)X_{in}(s) &\Rightarrow x_{out}(t) = \mathcal{L}^{-1}\left[H(s)X_{in}(s)\right] \\ &= \int_0^t h(\tau)x_{in}(t-\tau)d\tau = \int_0^t x_{in}(\tau)h(t-\tau)d\tau,\ 0 \equiv -0, \end{aligned} \tag{3.1-30}$$

i.e. with 0 including the origin. Applying this theorem to the transform of the state-space system response

$$Y(s) = [C(sI-A)^{-1}]x(0) + [C(sI-A)^{-1}B]U(s)$$

yields

$$y(t) = Ce^{At}x(0) + \int_0^t Ce^{A(t-\tau)}Bu(\tau)d\tau,\ e^{At} = \mathcal{L}^{-1}\left[(sI-A)^{-1}\right]. \tag{3.1-31}$$

3.1.6.1 *Embedding initial conditions into input/output framework through convolution*

Considering the first order system $\dot{x} = ax + u,\ y = x,\ x(0)$ — initial condition, e.g., mass/damper/clamp system with the force input u applied to the mass and the mass velocity x taken as the output yields

$$y(t) = \underset{\substack{\nearrow \\ \text{Initial velocity (at } t=0)}}{e^{at}x(0)} + \int_0^t e^{a(t-\tau)}u(\tau)d\tau = \int_0^t e^{a(t-\tau)}\underbrace{[\underset{\substack{\nearrow \\ \text{Velocity jump at } t=0}}{x(0)\delta(\tau)} + u(\tau)]}_{u_1(\tau)}d\tau = \int_0^t e^{a(t-\tau)}u_1(\tau)d\tau. \tag{3.1-32}$$

Expression (3.1-32) shows that the effect of the initial condition $x(0)$ can be embedded into the first order system input/output representation by redefining the input to include the delta-function. Then, *the response of the system at rest to impulsive input becomes identical to the response of the system to initial conditions under zero input*. With the appropriate modifications this extends to any LTI system with no hidden modes. For example, as seen from the impulse response of a simple mass in Figure 3.1-6, the effect of impulse on this system resets the initial velocity from 0 to v_0. Similar resetting of initial velocity will take place in MSD system.

Unlike the simple mass and MSD system, in MS2D system the input is not the force, but the displacement, and it is differentiated. Therefore, when displacement is an impulse, the mass is affected by the force that is the derivative of this impulse, referred to as ***doublet***, $\dot{\delta}(t)$, making the mass velocity infinite at the instant of impact, and the mass position — system output — discontinuous.

The impulse response of MS2D system is then seen to be equivalent to doublet response of MSD system, showing in (3.1-29) and Figure 3.1-13 that not only the velocity, but also the position in the latter system can be reset to nonzero value through an appropriate input. Namely, since superposition is valid for the MSD system due to the linearity of the latter, adding simultaneous impulse and doublet to form an input and applying it to MSD system simultaneously resets both system velocity and position. This shows that *initial conditions can be easily embedded into an input/output framework, and there is no loss of generality in dropping initial conditions from consideration if the hidden modes are of no interest and decay on their own.*

Example 3.1-2: Consider the system with a causal impulse response $h(t) = \delta(t)\text{-}e^{-t}u(t)$ where $u(t)$ is the unit step. Derive analytically the response, $y(t)$, of this system to a causal input given by the unit step $u(t)$ using the ***noncausal convolution integral*** $y(t)=\int_{-\infty}^{\infty} h(\tau)u(t-\tau)d\tau$. Show graphically, using folding and shifting, how the integrand product defines the integration limits. Sketch the resulting response.

Solution: The signals $h(\tau)$, $u(t-\tau)$, and $h(\tau)u(t-\tau)$ are shown in Figure 3.1-14.

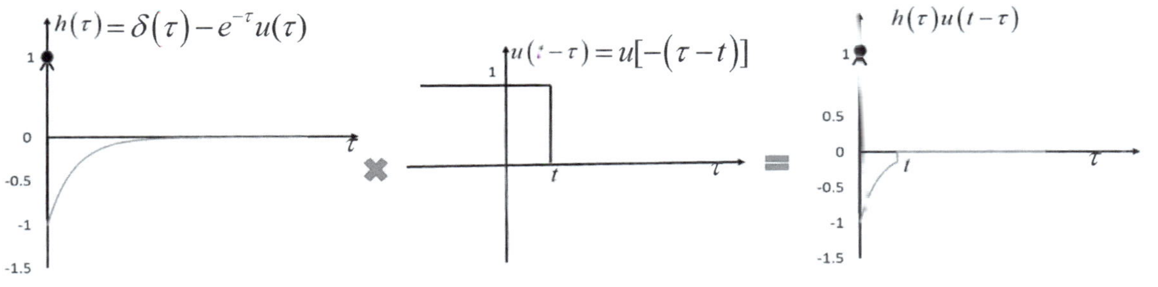

Fig. 3.1-14. Setting up the limits of integration.

The product $h(\tau)u(t-\tau)$ imposes the [0, t] integration interval, yielding ***causal convolution***, and hence causal response, in which any noncausal function of time needs to be multiplied by $u(t)$, as shown below.

$$y(t)=\int_{-\infty}^{\infty} h(\tau)u(t-\tau)d\tau=\int_{-\infty}^{\infty} h(\tau)u[-(\tau-t)]d\tau=\int_0^t[\delta(\tau)-e^{-\tau}]d\tau=\int_0^t\delta(\tau)d\tau-\int_0^t e^{-\tau}d\tau=$$

$$u(t)-[(-e^{-\tau})\big|_0^t]u(t)=u(t)-(1-e^{-t})u(t)=u(t)-u(t)+e^{-t}u(t)=e^{-t}u(t),\quad -\infty\le t\le\infty.$$

The response $y(t) = e^{-t}u(t), \; -\infty \le t \le \infty,$ is given below in Figure 3.1-15.

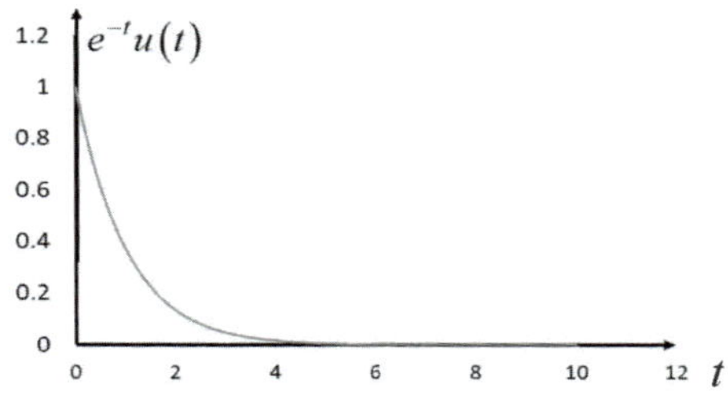

Fig. 3.1-15. System response to a unit step.

3.1.6.2 *Convolution integral as a filter*

Convolution integral provides an explicit calculation that performs filtering of the input signal. To see this, consider a causal input $u(t)$ and a causal infinite duration impulse response, often referred to as ***infinite impulse response*** **(*IIR*)** shown in Figures 3.1-16 and 3.1-17, respectively:

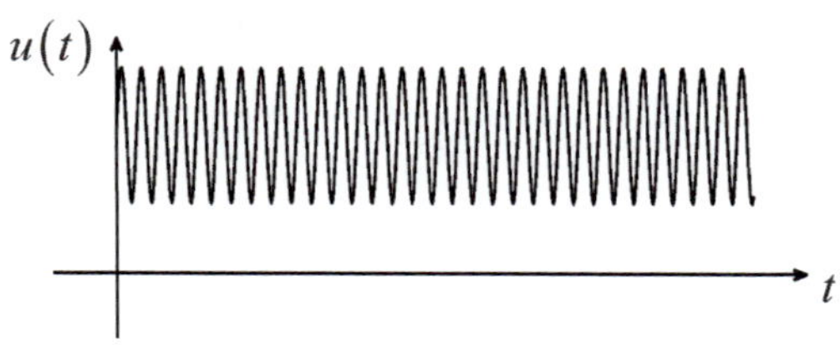

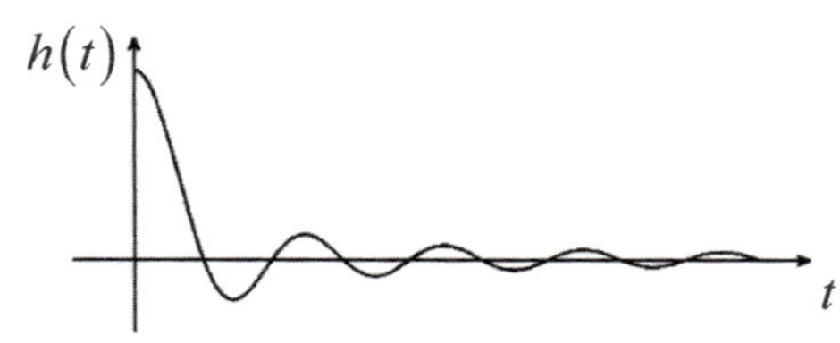

Fig. 3.1-16. Causal input signal $u(t)$.

Fig. 3.1-17. Causal infinite impulse response (IIR) $h(t)$.

Then, convolution of a causal input signal with a causal IIR yields a causal IIR filter given by

$$y(t) = \int_0^t u(\tau)h(t-\tau)d\tau = \int_0^t u(\tau)h[-(\tau-t)]d\tau \tag{3.1-33}$$

and shown in Figures 3.1-18 and 3.1-19, respectively:

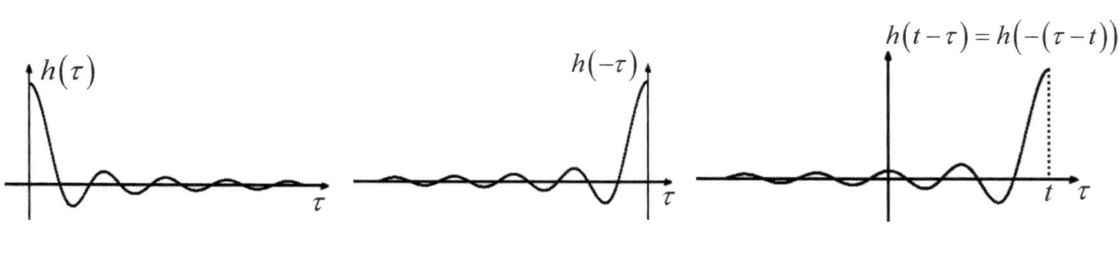

Fig. 3.1-18. Converting $h(\tau)$ into $h(t\text{-}\tau)$ using folding and shifting.

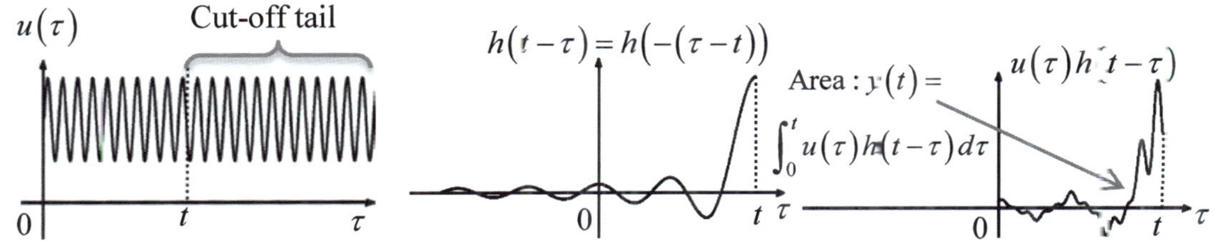

Fig. 3.1-19. Graphical convolution representation of the IIR filter.

From Figure 3.1-19 it is seen that filter action becomes more accurate as t moves to the right, i.e. increases, bringing out more of the impulse response to play a bigger role in shaping its product with the input signal. The drawback, however, is that the calculation time interval [0, t] goes to infinity, making calculation size grow without bound.

The latter problem can be addressed by noting that filters are stable dynamical systems with decaying to zero impulse responses. This permits truncating causal infinite impulse response at $t = t_{tr}$ for a large enough t_{tr} after the impulse response decays well enough to ensure that the error caused by truncation will be insignificant. This yields a finite duration impulse response $h_t(t)$, referred to as ***finite impulse response*** (FIR), shown in Figure 3.1-20, where the subscript t indicates truncation:

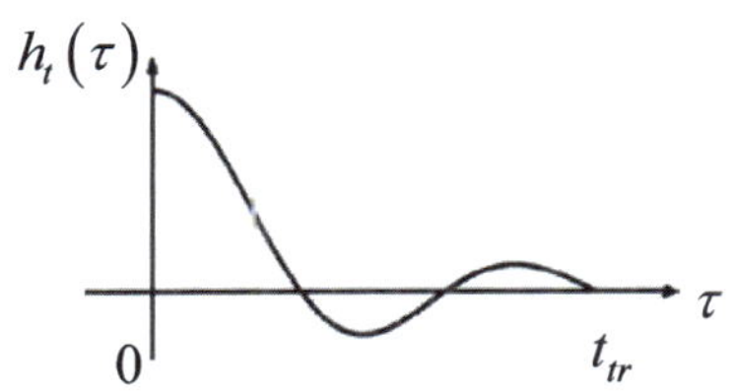

Fig. 3.1-20. Truncated (finite) impulse response.

Convolution with FIR is graphically shown in Figures 3.1-21 and 3.1-22. The expression of the IIR filter (3.1-21) is seen to change for the FIR filter to

$$y(t)=\int_{t-t_{tr}}^{t} u(\tau)h_{tr}(t-\tau)d\tau=\int_{t-t_{tr}}^{t} u(\tau)h_{tr}\left[-(\tau-t)\right]d\tau. \quad (3.1\text{-}34)$$

The impulse response truncation is seen to result in the input signal being weighted by the filter impulse response only over the ***moving, or sliding, window*** of fixed length **[*t, t-t_{tr}*]**. Therefor this filter is also referred to as the ***moving average* (*MA*)** filter. As it is seen, the amount of calculation if discretized, stays the same, with accuracy depending on the size of the moving window.

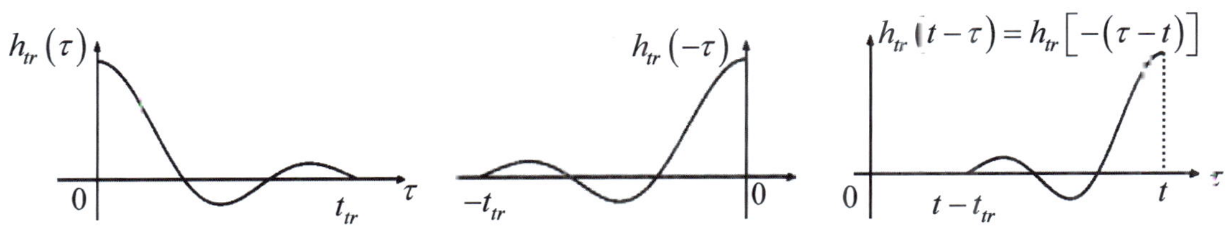

Fig. 3.1-21. Converting $h_{tr}(\tau)$ into $h_{tr}(t\text{-}\tau)$ using folding and shifting.

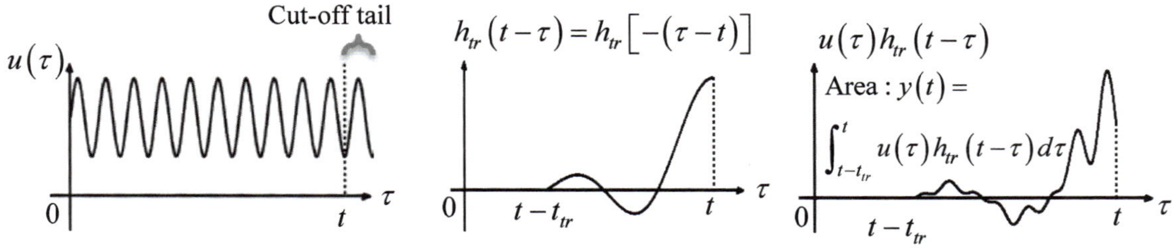

Fig. 3.1-22. Graphical convolution representation of the FIR filter.

3.1.7 *Pole-zero map: a look inside the dynamic system behavior*

The pole-zero map acts as a visual representation of the transfer function. Since transfer function is defined by the roots of numerator and denominator polynomials, which are, respectively, system zeros and poles, the location of both uniquely determines system properties. Pole/zero map permits quick visualization of these properties. For example, pole location shows lack or degree of stability, susceptibility to resonances, and response type, and zero location determines the step response direction (so called inverse response — going initially the other way) as well as filtering properties in terms of the frequency response when zeros at infinity are included as well. Poles are typically discussed in dynamics courses and zeros are paid as much attention at in filtering and control.

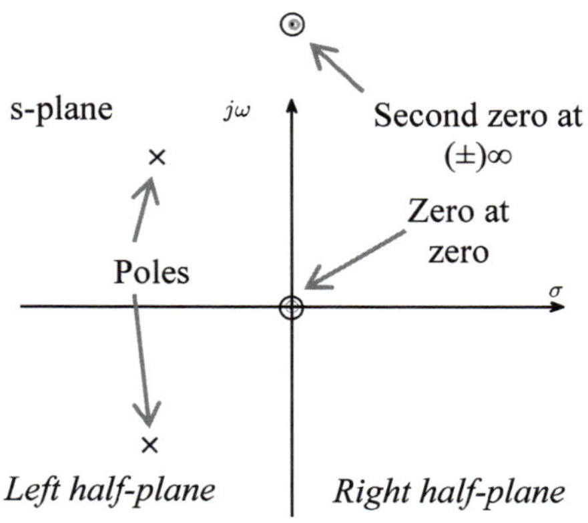

Fig. 3.1-23. Pole-zero map of the MS2D system.

For the right and the left half-planes, we have the following terminology:

- *open half-plane of s-plane:* does not include the $j\omega$-axis,
- *closed half-plane of s-plane:* does include the $j\omega$-axis.

If all poles are in the open left half-plane, the system is stable; complex conjugate poles imply that the system is underdamped, causing large magnitude response to excitation at resonance; a zero at the origin blocks constant inputs, a zero at ∞ blocks high frequency inputs.

Derivative on input becomes a zero at $s = 0$: a "blocking zero" that blocks any effect of constant (i.e. zero frequency) inputs on outputs: $H(j\omega) = 0$ for $\omega = 0$.

The latter blocking zero has been introduced into MS2D system transfer function through the MSD system redesign — addition of a damper at the input. The pole/zero map of the resulting transfer function (3.1-6) shown in Figure 3.1-23 is seen to have more poles than finite zeros. But considering zeros at "∞" equalizes the number of zeros and poles:

"missing" zero at $1/0 = \infty$

$$\lim_{|s|\to\infty}\left|\frac{X_{out}(s)}{X_{in}(s)}\right| = \lim_{|s|\to\infty}\left|\frac{b_2}{m}\frac{s}{s^2+\frac{(b_1+b_2)}{m}s+\frac{k}{m}}\right| = \frac{b_2}{m}\lim_{|s|\to\infty}\frac{1}{|s|} = 0, H(s) = \frac{b_2}{m}\frac{\overbrace{s(1-0\cdot s)}}{s^2+\frac{(b_1+b_2)}{m}s+\frac{k}{m}}.$$

3.1.8 *Frequency response: calculation and use*

System frequency response characterizes how a system modifies the parameters of the input sinusoid — a *noncausal* signal — in producing the output, also periodic noncausal signal. The integrative system portrait in Figure 3.1-1 is at this point, however, restricted only to causal systems, i.e. the ones with causal impulse response, which guarantees the existence of *causal* (single-sided) Laplace transform. We will therefore find system frequency response by simply substituting s by $j\omega$ in system transfer function given by this Laplace transform and calculating magnitude and phase of the resulting expression as a function of ω.

We cannot show why this works in the time domain using convolution theorem, since causal Laplace transform is not defined for a noncausal input, and taking the inverse Laplace transform of $H(s)[\omega/((\sigma+s)^2+\omega^2)]$ will simply yield a convolution with a causal input $cos(\omega t)u(t)$. Therefore, we will show this on the MS2D example.

The full noncausal development of frequency response will be carried out in Chapter 6 through noncausal tools — convolution and Fourier transform, with much of the chapter devoted to the various forms of the latter to carry out spectral analysis.

That's where we will show why substituting s by $j\omega$ works in the general case. Instead of a product of causal Laplace transforms, through a noncausal convolution of causal impulse response with noncausal input, we will obtain a *noncausal time-domain* expression where the complex gain $H(j\omega)$ multiplies periodic (noncausal) input function.

3.1.8.1 *Noncausal input*

First, we find solution of ODE (3.1-2) with $b_1 = b_2 = k = 0.5$ and $m = 1$ for the true sinusoidal (non-causal) input by direct substitution:

$$\ddot{x}_{out} + \dot{x}_{out} + 0.5x_{out} = 0.5\dot{x}_{in},\ x_{in} = \cos(\omega t).$$

Assume solution has the same format: $x_{out}(t) = A\cos(\omega t) + B\sin(\omega t)$.

Substitute the formula into the ODE, combine similar terms:

$$
\begin{aligned}
&\left(-\omega^2 A\cos(\omega t)-\omega^2 B\sin(\omega t)\right) \longleftarrow \ddot{x}_{out} \\
&+\left(-\omega A\sin(\omega t)+\omega B\cos(\omega t)\right) \longleftarrow \dot{x}_{out} \qquad\qquad \dot{x}_{in} \\
&+0.5\left(A\cos(\omega t)+B\sin(\omega t)\right) \longleftarrow x_{out} \qquad\qquad \downarrow \\
&=\left(-\omega^2 A+\omega B+0.5A\right)\cos(\omega t)+\left(-\omega^2 B-\omega A+0.5B\right)\sin(\omega t)=-0.5\omega\sin(\omega t).
\end{aligned}
$$

Solve the system of equations:

$$
\left.\begin{aligned}-\omega^2 A+\omega B+0.5A=0\\ -\omega^2 B-\omega A+0.5B=-0.5\omega\end{aligned}\right\}\Rightarrow
\left.\begin{aligned}A=\frac{0.5\omega^2}{\omega^4+0.25}\\ B=\frac{0.5\omega^3-0.25\omega}{\omega^4+0.25}\end{aligned}\right\}\Rightarrow
\begin{aligned}x_{out}(t)=\frac{0.5\omega^2}{\omega^4+0.25}\cos(\omega t)\\ +\frac{0.5\omega^3-0.25\omega}{\omega^4+0.25}\sin(\omega t).\end{aligned}
$$

Add sinusoids with the same frequency in

$$x_{out}(t)=\frac{0.5\omega^2}{\omega^4+0.25}\cos(\omega t)+\frac{0.5\omega^3-0.25\omega}{\omega^4+0.25}\sin(\omega t).$$

For this purpose, manipulating the cosine addition formula

$$\cos(\varphi)\cos(\theta)-\sin(\varphi)\sin(\theta)=\cos(\varphi+\theta)$$

to obtain (we will review this in more detail later)

$$A\cos(\theta)+B\sin\theta=\sqrt{A^2+B^2}\cos\left(\theta+\operatorname{atan}(-B/A)\right).$$

Therefore, noncausal system response takes the form

$$x_{out}(t)=\frac{0.5\omega^2}{\omega^4+0.25}\cos(\omega t)+\frac{0.5\omega^3-0.25\omega}{\omega^4+0.25}\sin(\omega t)=\frac{0.5\omega}{\sqrt{\omega^4+0.25}}\cos\left(\omega t+\operatorname{atan}\left(\frac{0.5-\omega^2}{\omega}\right)\right).$$

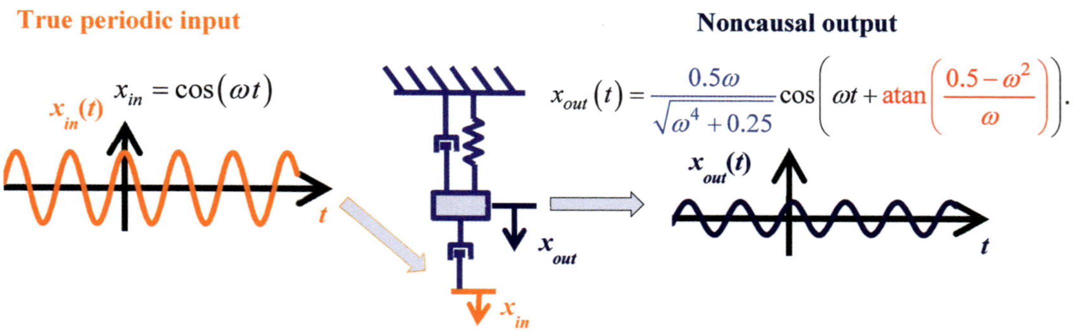

Fig. 3.1-24. Response of a causal system to periodic input.

The entire noncausal input-output setting is shown in Figure 3.1-24.

3.1.8.2 *Causal input*

Let us now find solution to ODE (3.1-2) with $b_1 = b_2 = k = 0.5$ and $m = 1$ given by $\ddot{x}_{out} + \dot{x}_{out} + 0.5x_{out} = 0.5\dot{x}_{in}$, $x_{out}(0) = \dot{x}_{out}(0) = 0$ for causal input $x_{in} = \cos(\omega t)u(t)$ shown on the left in Figure 3.1-25 below using single-sided Laplace transform. The transfer function $H(s)$ of the system and the transform $X_{in}(s)$ of the input are, respectively, given by the following Laplace transform pairs (Appendix E, Sections E.3, E.4):

$$\ddot{x}_{out} + \dot{x}_{out} + 0.5x_{out} = 0.5\dot{x}_{in},\ x_{out}(0) = \dot{x}_{out}(0) = 0 \overset{\mathcal{L}}{\rightleftharpoons} H(s) = X_{out}(s)/X_{in}(s) = \frac{0.5s}{s^2+s+0.5},$$

$$x_{in} = \cos(\omega t)u(t) \overset{\mathcal{L}}{\rightleftharpoons} X_{in}(s) = \frac{s}{s^2+\omega^2}.$$

Since $H(s)$ is seen to be underdamped ($\zeta \approx 0.7$), the system output transform $X_{out}(s)$ is simplified below using completion of squares in its denominator along with partial fractional expansions (3.1-27) and (3.1-28). First, setting

$$X_{out}(s) = H(s)X_{in}(s) = \frac{0.5s}{s^2+s+0.5}\cdot\frac{s}{s^2+\omega^2} = \frac{0.5s^2}{\left(s^2+\omega^2\right)\left(s^2+s+0.5\right)} = \frac{As+B}{s^2+\omega^2} + \frac{Cs+D}{s^2+s+0.5},$$

the expressions for A, B, C, and D are obtained as follows:

$$0.5s^2 = (As+B)\left(s^2+s+0.5\right) + (Cs+D)\left(s^2+\omega^2\right) =$$
$$(A+C)s^3 + (B+A+D)s^2 + \left(0.5A+B+\omega^2 C\right)s + \left(0.5B+\omega^2 D\right)$$

$$\Rightarrow \begin{cases} A+C=0,\ A+B+D=0.5, \\ 0.5A+B+\omega^2 C = 0, \\ 0.5B+\omega^2 D = 0. \end{cases} \Rightarrow \begin{cases} A+B+D=0.5, \\ \left(0.5-\omega^2\right)A + B = 0, \\ 0.5B+\omega^2 D = 0. \end{cases} \Rightarrow \begin{cases} A+B+D=0.5, \\ A = [1/(\omega^2-0.5)]B, \\ D = -(0.5/\omega^2)B. \end{cases}$$

This yields

$$\left(\frac{1}{\omega^2-0.5}+1-\frac{0.5}{\omega^2}\right)B = 0.5 \Rightarrow B = \frac{0.5\omega^4-0.25\omega^2}{\omega^4+0.25},\ A = \frac{1}{\omega^2-0.5}B = \frac{1}{\omega^2-0.5}\frac{0.5\omega^2\left(\omega^2-0.5\right)}{\omega^4+0.25}$$

$$= \frac{0.5\omega^2}{\omega^4+0.25},\ C = -A = \frac{-0.5\omega^2}{\omega^4+0.25},\ D = -\frac{0.5}{\omega^2}B = -\frac{0.5}{\omega^2}\frac{0.5\omega^2\left(\omega^2-0.5\right)}{\omega^4+0.25} = -\frac{0.25\omega^2-0.125}{\omega^4+0.25}$$

Substituting A, B, C, D into the expansion for $X_{\text{out}}(s)$ yields

$$X_{out}(s)=\frac{0.5s^2}{\left(s^2+\omega^2\right)\left(s^2+s+0.5\right)}=\frac{0.5\omega^2}{\omega^4+0.25}\frac{s}{s^2+\omega^2}+\frac{0.5\omega^4-0.25\omega^2}{\omega^4+0.25}\frac{1}{s^2+\omega^2}$$
$$+\frac{-0.5\omega^2}{\omega^4+0.25}\frac{s}{s^2+s+0.5}-\frac{0.25\omega^2-0.125}{\omega^4+0.25}\frac{1}{s^2+s+0.5}.$$

Noting that $s^2+s+0.5=s^2+2\cdot 0.5\cdot s+0.25+0.25=\left(s+0.5\right)^2+0.5^2$ and setting $s=s+0.5-0.5$ in the numerator of the third term of the expansion above yields

$$X_{out}\left(s\right)=\frac{0.5\omega^2}{\omega^4+0.25}\cdot\frac{s}{s^2+\omega^2}+\frac{0.5\omega^3-0.25\omega}{\omega^4+0.25}\cdot\frac{\omega}{s^2+\omega^2}-\frac{0.5\omega^2}{\omega^4+0.25}\cdot\frac{s+0.5}{\left(s+0.5\right)^2+0.5^2}$$
$$-\frac{-0.25\omega^2+0.25\omega^2-0.125}{\omega^4+0.25}\frac{1}{\left(s+0.5\right)^2+0.5^2}=\frac{0.5\omega^2}{\omega^4+0.25}\cdot\frac{s}{s^2+\omega^2}+\frac{0.5\omega^3-0.25\omega}{\omega^4+0.25}\cdot\frac{\omega}{s^2+\omega^2}$$
$$-\frac{0.5\omega^2}{\omega^4+0.25}\cdot\frac{s+0.5}{\left(s+0.5\right)^2+0.5^2}+\frac{0.25}{\omega^4+0.25}\cdot\frac{0.5}{\left(s+0.5\right)^2+0.5^2},$$

producing the desired inverse Laplace transform

$$x_{out}\left(t\right)=\frac{0.5\omega^2}{\omega^4+0.25}\cos\left(\omega t\right)u\left(t\right)+\frac{0.5\omega^3-0.25\omega}{\omega^4+0.25}\sin\left(\omega t\right)u\left(t\right)+e^{-0.5t}\left(-\frac{0.5\omega^2}{\omega^4+0.25}\cos\left(0.5t\right)\right.$$
$$\left.+\frac{0.25}{\omega^4+0.25}\sin\left(0.5t\right)\right)u\left(t\right)=\frac{0.5\omega}{\sqrt{\omega^4+0.25}}\cos\left(\omega t+\text{atan}\left(\frac{0.5-\omega^2}{\omega}\right)\right)u(t)+$$
$$e^{-0.5t}\left(-\frac{0.5\omega^2}{\omega^4+0.25}\cos\left(0.5t\right)+\frac{0.25}{\omega^4+0.25}\sin\left(0.5t\right)\right)u\left(t\right).$$

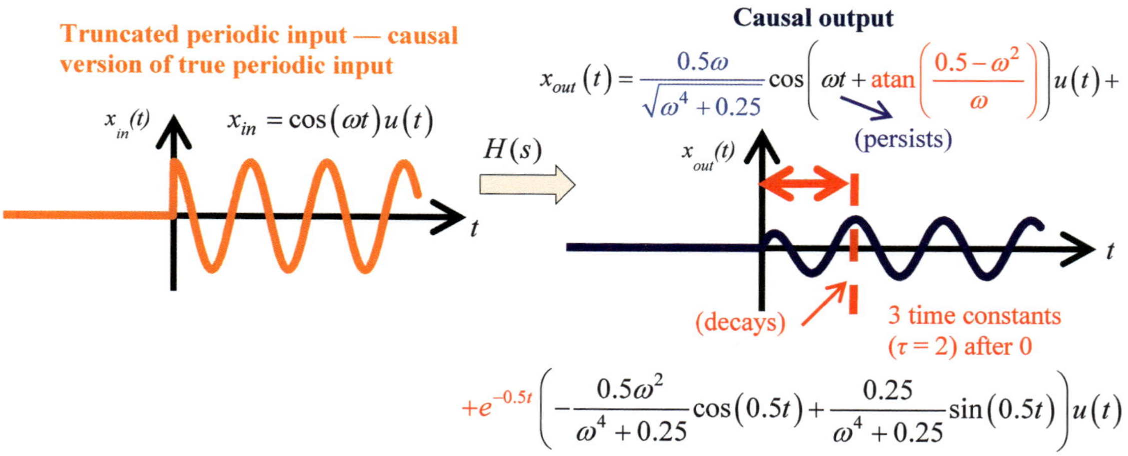

Fig. 3.1-25. Response of a causal system, MS2D, to truncated periodic input.

This causal response is depicted in Figure 3.1-25.

3.1.8.3 *Frequency response of a causal system: convergence of causal response to non-causal one*

Let us now compare the two cases above. As seen from pole/zero map (Figure 3.1-23), MS2D system is stable. Therefore, the system response to the causal version of a periodic input converges at $t = \infty$ in theory and after about 3τ in practice to the response of the same system to the true (noncausal) periodic input as the initial transient dies out due to $e^{-t/\tau}u(t)$ multiplier, as shown on the right in Figure 3.1-25. This makes system transfer function a legitimate tool for finding frequency response through looking at the stationary part of the response in time domain.

This is also the basis for finding system frequency response in practice using long enough real data records: output data to sine excitation are collected *until magnitude and phase stop changing over several cycles*, indicating that the transients died out.

Calculation of $|H(j\omega)|$ and $\angle H(j\omega)$ (complex variables will be reviewed later in the book) yields exact match of the time domain response magnitude and phase of the noncausal response in Figure 3.1-25:

$$|H(j\omega)| = \left|\frac{0.5\omega}{\sqrt{(0.5-\omega^2)^2+\omega^2}}\right| = \left|\frac{0.5\omega}{\sqrt{\omega^4+0.25}}\right|$$

$$H(j\omega) = \frac{0.5(j\omega)}{(j\omega)^2+(j\omega)+0.5} = \frac{j0.5\omega}{(0.5-\omega^2)+j\omega}$$

$$\angle H(j\omega) = \angle(0.5j\omega) - \angle\left((0.5-\omega^2)+j\omega\right) = \frac{\pi}{2} - \text{atan}\left(\frac{\omega}{0.5-\omega^2}\right) = \text{atan}\left(\frac{0.5-\omega^2}{\omega}\right).$$

Therefore, noncausal time domain response of MS2D system can be expressed in terms of its transfer function as: $x_{in} = \cos(\omega t) \to x_{out}(t) = |H(j\omega)|\cos(\omega t + \angle H(j\omega))$, indicating that $\{|H(j\omega)|, \angle H(j\omega)\}$ vesus ω shown in Figure 3.1-26 is the frequency response of MS2D system with $b_1 = b_2 = k = 0.5$ and $m = 1$.

To rigorously justify the use of the transfer function for determining the frequency response in the general case, we will introduce the Fourier Transform through convolution later in the course. We will arrive at the noncausal version of $H(j\omega)$ directly, rather than through heuristic substitution of s by $j\omega$ in the transfer function. This will also justify obtaining frequency response from transfer function through complex variable phasor analysis.

3.1.8.4 *What does system frequency response tell?*

MS2D system frequency response shows that

- input signal magnitude is multiplied by $|H(j\omega)|$: $|X_{out}(j\omega)| = |H(j\omega)||X_{in}(j\omega)|$,
- input signal phase is shifted by $\angle H(j\omega)$: $\angle X_{out}(j\omega) = \angle X_{in}(s) + \angle H(j\omega)$.

Let's take a closer look at the MS2D system properties brought out by its magnitude response in Figure 3.1-26:

$$\left|H(j\omega)\right| = \left|\frac{X_{out}(s)}{X_{in}(s)}\right| = \left|\frac{b_2}{m}\frac{j\omega}{(j\omega)^2 + \frac{(b_1+b_2)}{m}j\omega + \frac{k}{m}}\right|$$
$$\Rightarrow \left|X_{out}(j\omega)\right| = \left|\frac{b_2}{m}\frac{1}{(j\omega)^2 + \frac{(b_1+b_2)}{m}j\omega + \frac{k}{m}}\right|\left|\omega\right|\left|X_{in}(j\omega)\right| \quad (3.1\text{-}35)$$

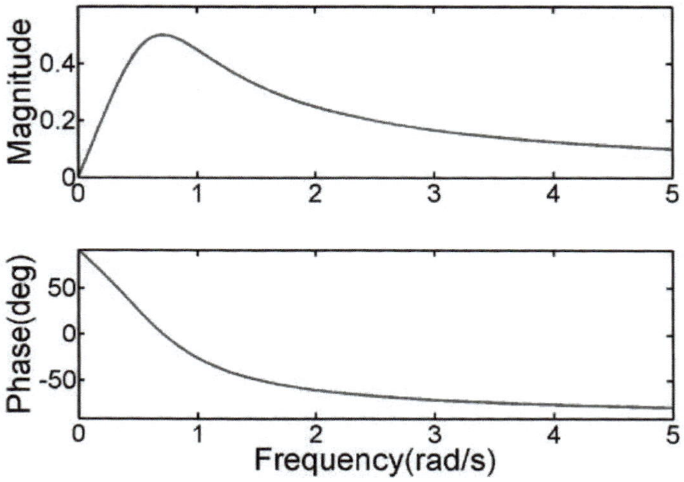

Fig. 3.1-26. Frequency response: magnitude and phase.

The *blocking zero at* $\omega = 0$ causes system frequency response to multiply magnitude of constant ($\omega = 0$) input signals by 0, thereby setting output signal magnitude to 0. Because of the frequency response continuity, this zero also causes output signal magnitudes for very low frequency inputs to be close to 0.

The *blocking zero at* $\omega = \infty$ causes filter magnitude to look like $1/\omega$ for large ω. Because of the frequency response continuity, this zero causes output signal magnitudes for very high frequency inputs to be close to 0.

The *system poles* are assumed to be complex, and consequently *underdamped*, creating a pronounced *resonance peak* in system magnitude response.

3.1.9 *Bode plots*

Linear scale in frequency response graphs focuses on their ***local*** details, such as magnitude ripple, and is therefore extensively used in filter design to visualize the magnitude ripple/phase lag tradeoffs. In control, however, the ***asymptotic*** frequency domain properties of LTI systems are often of more interest. The latter are brought into sharper focus by Bode plots, where the frequency axis scale shows the "orders of magnitude" changes, well represented by a ***logarithmic scale*** and ***graph asymptotes***. Magnitude $|H(j\omega)|$ axis is scaled in a Bode magnitude plot in decibels (dB):

$$|H(j\omega)| \text{ in dB} \triangleq 20\log_{10}|H(j\omega)| = 20\log_{10}\left[k\frac{(s-z_1)(s-z_2)\cdots(s-z_m)}{(s-p_1)(s-p_2)\cdots(s-p_n)}\right]\Bigg|_{s=j\omega} = 20\log_{10}k$$

$$+20\log_{10}|(s-z_1)| + 20\log_{10}|(s-z_2)| + \cdots - 20\log_{10}|(s-p_1)| - 20\log_{10}|(s-p_2)| - \cdots\Big|_{s=j\omega},$$

making the total magnitude response the sum of individual responses, while phase stays within small range for which $\measuredangle H(j\omega)$ axis scaled linearly is sufficient. The first and the second order systems frequency response graphs obtained as the result of such rescaling are reviewed next.

3.1.9.1 *1st order system*

For the first order system, we have $H(s) = k/(s-p_1)$.

Assuming stable system yields $p_1 = -a,\ a = const > 0$, so that $H(s) = k/(s+a)$.

Assuming gain at zero frequency $s = j\omega = 0$, referred to as DC gain, to be unity, i.e. $H(s) = k/a = 1$, yields $k = a$ and the frequency response function

$$H(s)\big|_{s=j\omega} = a/(s+a)\big|_{s=j\omega} = 1/[(j\omega/a)+1] = 1/[(j\omega/\omega_c)+1],$$

where ω_c is referred to as the cutoff frequency, discussed in more detail in Chapter 4.

This yields magnitude response

$$20\log_{10}|H(j\omega)| = 20\log_{10}|1| - 20\log_{10}|(j\omega/\omega_c)+1| = -20\log_{10}\left|\sqrt{(\omega/\omega_c)^2+1}\right|$$

and phase response

$$\measuredangle H(j\omega) = \measuredangle 1/[(j\omega/\omega_c)+1] = \measuredangle 1 - \measuredangle[(j\omega/\omega_c)+1] = 0 - \measuredangle[(j\omega/\omega_c)+1] = -\tan^{-1}(\omega/\omega_c).$$

Next, we turn to asymptotes. Letting $\omega << \omega_c$ we obtain

$$20\log_{10}|H(j\omega)| \approx -20\log_{10}\left|\sqrt{1}\right| = 0 \text{ and } \measuredangle H(j\omega) \approx -\tan^{-1}(0) = 0,$$

whereas letting $\omega >> \omega_c$ yields

$$20\log_{10}\left|H(j\omega)\right| \approx -20\log_{10}\left|\omega/\omega_c\right| = 20\log_{10}\omega_c - 20\log_{10}\omega = -(20\log_{10}\omega - 20\log_{10}\omega_c)$$

and

$$\measuredangle H(j\omega) \approx -\tan^{-1}(\infty) = -90^o.$$

The latter magnitude asymptote represents a line $-20\log_{10}\omega$ with a drop of 20 dB for every 10 times increase in ω (negative slope of 20 dB/decade) moved to a "zero place" $20\log_{10}\omega_c$ which determines where this line crosses zero.

Indeed, setting $\omega = \omega_c$ we obtain $-20\log_{10}\left|\omega_c/\omega_c\right| = -20\log_{10}1 = 0$, making both magnitude asymptotes connect at ω_c, and also yielding at this point

$$\measuredangle H(j\omega_c) \approx -\tan^{-1}(1) = -45^o.$$

The corresponding Bode plot is shown in Figure 3.1-27.

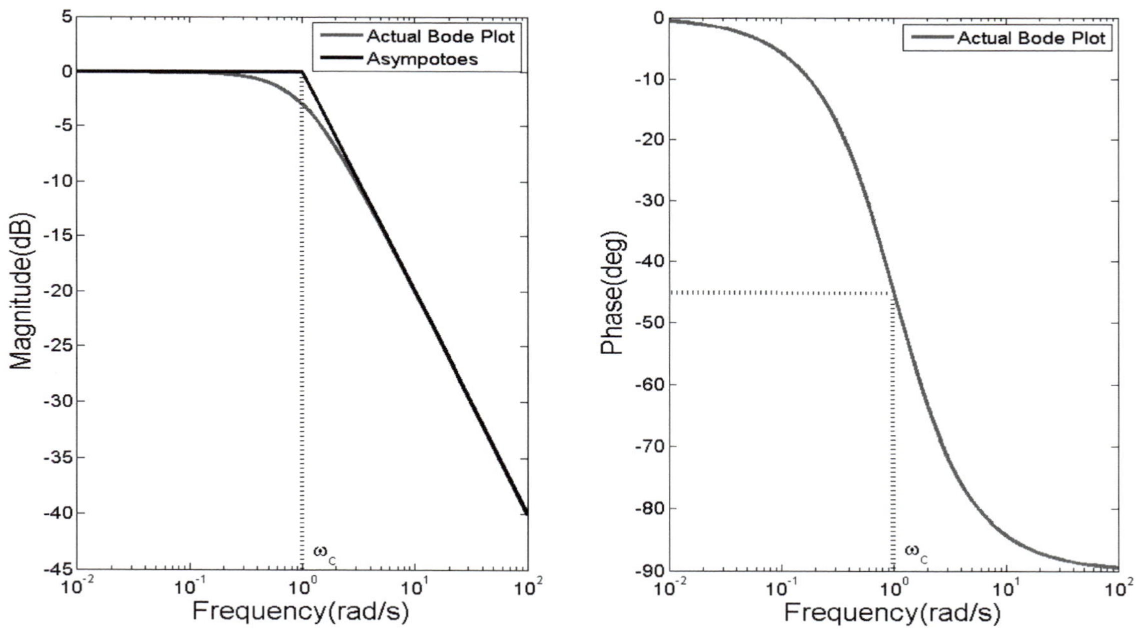

Fig. 3.1-27. The first order system Bode plot; asymptotes form a corner at "corner frequency" ω_c.

3.1.9.2 *2nd order MSD system*

For this system $H(s) = \omega_n^2/(s^2 + 2\zeta\omega_n s + \omega_n^2)$, and frequency and log-magnitude responses are

$$H(j\omega) = 1/[-(\omega/\omega_n)^2 + 2\xi j(\omega/\omega_n) + 1]$$

and

$$-20\log_{10}\left|\sqrt{[1-(\omega/\omega_n)^2]^2 + 4\xi^2(\omega/\omega_n)^2}\right|,$$

yielding, respectively, asymptotes

$$-20\log_{10}\left|\sqrt{1}\right| = 0,\ \omega << \omega_n,$$

and

$$-(40\log_{10}\omega - 40\log_{10}\omega_n), \omega >> \omega_n,$$

corner frequency $\omega_c = \omega_n$, and phase response

$$\measuredangle H(j\omega) = -\tan^{-1} 2\zeta(\omega/\omega_n)/[1-(\omega/\omega_n)^2], \measuredangle H(j\infty) = -\pi.$$

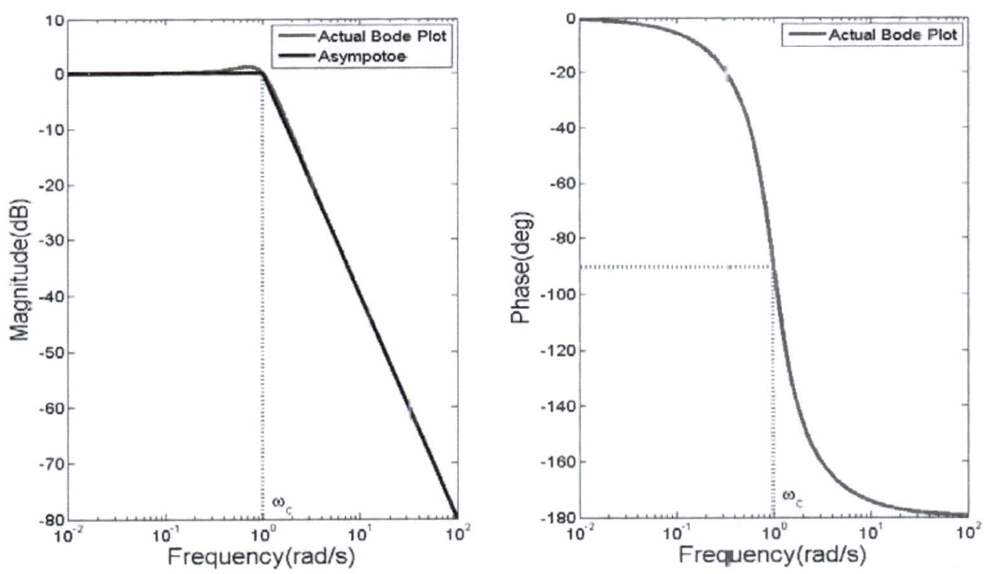

Fig. 3.1-28. MSD system Bode plot; asymptotes form a corner at "corner" (or break) frequency ω_n.

3.1.9.3 *2nd order MS2D system*

For this system $H(s) = \omega_n^2 s/(s^2 + 2\zeta\omega_n s + \omega_n^2)$, and frequency and log-magnitude responses are

$$H(j\omega) = j\omega / \left[-\left(\frac{\omega}{\omega_n}\right)^2 + 2\xi j\left(\frac{\omega}{\omega_n}\right) + 1 \right] \text{ and}$$

$$20\log_{10}\omega - 20\log_{10}\left|\sqrt{[1-(\omega/\omega_n)^2]^2 + 4\xi^2(\omega/\omega_n)^2}\right|,$$

respectively, yielding asymptotes

$$20\log_{10}\omega,\ \omega << \omega_n,$$

and

$$-(20\log_{10}\omega - 40\log_{10}\omega_n),\ \ \omega >> \omega_n,$$

corner frequency $\omega_c = \omega_n$ where asymptotes connect, and phase response

$$\measuredangle H(j\omega) = 90^o - \tan^{-1}2\zeta(\omega/\omega_n)/[1-(\omega/\omega_n)^2],\ \measuredangle H(j\infty) = -\pi/2.$$

On the left and the right of the corner frequency magnitude graph drops 20 dB/decade as, respectively, $\omega \to 0$, and $\omega \to \infty$, clearly showing the effect of blocking zeros in the vicinity of very low and very high frequencies.

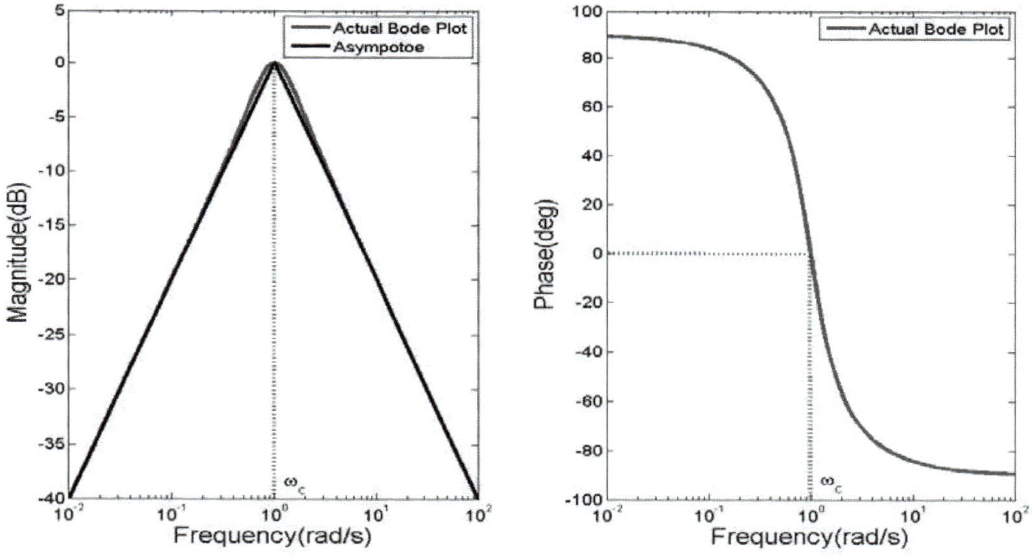

Fig. 3.1-29. MS2D system Bode plot; asymptotes clearly show blocking on both sides of $\omega_c = \omega_n$.

3.1.10 *Resonance in the second order systems*

Since the mold velocity distortion is caused by the beam resonance, and the second order models with various zero configurations provide good approximations of the beam behavior around the resonance frequency that are suitable for controller design, we will review the resonance calculation for basic second order systems.

3.1.10.1 *Resonance frequency and peak of MSD system*

Computing the magnitude of MSD transfer function from its frequency response yields

$$H(j\omega)=\frac{1}{-\left(\frac{\omega}{\omega_n}\right)^2+2\xi j\frac{\omega}{\omega_n}+1}\Rightarrow \left|H(j\omega)\right|=\frac{1}{\sqrt{\left(1-\left(\frac{\omega}{\omega_n}\right)^2\right)^2+\left(2\xi\frac{\omega}{\omega_n}\right)^2}}=\frac{1}{\sqrt{\left(1-\frac{\omega^2}{\omega_n^2}\right)^2+4\xi^2\frac{\omega^2}{\omega_n^2}}}.$$

Denote:

$$h(\omega)=\left(1-\frac{\omega^2}{\omega_n^2}\right)^2+4\xi^2\frac{\omega^2}{\omega_n^2}=\frac{\omega^4}{\omega_n^4}-2\frac{\omega^2}{\omega_n^2}-1+4\xi^2\frac{\omega^2}{\omega_n^2}=\frac{\omega^4}{\omega_n^4}-2\left(1-2\xi^2\right)\frac{\omega^2}{\omega_n^2}+1+\left(1-2\xi^2\right)^2$$

$$-\left(1-2\xi^2\right)^2=\left[\frac{\omega^2}{\omega_n^2}-\left(1-2\xi^2\right)\right]^2+1+-1+4\xi^2-4\xi^4=\left[\frac{\omega^2}{\omega_n^2}-\left(1-2\xi^2\right)\right]^2+4\xi^2\left(1-\xi^2\right).$$

Then, when $h(\omega)$ reaches its minimum value, $\left|H(j\omega)\right|=1/\sqrt{h(\omega)}$ reaches its maximum. This takes place when the term in brackets is set to zero: $\left[(\omega^2/\omega_n^2)-\left(1-2\xi^2\right)\right]^2=0,$ yielding the resonance frequency $\omega_{res}=\omega_n\sqrt{1-2\xi^2}$ and the maximum magnitude, or the resonance peak, M_r

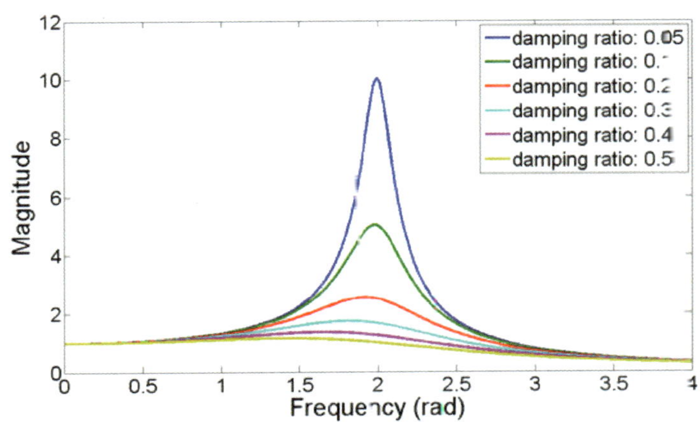

Fig. 3.1-30. Magnitude response of MSD system.

$$M_r = |H(j\omega)|_{\max} = |H(j\omega_{res})| = \frac{1}{\sqrt{h(\omega_{res})}} = \frac{1}{\sqrt{4\xi^2(1-\xi^2)}} = \frac{1}{2\xi\sqrt{1-\xi^2}}.$$

Figure 3.1-30 presents the magnitude response of MSD system, showing the relation of the resonance frequency and the maximum amplitude to the damping ratio.

3.1.10.2 *Resonance frequency and peak of MS2D system*

The transfer function of MS2D system is

$$H(s) = \frac{b_2 s}{ms^2 + (b_1 + b_2)s + k} = \frac{\frac{b_2}{m}s}{s^2 + \frac{(b_1+b_2)}{m}s + \frac{k}{m}} = \frac{as}{s^2 + 2\xi\omega_n s + \omega_n^2}$$

where a stands for b_2/m to simplify the notation. Then

$$|H(j\omega)| = \left|\frac{aj\omega}{(j\omega)^2 + 2\xi\omega_n(j\omega) + \omega_n^2}\right| = \left|\frac{aj\omega}{-\omega^2 + 2\xi\omega_n j\omega + \omega_n^2}\right| = \frac{aj}{-\omega + 2\xi\omega_n j + \frac{\omega_n^2}{\omega}} = \frac{|a|}{\sqrt{\left(\frac{\omega_n^2}{\omega} - \omega\right)^2 + 4\xi^2\omega_n^2}}.$$

Denote: $h(\omega) = \left(\frac{\omega_n^2}{\omega} - \omega\right)^2 + 4\xi^2\omega_n^2$. Again, $h(\omega)$ reaches its minimum value

$h(\omega)_{\min} = 4\xi^2\omega_n^2$, when $\left[\frac{\omega_n^2}{\omega} - \omega\right]^2 = 0 \Rightarrow \omega_{res} = \omega_n$ and $M_r = |H(j\omega_{res})| = \frac{|a|}{\sqrt{4\xi^2\omega_n^2}} = \frac{|a|}{2\xi\omega_n}.$

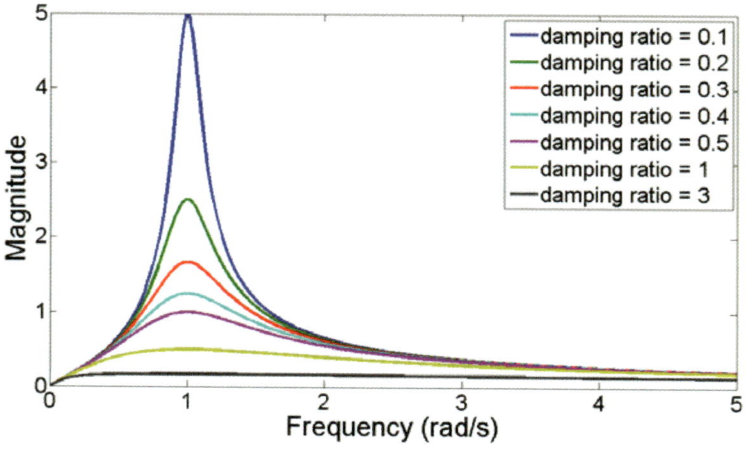

Fig. 3.1-31. Magnitude response of MS2D system.

The magnitude response of $H(s)=\dfrac{s}{s^2+2\xi s+1}$ is shown in Figure. 3.1-31.

Resonance frequency and peak of the second order system with 2 zeroes at the origin. Second order system with 2 zeroes at the origin has the form:

$$H(s)=\frac{as^2}{s^2+2\xi\omega_n s+\omega_n^2},$$

$$|H(j\omega)|=\frac{a}{1+2\xi\omega_n\frac{1}{s}+\frac{\omega_n^2}{s^2}}=\frac{a}{1-2\xi j\frac{\omega_n}{\omega}-\frac{\omega_n^2}{\omega^2}}=\frac{|a|}{\sqrt{\left(\frac{\omega_n^2}{\omega^2}-1\right)^2+4\xi^2\frac{\omega_n^2}{\omega^2}}}.$$

Similar to MSD system, when $\omega\equiv\omega_{res}=\dfrac{\omega_n}{\sqrt{1-2\xi^2}}$, we have $M_r=|H(j\omega_{res})|=\dfrac{1}{2\xi\sqrt{1-\xi^2}}$.

The magnitude response of $H(s)=\dfrac{s^2}{s^2+2\xi s+1}$ is shown in Figure. 3.1-32.

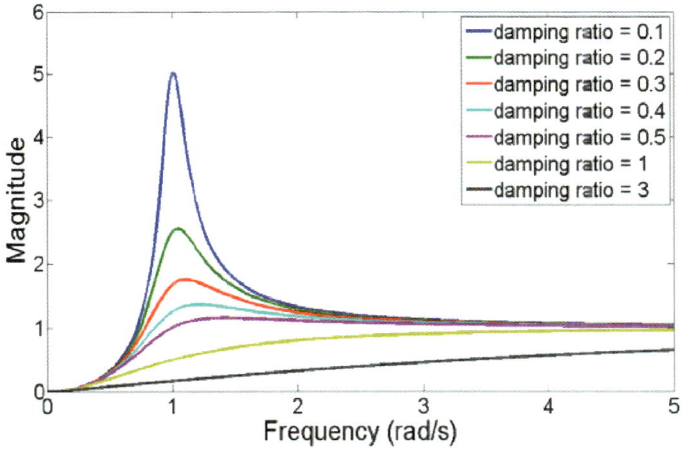

Fig. 3.1-32. Magnitude response of second order system with 2 zeros at origin.

Resonance frequency and peak of the beam system. From system identification to be discussed in more detail in Chapter 4, the 2nd order transfer function representing the best fit to the test data around the resonance frequency for the hardware testbed beam has the following form:

$$H(s)=\frac{bs+c}{s^2+2\xi\omega_n s+\omega_n^2}.$$

The magnitude of above transfer function is

$$|H(j\omega)| = \frac{\sqrt{b^2\omega^2 + c^2}}{\sqrt{(\omega_n^2 - \omega^2)^2 + 4\xi^2\omega_n^2\omega^2}} = \sqrt{\frac{b^2\omega^2 + c^2}{(\omega_n^2 - \omega^2)^2 + 4\xi^2\omega_n^2\omega^2}}.$$

Let's denote $x = \omega^2$. Then

$$\frac{b^2\omega^2 + c^2}{(\omega_n^2 - \omega^2)^2 + 4\xi^2\omega_n^2\omega^2} \equiv h(\omega^2) \Rightarrow h(x) = \frac{b^2 x + c^2}{(\omega_n^2 - x)^2 + 4\xi^2\omega_n^2 x}.$$

When $dh(x)/dx = 0$, $h(x)$ attains its maximum value. Computing the latter derivative:

$$h'(x) = \frac{b^2\left((\omega_n^2 - x)^2 + 4\xi^2\omega_n^2 x\right) - \left(2(\omega_n^2 - x)(-1) + 4\xi^2\omega_n^2\right)(b^2 x + c^2)}{P(x)}$$
$$= \frac{-b^2x^2 - 2c^2x + b^2\omega_n^4 - (4\xi^2 - 2)\omega_n^2 c^2}{P(x)}$$

and setting $h'(x) = 0$ yields

$$-b^2x^2 - 2c^2x + 2\omega_n^2c^2 + b^2\omega_n^4 - 4\xi^2\omega_n^2c^2 = 0.$$

Solving the above equality with respect to x gives:

$$x = \frac{c^2 \pm \sqrt{c^4 - (-b^2)(2\omega_n^2c^2 + b^2\omega_n^4 - 4\xi^2\omega_n^2c^2)}}{-b^2} = \omega_{res}^2.$$

The best transfer function fit to test data yields

$$H(s) = \frac{4.9161s + 2918}{s^2 + 1.482s + 3272} \text{ with } \omega_n = 57.2014,\ \xi = 0.013, b = 4.9161, \text{ and } c = 2981.$$

Substituting the latter values into the above equation for x and selecting the root $x = 3271$ yields

$$\omega_{res} = \sqrt{x} = 57.2\ rad/s = 9.1\ Hz,$$

which is very close to 9.2 Hz — the resonance frequency observed in experimental magnitude response.

3.1.11 *What each representation is good for:*

I) ***System layout*** provides quick overview of system components and interconnections necessary to understand system functioning. It also helps with sensor and actuator placement, component sizing, design modifications, such as addition of damper to MSD system to create blocking zero, and development of control system topology most suitable for the problem.

II) ***System physics*** brings the necessary insights into system behavior, permitting solution of system integration and controller design problems through model building and diagnostics.
III) LCCODEs provide the form suitable for analyzing system properties and finding response to input signals and initial conditions.
IV) ***Transfer function*** permits system identification from input/output data, system analysis and diagnostics, filter design, control system design, and system integration. Transfer function matrix is also the main tool for designing robust controllers for systems with uncertain and varying dynamics.
V) ***LCCODEs state space representation*** provides the main tool for system modeling and simulation, development and implementation of filters, including the advanced ones, like Kalman filters (not covered in this book), for systems with nonstationary noisy measurements, and design and implementation of robust controllers for systems with uncertain and varying dynamics.
VI) ***Impulse response*** provides testing basis for experimental system modeling and discrete/continuous spectral diagnostic tools to be introduced later in the book.
VII) ***Convolution integral*** gives computational algorithms for IIR and FIR filtering. In subsequent chapters it will be shown that it also provides fundamental insights into the behavior of the ideal filters as noncausal systems and permits rigorous definition of frequency response through noncausal Fourier transform.
VIII) ***Pole-zero map*** shows system complexity, filtering properties, stability, and system response properties. It is also extensively used in the control system design.
IX) ***Frequency response*** permits filter design and data-based modeling of otherwise unmodelable systems. It is also one of the main tools for designing robust controllers for systems with uncertain and varying dynamics.

3.2 Review of Parametric System Modeling, Subsystem Commutativity, Firsts Look at System Integration — Subsystem Loading, Output-Dependent System Properties

Parametric system modeling and interconnection: Representing systems or their elements through fundamental physical laws yields their parametric models in the form of ODEs, like (3.1-2) for MS2D system or partial differential equations (PDEs), like (1.5-1)–(1.5-4) for beam equation, where parameters represent physically motivated quantities. To model a system like a mold oscillator hardware testbed, ODEs and PDEs, generally nonlinear, are interconnected through inputs and outputs that in the case of PDEs take the form of the boundary conditions, such as (1.5-5)–(1.5-7). An example of such modeling — software testbed — has been described in Chapter 1, with details given in Appendices G and H.

To better understand the properties of the software testbed, Appendix G contains detailed derivation of linear Euler-Bernoulli and Timoshenko beam PDEs obtained from Lagrangian and Hamilton's principle, their modal decompositions, and recommendation on parameter adjustment for matching the desired resonance frequency. However, to bring out the basic pitfalls of system integration we will start with simple linear electrical circuits, first briefly reviewing their transfer function derivation.

3.2.1 *Passive electrical circuits*

3.2.1.1 *Transfer functions of passive electrical circuits:*

System component with no external power source will be called ***passive component***. Circuits with passive electrical components (resistor, capacitor and inductor) can be modeled using the equivalent (complex) ***impedance***, as outlined below

Complex impedance is defined as $Z(s)=V(s)/I(s)=$ potential difference/current, and for main passive components is given by:

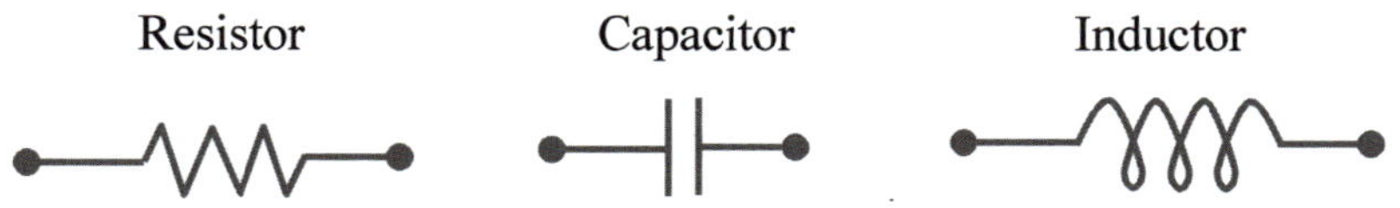

$Z_R(s)=R$ $\qquad$ $Z_C(s)=\dfrac{1}{Cs}$ $\qquad$ $Z_L(s)=Ls$

Fig. 3.2-1. Circuit representation and impedances of common passive components.

Series and parallel connections of components are, respectively, described by:

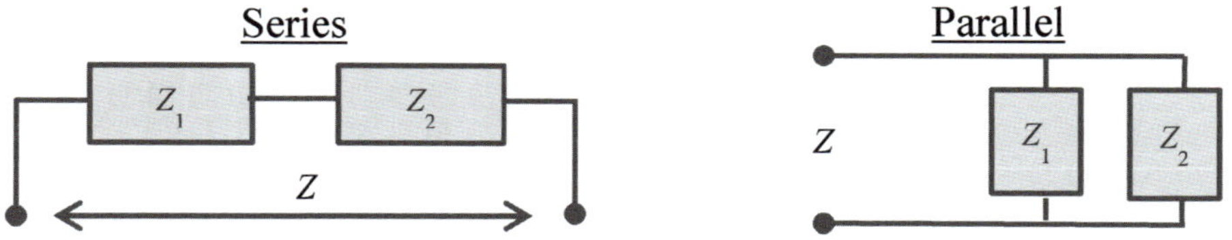

$Z=Z_1+Z_2$

$$\frac{1}{Z}=\frac{1}{Z_1}+\frac{1}{Z_2}\Rightarrow Z=\frac{1}{\dfrac{1}{Z_1}+\dfrac{1}{Z_2}}$$

Fig. 3.2-2. Complex impedance elements connected in series and in parallel.

Drawing, interpreting, and simplifying a circuit diagram. Working with real circuits typically involves four steps, three of which are depicted in Figures 3.2-3–3.2-5 below:

1) Drawing current loops and calculating voltage drops using Kirchhoff's Voltage Law:

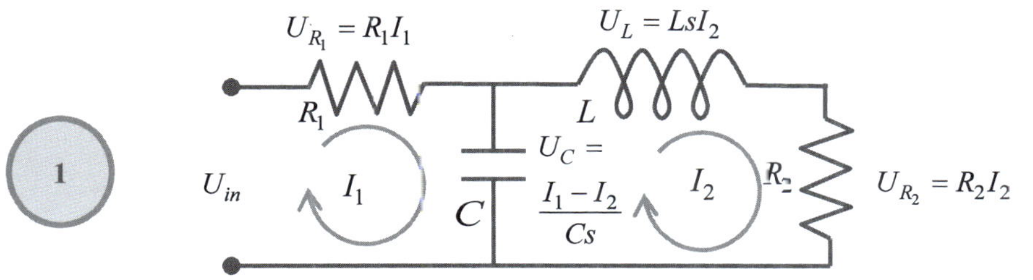

Fig. 3.2-3. Current loops with corresponding voltage drops.

2) Locating the input, the output, and the common ground in the actual physical circuit:

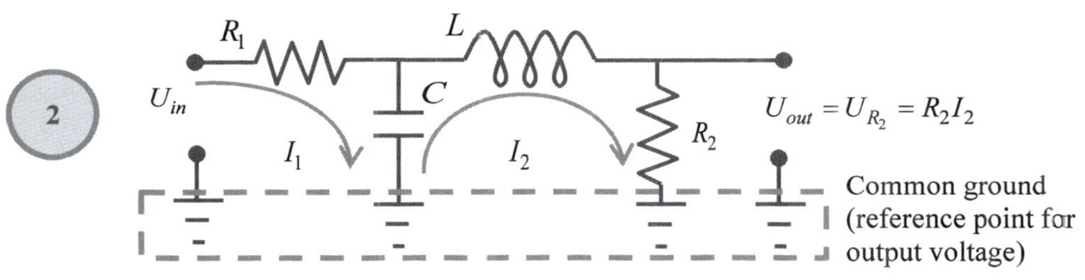

Fig. 3.2-4. Circuit input, output, and common ground.

3) Simplifying circuit by adding impedances in series and parallel:

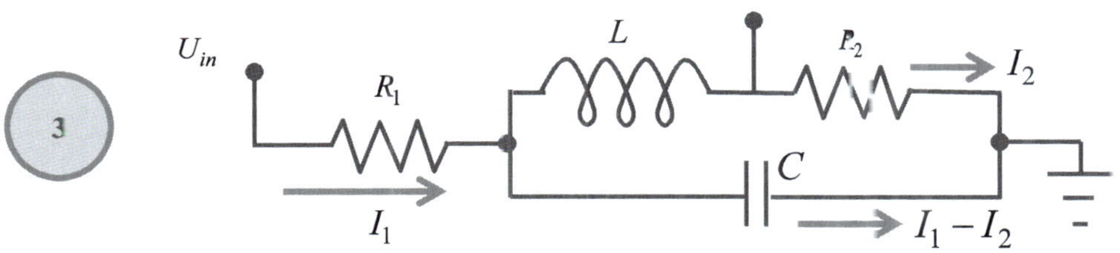

Fig. 3.2-5. Identifying impedances connected in series and parallel and adding them.

4) Simulating the circuit performance — this will be discussed in example solution to Homework 3.

Example 3.2-1: find the transfer function of the following circuit:

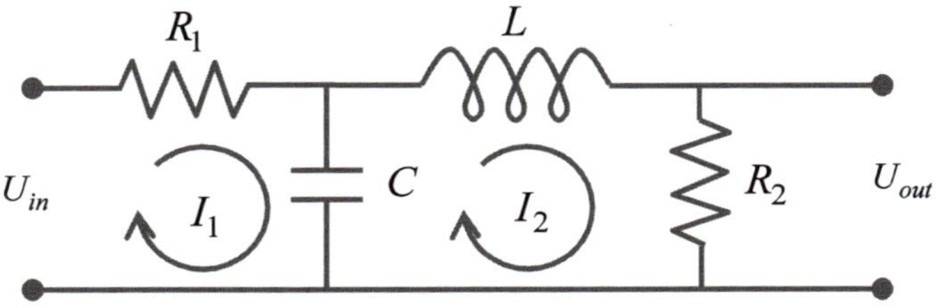

Fig. 3.2-6. Circuit topology for Example 3.2-1.

Solution: Defining impedances: $Z_1 = R_1,\ \ Z_2 = R_2,\ \ Z_L = Ls,\ \ Z_C = 1/Cs$, Kirchhoff's Voltage Law yields:

$$U_{in} = Z_1 I_1 + Z_C\left(I_1 - I_2\right) \tag{3.2-1}$$

$$0 = Z_L I_2 + Z_2 I_2 + Z_C\left(I_2 - I_1\right) \tag{3.2-2}$$

with output given by $U_{out} = Z_2 I_2$.

Using (3.2-2), current in the first loop is obtained in terms of circuit output as follows:

$$0 = Z_L I_2 + Z_2 I_2 + Z_C\left(I_2 - I_1\right) = \frac{Z_L}{Z_2} Z_2 I_2 + Z_2 I_2 + \frac{Z_C}{Z_2} Z_2 I_2 - Z_C I_1 = \frac{Z_L}{Z_2} U_{out} + U_{out} + \frac{Z_C}{Z_2} U_{out} - Z_C I_1$$

$$\Rightarrow \left(\frac{Z_L}{Z_2} + 1 + \frac{Z_C}{Z_2}\right) U_{out} = Z_C I_1 \Rightarrow I_1 = \frac{Z_L / Z_2 + 1 + Z_C / Z_2}{Z_C} U_{out}.$$

Now, using (3.2-1), input is expressed in terms of output as follows:

$$U_{in} = Z_1 I_1 + Z_C\left(I_1 - I_2\right) = \left(Z_1 + Z_C\right) I_1 - \frac{Z_C}{Z_2} Z_2 I_2 = \left(Z_1 + Z_C\right) \frac{Z_L / Z_2 + 1 + Z_C / Z_2}{Z_C} U_{out} - \frac{Z_C}{Z_2} U_{out}$$

$$= \left[\left(Z_1 + Z_C\right) \frac{Z_L / Z_2 + 1 + Z_C / Z_2}{Z_C} - \frac{Z_C}{Z_2}\right] U_{out}.$$

This yields transfer function

$$\frac{U_{out}}{U_{in}} = \frac{1}{\left(Z_1 + Z_C\right) \dfrac{Z_L/Z_2 + 1 + Z_C/Z_2}{Z_C} - \dfrac{Z_C}{Z_2}} = \frac{1}{\left(R_1 + \dfrac{1}{Cs}\right) \dfrac{Ls/R_2 + 1 + (1/Cs)/R_2}{1/Cs} - \dfrac{1/Cs}{R_2}}$$

$$= \frac{1}{\left(R_1 + \dfrac{1}{Cs}\right)\left(\dfrac{Ls}{R_2} + 1 + \dfrac{1}{R_2 Cs}\right) Cs - \dfrac{1}{R_2 Cs}} = \frac{1}{\left(R_1 Cs + 1\right)\left(\dfrac{Ls \cdot Cs + R_2 Cs + 1}{R_2 Cs}\right) - \dfrac{1}{R_2 Cs}}$$

$$=\frac{R_2Cs}{(R_1Cs+1)(LCs^2+R_2Cs+1)-1}=\frac{R_2Cs}{(R_1Cs+1)(LCs^2+R_2Cs+1)-1}$$

$$=\frac{R_2Cs}{R_1LC^2s^3+LCs^2+R_1R_2C^2s^2+R_2Cs+R_1Cs+1-1}=\frac{R_2C}{R_1LC^2s^2+(LC+R_1R_2C^2)s+(R_2C+R_1C)}$$

$$=\frac{R_2C}{R_1LC^2s^2+(L+R_1R_2C)Cs+(R_1+R_2)C}=\frac{R_2}{R_1LCs^2+(L+R_1R_2C)s+(R_1+R_2)}.$$

$$\therefore \frac{U_{out}}{U_{in}}=\frac{R_2}{(R_1+R_2)}\cdot\frac{1}{\frac{R_1LC}{R_1+R_2}s^2+\frac{L+R_1R_2C}{R_1+R_2}s+1}=\frac{R_2}{R_1LC}\cdot\frac{1}{s^2+\frac{L+R_1R_2C}{R_1LC}s+\frac{R_1+R_2}{R_1LC}}.$$

Setting $s=0$, $R_2/(R_1+R_2)$ is seen to be the ***DC gain*** of the circuit. R_2/R_1LC is a ***monic gain***, due to numerator and denominator polynomials being monic (coefficient at the highest degree of s is 1).

3.2.1.2 *Properties of passive circuits*

Lack of commutativity — Example 3.2-2: Let's connect two passive circuits as follows:

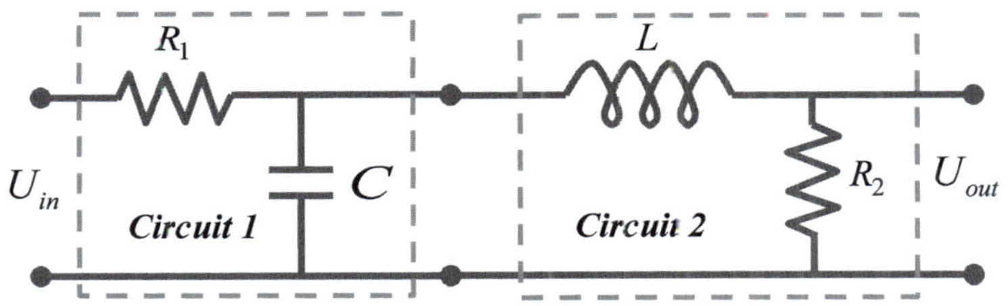

$$\frac{U_{out}}{U_{in}}=\frac{\frac{R_2}{R_1LC}}{s^2+\frac{L+R_1R_2C}{R_1LC}s+\frac{R_1+R_2}{R_1LC}}$$

Fig. 3.2-7. Interconnection of two passive systems.

We calculated the transfer function for this circuit in Example 3.2-1. Let's now reverse the interconnection order:

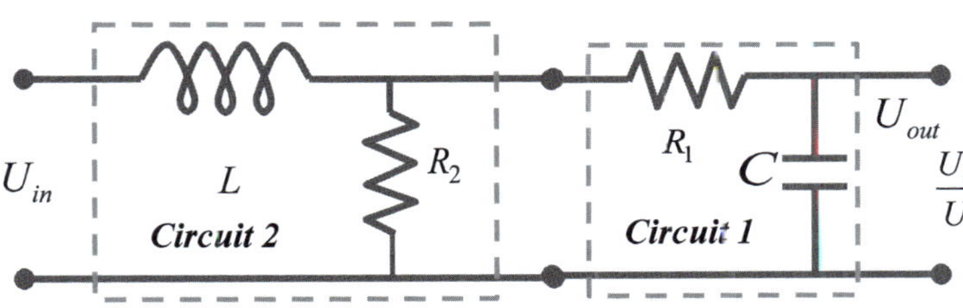

$$\frac{U_{out}}{U_{in}}=\frac{\frac{R_2}{(R_1+R_2)LC}}{s^2+\frac{L+R_1R_2C}{(R_1+R_2)LC}s+\frac{R_2}{(R_1+R_2)LC}}$$

Fig. 3.2-8. Changing the order of passive systems interconnection.

We see that switching the order gives a completely different transfer function. Passive subsystems do not commute!

Loading — Example 3.2-3: Passive circuits do not retain their properties if other circuits are attached to them. A seen in Figure 3.2-9, attaching Circuit 2 changes the behavior of Circuit 1 or, said another way, Circuit 2 ***loads*** Circuit 1.

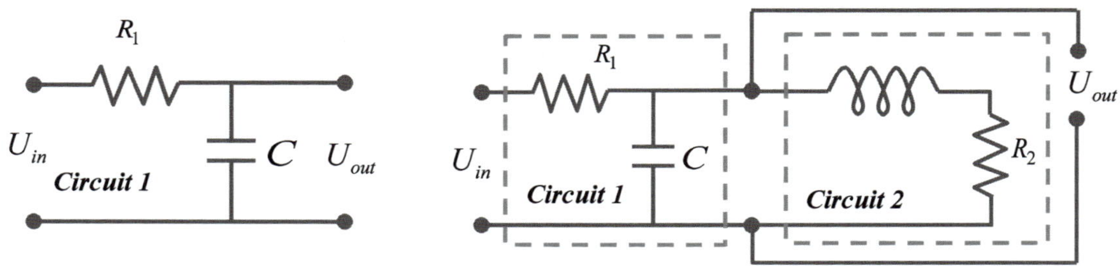

$$\frac{U_{out}}{U_{in}} = \frac{1}{R_1C} \Big/ \left(s + \frac{1}{R_1C} \right) \qquad \frac{U_{out}}{U_{in}} = \frac{R_2}{R_1LC}\left(\frac{L}{R_2}s+1\right) \Big/ \left(s^2 + \frac{L+R_1R_2C}{R_1LC}s + \frac{R_1+R_2}{R_1LC} \right)$$

Fig. 3.2-9. Illustration of loading.

3.2.2 *System integration problems*

3.2.2.1 *Actuator sizing: hydraulic valve selection*

Actuator sizing is central to load handling: reduction of distortion under load is typically addressed through selecting more powerful actuator. However, as discussed in Chapter 1, this is not always possible, requiring bigger actuator size and power unit as shown in Figures 3.2-10 and 3.2-11.

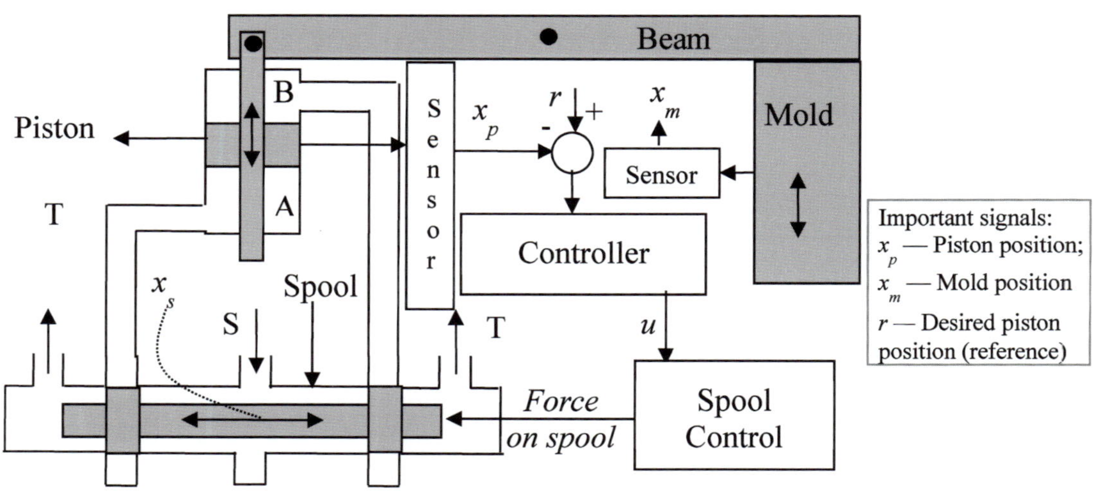

Fig. 3.2-10. Distortion reduction in operation of the servo-beam structure requires bigger actuator.

Fig. 3.2-11. Bigger actuator requires more space in the load assembly and bigger power unit.

3.2.2.2 *Actuator loading*

Even with proper sizing, the testbed also distorts the mold velocity, as seen in Figure 3.2-12.

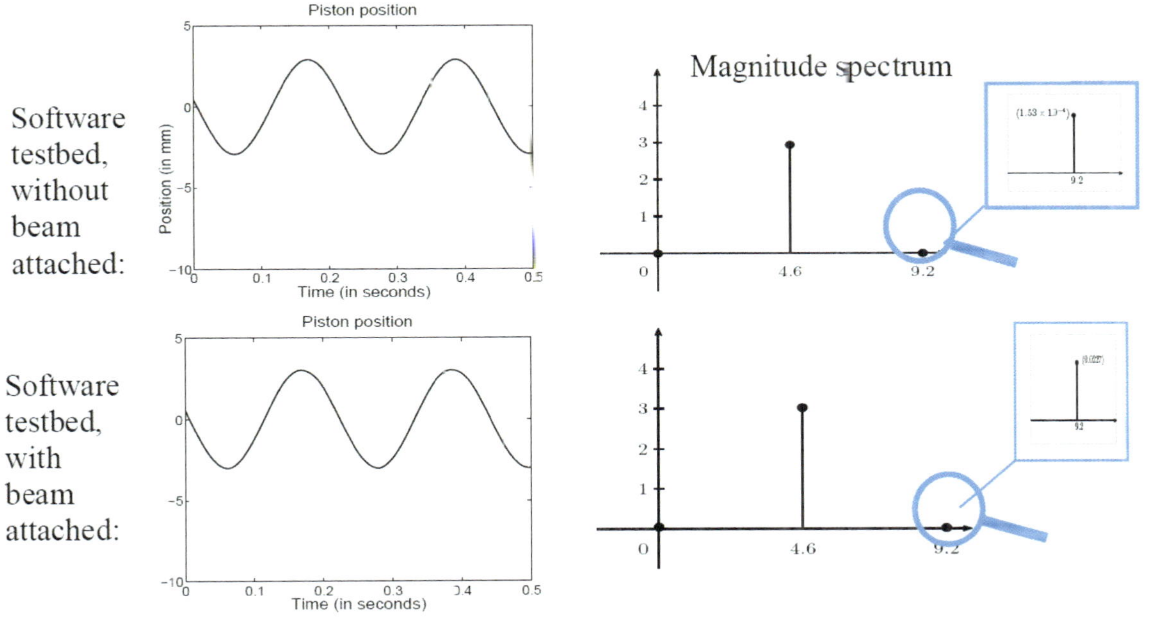

Fig. 3.2-12. Actuator loading introduces small-scale periodic components.

In this figure, the magnitude of the resonance frequency component is seen to be scaled-up (amplified) by 2 orders of magnitude when the piston is driving (actuating) the beam!

Clearly, loading effect takes place in spite of the fact that valve is externally powered, i.e. is *active* system component, as shown in Figure 3.2-13.

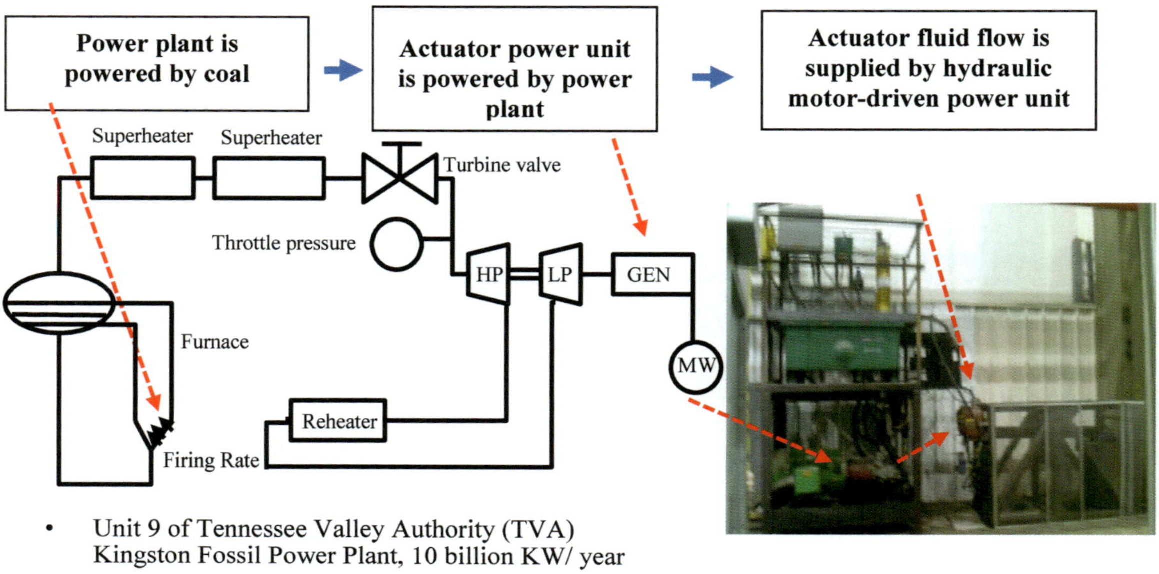

Fig. 3.2-13. Valve external power flow.

Why did testbed valve distort? Actuator is referred to as the ***ideal actuator*** when it does not introduce its own dynamics into the system. Generalizing this concept, we will call a subsystem that doesn't exhibit loading effect an ***ideal active subsystem***.

Testbed and productions unit valves are active components, but *non-ideal* ones. In software testbed, valve changes its behavior under load by increasing the second harmonic of driving frequency by two orders of magnitude from 1.5×10^{-4} to 2.27×10^{-2}. This is just enough to reproduce the production unit problem: more powerful actuator could have smaller loading effect, but it would be too big which is not practical. This is a generic problem — real actuators are non-ideal!

3.2.2.3 *Coupling and decoupling*

Noncommutativity and loading are manifestations of the effect of ***subsystem coupling***. Two subsystems comprising a dynamical system are said to be ***coupled*** if the state of one of them affects the state of the other one. We will call the coupling ***unidirectional***, or directed (a convenient oxymoron), if only one of the systems is affected, and ***bidirectional***, or undirected, if both are. The latter case will be also called ***cross-coupling***. When system consists of more than two interacting subsystems the term "coupling" is still used.

To illustrate the effect of coupling let us try to create a mechanical differentiator with the force at its right end proportional to the velocity of its left end using the damper with coefficient b_2:

$$F_{out,b_2}(s) / X_{in,b_2}(s) = H_{b_2}(s) = b_2 s$$

and apply this force to the mass of the MSD system to obtain transfer function

$$X_{out,\,MSD}(s) / X_{in,b_2}(s) = H_{b_2} H_{MSD} = b_2 s / (ms^2 + b_1 s + k).$$

First, we assume that both ends of the damper are free and attach its right end to the MSD system mass. The force generated by the damper and its Laplace transform are given by the equations on the left in Figure 3.2-14. When the right damper end is attached to the MSD system mass, the damper and the MSD system become cross-coupled, with the MSD system mass motion affecting the force generated by the damper through the term $-b_2 \dot{x}_{out}(t)$ in equation (3.1-1). This yields MS2D system transfer function.

$$X_{out,\,MS2D}(s) / X_{in,b_2}(s) = b_2 s / [ms^2 + (b_1 + b_2)s + k] \equiv H_{MS2D} \neq H_{b_2} H_{MSD} = b_2 s / (ms^2 + b_1 s + k).$$

In this case motions at both damper ends are its inputs and the system represented by a damper is two-input/one-output (force), not lending itself to acting as input differentiator, as shown in Figure 3.2-14.

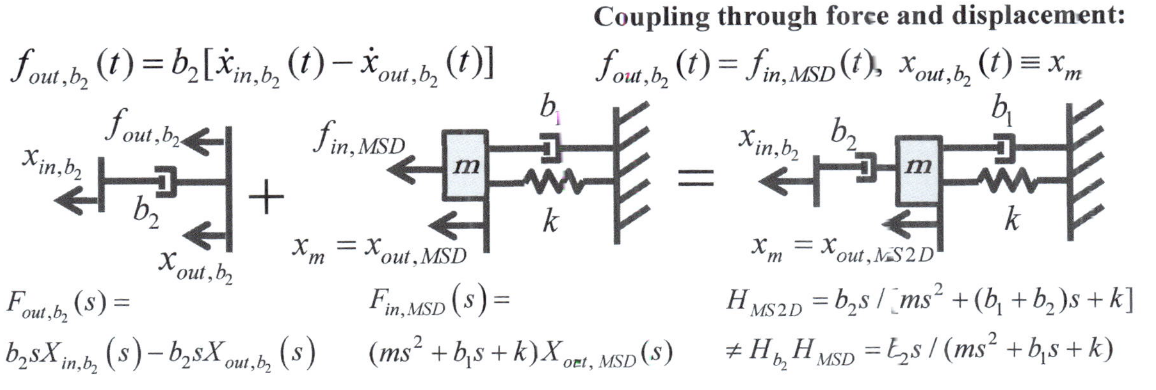

Fig. 3.2-14. Differentiator coupling — state of driven MSD system affects differentiator output force.

3.2.2.4 *Buffer — dynamic decoupling*

To convert the b_2 damper into differentiator, we can straightjacket the motion of the right end of damper b_2 by stiffening up MSD system, e.g. by drastically increasing spring constant. This would make ***MSD mechanical input impedance close to infinity***, as if the b_2 damper is acting against a stiff vertical beam as shown in Figure 3.2-15 (left). This infinite resistance to motion essentially ***decouples*** the damper from the state of MSD system by suppressing the velocity-induced coupling term $-b_2 \dot{x}_{out,b_2}(t)$.

If we now, keeping infinite input impedance, transfer the force acting on a beam to the mass of another MSD system that is not stiff, we can retain the differentiator action if we ensure that the latter system can move freely in response to this force. This implies that the transmitter of this force to the second MSD system must have ***zero output mechanical impedance***. The device that satisfies both of these input/output properties, namely, infinite input impedance/zero output impedance, is called ***buffer*** because it eliminates dynamic system cross-coupling, or loading, while retaining unidirectional coupling through transmission of information about the force and recreation of the latter in driving the load.

Coupling through force only:

$f_{out,b_2}(t) = b_2\dot{x}_{in,b_2}(t), x_{out,b_2} = \dot{x}_{out,b_2}(t) = 0$ $\qquad$ $f_{out,b_2}(t) = f_{in,MSD}(t)$

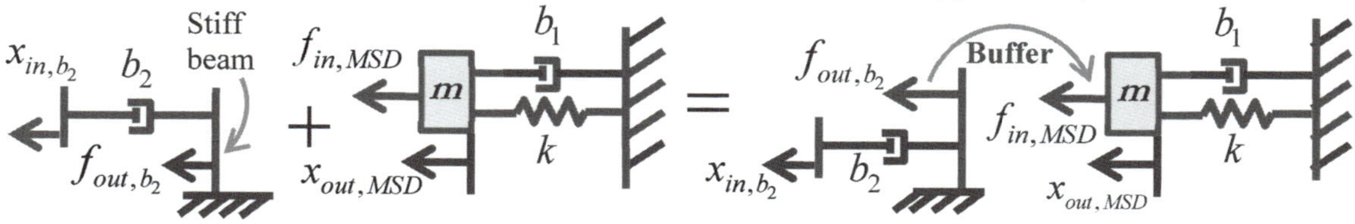

$F_{out,b_2}(s)/X_{in,b_2}(s) = H_{b_2}(s) = b_2 s$ $\qquad$ $X_{out,MSD}(s)/F_{in,MSD}(s) = H_{MSD} = 1/(ms^2 + b_1 s + k)$ $\qquad$ $X_{out,MSD}(s)/X_{in,b_2}(s) = H_{b_2,MSD} = H_{b_2}H_{MSD} = b_2 s/(ms^2 + b_1 s + k) \neq H_{MS2D}$

Fig. 3.2-15. Differentiator decoupling — driven MSD system does not affect differentiator output force and is not constrained in displacement and velocity in its motion under this force.

Building a buffer — transduction. Such force information transfer and recreation, problematic to achieve by purely mechanical rigid body means, can be easier accomplished through ***transduction — conversion of one form of energy into another***.

Transduction includes ***sensing — reading off physical variables and converting the result into an energy form convenient for further processing*** (typically, voltage), and ***actuation — application to a system of a signal received in one energy form and converted into another energy form capable of affecting system behavior***.

As already seen in the hardware testbed and the mold oscillator system description, sensing and further processing of sensor signal for decision making typically involves small magnitude signals, while actuation (decision implementation) — conversion of small magnitude signals of one energy form into large magnitude ones in another energy form, often requiring external power supply to inforce action dictated by a small signal.

Applying transduction to system in Figure 3.2-15, the buffer can be created as follows:

- force sensing could be accomplished through a strain gage force sensor attached to a stiff beam, as described in Section 3.4, with the sensor signal fed into signal processor input,
- force application can be accomplished through feeding signal processor output into either hydraulic actuator, like the one described in Section 1.5 of Chapter 1, but with piston attached to the mass, or electrical amplifier with voice coil actuator attached to the mass.

Then, using force-pressure-voltage and displacement-(fluid flow)-current analogies, the electrical amplifier must have sufficiently high current (very low output impedance) to allow unobstructed electromagnetically induced motion. Similarly, there should be little limitation on fluid flow rate in the piston cylinder to avoid impeding piston motion. Buffer based on the latter actuation is shown below.

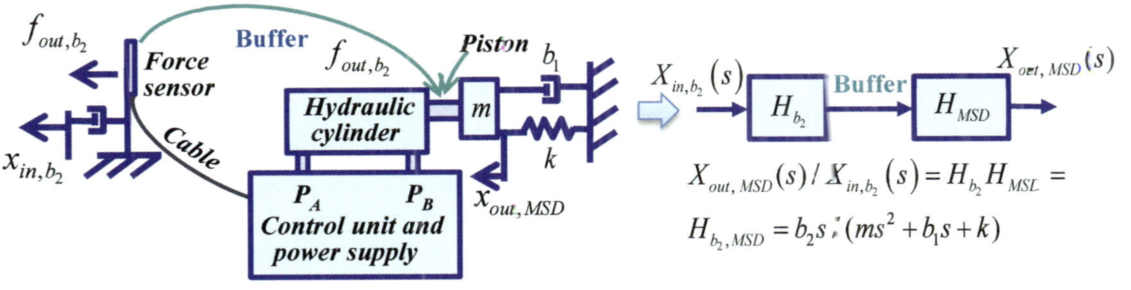

Fig. 3.2-16. Differentiator decoupling — buffer is implemented through force sensor, electronic control unit (signal processor), and hydraulic actuator with external power supply.

Neglecting piston damping and loading yields direct correspondence between hydraulic pressure on the piston (a_p — scaled pressure difference $P_A - P_B$) and force acting on the mass at the piston attachment point:

$$f_{out,b_2}(t)=b_2\dot{x}_{in,b_2}(t)=\left(P_A-P_B\right)a_p=m\ddot{x}_{out}+b_1\dot{x}_{out}+kx_{out}.$$

3.2.2.5 *Effect of output selection on system properties*

One of the key system properties extensively used in filter design is pick-off point relocation for otherwise unchanged system layout, seen in Figure 3.2-17 to dramatically affect system filtering properties, changing system zeros, as well as system topology from the input/output standpoint.

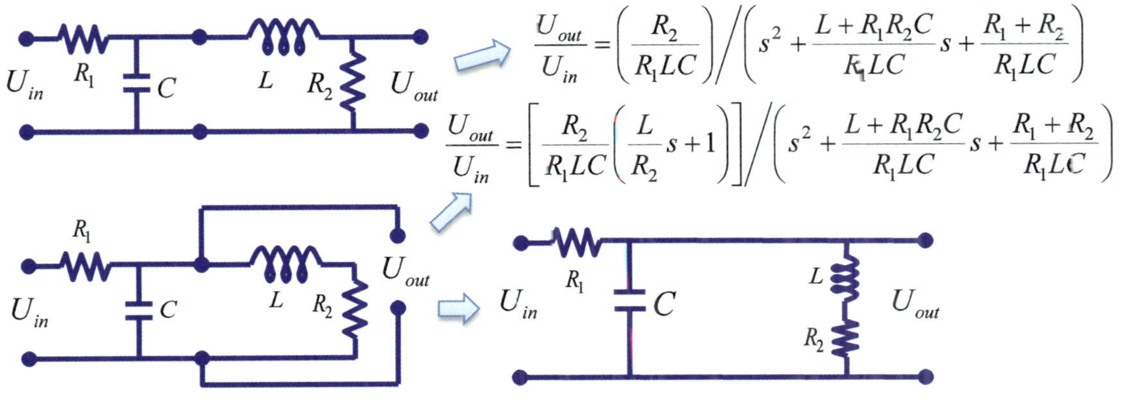

Fig. 3.2-17. Change of system filtering topology and filtering action through pick-off points' relocation.

3.3 First Look at Filtering: Basic Instrumentation Techniques — AC versus DC Coupling. Active (Op Amp) Circuits, Loading Mitigation, Logic Components

3.3.1 *Basic filtering: AC versus DC coupling*

Elimination of the DC offset, while passing high frequency signals, or elimination of the high frequency noise, while passing the low frequency signals are referred to, respectively, as the AC and the DC coupling, and are accomplished in their simplest form by the filter circuits shown in Figures 3.3-1 and 3.3-3, respectively.

3.3.1.1 *AC coupling*

In this case, the complex impedance method yields

$$V_{out} = RI \Rightarrow V_{in} = \frac{1}{Cs}I + RI = \frac{1}{RCs}RI + RI = \frac{1}{RCs}V_{out} + V_{out} = \left(\frac{1}{RCs}+1\right)V_{out} = \left(\frac{1+RCs}{RCs}\right)V_{out}$$

$$= \frac{s+(1/RC)}{s}V_{out} \Rightarrow H_{AC} = \frac{V_{out}(s)}{V_{in}(s)} = \frac{s}{s+1/RC} \Rightarrow \frac{d}{dt}v_{out}(t) + \frac{1}{RC}v_{out}(t) = \frac{d}{dt}v_{in}(t).$$

From above transfer function it is seen that $|H_{AC}(0)| = 0$ and $|H_{AC}(\infty)| = 1$, i.e. H_{AC} has blocking zero at zero that appears due to the blocking capacitor, which generates the factor s in the numerator of the transfer function. The latter corresponds to the derivative at the input, blocking DC signal.

As seen from the magnitude response shown in Figure 3.3-2, the signals with frequencies in the vicinity of zero frequency are attenuated as well, while high frequency signals are passed almost undistorted. This action is therefore referred to as the ***high-pass filtering***.

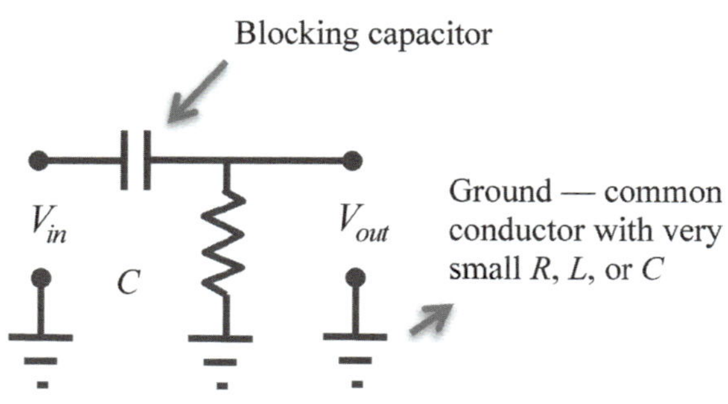

Fig. 3.3-1. AC coupler.

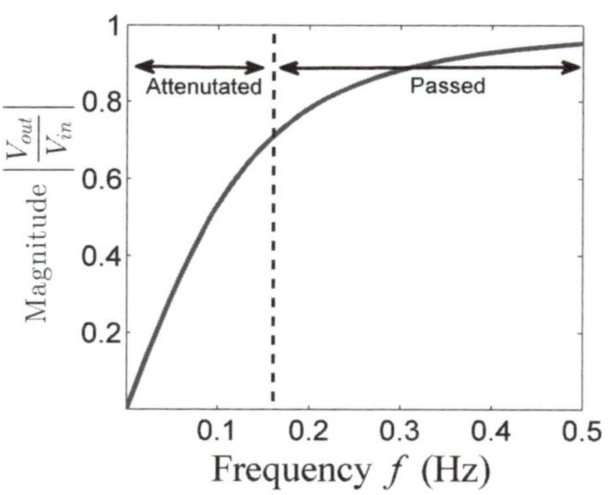

Fig. 3.3-2. High-pass filter magnitude response.

3.3.1.2 *DC coupling*

Using the complex impedance techniques yields

$$V_{out} = \frac{1}{Cs} I \Rightarrow V_{in} = RI + \frac{1}{Cs} I = RCs\frac{1}{Cs} I + \frac{1}{Cs} I = (RCs+1)V_{out} \Rightarrow H_{DC} = \frac{V_{out}}{V_{in}} = \frac{1}{RCs+1} = \frac{1/RC}{s+1/RC}.$$

From above transfer function it is seen that $|H_{DC}(0)| = 1$ and $|H_{DC}(\infty)| = 0$, i.e. H_{DC} has blocking zero at infinity, blocking high frequency, or AC signals. This takes place due to the strictly proper transfer function, which has the order of denominator bigger than the order of numerator.

As seen from the magnitude response shown in Figure 3.3-4, the signals with high frequencies are attenuated, while the low frequency signals are passed relatively undistorted. This action is therefore referred to as the ***low-pass filtering***.

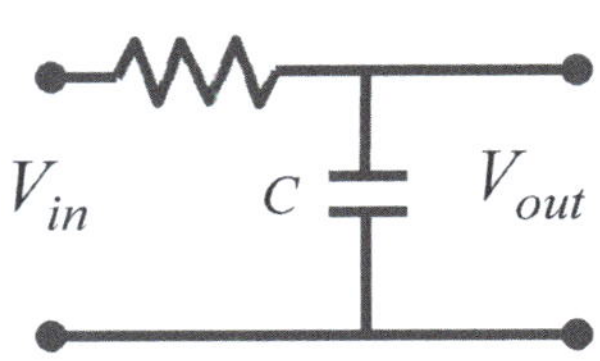

Fig. 3.3-3. DC coupler.

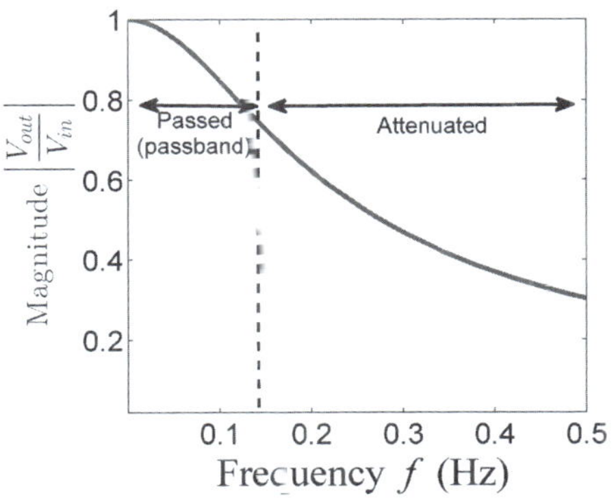

Fig. 3.3-4. Low-pass filter magnitude response.

The result of changing the oscilloscope coupling from DC to AC in displaying low frequency signal is shown below in Figure 3.3-5 — the signal is eliminated.

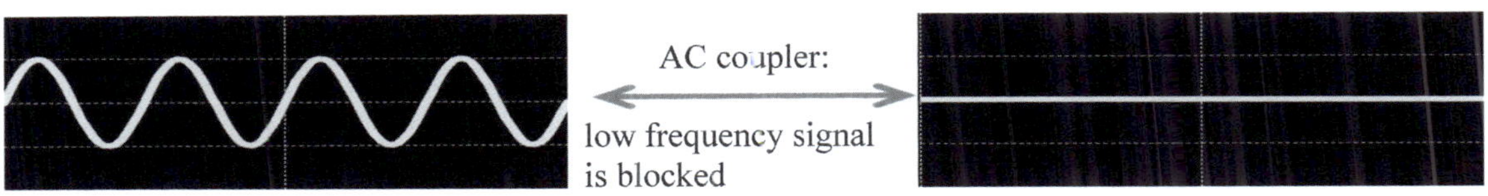

Fig. 3.3-5. Scope display of low frequency signal for coupling switched from DC to AC.

3.3.2 *Op amp based active circuits*

3.3.2.1 *Operational amplifiers*[1,2] *(op amps)*

Op amp is one of the key signal processing devices. It is an active element due to presence of an external power supply. The op amp schematics and basic properties are shown in Figure 3.3-6.

---

[1] www.analog.com/media/en/technical-documentation/data-sheets/AD8557.pdf
[2] www.ti.com/lit/an/sboa092a/sboa092a.pdf

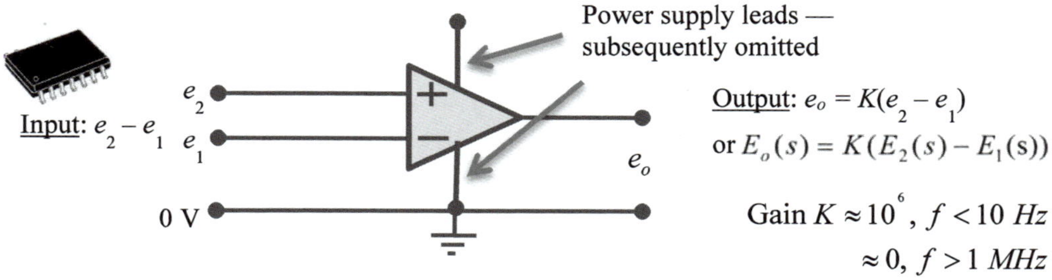

Fig. 3.3-6. Operational amplifier.

The main assumption about an ideal op amp is that it has the buffer properties:

- ∞ input impedance $\Rightarrow$ 0 input current
- 0 output impedance $\Rightarrow$ output voltage not affected by load.

Additional ideal op amp assumptions are

- ∞ voltage gain and slew rate (i.e. step input produces step output with no distortion).

Op amp serves as the basic building block for a variety of circuits that enable many indispensable operations with signals. Some of these circuits are briefly reviewed next.

3.3.2.2 *Inverting amplifier*

The schematics of the inverting amplifier is shown in Figure 3.3-7. In this circuit, the op amp shown in Figure 3.3-6 is "flipped", and its positive lead is connected to the ground, so that $e_2 = 0$.

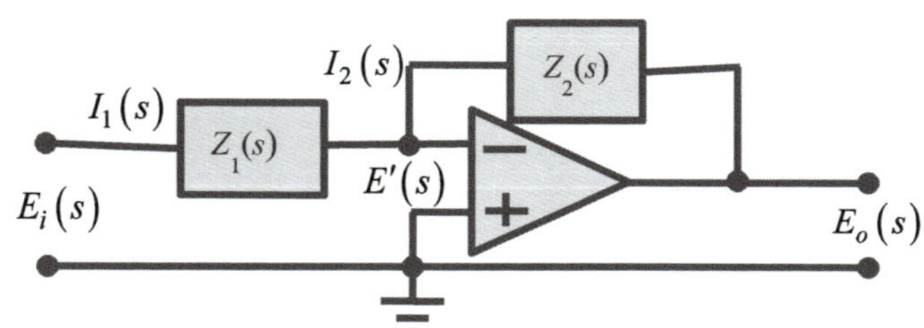

Fig. 3.3-7. Inverting amplifier.

The transfer function of the inverting amplifier can be derived as follows:

$$I_1 = \frac{E_i - E'}{Z_1} = \frac{E_i}{Z_1} - \frac{E'}{Z_1} = I_2 = \frac{E' - E_o}{Z_2} = \frac{E'}{Z_2} - \frac{E_o}{Z_2} \Rightarrow \frac{E_i}{Z_1} + \frac{E_o}{Z_2} = \left(\frac{1}{Z_1} + \frac{1}{Z_2}\right) E' \text{ or } E'\left(\frac{1}{Z_1} + \frac{1}{Z_2}\right) = \frac{E_i}{Z_1} + \frac{E_o}{Z_2},$$

but $E_o = K(E_2 - E_1) = -KE'$, therefore $E' = -E_o / K$ and

$$-\frac{E_o}{K}\left(\frac{1}{Z_1} + \frac{1}{Z_2}\right) = \frac{E_i}{Z_1} + \frac{E_o}{Z_2} \Rightarrow -E_o\left(\frac{1}{KZ_1} + \frac{1}{KZ_2} + \frac{1}{Z_2}\right) = \frac{E_i}{Z_1} \Rightarrow \frac{E_o}{E_i} = -\frac{1}{Z_1} \Big/ \left(\frac{1}{KZ_1} + \frac{1}{KZ_2} + \frac{1}{Z_2}\right) =$$

$$-\frac{Z_2}{Z_1}\bigg/\left(\frac{Z_2}{KZ_1}+\frac{1}{K}+1\right)=-\frac{Z_2}{Z_1}/\left(\left(\frac{Z_2}{Z_1}+1\right)\bigg/K+1\right)\Rightarrow\frac{E_o}{E_i}\approx-\frac{Z_2}{Z_1}\text{ for }\left|\frac{Z_2}{Z_1}+1\right|<<K\Rightarrow E_o\approx-\frac{Z_2}{Z_1}E_i. \quad (3.3\text{-}1)$$

This also implies that, with K infinite, $E_o=-(Z_2/Z_1)E_o$ and $E'=-E_o/K$ is kept at zero, quite unlike the passive circuit that sets $E'=I_2Z_2+E_o E'=I_2Z_2+E_o$ with op amp removed. Equation (3.3-1), thus, shows that this circuit inverts and amplifies input signal, making it inverting amplifier.

3.3.2.3 *Inverting summer*

Adding inputs to the circuit above leads to the configuration of the form

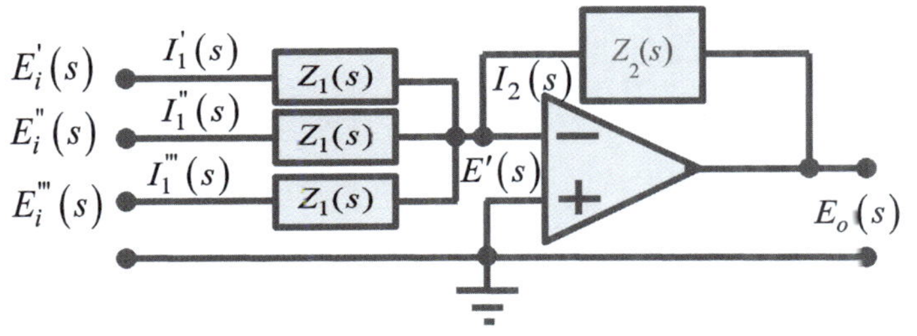

Fig. 3.3-8. Inverting summer.

$$\frac{E_o}{E_i}=-\frac{Z_2}{Z_1}\Rightarrow\ E_o=-\frac{Z_2}{Z_1}E_i'-\frac{Z_2}{Z_1}E_i''-\frac{Z_2}{Z_1}E_i'''=-\frac{Z_2}{Z_1}[E_i'+E_i''+E_i''']. \quad (3.3\text{-}2)$$

Equation (3.3-2) shows addition summing action of the latter circuit, making it "inverting summer".

3.3.2.4 *Inverting differentiator*

Setting in an inverting amplifier $Z_1=1/Cs$ and $Z_2=R$ yields configuration

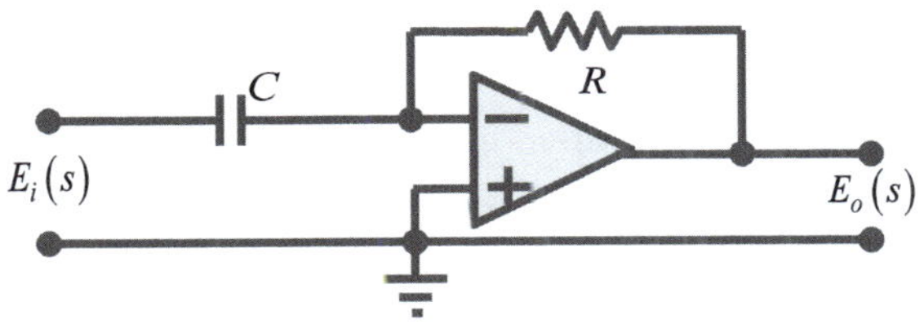

Fig. 3.3-9. Inverting differentiator.

and the input/output relation

$$\frac{E_o}{E_i}=-\frac{Z_2}{Z_1}=-\frac{R}{1/Cs}=-RCs\Rightarrow E_o(s)=-RCsE_i(s)\Rightarrow e_o(t)=-RC\frac{d}{dt}e_i(t). \quad (3.3\text{-}3)$$

3.3.2.5 *Inverting integrator*

Setting in an inverting amplifier $Z_1 = R$ and $Z_2 = 1/Cs$ yields configuration

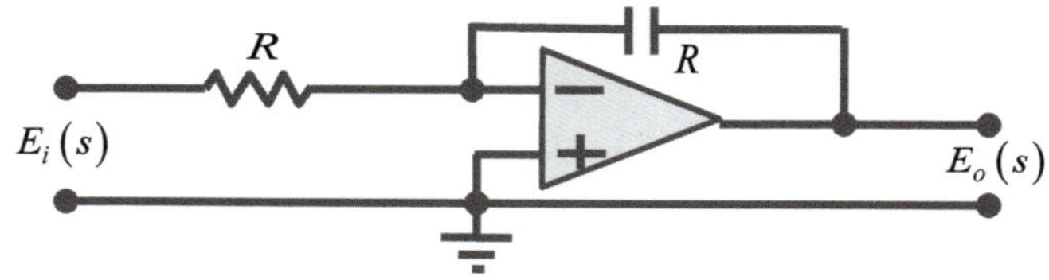

Fig. 3.3-10. Inverting integrator.

and the input/output relation

$$\frac{E_o}{E_i} = -\frac{Z_2}{Z_1} = -\frac{1/Cs}{R} = -\frac{1}{RCs} \Rightarrow E_o(s) = -\frac{1}{RCs}E_i(s) \Rightarrow e_o(t) = e_o(0) - \frac{1}{RC}\int_0^t e_i(\tau)d\tau. \quad (3.3\text{-}4)$$

3.3.2.6 *Current-to-voltage converter*

This circuit is used when sensor is a current source, as, for example, a photodiode light sensor.

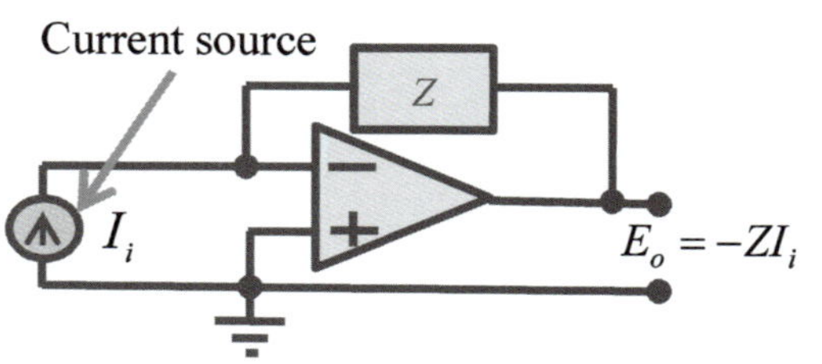

$$\frac{E_o}{E_i} = -\frac{Z_2}{Z_1} \Rightarrow E_o = -E_i\frac{Z_2}{Z_1}$$

$$= -\frac{E_i}{Z_1}Z_2 = -I_iZ_2 \equiv -I_iZ.$$

Fig. 3.3-11. Current-to-voltage converter.

3.3.2.7 *Comparator — a precursor to an analog-to-digital converter*

This device, shown in Figure 3.3-12, is drawn like a regular op amp, but it functions differently: it compares two voltages (or currents) and switches its output when the voltage between its inputs crosses zero to indicate which is larger. Therefore, it converts an analog input signal into an output signal with only two possible states — higher or lower, which can be associated with a single bit on (1) or off (0), respectively. It can also compare a varying voltage and a stable reference. *If reference e_1 is zero, then just the analog signal e_2 sign will designate digital* 1 *(e.g., for* +*) or* 0 *(for* −*).*

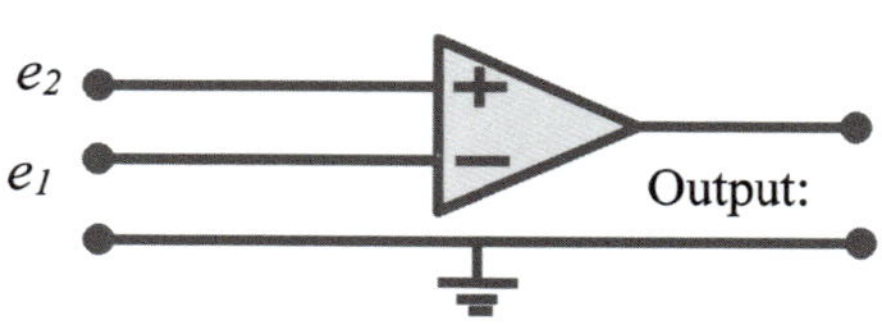

$$v_{out} = A(e_2 - e_1) = \begin{cases} +v_{max}, e_2 > e_1 \Rightarrow 1 \\ -|v_{min}|, e_2 < e_1 \Rightarrow 0 \end{cases}$$

Fig. 3.3-12. Comparator.

3.3.2.8 *Noninverting amplifier*

Here, the op amp shown in Figure 3.3-6 retains position of its leads in relation to the ground.

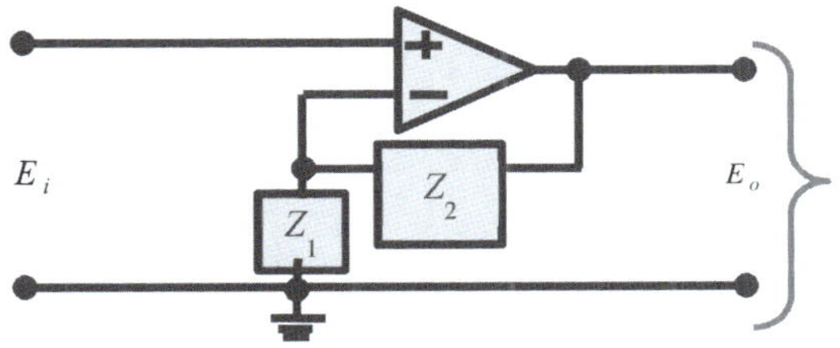

$$\frac{E_o}{E_i} = \frac{Z_1 + Z_2}{Z_1} \Rightarrow E_o = \frac{Z_1 + Z_2}{Z_1} E_i. \qquad (3.3\text{-}5)$$

Fig. 3.3-13. Noninverting amplifier.

3.3.2.9 *Noninverting amplifier as buffer*

Buffer introduced in Section 3.2 can be viewed as a one-directional signal isolator that, at the stage of system integration, enables retention of the properties of the individual system components after their interconnection.

For example, given two filters implemented as passive circuits with transfer functions $H_1(s)$ and $H_2(s)$, respectively, simply connecting these two circuits in series will not produce a circuit with the transfer function $H_3 = H_1 H_2$, if the latter is desired.

Problem: H_2 exerts ***load*** on H_1. Solution: providing load isolation through ***buffer, or follower,*** as shown in Figure 3.3-14.

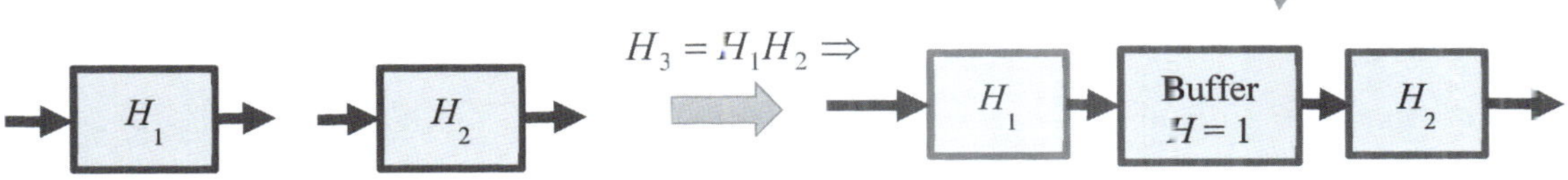

Fig. 3.3-14. Load isolation through buffer.

Buffer can be obtained from noninverting amplifier given in Figure 3.3-13 by breaking the path with Z_1 and shorting the path with Z_2. This yields the circuit as shown in Figure 3.3-15 below:

$$Z_1 = \infty,\ Z_2 = 0$$
$$\Rightarrow \frac{E_o}{E_i} = 1.$$

Fig. 3.3-15. Noninverting unity gain buffer, also referred to as follower.

3.3.3 *Other types of signal isolators*

Another widely used load isolator is optical isolator that provides a unidirectional information. Magnetic transformer serves as an electrical isolator, however it does not eliminate loading due to back EMF. Both isolators are depicted in Figure 3.3-16 below, with their main features indicated.

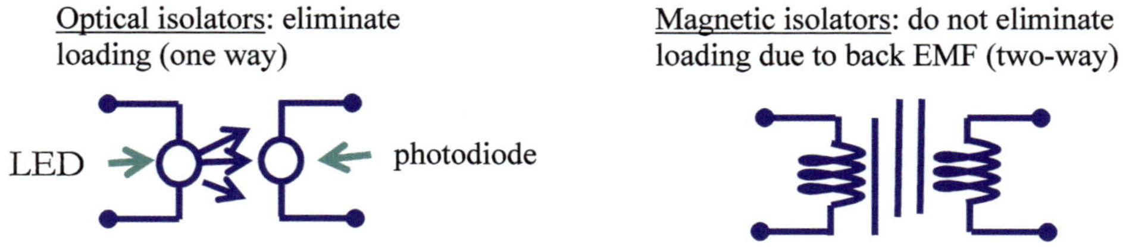

Fig. 3.3-16. Noninverting unity gain buffer.

3.3.4 *Two logic components used in signal processing: the 1-bit analog-to-digital converter (ADC) and the 1-bit digital-to-analog converter (DAC)*

3.3.4.1 *The comparator followed by the clocked type-D rising-edge-triggered flip-flop (latch) yields the latched comparator — the 1-bit analog-to-digital converter (ADC)*

At the time moment when the clock edge rises, the D (data) latch input connected to the comparator output captures the output sign of the latter and passes it as a 1 bit information to the latch output Q (1 for +, 0 for −). It is then held there for one clock cycle. The next rising edge resets the latch. Thus, comparator with grounded "−" lead followed by the latch forms *clocked latched comparator*

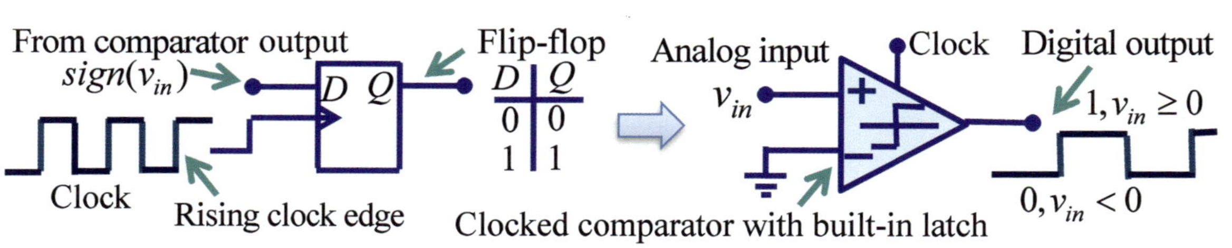

Fig. 3.3-17. Rising-clock-edge-triggered flip-flop built into comparator yields the latched comparator — the 1-bit ADC.

3.3.4.2 *The single-bit digital-to-analog converter*

The 1-bit digital-to-analog converter (DAC) shown below acts as a switch with digital input and analog output. When the value of 1 appears at the input, this switch connects its analog output lead to the positive voltage rail fixed at $+V_{DR}$, and, conversely, when the value of 0 appears at the input, the switch connects its analog output lead to the negative voltage rail fixed at $-V_{DR}$. The output value is kept until the next bit arrives, causing the switch to connect to the rail as set by bit value.

$$\left.\begin{array}{c}\text{Analog}\\ \text{output}\end{array}\right\} = \begin{cases} +V_{DR} & \text{for input bit value 1,} \\ -V_{DR} & \text{for input bit value 0,} \end{cases}$$

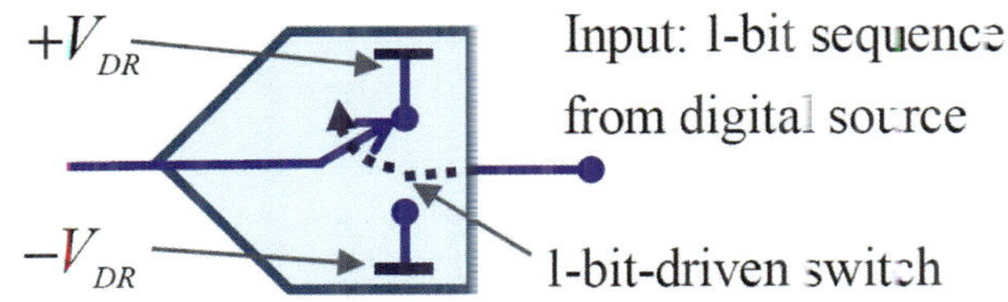

Fig. 3.3-18. Single-bit digital-to-analog converter.

3.3.5 *Voltage dividers, potentiometers, and analog-to-digital converter-encoder*

3.3.5.1 *Voltage divider and its loading, displacement sensor, potentiometer*

Consider a simple voltage divider in Figure 3.3-19. This circuit can be used to cut voltage in half, if a lower voltage is needed.

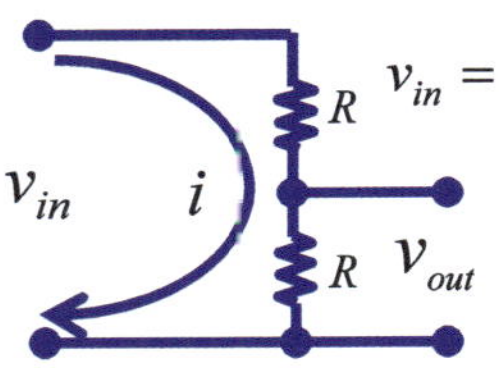

$$v_{in} = Ri + Ri = 2Ri,$$
$$v_{out} = Ri = \frac{1}{2}2Ri = \boxed{\frac{1}{2}v_{in}.}$$

Fig. 3.3-19. Simple voltage divider.

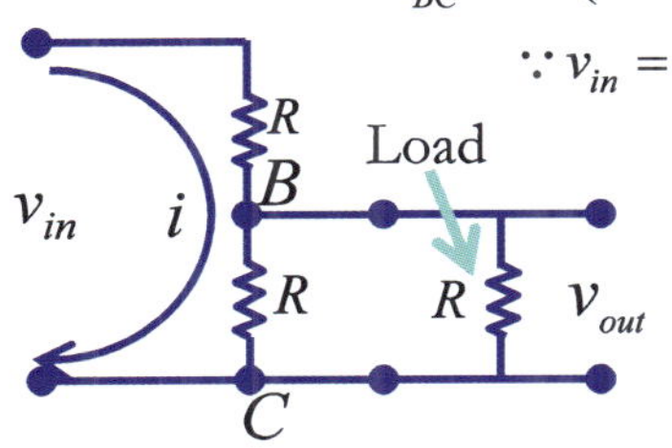

$$R_{BC} = 1/(1/R + 1/R) = 1/(2/R) = R/2$$
$$\because v_{in} = iR + iR/2 = i \cdot 2R_{BC} + iR_{BC}$$
$$= 3iR_{BC} = 3v_{out} \Rightarrow$$
$$\boxed{v_{out} = iR_{BC} = \frac{1}{3}v_{in}.}$$

Fig. 3.3-20. Voltage divider with load attached — the division function is lost

But, suppose we try to use this divider to power a load given by a simple resistor. As seen from calculation in Figure 3.3-20, in this case, voltage divider stops working: it is a passive circuit, therefore attaching a passive circuit (resistor) to it **loads** it, so that it no longer functions correctly.

Connecting *N* identical resistors in series forms an *N*-step voltage divider shown in Figure 3.3-21.

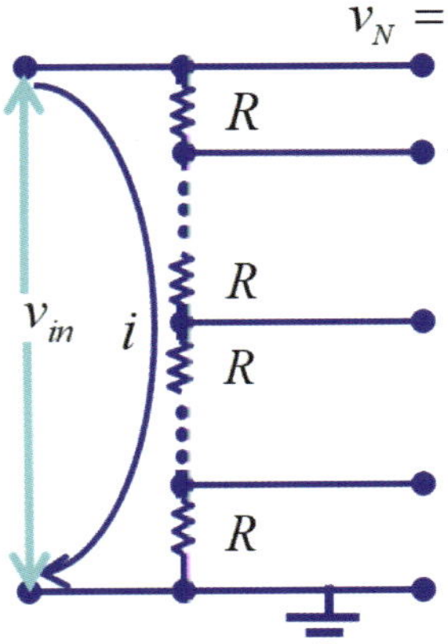

$$v_N = (N/N)v_{in} = v_{in}$$
$$v_{N-1} = \frac{N-1}{N}v_{in}$$
$$v_n = \frac{n}{N}v_{in}$$
$$v_1 = \frac{1}{N}v_{in}$$

Fig. 3.3-21. The *N*-steps voltage divider: *N* identical resistors connected in series form a voltage ladder.

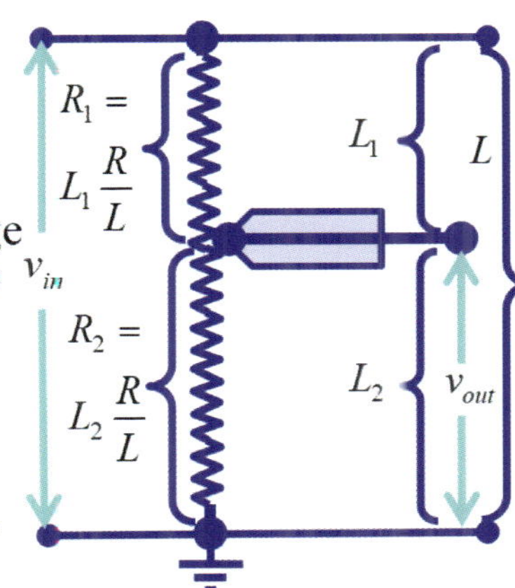

$$R_1 = L_1\frac{R}{L}, \quad R_2 = L_2\frac{R}{L}$$

$$v_{out} = \frac{R_2}{R_1 + R_2}v_{in} = \frac{L_2\frac{R}{L}}{L_1\frac{R}{L} + L_2\frac{R}{L}}v_{in} = \frac{L_2}{L_1 + L_2}v_{in} = \frac{L_2}{L}v_{in}$$
$$\Rightarrow L_2 = L\frac{v_{out}}{v_{in}},$$

Fig. 3.3-22. A continuous resistor voltage divider acting as the displacement sensor.

Decreasing each resistor value, while increasing the number of steps to keep the total resistance constant converts the N-step divider into a continuous voltage divider, as shown in Figure 3.3-22.

The resistance is then directly proportional to the length. Using this relation, movable voltage probe (wiper) picks off voltage that is proportional to distance from the grounded lead: i.e., the *voltage* measurement can be used to find the *distance* from the probe to the end of the resistor if current through both R_1 and R_2 is identical.

By curling the one long resistor into a circle, we can turn the measurement of the length into the measurement of the *angle* of the probe position as shown in Figure 3.3-23.

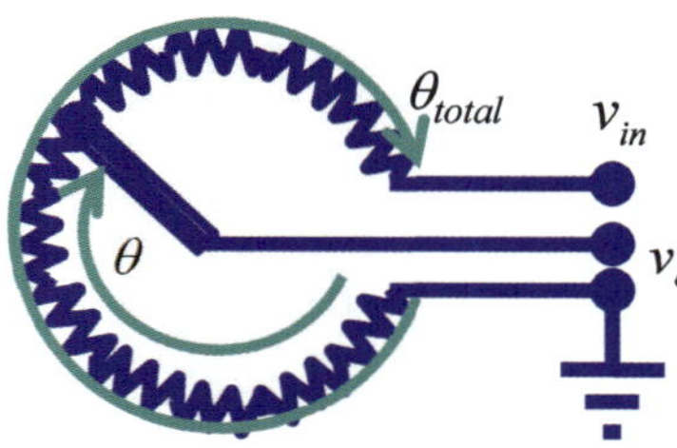

$$v_{out} = \frac{R_2}{R_1 + R_2} v_{in}$$
$$= \frac{\theta}{\theta_{total}} v_{in} \Rightarrow$$
$$\boxed{\theta = \frac{\theta_{total}}{v_{in}} v_{out}}$$

Fig. 3.3-23. Potentiometer.

Fig. 3.3-24. Potentiometers used in a guitar.

The result is called a **potentiometer**. It is used as volume control and angular displacement sensor. To avoid its loading, *buffer should be placed between the pot and the load.*

3.3.5.2 *Flash ADC — analog signal quantizer and binary encoder of comparator state*

In Figure 3.3-25 an 8R resistor string forms a reference voltage ladder, with each R forming a step:

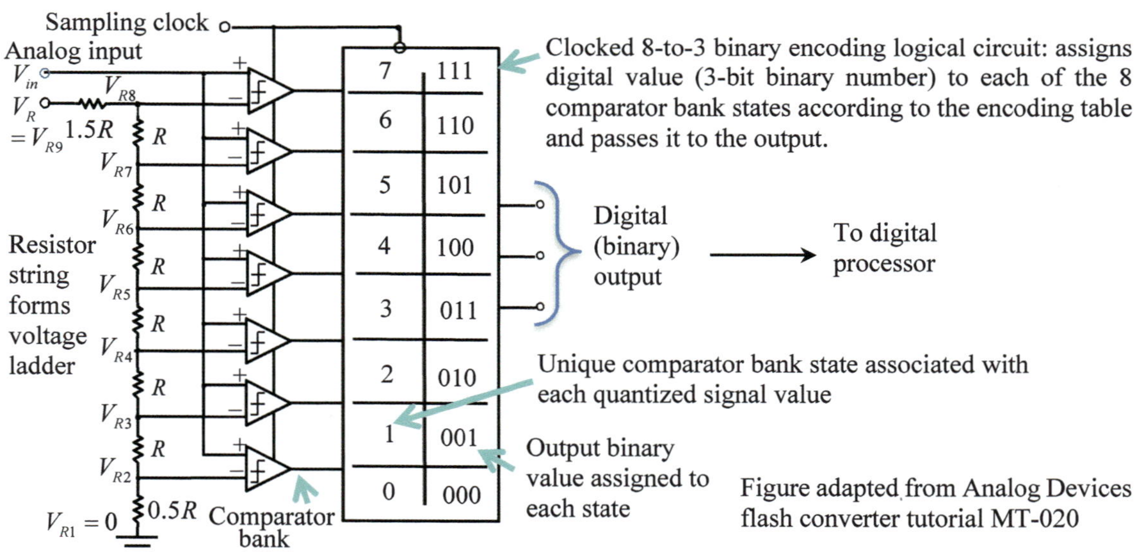

Fig. 3.3-25. Flash ADC — analog input and each ladder step reference voltage are fed into the corresponding comparator.

corresponds to the $V_R/8$ voltage drop from V_R ($\equiv V_{R9}$) > 0 towards V_{R1}, yielding the $V_\Delta = V_R/8$ voltage *quantization resolution*. This makes the voltages of the 1st, the ith, $i = 2, \ldots, 7$, and the 8th reference intervals lie, respectively, in the $[0, V_\Delta/2)$ (upper bound not included), $[V_\Delta/2 + (i\text{-}2)V_\Delta, V_\Delta/2 + (i\text{-}1)V_\Delta)$, and $[V_\Delta/2 + 6V_\Delta, V_R = 8V_\Delta]$ voltage ranges. Expressing the number of each of the resulting $2^3 = 8$ reference voltage quantization levels in a binary form yields the above 3-bit binary encoding table.

A very fast parallelized sampling/quantization *flash ADC* architecture is then formed by adding a bank of 7 stand-alone nonloading op-amp-based latched clocked comparators and connecting the negative pin of each of them to each of the 7 links between two adjacent resistors. The input voltage is connected to the positive pins of all comparators. When on the rising clock cycle edge $V_{Rk} \leq V_{in} < V_{R(k+1)}$, $k = 1, \ldots, 8$, the lower k-1 comparators latch up, and the encoder counts this number and puts it into binary form, yielding the so-called *Pulse Code Modulation*.

3.4 Basic Motion Sensing: Strain Gauge, Basic Instrumentation Techniques: Bridge Circuit, Instrumentation Amplifier, and Common Mode Rejection Ratio (CMRR)

3.4.1 *Strain Gauge*

3.4.1.1 *Strain gauge construction and use*

Strain gauge I one of the key sensors used in industrial as well as medical applications. Strain ε, or tiny elongation in elastic material can be measured to detect changes in force, temperature, and acceleration. Strain gauges measure elongation or compression via the change in resistance of thin-film resistors (metal or semiconductor) caused by their deformation:

$$\frac{\Delta R}{R} = \mathrm{GF}\varepsilon.$$

Where, GF is sensitivity, or gauge factor: 2–6 for metallic resistors, 40–200 for semiconductor resistors. The typical strain gauge configuration is shown in Figure 3.4-1.

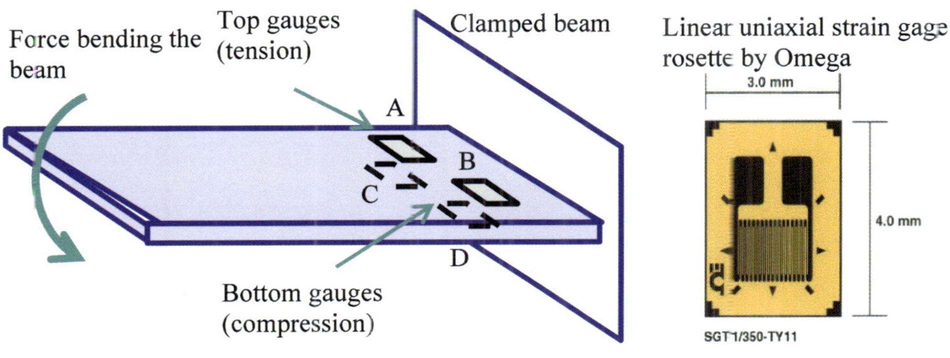

Fig. 3.4-1. Strain gauge as a load cell — force sensor.

3.4.1.2 *Bridge circuit for difference measurement*

The extraction of the strain gauge sensor signal is typically carried out using the bridge circuit shown in Figure 3.4-2.

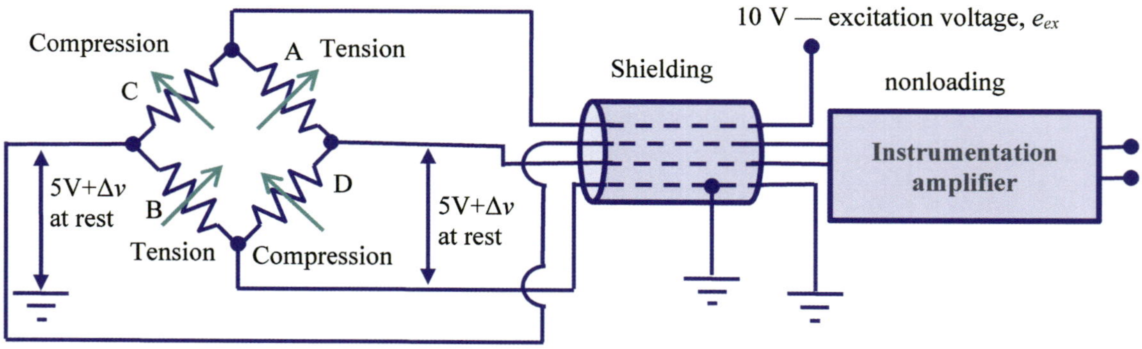

Fig. 3.4-2. Bridge circuits.

When the system is at rest, $e_1' = e_2' = e_{CM}$ (*common mode* voltage) — signal common to both strain gauge pick-off points. Since strain gauge is a little antenna, it generates at all times EMF induced parasitics, denoted further as Δv. Deformation signal in each gauge output wire is "riding" on (fluctuating around) common mode voltage (here e_{CM} = offset voltage 5 V+ parasitics Δv, with respect to ground). The ideal instrumentation amplifier should cancel out the common mode and pass to the output only the amplified *voltage difference.*

3.4.2 *Ideal instrumentation amplifier*

Ideal instrumentation (difference) amplifier in its simplest form is shown in Figure 3.4-3. In this circuit, the resistors are assumed to be matched perfectly. The circuit is also assumed to have

- 0 voltage offset — output voltage is zero when voltages at both inputs are identical
- insensitivity of output voltage to simultaneous changes in both input voltages by a "common", i.e. identical at both inputs, voltage value.

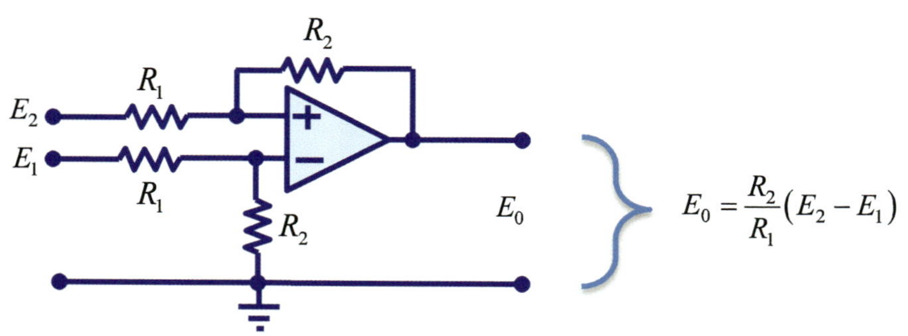

Fig. 3.4-3. Ideal instrumentation amplifier.

The description of useful signals extraction by this circuit is given next.

Strain gauge signal at instrumentation amplifier input, normal mode signal. For the strain gauge example (Figure 3.4-3), as well as for other sensors, the input signals to instrumentation amp are not just the sensing element outputs, but include other signals:

$$e_1' = e_{CM} + e_1 - \Delta v, \quad e_2' = e_{CM} + e_2 + \Delta v.$$

Here e_{CM} is due to excitation voltage, e_2 is due to sensing element, and Δv is EMF induced voltage parasitics, since strain gage is a little antenna. Therefore

$$e_2' - e_1' = (e_{CM} + e_2 + \Delta v) - (e_{CM} + e_1 + \Delta v) = e_2 - e_1 \equiv e_{inp}, \ |e_{inp}| = |e_2| + |e_1|$$

where $e_2' - e_1'$ is the actual signal, and e_{inp} is referred to as the ***normal mode*** or ***normal mode signal***. The latter is the actual signal (in millivolts) generated by the sensor, and it can take positive or negative values.

Normal mode gain. The amplifier gain that relates $e_2 - e_1$ to e_o is called the normal mode gain, G_{NM}, i.e.

$$e_o = G_{NM}(e_2 - e_1). \tag{3.4-1}$$

3.4.3 *Real instrumentation amplifier*

3.4.3.1 *Common mode rejection ratio*

Real instrumentation amplifiers are characterized by nonidealities arising from mismatch in characteristics of the identically labeled components, such as resistance values and/or op amp gains, caused by manufacturing tolerances, as well as circuit design features such as offsets. Both result in distorting the ideal amplifier input/output relation (3.4-1). Attaining the required amplification precision in spite of this distortion is accomplished through an analytical tool referred to as the Common Mode Rejection Ratio (CMRR). This tool guides component selection and the choice of circuit topology under the constraint of providing the necessary input impedance, as described next.

Many more details on the concepts and the circuit given below can be found in 'The Art of Electronics'[3].

Common mode gain. For $e_1 = e_2 = 0$ (and therefore $e_1' = e_2' = e_{CM}$) real instrumentation amplifier, due to imperfect e_{CM} subtraction, has $E_o = e_{lk} \neq 0$, which we will refer to as the leakage output voltage. This leakage voltage can be related to e_{CM} at the input leads by modeling it as a product:

$$|e_{lk}| = G_{CM} e_{CM} \tag{3.4-2}$$

where the gain G_{CM} relating $|e_{lk}|$ to e_{CM} is called the common mode gain.

---

[3]Horowitz and Hill, The Art of Electronics, 2nd edition, Cambridge University Press, 1989 and 3rd edition, 2015.

Then amplifier equation in its simplest form (without the offset voltage) is given by:

$$e_o = G_{NM}\left(e_2' - e_1'\right) + e_{lk} = G_{NM}\left(e_2 - e_1\right) + e_{lk}.$$

The output now is seen to consist of two parts. The amplified version $G_{NM}(e_2 - e_1)$ of the actual (normal mode) signal $(e_2 - e_1)$ we are trying to measure and "leakage" voltage e_{lk}. In $|e_{lk}| = G_{CM}e_{CM}$ the change in the excitation voltage, and hence, in e_{CM} would result in the change in both G_{CM} and e_{lk}. From this equation it is seen that e_{lk} acts as an offset to the measurement: when $|G_{NM}(e_2 - e_1)|$ is smaller than $|e_{lk}|$, the normal mode signal (its amplified version) can have its sign reversed at amp output by the leakage voltage.

Instrumentation amplifier lower resolution bound. Considering e_{lk} to be the leakage floor voltage at the output, set $|(e_2 - e_1)|G_{NM} = |e_{lk}|$, then,

$$\frac{|e_{lk}|}{G_{NM}} = e_{lb} \tag{3.4-3}$$

is a lower bound on the input voltage. As shown in Figure 3.4-4:

- if $|e_2 - e_1| < e_{lb}$, the output signal from the sensor is drowned in the leakage (can have its sign reversed at amp output by the leakage voltage),
- if $|e_2 - e_1| > e_{lb}$, the output voltage generated by the normal mode signal due to the sensor can be seen.

Fig. 3.4-4. Instrumentation amplifier lower resolution bound.

Common mode rejection ratio (CMMR). Set $e_{lb}\,G_{NM}\,(= |e_{lk}|) = e_{CM}\,G_{CM}$. Then,

$$e_{lb}\frac{G_{NM}}{G_{CM}} = e_{CM} \text{ or } \frac{G_{NM}}{G_{CM}} = \frac{e_{CM}}{e_{lb}}$$

where e_{CM} is common mode voltage (fixed) and e_{lb} is minimum resolution for signal, which needs to be specified. The smaller the G_{CM} and/or the bigger the G_{NM}, the smaller the e_{lb} is. As $G_{NM}/G_{CM} \to \infty$, $e_{lb} \to 0$, i.e. common mode interference is rejected. Therefore G_{NM}/G_{CM} is called the common mode rejection ratio, or in dB,

$$CMRR = 20\log_{10}\left(\frac{G_{NM}}{G_{CM}}\right)\text{dB}\left(\approx\ 120\ dB,\ i.e.\ 10^6\ is\ a\ good\ value.\right)$$

3.4.3.2 *Circuit topology*

Classical three amp topology. The amplifier circuit shown in Figure 3.4-2 has limited precision capabilities due to the need to perfectly match resistors, which is practically impossible. It can also suffer from insufficient input impedance, resulting in some loading. Therefore, the amplifiers with large gain, large input impedance, and large CMRR, such as AD8557, are typically based on the three op amps topology shown in Figure 3.4-5 below.

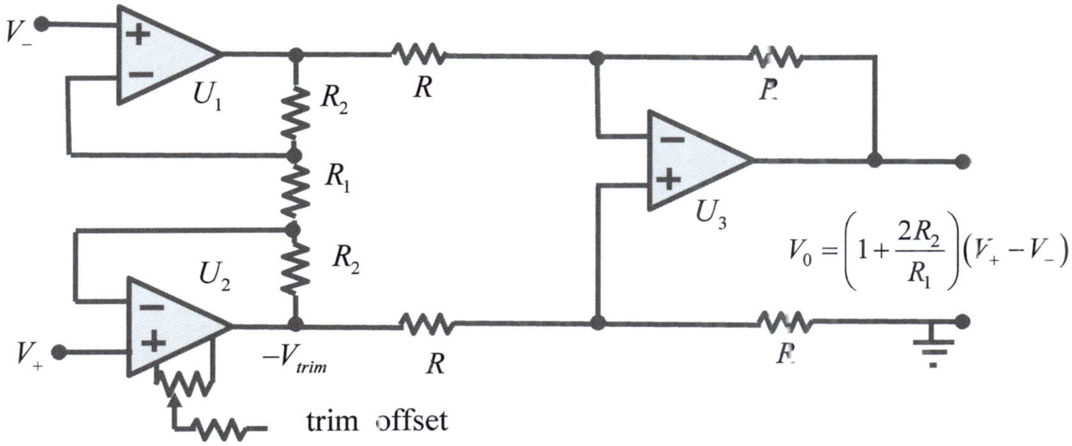

Fig. 3.4-5. Instrumentation amplifier: classical three op amps topology. Resistors, for example, can be selected as follows: $R = R_2 = 22.5\ k\Omega$, $R_1 = 365\ \Omega$.

The input op amps U_1 and U_2 must have high CMRR, however U_3 is not required to have high CMRR and resistors R in its circuit portion do not need precise matching.

Taking into account the offset voltage, the amplifier equation used for it is typically given by

$$v_{out} = G_{NM}\left(v_+ - v_-\right) + G_{CM}\frac{v_+ + v_-}{2} + v_{offset}.$$

Increasing the normal mode gain without bound is seen below to completely reject both the common mode and the offset voltage:

$$G_{NM} \to \infty \Rightarrow \left(v_+ - v_-\right) + \frac{G_{CM}\frac{v_+ + v_-}{2} + v_{offset}}{G_{NM}} = \left(e_+ - e_-\right) + \frac{G_{CM}}{G_{NM}}e_{CM} + \frac{v_{offset}}{G_{NM}} \to \left(e_+ - e_-\right).$$

This is an affine looking (linear with offset) model of distortion. To relate the latter model to a simpler affine model of distortion used earlier we note that i) $(v_+ + v_-)/2$ is the same as the common mode voltage for the strain gauge, ii) $G_{CM}(v_+ + v_-)/2$ is a representation of leakage in terms of half of excitation voltage, and iii) the offset voltage was considered as part of the leakage.

Instrumentation amplifier AD620. The topology given in Figure 3.4-5 is used as the basis in a number of commercially available instrumentation amplifiers. One of the most popular ones is Analog Devices AD620 whose top view packaging is pictured below.

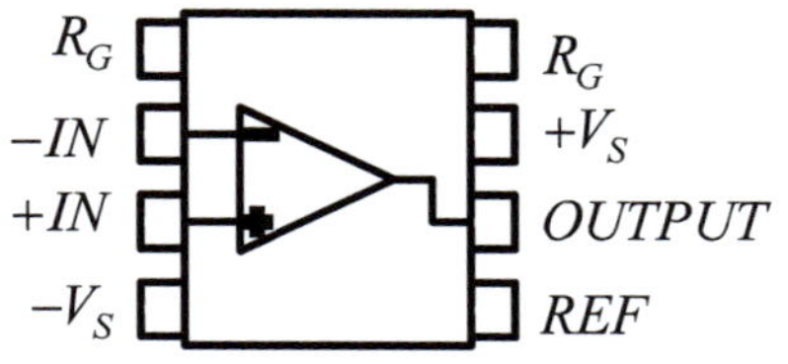

Fig. 3.4-6. Packaging schematics of instrumentation amplifier AD620.

Its circuitry is given in data sheets that can be found online[4] (pp. 15 and 16 are on RF interference, CMR, and grounding).

3.4.4 *Total Harmonic Distortion (THD): a measure of the distortion of a transducer/amplifier response to a pure sine input*

The nonlinearity of a non-ideal transducer/amplifier can be quantified in terms of its Total Harmonic Distortion (THD): given a pure sine input, an amplifier THD is the square root of the ratio, in %, of the sum of powers, $P_2 + \ldots + P_n$, of all harmonic multiples in its output to the power P_1 of the sine output component which has the input frequency (%RMS of the relative distortion):

$$\sqrt{\left(\sum_{i=2}^{n} P_i\right) / P_1} \times 100\% = \sqrt{\left(\sum_{i=2}^{n} A_i^2\right) / A_1} \times 100\%.$$

Here A_i is the magnitude of the i'th pure sinusoidal output component, as shown in Figure 3.4-7.

The hardware testbed Rexroth actuator THD under load is seen to be $A_2/A_1 \times 100\% = (0.035/3)100\% \approx 1\%$ — quite small, as seen in Figure 3.4-7 piston position graph, but still not small enough to prevent large resonance-induced motion distortion of the beam/mold assembly, as seen in Figure 3.4-7 mold position graph below. Ideally, this application requires attaining a 0.01% THD. Such THD reduction is achieved in Chapter 9 through control techniques, solving the distortion problem.

[4] http://www.analog.com/static/imported-files/data_sheets/AD620.pdf

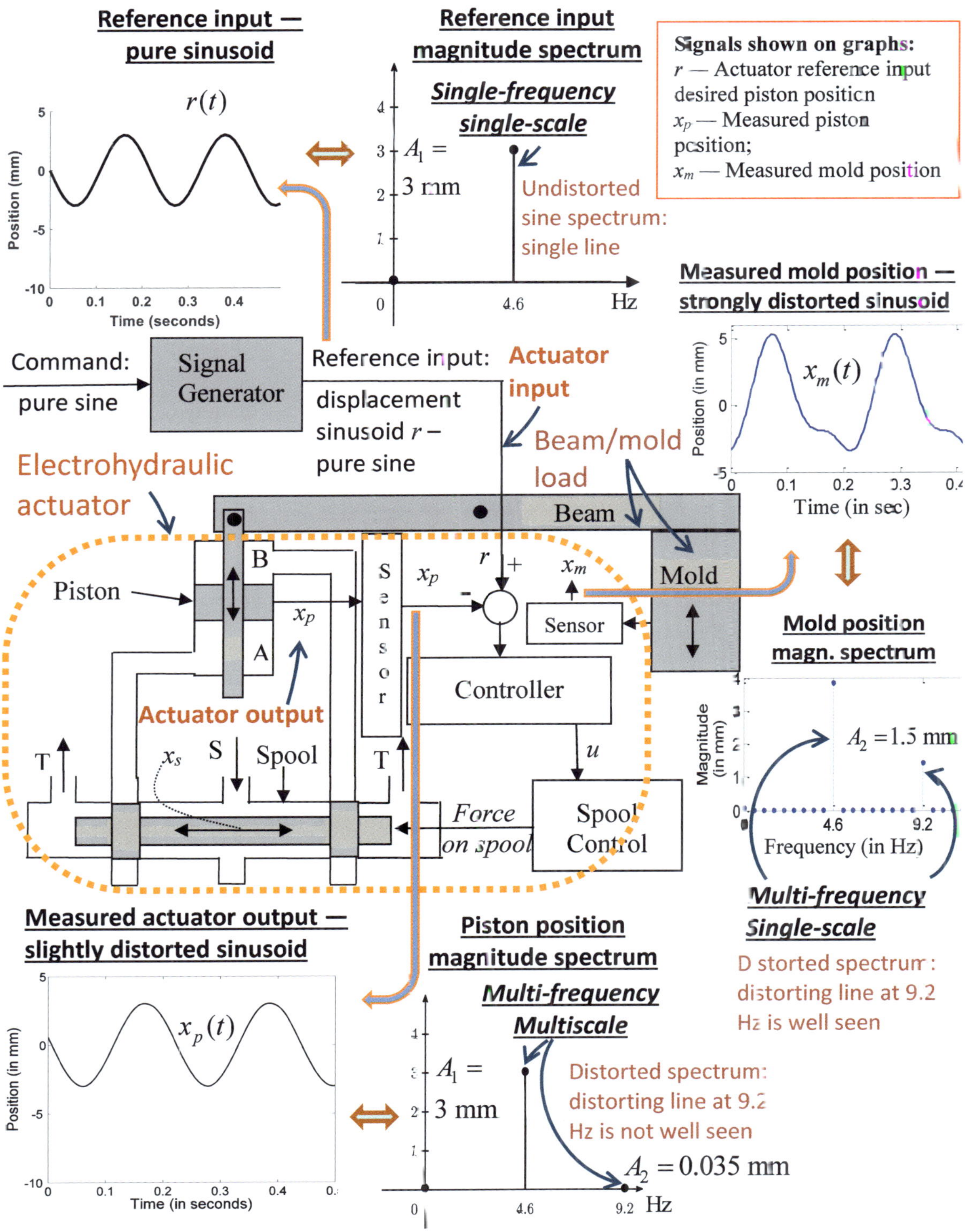

Fig. 3.4-7. Aggregated view of harmonic distortion induced by the actuator hydraulics under load in mold oscillation.

Homework 4: Solved Examples and Solution Guide

Problem 1: Let $R_1 = \ldots = R_5 = C_1 = C_2 = L_1 = L_2 = 1$ in Figure HW4-1.

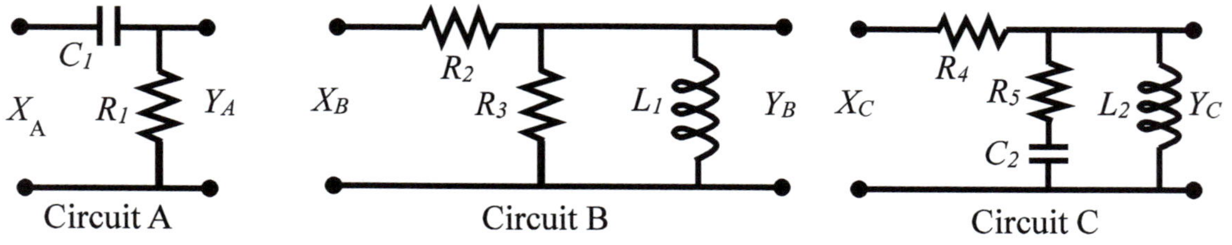

Fig. Hw4-1. Passive circuits for illustrating passive subsystems interconnection properties.

(a) For each circuit in Figure HW4-1, derive the transfer function from input voltage, *x*, to output voltage, *y*, using the equivalent impedance method and impulse responses using partial fraction expansion and inverse Laplace transform.
(b) Connect the output leads of Circuit A directly to the input leads of circuit B, and call the result Circuit D. Find the transfer function and impulse response of the latter.
(c) Insert a buffer between the output leads of Circuit A and the input leads of circuit B, and call the result Circuit E. Find the transfer function and impulse response.

Solution:

(a) For Circuit A, shown in Figure Hw4-2, there is only one current to consider. Noting that the impedance of the resistor is $Z_{R_1} = 1$ and the impedance of the capacitor is $Z_{C_1} = 1/s$, the voltage drops are

$$X_A = Z_{R_1} I + Z_{C_1} I = \left(Z_{R_1} + Z_{C_1}\right) I = \left(1 + 1/s\right) I \text{ and } Y_A = Z_{R_1} I = I.$$

Taking the ratio, $\dfrac{Y_A}{X_A} = \dfrac{I}{\left(1 + (1/s)\right) I} = \dfrac{s}{s+1}$ yields the first-order transfer function, which permits finding the impulse response by taking its inverse Laplace transform:

$$h(t) = \mathcal{L}^{-1}\left\{\frac{s}{s+1}\right\} = \mathcal{L}^{-1}\left\{1 - \frac{1}{s+1}\right\} = \delta(t) - e^{-t}u(t).$$

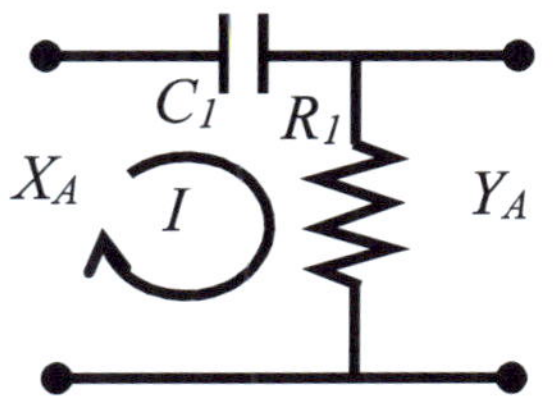

Fig. Hw4-2. Circuit A.

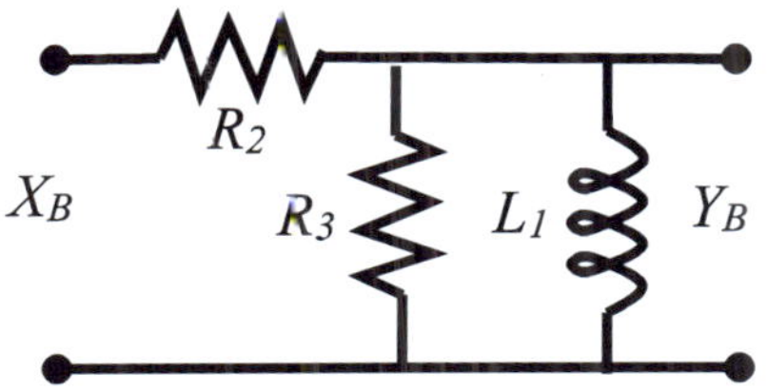

Fig. Hw4-3. Circuit B.

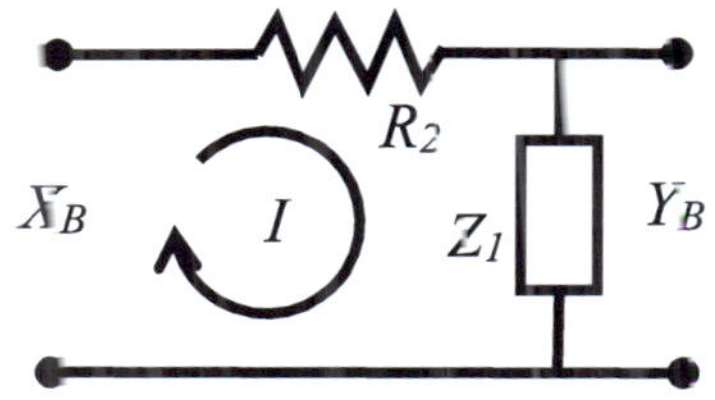

Fig. Hw4-4. Modified Circuit B.

Finding impulse response for Circuit B which has two loops is slightly more complicated. First, we note that the resistors and the inductor impedances are $Z_{R_2} = Z_{R_3} = 1$ and $Z_{L_1} = s$, respectively. Because R_3 and L_1, as seen in Figure Hw4-3, are connected in parallel, we can combine them to obtain the new circuit shown in Figure Hw4-4.

Since $Z_{R_3} = R_3 = 1,\ Z_{L_1} = L_1 s = s$, the combined impedance is

$$Z_1 = \frac{1}{1/Z_{R_3} + 1/Z_{L_1}} = \frac{1}{1/R_3 + 1/L_1 s} = \frac{R_3 L_1 s}{L_1 s + R_3} = \frac{s}{s+1}.$$

Therefore, the voltage drops and the circuit transfer function are computed as follows:

$$X_B = Z_{R_2} I + Z_1 I = \left(Z_{R_2} + Z_1\right) I = \left(1 + \frac{s}{s+1}\right) I = \frac{2s+1}{s+1} I,\ Y_B = Z_1 I = \frac{s}{s+1} I,$$

$$\Rightarrow \frac{Y_B}{X_B} = \left(\frac{s}{s+1} I\right) / \left(\frac{2s+1}{s+1}\right) I = \frac{s}{2s+1}.$$

Now, to obtain the impulse response, we take the inverse Laplace transform

$$h(t) = \mathcal{L}^{-1}\left\{\frac{s}{2s+1}\right\} = \mathcal{L}^{-1}\left\{\frac{1}{2} - \frac{0.5}{2s+1}\right\} = \frac{1}{2}\delta(t) - 0.5e^{-0.5t}u(t).$$

For Circuit C, the impendences are $Z_{R_4} = Z_{R_5} = R_4 = R_5 = 1,\ Z_{L_2} = L_2 s = s,\ Z_{C_2} = 1/(C_2 s) = 1/s.$

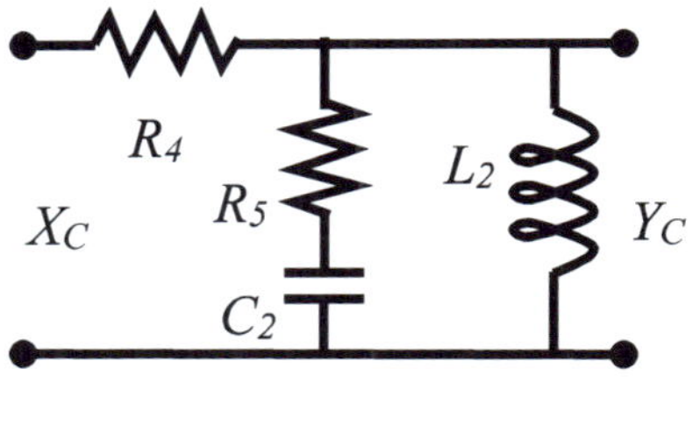

Fig. Hw4-5. Circuit C.

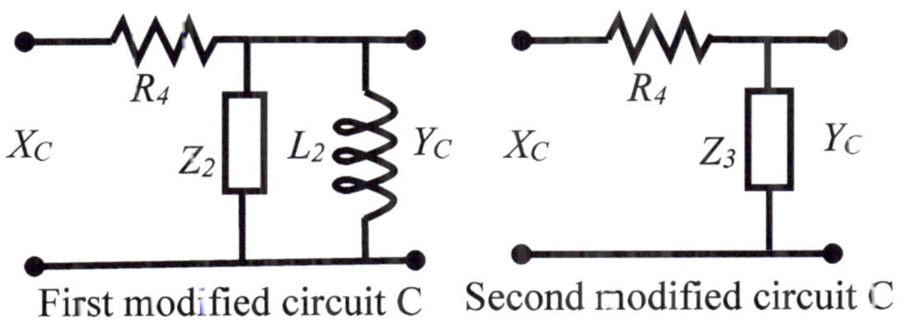

Fig. Hw4-6. Modified circuit C.

Since R_5 and C_2 are connected in series, their combined impedance, Z_2, is given by $Z_2 = Z_{R_5} + Z_{C_5} = 1 + 1/s = (s + 1)/s$. Now, Z_2 and Z_{L_2} are parallel impedances, therefore we can combine them as shown below, with the corresponding simplified circuits given in Figure Hw4-6.

$$\frac{1}{Z_3}=\frac{1}{Z_2}+\frac{1}{Z_{L_2}}=\frac{s}{s+1}+\frac{1}{s}=\frac{s^2+s+1}{s(s+1)}\Rightarrow Z_3=\frac{s(s+1)}{s^2+s+1}.$$

Calculating the voltage drop across Z_3 in the second modified Circuit C yields

$$\frac{Y_C}{X_C}=\frac{Z_3}{R_4+Z_3}=\left(\frac{s(s+1)}{s^2+s+1}\right)\Bigg/\left(1+\frac{s(s+1)}{s^2+s+1}\right)=\frac{s(s+1)}{2s^2+2s+1}.$$

To take the inverse Laplace transform of the resulting proper transfer function, we use polynomial division to decompose it into a constant and a strictly proper transfer function and complete the squares in the denominator of the latter. This yields

$$\frac{s(s+1)}{2s^2+2s+1}=\frac{1}{2}-\frac{0.5}{2s^2+2s+1}=\frac{1}{2}-\frac{0.5}{2(s+0.5)^2+0.5}=\frac{1}{2}-\frac{1}{2}\cdot\frac{0.5}{(s+0.5)^2+0.25}\Rightarrow$$

$$h(t)=\mathcal{L}^{-1}\left\{\frac{1}{2}-\frac{1}{2}\cdot\frac{0.5}{(s+0.5)^2+0.25}\right\}=\frac{1}{2}\delta(t)-0.5e^{-0.5t}\sin(0.5t)u(t).$$

(b) Circuit D will take the form

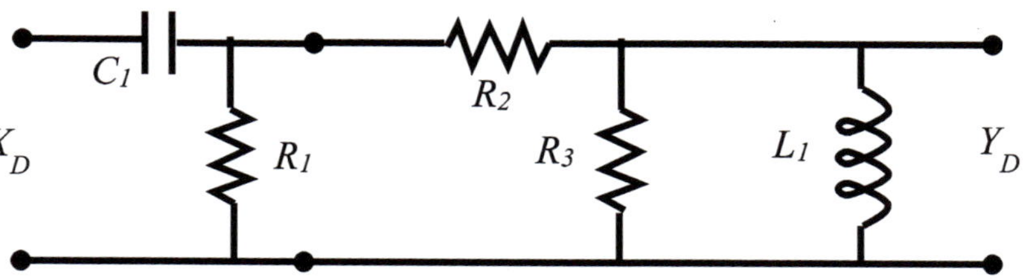

Fig. Hw4-7. Circuit D.

Again, combining R and L connected in parallel yields the following modified circuit:

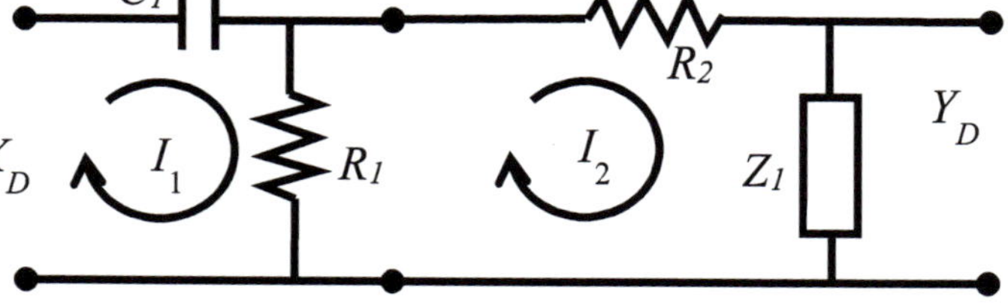

Fig. Hw4-8. Modified circuit D.

The impedances are the same as in (a) and (b), $Z_{R_1}=1$, $Z_{C_1}=1/s$, and $Z_1=\frac{s}{s+1}$. Now we have a more complicated circuit, but we can follow the same procedure as before.

For the output, and the left and the right loops we, respectively, have

$$Y_D = Z_1 I_2 = \frac{s}{s+1} I_2,\quad Z_{C_1} I_1 + (I_1 - I_2) Z_{R_1} = X_D.$$

This yields

$$I_2\left(Z_{R_2}+Z_1\right)=\left(I_1-I_2\right)Z_{R_1} \Rightarrow I_2\left(1+\frac{s}{s+1}\right)=\left(I_1-I_2\right)\Rightarrow I_1=\frac{3s+2}{s+1}I_2.$$

Substituting the relationships above for X_D and I_1 yields

$$X_D = Z_{C_1}I_1+\left(I_1-I_2\right)Z_{R_1}=\frac{1}{s}\cdot\frac{3s+2}{s+1}I_2+\frac{2s+1}{s+1}I_2=\frac{2(s+1)^2}{s(s+1)}I_2=\frac{2(s+1)}{s}I_2 \Rightarrow$$

$$\frac{Y_D}{X_D}=H_D(s)=\frac{\frac{s}{s+1}I_2}{\frac{2(s+1)}{s}I_2}=\frac{s^2}{2(s+1)^2}=\frac{s^2}{2s^2+4s+2}=\frac{1}{2}-\frac{2s+1}{2s^2+4s+2}=$$

$$\frac{1}{2}-\left(\frac{2(s+1)}{2s^2+4s+2}-\frac{1}{2s^2+4s+2}\right)=\frac{1}{2}-\left(\frac{1}{s+1}-\frac{1}{2(s+1)^2}\right)=\frac{1}{2}-\frac{1}{s+1}+\frac{1}{2}\frac{1}{(s+1)^2},$$

which yields the inverse Laplace transform

$$h_D(t)=\mathcal{L}^{-1}\left\{H_D(s)\right\}=\mathcal{L}^{-1}\left\{\frac{1}{2}-\frac{1}{s+1}+\frac{1}{2}\frac{1}{(s+1)^2}\right\}=\frac{1}{2}\delta(t)-\left(e^{-t}+\frac{1}{2}te^{-t}\right)u(t).$$

Simulation check. This was a more complicated problem, and it might help to double-check the answer. This can be done using Simulink, and the "Simscape" modelling tools found in the Simulink library. The "Electrical" section of the "Foundation Library" includes all the electrical elements needed to create this circuit. For example, the component "capacitor" can be found by searching for "capacitor" in the Simulink Library as shown in Figure HW4-9a.

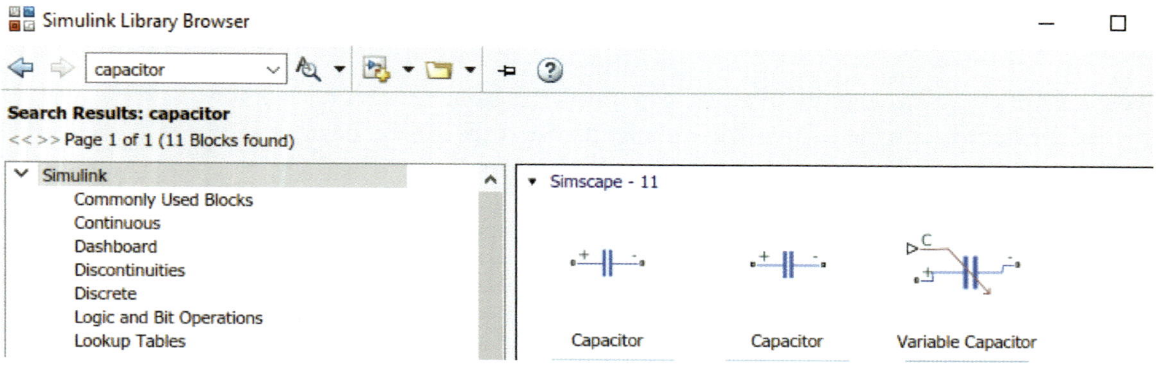

Fig. Hw4-9a. Simulink Library for capacitor.

The resistors, inductor, capacitor, and ground are all in the "Electrical Elements" library. After getting all the resistors, inductors, capacitors needed, we can construct the circuit in Figure HW4-7 in Simulink as shown in Figure Hw4-9b. Note that the values of a component can be set by double clicking the corresponding block.

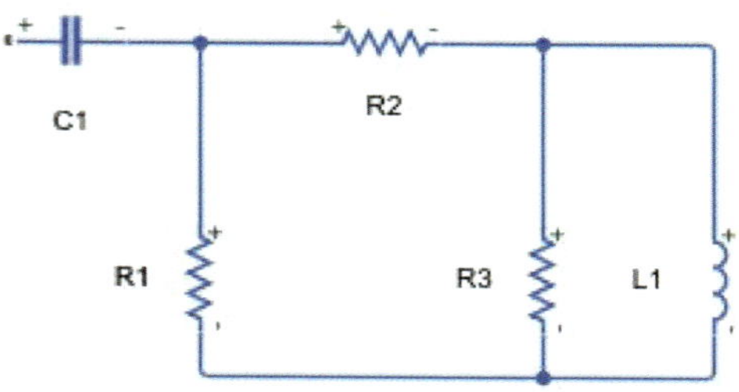

Fig. Hw4-9b. Simulink block diagram of electrical components in circuit D.

Then the voltage sensor (in the "Electrical Sensors" library), and the controlled voltage source (in the "Electrical Sources" library) are needed for the circuits. All of these elements must be attached as they would in the actual physical system. For example, voltage sensors should be attached in parallel and current sensors — in series, as shown in Figure Hw4-9c.

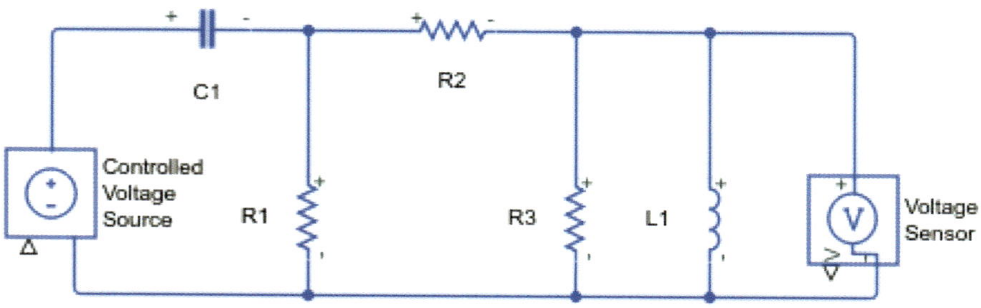

Fig. Hw4-9c. Simulink block diagram of electrical components in circuit D with sensor and source.

Once this is done, a few other additions are necessary for the model to work. First, a "Solver Configuration" block must be attached to the circuit. Second, the signals connecting the Simscape elements are not of the same data type as those for the usual signals in Matlab. To convert them to normal signals and back, one would need, respectively, "PS-Simulink Converter" and "Simulink-PS Converter" blocks from the "Utilities" library in Simscape. These should go between the input (the step block) and the output (the scope). The complete block diagram is shown in Figure Hw4-9d, where the electrical circuit part is given in blue color, while the normal Simulink signal is in black.

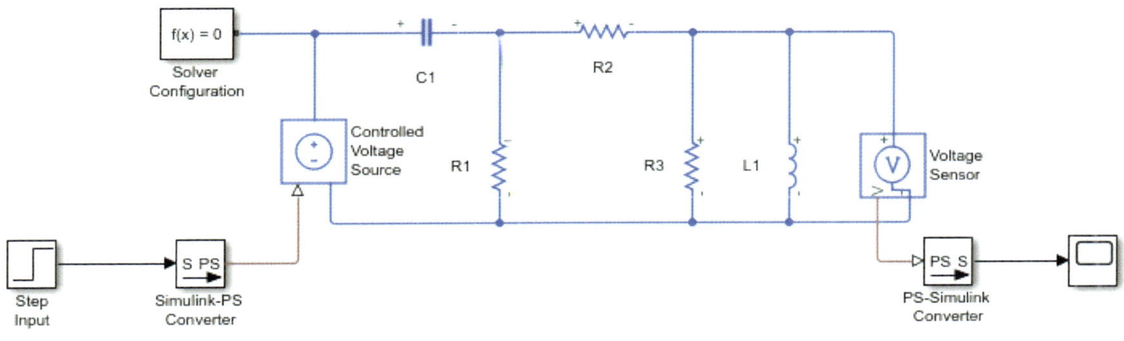

Fig. Hw4-9d. Simulink block diagram of electrical components in circuit D.

From the previous calculation we obtained the transfer function of circuit D given by

$$H(s)=\frac{s^2}{2s^2+4s+2}.$$

We can add this function to Simulink model and compare its response with that of the circuit, as shown in Figure Hw4-9e.

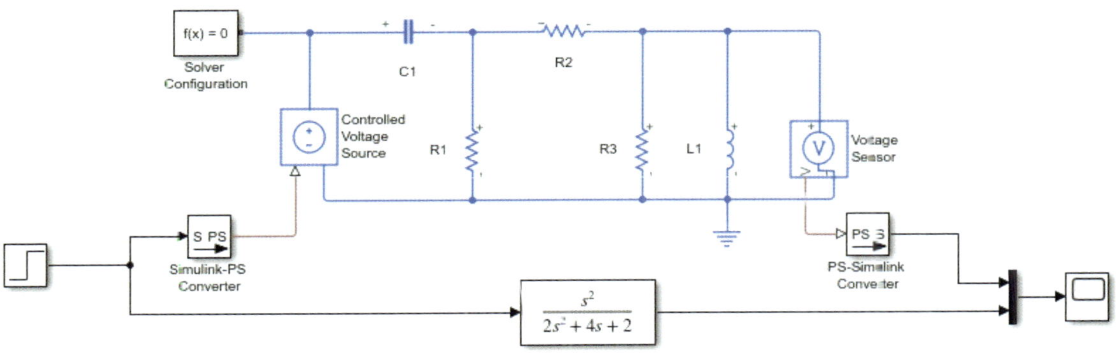

Fig. Hw4-9e. Simulink block diagram comparing the response of circuit D with that of its transfer function.

This response of the step input is shown in Figure Hw4-10. From the scope we can see that the outputs of the actual circuit and the transfer function in response to step input match.

Finally, Simulink is unable to have an impulse input. The impulse response can then be approximated by taking the derivative of the step response. In Simulink, we can search for the block we needed directly by double clicking at a white space and type in the block you needed, as shown in Figure Hw4-9f, and the final block diagram is shown in Figure Hw4-9g.

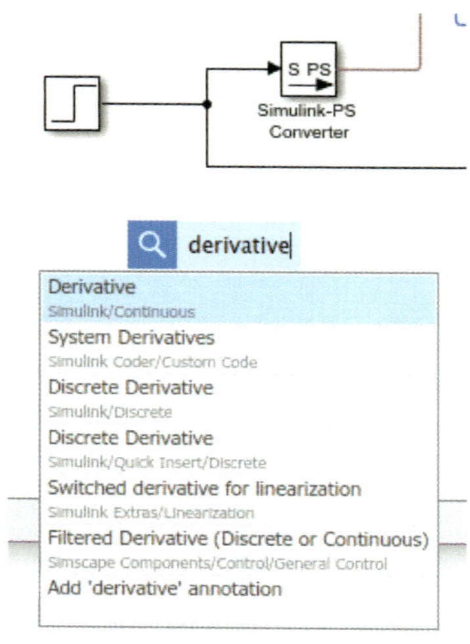

Fig. Hw4-9f. Searching for derivative block in Simulink.

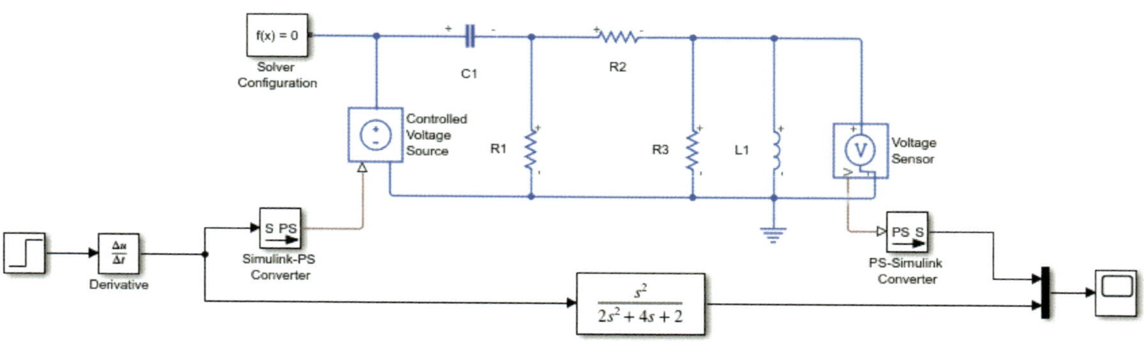

Fig. Hw4-9g. Simulink block diagram comparing the response of circuit D with that of its transfer function.

This output is shown below. From the scope we can see that the outputs of the actual circuit and the transfer function in response to step input match.

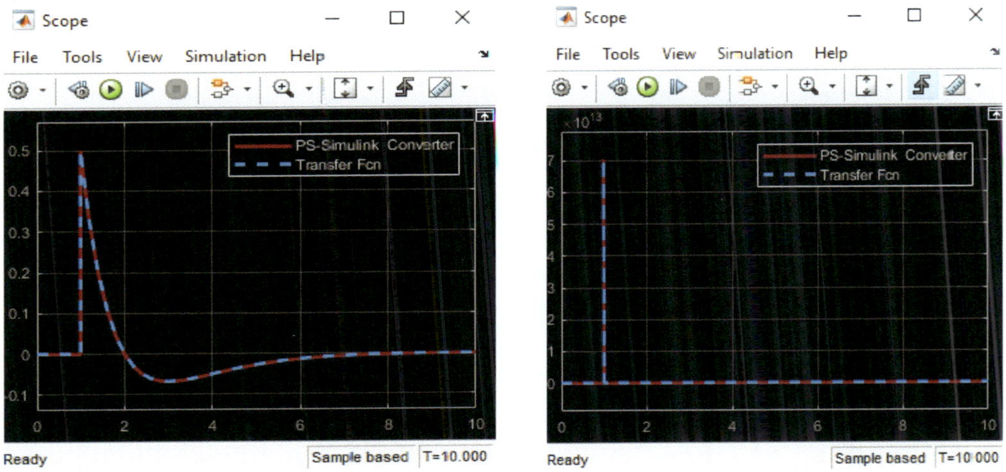

Fig. Hw4-10. Scope output from circuit D Simulink model with step input (left) and impulse input (right).

(c) Circuit E takes the form

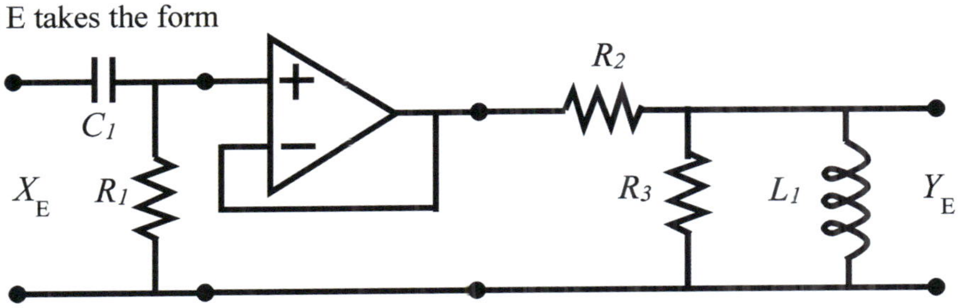

Fig. Hw4-11. Circuit E.

The effect of the buffer, or follower, is that the voltage is the same on either side, but there is no current connection between the two. That is, we can look at this interconnection as at two separate circuits,

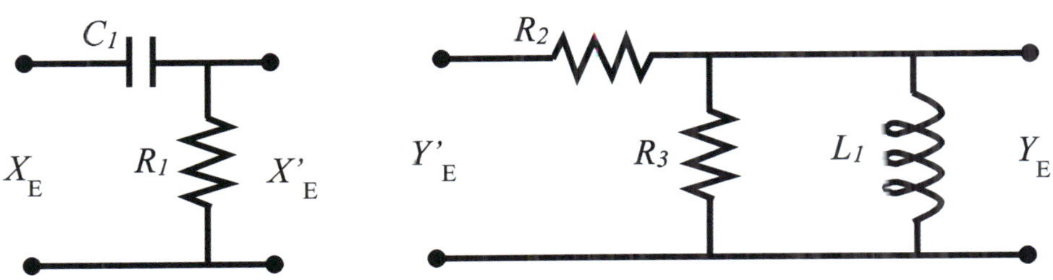

Fig. Hw4-12. Two passive sub-circuits of circuit E.

with the condition that $X'_E = Y'_E$. However, these circuits are the ones for which we obtained transfer function in part (a):

$$\frac{X'_E}{X_E} = \frac{Y_A}{X_A} = \frac{s}{s+1} \text{ and } \frac{Y_E}{Y'_E} = \frac{Y_B}{X_B} = \frac{s}{2s+1}.$$

Multiplication of these transfer functions yields

$$\frac{Y_E}{X_E} = \frac{Y_E}{X_E}\cdot\frac{X'_E}{X'_E} = \frac{Y_E}{X_E}\cdot\frac{X'_E}{Y'_E} = \frac{Y_E}{Y'_E}\cdot\frac{X'_E}{X_E} = \frac{s}{2s+1}\cdot\frac{s}{s+1} = \frac{s^2}{2s^2+3s+1}.$$

The resulting transfer function has two real roots, so we will use partial fraction expansion to find the impulse response. Setting

$$\frac{s^2}{2s^2+3s+1} = \frac{s^2}{(2s+1)(s+1)} = \frac{A}{2s+1} + \frac{B}{s+1} + C$$

and multiplying both sides by $\left(2s^2+3s+1\right)$ gives

$$s^2 = A(s+1) + B(2s+1) + C\left(2s^2+3s+1\right) = 2Cs^2 + (A+2B+3C)s + (A+B+C)$$

$$\because C = 1/2,\ A+2B = -3/2, \text{ and } A+B = -1/2.$$

The solution to this set of equations is A = 1/2 and $B = -1$. So,

$$\frac{Y_E}{X_E} = \frac{1}{4}\frac{1}{s+\frac{1}{2}} - \frac{1}{s+1} + \frac{1}{2} \Rightarrow h(t) = \mathcal{L}^{-1}\left\{\frac{1}{2} + \frac{1}{4}\frac{1}{s+1/2} - \frac{1}{s+1}\right\} = \frac{1}{2}\delta(t) - \left(e^{-t} + \frac{1}{4}e^{-\frac{1}{2}t}\right)u(t).$$

Side Note. Once again, this can be checked using Simulink. The Op-Amp block is also in the "Electrical Elements" library in Simscape.

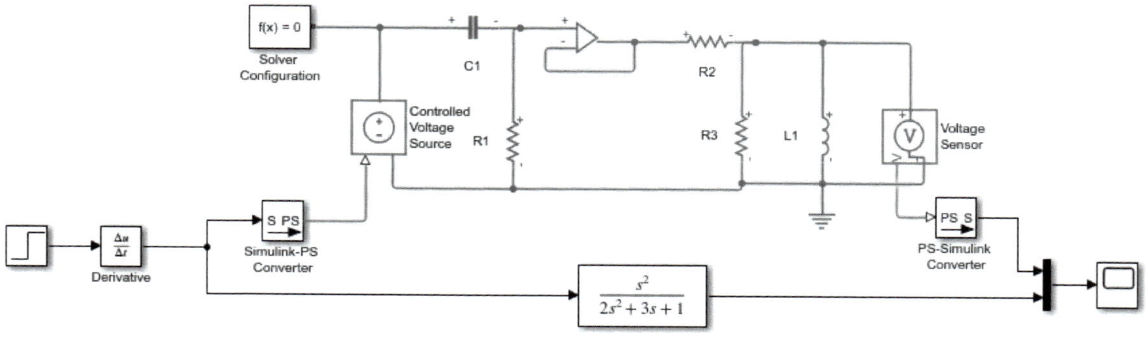

Fig. Hw4-13. Simulink block diagram of circuit E.

Problem 2 (ideal op amps):
(a) Explain why an ideal op amp does not load the preceding circuit;
(b) Show how an op amp can serve as a buffer, isolating circuits and allowing them to be connected in cascade;
(c) Show how any number of op amps can be connected in cascade.

Solution:
(a) No current can flow through the ideal op amp; it has infinite input impedance. Therefore, it cannot load any circuit before it.
(b) The input infinite impedance means that no current from the circuit in front of an op amp can affect the circuit after it. The zero output impedance means that any current required by the voltage transmitted through the op amp will be supplied without voltage distortion. This allows the two circuits to be cascaded in an unloading manner, i.e. without altering the transfer function of each circuit.
(c) If op amps are connected in cascade, no current can pass from one to the other, so that each op amp does not load the adjacent op amps.

Problem 3: Action of unloaded potentiometer. In Figure Hw4-16, let the resistors R_1 and R_2 represent the variable resistances of the corresponding resistance winding of the pot. Show that if there is infinite impedance between the wiper and terminal A (output of pot), the output voltage will be proportional to shaft angle (see Figure Hw4-15).

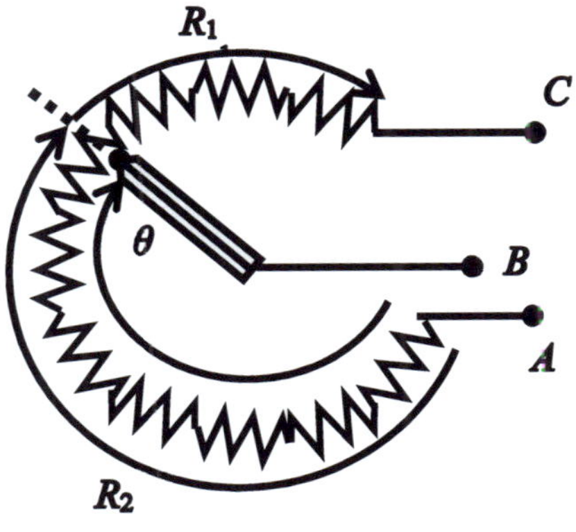

Fig. Hw4-15. Rotary potentiometer.

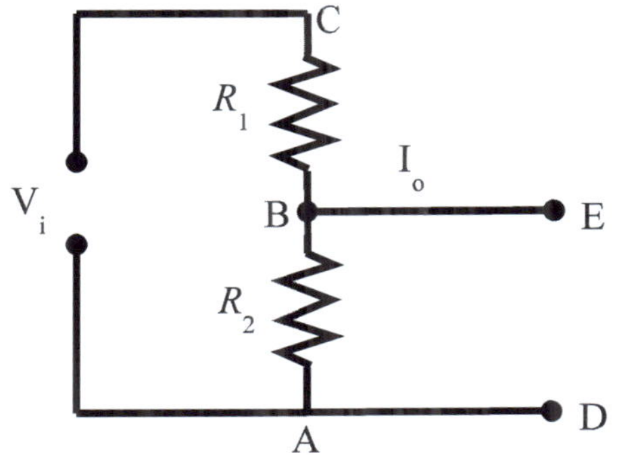

Fig. Hw4-16. Rotary potentiometer as variable voltage divider.

R_2 varies with shaft angle θ, because the wiper moves along the resistive element altering the length of the conductive part. If impedance between D and E is infinite, the current through R_2 and R_1 is the same. By Ohm's law, the output voltage across A and B will vary with R_2 proportional to shaft angle.

A voltage loop yields:

$$V_i - R_1 i - R_2 i = 0,\ \ R_2 i = V_o \Rightarrow i = \frac{V_o}{R_2},$$

$$V_i - V_o\left(\frac{R_1}{R_2}\right) - V_o = 0 \Rightarrow V_i = V_o\left(\frac{R_1}{R_2}+1\right),\ \ V_o = V_i\left(1\Big/\left(\frac{R_1}{R_2}+1\right)\right) = V_i\left(\frac{R_2}{R_1+R_2}\right) = V_i\frac{\theta}{2\pi}.$$

If the impedance between E and D is not infinite, the expression above will be distorted, the phenomenon referred to as ***loading of a potentiometer*** — the smaller the impedance between E and D, the larger the distortion.

Problem 4. *Loading in nonlinear systems.* Try to visualize the hydraulic actuator loading in mold oscillator system as follows:

1) Simulate hydraulic actuator without load (beam/mass load is not attached — refer to Chapter 2, Problem 12 solution guide, Part C, pp. 101 and 102) — check if its response (piston displacement) to 4.6 Hz sinusoidal reference input has sinusoidal components at integer multiple frequencies of the reference frequency, possibly with very tiny magnitudes (if so, this is manifestation of a small actuator nonlinearity even under no load). Use P-controller, not tracking controller.
2) Repeat simulation, now with beam/mass load attached, and check if the magnitudes of the sinusoidal components in piston displacement at integer multiple frequencies of the reference frequency are noticeably bigger. If the latter takes place, this means that the actuator nonlinearity increases — it is manifestation of the effect of the load on actuator performance. Actuator selection for a particular load is referred to in industry as actuator sizing, and loss of actuator performance under load usually indicates poor actuator sizing, but sometimes also fundamental system integration problems, like nonlinearity of a well sized actuator exciting the beam resonance in mold oscillator. Use P-controller, not tracking controller.
3) If you couldn't see anything by observing piston displacement waveforms in 1) and 2), use the brute force way: generate numerically pure sinusoid with the magnitude of piston displacement waveform and frequency 4.6 Hz, subtract it from each of the piston displacement waveforms, plot the differences and comment on their relative size. Do your best, and give any results you can come up with, but don't over-exert yourself trying to get an exact decomposition. The point here is just to see how difficult the problem is, and to contrast this with tools presented later in the book.

In each graph clearly label axes indicating the values and the units (e.g., sec).

Chapter 4

Sampling Basics, Harmonic Signals, and Signal Spectrum

4.1 Sampling Basics

4.1.1 *Discrete time sinusoid and digital frequency*

Signals available for processing — analog position and velocity sensor output. Solution of Nucor mold oscillator problem is based on sensing rigid body position (displacement). The Temposonic sensor: R-Series, Model RH, used for this purpose, provides position and velocity measurements with 1 μm and 0.1 mm/s resolution, respectively, as well as analog outputs: voltage: +10 to −10 Vdc (load > 5kOhms) and current: 4 to 20 mA (load 0/500 Ohms).

The sensing action employs ***magnetostriction*** — an elastic deformation under MF. Namely, electric pulse is sent through a ferromagnetic sensing tube. Interaction with permanent magnetic field of moving magnet causes in the tube the torsional strain that travels back. By comparing timing of the electrical pulse and the strain travel time, the position of a magnet can be determined. Thus, the sensor output is an analog signal sampled by the ***data acquisition system (DAC)***.

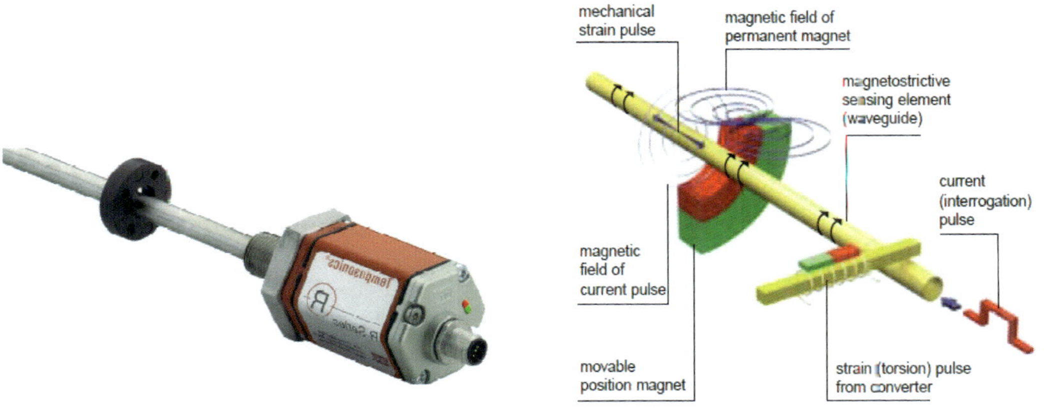

Fig. 4.1-1. Temposonic position sensor, could be screwed into piston bottom for piston position measurement. (Image: MTS https://www.mtssensors.com/portals/mts_sensors/documents/R-Series/Catalog_R-Series_551303_EN.pdf p.4).

4.1.1.1 *Sampling of continuous time (CT) sinusoid*

Sampling process. Consider $x(t) = \cos(2\pi f_1 t)$, $f_1 = 1/T_1$. Subscript 1 is used to denote the fundamental period and the corresponding first harmonic. We will occasionally drop this subscript for brevity. Recall that for the kth harmonic of fundamental frequency f_1 ($A_k = 1$, $\theta_k = 0$) we have:

$$x(t) = \cos(2\pi k f_1 t), \text{ where } f_k = kf_1 \Rightarrow T_k = \frac{1}{f_k} = \frac{1}{kf_1} = \frac{1}{k}T_1. \tag{4.1-1}$$

The first key element of the sampling process is decomposing the continuous independent variable (argument) t in $x(t)$ into a set of discrete points, $t = nt_s$, and continuous intervals, $t \neq nt_s$,

$$t \to nt_s = n(1/S_F) \tag{4.1-2}$$

where t_s is ***the sampling interval***, and $S_F = 1/t_s$ **Hz** is ***the sampling rate*** or ***sampling frequency*** — the inverse of t_s, and n is an integer used in mapping a continuum t into a countable set of points nt_s, as shown in Figure 4.1-2.

The second key sampling element, also shown in Figure 4.1-2, is retaining the values, $x(nt_s)$, of signal $x(t)$ at discrete argument values nt_s, and replacing signal $x(t)$ at all other values of t by zero. This yields a signal $x_s(t)$, which is defined for all t and referred to as the ***sampled-data signal***. The latter is shown in Figure 4.1-2 in the lower graph, assuming 1/6th of a cycle per sampling interval t_s.

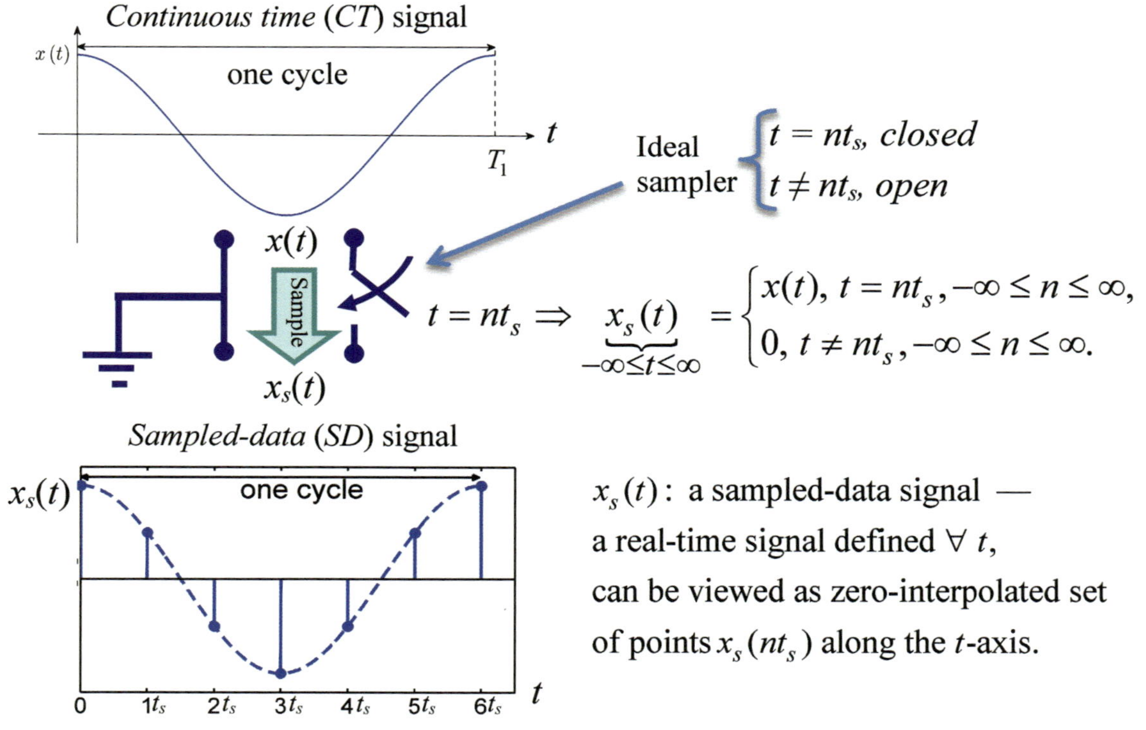

Fig. 4.1-2. Sampling continuous time signal $x(t)$ to obtain sampled-data signal $x_s(t)$ and discrete time signal $x_s[n]$.

Digital frequency. Assuming period T_1 in Figure 4.1-2 to be equal to 6 seconds, i.e. frequency f_1 = 1/6 Hz, or 1/6 cycle/second, let's set the sampling interval t_s of the sampler shown in this figure to 1 second. Then, the ***DT signal***, corresponding the sampled-data signal $x_s(t)$, is given by the function $x(nt_s) \equiv x_s[n]$ in Figure 4.1-3 — a non-real-time sampled version of $x(t)$ indexed only by n.

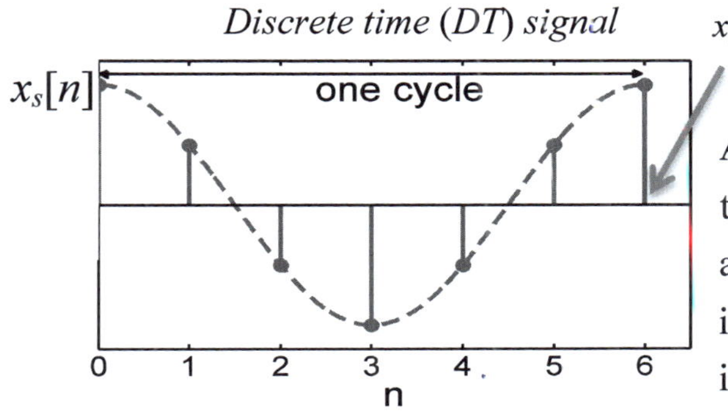

$x_s[n+6] = x_s[n] = x_s[n \pm kN_1], k = 0,1,2,\ldots$

$N_1 = 6 =$ ***period*** *(integer)*

A *DT* signal $x_s[n]$ is a non-real-time sampled version of $x(t)$ defined as an array of numbers $x_s(nt_s) \equiv x_s[n]$ indexed only by n and carrying no information about the value of t_s.

Fig. 4.1-3. Discrete time signal $x_s[n]$ — a non-real-time representation of the SD signal $x(nt_s)$ in Figure 4.1-2 indexed by integer n.

Since $T_1 = 6t_s$, we have $T_1/t_s = 6$, i.e. 6 samples per cycle, or 1/6th of a ***cycle per sample***, rather than per second. The latter, referred to as the ***digital frequency*** of $x_s[n]$, is a discrete time counterpart of continuous frequency cycle/sec. Also, since $x_s[n]$ repeats itself after 6 samples, the value $N_1 = T_1/t_s = 6$ represents ***period*** of $x_s[n]$, so that $x_s[n+6] = x_s[n] = x_s[n \pm kN_1]$, $k = 0, 1, 2\ldots$

Thus, for any integer $N_1 = T_1/t_s \equiv N$ sampled signal $x_s[n]$ is an N-periodic DT signal (DT sinusoid) with fundamental frequency given by

- $F = \frac{1}{N}\frac{\text{cycles}}{\text{sample}}$, F is ***digital frequency in cycles per sample***, or fraction of a cycle:

$$N \cdot t_s = T \Rightarrow N = \frac{T}{t_s},\ F = \frac{1}{N} = \frac{t_s}{T},\ \text{or}$$

- $\Omega = 2\pi F = \frac{2\pi}{N}\frac{\text{radians}}{\text{sample}}$, Ω is ***digital frequency in radians per sample,***

$$\because x_s[n] = \cos(2\pi \underbrace{f}_{\frac{1}{T}} \underbrace{nt_s}_{\text{Instead of } t}) = \cos\left(2\pi \frac{t_s}{\underbrace{T}_{T=Nt_s}} n\right) = \cos\left(2\pi \underbrace{\frac{1}{N}}_{F} n\right) = \cos\left(\underbrace{2\pi \frac{1}{N}}_{\Omega} n\right) = \underbrace{\cos(2\pi F n) = \cos(\Omega n)}_{\textit{Discrete time sinusoid}}.$$

The function $\cos(2\pi F n)$ is thus seen to be the discrete time sinusoid with period N.

We also note that $F = \frac{t_s}{T} = \frac{1/T}{1/t_s} = \boxed{\frac{f}{S_F} = \frac{\text{CT frequency}}{\text{sampling frequency}}}.$

4.1.1.2 *Reconstruction of CT periodic signals from DT counterparts*

Given sampling interval t_s, the reconstruction of a CT sinusoid corresponding to a DT sinusoid $\cos(2\pi Fn)$ is carried out through the inverse of the sampling map (4.1-2) given by a map

$$n \to t/t_s = t/(1/S_F) = S_F t, \tag{4.1-3}$$

yielding

$$\cos(2\pi Fn) \underset{n \to S_F t}{\rightarrow} \cos(2\pi(f/S_F)S_F t) = \cos(2\pi f t). \tag{4.1-4}$$

4.1.1.3 *Aliasing*

Example 4.1-1: Let's sample $x(t) = \cos\left(2\pi f\left(t - \frac{T}{4}\right)\right) = \cos\left(2\pi f t - 2\pi \frac{1}{T}\frac{T}{4}\right) = \cos\left(2\pi f t - \frac{\pi}{2}\right)$ by taking

$$t_s = \frac{1}{2}T \Rightarrow S_F = \frac{1}{t_s} = \frac{1}{(1/2)T} = 2\frac{1}{T} = 2f, \ \therefore F = \frac{f}{S_F} = \frac{f}{2f} = \frac{1}{2}.$$

This results in the signals shown in Figure 4.1-4 below:

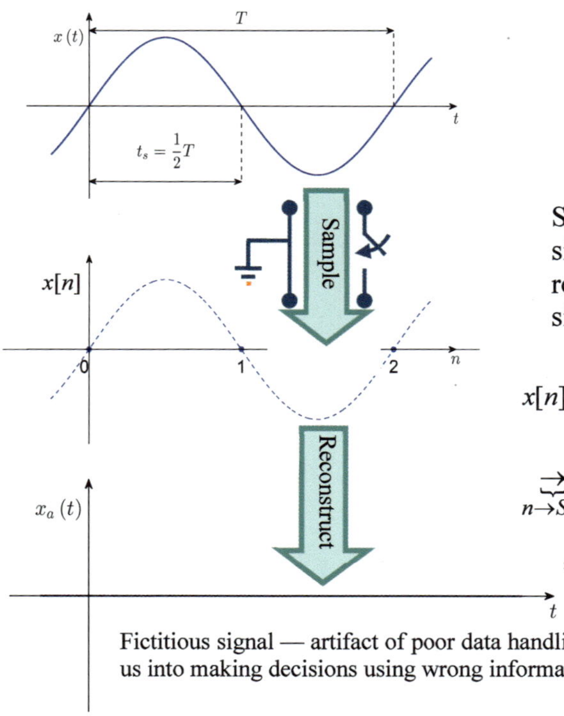

Since we have constant DT signal, for CT signal reconstruction we pick DT sinusoid with zero frequency:

$$x[n] = \cos\left(2\pi Fn - \frac{\pi}{2}\right) = 0,\ F = 0,\ S_F = 2/T,$$

$$\underset{n \to S_F t}{\rightarrow} \cos\left(2\pi(0/S_F)(S_F)t - \frac{\pi}{2}\right)$$

$$= \cos\left(2\pi 0 t - \frac{\pi}{2}\right) = 0 \text{ for all } t$$

Fig. 4.1-4. The generation of alias $x_a(t)$ of the original signal $x(t)$.

Comparing the original signal $x(t)$ shown on top in Figure 4.1-4 with the sampled signal shown in the middle and seen to contain only the zero values of $x(t)$, we discover that we have "lost" $x(t)$! If, not realizing this, we reconstruct $x(t)$ from $x_s[n]$, it would look like the bottom graph, and we would think that this is how the original signal looked.

This new $x_a(t)$ is called an ***alias*** of $x(t)$, and $x(t)$ is said to be ***aliased.*** The change in the reconstructed signal compared to the original is called ***aliasing*** (acquisition of a mistaken identity).

4.1.1.4 *Periodicity of DT signal*

If the frequency of a continuous time sinusoid signal is f Hz, then the period of this signal is clearly $T = 1/f$ seconds. Then, analogy suggests that if the frequency of a discrete time sinusoid signal is F cycles/sample, then the period of this signal is $N = 1/F$ samples. ***This is not true in general!***
Example 4.1-2: Consider $x[n] = \cos(2\pi \cdot 0.8 \cdot n) \Rightarrow F = 0.8 \Rightarrow 1/F = 1.25$.

The latter value is not even an integer, so cannot be N. Is this $x[n]$ periodic? Figure 4.1-4 shows that this signal looks periodic with period $N = 5$.

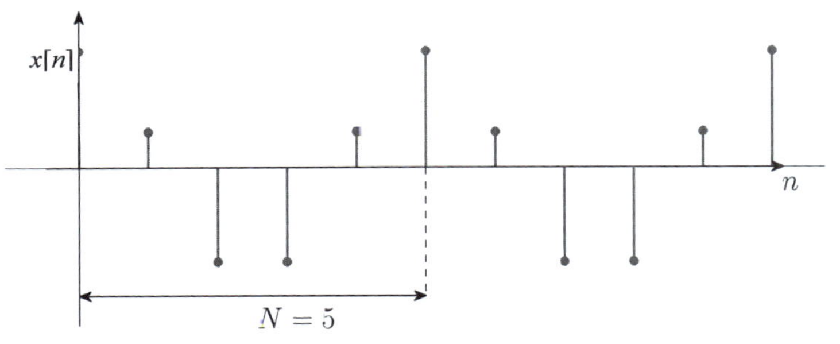

Fig. 4.1-5. $x[n]$ for Example 4.1-2.

Let's check analytically if this is true:

$$x[n] = \cos\left(2\pi \cdot 0.8 \cdot (n+5)\right) = \cos\left(2\pi \cdot 0.8 \cdot n + 2\pi \cdot 0.8 \cdot 5\right)$$
$$= \cos\left(2\pi \cdot 0.8 \cdot n + 2\pi \cdot 4)\right) = \cos\left(2\pi \cdot 0.8 \cdot n\right).$$

So $x[n]$ indeed has period $N = 5$ and we also note that $FN = 0.8{\cdot}5 = 4$.

Let's generalize this example: by the definition of the period, N is an integer such that, for any n, $x[n + N] = x[n]$. For which F is this true? We have

$$\cos\left(2\pi F\left(n+N\right)\right) = \cos\left(2\pi Fn\right) \Rightarrow \cos\left(2\pi Fn + 2\pi FN\right)$$
$$= \cos\left(2\pi Fn\right) \Rightarrow \cos\left(\alpha + 2\pi FN\right) = \cos\left(\alpha\right). \tag{4.1-5}$$

The latter holds if and only if $FN = k$ (some integer).

Therefore, for $x[n]$ to be periodic F must be a ratio of two integers: $F = k/N$

$$\Rightarrow F = \frac{k}{N} = \frac{f_1}{S_F} = \frac{1/T}{1/t_s} = \frac{t_s}{T} = \frac{kt'_s}{Nt'_s}. \tag{4.1-6}$$

Example 4.1-3: Consider $x[n] = \cos(2 \cdot n)$:

$$x[n] = \cos(2 \cdot n) = \cos(2\pi \cdot 1/\pi \cdot n) \Rightarrow F = \frac{1}{\pi} \neq \text{ rational.}$$

Therefore this signal is not periodic. Thus, given CT frequency f, both k and N depend on S_F.

If F is not the ratio of two integers (i.e. rational), then the signal has no DT period, but the envelope is still periodic in CT, as shown in Figure 4.1-5. In general, period N is the integer in the denominator of F after the common factors in numerator and denominator of F have been cancelled, as shown in (4.1-6).

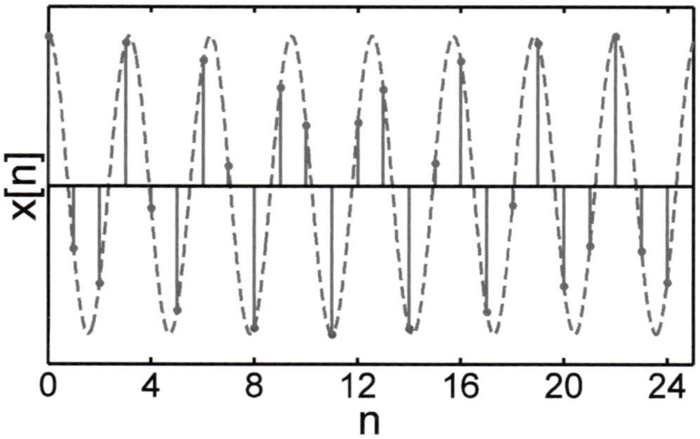

Fig. 4.1-6. DT non-sinusoid plot — the signal never exactly repeats itself.

4.2 Frequency of DT Sinusoid, Principal Range, Aliasing as Folding, Sampling Theorem

4.2.1 *Frequency of DT sinusoid*

Let us now get back to Example 4.1-2, where $N = 5$ and $F = 0.8$, and take a look at how many cycles this F gives per 5 samples:

$$F = 0.8 = \frac{8}{10} = \frac{4}{5}.$$

Thus, 5 samples (period) of DT sinusoid were sampling four cycles of CT sinusoid. Enveloping four cycles of CT sinusoid over one cycle of this DT sinusoid gives the following plot:

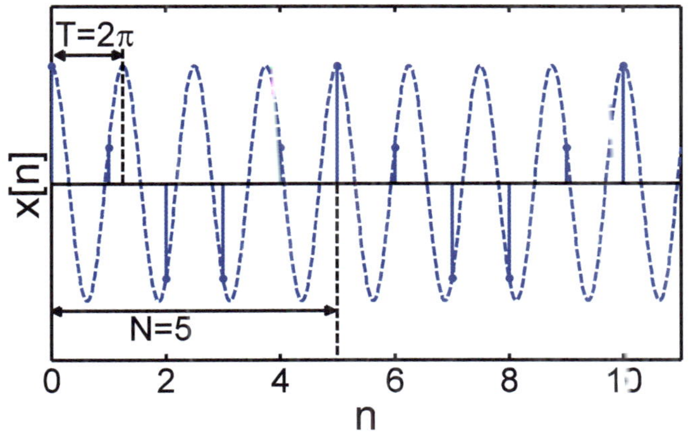

Fig. 4.2-1. The actual CT sinusoid plot.

But taking a closer look at the graph of $x[n]$, it is seen that a single cycle of CT sinusoid also exactly envelopes $x[n]$ over period 5 of the latter, as shown on the graph below:

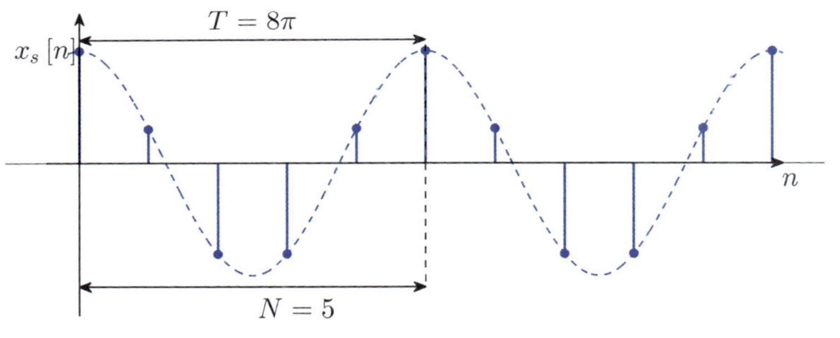

Fig. 4.2-2. The aliased CT sinusoid plot.

This means that *the same* DT signal $x[n]$ with period $N = 5$ can have digital frequency $F = 1/5$, as well as another digital frequency $F = 4/5$! What's going on?

To answer this question, consider the following equation:

$$x[n] = \cos\left(2\pi \underbrace{\frac{k}{N}}_{F} n\right) = \cos(2\pi F n), \text{ where } F = \frac{k}{N}.$$

Change F by adding to it an integer m: $F \to F + m$:

$$\cos(2\pi(F+m)n) = \cos(2\pi Fn + 2\pi mn) = \cos(2\pi Fn).$$

That is, integer change in F does not change digital frequency. What does?

Change of k in $F = k/N$: take $F_1 = 1/N$ as the fundamental frequency, then $F_k = kF_1 = k/N$, $k = 0, 1, 2, \ldots, N\text{-}1$, and $\cos(2\pi F_k n)$ is seen to be the kth harmonic with respect to $\cos(2\pi F_1 n)$.

4.2.1.1 *DT harmonics*

Let's now note that for any integer m:

$$\cos(2\pi F_{mN+k} n) = \cos\left(2\pi \frac{mN+k}{N} n\right) = \cos\left[\left(2\pi m + 2\pi \frac{k}{N}\right) n\right] = \cos\left(2\pi \frac{k}{N} n\right). \quad (4.2\text{-}1)$$

So, it looks like DT sinusoid of period N gives rise to only N harmonics: $k = 0, 1, \ldots, N\text{-}1$, compared to a CT sinusoid which gives rise to an infinite number of harmonics. Note that each DT harmonic with $F_k = k/N$ lies in the [0, 1] range. The actual periods corresponding to each F_k vary from harmonic to harmonic.

Example 4.2-1: for $N = 10$: $F_1 = \frac{1}{10}$: Fundamental frequency

$$k = 0,\quad x_0[n] = A_0 \cos\left(2\pi \cdot \frac{0}{10} \cdot n + \theta_0\right);\quad k = 1,\quad x_1[n] = A_1 \cos\left(2\pi \cdot \frac{1}{10} \cdot n + \theta_1\right);$$

$$k = 2,\quad x_2[n] = A_2 \cos\left(2\pi \cdot \frac{2}{10} \cdot n + \theta_2\right);\quad k = 3,\quad x_3[n] = A_3 \cos\left(2\pi \cdot \frac{3}{10} \cdot n + \theta_3\right);$$

Period = 5 Period = 10

…

$$k = 9,\quad x_9[n] = A_9 \cos\left(2\pi \cdot \frac{9}{10} \cdot n + \theta_9\right).$$

Period = 10

4.2.2 *Digital frequency folding*

Digital frequency bounds. We also note that in Example 4.2-1 harmonic for $k = 10$, i.e. for $F_{10} = 1$,

$$x_{10}[n] = A_{10} \cos\left(2\pi \cdot \frac{10}{10} \cdot n + \theta_{10}\right) = A_{10} \cos(2\pi n + \theta_{10}) = A_{10} \cos(\theta_{10}) = const$$

reduces to that for $k = 0$, i.e. for $F_0 = 0$:

$$A_{10} \cos(\theta_{10}) = A_{10} \cos(2\pi \cdot 0 \cdot n + \theta_{10}) \equiv A_0 \cos(2\pi \cdot 0 \cdot n + \theta_0) = A_0 \cos(\theta_0).$$

Further on, $F_5 = 1/2$ is also similar to $F_0 = 0$, yielding constant signal

$$x_5[n] = A_5 \cos\left(2\pi \cdot \frac{5}{10} \cdot n + \theta_5\right) = A_5 \cos\left(2\pi \cdot \frac{1}{2} \cdot n + \theta_5\right) \underset{\theta_5 = -\pi/2}{=} A_5 \sin(\pi n) = 0.$$

Clearly, the values $F = 0$, $F = 1/2$, and $F = 1$ must represent some frequency bounds.

Folding and principal range. Are the rest of the N-1 harmonics distinct? Getting back to Example 4.2-1 with $F = F_4 = 4/5$ we obtain:

$$\cos\left(2\pi\cdot\underbrace{\frac{4}{5}}_{F_4}\cdot n\right)=\cos\left(2\pi\cdot\left(\overbrace{-\frac{1}{5}}^{-F_1}+\overbrace{1}^{\text{Integer}}\right)\cdot n\right)=\cos\left(2\pi\cdot\left(-\frac{1}{5}\right)\cdot n+2\pi\cdot 1\cdot n\right)=\cos\left(2\pi\cdot\underbrace{\left(-\frac{1}{5}\right)}_{}\cdot n\right).$$

$F_4 \longrightarrow$ Frequency F_4 is folded to $-F_1$ $\longrightarrow$

By adding and subtracting integers we can always transform the frequency F of a DT sinusoid into an equivalent frequency within [−½, ½] — the **principal range.**

Further on, by phase shift, we can map F to the range [0, ½]. Adding phase θ_4 in the example to see this clearer yields:

$$\cos\left(2\pi\cdot\left(-\frac{1}{5}\right)\cdot n+\theta_4\right)=\cos\left(-2\pi\cdot\frac{1}{5}\cdot n+\theta_4\right)=\cos\left(-\left(2\pi\cdot\frac{1}{5}\cdot n-\theta_4\right)\right)=\cos\left(2\pi\cdot\frac{1}{5}\cdot n-\theta_4\right)$$

$\cos(\phi)=\cos(-\phi)$, i.e. cosine is even

Thus, $F_n<\frac{1}{2}\Rightarrow\frac{k}{N}<\frac{1}{2}\Rightarrow k<\frac{N}{2}$.

From above equations we can see that frequency F_4 has turned into frequency F_1. This transition is referred to as ***folding*** due to its symmetry around 1/2 if viewed graphically.

Denoting by *int*[N/2] the integer part of the value in brackets, folding implies that the DT sinusoid of even period N has *int*[N/2] distinct harmonics, k = 0, 1, …, *int*[N/2]-1, and of odd period N has *int*[N/2] + 1 distinct harmonics, k = 0, 1, …, *int*[N/2]. This is seen in the example shown below.

Example 4.2-2: $N = 10 \Rightarrow \text{int}[N/2] = 5 \Rightarrow F_0 = 0, F_1 = 1/10, F_2 = 2/10, F_3 = 3/10, F_4 = 4/10$, $N = 9 \Rightarrow \text{int}[N/2] + 1 = 5 \Rightarrow F_0 = 0, F_1 = 1/9, F_2 = 2/9, F_3 = 3/9, F_4 = 4/9\ (< 1/2)$.

4.2.2.1 *Aliasing as folding*

Let us now explicitly relate folding to aliasing. Consider a sampled CT sinusoid:

$$\cos(2\pi f_1 t+\theta_1)\xrightarrow{t\to nt_s,\,\theta\equiv\theta_k}\cos\left(2\pi\overbrace{\frac{f_1}{s_F}}^{F_k}n+\theta_k\right). \tag{4.2-2}$$

Since the frequency, F_k, (and harmonic number k) of the sampled signal (4.2-2) is determined by the sampling rate S_F as $F_k = f_1/S_F$, if $|F_k| > ½$, the DT signal folds to an identical one with frequency $0 \le F < ½$. As we have seen, only $0 \le F_k < ½$ ensures that sampled signal accurately represents the original signal:

$$F_k=\frac{f_1}{S_F}<\frac{1}{2}\Rightarrow S_F>2f_1. \tag{4.2-3}$$

The frequency $F = ½$, or $f = (½)S_F$ is called the ***folding frequency***. If $F_k > ½$, it is folded to F_a in the range [0, ½], producing the CT alias $x_a(t)$ of $x(t)$ with frequency $f_a = F_a S_F < f_1$.

Example 4.2-3: Assume that the original analog signal $x(t)$ has frequency f_1. Figure 4.2-3 shows sampling of this signal with S_F that yields five samples over four cycles.

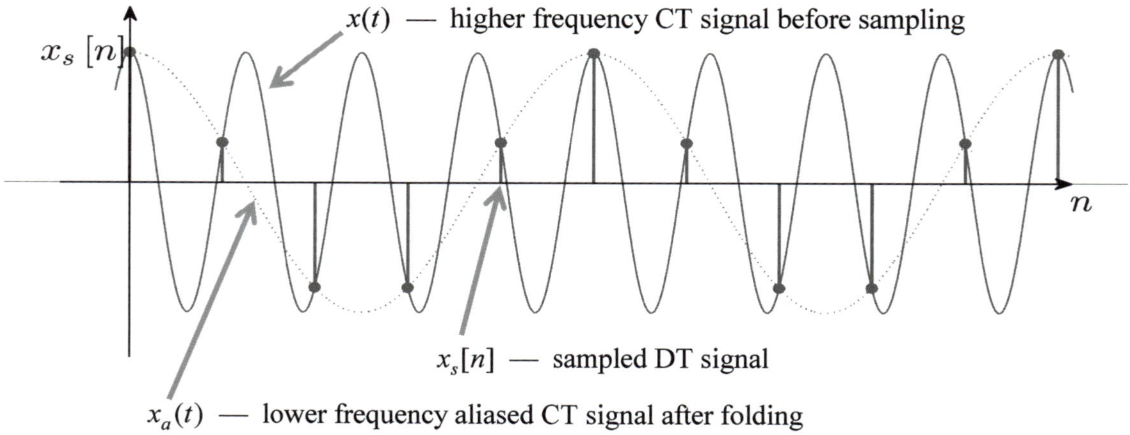

Fig. 4.2-3. Original sinusoid signal $x(t)$, DT signal $x_s[n]$ produced by sampling rate 5 samples per 4 cycle, and CT signal $x_a(t)$ reconstructed from DT sampling signal.

This sampling produces the sampled signal with digital frequency $F_4 = 4/5 > 1/2$, so that $f_1 = F_4 S_F$.

After integer subtraction and phase shift, F_4 folds into $F_1 = |4/5-1| = 1/5 \equiv F$a. Reconstruction of the analog signal using the latter digital frequency and the original sampling frequency yields $f_a = F_a S_F$, so that

$$\frac{f_a}{f_1} = \frac{F_a S_F}{F_4 S_F} = \frac{F_1 S_F}{F_4 S_F} = \frac{(1/5)S_F}{(4/5)S_F} = \frac{1}{4}.$$

Thus, the alias $x_a(t)$ has 4 times lower frequency than $x(t)$.

4.2.2.2 *Sampling theorem*

We are now ready to state the key result of sampling.

Definition: If the highest frequency in the signal is f_B, the signal is said to be ***bandlimited to*** f_B.

Sampling Theorem (Nyquist-Shannon-Kotelnikov): To avoid aliasing, a bandlimited to f_b signal should be sampled at a rate $S_F > 2f_B$. The reconstructed signal is then given by,

$$x(t) = \sum_{n=-\infty}^{\infty} x[n]\operatorname{sinc}\left(\frac{t - nt_s}{t_s}\right). \tag{4.2-4}$$

The critical sampling rate $S_N = 2f_B$ is called the ***Nyquist rate***.

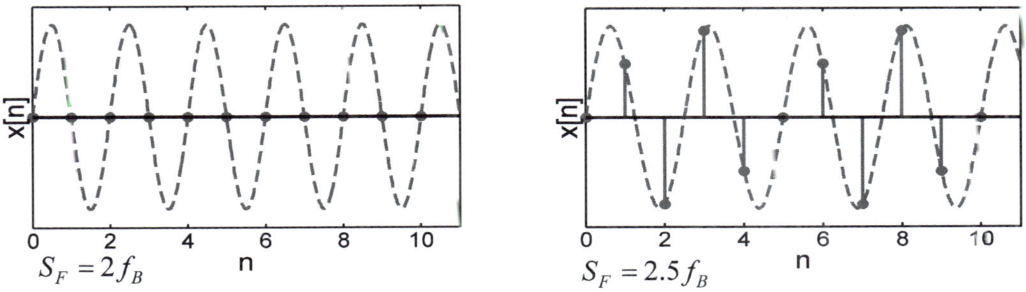

Fig. 4.2-4. Left graph — the original analog signal (dashed curve) and the sampled signal produced by sampling at $2f_B$ making it unsuitable for the original signal reconstruction. Right graph — sampling at $2.5f_B$ admits reconstruction.

4.3 Sampling Basics: Analog-to-Digital (A/D) and Digital-to-Analog (D/A) Conversion. Harmonic Signals: Review of Complex Numbers, Complex Variables, and Complex Functions

4.3.1 *Signal conversion*

4.3.1.1 *Analog-to-digital (A/D) conversion: two approaches — PCM and PDM*

A/D conversion is modulation of a digital carrier signal through analog signals characteristics. To perform A/D conversion we first carry out ***sampling***: i) select the sampling frequency S_F as dictated by application: rule of thumb is $S_F = 10$ (highest frequency of interest), and ii) bandlimit the analog signal to $f_B < S_F/2$ via low-pass (***antialiasing***) filter to reduce aliasing. Next, we encode the sampled signal in terms of either its 1) magnitude, referred to as PCM (Pulse Code Modulation) depicted in Figure 4.3-1, or 2) slope, referred to as PDM (Pulse Density Modulation), shown Figure 4.3-2.

PCM employs magnitude ***quantization***: for each sample, selection of the binary word with the closest to sample value admitted by an L-level quantizer, L = power of 2, i.e. $L = 2^B$, where integer B is ***bit word size (depth)***. For example, a 12-bit quantizer has 2^{12}, or 4096 levels. Figure 4.3-1 shows the sampled signal with sampling interval $t_s = 1/S_F$ and digital signal with resolution = (max-min)/L.

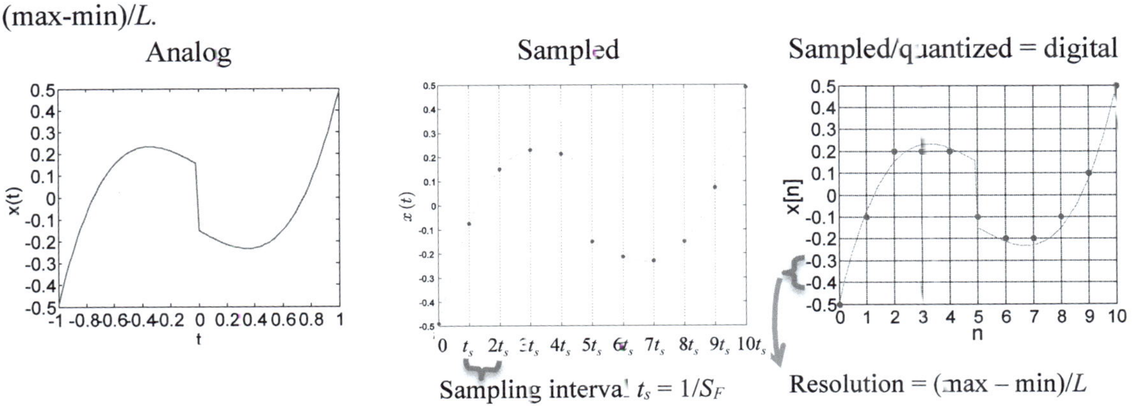

Fig. 4.3-1. Analog to digital conversion in Pulse Code Modulation.

Example 4.3-1, PCM audio: The ***dynamic range*** in audio applications is a signal-to-noise ratio measured as *the ratio in dB of the loudest undistorted signal to the quietest audible signal*. For PCM it is specified through the bit depth, denoted here by B. The standard CD recording has 44.1/16 format, i.e. $S_F = 44.1$ kHz (frequency range under 22 kHz) and $B = 16$, yielding $L = 2^{16}$ levels and theoretical dynamic range of 96 dB. The vinyl record has frequency range from 7 Hz to over 50 kHz and dynamic range up to 70 dB (theoretical limit 110 dB), so CDs have been a far cry from the all-analog vinyl records of the sixties. However, present digitally mastered vinyl is quite poor.

Modern high definition ("Hi-Res") PCM recordings have addressed the CD format drawbacks by raising bit depth to 24 (dynamic range of 144 dB) and S_F up to 352 kHz (8 times 44.1 Hz). This has yielded PCM formats of 96/24 (10–20 MB/song), 192/24 (50–150 MB/song), and 352/24 (DXD — digital extreme definition, 200–300 MB/song, approaches good all-analog vinyl record quality). To hear the sound difference, compare 44.1/16 and 96/24 with 192/24 in the following tracks: https://www.hdtracks.com/#/album/5df14cc30bee25c09bc16893 The Allman Brothers Band — track 4: guitar sound at the clip's end — 96/24 — slightly smothered, 192/24 — really sings due to a *higher sampling rate* better capturing overtones.
https://www.hdtracks.com/#/album/5de33ce33f6dd7ed9300b1c9 Wilson Pickett — track 2: guitar chords during singing — 44.1/16 — barely heard, 192/24 — chords sound crisp mostly due to a *higher dynamic range* of 24 vs 16 bits depth in magnitude encoding.

Example 4.3-2, PDM audio: PDM encodes the analog signal's rate of change into the relative density of the pulses (1's) — a faster signal value change generates more 1's per time interval. Therefore, the result is not the sequence of binary B-length words at the PCM sampling rates, but the stream of clocked 0's and 1's, referred to as the Direct Stream Digital (DSD) sequence at much higher sampling rates. The encoding technique used is ***delta-sigma modulation***, considered in detailed in Chapter 6. The DSD formats are: i) single rate, or DSD64 (2.8 mHz — 64 times 44.1 kHz), roughly equivalent to 192/24 PCM, 2) double rate DSD, or DSD128 (5.6 mHz — 128 times 44.1 kHz), like DXD — 352/24 PCM, but better, 3) quad DSD, or DSD256 (11.2 mHz — 256 times 44.1 kHz), like 764/32 PCM — double DXD, not yet available, and finally DSD512 — the top! DSD1024 is used for digital masters. Compare now 192/24 PCM with 11.2896 DSD in Jazz At The Pawnshop: Late Night — track 5 at https://www.prostudiomasters.com/search?cs=1&q=Jazz+at+the+pawnshop#quickview/album/5713.
We thought 192/24 PCM is pretty good, but 11.2 DSD is way better!

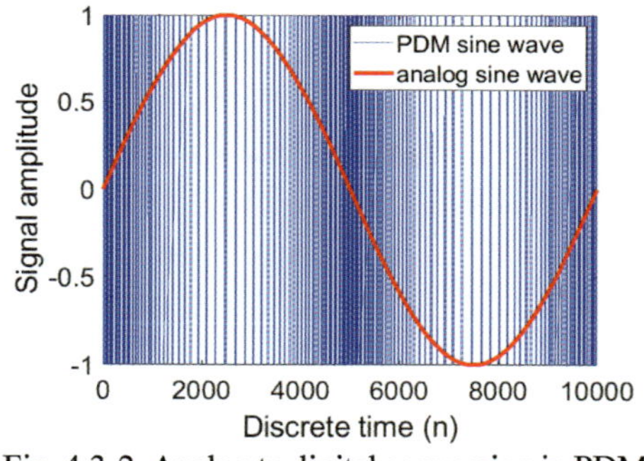

Fig. 4.3-2. Analog to digital conversion in PDM. Dark lines are 1's, white lines are 0's.

4.3.1.2 *Digital-to-Analog (D/A) conversion*

This conversion is also known as signal reconstruction. It requires interpolating between sampled data points, as shown in Figure 4.3-3. The more samples used in a single fit, the better.

For example: step interpolation (zero-order hold), linear interpolation (first-order hold), ideal interpolation: via sinc function.

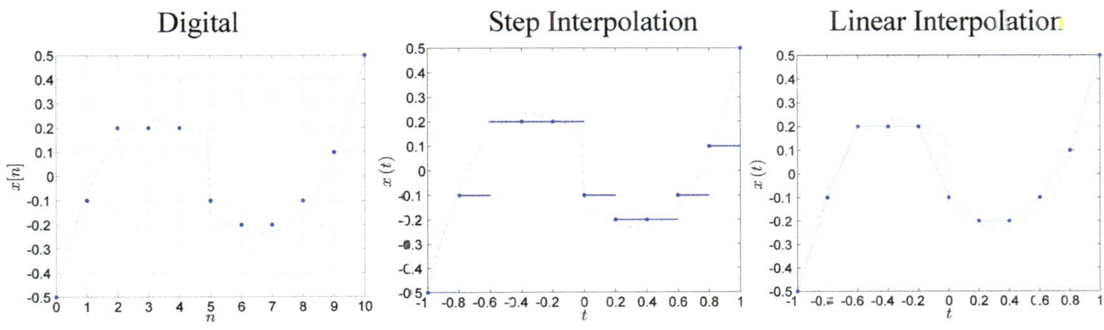

Fig. 4.3-3. Digital to analog conversion.

4.3.2 *Harmonic signals*

Motivation:

- ***Diagnostics:*** determining signal composition and on its basis developing understanding of system behavior.
- ***System identification:*** finding system description from input and output signals, and on its basis developing analytical model and characterizing system uncertainty.
- ***Control:*** designing a control system that performs complex oscillatory motion through linear actuation.
- ***Communication*** — understanding the concept of bandwidth in data transmission over the Internet and in wireless command and control.

Fig. 4.3-4: Application of harmonic signals.

4.3.3 *Review of complex numbers, variables, and functions*

4.3.3.1 *Fundamentals of complex variable*

Recall Taylor series for e^x, $sin(x)$, and $cos(x)$:

$$e^x = 1 + x + \frac{1}{2}x^2 + \frac{1}{3!}x^3 + \ldots,$$

$$\cos(x) = 1 - \frac{1}{2}x^2 + \frac{1}{4!}x^4 - \frac{1}{6!}x^6 \ldots, \quad \sin(x) = x - \frac{1}{3!}x^3 + \frac{1}{5!}x^5 - \frac{1}{7!}x^7 + \ldots .$$

Therefore, for any real number, θ, for $x = j\theta$, $j = \sqrt{-1}$ we have **Euler's Formula**:

$$e^{j\theta} = 1 + j\theta + \frac{1}{2!}(j\theta)^2 + \frac{1}{3!}(j\theta)^3 + \ldots = 1 + j\theta - \frac{1}{2!}\theta^2 - j\frac{1}{3!}\theta^3 + \ldots$$

$$= \underbrace{\left(1 - \frac{1}{2!}\theta^2 + \frac{1}{4!}\theta^4 - \ldots\right)}_{\cos(\theta)} + j\underbrace{\left(\theta - \frac{1}{3!}\theta^3 + \frac{1}{5!}\theta^5 - \ldots\right)}_{\sin(\theta)} = \cos\theta + j\sin\theta.$$

Homework Tip: Euler's formula can be used to derive any of the useful trigonometric identities. In homeworks, projects, or exams, if it is difficult to figure out how to integrate a trigonometric function, like $\sin(t)\cos(t)$ and one doesn't remember the double-angle formulas, writing it in terms of e helps in working it out much easier.

Using Euler's formula, any complex number, z, can be written in "**polar form**" or "**exponential form**" in terms of its **magnitude**, $|z|$, and **phase**, θ:

Rectangular form: $z = x + jy$ polar form: $z = |z|\angle z$ exponential form: $z = |z|e^{j\theta}$ where, $x = \operatorname{Re} z = |z|\cos\theta$, $y = \operatorname{Im} z = |z|\sin\theta$, $|z| = \left|\sqrt{x^2 + y^2}\right|$, and $\angle z \equiv \theta = \tan^{-1}\left(\frac{y}{x}\right)$.

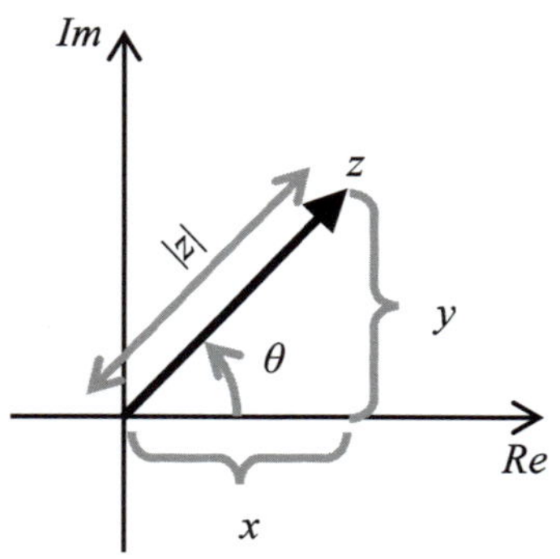

Fig. 4.3-5. Complex number.

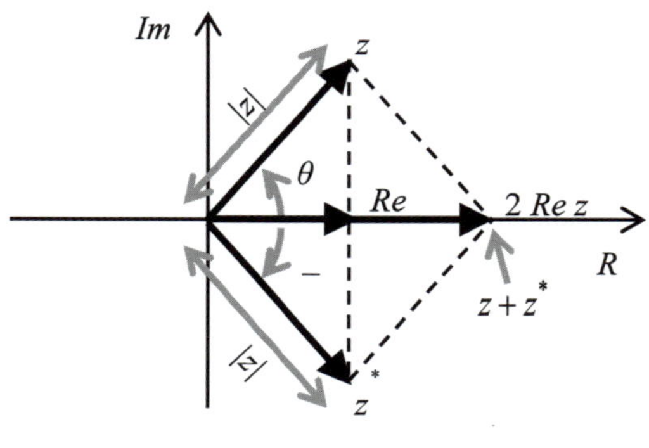

Fig. 4.3-6. Sum of complex conjugate numbers give real number.

The **complex conjugate** of z is:

$$z^* \equiv x - jy = |z|(\cos\theta - j\sin\theta) = |z|(\cos(-\theta) + j\sin(-\theta)) \Rightarrow \boxed{z^* = |z|e^{-j\theta}.}$$

From Figure 4.3-6 we have the following relations:

$$z+z^* = (x+jy)+(x-jy) = 2x \Rightarrow \boxed{\operatorname{Re} z = \frac{z+z^*}{2}},$$

$$z-z^* = (x+jy)-(x-jy) = j2y \Rightarrow \boxed{\operatorname{Im} z = \frac{z-z^*}{j2}}.$$

It is much easier to multiply complex numbers when they are in exponential form:

$$z_1 \cdot z_2 = |z_1| e^{j\theta_1} \cdot |z_2| e^{j\theta_2} = |z_1| \cdot |z_2| e^{j(\theta_1+\theta_2)}.$$

In particular

$$z \cdot z^* = |z| e^{j\theta} \cdot |z| e^{-j\theta} = |z||z| e^{j(\theta-\theta)} = |z|^2 \quad \Rightarrow \boxed{|z| = \left|\sqrt{z \cdot z^*}\right|}.$$

Unit circle: A useful analytical/graphical tool is the unit circle in the complex plane:

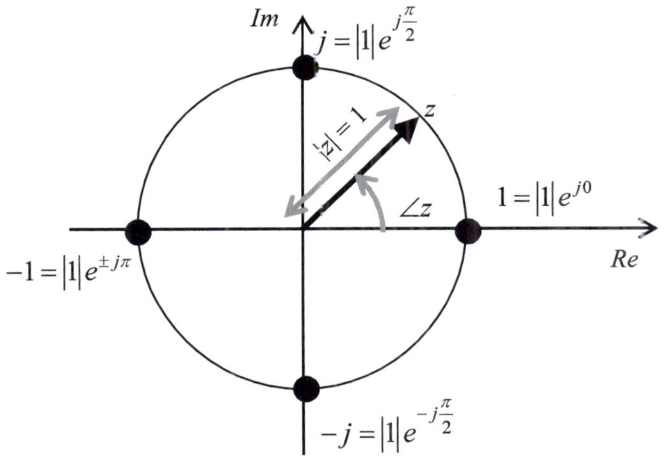

Fig. 4.3-7. Unit circle in a complex plane.

For example, multiplying $z_1 \cdot j = |z| e^{j\theta_1} \cdot e^{j\pi/2} = |z| e^{j(\theta_1+\pi/2)}$ adds positive value $\pi/2$ to θ, i.e. rotates z counterclockwise by $\pi/2$.

4.3.3.2 *Functions of complex variables*

If x and y are varying, z is called a complex variable.

Example 4.3-3: Laplace transformation is in terms of $s=\sigma+j\omega$, where σ and ω (both real) may vary. We can also have functions of complex variables, or complex functions

$$F(z)=\operatorname{Re}F+\operatorname{Im}F,\ |F(z)|=\left|\sqrt{(\operatorname{Re}F)^2+(\operatorname{Im}F)^2}\right|,\ \angle F=\tan^{-1}\left(\frac{\operatorname{Im}F}{\operatorname{Re}F}\right).$$

Example 4.3-4: Transfer functions

$$H(s)\to H(j\omega)=\operatorname{Re}(H(j\omega))+j\operatorname{Im}(H(j\omega))\Rightarrow$$

$$|H(j\omega)|=\left|\sqrt{(\operatorname{Re}(H(j\omega)))^2+(\operatorname{Im}(H(j\omega)))^2}\right|,\qquad \angle H(j\omega)=\tan^{-1}\left(\frac{\operatorname{Im}H(j\omega)}{\operatorname{Re}H(j\omega)}\right),$$

where $H(j\omega)$ is the frequency response, $|H(j\omega)|$ is the magnitude response, and $\angle H(j\omega)$ is phase response.

Derivatives and integrals of complex exponential: Complex exponential is the workhorse of modern signal processing due to the fact that differentiation and integration do not change the function $e^{j\omega t}$, they just shift its phase, or time, and rescale its magnitude:

$$\frac{d}{dt}e^{j\omega_1 t}=j\omega_1 e^{j\omega_1 t}=\omega_1\underbrace{e^{j\pi/2}}_{j}e^{j\omega_1 t}=\omega_1 e^{j(\underbrace{\omega_1 t+\pi/2}_{\theta_1})}=\omega_1 e^{j\omega_1[t\underbrace{-(-\pi/2\omega_1)}_{t_1}]},$$

Phase lead by $\pi/2$, or time advance by $-\pi/2\omega_1$

$$\int e^{j\omega_1 t}dt=\frac{1}{j\omega_1}e^{j\omega_1 t}=-\frac{1}{\omega_1}je^{j\omega_1 t}=\frac{1}{\omega_1}\underbrace{e^{j(-\pi/2)}}_{-j}e^{j\omega_1 t}=\frac{1}{\omega_1}e^{j(\underbrace{\omega_1 t-\pi/2}_{\theta_1})}=\frac{1}{\omega_1}e^{j\omega_1[t\underbrace{-\pi/2\omega_1}_{t_1}]}.$$

Phase lag by $-\pi/2$, or time delay by $\pi/2\omega_1$

Therefore, $e^{j\omega t}$ (or, more generally, e^{jst}, $s=\sigma+j\omega$) can be thought of as "intimately related" to these operations. Differentiating or integrating it several times yields iterative application of the same property. Recalling eigenvectors and eigenvalues from linear algebra, we can similarly describe $j\omega_1$ as an ***eigenvalue*** of the differential operator *d*/*dt*, and $e^{j\omega_1 t}$ as the corresponding ***eigenfunction***.

Function $e^{j\omega_1 t}$ is, then, also seen to be intimately related to LCCODEs, since the latter are formed by sums of derivatives, and, hence, to linear time-invariant filters, since we can think of linear filters as linear differential equations.

These eigenfunction properties can be used for describing the filtering action as follows: 1) decompose the input into exponential functions, 2) pass each individual component through the filter, 3) reassemble rescaled and phase-shifted components at the output and obtain filtering action.

4.4 Complex Harmonic Signals. Discrete Double-Sided Spectrum of CT and DT Periodic Signals. First Look at the Exponential Fourier Series

4.4.1 *From real harmonics to complex ones. Discrete double-sided spectrum of periodic signals*

So far, we've taken a detailed look at harmonic signals like $A\cos(2\pi kft + \theta)$. This is a real function of t (i.e. it gives a real number for each fixed t). Now, we want to put this function in a form that lends itself to: 1) easy analysis of filtering action and 2) easy numerical processing for extracting the key signal features.

How can this be achieved? We use the fact that a real part of a complex number = half (complex number + conjugate of the same complex number), or $Re(z) = (z + z^*)/2$. But which complex number, z, to use? Complex exponential $e^{j\omega t}$. Why?

As indicated earlier, complex exponentials are processed by a filter in a completely transparent way, since taking the derivative of a complex exponential $e^{j\omega_1 t}$ just multiplies $e^{j\omega_1 t}$ by $\omega_1 j = \omega_1 e^{j(\pi/2)}$, which

- advances $e^{j\omega_1 t}$ in time by $-\pi/2\omega_1$, or introduces phase lead $\pi/2$, and
- rescales magnitude of $e^{j\omega_1 t}$ by ω_1.

In the rest of the book, complex exponentials will be used as the basis for developing a variety of signal processing techniques generally referred to as ***Fourier, or harmonic, analysis***.

4.4.1.1 *The k'th complex CT harmonic. Complex representation of the k'th CT real harmonic*

CT harmonic signal $x_k(t) = A_k e^{j(2\pi k f_1 t + \theta_k)}$, where $f_k = kf_1$, will be referred to as ***the kth complex harmonic (signal)*** with respect to fundamental frequency f_1. Henceforth, we will use subscript p for denoting CT or DT real periodic signals to contrast them, as necessary, against complex and/or nonperiodic signals. Then, CT time real harmonic representation in terms of the CT complex harmonic can be given as

$$x_{kp}(t) = \operatorname{Re} x_k(t) = A_k\cos(2\pi f_k t + \theta_k) = \frac{1}{2}x_k(t) + \frac{1}{2}x_k^* = \frac{A_k}{2}e^{j(2\pi f_k t + \theta_k)} + \frac{A_k}{2}e^{-j(2\pi f_k t + \theta_k)}$$

$$= \frac{A_k}{2}e^{j(2\pi f_k t + \theta_k)} + \frac{A_k}{2}e^{j(2\pi(-f_k)t - \theta_k)} = \frac{A_k}{2}e^{j\theta_k}e^{j2\pi f_k t} + \frac{A_{-k}}{2}e^{j(-\theta_k)}e^{j2\pi(-f_k)t}.$$

CT k'th real harmonic (pointing to $x_{kp}(t)$)

$$\frac{A_{-k}}{2} \triangleq \frac{A_k}{2} \qquad \theta_{-k} \triangleq -\theta_k \qquad f_{-k} \triangleq -f_k = (-k)f_1$$

<u>Notation:</u> we will use subscript p as needed *to emphasize that signal is real and periodic* or to *designate for brevity the variables related to piston motion*. When the distinction between the two is unclear from the context, subscript *pst* will be used for the latter.

As $t\uparrow$ (t increases), $(1/2)x_k(t)$ and $(1/2)x_k^*(t)$ rotate with the same angular velocity in the opposite directions, so that their sum $x_{kp}(t)$ oscillates only along the real axis:

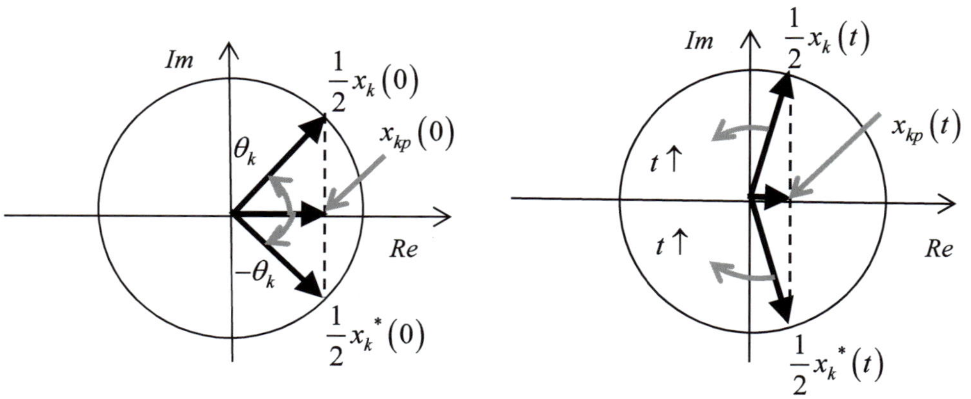

Fig. 4.4-1. Addition of CT complex conjugate harmonics produces real harmonic.

4.4.1.2 *Two forms of discrete spectrum of a real periodic signal*

Discrete spectrum of the k'th CT real harmonic. ***The one-side, or single-sided, and two-sided, or double-sided, spectrum of*** $x_p(t) = A_k \cos(2\pi f_k t + \theta_k)$ are shown, respectively, on the left and on the right in Figure 4.4-2.

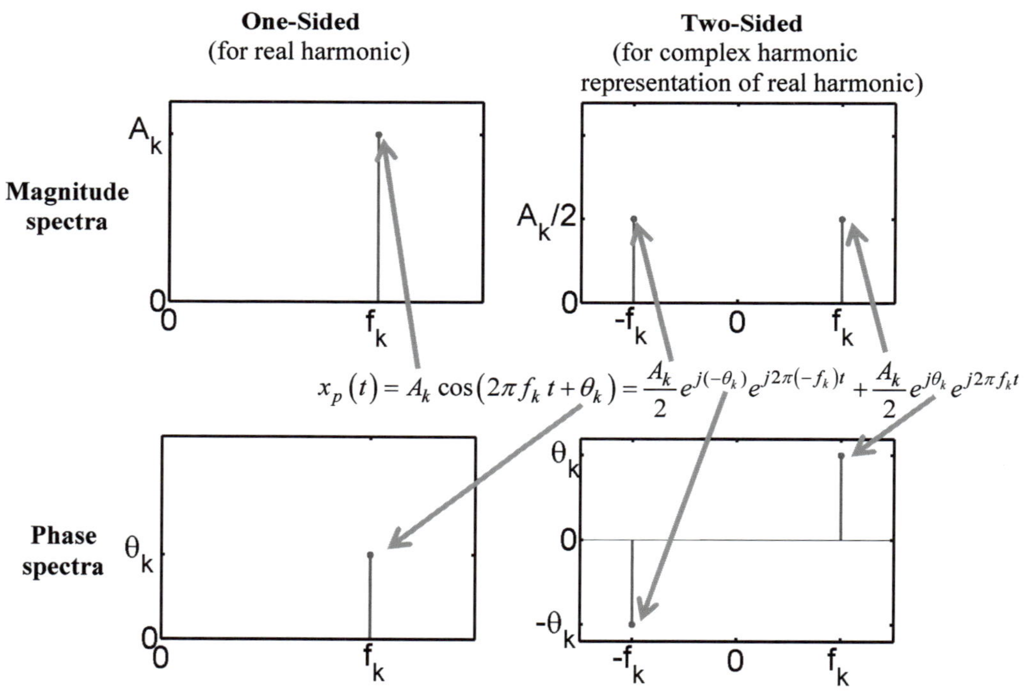

Fig. 4.4-2. Discrete spectrum of the *k*'th real CT harmonic signal.

Discrete two-sided spectrum of a real periodic CT signal. These spectrum are formed through representing real harmonics in terms of complex ones and following the representation of Figure 4.4-2 for every harmonic. For example, for a signal bandlimited to $f_B = f_3$ we have

$$x_p(t) = x_{0p}(t) + x_{1p}(t) + x_{2p}(t) + x_{3p}(t) = A_0 + A_1\cos(2\pi f_1 t + \theta_1) + A_2\cos(2\pi(2f_1)t + \theta_2)$$
$$+A_3\cos(2\pi(3f_1)t + \theta_3) = A_0 + \frac{A_1}{2}e^{j(-\theta_1)}e^{j2\pi(-f_1)t} + \frac{A_1}{2}e^{j\theta_1}e^{j2\pi f_1 t}$$
$$+\frac{A_2}{2}e^{j(-\theta_2)}e^{j2\pi(-2f_1)t} + \frac{A_2}{2}e^{j\theta_2}e^{j2\pi(2f_1)t} + \frac{A_3}{2}e^{j(-\theta_3)}e^{j2\pi(-3f_1)t} + \frac{A_3}{2}e^{j\theta_3}e^{j2\pi(3f_1)t}$$

with the corresponding spectra

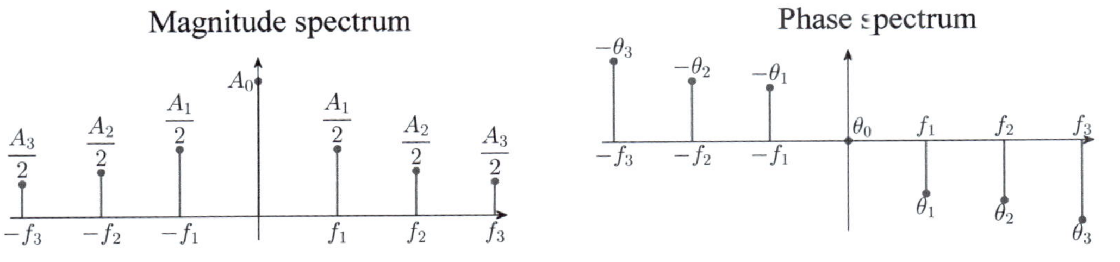

Fig. 4.4-3. Two-sided discrete spectrum for a CT harmonic signal bandlimited to $f_B = f_3$.

From Figure 4.4-3 we can see that two-sided magnitude spectrum has even symmetry and two-sided phase spectrum has odd symmetry.

Example 4.4-1: CT complex harmonic signal decomposition:

$$x_p(t) = 4 + 5\cos\left(20\pi t + \frac{\pi}{4}\right) + 3\cos\left(40\pi t + \frac{\pi}{5}\right)$$
$$= 4\cos(2\pi\cdot 10\cdot 0\cdot t) + 5\cos\left(2\pi\cdot 10\cdot 1\cdot t + \frac{\pi}{4}\right) + 3\cos\left(2\pi\cdot 10\cdot 2\cdot t + \frac{\pi}{5}\right)$$
$$= \underbrace{2e^{j0}}_{X_s[0]/2} e^{j2\pi\cdot 10\cdot 0\cdot t} + \underbrace{2e^{j(-0)}}_{X_s[-0]/2} e^{j2\pi\cdot 10\cdot(-0)\cdot t} + \underbrace{2.5e^{j\frac{\pi}{4}}}_{X_s[1]} e^{j2\pi\cdot 10\cdot 1\cdot t} + \underbrace{2.5e^{j(-\frac{\pi}{4})}}_{X_s[-1]} e^{j2\pi\cdot 10\cdot(-1)\cdot t}$$
$$+\underbrace{1.5e^{j\frac{\pi}{5}}}_{X_s[2]} e^{j2\pi\cdot 10\cdot 2\cdot t} + \underbrace{1.5e^{j(-\frac{\pi}{5})}}_{X_s[-2]} e^{j2\pi\cdot 10\,(-2)\cdot t} = \sum_{k=-2}^{2} X_s[k]e^{j2\pi k10t}.$$

$\sum_{k=-2}^{2} X_s[k]e^{j2\pi k10t}$ is the **exponential Fourier series representation** of $x_p(t)$ shown in Figure 4.4-4, where $\{X_s[-2], X_s[-1], X_s[0], X_s[1], X_s[2]\} = \{X_s[k]\}_{k=-2}^{2}$, $X_s[0] = X_s[0]/2 + X_s[-0]/2$, are the **exponential Fourier series coefficients**.

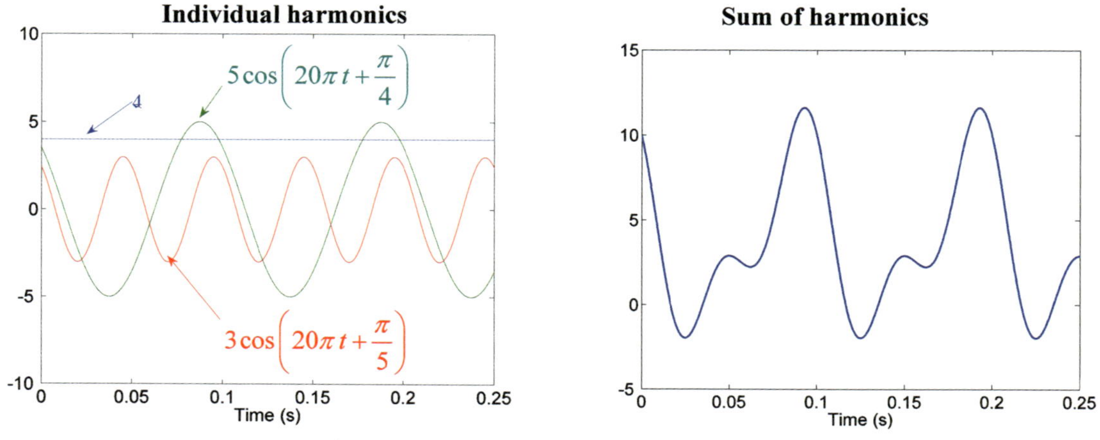

Fig. 4.4-4. CT complex harmonics form a real CT signal.

4.4.2 *DT complex harmonic signals*

Sampling of CT complex harmonic signal with sampling interval t_s yields:

$$x_1(t) = A_1 e^{j(2\pi f_1 t+\theta_1)} \xrightarrow[t\to nt_s]{} x_{t_s}[n] = A_1 e^{j(2\pi \underbrace{f_1 n t_s}_{1/S_F} +\theta_1)}$$

$$= A_1 e^{j\left(2\pi \frac{f_1}{S_F} n+\theta_1\right)} \underset{\frac{f_1}{S_F}=\frac{k}{N}=F_k}{=} \overbrace{A_k e^{j(2\pi F_k n+\theta_k)}}^{\textbf{Discrete-time complex harmonic}},$$

where DT $A_k \equiv$ CT A_1, DT $\theta_k \equiv$ CT θ_1, and $t_s = \frac{1}{S_F}$

$$\operatorname{Re} x_{t_s}[n] = x_1[n] = \underbrace{A_1 \cos(2\pi F_1 n + \theta_1)}_{\textbf{Discrete-time real harmonic}}.$$

Example 4.4-2: DT complex harmonic signal decomposition: Let us sample the waveform in the Example 4.4-1 as follows:

$$S_F = 45 \text{ Hz} \Rightarrow F = \frac{f_1}{S_F} = \frac{2}{9} \Rightarrow F_1 = \frac{1}{9},\ F_k = \frac{k}{9}.$$

This yields DT periodic function

$$x_{sp}[n] = 4 + 5\cos\left(2\pi\cdot\frac{2}{9}\cdot n + \frac{\pi}{4}\right) + 3\cos\left(2\pi\cdot\frac{4}{9}\cdot n + \frac{\pi}{5}\right) = \underbrace{4}_{X[0]} + \underbrace{2.5e^{j\frac{\pi}{4}}}_{X[2]} e^{j2\pi\cdot 2\cdot\frac{1}{9}\cdot n} +$$

$$\underbrace{2.5e^{j(-\frac{\pi}{4})}}_{X[-2]} e^{j2\pi\cdot\frac{1}{9}\cdot(-2)\cdot n} + \underbrace{1.5e^{j\frac{\pi}{5}}}_{X[4]} e^{j2\pi\cdot 4\cdot\frac{1}{9}\cdot n} + \underbrace{1.5e^{j(-\frac{\pi}{5})}}_{X[-4]} e^{j2\pi\cdot(-4)\cdot\frac{1}{9}\cdot n} = \sum_{k=-4}^{4} X[k] e^{j2\pi k\frac{1}{9}n},$$

where the last line is the **discrete Fourier series representation** of $x_{sp}[n]$, and $\{X_s[-4], X_s[-2], X_s[0], X_s[2], X_s[4]\} = \{X_s[k]\}_{k=-4}^{4}$, $X_s[-3] = X_s[-1] = X_s[1] = X_s[3] = 0$ are the **discrete (exponential) Fourier series coefficients**. This signal is shown in Figure 4.4-5.

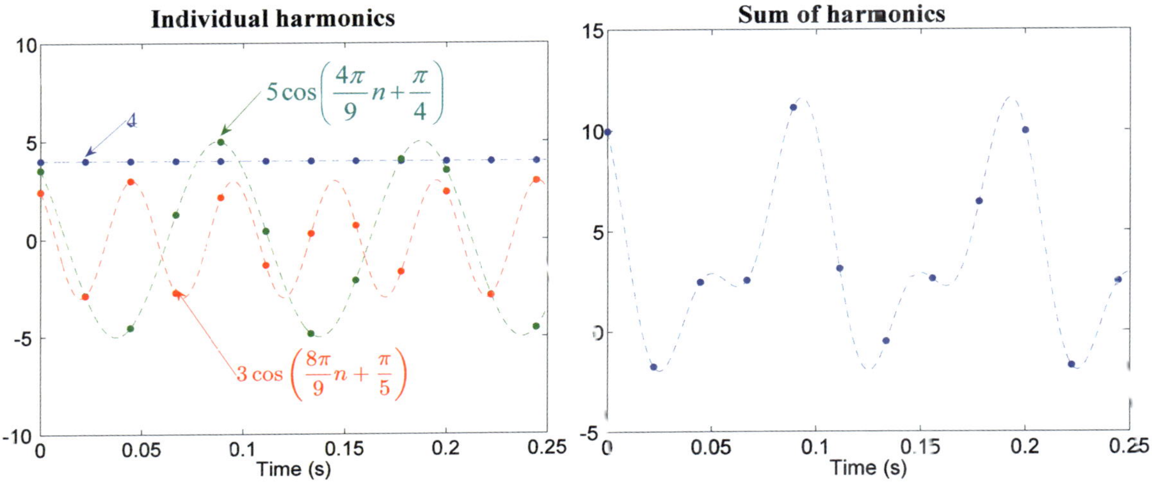

Fig. 4.4-5. CT complex harmonics form a real CT signal.

4.4.3 *Adding harmonics after folding and overlap*

Suppose we sample a signal bandlimited to $f_b = 5f_1$, i.e. with Nyquist frequency $S_N = 10f_1$, at $S_F = 6f_1 < S_N$. This makes the sampled fifth CT harmonic fold onto the first DT harmonic:

$S_F = 6f_1 > 2f_1 \rightarrow$ not aliased

$S_F = 6f_1 < S_N = 2f_b = 10f_1 \rightarrow$ aliased: $5f_1/6f_1 = 5/6$ is folded to $F_1 = 1/6$

$$x_p(t) = A_1\cos(2\pi f_1 t + \theta_1) + A_5\cos(2\pi\cdot 5f_1 t + \theta_5) \xrightarrow{\text{sampled at } S_F = 6f_1}$$

$$x_{al,p}[n] = A_{al}\cos(2\pi F_1 n + \theta_{al}) = A_1\cos(2\pi F_1 n + \theta_1) + A_5\cos(2\pi F_1 n - \theta_5).$$

We can add the aliased harmonic to the correctly-sampled harmonic analytically using the following figure:

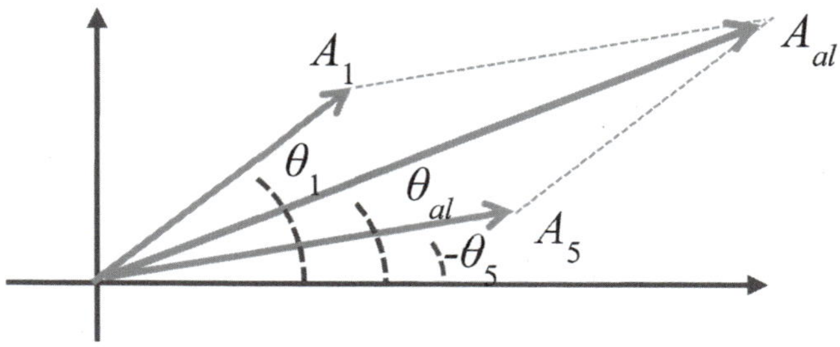

Fig. 4.4-6. Sum of harmonics.

4.4.3.1 *Adding harmonics graphically*

Example 4.4-3:

$$x_p(t) = 5\cos\left(2\pi\cdot\overbrace{10}^{f_1}\cdot t + \frac{\pi}{3}\right) + 4\cos\left(2\pi\cdot\overbrace{50}^{f_5}\cdot t \overbrace{-\frac{\pi}{5}}^{\theta_5}\right)$$ sampled at $S_F = 60$ Hz. What is the resulting signal?

First, find sampled DT signal and apply aliasing

$$x_{al,p}[n] = 5\cos\left(2\pi\cdot\frac{10}{60}\cdot n + \frac{\pi}{3}\right) + 4\cos\left(2\pi\cdot\frac{50}{60}\cdot n - \frac{\pi}{5}\right) = 5\cos\left(2\pi\cdot\frac{1}{6}\cdot n + \frac{\pi}{3}\right) + 4\cos\left(2\pi\cdot\frac{5}{6}\cdot n - \frac{\pi}{5}\right)$$
$$= 5\cos\left(2\pi\cdot\frac{1}{6}\cdot n + \frac{\pi}{3}\right) + 4\cos\left(2\pi\cdot\frac{-1}{6}\cdot n - \frac{\pi}{5}\right) = 5\cos\left(2\pi\cdot\frac{1}{6}\cdot n + \frac{\pi}{3}\right) + 4\cos\left(2\pi\cdot\frac{1}{6}\cdot n + \underbrace{\frac{\pi}{5}}_{-\theta_5}\right).$$

After folding, the second signal overlaps (is folded on top of) the first one:

$$x_{al,p}[n] = 5\cos\left(2\pi\cdot\frac{1}{6}\cdot n + \frac{\pi}{3}\right) + 4\cos\left(2\pi\cdot\frac{1}{6}\cdot n + \frac{\pi}{5}\right).$$

That is, we want to find $C \equiv A_{al}$ and $\theta \equiv \theta_{al}$ that satisfy

$$x_{al,p}[n] = C\cos\left(2\pi\cdot\frac{1}{6}\cdot n + \theta\right) = 5\cos\left(2\pi\cdot\frac{1}{6}\cdot n + \frac{\pi}{3}\right) + 4\cos\left(2\pi\cdot\frac{1}{6}\cdot n + \frac{\pi}{5}\right).$$

We can carry this out geometrically or algebraically as shown in Figures 4.4-7–4.4-8 below.

Set $n = 0$. Then, recalling that

$A\cos(\theta) = \mathrm{Re}[A\cos(\theta) + jA\sin(\theta)] = \mathrm{Re}(Ae^{j\theta})$, we have

$$x_{al,p}[0] = 5\cos\left(\frac{\pi}{3}\right) + 4\cos\left(\frac{\pi}{5}\right) = \mathrm{Re}\left(5e^{j\frac{\pi}{3}} + 4e^{j\frac{\pi}{5}}\right)$$

$$= \mathrm{Re}\left(5\left(\cos\left(\frac{\pi}{3}\right) + j\sin\left(\frac{\pi}{3}\right)\right) + 4\left(\cos\left(\frac{\pi}{5}\right) + j\sin\left(\frac{\pi}{5}\right)\right)\right)$$

$$= \mathrm{Re}\left(5\cos\left(\frac{\pi}{3}\right) + 4\cos\left(\frac{\pi}{5}\right) + j\left(5\sin\left(\frac{\pi}{3}\right) + 4\sin\left(\frac{\pi}{5}\right)\right)\right)$$

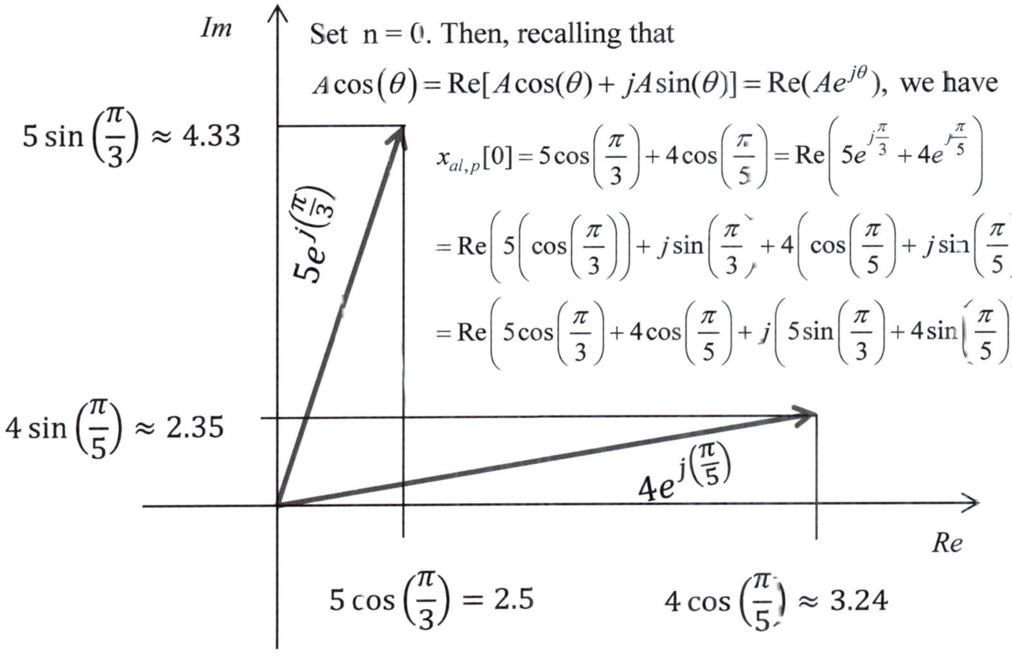

$$= \mathrm{Re}(C(\cos\theta + j\sin\theta))$$

$$= \mathrm{Re}\,Ce^{j\theta} = \mathrm{Re}(8.81e^{j0.861})$$

$$= C\cos\theta = 8.81\cos(0.861)$$

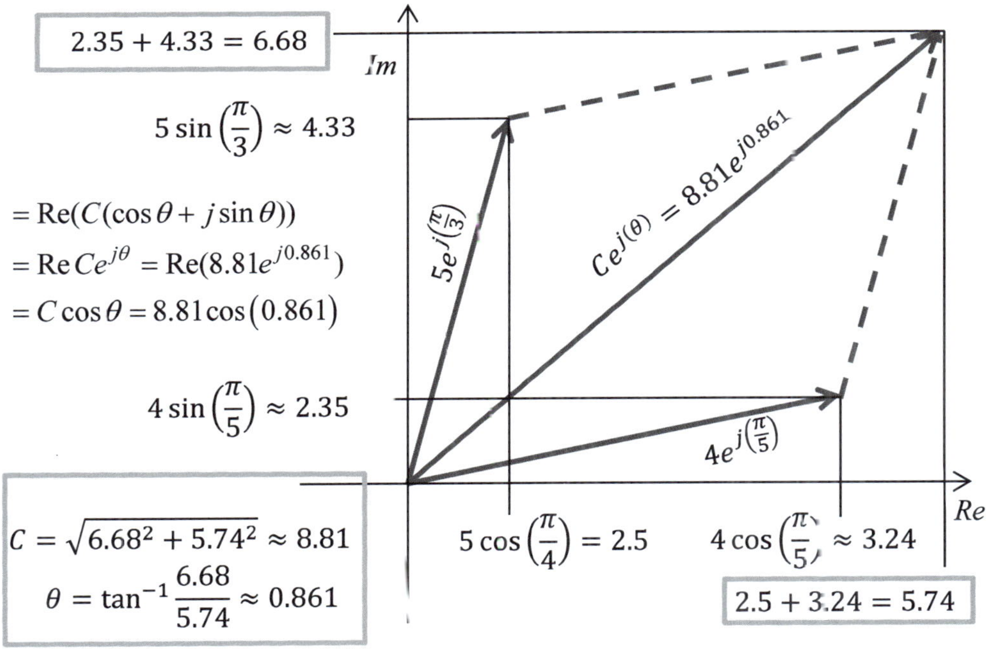

$$C = \sqrt{6.68^2 + 5.74^2} \approx 8.81$$

$$\theta = \tan^{-1}\frac{6.68}{5.74} \approx 0.861$$

Fig. 4.4-7. Adding initial condition vectors.

Putting back $2\pi\frac{1}{6}n$ yields:

$x_{al,p}[n] = \mathrm{Re}\,Ce^{j\theta}e^{j2\pi F_1 n} =$

$\mathrm{Re}(8.81e^{j0.861}e^{j2\pi\frac{1}{6}n})$

$= C\cos(2\pi F_1 n + \theta)$

$$= 8.81\cos\left(2\pi\cdot\frac{1}{6}\cdot n + 0.861\right)$$

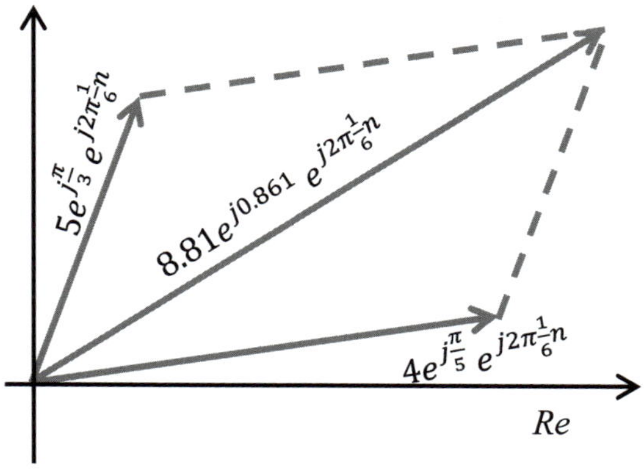

Fig. 4.4-8. Augmenting initial condition vectors with time evolution variables.

4.4.3.2 *Adding harmonic algebraically*

Example 4.4-4: Consider $x_{al,p}[n] = 5\cos\left(2\pi\cdot\frac{1}{6}\cdot n + \frac{\pi}{3}\right) + 4\cos\left(2\pi\cdot\frac{1}{6}\cdot n + \frac{\pi}{5}\right)$. Using cosine angle addition formula: $\cos(\alpha+\beta) = \cos(\alpha)\cos(\beta) - \sin(\alpha)\sin(\beta)$ yields

$$x_{al,p}[n] = 5\cos\left(2\pi\cdot\frac{1}{6}\cdot n + \frac{\pi}{3}\right) + 4\cos\left(2\pi\cdot\frac{1}{6}\cdot n + \frac{\pi}{5}\right) = 5\left(\cos\left(2\pi\cdot\frac{1}{6}\cdot n\right)\cos\left(\frac{\pi}{3}\right)\right.$$
$$\left.-\sin\left(2\pi\cdot\frac{1}{6}\cdot n\right)\sin\left(\frac{\pi}{3}\right)\right) + 4\left(\cos\left(2\pi\cdot\frac{1}{6}\cdot n\right)\cos\left(\frac{\pi}{5}\right) - \sin\left(2\pi\cdot\frac{1}{6}\cdot n\right)\sin\left(\frac{\pi}{5}\right)\right)$$

(combining like terms) $= \cos\left(2\pi\cdot\frac{1}{6}\cdot n\right)\left(5\cos\left(\frac{\pi}{3}\right) + 4\cos\left(\frac{\pi}{5}\right)\right) - \sin\left(2\pi\cdot\frac{1}{6}\cdot n\right)\left(5\sin\left(\frac{\pi}{3}\right)\right.$ $\left.+4\sin\left(\frac{\pi}{5}\right)\right) \approx \cos\left(2\pi\cdot\frac{1}{6}\cdot n\right)\cdot 5.74 - \sin\left(2\pi\cdot\frac{1}{6}\cdot n\right)\cdot 6.68 =$ (applying the angle addition formula again)

$$\approx \cos\left(2\pi\cdot\frac{1}{6}\cdot n\right)\cdot 5.74 - \sin\left(2\pi\cdot\frac{1}{6}\cdot n\right)\cdot 6.68 = C\left(\cos\left(2\pi\cdot\frac{1}{6}\cdot n\right)\cos(\theta) - \sin\left(2\pi\cdot\frac{1}{6}\cdot n\right)\sin(\theta)\right)$$

$\Rightarrow 5.74 = C\cos(\theta)$, $6.68 = C\sin(\theta)$. Finally, solving the latter for C and θ yields

$$\left.\begin{aligned} C &= \sqrt{5.74^2 + 6.68^2} \approx 8.81 \\ \theta &= \tan^{-1}\left(\frac{6.68}{5.74}\right) \approx 0.861 \end{aligned}\right\} \Rightarrow x_{al,p}[n] = 5\cos\left(2\pi\cdot\frac{1}{6}\cdot n + \frac{\pi}{3}\right) + 4\cos\left(2\pi\cdot\frac{1}{6}\cdot n + \frac{\pi}{5}\right)$$

$$\approx 8.81\cos\left(2\pi\cdot\frac{1}{6}\cdot n + 0.861\right).$$

4.4.3.3 *Adding harmonics with different magnitude scales*

Let's now consider the sum vector, z_3, of vectors z_2 and z_1, and another sum vector, z_4, of vector z_2 and a shorter version of z_1, as shown in Figure 4.4-9. Assume that the magnitude A_1 of vector z_1 is increasing to infinity while the magnitude A_2 of vector z_2 is staying constant. Then, the relative contribution of the projections of the vector z_2 with fixed magnitude A_2 to the sums z_3 and z_4 is seen to become smaller as A_1 increases. This makes the sum vector angle and magnitude monotonically converge to those of z_1, i.e the sum vector converge to z_1 itself, as seen by comparing z_4 with z_3. This relation can be formally expressed as follows: for any three complex numbers z_1, z_2, and z_3, where z_1 is varying, z_2 is fixed, and $z_3 = z_1 + z_2$:

$$|z_1| \to \infty \Rightarrow \begin{Bmatrix} \angle(z_3 = z_1 + z_2) \to \angle z_1 \\ |z_3 = z_1 + z_2| \to |z_1| \end{Bmatrix} \Rightarrow z_3 = z_1 + z_2 \to z_1. \qquad (4.4\text{-}1)$$

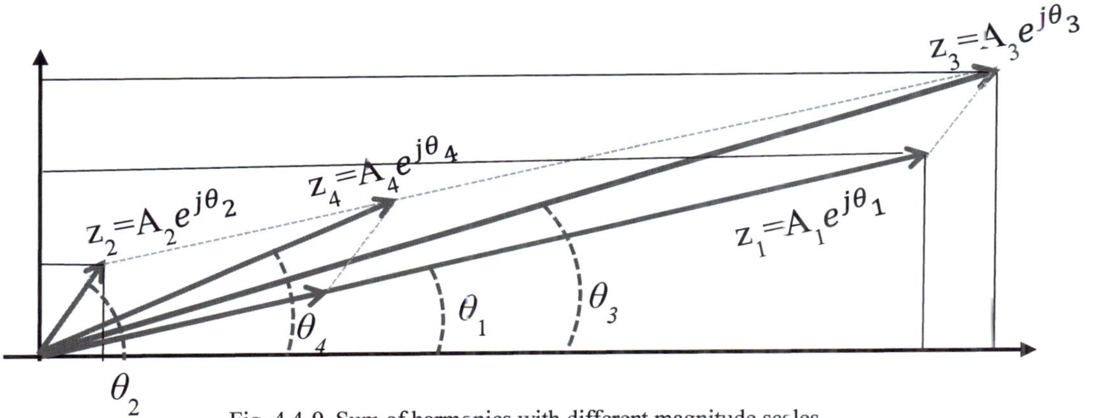

Fig. 4.4-9. Sum of harmonics with different magnitude scales.

Homework 5: Solved Examples and Solution Guide

Problem 1: Determine if the following signals are periodic, and compute the fundamental period *N* for the periodic ones:

(a) $x[n] = \cos(n/17)$,
(b) $x[n] = -\sin(\pi n/6) + |3\cos(\pi n/9)|$,
(c) $x[n] = 6 + 3\sin(9\pi n/7)$,
(d) $x[n] = \sin(3\pi n/7) + \cos(5\pi n/11)$,
(e) $x[n] = e^{-j0.6\pi n}$,
(f) $x[n] = \left|e^{-j0.6\pi n}\right|$,
(g) $x[n] = 4e^{j0.4\pi n} + 4e^{-j0.5\pi n}$,
(h) $x[n] = (-1/j)^{n/15}$.

Solution:

(a) $\cos(2\pi nF) = \cos(n/17)$, $\because 2\pi F = \frac{1}{17} \Rightarrow F = \frac{1}{34\pi}$.
Since F is not a ratio of integers, the signal is non-periodic.

(b) For $-\sin(\pi n/6)$, similarly to (a) above, $2\pi F_1 = \frac{\pi}{6} \Rightarrow F_1 = \frac{1}{12}$; for $|3\cos(\pi n/9)|$, $\pi F_2 = \frac{\pi}{9} \Rightarrow F_2 = \frac{1}{9}$. Both are ratio of integers. $N_1 = 1/F_1 = 12, N_2 = 1/F_2 = 9$.
Common period: $lcm(N_1, N_2) = N = 36$.

(c) $2\pi F = \frac{9\pi}{7} \Rightarrow F = \frac{9}{14}$, $N = \frac{14}{9}$. Common period: $lcm(N,1) = 14$

(d) $2\pi F_1 = \frac{3\pi}{7}, 2\pi F_2 = \frac{5\pi}{11} \Rightarrow F_1 = \frac{3}{14}, F_2 = \frac{5}{22}$. Common period: $lcm(N_1, N_2) = 154$.

(e) Using Euler's formula $x[n] = \cos(0.6n\pi) - j\sin(0.6n\pi)$,
$\because 2\pi F = 0.6\pi \Rightarrow F = 0.6/2 = 3/10$. Common period: $lcm(N,1) = 10$.

(f) $x[n] = \left|e^{-j0.6\pi n}\right| = |\cos(-0.6n\pi) + j\sin(-0.6n\pi)| = \left|\sqrt{\cos^2(-0.6n\pi) + \sin^2(-0.6n\pi)}\right| = 1$.
Therefore, the signal is constant, and hence periodic with arbitrary period.

(g) $x[n] = 4\cos(0.4n\pi) + 4j\sin(0.4n\pi) + 4\cos(0.5n\pi) - 4j\sin(0.5n\pi)$.
$F_1 = 1/5, F_2 = 1/4$. Common period: $lcm(N_1, N_2) = 20$.

(h) $x[n] = (-1/j)^{n/15} = (j)^{n/15} = (e^{j\pi/2})^{n/15} = e^{jn\pi/30} = \cos(n\pi/30) + j\sin(n\pi/30)$.
$F = 1/60$. Common period: $lcm(N,1) = 60$.

Problem 2: Express each signal in terms of digital frequency i) $|F| < 0.5$, ii) $3 < F < 4$:

(a) $x[n] = \sin(7n\pi/5)$,
(b) $x[n] = \cos(6\pi n/5) - 6\sin(9\pi n/5)$.

Solution:

(a) $2\pi F = \frac{7}{5}\pi \Rightarrow F = 0.7 > 0.5$, $\because$ for $|F| < 0.5$, $F = 0.7 - 1 = -0.3$, so that
$x[n] = \sin(-0.6n\pi) = -\sin(0.6n\pi)$.
For $3 < F < 4$, $F = 4 - 0.3 = 3.7$, so that $x[n] = \sin(7.4n\pi)$.

(b) $F_1 = 0.6 > 0.5, F_2 = 0.9 > 0.5$, for $|F| < 0.5, F_1 = 0.6 - 1 = -0.4, F_2 = 0.9 - 1 = -0.1$.
$\because x[n] = \cos(0.8n\pi) + 6\sin(0.2n\pi)$.
For $3 < F < 4$, $F_1 = 4 - 0.4 = 3.6$, $F_2 = 4 - 0.1 = 3.9$. $x[n] = \cos(7.2n\pi) - 6\sin(7.8n\pi)$.

Problem 3: Assume that for a given reference input, under Bosch controller, the software testbed mold displacement signal is:

a) $x_{p1}(t) = 0.08 + 3.6\cos(2\pi 4.6t - 5\pi/8) + 1.3\cos(2\pi 9.2t + 7\pi/9)$,

the hardware testbed mold displacement signal is:

b) $x_{p2}(t) = -0.1 + 3.8\cos(2\pi 4.6t - 3\pi/5) + 1.2\cos(2\pi 9.2t + 5\pi/8)$,

and the production unit mold displacement signal is:

c) $x_{p3}(t) = 0.6 + 3.1\cos(2\pi 4.4t - 6\pi/11) + 0.6\cos(2\pi 8.8t + \pi/5) + 1.2\cos(2\pi 13.2t + \pi/4)$.

Determine if aliasing takes place when the above signals are sampled at a sampling rate $S_F =$ 12Hz **and express each sampled signal in terms of digital frequency in the principal range** $|F| \le 0.5$.

Solution:

(a) $2\pi f_1 = 2\pi \cdot 4.6,\, 2\pi f_2 = 2\pi \cdot 9.2 \Rightarrow f_1 = 4.6,\, f_2 = 9.2$,

$$F_1 = \frac{f_1}{S_F} = \frac{4.6}{12} = \frac{23}{60} < 0.5,\; F_2 = \frac{f_2}{S_F} = \frac{9.2}{12} = \frac{23}{30} > 0.5,$$

Therefore, aliasing takes place. For $|F| \le 0.5$:

$$F_1' = F_1 = \frac{23}{60},\; F_2' = F_2 - 1 = \frac{23}{30} - 1 = -\frac{7}{30},$$

$$x_{p1}[n] = 0.08 + 3.6\cos\left(2\pi F_1' n - \frac{5\pi}{8}\right) + 1.3\cos\left(2\pi F_2' n + \frac{7\pi}{9}\right)$$

$$= 0.08 + 3.6\cos\left(2\pi \frac{23}{60} n - \frac{5\pi}{8}\right) + 1.3\cos\left(-2\pi \frac{7}{30} n + \frac{7\pi}{9}\right)$$

$$= 0.08 + 3.6\cos\left(\frac{23}{30}\pi n - \frac{5\pi}{8}\right) + 1.3\cos\left(\frac{7}{15}\pi n - \frac{7\pi}{9}\right)$$

(b) Since the hardware testbed mold displacement signal frequencies are the same as those in part (a), just with different magnitudes and phases, aliasing takes place:

$$x_{p2}[n] = -0.1 + 3.8\cos\left(2\pi \frac{23}{60} n - \frac{3\pi}{5}\right) + 1.2\cos\left(-2\pi \frac{7}{30} n + \frac{5\pi}{8}\right)$$

$$= -0.1 + 3.8\cos\left(\frac{23}{30}\pi n - \frac{3\pi}{5}\right) + 1.2\cos\left(\frac{7}{15}\pi n - \frac{5\pi}{8}\right).$$

(c) $2\pi f_1 = 2\pi \cdot 4.4,\; 2\pi f_2 = 2\pi \cdot 8.8,\; 2\pi f_3 = 2\pi \cdot 13.2 \Rightarrow f_1 = 4.4,\; f_2 = 8.8,\; f_3 = 13.2$,

$$\because F_1 = \frac{f_1}{S_F} = \frac{4.4}{12} = \frac{11}{30} < 0.5,\; F_2 = \frac{f_2}{S_F} = \frac{8.8}{12} = \frac{11}{15} > 0.5,\; F_3 = \frac{f_3}{S_F} = \frac{13.2}{12} = \frac{11}{10} > 0.5,$$

therefore, aliasing takes place. For $|F| \le 0.5$:

$$F_1' = F_1 = \frac{11}{30},\ F_2' = F_2 - 1 = \frac{11}{15} - 1 = -\frac{4}{15},\ F_3' = F_3 - 1 = \frac{11}{10} - 1 = \frac{1}{10},\ \text{yielding}$$

$$x_{p1}[n] = 0.6 + 3.1\cos\left(2\pi F_1' n - \frac{6\pi}{11}\right) + 0.6\cos\left(2\pi F_2' n + \frac{\pi}{5}\right) + 1.2\cos\left(2\pi F_2' n + \frac{\pi}{4}\right)$$

$$= 0.6 + 3.1\cos\left(2\pi \frac{11}{30} n - \frac{6\pi}{11}\right) + 0.6\cos\left(-2\pi \frac{4}{15} n + \frac{\pi}{5}\right) + 1.2\cos\left(2\pi \frac{1}{10} n + \frac{\pi}{4}\right)$$

$$= 0.6 + 3.1\cos\left(\frac{11}{15}\pi n - \frac{6\pi}{11}\right) + 0.6\cos\left(\frac{8}{15}\pi n - \frac{\pi}{5}\right) + 1.2\cos\left(\frac{1}{5}\pi n + \frac{\pi}{4}\right).$$

Problem 4: The simplest analog method for changing a sound pitch is to record a sound waveform and replay it at a speed that differs from the one used in its recording. The digital counterpart of that is to reconstruct a sampled signal at a different sampling rate. Suppose we sample an analog signal $x(t) = \cos(12600\pi t + \pi/4)$ at a sampling rate of 18 kHz;

(a) Determine the resulting sampled signal.

(b) Determine the reconstructed analog signal frequency for the reconstruction rate S_R = 10 kHz.

(c) Determine the reconstruction rate S_R for raising the original signal pitch (frequency) 1.5 times.

(d) Determine the reconstruction rate S_R for lowering the original signal pitch 1.5 times.

Solution:

(a) $f = \frac{12600}{2} = 6300 \Rightarrow F = \frac{f}{S} = \frac{6300}{18000} = \frac{7}{20}$. $x[n] = \cos\left(2\pi n \frac{7}{20} + 0.25\pi\right) = \cos\left(\frac{7\pi n}{10} + 0.25\pi\right)$.

(b) $f_R = FS_R = \left(\frac{7}{20}\right)(10000) = 3500 Hz.$

(c) $f_R = FS_R = \left(\frac{7}{20}\right)S_R = 6300 \times 1.5 Hz \Rightarrow S_R = 27000 Hz.$

(d) $f_R = FS_R = \left(\frac{7}{20}\right)S_R = \frac{6300}{1.5} Hz \Rightarrow S_R = 12000 Hz.$

Problem 5: The normal blood pressure range is — lower — 80 mmHg, upper — 120 mmHg. We will limit the range to the normal one for this problem. Monitoring any other range can be assigned as a homework problem. Consider a full bridge blood pressure monitor with maximum difference output voltage under full load of 120 mmHg to be 4 mV and calibration accurate to 0.5 mmHg. The output of the amplifier should be 10 V under full load. Draw layout and schematics. Indicate the composition of the signal at the output of the strain gauge in terms of common mode, normal mode, and parasitic (noise) signal voltages, indicate analytically how instrumentation amp extracts normal mode signal. Sketch a layout of a full

bridge strain gauge wrapped around an artery inside the human body during minimally invasive surgery[1].

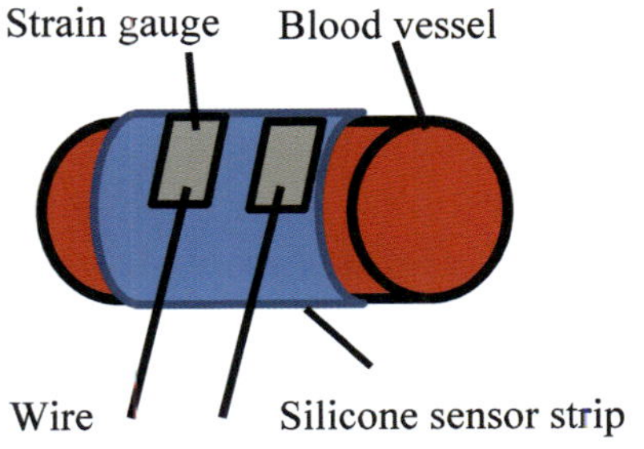

Fig. Hw5-1. Silicone sensor strip with strain gauge wrapped around the blood vessel.

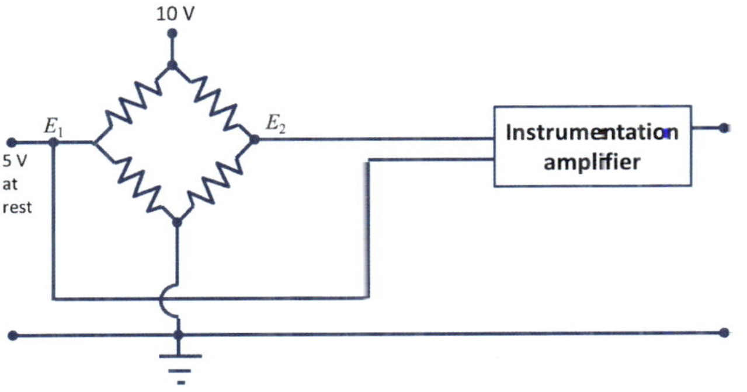

Fig. Hw5-2. Circuit representation of the strain gauge.

Solution:

Max output under full load = 4 mV. Max output of the amplifier = 10 V. Gain required: $V_{out}/V_{in} = 10V/0.004V = 2500$ (minimum gain). Common mode voltage is still $E_{CM} = 5$ V Minimum measurement: 80 mmHg–0 mV. Maximum measurement: 120 mmHg–4 mV. Calibration accuracy (resolution) to 0.5 mmHg.

So, the normal mode lower bound signal voltage is $e_{lb} = \dfrac{0.5\ mmHg}{(120-80)\,mmHg} \cdot 0.004V = 5 \cdot 10^{-5}\,V$. This is the minimum size of strain gauge signal we want to be able to resolve.

$$CMRR = \frac{G_{NM}}{G_{CM}} = \frac{E_{CM}}{e_{lb}} = \frac{5V}{5 \cdot 10^{-5}V} = 10^5 \text{ or } 20\log_{10}(CMRR) = 100\ dB.$$

The compositions of E_1 and E_2 are, respectively, $E_1 = E_{CM} + e_1 + \Delta V$ and $E_2 = E_{CM} + e_2 + \Delta V$ where E_{CM} is the common mode voltage (5 V, in this case), $e_1 - e_2$ is the normal mode output of the strain gauge, and ΔV contains the parasitics. The instrumentation amplifier (shown in Figure 3.4-5) has a transfer function

$$E_o = \left(1+\frac{2R_2}{R_1}\right)(E_1 - E_2) = \left(1+\frac{2R_2}{R_1}\right)\left((E_{CM}+e_1+\Delta V)-(E_{CM}+e_2+\Delta V)\right) = \left(1+\frac{2R_2}{R_1}\right)(e_1 - e_2).$$

Problem 6:

(a) $x(t) = 5 - 2\sin(10\pi t) + 3\sin(15\pi t)$; **(b)** $x(t) = 3\sin(20\pi t - \pi/4) + 3\sin(30\pi t + \pi/6)$;

---

[1]*Biomed Microdevices. 2012 Jun; 14(3):573–81. doi: 10.1007/s10544-012-9636-9. Highly flexible capacitive strain gauge for continuous long-term blood pressure monitoring. Bingger P[1], Zens M, Woias P.

i) Represent the above functions in the forms that explicitly show magnitudes and phases for their one-sided and two-sided magnitude and phase spectra; (ii) For each waveform indicate fundamental frequency and harmonic number of each component, and draw the corresponding magnitude and phase spectrum; (iii) Now, assume the functions above are passed through a sampler. Choose a sampling frequency in such a manner that the lowest discrete time harmonic in each corresponding sampled function has eight samples per cycle. Rewrite the sampled functions to explicitly show magnitudes and phases for their discrete time one-sided and two-sided magnitude and phase spectra, indicate the harmonic number of each component and draw the corresponding discrete time magnitude and phase spectra.

Solution: (a) (i) Using $\cos(\alpha+\pi/2)=-\sin(\alpha)$ and $\cos(\alpha-\pi/2)=\sin(\alpha)$, $x(t)$ can be transformed into the following sum of cosines:

$$x(t)=5-2\sin(10\pi t)+3\sin(15\pi t)=5+2\cos\left(2\pi t\left(\underbrace{5}_{k_1f_1}\right)+\frac{\pi}{2}\right)+3\cos\left(2\pi t\left(\underbrace{\frac{15}{2}}_{k_2f_2}\right)-\frac{\pi}{2}\right)$$

$$=5+e^{j(-\frac{\pi}{2})}e^{j2\pi(-5)t}+e^{j\frac{\pi}{2}}e^{j2\pi(5)t}+1.5e^{j\frac{\pi}{2}}e^{j2\pi(-7.5)t}+1.5e^{j(-\frac{\pi}{2})}e^{j2\pi(7.5)t}$$

$$=5+e^{j(-\frac{\pi}{2})}e^{j2\pi(-k_1f_1)t}+e^{j\frac{\pi}{2}}e^{j2\pi(k_1f_1)t}+1.5e^{j\frac{\pi}{2}}e^{j2\pi(-k_2f_1)t}+1.5e^{j(-\frac{\pi}{2})}e^{j2\pi(k_2f_1)t}.$$

(ii) Here the harmonic numbers to be determined are denoted as k_i, where the subscript i designates the order of the component under consideration. Then, $k_1f_1=5$ and $k_2f_1=15/2=7.5$, where the fundamental frequency f_1 is given by the greatest common divisor, 2.5 Hz; this yields harmonic numbers $k_1=2$ and $k_2=3$.

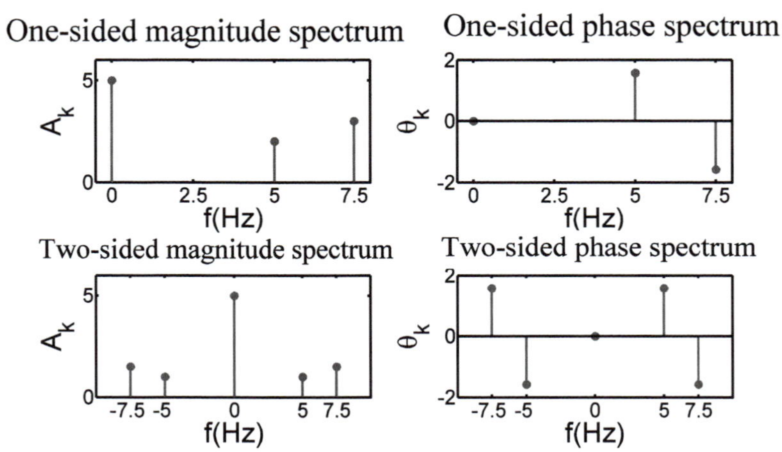

Fig. Hw5-3. CT function magnitude and phase spectrum for Problem 6(a).

(iii) Discrete Time: the lowest frequency in $x(t)$ is 5 Hz, so $f_{low} = 5$ Hz. The DT counterpart of $x(t)$ has the form

$$x[n] = 5 + 2\cos\left(2\pi k_1 F_1 n + \frac{\pi}{2}\right) + 3\cos\left(2\pi k_2 F_1 n - \frac{\pi}{2}\right).$$

According to the given task the lowest digital frequency needs to have 8 samples per cycle, therefore

$F_{low} = \dfrac{f_{low}}{S_f} = \dfrac{1}{8}$ cycle/sample. $S_f = f_{low}(8) = (5)(8) = 40\ Hz$, and

$$x[n] = 5 + 2\cos\left(2\pi\frac{5}{40}n + \frac{\pi}{2}\right) + 3\cos\left(2\pi\frac{15}{80}n - \frac{\pi}{2}\right)$$

$$= 5 + 2\cos\left(2\pi\underbrace{(1/8)}_{k_1F_1}n + \frac{\pi}{2}\right) + 3\cos\left(2\pi\underbrace{(3/16)}_{k_2F_1}n - \frac{\pi}{2}\right) = 5 + 2\cos\left(2\pi\frac{2}{16}n - \frac{\pi}{2}\right) + 3\cos\left(2\pi\frac{3}{16}n - \frac{\pi}{2}\right)$$

$$= 5 + e^{j\left(-\frac{\pi}{2}\right)}e^{j2\pi\left(-\frac{2}{16}\right)n} + e^{j\frac{\pi}{2}}e^{j2\pi\left(\frac{2}{16}\right)n} + 1.5e^{j\frac{\pi}{2}}e^{j2\pi\left(-\frac{3}{16}\right)n} + 1.5e^{j\left(-\frac{\pi}{2}\right)}e^{j2\pi\left(\frac{3}{16}\right)n}$$

Therefore, we have $F_1 = \dfrac{1}{16}$; $k_1 = 2$; $k_2 = 3$.

In this particular case CT and DT harmonic numbers came out to be the same, $k_1 = 2$ and $k_2 = 3$.

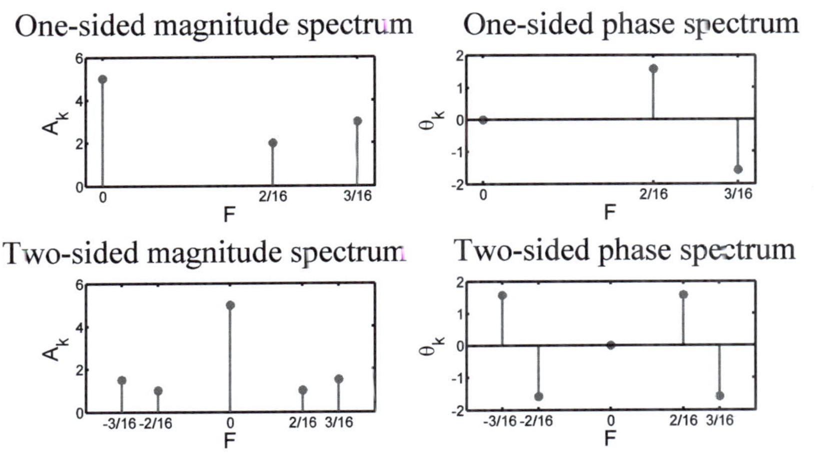

Fig. Hw5-4. DT function magnitude and phase spectrum for Problem 6(a).

(b) $x(t) = 3\sin\left(20\pi t - \frac{\pi}{4}\right) + 3\sin\left(30\pi t + \frac{\pi}{6}\right)$

(i) Convert $x(t)$ from sines to cosines by formula $\sin(2\pi k_1 f_1 t + \theta) = \cos(2\pi k f_1 t + \theta - \pi/2)$:

$$x(t)=3\cos\left(2\pi t(10)-\frac{\pi}{4}-\frac{\pi}{2}\right)+3\cos\left(2\pi t(15)+\frac{\pi}{6}-\frac{\pi}{2}\right)$$

$$=3\cos\left(2\pi t\left(\underbrace{10}_{k_1 f_1}\right)-\frac{3\pi}{4}\right)+3\cos\left(2\pi t\left(\underbrace{15}_{k_2 f_1}\right)-\frac{\pi}{3}\right)$$

$$=1.5e^{j\frac{3\pi}{4}}e^{j2\pi(-10)t}+1.5e^{-j\frac{3\pi}{4}}e^{j2\pi(10)t}+1.5e^{j\frac{\pi}{3}}e^{j2\pi(-15)t}+1.5e^{-j\frac{\pi}{3}}e^{j2\pi(15)t}$$

$$=1.5e^{j\frac{3\pi}{4}}e^{j2\pi(-k_1f_1)t}+1.5e^{-j\frac{3\pi}{4}}e^{j2\pi(k_1f_1)t}+1.5e^{j\frac{\pi}{3}}e^{j2\pi(-k_2f_1)t}+1.5e^{-j\frac{\pi}{3}}e^{j2\pi(k_2f_1)t}.$$

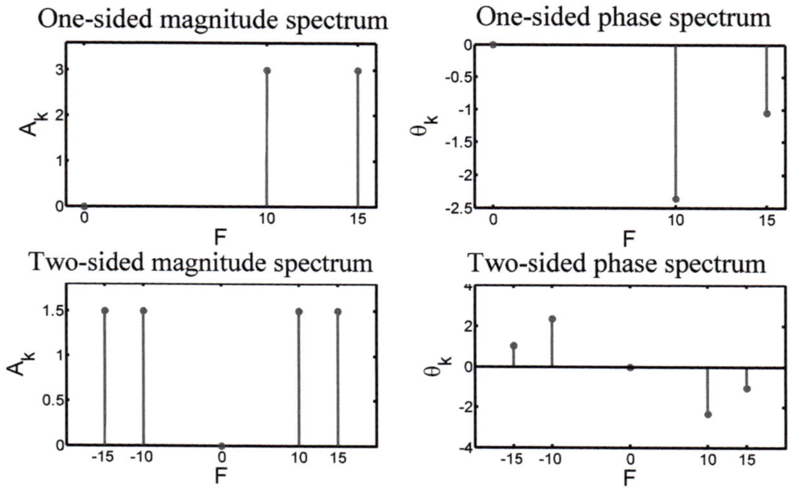

Fig. Hw5-5. CT function magnitude and phase spectrum for Problem 6(b).

(ii) Here the harmonic numbers to be determined are denoted as k_i, where the subscript *i* designates the order of the component under consideration. Then, $k_1 f_1 = 10$ and $k_2 f_1 = 15$, where the fundamental frequency f_1 is given by the greatest common divisor, 5 Hz; this yields harmonic numbers $k_1 = 2$ and $k_2 = 3$.

(iii) Discrete Time: the lowest frequency in *x*(*t*) is 10 Hz, so $f_{low} = 10$ Hz. The DT counterpart of *x*(*t*) has the form $x[n] = 3\cos(2\pi k_1 F_1 n - 3\pi/4) + 3\cos(2\pi k_2 F_1 n - \pi/3)$.

According to the given task the lowest digital frequency needs to have 8 samples per cycle, therefore

$F_{low} = \dfrac{f_{low}}{S_f} = \dfrac{1}{8}$ cycle/sample. $S_f = f_{low}(8) = (10)(8) = 80\ Hz$, and

$$x[n]=3\cos\left(2\pi(10)\left(\frac{1}{80}\right)n-\frac{3\pi}{4}\right)+3\cos\left(2\pi(15)\left(\frac{1}{80}\right)n-\frac{\pi}{3}\right)=3\cos\left(2\pi\left(\underbrace{1/8}_{k_1F_1}\right)n-\frac{3\pi}{4}\right)+$$

$$3\cos\left(2\pi\left(\underbrace{3/16}_{k_2F_1}\right)n-\frac{\pi}{3}\right)=3\cos\left(2\pi\left(\frac{2}{16}\right)n-\frac{3\pi}{4}\right)+3\cos\left(2\pi\left(\frac{3}{16}\right)n-\frac{\pi}{3}\right)=1.5e^{j\left(\frac{3\pi}{4}\right)}e^{j2\pi\left(-\frac{2}{16}\right)n}+$$

$$1.5e^{j\left(-\frac{3\pi}{4}\right)}e^{j2\pi\left(\frac{2}{16}\right)n}+1.5e^{j\left(\frac{\pi}{3}\right)}e^{j2\pi\left(-\frac{3}{16}\right)n}+1.5e^{j\left(-\frac{\pi}{3}\right)}e^{j2\pi\left(\frac{3}{16}\right)n}\quad\therefore\ F_1=1/16;\ k_1=2;\ k_2=3.$$

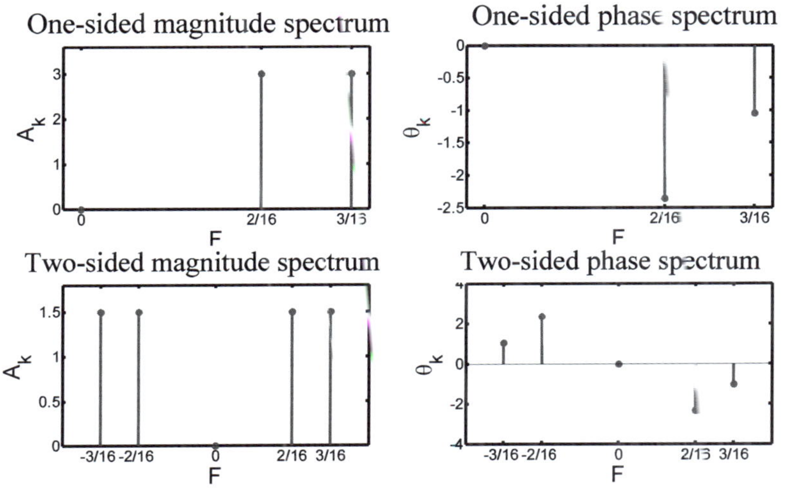

Fig. Hw5-6. DT function magnitude and phase spectrum for Problem 6(b).

Problem 7. Computing system frequency response using DT step interpolation.

The signal from digital controller to the plant actuator is typically passed through zero-order hold, or step interpolation. If the latter is low resolution, carrying out the planned tasks becomes impossible. Assume that the reference signal $r(t) = a\cos(2\pi ft)$ with f = 4.6 Hz and a = 3.0 mm is converted into DT sinusoid $a\cos(2\pi Fn)$ using the sampling interval to obtain digital frequency $F = 0.1$ cycles/sample by passing it through the A/D converter. Pass the resulting DT reference signal through the D/A converter represented by the zero-order hold using the same sampling interval to obtain the analog stair-step (zero-order-hold) approximation of the original analog reference input. A) Plot the following versions of the reference input: i) the original, $r(t)$, ii) the sampled-data, or discrete-continuous, $r_s(t)$, iii) the DT, $r_s[n]$, and iv) the step-interpolated, $r_{zoh}(t)$. B) using this step-interpolated signal, excite the beam of Hw1, with no mass attached to it, at $t = 0$; compute and plot piston and mold positions under P-controller only, comment verbally and indicate on the graph if there is any transient, if and when this transient dies out, if there is the remaining stationary part of the response, and if it is distorted. Comparing the reference input and the stationary part of the response graphs, compute system magnitude and phase responses at 4.6 Hz using the heuristic approach of Hw 3. C) repeat tasks of B) with mass attached to the beam. In B and C tasks above, provide both input and response graphs in an easily comparable format to support your answers. Put your name in each graph and clearly label axes indicating the values and the units (e.g., sec).

Solution guide: A) With digital frequency F = 0.1 cycles/sample, the sampling frequency for CT to DT signal conversion is 46 Hz. Then, zero order hold interpolation generates the continuous-time input signal $u(t)$ by holding each sample value $u[k]$ constant over one sampling interval. The zero-order hold signal output from the CT signal can be directly obtained through Zero-Order Hold block in Simulink, found in Simulink library 'Discrete'. For this problem the "sample time" for the Zero-Order Hold block would be 1/46, as shown below on the left in Figure Hw5-7, resulting in signals from using Simulink shown in Figure Hw5-8.

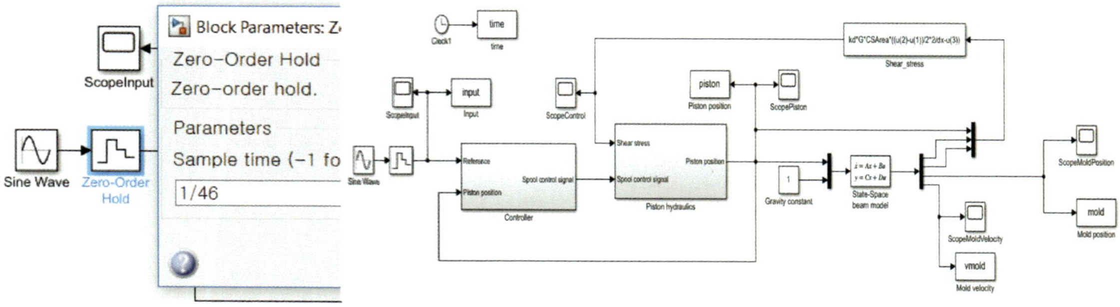

Fig. Hw5-7. Sinusoidal input enters a zero-order hold sampling block (left) in Testbed_simulation_slx diagram (right).

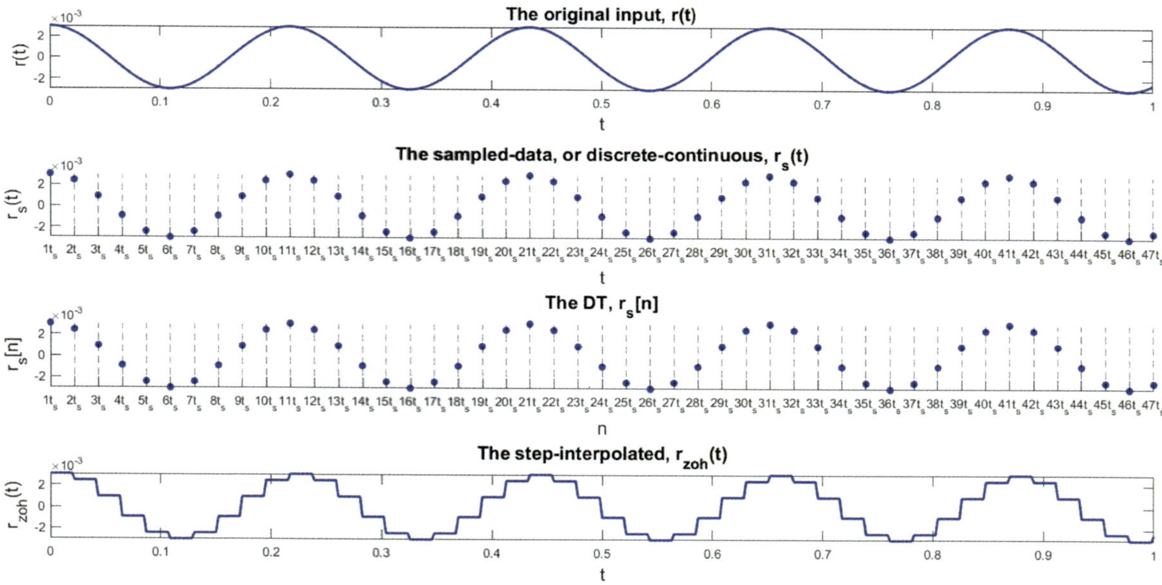

Fig. Hw5-8. Plots of the different versions of the reference input: i) the original, $r(t)$, ii) the sampled-data, or discrete-continuous, $r_s(t)$, iii) the DT, $r_s[n]$, and iv) the step-interpolated, $r_{zoh}(t)$. where zoh stands for the zero-order hold.

A beam response to a step-interpolated DT signal with no mass attached computed by *Testbed_simulation.slx* under P-controller is shown in Figures Hw5-9 and Hw5-10. The stair-step shape of the input is seen to have no visible effect on the response shape, demonstrating that the input approximation accuracy is adequate.

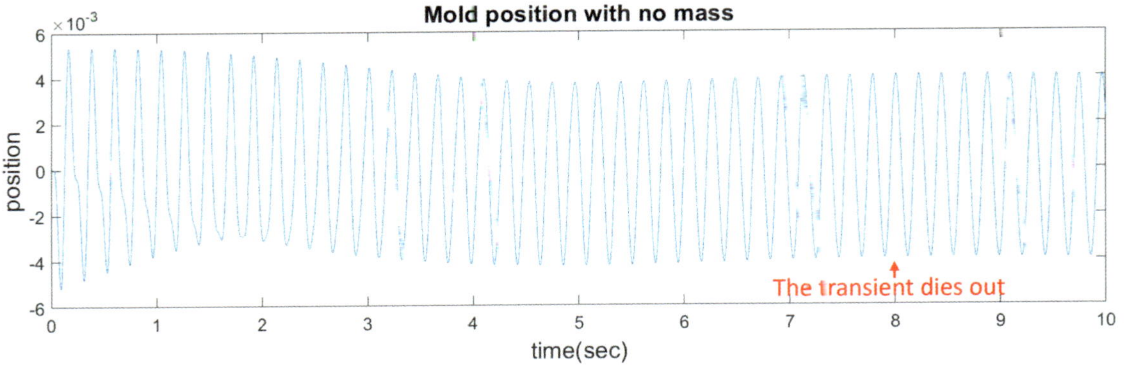

Fig. Hw5-9. Beam response with no mass to the step-interpolated DT signal. The transient that dies out around $t = 8$ sec.

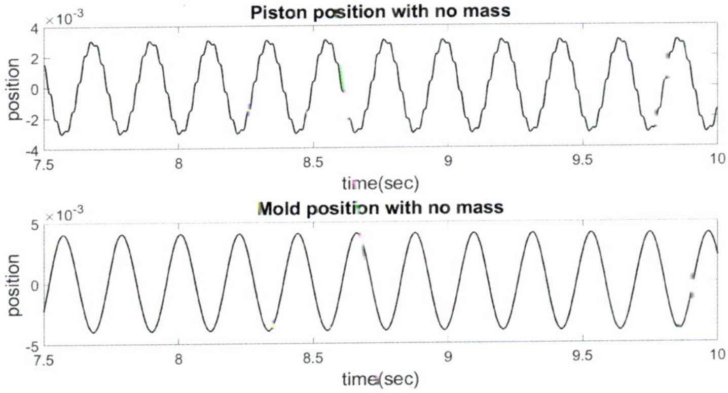

Fig. Hw5-10. The input graph and the stationary part of the response graph with no mass under P-controller.

The response of the system with mass and with P-controller only shows mold position with distortion

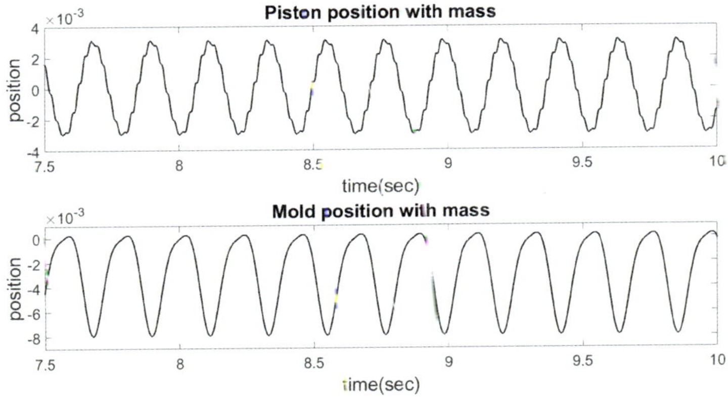

Fig. Hw5-11. The input graph and the stationary part of the response graph with mass under P-controller.

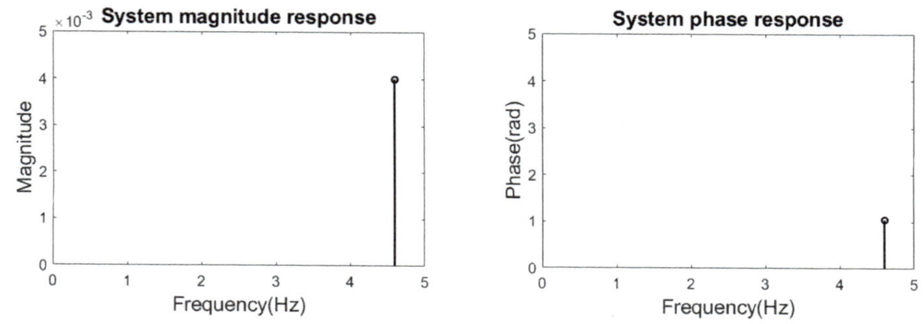

Fig. Hw5-12. The single-frequency, 4.6 Hz, magnitude and phase responses of the system with mass under P-controller.

4.5 Use of Spectrum for Diagnostics and System Identification. Ideal and Real Signal Reconstruction. The Need for Oversampling

4.5.1 *Use of spectrum for diagnostics*

4.5.1.1 *Finding spectrum from data*

So far, we've talked about spectrum as a theoretical tool: we can draw one only if we specifically know the frequencies in the signal. If we don't know the signal harmonic composition, we can try to find it by brute force approach: estimate fundamental harmonic magnitude and frequency, subtract it from the signal, plot the residual and estimate its magnitude and frequency. It might work fine for the testbed mold position, where only the second harmonic is of interest, but not so well for the production unit, where both the second and the third harmonics are of interest. And it would do much worse in catching the tiny harmonic multiples of the operating frequency in the piston position caused by loading effect due to actuator nonlinearity from data, such as that given by the identically looking plots in Figure 4.5-1. Hence, missing the reason for mold velocity distortion.

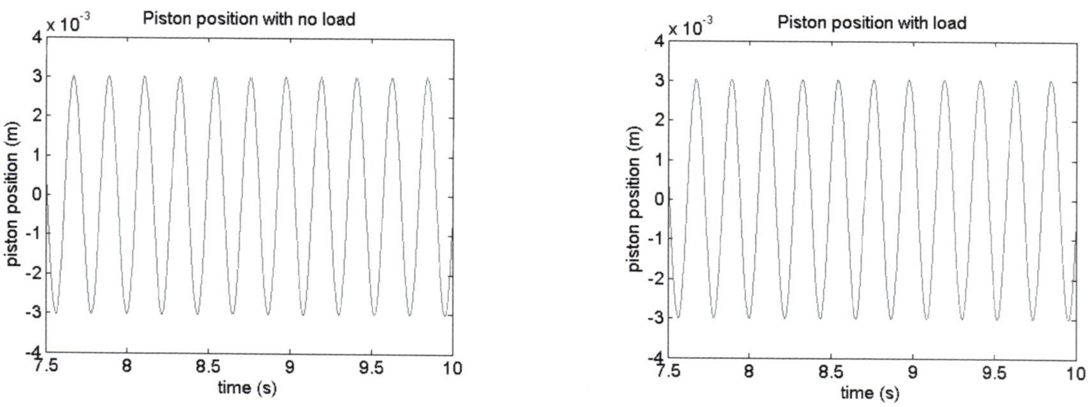

Fig. 4.5-1. Homework 4 plots — simulation of piston position without load (left) and with load (right).

Is it possible to *find the spectrum of a signal from data* ***directly and with high resolution***? It turns out the answer is yes! The key tool for doing this is called the ***Fourier analysis***.

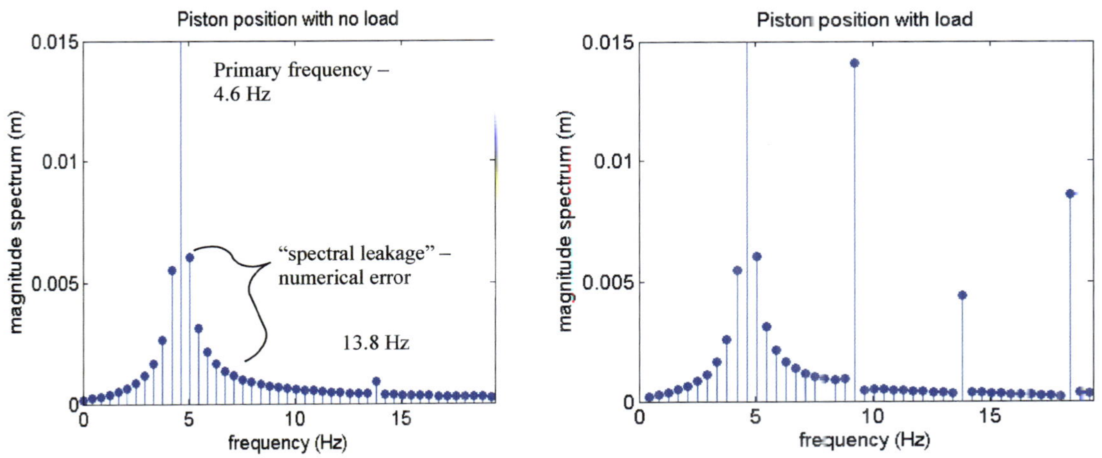

Fig. 4.5-2. Zoomed-in piston magnitude spectrum.

These plots were calculated using only built-in Matlab functions. Magnitudes at higher harmonics increased under load by 2 orders of magnitude, but were still drowned out by the primary harmonic when plotted against time, instead of frequency.

4.5.1.2 *Finding transfer function from data*

Let's recall that a transfer function is a general way of writing the input-output relationship for a linear time invariant physical system that yields frequency domain system properties:

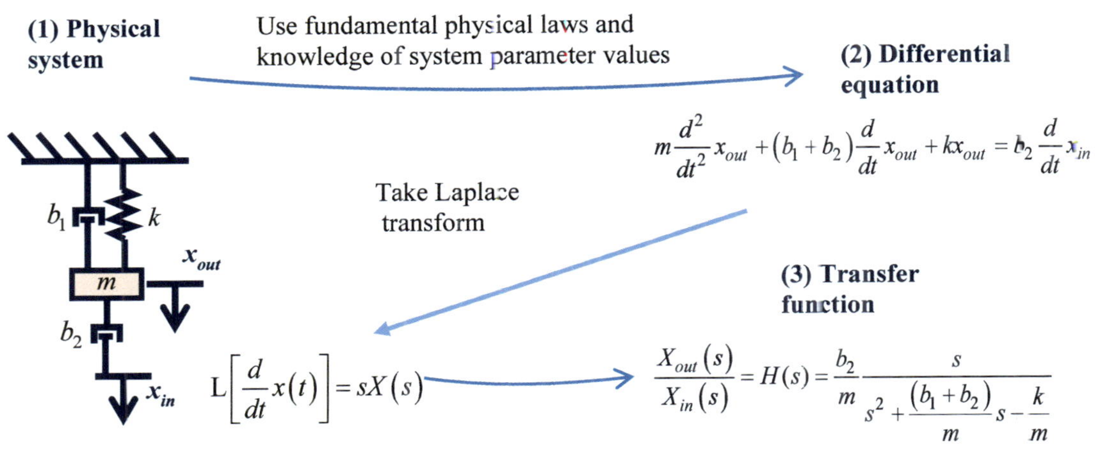

Fig. 4.5-3. MS2D system parametric modeling.

To write differential equation (2) in the figure above we need to know physical laws governing system behavior and also system parameter values. Useful shortcuts, like the complex impedance method, make it easy to go from (1) in the figure above straight to (3), skipping the equation writing.

So far, we have been using the transfer function for solving the differential equation given an input that has Laplace transform: take Laplace transform of input, multiply by transfer function, take inverse Laplace transform of result. But what if we don't know system structure or its physical laws or its parameters, and still would like to obtain system transfer function? Can we do it? Yes, using data.

One of the simplest ways to obtain system transfer function $H(s)$ is to first find system frequency response $H(j\omega)$ and then recover $H(s)$ from $H(j\omega)$, which is often possible.

To find $H(j\omega)$ we recall that a stable linear system turns sinusoidal input $x_{in}(t)$ at a constant frequency into a sinusoidal output $x_{out}(t)$ *at the same frequency*, but with changed magnitude and phase as follows:

- ***The magnitude of the input is multiplied by $|H(j\omega)|$ — the magnitude of $H(j\omega)$.***
- ***The phase of the input is increased by $\angle H(j\omega)$ — the phase of $H(j\omega)$.***

This is depicted in Figure 4.5-4 below:

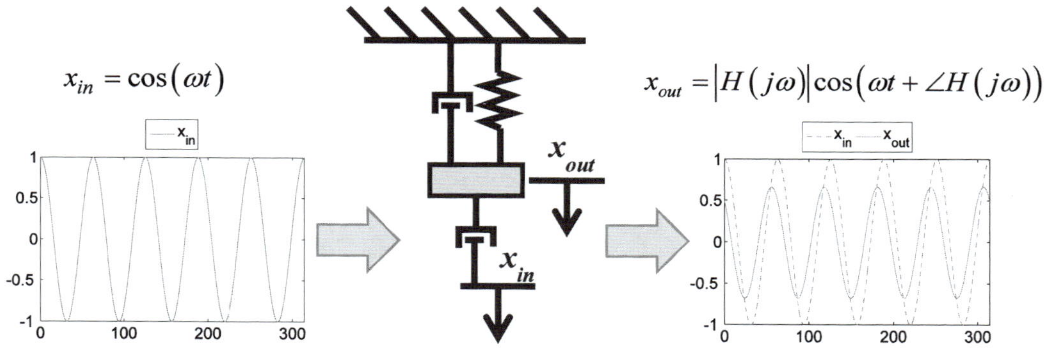

Fig. 4.5-4. Input and output of an MS2D system shows system gain and phase at a given frequency.

Thus, calculating the change in magnitude and phase by directly comparing the input and output signals for every ω, we obtain the system frequency response.

For every fixed ω this is a complex number, which has magnitude and phase:

$$H(j\omega) = \underbrace{|H(j\omega)|}_{\text{Magnitud}} e^{\overbrace{j\angle H(j\omega)}^{\text{Phase}}}.$$

4.5.1.3 *Black box versus grey box identification*

Transfer function identification has two main scenarios:

- We don't know the physical laws governing the system, so can't express $H(s)$ in terms of some physical parameters. This is called ***black box identification.*** This scenario is conceptually depicted in Figure 4.5-5.
- We can write system transfer function in terms of physical parameters with unknown values. This is called ***grey box identification.*** This scenario is conceptually depicted in Figure 4.5-6.

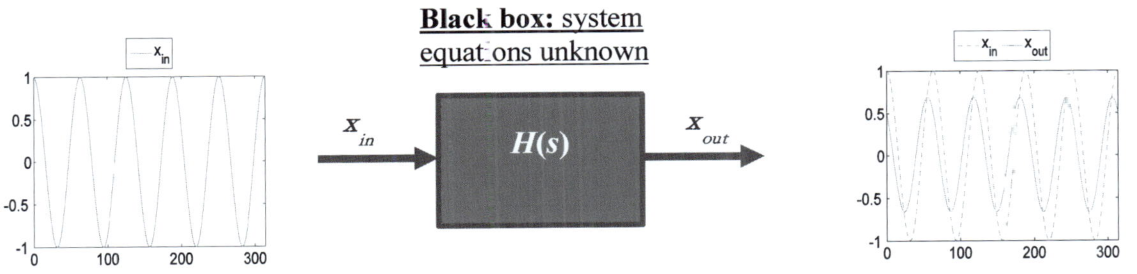

Fig. 4.5-5. Black box system and its input and output.

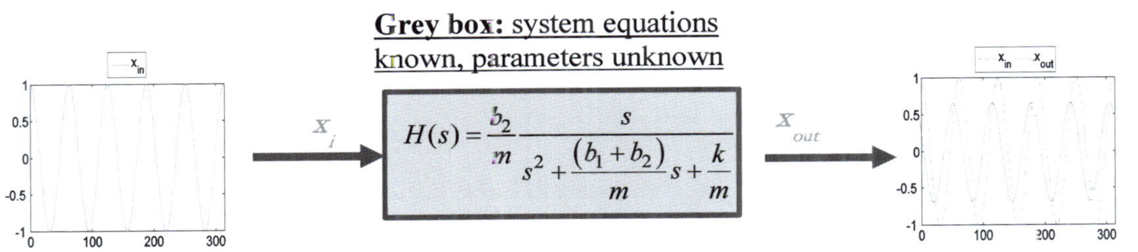

Fig. 4.5-6. Grey box system and its input and output

System frequency response works for both cases.

Example 4.5-1: Grey box system identification from input-output data. Consider a circuit

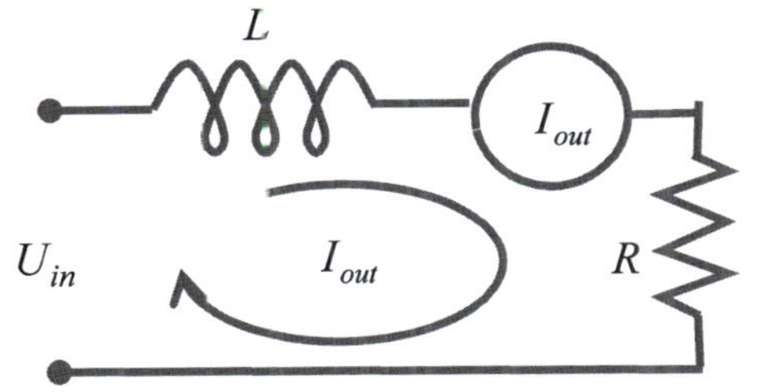

Fig. 4.5-7. Circuit layout.

where the values of L and R are unknown. Impedances are: $Z_R = R$, $Z_L = Ls$, $Z = Z_R + Z_L = R + Ls$. Let's take ammeter measurement I_{out} and apply Kirchhoff's Voltage Law.

Then

$$U_{in} = ZI_{out} \Rightarrow I_{out} = \frac{1}{Z}U_{in} \Rightarrow H(s) = \frac{I_{out}}{U_{in}} = \frac{1}{Z}$$

$$\because H(s) = \frac{1}{R+sL} \Rightarrow H(j\omega) = \frac{1}{R+jL\omega} = \frac{1}{\left|\sqrt{R^2+L^2\omega^2}\right| e^{j\tan^{-1}\frac{L\omega}{R}}} = \frac{1}{\left|\sqrt{R^2+L^2\omega^2}\right|} e^{-j\tan^{-1}\frac{L\omega}{R}}.$$

Choose frequency ω_1 and apply $u_{in} = A_{in}\cos(\omega_1 t)$. Obtain $I_{out} = A_{out}\cos(\omega_1 t + \theta)$.

Set: $A_{out} = \frac{1}{\left|\sqrt{R^2+L^2\omega_1^2}\right|}A_{in}$, $\theta_1 = -\tan^{-1}\frac{L\omega_1}{R}$, calculate L and R — *identify system from input/output data.*

Alternatively, instead of using phase, repeat magnitude relation for another frequency. Similar procedure is used at present in production at Nucor Steel — after every bi-weekly mold replacement, resonance frequency value of the mold oscillator is updated through input/output data collection to tune the controller.

4.5.2 *Signal reconstruction*

Suppose we sample a signal. Can we reconstruct the full CT signal from the DT sampled data?
Example 4.5-2: Consider signal $x(t) = \cos(2\pi t) + 0.5\cos(2\pi \cdot 0.3 \cdot t) - 0.5\cos(2\pi \cdot 1.65 \cdot t)$:

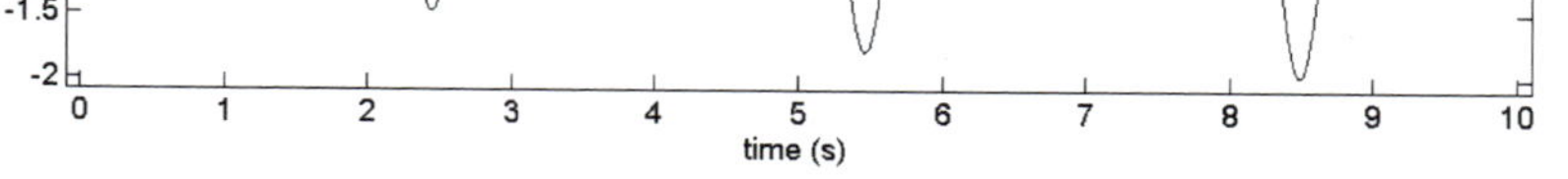

Fig. 4.5-8. Original signal with highest frequency: $f_B = 1.65$ Hz.

4.5.2.1 *Ideal signal reconstruction*

It turns out, **yes.** In theory, we can reconstruct the signal *exactly* as long as the sampled signal is *not aliased.* As an example, consider sampling of a signal given in Figure 4.5-8:

$$x[n] = x(nt_s) = x\left(\frac{n}{S_F}\right).$$

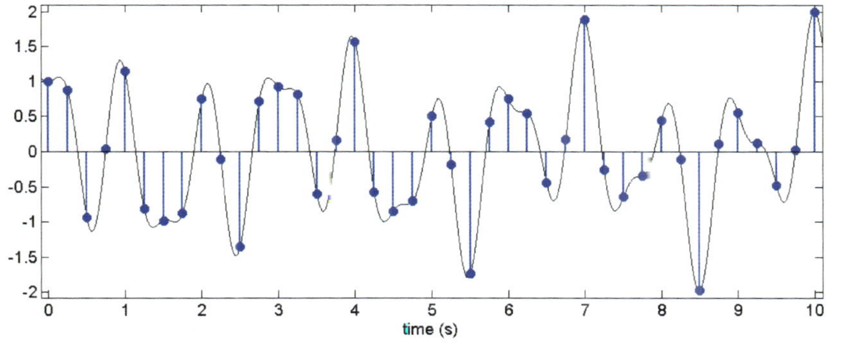

Fig. 4.5-9. Ideal signal reconstruction.

Since the highest signal frequency is $f_B = 1.65$ Hz, taking sampling rate: $S_F = 4$ Hz yields the highest digital frequency $F = f_B / S_F = 1.65 / 4 = 0.4125 < 0.5$, therefore there is no aliasing.

The Nyquist-Shannon-Kotelnikov sampling theorem says that, with no aliasing, the exact signal reconstruction is carried out using sinc function as follows:

$$x(t) = \sum_{n=-\infty}^{\infty} x[n]\text{sinc}\left(\frac{t - nt_s}{t_s}\right).$$

Intuition behind this theorem is illustrated in Figure 4.5-10.

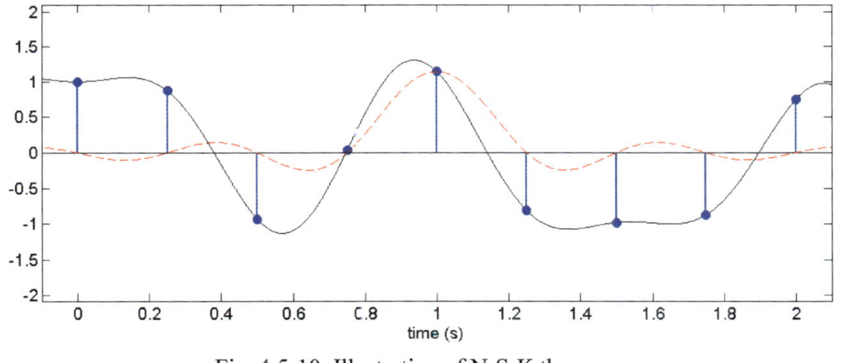

Fig. 4.5-10. Illustration of N-S-K theorem.

The sinc functions employed in the reconstruction are equal to 1 at the given sampled data point and to 0 at all others:

$$\text{sinc}\left(\frac{t - nt_s}{t_s}\right)_{t=nt_s} = \text{sinc}(0) = 1, \text{sinc}\left(\frac{t - nt_s}{t_s}\right)_{t=mt_s, m\neq n}$$

$$= \text{sinc}(m-n)_{m\neq n} = 0,\ 0 < \text{sinc}\left(\frac{t - nt_s}{t_s}\right)_{t\neq mt_s} < 1,$$

but their infinite summation fills out (ideally interpolates) all the empty spaces between data points.

Adding together a sinc for each data point gives the original signal back *in its entirety*, as shown in Figure 4.5-11 where all peaks and valleys between samples are reconstructed!

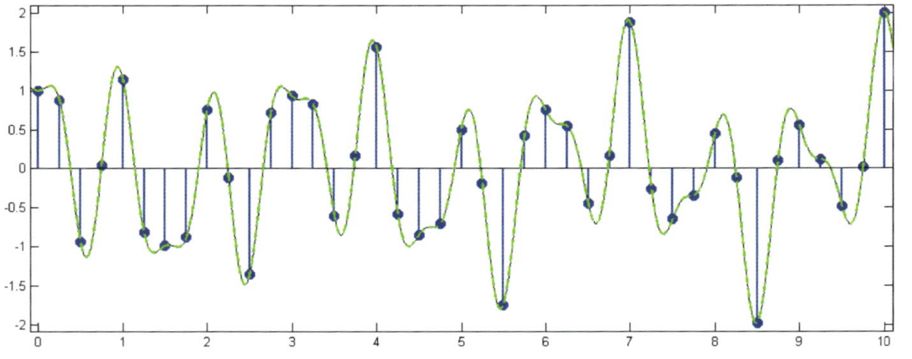

Fig. 4.5-11. Fourier reconstruction.

This technique is sometimes referred to as "Fourier reconstruction". It requires an infinite number of samples to fully reconstruct the signal. As a consequence, it is *non-causal* and it shows that attaining 100% reconstruction accuracy is computationally infeasible.

However, if the sampled signal is aliased, the reconstruction will not be accurate even with infinite number of samples, as seen in Figure 4.5-12, where for the highest frequency $f = 1.65$ Hz, the sampling rate: $S_F = 0.8$ Hz yields the highest digital frequency $F = f / S_F = 1.65 / 4 = 2.0625 > 0.5$, resulting in aliasing.

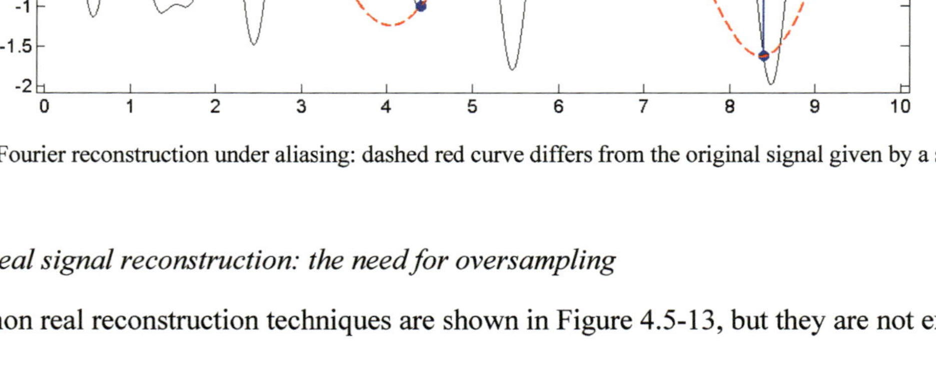

Fig. 4.5-12. Fourier reconstruction under aliasing: dashed red curve differs from the original signal given by a solid grey curve.

4.5.2.2 *Real signal reconstruction: the need for oversampling*

The common real reconstruction techniques are shown in Figure 4.5-13, but they are not exact.

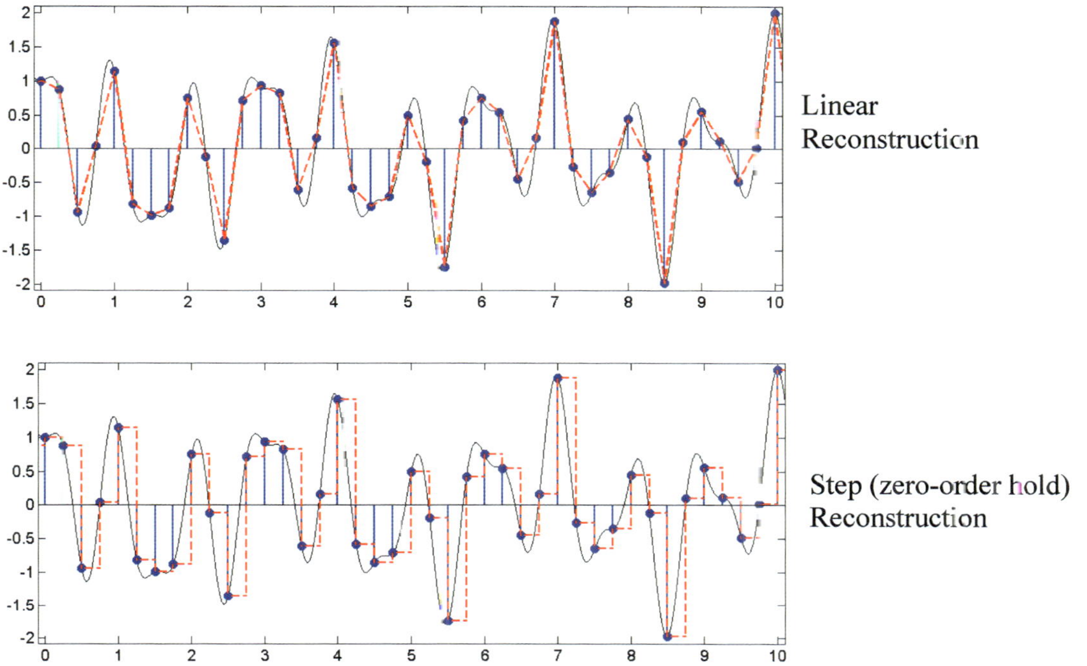

Fig. 4.5-13. Real signal reconstruction using linear and zero-order hold (step) interpolation.

What can one do to improve the reconstruction accuracy? Oversample: choose much higher sampling rate than twice the maximum frequency of interest, e.g. 1024 times oversampling for high quality studio master record: higher oversampling permits better reconstruction — that's why we need $S_F = 352$ KHz for high definition audio, even though we can hear only up to 20 KHz.

Example 4.5-3: Lack of knowledge of noise properties brings in the need for oversampling. Indeed, putting an electromagnet next to a thermocouple and sampling the resulting signal can produce a reading given by dashed red curve in Figure 4.5-14:

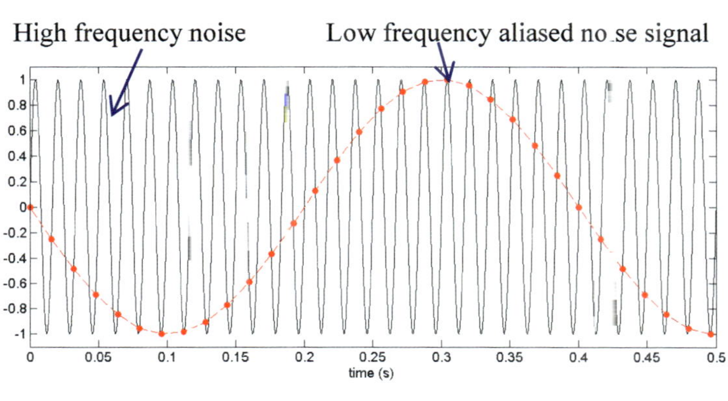

Fig. 4.5-14. Aliasing example from thermocouple reading.

The single frequency 60 Hz noise is aliased into a reasonably looking low frequency, but completely fake, thermocouple oscilloscope reading. Aliasing occurred because the frequency range of the disturbance was unknown, but the sampling frequency was chosen to sample slow thermocouple reading, and therefore turned out to be too low to correctly sample the distorted sensor signal. Thus, noise frequency uncertainty is another reason for oversampling of sensor data.

4.5.3 *DT spectrum plots*

4.5.3.1 *Building DT spectrum plots for an under- and over-sampled CT periodic signal*

To avoid misinterpreting spectrum data, let's get comfortable with the details of the spectrum graphs through putting together DT spectrum plots for a simple CT periodic signal:

Bandlimit

CT signal: $$x_p(t)=5\cos\left(2\pi\cdot\underbrace{10}_{f_1}\cdot t+\frac{\pi}{4}\right)+3\cos\left(2\pi\cdot\underbrace{10\cdot 2}_{f_2=2f_1}\cdot t+\frac{\pi}{5}\right)+2\cos\left(2\pi\cdot\underbrace{10\cdot 3}_{f_3=3f_1=f_B}\cdot t+\frac{\pi}{6}\right). \quad (4.5\text{-}1)$$

Thus, we have: $f_1=10Hz,\ f_2=2f_1=20Hz,\ f_3=3f_1=30Hz=f_B\leftarrow$ bandlimit.

To bring out the effect of aliasing on spectrum plots we will specify two sampling rates:

DT signal 1: $S_F=45$ Hz $< S_N=2f_B=60$ Hz - $S_N/S_F=60/45=$ **1.33 times undersampling**

We first check the digital frequency of each harmonic:

$F_2=\dfrac{f_1}{S_F}=\dfrac{2}{9},\ F_4=\dfrac{f_2}{S_F}=\dfrac{4}{9},\ F_6=\dfrac{f_3}{S_F}=\dfrac{6}{9}>\dfrac{1}{2}$. The third harmonic clearly has aliasing, yielding

$$x_{p1}[n]=5\cos\left[2\pi\cdot\frac{2}{9}\cdot n+\frac{\pi}{4}\right]+3\cos\left[2\pi\cdot\frac{4}{9}\cdot n+\frac{\pi}{5}\right]+2\cos\left[2\pi\cdot\frac{6}{9}\cdot n+\frac{\pi}{6}\right]$$

Folding (4.5-2)

$$=5\cos\left[2\pi\cdot\frac{2}{9}\cdot n+\frac{\pi}{4}\right]+3\cos\left[2\pi\cdot\frac{4}{9}\cdot n+\frac{\pi}{5}\right]+2\cos\left[2\pi\cdot\left(\frac{3}{9}\right)\cdot n-\frac{\pi}{6}\right].$$

Folded frequency

DT signal 2: $S_F=90$ Hz $> S_N=2f_B=60$ Hz - $S_F/S_N=90/60=$ 1.5 times oversampling

Checking the digital frequency of each harmonic:

$F_2=\dfrac{f_1}{S_F}=\dfrac{1}{9},\ F_4=\dfrac{f_2}{S_F}=\dfrac{2}{9},\ F_6=\dfrac{f_3}{S_F}=\dfrac{3}{9}<\dfrac{1}{2}$ shows that the second DT signal has no aliasing:

$$x_{p2}[n]=5\cos\left[2\pi\cdot\frac{1}{9}\cdot n+\frac{\pi}{4}\right]+3\cos\left[2\pi\cdot\frac{2}{9}\cdot n+\frac{\pi}{5}\right]+2\cos\left[2\pi\cdot\frac{3}{9}\cdot n+\frac{\pi}{6}\right]. \quad (4.5\text{-}3)$$

The CT signal above, along with its sampled and reconstructed signals is shown in Figure 4.5-15.

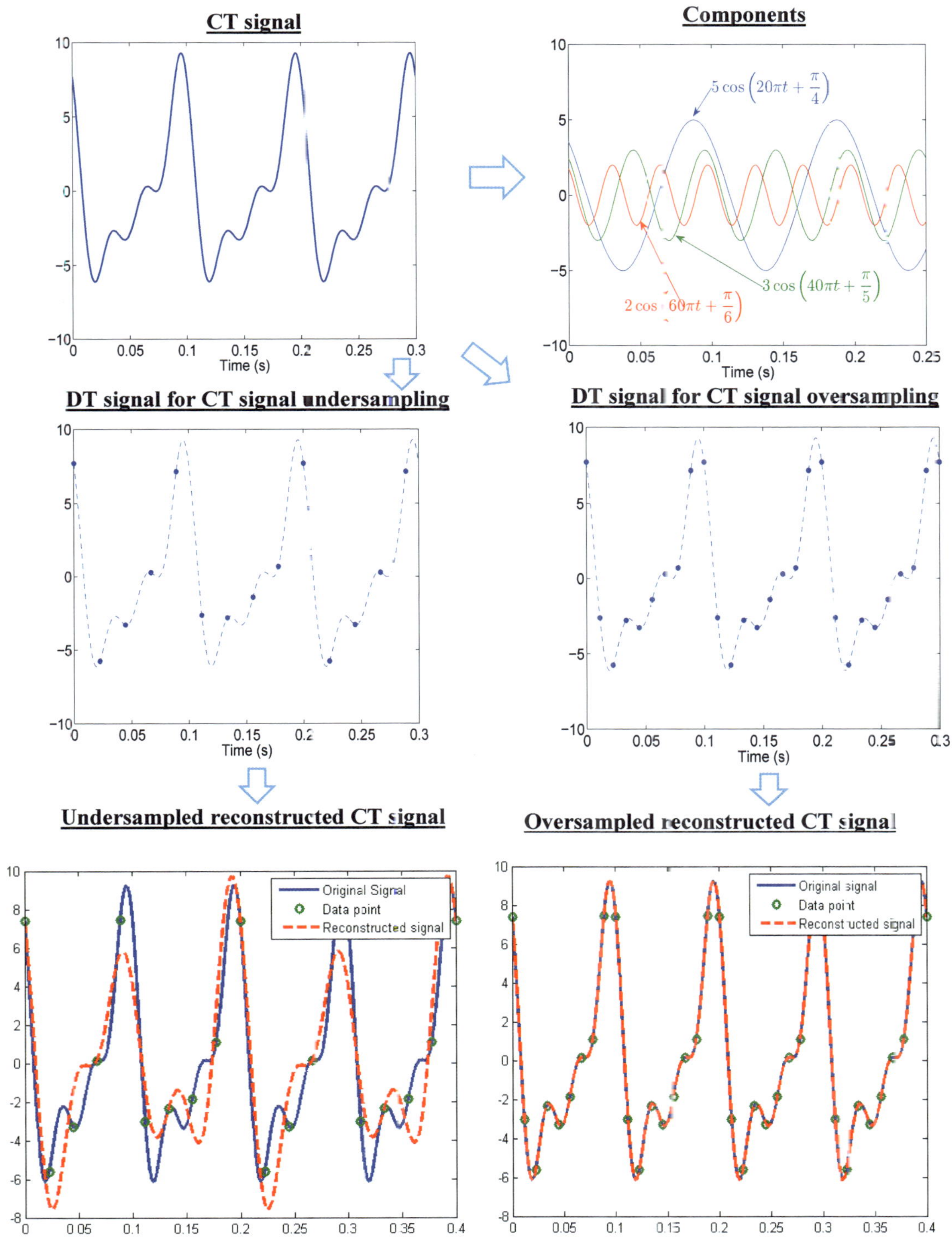

Fig. 4.5-15. Undersampling and oversampling of a CT signal (4.5-1). Undersampling halved the highest frequency.

4.5.3.2 *Spectral image folding and repetition on digital frequency axis*

The frequency content of a CT signal is typically visualized with the help of the spectrum (or spectral) plots of the DT signal version. To bring out the pitfalls of this visualization, let's first consider plotting the DT signal 1 magnitude spectrum, as given in (4.5-2), corresponding to the first harmonic of the CT signal (4.5-1).

There are three key steps in doing so. The first one is to mark the sampling rate and its half, given here as 45 Hz and 22.5 Hz, respectively, on the CT frequency axis. Then the spectrum line location with respect to the $S_F/2$ immediately shows whether or not the CT spectrum is within the Nyquist frequency range. In Figure 4.5-16 the first CT harmonic is within this range: $f_1 < S_F / 2$.

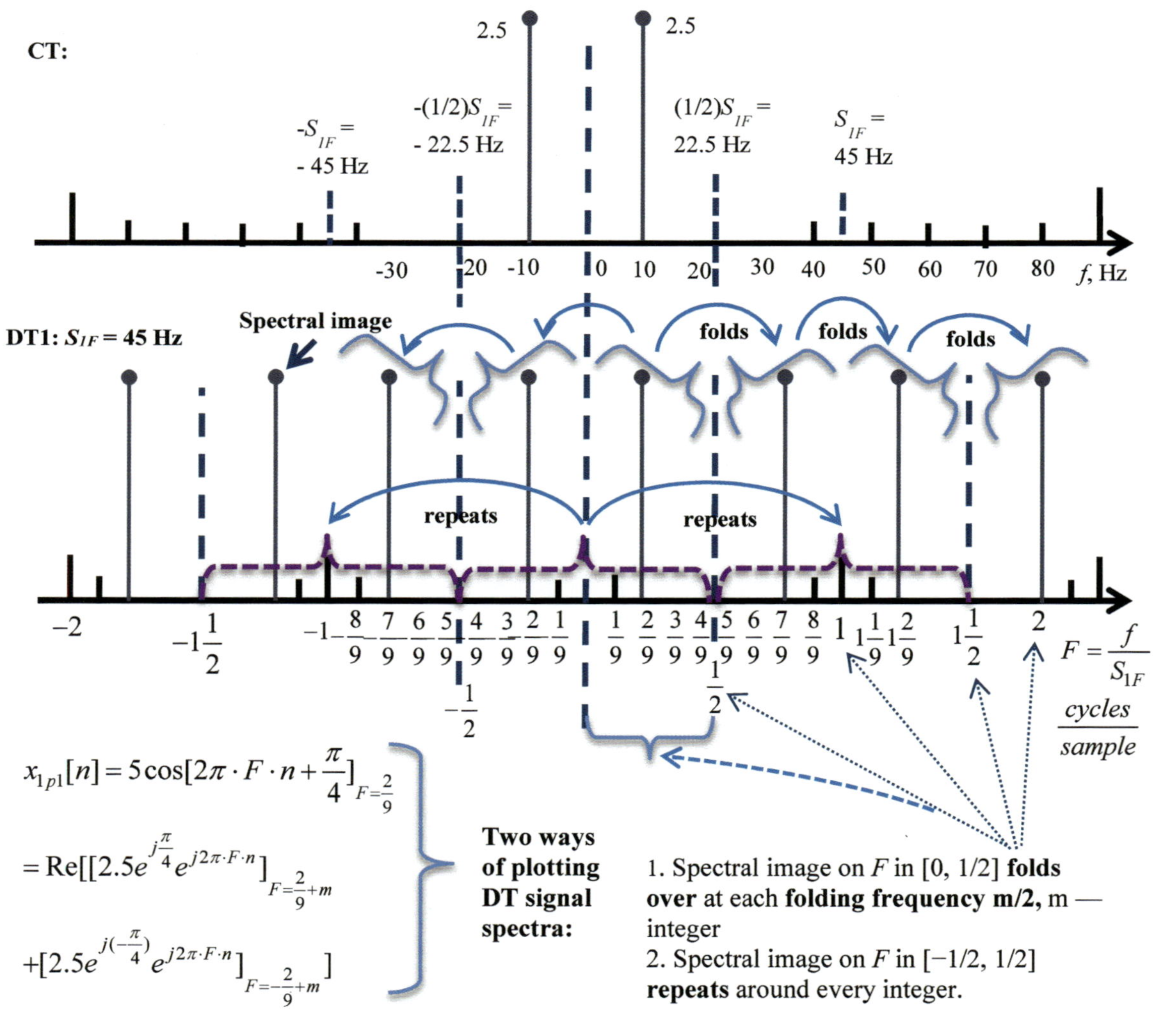

Fig. 4.5-16. DT signal spectrum folding and repetition.

The next step is to draw the digital frequency axis. This is done by simply rescaling the CT frequency axis by S_F as $F = f/S_F$. Then, the DT counterparts of the CT axis points $S_F = 45$ Hz and $S_F/2 = 22.5$ Hz will be simply F = 1 and F = 1/2, respectively, clearly delineating the principal range [0, 1/2] on the digital frequency axis.

Finally the CT spectral lines are transferred to the DT spectral plot at the digital frequency points $F_2 = f_1/S_F = 2/9$ and $-F_2 = -2/9$, but because the addition of integer does not change the DT function, spectral lines repeat around every integer or fold around every half-integer along the digital frequency axis as shown in Figure 4.5-13. This is why $F = 1/2$ is called the *folding frequency*.

As an exercise, try to unbundle folding from repetition, shown together in Figure 4.5-16, by sketching separate DT1 graph for each.

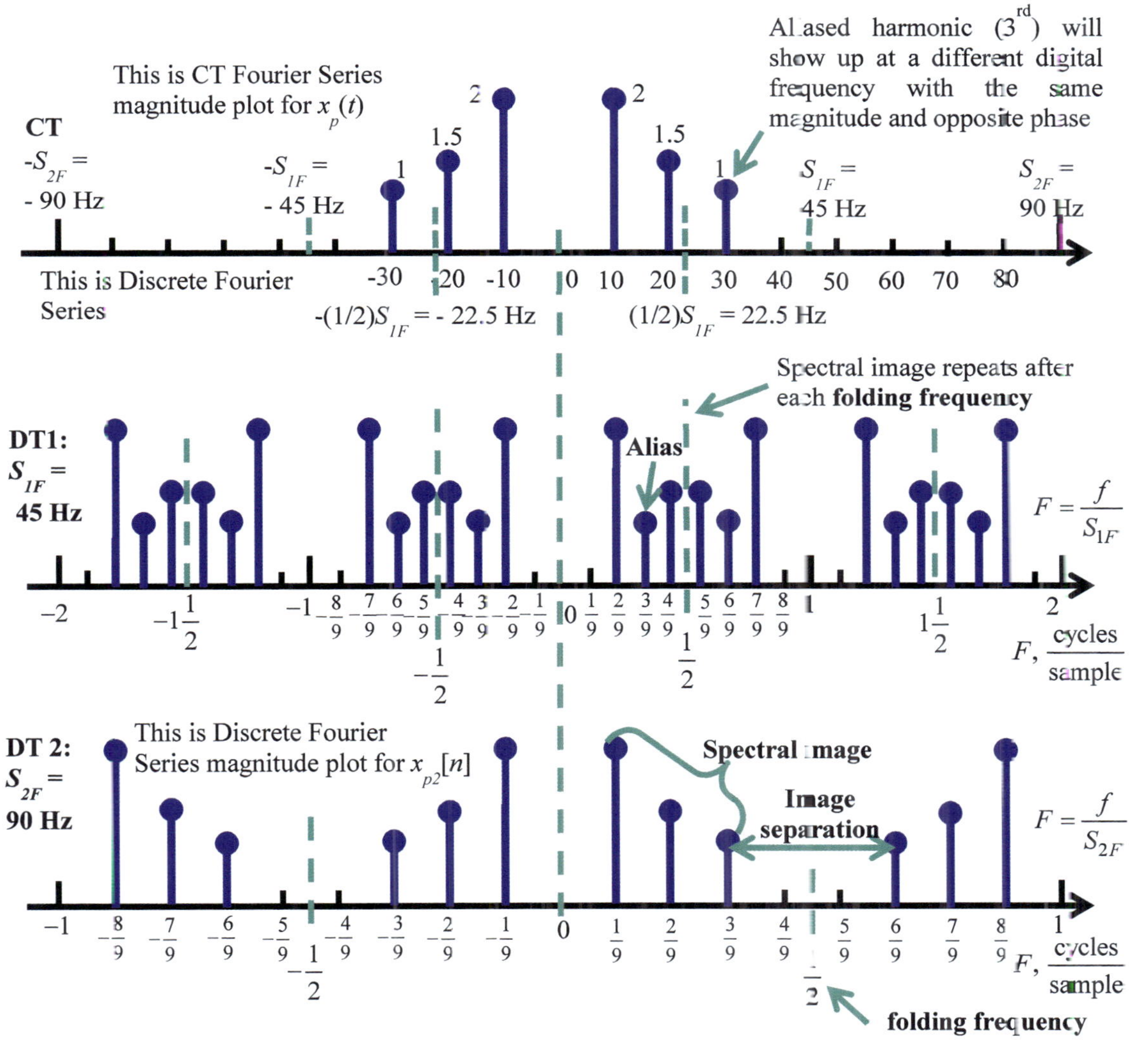

Fig. 4.5-17. Two-sided magnitude spectrum of CT signal (4.5-1) and its aliased and not aliased DT versions.

Figure 4.5-17 shows DT magnitude spectrum of (4.5-1) with all the components, clearly contrasting the aliasing of the third harmonic for $S_F = 45$ Hz in the middle plot against its non-aliased version for $S_F = 90$ Hz in the lower plot.

4.5.3.3 *Aliasing as spectral image overlap*

Another way to think of aliasing is in terms of the overlapping spectral images, as shown in Figure 4.5-18. Empty space between the images is called **image separation**, which relaxes sharp cutoff of recovery filters that reconstruct analog signals from samples, and reduces distortion caused by aliasing of the low-magnitude harmonics beyond f_B passed by real anti-aliasing (low-pass) filter.

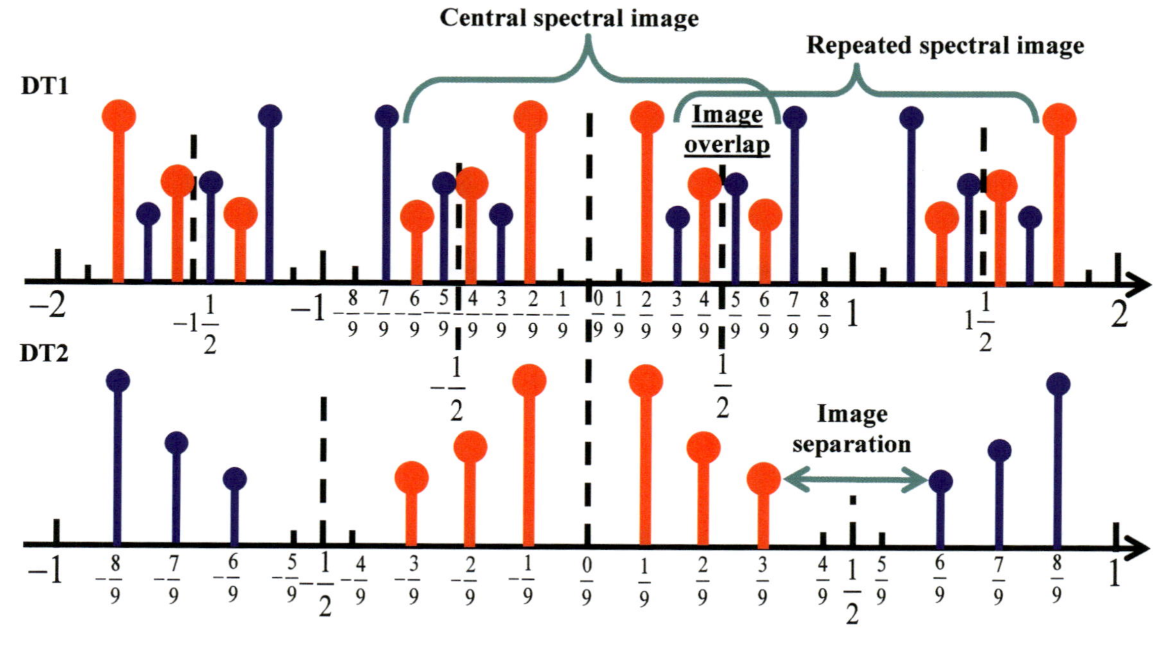

Fig. 4.5-18. DT aliasing interpreted as image overlap.

Further increase in sampling rate makes central spectral image cluster tighter around 0, producing bigger image separation and permitting carrying out further reduction of the undesirable high frequency magnitudes by the digital filters with less phase distortion in the frequency range of interest.

Homework 6: Solved Examples and Solution Guide

Problem 1: Determine if the signal $x(t) = e^{j2\pi f_1 t}$ is periodic; if it is, find its period. Using this result determine if the following signals are periodic, and if they are, find their periods:
(a) $y_1(t) = (3/j)x(0.5-5t) - 3(-2/j)x(0.3t)$,

(b) $y_2(t) = 3e^{j8\pi t} + 5e^{-6\pi t}$,

(c) $y_3(t) = y_2((3/(-j2\pi))t)$,

(d) $y_4(t) = y_2((-1/3\pi)t)$,

(e) $y_5(t) = \sum_{k=-\infty}^{\infty} X[k]e^{j2\pi k f_1 t}$,

(f) $y_6(t) = y_5((3/(-j2\pi))t)$,

(g) $y_7(t) = y_5((-1/3\pi)t)$.

Solution: Euler's identity: $e^{\pm j\alpha} = \cos\alpha \pm j\sin\alpha$.

Thus, $x(t) = e^{j2\pi f_1 t} = \cos(2\pi f_1 t) + j\sin(2\pi f_1 t)$ or $x(t) = e^{j\omega_1 t} = \cos(\omega_1 t) + j\sin(\omega_1 t)$.

Since each term has period $T_1 = \frac{1}{f_1}$, $x(t)$ is T_1 periodic.

(a) For the signal $y_1(t) = (3/j)x(0.5-5t) - 3(-2/j)x(0.3t)$,

$$y_1(t) = (3/j)e^{j2\pi f_1(0.5-5t)} + (6/j)e^{j2\pi f_1(0.3t)}$$
$$= \frac{3}{j}\left[\cos(2\pi f_1(0.5-5t)) + j\sin(2\pi f_1(0.5-5t))\right] + \frac{6}{j}\left[\cos(2\pi f_1(0.3t)) + j\sin(2\pi f_1(0.3t))\right]$$
$$= 3\left[-j\cos(2\pi f_1(0.5-5t)) + \sin(2\pi f_1(0.5-5t))\right] + 6\left[-j\cos(2\pi f_1(0.3t)) + \sin(2\pi f_1(0.3t))\right],$$

$$F_1 = \frac{10\pi f_1}{2\pi} = 5f_1;\ F_2 = \frac{0.6\pi f_1}{2\pi} = 0.3f_1 \Rightarrow T_1 = \frac{1}{5f_1};\ T_2 = \frac{10}{3f_1}.$$

Fundamental frequency $\frac{f_1}{10} \Rightarrow T = \frac{10}{f_1}$, so $y_1(t)$ is periodic with $T = \frac{10}{f_1}$.

(b) For the signal $y_2(t) = 3e^{j8\pi t} + 5e^{-6\pi t}$,

by Euler's formula, we have: $y_2(t) = 3(\cos(8\pi t) + j\sin(8\pi t)) + \underbrace{5e^{-6\pi t}}_{\text{nonperiodic}}$.

Since the second term is a decaying exponential, $y_2(t)$ is not periodic.

(c) For the signal $y_3(t) = y_2((3/(-j2\pi))t)$,

$$y_3(t) = y_2((3/(-j2\pi))t) = 3e^{j8\pi(3/(-j2\pi))t} + 5e^{-6\pi(3/(-j2\pi))t} = \underbrace{3e^{-12t}}_{\text{nonperiodic}} + 5e^{-j9t}.$$

Since the first term is a decaying exponential, $y_3(t)$ is not periodic.

(d) For the signal $y_4(t) = y_2((-1/3\pi)t)$,

$$y_4(t) = y_2((-1/3\pi)t) = 3e^{j8\pi(-1/3\pi)t} + 5e^{-6\pi(-1/3\pi)t} = 3e^{-j(8/3)t} + \underbrace{5e^{2t}}_{\text{nonperiodic}}.$$

Since the second term is a growing exponential, $y_4(t)$ is not periodic.

(e) For the signal $y_5(t)=\sum_{k=-\infty}^{\infty} X[k]e^{j2\pi kf_1t}$,

$$y_5(t)=\sum_{k=-\infty}^{\infty} X[k]e^{j2\pi kf_1t}=\sum_{k=-\infty}^{\infty} X[k]\left(\cos(2\pi kf_1t)+j\sin(2\pi kf_1t)\right).$$

Since $X[k]$ is just a coefficient not dependent on *t*, $y_5(t)$ is periodic with fundamental frequency f_1.

Therefore, $y_5(t)$ is periodic with period $T=\frac{1}{f_1}$.

(f) For the signal $y_6(t)=y_5((3/(-j2\pi))t)$,

$$y_6(t)=\sum_{k=-\infty}^{\infty} X[k]e^{j2\pi kf_1(3/(-j2\pi))t}=\sum_{k=-\infty}^{\infty} X[k]e^{-3kf_1t}.$$

Since every term of $y_6(t)$ is exponential decaying, $y_6(t)$ is not periodic and will eventually decay to 0.

(g) For the signal $y_7(t)=y_5((-1/3\pi)t)$,

$$y_7(t)=\sum_{k=-\infty}^{\infty} X[k]e^{j2\pi kf_1(-1/3\pi)t}=\sum_{k=-\infty}^{\infty} X[k]e^{-j(2/3)kf_1t}$$

$$=\sum_{k=-\infty}^{\infty} X[k]\left(\cos(2\pi k(f_1/3\pi)t)-j\sin(2\pi k(f_1/3\pi)t)\right).$$

Since $X[k]$ is a coefficient independent on *t*, $y_7(t)$ is periodic with fundamental frequency $f_1/3\pi$.

Therefore, $y_7(t)$ is periodic with period $T=3\pi/f_1$.

Problem 2: Each value of complex-valued signal $x[n]$ can be represented by its polar form or rectangular form as a function of *n*. (a) Draw both representations for a signal $x[n]=\left\{5,\underset{\uparrow}{-1},-3+4j,4-3j,-5j\right\}$; (b) Using Matlab, plot both representations for a signal $x[n]=e^{-j0.8n\pi}$, check if *x*[*n*] periodic, and if it is, establish its period from the Matlab plots. Indicate which plots clearly show the period of this function.

Solution:

(a) $x[n]=\left\{5,\underset{\uparrow}{-1},-3+4j,4-3j,-5j\right\}$; $x[n]=RE+jIM$; $\Rightarrow$

$$x[n]_{magnitude}=\sqrt{RE^2+IM^2}=\{5,1,5,5,5\},\quad x[n]_{phase}=\tan^{-1}\left(\tfrac{IM}{RE}\right)=\left\{0,-\pi,2.2143,-0.6435,-\frac{\pi}{2}\right\}.$$

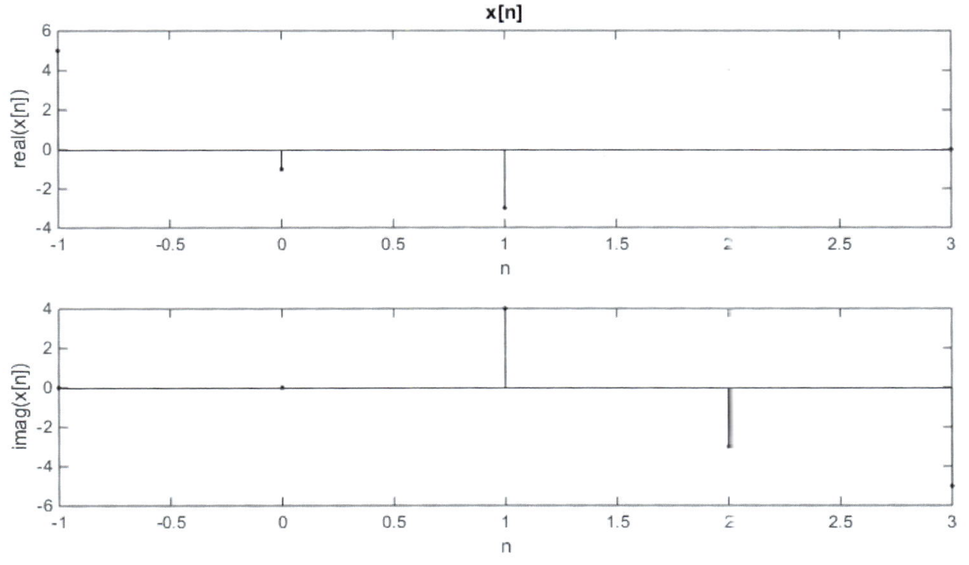

Fig. Hw6-1. Real and imaginary part of x[n] for Problem 2(a).

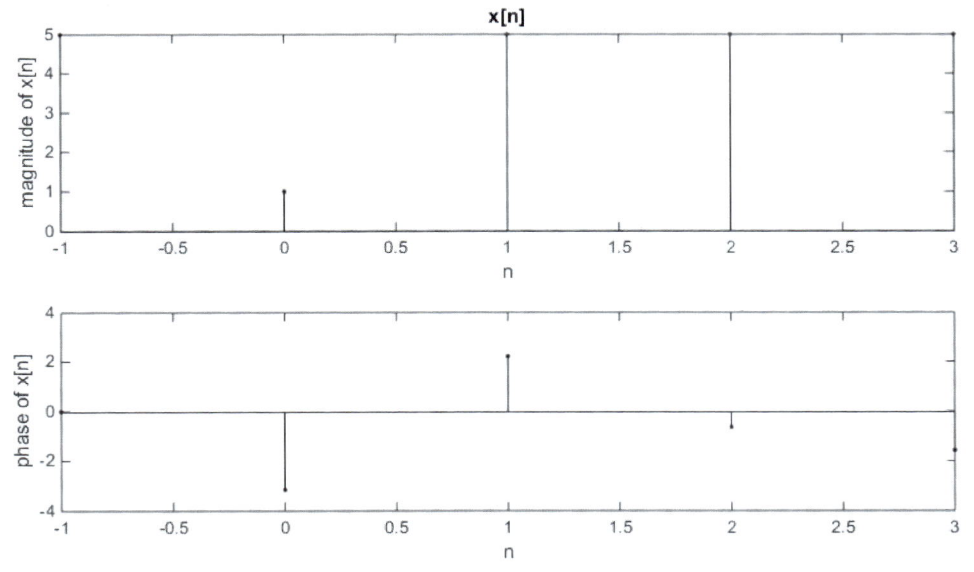

Fig. Hw6-2. Magnitude and phase of x[n] for Problem 2(a).

(b) $x[n]=e^{-j0.8n\pi}$; $e^{-j0.8\pi n}=e^{j(-0.8)n\pi}=e^{j2\pi(-F)n}$ $\therefore$ $F=\dfrac{0.8}{2}=0.4$, $N=\dfrac{5}{2}$, but signal $x(n+N)$ is defined only for integer argument, therefore period has to be integer, so $N=5$. From the two plots, we can detect $x[n]=e^{-j0.8n\pi}$ is periodic, with period $N=5$. Compare the two plots, we can conclude the magnitude and phase spectrum shows the period of the function more clearly.

MATLAB commands:

Figure HW6-3:

```
n=-10:1:10;x=exp(-j*0.8*n*pi);
subplot(2,1,1),stem(n,real(x)),subplot(2,1,2),stem(n,imag(x));
```

Figure HW6-4:

```
n=-10:1:10;x=exp(-j*0.8*n*pi);
subplot(2,1,1),stem(n,abs(x)),subplot(2,1,2),stem(n,angle(x));
```

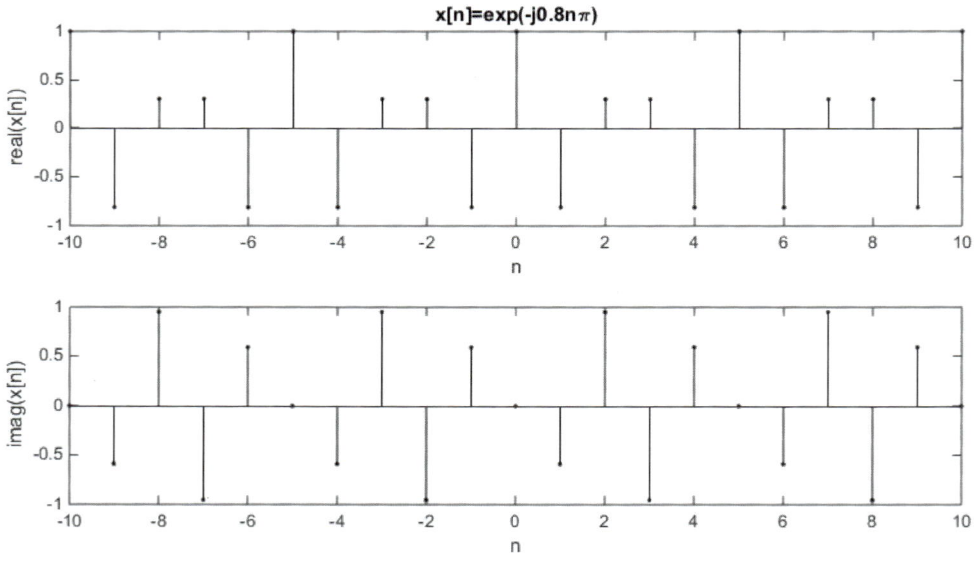

Fig. Hw6-3. Real and imaginary part for Problem 2(b).

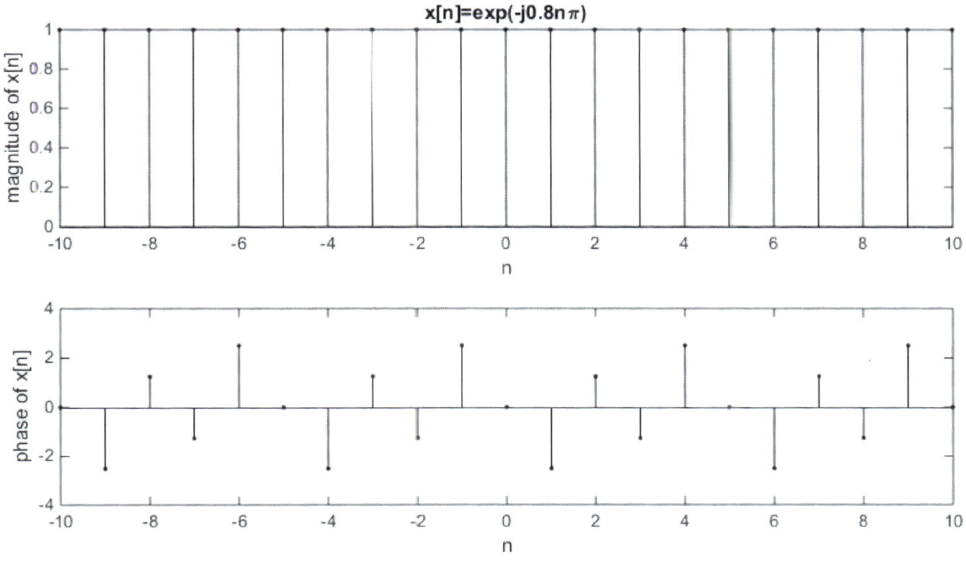

Fig. Hw6-4. Magnitude and phase spectrum for Problem 2(b).

Problem 3 (Grey box system identification via frequency response): An electric motor in a blender can be roughly represented by a series connection of R and L.

(a) Determine R and L values by making the following tests:

i) AC: $V = 110\ V_{(RMS)}$, $f = 60\ Hz$, $\Rightarrow |I_{ac}| = 8\ A$,

ii) DC: $V = 5\ V$, $f = 0\ Hz$, $\Rightarrow I_{dc} = 8A$.

(b) Now, taking the voltage drops across R and L as two separate outputs, find the corresponding input/output transfer functions and plot the input (both magnitude and phase) voltage spectra to the motor and the corresponding voltage spectra for each output for both i) and ii) of part (a).

Solution:

(a) For DC current, $L \to 0$. Therefore, we can simply use Ohm's law to find the resistance, R, in series with the voltage supply, V.

$$V = IR \Rightarrow R = \frac{V}{I} = \frac{5V}{8A} = 0.625\Omega.$$

For AC current,

$$|V| = |110/0.7071| = |I||Z_{eq}| = I_{ac}\ R + Z_L| = I_{ac}|R + j\omega L| = I_{ac}\left|\sqrt{R^2 + (\omega L)^2}\right|,$$

$110/0.7071 = (8)\left|\sqrt{(0.625)^2 + (2\pi \cdot 60 \cdot L)^2}\right|$. Solve for L: $L = 0.0516\ \text{H}$.

(b) Denote the voltage drops across R and L by V_R and V_L, using complex impedances, we can write,

$$G_R(s) = \frac{V_R}{V} = \frac{IZ_R}{I(Z_R + Z_L)} = \frac{Z_R}{Z_R + Z_L} = \frac{R}{R + Ls} \approx \frac{0.625}{0.625 + 0.0516s},$$

$$G_L(s) = \frac{V_L}{V} = \frac{IZ_L}{I(Z_R + Z_L)} = \frac{Z_L}{Z_R + Z_L} = \frac{Ls}{R + Ls} \approx \frac{0.0364s}{0.625 + 0.0516s}.$$

For both cases, assume the input signal has phase 0, then the magnitudes and phases of the outputs can be calculated using the transfer functions evaluated at $j\omega$.
For case i) in (a), $\omega = 120\pi$ rad/s, then

$$G_R(s) \approx \frac{0.625}{0.625 + 0.0516 \cdot 120\pi j} = \frac{0.625}{0.625 + 19.453j},$$

$$G_L(s) \approx \frac{0.0516 \cdot 120\pi j}{0.625 + 0.0516 \cdot 120\pi j} = \frac{19.453j}{0.625 + 19.453j}.$$

This means,

$$|V_R| = |V||G_R| \approx (110/0.7071)\left|\frac{0.625}{0.625 + 19.453j}\right| = 155.565\frac{0.625}{\sqrt{0.625^2 + 19.453^2}} = 4.996,$$

$$\angle V_R = \angle V + \angle G_R \approx 0 + \angle\frac{0.625}{0.625 + 19.453j} = \angle 0.625 - \angle(0.625 + 19.453j) = -1.539$$

and

$$|V_L| = |V||G_L| \approx (110/0.7071)\left|\frac{19.453j}{0.625 + 19.453j}\right| = 155.565\frac{19.453}{\sqrt{0.625^2 + 19.453^2}} = 155.485,$$

$$\angle V_L = \angle V + \angle G_L \approx 0 + \angle\frac{19.453j}{0.625 + 19.453j} = \angle 19.453j - \angle(0.625 + 19.453j) = 0.0321.$$

For case ii) in (a), $\omega = 0$ rad/s, then

$$G_R(s) = \frac{0.625}{0.625 + 0.0516 \cdot 0j} = 1,, \; G_L(s) = \frac{0.0516 \cdot 0j}{0.625 + 0.0516 \cdot 0j} = 0.$$

This means

$$|V_R| = |V||G_R| = 5|1| = 5, \; \angle V_R = \angle V + \angle G_R = 0 + \angle 1 = 0$$

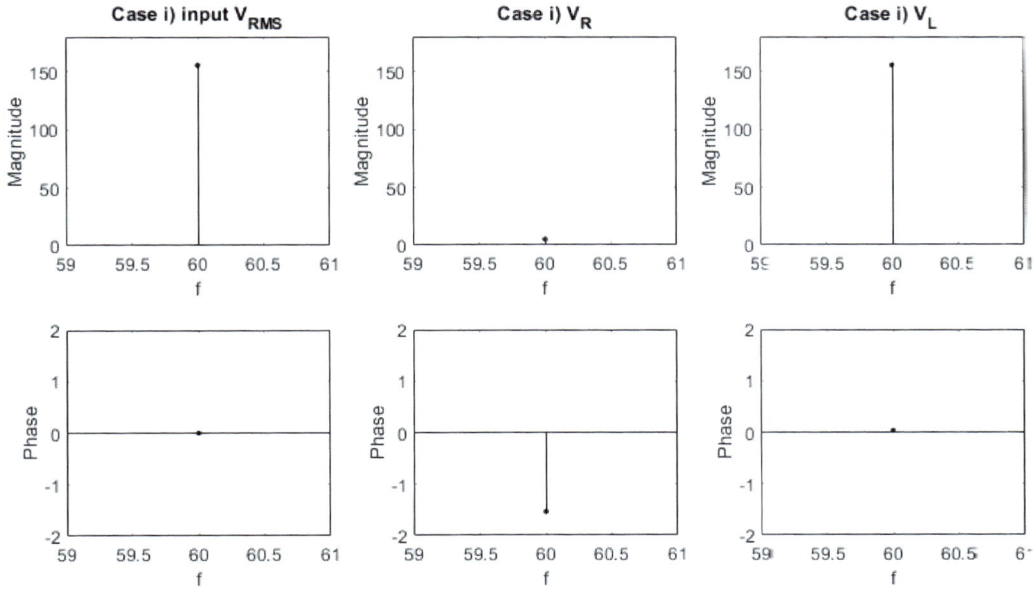

Fig. Hw6-5. Magnitude and phase spectrum for case i) of Problem 3.

and $|V_L| = |V||G_L| = 5|0| = 0$, $\angle V_L = \angle V + \angle G_L = 0 + \angle 0 = 0 = 0$.

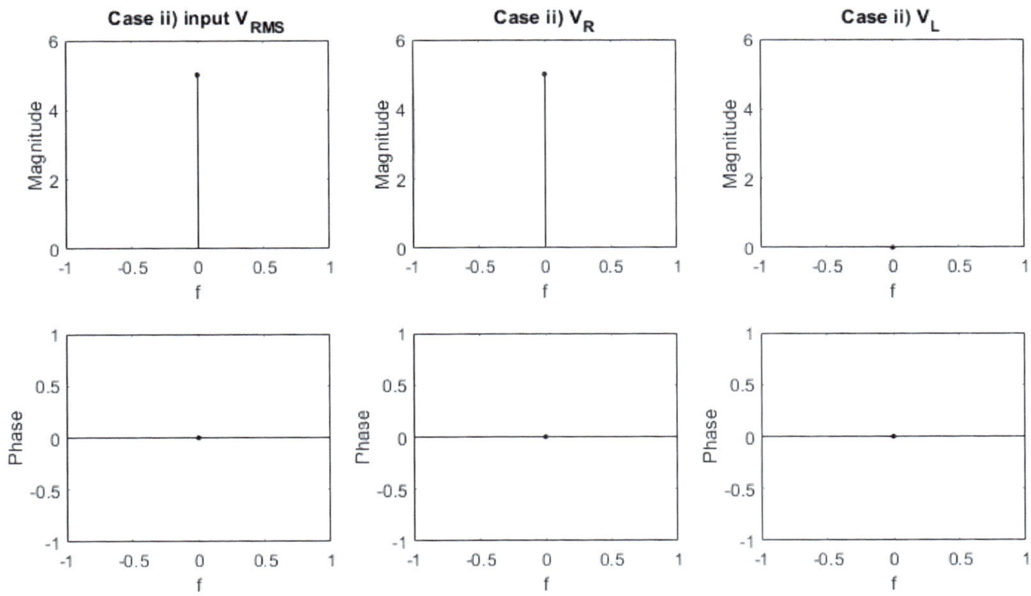

Fig. Hw6-6. Magnitude and phase spectrum for case ii) of Problem 3.

Problem 4: Consider a blender with the model identified in Problem 3 and a coffee/tea immersion heater modeled as a $12\,\Omega$ resistor. Determine the current magnitude for:
(a) the blender for $|V|=90\ V,\ \ f=60\ Hz$;
(b) the heater for $|V|=10\ V,\ \ f=600\ Hz$;
(c) the heater for $|V|=100\ V,\ \ f=60\ Hz$.

Now, taking the voltage as the input and the current as the output, derive the corresponding input/output transfer function of each appliance and plot the input (both magnitude and phase) spectra and the corresponding current output spectra for (a), (b), and (c).

Solution:

(**a**) For the blender, where $|V|=90V$ and $\omega=2\pi f=120\pi$ rad/sec,

$$|I|=\frac{|V|}{|Z|}=\frac{|V|}{|R+j\omega L|}=\frac{90}{|0.625+j(120\pi)(0.0516)|}=\frac{90}{\sqrt{0.625^2+(120\pi)^2(0.0516)^2}}=4.6242A.$$

Using complex impedances, we can write

$$G_{IV}(s)=\frac{I}{V}=\frac{I}{I(Z_R+Z_L)}=\frac{1}{Z_R+Z_L}=\frac{1}{R+Ls}.$$

Assume the phase of the input V is 0, then the phase of I is

$$\angle I=\angle V+\angle G_{IV}=0+\angle\frac{1}{0.625+j0.0516\cdot 120\pi}=-1.5387.$$

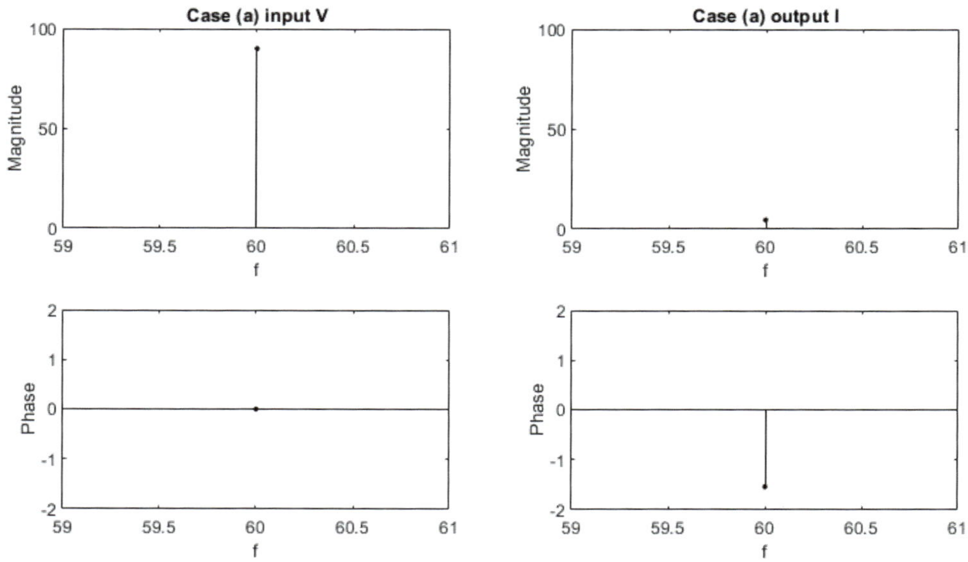

Fig. Hw6-7. Magnitude and phase spectrum for case (a) of Problem 4.

(b) For the heater, pure resistance, $R = 12\ \Omega$, where $|V| = 10V$ and $f = 600$ Hz,

$$|I| = \frac{|V|}{|R|} = \frac{10}{12} = 0.8333A.$$

Using complex impedances, we can write

$$G_{IV}(s) = \frac{I}{V} = \frac{I}{IZ_R} = \frac{1}{Z_R} = \frac{1}{R}.$$

Assume the phase of the input V is 0, then the phase of I is

$$\angle I = \angle V + \angle G_{IV} = 0 + \angle \frac{1}{0.625} = 0.$$

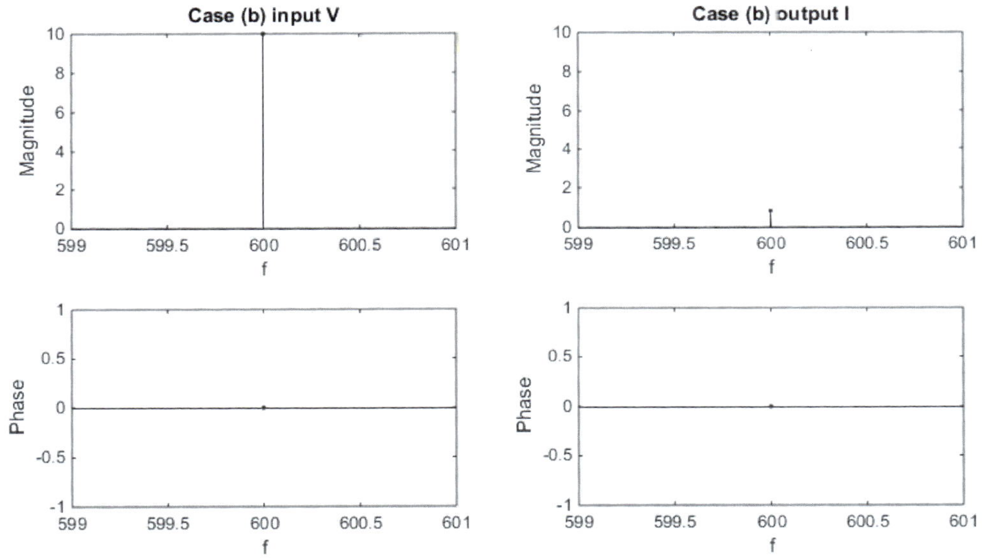

Fig. Hw6-8. Magnitude and phase spectrum for case (b) of Problem 4.

(c) For the heater, pure resistance, $R = 12\ \Omega$, where $|V| = 100V$ and $f = 60$ Hz,

$$|I| = \frac{|V|}{|R|} = \frac{100}{12} = 8.333A.$$

Using complex impedances, we can write

$$G_{IV}(s) = \frac{I}{V} = \frac{I}{IZ_R} = \frac{1}{Z_R} = \frac{1}{R}.$$

Assume the phase of the input V is 0, then the phase of I is

$$\angle I = \angle V + \angle G_{IV} = 0 + \angle \frac{1}{0.625} = 0.$$

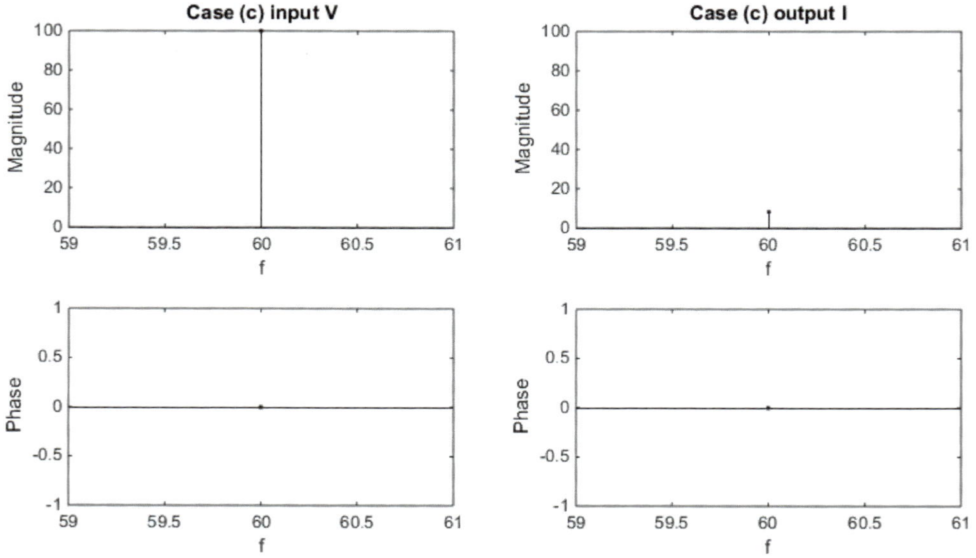

Fig. Hw6-9. Magnitude and phase spectrum for case (c) of Problem 4.

Problem 5. Assume that for a given reference input, under Bosch controller, the production unit mold position signal is:

$$x_p(t) = 0.3 + 3.1\cos(2\pi 4.4t - 3\pi/8) + 0.12\cos(2\pi 8.8t - \pi/5) + 0.23\cos(2\pi 13.2t + \pi/7).$$

Represent this signal in terms of CT complex harmonics, write Exponential Fourier Series coefficients for this signal. Using MATLAB, plot single-sided and double-sided magnitude and phase spectra for this signal, and indicate in detail how Exponential Fourier Series coefficients of this signal are related to its spectra.

Solution: None given. The problem is left for an independent study.

Problem 6: Assume that $x(t)=\cos\left(2\pi f_1 t+\frac{\pi}{4}\right)$ is sampled as follows, starting at $t = 0$: (a) 3 samples over 1 period; (b) 16 samples over 5 periods; (c) 16 samples over 9 periods. Find the DFS coefficients of the sampled signal in each case.

Solution:

$$x(t)=\sin\left(2\pi f_1 t+\tfrac{\pi}{4}\right)=\cos\left(2\pi f_1 t+\tfrac{\pi}{4}-\tfrac{\pi}{2}\right)=\cos\left(2\pi f_1 t-\tfrac{\pi}{4}\right),$$

(a) 3 samples over 1 period

$$\frac{3}{1/f_1}=3f_1=S_f \text{ (no aliasing),}$$

$$N_1=3,\ k=1 \Rightarrow x(n)=\cos\left(2\pi\frac{1}{3}n-\frac{\pi}{4}\right)=\frac{1}{2}e^{j\frac{\pi}{4}}e^{j2\pi\left(-\frac{1}{3}\right)n}+\frac{1}{2}e^{-j\frac{\pi}{4}}e^{j2\pi\frac{1}{3}n}.$$

Nonzero coefficients at $k=-1$ and $k=1$ in the principal range, or since coefficients repeat every N_1 harmonic numbers, $k=1$ and $k=2$ in the range $0\le F\le 1$,

$$\{X_{DFS}[k]\}=\{0,\tfrac{1}{2}e^{-j\frac{\pi}{4}},\tfrac{1}{2}e^{j\frac{\pi}{4}}\}.$$

(b) 16 samples over 5 periods

$$\frac{16}{5}f_1=S_f \text{ (no aliasing), } N_1=16,\ k=5,\ x(n)=\cos\left(2\pi\tfrac{5}{16}n-\tfrac{\pi}{4}\right).$$

Nonzero coefficient at $k = 5$ and $k = N_1\text{-}5 = 11$ for range of F in [0,1],

$$\{X_{DFS}[k]\}=\{0,0,0,0,0,\tfrac{1}{2}e^{-j\frac{\pi}{4}},0,0,0,0,0,\tfrac{1}{2}e^{j\frac{\pi}{4}},0,0,0,0\}.$$

(c) 16 samples over 9 periods

$$\frac{16}{9}f_1=S_f \text{ (aliasing occurs!), } N_1=16,\ k=9,\ x(n)=\cos\left(2\pi\tfrac{9}{16}n-\tfrac{\pi}{4}\right).$$

Nonzero coefficient at $k = 9$ and $k = 16\text{-}9 = 7$:

$$\{X_{DFS}[k]\}=\{0,0,0,0,0,0,0,\tfrac{1}{2}e^{j\frac{\pi}{4}},0,\tfrac{1}{2}e^{-j\frac{\pi}{4}},0,0,0,0,0,0\}.$$

Problem 7: Assume that the production unit mold position signal is

$$x(t)=5\cos\left(2\pi(10)t+\tfrac{\pi}{4}\right)+3\cos\left(2\pi(20)t+\tfrac{\pi}{5}\right)+2\cos\left(2\pi(30)t+\tfrac{\pi}{6}\right).$$

Compute a) 2 times undersampled, and b) 3 and 6 times oversampled DT versions of this function in polar form. Determine if aliasing takes place and fold the aliased signal into the principal range (don't leave terms with the same digital frequency as separate). Indicate DFS of each sampled function.

Solution:

a) Compute 2 times undersampled (aliasing occurs)

By observation, $f_0 = 10$; f_b (bandlimit) $= 3 f_0 = 30$; (highest frequency = 30 Hz)

$S_N = 2 f_b = 60$ Hz (critical sampling rate).

For 2 times undersampling, $S_f = (60)/2 = 30$ Hz

$F_1 = \frac{10}{30} = \frac{1}{3}; F_2 = \frac{20}{30} = \frac{2}{3} \to -\frac{1}{3}; F_3 = \frac{30}{30} = 1 \to 0$ **(into the principal range, i.e. folding)**

$$x[n] = 5\cos\left(2\pi\tfrac{1}{3}n + \tfrac{\pi}{4}\right) + 3\cos\left(-2\pi\tfrac{1}{3}n + \tfrac{\pi}{5}\right) + 2\cos\left(2\pi 0n + \tfrac{\pi}{6}\right)$$
$$= 5\cos\left(2\pi\tfrac{1}{3}n + \tfrac{\pi}{4}\right) + 3\cos\left(2\pi\tfrac{1}{3}n - \tfrac{\pi}{5}\right) + 2\cos\left(2\pi\tfrac{0}{3}n + \tfrac{\pi}{6}\right).$$

Since the first and second terms have the same digital frequency, then we continue the transform,

$$x[n] = 5\cos\left(2\pi\tfrac{1}{3}n + \tfrac{\pi}{4}\right) + 3\cos\left(2\pi\tfrac{1}{3}n - \tfrac{\pi}{5}\right) + 2\cos\left(2\pi\tfrac{0}{3}n + \tfrac{\pi}{6}\right)$$
$$= 5\left[\cos\left(2\pi\tfrac{1}{3}n\right)\cos\left(\tfrac{\pi}{4}\right) - \sin\left(2\pi\tfrac{1}{3}n\right)\sin\left(\tfrac{\pi}{4}\right)\right] + 3\left[\cos\left(2\pi\tfrac{1}{3}n\right)\cos\left(\tfrac{\pi}{5}\right) + \sin\left(2\pi\tfrac{1}{3}n\right)\sin\left(\tfrac{\pi}{5}\right)\right]$$
$$+ 2\cos\left(2\pi\tfrac{0}{3}n + \tfrac{\pi}{6}\right)$$
$$= \left(5\cos\left(\tfrac{\pi}{4}\right) + 3\cos\left(\tfrac{\pi}{5}\right)\right)\cos\left(2\pi\tfrac{1}{3}n\right) + \left(3\sin\left(\tfrac{\pi}{5}\right) - 5\sin\left(\tfrac{\pi}{4}\right)\right)\sin\left(2\pi\tfrac{1}{3}n\right) + 2\cos\left(2\pi\tfrac{0}{3}n + \tfrac{\pi}{6}\right)$$
$$= 5.9626\cos\left(2\pi\tfrac{1}{3}n\right) - 1.7722\sin\left(2\pi\tfrac{1}{3}n\right) + 2\cos\left(2\pi\tfrac{0}{3}n + \tfrac{\pi}{6}\right)$$
$$= \sqrt{5.9626^2 + (-1.7722)^2}\cos\left(2\pi\tfrac{1}{3}n + \tan^{-1}\left(\frac{-1.7722}{5.9626}\right)\right) + 2\cos\left(2\pi\tfrac{0}{3}n + \tfrac{\pi}{6}\right)$$
$$= 6.2204\cos\left(2\pi\tfrac{1}{3}n\text{-}0.2889\right) + 2\cos\left(2\pi\tfrac{0}{3}n + \tfrac{\pi}{6}\right).$$

For DFS:

$$x[n] = (3.1102e^{-j0.2889}e^{j2\pi\frac{1}{3}n} + 3.1102e^{j0.2889}e^{-j2\pi\frac{1}{3}n}) + (e^{j\frac{\pi}{6}}e^{j2\pi\frac{0}{3}n} + e^{-j\frac{\pi}{6}}e^{-j2\pi\frac{0}{3}n})$$

since $e^{j2\pi n} = 1$, then,

$$x[n] = (3.1102e^{-j0.2889})e^{j2\pi\frac{1}{3}n} + (3.1102e^{j0.2889})e^{j2\pi\frac{2}{3}n} + (e^{j\frac{\pi}{6}} + e^{-j\frac{\pi}{6}})e^{j2\pi\frac{0}{3}n}$$
$$= (\sqrt{3}e^{j0})e^{j2\pi\frac{0}{3}n} + (3.1102e^{-j0.2889})e^{j2\pi\frac{1}{3}n} + (3.1102e^{j0.2889})e^{j2\pi\frac{2}{3}n}.$$

DFS coefficients: $\{X_{DFS}[0], X_{DFS}[1], X_{DFS}[2]\} = \{(\sqrt{3}e^{j0}), (3.1102e^{-j0.2889}), (3.1102e^{j0.2889})\}$ and

$x[n] = \sum_{k=0}^{N_1-1} X_{DFS}[k]e^{j2\pi\frac{k}{N_1}n}$ where $N_1 = 3$.

b) Compute 3 times oversampled (aliasing doesn't occur)

$$S_f = (60) * 3 = 180 \text{ Hz}, \ F_1 = \tfrac{10}{180} = \tfrac{1}{18}; F_2 = \tfrac{20}{180} = \tfrac{2}{18}; F_3 = \tfrac{30}{180} = \tfrac{3}{18},$$
$$x[n] = 5\cos\left(2\pi\tfrac{1}{18}n + \tfrac{\pi}{4}\right) + 3\cos\left(2\pi\tfrac{2}{18}n + \tfrac{\pi}{5}\right) + 2\cos\left(2\pi\tfrac{3}{18}n + \tfrac{\pi}{6}\right).$$

For DFS:

$$
\begin{aligned}
x[n] &= (2.5e^{j\frac{\pi}{4}}e^{j2\pi\frac{1}{18}n} + 2.5e^{-j\frac{\pi}{4}}e^{-j2\pi\frac{1}{18}n}) + (1.5e^{j\frac{\pi}{5}}e^{j2\pi\frac{2}{18}n} + 1.5e^{-j\frac{\pi}{5}}e^{-j2\pi\frac{2}{18}n}) \\
&\quad + (e^{j\frac{\pi}{6}}e^{j2\pi\frac{3}{18}n} + e^{-j\frac{\pi}{6}}e^{-j2\pi\frac{3}{18}n}) \\
&= (2.5e^{j\frac{\pi}{4}}e^{j2\pi\frac{1}{18}n} + 2.5e^{-j\frac{\pi}{4}}e^{-j2\pi\frac{1}{18}n}\cdot e^{j2\pi n}) + (1.5e^{j\frac{\pi}{5}}e^{j2\pi\frac{2}{18}n} + 1.5e^{-j\frac{\pi}{5}}e^{-j2\pi\frac{2}{18}n}\cdot e^{j2\pi n}) \\
&\quad + (e^{j\frac{\pi}{6}}e^{j2\pi\frac{3}{18}n} + e^{-j\frac{\pi}{6}}e^{-j2\pi\frac{3}{18}n}\cdot e^{j2\pi n}) \\
&= 2.5e^{j\frac{\pi}{4}}e^{j2\pi\frac{1}{18}n} + 2.5e^{-j\frac{\pi}{4}}e^{j2\pi\frac{17}{18}n} + 1.5e^{j\frac{\pi}{5}}e^{j2\pi\frac{2}{18}n} + 1.5e^{-j\frac{\pi}{5}}e^{j2\pi\frac{16}{18}n} - e^{j\frac{\pi}{6}}e^{j2\pi\frac{3}{18}n} + e^{-j\frac{\pi}{5}}e^{j2\pi\frac{15}{13}n}.
\end{aligned}
$$

DFS coefficients:

$$\{X_{DFS}[0], X_{DFS}[1], X_{DFS}[2], X_{DFS}[3], X_{DFS}[4], \ldots, X_{DFS}[14], X_{DFS}[15], X_{DFS}[16], X_{DFS}[17]\}$$

$$= \left\{0, 2.5e^{j\frac{\pi}{4}}, 1.5e^{j\frac{\pi}{5}}, e^{j\frac{\pi}{6}}, 0, \ldots, 0, e^{-j\frac{\pi}{6}}, 1.5e^{-j\frac{\pi}{3}}, 2.5e^{-j\frac{\pi}{4}}\right\} \quad \text{and} \quad x[n] = \sum_{k=0}^{N_1-1} X_{DFS}[k]e^{j2\pi\frac{k}{N_1}n} \quad \text{where}$$

$N_1 = 18$.

Compute 6 times oversampled (aliasing doesn't occur)

$$S_f = (60) * 6 = 360 \text{ Hz}, \; F_1 = \tfrac{10}{360} = \tfrac{1}{36}; F_2 = \tfrac{20}{360} = \tfrac{2}{36}; F_3 = \tfrac{30}{180} = \tfrac{3}{36},$$

$$x[n] = 5\cos\left(2\pi\tfrac{1}{36}n + \tfrac{\pi}{4}\right) + 3\cos\left(2\pi\tfrac{2}{36}n + \tfrac{\pi}{5}\right) + 2\cos\left(2\pi\tfrac{3}{36}n + \tfrac{\pi}{6}\right).$$

For DFS:

$$
\begin{aligned}
x[n] &= (2.5e^{j\frac{\pi}{4}}e^{j2\pi\frac{1}{36}n} + 2.5e^{-j\frac{\pi}{4}}e^{-j2\pi\frac{1}{36}n}) + (1.5e^{j\frac{\pi}{5}}e^{j2\pi\frac{2}{36}n} + 1.5e^{-j\frac{\pi}{5}}e^{-j2\pi\frac{2}{36}n}) \\
&\quad + (e^{j\frac{\pi}{6}}e^{j2\pi\frac{3}{36}n} + e^{-j\frac{\pi}{6}}e^{-j2\pi\frac{3}{36}n}) \\
&= (2.5e^{j\frac{\pi}{4}}e^{j2\pi\frac{1}{36}n} + 2.5e^{-j\frac{\pi}{4}}e^{-j2\pi\frac{1}{36}n}\cdot e^{j2\pi n}) + (1.5e^{j\frac{\pi}{5}}e^{j2\pi\frac{2}{36}n} + 1.5e^{-j\frac{\pi}{5}}e^{-j2\pi\frac{2}{36}n}\cdot e^{j2\pi n}) \\
&\quad + (e^{j\frac{\pi}{6}}e^{j2\pi\frac{3}{36}n} + e^{-j\frac{\pi}{6}}e^{-j2\pi\frac{3}{36}n}\cdot e^{j2\pi n}) \\
&= 2.5e^{j\frac{\pi}{4}}e^{j2\pi\frac{1}{36}n} + 2.5e^{-j\frac{\pi}{4}}e^{j2\pi\frac{35}{36}n} + 1.5e^{j\frac{\pi}{5}}e^{j2\pi\frac{2}{36}n} + 1.5e^{-j\frac{\pi}{5}}e^{j2\pi\frac{34}{36}n} + e^{j\frac{\pi}{6}}e^{j2\pi\frac{3}{36}n} + e^{-j\frac{\pi}{6}}e^{j2\pi\frac{33}{36}n}.
\end{aligned}
$$

DFS coefficients:

$$\{X_{DFS}[0], X_{DFS}[1], X_{DFS}[2], X_{DFS}[3], X_{DFS}[4], \ldots, X_{DFS}[32], X_{DFS}[33], X_{DFS}[34], X_{DFS}[35]\}$$

$$= \left\{0, 2.5e^{j\frac{\pi}{4}}, 1.5e^{j\frac{\pi}{5}}, e^{j\frac{\pi}{6}}, 0, \ldots, 0, e^{-j\frac{\pi}{6}}, 1.5e^{-j\frac{\pi}{5}}, 2.5e^{-j\frac{\pi}{4}}\right\} \text{ and } x[n] = \sum_{k=0}^{N_1-1} X_{DFS}[k]e^{j2\pi\frac{k}{N_1}n}, \; N_1 = 36.$$

Problem 8: Compute and plot two-sided magnitude spectrum of the CT function of Problem 7 above vs. magnitude spectra of its 2 times undersampled and 3 and 6 times oversampled DT versions. Compute and plot the corresponding phase spectra. On all plots clearly indicate the image folding, repetition, overlap, and separation or the lack of thereof, as relevant. Comment on the size of magnitude spectrum image separation as a function of sampling rate and support your comment with specific numbers for the function sampled in this problem.

Solution:

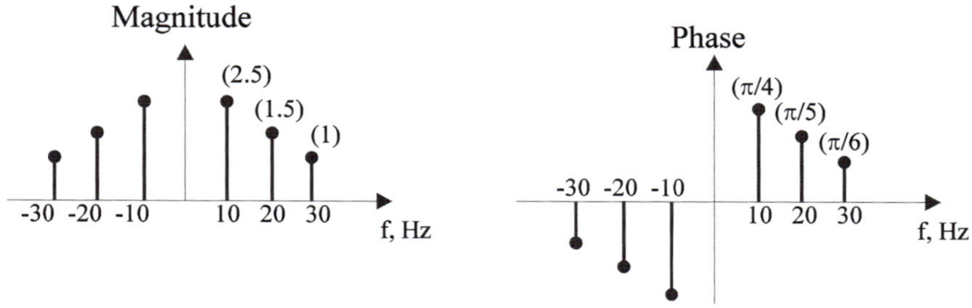

Fig. Hw6-10. CT signal magnitude and phase spectra for Problem 8.

2 times undersampling:

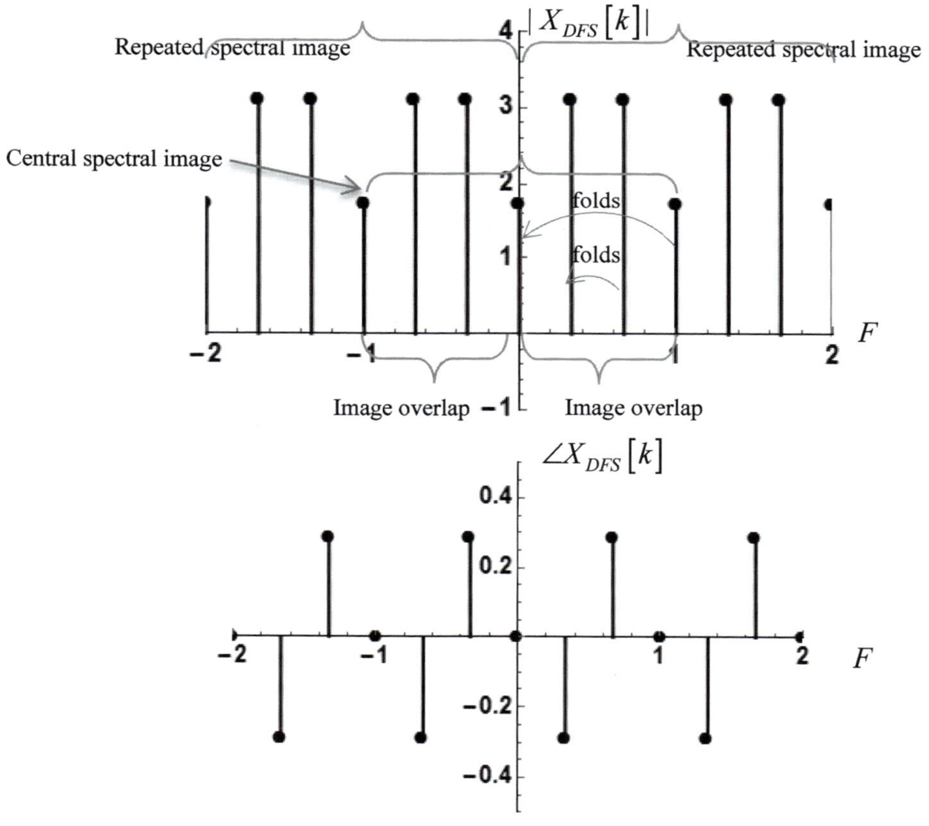

Fig. Hw6-11. DT signal magnitude and phase spectrum for 2 times undersampling.

3 times oversampling

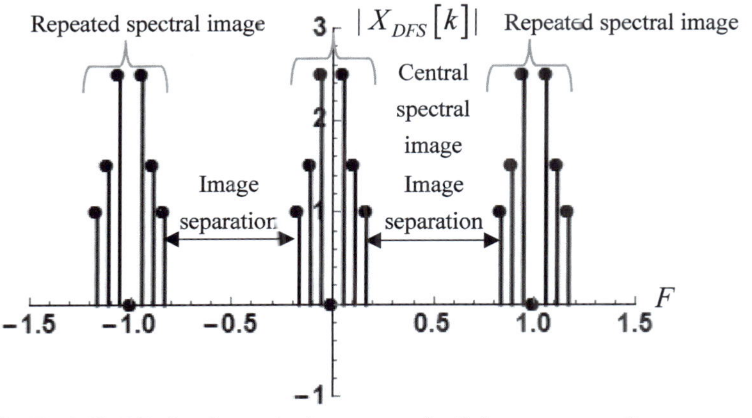

Fig. Hw6-12. DT signal magnitude spectrum for 3 times oversampling.

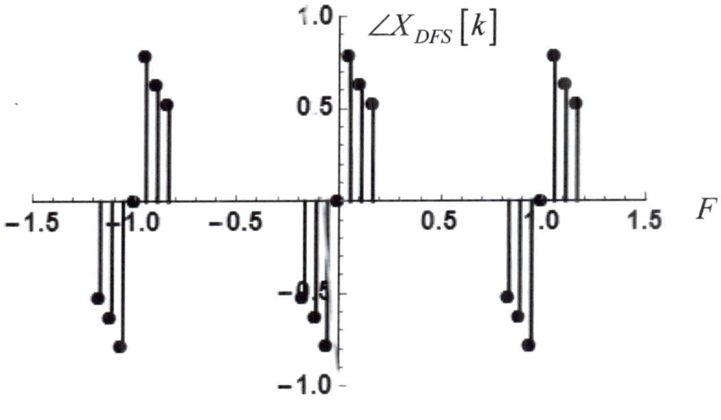

Fig. Hw6-13. DT signal phase spectrum for 3 times oversampling.

6 times oversampling:

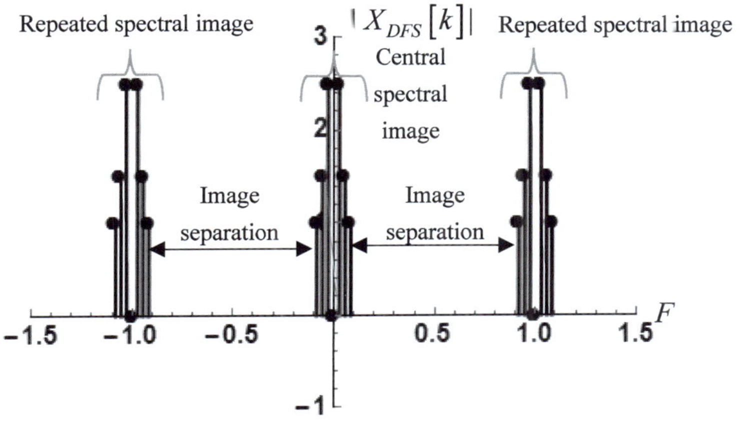

Fig. Hw6-14. DT signal phase spectrum for 3 times oversampling.

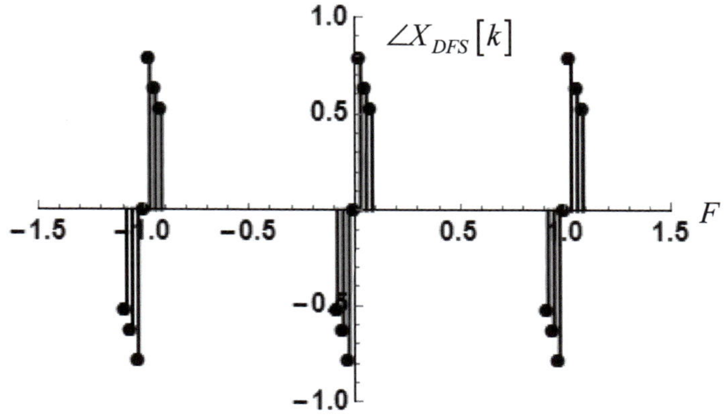

Fig. Hw6-15. DT signal magnitude and phase spectra for 6 times oversampling.

Problem 9: Write analytical expressions (specific ones, with the formulas explicitly related to the original function) for the reconstruction of the original CT function from its 3 and 6 times oversampled DT versions obtained in Problem 7 using ideal interpolation. Reconstruct numerically the original CT function from its 3 and 6 times oversampled DT versions using ideal, linear, and step interpolation and plot the reconstructed CT versions. For each interpolation type, comment on the reconstruction quality of the original function using 3 times oversampled DT version versus that using 6 times oversampled DT version.

Solution:

We have the following formula for ideal reconstruction of a sampled signal:

$$x(\mathrm{t}) = \sum_{n=-\infty}^{\infty} x[\mathrm{n}]\mathrm{sinc}((t - nt_s)/t) = \sum_{n=-\infty}^{\infty} x[\mathrm{n}]\mathrm{sinc}(S_F t - n).$$

All we need to do for this problem is using formula for $x[n]$ from Problem 8 for 2 different sampling rates.

For 3 times oversampling ($S_F = 180Hz$)

$$x_{sp}[n] = 5\cos\left(2\pi\tfrac{1}{18}n + \tfrac{\pi}{4}\right) + 3\cos\left(2\pi\tfrac{2}{18}n + \tfrac{\pi}{5}\right) + 2\cos\left(2\pi\tfrac{3}{18}n + \tfrac{\pi}{6}\right)$$

$$x(\mathrm{t}) = \sum_{n=-\infty}^{\infty} [5\cos\left(2\pi\tfrac{1}{18}n + \tfrac{\pi}{4}\right) + 3\cos\left(2\pi\tfrac{2}{18}n + \tfrac{\pi}{5}\right) + 2\cos\left(2\pi\tfrac{3}{18}n + \tfrac{\pi}{6}\right)]\mathrm{sinc}(180t - n).$$

For 6 times oversampling (($S_F = 360Hz$)

$$x_{sp}[n] = 5\cos\left(2\pi\tfrac{1}{36}n + \tfrac{\pi}{4}\right) + 3\cos\left(2\pi\tfrac{2}{36}n + \tfrac{\pi}{5}\right) + 2\cos\left(2\pi\tfrac{3}{36}n + \tfrac{\pi}{6}\right)$$

$$x(\mathrm{t}) = \sum_{n=-\infty}^{\infty} [5\cos\left(2\pi\tfrac{1}{36}n + \tfrac{\pi}{4}\right) + 3\cos\left(2\pi\tfrac{2}{36}n + \tfrac{\pi}{5}\right) + 2\cos\left(2\pi\tfrac{3}{36}n + \tfrac{\pi}{6}\right)]\mathrm{sinc}(360t - n).$$

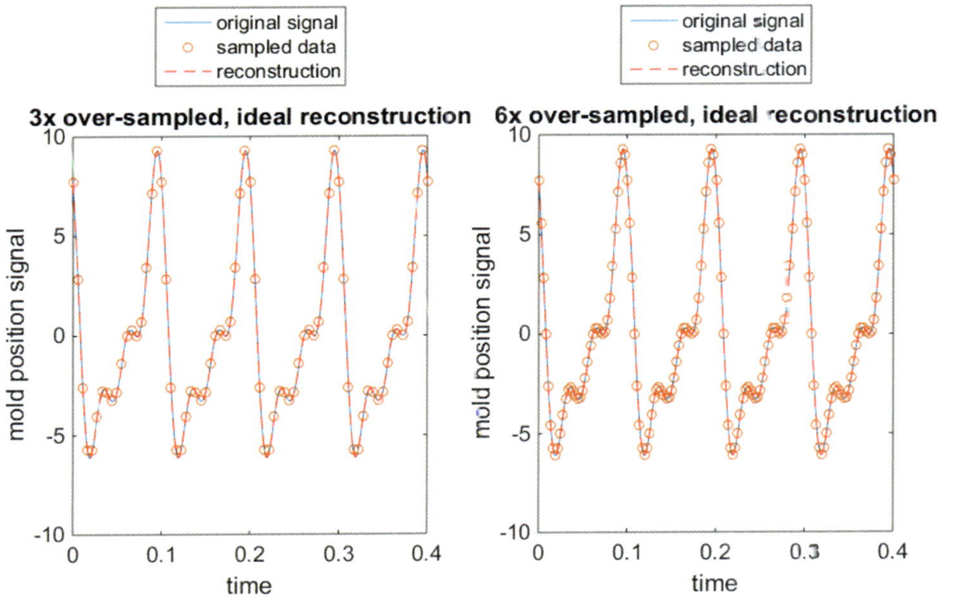

Fig. Hw6-15. Signal reconstruction using ideal interpolation.

For signal reconstruction using ideal interpolation, both 3x and 6x oversampled cases provide exact reconstruction of the original signal, whereas linear interpolation gives closer reconstruction of the original signal for 6x oversampled case than for 3x oversampled case:

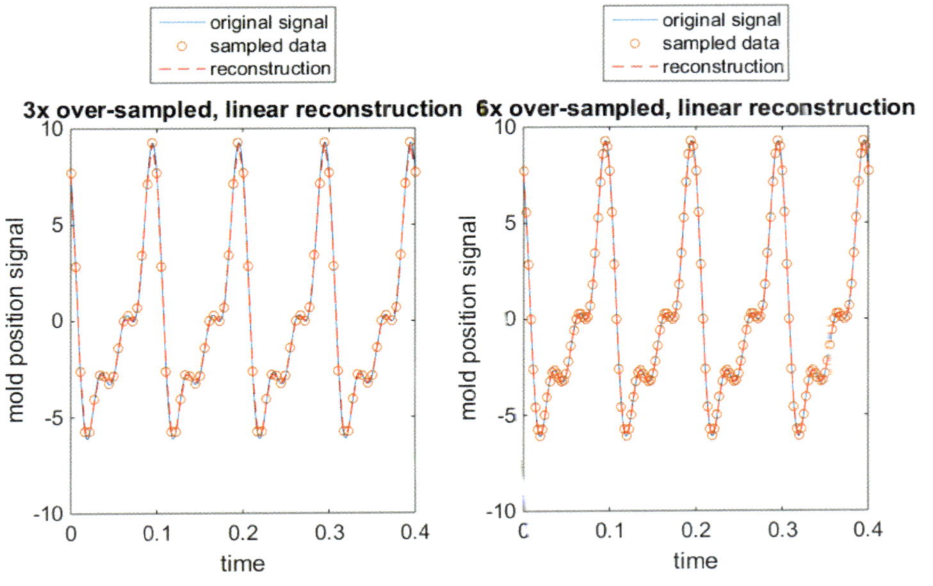

Fig. Hw6-16. Signal reconstruction using linear interpolation.

Signal reconstruction using step interpolation gives closer reconstruction of the original signal for 6x oversampling than for 3x oversampling, as well:

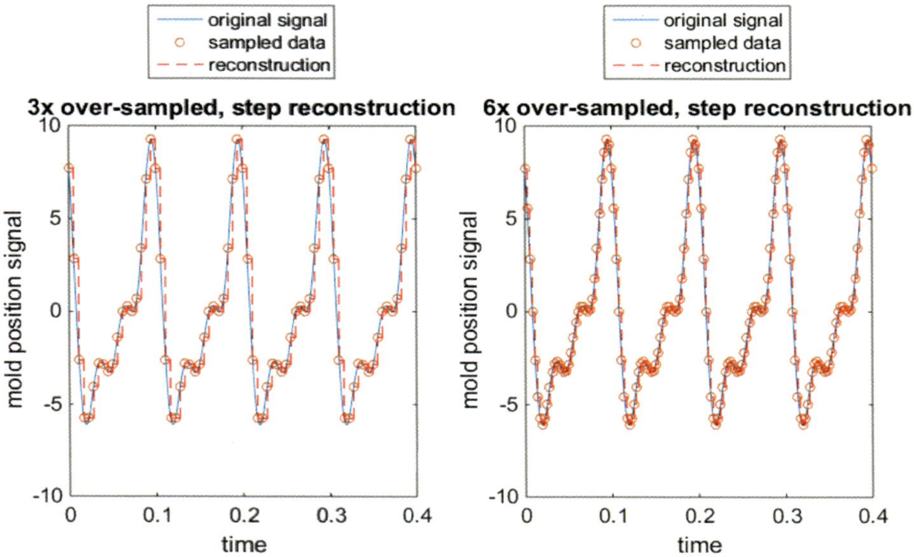

Fig. Hw6-17. Signal reconstruction using step interpolation.

Chapter 5

Function Projection and Fourier Series

5.1 Projection of a Signal, Continuous Time and Discrete Time Harmonic Bases, Continuous Time Fourier Series

5.1.1 *Continuous time harmonic basis*

Recall from linear algebra: the ***inner, dot, or scalar product*** of two vectors $\vec{e}_1$ and $\vec{e}_2$ with the same point of origin is written as $\vec{e}_1 \cdot \vec{e}_2$ and defined as $|\vec{e}_1||\vec{e}_2|\cos\theta$, where $|\vec{e}_1|$ and $|\vec{e}_2|$ are their respective lengths and θ is the angle between them. Vectors that are orthogonal to each other, i.e. with $\vec{e}_1 \cdot \vec{e}_2 = 0$ can be taken as ***basis vectors*** that form a ***vector space***.

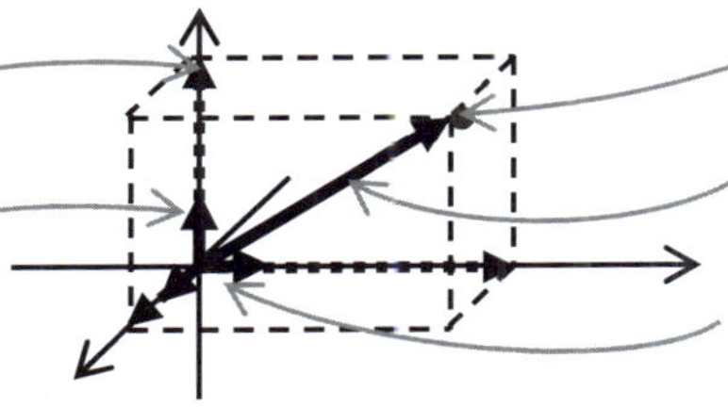

Fig. 5.1-1. Basis vectors in vector space.

Further on, these vectors can be chosen to be orthonormal, i.e. unit vectors: $\vec{e}_1 \cdot \vec{e}_1 = \vec{e}_2 \cdot \vec{e}_2 = 1$. E.g., orthonormal vectors $\vec{i}, \vec{j}, \vec{k}$ form a Cartesian space. A point x in the latter space is then associated with a vector $\vec{x}$ from the origin to this point.

So, the usual practice of writing a vector as an ordered triple of numbers that are Cartesian coordinates is as either $\vec{x} = (x_1, x_2, x_3)$, referred to as geometric, or coordinate, notation, or $\vec{x} = [x_1, x_2, x_3]^T$, referred to as algebraic, or vector/matrix notation, both meaning $\vec{x} = x_1\vec{i} + x_2\vec{j} + x_3\vec{k}$ where the coefficients (weights) x_i are calculated using the dot product: $x_1 = \vec{x} \cdot \vec{i}$, $x_2 = \vec{x} \cdot \vec{j}$, and $x_3 = \vec{x} \cdot \vec{k}$.

Then, the dot product between two vectors $\vec{x} = x_1\vec{i} + x_2\vec{j} + x_3\vec{k}$ and $\vec{y} = y_1\vec{i} + y_2\vec{j} + y_3\vec{k}$ with y_i, $i = 1, 2, 3$, calculated similarly can be written as

$$\vec{x} \cdot \vec{y} = x_1 y_1 + x_2 y_2 + x_3 y_3 = \sum_{i=1}^{3} x_i y_i = \begin{bmatrix} x_1 \\ x_2 \\ x_3 \end{bmatrix}^T \begin{bmatrix} y_1 \\ y_2 \\ y_3 \end{bmatrix}, \tag{5.1-1}$$

with the norm (length) of the vector given by

$$\left|\vec{x}\right| = \sqrt{\vec{x}\cdot\vec{x}} = \sqrt{x_1^2 + x_2^2 + x_3^2}. \tag{5.1-2}$$

5.1.1.1 *Projections*

Recall from linear algebra and Euclidean geometry: the following are two equivalent ways to describe a projection:

(1) Decompose the vector as follows: $\vec{x}$ = (a vector along the line generated by $\vec{y}$) + (a vector perpendicular to this line). The vector on the line is ***the projection of*** $\vec{x}$ ***onto*** $\vec{y}$. This decomposition is unique.

(2) The projection of $\vec{x}$ onto $\vec{y}$ is the vector along the same line as $\vec{y}$ that is closest to $\vec{x}$.

Both of these definitions transfer to signals using the inner product to be introduced shortly.

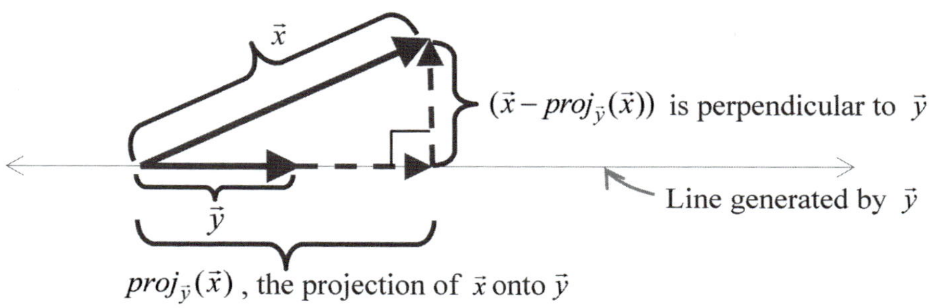

Fig. 5.1-2. Projection.

5.1.1.2 *General orthogonal basis*

Now, suppose we want to write a vector as a sum of vectors that are perpendicular, but not unit vectors. For example, instead of $\vec{i} = (1,0,0)$, $\vec{j} = (0,1,0)$ and $\vec{k} = (0,0,1)$, suppose we want to write:

$$\vec{x} = x_1\vec{b_1} + x_2\vec{b_2} + x_3\vec{b_3}$$

where $\vec{b_1} = (1,0,0)$, $\vec{b_2} = (0,1,1)$, and $\vec{b_3} = (0,1,-1)$. These are clearly perpendicular to each other, but the last two are not unit vectors. They are, however, easily turned into unit vectors by normalizing them: $\vec{e_i} = \vec{b_i}/\left|\vec{b_1}\right|$. Then, $\vec{x}$ can be decomposed in terms of these unit vectors as

$$\vec{x} = \hat{x}_1\vec{e_1} + \hat{x}_2\vec{e_2} + \hat{x}_3\vec{e_3}$$

where $\hat{x}_i = \vec{x}\cdot\vec{e_i} = \vec{x}\cdot\left(\vec{b_i}/\left|\vec{b_i}\right|\right) = \left(\vec{x}\cdot\vec{b_i}\right)/\left|\vec{b_i}\right|$. This quantity $\hat{x}_i$ is called the ***scalar projection*** of $\vec{x}$ onto $\vec{b_i}$. Now, following through with the calculation,

$$\vec{x}=\left(\vec{x}\cdot\vec{e_1}\right)\vec{e_1}+\left(\vec{x}\cdot\vec{e_2}\right)\vec{e_2}+\left(\vec{x}\cdot\vec{e_3}\right)\vec{e_3}=\left(\vec{x}\cdot\frac{\vec{b_1}}{\left|\vec{b_1}\right|}\right)\frac{\vec{b_1}}{\left|\vec{b_1}\right|}+\left(\vec{x}\cdot\frac{\vec{b_2}}{\left|\vec{b_2}\right|}\right)\frac{\vec{b_2}}{\left|\vec{b_2}\right|}+\left(\vec{x}\cdot\frac{\vec{b_3}}{\left|\vec{b_3}\right|}\right)\frac{\vec{b_3}}{\left|\vec{b_3}\right|}$$
$$=\frac{\vec{x}\cdot\vec{b_1}}{\left|\vec{b_1}\right|^2}\vec{b_1}+\frac{\vec{x}\cdot\vec{b_2}}{\left|\vec{b_2}\right|^2}\vec{b_2}+\frac{\vec{x}\cdot\vec{b_3}}{\left|\vec{b_3}\right|^2}\vec{b_3}=\frac{\vec{x}\cdot\vec{b_1}}{\vec{b_1}\cdot\vec{b_1}}\vec{b_1}+\frac{\vec{x}\cdot\vec{b_2}}{\vec{b_2}\cdot\vec{b_2}}\vec{b_2}+\frac{\vec{x}\cdot\vec{b_3}}{\vec{b_3}\cdot\vec{b_3}}\vec{b_3}. \quad (5.1\text{-}3)$$

The vector $\left(\left(\vec{x}\cdot\vec{b_i}\right)/\left(\vec{b_i}\cdot\vec{b_i}\right)\right)\vec{b_i}$ is called the ***vector projection*** of $\vec{x}$ onto $\vec{b_i}$.

Using the three vectors above,

$$\begin{bmatrix}x_1\\x_2\\x_3\end{bmatrix}=\left(\begin{bmatrix}x_1\\x_2\\x_3\end{bmatrix}^T\cdot\begin{bmatrix}1\\0\\0\end{bmatrix}\right)\begin{bmatrix}1\\0\\0\end{bmatrix}+\frac{1}{2}\left(\begin{bmatrix}x_1\\x_2\\x_3\end{bmatrix}^T\cdot\begin{bmatrix}0\\1\\1\end{bmatrix}\right)\begin{bmatrix}0\\1\\1\end{bmatrix}+\frac{1}{2}\left(\begin{bmatrix}x_1\\x_2\\x_3\end{bmatrix}^T\cdot\begin{bmatrix}0\\1\\-1\end{bmatrix}\right)\begin{bmatrix}0\\1\\-1\end{bmatrix}$$
$$=\left(x_1\right)\begin{bmatrix}1\\0\\0\end{bmatrix}+\frac{1}{2}\left(x_2+x_3\right)\begin{bmatrix}0\\1\\1\end{bmatrix}+\frac{1}{2}\left(x_2-x_3\right)\begin{bmatrix}0\\1\\-1\end{bmatrix}.$$

5.1.1.3 *Orthogonal basis for signals. Vector projection for functions*

Let's now associate, respectively, the elements of each of two vectors in dot product (5.1-1) with the samples of each of two CT T-periodic signals $y_i(t)$ and $y_j(t)$ over T, produced by the t_s sampling interval. Then, if $t_s \to 0$, the number of samples in the summation $\to \infty$, and in the limit summation becomes integration under some reasonable assumptions. The concept of projection can then be extended to CT T-periodic signals using the resulting representation of the inner (dot) product:

$$y_i(t)\cdot y_j(t)=\frac{1}{T}\int_T y_i(t)\,y_j^*(t)\,dt. \quad (5.1\text{-}4)$$

Then, the function norm can be defined as

$$\left\|y_i(t)\right\|=\left|\sqrt{P_{y_i}}\right|=RMS_{y_i}=\left|\sqrt{y_i(t)\cdot y_i(t)}\right|=\left|\sqrt{\frac{1}{T}\int_T y_i(t)\,y_i^*(t)\,dt}\right|.$$

This is the definition of RMS (root-mean-square) value that we gave in Chapter 2. The reason we divide the integrated quantity by the period T is to make the constant periodic function $y(t) = 1$ a unit function. Multiplication by the complex conjugate of $y_i(t)$ permits the use of complex functions like e^{jt} for forming a basis.

Then $y_i(t)$ and $y_j(t)$ can be called ***orthogonal*** if

$$y_i(t)\cdot y_j(t)=\frac{1}{T}\int_T y_i(t)\,y_j^*(t)\,dt=0. \quad (5.1\text{-}5)$$

Functions $y_i(t)$, $i \in [-\infty, \infty]$, satisfying (5.1-5) form an ***orthogonal basis***. The "vector," or normalized, projection $p_{y_i}(x(t))$ of $x(t)$ onto $y_i(t)$ can be computed as

$$p_{y_i}(x) = p_i y_i(t) = \frac{x(t)\cdot y_i(t)}{y_i(t)\cdot y_i(t)} y_i(t) = \frac{\frac{1}{T}\int_T x(t) y_i^*(t)\,dt}{\frac{1}{T}\int_T y_i(t) y_i^*(t)\,dt} y_i(t) = \underbrace{\left(\frac{\frac{1}{T}\int_T x(t) y_i^*(t)\,dt}{\|y_i(t)\|^2} \right)}_{p_i} y_i(t), \quad (5.1\text{-}6)$$

where p_i is the projection coefficient (weight) for the projection of $x(t)$ onto $y_i(t)$ basis function. Then, given a vector $\vec{x}$ expressed through vector projection (5.1-3) as

$$\vec{x} = \frac{\vec{x}\cdot\vec{b_1}}{\left|\vec{b_1}\right|^2}\vec{b_1} + \frac{\vec{x}\cdot\vec{b_2}}{\left|\vec{b_2}\right|^2}\vec{b_2} + \frac{\vec{x}\cdot\vec{b_3}}{\left|\vec{b_3}\right|^2}\vec{b_3},$$

its counterpart for T-periodic $x(t)$ with three components expressed through harmonically related T-periodic basis functions $y_i(t)$ is given by

$$x(t) = \frac{x(t)\cdot y_1(t)}{\|y_1(t)\|^2} y_1(t) + \frac{x(t)\cdot y_2(t)}{\|y_2(t)\|^2} y_2(t) + \frac{x(t)\cdot y_3(t)}{\|y_3(t)\|^2} y_3(t).$$

If we pick $y_i(t)$ so that

$$y_i(t)\cdot y_i(t) = 1\ \forall i, \quad (5.1\text{-}7)$$

i.e. turn $y_i(t)$ into a "unit vector" (RMS = 1), the projection $p_{y_i(t)}(x(t))$ of $x(t)$ onto unit function $y_i(t)$ simplifies to

$$p_{y_i}(x(t)) = [x(t)\cdot y_i(t)] y_i(t) = p_i y_i(t) = \left[\frac{1}{T}\int_T x(t) y_i^*(t)\,dt\right] y_i(t).$$

Functions satisfying (5.1-5) and (5.1-7) are called ***orthonormal***, and their set $y_i(t)$, $i \in [-\infty, \infty]$, forms ***orthonormal basis***.

Similarly, a dot product for two DT N-periodic signals $y_i[n]$ and $y_j[n]$ is given by

$$y_i[n]\cdot y_j[n] = \frac{1}{N}\sum_{n=0}^{N-1} y_i[n]\cdot y_j^*[n]$$

with the corresponding subsequent definitions.

Example 5.1-1: CT complex harmonic basis. Let's now check the properties of $y_i(t) = e^{ji\omega_1 t}$. Take $y_m(t) = e^{jm\omega_1 t}$ and $y_n(t) = e^{jn\omega_1 t}$. Then, since their common period is $T_1 = 2\pi/\omega_1$ and inner product is given by

$$p_n = y_m(t)\bullet y_n(t) / y_n(t)\bullet y_n(t) = \left(\frac{1}{T_1}\int_{T_1} y_m(t)y_n^*(t)dt\right) \Big/ \underbrace{\left(\frac{1}{T_1}\int_{T_1} y_n(t)y_n^*(t)dt\right)}_{=1 \text{ – to be shown below}} = \tag{5.1-8}$$

$$y_m(t)\bullet y_n(t) = \frac{1}{T_1}\int_{T_1} y_m(t)y_n^*(t)dt = \frac{1}{T_1}\int_0^{T_1} e^{jm\omega_1 t}e^{-jn\omega_1 t}dt = \frac{1}{T_1}\int_0^{T_1} e^{j(m-n)\omega_1 t}dt.$$

Checking (5.1-8) for unit size (5.1-7) and orthogonality (5.1-5) yields, respectively,

$$m = n \Rightarrow y_m(t)\cdot y_n(t) = \frac{1}{T_1}\int_0^{T_1} e^{j(m-n)\omega_1 t}dt = \frac{1}{T_1}\int_0^{T_1} e^{j\cdot 0\cdot\omega_1 t}dt = \frac{1}{T_1}\int_0^{T_1} 1dt = 1$$

← CT complex harmonic functions are "unit vectors" for signals

and

$$m \neq n \Rightarrow y_m(t)\cdot y_n(t) = \frac{1}{T_1}\int_0^{T_1} e^{j(m-n)\omega_1 t}dt = \frac{1}{(m-n)T_1\omega_1}\underbrace{e^{j(m-n)\omega_1 t}}_{T_1\text{-periodic}}\Big|_0^{T_1} = 0.$$

← CT complex harmonic functions are mutually orthogonal

Consequently, CT complex harmonic functions are orthonormal, and they form (***span***) a ***function space*** with a unit basis:

$$\{e^{j2\pi k\frac{1}{T_1}t}\}_{k=-\infty}^{\infty}. \tag{5.1-9}$$

Example 5.1-2: DT complex harmonic basis. Take $y_k[n] = e^{j2\pi kF_1 n}$, $y_m[n] = e^{j2\pi mF_1 n}$, then

$$p_m := \frac{1}{N_1}\sum_{n=0}^{N_1-1} y_k[n]y_m^*[n] = \frac{1}{N_1}\sum_{n=0}^{N_1-1} e^{j2\pi kF_1 n}e^{-j2\pi mF_1 n} = 0, \text{ if m} \neq \text{k},$$

i.e. $y_k[n]$ is orthogonal to $y_m[n]$ (show this in the homework).

Therefore, DT harmonic functions form a function space with a unit basis

$$\{e^{j2\pi k\cdot\frac{1}{N_1}\cdot n}\}_{k=0}^{N_1-1}. \tag{5.1-10}$$

5.1.2 *Continuous time Exponential Fourier series*

Using the above definitions, an arbitrary CT T_1-periodic signal can be thought of as a point (vector) in a function space spanned by CT harmonic basis (5.1-9).

The ***Fourier series coefficients*** of an arbitrary CT periodic signal can then be viewed as a set of the coefficients (weights) of projections onto this basis, with the kth projection coefficient given by:

$$X_s[k] = |X_s[k]|e^{j\theta_k} \longleftarrow X_s[k] = \frac{1}{T_1}\int_{T_1} x_p(t)e^{-jk\omega_1 t}dt. \tag{5.1-11}$$

- a complex number

Using the projections with the weights (5.1-11) yields real periodic signal representation:

$$x_p(t) = \sum_{k=-\infty}^{\infty} X_s[k] e^{jk\omega_1 t}, \tag{5.1-12}$$

referred to as the ***Exponential Fourier series*** (***EFS***).

5.1.2.1 *Going from the exponential form to the polar and the trigonometric forms*

In addition to the Exponential Fourier series, two other forms, ***polar***, also referred to as ***magnitude and phase***, and ***trigonometric***, are widely used — polar because it represents the actual physical waveforms in terms of magnitudes and phases, and trigonometric because it uses for representation familiar sines and cosines.

To arrive at these forms for the real periodic signal starting from the exponential form we rewrite summation in (5.1-12) to range from 0 to ∞, instead of $-\infty$ to ∞ as

Recall that $X_s[k]$ and $X_s[-k]$ for $k = 1, 2, \ldots$ are complex conjugates, so that replacing k in this term with $-k$ for $k = 1, 2, \ldots$ gives $X_s[k]^*$ - a complex conjugate of $X_s[k]$

$$\begin{aligned} x_p(t) &= \sum_{k=-\infty}^{\infty} X_s[k] e^{jk\omega_1 t} = \sum_{k=-\infty}^{\infty} |X_s[k]| e^{j\angle X_s[k]} e^{jk\omega_1 t} = \\ &\sum_{k=-1}^{-\infty} X_s[k] e^{jk\omega_1 t} + X_s[0] + \sum_{k=1}^{\infty} X_s[k] e^{jk\omega_1 t} = \\ &\sum_{k=1}^{\infty} \underbrace{X_s[-k]}_{X_s[k]^*} e^{-jk\omega_1 t} + X_s[0] + \sum_{k=1}^{\infty} X_s[k] e^{jk\omega_1 t}. \end{aligned} \tag{5.1-13}$$

Polar, or magnitude and phase form. Expression (5.1-13) can be rewritten as

$$x_p(t) = \underbrace{X_s[0]}_{\mathrm{Re}\, z_0} + \sum_{k=1}^{\infty} \Big(\underbrace{|X_s[k]| e^{j(k\omega_1 t + \angle X_s[k])}}_{z_k} + \underbrace{\overbrace{|X_s[k]|}^{=|X_s[-k]|} e^{-j(k\omega_1 t + \angle X_s[k])}}_{z_k^*} \Big). \tag{5.1-14}$$

Applying Euler's formula to (5.1-14), the polar form is calculated as

$$x_p(t) = \mathrm{Re}\, z_0 + \sum_{k=1}^{\infty} 2\, \mathrm{Re}\, z_k \quad \leftarrow \mathrm{Re}\, z_k = \mathrm{Re}\, z_k^*$$

$$= \underbrace{\underbrace{|X_s[0]|}_{c_0} \cos\Big(\underbrace{\angle X_s[0]}_{\theta_0 = \tan^{-1} \frac{\mathrm{Im} X_s[0]}{\mathrm{Re} X_s[0]}} \Big) + \sum_{k=1}^{\infty} \underbrace{2|X_s[k]|}_{c_k = 2\left|\sqrt{(\mathrm{Im} X_s[k])^2 + (\mathrm{Re} X_s[k])^2}\right|} \cos\Big(k\omega_1 t + \underbrace{\angle X_s[k]}_{\theta_k = \tan^{-1} \frac{\mathrm{Im} X_s[k]}{\mathrm{Re} X_s[k]}} \Big)}_{\text{Polar (magnitude and phase) form}}$$

$$= \sum_{k=0}^{\infty} c_k \cos(k\omega_1 t + \theta_k).$$

Trigonometric form. Expression (5.1-13) can also be rewritten in the form

$$x_p(t) = X_s[0] + \sum_{k=1}^{\infty}\left(X_s[k]e^{jk\omega_1 t} + X_s[k]^* e^{-jk\omega_1 t}\right). \tag{5.1-15}$$

Then, applying Euler's formula to (5.1-15), the trigonometric form is calculated as

$$\begin{aligned}
x_p(t) &= X_s[0] + \sum_{k=1}^{\infty}\left(X_s[k]e^{jk\omega_1 t} + X_s[k]^* e^{-jk\omega_1 t}\right) \\
&= X_s[0] + \sum_{k=1}^{\infty}\left(X_s[k]\left(\cos(k\omega_1 t) + j\sin(k\omega_1 t)\right) + X_s[k]^*\left(\cos(k\omega_1 t) - j\sin(k\omega_1 t)\right)\right) \\
&= \underbrace{\underbrace{X_s[0]}_{a_0} + \sum_{k=1}^{\infty}\Big(\underbrace{X_s[k] + X_s[k]^*}_{a_k=2\operatorname{Re}X_s[k]}\Big)\cos(k\omega_1 t) + \underbrace{j(\overbrace{X_s[k] - X_s[k]^*}^{j2\operatorname{Im}X_s[k]})}_{b_k=-2\operatorname{Im}X_s[k]}\sin(k\omega_1 t)}_{\text{Trigonometric form}} \\
&= \sum_{k=0}^{\infty}\left[a_k\cos(k\omega_1 t) + b_k\sin(k\omega_1 t)\right].
\end{aligned}$$

Direct calculation of the trigonometric form coefficients using sines and cosines. Since the latter are orthogonal (perpendicular) in the same sense as complex exponential functions, but they are not unit functions, formula (5.1-6) for projection calculation would need to be used:

$$\frac{1}{T_1}\int_0^{T_1}\sin(m\omega_1 t)\cos(n\omega_1 t)\,dt = 0,$$

$$\frac{1}{T_1}\int_0^{T_1}\cos(n\omega_1 t)\cos(m\omega_1 t)\,dt = \begin{cases} 0, & \text{if } m \neq n, \\ \dfrac{1}{2}, & \text{if } m = n \neq 0, \\ 1, & \text{if } m = n = 0, \end{cases}$$

$$\frac{1}{T_1}\int_0^{T_1}\sin(n\omega_1 t)\sin(m\omega_1 t)\,dt = \begin{cases} 0, & \text{if } m \neq n, \\ \dfrac{1}{2}, & \text{if } m = n \neq 0, \\ 0, & \text{if } m = n = 0. \end{cases}$$

These are not equal to 1, like they were for complex exponentials.

These integrals can be taken using the trigonometric identities for $sin(x-y)$ and $cos(x+y)$. Then the trigonometric form coefficients are given by:

$$a_0 = X_s[0] = \frac{1}{T_1}\int_0^{T_1}x_p(t)\,dt$$

$$a_k = \frac{\frac{1}{T_1}\int_0^{T_1} x_p(t)\cos(k\omega_1 t)dt}{\frac{1}{T_1}\int_0^{T_1}\cos(k\omega_1 t)\cos(k\omega_1 t)dt} = \frac{\frac{1}{T_1}\int_0^{T_1} x_p(t)\cos(k\omega_1 t)dt}{\frac{1}{2}} = \frac{2}{T_1}\int_0^{T_1} x_p(t)\cos(k\omega_1 t)dt,$$

$$b_k = \frac{\frac{1}{T_1}\int_0^{T_1} x_p(t)\sin(k\omega_1 t)dt}{\frac{1}{T_1}\int_0^{T_1}\sin(k\omega_1 t)\sin(k\omega_1 t)dt} = \frac{\frac{1}{T_1}\int_0^{T_1} x_p(t)\sin(k\omega_1 t)dt}{\frac{1}{2}} = \frac{2}{T_1}\int_0^{T_1} x_p(t)\sin(k\omega_1 t)dt. \quad (5.1\text{-}16)$$

5.1.2.2 *Summary: three forms of continuous time Fourier series signal representation*

We can represent the original signal in terms of its projections onto CT harmonic basis as

$$x_p(t) = \sum_{k=-\infty}^{\infty} \underbrace{X_s[k]e^{jk\omega_1 t}}_{k\text{'th projection of } x_p(t) \text{ onto } k\text{'th harmonic}} \quad \longleftarrow \textbf{Exponential form of Fourier Series}$$

$$= \sum_{k=0}^{\infty} c_k \cos(k\omega_1 t + \theta_k) \quad \longleftarrow \textbf{Polar, or magnitude and phase, form}$$

$$= \sum_{k=0}^{\infty} \left[a_k \cos(k\omega_1 t) + b_k \sin(k\omega_1 t)\right] \quad \longleftarrow \textbf{Trigonometric form}$$

$k=0, X_s[0]$ always real:	$k \neq 0$: **Trigonometric form**	**Polar form**
$a_0 = X_s[0],\ c_0 = \lvert X_s[0]\rvert,$	$a_k = 2\,\mathrm{Re}\,X_s[k],\ \mathrm{Re}\,X_s[k] = \frac{a_k}{2},$	$c_k = 2\lvert X_s[k]\rvert,$
$\theta_0 = \tan^{-1}\frac{\mathrm{Im}\,X[0]}{\mathrm{Re}\,X[0]} = \begin{cases} 0, & X_s[0] \geq 0, \\ \pm\pi, & X_s[0] < 0. \end{cases}$	$b_k = -2\,\mathrm{Im}\,X_s[k], \mathrm{Im}\,X_s[k] = -\frac{b_k}{2}.$	$\theta_k = \tan^{-1}\frac{\mathrm{Im}\,X[k]}{\mathrm{Re}\,X[k]}.$

Using trigonometric form to calculate exponential form. Computing for each k two integrals (separately for cosine and sine), we can calculate exponential form as follows:

$$X_s[k] = \mathrm{Re}\,X_s[k] + j\mathrm{Im}X_s[k] = a_k/2 - jb_k/2 = \left|X_s[k]\right|e^{j\angle X_s[k]} = \left|\sqrt{(a_k^2/4)+(b_k^2/4)}\right|e^{j\tan^{-1}(-b_k/a_k)}. \quad (5.1\text{-}17)$$

One should be able to recognize the signals that are *already in one of the FS forms* (as in Example 5.1-3), making FS coefficients calculation irrelevant. If, given EFS signal representation $\sum_{k=-\infty}^{\infty} X_s[k]e^{jk\omega_1 t}$, computation of its *m'th* projection coefficient $X_s[m]$ is still carried out, the result naturally yields the same coefficient:

$$\frac{1}{T_1}\int_{T_1} x_p(t)e^{-jm\omega_1 t}dt = \frac{1}{T_1}\int_{T_1}\left(\sum_{k=-\infty}^{\infty} X_s[k]e^{jk\omega_1 t}\right)e^{-jm\omega_1 t}dt$$

$$= \frac{1}{T_1}\sum_{k=-\infty}^{\infty} X_s[k]\underbrace{\int_{T_1} e^{jk\omega_1 t}e^{-jm\omega_1 t}dt}_{=\begin{smallmatrix} T_1,\ m=k \\ 0,\ m\neq k\end{smallmatrix}} = \frac{1}{T_1}X_s[m]T_1 = X_s[m].$$

We will give several examples of FS calculation: 1) *numerical CT* (this section, using CT EFS), 2) *numerical DT* (next section, using Discrete FS), both for the same software testbed data record, and 3) *analytical CT* (section 5.3, using CT EFS) for six basic analytically given waveforms.

5.1.2.3 *Fourier series and discrete spectra*

Fourier series coefficients of a signal yield its discrete spectra, or spectral graphs, as follows:

- $|X_s(k)|$ versus $k\omega_1$ — discrete 2-sided magnitude spectrum
- θ_k versus $k\omega_1$ — discrete 1- or 2-sided phase spectrum
- c_k versus $k\omega_1$ — discrete 1-sided magnitude spectrum.

These graphs permit quick visualization of the main signal characteristics, like signal frequency content, providing the key system diagnostic tool as shown in the next example, and bandwidth to be discussed in Section 6.

Example 5.1-3: calculating CT harmonics numerically using software testbed data.
Problem 4 in Homework 4 solution guide asked to check the magnitudes of the sinusoidal components of the software testbed piston position with the weight of the beam and the mold and without it. No consistent methodology, however, was suggested for carrying out this task.

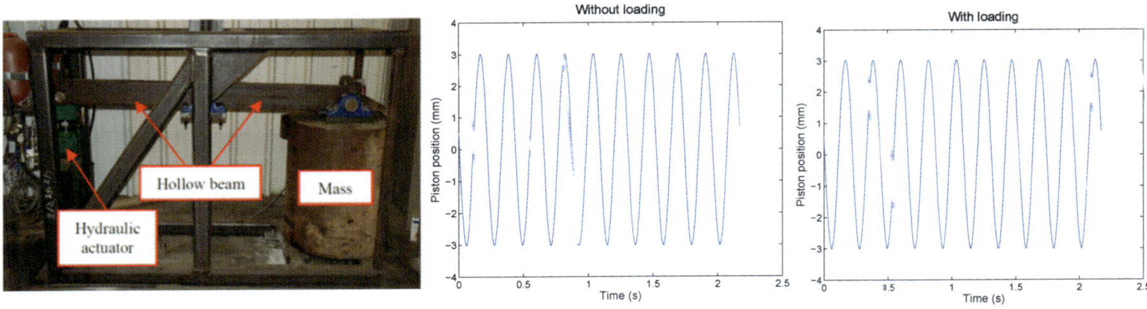

Fig. 5.1-3. Testbed piston position graphs without loading (left) and with it look indistinguishable.

The corresponding piston position graphs are given in Figure 5.1-3 and on visual inspection look identical. Let us now apply Fourier series to examine them.

The main component frequency for both signals is 4.6 Hz — that of the desired reference sinusoid the piston is trying to track. Let's try to remove the main sinusoidal part and see what is left.

The data from the first 10 cycles at 4.6 Hz collected from Simulink model is shown below.

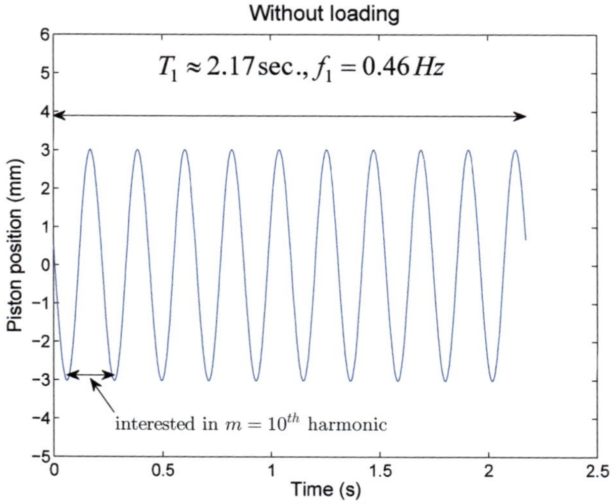

Fig. 5.1-4. CT data record of piston position without load.

Collecting data set over 10 periods and defining its duration as $T_1 = 2.17$ so that $f_1 = 0.46$ Hz, makes 4.6 Hz to be the 10th harmonic. Calculating CT harmonic coefficients $X_s[10]$ and $X_s[-10]$:

$$X_s[10] = \frac{1}{T_1}\int_{T_1} x_p(t)e^{-j10\cdot 2\cdot\pi\cdot 0.46t}dt,\ X_s[-10] = \frac{1}{T_1}\int_{T_1} x_p(t)e^{-j(-10)\cdot 2\cdot\pi\cdot 0.46t}dt = \left(X_s[10]\right)^*,$$

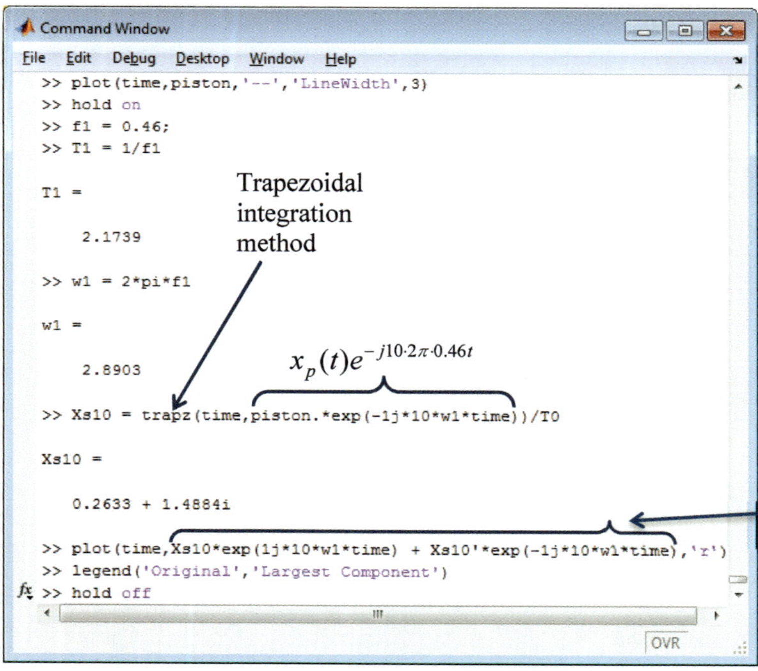

through approximate numerical integration as shown in Figure 5.1-5 yields

$$X_s[10] = 0.2633 + j1.4884 \Rightarrow$$
$$X_s[-10] = 0.2633 - j1.4884,$$
$$c_{10} = 3.0231,\ \theta_{10} = 1.3957,$$
$$a_{10} = 0.5265,\ b_{10} = 2.9769,$$

which permits computing the 10th harmonic at 4.6 Hz as a function of time:

$$x_{p10}(t) = X_s[10]e^{j10\omega_1 t}$$
$$+X_s[-10]e^{-j10\omega_1 t} =$$
$$a_{10}\cos(10\omega_1 t) + b_{10}\sin(10\omega_1 t)$$
$$= c_{10}\cos(10\omega_1 t + \theta_{10}).$$

Fig. 5.1-5. Matlab commands for calculating and plotting CT harmonics from the data record.

As seen from the upper two graphs in Figure 5.1-6, both signals look identical due to the strongly dominant component at 4.6 Hz. However, subtracting the latter from the original signal $x_p(t)$, the high-frequency residuals are seen in the lower two graphs in Figure 5.1-6 to be an order of magnitude larger for the loaded piston.

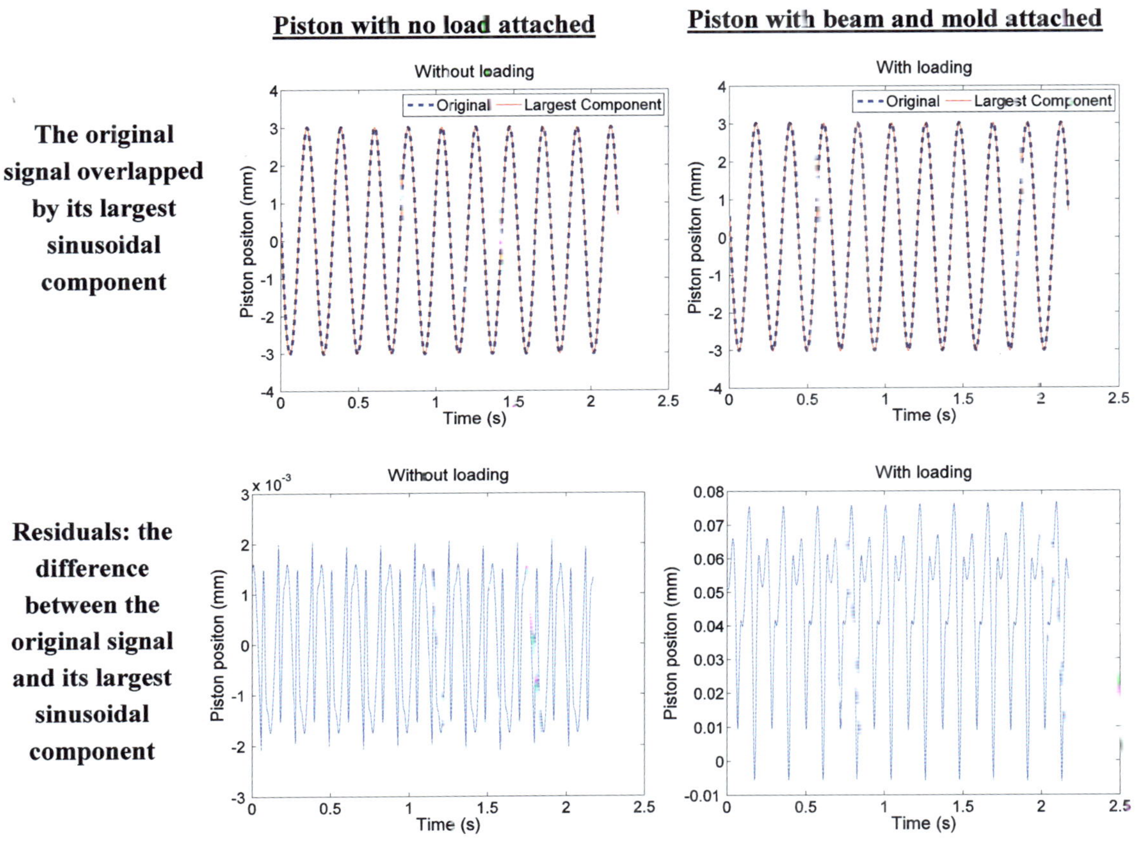

Fig. 5.1-6. Piston position with load and without it — calculation of CT residuals.

5.2 Discrete Fourier Series, Power of Periodic CT and DT Signals — Parseval's Relationship

5.2.1 *Discrete Fourier Series (DFS)*

Similarly to the continuous time case, an arbitrary DT N_1-periodic signal can be thought of as a point (vector) in a function space spanned by DT harmonic basis (5.1-10).

Then, Discrete Fourier Series Coefficients, or Discrete Fourier Series (DFS) are given by:

$$x_p(t) \xrightarrow[t \to nt_s]{T_1 = N_1 t_s} x_p[n] \Rightarrow X_{DFS}[k] = \frac{1}{N_1} \sum_{n=0}^{N_1 - 1} x_p[n] e^{-j2\pi \frac{k}{N_1} n}, k = 0, \ldots, N_1 - 1 \qquad (5.2\text{-}1)$$

$X_{DFS}[k]$ — coefficient of projection $X_{DFS}[k]e^{j2\pi\frac{k}{N_1}n}$ of $x_p[n]$ onto k^{th} basis function: $e^{j2\pi\frac{k}{N_1}n}$.

- The DFS is N_1-periodic (repeats itself every N_1 samples)
- For real $x_p[n]$, the DFS is symmetric about 0 on [−½, ½]
  - Magnitude has even symmetry
  - Phase has odd symmetry

$$x_p[n] = \sum_{k=0}^{N_1-1} X_{DFS}[k]e^{j2\pi\frac{k}{N_1}n} = \sum_{k=0}^{N_1-1} X_{DFS}[k]e^{j\Omega_k n}. \tag{5.2-2}$$

Example 5.2-1: calculating DT harmonics numerically: software testbed data sampled. To rework Problem 4 in DT, assume we have sampled the CT data over $T_1 = 2.17$ sec with the ts = 0.002 sec, producing $N_1 = 1087$ data points and digital frequency $F_1 = 1$ cycle/1087 samples:

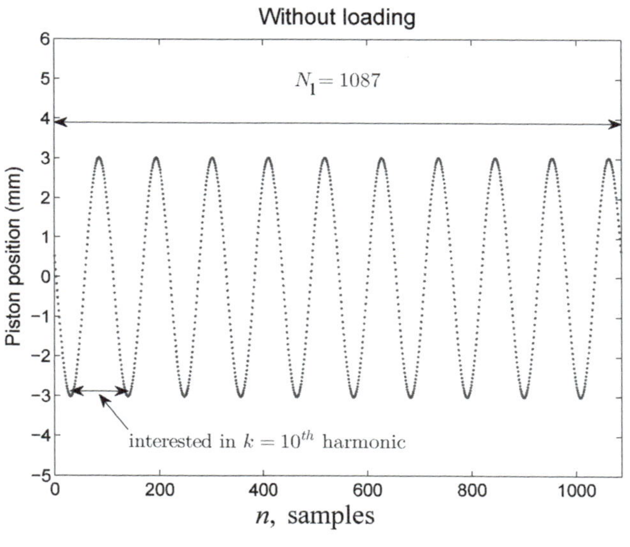

Fig. 5.2-1. DT data record of piston position without load.

The harmonic of interest is now the 10th harmonic $X_{DFS}[10]$ since we have sampled 10 cycles over 1087 samples:

$$X_{DFS}[10] = \frac{1}{N_1}\sum_{n=0}^{N_1-1} x_p[n]e^{-j\cdot 2\pi\frac{10}{N_1}n},\ X_{DFS}[-10] = \frac{1}{N_1}\sum_{n=0}^{N_1-1} x_p[n]e^{-j\cdot 2\pi\frac{-10}{N_1}n} = X^*_{DFS}[10].$$

Computation of this harmonic and the residuals is shown in Figures 5.2-2 and 5.2-3, respectively.

```
>> plot(piston,'o')
>> N = length(piston)

N =

        1087

>> n = (0:(N-1))';
>> X10 = sum(piston.*exp(-j*2*pi*(10/N)*n))/N

X10 =

   0.2568 + 1.4855i

>> hold on
>> plot(X10*exp(j*2*pi*(10/N)*n) + X10'*exp(-j*2*pi*(10/N)*n),'r.')
>> legend('Original','Largest Component')
>> hold off
```

$$x_p[n]e^{-j2\cdot\pi\cdot\frac{10}{N_1}n}$$

$$X_{DFS}[10] = 0.2568 + j1.4855 \Rightarrow$$

$$X_{DFS}[-10] = 0.2568 - j1.4855,$$

$$\Downarrow$$

$$x_{p10}[n] = X_{DFS}[10]e^{j2\pi\frac{10}{N_1}n} + X_{DFS}[-10]e^{-j2\pi\frac{10}{N_1}n}.$$

Fig. 5.2-2. Matlab commands for calculating and plotting the 10th DT harmonic $x_{p10}[n]$ from DT data record $x_p[n]$.

Subtracting $x_{p10}[n]$ from the original DT data record $x_p[n]$, the high-frequency residuals are seen in the lower two graphs in Figure 5.2-3 to be an order of magnitude larger for the loaded piston.

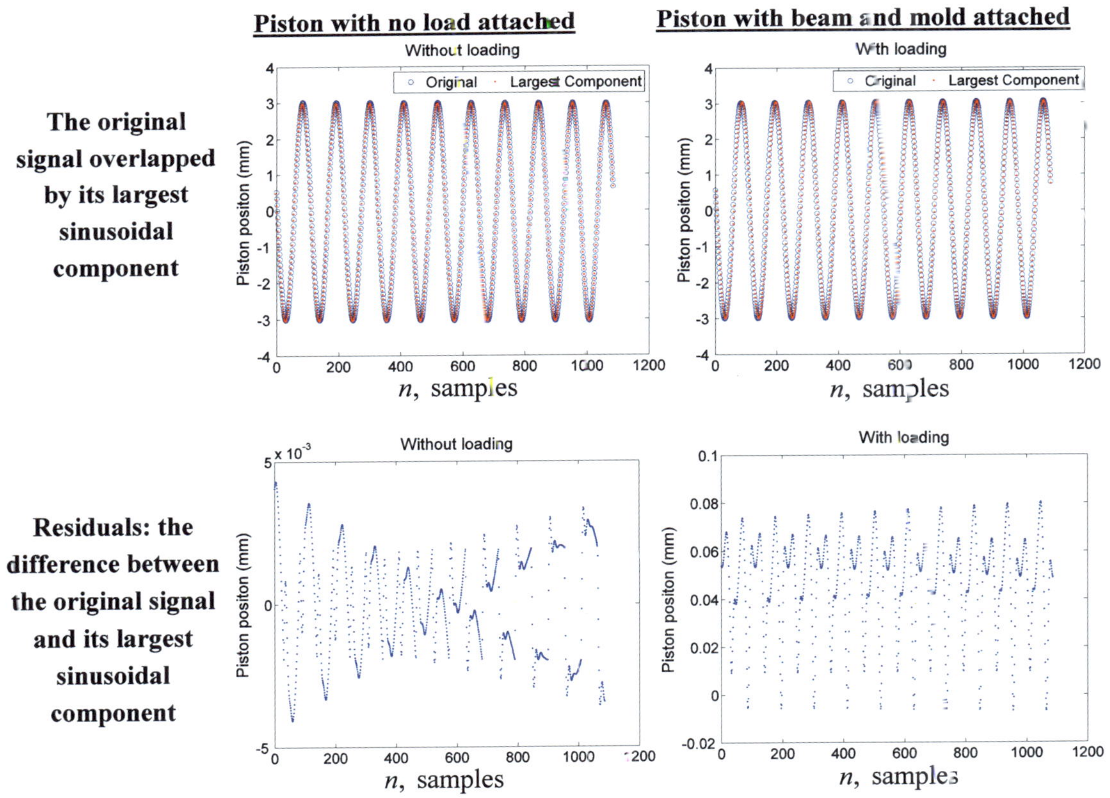

Fig. 5.2-3. Piston position with load and without it — calculation of DT residuals

Note that in the absence of aliasing ($S_F > S_N$), neglecting minor mismatch due to the computational error, ***CT and DT discrete spectra have the same magnitudes and phases, i.e. are identical:*** e.g. $X_s[10]=0.2633+j1.4884=1.5115e^{j1.3957} \approx X_{DFS}[10]=0.2568+j1.4855=1.5075e^{j1.3996}$.

5.2.2 *Power of periodic signals: Parseval's relationship*

Power of a periodic signal, CT. Power of $x_p(t)$ can be calculated using its Fourier series expansion:

$$P_x=\frac{1}{T_1}\int_{T_1}\left|x_p(t)\right|^2 dt=\frac{1}{T_1}\int_{T_1}x_p(t)x_p^*(t)dt=\frac{1}{T_1}\int_{T_1}\underbrace{\sum_{k=-\infty}^{\infty}X_s[k]e^{jk\omega_1 t}}_{x_p(t)}\underbrace{\sum_{m=-\infty}^{\infty}X_s^*[m]e^{-jm\omega_1 t}}_{x_p^*(t)}dt$$

$$=\sum_{k=-\infty}^{\infty}X_s[k]X_s^*[k]\underbrace{\frac{1}{T_1}\int_{T_1}e^{jk\omega_1 t}e^{-jk\omega_1 t}dt}_{1}=\sum_{k=-\infty}^{\infty}\left|X_s[k]\right|^2=\sum_{k=-\infty}^{\infty}P_s[k],$$

or, in terms of the polar Fourier series coefficients, c_k:

$$\left.\begin{array}{l}\left|X_s[0]\right|=c_0\\ \left|X_s[k]\right|=\frac{1}{2}c_k\\ \left|X_s[-k]\right|=\frac{1}{2}c_k\end{array}\right\}\xrightarrow[x_p(t)]{\text{For real}}\quad P_x=\sum_{k=-\infty}^{\infty}\left|X_s[k]\right|^2=\sum_{k=-\infty}^{-1}\left|X_s[k]\right|^2+\left|X_s[0]\right|^2+\sum_{k=1}^{\infty}\left|X_s[k]\right|^2=$$

$$\left|X_s[0]\right|^2+\sum_{k=1}^{\infty}2\left|X_s[k]\right|^2=c_0^2+\sum_{k=1}^{\infty}\frac{1}{2}c_k^2=\overbrace{c_0^2}+\overbrace{\sum_{k=1}^{\infty}2P_s[k]}^{\text{Actual power of }k^{th}\text{ component}}$$

$$\because\ 2P_s[k]=\frac{1}{2}c_k^2\Rightarrow P_s[k]=\left|X_s[k]\right|^2=\frac{1}{4}c_k^2.$$

The above fact is ***Parseval's Theorem:*** *the total power of a periodic signal is the sum of the powers of its harmonics.* In addition to magnitude and phase spectra, we can plot power spectrum:

- Two-sided: $P_s[k]$ versus $k\omega_1$ — just square of magnitude spectrum,
- One-sided: $2P_s[k]$ versus $k\omega_1$.

Power of a periodic signal: DT. In this case, the same relationship holds:

$$P_x=\frac{1}{N_1}\sum_{n=0}^{N_1-1}\left|x_p[n]\right|^2=\sum_{k=0}^{N_1-1}\left|X_{DFS}[k]\right|^2=\sum_{k=0}^{N_1-1}P_{DFS}[k],\ \text{DT power spectrum}: \tag{5.2-3}$$

$P_{DFS}[k]$ versus k repeats every N_1 steps: $P_{DFS}[k]=\left|X_{DFS}[k]\right|^2,\ 0\le k\le N_1-1$.

Extension of vector space techniques to function space for periodic signals. The latter extension is carried out by applying to these signals the counterparts of vector space operations given below.

Table 5.2-1. Function space counterparts of vector space representation.

Type of vector	(real) 3D vectors	real T_1-periodic CT signals	real N_1-periodic DT signals
Notation	$x=\begin{pmatrix}x_1 & x_2 & x_3\end{pmatrix}^T$	$x(t)$	$x[n]$
Dot product, (Signal power)	$x^T y = x_1y_1 + x_2y_2 + x_3y_3$	$\frac{1}{T_1}\int_{T_1} x(t)y^*(t)dt$	$\frac{1}{N_1}\sum_{n=0}^{N_1-1} x[n]y^*[n]$
Absolute value, RMS	$\lvert x\rvert = \left\lvert\sqrt{x^T x}\right\rvert$	$RMS = \left\lvert\sqrt{P_x}\right\rvert = \left\lvert\sqrt{\frac{1}{T_1}\int_{T_1} x(t)x^*(t)dt}\right\rvert$	$RMS = \left\lvert\sqrt{P_x}\right\rvert = \left\lvert\sqrt{\frac{1}{N_1}\sum_{n=0}^{N_1-1} x[n]x^*[n]}\right\rvert$
Projection	$proj_y(x) = \frac{x^T y}{y^T y} y$	$proj_y(x) = \frac{\frac{1}{T_1}\int_{T_1} x(t)y^*(t)dt}{\frac{1}{T_1}\int_{T_1} y(t)y^*(t)dt} y(t)$	$proj_y(x) = \frac{\frac{1}{N_1}\sum_{n=0}^{N_1-1} x[n]y^*[n]}{\frac{1}{N_1}\sum_{n=0}^{N_1-1} y[n]y^*[n]} y[n]$
Unit basis	$\begin{bmatrix}1\\0\\0\end{bmatrix}, \begin{bmatrix}0\\1\\0\end{bmatrix}, \begin{bmatrix}0\\0\\1\end{bmatrix}$	$\{e^{j2\pi k\cdot\frac{1}{T_1}\cdot t}\}_{k=-\infty}^{\infty}$	$\{e^{j2\pi k\cdot\frac{1}{N_1}\cdot n}\}_{k=0}^{N_1-1}$

Homework 7: Solved Examples and Solution Guide

Problem 1: Show that DT harmonics form an orthogonal basis.

Solution:

We need to show that $e^{j\Omega_1 kn} \bullet e^{j\Omega_1 mn} = \begin{cases} 0, & k \neq m, \\ \text{some real number}, & k = m. \end{cases}$

Let $x[n] = e^{j\Omega_1 kn}$, $y[n] = e^{j\Omega_1 mn}$, $\Omega_1 = \frac{2\pi}{N_1}$, and denote $P \doteq \sum_{n=0}^{N_1-1} x[n]y^*[n]$.

$$\therefore P = \sum_{n=0}^{N_1-1} x[n]y^*[n] = \sum_{n=0}^{N_1-1} e^{j\Omega_1 kn}e^{-j\Omega_1 mn} = \sum_{n=0}^{N_1-1} e^{j\Omega_1(k-m)n} = \sum_{n=0}^{N_1} e^{j\Omega_1(k-m)n} - e^{j\Omega_1(k-m)N_1}$$

$$= \sum_{n=0}^{N_1}\left(e^{j\Omega_1(k-m)}\right)^n - e^{j\Omega_1 N_1} \text{ (notice: when } k = m\text{, } e^{j\Omega_1(k-m)} = 1\text{).}$$

Apply identity: $\sum_{n=0}^{N}\alpha^n = \begin{cases}\frac{1-\alpha^{N+1}}{1-\alpha} & \text{for } \alpha \neq 1,\\ N+1 & \text{for } \alpha = 1,\end{cases}$ to the term with summation yields

$$P=\begin{cases}\dfrac{1-\left(e^{j\Omega_1(k-m)}\right)^{N_1+1}}{1-e^{j\Omega_1(k-m)}}-e^{j\Omega_1(k-m)N_1}=0, & k\neq m,\\ N_1+1-e^{j\Omega_1(k-m)N_1}=N_1+1-1=N_1, & k=m.\end{cases}$$ The latter follows from

$$\frac{1-\left(e^{j\Omega_1(k-m)}\right)^{N_1+1}}{1-e^{j\Omega_1(k-m)}}-e^{j\Omega_1(k-m)N_1}=\frac{1-e^{j\Omega_1(k-m)(N_1+1)}}{1-e^{j\Omega_1(k-m)}}-\frac{e^{j\Omega_1(k-m)N_1}\left(1-e^{j\Omega_1(k-m)}\right)}{1-e^{j\Omega_1(k-m)}}$$
$$=\frac{1-e^{j\Omega_1(k-m)(N_1+1)}-e^{j\Omega_1(k-m)N_1}+e^{j\Omega_1(k-m)(N_1+1)}}{1-e^{j\Omega_1(k-m)}}=\frac{1-e^{j\Omega_1(k-m)N_1}}{1-e^{j\Omega_1(k-m)}}=\frac{1-e^{j\frac{2\pi}{N_1}(k-m)N_1}}{1-e^{j\Omega_1(k-m)}}$$
$$=\frac{1-e^{j2\pi(k-m)}}{1-e^{j\Omega_1(k-m)}}=\frac{1-1}{1-e^{j\Omega_1(k-m)}}=0.$$

Problem 2: Represent the following periodic signals in terms of exponential, polar, and trigonometric Fourier series, indicate the fundamental period and the harmonic numbers of all the terms, and calculate the signal power and the rms value;

(a) $x(t)=\sum_{k=-3}^{3}\frac{4}{k^2}\cos\left(\frac{k\pi}{4}\right)e^{-jk\pi/5}e^{j7k\pi t}$;

(b) $x(t)=3+5\sin\left(6\pi t-\frac{\pi}{5}\right)-7\cos\left(14\pi t+\frac{\pi}{4}\right)-3\cos(16\pi t)$;

(c) $x(t)=\left[\sin(2t)-\cos(3t)\right]\left[\cos(2t)-\sin(3t)\right]$.

Solution: (a) $x(t)=\sum_{k=-3}^{3}\frac{4}{k^2}\cos\left(\frac{k\pi}{4}\right)e^{-jk\pi/5}e^{j7k\pi t}$.

Notice that $\cos\left(\frac{k\pi}{4}\right)=0$ for $k=\pm 2$, so that $x(t)=\frac{4}{9}\cos\left(-\frac{3\pi}{4}\right)e^{j3\pi/5}e^{j(-3)7\pi t}+\frac{4}{1}\cos\left(-\frac{1\pi}{4}\right)e^{j1\pi/5}e^{j(-1)7\pi t}$
$+\frac{4}{1}\cos\left(\frac{1\pi}{4}\right)e^{j-1\pi/5}e^{j(1)7\pi t}+\frac{4}{9}\cos\left(\frac{3\pi}{4}\right)e^{-j3\pi/5}e^{j(3)7\pi t}$.

$x(t)=-0.3143e^{j3\pi/5}e^{-j21\pi t}+2.828e^{j1\pi/5}e^{-j7\pi t}+2.828e^{-j1\pi/5}e^{j7\pi t}-0.3143e^{-j3\pi/5}e^{j21\pi t}$ (exponential form),
$x(t)=-0.6286\cos(3\pi/5-21\pi t)+5.656\cos(\pi/5-7\pi t)=0.6286\cos(21\pi t-3\pi/5+\pi)+5.656\cos(7\pi t-\pi/5)$
$=0.6286\cos(21\pi t+2\pi/5)+5.656\cos(7\pi t-\pi/5)$ (polar form)
$=-0.6286(\cos(3\pi/5)\cos(21\pi t))-0.6286(\sin(3\pi/5)\sin(21\pi t))$
$+5.656(\cos(\pi/5)\cos(7\pi t))+5.656(\sin(\pi/5)\sin(7\pi t))$
$=0.1942\cos(21\pi t)-0.5978\sin(21\pi t)+4.5758\cos(7\pi t)+3.3245\sin(7\pi t))$ (trigonometric form).
Frequencies present: 10.5 Hz and 3.5Hz. So, fundamental frequency $f_1=GCD(10.5,3.5)=3.5\,Hz$.
So, harmonics present: $k=1$ (fundamental) and $k=3$ (3rd harmonic).
From Exp. form, $P=2(2.828)^2+2(0.3143)^2=16.2$ and $x_{rms}=\sqrt{P}=4.02$.

(b) $x(t)=3+5\sin\left(6\pi t-\frac{\pi}{5}\right)-7\cos\left(14\pi t+\frac{\pi}{4}\right)-3\cos(16\pi t)$.

Notice that: $\cos(\alpha)=-\cos(\alpha-\pi)$ and $\sin(\alpha)=\cos\left(\alpha-\frac{\pi}{2}\right)$. Therefore,

$x(t)=3+5\cos\left(6\pi t-\frac{7\pi}{10}\right)+7\cos\left(14\pi t-\frac{3\pi}{4}\right)-3\cos(16\pi t)$ (polar form).

$x(t)=3+\frac{5}{2}e^{-j\frac{7\pi}{10}}e^{j6\pi t}+\frac{5}{2}e^{j\frac{7\pi}{10}}e^{-j6\pi t}+\frac{7}{2}e^{-j\frac{3\pi}{4}}e^{j14\pi t}+\frac{7}{2}e^{j\frac{3\pi}{4}}e^{-j14\pi t}-\frac{3}{2}e^{j\text{-}16\pi t}-\frac{3}{2}e^{j16\pi t}$ (exponential form) or

$x(t)=3+5\cos\left(6\pi t\right)\cos\left(\frac{7\pi}{10}\right)+5\sin\left(6\pi t\right)\sin\left(\frac{7\pi}{10}\right)+$

$7\cos\left(14\pi t\right)\cos\left(\frac{3\pi}{4}\right)+7\sin\left(14\pi t\right)\sin\left(\frac{3\pi}{4}\right)-3\cos(16\pi t)$

$=3-2.9389\cos\left(6\pi t\right)+4.045\sin\left(6\pi t\right)-4.9498\cos\left(14\pi t\right)+4.9498\sin\left(14\pi t\right)-3\cos(16\pi t)$

(trigonometric form).

Frequencies present: 3 Hz and 7 Hz. So, fundamental freq. $f_1=GCD(3,7)=1\ Hz$.

So, harmonics present: $k=3$ (3rd harmonic) and $k=7$ (7th harmonic).

From polar form, $P=(3)^2+0.5\left[5^2+7^2+3^2\right]=50.5$ and $x_{rms}=\sqrt{P}=7.106$.

(c) $x(t)=\left[\sin(2t)-\cos(3t)\right]\left[\cos(2t)-\sin(3t)\right]$

$=\sin(2t)\cos(2t)-\sin(2t)\sin(3t)-\cos(3t)\cos(2t)+\cos(3t)\sin(3t)$

$=1/2(\sin(4t))+1/2(\sin(6t))-(\cos(2t-3t))$

$=1/2(\sin(4t))+1/2(\sin(6t))-(\cos(t))$ (trigonometric form)

$=1/2(\cos(4t-\pi/2))+1/2(\cos(6t-\pi/2))+(\cos(t-\pi))$ (polar form)

$=\frac{1}{4}(e^{-j\pi/2}e^{j4t}+e^{j\pi/2}e^{-j4t})+\frac{1}{4}(e^{-j\pi/2}e^{j6t}+e^{j\pi/2}e^{-j6t})+\frac{1}{2}(e^{-j\pi}e^{jt}+e^{j\pi}e^{-jt})$ (exponential form).

Matching $e^{j\omega t}$ to the corresponding terms, frequencies present are 4, 6, and 1 rad/s. So, fundamental freq. $\omega_1=GCD(4,6,1)=1$ rad/s and harmonics present are $k=[1,4,6]$.

From polar form: $P=0.5\left[(0.5)^2+(0.5)^2+(1)^2\right]=0.75$ and $x_{rms}=\sqrt{P}=0.866$.

Problem 3: Find the coefficients of the exponential, polar, and trigonometric Fourier series representation for $x(t)=\sum_{k=0}^{\infty}\frac{4}{k^2}\sin^2\left(\frac{k\pi}{4}\right)\cos\left(400k\pi t\right)$; indicate the dc term, the fundamental frequency f_1, and the fundamental period T_1.

Solution: Look at trigonometric form: $x_p(t) = a_o + \sum_{k=1}^{\infty}\left[a_k \cos(2\pi k f_1 t) + b_k \sin(2\pi k f_1 t)\right]$. Then

$$a_o = \frac{4}{k^2}\sin^2\left(\tfrac{k\pi}{4}\right)\Big|_{k=0} = \lim_{k\to 0}\frac{d(4\sin^2\left(\tfrac{k\pi}{4}\right))/dk}{dk^2/dk} = \lim_{k\to 0}\frac{4\cdot 2\sin\left(\tfrac{k\pi}{4}\right)\cos\left(\tfrac{k\pi}{4}\right)\cdot\tfrac{\pi}{4}}{2k} = \lim_{k\to 0}\frac{\pi\sin\left(\tfrac{k\pi}{4}\right)}{k}$$

$$= 4\lim_{k\to 0}\frac{\pi d\sin\left(\tfrac{k\pi}{4}\right)/dk}{dk/dk} = \lim_{k\to 0}\frac{\pi\cdot\tfrac{\pi}{4}\cos\left(\tfrac{k\pi}{4}\right)}{1} = \frac{\pi^2}{4};\ a_k = \frac{4}{k^2}\sin^2\left(\frac{k\pi}{4}\right);\ b_k = 0;\ f_1 = 200\text{ Hz.}$$

(a) Identify the fundamental frequency f_1 and time period T_1; $f_1 = 200$ Hz and $T_1 = 1/f_1 = 5$ ms.

(b) Identify the Fourier series coefficients c_k, θ_k, and $X[k]$.

Look at the polar form: $x_p(t) = c_o + \sum_{k=1}^{\infty} c_k \cos(2\pi k f_1 t + \theta_k)$, $c_k \ge 0$, $c_o \ge 0$,

$$c_o = a_o = \frac{\pi^2}{4};\ c_k = \frac{4}{k^2}\sin^2\left(\tfrac{k\pi}{4}\right) \ge 0;\ \theta_k = 0;\ f_1 = 200\text{ Hz.}$$

Look at the exponential form: $x_p(t) = \sum_{k=-\infty}^{\infty} X_s[k]e^{j2\pi k f_1 t}$. Then, since $\sin(-x) = -\sin(x)$, we have

$$X_s[k] = \begin{cases} \frac{2}{k^2}\sin^2\left(\tfrac{k\pi}{4}\right), & k \ge 1, \\ \pi^2/4, & k = 0, \\ \frac{2}{k^2}\sin^2\left(\tfrac{k\pi}{4}\right), & k \le -1. \end{cases}$$

5.3 Analytical CT Fourier Series Calculation for Six Basic Waveforms

5.3.1 *Square wave Exponential Fourier series representation and spectra calculation*

5.3.1.1 *Fourier series representation of the square wave signal*

Recall periodic pulse train (square wave) from Section 2.2:

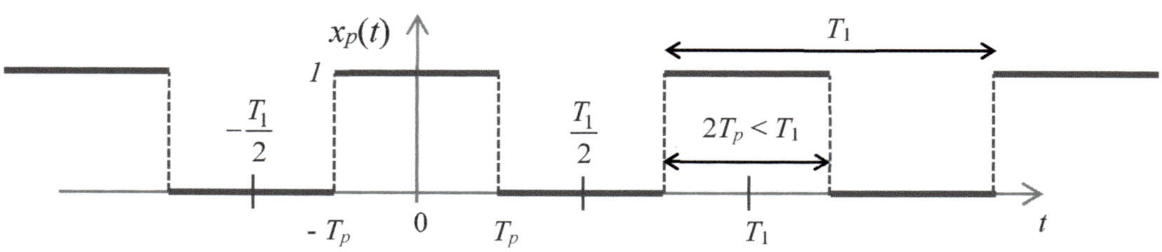

Fig. 5.3-1. Periodic pulse train signal.

and its representation in terms of rectangle signals:

$$x_p(t) = ... + \text{rect}(t/2T_p) + \text{rect}[(t-T)/2T_p] + ... = \sum_{n=-\infty}^{\infty} \text{rect}[(t-nT_1)/2T_p],\ T_p < T_1/2. \quad (5.3\text{-}1)$$

The fundamental period and frequency of this signal are, respectively T_1 and $\omega_1 = 2\pi/T_1$.

Let's find the Fourier series coefficients for this signal to obtain its Fourier series representation, often simply called Fourier series

$$x_p(t) = \sum_{k=-\infty}^{\infty} X_s[k] e^{jk\omega_1 t}. \quad (5.3\text{-}2)$$

Let's use $[-T_1/2, T_1/2]$ as the integration period. This makes the calculation easier, as we will see shortly. Then

$$X_s[k] = \frac{1}{T_1}\int_{-\frac{T_1}{2}}^{\frac{T_1}{2}} \underbrace{\text{rect}\left(\frac{t}{2T_p}\right)}_{1,\ \text{when } |t| \le T_p,\ 0,\ \text{otherwise}} e^{-jk\frac{2\pi}{T_1}t} dt = \frac{1}{T_1}\int_{-T_p}^{T_p} e^{-\frac{j2k\pi}{T_1}t} dt.$$

For $k = 0$,

$$X_s[0] = \frac{1}{T_1}\int_{-T_p}^{T_p} e^{-\frac{j2\cdot 0\cdot\pi}{T_1}t} dt = \frac{1}{T_1}\int_{-T_p}^{T_p} e^0 dt = \frac{1}{T_1}\int_{-T_p}^{T_p} 1dt = \frac{2T_p}{T_1}. \quad (5.3\text{-}3)$$

For $k \neq 0$, complex exponentials integrate just like regular exponentials. This is why we chose complex exponentials as the "basis" functions:

$$X_s[k] = \frac{1}{T_1}\int_{-T_p}^{T_p} e^{-\frac{j2k\pi}{T_1}t} dt = \frac{1}{T_1}\left(-\frac{T_1}{j2k\pi}\right) e^{-\frac{j2k\pi}{T_1}t}\Bigg|_{-T_p}^{T_p} = \frac{1}{j2k\pi}[-(e^{-\frac{j2k\pi}{T_1}T_p} - e^{-\frac{j2k\pi}{T_1}(-T_p)})] =$$

$$\frac{1}{k\pi}\cdot\frac{1}{j2}[e^{j\frac{2k\pi T_p}{T_1}} - e^{-j\frac{2k\pi T_p}{T_1}}] = \frac{1}{k\pi}\cdot\frac{1}{j2}\left[\underbrace{\cos\left(\frac{2k\pi T_p}{T_1}\right) + j\sin\left(\frac{2k\pi T_p}{T_1}\right)}_{\text{Euler's formula}} - \underbrace{\left(\cos\left(\frac{2k\pi T_p}{T_1}\right) - j\sin\left(\frac{2k\pi T_p}{T_1}\right)\right)}_{\text{Euler's formula}}\right]$$

$$= \frac{1}{k\pi}\cdot\frac{1}{j2}\left[2j\sin\left(\frac{2k\pi T_p}{T_1}\right)\right] = \frac{1}{k\pi}\cdot\sin\left(\frac{2k\pi T_p}{T_1}\right) = \frac{2T_p}{T_1}\frac{\sin\left(\pi k\frac{2T_p}{T_1}\right)}{\pi k\frac{2T_p}{T_1}} = \frac{2T_p}{T_1}\text{sinc}\left(k\frac{2T_p}{T_1}\right).$$

In summary:

$$\boxed{X_s[k] = \frac{2T_p}{T_1}\text{sinc}\left(k\frac{2T_p}{T_1}\right)} \Rightarrow x_p(t) = \sum_{k=-\infty}^{\infty} \underbrace{\left(\frac{2T_p}{T_1}\right)}_{\text{Duty ratio}} \text{sinc}\left(k\frac{2T_p}{T_1}\right) e^{jk\frac{2\pi}{T}t}. \quad (5.3\text{-}4)$$

For $k = 0$, the value of $\text{sinc}(0) = 1$.

Note that all $X_s[k]$ are real (positive or negative), yielding the following three values:

$$\text{Magnitude: } \left|X_s[k]\right| = \frac{2T_p}{T_1}\left|\text{sinc}\left(k\frac{2T_p}{T_1}\right)\right|, \text{ Phase: } \angle X_s[k] = \begin{cases} 0, & X_s[k] \geq 0 \\ 1, & X_s[k] < 0 \end{cases}, \tag{5.3-5}$$

$$\text{Power: } P_s[k] = \left|X_s[k]\right|^2 = \left(\frac{2T_p}{T_1}\text{sinc}\left(k\frac{2T_p}{T_1}\right)\right)^2. \tag{5.3-6}$$

5.3.1.2 *Plotting square wave spectra: Fourier coefficients as functions of frequency*

Take $T_1 = 8T_p$. Recalling that sinc function has the form

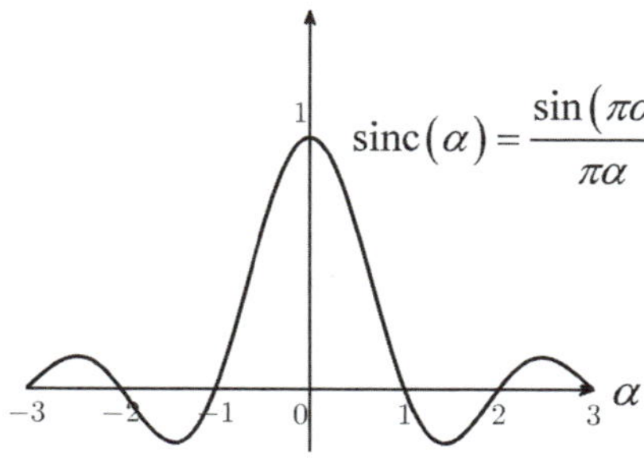

Fig. 5.3-2. Sinc function analytical expression and graph.

yields

$$X_s[k] = \frac{2T_p}{T_1}\text{sinc}\left(k\frac{2T_p}{T_1}\right) = \frac{1}{4}\text{sinc}\left(\frac{k}{4}\right) =$$

$$\frac{1}{4}\text{sinc}\left(\frac{1}{4\omega_1}k\omega_1\right) = \begin{cases} \overbrace{\frac{1}{4}\left|\text{sinc}\left(\frac{1}{4\omega_1}k\omega_1\right)\right|}^{\left|X_s[k]\right|} e^{j0}, & \text{sinc}\left(\frac{1}{4\omega_1}k\omega_1\right) \geq 0, \\ \underbrace{\frac{1}{4}\left|\text{sinc}\left(\frac{1}{4\omega_1}k\omega_1\right)\right|}_{\left|X_s[k]\right|} e^{j(\pm\pi)}, & \text{sinc}\left(\frac{1}{4\omega_1}k\omega_1\right) < 0. \end{cases}$$

Since sinc(α) takes 0 values at integer values of α, sinc($k/4$) takes zero values at k = integer multiples of 4, and, hence, sinc($k\omega_1/4k\omega_1$) — at $k\omega_1$ = *integer* multiples of $4\omega_1$. This yields spectra of the pulse train given in Figure 5.3-3.

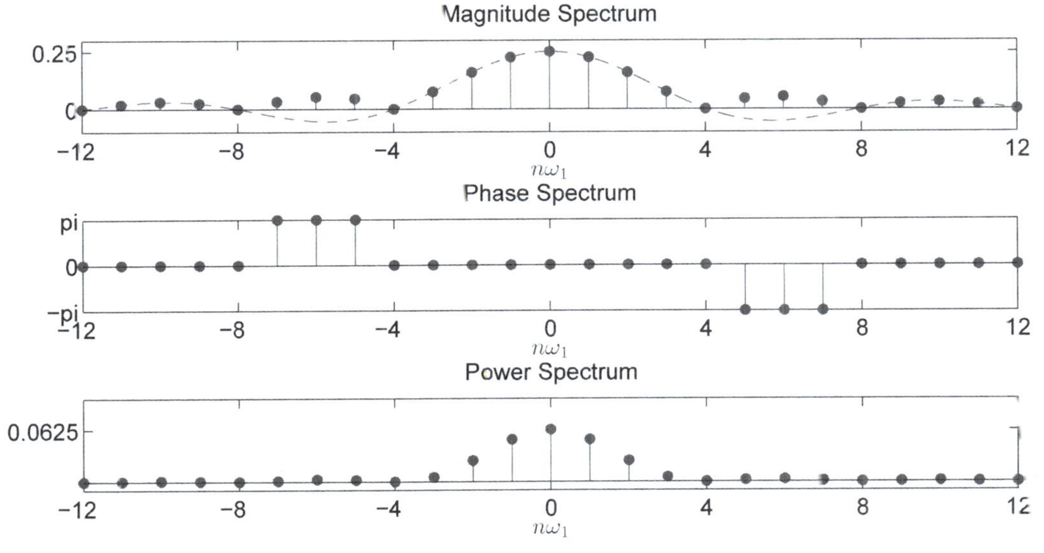

Fig. 5.3-3. Square wave magnitude, phase, and power spectra.

We note that for the phase spectrum, π and $-\pi$ is the same point, making the choice of the phase spectrum lines going up on the right and down on the left arbitrary (in line with odd symmetry for real periodic signal phase spectra).

5.3.2 *Exponential Fourier series coefficients calculation for five other basic waveforms*

5.3.2.1 *Exponential Fourier series coefficients calculation for the impulse train*

From Figure 5.3-4, and by the sifting property of impulses, we have:

$$X[k]=\frac{1}{T_1}\int_{-\frac{T_1}{2}}^{\frac{T_1}{2}}\delta(t)e^{-j2\pi kf_1t}dt=\frac{1}{T_1}\cdot e^{-j2\pi kf_1t}\Big|_{t=0}=\frac{1}{T_1}\cdot 1=\frac{1}{T_1}\ \forall k.$$

This shows that all the Fourier series coefficients of the impulse train are $1/T_1$, i.e. don't decay with frequency increase. This means that impulse train has infinite power, i.e. is not a power signal.

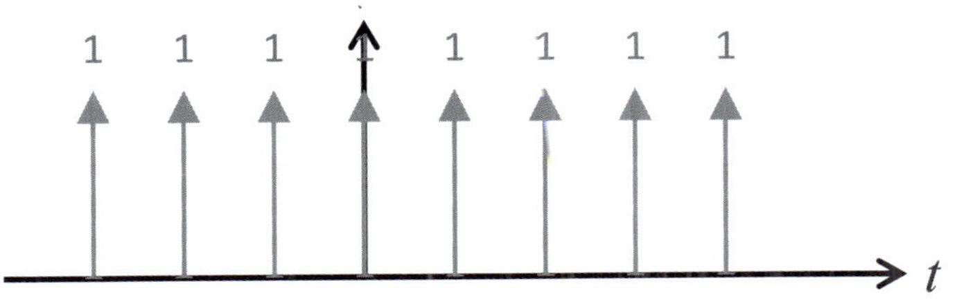

Fig. 5.3-4. Impulse train.

5.3.2.2 *Exponential Fourier series coefficients calculation for the sawtooth waveform*

From Figure 5.3-5, $T_1 = 2,\ f_1 = \frac{1}{2}$ and $x_p(t) = t, 0 \le t \le 2$ we have

$$a_0 = \frac{1}{2}\int_0^2 t dt = 1,$$

$$a_k = \frac{2}{2}\int_0^2 t\cos\left(2\pi k \frac{1}{2} t\right)dt = \int_0^2 t\cos(\pi k t)dt = t\frac{1}{\pi k}\sin(\pi k t)\Big|_0^2 - \int_0^2 \frac{1}{\pi k}\sin(\pi k t)dt$$

$$= \frac{1}{\pi k}t\sin(\pi k t) + \frac{1}{(\pi k)^2}\cos(\pi k t)\Big|_0^2 = 0,$$

$$b_k = \frac{2}{2}\int_0^2 t\sin\left(2\pi k \frac{1}{2} t\right)dt = \int_0^2 t\sin(\pi k t)dt = t\left(-\frac{1}{\pi k}\cos(\pi k t)\right)\Big|_0^2 - \int_0^2 -\frac{1}{\pi k}\cos(\pi k t)dt$$

$$= -\frac{1}{\pi k}t\cos(\pi k t) + \frac{1}{(\pi k)^2}\sin(\pi k t)\Big|_0^2 = -\frac{2}{\pi k}.$$

Therefore, $X[0] = 1,\ X[k] = \frac{j}{\pi k}, (k \ne 0).$

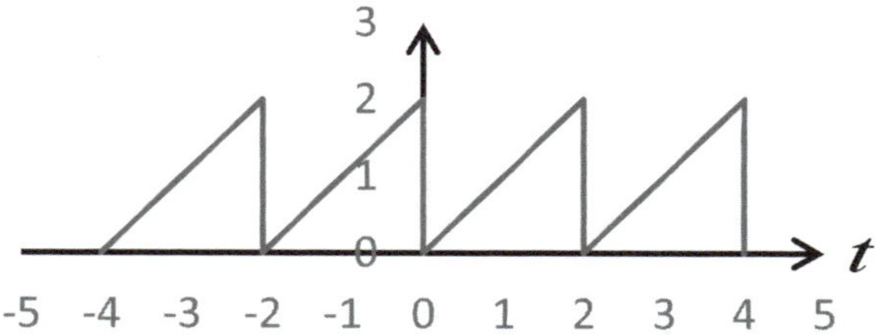

Fig. 5.3-5. Sawtooth signal.

We note that from Figure 5.3-5 $x_p(t)$ is seen to have odd symmetry if the constant term is adjusted accordingly, which also yields $a_k = 0,\ k \ne 0$.

5.3.2.3 *Exponential Fourier series coefficients calculation for triangular pulse train*

For the triangular pulse train showed in Figure 5.3-6, we have $T_1 = 2,\ f_1 = \frac{1}{2}$ and

$$x_p(t) = \begin{cases} t, 0 \le t \le 1, \\ 2 - t, 1 \le t \le 2. \end{cases}$$

Then we can compute the Fourier series coefficients as follows:

$$a_0 = \frac{1}{2}\left(\int_0^1 t\,dt + \int_1^2 (2-t)\,dt\right) = \frac{1}{2},$$

$$a_k = \frac{2}{2}\left(\int_0^1 t\cos\left(2\pi k\frac{1}{2}t\right)dt + \int_1^2 (2-t)\cos\left(2\pi k\frac{1}{2}t\right)dt\right) = \begin{cases} 0, & k\ even, \\ \dfrac{-4}{(\pi k)^2}, & k\ odd, \end{cases}$$

$$b_k = \frac{2}{2}\left(\int_0^1 t\sin\left(2\pi k\frac{1}{2}t\right)dt + \int_1^2 (2-t)\sin\left(2\pi k\frac{1}{2}t\right)dt\right) = 0.$$

Therefore, we have:

$$X[0] = \frac{1}{2},\ X[k] = -\frac{2}{(\pi k)^2}, (k\ odd).$$

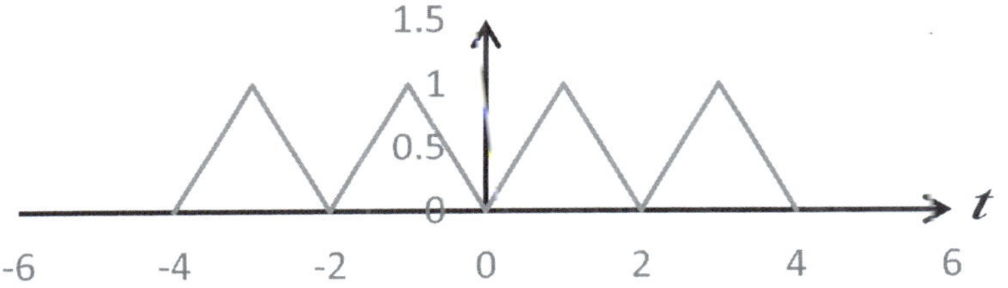

Fig. 5.3-6. Triangular pulse train.

We also note that from Figure 5.3-6 $x_p(t)$ is seen to have even symmetry, yielding $b_k = 0,\ k \neq 0$.

5.3.2.4 *Exponential Fourier series coefficients calculation for the half-wave rectified sine*

From Figure 5.3-7, we have $T_1 = 2,\ f_1 = \frac{1}{2}$ and $x_p(t) = \sin(\pi t), 0 \le t \le 1$.

Then we compute the coefficients as follows:

$$a_0 = \frac{1}{2}\int_0^1 \sin(\pi t)\,dt = \frac{1}{2}\cdot\frac{1}{\pi}\left(-\cos(\pi t)\right)\Big|_0^1 = \frac{1}{\pi},$$

$$a_k = \frac{2}{2}\int_0^1 \sin(\pi t)\cos(\pi k t)\,dt = \frac{1}{2}\int_0^1 \left\{\sin\left[(1+k)\pi t\right] + \sin\left[(1-k)\pi t\right]\right\}dt.$$

Upon simplification, using $\cos(1 \pm k)\pi = -\cos(k\pi)$, this yields

$$a_k = \frac{-1}{2\pi}\left[\frac{\cos\left[(1+k)\pi t\right]}{1+k} + \frac{\cos\left[(1-k)\pi t\right]}{1-k}\right]\Bigg|_0^1 = \frac{1+\cos(k\pi)}{\pi(1-k^2)} = \frac{2}{\pi(1-k^2)}\ (k\ even).$$

Similarly,

$$b_k = \frac{2}{2}\int_0^1 \sin(\pi t)\sin(\pi k t)\,dt = \frac{1}{2}\int_0^1 \{\cos[(1+k)\pi t] - \cos[(1-k)\pi t]\}\,dt$$

$$= \frac{1}{2\pi}\left[\frac{\sin[(1+k)\pi t]}{1+k} + \frac{\sin[(1-k)\pi t]}{1-k}\right]\Bigg|_0^1 = 0.$$

Now, we encounter a little problem: if $b_k = 0$, the signal $x_p(t)$ must be even symmetric. However, it is not! The catch is that for $k = 1$ the denominator $1 - k = 0$, making the above equation for b_k undefined. Thus we calculate b_1 separately:

$$b_1 = \frac{2}{2}\int_0^1 \sin(\pi t)\sin(\pi t)\,dt = \frac{1}{2}\int_0^1 [1 - \cos(2\pi t)]\,dt = \frac{1}{2}.$$

Thus, $b_1 = \frac{1}{2}$ and $b_k = 0, k > 1$. Computing a_1 separately yields $a_1 = 0$. Therefore,

$$X[0] = \frac{1}{\pi},\ X[1] = \frac{-j}{4},\ X[k] = \frac{1}{\pi(1-k^2)},\ (k\ even).$$

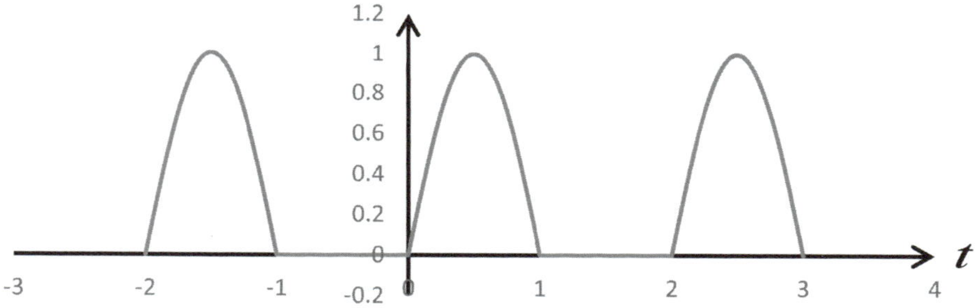

Fig. 5.3-7. Half-wave rectified sine.

5.3.2.5 *Exponential Fourier series coefficients calculation for the truncated periodically extended exponential signal*

From Figure 5.3-8, we have $T_1 = 2, f_1 = \frac{1}{2}$ and $x_{tp}(t) = e^t, 0 \le t \le 1$, repeats every T_1 sec, $T_1 \ge 1$. Its exponential Fourier series coefficients are given by

$$X[0]=\frac{1}{T_1}\int_{T_1}x(t)dt=\frac{1}{2}\int_0^1 e^t dt=\frac{e-1}{2},$$

$$X[k]=\frac{1}{T_1}\int_{T_1}x(t)e^{-j2\pi kf_1t}dt=\frac{1}{2}\int_0^1 e^te^{-j2\pi k\frac{1}{2}t}dt=\frac{1}{2}\int_0^1 e^{t(1-jk\pi)}dt$$

$$=\frac{0.5e^{t(1-jk\pi)}}{1-jk\pi}\bigg|_0^1=\frac{0.5\left(e^{1-jk\pi}-1\right)}{1-jk\pi}.$$

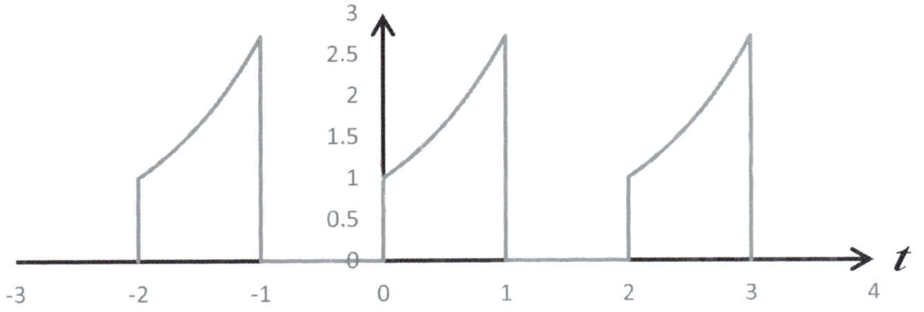

Fig. 5.3-8. Truncated and periodically extended exponential signal.

5.4 Signal Speed and Bandwidth, Gibbs Phenomenon, Windowing, Filtering as Complex Windowing

5.4.1 *Link between time and frequency domains: signal speed*

We are writing signals as the sums of sinusoids. What about the derivatives of the latter?

$$\frac{d}{dt}\cos(\omega_i t)=-\omega_i\sin(\omega_i t)\ \Rightarrow\ \max_t[-\omega_i\sin(\omega_i t)]=\omega_i\max_t[-\sin(\omega_i t)]$$
$$=\omega_i[-\sin(\omega_i t)]_{t=(3/4)T_i}=\omega_i\cdot 1=\omega_i,$$

i.e. sinusoid of frequency ω_i has the maximal rate of change, i.e. the steepest slope, also ω_i! Let's call this value the ***signal speed***, keeping in mind that although it is not the speed of signal propagation through media, it directly relates to the *speed of information propagation* through media.

Example 5.4-1: Signal speed and its relation to signal frequency.
Figure 5.4-1 shows three harmonics of a fundamental frequency ω_1: increase in frequency, or decrease in period, is seen to proportionally increase the maximal signal slope:

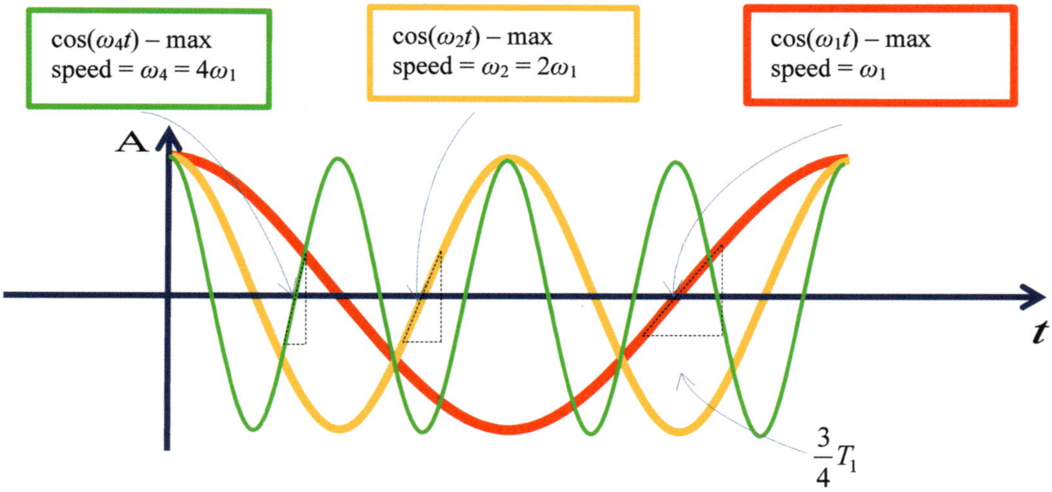

Fig. 5.4-1. Three harmonics of a signal with the indicated signal speed.

This figure makes it explicitly clear that capturing ***faster changing*** signals can be accomplished only through the inclusion of the ***higher frequency*** components. This requires more bandwidth, higher sampling rates, and faster computation.

Example 5.4-2: Speed and bandwidth of the square wave.

Consider square wave, such as that given in Figure 5.3-4,

$$x_p(t) = \sum_{n=-\infty}^{\infty} \text{rect}\left(\frac{t-nT_1}{2T_p}\right) = \sum_{k=-\infty}^{\infty} \frac{\omega_1 T_p}{\pi} \text{sinc}\left(\frac{\omega_k T_p}{\pi}\right) e^{j\omega_k t}. \tag{5.4-1}$$

Its maximum speed is $dx_p(t)/dt = \infty = lim_{k\to\infty}\omega_k$.

Squire wave has an infinitely fast rate of change in the signal, so we need infinitely high frequency components to capture it. The first equality in (5.4-1) is valid ***pointwise***, but how about the second one? Does it really work out to build such abruptly changing (infinite speed) signals through smooth basis functions?

5.4.2 *Convergence of Fourier series: building abruptly changing (infinite speed) signals through smooth basis functions*

So far, it looked like we mean pointwise convergence when we represent a signal in terms of its Fourier series projections:

$$x_p(t) = \sum_{k=-\infty}^{\infty} X_s[k] e^{jk\omega_1 t} = \sum_{k=-\infty}^{\infty} X_s[k] e^{jk\omega_1 t}. \tag{5.4-2}$$

This, however, has been an abuse of notation: the convergence, as shown in Chapter 10, is only in the ***"energy" sense***: the power of the partial-sum error $x_p(t)-\sum_{k=-N}^{N} X_s[k]e^{jk\omega_1 t} \rightarrow 0$ as $N \rightarrow \infty$, where N is a finite number. Convergence of sequences of functions is a tricky property to explore, so we will look only at some aspects of "energy" convergence. For example, knowing that the individual terms, $e^{jk\omega_1 t}$, are all continuous, what happens when $x_p(t)$ is discontinuous?

Example 5.4-3. Square wave convergence.

Let's take $T_1 = 1 = 8T_p$, then

$$x_p(t)=\sum_{k=-\infty}^{\infty} X_s[k]e^{jk\omega_1 t}=\sum_{k=-\infty}^{\infty}\frac{1}{4}\text{sinc}\left(\frac{k}{4}\right)e^{jk\frac{2\pi}{T_1}t}=\sum_{k=-\infty}^{\infty}\frac{1}{4}\text{sinc}\left(\frac{k}{4}\right)e^{j2k\pi t}. \qquad (5.4\text{-}3)$$

Let's graph this summation for increasing values of N, to see what it means for the sum to converge in the "energy" sense.

Sum from −1 to 1: $\sum_{k=-1}^{1}\frac{1}{4}\text{sinc}\left(\frac{k}{4}\right)e^{j2\pi kt}$

Sum from −10 to 10: $\sum_{k=-10}^{10}\frac{1}{4}\text{sinc}\left(\frac{k}{4}\right)e^{j2\pi kt}$

Sum from −100 to 100: $\sum_{k=-100}^{100}\frac{1}{4}\text{sinc}\left(\frac{k}{4}\right)e^{j2\pi kt}$

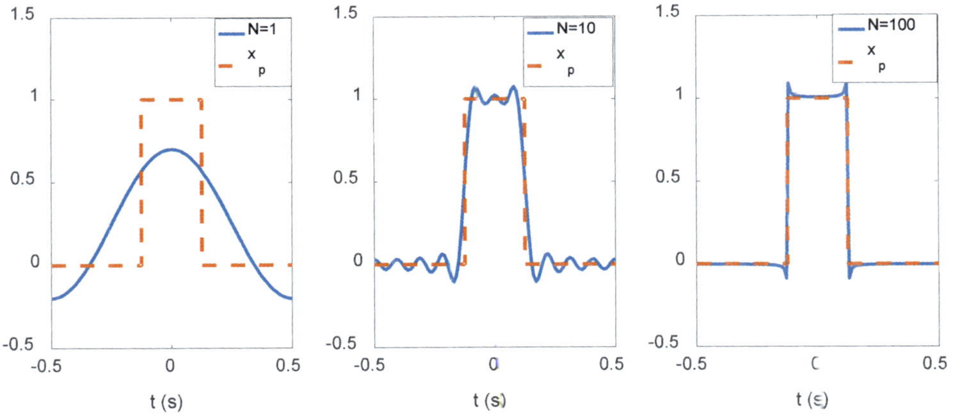

Fig. 5.4-2. Sum of Fourier series for the square wave.

Figure 5.4-2 shows that convergence in the "energy" sense does not imply pointwise convergence. The power of the high frequency errors at the jump discontinuities of the original signal goes to 0 as $N \rightarrow \infty$, but the magnitude does not. The latter fact is referred to as *ringing*, or *Gibbs effect*. But we'd still like to get rid of it. How? One option is *windowing* discussed next.

5.4.3 *Spectral windowing*

Windowing (filtering) invokes the link between the time and the frequency domains to smooth out the errors at discontinuities by applying a **spectral window** that reshapes and/or gets rid of high-frequency components. Windowing consists in approximating the original function $x_p(t)$ as

$$x_p(t) \approx \sum_{k=-M}^{M} W[k] X_s[k] e^{jk\omega_1 t}.$$

Finite sum ↗ (the summation) — Each component is modified by the corresponding real number $W[k]$

Passing a square wave through the Hamming spectral window is shown in Figure 5.4-3.

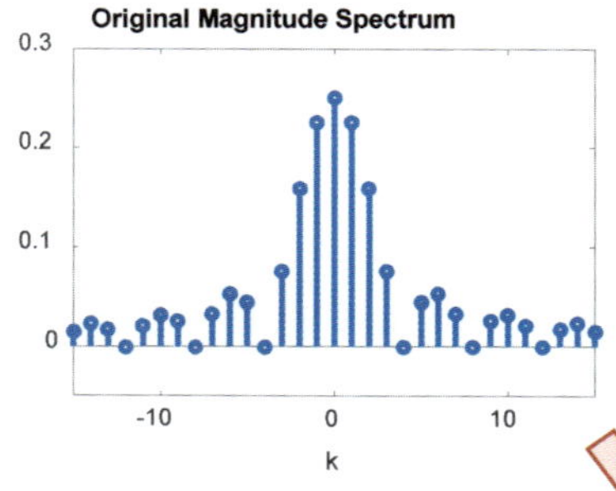

Window Type	Expression for $W[k]$ Over $-N \le k \le N$
Boxcar	$\text{rect}(k/2N) = 1$
Bartlett	$\text{tri}(k/N) = 1 - \lvert k \rvert / N$
Von Hann (Hanning)	$0.5 + 0.5\cos(k\pi/N)$
Hamming	$0.54 + 0.46\cos(k\pi/N)$
Cosine	$\cos(k\pi/(2N))$
Riemann	$\text{sinc}(k\pi/N)$
Blackman	$0.42 + 0.5\cos(k\pi/N) + 0.08\cos(2k\pi/N)$

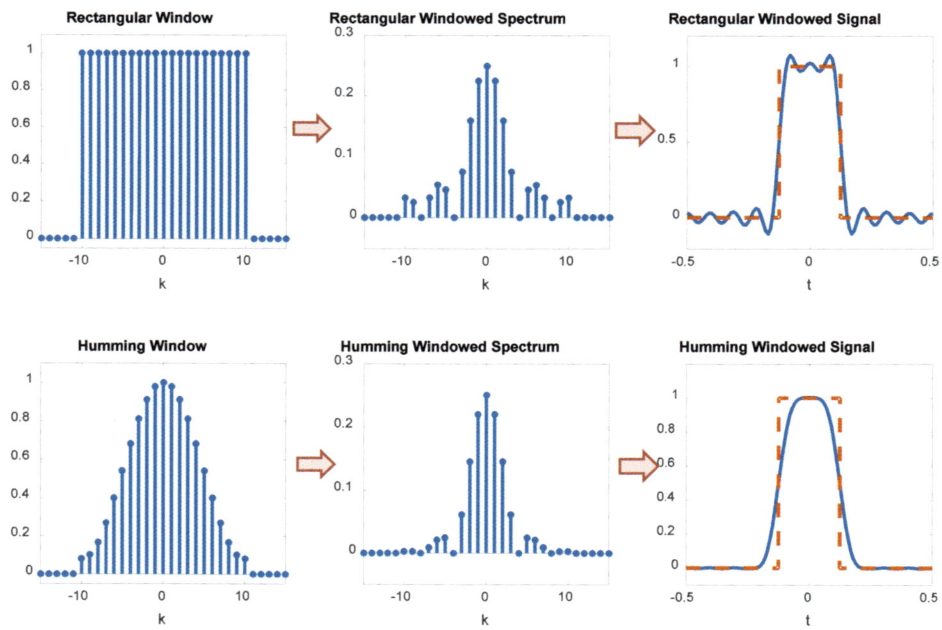

Fig. 5.4-3. Table of the main spectral windows and Illustration of rectangular and Hamming windowing.

It is seen that this window multiplies each term in the Fourier series by a specific real number, Hamming weight, making windowed signal to be a particular weighted local average of the original signal. This action implements one of the key ideas in signal processing and control: ***multiplying the frequency components by numbers, in general complex ones, performs time-domain filtering***.

So far, our discussion has been limited to ***spectral windowing*** applied to the Fourier series Windows can also be applied to the sampled data itself through convolution, referred to as ***time-domain windowing***. A more general filtering version of the latter will be considered in Chapter 7.

Carrying out spectral windowing is difficult: it is not always clear what shape and size will give the best results. For example, tapered windows are usually better suited for avoiding Gibb's effect. There are many windows shapes: Rectangular, Bartlett, Hamming, Blackman, and others. The main spectral windows are given in the table form in Figure 5.4-3.

5.4.4 *What are we getting through bandwidth, i.e. means for handling almost abruptly changing (very high speed) signals*

Techno-social revolution: information technology (IT) — Internet, wireless, audiovisual systems, industrial and health monitoring and control, pervasive real-time robotic systems and drones, globalization — round-the-clock design, social networking, internet of things, machine learning.

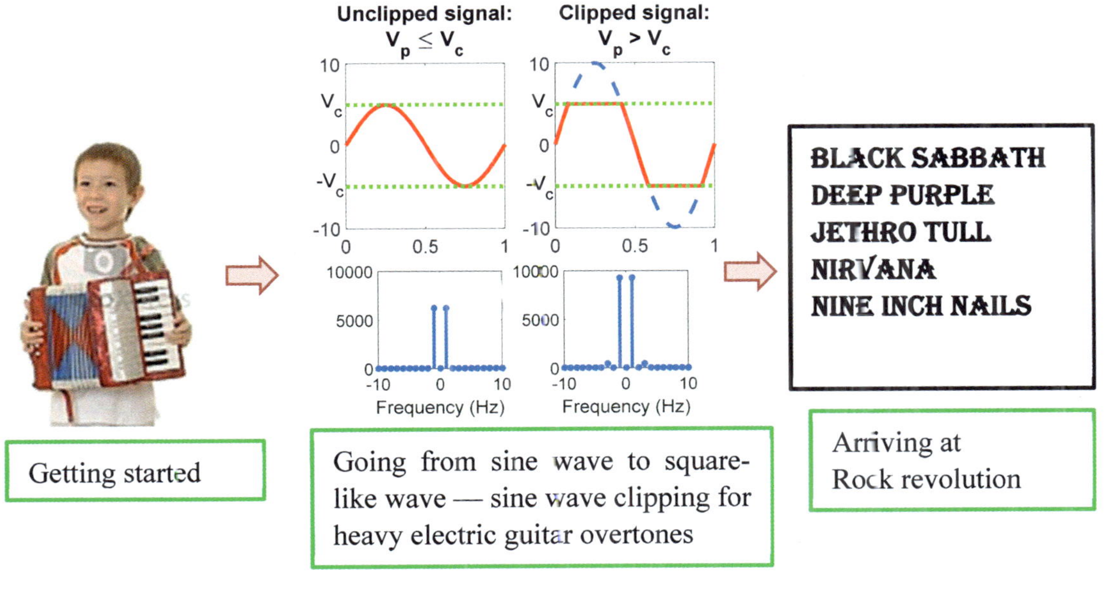

Fig. 5.4-4. Example of bandwidth utilization.

Psycho-acoustic revolution: jazz, rock, pop, rap, and the accompanying social upheaval.

5.5 Sampling Continuous Signal Spectra: Computation of $H(j\omega)$ from Fourier Series

So far, we've discussed how to make the time axis accessible to digital signal processing (DSP): bandlimit (low-pass filter) the signal to pass frequencies in the range of interest with minimal distortion, but minimize magnitude of potentially aliased high frequency signal part if it gets into the principal range. Oversample (at least 10 times) and quantize to generate a data file, stopping at some point to keep the data set small enough to store.

Now, we're going to focus on convenient computational access to the frequency axis — discretizing the latter and then approximating continuous spectra of nonperiodic signals, like impulse responses, by discrete spectra. Having both discrete time and discrete frequency permits application of Fast Fourier Transform (FFT) algorithms — the workhorse of modern technology.

We will start with calculating the transfer function $H(s)$, or at least $H(j\omega)$ from the impulse response $h(t)$ that is ***experimentally obtained not as a CT function, but as a finite duration sampled data record.*** Previously, we've done this by finding characteristics of the response (e.g. decay rate) and fitting it to a system model (e.g. first order or second order). For more complex motions this is not possible, so we want to have a computer do this, and for that we need a systematic method.

Example 5.5-1: Impulse response of a system whose continuous signal spectra are sampled. Let us sample frequency response of RC circuit (low-pass filter) through its impulse response $h(t)$:

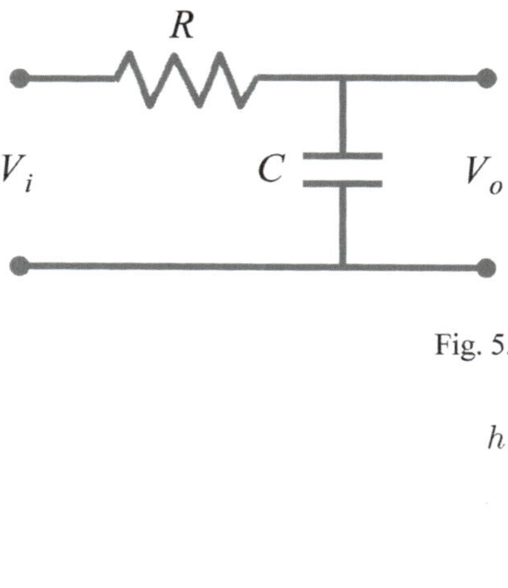

$$H(s) = \frac{V_o(s)}{V_i(s)} = \frac{IZ_C}{I(Z_R + Z_C)} = \frac{1/Cs}{R + \frac{1}{Cs}} = \frac{1}{\underbrace{RC}_{\tau}s + 1} = \frac{1/\tau}{s + \frac{1}{\tau}} \Rightarrow$$

$$h(t) = \mathcal{L}^{-1}\left[H(s)\right] = \mathcal{L}^{-1}\left[\frac{1/\tau}{s + (1/\tau)}\right] = \begin{cases} \frac{1}{\tau}e^{-\frac{t}{\tau}}, t \geq 0, \\ 0, t < 0. \end{cases}$$

Fig. 5.5-1. RC circuit and its impulse response.

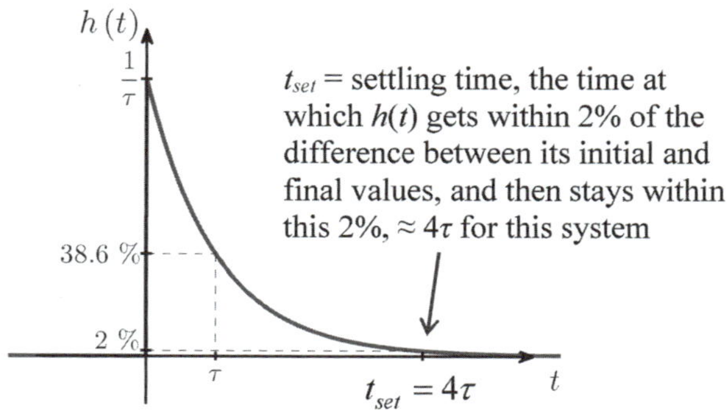

Fig. 5.5-2. Impulse response plot of the RC circuit in Figure 5.5-1.

As indicated in Chapter 3, we refer to this signal as the (CT) *infinite duration impulse response* or simply, IIR, and the filters with such impulse responses are referred to as ***CT or DT infinite impulse response, or IIR, filters***.

(1) Truncate $h(t)$ in time at some t_{tr}, say $t_{tr} = t_{set}$ to obtain truncated signal $h_t(t)$ that has finite duration:

$$h_t(t) = \begin{cases} h(t), & -\infty \le t \le t_{tr} = t_{set} \\ 0, & t > t_{tr} \end{cases} \Rightarrow$$

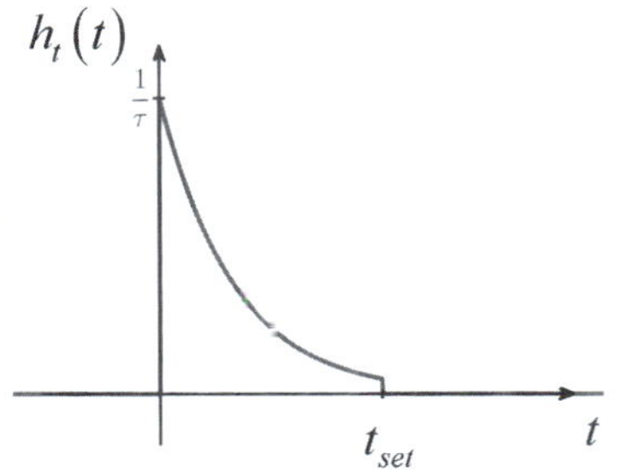

Fig. 5.5-3. CT signal truncation.

As indicated in Chapter 3, filters with such finite duration impulse response are referred to as ***finite impulse response, or FIR, filters***.

(2) Set $T_1 \ge t_{set}$, and define $h_{tp}(t)$ to be the T_1-periodic extension of $h_t(t)$:

$$h_{tp}(t) = \sum_{n=-\infty}^{\infty} h_t(t - nT_1) \Rightarrow$$

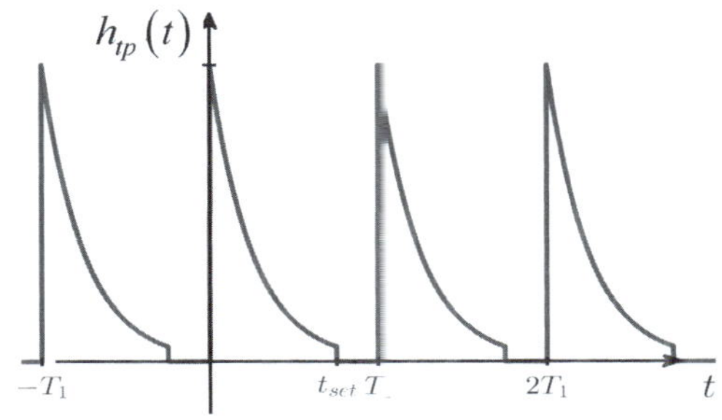

Fig. 5.5-4. Periodic extension of the signal in Figure 5.5-3.

(3) Now we've got a periodic function with jumps similar to the square wave. Although this step at first glance looks innocuous, it actually constitutes a tectonic shift in signal representation: *continuous spectra associated with system magnitude and phase response have now turned into discrete spectra due to signal time domain periodicity*. Indeed, taking fundamental period as T_1, so that $\omega_1 = 2\pi/T_1$, yields the Fourier Series of $h_{tp}(t)$:

$$H_s[k] = \frac{1}{T_1}\int_0^{T_1} h_{tp}(t) e^{-jk\frac{2\pi}{T_1}t}\, dt = \frac{1}{T_1}\int_{-\infty}^{\infty} h_t(t) e^{-jk\frac{2\pi}{T_1}t}\, dt = \frac{1}{T_1}\int_0^{t_{set}} h(t) e^{-jk\frac{2\pi}{T_1}t}\, dt. \quad (5.5\text{-}1)$$

↑ Since $h_t(t)$ is 0, except on $[0, t_{set}]$

For the RC circuit:

$$H_s[k]=\frac{1}{T_1}\int_0^{t_{set}}\frac{1}{\tau}e^{-\frac{t}{\tau}}e^{-jk\frac{2\pi}{T_1}t}dt=\frac{1}{\tau T_1}\int_0^{t_{set}}e^{-\left(jk\frac{2\pi}{T_1}+\frac{1}{\tau}\right)t}dt$$

$$=\frac{1}{T_1}\frac{-1}{j\frac{2k\pi\tau}{T_1}+1}e^{-\left(jk\frac{2\pi}{T_1}+\frac{1}{\tau}\right)t}\Bigg|_0^{t_{set}}=\frac{1}{T_1}\frac{1-e^{-\left(jk\frac{2\pi}{T_1}+\frac{1}{\tau}\right)t_{set}}}{j\frac{2k\pi\tau}{T_1}+1}.$$

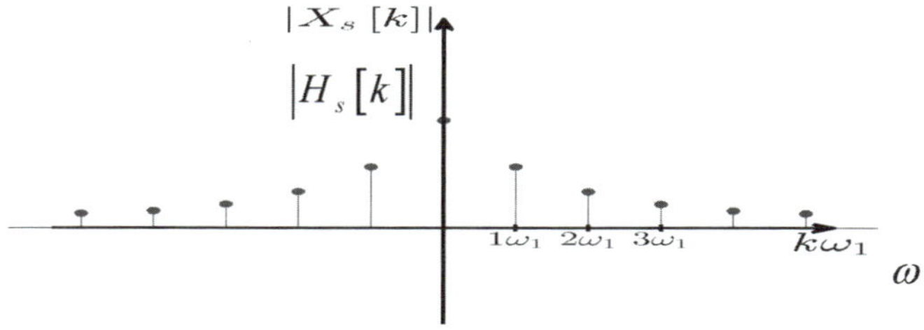

Fig. 5.5-5. Sampling continuous spectra.

(4) Take $T_1 \to \infty$ to single out $h_t(t)$ from $h_{tp}(t)$. Then:

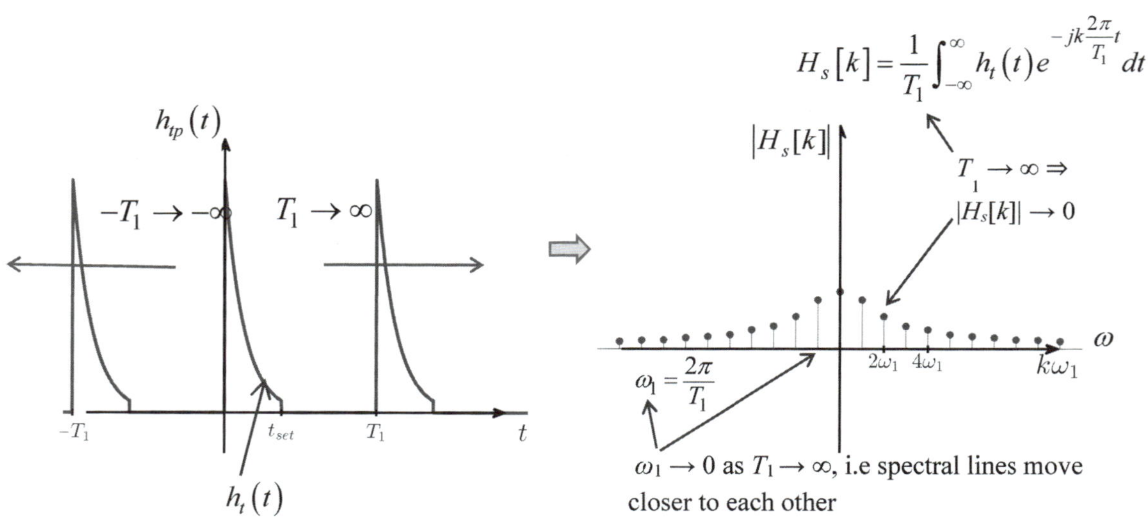

Fig. 5.5-6. Periodic extension removal.

Thus, $|H_s[k]| \to 0$ as $T_1 \to \infty$.

(5) Multiply $H_s[k]$ by T_1, then

$$T_1 H_s[k] = \int_{-\infty}^{\infty} h_t(t) e^{-jk\overbrace{\frac{2\pi}{T_1}}^{\omega_1}} dt = \left(\underbrace{\int_{-\infty}^{\infty} h_t(t) e^{-j\omega t} dt}_{H_t(j\omega)} \right)\Bigg|_{\omega = k\omega_1} = H_t(j\omega)\big|_{\omega=k\omega_1} = H_t(jk\omega_1), \tag{5.5-2}$$

where subscript t indicates distortion due to truncation, and $H_t(jk\omega_1)$ no longer $\to 0$ as $T_1 \to \infty$, since the envelope $H_t(j\omega)$ of $T_1 H_s[k]$ becomes independent of T_1 and depends only on $h_t(t)$, and so its magnitude $|H_t(j\omega)|$ no longer $\to 0$ as $T_1 \to \infty$. However, since $H_s[k] = \frac{1}{T_1} H_t(jk\omega_1)$, i.e. scaled by T_1 FS coefficients $T_1 H_s[k]$ of periodic extension $h_{tp}(t)$ provide samples of their envelope $H_t(j\omega)$ at $\omega = k\omega_1$, the envelope can be revealed clearer by making samples $H_t(jk\omega_1)$, $-\infty < k < \infty$, move closer to each other through decreasing ω_1, i.e. increasing T_1, as shown below.

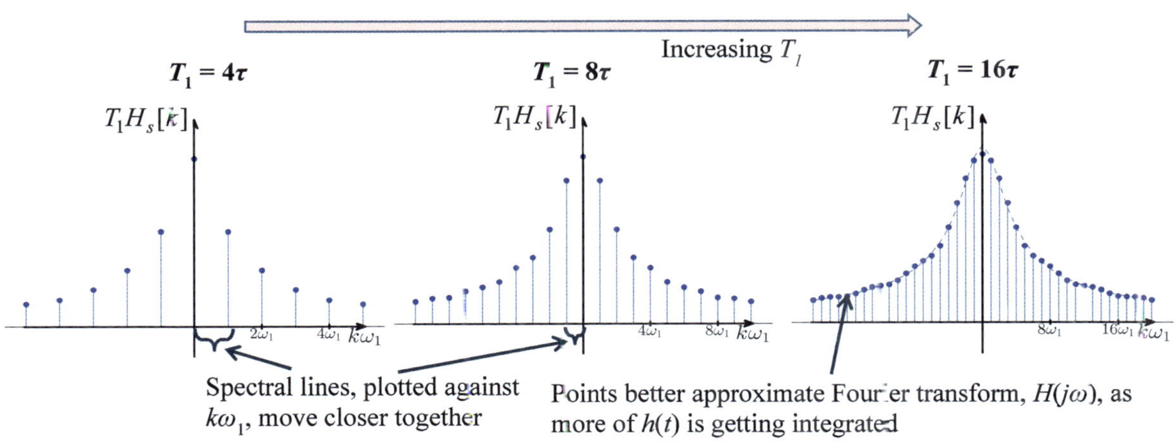

Fig. 5.5-7. Spectra with different T_1.

What does multiplying by T_1 do: transforming power signal into energy signal. $h_{tp}(t)$ is seen to be a power signal, not an energy signal. But if we let $T_1 \to \infty$, then, since energy of $h_{tp}(t)$ in each period is fixed and independent of period T_1, power of $h_{tp}(t)$, i.e. energy of $h_{tp}(t)$ per period size $\to 0$:

$$P_{h_{tp}} = \frac{1}{T_1} \int_{T_1} |h_{tp}(t)|^2 dt = \sum_{k=-\infty}^{\infty} |H_s[k]|^2 \underset{T_1 \to \infty}{\to} 0. \tag{5.5-3}$$

Therefore, as $T_1 \to \infty$, power of each harmonic $|X_s[k]|^2 \to 0$ and, hence, each Fourier coefficient magnitude $|X_s[k]| \to 0$.

Multiplying $P_{h_{tp}} = \frac{1}{T_1}\int_{T_1} |h_{tp}(t)|^2 dt = \sum_{k=-\infty}^{\infty} |H_s[k]|^2 \underset{T_1\to\infty}{\to} 0$ by T_1 as $T_1 \to \infty$ yields

$E_x = \int_{T_1} |h_t(t)|^2 dt \approx \int_{-\infty}^{\infty} |h(t)|^2 dt$ where $h_t(t) \approx h(t)$ is an energy signal.

5.5.1.1 *Sampling continuous signal spectra: computation of H(jω) from Fourier series — Fourier transform*

Thus, since $\omega_1 \to 0$ as $T_1 \to \infty$, we can pick k so that $k\omega_1 \approx \omega$ for any frequency ω. The complex function

$$H_t(j\omega) = \int_{-\infty}^{\infty} h_t(t)e^{-j\omega t}dt \underbrace{=}_{\text{for causal response}} \int_0^{\infty} h_t(t)e^{-j\omega t}dt \tag{5.5-4}$$

is called **Fourier transform** of $h_t(t)$. Removing truncation, i.e. letting $t_{tr} \to \infty$, we obtain:

$$H(j\omega) = \int_{-\infty}^{\infty} h(t)e^{-j\omega t}dt. \tag{5.5-5}$$

Thus, $H_t(j\omega)$ is an approximation of $H(j\omega)$ with $H_t(j\omega) \to H(j\omega)$ as $t_{tr} \to \infty$.

Therefore, increasing T_1 we obtain more spectral lines per any fixed interval on frequency axis. If, while increasing T_1, we keep picking up more of $h(t)$, then we reduce the error between the actual $H(j\omega)$ and its approximate samples $T_1H_s[k]$ at $k\omega_1$. If we retain $h_t(t)$, we still get denser samples, but do not reduce the error.

5.5.1.2 *Relation between Laplace transform and Fourier transform and their existence for causal signals*

Distinguishing between Fourier transforms of causal and noncausal signals is centrally important for understanding of filtering, since impulse responses of real filters are causal and those of ideal filters are noncausal. Therefore, frequency responses of real and ideal filters will be Fourier transforms of causal and noncausal signals, respectively. For ***causal square integrable*** signal, i.e. for $h(t)$ being causal energy signal, frequency response (transfer function) given by Fourier transform $H(j\omega)$ of $h(t)$ exists and is Laplace transform of $h(t)$ with $s = j\omega$:

$$H(s)\Big|_{s=j\omega} = \int_{-\infty}^{\infty} h(t)e^{-st}dt\Big|_{s=j\omega} = \int_{-\infty}^{\infty} h(t)e^{-(\sigma+j\omega)t}dt\Big|_{\sigma=0} = \int_{-\infty}^{\infty} h(t)e^{-j\omega t}dt = H(j\omega). \tag{5.5-6}$$

For ***causal nonsquare integrable*** signal $h(t)$ with exponential growth, exponential weighting factor $e^{-\sigma t}$ with σ positive and large enough makes the product $h_w(t) \equiv h(t)e^{-\sigma t}$ square integrable. Then

Fourier transform of $h(t)e^{-\sigma t}$ exists and it becomes Laplace transform of $h(t)$:

$$H_{h_w}(j\omega)=\int_0^\infty \overbrace{h(t)e^{-\sigma t}}^{h_w(t)} e^{-j\omega t}dt=\int_0^\infty h(t)e^{-\overbrace{(\sigma+j\omega)}^{s}t}dt=\int_0^\infty h(t)e^{-st}dt=H(s). \qquad (5.5\text{-}7)$$

For ***noncausal square integrable*** signal, i.e. for $h(t)$ being noncausal energy signal, frequency response (transfer function) given by Fourier transform $H(j\omega)$ of $h(t)$ exists and is still Laplace transform, now two-sided, of $h(t)$ with $s=j\omega$:

$$H(s)\Big|_{s=j\omega}=\int_{-\infty}^{\infty} h(t)e^{-st}dt\Big|_{s=j\omega}=\int_{-\infty}^{\infty} h(t)e^{-(\sigma+j\omega)t}dt\Big|_{\sigma=0}=\int_{-\infty}^{\infty} h(t)e^{-j\omega t}dt=H(j\omega). \qquad (5.5\text{-}8)$$

For ***noncausal nonsquare integrable*** signal $h(t)$ with exponential growth, exponential weighting factor $e^{-\sigma t}$ with σ positive and large makes the product $h_w(t)\equiv h(t)e^{-\sigma t}$ exponentially growing as $t\rightarrow-\infty$, i.e. not square integrable over $t\rightarrow(-\infty,\infty)$. Therefore, both Fourier transform of $h(t)e^{-\sigma}$ and Laplace transform of $h(t)$ do not exist:

$$H_{h_w}(j\omega)=\int_{-\infty}^{\infty} \overbrace{h(t)e^{-\sigma t}}^{h_w(t)} e^{-j\omega t}dt=\int_{-\infty}^{\infty} h(t)e^{-\overbrace{(\sigma+j\omega)}^{s}t}dt=\int_{-\infty}^{\infty} h(t)e^{-st}dt=H(s)=\infty. \qquad (5.5\text{-}9)$$

5.6 Computing System Frequency Response from Finite Duration Sampled Data Record: Discrete Fourier Transform (DFT)

5.6.1 *Sampling CF spectra: computation of $H(j\omega)$ from Fourier series — summary*

Figure 5.6-1 depicts relations among signals and their spectra, the first — just shown, and the rest — to be shown in Chapter 6. CF and DF stand for continuous and discrete frequency, respectively.

A. CT time domain periodicity implies DF nonperiodic frequency spectrum, and vice versa. To use this property, we make any finite duration signal time-periodic through its periodic extension.

B. The dual is also true (cf. Chapter 6). Namely, *CF frequency domain periodicity implies DT nonperiodic time domain signal, and vice versa.*

C. Discrete-time/discrete-frequency (DT/DF) setting requires periodicity in both domains. This fact will be shown in Chapter 6. Employing the latter setting, our ultimate goal — detailed understanding of how to analyze noisy finite duration DT data record will be accomplished in Chapter 8 using FFT.

Returning to Section 5.5 summary, computation of IIR filter frequency domain representation $H(j\omega)$ from its impulse response involves the following steps assembled in Figure 5.6-2:

(1) Convert IIR filter $H(j\omega)$ with impulse response $h(t)$ to FIR filter $H_t(j\omega)$ with impulse response $h_t(t)$ by truncating $h(t)$ at $t_{tr}=t_{set}$.

(2) Take periodic extension of $h_t(t)$ with period $T_1>t_{set}$, call the extension $h_{tp}(t)$.

(3) Calculate Fourier Series, $H_s[k]$ of periodic extension $h_{tp}(t)$.

(4) Improve sampling resolution $\omega_1 = 2\pi/T_1$ by increasing T_1 so that ω_1 decreases. This, however, makes $h_{tp}(t)$ sparser, with signal power and, hence magnitude spectrum $|H_s[k]|$ going to zero as $T_1 \to \infty$, making resulting spectra completely diverge from $H_t(j\omega)$.

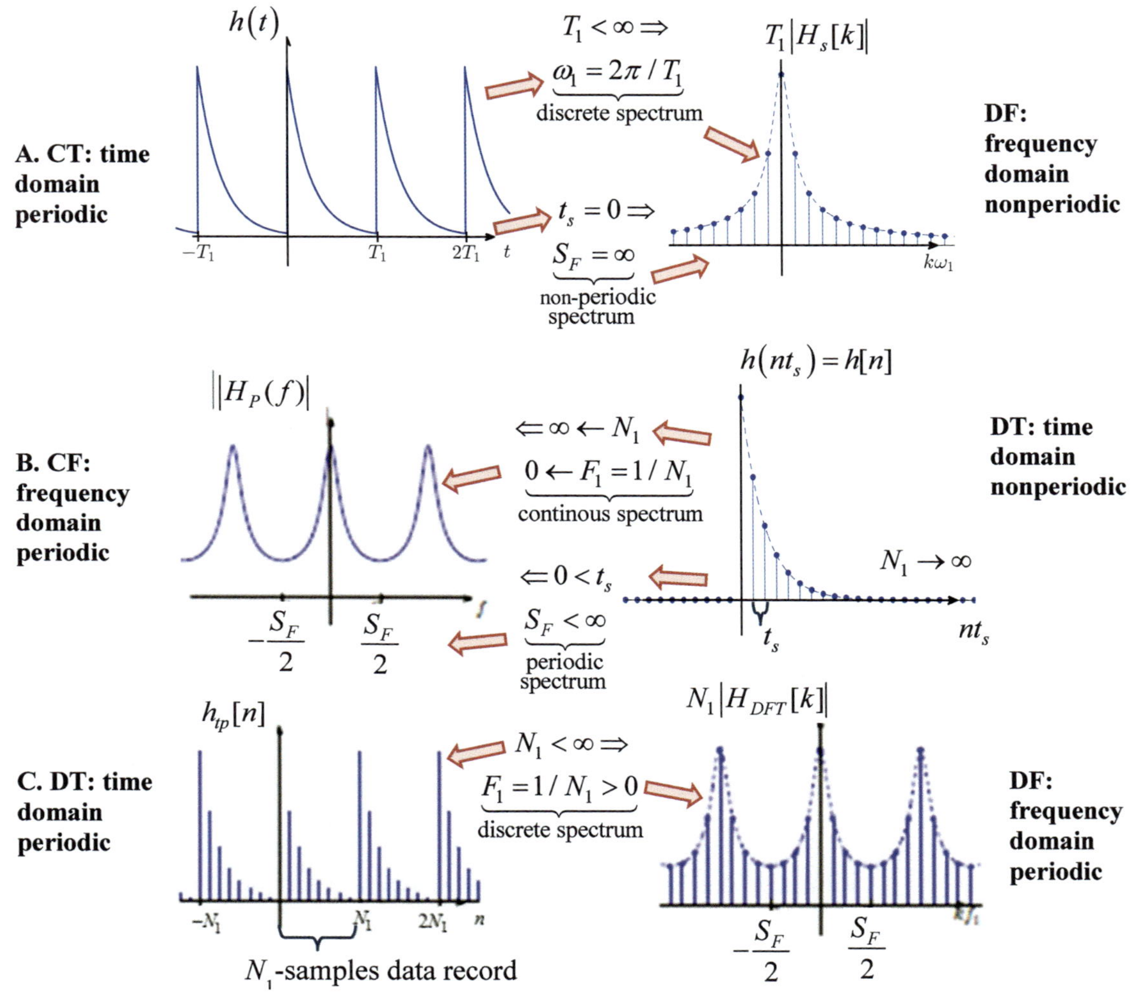

Fig. 5.6-1. The key relations among signals and their spectra: ***continuity of spectra implies absence of N_1 or T_1, and hence, non-periodicity of CT or DT signal***, and, conversely, ***continuity of signal implies absence of t_s, and therefore S_F, and, hence, non-periodicity of spectra***.

(5) Relate $H_s[k]$ to $H_t(j\omega)$ by multiplying $H_s[k]$ by T_1 to obtain $T_1H_s[k] = H_t(jk\omega_1)$ — the exact samples of the "envelope" FIR transfer function $H_t(j\omega)$ and the approximate samples of the "envelope" transfer function $H(j\omega)$, *both not bandlimited*. As $T_1 \to \infty$ ($\omega_1 \to 0$), $T_1H_s[k]$ give more points closer to each other, and if $t_{tr} \to \infty$, also closer to $H(j\omega)$.

Thus, CT FIR filter has its frequency response continuous w.r.t. frequency axis, but admits ***exact*** sampling of its frequency response through periodic extension and discrete spectrum calculation $T_1H_s[k]$, whereas CT IIR filter admits only ***approximate*** sampling of its frequency response.

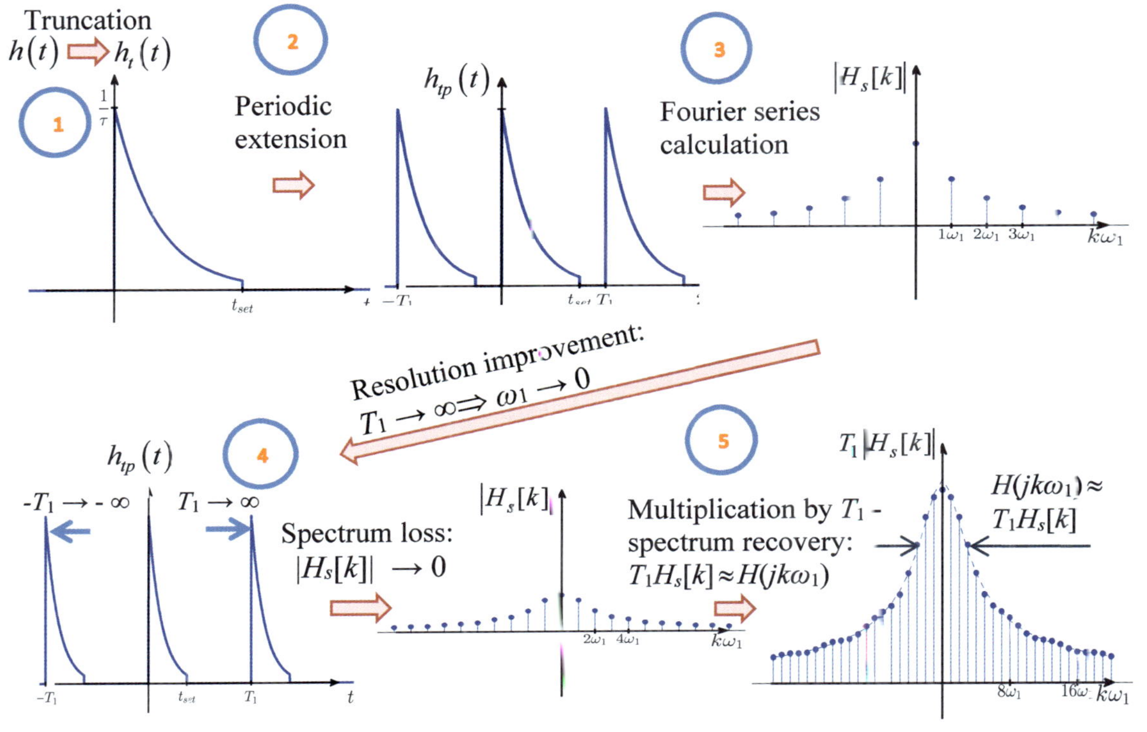

Fig. 5.6-2. Sampling continuous signal spectra.

Until now, we've been assuming that we know the exact analytical form of $h(t)$. But, in reality, we usually obtain a DT finite duration data record from experimental measurements. So, we need step 6: ***bandlimiting and sampling***.

5.6.2 *Sampling CF spectra: computation of H(jω) from Fourier series using DT data*

(6) Bandlimit (by low pass filtering) $h_{tp}(t)$ to have maximum frequency f_B, call it $h_{tpb}(t)$. Sample it at the rate $S_F > 2f_B$ (to prevent aliasing), take $T_1 = t_s N_1$. Then:

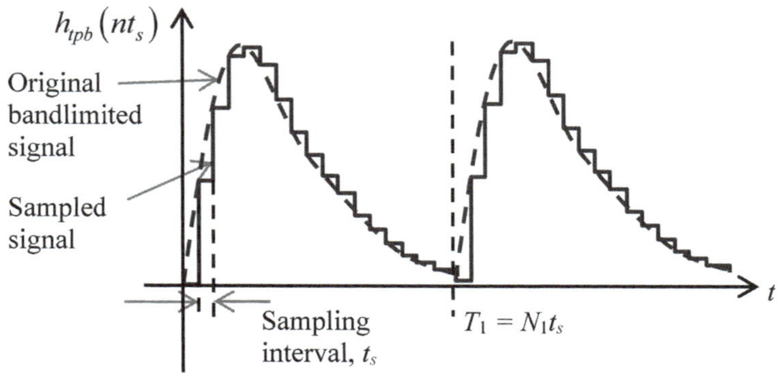

Fig. 5.6-3. N_1-samples DT data record $h_{tpb}(nt_s) \equiv h_{tpb}[n]$.

$$\underbrace{T_1 H_s[k]}_{\substack{\text{Not}\\ \text{bandlimited}\\ h_{tp}}} \approx \underbrace{T_1 \frac{1}{T_1}\int_0^{T_1} h_{tpb}(t) e^{-jk\frac{2\pi}{T_1}t} dt}_{\substack{\text{Bandlimited CT Fourier}\\ \text{series coefficients } H_{sb}[k]}} = t_s N_1 \underbrace{\frac{1}{N_1}\sum_{n=0}^{N_1-1} \overbrace{h_{tpb}(nt_s)}^{h_{tpb}[n]} e^{-jk\frac{2\pi}{N_1 t_s}nt_s}}_{\substack{\text{Discrete Fourier Series } H_{DFS}[k]:\\ \text{exactly equal to } H_{sb}[k]}} = t_s \overbrace{\sum_{n=0}^{N_1-1} \underbrace{h_{tpb}[n]}_{\substack{\text{The actual } N_1\text{-}\\ \text{samples data record}\\ \text{for } n\in[0,\, N_1-1]}} e^{-j2\pi k\frac{1}{N_1}n}}^{H_{DFT}[k]}. \quad (5.6\text{-}1)$$

Thus, $N_1 \cdot H_{DFS}[k]$ computed from real impulse response finite duration data record $h_{tpb}[n]$, $n\in[0, N_1-1]$, which can be viewed as one period of truncated, periodically extended, bandlimited, sampled version $h_{tpb}(nt_s)$ of the actual impulse response $h(t)$, for even N_1 gives $N_1/2$ and for odd N_1 gives $\text{int}(N_1/2) + 1$ approximate sample points $t_s\, N_1 \cdot H_{DFS}[k]$ of the actual frequency response $H(j\omega)$, i.e. of the Fourier transform of $h(t)$, over the frequency range $\omega\in[0, 2\pi/(2t_s)]$, i.e.

$$H(jk\omega_1) \underbrace{\approx}_{\substack{\text{Truncating } h(t),\\ \text{obtaining } h_t(t)}} H_t(jk\omega_1) \underbrace{=}_{\substack{\text{Periodically}\\ \text{extending } h_t(t),\\ \text{obtaining } h_{tp}(t)}} T_1 H_s[k] \underbrace{\approx}_{\substack{\text{Bandlimiting } h_{tp}(t),\\ \text{obtaining } h_{tpb}(t):}} T_1 H_{sb}[k] \underbrace{=}_{} t_s \overbrace{N_1 \cdot H_{DFS}[k]}^{H_{DFT}[k]} \equiv t_s H_{DFT}[k], \quad (5.6\text{-}2)$$

Sampling $h_{tpb}(t)$, obtaining $h_{tpb}[n]$, $n\in[0,\ N_1-1]$: ***DT finite duration (N_1 – samples) data record!***

$k = 0, \ldots, (N_1-1)/2$, N_1-odd, or $k = 0, \ldots, \text{int}[(N_1-1)/2]$, N_1-even.

Thus $H(j\omega)\big|_{\omega=k\omega_1} \approx t_s N_1 H_{DFS}[k] \doteq t_s H_{DFT}[k]$. That's why $N_1 \cdot X_{DFS}[k] \equiv X_{DFT}[k]$ is called ***Discrete Fourier Transform.***

5.6.3 *Discrete Fourier Transform*

Thus, the ***Discrete Fourier Transform*** (DFT) for a DT signal $x[n]$ is defined as:

$$X_{DFT}[k]=\sum_{n=0}^{N_1-1}x[n]e^{-j2\pi k\frac{1}{N_1}n}. \qquad (5.6\text{-}3)$$

From this definition, we can see that the Discrete Fourier Series coefficients and the Discrete Fourier Transform have the relationship:

$$X_{DFT}[k]=N_1\cdot X_{DFS}[k]. \qquad (5.6\text{-}4)$$

Therefore, the Discrete Fourier Transform is N_1-periodic, i.e. $X_{DFT}[k] = X_{DFT}[k + N_1]$, like the Discrete Fourier Series, and symmetric about $N_1/2$ for even N_1 and int$(N_1/2) + 1$ for odd N_1. Its magnitude has even symmetry and phase has odd symmetry. And, the Inverse Discrete Fourier Transform (IDFT) is similar to the Inverse Discrete Fourier Series:

$$x[n]=\frac{1}{N_1}\sum_{n=0}^{N_1-1}X_{DFT}[k]e^{j2\pi k\frac{1}{N_1}n}. \qquad (5.6\text{-}5)$$

Bandlimiting. Just as with the Discrete Fourier Series, the Discrete Fourier Transform is N_1-periodic. This means that aliasing is still a potential problem. The highest number of digital frequency samples of the analog spectrum the DFT can provide is $N_1/2$ for even N_1 and int$(N_1/2) + 1$ for odd N_1. So, to correctly represent higher analog frequencies we need larger number of sampling points N_1 and faster sampling rate S_F. Moreover, CT non-periodic causal signals have Fourier transforms that are non-zero on the entire $\omega\in(-\infty, \infty)$ interval, so we need to cut off the frequency at a certain point for CT signals as well which is possible only approximately.

Thus, we should bandlimit the signal when computing the Fourier transform (continuous or discrete) well enough, and whatever cannot be bandlimited should be oversampled to the reasonable extent to minimize aliasing.

5.7 DFS and DFT as Diagnostic Tools for Catching Low Magnitude Harmonics. Fast Fourier Transform and its Use

We will now address spectra calculation of residuals such as those of Examples 5.1-4 and 5.2-1, noting that due to relation (5.6-2), DFS and DFT calculation for N_1-data record yields values that differ only by a factor N_1.

Example 5.7-1: Calculating DFS magnitude spectra from data — some pitfalls.
Let's use the data from Example 5.2-1, which is a sampled data record from Problem 4 of Homework 4 solution guide: assume we have sampled in real time the position of a piston, driven with the reference oscillation frequency f_1 = 4.6 Hz, in meters with sampling interval 0.002 sec, and collected now N_1 = 1250 data points that are plotted in Figure 5.7-1, instead of 1087 data points. Looking at the data, we might expect the results to be large at the piston frequency, and very small everywhere else, but that's not what we are getting, as seen from calculation below Figure 5.7-2:

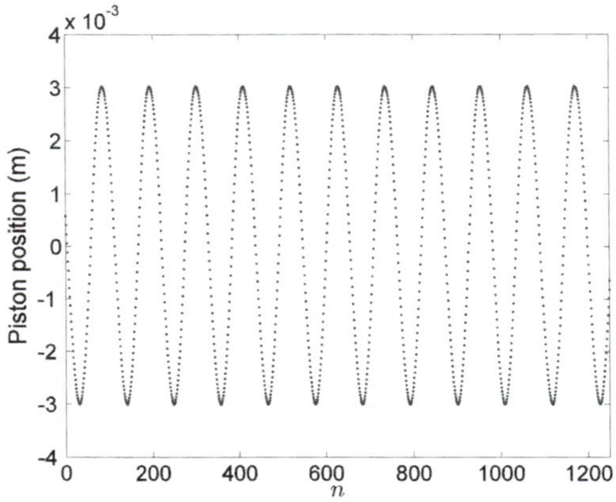

Fig. 5.7-1. Data from Problem 4 of Homework 4 solution guide plotted for 1250 data points.

MATLAB COMMAND 5.7-2

```
>> load model_piston_data.mat
>> N1 = length(piston);  ← N1 = 1250
>> n = 0:(N1-1);
>> figure
>> plot(n,piston,'.')
>> XDFS = ZEROS(1,N1);
>> for k = 0:(N1-1)
       XDFS(k + 1) = (1/N1)*sum(piston'.*exp(-1j*2*pi*k*n/N1))
>> end;
>> figure
>> stem(n,abs(XDFS))
```

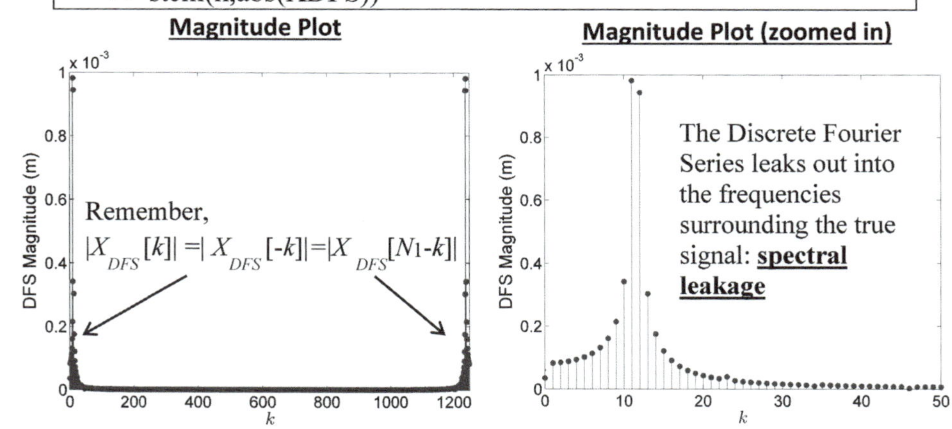

Fig. 5.7-2. MATLAB COMMAND 5.7-2 calculation result shows DFS with spectral leakage.

5.7.1.1 *Spectral leakage and its avoidance*

What happened? Briefly examining the plot in Figure 5.7-1, we find two ways in which it differs from that in Figure 5.2-1 of Example 5.2-1: 1) it contains 11 cycles instead of 10, and 2) it also has an additional part, which is a bit less than half-cycle. Thus, approximately eleven periods were acquired to work with the original single period. This suggests that the latter now be viewed as the eleventh harmonic relative to new fundamental period.

There is a problem, however, with this view. When the data is collected so that it does not contain an integer number of periods of interest, but the entire data record length is taken as the single, and hence, the fundamental period, there is no digital (and hence underlying analog) frequency clearly corresponding to an integer submultiple (a harmonic) of this fundamental period.

To see this for the Example 5.7-1 data, we simply note that $f_{11} = 4.6$ Hz, $T_1 = 11T_{11} = 11/f_{11} = 11/4.6 = 2.391$ sec $\Rightarrow N_1 = 2.391/0.002 = 1195$ data points (with small round-off error). However, 1250 data points length corresponds to $T_1 = 1250 \cdot 0.002 = 2.5$ sec, rather than 2.391 sec, so that $f_{11} = 1/(2.5/11) = 11/2.5 = 4.4$ Hz. The latter frequency, however, is not supported by the oscillatory curve shape, which results in summing up samples over the time interval unrelated to the period of interest.

The result of the above mismatch manifests itself in appearance of several surrounding terms of the Discrete Fourier Series around single frequency component spectrum at 4.6 Hz, seen in Figure 5.7-2 and referred to as **spectral leakage**.

Example 5.7-2: Removing leakage. So what if we select an integer number of periods, for example 11 by taking 1195 data points as shown in Figure 5.7-3, and discarding the rest?

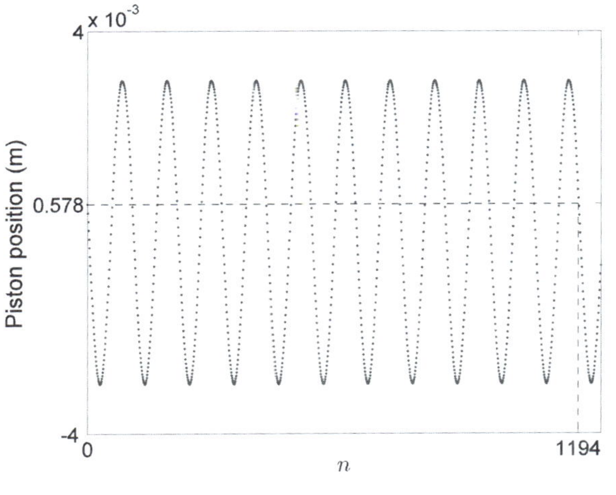

Fig. 5.7-3. Limiting data to $N_1 = 1195$ data points from $n = 0$ to $n = N_1\text{-}1 = 1194$.

As seen in Figure 5.7-4, the leakage is almost completely gone, with the remaining small-scale zoomed leakage caused by other, lesser, factors, such as the choice of sampling rate.

MATLAB COMMAND 5.7-4

```
>> load model_piston_data.mat
>> N1 = 1195;
>> n = 0:(N1-1);
>> figure
>> plot(n,piston,'.')
>> XDFS = ZEROS(1,N1);
>> for k = 0:(N1-1)
      XDFS(k + 1) = (1/N1)*sum(piston'.*exp(-1j*2*pi*k*n/N1))
>> end;
>> figure
>> stem(n,abs(XDFS))
```

Selecting only $N_1 = 1195$ data points now

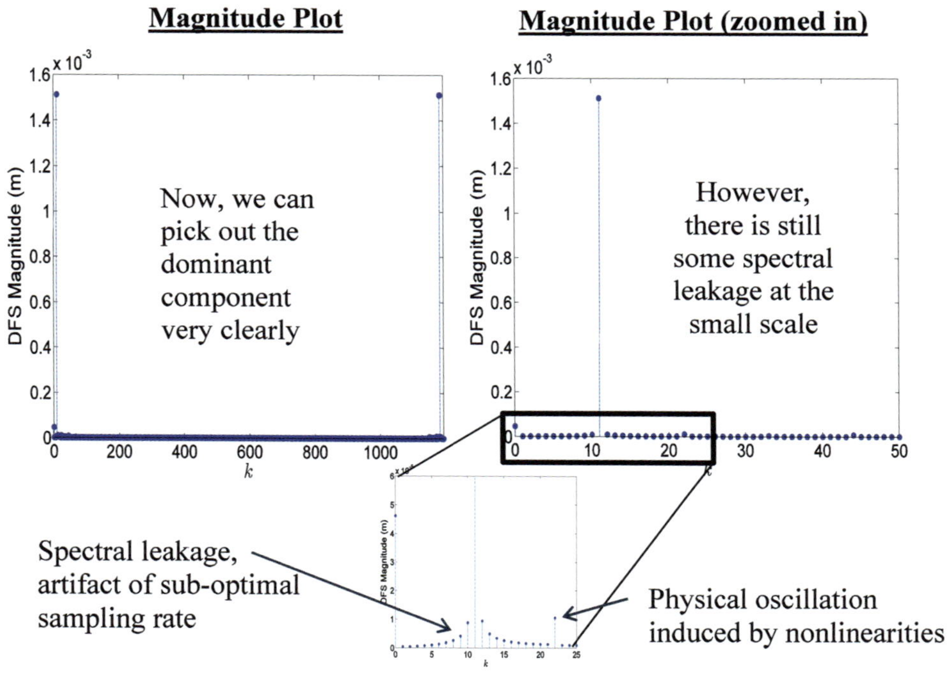

Fig. 5.7-4. MATLAB COMMAND 5.7-4 result shows DFS magnitude spectrum with no leakage, based on 1195 data points.

5.7.1.2 *Discrete Fourier Transform of periodic signals*

Now that we understand the importance of the correct data length selection for plotting DFS spectra, let us assume that this selection is done perfectly and consider plotting DFT spectra.

For this purpose, we will focus on sampling of a single cycle of a CT T_1-periodic sinusoid $x(t)=\cos(2\pi f_1 t+\theta)$, $f_1 = 1/T_1$.

We recall that its Fourier series coefficients are, respectively $0.5e^{j\theta}$ at $f = f_1$ and $0.5e^{-j\theta}$ at $f = -f_1$, i.e. $x(t) = \cos(2\pi f_1 t + \theta) = 0.5e^{j\theta}e^{j2\pi f_1 t} + 0.5e^{-j\theta}e^{j2\pi(-f_1)t}$.

Case 1: What happens if we take DFT of sampled sinusoid as sampling rate increases? As t_s changes, we choose it to stay as integer submultiple of T_1, so that as t_s decreases, $N_1 = T_1/t_s$ increases, and $F_1 = 1/N_1$ decreases. This yields:

$$x(t) = \cos(2\pi f_1 t + \theta) = 0.5e^{j\theta}e^{j2\pi f_1 t} + 0.5e^{-j\theta}e^{j2\pi(-f_1)t} \underset{\substack{t=nt_s\\T_1=N_1t_s}}{\underbrace{\Rightarrow}}$$

$$\underbrace{0.5}_{|X_{DFS}[1]|} e^{j\theta}e^{j2\pi\frac{1}{N_1}n} + \underbrace{0.5}_{|X_{DFS}[-1]|} e^{-j\theta}e^{-j2\pi\frac{1}{N_1}n} \Rightarrow \tag{5.7-1}$$

$$\underbrace{N_1 \underbrace{0.5}_{|X_{DFS}[1]|}}_{|X_{DFT}[1]|} e^{j\theta}e^{j2\pi\frac{1}{N_1}n} + \underbrace{N_1 \underbrace{0.5}_{|X_{DFS}[-1]|}}_{|X_{DFT}[-1]|} e^{-j\theta}e^{-j2\pi\frac{1}{N_1}n} \underset{\substack{t_s\to0\Rightarrow\\N_1\to\infty}}{\underbrace{\Rightarrow}} \begin{cases} \underbrace{N \underbrace{0.5}_{|X_{DFS}[1]|}}_{|X_{DFT}[1]|\to\infty}, \\ \underbrace{N_1 \underbrace{0.5}_{|X_{DFS}[-1]|}}_{|X_{DFT}[-1]|\to\infty}. \end{cases}$$

Thus, as the sampling rate of CT sinusoid increases, DT version of the latter that was initially mapped into the first DFS harmonic, stays as the first DFS harmonic with the same magnitude, but gets more samples per one cycle, however its corresponding DFT magnitude goes to infinity!

Case 2: What happens if we take DFT of sampled sinusoid as data record size increases? Viewing a single cycle as the ***k'th cycle*** where index k in $T_k = t_sN_k$ can vary, but the values of t_s, T_k, N_k, and hence digital frequency $1/N_k$ stay the same irrespectively of the value of k, we increase CT data length T_1 to include more CT and hence more DT cycles, $T_1 = kT_k \Rightarrow N_1 = kN_k$ (add full cycles to avoid leakage). This yields:

$$x(t) = \cos(2\pi f_k t + \theta) = 0.5e^{j\theta}e^{j2\pi f_k t} + 0.5e^{-j\theta}e^{j2\pi(-f_k)t} \underset{\substack{t=nt_s\\T_1=kT_k=kN_kt_s}}{\underbrace{\Rightarrow}}$$

$$\underbrace{0.5}_{|X_{DFS}[k]|} e^{j\theta}e^{j2\pi\frac{k}{N_1}n} + \underbrace{0.5}_{|X_{DFS}[-k]|} e^{-j\theta}e^{-j2\pi\frac{k}{N_1}n} \Rightarrow$$

$$\underbrace{N_1 \underbrace{0.5}_{|X_{DFS}[k]|}}_{|X_{DFT}[k]|} e^{j\theta}e^{j2\pi\frac{k}{N_1}n} + \underbrace{N_1 \underbrace{0.5}_{|X_{DFS}[-k]|}}_{|X_{DFT}[-k]|} e^{-j\theta}e^{-j2\pi\frac{k}{N_1}n} \underset{\substack{k\to\infty\Rightarrow\\N_1=kN_k\to\infty}}{\underbrace{\Rightarrow}} \begin{cases} \underbrace{N_1 \underbrace{0.5}_{|X_{DFS}[k]|}}_{|X_{DFT}[k]|\to\infty}, \\ \underbrace{N_1 \underbrace{0.5}_{|X_{DFS}[-k]|}}_{|X_{DFT}[-k]|\to\infty}. \end{cases} \tag{5.7-2}$$

Thus, as the data record of CT sinusoid sampled with fixed sampling rate increases, DT sinusoid keeps the same number of samples per one cycle and its DFS keeps the same magnitude, however its harmonic number k and its DFT magnitude go to infinity!

Example 5.7-3, Case 2: The signal $x(t) = 2\cos(10\pi t)$ is 2 times oversampled and data is collected for a) 1 period, b) 2 periods, and c) 4 periods. Find and sketch its DFT for data in a), b), and c).

The frequency of $x(t)$ is 5 Hz, therefore Nyquist rate is $S_N = 10$ Hz, and twice oversampling gives the sampling rate $S_F = 20\ Hz$ and the digital frequency $F = 50/20 = 1/4 = 2/8 = 4/16$. Referring to (5.7-2), for the data collected during one, two, and four cycles, the index k takes, respectively, the values of 1, 2, and 4. The size, N_1, of data record (as a function of n) correspondingly increases from $N_1 = 4$ for $k = 1$ to $N_1 = 2N_2 = 2.4 = 8$ for $k = 2$, and to $N_1 = 4N_4 = 4.4 = 16$ for $k = 4$. Consequently, for $k = 1, 2, 4$, $|X_{DFS}[k]| = 1$, and $|X_{DFT}[k]| = kN_k|X_{DFS}[k]|$ yields, respectively, $|X_{DFT}[1]| = 1 \cdot 4 \cdot |X_{DFS}[1]| = 1 \cdot 4 \cdot 1 = 4$, $|X_{DFT}[2]| = 2 \cdot 4 \cdot |X_{DFS}[2]| = 2 \cdot 4 \cdot 1 = 8$, and $|X_{DFT}[4]| = 4 \cdot 4 \cdot |X_{DFS}[4]| = 4 \cdot 4 \cdot 1 = 16$. The two nonzero DFT values will, respectively, appear at $k = 1$ and $k = 4 - 1 = 3$, at $k = 2$ and $k = 8 - 2 = 6$, and at $k = 4$ and $k = 16 - 4 = 12$, as shown in Figure 5.7-5.

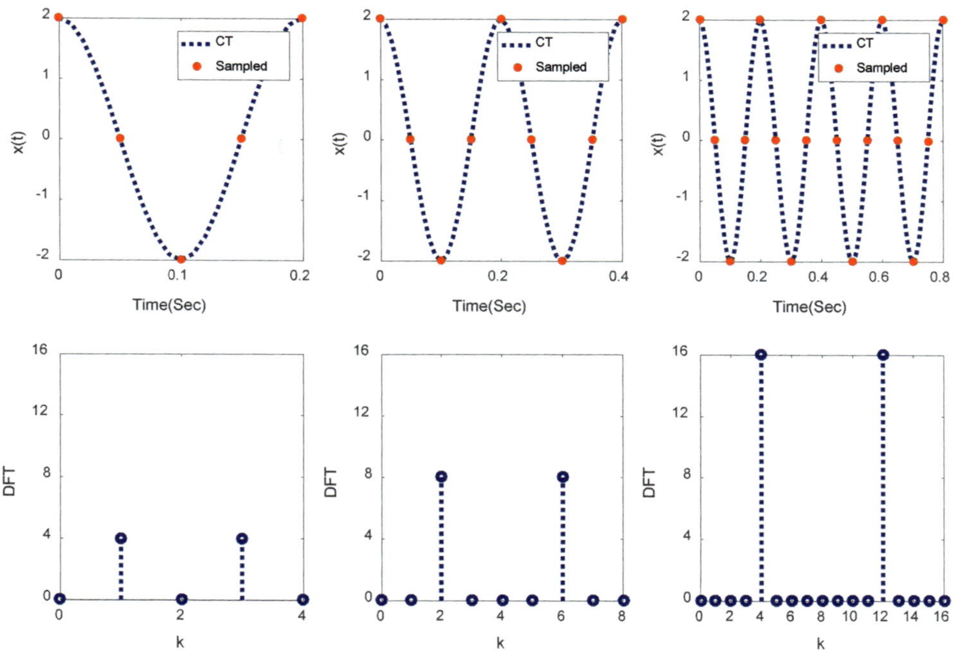

Fig. 5.7-5. The signal $x(t) = 2\cos(10\pi t)$ of Example 5.7-3, its three sampled data records corresponding, respectively, to k = 1, 2, 4, and the DFT plots corresponding to each data record.

Thus, DFT magnitude of periodic sampled signal goes to infinity at harmonics with nonzero magnitude as the data record increases, while keeping the same sampling rate.

We will develop an underlying CT analytical framework for this behavior in Chapter 8.

To do calculations fast we use FFT — Fast Fourier Transform, briefly described next.

5.7.2 *Fast Fourier Transform (FFT)*

The fast Fourier transform (FFT) is an algorithm that quickly calculates discrete Fourier transform (DFT). FFT is widely used for many applications in engineering, science, and mathematics. A well-known mathematician G. Strang called it "the most important numerical algorithms of our lifetime." It is executed billions of times every second throughout the planet and outside of it in communication applications.

There are many different FFT algorithms involving wide range of mathematics, but the most commonly used FFT has been Cooley-Tukey FFT algorithm, named after J.W. Cooley and John Tukey. The MATLAB 'fft' command is also based on this algorithm.

The simplest and most common idea of the Cooley-Tukey FFT algorithm is breaking up a transform of even length N into two transforms of length $N/2$ using the following identity:

$$\sum_{n=0}^{N-1} x[n]e^{-j2\pi k\frac{1}{N}n} = \underbrace{\sum_{m=0}^{N/2-1} x[2m]e^{-j2\pi k\frac{1}{N}(2m)}}_{\text{DFT of even-indexed part of } x[m]} + \underbrace{\sum_{m=0}^{N/2-1} x[2m+1]e^{-j2\pi k\frac{1}{N}(2m+1)}}_{\text{DFT of odd-indexed part of } x[m]}$$

$$= E_k + e^{-j2\pi k\frac{1}{N}} O_k$$

where, DFT of the even-indexed $x[2m]$ is denoted by E_k, and DFT of the odd-indexed $x[2m+1]$ is denoted by O_k.

More generally, Cooley-Tukey FFT algorithm re-expresses the discrete Fourier transform (DFT) of an arbitrary composite size $N = N_1N_2$ in terms of smaller DFTs of sizes N_1 and N_2. Then this algorithm first computes N_1 DFTs of size N_2, then computes N_2 DFTs of size N_1.

FFT reduces the DFT number of computations, N^2, to $NlogN$. The computational speedup due to this reduction is very significant: for example, for N = 4096 the speedup factor is 683.

MATLAB has a direct command for FFT — 'fft'. For example, Y = fft(x) returns the discrete Fourier transform (DFT) of vector x, computed with FFT algorithm.

5.7.2.1 *Calculation of DFS using FFT*

Example 5.7-4: Let us now use fft command to calculate the DFT of sampled data record of length N_1, which then also permits calculating DFS of sampled data.

MATLAB COMMAND 5.7-6
>>N1 = length(piston) >> n = 0:(N1-1); >> figure >> plot(n,piston,'.') >> XDFT = fft(piston); >> XDFS = XDFT/N1; >> figure >> stem(n,abs(XDFS))

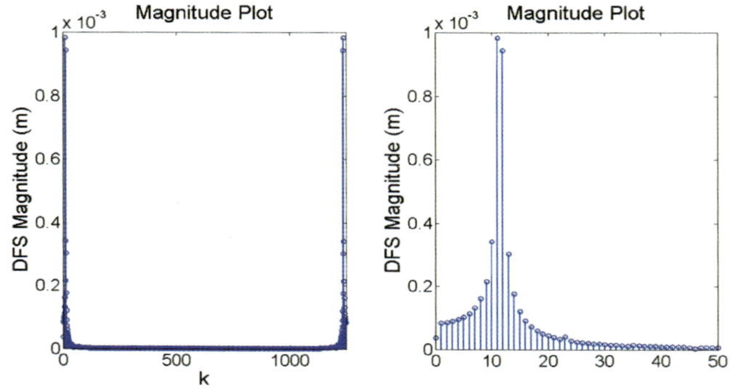

Fig. 5.7-6. DFS calculated through MATLAB COMMAND 5.7-6 using FFT shows leakage.

Just as in the DFS calculation of sampled data of length 1250, we have spectral leakage for FFT, as well. Since 'fft' command gives the vector of DFT of piston data with the same length, if we want to avoid the leakage problem, we need to set up a new vector of piston data with the length of 1195. Then using the exact same command as MATLAB COMMAND 5.7-6 we obtain:

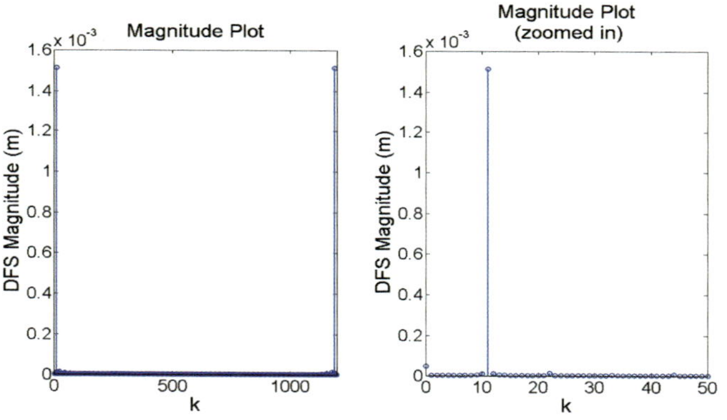

Fig. 5.7-7. DFS calculated through FFT without leakage.

Comparing Figures 5.7-1–5.7-4 with the latter ones, the FFT gives the exact same result as DFT, but with significantly reduced computational load.

5.7.3 *Developing DFT-based magnifying glass*

Now we are ready to carry out diagnostics on the mold oscillation system. We will employ FFT for both DFS and DFT based diagnostics.

5.7.3.1 *Catching the culprit based on our hypothesis*

In the search for the beam resonance excitation driver, first we are going to plot DFS magnitude spectrum and zoom on the very small magnitudes not seen well on the regular plot. The zooming is motivated by our hypothesis that higher frequency harmonic multiples of the reference oscillation frequency can be generated at the input to the beam due to the actuator nonlinearity.

Example 5.7-5: Run Nucor software testbed with periodic reference signal with magnitude 3 mm and frequency 4.6 Hz under 50% load under Bosch controller for full 10 periods after transients die out. Discretize data with $S_F = 460\ Hz$, generate its DFS plot that includes the first multiple at 9.2 Hz.

The input signal period is 1/4.6 sec. Since we will collect 10 periods, we will label it as T_{10}.

Since $S_F = 460\ Hz$, $t_s = 1/S_F = 1/460$ sec, and the CT time data collection time interval, denoted as T_1, is full 10 periods, the discrete data record length is:

$$N_1 = \frac{T_1}{t_s} = \frac{10T_{10}}{t_s} = \frac{10/4.6}{1/460} = 1000.$$

We make these changes to the data collection blocks in the Simulink model:

MATLAB COMMAND 5.7-8

```
>> N1 = length(piston)
>> T1 = 10*1/4.6;
>> XDFT = fft(piston);
%The following 3 lines are used to create the
two-sided XDFT from the single side of XDFT
>> XDFT_positive = XDFT(2:N1);
>> XDFT_negative = flipud(XDFT_positive);
>> XDFT_ts = [XDFT_negative;XDFT];
>> XDFS = XDFT_ts/N1;
>> k = -(N1-1):(N1-1);
>> f = k/T1; %f = k/T
>> figure;
>> stem(f,abs(XDFS));
```

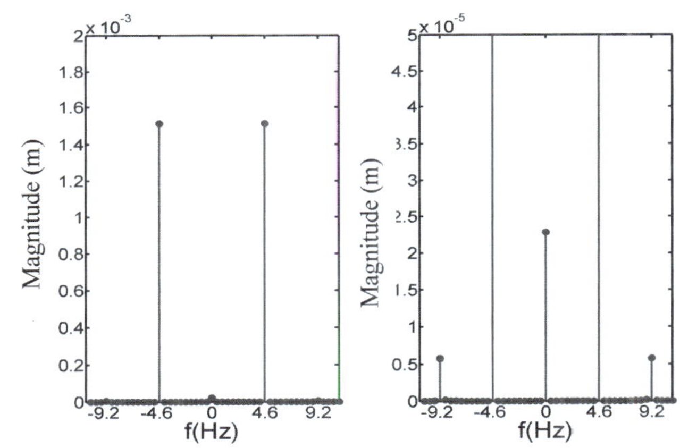

Fig. 5.7-8. DFS magnitude spectra of piston data record with data duration of 10 seconds: (left) scaled to fit the highest magnitude harmonic, (right) zoomed in to see the conjectured low magnitude harmonic at 9.2 Hz.

The DFS magnitude plot in Figure 5.7-8 fully shows the 4.6 Hz harmonic magnitude, but fails to display anything else. Due to our conjecture, however, that piston nonlinearity will yield a spike at 9.2 Hz we zoom in at 9.2 Hz. When the latter spectral line is revealed, the magnitude of the first harmonic goes far outside of the plot (off-scale), pointing to a multiscale signal composition.

5.7.3.2 *Catching the culprit with no hypothesis using the properties of DFT of periodic signals*

But what if we haven't had such a hypothesis? Would we still have had a chance to find the resonance driver? Yes, through several repetitive calculations of DFT magnitude spectra of the data according to either Case 1 or Case 2 given, respectively, by formulas (5.7-1) and (5.7-2), as presented next.

Example 5.7-6: Rework Example 5.7-5 with DFT magnitude spectrum computation using the software testbed. Keep decreasing t_s for the same CT data duration without changing the vertical scale of the plot and explore what happens with DFT magnitude spectra.

MATLAB COMMAND 5.7-9

```
>> N1 = length(piston)
>> T1 = 10*1/4.6;
>> XDFT = fft(piston);
%The following 3 lines are used to create the two-side XDFT from the single side of XDFT
>> XDFT_positive = XDFT(2:N1);
>> XDFT_negative = flipud(XDFT_positive);
>> XDFT_ts = [XDFT_negative;XDFT];
>> k = -(N1-1):(N1-1);
>> f = k/T1; %f = k/T
>> figure;
>> stem(f,abs(XDFT_ts));
```

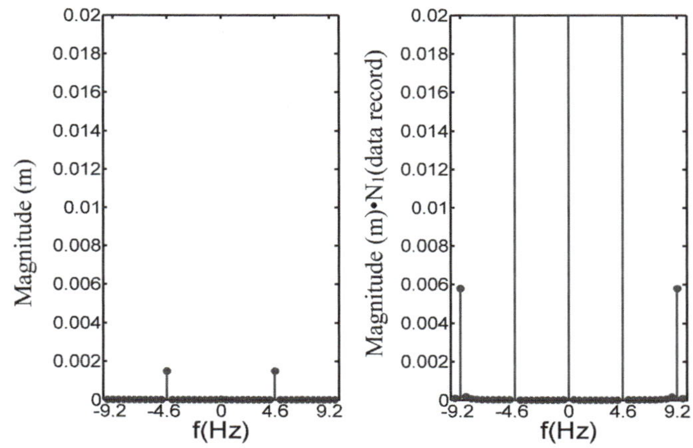

Fig. 5.7-9. DFS (left) and DFT (right) magnitude spectra of piston data record with data duration of 10 seconds.

In Figure 5.7-9, the DFS magnitude at 9.2 Hz is too small to be seen, and we don't zoom in on it in the absence of harmonic multiple hypothesis. So we completely miss this harmonic. In contrast to that, for the same t_s, the DFT magnitude spectrum shows the spectral line at 9.2 Hz well. This takes place because DFT multiplies DFS magnitudes by $N_1 = T_1/t_s$, amplifying them and making them go to infinity for the same T_1 as t_s decreases. So, if there is a small magnitude DFS harmonic, and we keep the same scale of the vertical axis while decreasing t_s, DFT will "amplify" both magnitudes like a magnifying glass: magnitude at 4.6 Hz will go outside of the plot and magnitude at 9.2 Hz will start being visible.

This property of DFT can be interpreted as mapping a multiscale DFS signal representation into a single-scale one, with all the nonzero harmonics magnitudes scaled up to infinity, e.g. like $\delta(f = 9.2\ Hz)$.

Although DFT of a sampled signal represents an approximation of the CT Fourier transform of the original CT signal, we haven't yet rigorously linked the two for a sinusoid. This will be carried out in Chapter 8 where we will address detection of sinusoids in the presence of noise.

Homework 8: Solved Examples and Solution Guide

Problem 1. Using software testbed generate and plot a continuous time piston position data record for periodic reference signal with magnitude 3 mm and frequency 4.6 Hz at 50% load

under Bosch controller for full 10 periods after transients die out. Discretize data with $S_{F1} = 460\ Hz$ **and** $S_{F2} = 920\ Hz$ **and plot DT data record.**

Solution:

This problem has been worked out in part in Example 5.7-5. We will repeat the latter as needed, but supplement it with step-by-step explanations.

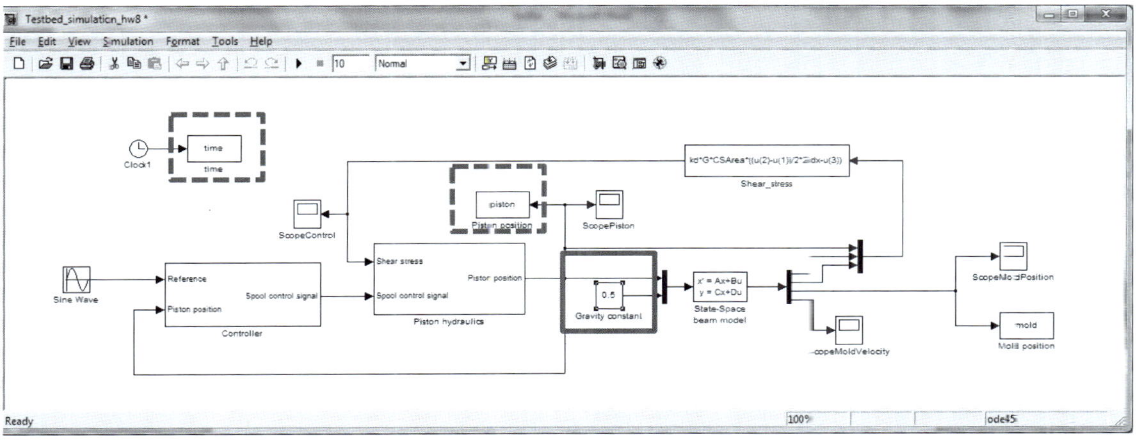

Fig. Hw8-1. Simulink model of the testbed.

In order to obtain the desired data, we have to make a few changes to the simulation program.

First, in order to limit the load to 50% of the nominal value, we can change the "gravity constant" in the model. This is not identical to changing the load, since it also affects the weight of the beam, but the error is small. A gravity constant of 1 would be 100% load, so 50% load would require a gravity constant of 0.5.

Second, we need to discretize the data at 460 Hz and collect 10 periods at 4.6 Hz. (When calculating a DFS or DFT, it is best to have an integer number of fundamental periods, or else as long a data set as possible, in order to prevent *spectral leakage*.) This is equivalent to a sampling time of: $t_s = 1/S_F = 1/460$ seconds.

The input signal period is 1/4.6 sec. Since we will collect 10 periods, we will label it as T_{10}. Since $S_F = 460\ Hz$, $t_s = 1/S_F = 1/460$ sec, and the CT time data collection time interval, denoted as T_1, is full 10 periods, the discrete data record length is:

$$N_1 = \frac{T_1}{t_s} = \frac{10T_{10}}{t_s} = \frac{10/4.6}{1/460} = 1000.$$

We need to make these changes to the data collection blocks in the Simulink model.

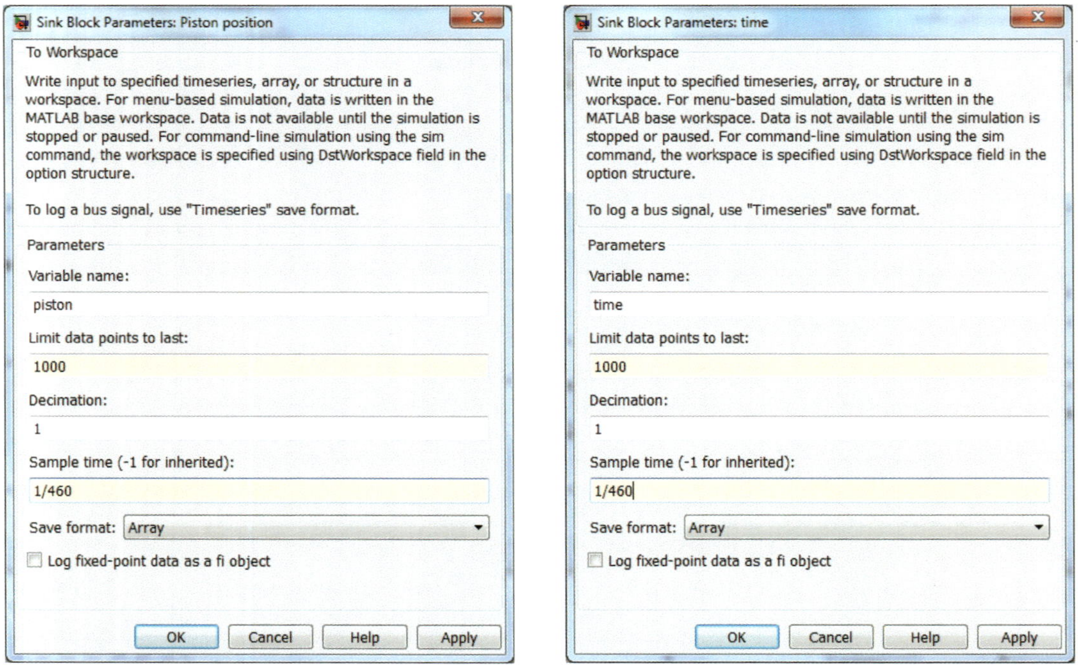

Fig. Hw8-2. Workspace parameters in testbed model.

Now, we can run the simulation and acquire 10 periods after the transient died down. Since the transient does not show up in the piston, we need to look at the mold output to make sure that we have collected enough stationary data.

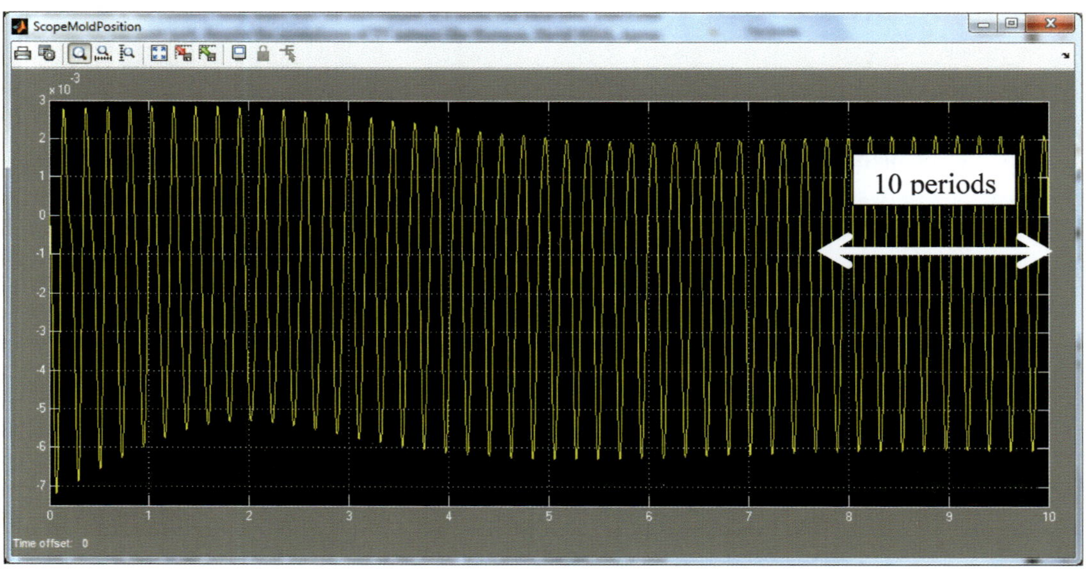

Fig. Hw8-3. Scope result of Simulink model for mold displacement.

The double arrow (>>) lines in the rest of this example indicate Matlab commands used for this problem. To make sure we collected the data correctly, we can plot the piston data:

>> plot(time,piston)

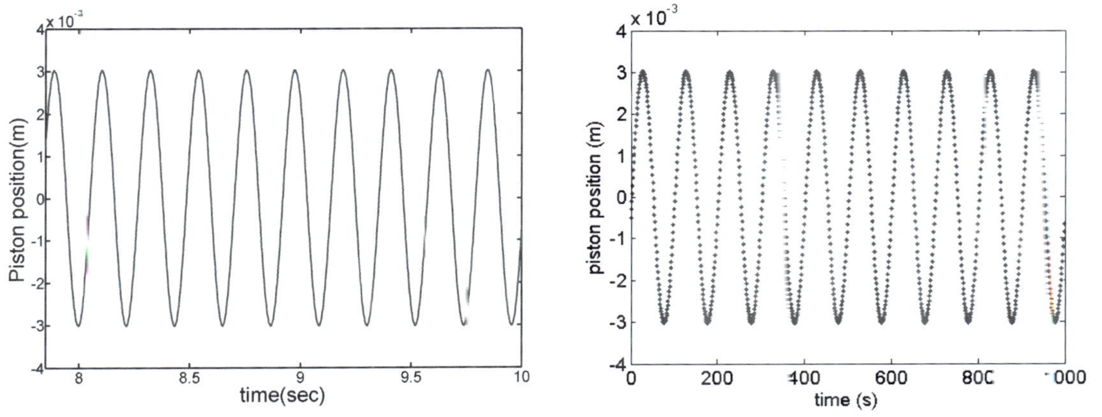

Fig. Hw8-4. Continuous (left)) and discrete (right) piston displacement result from simulation.

Now, to plot the data as a DT signal we use:

>> N_1 = 1000;

>> n = (0:(N_1-1))';

The transpose here is to make *n* a column vector, like *piston* is.

>> plot(n,piston,'.')

While in most modern programming languages the data is numbered from 0 to N-1, in this book for clarity we number the data from 0 to N_1-1. Matlab, however, indexes arrays from 1 to N (or from 1 to N_1).

a) Compute, using Matlab, and plot the data record DFS, give analytical expression of this DFS, and comment on the presence and the magnitude of any harmonics other than the fundamental one.

Now we can calculate the DFS of this truncated data set. Recall the formula for the DFS coefficients for periodic signal $x_p[n]$:

$$X_{DFS}\left[k\right] = \frac{1}{N_1} X_{DFT} = \frac{1}{N_1} \sum_{n=0}^{N_1-1} x_p\left[n\right] e^{-j2\pi k \frac{1}{N} n}.$$

In Matlab, this can be done by the following commands:

>> x = piston;

n, samples

>> XDFS = zeros(N_1,1);

>> for k = 0:(N_1-1); XDFS(k + 1) = (1/N_1)*sum(x'.*exp(-1j*2*pi*(k/N_1)*n)'); end;

>> k = (0:(N_1-1))';

>> F = k/N_1;

>> subplot(1,2,1);

```
>> stem(F,abs(XDFS));
>> subplot(1,2,2);
>> stem(F,angle(XDFS));
```

Since the data size is $N_1 = 1000$, the corresponding digital frequency is $F_1 = 1/N_1 = 1/1000$ cycles/sample, and at each harmonic k the digital frequency is $k/F_1 = k/1000$. This permits us to plot the DFS spectrum with respect to the digital frequency axis.

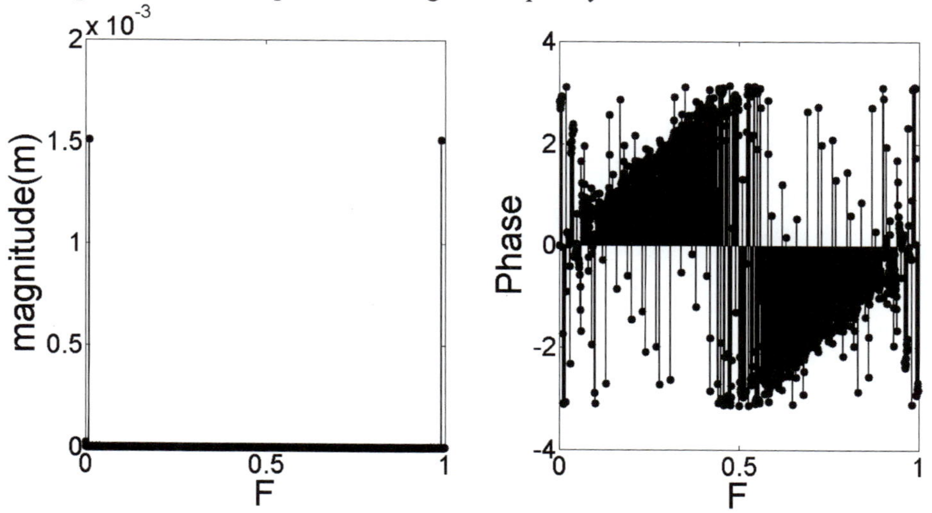

Fig. Hw8-5. DFS of piston data plotted with respect to digital frequency.

The question then asks to write the DFS analytically. Most of the terms appear to have magnitude close to 0, so we don't need to report them. There are two obviously non-zero terms, which correspond to the fundamental frequency of the piston. Zooming in on the lower left-hand corner of the magnitude plot yields

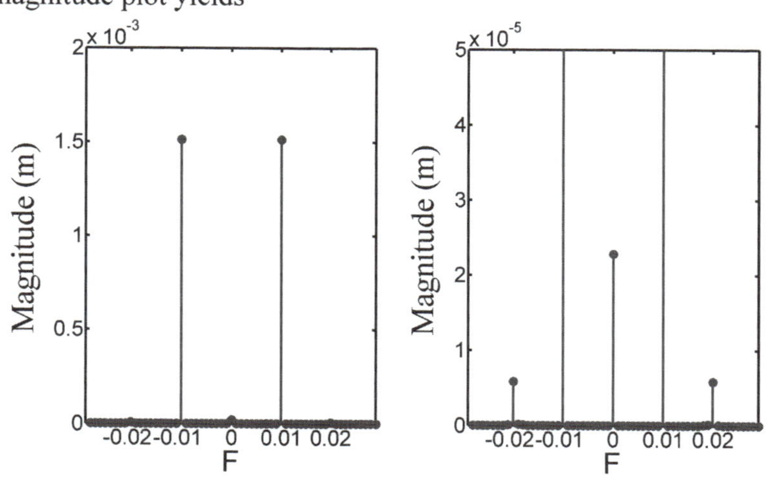

Fig. Hw8-7. Zoom in of magnitude DFS plotted with respect to digital frequency.

Which terms we include in the DFS is now a bit of a judgment call. There's a significant component at $k = 0$ (DC component) that should be included. We will also include the terms at k = 20, 30, and 40 — all integer multiples of k =10 — driving digital frequency. Also, due to symmetry, all $X_{DFS}[N_1\text{-}k]$ should be complex conjugate of $X_{DFS}[k]$: $X_{DFS}[N_1\text{-}k] = X_{DFS}[k]^*$, so if we include $X_{DFS}[k]$ in the DFS we need to include $X_{DFS}[N_1\text{-}k]$.

Also, keeping in mind that in our notation $X_{DFS}[0]$ is the first term, but in Matlab, $X_{DFS}[1]$ is the first term, we adjust indices accordingly.

So, the DFS is:

$$\left\{\begin{matrix} X_{DFS}[0], X_{DFS}[10], , X_{DFS}[20], X_{DFS}[30], X_{DFS}[40], \\ X_{DFS}[960], X_{DFS}[970], X_{DFS}[980], X_{DFS}[990] \end{matrix}\right\}$$
$$= \left\{\begin{matrix} 2.28\cdot 10^{-5}, (-0.25 - j1.49)\cdot 10^{-3}, (3.58 - j4.57)\cdot 10^{-6}, (-2.46 - j2.63)\cdot 10^{-6}, (-2.28 + j2.71)\cdot 10^{-6}, \\ (-2.28 - j2.71)\cdot 10^{-6}, (-2.46 + j2.63)\cdot 10^{-6}, (3.58 + j4.57)\cdot 10^{-6}, (-0.25 + j1.49)\cdot 10^{-3} \end{matrix}\right\}.$$

There are small, but noticeable components away from the fundamental frequency. However, they are all at harmonics of the fundamental frequency. These are likely caused by nonlinearity in the piston.

b) Give analytical expression of the DT signal (write IDFS) in terms of its DFS, simulate it and check if it matches the data record.

The general formula for the inverse DFS (IDFS) is

$$x_p[n] = \sum_{k=0}^{N_1-1} X_{DFS}[k] e^{j2\pi \frac{k}{N_1} n}.$$

Using the DFS written above,

$$\begin{aligned} x_p[n] &= X_{DFS}[0] e^{j2\pi \frac{0}{N} n} + X_{DFS}[10] e^{j2\pi \frac{10}{N} n} + X_{DFS}[20] e^{j2\pi \frac{20}{N} n} + X_{DFS}[30] e^{j2\pi \frac{30}{N} n} \\ &+ X_{DFS}[40] e^{j2\pi \frac{40}{N} n} + X_{DFS}[960] e^{j2\pi \frac{960}{N} n} + X_{DFS}[970] e^{j2\pi \frac{970}{N} n} + X_{DFS}[980] e^{j2\pi \frac{980}{N} n} \\ &+ X_{DFS}[990] e^{j2\pi \frac{990}{N} n} = 2.28\cdot 10^{-5} e^{j2\pi \frac{0}{1000} n} + (-0.25 - j1.49)\cdot 10^{-3} e^{j2\pi \frac{10}{1000} n} \\ &+ (3.58 - j4.57)\cdot 10^{-6} e^{j2\pi \frac{20}{1000} n} + (-2.46 - j2.63)\cdot 10^{-6} e^{j2\pi \frac{30}{1000} n} + (-2.28 + j2.71)\cdot 10^{-6} e^{j2\pi \frac{40}{1000} n} \\ &+ (-2.28 - j2.71)\cdot 10^{-6} e^{j2\pi \frac{960}{1000} n} + (-2.46 + j2.63)\cdot 10^{-6} e^{j2\pi \frac{970}{1000} n} + (3.58 + j4.57)\cdot 10^{-6} e^{j2\pi \frac{980}{1000} n} \\ &+ (-0.25 + j1.49)\cdot 10^{-3} e^{j2\pi \frac{990}{1000} n}. \end{aligned}$$

In Matlab:

```
>> xIDFS = XDFS(0+1)*exp(1j*2*pi*(0/N1)*n) +
   XDFS (10+1)*exp(1j*2*pi*(10/N1)*n) +
   XDFS (20+1)*exp(1j*2*pi*(20/N1)*n) +
```

```
    XDFS (30+1)*exp(1j*2*pi*(30/N1)*n) +
    XDFS (40+1)*exp(1j*2*pi*(40/N1)*n) +
    XDFS (N1-40+1)*exp(1j*2*pi*((N1-40)/N1)*n) +
    XDFS (N1-30+1)*exp(1j*2*pi*((N1-30)/N1)*n) +
    XDFS (N1-20+1)*exp(1j*2*pi*((N1-20)/N1)*n) + .
    XDFS (N1-10+1)*exp(1j*2*pi*((N1-10)/N1)*n);
>> plot(n,x,'.',n,xIDFS,'o')
```

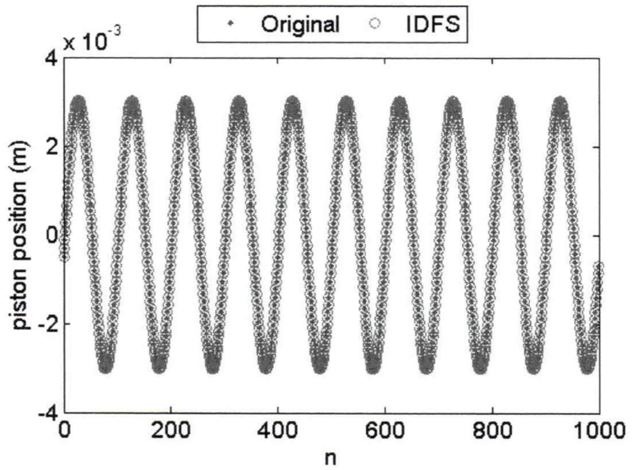

Fig. Hw8-8. Original data record and IDFS.

The IDFS matches well the original data.

c) Taking the DT signal written in terms of its DFS in b) and using sampling frequency S_{F1} = 460 Hz, give analytical expression of (reconstruct) the corresponding CT signal and check if it matches the DT data record obtained with S_{F1} = 460 Hz at nt_s data points.

The programming convention for the DFS is to use frequencies between 0 and (N_1-1)/N_1, so, first, we have to rewrite the IDFS so that all frequencies shift into the principal range, [−½, ½]:

$$x_p[n] = 2.28\cdot10^{-5}e^{j2\pi\frac{0}{1000}n} + (-0.25-j1.49)\cdot10^{-3}e^{j2\pi\frac{10}{1000}n} + (3.58-j4.57)\cdot10^{-6}e^{j2\pi\frac{20}{1000}n}$$
$$+(-2.46-j2.63)\cdot10^{-6}e^{j2\pi\frac{30}{1000}n} + (-2.28+j2.71)\cdot10^{-6}e^{j2\pi\frac{40}{1000}n}$$
$$+(-2.28-j2.71)\cdot10^{-6}e^{j2\pi\left(-\frac{40}{1000}\right)n} + (-2.46+j2.63)\cdot10^{-6}e^{j2\pi\left(-\frac{30}{1000}\right)n}$$
$$+(3.58+j4.57)\cdot10^{-6}e^{j2\pi\left(-\frac{20}{1000}\right)n} + (-0.25-j1.49)\cdot10^{-3}e^{j2\pi\left(-\frac{10}{1000}\right)n}.$$

As described in Section 4.1 of Chapter 4, for the sampling frequency S_F = 460 samples/second given in the data file, we reconstruct the analytical CT signal through the map (4.1-3): $n \to t/t_s = t/(1/S_F) = S_F t = 460t$ substituted into the expression above. This yields

$$x_p(t) = 2.28\cdot 10^{-5} e^{j2\pi\frac{0\cdot 460}{1000}t} + (-0.25 - j1.49)\cdot 10^{-3} e^{j2\pi\frac{10\cdot 460}{1000}t} + (3.58 - j4.57)\cdot 10^{-6} e^{j2\pi\frac{20\cdot 460}{1000}t}$$

$$+(-2.46 - j2.63)\cdot 10^{-6} e^{j2\pi\frac{30\cdot 460}{1000}t} + (-2.28 + j2.71)\cdot 10^{-6} e^{j2\pi\frac{40\cdot 460}{1000}t}$$

$$+(-2.28 - j2.71)\cdot 10^{-6} e^{j2\pi\left(-\frac{40\cdot 460}{1000}\right)t} + (-2.46 + j2.63)\cdot 10^{-6} e^{j2\pi\left(-\frac{30\cdot 460}{1000}\right)t}$$

$$+(3.58 + j4.57)\cdot 10^{-6} e^{j2\pi\left(-\frac{20\cdot 460}{1000}\right)t} + (-0.25 - j1.49)\cdot 10^{-3} e^{j2\pi\left(-\frac{10\cdot 460}{1000}\right)t} =$$

$$2.28\cdot 10^{-5} + (-0.25 - j1.49)\cdot 10^{-3} e^{j2\pi 4.6t} + (3.58 - j4.57)\cdot 10^{-6} e^{j2\pi 9.2t} +$$

$$(-2.46 - j2.63)\cdot 10^{-6} e^{j2\pi 13.8t} + (-2.28 + j2.71)\cdot 10^{-6} e^{j2\pi 18.4t} + (-2.28 - j2.71)\cdot 10^{-6} e^{j2\pi(-18.4)t} +$$

$$(-2.46 + j2.63)\cdot 10^{-6} e^{j2\pi(-13.8)t} + (3.58 + j4.57)\cdot 10^{-6} e^{j2\pi(-9.2)t} + (-0.25 - j1.49)\cdot 10^{-3} e^{j2\pi(-4.6)t}.$$

In Matlab:

```
>> SF = 460;
>> ts = 1/SF;
>> t = 0:0.0001:(N1-1)*ts;
>> xCT = XDFS(0+1)*exp(1j*2*pi*(0/N1)*SF*t) + ...
   XDFS (10+1)*exp(1j*2*pi*(10/N1)*SF*t) + ...
   XDFS (20+1)*exp(1j*2*pi*(20/N1)*SF*t) + ...
   XDFS (30+1)*exp(1j*2*pi*(30/N1)*SF*t) + ...
   XDFS (40+1)*exp(1j*2*pi*(40/N1)*SF*t) + ...
   XDFS (N-40+1)*exp(1j*2*pi*((-40)/N1)*SF*t) + ...
   XDFS (N-30+1)*exp(1j*2*pi*((-30)/N1)*SF*t) + ...
   XDFS (N-20+1)*exp(1j*2*pi*((-20)/N1)*SF*t) + ...
   XDFS (N-10+1)*exp(1j*2*pi*((-10)/N1)*SF*t);
>> plot(n*ts,x,'.',t,xCT)
```

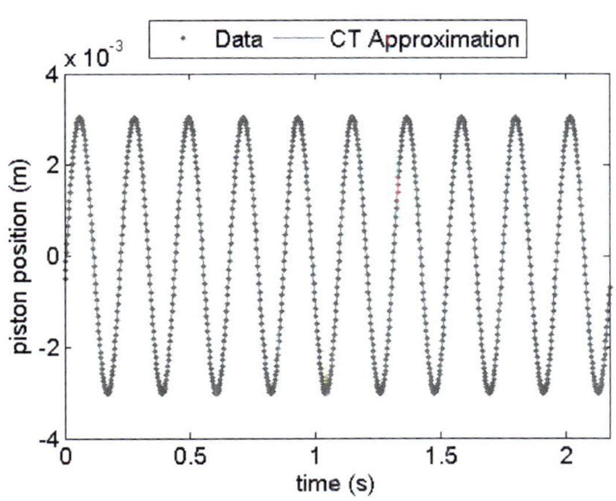

Fig. Hw8-9. Original data record and CT approximation.

The analytical CT calculation matches well with the original data.

d) Compute, using Matlab FFT algorithm, DFT of the data record obtained with S_{F1} = 460Hz and give analytical expression of this DFT. Plot magnitude spectrum of this DFT and on its basis comment on the presence and magnitude of any harmonics other than the fundamental one.

Since the data size is $N_1 = 1000$, the corresponding digital frequency is $F_1 = 1/N_1 = 1/1000$ cycles/sample. Therefore, for each harmonic k the digital frequency is $F_k = k/F_1 = k/1000$. This permits plotting the spectrum along digital frequency axis F at discrete points F_k.

Also, for the 4.6 Hz harmonic we have $F_k = 4.6 / S_F = 4.6 / 460 = 1 / 100 = 10(1 / 1000)$. Therefore, this harmonic is the 10th harmonic with respect to fundamental harmonic of the data record DT periodic extension.

Rescaling the digital frequency axis F to analog frequency f for the corresponding discrete set of points then takes the form $f_k = F_k S_F = (k / 1000)460 = 4.6k / 100$ Hz.

Now, following the MATLAB COMMAND 5.7-9 in Example 5.7-6 yields the plot shown below in Figure HW8-10.

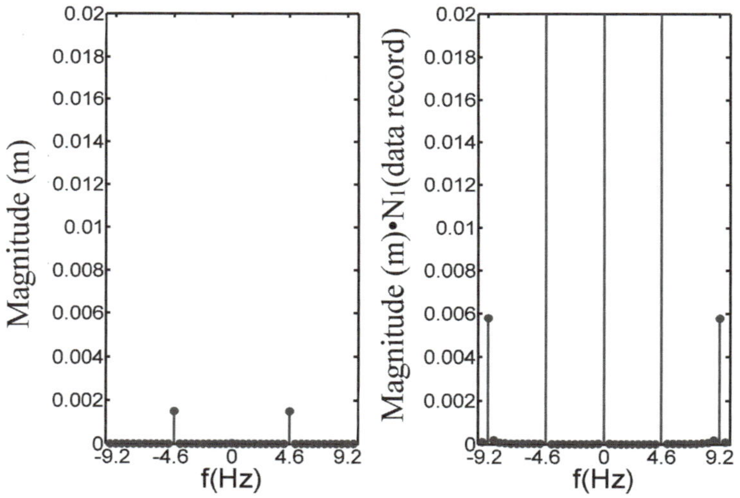

Fig. Hw8-10. Comparison of DFS (left) and DFT (right) magnitude spectra of the data record.

By inspection, most of the terms in the computed DFT are approximately 0. The clearly non-zero terms are:

$$\{X[0], X[10], X[20], X[30], X[40], X[960], X[970], X[980], X[990]\}$$
$$= \left\{ \begin{array}{l} 0.0228, (-0.2529 - j1.4919), (0.0036 - j0.0046), (-0.0025 - j0.0026), (-0.0023 + j0.0027), \\ \quad (-0.0023 - j0.0027), (-0.0025 + j0.0026), (0.0036 + j0.0046), (-0.2529 + j1.4919) \end{array} \right\}.$$

e) Perform numerical computations of d) for the following DT data records:

A) 10 CT periods discretized with $S_{F1} = 460$ Hz,

B) 5 CT periods discretized with $S_{F2} = 460$ Hz,

C) 5 CT periods discretized with $S_{F3} = 920$ Hz.

Comment on the behavior of DFT magnitude spectra in detecting small periodic components for two cases:

1) decreasing t_s for the same CT data record by comparing B) and C) spectra, and

2) increasing the CT data length for the same t_s by comparing A) and B) spectra.

Following the same procedure as in the previous questions we have:

A) 10 CT periods discretized with $S_{F1} = 460$ Hz

This case has been considered in the previous questions, so we don't need to rerun the testbed Simulink file and just use the previous data.

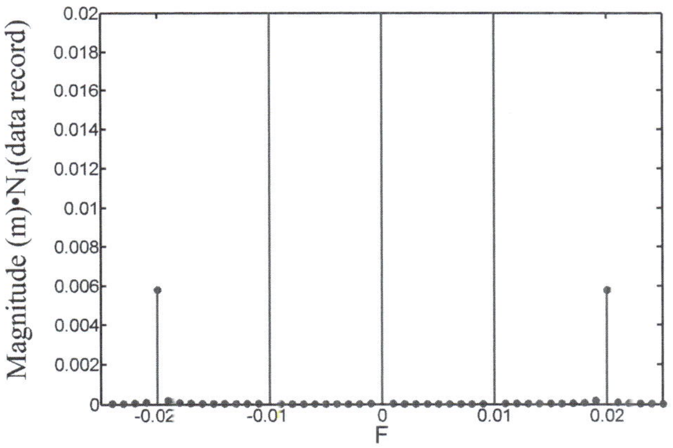

Fig. Hw8-11. Magnitude spectrum of DFT for 10 CT periods and 460 Hz sampling frequency.

$$\{X[0], X[10], X[20], X[30], X[40], X[960], X[970], X[980], X[990]\}$$
$$= \left\{ \begin{matrix} 0.0028, (-0.2529 - j1.4919), (0.0036 - j0.0046), (-0.0025 - j0.0026), (-0.0023 + j0.0027), \\ (-0.0023 - j0.0027), (-0.0025 + j0.0026), (0.0036 + j0.0046), (-0.2529 + j1.4919) \end{matrix} \right\}.$$

B) 5 CT periods discretized with $S_{F1} = 460$ Hz.

In this case we have different number of CT periods required, so we need to rerun the testbed Simulink file. With different CT data record size we have

$$N_1 = \frac{5 \cdot 1/4.6}{1/460} = 500.$$

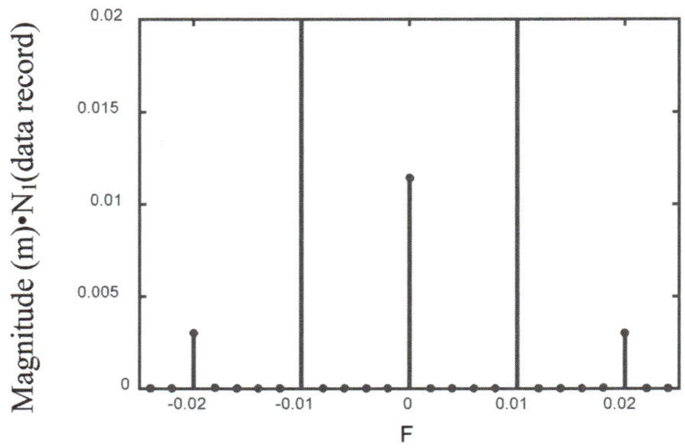

Fig. Hw8-12. Magnitude spectrum of DFT under 5 CT periods and 460 Hz sampling frequency.

$$\{X[0],\, X[5],\, X[10],\, X[15],\, X[20],\, X[480],\, X[485],\, X[490],\, X[495]\}$$
$$=\left\{\begin{matrix} 0.0114,\, (-0.1265-j0.7459),\, (0.0018-j0.0024),\, (-0.0012-j0.0013), (-0.0012+j0.0013), \\ (-0.0012-j0.0013),\, (-0.0012+j0.0013),\, (0.0018+j0.0024),\, (-0.1265+j0.7459) \end{matrix}\right\}.$$

C) 5 CT periods discretized with $S_{F2} = 920$ Hz.

In this case we have different number of CT periods and different time step t_s required, so we need to rerun the testbed Simulink file. With different CT data record and different t_s we have

$$t_s = 1/920, \quad N_1 = \frac{5 \cdot 1/4.6}{1/920} = 1000.$$

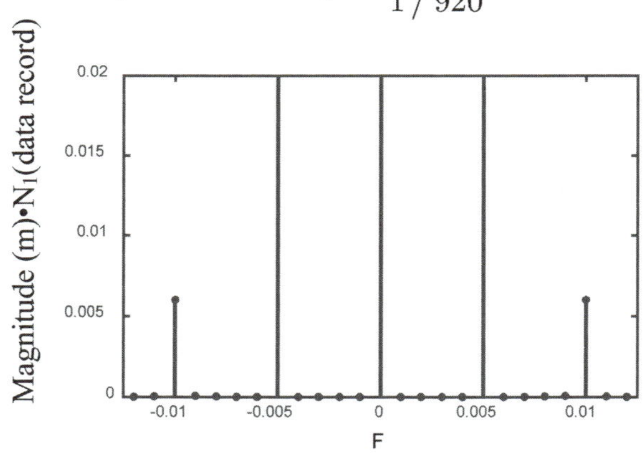

Fig. Hw8-13. Magnitude spectrum of DFT under 5 CT periods and 920 Hz sampling frequency.

$$\{X[0], X[5], X[10], X[15], X[20], X[980], X[985], X[990], X[995]\}$$
$$=\left\{\begin{matrix}0.00229, (-0.2997-j1.4832), (0.0033-j0.0050), (-0.0027-j0.0024), (-0.0020+j0.0029),\\ (-0.0020-j0.0029), (-0.0027+j0.0024), (0.0033+j0.0050), (-0.2997+j1.4832)\end{matrix}\right\}.$$

From A and B, we can conclude that with the same t_s, if the data length increased, the magnitude of the harmonics will increase too; From B) and C), we can conclude that with same CT data record, reducing the t_s increases the magnitudes of the harmonics. Therefore, in order to identify the harmonics buried in noise we can increase the CT data record length and/or sampling frequency.

Problem 2: Draw several periods of a periodic signal defined over one period as $x(t) = e^{-2t}$ $0 \le t \le 2$, with the period $T_1 = 2$, and find the exponential Fourier series coefficients. Derive analytical expressions for, and plot, the single and double-sided magnitude, phase, and power spectra. Indicate how the resulting double-sided magnitude and phase spectra can be used to approximately reveal the frequency response of the system with impulse response given by $h(t) = e^{-2t}, 0 \le t \le \infty$.

Solution:

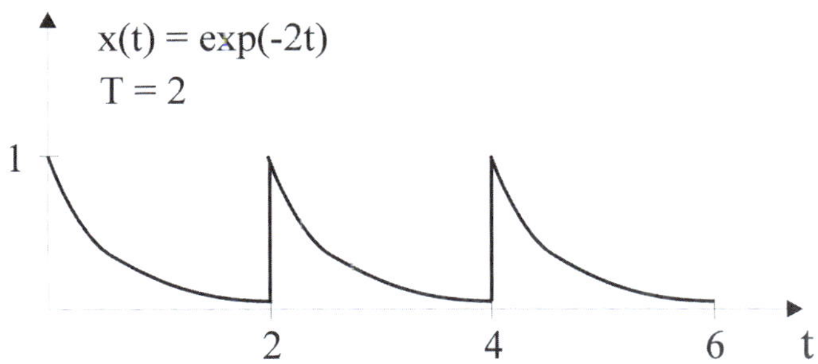

Fig. Hw8-14. Signal from Problem 2.

Since the fundamental period $T_1 = 2$ then the fundamental frequency are $f_1 = \frac{1}{T_1} = \frac{1}{2}$ and $\omega_1 = 2\pi f_1 = \pi$. Since $e^{j2\pi k} = 1$, we have:

$$H_s[k] = \frac{1}{T_1}\int_0^{T_1} x(t)e^{-jk\omega_1 t}dt = \frac{1}{T_1}\int_0^{T_1} h(t)e^{-jk\omega_1 t}dt = \frac{1}{2}\int_0^2 e^{-2t}e^{-jk\pi t}dt = \frac{1}{2}\int_0^2 e^{-(2+jk\pi)t}dt$$
$$= \frac{1}{2}\left[\frac{-e^{-(2+jk\pi)t}}{2+jk\pi}\Bigg|_0^2\right] = \frac{1}{2}\left[\frac{-e^{-4-j2\pi k}+1}{2+jk\pi}\right] = \frac{1}{2}\left[\frac{1-e^{-4}e^{-j2\pi k}}{2+jk\pi}\right] = \frac{1-e^{-4}}{4+j2\pi k}.$$

Analytical expressions for magnitude and phase spectrum:

$$H_s[k] = \frac{1-e^{-4}}{4+j2\pi k} = \frac{1-e^{-4}}{4+j2\pi k}\cdot\frac{4-j2\pi k}{4-j2\pi k} = \frac{\left(4-4e^{-4}\right)+j\left(e^{-4}2\pi k - 2\pi k\right)}{4^2+\left(2\pi k\right)^2}$$

$$= \overbrace{\frac{4-4e^{-4}}{4^2+\left(2\pi k\right)^2}}^{RE} + j\overbrace{\left(\frac{e^{-4}2\pi k - 2\pi k}{4^2+\left(2\pi k\right)^2}\right)}^{IM}.$$

Magnitude: $\left|H_s[k]\right| = \sqrt{RE^2+IM^2} = \left|\frac{1-e^{-4}}{4+j2\pi k}\right| = \frac{\left|1-e^{-4}\right|}{\left|4+j2\pi k\right|} = \frac{\left|1-e^{-4}\right|}{\sqrt{4^2+\left(2\pi k\right)^2}}.$

Phase:

$$\phi_{rad} = \tan^{-1}\left(\frac{IM}{RE}\right) = \tan^{-1}\left(\frac{e^{-4}2\pi k - 2\pi k}{4-4e^{-4}}\right) = \tan^{-1}\left(\frac{-2\pi k\left(1-e^{-4}\right)}{4\left(1-e^{-4}\right)}\right) = -\tan^{-1}\left(\frac{\pi k}{2}\right).$$

For 1st to 5th harmonics, the magnitude, phase and power spectra coefficients are listed below.

k	$\left\|H_s[k]\right\|$	$\phi(rad/s)$	$\phi(\text{degree})$	$\left\|H_s[k]\right\|^2$
−5	0.0310	1.4442	82.74	0.0010
−4	0.0386	1.4130	80.96	0.0015
−3	0.0509	1.3617	78.02	0.0026
−2	0.0744	1.2626	72.34	0.0055
−1	0.1318	1.0039	57.52	0.0174
0	0.2454	0	0	0.0602
1	0.1318	−1.0039	−57.52	0.0174
2	0.0744	−1.2626	−72.34	0.0055
3	0.0509	−1.3617	−78.02	0.0026
4	0.0386	−1.4130	−80.96	0.0015
5	0.0310	−1.4442	−82.74	0.0010

The resulting two-sided magnitude, phase, and power spectra take the form:

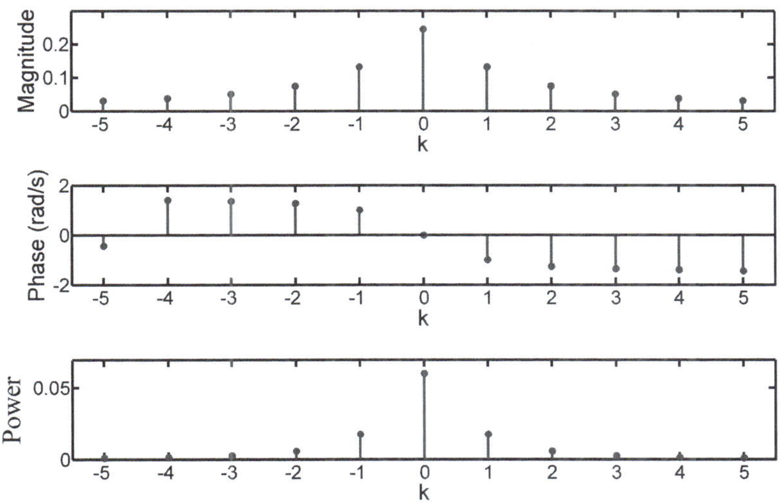

Fig. Hw8-15. Two-sided magnitude, phase, and power spectra.

Let's now multiply each of the magnitude spectrum components by $T_s = 2$ and take phase spectrum unchanged. Then, the resulting spectra yield, respectively, the approximate samples of the magnitude $|H(j\omega)|$ and the phase $\angle H(j\omega)$ of the Fourier transform $H(j\omega)$ of $h(t) = e^{-2t}$, $0 \le t \le \infty$. The envelopes of the latter spectra approximate the magnitude and the phase responses of the system with the impulse response $h(t) = e^{-2t}, 0 \le t \le \infty$.

For single-sided spectrum we have: $c_0 = H_s[0]$, $c_k = 2|H_s[k]|$, $p_k = 2|H_s[k]|^2$

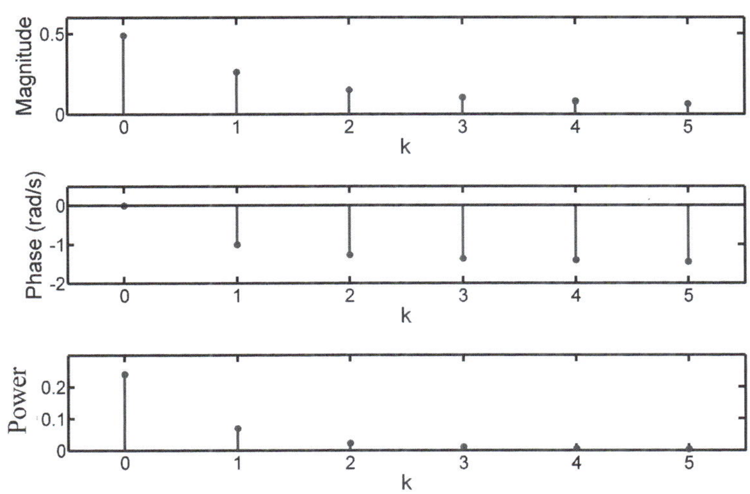

Fig. Hw8-16. Single-sided magnitude, phase, and power spectra.

Problem 3: Assume the system has a noncausal impulse response: $h(t) = \mathrm{rect}(t / 2T_p)$**,**

(a) Derive system's frequency response;

(b) Evaluate system's frequency response via Fourier series taking (i) $T_1 = 4T_p$ **and (ii)** $T_1 = 8T_p$ **and plot corresponding magnitude spectra;**

(c) Bandlimit each Fourier Series evaluation of the frequency response to the central "lobe", choose 4 times oversampling for both (i) and (ii), write down DFS and DFT for (i) and (ii), plot DFT and DFS magnitude spectra for (i) and (ii).

Solution: The figure of $h(t)$ is shown in Figure HW8-14.

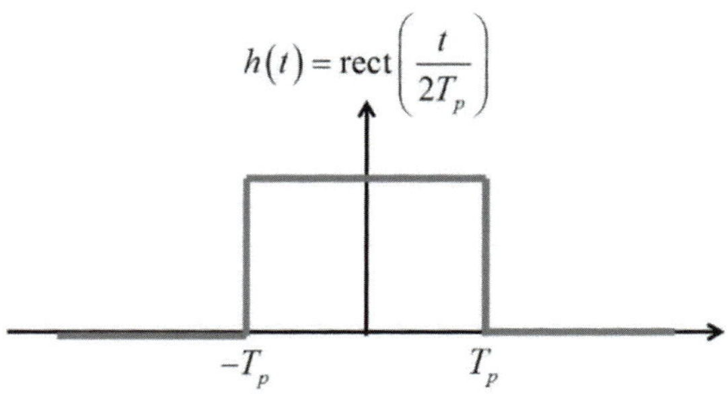

Fig. Hw8-17. Noncausal signal $h(t)$.

(a) Frequency Response:

$$H(j\omega) = \int_{-\infty}^{\infty} h(t)e^{-j\omega t}dt = \int_{-T_p}^{T_p} e^{-j\omega t}dt = \frac{-1}{j\omega}\left(e^{-j\omega t}\right)\Big|_{-T_p}^{T_p} = \frac{-1}{j\omega}\left(e^{-j\omega T_p} - e^{j\omega T_p}\right)$$

$$= \frac{e^{j\omega T_p} - e^{-j\omega T_p}}{j\omega} = \frac{2}{\omega}\frac{e^{j\omega T_p} - e^{-j\omega T_p}}{j2} = \frac{2}{\omega}\sin\left(\omega T_p\right) = \frac{2T_p}{\pi\frac{\omega}{\pi}T_p}\sin\left(\pi\frac{\omega}{\pi}T_p\right) =$$

$$2T_p\,\mathrm{sinc}\left(\frac{\omega}{\pi}T_p\right) = 2T_p\,\mathrm{sinc}\left(2fT_p\right).\ \text{Note: } \frac{e^{j\theta} - e^{-j\theta}}{j2} = \sin(\theta) \text{ and } \frac{\sin(\pi\phi)}{\pi\phi} = \mathrm{sinc}(\phi).$$

(b)

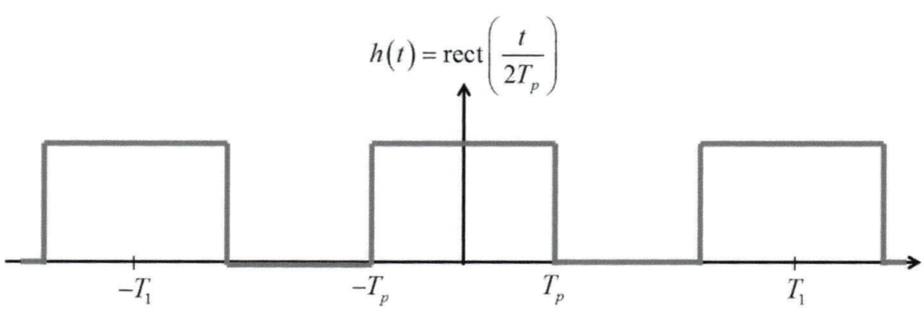

Fig. Hw8-18. Periodic expansion of signal $h(t)$.

Define $h_p(t)$ to be T_1-periodic continuous-time extension of $h(t)$, then $h_p(t)$ has Fourier series representation $h_p(t) = \sum_{n=-\infty}^{\infty} H_s[k]e^{jk\omega_1 t} = \sum_{n=-\infty}^{\infty} \text{rect}\left[\frac{t - nT_1}{2T_p}\right]$ and system's frequency response can be evaluated by $H(j\omega)\big|_{\omega=k\omega_1} = T_1 H_s[k]$. Therefore, we need to find Fourier Series coefficients $H_s[k]$:

$$H_s[0] = \frac{1}{T_1}\int_{-T_p}^{T_p} 1dt = \frac{2T_p}{T_1},$$

$$H_s[k] = \frac{1}{T_1}\int_{-T_p}^{T_p} e^{-jk\omega_1 t}dt = \frac{-1}{jk\omega_1 T_1}\left(e^{-jk\omega_1 T_p} - e^{jk\omega_1 T_p}\right) = \frac{e^{jk\omega_1 T_p} - e^{-jk\omega_1 T_p}}{jk\omega_1 T_1}.$$

Note: $\dfrac{e^{j\theta} - e^{-j\theta}}{j2} = \sin(\theta)$

$$\Rightarrow H_s[k] = \frac{2}{k\omega_1 T_1}\sin(k\omega_1 T_p);\ \text{note: } \omega_1 T_1 = 2\pi$$

$$\Rightarrow H_s[k] = \frac{2}{2\pi k}\sin\left(2\pi k\frac{T_p}{T_1}\right) = \frac{2T_p / T_1}{2\pi k T_p / T_1}\sin\left(2\pi k\frac{T_p}{T_1}\right);\ \text{note: } \frac{\sin(\pi\phi)}{\pi\phi} = \text{sinc}(\phi)$$

$$\Rightarrow H_s[k] = 2\frac{T_p}{T_1}\text{sinc}\left(2k\frac{T_p}{T_1}\right) \text{ and}$$

$$H(j\omega)\big|_{\omega=k\omega_1} = T_1 H_s[k] = 2T_p\,\text{sinc}\left(2k\frac{T_p}{T_1}\right) = 2T_p\,\text{sinc}\left(k\frac{2\pi}{T_1}\frac{T_p}{\pi}\right) = 2T_p\,\text{sinc}\left(\omega\frac{T_p}{\pi}\right)\Bigg|_{\omega=k\omega_1}.$$

Fourier series:

(i) $T_1 = 4T_p$,

$$H_s[k] = 2\left(\frac{1}{4}\right)\text{sinc}\left(2k\left(\frac{1}{4}\right)\right) = \frac{1}{2}\text{sinc}\left(\frac{k}{2}\right) = \frac{1}{2}\text{sinc}\left(\frac{1}{2\omega_1}k\omega_1\right),\ \left|H(j\omega)\right|_{\omega=k\omega_1} = T_1\left|H_s[k]\right|.$$

(ii) $T_1 = 8T_p$, $H_s[k] = 2\left(\frac{1}{8}\right)\text{sinc}\left(2k\left(\frac{1}{8}\right)\right) = \frac{1}{4}\text{sinc}\left(\frac{k}{4}\right) = \frac{1}{4}\text{sinc}\left(\frac{1}{4\omega_1}k\omega_1\right)$.

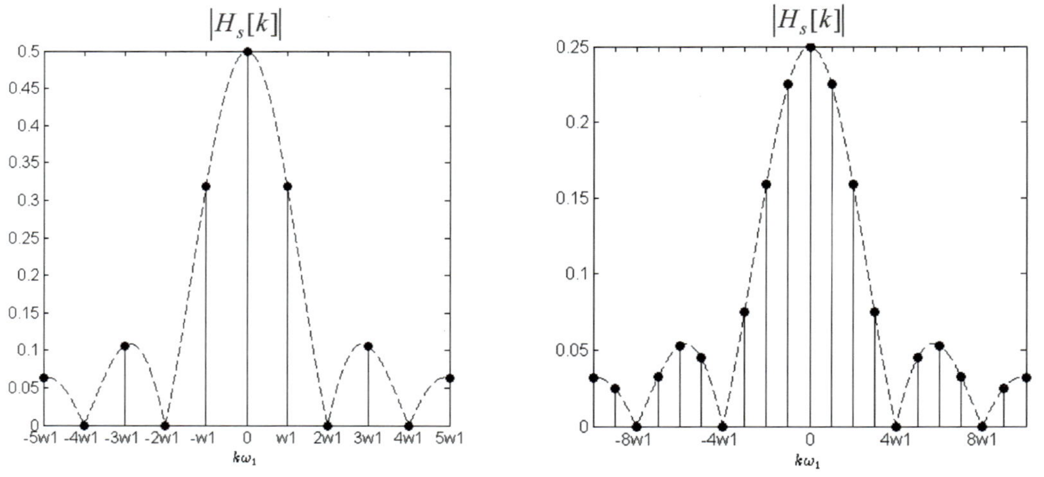

Fig. Hw8-19. Magnitude spectra for (a) $T_1 = 4T_p$ (b) $T_1 = 8T_p$.

Note: An increase in the period, T_1, of the periodic extension $h_p(t)$ of $h(t)$ increases the density (fidelity) of the representation of $H(j\omega)$ (continuous function of ω) via Fourier series coefficients of $h_p(t)$.

(c) Discrete time — sampling:

We cannot sample $h(t) = \mathrm{rect}\left(\dfrac{t}{2T_p}\right)$ without aliasing because $h(t)$ is not bandlimited.

Therefore we bandlimit $h_p(t)$ to central lobe and then set sampling frequency S_F with respect to the central lobe band limit for each evaluation and sample the bandlimited periodic extension.

(i) $T_1 = 4T_p$,

$f_b = 2\omega_1$ for band limiting to center lobe, i.e. k is restricted to $[-2,-1,0,1,2]$.

For 4x oversampling: $S_F = 4(2f_b) = 16\omega_1$ and $N = 16$. Since there is no aliasing,

$$H_s[k] = H_{DFS}[k] = \frac{1}{2}\mathrm{sinc}\left(\frac{k}{2}\right) = \frac{1}{2}\mathrm{sinc}\left(\frac{1}{2/16}\frac{k}{16}\right),\ k = -2,-1,0,1,2.\ \ |H_{DFS}[k]|,\ T_1 = 4T_p,$$

$$H_{DFT}[k] = N \cdot H_{DFS}[k], T_1 = t_s N = \frac{1}{S_F}N,$$

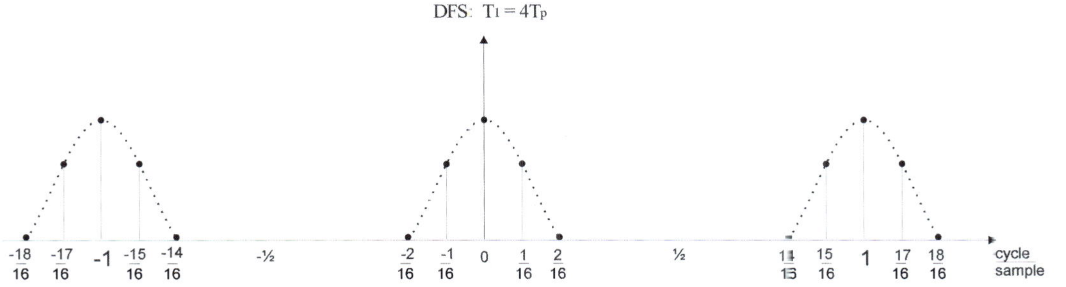

Fig. Hw8-20. DFS magnitude spectrum for $T_1 = 4T_p$ with 4 times oversampling.

$$H(j\omega)\big|_{\omega=k\omega_1} \approx T_1 X_s[k] = T_1 X_{DFS}[k] = t_s N X_{DFS}[k] = t_s X_{DFT}[k].$$

(ii) $T_1 = 8T_p$,

$f_b = 4\omega_1$ for bandlimiting to center lobe, i.e. k is restricted to $[-4, -3, -2, -1, 0, 1, 2, 3, 4]$.

For 4x oversampling: $S_F = 4(2f_b) = 32\omega_1$ and $N = 32$.

Since no aliasing occurs, $H_s[k] = H_{DFS}[k] = \dfrac{1}{4}\operatorname{sinc}\left(\dfrac{k}{4}\right) = \dfrac{1}{4}\operatorname{sinc}\left(\dfrac{1}{4/32}\dfrac{k}{32}\right)$, $k = [-4, -3, -2, -1, 0, 1, 2, 3, 4]$.

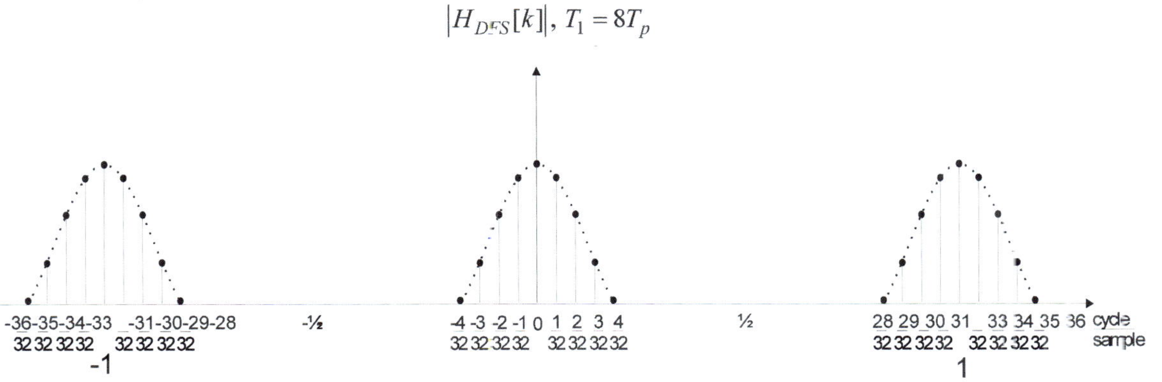

Fig. Hw8-21. DFS magnitude spectrum for $T_1 = 8T_p$ with 4 times oversampling.

So, $H(j\omega)$ can be evaluated by sampling $h(t)$, computing DFT of sampled $h(t)$ and multiplying it by sampling interval t_s.

Summary of formulas for HW #8

1) Fourier Series (FS): used for continuous periodic signal $x_p(t)$ with period T_1

$$X[k] = \frac{1}{T}\int_T x_p(t)e^{-j(2\pi k/T_1)t}dt\,,\ x_p(t) = \sum_{k=-\infty}^{\infty} X[k]e^{j(2\pi k/T_1)t}\,.$$

2) Discrete Fourier series (DFS): used for discrete periodic sequence $x_p[n]$ with period N_1

$$X_{DFS}[k] = \frac{1}{N_1}\sum_{n=0}^{N_1-1} x_p[n]e^{-j(2\pi k/N_1)n},\ x_p[n] = \sum_{k=0}^{N_1-1} X_{DFS}[k]e^{j(2\pi k/N_1)n}. \qquad (*)$$

3) Discrete Fourier transform (DFT): used for discrete usually finite sequence $x[n]$ with length N_1, can be interpreted as a frequency domain sampled version of DTFT (of finite sequence $x[n]$ with length N_1) with interval $2\pi / N_1$ over $0 \le \omega \le 2\pi$

$$X_{DFT}[k] = \sum_{n=0}^{N_1-1} x[n]e^{-j(2\pi k/N_1)n}, \qquad k = 0,\ldots,N_1-1, \qquad (**)$$

$$x[n] = \frac{1}{N_1}\sum_{k=0}^{N_1-1} X_{DFT}[k]e^{j(2\pi k/N_1)n}, \qquad n = 0,\ldots N_1-1.$$

4) If N_1 point DFT is calculated for a periodic sequence of period N_1, the sequence over only $n = 0,\ldots,N_1-1$ should be considered:

$$(*) \text{ and } (**) \Rightarrow X_{DFT}[k] = N_1 \cdot X_{DFS}[k], \qquad k = 0,\ldots,N_1-1.$$

Chapter 6

Linear System Characteristics, Fourier Transform and Data Acquisition, Introduction to Filters

6.1 Linear Systems: Parametric versus Nonparametric Models, Remaining Objectives and Key Relations, Impulse Response of DT System

So far we have been focusing on the well-structured analytical systems and signals descriptions such as, respectively, LCCODEs and Fourier series. Real systems and data, however, are messy, i.e. much less structured. To better handle this increased complexity, we are going to expand our capabilities in four directions: i) system modeling, ii) signal conditioning, iii) diagnostics with the relevant signal processing, and iv) system dynamics modification. This will be accomplished by introducing, respectively, nonparametric system identification and representation and filtering in the present chapter, filter design in Chapter 7, correlation and discrete-continuous spectral analysis in Chapter 8, and controller design in Chapter 9. We will proceed by first reviewing the familiar techniques and then bringing them closer to addressing insufficiently structured systems and data. Unless otherwise indicated, we will use the abbreviations $H(\omega) \equiv H(j\omega)$ and $H(f) \equiv H(j2\pi f)$

6.1.1 *Parametric system models based on physical laws*

The physical systems considered in Chapter 3 can be ***parameterized by physical quantities*** (e.g., mass) and described by differential equations that use these quantities as ***coefficients (parameters)***.

For example, systems represented in terms of components with well-defined dynamics and clear interconnection rules, such as a mass-spring-damper system in Figure 6.1-1) and a resistor-inductor-capacitor (RLC) circuit (Figure 6.1-2) give rise to ***parametric models*** represented by ODEs and/or PDEs. The techniques in this book are applicable to linear time invariant (LTI) systems and nonlinear systems linearizable in some operating ranges. Let us briefly review the properties of LCCODEs.

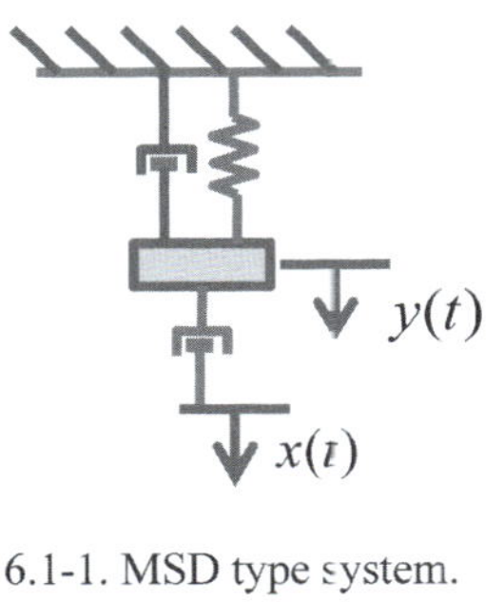

Fig. 6.1-1. MSD type system.

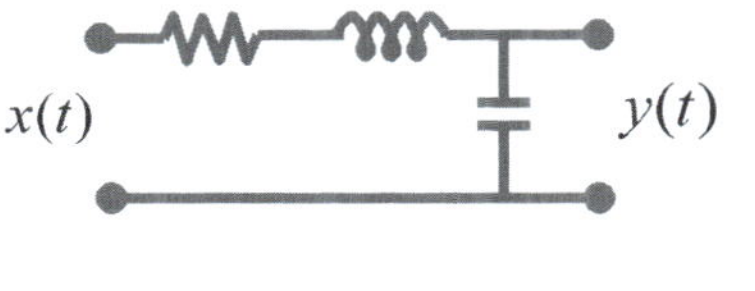

Fig. 6.1-2. Resistor-inductor-capacitor (RLC) circuit.

6.1.1.1 *Parametric system models: linear constant coefficient ODEs*

Consider a linear, constant-coefficient ordinary differential equation (LCCODE) with input $x(t)$ and zero initial conditions:

$$\begin{gathered} y^{(n)}(t)+a_{n-1}y^{(n-1)}(t)+\ldots+a_2y''+a_1y'(t)+a_0y(t) \\ =b_nx^{(n)}(t)+\ldots+b_2x''(t)+b_1x'(t)+b_0x(t),\ y^{(i)}=0,\ i=1,\ldots,n-1. \end{gathered} \tag{6.1-1}$$

Here coefficients a_i*'s* and b_i*'s* are *system parameters*. Let's step back for a moment and look at this ODE as a "black box." We provide an input $x(t)$ to the LCCODE and receive a response $y(t)$.

Even if we don't know the actual equation behavior, we know that the equation linearity and time invariance provide some nice properties reviewed next.

6.1.1.2 *Parametric system models: LCCODE linearity and time invariance*

The ODE ***linearity*** implies that i) ***multiplication by a constant*** is passed through:

and ii) ***sums*** are passed through:

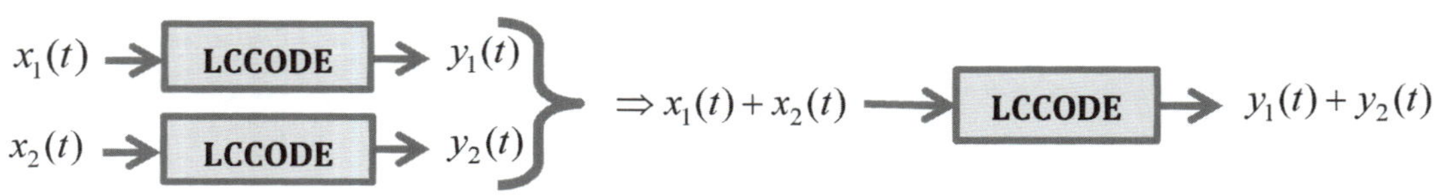

So, instead of solving the ODE with a given input directly, we can break the input up into pieces, sinusoids or delta-functions, for which the ODE is easier to solve.

For example, in Chapter 4 we developed analytical representation of periodic functions in terms of ***Fourier series*** — in general infinite ***sums of sinusoids*** — using Euler's formula and complex exponentials:

$$x_p(t)=\sum_{k=-\infty}^{\infty}X_s[k]e^{jk\omega_1 t}\Leftrightarrow X_s[k]=\frac{1}{T_1}\int_{-\frac{T_1}{2}}^{\frac{T_1}{2}}x_p(t)e^{-jk\omega_1 t}dt. \tag{6.1-2}$$

Due to linearity, this representation can be employed for obtaining particular solutions to a LCCODE by passing through the latter each term comprising the input, and then adding the outputs together:

$$e^{jk\omega_1 t} \rightarrow \boxed{\textbf{LCCODE}} \rightarrow y_k(t)$$

$$\sum_{k=-\infty}^{\infty} X_s[k]e^{jk\omega_1 t} \rightarrow \boxed{\textbf{LCCODE}} \rightarrow \sum_{k=-\infty}^{\infty} X_s[k]y_k(t)$$

The ODE ***time-invariance***, i.e. ODE coefficients being constant, implies that ***time delays*** are passed through.

$$x(t-T) \longrightarrow \boxed{\textbf{LCCODE}} \longrightarrow y(t-T)$$

6.1.1.3 *Parametric system models: LCCODE and rational transfer functions*

In Chapter 3 we reviewed the use of ***single-sided Laplace transform*** as a convenient way to solve LCCODEs and find the behavior of a physical LTI system. We proceeded by converting real functions of real variable, t, (time) into complex functions of complex variable s:

$$X(s) = \int_0^\infty x(t)e^{-st}dt, \qquad s = \sigma + j\omega. \tag{6.1-3}$$

This turns derivatives in time domain into multiplication by s in Laplace domain:

$$L\left[\frac{d}{dt}x(t)\right] = sX(s) - x(0). \tag{6.1-4}$$

The latter permits obtaining input/output LTI system description in terms of ***rational transfer function*** — algebraic relation between input and output transforms in terms of polynomials *parametrized by LCCODE coefficients*:

$$x(t) \longrightarrow \boxed{\textbf{LCCODE}} \longrightarrow y(t)$$

$$y^{(n)} + a_{n-1}y^{(n-1)} + \ldots + a_1 y' + a_0 y = b_n x^{(n)} + b_{n-1}x^{(n-1)} + \ldots + b_1 x' + b_0 x$$

$$\Big\Downarrow \quad \textbf{Laplace Transform}$$

$$s^n Y + a_{n-1}s^{n-1}Y + \ldots + a_1 s^{n-1} s Y + a_0 Y = b_n s^n X + b_{n-1}s^{n-1}X + \ldots + b_1 sX + b_0 X$$

$$Y = \underbrace{\frac{b_n s^n + b_{n-1}s^{n-1} + \ldots + b_1 s + b_0}{s^n + a_{n-1}s^{n-1} + \ldots + a_1 s^{n-1} s + a_0}}_{H(s)\ \longleftarrow\ \textbf{Parametric transfer function}} X \tag{6.1-5}$$

Solving an ODE then amounts to taking the Laplace transform of the input, multiplying it by the transfer function, and then taking the inverse Laplace transform of the resulting product. This technique presumes availability of LCCODEs. Now, we are going to move beyond this setting.

6.1.1.4 *Working with field test data: motivation for Fourier series/transform computation (spectral analysis)*

In the course of the mold oscillator distortion control project several modeling efforts have been made. The results are summed up in Figure 6.1-3.

<u>Hardware testbed model:</u> ODE/PDE for simulation? <u>Yes!</u>
From physical laws, geometric measurements, material properties: *very time consuming, requires advanced modeling background – <u>that's our software testbed!</u>*

<u>Real system model:</u> ODE/PDE for simulation? From physical laws? From finite element code?
<u>No way!</u>
<u>Only nonparametric from test data!</u>

Fig. 6.1-3. Hardware testbed system and real system modeling summary.

This experience tells us that finding system dynamics and changes in the latter over time and carrying out system diagnostics, all based on data, could often be the only way to deal with the real physical system complexity. With this motivation, we will focus on the following response types and related tools capable of yielding nonparametric models and providing their diagnostics:

(1) *Impulse response*, convolution, CT Fourier transform
(2) *Sine response* (*frequency response*) at distinct frequencies, CT Fourier series and transform
(3) *Random excitation response,* correlation, CT Fourier transform, DFT
(4) *Noisy response with sinusoids*, correlation, CT Fourier transform, DFT, mixed spectra.

6.1.1.5 *From parametric to nonparametric models: impulse response and convolution for general linear time-invariant systems*

Convolution integral provides response to an ***arbitrary input*** in terms of ***the system impulse response*** not only for LCCODEs, but for arbitrary linear time-invariant (LTI) systems:

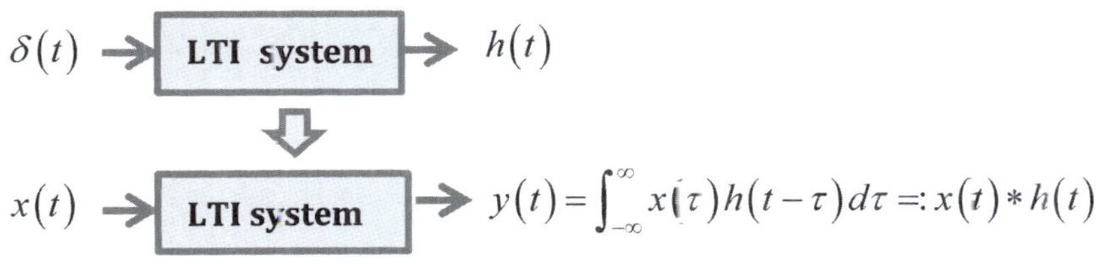

Since Fourier transform is an LTI map itself, convolution in the time domain is mapped into multiplication of the constituent signal transforms in the frequency domain for any LTI system:

$$L[x(t) * h(t)] = X(\omega)H(\omega) \qquad (6.1\text{-}6)$$

where $H(j\omega)$ is Fourier transform of system impulse response, or system transfer function:

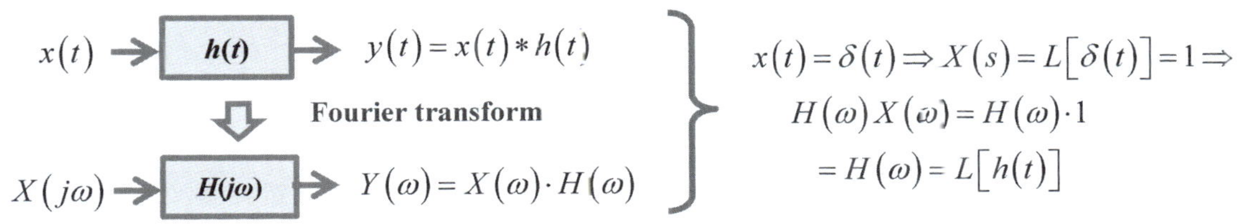

This implies that applying impulse input and computing Fourier transform of the response yields ***nonparametric*** LTI system transfer function, i.e. the one not parametrized by any physical quantity.

6.1.1.6 *From parametric to nonparametric models: LTI system frequency response*

Another LTI system property key to nonparametric modeling is system action on a periodic signal:

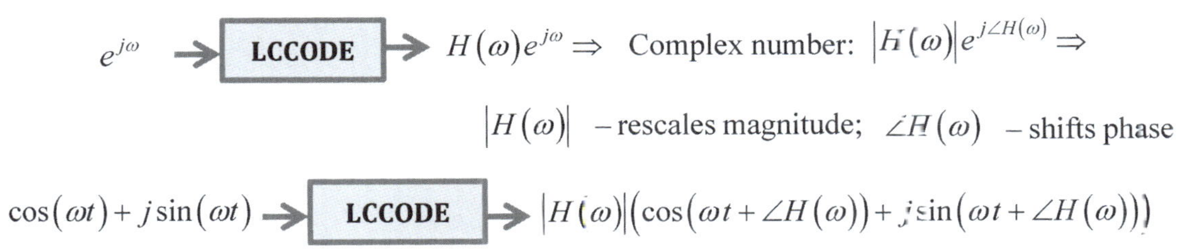

This action, referred to as ***frequency response***, was obtained for a LLCCODE analytically by evaluating the corresponding Laplace transform based transfer function on the $j\omega$-axis:

$$H(\omega) = \int_0^{\infty} h(t)e^{-j\omega t}dt = H(s)\big|_{s=j\omega}. \qquad (6.1\text{-}7)$$

However, this response can be obtained for any LTI system experimentally through making a table of the physical system ***filtering*** properties: amplification and attenuation gains along with the phase shifts at a range of frequencies, yielding a nonparametric system model.

6.1.2 *Working with noisy data and poorly parametrizable systems: key objectives*

So far we have introduced the following tools for signal processing and system representation:

i) *convolution integral* — calculates system response to arbitrary input through impulse response,

ii) *continuous and discrete Fourier series and transforms* — 1) calculate frequency content of signals with discrete spectra, for example, for capturing small magnitude periodic components, such as those in Examples 5.7-1–5.7-6, 2) calculate and approximate continuous spectra and frequency responses of systems, for example, to find and quantify resonance properties of nonparametric models, such as mold oscillator frequency response in Figure 1.6-8.

Thus, these tools provide capability for analyzing multiscale data and using deterministic signals, impulsive or periodic, to analyze, identify, and diagnose the time domain behavior of an LTI system not necessarily well parametrizable though fundamental physical laws. However, in real systems the test data are combinations of periodic, non-periodic, and random signals.

To enhance the attained capability in the presence of these signals we are going to focus on:

- ***analog and digital filter design*** for acquiring good field test data
- ***discrete/continuous spectral analysis*** for field test data *features extraction* to bring out signal composition, diagnose systems, and identify nonparametric system models
- ***feedback/feedforward controller design*** for shaping the filtering action and the time domain behavior of a real physical system with available model or without it.

To consistently address these topics we will systematize Fourier transform techniques and build analytical tools for analog and digital filter design.

6.1.2.1 *Key relations: nonparametric (only data based) modeling tool*

Problem: obtain frequency response from test data. We can use the following approaches.

i) Impulse input, nonparametric model. Compute Fourier transform of impulse response:

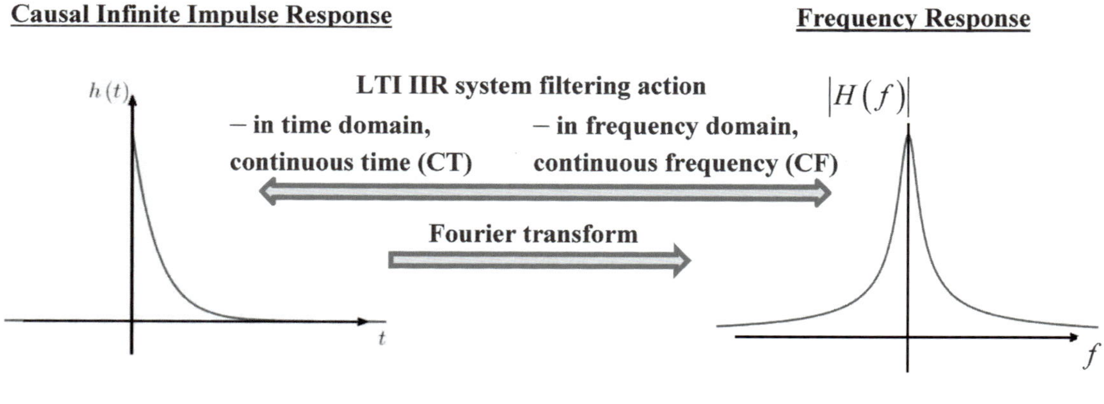

Fig. 6.1-4. Relation between CT LTI system impulse response and frequency response.

ii) Impulse input, parametric model, simple dynamics: fit a model to the impulse response, and calculate the corresponding transfer function. This method requires knowing beforehand the general model of the system, e.g. underdamped second-order one, as shown in Figure 6.1-5.

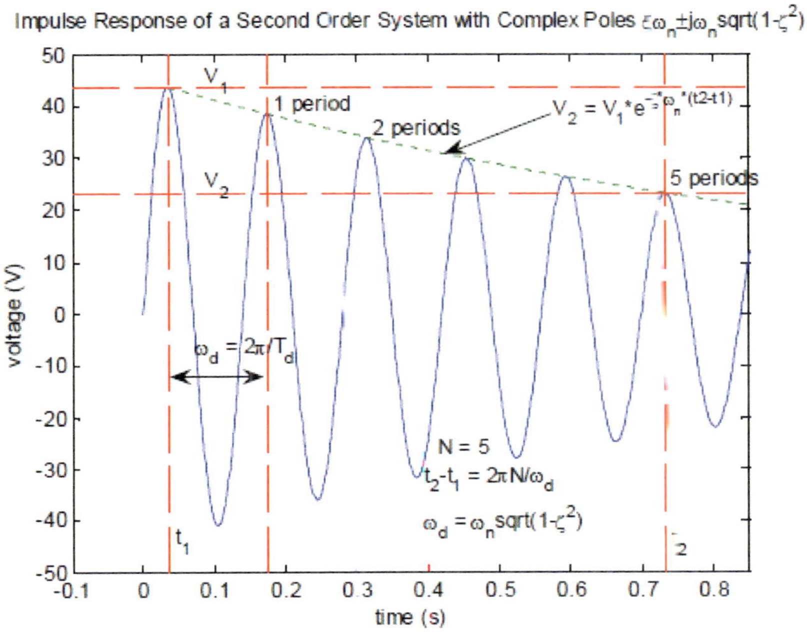

Fig. 6.1-5. Impulse response of a damped second order system.

iii) Sine input (sine-sweep): Provide to the system a sinusoidal input, and slowly increase its frequency. Measure the output magnitude and phase as shown in Figure 6.1-6. This method is useful, but time-consuming. Also, system excitation at the frequencies of interest can cause damage. This method was used to build the beam assembly waterfall plot, and is currently used in production resonance controller tuning, but care is exercised to avoid beam damage at resonance.

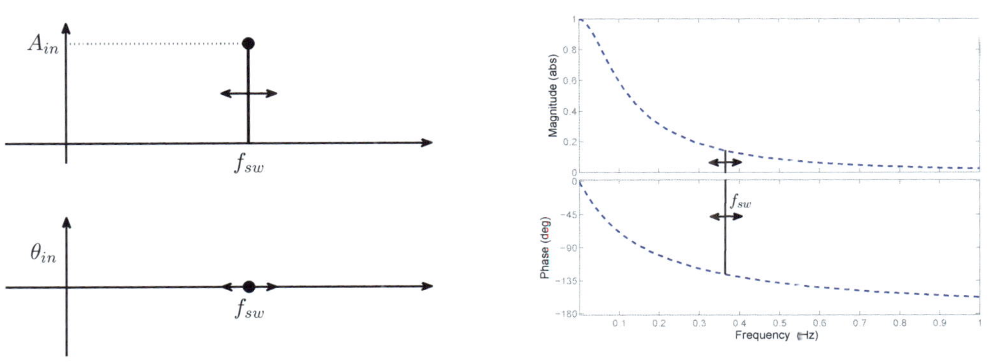

Fig. 6.1-6. Frequency response - nonparametric LTI system model identified through swept-sine input.

6.1.2.2 *The key signal processing and system modeling relations: calculating nonparametric model from impulse response*

Since data record is finite, and, hence, impulse response is truncated, we take FIR approach — define the T_1-periodic extension of the truncated impulse response and calculate Fourier series for fundamental frequency $\omega_1 = 2\pi/T_1 = 2\pi f_1$. The Fourier series coefficients can then be used to find the "sampled" values of the Fourier Transform: $H(k\omega_1) \equiv H(kf_1) \approx T_1 H_s[k] = T_1 H_s[(1/f_1)kf_1] \equiv H_t(kf_1)$, approximate due to truncation (hence subscript t in $H_t(kf_1)$), as shown in Figure 6.1-7.

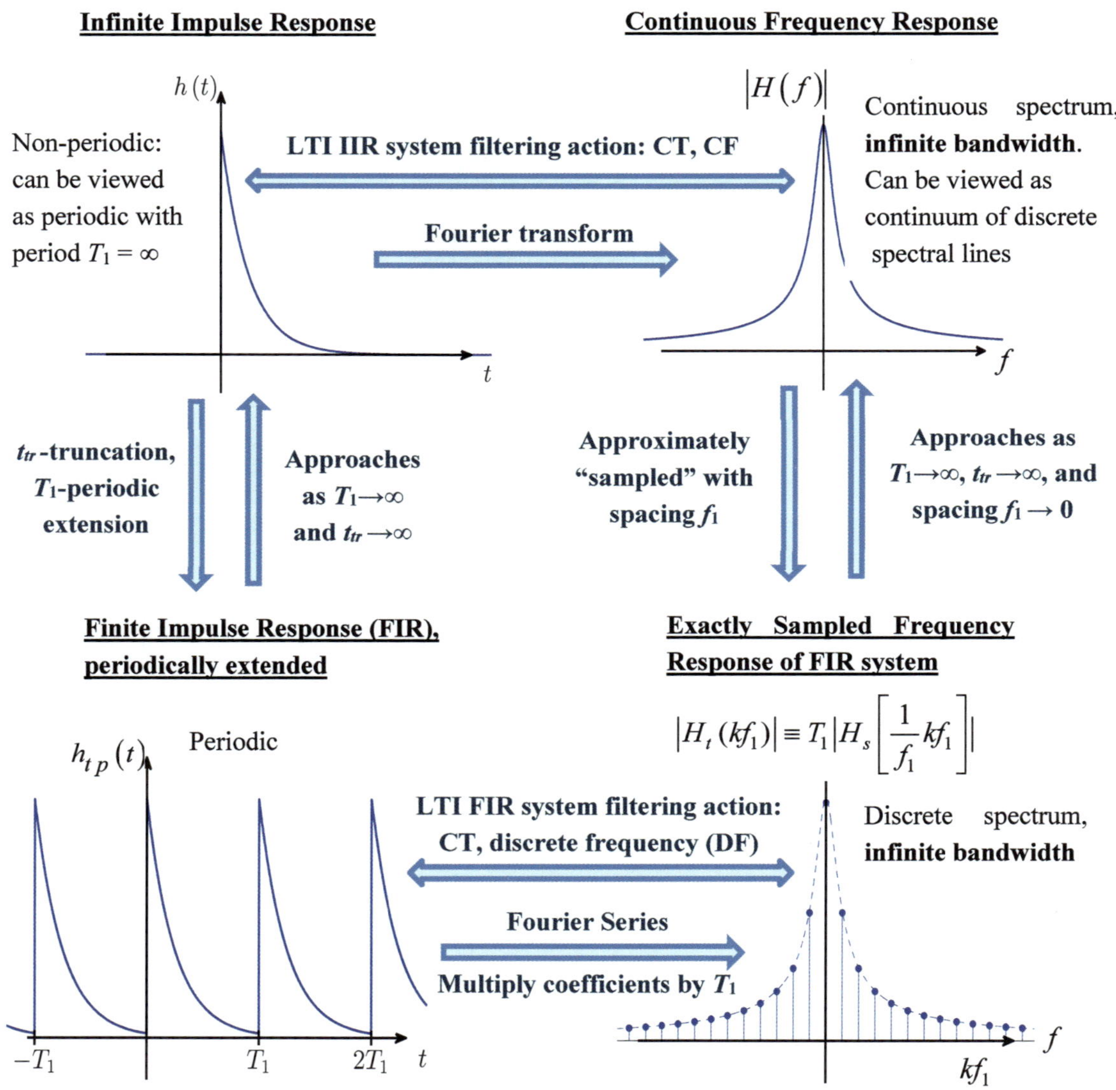

Fig. 6.1-7. Frequency response — nonparametric LTI model identified through impulse response and the corresponding Fourier series and transform calculations.

6.1.2.3 *Key relations: recap*

Both CT and DT Fourier series and transforms are based on writing a periodic or nonperiodic signal as a sum of discrete frequency sinusoids or integral of a continuum of sinusoids. For truncated periodically extended LTI system impulse response the key relations are summarized below.

Parenthesis and brackets indicate that the variables are defined on the continuous and integer value sets, respectively.

Fourier Series:

$$H_s[k]=\frac{1}{T_1}\int_0^{T_1} h_{tp}(t)e^{-jk2\pi\frac{1}{T_1}t}dt .$$

$$h_{tp}(t)=\sum_{k=-\infty}^{\infty} H_s[k]e^{jk2\pi\frac{1}{T_1}t} .$$

- $h_{tp}(t)$ is periodic with period T_1
- The Fourier series may have an infinite number of non-zero terms
- If the Fourier series has a finite number of non-zero terms, we say $h_{tp}(t)$ is "bandlimited"

Discrete Fourier Series:

$$H_{DFS}[k]=\frac{1}{N_1}\sum_{n=0}^{N_1-1} h_{tp}[n]e^{-jk2\pi\frac{1}{N_1}n},$$

$$0\le k\le N_1-1.$$

$$h_{tp}[n]=\sum_{k=0}^{N_1-1} H_{DFS}[k]e^{jk2\pi\frac{1}{N_1}n},$$

$$0\le n\le N_1-1.$$

- $h_{tp}[n]$ be periodic with period N
- The Discrete Fourier Series will only have a finite number of terms, because of aliasing

Fourier Transform:

$$H(\omega)=\int_0^{\infty} h(t)e^{-j\omega t}dt=H(s)\big|_{s=j\omega}$$

or

$$H(f)=\int_0^{\infty} h(t)e^{-j2\pi ft}dt=H(s)\big|_{s=j2\pi f},$$

$h(t)$ is finite or infinite duration.

Discrete Fourier Transform:

$$H_{DFT}[k]=\sum_{n=0}^{N_1-1} h_{tp}[n]e^{-jk2\pi\frac{1}{N_1}n},$$

$$0\le k\le N_1-1.$$

$h_{tp}[n]$ is finite duration of length N_1.

We can now perform DT *finite duration* data spectral calculation through DFS and DFT.

But how about a DT system with *infinite duration* impulse response. Is there a counterpart of a CT IIR filter in DT? What is its frequency response? To answer these questions, assume that the DT impulse response comes from sampling of a CT impulse response, as shown in Figure 6.1-8.

Then, as Figure 6.1-8 indicates, the corresponding DT frequency response will be fundamentally distorted with respect to its CT counterpart due to aliasing. The latter is caused by sampling the infinite bandwidth CT impulse response with finite sampling frequency.

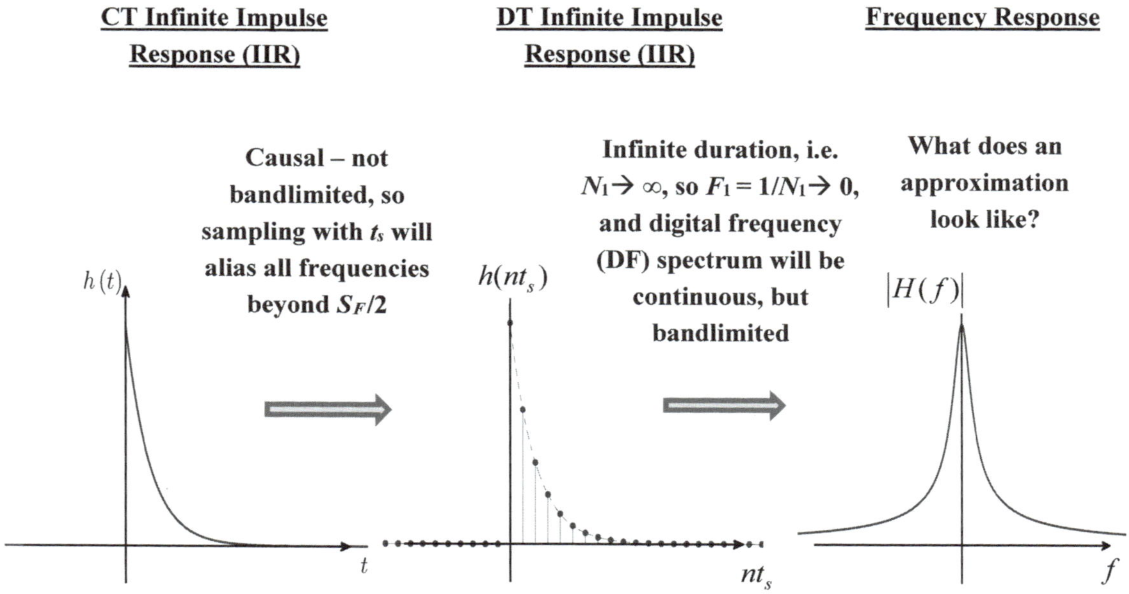

Fig. 6.1-8. Obtaining DT IIR system frequency response from sampled CT infinite impulse response.

Because DT IIR has infinite duration in n, and so can't be periodically extended in DT, any variant of Fourier Transform of this DT IIR will have ***continuous digital frequency F and map into continuous spectrum in [-$S_F/2$, $S_F/2$] on analog frequency axis f.***

6.2 Discrete Time Fourier Transform (DTFT), Fourier Transform: Theory and Practice. Modern PCM and PDM Based Data Acquisition (DAQ) Systems

6.2.1 *Key relations: frequency response of DT system and its relation to frequency response of CT system*

To find frequency response of DT systems with infinite duration impulse response, let's try, in DT, the approximation we used to obtain the CT Fourier Transform, and see what comes out.

Not bandlimited

1. CT infinite duration impulse response:

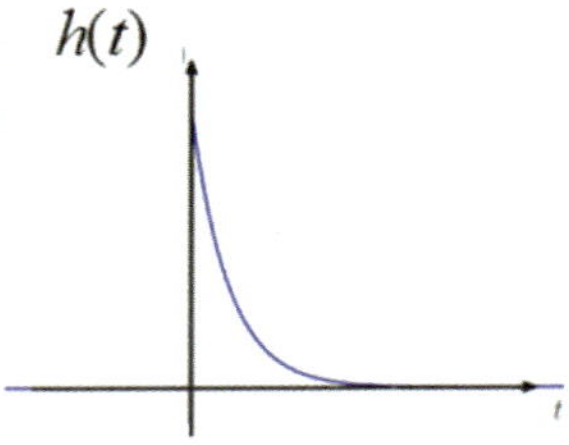

2. Its truncation to t_{tr} and T_1-periodic extension,

$t_{tr} \le T_1$:

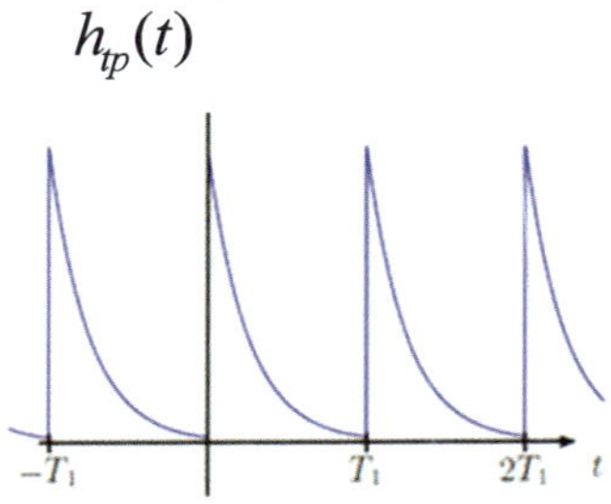

3. Fourier series of periodic extension:

$$H_s[k]=\frac{1}{T_1}\int_0^{T_1} h_{tp}(t)e^{-j2\pi\frac{k}{T_1}t}dt,$$ **multiplying by T_1 :**

$$|H_t(kf_1)|=T_1\left|H_s\left[\frac{1}{f_1}kf_1\right]\right|$$

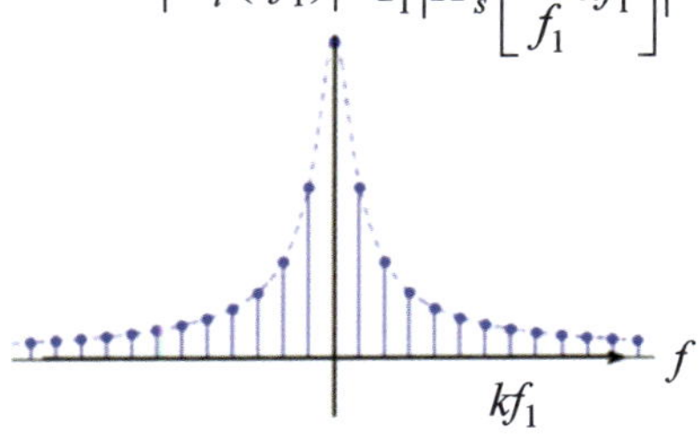

4. Retracting periodic extension: letting $T_1\to\infty$, $t_{tr}\to\infty$, $kf_1\to f$:

$$T_1H_s[k]\to\int_{-\infty}^{\infty}h(t)e^{-j2\pi ft}dt=H(f)$$

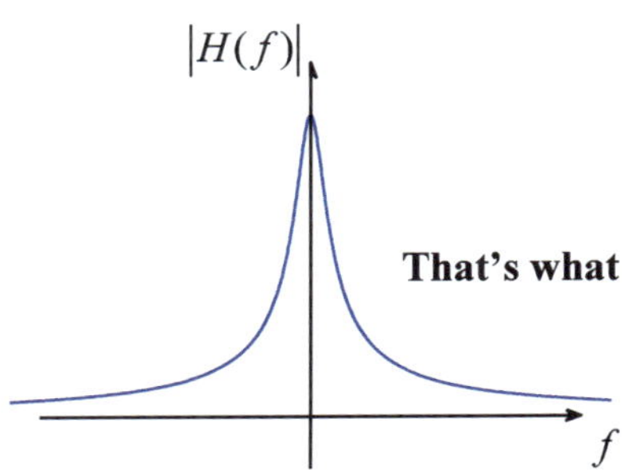

Sampled, bandlimited, aliased

1. DT infinite duration impulse response:

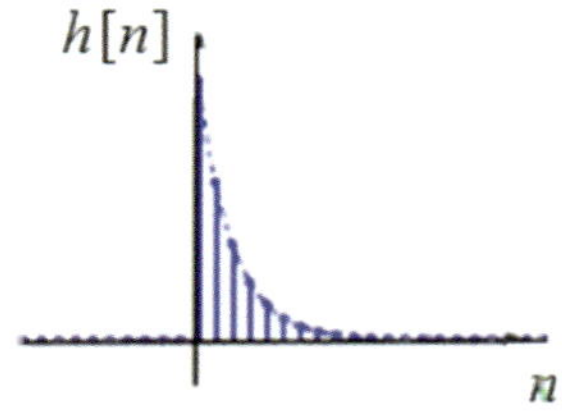

2. Its truncation to n_{tr} and N_1-periodic extension,

$n_{tr} \le N_1$:

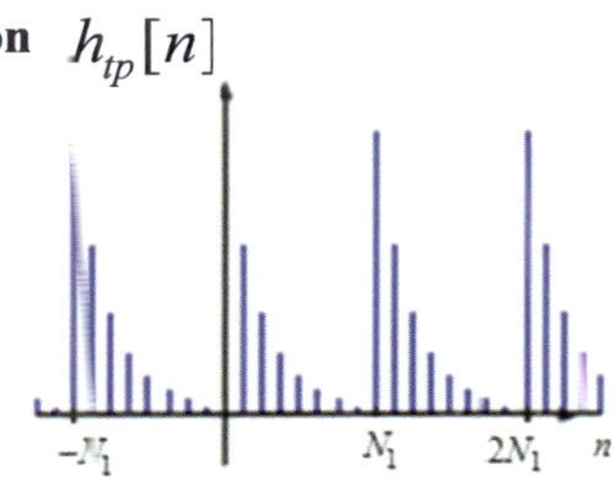

3. Discrete Fourier series of periodic extension:

$$H_{DFS}[k]=\frac{1}{N_1}\sum_{n=0}^{N_1-1}h_{tp}[n]e^{-j2\pi\frac{k}{N_1}n},$$ **multiplying by N_1:**

$$\left|H_{tp}\left(kF_1=k\frac{f_1}{S_F}\right)\right|=N_1\left|H_{DFS}\left[\frac{1}{f_1}kf_1\right]\right|$$

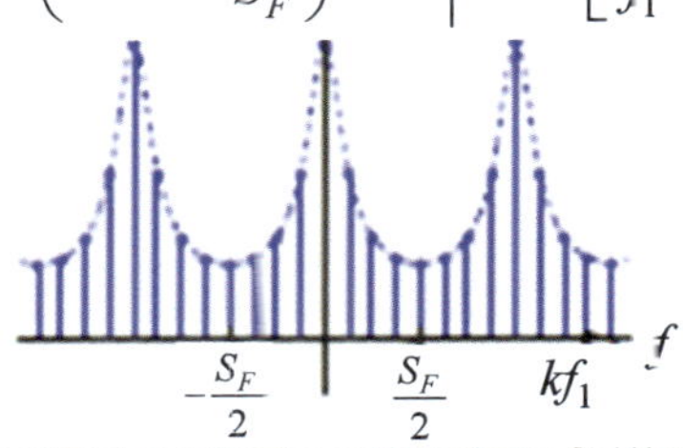

4. Retracting periodic extension: letting $N_1\to\infty$, $n_{tr}\to\infty$, $kF_1\to F$:

$$N_1H_{DFS}[k]\to\sum_{n=-\infty}^{\infty}h[n]e^{-j2\pi Fn}=H_p(F)$$

$$\left|H_p\left(F=\frac{f}{S_F}\right)\right|$$

That's what approximation looks like!

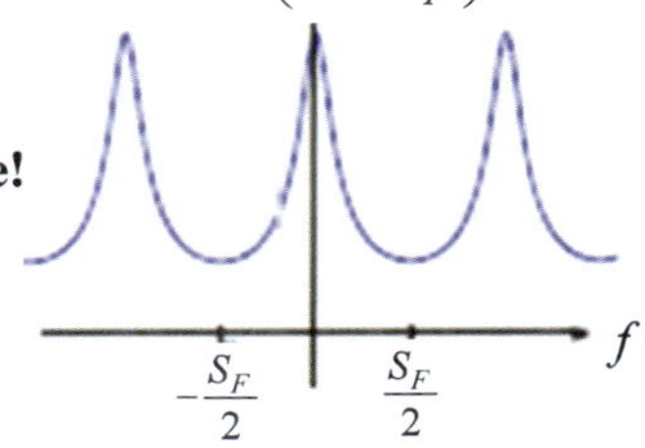

Fig. 6.2-1. CT system frequency response approximation through the use of infinite duration sampled impulse response.

6.2.2 *Discrete Time Fourier transform (DTFT) from DFS*

The final relation in Figure 6.2-1 is called the **Discrete Time Fourier Transform** (DTFT):

$$X_p(F) \equiv X_p\left(e^{j2\pi F}\right) = \sum_{n=-\infty}^{\infty} x[n]e^{-j2\pi Fn}. \tag{6.2-1}$$

Note that just like the Fourier transform, the Discrete Time Fourier transform is defined for the continuum of frequencies, but in [−1/2, 1/2] interval on digital frequency axis $F = f/S_F$ which can be mapped into $[-S_F/2, S_F/2]$ on analog frequency axis f.

Since $e^{j2\pi(F+1)n} = e^{j2\pi Fn}e^{j2\pi n} = e^{j2\pi Fn}$, the Discrete Time Fourier Transform is periodic (hence subscript p) with period 1 on digital frequency axis F (aliasing, again).

Example 6.2-1: Derive the DTFT of the signal shown in Figure 6.2-2.

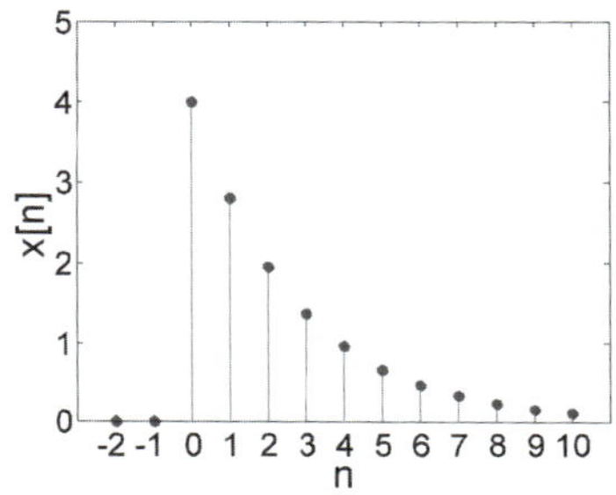

$$x[n] = \alpha^n u[n], \quad |\alpha| < 1$$

Fig. 6.2-2. DT signal $x[n]$.

Remember from Lecture 3 example, this is the sampled version of $e^{-\sigma t}$ for $\sigma > 0$. Its DFFT is

$$X_p(F) = \sum_{n=0}^{\infty} \alpha^n e^{-j2\pi nF} = \sum_{n=0}^{\infty}\left(\alpha e^{-j2\pi F}\right)^n = \frac{1}{1-\alpha e^{-j2\pi F}}. \tag{6.2-2}$$

This is also the impulse response of the first order constant coefficient discrete time linear system (to be looked at in the next chapter).

6.2.2.1 *Discrete Time Fourier Transform: relation to Continuous Time Fourier Transform and DFT*

Let's now take a look at Fourier Transform:

$$X(\omega) \equiv X(j\omega) = \int_{-\infty}^{\infty} x(t)e^{-j\omega t}dt \text{ or } X(f) \equiv X(j2\pi f) = \int_{-\infty}^{\infty} x(t)e^{-j2\pi ft}dt. \tag{6.2-3}$$

Let's bandlimit $x(t)$ (can be done only approximately), sample it (neglecting small aliasing), and compute the Fourier Transform of the sampled signal:

$$X(f)=\int_{-\infty}^{\infty} x(t)e^{-j2\pi ft}dt \xrightarrow[\substack{f_{\max}\le f_b,\, t\to nt_s,\, S_F=1/t_s,\\ \int\to\Sigma,\ dt\to t_s,\ x(t)\to x(nt_s),\\ f\in(-\infty,\infty)\to F=\frac{f}{S_F}\in(-1/2,1/2)}]{} t_s\sum_{n=-\infty}^{\infty} x(nt_s)e^{-j2\pi fnt_s}$$

$$=t_s\sum_{n=-\infty}^{\infty} x(nt_s)e^{-j2\pi\frac{f}{S_F}n}=t_s\sum_{n=-\infty}^{\infty} x(nt_s)e^{-j2\pi Fn}=t_sX_p(F)\Rightarrow$$

truncate $x(nt_s)$, periodically extend truncated signal, get discrete spectrum to sample $X_p(F)$:

$$t_sX_p(F)_{F=k/N_1}\approx t_sX_{tp}(F)_{F=k/N_1}=t_sX_{DFT}[k]\approx X\left(k\frac{1}{N_1t_s}\right)=X\left(k\frac{1}{T_1}\right)=X(kf_1). \quad (6.2\text{-}4)$$

distorted due to truncation (first $\approx$) — distorted due to bandlimiting and aliasing (second $\approx$)

6.2.3 *Key time-domain/frequency-domain relations for processing test data*

6.2.3.1 *Interrelating DT system impulse and frequency responses using DTFT and DFT*

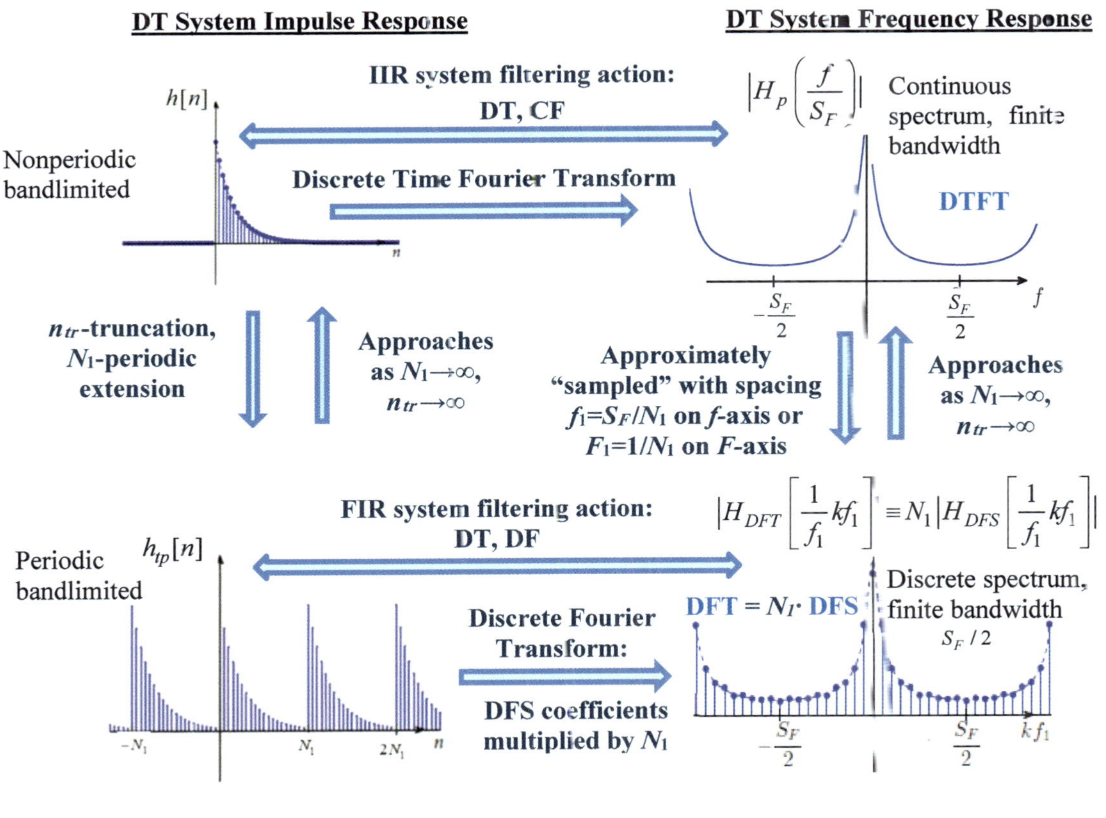

Fig. 6.2-3. Obtaining DT filter frequency response approximation through DFT.

6.2.3.2 *Estimating the CT system frequency response (FT) from its sampled IIR and FIR impulse responses using DTFT and DFT with small truncation error*

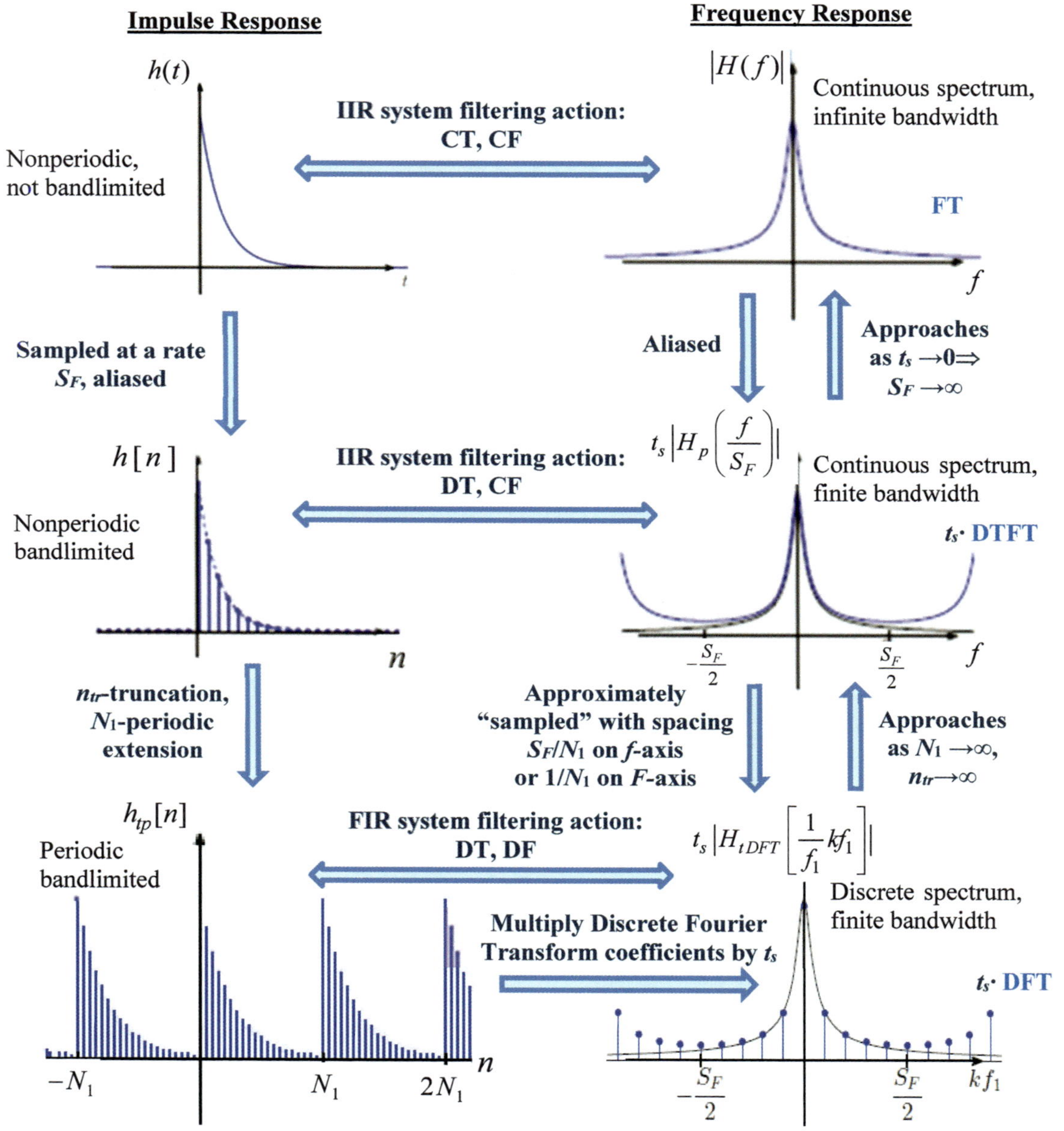

Fig. 6.2-4. Relation between CT IIR and DT IIR and FIR frequency responses in terms of FT, DTFT, and DFT with small truncation error.

6.2.3.3 *Estimating the CT system frequency response (FT) from its sampled IIR and FIR impulse responses using DTFT, distorted DTFT, and DFT with large truncation error*

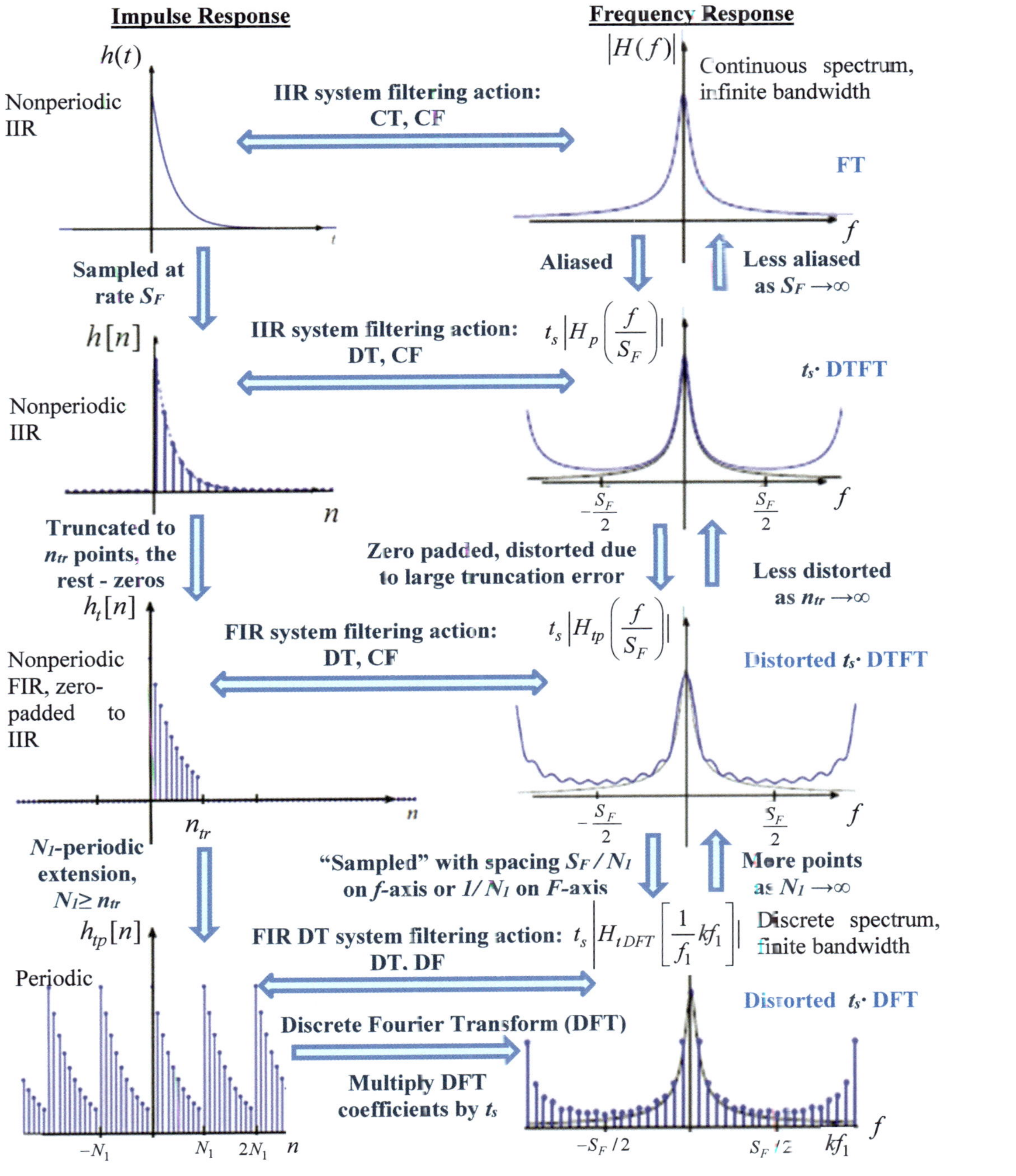

Fig. 6.2-5. Relation between CT IIR and DT IIR and FIR frequency responses in terms of FT, DTFT, distorted DTFT, and DFT with large truncation error.

6.2.3.4 *Putting FT, DTFT, distorted DTFT, and DFT together: analytical portrait*

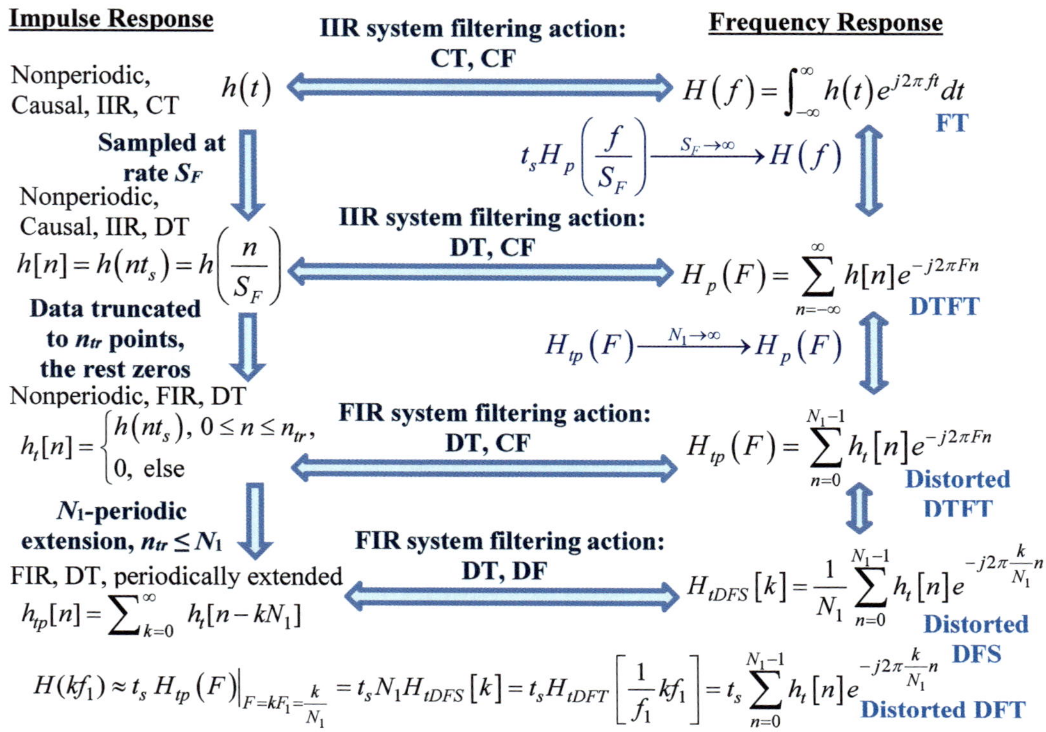

6.2.3.5 *From impulse response to frequency response: practical shortcut*

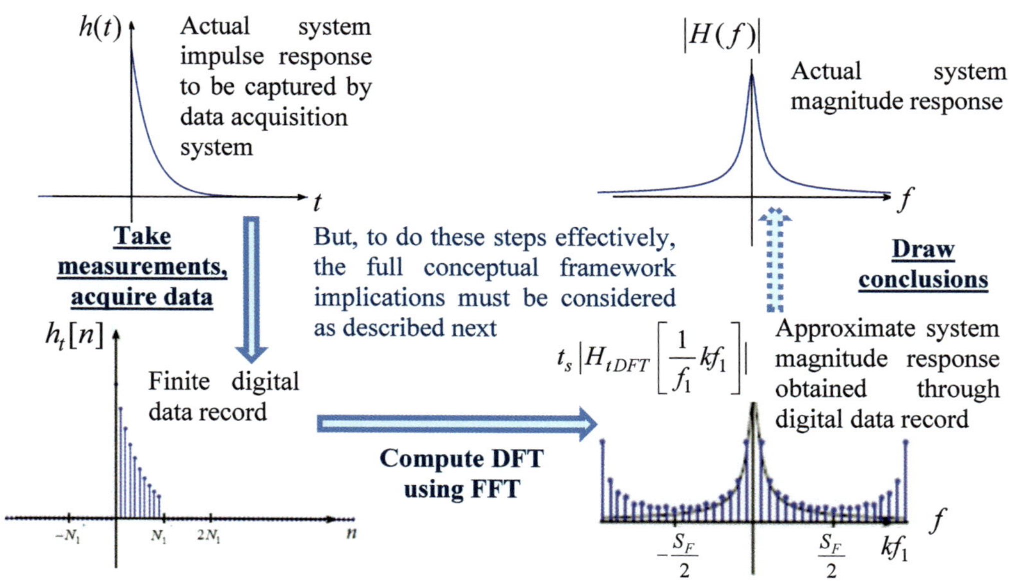

Fig. 6.2-6. Estimating IIR CT system frequency response in terms of DFT: computational framework.

6.2.3.6 *From impulse response to frequency response: practical issues*

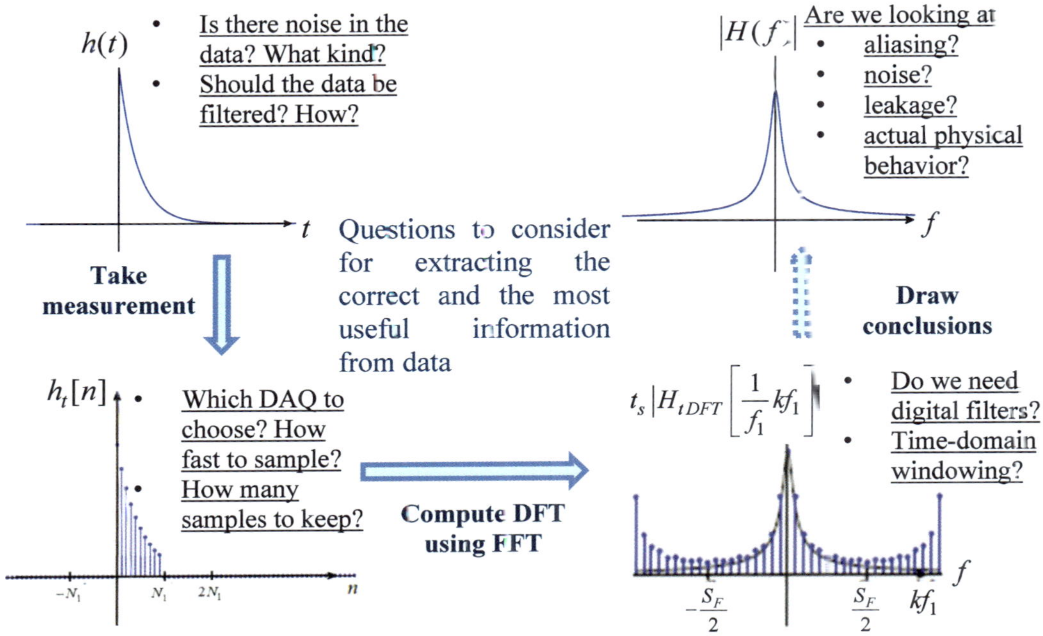

Fig. 6.2-7. Estimating IIR CT system frequency response in terms of DFT: practical issues.

6.2.3.7 *PCM versus PDM based data acquisition system (DAS, DAQ) selection*

The main criterion for a DAQ selection is the frequency range of interest, the fidelity required, the data size, and the application. They determine the sampling frequency $S_F = 1/t_s$ and the antialiasing filter type and order. The latter are based on the relation of the magnitude spectral density of the original CT signal (the upper left graph) to those of the scaled by t_s DTFTs of the DT signals (cf. Figure 6.2-5) for different S_F values, as sketched in Figure 6.2-8. In this figure, the upper bound on the frequency range of interest and the signal bandlimit (theoretically nonexistent) are denoted as $f_{\max}$ Hz, and f_B Hz, respectively. The CT signal is first passed through an antialiasing low pass filter to attenuate the high frequencies beyond $f_{\max}$, and only then the low-pass-filtered signal is sampled with the frequency $S_F = 1/t_s > 2\,f_{\max}$. The upper right plot depicts aliasing for $2f_B > S_F > 2\,f_{\max}$.

The figure shows how to ease the real low pass filter drawbacks: 1) a sharper roll off leading to a higher filter order and a larger passband phase distortion (Chapter 7), 2) inability to completely block the noise, and 3) passband distortion. The lower two graphs in Figure 6.2-8 show that a larger oversampling with $S_F >> 2\,f_{\max}$ enables the use of filters with a larger passband, a much more gradual roll-off, and a much larger signal attenuation at the sampling frequency, all resulting in a significantly lower distortion of the sampled signal in the frequency range of interest. Linear-phase Bessel filters (Chapter 7) provide further phase distortion reduction, desirable in audio applications, whereas for mechanical measurements magnitude is central, so Cauer filters are fine.

In PCM, DAQ data volume sharply increases with the sampling rate and bit depth increase.

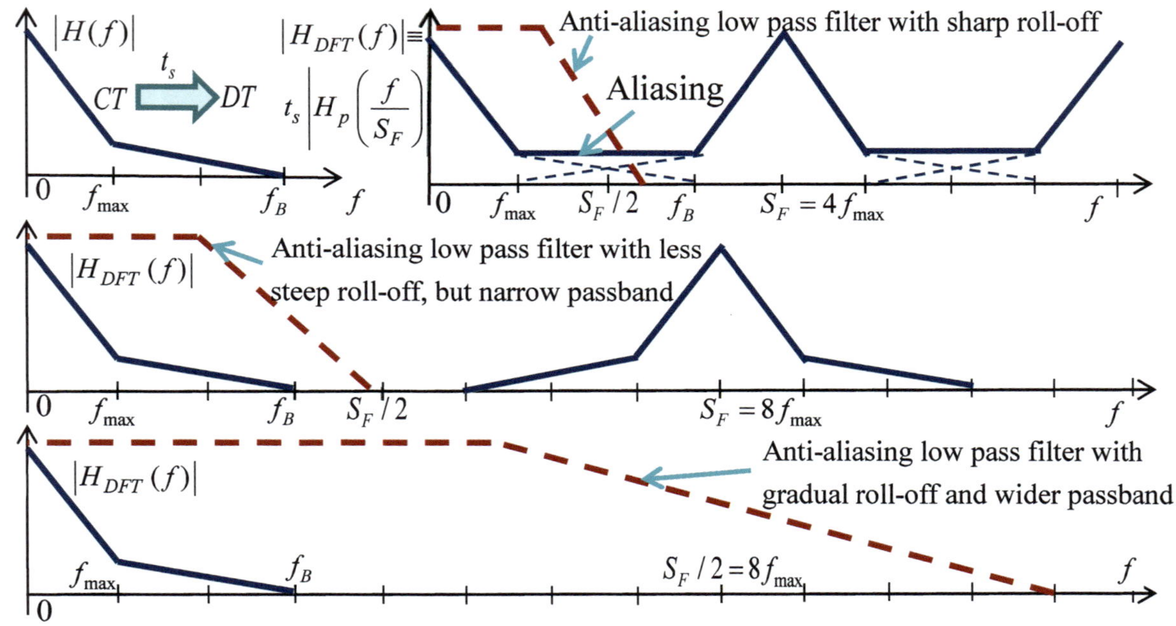

Fig. 6.2-8. The interplay between the sampling frequency and the antialiasing filter selection.

PDM partially alleviates this problem through the negative feedback based delta-sigma modulation (DSM). Figure 6.2-9 shows the key DSM components: 1. Differencer (differencing op amp) with a positive lead (pin), to which the input signal voltage is applied, and a negative lead, to which an output of the feedback return path is applied. 2. Integrator of the differencer output. 3. Clocked latched comparator generating 0 or 1 depending on the sign of the difference between the input and the reference voltages, and switching the data flip-flop (DFF) to latch the value until the next change. 4. 1-bit DAC (digital-to-analog converter) matching output $-V_{DR}$ to input 0 and $+V_{DR}$ to input 1. This topology makes DSM work by bouncing around the analog input signal as follows.

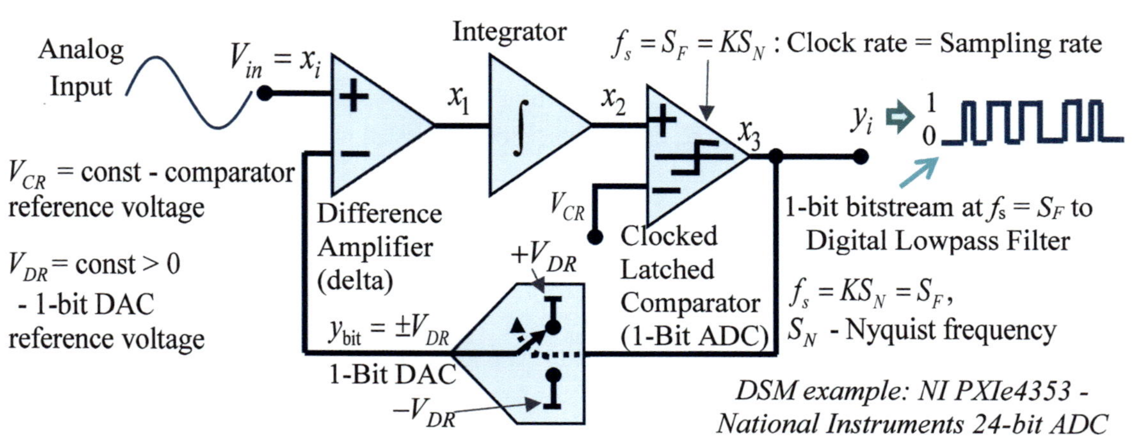

Fig. 6.2-9. Delta-sigma ADC modulator negative feedback topology and its components. Oversampling factor $K >> 1$.

For simplicity, set $V_{CR} = 0$ and limit V_{in} to $\pm V_{DR}$. Also, set the integrator gain to reach output $x_2 = V_{DR}$ for V_{DR} const. input during one clock cycle starting from integrator zero initial condition (i.c.). DSM then works as follows. 1. At the clock cycle rising edge, 1-bit DAC takes the current bitsteam output value y_i and converts it to the corresponding analog value $y_{bit} = \pm V_{DR}$. 2 y_{bit} is then subtracted by the differencer from V_{in} to obtain output x_1. 3. For $f_s >> S_N$, x_1 is near constant during one cycle, so integrator generates an up (for positive x_1) or down (for negative x_1) ramp, and after one cycle reproduces x_1 and adds it to its i.c., that resulted from the previous integration step, to form x_2. 4. At the end of the latter cycle, a comparator latches y_i as logic 1 if x_2 is greater than or equal to zero volts and as logic 0 otherwise. 5. At the next rising clock edge, the latter y_i is again fed to a 1-bit DAC, making the differencer reduce or increase the incoming signal V_{in} by V_{DR} for, respectively, logic 1 and logic 0 at 1-bit DAC input, often in the direction opposite to that of the previous cycle.

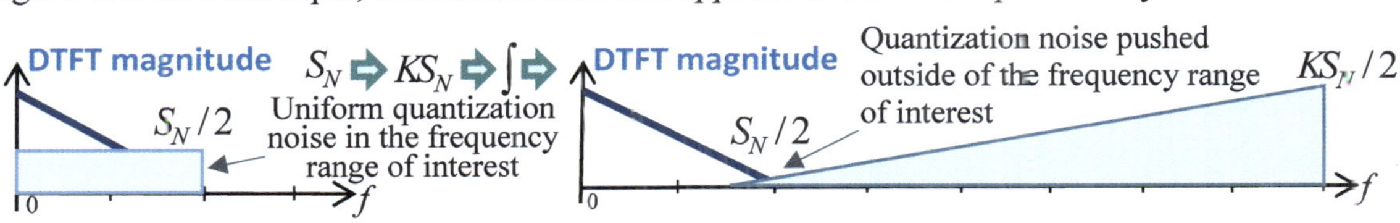

Fig. 6.2-10. Oversampling S_N by a factor of K and integrating pushes the quantization noise out towards $KS_N/2$.

$\pm V_{DR}$ bouncing generates large uniform quantization noise with similar number and size of positive and negative quantization errors across the frequency range of interest [0, $S_N/2$] Hz, which add up to much smaller values in this range when passed though the integrator as shown in Figure 6.2-10. The Simscape DSM ADC schematics is given below. The *PDM and the PCM digital-to-analog converters (DACs)* structures shown below are essentially the *reverses of the corresponding ADCs*.

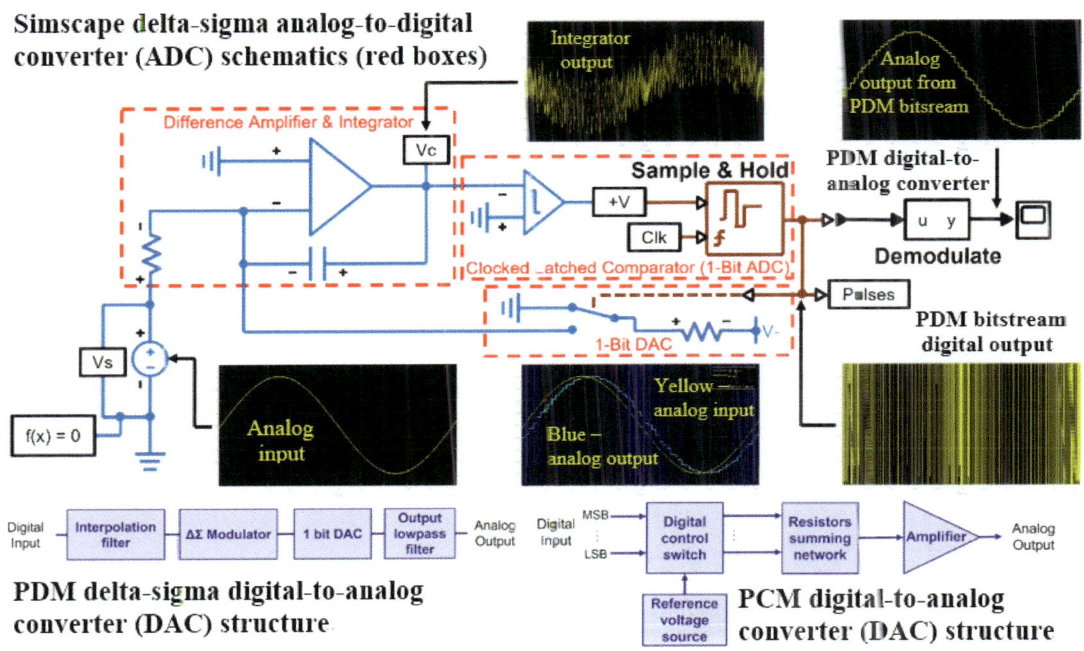

6.3 Introduction to Nonparametric System Modeling: Convolution as Both Impulse and Frequency Response Based Filtering of CT and DT Signals

6.3.1 *Working with field test data: linearity and time-invariance admit standard response-based signal processing techniques for finding nonparametric models*

In industrial applications system representation, changes in dynamics over time, and signal characteristics can be often obtained only though collecting field test data, as indicated in Figure 6.3-1. Often system linearity and time-invariance permit application of standard tools as follows.
1) *Impulse response* is processed through Fourier Transform to obtain transfer function.
2) *Sine response (frequency response)* at a number of frequencies, or, if available, swept-sine response, is processed through Fourier series to obtain envelope that represents Fourier Transform, from which we obtain transfer function. Fourier series also reveals signal distortion.
3) *Random excitation response* is processed through correlation and Fourier Transform.
4) *Noisy response* is processed by correlation and Fourier Transform to reveal possible periodic components in noise, leading to mixed spectral analysis.

Showing how to do all the above is the focus of this book. In mechanical systems we can often apply impulsive input, whereas in process control the test input is typically a step. Sine and random signal excitation are extensively used in both types of systems.

- **These real physical systems are not described best by LCCODEs.**
- **The left one is best described by a system of coupled ODEs and PDEs. The right one can't be described by equations at all!**
- **Both can be said to be linear-like in response to input: scaling of input gives about the same scaling of the output in the operating range of oscillation magnitudes.**
- **Both systems don't change much over time, so time shift in input produces the same response, shifted by the same value.**
- **Systems like this are called Linear Time Invariant, and all the response-based techniques indicated above are applicable to them.**

Fig. 6.3-1. Field test data and system linearity and time-invariance permit modeling of systems not modelable through finite element techniques and/or fundamental physical laws by applying standard signal processing techniques.

How to get field test data: data acquisition system at Nucor

Marketplace offers a great variety of data acquisition systems. Nucor system looks as follows:

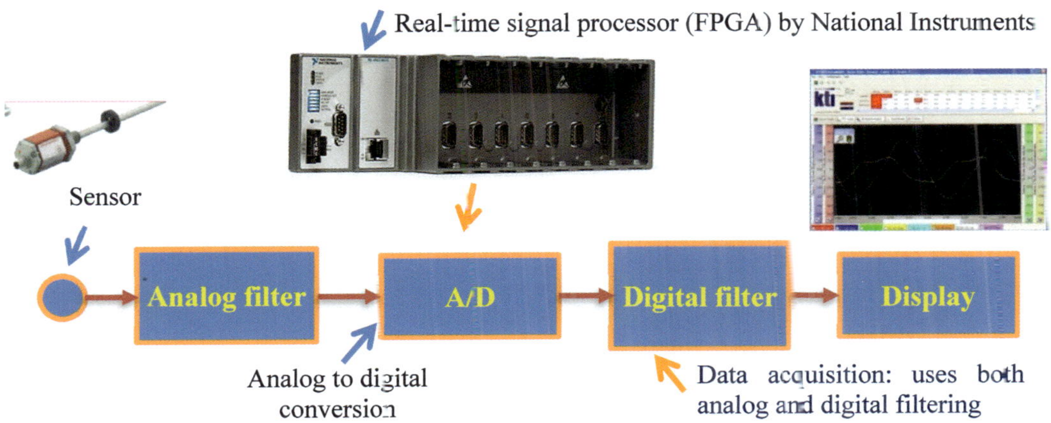

Fig. 6.3-2. Schematic layout of data acquisition system at Nucor.

6.3.2 *Building a noncausal convolution integral*

The **Fourier Transform** is a useful tool to describe and **analyze *signals (spectrum computation) and characterize systems (system identification)***. We were able to use it for finding small periodic components hidden in the piston output. The **Fourier Transform** also gives us a useful tool to describe and analyze the action of ***systems*** (physical systems, filters, controllers, etc.) **on *signals (filter design)***. First, we recall a key relationship between the input and the output of a *causal* linear, time-invariant system: **The solution of a linear, constant-coefficient ordinary differential equation with an arbitrary input *u*(*t*) is the convolution of the ODE's impulse response with *u(t)*.**

To derive this relationship for the general noncausal signals and systems, *necessary for the understanding of filtering*, will define the general noncausal input/output pairs of weighted impulses and the corresponding system responses and in the limit make these pairs form input/output continua representing response of a noncausal system to a noncausal input.

For this purpose let's first recall how to express a continuous function $x(t)$ such as that shown in Figure 6.3-3 by a continuum of impulses. This will include two steps, 1) approximation of a CT signal by a piecewise constant function, and 2) taking the limit of the latter to obtain the continuum of weighted impulses that through integration build up to the original continuous function.

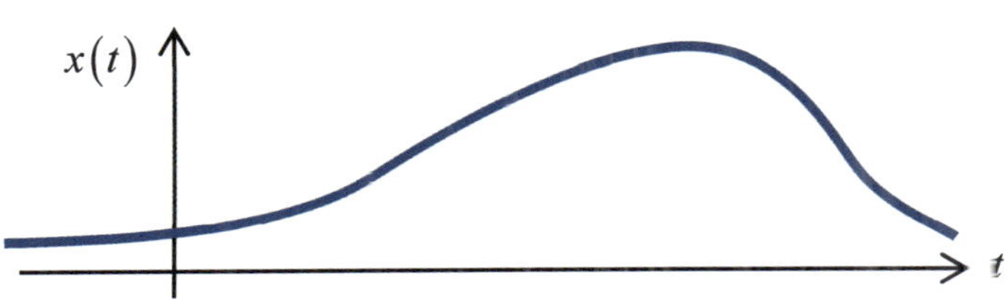

Fig. 6.3-3. Original CT signal.

6.3.2.1 *Piecewise constant approximation of a continuous function*

First, we recall that through shifting, stretching, and scaling of rectangles we can obtain an arbitrary rectangle as shown below:

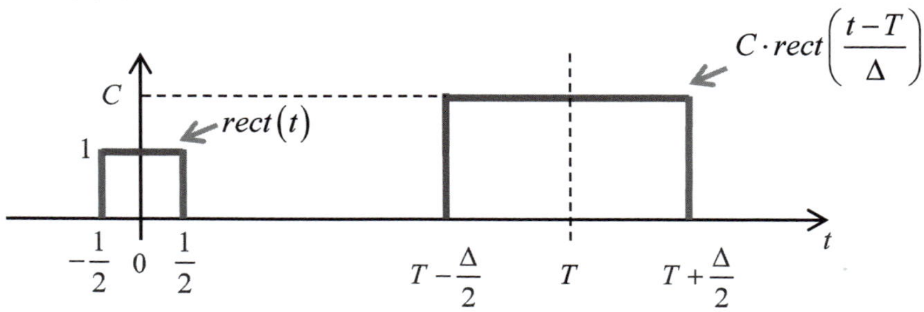

Fig. 6.3-4. Shifting, stretching, and scaling of a rectangle.

Next, we recall that a **pulse** signal of width Δ, denoted as $p_\Delta(t)$, is a rectangle signal with width Δ and height $1/\Delta$, i.e. unit area, so that the CT impulse, $\delta(t)$, can be thought of as the limit of a pulse whose width goes to *0*, with the area under it remaining 1:

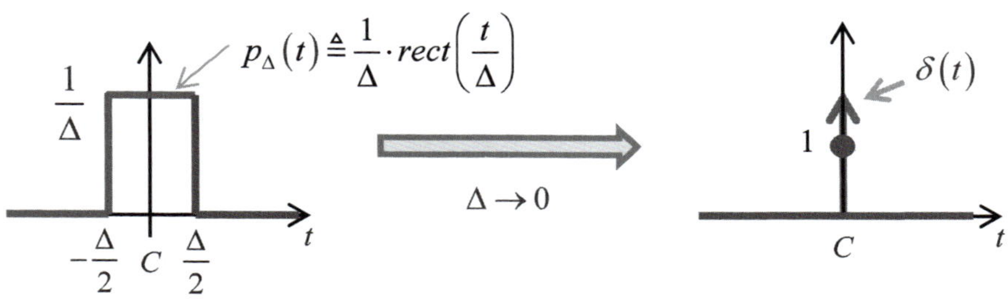

Fig. 6.3-5. CT impulse from pulse.

Multiplying this pulse by constant C rescales area to C.

Finally, we define a point $x(n\Delta)$ on the original function and associate with this point a rectangle placed at $t = n\Delta$ with width Δ and height $x(n\Delta)$:

$$x(n\Delta) \cdot rect\left(\frac{t-n\Delta}{\Delta}\right).$$

Summing up the weighted rectangles yields:

$$x_\Delta(t) \triangleq \sum_{n=-\infty}^{\infty} x(n\Delta) \cdot \mathrm{rect}\left(\frac{t-n\Delta}{\Delta}\right). \tag{6.3-1}$$

The application of the latter to the function shown in Figure 6.3-3 gives an approximation $x_\Delta(t)$ shown below:

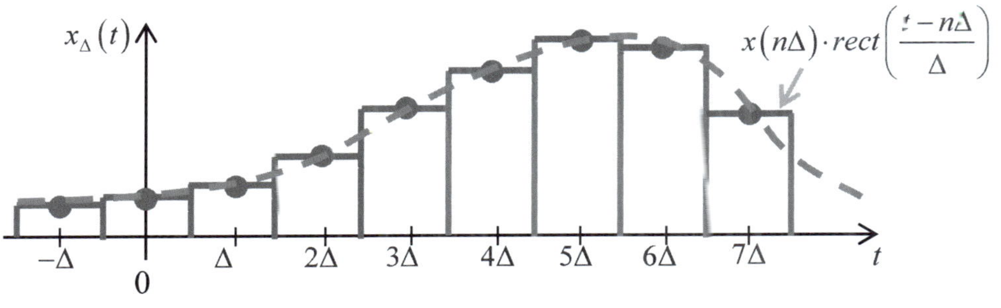

Fig. 6.3-6. Piecewise continuous (stair-step) approximation $x_\Delta(t)$ of function $x(t)$.

6.3.2.2 *Exact expression of continuous function in terms of the continuum of impulses*

Since the impulse $C\delta(t)$ is the limit of a constant area C rectangle with width $\rightarrow 0$, we have:

$$C\delta(t) = C\lim_{\Delta\to 0}(1/\Delta)\cdot \mathrm{rect}(t/\Delta) \Rightarrow x_\Delta(t) = \sum_{n=-\infty}^{\infty} x(n\Delta)\cdot \mathrm{rect}\left(\frac{t-n\Delta}{\Delta}\right) =$$

$$\sum_{n=-\infty}^{\infty} x(n\Delta)\cdot p_\Delta(t-n\Delta)\cdot\Delta \underset{\substack{\Delta\to d\tau\\ n\Delta\to\tau\\ \Sigma\to\int}}{\rightarrow} \int_{-\infty}^{\infty} x(\tau)\delta(t-\tau)d\tau = \int_{-\infty}^{\infty} x(\tau)\delta(\tau-t)d\tau = x(t).$$

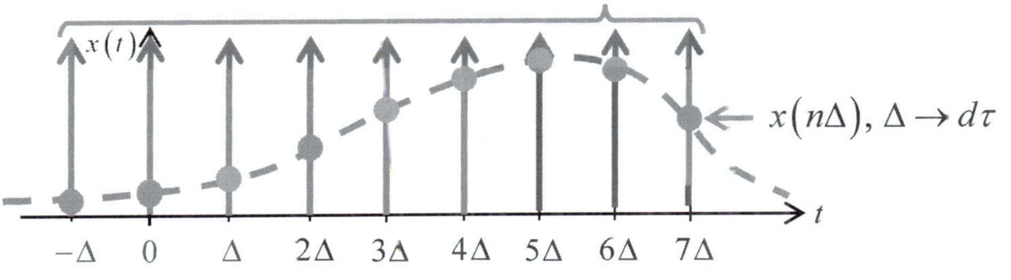

Fig. 6.3-7. Exact continuous function representation through a continuum of infinitely closely spaced weighted impulses.

6.3.3 *CT: impulse response*

Now, let's take a LTI system, and input the Δ-width pulse:

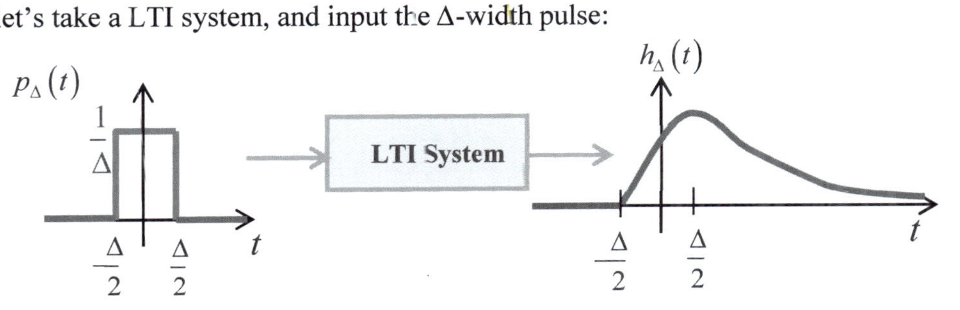

Fig. 6.3-8. Impulse response of a LTI system.

Because the system is linear and constant coefficient, shifting the input in time and stretching it in magnitude shifts and stretches the output, respectively, by the same amounts:

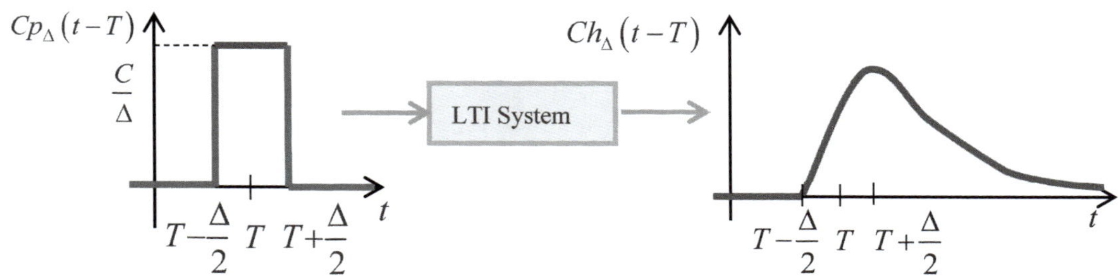

Fig. 6.3-9. Shifting and stretching - impulse response of a LTI system.

As $\Delta \rightarrow 0$, $p_\Delta(t) \rightarrow \delta(t)$ (the CT impulse), and $h_\Delta(t) \rightarrow h(t)$ (the CT **impulse response** of the system):

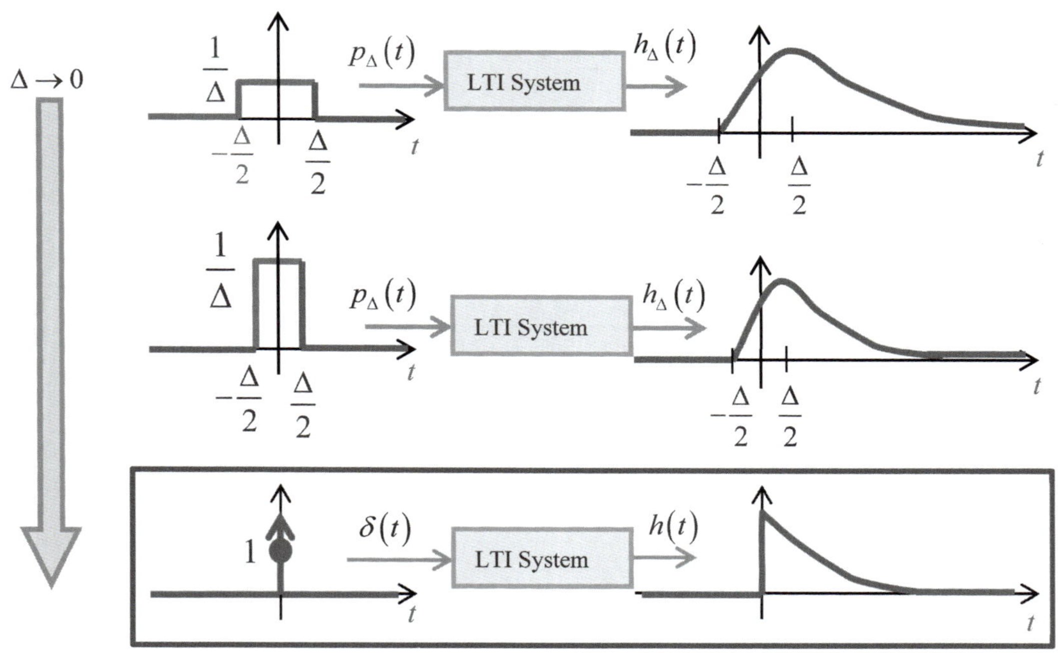

Fig. 6.3-10. CT impulse response of a LTI system as $\Delta \rightarrow 0$.

6.3.3.1 *CT: finding LTI system output via impulse approximation*

For a LTI system, the sum of pulses forming this pulse-train will pass through as the sum (superposition) of identically scaled pulse responses:

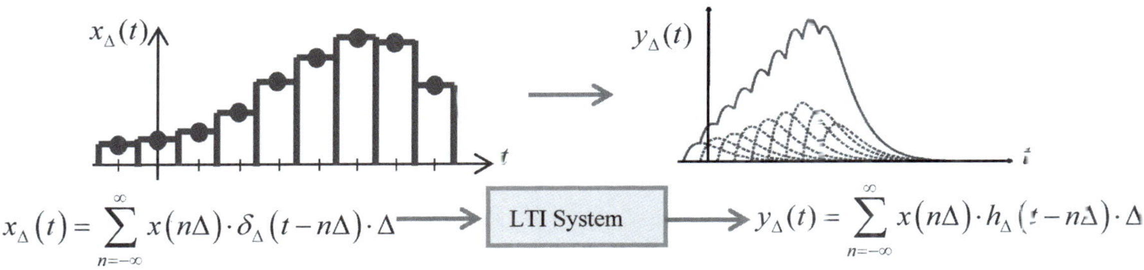

Fig. 6.3-11. Approximation of the output signal through the pulse response summation.

As $\Delta\to 0$, the summations become integrals:

$$\sum\Delta\to\int d\tau, n\Delta\to\tau, p_\Delta(t)\to\delta(t), h_\Delta(t)\to h(t), x_\Delta(t)\to x(t), y_\Delta(t)\to y(t),$$

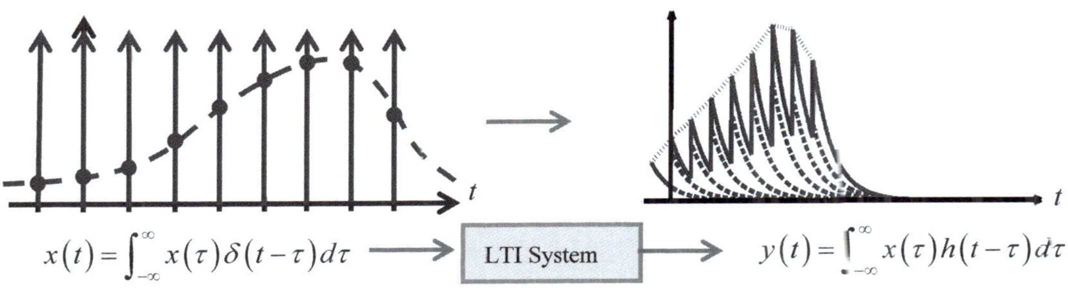

Fig. 6.3-12. Exact continuous system output representation in terms of a continuum of infinitely closely spaced weighted impulse responses.

6.3.3.2 *CT: finding the LTI system output using its impulse response*

Thus, knowing the noncausal system impulse response:

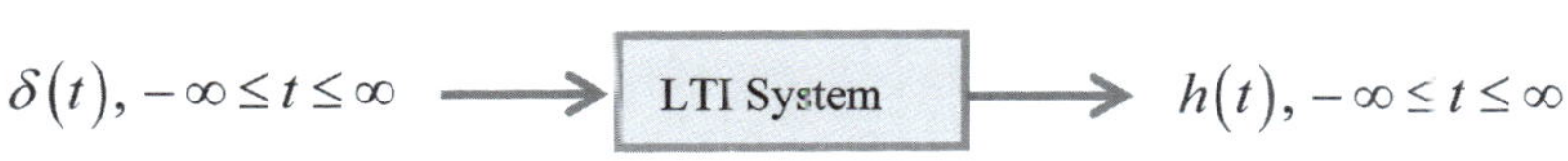

we can express an arbitrary <u>noncausal</u> input $x(t)$ as an integral of weighted and shifted impulses, and, noting that <u>causal zero initial conditions are represented by integration from $-\infty$ to 0</u>, find the system response $y(t)$ as:

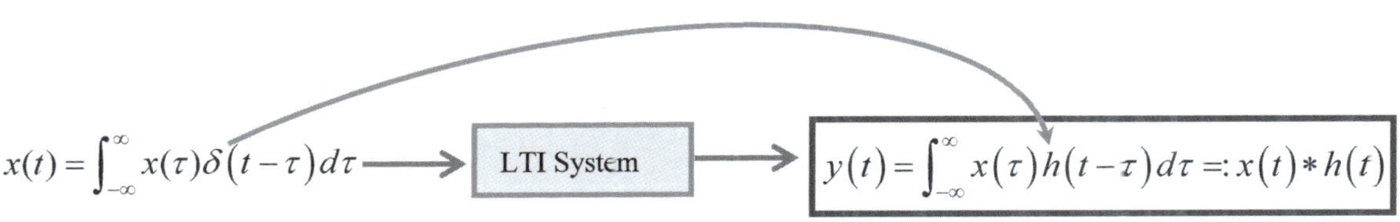

Thus, we have derived the generally noncausal ***superposition or convolution integral*** that performs operation on $u(t)$ and $h(t)$ called **noncausal convolution**:

$$\boxed{x(t)*h(t):=\int_{-\infty}^{\infty}x(\tau)h(t-\tau)d\tau,\ -\infty\le t\le\infty.} \tag{6.3-2}$$

Some Useful Convolution Properties:

Symmetry (commutativity): $x*h=h*x$ (6.3-3)

Linearity: $(x_1+x_2)*h=x_1*h+x_2*h,\ (Cx)*h=C(x*h)$ (C is a constant) (6.3-4)

Time shift: $x(t-T)*h(t)=(x*h)(t-T)$ (6.3-5)

Derivative: $x'*h=(x*h)'$ (6.3-6)

Impulse — identity operator: $x*\delta=x$ (6.3-7)

6.3.3.3 *DT: linear constant coefficient difference equations*

Linear, time-invariant systems in Discrete Time are usually described by **difference equations**

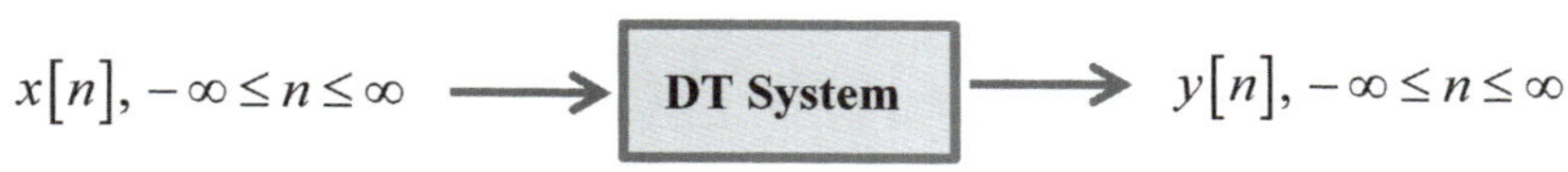

$$y[n]+A_1y[n-1]+\ldots+A_ky[n-k]=B_0x[n]+B_1x[n-1]+\ldots+B_kx[n-k]. \tag{6.3-8}$$

A linear, time-invariant Discrete Time system has the same superposition properties as a linear, Continuous-Time system. So, we can again find the response of this system to any signal using the impulse response, $h[n]$,

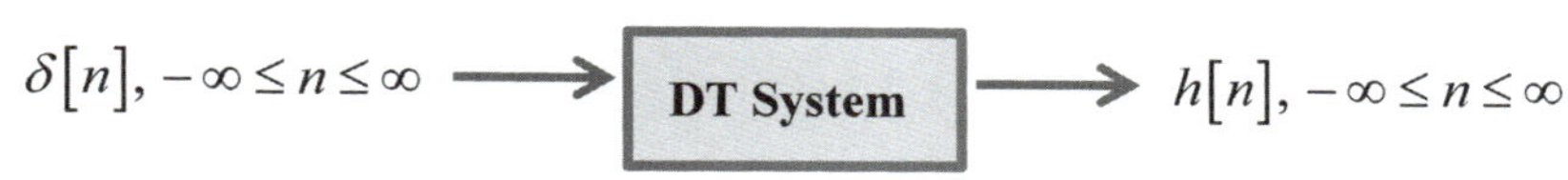

recalling that in DT: $\delta[n]=\begin{cases}1, & n=0,\\ 0, & \text{otherwise.}\end{cases}$

6.3.3.4 *DT: convolution*

If the DT system is linear and time invariant, its response $y[n]$ to an arbitrary input $x[n]$,

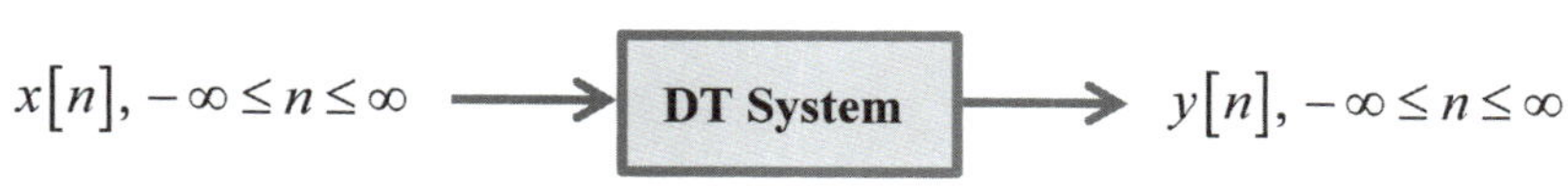

can be found using representation of an arbitrary input $x[n]$ in terms of $\delta[n]$:

$$x[n]=\sum_{k=-\infty}^{\infty}x[k]\delta[n-k],\ -\infty\le n\le\infty, \tag{6.3-9}$$

as superposition of impulse responses:

$$y[n]=\sum_{k=-\infty}^{\infty} x[k]h[n-k],\ -\infty \le n \le \infty. \tag{6.3-10}$$

This sum is called the **noncausal discrete time convolution**:

$$x[n]*h[n]:=\sum_{k=-\infty}^{\infty} x[k]h[n-k],\ -\infty \le n \le \infty. \tag{6.3-11}$$

Some useful convolution properties:

Symmetry (commutativity): $x*h=h*x$ (6.3-12)

Linearity: $(x_1+x_2)*h=x_1*h+x_2*h,\ (Cx)*h=C(x*h)$ (C is a constant). (6.3-13)

Time shift: $x[n-N]*h[n]=(x*h)[n-N]$. (6.3-14)

Difference: $(x[n]-x[n-1])*h=(x*h)[n]-(x*h)[n-1]$. (6.3-15)

Impulse — identity operator: $x*\delta=x$. (6.3-16)

6.3.4 *CT and DT system response*

6.3.4.1 *CT: system response to periodic input — a noncausal power signal*

Having in mind the convolution for finding system response

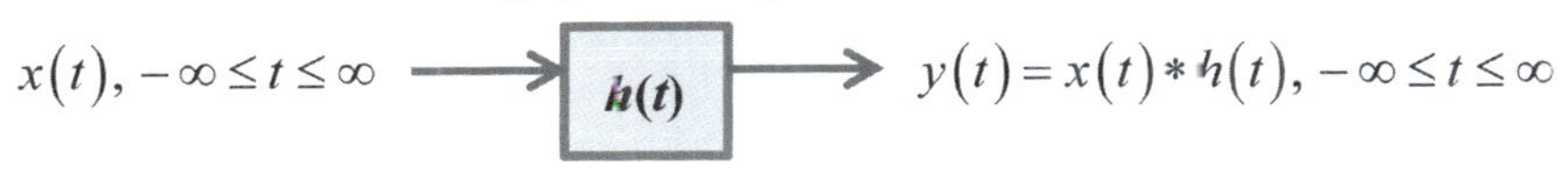

suppose $x(t)$ is a periodic, i.e. noncausal, with fundamental frequency ω_1 and Fourier Series:

$$x_p(t)=\sum_{k=-\infty}^{\infty} X_s[k]e^{jk\omega_1 t}. \tag{6.3-17}$$

Then, assuming not necessarily causal impulse response $h(t)$ that is an energy signal and using the noncausal convolution to calculate the output yields:

$$y(t)=x*h=h*x=\int_{-\infty}^{\infty} h(\tau)x_p(t-\tau)d\tau=\int_{-\infty}^{\infty} h(\tau)\sum_{k=-\infty}^{\infty} X_s[k]e^{jk\omega_1(t-\tau)}d\tau$$

$$=\sum_{k=-\infty}^{\infty}\left[X_s[k]e^{jk\omega_1 t}\int_{-\infty}^{\infty} h(\tau)e^{-jk\omega_1\tau}d\tau\right].$$

The integral in brackets represents the **two-sided Fourier Transform** of $h(t)$, denoted as $H(\omega)\equiv H(j\omega)=\int_{-\infty}^{\infty} h(t)e^{-j\omega t}dt$ and evaluated at $\omega = k\omega_1$, so that

$$y_p(t) = \sum_{k=-\infty}^{\infty} \underbrace{H(k\omega_1)X_s[k]}_{Y_s[k]} e^{jk\omega_1 t} = \sum_{k=-\infty}^{\infty} Y_s[k]e^{jk\omega_1 t}. \qquad (6.3\text{-}18)$$

Thus, the output is a power signal, **periodic with fundamental frequency ω_1 and Fourier series $Y_s[k] = H(k\omega_1)X_s[k]$.**

6.3.4.2 *CT: frequency response*

Expression (6.3-18) reveals the analytical details of an LTI system filtering action:

$$x_p(t) = \sum_{k=-\infty}^{\infty} X_s[k]e^{jk\omega_1 t} \xrightarrow{\quad} \boxed{H(\omega)} \xrightarrow{\quad} y_p(t) = \sum_{k=-\infty}^{\infty} H(k\omega_1)X_s[k]e^{jk\omega_1 t} = \sum_{k=-\infty}^{\infty} Y_s[k]e^{jk\omega_1 t}$$

$$= c_0 H(0) + \sum\nolimits_{k=1}^{\infty} c_k H_k \cos(k\omega_1 t + \theta_k + \varphi_k) \Rightarrow$$

$X_s[k]$ and $H(k\omega_1)$ are complex numbers:

$$X_s[k] = \underbrace{|X_s[k]|}_{X_k} e^{j\theta_k} \qquad\qquad H(k\omega_1) = \underbrace{|H(k\omega_1)|}_{H_k} e^{j\varphi_k}$$

$$\Rightarrow Y_s[k] = H(k\omega_1)X_s[k] = \overbrace{H_k X_k}^{|Y_s[k]|} e^{j\overbrace{(\theta_k+\varphi_k)}^{\angle Y_s[k]}} = |Y_s[k]|e^{j\angle Y_s[k]}.$$

In the output, the **frequencies don't change**, the **magnitude gets multiplied by $|H(k\omega_1)|$**, and the **phase gets shifted by φ_k**, i.e. the phase of $H(k\omega_1)$. Therefore, $H(\omega) = \int_{-\infty}^{\infty} h(t)e^{-j\omega t}dt$, the two-sided Fourier Transform of the impulse response, $h(t)$, is seen to represent the **frequency response** or the **transfer function** of the system. The impulse response $h(t)$ can have an arbitrary form, so that $H(j\omega)$ gives the **nonparametric LTI system model.**

6.3.4.3 *DT: system response in frequency domain*

Similarly, the DT convolution

$$x[n],\ -\infty \le n \le \infty \xrightarrow{\quad} \boxed{h[n]} \xrightarrow{\quad} y[n] = \sum_{k=-\infty}^{\infty} x[k]h[n-k],\ -\infty \le n \le \infty$$

can be used to obtain DT system frequency response by taking the DTFT as follows. Assume that $x[n]$ and $h[n]$ are noncausal, monotonic, and square summable. Then $x[\pm\infty] = h[\pm\infty] = 0$, and

$$Y_p(F) = \sum_{n=-\infty}^{\infty} y[n]e^{-j2\pi Fn} = \sum_{n=-\infty}^{\infty} \Big(\sum_{k=-\infty}^{\infty} x[k]h[n-k]\Big) \underbrace{e^{-j2\pi Fn}}_{\text{move into inner summation}} =$$

$$\sum_{n=-\infty}^{\infty}(\sum_{k=-\infty}^{\infty}x[k]\underbrace{e^{-j2\pi kF}e^{j2\pi kF}}_{=1}h[n-k]e^{-j2\pi Fn})=\sum_{k=-\infty}^{\infty}(x[k]e^{-j2\pi Fk}\sum_{n=-\infty}^{\infty}h[n-k]e^{-j2\pi F(n-k)})$$

Change the order of summation (Fubini's theorem), take out functions of k, set $m = n-k$, sum up as $H_p(F)$, take out of k-summation

$$=\sum_{k=-\infty}^{\infty}(x[k]e^{-j2\pi Fk}\sum_{m=-\infty}^{\infty}h[m]e^{-j2\pi Fm})=X_p(F)H_p(F).$$

DT convolution is equivalent to multiplication of the DTFT:

$$y[n]=\sum_{k=-\infty}^{\infty}x[k]h[n-k]\Leftrightarrow Y_p(F)=H_p(F)X_p(F),\ -\infty\le n\le\infty. \quad (6.3\text{-}19)$$

6.3.4.4 *DT: transfer functions*

To derive a DT system transfer function, let's start with taking the DTFT of $x[n-1]$:

$$\sum_{n=-\infty}^{\infty}x[n-1]e^{-j2\pi Fn}=e^{-j2\pi F}\sum_{n=-\infty}^{\infty}x[n-1]e^{-j2\pi F(n-1)}=e^{-j2\pi F}X_p(F).$$

Using the latter, and taking the DTFT of a difference equation yields DT transfer function as follows:

$$\begin{aligned}&y[n]+A_1y[n-1]+\ldots+A_ky[n-k]=B_0x[n]+B_1x[n-1]+\ldots+B_kx[n-k]\Rightarrow\\&Y_p(F)+A_1e^{-j2\pi F}Y_p(F)+\ldots+A_ke^{-j2\pi kF}Y_p(F)=B_0eX_p(F)+B_1e^{-j2\pi F}X_p(F)+\ldots\\&+B_ke^{-j2\pi kF}X_p(F)\Rightarrow\frac{Y_p(F)}{X_p(F)}=\boxed{H_p(F)=\frac{B_0+B_1e^{-j2\pi F}+\ldots+B_ke^{-j2\pi kF}}{1+A_1e^{-j2\pi F}+\ldots+A_ke^{-j2\pi kF}}}.\end{aligned} \quad (6.3\text{-}20)$$

Example 6.3-1: Find the impulse response and the transfer function of the system

$$y[n]=\alpha y[n-1]+x[n],\ |\alpha|<1. \quad (6.3\text{-}21)$$

Setting initial condition $y[n-1] = 0$, iterative calculation of the response of (6.3-21)) to $\delta[n]$, or impulse response, yields

$$\begin{aligned}&n=0: y[0]=\alpha y[-1]+\delta[0]=\alpha\cdot 0+1=1;\ n=1: y[1]=\alpha y[1-1]+\delta[1]=\alpha y[0]+0\\&=\alpha\cdot 1=\alpha;\ n=2: y[2]=\alpha y[2-1]+\delta[2]=\alpha y[1]+0=\alpha\cdot\alpha=\alpha^2\Rightarrow h[n]=\alpha^n u[n].\end{aligned} \quad (6.3\text{-}22)$$

The transfer function can be computed as follows:

$$y[n]-\alpha y[n-1]=x[n]\Rightarrow(1-\alpha e^{-j2\pi F})Y_p(F)=X_p(F)\Rightarrow H_p(F)=\frac{1}{1-\alpha e^{-j2\pi F}}, \quad (6.3\text{-}23)$$

yielding, through the use of Euler's formula, the magnitude and the phase responses

$$|H_p(F)| = \frac{1}{|1-\alpha\cos(2\pi F)+j\alpha\sin(2\pi F)|} = \frac{1}{\left|\sqrt{1-2\alpha\cos(2\pi F)+\alpha^2\cos^2(2\pi F)+\alpha^2\sin^2(2\pi F)}\right|}$$

$$= \frac{1}{\left|\sqrt{1-2\alpha\cos(2\pi F)+\alpha^2}\right|},\quad \angle H_p(F) = \varphi_p(F) = -\tan^{-1}\left(\frac{\alpha\sin(2\pi F)}{1-\alpha\cos(2\pi F)}\right).$$

System impulse and frequency responses are given, respectively, in Figures 6.3-13 and 6.3-14.

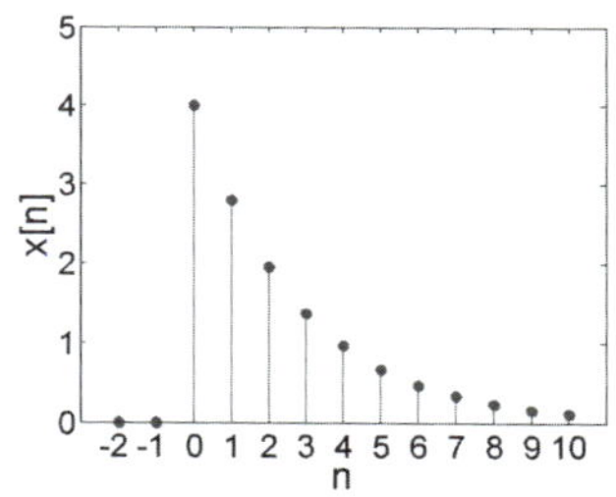

Fig. 6.3-13. Signal $x[n]$.

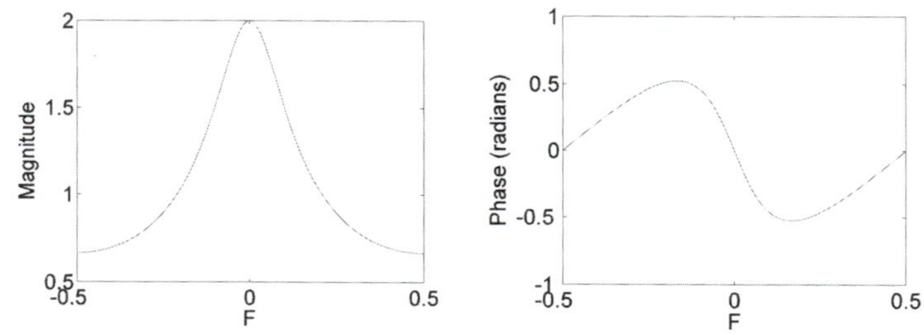

Fig. 6.3-14. System frequency response for $\alpha = 0.5$.

6.3.4.5 *DT: frequency response*

A DT system responds to a DT harmonic the same way a CT system responds to a CT harmonic:

$$x_p[n] = a\cos(2\pi Fn+\theta) \longrightarrow \boxed{H_p(F)} \longrightarrow y_p[n] = a|H_p(F)|\cos(2\pi Fn+\theta+\varphi_p(F))$$

i.e., it scales the magnitude and shifts the phase of the input DT sinusoid. Thus, the DT transfer function is seen to be the frequency response in exactly the same way as the CT transfer function.
Example 6.3-2: Compute the response of the DT system of the previous example with $\alpha = 0.5$ to a DT sinusoid. DT system in this case is given by

$$y[n] = 0.5y[n-1]+x[n] \Leftrightarrow H_p(F) = \frac{1}{1-0.5e^{-j2\pi F}}.$$

The response to a sinusoid $x_p[n] = 10\cos[0.5\pi n+(\pi/3)]$ is then computed as follows:

$$x_p[n] = 10\cos(2\pi\underbrace{\frac{1}{4}}_{F_1}n+\frac{\pi}{3}) \Rightarrow H_p(F) = \frac{1}{1-0.5\underbrace{e^{-j\frac{\pi}{2}}}_{-j}} = \frac{1}{1+j0.5} \approx 0.894e^{-j0.464}$$

$$\Rightarrow y_p[n] \approx 10\cdot 0.894\cos\left(2\pi\frac{1}{4}n+\frac{\pi}{3}-0.464\right) \approx 8.94\cos\left(2\pi\frac{1}{4}n+0.584\right).$$

6.3.4.6 *Convolution for two power signals: periodic convolution*

Application of convolution considered above to power signals, such as sinusoids, leads to infinite growth of the result. Therefore, convolving these functions over one period and normalizing the results through division by the period could be specified instead. The latter is called ***periodic convolution***:

$$y_p(t) = x_p(t) \circledast h_p(t) \triangleq \frac{1}{T_1}\int_0^{T_1} x_p(\tau)h_p(t-\tau)d\tau,\ 0 \le t \le T_1, \text{ for periodic power signals.}$$

For the nonperiodic signals or periodic noncommensurable signals, the integration interval and normalizing value are extended to infinity:

$$y(t) = x(t) \circledast h(t) \triangleq \lim_{T_1 \to \infty} \frac{1}{T_1}\int_{T_1} x(\tau)h(t-\tau)d\tau, \text{ for nonperiodic power signals.}$$

If $x_p(t) = \sum_{k=-\infty}^{\infty} X_s[k]e^{jk\omega_1 t}$ and $h_p(t) = \sum_{k=-\infty}^{\infty} H_s[k]e^{jk\omega_1 t}$ for the same ω_1, then

$$y_p(t) = x_p(t) \circledast h_p(t) = \sum_{k=-\infty}^{\infty} Y_s[k]e^{jk\omega_1 t} = \sum_{k=-\infty}^{\infty} X_s[k]H_s[k]e^{jk\omega_1 t} \Rightarrow \overbrace{x_p \circledast h_p \Leftrightarrow X_s[k]H_s[k]}^{\text{Fourier series pair}}. \quad (6.3\text{-}24)$$

Periodic convolution for DT is typically defined without normalizing factor $1/N_1$ to conveniently relate it to DFTs of its constituent functions. Namely, given two N_1 length sequences $x[n]$ and $h[n]$ (zero padded, if necessary) and their respective DFTs $X_{DFT}[k]$ and $H_{DFT}[k]$, the N_1-periodic extensions $x_p[n]$ and $h_p[n]$ of $x[n]$ and $h[n]$ due to N_1 periodicity of DFT yield (prove this)

$$y_p[n] = x_p[n] \circledast h_p[n] = \sum_{m=0}^{N-1} x_p[m]h_p[n-m] \overset{DFT}{\Rightarrow} Y_{pDFT}[k] \equiv Y_{DFT}[k] = X_{DFT}[k]H_{DFT}[k]. \quad (6.3\text{-}25)$$

The relations (6.3-24) and (6.3-25) will be used in Chapter 8 for building diagnostic tools.

6.4 Ideal and Real Filters: Continuous and Discrete Time

Frequency response as filter. Recall the response of an LTI system to periodic input:

$$x_p(t) = \sum_{k=-\infty}^{\infty} X_s[k]e^{jk\omega_1 t} \longrightarrow \boxed{H(\omega)} \longrightarrow y_p(t) = \sum_{k=-\infty}^{\infty} H(k\omega_1)X_s[k]e^{jk\omega_1 t}$$

If $|H(k\omega_1)|$ that multiplies the magnitude of the kth harmonic is small (<< 1), harmonic gets ***attenuated*** (suppressed), i.e. filtered out so that it is not present in the output. If $|H(k\omega_1)|$ is large (> 1), harmonic gets ***amplified***. So we can think of LTI systems as **filters** acting on the input.

6.4.1 *Continuous time ideal filters*

6.4.1.1 *CT: magnitude responses of ideal filters*

For ideal filters, magnitude responses are 1 at the passband, 0 at the stopband, and the phase shift is 0 (no time delay):

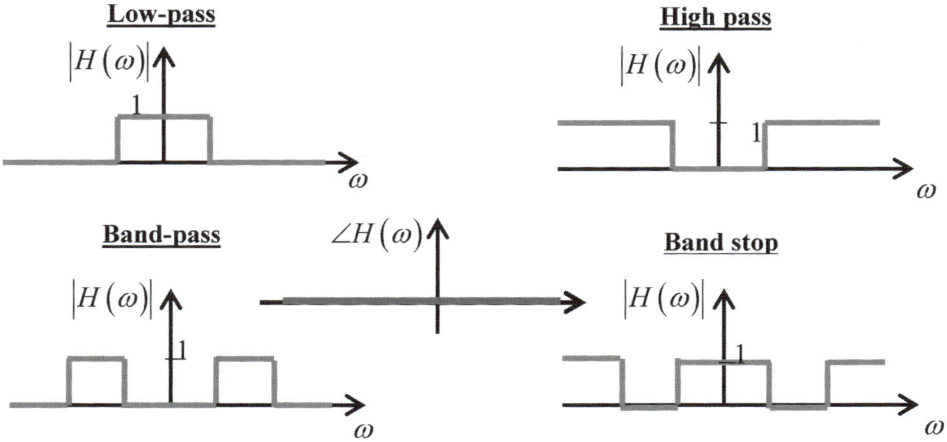

Fig. 6.4-1. Magnitude responses of ideal filters for the basic filter types.

6.4.1.2 *CT: Ideal Low Pass Filter (ILPF)*

Example 6.4-1: Ideal low-pass filter passing square wave:

$x_p(t) = \sum_{k=-\infty}^{\infty} X_s[k] e^{jk\omega_1 t}$ → $H(\omega)$ → $y_p(t) = \sum_{k=-\infty}^{\infty} H(k\omega_1) X_s[k] e^{jk\omega_1 t} = \sum_{k=-\infty}^{\infty} X_k H_k e^{j\theta_k} e^{jk\omega_1 t}$, i.e. ideal filters become rectangular spectral windows

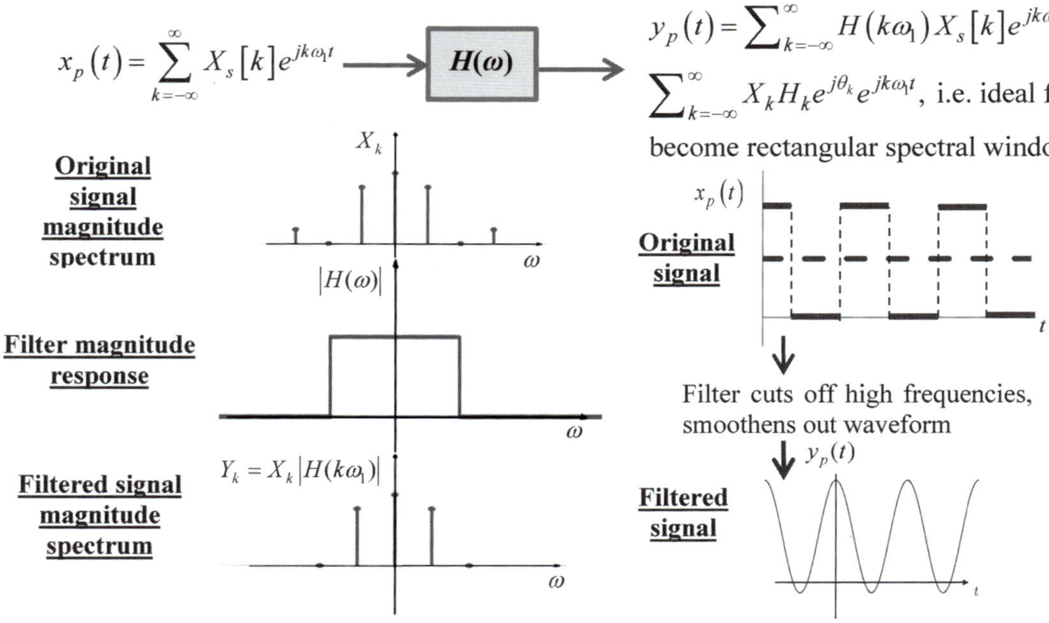

Fig. 6.4-2. Ideal low pass filter acting on the square wave input.

CT: ideal low pass filtering in detail — filtering input magnitude spectrum: $Y_k = X_k H_k$.

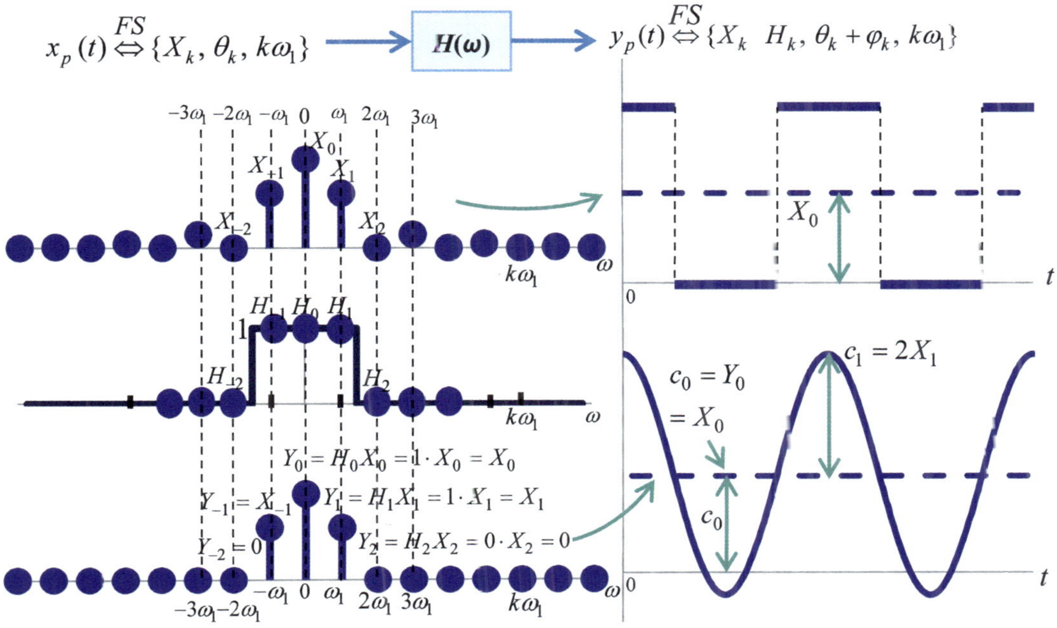

Fig. 6.4-3. ILPF Filtering square wave magnitude spectrum.

CT: ideal low pass filtering in detail — filtering input phase spectrum: $\angle Y_s[k] = \theta_k + \varphi_k$.

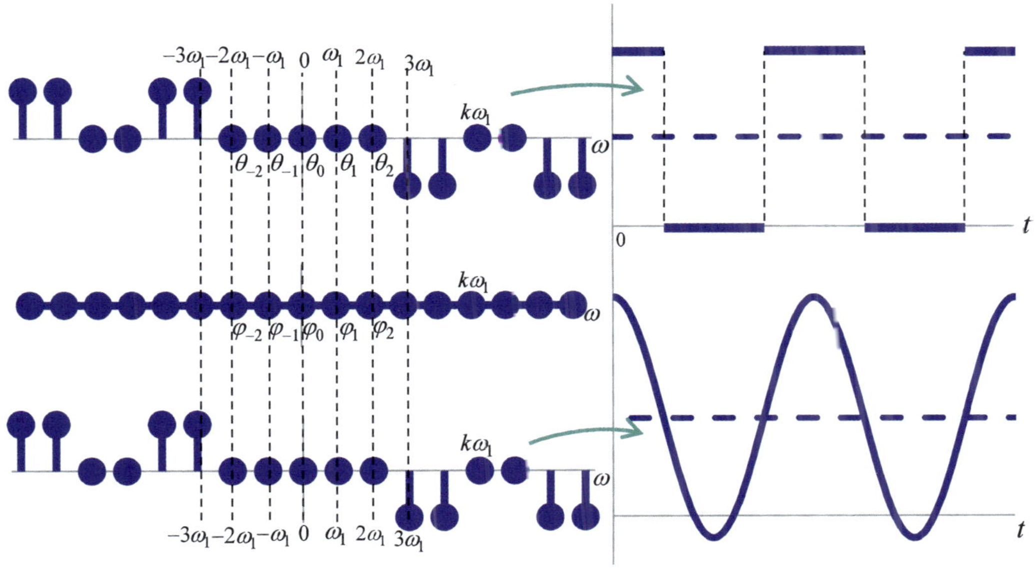

Fig. 6.4-4. ILPF filtering square wave phase spectrum.

6.4.1.3 *CT: ideal low-pass filter frequency response*

As shown in Figure 6.4-1, an ideal low-pass filter has a rectangle magnitude response of height 1 and a flat 0 for phase response, making the response bandlimited to ω_c — the cutoff frequency:

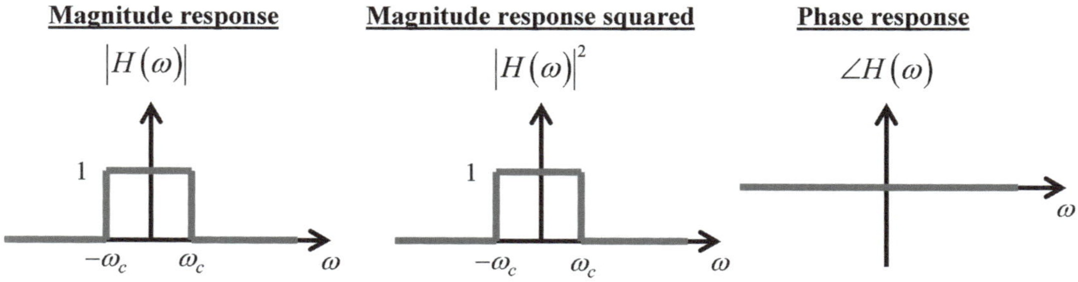

Fig. 6.4-5. ILPF frequency response.

$$\text{Therefore, } H(\omega) = \text{rect}\left(\frac{\omega}{2\omega_c}\right)e^{j0} = \text{rect}\left(\frac{\omega}{2\omega_c}\right). \tag{6.4-1}$$

Example 6.4-2: Compute the response of the ILPF with the cutoff frequency $\omega_c = 10\pi$ to signals 1) $x_{in}(t) = 4\cos(7\pi t - \pi/6)$ and 2) $x_{in}(t) = 4\cos(14\pi t - \pi/6)$.
The response is given as follows:

$$\text{1) } x_{in}(t) = 4\cos\left(\underbrace{7\pi}_{\omega_1<\omega_c} t - \frac{\pi}{6}\right) \Rightarrow \begin{matrix} |H(\omega_1)| = 1 \\ \varphi(\omega_1) = 0 \end{matrix} \Rightarrow x_{out}(t) = 4\cos\left(7\pi t - \frac{\pi}{6}\right),$$

$$\text{2) } x_{in}(t) = 4\cos\left(\underbrace{14\pi}_{\omega_2>\omega_c} t - \frac{\pi}{6}\right) \Rightarrow \begin{matrix} |H(\omega_2)| = 0 \\ \varphi(\omega_1) = 0 \end{matrix} \Rightarrow x_{out}(t) = 0.$$

6.4.1.4 *CT: ideal low-pass filter impulse response*

Taking the inverse Fourier Transform of the ILPF transfer function $H(\omega) = \text{rect}(\omega/2\omega_c)$ yields

$$h(t) = \frac{1}{2\pi}\int_{-\infty}^{\infty} H(\omega)e^{j\omega t}d\omega = \frac{1}{2\pi}\int_{-\infty}^{\infty} \text{rect}\left(\frac{\omega}{2\omega_c}\right)e^{j\omega t}d\omega = \frac{1}{2\pi}\int_{-\omega_c}^{\omega_c} e^{j\omega t}d\omega$$

$$= \frac{1}{2\pi}\frac{1}{jt}\left(e^{j\omega_c t} - e^{-j\omega_c t}\right) = \frac{1}{\pi t}\sin(\omega_c t) = \frac{\omega_c}{\pi}\frac{\sin\left(\pi\frac{\omega_c t}{\pi}\right)}{\left(\pi\frac{\omega_c t}{\pi}\right)} \boxed{= \frac{\omega_c}{\pi}\text{sinc}\left(\frac{\omega_c}{\pi}t\right)} \tag{6.4-2}$$

shown in Figure 6.4-6.

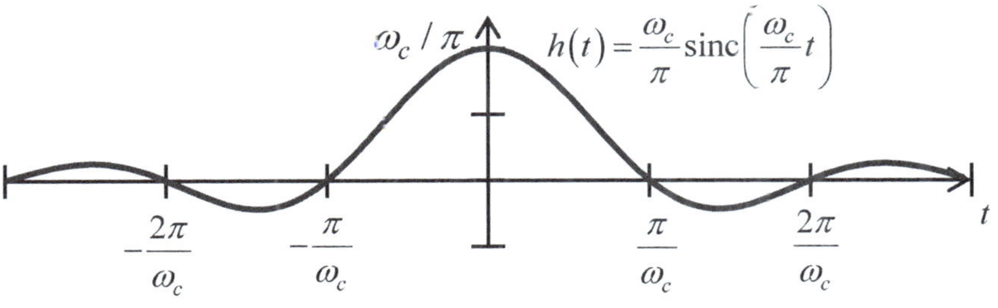

Fig. 6.4-6. ILPF Impulse Response.

Note that the impulse response is noncausal, and in fact extends all the way from $-\infty$ to ∞. That is, the output of the ideal filter depends on the entire signal, from $-\infty$ to ∞.

This output is clearly impossible to compute in practice, or even approximate well if we need to obtain it in real-time: if we shift the impulse response far to the right to make it causal-like, we will get the similar magnitude response, but enormous phase distortion, or time lag, so that we will have to wait for a long time for anything to show up at the output.

Thus, the ideal filters are noncausal, and hence impossible to implement.

6.4.2 *Continuous time real filters*

Real filters are characterized by a causal impulse response and in their simplest form are represented by the linear, time-invariant systems (LCCODEs), and their filtering action is described by their **frequency response** — response to periodic inputs.

Example 6.4-3: Recall RC circuit that performed DC coupling in Chapter 4:

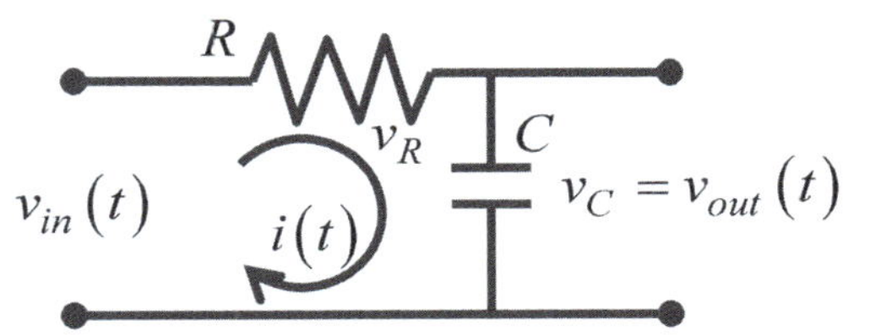

$$H(s) = \frac{V_{out}}{V_{in}} = \frac{V_C}{V_R + V_C} = \frac{(1/Cs)I}{RI + (1/Cs)I} = \frac{1}{1 + RCs} = \frac{1}{1 + \tau s},$$

Time constant: $\tau = RC$

Fig. 6.4-7. RC circuit and its transfer function.

with impulse response:

$$h(t) = L^{-1}\left\{\frac{1}{1 + \tau s}\right\} = \frac{1}{\tau} e^{-\frac{t}{\tau}} u(t), \tag{6.4-3}$$

and frequency response:

$$H(\omega) = \frac{1}{1 + j\omega\tau} = \frac{1}{1 + j\omega\tau} \cdot \frac{1 - j\omega\tau}{1 - j\omega\tau} = \frac{1 - j\omega\tau}{1 + \omega^2\tau^2} = \underbrace{\frac{1}{1 + \omega^2\tau^2}}_{\text{Re}\,H(\omega)} + j\underbrace{\frac{-\omega\tau}{1 + \omega^2\tau^2}}_{\text{Im}\,H(\omega)} = |H(j\omega)| e^{j\varphi(\omega)}, \tag{6.4-4}$$

where

$$|H(\omega)| = \left|\sqrt{\mathrm{Re}^2 H(\omega) + \mathrm{Im}^2 H(\omega)}\right| = \left|\sqrt{\frac{1}{\left(1+\omega^2\tau^2\right)^2} + \frac{\omega^2\tau^2}{\left(1+\omega^2\tau^2\right)^2}}\right| = \frac{1}{|\sqrt{1+\omega^2\tau^2}\,|}, \quad (6.4\text{-}5)$$

$$\varphi(\omega) = \tan^{-1}\frac{\mathrm{Im}\,H(j\omega)}{\mathrm{Re}\,H(j\omega)} = \tan^{-1}(-\omega\tau) = -\tan^{-1}(\omega\tau). \quad (6.4\text{-}6)$$

6.4.2.1 *Energy spectral density and magnitude spectral density*

We will now introduce the spectral counterparts of a periodic (power) signal that are used to characterize the continuous energy signal, using Example 6.4-3.

Given impulse response (6.4-3) and its Fourier transform (6.4-5), the energy of this impulse response is given by:

$$E_h = \int_{-\infty}^{\infty} |h(t)|^2\, dt = \int_0^{\infty} \frac{1}{\tau^2} e^{-\frac{2t}{\tau}}\, dt = -\frac{1}{2\tau} e^{-\frac{2t}{\tau}}\Big|_0^{\infty} = \frac{1}{2\tau}.$$

And integral of the magnitude response squared is given by

$$\int_{-\infty}^{\infty} |H(\omega)|^2\, d\omega = \int_{-\infty}^{\infty} \frac{1}{1+\omega^2\tau^2}\, d\omega = \frac{1}{\tau}\tan^{-1}(\omega\tau)\Big|_{-\infty}^{\infty} = \frac{1}{\tau}\left(\frac{\pi}{2} - \left(-\frac{\pi}{2}\right)\right) = \frac{\pi}{\tau},$$

where we used an indefinite integral formula from Appendix A. Therefore,

$$E_h = \int_{-\infty}^{\infty} |h(t)|^2\, dt = \frac{1}{2\tau} = \frac{1}{2\pi}\frac{\pi}{\tau} = \frac{1}{2\pi}\int_{-\infty}^{\infty} |H(\omega)|^2\, d\omega$$

$$= \int_{-\infty}^{\infty} \underbrace{[(1/\sqrt{2\pi})|H(\omega)|]^2}_{\substack{\text{Energy Spectral Density}\\ \text{as a function of } \omega}}\, d\omega = \int_{-\infty}^{\infty} \underbrace{|H(f)|^2}_{\substack{\text{Energy Spectral Density}\\ \text{as a function of } f}}\, df,$$

i.e., the energy of a signal $h(t)$ is given by

— the integral of the squared magnitude $|H(f)|^2 \equiv |H(j2\pi f)|^2$ of its Fourier transform (transfer function) with respect to f, or

— the integral of the scaled squared magnitude response $(1/2\pi)|H(\omega)|^2 = [(1/\sqrt{2\pi})|H(\omega)|]^2 \equiv |H(\omega)|_{MD}^2$ with respect to ω, where $|H(\omega)|_{MD}$ will be defined below.

This indicates that the latter quantities represent **energy (spectral) density**. Typically, only the first one — a function of f — is used for plotting the energy density graphs, however there is nothing wrong with also using the second one — a function of ω — if the scaling factor is shown on the graph:

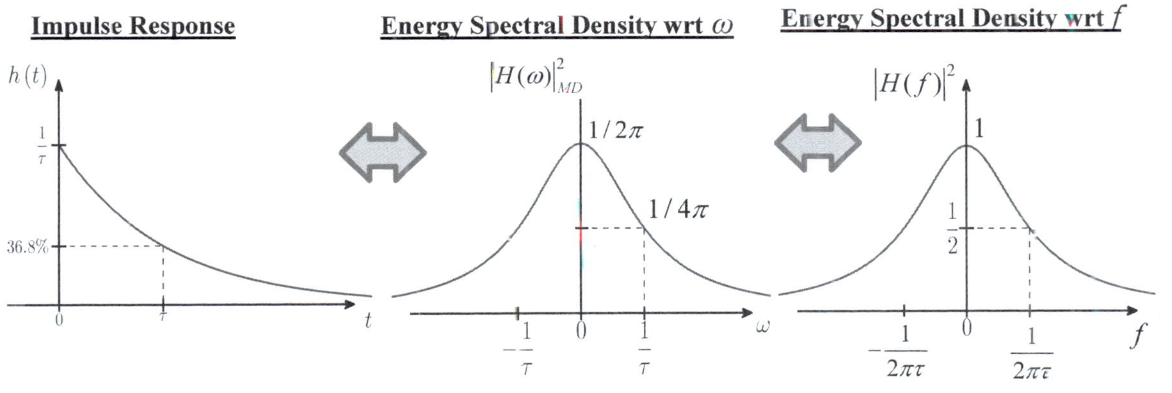

Fig. 6.4-8. Impulse response and the corresponding Energy Spectral Density plotted with respect to ω and f.

In power signals with discrete spectra, the power of a harmonic is the square of the harmonic magnitude. By analogy, since an energy signal Fourier transform is a continuous function of frequency: $H(f) \equiv H(j2\pi f)$, and the square of its magnitude $|H(f)|^2$ is the signal energy spectral density, the Fourier transform magnitude $|H(f)| \equiv |H(j2\pi f)|$ is called signal **magnitude (spectral) density** (as a function of f).

What, then, about the familiar **magnitude response** graph $|H(\omega)|$? If it is scaled by $1/\sqrt{2\pi}$, it becomes a **magnitude spectral density** graph as a function of ω: $|H(\omega)|_{MD} \triangleq (1/\sqrt{2\pi})|H(\omega)|$:

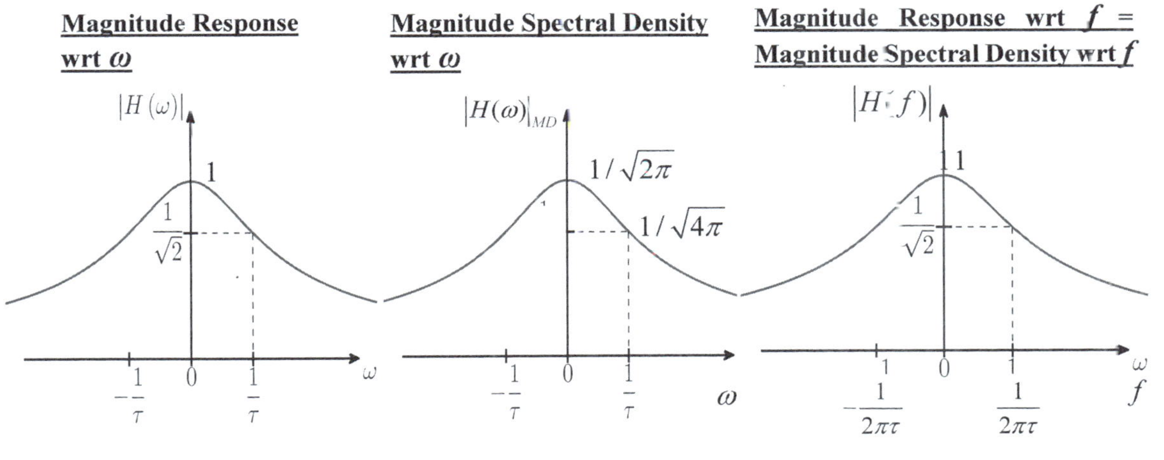

Fig. 6.4-9. Magnitude response and the corresponding Magnitude Spectral Density plotted with respect to ω and f.

The **break frequency,** $\omega_B = 2\pi f_B$, or the half-power frequency is defined as that at which the power of a filtered periodic signal drops to 1/2 of its input value:

$$\left|H(\omega)\right| = \frac{1}{\sqrt{1+\omega_B^2\tau^2}} = \frac{1}{\sqrt{2}} \Rightarrow 1+\omega_B^2\tau^2 = 2 \Rightarrow \omega_B = \frac{1}{\tau} \Rightarrow f_B = \frac{1}{2\pi\tau}. \qquad (6.4\text{-}8)$$

Note that time constant τ is given in sec/rad, since $\omega_B = 1/\tau$ is given in radians/second. Unlike the ideal LPF, the real LPF is seen to be *not bandlimited*, so, the *break frequency*, although imperfect, is used as a measure of the real LPF filter *bandwidth*.

6.4.2.2 *CT: from real low-pass filter (LPF) to all-pass filter (APF)*

The changes in the time constant τ and bandwidth $\omega_B = 1/\tau$ of the pair

$$H(\omega) = \frac{1}{1+j\omega\tau} = \frac{1/\tau}{(1/\tau)+j\omega} \Longleftrightarrow h(t) = \frac{1}{\tau}e^{-\frac{t}{\tau}}u(t)$$

fundamentally affect the LPF filter properties. Indeed, noting that

$$\int_{-\infty}^{\infty} h(t)\,dt = \int_0^{\infty} \frac{1}{\tau}e^{-\frac{t}{\tau}}dt = \frac{1}{\tau}\left(-\tau e^{-\frac{t}{\tau}}\Big|_0^{\infty}\right) = -e^{-\frac{t}{\tau}}\Big|_0^{\infty} = 1,$$

if $\tau \to 0$ so that $\omega_B = \frac{1}{\tau} \to \infty$, then $h(t) \to \delta(t)$ and $H(j\omega) = \frac{1}{1+j\omega\tau} \to 1$, as shown below:

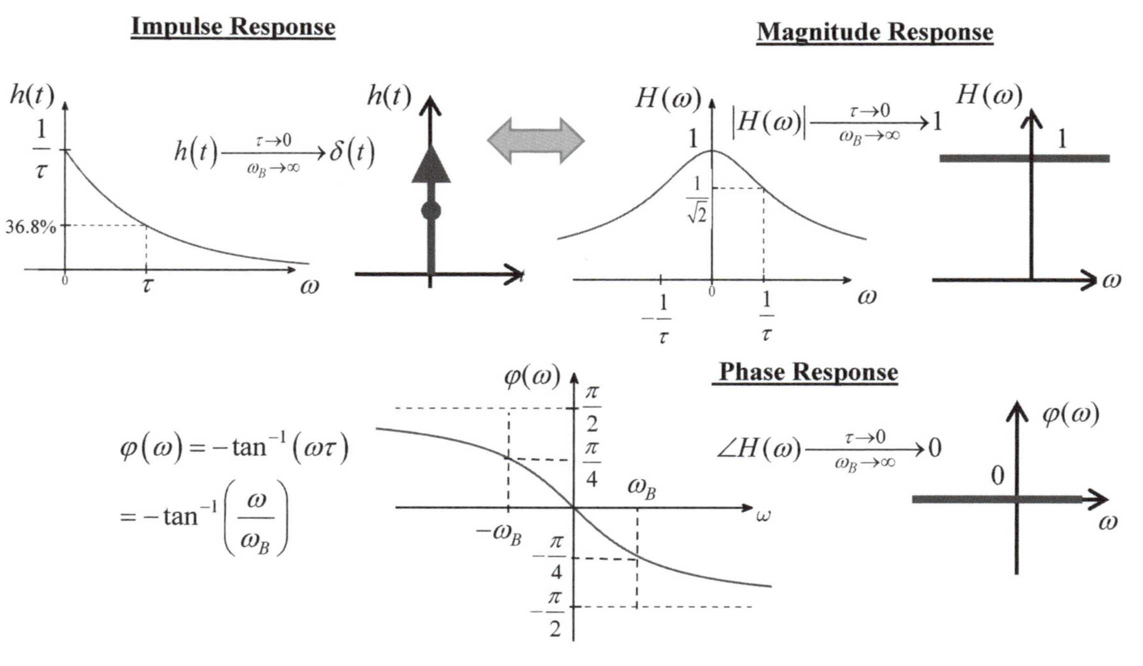

Fig. 6.4-10. Convergence of the real LPF impulse, magnitude, and phase responses to those of the ideal all-pass filter.

The responses of the filter obtained in the limit show that this filter *passes all frequencies with no distortion*, representing, therefore, the **ideal all-pass filter**. Approaching the latter is extremely expensive in practice: Nordost Odin Supreme eight-foot speaker cable pair costs $28,999.

As shown in Figure 6.4-11, the time-constant τ governs the decay rate of the impulse response — a smaller τ corresponds to a faster decay and a higher filter **bandwidth**.

The signal is never **bandlimited**: the corresponding magnitude response never reaches 0 for any finite ω, i.e. this filter has no actual cutoff frequency ω_C, and when ω_B is called ω_C, the subscript c stands for the **corner frequency** in reference to the corresponding Bode plot.

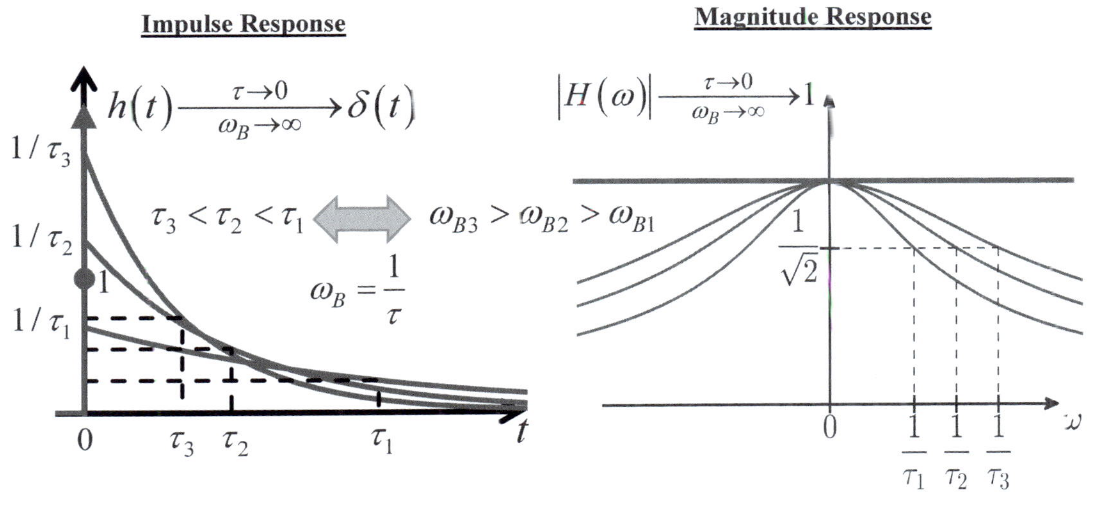

Fig. 6.4-11. Impulse response and magnitude response of a real LPF with different time constants.

Example 6.4-4: Let us pass a periodic signal through a real LPF:

$$v_{in}(t) = 4\sin\left(20\pi t - \frac{\pi}{6}\right) = 4\cos\left(20\pi t - \frac{\pi}{6} - \frac{\pi}{2}\right) = 4\cos\left(\underbrace{20\pi}_{\omega_1} t - \frac{2}{3}\pi\right),$$

$$|H(\omega_1)| = \frac{1}{\sqrt{1+\omega_1^2\tau^2}} = \frac{1}{\sqrt{1+400\pi^2\tau^2}}, \qquad \varphi(\omega_1) = -\tan^{-1}(\omega_1\tau) = -\tan^{-1}(20\pi\tau),$$

$$v_{out}(t) = 4\underbrace{\frac{1}{\sqrt{1+400\pi^2\tau^2}}}_{|H(\omega_1)|}\cos\left(20\pi t - \frac{2}{3}\pi \underbrace{-\tan^{-1}(20\pi\tau)}_{\varphi(\omega_1)}\right)$$. Similarly,

$$v_{in}(t) = \sum_{k=1}^{N} c_k \cos(k\omega_1 t + \theta_k) \Rightarrow \begin{matrix} H_k = |H(k\omega_1)| \\ \varphi_k = \varphi(k\omega_1) \end{matrix} \Rightarrow v_{out}(t) = \sum_{k=1}^{N} H_k c_k \cos(k\omega_1 t + \theta_k + \varphi_k).$$

6.4.3 *Discrete time filters*

6.4.3.1 *Ideal DT filters*

Recalling that the DT spectrum is periodic with period 1 in digital frequency F, the ideal DT filters cut off portions of the DT spectrum analogously to the ideal CT filters:

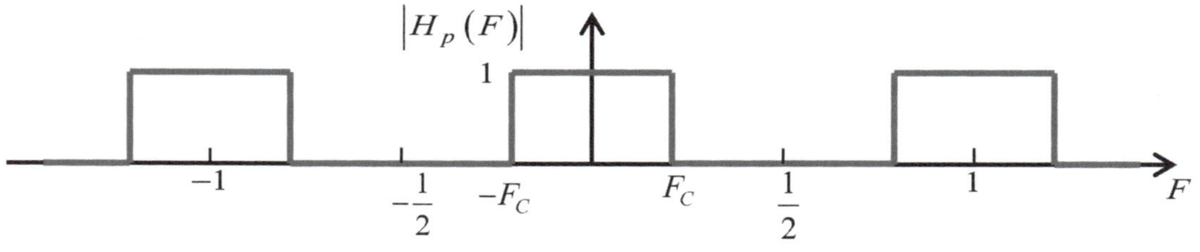

Fig. 6.4-12. DTFT magnitude yields ILPF magnitude spectral density, or frequency response.

DT filter response is computed through DT convolution:

$$y[n]=\sum_{k=-\infty}^{\infty}x[k]h[n-k]\Leftrightarrow Y_p(F)=H_p(F)X_p(F). \tag{6.4-9}$$

The IDTFT impulse response calculation of the DT ILPF yields a noncausal function of F:

$$\begin{aligned}h[n]&=\int_{-\frac{1}{2}}^{\frac{1}{2}}H_p(F)e^{j2\pi nF}dF=\int_{-F_C}^{F_C}e^{j2\pi nF}dF=\frac{1}{\pi n}\frac{1}{j2}\left(e^{j2\pi nF_C}-e^{-j2\pi nF_C}\right)\\&=\frac{\sin(2\pi nF_C)}{\pi n}=2F_C\operatorname{sinc}(2F_Cn).\end{aligned} \tag{6.4-10}$$

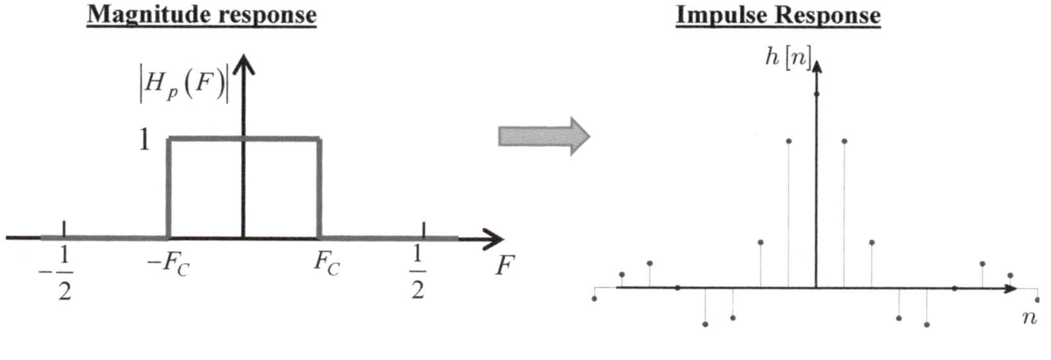

Fig. 6.4-13. Impulse response of a DT ILPF.

Applying to a DT signal a low-pass filter that cuts off DT signal frequency at Fc is equivalent to convolving this signal with the filter impulse response shown in Figure 6.4-13 for $-\infty \leq n \leq \infty$.

Since this convolution is noncausal — it involves a noncausal impulse response as well as the future values of the input signal — the ideal DT filters cannot be implemented for real-time computation.

The magnitude responses of the ideal CT filters are obtained by simply shifting and/or subtracting from 1 the ideal low-pass filter magnitude response:

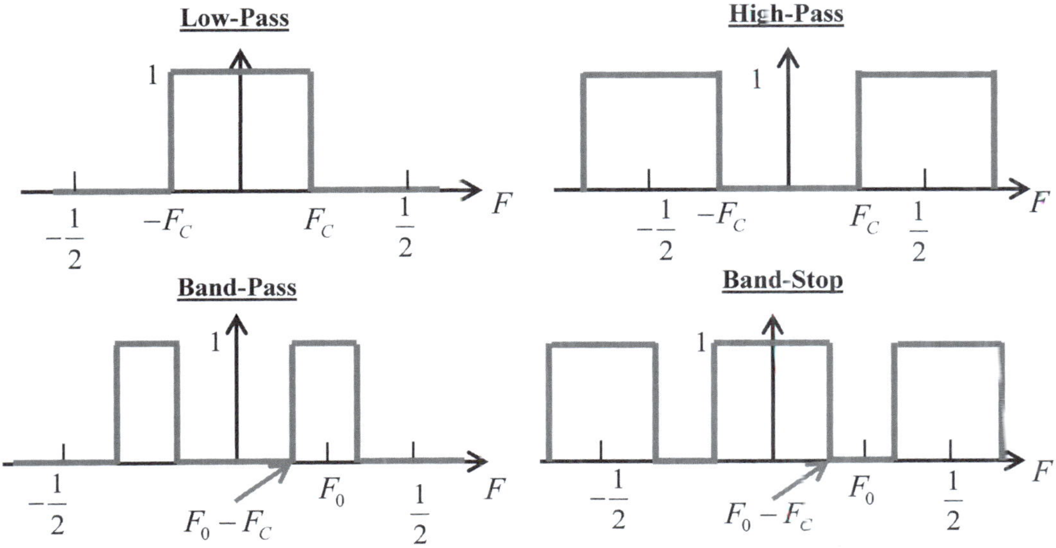

Fig. 6.4-14. DT magnitude spectral densities if the ideal filters.

6.4.3.2 *Real DT filters*

As real DT filter can be defined in a variety of ways: through a difference equation, a causal infinite impulse response (IIR), and a causal finite impulse response (FIR). Unlike a CT real filter, a DT real filter is characterized by the nondecaying to zero frequency response.

Example 6.4-5: First order real linear DT LPF is given by

$$y[n] = \alpha y[n-1] + x[n]. \tag{6.4-11}$$

Its impulse response and the transfer function (computed in (6.3-24)) are, respectively, given by

$$h[n] = \alpha^n u[n] \text{ and } H_p(F) = \frac{1}{1-\alpha e^{-j2\pi F}},$$

yielding magnitude and phase responses given, respectively, by

$$|H_p(F)| = \frac{1}{\sqrt{1-2\alpha\cos(2\pi F)+\alpha^2}} \quad \text{and} \quad \varphi_p(F) = -\tan^{-1}\left(\frac{\alpha\sin(2\pi F)}{1-\alpha\cos(2\pi F)}\right).$$

The corresponding magnitude and impulse response graphs are shown in Figure 6.4-15:

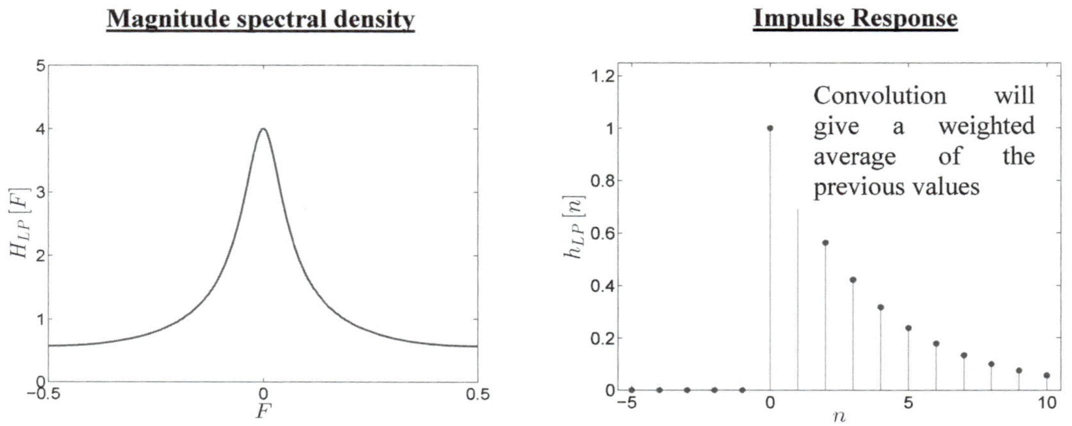

$$H_{LP}(F) = \frac{1}{1-\alpha e^{-j2\pi F}} \qquad h_{LP}[n] = \alpha^n u[n]$$

Fig. 6.4-15. DT real low pass filter.

Example 6.4-6: Given the first order real linear DT LPF, obtain the first order real linear DT HPF.

From the LPF of the previous example, we can quickly construct a HPF through the following simple consideration: instead of passing signals with frequencies only near 0, i.e. in the vicinity of the DC value, we want to pass signals with frequencies near the highest possible digital frequency, which is 0.5.

For this purpose, we shift the "zero place" $F = 0$ of the first order DT ILPF in the frequency axis F to F = 0.5 as follows:

$$H_{HP}(F) = H_{LP}(F+0.5) = \frac{1}{1-\alpha e^{-j2\pi(F+0.5)}} = \frac{1}{1-\alpha e^{-j2\pi F} e^{-j\pi}} = \frac{1}{1+\alpha e^{-j2\pi F}}. \quad (6.4\text{-}12)$$

As seen from (6.4-12), this leads to the α sign change, yielding

difference equation $$y[n+1] = -\alpha y[n] + x[n], \quad (6.4\text{-}13)$$

impulse response $$h[n] = (-\alpha)^n u[n], \quad (6.4\text{-}14)$$

magnitude response $$|H_p(F)| = \frac{1}{\sqrt{1+2\alpha\cos(2\pi F)+\alpha^2}}, \quad (6.4\text{-}15)$$

and phase response $$\varphi_p(F) = -\tan^{-1}\left(\frac{-\alpha\sin(2\pi F)}{1+\alpha\cos(2\pi F)}\right). \quad (6.4\text{-}16)$$

The first order DT HPF magnitude and impulse responses are shown in Figure 6.4-17.

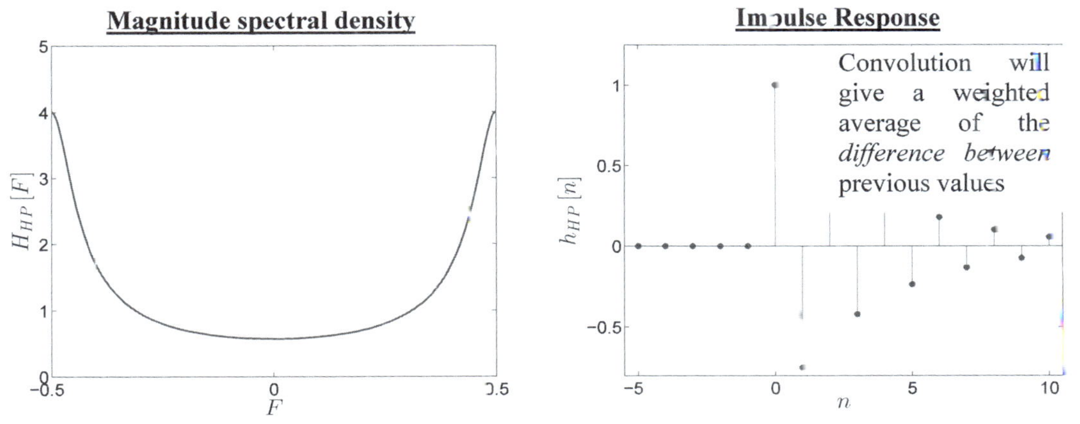

Fig. 6.4-16. DT first order real high pass filter magnitude and impulse responses.

Project 1: Solved Examples and Solution Guide

This project combines sampling, aliasing, and ideal and real filtering to build a conceptual and analytical framework for understanding the working of the front end of a data acquisition system.

Problem 1: A periodic half-wave rectified sine waveform $x_p(t)$ needs to be sampled with data acquisition system;

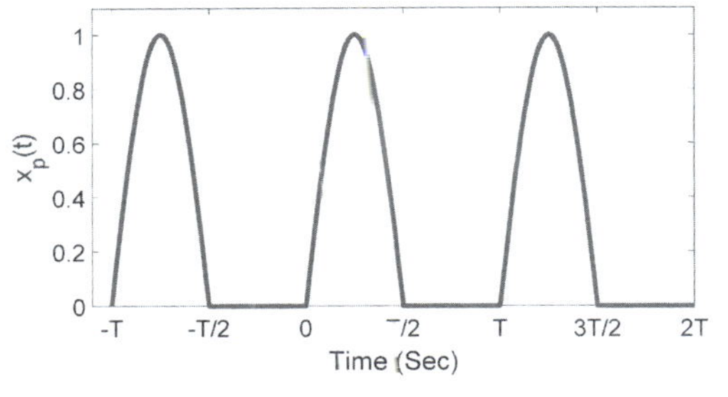

Fig. P1-1. Half-wave rectified sine waveform.

(a) For T = 0.1 sec., can the sampling period t_s of data acquisition system be selected to avoid aliasing? Explain, using the relationship between time domain and frequency domain for periodic signals.

Solution: It is impossible to avoid aliasing, since this is a non-smooth function. At each corner point (…, $-T/2$, 0, $T/2$, T, …), there is a step change in the first derivative (and higher derivatives, also, but the change in the first derivative is the obvious one). Step changes, in a function itself or its derivatives, cannot be achieved with finitely many Fourier Series terms that are infinitely continuously differentiable.

This means the signal is not bandlimited in the frequency domain. Therefore, it would require an infinitely large sampling rate to avoid aliasing, which is not possible. There is no upper bound, f_b, on the signal, and the inequality, $S_f > 2f_b$, can never be satisfied.

(b) Derive a polar Fourier series representation of $x_p(t)$ given above for $T = 0.1$ sec. and draw its single-sided magnitude, phase, and power spectra for the dc component plus the first seven harmonics.

Solution:

$$a_o = \tfrac{1}{T}\int_T x(t)\,dt = \tfrac{1}{0.1}\int_0^{\frac{T}{2}} \sin(2\pi 10t)\,dt = 10\int_0^{.05} \sin(20\pi t)\,dt$$
$$= 10\left(-\tfrac{1}{20\pi}\cos(20\pi t)\right)\Big|_0^{.05} = \tfrac{1}{2\pi}\left(-\cos(\pi) + \cos(0)\right) = \tfrac{1}{2\pi}(2) = \tfrac{1}{\pi},$$

$$a_k = \tfrac{2}{T}\int_T x(t)\cos(2\pi k f_1 t)\,dt = \tfrac{2}{0.1}\int_0^{0.05} \sin(20\pi t)\cos(20\pi kt)\,dt = 20\int_0^{0.05} \tfrac{1}{2}[\sin(20\pi t - 20\pi kt)$$
$$+ \sin(20\pi t + 20\pi kt)]dt = 10\int_0^{0.05}\left[\sin(20\pi t(1-k)) + \sin(20\pi t(1+k))\right]dt$$
$$= 10\left[\int_0^{0.05} \sin(20\pi t(1-k))\,dt + \int_0^{0.05} \sin(20\pi t(1+k))\,dt\right]$$
$$= 10\left[\tfrac{1}{20\pi(1-k)}\left(-\cos(20\pi t(1-k))\right)\right]_0^{0.05} + 10\left[\tfrac{1}{20\pi(1+k)}\left(-\cos(20\pi t(1+k))\right)\right]_0^{0.05}$$
$$= \tfrac{1}{2\pi}\left[\tfrac{1}{(1-k)}\left(-\cos(\pi(1-k)) + \overbrace{\cos(0)}^{1}\right) + \tfrac{1}{(1+k)}\left(-\cos(\pi(1+k)) + \overbrace{\cos(0)}^{1}\right)\right]$$

[using $\cos(\pi(1 \pm k)) = \cos(k\pi \pm \pi) = -\cos(k\pi)$]

$$= \tfrac{1}{2\pi}\left[\tfrac{1}{(1-k)}\left(\cos(k\pi)+1\right) + \tfrac{1}{(1+k)}\left(\cos(k\pi)+1\right)\right] = \tfrac{1}{2\pi}\left(\tfrac{1}{(1-k)} + \tfrac{1}{(1+k)}\right)\left(\cos(k\pi)+1\right)$$
$$= \tfrac{1}{2\pi}\left(\tfrac{(1+k)+(1-k)}{(1+k)(1-k)}\right)\left(\cos(k\pi)+1\right) = \tfrac{1}{2\pi}\tfrac{2}{\left(1-k^2\right)}\left(\cos(k\pi)+1\right) = \tfrac{1}{\pi\left(1-k^2\right)}\left(\cos(k\pi)+1\right).$$

Therefore we have: $a_k = \begin{cases} \frac{2}{\pi\left(1-k^2\right)}, & \text{even k,} \\ 0, & \text{odd k.} \end{cases}$

$$b_k = \tfrac{2}{T}\int_T x(t)\sin(2\pi k f_1 t)\,dt = 20\int_0^{0.05} \sin(20\pi t)\sin(20k\pi t)\,dt$$
$$= 20\int_0^{0.05} \tfrac{1}{2}\left[\cos(20\pi t - 20\pi kt) - \cos(20\pi t + 20\pi kt)\right]dt$$
$$= 10\left[\int_0^{0.05} \cos(20\pi t(1-k))\,dt - \int_0^{0.05} \cos(20\pi t(1+k))\,dt\right]$$
$$= 10\left[\tfrac{1}{20\pi(1-k)}\sin(20\pi t(1-k))\right]_0^{0.05} - 10\left[\tfrac{1}{20\pi(1+k)}\sin(20\pi t(1+k))\right]_0^{0.05}$$

$$=\tfrac{1}{2\pi}\left(\tfrac{1}{1-k}-\tfrac{1}{1+k}\right)[(\overbrace{\sin\left(\pi(1-k)\right)}^{0}-\overbrace{\sin(0)}^{0})-(\overbrace{\sin\left(\pi(1+k)\right)}^{0}-\overbrace{\sin(0)}^{0})]=0.$$

This calculation leaves b_1 undefined due to $1-k$ in the denominator, so it should be calculated separately. Starting from $b_k=\frac{2}{T}\int_T x(t)\sin(2\pi k f_1 t)\,dt$ and substituting $k=1$ yields:

$$\begin{aligned}b_1&=10\left[\int_0^{0.05}\cos\left(20\pi t(1-1)\right)dt-\int_0^{0.05}\cos\left(20\pi t(1+1)\right)dt\right]\\&=10\left[\int_0^{0.05}\cos(0)\,dt-\int_0^{0.05}\cos(40\pi t)\,dt\right]=10\left[\int_0^{0.05}1\,dt-\int_0^{0.05}\cos(40\pi t)\,dt\right]\\&=10\cdot 0.05-10\left[\tfrac{1}{40\pi}\sin(40\pi t)\right]_0^{0.05}=0.5-10\left[\sin(2\pi)-\sin(0)\right]=0.5.\end{aligned}$$

Therefore we have: $b_k=\begin{cases}\frac{1}{2}, & k=1,\\ 0, & \text{else.}\end{cases}$

Now transform to Polar form:

$$c_1=\sqrt{a_1^2+b_1^2}=\tfrac{1}{2};\ \theta_1=\tan^{-1}\left(-\tfrac{b_1}{a_1}\right)=-\tan^{-1}\left(\tfrac{0.5}{0}\right)=-\tfrac{\pi}{2};\ c_2=\left|\frac{2}{\pi\left(1-k^2\right)}\right|_{k=2}=\tfrac{2}{3\pi};\ \theta_2=\pi;$$

$$c_3=0;\ \theta_3=0;\ \ c_4=\left|\frac{2}{\pi\left(1-k^2\right)}\right|_{k=4}=\tfrac{2}{15\pi}\ ;\ \theta_4=\pi;\ c_5=0;\ \theta_5=0;$$

$$c_6=\left|\frac{2}{\pi\left(1-k^2\right)}\right|_{k=6}=\tfrac{2}{35\pi}\ ;\ \theta_6=\pi;\ \ c_7=0;\ \theta_7=0.$$

So, $x_p(t)=c_o+\sum_{k=1}^{7}c_k\cos\left(2\pi k f_1 t+\theta_k\right)$

$$=\tfrac{1}{\pi}+\tfrac{1}{2}\cos\left(2\pi 10t+\tfrac{\pi}{2}\right)+\tfrac{2}{3\pi}\cos\left(2\pi 20t+\pi\right)+\tfrac{2}{15\pi}\cos\left(2\pi 40t+\pi\right)+\tfrac{2}{35\pi}\cos\left(2\pi 60t+\pi\right).$$

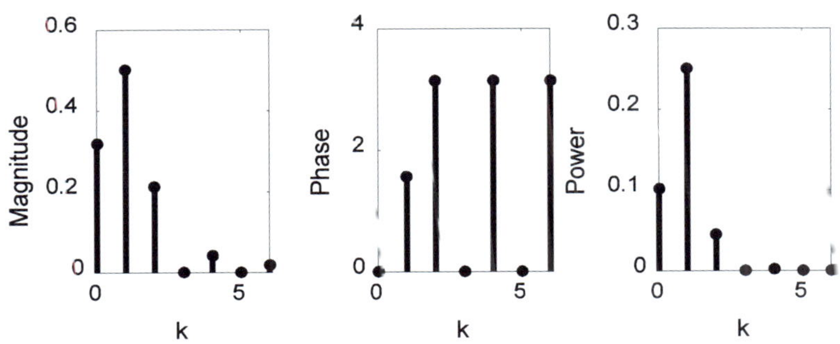

Fig. P1-2. Single-sided magnitude, phase, and power spectra.

(c) Find the sampled version of the first seven harmonics of $x_p(t)$ in (b) with digital frequency $|F| \le 1/2$ if the sampling period is fixed at $t_s = \frac{1}{70}$ sec.

Solution:

$$t_s = \tfrac{1}{70}\text{ sec} \Rightarrow S_f = 70\text{ Hz} \Rightarrow F_1 = \tfrac{f_o}{S_f} = \tfrac{10}{70} = \tfrac{1}{7},\quad x(n) = c_k\cos(2\pi k F_1 n + \theta_k),$$

$$k=1:\ x_1(n) = \tfrac{1}{2}\cos\left(2\pi\tfrac{1}{7}n + \tfrac{\pi}{2}\right),\quad k=2:\ x_2(n) = \tfrac{2}{3\pi}\cos\left(2\pi\tfrac{2}{7}n + \pi\right),$$

$$k=4:\ x_4(n) = \tfrac{2}{15\pi}\cos\left(2\pi\tfrac{4}{7}n + \pi\right) \Rightarrow \text{ folds to: } \tfrac{2}{15\pi}\cos\left(2\pi\left(\tfrac{4}{7}-1\right)n + \pi\right) = \tfrac{2}{15\pi}\cos\left(2\pi\tfrac{3}{7}n - \pi\right),$$

$$k=6:\ x_6(n) = \tfrac{2}{35\pi}\cos\left(2\pi\tfrac{6}{7}n + \pi\right) \Rightarrow \text{ folds to: } \tfrac{2}{35\pi}\cos\left(2\pi\left(\tfrac{6}{7}-1\right)n + \pi\right) = \tfrac{2}{35\pi}\cos\left(2\pi\tfrac{1}{7}n - \pi\right).$$

The 4th, 5th, 6th, and 7th harmonics get aliased (i.e. mapped to lower frequency) and fold onto the 1st, the 2nd, the 3rd harmonics, and the dc component, respectively. However, the 5th and the 7th harmonics have zero magnitude and phase (from part (b)) and therefore do not contribute to the aliasing.

$$x_p(t) = \tfrac{1}{\pi} + \tfrac{1}{2}\cos\left(2\pi 10t + \tfrac{\pi}{2}\right) + \tfrac{2}{3\pi}\cos(2\pi 20t + \pi) + \tfrac{2}{15\pi}\cos(2\pi 40t + \pi) + \tfrac{2}{35\pi}\cos(2\pi 60t + \pi)$$

$$x_p[n] = \tfrac{1}{\pi} + \tfrac{1}{2}\cos\left(2\pi\tfrac{1}{7}n + \tfrac{\pi}{2}\right) + \tfrac{2}{35\pi}\cos\left(2\pi\tfrac{1}{7}n - \pi\right) + \tfrac{2}{3\pi}\cos\left(2\pi\tfrac{2}{7}n + \pi\right) + \tfrac{2}{15\pi}\cos\left(2\pi\tfrac{3}{7}n + \pi\right)$$

$$= \tfrac{1}{\pi} + \overbrace{2\sqrt{\left(\tfrac{1}{2}\right)^2 + \left(\tfrac{2}{35\pi}\right)^2}}^{0.5003}\cos\left[2\pi\tfrac{1}{7}n + \pi - \tan^{-1}\left(\tfrac{35\pi}{4}\right)\right] + \tfrac{2}{3\pi}\cos\left(2\pi\tfrac{2}{7}n + \pi\right) + \tfrac{2}{15\pi}\cos\left(2\pi\tfrac{3}{7}n + \pi\right),$$

where the first harmonic has been computed by folding the aliased 6th harmonic on top of the nonaliased one:

$$\tfrac{1}{2}\cos\left(2\pi\tfrac{1}{7}n + \tfrac{\pi}{2}\right) + \tfrac{2}{35\pi}\cos\left(2\pi\tfrac{1}{7}n - \pi\right) = \tfrac{1}{4}e^{-j\frac{\pi}{2}}e^{-j2\pi\frac{1}{7}n} + \tfrac{1}{35\pi}e^{j\pi}e^{-j2\pi\frac{1}{7}n} + \tfrac{1}{4}e^{j\frac{\pi}{2}}e^{j2\pi\frac{1}{7}n}$$
$$+ \tfrac{1}{35\pi}e^{-j\pi}e^{j2\pi\frac{1}{7}n} = \left(\tfrac{1}{4}e^{-j\pi/2} + \tfrac{1}{35\pi}e^{j\pi}\right)e^{-j2\pi\frac{1}{7}n} + \left(\tfrac{1}{4}e^{j\pi/2} + \tfrac{1}{35\pi}e^{-j\pi}\right)e^{j2\pi\frac{1}{7}n}$$

— decompose further using Euler's formula; note that only the expression in the parenthesis in the first term needs to be reduced, since the expression in the parenthesis in the second term will be its complex conjugate:

$$= [\tfrac{1}{4}(\cos\pi/2 - j\sin\pi/2) + \tfrac{1}{35\pi}(\cos\pi + j\sin\pi)]e^{-j2\pi\frac{1}{7}n} + [\tfrac{1}{4}(\cos\pi/2 + j\sin\pi/2)$$
$$+ \tfrac{1}{35\pi}(\cos\pi - j\sin\pi)]e^{j2\pi\frac{1}{7}n} = \left(-\tfrac{1}{4}j - \tfrac{1}{35\pi}\right)e^{-j2\pi\frac{1}{7}n} + \left(\tfrac{1}{4}j - \tfrac{1}{35\pi}\right)e^{j2\pi\frac{1}{7}n}.$$

Now, establish the quadrant for the polar representation of the complex number in the first term parenthesis, and, respectively, the complex number in the second parenthesis, and put both into the polar form:

$$= \left[\sqrt{\left(\tfrac{1}{4}\right)^2 + \left(\tfrac{1}{35\pi}\right)^2}\,e^{-j\left(\pi - \tan^{-1}\frac{1/4}{1/35\pi}\right)}\right]e^{-j2\pi\frac{1}{7}n} + \left[\sqrt{\left(\tfrac{1}{4}\right)^2 + \left(\tfrac{1}{35\pi}\right)^2}\,e^{j\left(\pi - \tan^{-1}\frac{1/4}{1/35\pi}\right)}\right]e^{j2\pi\frac{1}{7}n}$$

$$= 2\sqrt{\left(\tfrac{1}{4}\right)^2 + \left(\tfrac{1}{35\pi}\right)^2} \cos\left[2\pi \tfrac{1}{7} n + \left(\pi - \tan^{-1} \tfrac{35\pi}{4}\right)\right].$$

(d) Suppose now that you can pass $x_p(t)$ in (b) through an ideal low pass filter before sampling. Select the maximum possible bandwidth of this ideal low pass filter to avoid aliasing when you sample its output with the sampling period of part (c). For the purpose of selection of the filter bandwidth assume nonzero harmonic magnitudes at all harmonic numbers $k = 1, 2, 3, 4, \ldots$ Draw the output of this filter: include both time domain and single-sided magnitude, phase, and power spectra plots.

Solution: From part (c), $S_f = 70$ Hz. Therefore, we bandlimit the CT signal to $f_c = \frac{S_f}{2} = 35$ Hz to avoid any aliasing after sampling. However, since there are no aliased components with frequencies less than 40 Hz (4th harmonic) we select the ideal filter bandwidth slightly less than 40 Hz, i.e. $40 - \epsilon$, where ϵ is a small number.

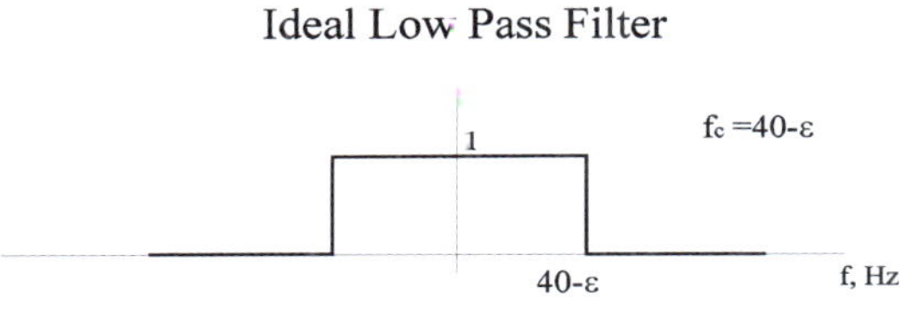

Fig. P1-3. Ideal low pass filter.

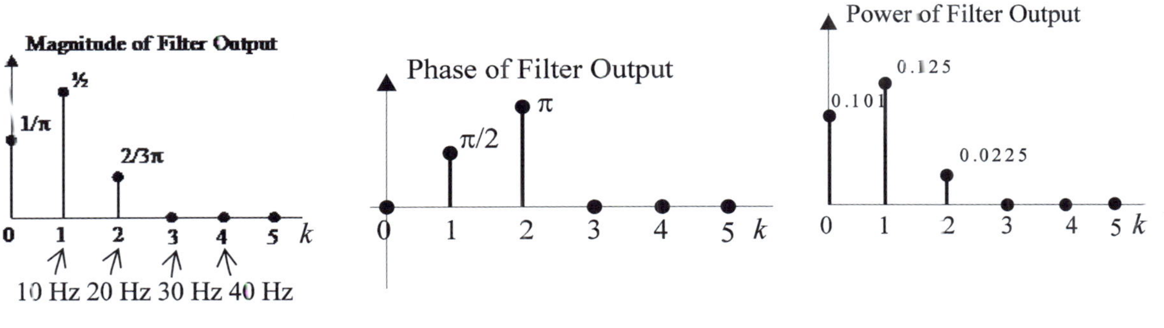

Fig. P1-4. Single side magnitude spectrum, phase spectrum, and power spectrum of filter output.

(e) Find the normalized power P_x of $x_p(t)$ of (b).

Solution: we note that according to Figure P1-1 we need to integrate $x_p^2(t)$ over half-period:

$$P_x = \frac{1}{T}\int_0^T x_p(t)^2\,dt = \frac{1}{T}\int_o^{T/2}\sin^2\left(2\pi\frac{1}{T}t\right)dt = \frac{1}{T}\int_o^{T/2}\frac{1}{2}\left(1-\cos\left(4\pi\frac{1}{T}t\right)\right)dt = \frac{1}{2T}\left[\int_o^{T/2}dt - \right.$$

$$\left.\int_0^{T/2}\cos\left(4\pi\frac{1}{T}t\right)dt\right] = \frac{1}{2T}\left[\left(\frac{T}{2}\right) - \frac{1}{4\pi\frac{1}{T}}\sin\left(4\pi\frac{1}{T}t\right)\Big|_0^{\frac{T}{2}}\right] = \frac{1}{4} - \frac{1}{8\pi}\left(\sin\left(4\pi\frac{1}{T}\frac{T}{2}\right) - \sin(0)\right) = \frac{1}{4}.$$

(f) What percentage of P_x does the output of the ideal filter of (d) contain? What can you say now about the accuracy of your sampled version even though it has no aliasing? Would smaller t_s help? Explain.

Solution:

P_x for filtered output: $P_{xf} = c_o^2 + \frac{1}{2}c_1^2 + \frac{1}{2}c_2^2 = \frac{1}{\pi^2} + \frac{1}{8} + \frac{2}{9\pi^2} = 0.2488 \Rightarrow$

$\frac{P_{xf}}{P_x} = \frac{0.2488}{0.25} = .9952 \quad \Rightarrow \quad 99.52\%$ of the original power recovered after filtering.

Very little signal is lost due to filtering. A smaller t_s would improve the accuracy of the sampled signal, capturing more of the harmonics with less loss after filtering.

(g) Now, we will make a conceptual transition to real filtering. Suppose you have selected in part (d) an ideal low pass filter with the bandwidth twice higher than that required to prevent aliasing for $t_s = 1/70$ sec. What percentage of P_x will correspond to components of $x_p(t)$ which will be aliased?

Solution: The ideal filter with the cutoff frequency twice higher than in (d) will have the passband $2f_c = 80 - 2\varepsilon$. So, this filter passes up to the 7th harmonic (70Hz) through. Then the filtered signal is

$$x_{pf}(t) = x_p(t) = c_o + \sum_{k=1,2,4,6} c_k\cos(2\pi kf_1 t + \theta_k)$$

$$= \frac{1}{\pi} + \frac{1}{2}\cos\left(2\pi 10t + \frac{\pi}{2}\right) + \frac{2}{3\pi}\cos(2\pi 20t + \pi) + \frac{2}{15\pi}\cos(2\pi 40t + \pi) + \frac{2}{35\pi}\cos(2\pi 60t + \pi)$$

and $c_4\cos(2\pi 40t + \theta_4)$ and $c_6\cos(2\pi 60t + \theta_6)$ will be aliased.

$$P_4 = \frac{1}{2}\left(\frac{2}{15\pi}\right)^2 = 9.01\cdot 10^{-4};\ P_6 = \frac{1}{2}\left(\frac{2}{35\pi}\right)^2 = 1.65\cdot 10^{-4}.$$

So the aliased components constitute:

$$\frac{(P_4 + P_6)}{P_x} = \frac{9.01\cdot 10^{-4} + 1.65\cdot 10^{-4}}{0.25} = 4.3\cdot 10^{-3} = 0.43\% \text{ of the total power passed.}$$

Problem 2: Suppose a periodic waveform $x_p(t)$ of Problem 1 ($T = 0.1$ sec) is sampled with $t_s = \frac{1}{70}$ sec. and prior to sampling $x_p(t)$ is passed through (anti-aliasing) low pass filter implemented as an RC circuit whose output is the capacitor voltage.

(a) Select the values of R and C of a low pass RC filter to place its break frequency (bandwidth) in the middle between the frequencies corresponding to the highest nonaliased and the lowest aliased harmonic frequencies. For the purpose of selection of the filter

bandwidth assume nonzero harmonic magnitudes at all harmonic numbers ..., −4, −3, −2, −1, 0, 1, 2, 3, 4, ... Write analytical expressions for the filter magnitude response wrt rad/sec, and magnitude density, phase, and energy spectral density wrt Hz, and draw the corresponding double-sided plots. Derive the impulse response of this filter and explain the relation between the filter bandwidth and the speed of decay of its impulse response.

Solution: Place the break, or cutoff, frequency between the aliased and non-aliased harmonic frequencies:

highest non-aliased frequency: 30 Hz, lowest aliased frequency: 40 Hz

break frequency: $f_c = \frac{30+40}{2} = 35$ Hz $\Rightarrow$ $\omega_c = 2\pi f_c = 70\pi = 219.9$ rad/sec.

Then, $\tau = \frac{1}{\omega_c} = \frac{1}{70\pi}$.

And for low-pass RC filter: $H(j\omega) = \frac{1}{1+j\omega RC} \Rightarrow \tau = RC = \frac{1}{70\pi}$.

So any combination of resistor/capacitor whose product is $\frac{1}{70\pi}$ will fulfill these requirements, e.g., $R = 1000\Omega$ and $C = 4.54\ \mu F$.

Magnitude response: $|H(j\omega)| = \frac{1}{\sqrt{1+(\omega\tau)^2}} = \frac{1}{\sqrt{1+\frac{\omega^2}{(70\pi)^2}}}$,

magnitude spectral density: $|H(f)| = \frac{1}{\sqrt{1+(2\pi f\tau)^2}} = \frac{1}{\sqrt{1+\frac{(2\pi f)^2}{(70\pi)^2}}} = \frac{1}{\sqrt{1+(\frac{f}{35})^2}}$,

phase: $\theta_\omega = -\tan^{-1}(2\pi f\tau) = -\tan^{-1}\left(\frac{f}{35}\right)$,

energy spectral density $|H(f)|^2 = \left|\frac{1}{\sqrt{1+(2\pi f\tau)^2}}\right|^2 = \frac{1}{1+(2\pi f\tau)^2} = \frac{1}{1+(\frac{f}{35})^2}$.

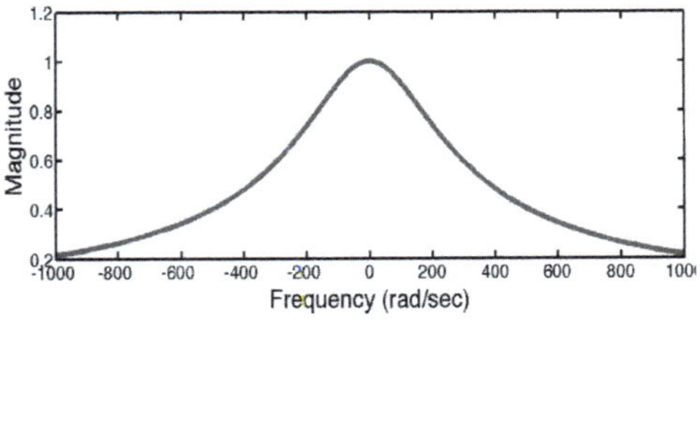

Fig. P1-5. Double sided magnitude response of the filter.

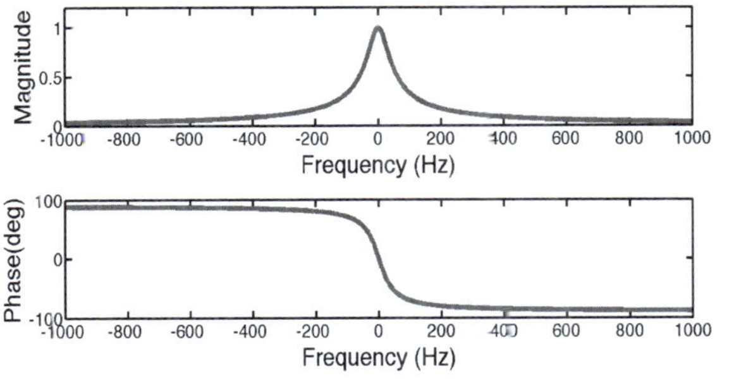

Fig. P1-6. Double sided magnitude spectral density and phase of the filter.

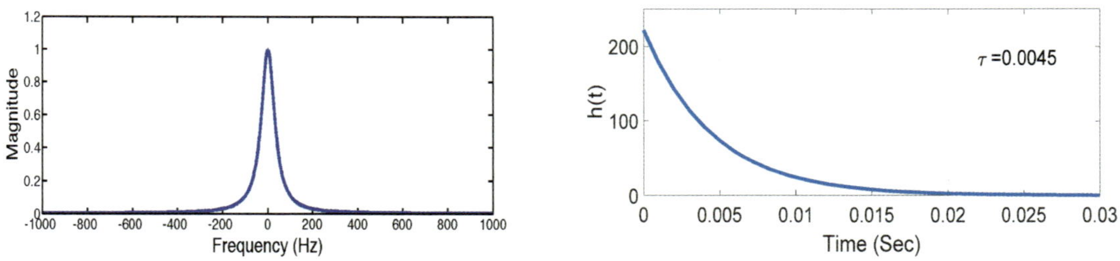

Fig. P1-7. Double sided energy spectral density and impulse response of the filter.

Impulse Response: $H(j\omega)=\frac{1}{1+j\omega\tau} \Rightarrow H(s)=\frac{1}{1+s\tau}=\frac{1}{\tau}\left(\frac{1}{s+\frac{1}{\tau}}\right) \overset{\text{inverse Laplace } \pounds^{-1}}{\Rightarrow} h(t)=\frac{1}{\tau}e^{-\frac{t}{\tau}},\ t\geq 0.$

Filter bandwidth: $\omega_b=2\pi f_b=70\pi=\frac{1}{\tau}=219.8$ rad/sec, or $f_b=35$ Hz.

As filter bandwidth increases, the filter time constant, τ, decreases, i.e. the system decays faster with increase in bandwidth (see the plot):

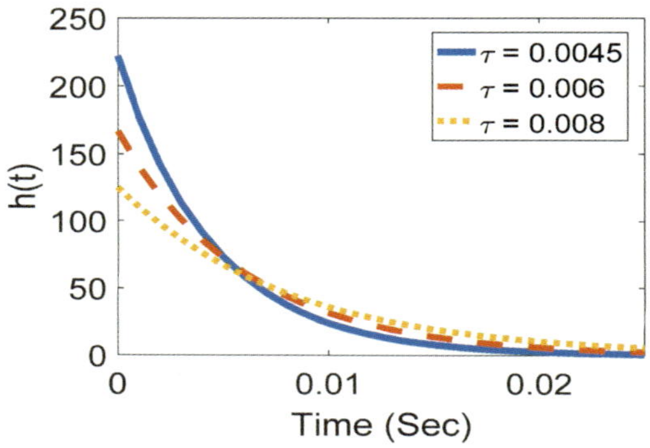

Fig. P1-8. Impulse response of the filter with different time constants.

(b) Compute and draw the double-sided magnitude spectrum of the output (take seven harmonics on each side) of the RC filter if the filter input is $x_p(t)$ defined above.

Solution: We can compute the filtering effect by using the expression for the periodic filtered signal

$$x_{pf}(t)=c_o+\sum_{k=1}^{6}c_k\left|H(jk\omega_1)\right|_{\omega=2\pi kf_1}\cos\left(2\pi kf_1t+\theta_k+\angle(H(jk\omega_1)_{\omega=2\pi kf_1}\right),$$

$$\left|H(j\omega)\right|=\frac{1}{\sqrt{1+(\omega\tau)^2}}=\frac{1}{\sqrt{1+\frac{\omega^2}{(70\pi)^2}}}.$$

Calculation of the double-sided spectrum of the output signal and the corresponding spectral plot are given in the Figure P1-9 below:

k	$k\omega_1$	$\lvert H(jk\omega_1)\rvert$	c_k	$\lvert X_s[k]\rvert$	$\lvert H(jk\omega_1)\rvert \times \lvert X_s[k]\rvert$
0	0	1	$1/\pi$	$1/\pi$	0.3183
±1	$\pm20\pi$	0.9615	$1/2$	$1/4$	0.2404
±2	$\pm40\pi$	0.8682	$2/3\pi$	$1/3\pi$	0.0921
±3	$\pm60\pi$	0.7593	0	0	0
±4	$\pm80\pi$	0.6585	$2/15\pi$	$1/15\pi$	0.0140
±5	$\pm100\pi$	0.5735	0	0	0
±6	$\pm120\pi$	0.5039	$2/35\pi$	$1/35\pi$	0.0046
±7	$\pm140\pi$	0.4472	0	0	0

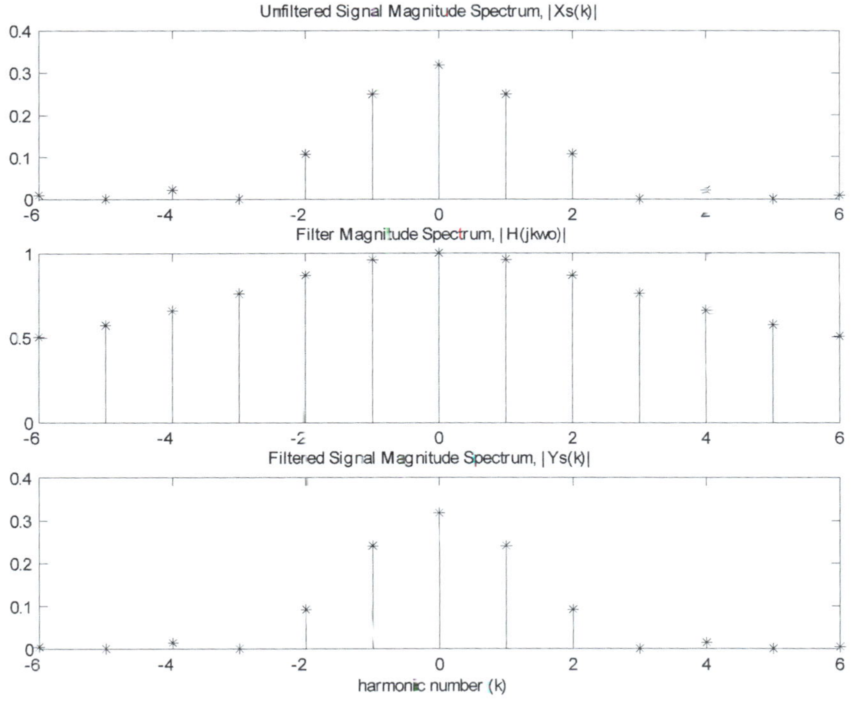

Fig. P1-9. Two-sided magnitude spectrum of filter output.

(c) Now we have noticed that the real filter works quite unlike a correctly selected ideal filter: it decreases the magnitude of the harmonics that are not going to be aliased after sampling and does not completely block harmonics which will be aliased after sampling, like in part (g) of Problem 1. That is why we need a certain amount of oversampling. Let us analyze these drawbacks quantitatively. Let us first consider the filter input signal $x_p(t)$. What is the percentage of the cumulative power of the three lowest frequency harmonics of $x_p(t)$ which will

be aliased after sampling in relation to the cumulative power of the harmonics of $x_p(t)$ which will not be aliased? Now, let us look at the output of our RC filter, (we will call it $y_p(t)$). What is the percentage of the cumulative power of the three lowest frequency harmonics of $y_p(t)$ which will be aliased after sampling in relation to the cumulative power of harmonics of $y_p(t)$ which will not be aliased? Does your RC filter do a good job to prevent aliasing? Explain. How much oversampling is needed to practically prevent aliasing? Explain.

Solution: What is the percentage of power of harmonics that will be aliased compared to that of harmonics that will not be aliased in the input signal?

Harmonics to be aliased: c_4, c_5, c_6, c_7:

$$P_{aliased} = \tfrac{1}{2}\left(\tfrac{2}{15\pi}\right)^2 + 0 + \tfrac{1}{2}\left(\tfrac{2}{35\pi}\right)^2 + 0 = 0.0011.$$

Harmonics to be not aliased: c_0, c_1, c_2:

$$P_{non-aliased} = \left(\tfrac{1}{\pi}\right)^2 + \tfrac{1}{2}\left(\tfrac{1}{2}\right)^2 + \tfrac{1}{2}\left(\tfrac{2}{3\pi}\right)^2 = 0.2488.$$

$$\frac{P_{aliased}}{P_{non-aliased}} = \frac{0.0011}{0.2488} = 0.0044 = 0.44\%.$$

From the filter output: (y_p) from the table in (b)

$$P_{aliased} = 2\left(1.397\cdot 10^{-2}\right)^2 + 0 + 2\left(4.582\cdot 10^{-3}\right)^2 + 0 = 4.32\times 10^{-4}.$$

$$P_{non-aliased} = \left(\tfrac{1}{\pi}\right)^2 + 2\left(0.2404\right)^2 + 2\left(9.212\cdot 10^{-2}\right)^2 = 0.2339.$$

$$\frac{P_{aliased}}{P_{non-aliased}} = \frac{4.32\times 10^{-4}}{0.2339} = 1.85\times 10^{-3} = 0.185\%.$$

So, the filter does reduce the contribution of the aliased values. However, the ideal filter does a perfect job because it eliminates everything outside of the range. 10x oversampling should effectively prevent aliasing.

Problem 3. Suppose the RC filter periodic waveform output signal $y_p(t)$ of Problem 2(b) is sampled with $S_F = 64$ Hz. and after sampling the sampled $y_p(nt_s)$ is passed (assume that only the dc component and the first six real harmonics have been sampled, and neglect the rest) through digital filter with pulse response that is obtained by first scaling the impulse response of RC filter from Problem 2(a) by multiplying it by 1/160, then sampling the result with $S_F = 160$ Hz.

(a) Compute frequency response of this digital filter. Write analytical expressions for magnitude density, phase, and energy spectral density of this filter and draw the corresponding double-sided plots. Explain the relation between the filter bandwidth and the speed of decay of its impulse response and classify this filter.

Solution: The impulse response for the RC filter was: $h(t) = 70\pi e^{-70\pi t}u(t)$.

The pulse response of the digital filter is given to be this signal, sampled with rate 160 Hz, and also scaled by 1/160. Writing this out analytically takes the form

$$h[n]=\frac{1}{160}h\left(\frac{n}{S_F}\right)=\frac{1}{160}70\pi e^{-70\pi\frac{n}{S_F}}u(t_s n)=\frac{7\pi}{16}e^{-\frac{70\pi}{160}n}u[n]=\frac{7\pi}{16}\left(e^{-\frac{7}{16}\pi}\right)^n u[n].$$

Since $0<e^{-\frac{7\pi}{16}}<1$, we can calculate the DTFT of this pulse response as:

$$H(F)=\sum_{n=-\infty}^{\infty}h[n]e^{-j2\pi Fn}=\sum_{n=-\infty}^{\infty}\left(\frac{7\pi}{16}(e^{-\frac{7\pi}{16}})^n u[n]\right)e^{-j2\pi Fn}=\frac{7\pi}{16}\sum_{n=0}^{\infty}(e^{-\frac{7\pi}{16}-j2\pi F})^n=$$

$$\frac{7\pi}{16}\frac{1}{1-e^{-\frac{7\pi}{16}}e^{-j2\pi F}}=\frac{7\pi}{16}\frac{1}{1-e^{-\frac{7\pi}{16}}\left(\cos(2\pi F)-j\sin(2\pi F)\right)}=$$

$$\frac{7\pi}{16}\frac{1}{1-e^{-\frac{7\pi}{16}}\cos(2\pi F)+je^{-\frac{7\pi}{16}}\sin(2\pi F)}\approx\frac{1.3744}{1-0.2530\cos(2\pi F)+j0.2530\sin(2\pi F)}.$$

This yields the magnitude response wrt F, which is also magnitude spectral density:

$$|H(F)|=\left|\frac{7\pi}{16}\frac{1}{1-e^{-\frac{7\pi}{16}}\cos(2\pi F)+je^{-\frac{7\pi}{16}}\sin(2\pi F)}\right|$$

$$=\frac{7\pi}{16}\cdot\frac{1}{\left|1-e^{-\frac{7\pi}{16}}\cos(2\pi F)+je^{-\frac{7\pi}{16}}\sin(2\pi F)\right|}=\frac{7\pi}{16}\cdot\frac{1}{\sqrt{[1-e^{-\frac{7\pi}{16}}\cos(2\pi F)]^2+[e^{-\frac{7\pi}{16}}\sin(2\pi F)]^2}}$$

$$=\frac{7\pi}{16}\cdot\frac{1}{\sqrt{1-2e^{-\frac{7\pi}{16}}\cos(2\pi F)+e^{-\frac{7\pi}{8}}\cos^2(2\pi F)+e^{-\frac{7\pi}{8}}\sin^2(2\pi F)}}$$

$$=\frac{7\pi}{16}\cdot\frac{1}{\sqrt{1-2e^{-\frac{7\pi}{16}}\cos(2\pi F)+e^{-\frac{7\pi}{8}}}}\approx\frac{1.3744}{\sqrt{1.0640-0.5060\cos(2\pi F)}},$$

the phase:

$$\angle H(F)=\angle\left(\frac{7\pi}{16}\right)-\angle\left(1-e^{-\frac{7\pi}{16}}\cos(2\pi F)+je^{-\frac{7\pi}{16}}\sin(2\pi F)\right)=0-$$

$$\tan^{-1}\left(\frac{e^{-\frac{7\pi}{16}}\sin(2\pi F)}{1-e^{-\frac{7\pi}{16}}\cos(2\pi F)}\right)=-\tan^{-1}\left(\frac{e^{-\frac{7\pi}{16}}\sin(2\pi F)}{1-e^{-\frac{7\pi}{16}}\cos(2\pi F)}\right)\approx-\tan^{-1}\left(\frac{0.2530\sin(2\pi F)}{1-0.2530\cos(2\pi F)}\right),$$

and the energy spectral density:

$$\left|H(F)\right|^2 = \frac{49\pi^2}{256} \cdot \frac{1}{1 - 2e^{-\frac{7\pi}{16}}\cos(2\pi F) + e^{-\frac{7\pi}{8}}} \approx \frac{1.8891}{1.0640 - 0.5060\cos(2\pi F)},$$

all periodic, with period 1. The corresponding principal range plots are given below:

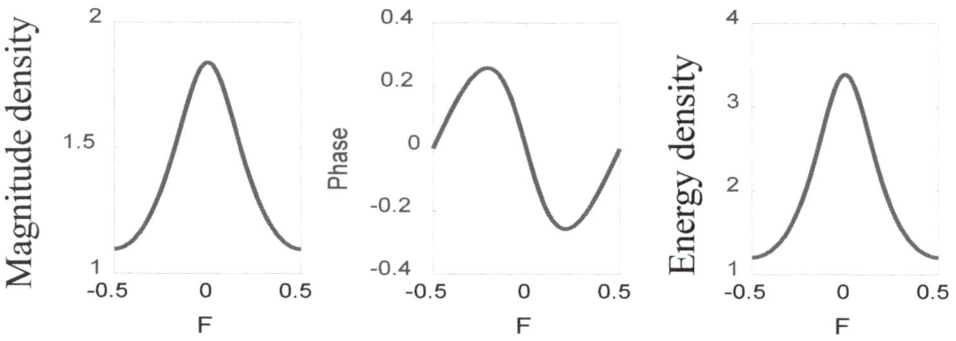

Fig. P1-10. Double-sided magnitude spectral density, phase, and energy spectral density (ESD) plots of a digital filter.

(b) Compute and draw the double-sided magnitude spectrum of the output if the filter input is $y_p(nt_s)$ defined above (with only dc component and the first six harmonics of the double-sided spectrum considered to have been sampled on both sides).

Solution: We can continue the table in Problem 2(b). Here, the input magnitudes are $\left|Y_s[k]\right| = \left|X_{sf}[k]\right|$, the filtered output magnitudes from Problem 2(b).

To find the effect of the digital filter, we first need to calculate the digital frequency, which can be found using the usual sampling relationships:

$$F = \frac{f}{S_f} = \frac{\omega t_s}{2\pi}.$$

In order to draw the double-sided magnitude spectrum correctly, we also have to take into account aliasing by making sure that each signal has digital frequency in the principal range [−1/2, 1/2]. These aliased frequencies are denoted as F_a in the table below.

Then, the magnitude response of the digital filter $\left|H(F)\right|$ can be calculated using the equations from part (a). Multiplying the input magnitudes by the filter magnitude gives the output magnitudes.

The magnitude spectrum of a DT signal repeats itself with period 1, so only the principal range needs to be plotted. The −1.5 to 1.5 range is plotted hear to illustrate the overlapping spectral images due to aliasing.

k	$\omega = k\omega_1$	$\lvert Y_s[k] \rvert$ (from 2b)	$F = \omega t_s/2\pi$	F_a	$\lvert H(F) \rvert$	$\lvert H(F) \rvert \times \lvert Y_s[k] \rvert$
0	0	0.3183	0	0	1.8399	0.5856
± 1	$\pm 20\pi$	0.2404	± 0.1563	± 0.1563	1.5534	0.3734
± 2	$\pm 40\pi$	0.0921	± 0.3125	± 0.3125	1.2256	0.1129
± 3	$\pm 60\pi$	0	± 0.4688	± 0.4688	1 1004	0
± 4	$\pm 80\pi$	0.0140	± 0.6250	∓ 0.3750	1.1527	0.0161
± 5	$\pm 100\pi$	0	± 0.7813	∓ 0.2188	1.3989	0
± 6	$\pm 120\pi$	0.0046	± 0.9375	∓ 0.0625	1.7795	0.0082
± 7	$\pm 140\pi$	0	± 1.0938	± 0.0938	1.7135	0

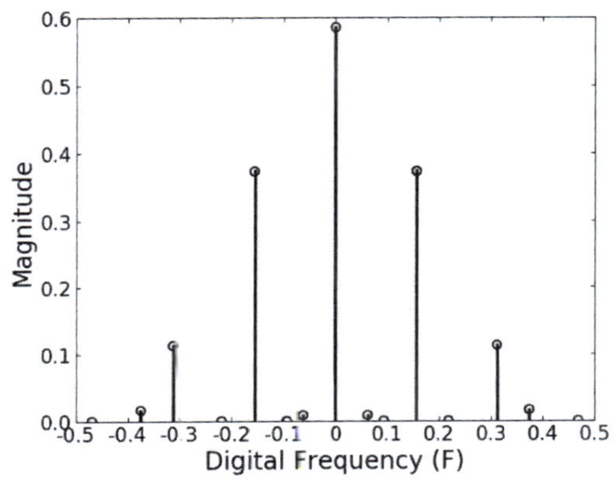

Fig. P1-11. The two-sided filter output signal magnitude spectrum.

Exam 1: Solved Examples and Solution Guide

Problem 1. Consider the filter (we will call it CR filter) shown in Figure 1 below with R = 1000 ohm and C = 0.001 F.

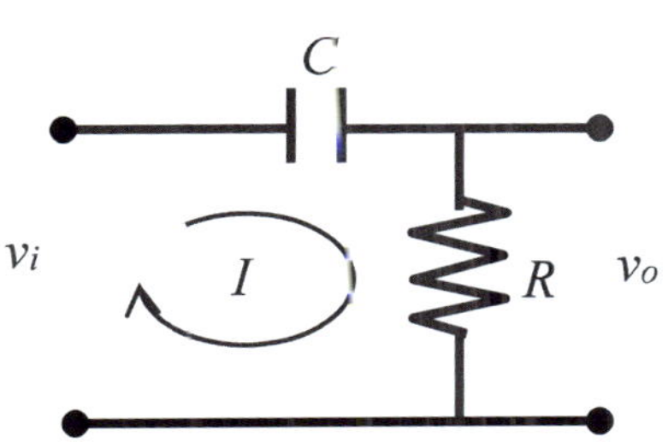

Fig. E1-1. CR circuit filter.

(a) Write analytical expressions for and draw sufficiently representative plots of the impulse, the magnitude, and the phase responses of this filter, the latter two with respect to ω.

Note: drawing plots requires obtaining a sufficiently representative, but not unnecessarily excessive number of data points first, and then connecting them. For example, in the present part, drawing the magnitude response requires computing the values of $|H(j\omega)|$ for several ω, marking the resulting points on the x-y plane, and drawing a curve to connect them. Also when drawing the plots, the correct labels and numbers along both axes should be used to avoid losing grade points.

Solution:

$$Z_R = R, Z_C = 1/Cs, \quad \left.\begin{array}{l} V_o = RI \\ V_i = (1/Cs + R)I \end{array}\right\} \to V_o = \frac{R}{1/Cs + R} V_i = \frac{RCs}{1 + RCs} V_i \Rightarrow$$

$$H(s) = \frac{V_o(s)}{V_i(s)} = \frac{RCs}{1 + RCs} = 1 - \frac{1}{RCs + 1} = 1 - \frac{1/RC}{s + 1/RC} \xrightarrow{\text{substitute } R,C} H(s) = 1 - \frac{1}{s+1}$$

$$\Rightarrow \begin{cases} h(t) = L^{-1}\{H(s)\} = \delta(t) - \dfrac{1}{RC}(e^{-t/RC})u(t) = \delta(t) - e^{-t}u(t), \\ H(j\omega) = \dfrac{RCj\omega}{1 + RCj\omega} = \dfrac{j\omega}{1 + j\omega}; \end{cases}$$

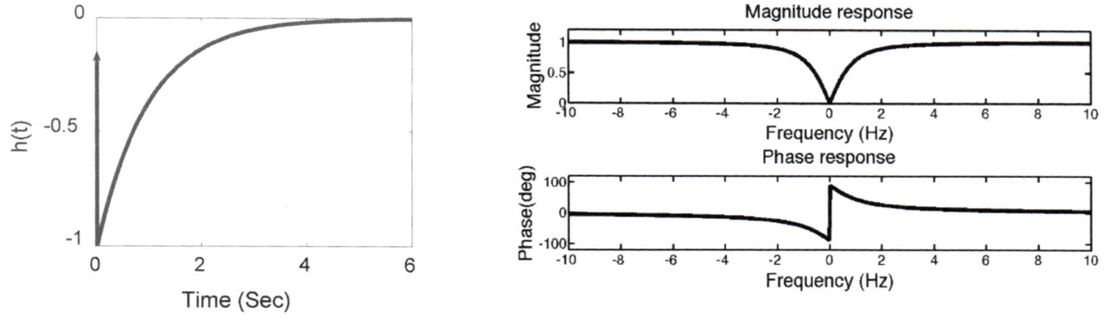

Fig. E1-2. The impulse, magnitude, and phase response of the CR filter.

(b) Classify the filter. Indicate where this filter is used in signal processing applications. Confirm your classification by taking as filter input $v_i(t) = 4 - 6\sin\left(6\pi t + \dfrac{\pi}{3}\right) - 4\cos\left(18\pi t - \dfrac{\pi}{6}\right)$**, computing the filter output** $v_o(t)$**, and examining the output frequency content.**

Solution: This is the high-pass filter that can be used as an AC-coupling circuit to filter out the DC component and attenuate the low frequency components in the DC vicinity.

To confirm the filter classification by passing through this filter a function

$$v_i(t) = 4 - 6\sin\left(6\pi t + \frac{\pi}{3}\right) - 4\cos\left(18\pi t - \frac{\pi}{6}\right)$$

we calculating the magnitudes and phases of the filter frequency response function at $\omega = 6\pi, 18\pi$ as follows:

$H(0) = 0,$

$$H(j6\pi) = \frac{j6\pi}{1 + j6\pi} \Rightarrow \ |H(j6\pi)| = \frac{6\pi}{\sqrt{1+(6\pi)^2}} \approx 1, \quad \angle H(j6\pi) = \frac{\pi}{2} - \tan^{-1}\left(\frac{6\pi}{1}\right) = 0.053(rad),$$

$$H(j18\pi) = \frac{j18\pi}{1 + j18\pi} \Rightarrow |H(j18\pi)| = \frac{18\pi}{\sqrt{1+(18\pi)^2}} \approx 1, \ \angle H(j18\pi) = \frac{\pi}{2} - \tan^{-1}\left(\frac{18\pi}{1}\right) = 0.018(rad).$$

$$\because v_o(t) = |H(0)|4 - |H(j6\pi)|6\sin\left(6\pi t + \frac{\pi}{3} + 0.053\right) - |H(j18\pi)|4\cos\left(18\pi t - \frac{\pi}{6} + 0.018\right)$$

$$= -6\sin(6\pi t + 1.1) - 4\cos(18\pi t - 0.506),$$

i.e. dc component is not passed and two high frequency components are passed with almost unattenuated magnitude.

(c) Choosing three different values of the RC product, show analytically and graphically how time domain and frequency domain filter characteristics are affected by the RC product value.

Solution: Let $RC = \tau$ for brevity. Then $h(t) = \delta(t) - (1/RC)(e^{-\frac{t}{RC}})u(t) = \delta(t) - (1/\tau)(e^{-\frac{t}{\tau}})u(t)$, and the graph corresponding to $\tau_1 < \tau_2 < \tau_3$ is given by

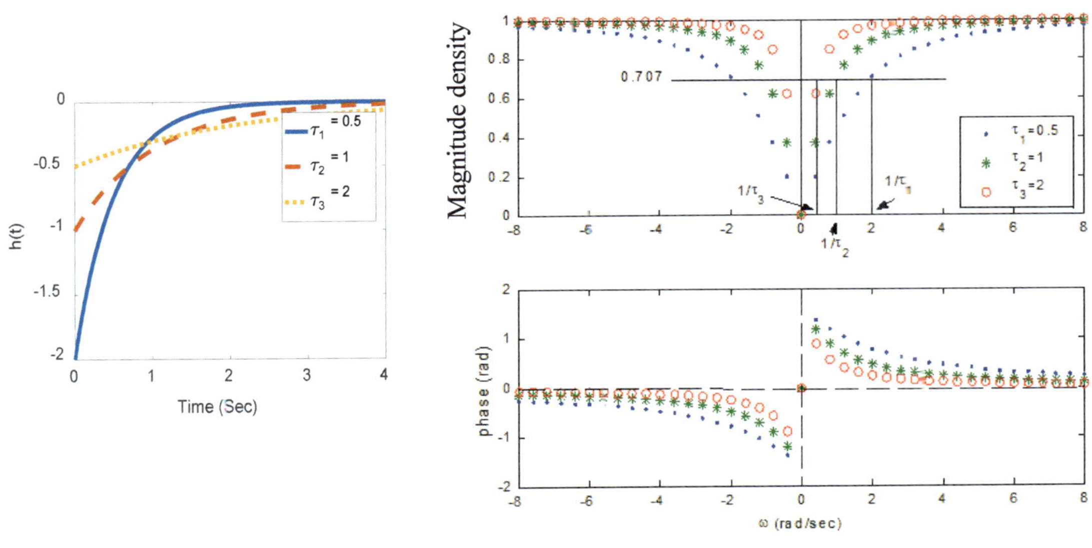

Fig. E1-3. Impulse, magnitude, and phase responses of the CR circuit with different values of the RC product.

As RC value increases, (τ increases), $1/\tau$ decreases, the exponential term drops out and the impulse response function becomes $\delta(t)$, which is an all-pass filter. Using

$$H(j\omega) = \frac{RCj\omega}{1 + RCj\omega} = \frac{\tau j\omega}{1 + \tau j\omega}$$

yields magnitude and phase response graphs in Figure E1-3. As RC value increases, (τ increases), the filter is seen to pass lower frequency components with decrease in phase angle.

(d) Draw a circuit that sequentially connects two filters in a non-loading configuration. Write the transfer function of this circuit and draw its magnitude spectral density.
Solution: The specified circuit is given in Figure E1-4.

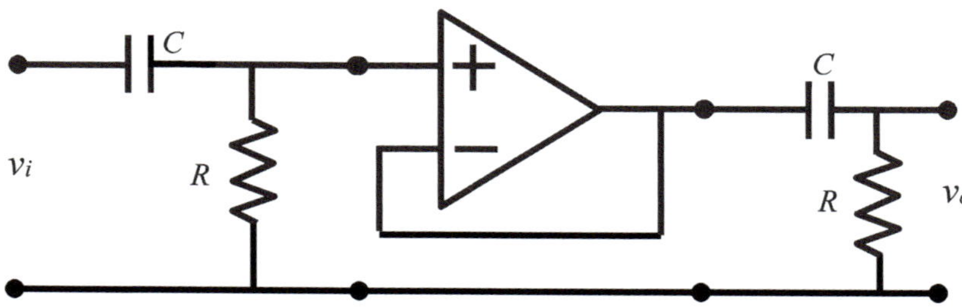

Fig. E1-4. Two RC circuits connected with buffer.

Because of nonloading, the transfer function of this circuit is given by the multiplication of two CR filter transfer functions: $H(s) = \frac{s}{1+s} \times \frac{s}{1+s} = \frac{s^2}{(1+s)^2}$. Magnitude response of the latter system is given in analytical and graphical forms as follows:

$$|H(j\omega)| = \frac{|(j\omega)^2|}{|(1+j\omega)^2|} = \frac{\omega^2}{\sqrt{(1-\omega^2)^2 + (2\omega)^2}},$$

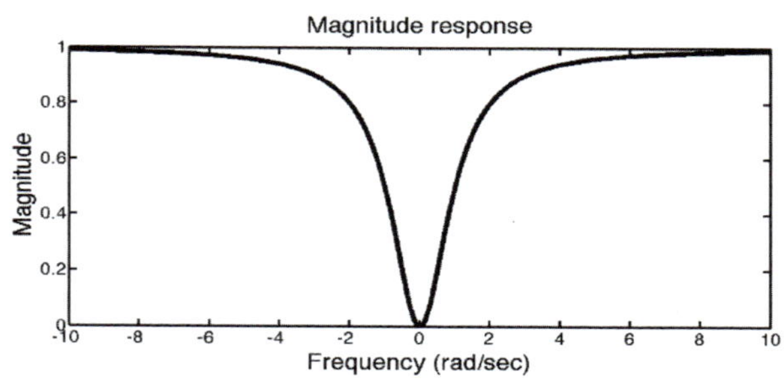

Fig. E1-5. Two-sided magnitude response of the circuit in Figure E1-6.

Problem 2: The following periodic function is sampled at $S_f = 12\,Hz$:

$$x_p(t) = 4 - 6\sin\left(6\pi t + \frac{\pi}{3}\right) - 4\cos\left(18\pi t - \frac{\pi}{6}\right).$$

(a) Write continuous time exponential Fourier series representation of this signal and draw its double-sided magnitude, phase, and power spectra. Indicate harmonic numbers in this representation and find the fundamental period of $x_p(t)$.

Solution: Recast a given function into polar Fourier series form:

$$\begin{aligned}
x_p(t) &= 4 - 6\sin\left(6\pi t + \frac{\pi}{3}\right) - 4\cos\left(18\pi t - \frac{\pi}{6}\right) \Rightarrow \text{transform sine to cosine}\\
&= 4 - 6\cos\left(6\pi t + \frac{\pi}{3} - \frac{\pi}{2}\right) - 4\cos\left(18\pi t - \frac{\pi}{6}\right) \Rightarrow \text{combine phase}\\
&= 4 - 6\cos\left(6\pi t - \frac{\pi}{6}\right) - 4\cos\left(18\pi t - \frac{\pi}{6}\right) \Rightarrow \text{transform to positive cosine}\\
&= 4 + 6\cos\left(6\pi t + \frac{5\pi}{6}\right) + 4\cos\left(18\pi t + \frac{5\pi}{6}\right). \quad (*)
\end{aligned}$$

Now, convert the latter to exponential Fourier series form and draw the graphs:

$$x_p(t) = 4 + 3e^{-j2\pi(3)t}e^{-j\frac{5\pi}{6}} + 3e^{j2\pi(3)t}e^{j\frac{5\pi}{6}} + 2e^{-j2\pi(9)t}e^{-j\frac{5\pi}{6}} + 2e^{j2\pi(9)t}e^{j\frac{5\pi}{6}},$$

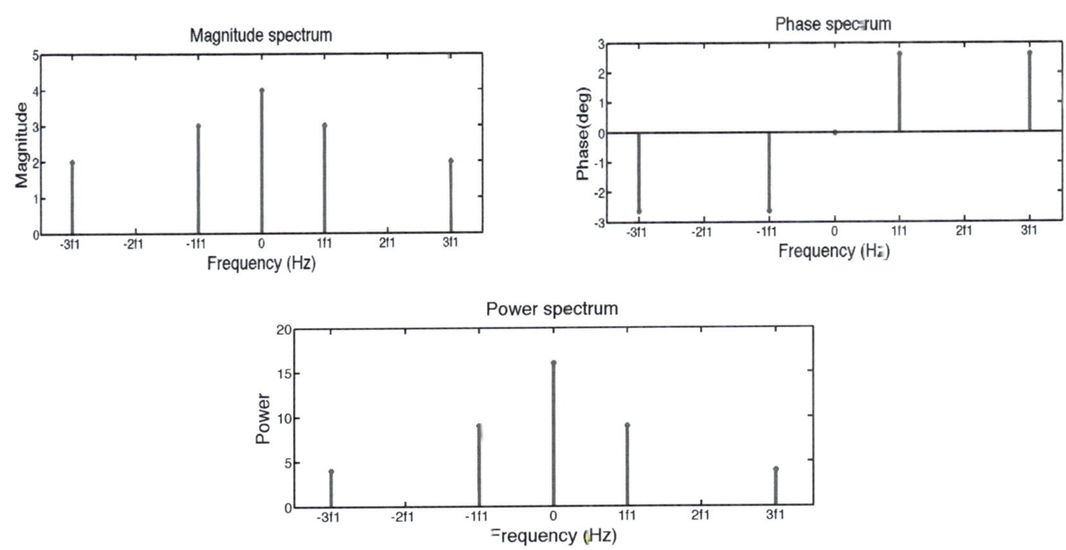

Fig. E1-6. Two-sided magnitude, phase and power spectra.

Fundamental frequency $f_1 = 3$ Hz. Fundamental period $T_1 = 1/3$ sec.

(b) Indicate which harmonics will be aliased by sampling and explain why. Give analytical expression of the discrete time sampled folded version $x_p[n]$ of $x_p(t)$ with digital frequencies in the $[0, 1/2)$ range.

Solution: Sampling frequency $S_f = 12\text{Hz} \Rightarrow f_b < S_f/2 = 6$ Hz, and since $f_1 = 3$ Hz, anything above $kf_1 = 6,\ k = 2$ will be aliased, i.e. the third harmonic will be aliased:

$$x_p[n] = 4 + 6\cos\left(2\pi\frac{3}{S_f}n + \frac{5\pi}{6}\right) + 4\cos\left(2\pi\frac{9}{S_f}n + \frac{5\pi}{6}\right) = 4 + 6\cos\left(2\pi\frac{1}{4}n + \frac{5\pi}{6}\right) + 4\cos\left(2\pi\frac{3}{4}n + \frac{5\pi}{6}\right)$$

$$= 4 + 6\cos\left(2\pi\frac{1}{4}n + \frac{5\pi}{6}\right) + 4\cos\left(2\pi\left(\frac{3}{4} - 1\right)n + \frac{5\pi}{6}\right) = 4 + 6\cos\left(2\pi\frac{1}{4}n + \frac{5\pi}{6}\right) + 4\cos\left(2\pi\frac{1}{4}n - \frac{5\pi}{6}\right)$$

$$= 4 + 3e^{-j2\pi(\frac{1}{4})n}e^{-j\frac{5\pi}{6}} + 3e^{j2\pi(\frac{1}{4})n}e^{j\frac{5\pi}{6}} + 2e^{-j2\pi(\frac{1}{4})n}e^{j\frac{5\pi}{6}} + 2e^{j2\pi(\frac{1}{4})n}e^{-j\frac{5\pi}{6}} = 4 + (3e^{j\frac{5\pi}{6}} + 2e^{-j\frac{5\pi}{6}})e^{j2\pi(\frac{1}{4})n}$$

$$+ (3e^{-j\frac{5\pi}{6}} + 2e^{j\frac{5\pi}{6}})e^{-j2\pi(\frac{1}{4})n} = 4 + 4.359e^{j3.03}e^{j2\pi\frac{1}{4}n} + 4.359e^{-j3.03}e^{-j2\pi\frac{1}{4}n} = 4 + 8.718\cos\left(2\pi\frac{1}{4}n + 3.03\right).$$

(c) Represent $x_p[n]$ of (b) by Discrete Fourier Series, draw DFS magnitude and phase spectra.

Solution: Converting the given function into DFS form has already been carried out above:

$$x_p[n] = 4 + 4.359e^{j3.03}e^{j2\pi\frac{1}{4}n} + 4.359e^{-j3.03}e^{-j2\pi\frac{1}{4}n}.$$

Drawing the graphs yields:

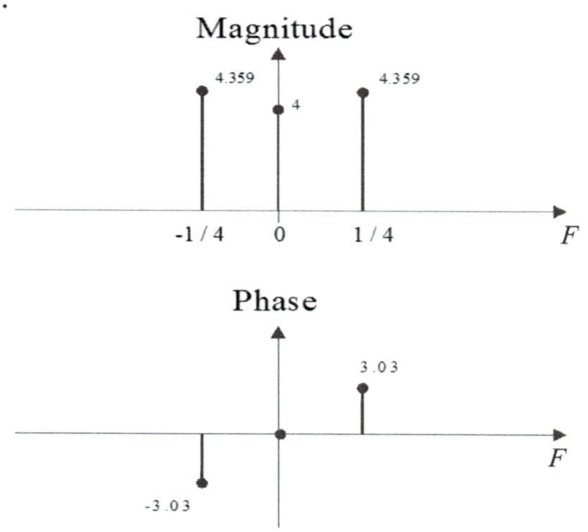

Fig. E1-7. Two-sided magnitude and phase spectra of the sampled function.

(d) Select S_f for 6x oversampling and draw the resulting DFS magnitude and phase spectra of the sampled signal.

Solution: For 6 times oversampling, $S_f = 6\cdot(2\cdot f_{\max}) = 6\cdot(2\cdot 9) = 108$ Hz, yielding the graphs

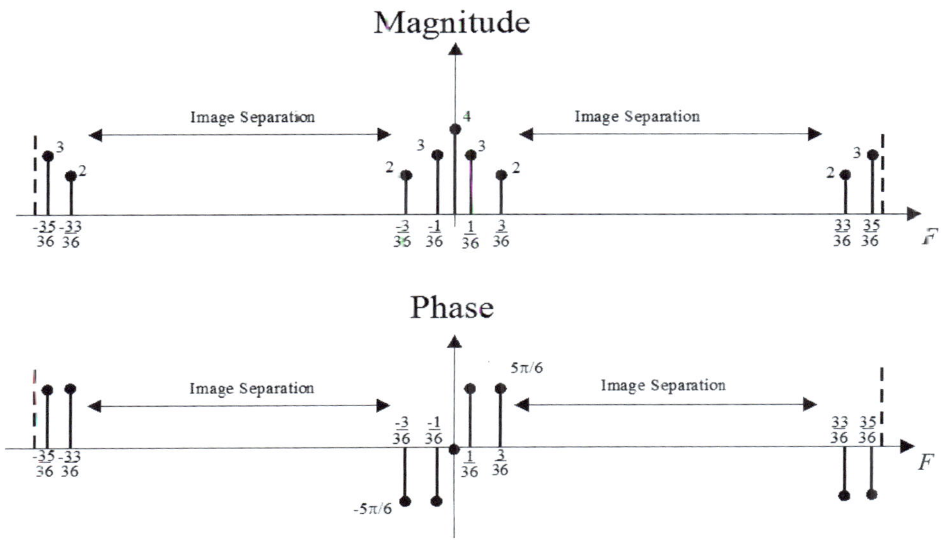

Fig. E1-8. Two-sided magnitude and phase spectra of the six times oversampled function.

(e) Explain the difference between magnitude spectra in (c) and (d) in terms of image separation.

Solution: There is no image separation in (c). In fact, since aliasing of the 3rd harmonic takes place, the magnitude and the phase of this harmonic are folded onto the first harmonic, misrepresenting the original spectra. In (d) there is the image separation, as indicated on the plot.

Problem 3: The periodic function of problem 2 is sampled at 12 Hz.
(a) Find the power of this signal using Parseval's Theorem.
Solution: Using equation (*) in Problem 2(a) which is already in polar Fourier series form yields

$$P = C_0^2 + \frac{1}{2}\sum_{k=1}^{\infty} C_k^2 = 4^2 + \frac{1}{2}(6)^2 + \frac{1}{2}(4)^2 = 42\,.$$

(b) Select an ideal low pass filter to block all of the components of $x_p(t)$ that will be aliased after sampling. Draw the magnitude and energy spectral densities and the phase spectrum.

Solution: To avoid aliasing, $\frac{S_f}{2} > k\cdot f_1 \Rightarrow k < \frac{S_f}{2f_1} = \frac{12}{2\cdot 3} = 2$, therefore, starting from the 2nd harmonic all higher harmonics will be aliased. Choose ideal filter cutoff frequency to be slightly less that the 2nd harmonic, i.e. $f_b = 6 - \varepsilon$ Hz, where ε is a small positive number:

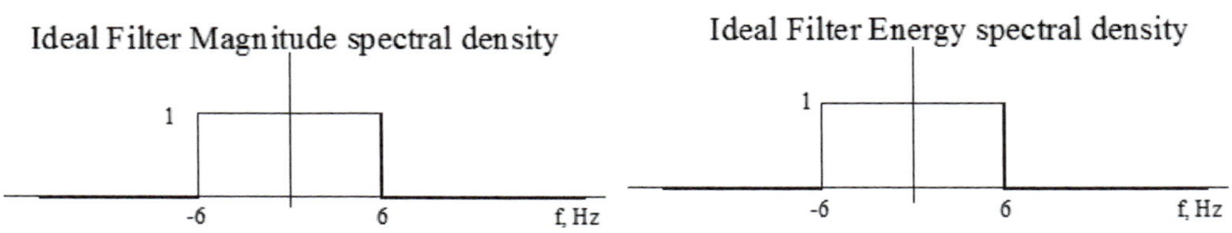

Fig. E1-9. Two side magnitude, phase and energy spectral density.

(c) Draw the magnitude and the phase spectra of the filter output of (b) if its input is $x_p(t)$. Calculate % power not passed through the filter.

Solution: Since only the DC component and the 1st harmonic are passed, the output signal spectra are:

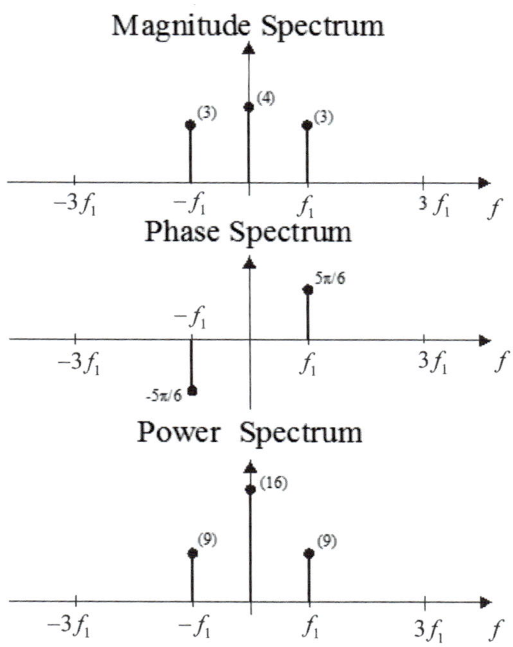

Fig. E1-10. Two-sided magnitude, phase and power spectra of the filtered signal.

The total power of the unfiltered signal: $P = 4^2 + \frac{1}{2}6^2 + \frac{1}{2}4^2 = 42,$

the total power of the filtered signal: $P = 4^2 + \frac{1}{2}6^2 = 34,$

% of total power not passed: $\frac{(42-34)}{42} \cdot 100 = 19.05\%.$

(d) Select the bandwidth of filter to be in the middle between the highest non-aliased harmonic and the lowest-aliased harmonic. Write the transfer function of this filter.

Solution: The highest non-aliased harmonic is 1, corresponding to $f = 3$ Hz, and the lowest aliased harmonic is 3, corresponding to $f = 9$ Hz.

Choosing $f_b = \dfrac{9+3}{2} = 6\text{ Hz} \Rightarrow \omega_b = 2\pi f_b = 12\pi$, and $\tau = RC = \dfrac{1}{\omega_b} = \dfrac{1}{12\pi}$

$$\therefore H(s) = \frac{1}{1+RCs} = \frac{1}{1+\tau s} = \frac{1}{1+\left(\dfrac{1}{12\pi}\right)s}.$$

(e) Show analytically and graphically on the magnitude and the phase spectral plots of the input, the filter, and the output how this low pass filter performs filtering action.

Solution: Analytically: using exponential Fourier series form:

$$x_p(t) = 4 + 3e^{-j2\pi(3)t}e^{-j\frac{5\pi}{6}} + 3e^{j2\pi(3)t}e^{j\frac{5\pi}{6}} + 2e^{-j2\pi(9)t}e^{-j\frac{5\pi}{6}} + 2e^{j2\pi(9)t}e^{j\frac{5\pi}{6}}$$

$$= \sum_{k=-3}^{3} X[k]e^{jk\omega_1 t} = \sum_{k=-3}^{3} |X[k]| \exp\left(jk\omega_1 t + \theta_k\right),$$

$$y_p(t) = \sum_{k=-3}^{3} |X[k]| \cdot |H(jk\omega_1)| \exp\left(jk\omega_1 t + \underbrace{\theta_k + \angle(H(jk\omega_1))}_{\phi_k} \right) = \sum_{k=-3}^{3} |Y[k]| \exp\left(jk\omega_1 t + \phi_k\right),$$

Magnitude | *Phase*

$$|H(jk\omega_1)| = \frac{1}{\sqrt{1+(k\omega_1\tau)^2}}, \qquad \angle(H(jk\omega_1)) = -\tan^{-1}(\omega\tau),$$

$$|H(jk\omega_1)|_{k=0} = \frac{1}{\sqrt{1+\left(\dfrac{0}{12\pi}\right)^2}} = 1, \qquad \angle(H(jk\omega_1))_{k=0} = 0\ rad,$$

$$|H(jk\omega_1)|_{k=\pm1} = \frac{1}{\sqrt{1+\left(\dfrac{6\pi}{12\pi}\right)^2}} = 0.8944, \qquad \angle(H(jk\omega_1))_{k=\pm1} = \mp\tan^{-1}\left(\frac{1}{2}\right) = \mp 0.4636\ rad,$$

$$|H(jk\omega_1)|_{k=\pm3} = \frac{1}{\sqrt{1+\left(\dfrac{18\pi}{12\pi}\right)^2}} = 0.5547, \qquad \angle(H(jk\omega_1))_{k=\pm3} = \mp\tan^{-1}\left(\frac{3}{2}\right) = \mp 0.9828\ rad,$$

$\therefore$ magnitude:

$$|Y(k)|_{k=0} = |H(jk\omega_1)|_{k=0} \cdot |X(k)|_{k=0} = 1 \cdot 4 = 4,$$
$$|Y(k)|_{k=\pm 1} = |H(jk\omega_1)|_{k=\pm 1} \cdot |X(k)|_{k=\pm 1} = 0.8944 \cdot 3 = 2.6832,$$
$$|Y(k)|_{k=\pm 3} = |H(jk\omega_1)|_{k=\pm 3} \cdot |X(k)|_{k=\pm 3} = 0.5547 \cdot 3 = 1.1094,$$

and phase:

$$(\phi_k)_{k=0} = \angle H(jk\omega_1)_{k=0} + \theta_0 = 0 + 0 = 0\ rad,$$
$$(\phi_k)_{k=\pm 1} = \angle H(jk\omega_1)_{k=\pm 1} + \theta_{\pm 1} = \mp 0.4636 \pm \left(\frac{5\pi}{6}\right) = \pm 2.154\ rad,$$
$$(\phi_k)_{k=\pm 3} = \angle H(jk\omega_1)_{k=\pm 3} + \theta_{\pm 3} = \mp 0.9828 \pm \left(\frac{5\pi}{6}\right) = \pm 1.64\ rad.$$

The corresponding graphs are:

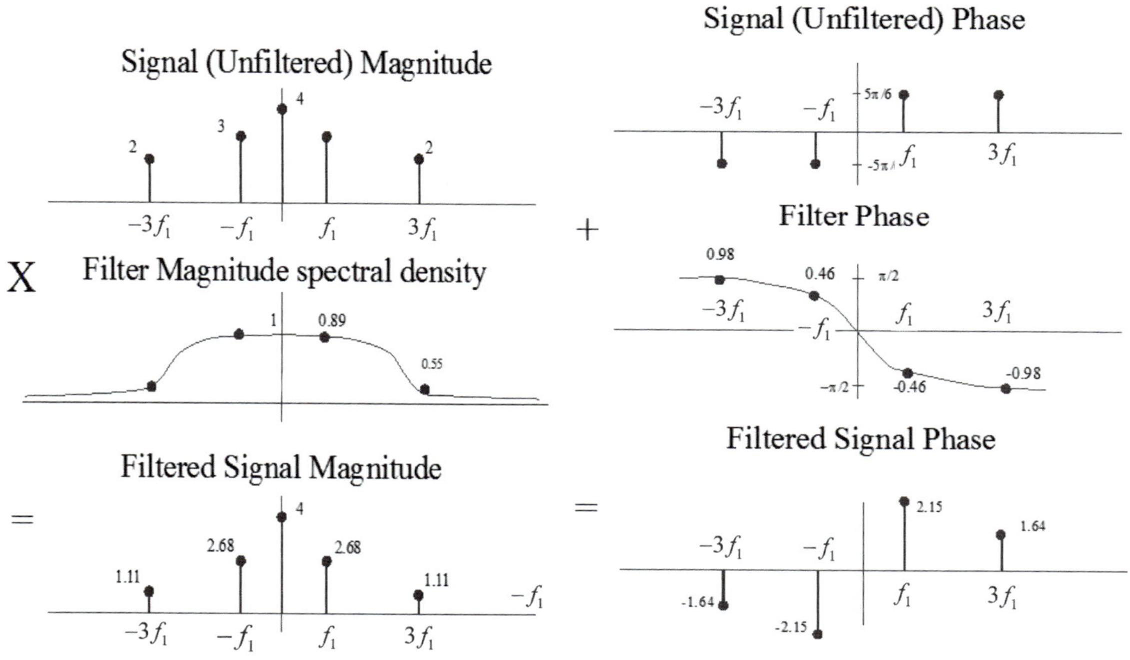

Fig. E1-11. Magnitude scaling and phase change of the input signal by the filter shown on the two-sided magnitude and phase spectra.

Problem 4: The periodic function of Problem 2

$$x_p(t) = 4 - 6\sin\left(6\pi t + \frac{\pi}{3}\right) - 4\cos\left(18\pi t - \frac{\pi}{6}\right)$$

is sampled at 36 Hz and the resulting DT $x_p[n]$ signal is passed through a digital high pass filter with transfer function $H_{hp}(F) = 1/(1 + 0.6e^{-j2\pi F})$.

a) Calculate the magnitudes of the DFS of the output signal.

Solution: using Euler's formula yields

$$x[n] = 4 - 6\sin\left(2\pi\frac{3}{36}n + \frac{\pi}{3}\right) - 4\cos\left(2\pi\frac{9}{36}n - \frac{\pi}{6}\right) = 4 + 6\cos\left(2\pi\frac{1}{12}n + \frac{5\pi}{6}\right) + 4\cos\left(2\pi\frac{3}{12}n + \frac{5\pi}{6}\right)$$

$$= 4 + 3\left(e^{2\pi\frac{1}{12}n}e^{\frac{5\pi}{6}} + e^{2\pi\left(-\frac{1}{12}\right)n}e^{-\frac{5\pi}{6}}\right) + 2\left(e^{2\pi\frac{3}{12}n}e^{\frac{5\pi}{6}} + e^{2\pi\left(-\frac{3}{12}\right)n}e^{-\frac{5\pi}{6}}\right).$$

For the filter

$$H(F) = \frac{1}{1+0.6e^{-j2\pi F}} = \frac{1}{1+0.6\cos(-2\pi F) - j0.6\sin(-2\pi F)},$$

$$|H(F)| = \left|\frac{1}{1+0.6\cos(-2\pi F) - j0.6\sin(-2\pi F)}\right| = \frac{1}{\sqrt{(1+0.6\cos(-2\pi F))^2 + (j0.6\sin(-2\pi F))^2}}$$

$$= \frac{1}{\sqrt{1+1.2\cos(2\pi F) + 0.36(\cos^2(2\pi F) + \sin^2(2\pi F))}} = \frac{1}{\sqrt{1.36+1.2\cos(2\pi F)}} \Rightarrow$$

k	$\lvert X_s[k]\rvert$	F	$\lvert H(F)\rvert$	$\lvert H(F)\rvert \times \lvert X_s[k]\rvert$
0	4	0	0.625	2.5
±1	3	±1/12	0.6456	1.94
±2	0	±2/12	0.7143	0
±3	2	±3/12	0.8575	1.7150

b) Show analytically and graphically on the magnitude and the magnitude density and phase spectral plots of the input $x_p[n]$, the filter, and the output $y_p[n]$ how this digital high pass filter performs filtering action. Use F cycles/sample as digital frequency axes units.

Solution: analytical expression for the magnitude is given in part a). For the phase we have

$$\angle H(F) = -\arctan\left(1+0.6e^{-j2\pi F}\right) = -\arctan\left(\frac{-0.6\sin(2\pi F)}{1+0.6\cos(2\pi F)}\right) \Rightarrow$$

k	$\theta_s[k]$	F	$\angle H(F)$	$\theta_s[k] + \angle H(F)$
0	0	0	0	0
±1	±2.618	±1/12	±0.1949	±2.8129
±2	0	±2/12	±0.3803	±0.3803
±3	±2.618	±3/12	±0.5404	±3.1584

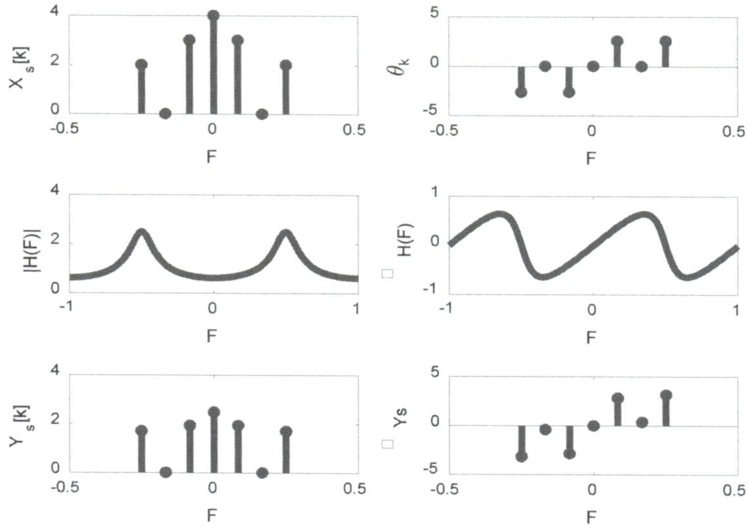

Fig. E1-12. Modification of the input signal phase by the filter is shown on the two-sided phase spectra.

c) Find (and explain how it was found) the transfer function of the high pass digital filer of the same form as that of $H_{hp}(F)$ given above, but with the filter bandwidth located between the two nonzero digital frequencies of the $x_p[n]$ obtained above in Problem 4.

Solution: the desired bandwidth is $\frac{2}{12}\times 2\times\pi=\frac{\pi}{3}$**.** The cut-off frequency is defined by $|H(\omega_c)|=\sqrt{2}/2$. The filter will have the form similar to that of $H_{hp}(F)$, yielding

$$H(\Omega)=\frac{1}{1+Ae^{-j\Omega}},|H(\Omega)|=\frac{1}{\sqrt{1+A^2+2A\cos(\Omega)}}.$$

Therefore, we have to solve the following problem:

$$|H(\Omega_1)|=|H(\Omega_2)|=\frac{\sqrt{2}}{2},|\Omega_1-\Omega_2|=\frac{\pi}{3}.$$

Solving the above equations yields $A=0.46$.

Chapter 7

CT and DT Filter Design and Implementation

Filtering is extraction of useful content from noisy measurements. In its simplest form, considered in the present book, is attenuates the signals outside of the frequency range of interest, as exemplified by the antialiasing filter. In its more complicated form, necessary for functioning of many modern systems and represented, for example, by Kalman filtering, it estimates system state on the basis of the system model and statistical noise properties.

This chapter focuses on the in-depth derivation of the most basic filter — analog Butterworth filter, and provides guidelines for basic low pass filter design as well as general CT and DT filter implementation. ***Digital filtering*** *is understood as DT filtering with quantization error when implemented in the digital signal processor*. The chapter also provides a glimpse at several filter types and gives an example of a practical filter circuit implemented in terms of op amps.

Modern data acquisition systems, such as that used in mold oscillator at Nucor include an analog antialiasing filter stage that precedes an A/D converter, and digital filter stage that follows the converter. In this chapter we will use the full arguments $H(j\omega)=H(j2\pi f)$ for the CT filter transfer function to ensure clarity in the derivations. For the DT filter transfer functions we will continue using $H(F)$ instead of $H(e^{j2\pi F})$ and $H(\Omega)$ instead of $H(e^{j\Omega})$.

7.1 Introduction to CT Filter Design: Motivation and Ideal Filter Approximation. Classical Filter Transfer Functions. Butterworth Filter: Existence

7.1.1 *Filter characteristics*

Example 7.1-1: Mold oscillator

The displacement of the mold is measured using four accelerometers, one located in each of the four corners of the mold. The mold is a shell formed by four walls subject to the turbulent steel flow that can induce small magnitude high frequency shell vibration. Let's model mold i) displacement, ii) velocity, and iii) acceleration, respectively, as follows:

i) displacement: $$x(t)=\underbrace{c_1\cos(\omega_1 t)+c_2\cos(2\omega_1 t)+c_3\cos(3\omega_1 t)}_{\text{Information of interest for control}}+\underbrace{\frac{c_1}{100}\cos(10\omega_1 t)}_{\text{Noise due to mold wall vibration}},\qquad(7.1\text{-}1)$$

ii) velocity: $$\underbrace{\dot{x}(t)=-c_1\omega_1\sin(\omega_1 t)-c_2 2\omega_1\sin(2\omega_1 t)-c_3 3\omega_1\sin(3\omega_1 t)}_{\text{Information of interest for production}}-\frac{c_1}{10}\omega_1\sin(10\omega_1 t),\qquad(7.1\text{-}2)$$

iii) acceleration: $\ddot{x}(t) = -c_1\omega_1^2\cos(\omega_1 t) - c_2 4\omega_1^2\cos(2\omega_1 t) - c_3 9\omega_1^2\cos(3\omega_1 t) - c_1\omega_1^2\cos(100\omega_1 t)$. (7.1-3)

What the corner sensors actually measure

Magnitude is equal to that of the reference signal

That is, even if the displacement noise is small in magnitude, if it is high frequency it may still be large in the measured signal. So, we need a low-pass filter to get rid of the noise.

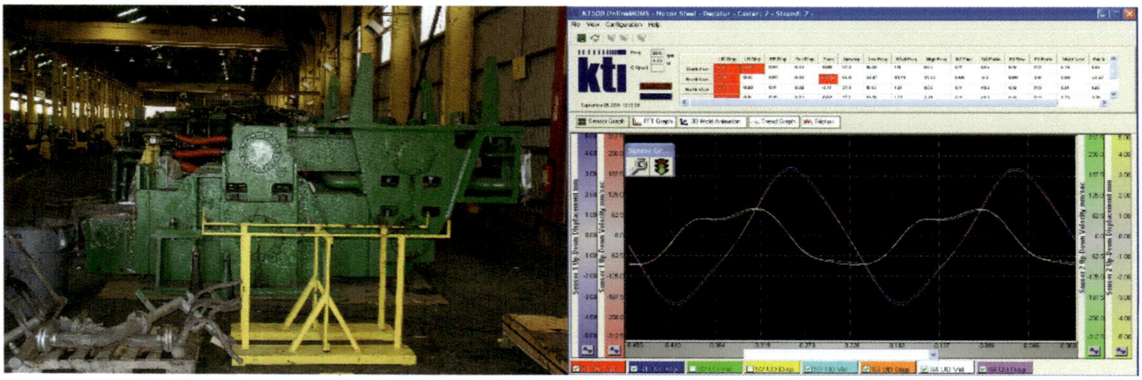

Fig. 7.1-1. Filtered accelerometer measurements converted to mold displacement and velocity.

The measured acceleration signal in the example above exhibits the following magnitude spectrum:

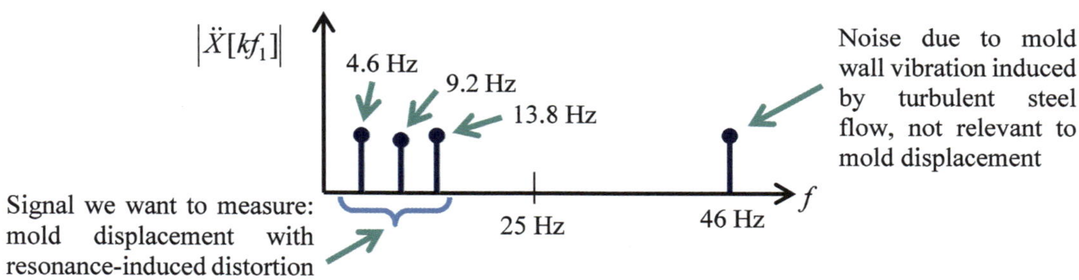

Fig. 7.1-2. Measured acceleration signal frequency characteristics.

Thus, the desired filter characteristic is of the form

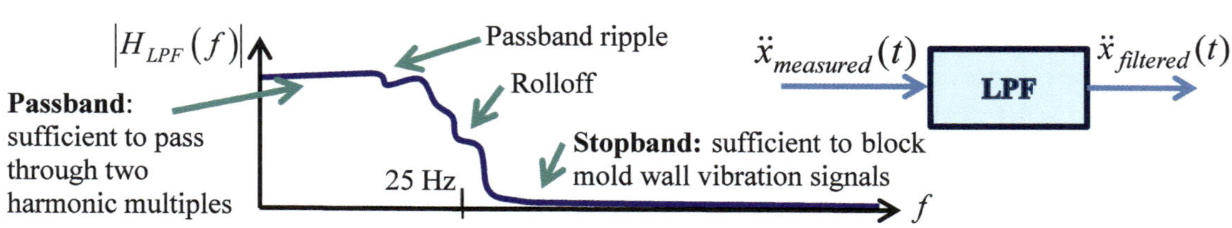

Fig. 7.1-3. Low-pass filter characteristics relevant to mold displacement measurement.

The passband and the stopband seen in Figure 7.1-3 are characterized, respectively, by ***filter gain***: $G = |H(f)|$, $G_{db} = 20\log|H(f)|$ and ***filter attenuation***: $A = 1/|H(f)|$, $A_{dB} = -20\log|H(f)|$.

Example 7.1-2: DC power supply.

One of the most basic filter applications is the AC/DC conversion. The rectifier circuit creates a waveform offset that induces the DC component, as shown below:

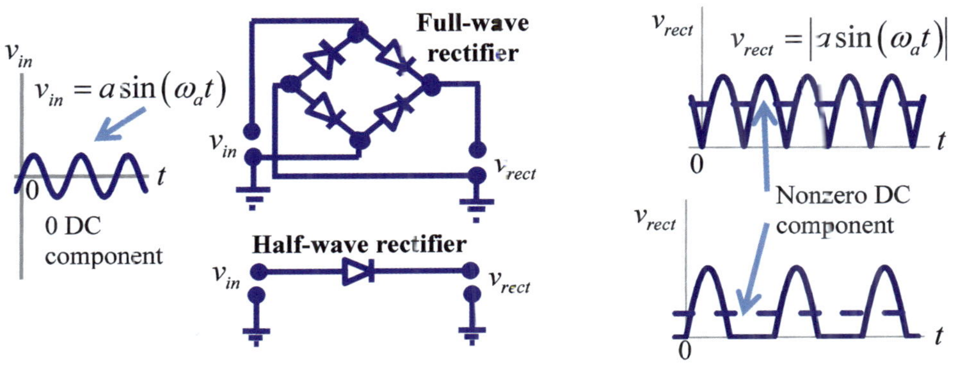

Fig. 7.1-4. DC power supply: rectifier block action.

However, only the DC component needs to be passed, while the rest filtered out. This is carried out by the low pass filter. Block diagram of a full-wave rectifier with low pass filter is shown below.

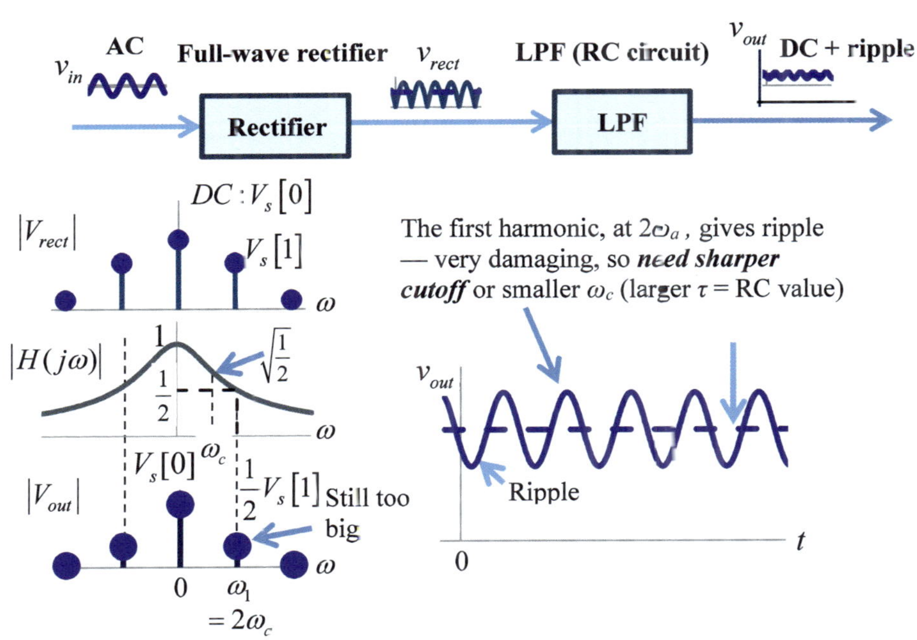

Fig. 7.1-5. DC power supply: low pass filter block action.

7.1.2 *Low pass filter*

7.1.2.1 *Single-pole low pass filter*

This filter has transfer function:

$$H(j\omega) = H_0 \frac{1}{1+j\omega\tau} = H_0 \frac{1/\tau}{j\omega + 1/\tau} = \left. \frac{H_0\omega_c}{s + \underbrace{\omega_c}_{1/\tau}} \right|_{s=j\omega} = H_0 \frac{1}{1 + j\dfrac{\omega}{\omega_c}} \tag{7.1-4}$$

where H_0 stands for DC gain and ω_c stands for break (cutoff) frequency. Phase shift:

$$\varphi(\omega) = -\tan^{-1}(\omega/\omega_c) \tag{7.1-5}$$

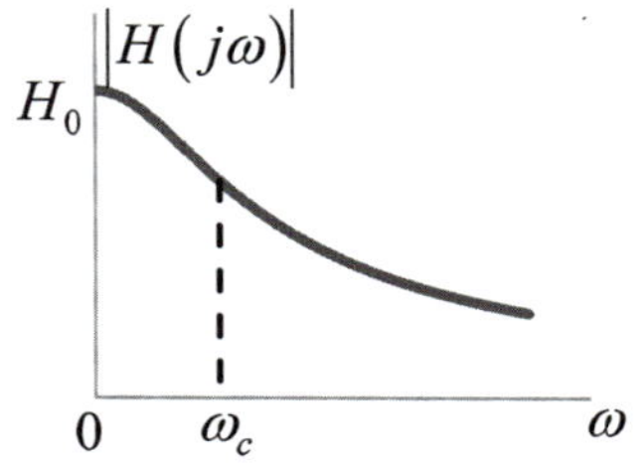

Fig. 7.1-6. Single-pole low-pass filter frequency response.

7.1.2.2 *Two-pole low pass filter*

The transfer function of this filter is

$$H(s) = \frac{H_0\omega_c^2}{s^2 + \alpha\omega_c s + \omega_c^2}, \tag{7.1-6}$$

where α is related to damping ζ coefficient as $\zeta = \alpha/2$. The quality of filter, "Q", is defined as

$$Q = 1/\alpha = 1/2\zeta. \tag{7.1-7}$$

Denoting the filter poles as $-a \pm jb$ yields

$$(s + a + jb)(s + a - jb) = s^2 + 2as + a^2 + b^2 \Rightarrow \omega_c^2 = a^2 + b^2 : 2a = \alpha\omega_c \Rightarrow \alpha = \frac{2a}{\sqrt{a^2 + b^2}}.$$

The magnitude response of the second order low pass filer is given in Figures 7.1-7 and 7.1-8, showing a flatter passband that that of the first order filter in Figure 7.1-6.

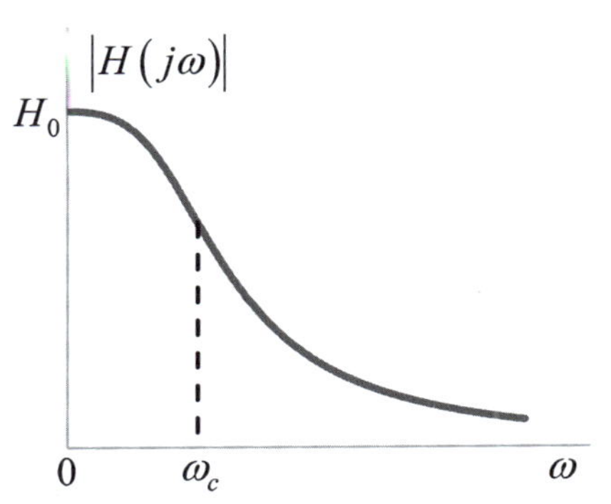

Fig. 7.1-7. Two-pole low-pass filter magnitude response.

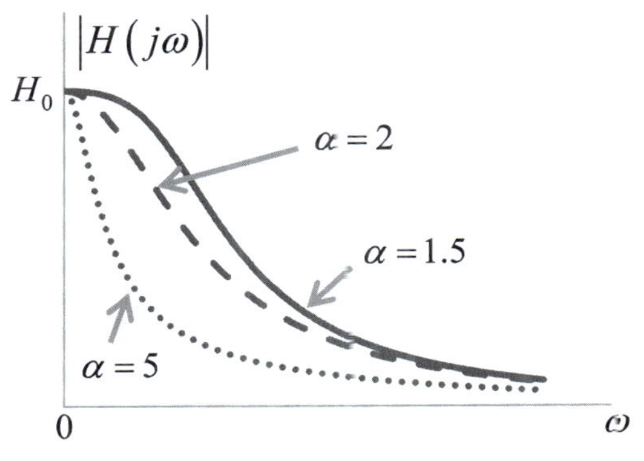

Fig. 7.1-8. Magnitude response with different damping.

Magnitude response falls off more quickly with the increase in α.

7.1.2.3 *Higher-order low pass filter*

As we have seen in Project 1, real first-order filters let through some components above the cutoff frequency, and distort some components below the cut-off frequency. The second order filters are seen in Figures 7.1-7 and 7.1-8 to have a flatter passband, while also showing a steeper drop-off, both desired properties, so it looks like the filter order matters. Can we then design higher order filters with even flatter passband and steeper rolloff to the stopband, as shown in Figure 7.1-9?

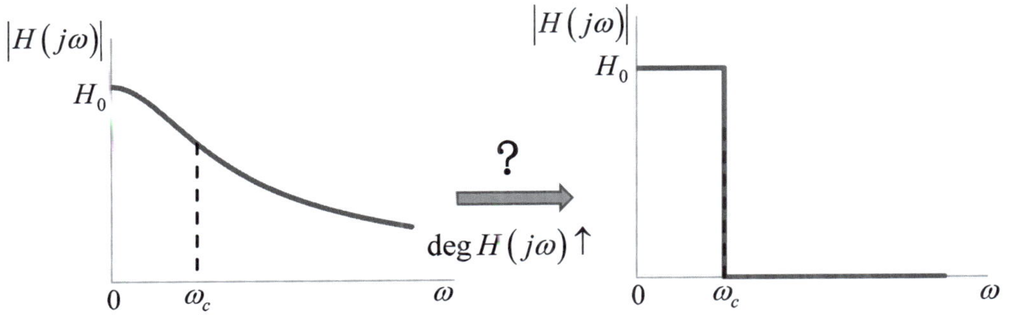

Fig. 7.1-9. Magnitude density: hypothesis — getting closer to ideal filter by increasing filter order.

It looks then like a reasonable thing to try building a high-order filter by cascading the first order filters. This is carried out next.

7.1.3 *Higher-order low-pass filters obtained by cascading lower order filters*

Let's cascade first order filters and check what happens with the magnitude response of the result. One single-pole filter:

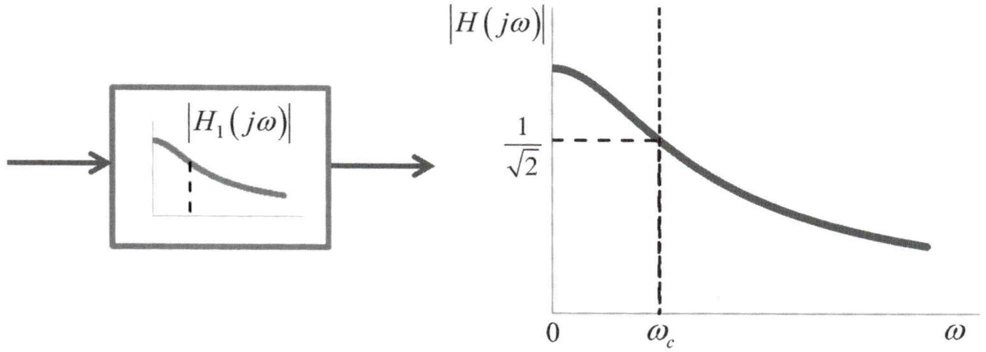

Fig. 7.1-10. One single-pole low-pass filter.

Two single-pole filters:

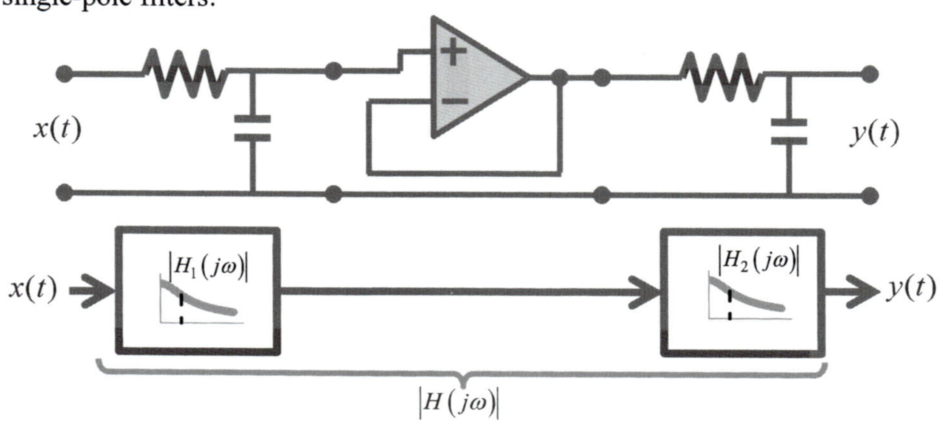

Fig. 7.1-11. Building a second-order filter by cascading two first order filters.

This yields

$$H(j\omega) = H_1(j\omega) \cdot H_2(j\omega) = |H_1(j\omega)| e^{j\varphi_1(\omega)} |H_2(j\omega)| e^{j\varphi_2(\omega)} = |H_1(j\omega)| |H_2(j\omega)| e^{j(\varphi_1(\omega)+\varphi_2(\omega))} \tag{7.1-8}$$

with the magnitude response

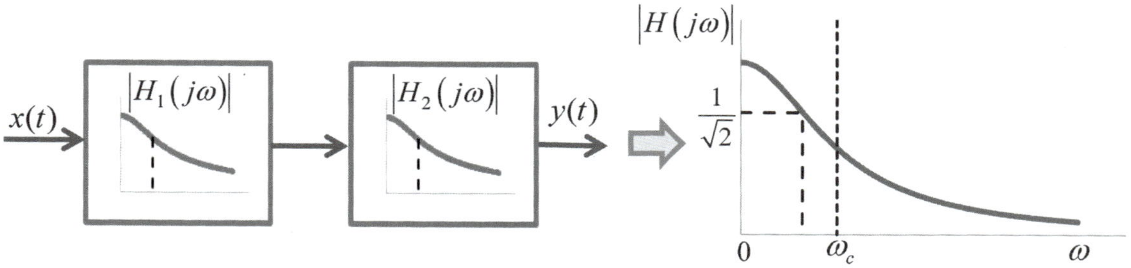

Fig. 7.1-12. Series connection of two single-pole low-pass filters.

Ten single-pole filters:

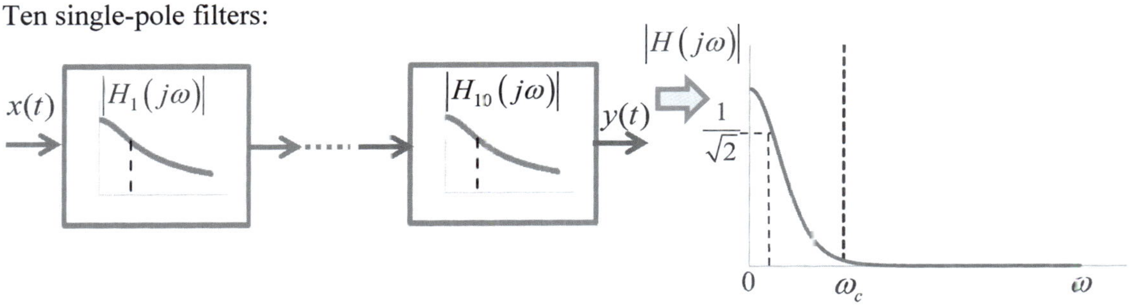

Fig. 7.1-13. Series connection of ten single-pole low-pass filters.

Obviously, this approach doesn't work, except for the DC power supply. Cascading single-pole filters together, i.e. arranging ***n* single-pole filters in series** to obtain

$$|H_1(j\omega)|^n = \left(\frac{1}{\sqrt{1+(\omega/\omega_c)^2}}\right)^n = \frac{1}{\sqrt{((1+(\omega/\omega_c)^2)^n}}, \tag{7.1-8}$$

is seen in the examples above to steepen the filter rolloff, but without making the rolloff stay at the desired cutoff frequency and forming the flat passband — as the filter order increases, the filter suppresses all frequencies except the DC signal — i.e. it does become ideal filter, but only for the DC signal.

This is rather obvious, since the first order filter magnitude response is less than 1 at all frequencies other than DC. Some of the filter transfer functions addressing this deficiency are presented in the next two sections, These filters are named after the scientists who discovered them.

7.2 Butterworth Filter: Magnitude Response, Existence, Transfer Function, Design and Implementation

7.2.1 *Butterworth filter magnitude response*

We would like an nth order filter to be steep near the cut-off point, and flat in the passband.

Stephen Butterworth (British physicist, 1885–1958) tried to find a filter with these properties by considering the magnitude response:

$$|H_n(j\omega)| = 1/\sqrt{1+(\omega/\omega_c)^{2n}} \tag{7.2-1}$$

where n is a design parameter (it will give filter order) chosen based on performance requirements.

The magnitude response shown in Figure 7.2-1 has the desired properties, getting steeper at the desired cutoff frequency, but does such filter exist and what is its form and phase response? The key tool to answer these questions is **spectral factorization** — the inverse of finding system energy spectral density: $|H_n(j\omega)|^2 = H_n(j\omega)\cdot H_n^*(j\omega) = H_n(j\omega)\cdot H_n(-j\omega)$, as will be seen below.

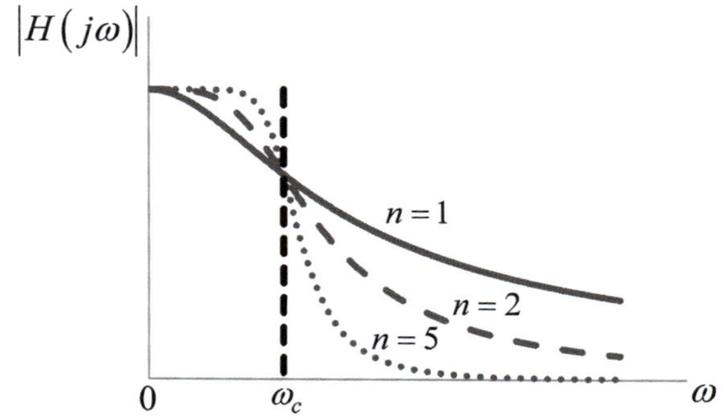

Fig. 7.2-1. n^{th} order Butterworth magnitude response.

7.2.2 *Butterworth filter: existence*

7.2.2.1 *Existence: a few preliminaries*

To answer these questions let's review some basic properties of polynomials and rational functions: **Polynomial factorization.** If we write the first, the second, and the third order polynomials as products of elementary factors

$$s - s_1,$$
$$(s-s_1)(s-s_2) = s^2 - s_1 s - s s_2 + s_1 s_2 = s^2 - (s_1+s_2)s + s_1 s_2,$$
$$(s-s_1)(s-s_2)(s-s_3) = s^3 - (s_1+s_2+s_3)s^2 + (s_1 s_2 + s_1 s_3 + s_2 s_3)s - s_1 s_2 s_3,$$

then by construction s_k, $k = 1$, ... are the roots of each polynomial, i.e., the values at which the value of a polynomial is equal to zero. By the main theorem of algebra, a polynomial of degree n.

$$s^n + a_{n-1}s^{n-1} + ... + a_0,$$

has exactly n roots s_k, $k = 1, ..., n$, and its set of roots $\{s_k, k = 1, ..., n,\}$ is unique.

This implies that the polynomial above has unique representation (factorization) in terms of its roots:

$$s^n + a_{n-1}s^{n-1} + ... + a_0 = (s-s_1)(s-s_2)...(s-s_n). \tag{7.2-2}$$

Unity DC gain filter form. From three examples we see that for a polynomial of degree n: $s^n + a_{n-1}s^{n-1} + ... + a_0$ with roots given by s_k, $k = 1, ..., n$, we have

$$a_0 = (-1)^n \prod_{k=1}^{n} s_k.$$

Denoting $D_n(s)$ to be the nth order denominator polynomial, this implies that an all-pole (no finite zeros) rational function with DC (static, i.e. at $s = 0$) gain 1 has the form

$$\frac{a_0}{D_n(s)} = \frac{a_0}{s^n + a_{n-1}s^{n-1} + \ldots + a_0} = \frac{(-1)^n \prod_{k=1}^{n} s_k}{(s-s_1)(s-s_2)\cdots(s-s_n)}. \tag{7.2-3}$$

Relations among the roots of a polynomial. If a second order polynomial is of the form $s^2 + a_0$, then from its factorization

$$(s-s_1)(s-s_2) = s^2 - s_1 s - s s_2 + s_1 s_2 = s^2 - (s_1 + s_2)s + s_1 s_2 \underset{s_2=-s_1}{=} s^2 + s_1 s_2 = s^2 + a_0,\ a_0 = s_1 s_2,$$

i.e. we must have $s_1 + s_2 = 0$, making its roots related as $s_2 = -s_1$. Then $s^2 + a_0$ can be represented as

$$s^2 + a_0 = s^2 + s_1 s_2 = (s-s_1)(s-s_2) \underset{s_2=-s_1}{=} (s-s_1)(s+s_1) = (-1)\underbrace{(s-s_1)}_{D_1(s)}\underbrace{(-s-s_1)}_{\tilde{D}_1(s)=D_1(-s)} = (-1)D_1(s)D_1(-s).$$

Where we defined $\tilde{D}_n(s) = D_n(-s)$ for general n^{th} degree polynomial Define also

$$\tilde{s}_k = -s_k.$$

Then

$$\begin{aligned} D_n(s) &= s^n + a_{n-1}s^{n-1} + \ldots + a_0 = (s-s_1)(s-s_2)\cdots(s-s_n) \Rightarrow \tilde{D}_n(s) \equiv D_n(-s) \\ &= (-s)^n + a_{n-1}(-s)^{n-1} + \ldots + a_0 = ((-s)-s_1)((-s)-s_2)\cdots((-s)-s_n) \\ &= ((-s)+\tilde{s}_1)((-s)+\tilde{s}_2)\cdots((-s)+\tilde{s}_n), \end{aligned}$$

i.e., if s_k is a root of $D_n(s)$, then $\tilde{s}_k = -s_k$ must be a root of $\tilde{D}_n(s) = D_n(-s)$. We will show that we can always compute such roots for a general polynomial the form $s^{2n} + s_0$, i.e. this pairwise pattern holds for $s^{2n} + s_0$ by uniqueness. Therefore, for an all-pole n^{th} order filter with static gain 1:

$$H_n(s) = \frac{a_0}{D_n(s)} = \frac{a_0}{s^n + a_{n-1}s^{n-1} + \ldots + a_0} \Rightarrow H_n(-s) = \frac{a_0}{D_n(-s)} =$$

$$\frac{a_0}{(-s)^n + a_{n-1}(-s)^{n-1} + \ldots + a_0} = \frac{a_0}{\tilde{D}_n(s)},\ a_0 = (-1)^n \prod_{k=1}^{n} s_k = \prod_{k=1}^{n}(-s_k) = \prod_{k=1}^{n}(\tilde{s}_k).$$

7.2.2.2 *Existence: finding transfer function form through spectral factorization*

Let's now try to find a filter with the transfer function that yields the magnitude response

$$\left|H_n(j\omega)\right| = \frac{1}{\sqrt{1+(\omega/\omega_c)^{2n}}}.$$

For notational simplification let's set $\omega_c = 1$: $\left|H_n(j\omega)\right| = \frac{1}{\sqrt{1+\omega^{2n}}}$, but keep track of the effect of having arbitrary value of ω_c.

First, let's assume that the filter is all-pole with static gain 1:

$$H_n(s) = \frac{a_0}{D_n(s)} = \frac{(-1)^n \prod_{k=1}^{n} s_k}{D_n(s)} \tag{7.2-4}$$

where s_k, $k = 1, \ldots, n$, are the roots of $D_n(s)$. Then, setting $s = j\omega$, calculating the square of the magnitude, and applying to it spectral factorization yields

$$\frac{1}{1+\omega^{2n}} = |H_n(j\omega)|^2 = H_n(j\omega)\cdot H_n^*(j\omega) =$$

$$H_n(j\omega)\cdot H_n(-j\omega) = \frac{a_0}{D_n(j\omega)}\cdot\frac{a_0}{D_n(-j\omega)} \Rightarrow \begin{cases} a_0^2 = 1 \Rightarrow a_0 = 1, \\ D_n(j\omega)\cdot D_n(-j\omega) = 1+\omega^{2n}. \end{cases}$$

7.2.2.3 *Existence: characteristic equation*

Since we replace s by $j\omega$ in $H_n(j\omega)$ when we calculate magnitude response, let's make a change of variables from $j\omega$ back to s (or from $j\omega/\omega_c$ to s/ω_c):

$$j\omega \to s \Rightarrow D_n(j\omega)\cdot D_n(-j\omega) = 1+(j\omega/j)^{2n} = 1+(j\omega\cdot(-j))^{2n} = 1+((-j)\cdot j\omega)^{2n} \underset{j\omega\to s}{\rightleftarrows}$$

$$D_n(s)\cdot D_n(-s) = 1+((-j)\cdot s)^{2n} = 1+(-j)^{2n}s^{2n} = 1+[(-j)^2]^n s^{2n} = 1+(-1)^n s^{2n}$$
$$= (-1)^n s^{2n} + 1 = (-1)^n s^{2n} + \underbrace{(-1)^n(-1)^n}_{1} = (-1)^n[s^{2n}+(-1)^n],$$

or, for general ω_c,

$$D_n(s/\omega_c)\cdot D_n(-s/\omega_c) = (-1)^n[(s/\omega_c)^{2n}+(-1)^n].$$

7.2.2.4 *Existence: denominator polynomial factorization*

Let's now factor product $D_n(s)D_n(-s)$ in terms of its roots s_k, $k = 1, \ldots, n$, as:

$$D_n(s)\cdot D_n(-s) = \underbrace{(s-s_1)(s-s_2)\cdots(s-s_n)}_{D_n(s)}\underbrace{(-s-s_1)(-s-s_2)\cdots(-s-s_n)}_{D_n(-s)}$$
$$= (-1)^n (s-s_1)(s-s_2)\cdots(s-s_n)(s-\underbrace{(-s_1)}_{s_{n+1}})(s-\underbrace{(-s_2)}_{s_{n+2}})\cdots(s-\underbrace{(-s_n)}_{s_{2n}})$$
$$= (-1)^n (s-s_1)(s-s_2)\cdots(s-s_{2n}), \text{ where } s_{k+n} = -s_k \text{ for } k = 1, \ldots, n,$$
$$= (-1)^n \underbrace{(s^{2n}+\ldots+\underbrace{(-1)^{2n}}_{1}\prod_{k=1}^{2n} s_k)}_{D_{2n}(s)} = (-1)^n \underbrace{(s^{2n}+\ldots+\prod_{k=1}^{2n} s_k)}_{D_{2n}(s)} = (-1)^n D_{2n}(s).$$

This yields

$$D_n(s)\cdot D_n(-s)=(-1)^n D_{2n}(s)=(-1)^n\left[s^{2n}+\ldots+\prod_{k=1}^{2n}s_k\right]=(-1)^n[s^{2n}+(-1)^n].$$

Now we note that we want to factor $D_{2n}(s)$ in terms of such s_k, $k=1, \ldots, 2n$, that

$$D_{2n}(s)=s^{2n}+a_{n-1}s^{2n-1}+\ldots+a_1 s+a_0=(s-s_1)(s-s_2)\ldots(s-s_{2n})=s^{2n}+(-1)^n,$$

or for general ω_c

$$D_{2n}(s/\omega_c)=(s/\omega_c)^{2n}+(-1)^n.$$

Substituting any such s_k into $D_{2n}(s)$ obviously yields $D_{2n}(s)=0$, so that

$$s_k^{2n}+(-1)^n=0,\ k=1,\ldots,2n.$$

Consequently, if we find s_k, $k=1, \ldots, 2n$, that satisfy polynomial equation

$$s^{2n}+(-1)^n=0,$$

the set of roots $\{s_k, k=1, \ldots, n\}$, due to its uniqueness, will then yield polynomial factorization

$$D_{2n}(s)=(s-s_1)(s-s_2)\ldots(s-s_{2n})=s^{2n}+\prod_{k=1}^{2n}s_k=s^{2n}+(-1)^n. \qquad (7.2\text{-}5)$$

7.2.2.5 *Existence: filter poles*

Thus, we need to find the roots of polynomial equation

$$s^{2n}+(-1)^n=0\Rightarrow s^{2n}=(-1)(-1)^n,$$

or for general ω_c

$$(s/\omega_c)^{2n}+(-1)^n=0\Rightarrow s^{2n}+\omega_c^{2n}(-1)^n=0\Rightarrow s^{2n}=\omega_c^{2n}(-1)(-1)^n.$$

Using Euler's formula for -1 and 1 yields:

$$s^{2n}=(-1)(-1)^n=\underbrace{e^{-j\pi}}_{-1}\underbrace{e^{j2k\pi}}_{1}(\underbrace{e^{j\pi}}_{-1})^n=e^{j(2k-1)\pi}e^{jn\pi}.$$

Taking the $(2n)^{\text{th}}$ root of both sides yields $2n$ roots of the form:

$$s_k=e^{j\frac{(2k-1)}{2n}\pi}e^{j\frac{\pi}{2}},\quad \text{or for general } \omega_c,\quad s_k=\omega_c e^{j\frac{(2k-1)}{2n}\pi}e^{j\frac{\pi}{2}}. \qquad (7.2\text{-}6)$$

How many distinct roots are there? Starting with $k = 1$ and incrementing k by one until $k = 2n$ gives a distinct root for each k, yielding $2n$ distinct roots total.

7.2.2.6 *Existence: no roots on the imaginary axis*

Rewriting (7.2-6) as $s_k = e^{j\frac{(2k-1)}{2n}\pi} e^{j\frac{\pi}{2}} = e^{j\frac{(2k-1)}{4n}2\pi} e^{j\frac{\pi}{2}}$, $k = 1, 2, \ldots, 2n$ gives $2n$ roots with the following properties:

- All the roots have the form $e^{j\theta}$ (or $\omega_c e^{j\theta}$), i.e. lie on the unit circle (or circle with radius ω_c) in the complex plane.
- For $k = 1, 2, \ldots, 2n$,

$$0 < \frac{1}{4n} \le \frac{(2k-1)}{4n} \le \frac{4n-1}{4n} < 1, \text{ and if } \frac{(2k-1)}{4n} = \frac{1}{2}, \text{ then } k = \frac{(2n+1)}{2} \Rightarrow \text{non-integer.}$$

Therefore $\frac{(2k-1)}{4n} \neq 0, \frac{(2k-1)}{4n} \neq \frac{1}{2}, \frac{(2k-1)}{4n} \neq 1$. This means that

$$s_k \neq \underbrace{e^{j\cdot 0\cdot 2\pi}}_{1} e^{j\frac{\pi}{2}} = \underbrace{e^{j\cdot 1\cdot 2\pi}}_{1} e^{j\frac{\pi}{2}} = e^{j\frac{\pi}{2}}, \text{ and } s_k \neq e^{j\cdot\frac{1}{2}\cdot 2\pi} e^{j\frac{\pi}{2}} = e^{j\pi} e^{j\frac{\pi}{2}} = e^{j\frac{3\pi}{2}},$$

i.e., ***there will be no roots on the imaginary axis* ($e^{j\pi/2}$ or $e^{j3\pi/2}$)** and hence no terms of the form $(s^2 + \omega_k^2)$ that would give rise to an oscillatory nondecaying signal.

Since

$$s_{k+1} = e^{j\frac{(2(k+1)-1)}{2n}\pi} e^{j\frac{\pi}{2}} = e^{j\frac{(2k+1)}{2n}\pi} e^{j\frac{\pi}{2}}, \quad k = 1, 2, \ldots, 2n,$$

$$\angle s_{k+1} - \angle s_k = \frac{(2k+1)}{2n}\pi + \frac{\pi}{2} - \left[\frac{(2k-1)}{2n}\pi + \frac{\pi}{2}\right] = \frac{(2k+1)-(2k-1)}{2n}\pi = \frac{2}{2n}\pi = \frac{\pi}{n},$$

i.e. the roots are spaced evenly with an angle of π/n between them.

This means that since $(\pi/n)2n = 2\pi$, there are n roots, $s_1, \ldots, s_n$, in the open left-half plane (stable poles) and n roots , $s_{n+1}, \ldots, s_{2n}$, in the open right-half plane (unstable poles).

The unstable poles are just the mirror images of the stable poles with respect to the origin along the lines running from them across the origin:

$$\begin{aligned} s_k &= e^{j\frac{(2k-1)}{2n}\pi} e^{j\frac{\pi}{2}} \Rightarrow s_{k+n} = e^{j\frac{(2(k+n)-1)}{2n}\pi} e^{j\frac{\pi}{2}} = e^{j\frac{(2k-1)}{2n}\pi} e^{j\frac{2n}{2n}\pi} e^{j\frac{\pi}{2}} \\ &= \overbrace{e^{j\pi}}^{-1} e^{j\frac{(2k-1)}{2n}\pi} e^{j\frac{\pi}{2}} = -e^{j\frac{(2k-1)}{2n}\pi} e^{j\frac{\pi}{2}} = -s_k, \quad k = 1, 2, \ldots, n, \end{aligned} \tag{7.2-7}$$

which by uniqueness of the roots proves the original assumption.

This means that the stable ($s_1, \ldots, s_n$) and the unstable ($s_{n+1}, \ldots, s_{2n}$) poles are, respectively, the poles of $H_n(s)$ (the roots of $D_n(s)$) and $H_n(-s)$ (the roots of $D_n(-s)$).

Do the roots we have found indeed yield

$$D_{2n}(s) = (s-s_1)(s-s_2)\ldots(s-s_{2n}) = s^{2n} + \prod_{k=1}^{2n} s_k = s^{2n} + (-1)^n \,?$$

Let's check for $n = 1$. In this case angle between roots is $\pi/n = \pi$, so

$$s_1 = e^{j\pi} = -1 \text{ and } s_2 = e^{j(\pi+\pi)} = e^{j2\pi} = 1 = -s_1.$$

Therefore

$$\begin{aligned} D_2(s) &= (s-s_1)(s-s_2) = (s-(-1))(s-1) = (s+1)(s-1) \\ &= s^2 + (1-1)s + (-1)(1) = s^2 + (-1)^1 = s^2 - 1. \end{aligned} \tag{7.2-8}$$

Similarly, this will hold for any n (that's certainly because we factored $D_{2n}(s)$ this way to start with) and the roots found yield the desired form of $D_{2n}(s)$ and, hence, the desired magnitude response.

Are we done? No. The existence of the roots that yield the desired transfer function does not ensure filter implementability — to allow that we still need to see if coefficients of $D(s)$ are real and $(-1)^n \prod_{k=1}^{n} s_k = \prod_{k=1}^{n} \tilde{s}_k = 1$.

7.2.2.7 *Existence: implementability*

The coefficients of $D_n(s)$ will be real if stable poles occur in complex conjugate pairs. To check if this is the case let's look at the complex conjugate of an arbitrary pole s_{k_1}:

$$\begin{aligned} s_{k_1}^* &= [e^{j\frac{2k_1-1}{2n}\pi} e^{j\frac{\pi}{2}}]^* = e^{j\frac{-(2k_1-1)}{2n}\pi} e^{-j\frac{\pi}{2}} = e^{-j\pi} e^{j\frac{-(2k_1-1)}{2n}\pi} e^{j\frac{\pi}{2}} = e^{j\pi} e^{j\frac{-(2k_1-1)}{2n}\pi} e^{j\frac{\pi}{2}} \\ &= e^{j\frac{2n}{2n}\pi} e^{j\frac{-2k_1+1}{2n}\pi} e^{j\frac{\pi}{2}} = e^{j\frac{2n-2k_1+1}{2n}\pi} e^{j\frac{\pi}{2}} = e^{j\frac{2n-2k_1+2-1}{2n}\pi} e^{j\frac{\pi}{2}} = e^{j\frac{2\overbrace{(n-k_1+1)}^{k_2}-1}{2n}\pi} e^{j\frac{\pi}{2}} \\ &= e^{j\frac{2k_2-1}{2n}\pi} e^{j\frac{\pi}{2}} \equiv s_{k_2},\ k_2 = n-k_1+1,\ k_1 = 1, 2, \ldots, n. \end{aligned}$$

Thus, for an arbitrary pole s_k we have

$$s_k^* = s_{n-k+1},\ k = 1, 2, \ldots, n; \text{ e.g. } s_1^* = s_n,\ s_2^* = s_{n-1}, \text{ etc.} \tag{7.2-9}$$

This shows that if n is even, both stable and unstable poles occur in complex conjugate pairs, and if n is odd, two additional roots appear that take values of -1 and 1, respectively, on the real axis (check for $n = 1$). This means that both $D_n(s)$ and $D_n(-s)$ are given by n-degree ***polynomials with real coefficients***, i.e. are physically implementable, and since $e^{j\theta}e^{-j\theta} = 1$,

$$(-1)^n \prod_{k=1}^{n} s_k = \prod_{k=1}^{n} \tilde{s}_k = 1 \text{ for } \omega_c = 1 \text{ or } (-1)^n \prod_{k=1}^{n} s_k = \prod_{k=1}^{n} \tilde{s}_k = \omega_c^n \text{ for arbitrary } \omega_c. \quad (7.2\text{-}10)$$

Thus, roots given by $s_k = e^{j\frac{(2k-1)}{2n}\pi} e^{j\frac{\pi}{2}}$, $k = 1, 2, \ldots, n$ satisfy all the requirements for the all-pole unity DC gain Butterworth filter realization

$$H_{n,\omega_c}(s) \triangleq H_n(s/\omega_c) = \frac{(-1)^n \prod_{k=1}^{n} s_k}{(s/\omega_c - s_1)(s/\omega_c - s_2)\cdots(s/\omega_c - s_{n-1})(s/\omega_c - s_n)} = \quad (7.2\text{-}11)$$
$$\frac{1}{(s/\omega_c - s_1)(s/\omega_c - s_2)\cdots(s/\omega_c - s_{n-1})(s/\omega_c - s_n)} = \frac{\omega_c^n}{(s-\omega_c s_1)(s-\omega_c s_2)\cdots(s-\omega_c s_{n-1})(s-\omega_c s_n)}.$$

Example 7.2-1: Let's take $n = 4$, $\omega_c = 1$. Then Butterworth filter poles are given by

$$s_k = e^{j\frac{(2k-1)}{2n}\pi} e^{j\frac{\pi}{2}}\Big|_{n=4} = e^{j\frac{2k-1}{2\cdot 4}\pi} e^{j\frac{4\pi}{8}} = e^{j\frac{2k-1+4}{8}\pi} = e^{j\frac{2k+3}{8}\pi}, k = 1, \ldots, 8 \Rightarrow$$

$$s_1 = e^{j\frac{5\pi}{8}},\ s_2 = e^{j\frac{7\pi}{8}},\ s_3 = e^{j\frac{9\pi}{8}} = e^{j(\frac{9\pi}{8}-2\pi)} = e^{j(\frac{9\pi}{8}-\frac{16\pi}{8})} = e^{-j\frac{7\pi}{8}} = s_2^*,\ s_4 = e^{j\frac{11\pi}{8}} = e^{-j\frac{5\pi}{8}} = s_1^*,$$
$$s_5 = e^{j\frac{13\pi}{8}} = e^{-j\frac{3\pi}{8}},\ s_6 = e^{j\frac{15\pi}{8}} = e^{-j\frac{\pi}{8}},\ s_7 = e^{j\frac{17\pi}{8}} = e^{j\frac{\pi}{8}} = s_6^*,\ s_8 = e^{j\frac{19\pi}{8}} = e^{j\frac{3\pi}{8}} = s_5^*$$

with the pole map given by

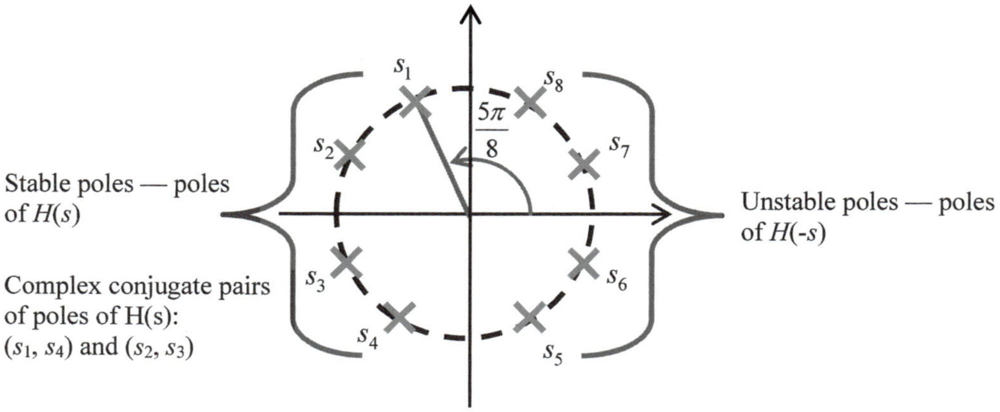

Fig. 7.2-2. Butterworth filter pole map for n = 4.

This pole map yields Butterworth filter transfer function

$$H_4(s)=\frac{(-1)^4\prod_{k=1}^{4}s_k}{(s-s_1)(s-s_2)(s-s_3)(s-s_4)}=\frac{s_1s_1^*s_2s_2^*}{(s-s_1)(s-s_1^*)(s-s_2)(s-s_2^*)}$$

$$=\frac{|s_1|^2|s_2|^2}{\left(s^2-2\operatorname{Re}s_1+|s_1|^2\right)\left(s^2-2\operatorname{Re}s_2+|s_2|^2\right)}=\frac{1}{s^4+2.6131s^3+3.4142s^2+2.16131s+1}.$$

The filter magnitude response is shown in Figure 7.2-3 (left) below.

Example 7.2-2: $n = 4$, ω_c = arbitrary. Rescaling s as s/ω_c yields

$$H_{4,\omega_c}(s)\triangleq H_4(s/\omega_c)=\frac{\omega_c^4}{(s-\omega_c s_1)(s-\omega_c s_2)(s-\omega_c s_3)(s-\omega_c s_4)}=\frac{\omega_c^4}{s^4+a_3s^3+a_2s^2+a_1s-\omega_c^4}.$$

Magnitude response of the resulting filter is shown in the right graph below

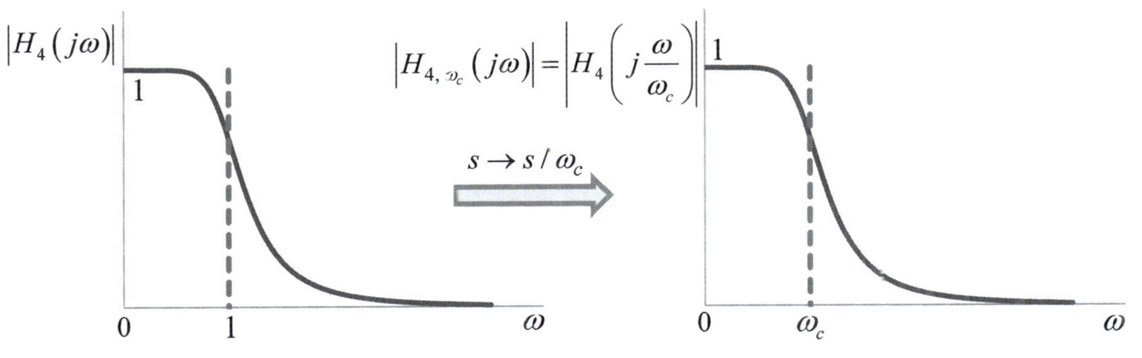

Fig. 7.2-3. 4th order Butterworth filter magnitude response. Left: $\omega_c = 1$, right: ω_c arbitrary.

7.2.3 *Butterworth filter: properties*

So why not take n = (a very large number) all the time? There are tradeoffs:

- Poles move closer to the imaginary axis, leaving smaller stability margin.
- Also, the higher the order, the larger the phase lag (Butterworth filter doesn't have the best phase response — other classical filters trade off some ripple in the passband and stopband, quite tolerable in most applications, for less phase distortion).

Tables 7.1 and 7.2 present, respectively, coefficients and factorization of the Butterworth filter required for its practical implementation.

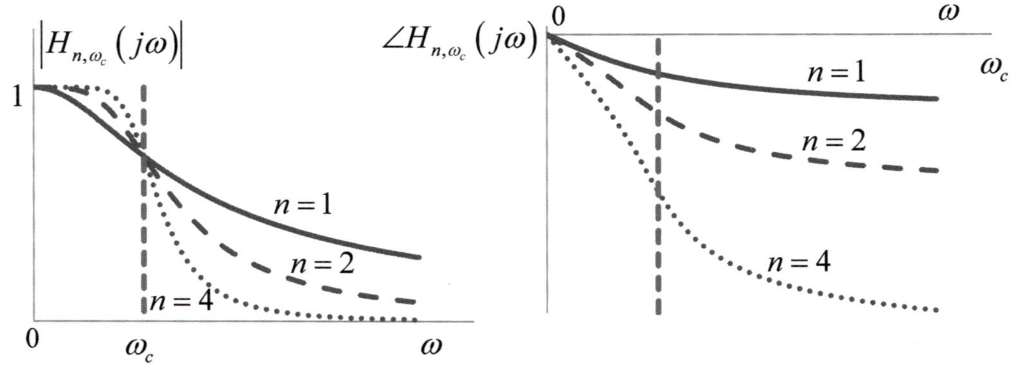

Fig. 7.2-4. N^{th} order Butterworth filter magnitude and phase responses — the higher the order, the larger the phase lag.

Table 7.1. Coefficients of Butterworth polynomials for filter $H_n(s) = 1/[s^n + a_{n-1}s^{n-1} + \ldots + a_1 s + 1]$.

n	a_1	a_2	a_3	a_4	a_5	a_6	a_7
2	1.4142						
3	2.0000	2.0000					
4	2.6131	3.4142	2.6131				
5	3.2361	5.2361	5.2361	3.2361			
6	3.8637	7.4641	9.1416	7.4641	3.8637		
7	4.4940	10.0978	14.5918	14.5918	10.0978	4.4940	
8	5.1528	13.1371	21.8462	25.6884	21.8462	13.1371	5.1528

Table 7.2. Factored Butterworth polynomials.

n	**Factored Polynomial**
1	$s+1$
2	$s^2+1.4142s+1$
3	$(s+1)(s^2+s+1)$
4	$(s^2+0.7654s+1)(s^2+1.8478s+1)$
5	$(s+1)(s^2+0.6180s+1)(s^2+1.6180s+1)$
6	$(s^2+0.5176+1)(s^2+1.4142s+1)(s^2+1.9319s+1)$
7	$(s+1)(s^2+0.4450+1)(s^2+1.2470s+1)(s^2+1.8019s+1)$
8	$(s^2+0.3902s+1)(s^2+1.1111s+1)(s^2+1.6639s+1)(s^2+1.9616s+1)$
9	$(s+1)(s^2+0.3473s+1)(s^2+s+1)(s^2+1.5321s+1)(s^2+1.8794s+1)$

7.2.4 *Filter gain and attenuation*

Assume that a signal $v_i(t)$ with Fourier transform $V_i(j\omega)$ is applied to an input of an LTI filter $H_n(j\omega)$, yielding an output signal $v_o(t)$ with Fourier transform $V_o(j\omega)$: $V_o(j\omega)=H_n(j\omega)V_i(j\omega)$, so that $|V_o|=|H_n(j\omega)||V_i|$, where $|V_i|=|V_i(j\omega)|$ and $|V_o|=|V_o(j\omega)|$ are the input and the output signal magnitudes (approximate magnitude densities, to be more precise), respectively.

Then, the ***filter gain*** and the ***filter attenuation*** at a particular frequency ω are, respectively, defined as the positive real scaling factors, G and A, by which the filtered, or output, signal magnitude is stretched or shrunk in comparison to that of the input signal:

$$G=|V_o|/|V_i|=|H_n(j\omega)| \text{ — simply, the filter magnitude response,}$$

$$A=|V_i|/|V_o|=1/|H_n(j\omega)| \text{ — simply, the inverse of the filter magnitude response,}$$

or, measured in ***decibels*** (dB):

$$G_{dB}=20\log_{10}|H(j\omega)|,\quad A_{dB}=20\log_{10}(1/|H(j\omega)|)=$$
$$-20\log_{10}|H(j\omega)|=20\log_{10}(|V_i|/|V_o|)=10\log_{10}(|V_i|^2/|V_o|^2).$$

7.2.5 *Butterworth filter — design basis*

With the increasing filter order n, the magnitude response $|V_o|/|V_i|=|H_n(j\omega)|$ of the filter in the stopband monotonically moves closer to zero. Therefore, given the lower bound L on periodic input magnitude attenuation $|V_i|/|V_o|$ at the frequency of interest $k\omega_c$, one can always find n so that

$$L\le(|V_i|/|V_o|)_{\omega=k\omega_c}=1/|H_n(jk\omega_c)|=1/\left.\frac{1}{\left|\sqrt{1+(\omega/\omega_c)^{2n}}\right|}\right|_{\omega=k\omega_c}=1/\frac{1}{\left|\sqrt{1+(k\omega_c/\omega_c)^{2n}}\right|}=\left|\sqrt{1+(k)^{2n}}\right|.$$

However, as the order increases, phase lag increases as well:

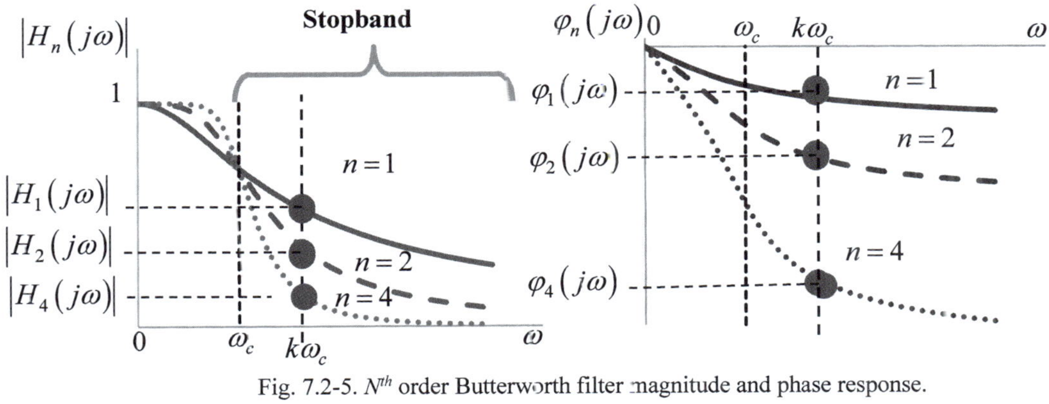

Fig. 7.2-5. N^{th} order Butterworth filter magnitude and phase response.

7.2.6 *Butterworth filter design example*

Goal: Find the transfer function of a Butterworth filter that has cutoff frequency of $\omega_c = 2.5 \times 10^3$ rad/s and attenuation of at least 10 dB at twice the cutoff frequency. We proceed through 4 steps.

1. **Use the Butterworth magnitude response to find the value of *n* that gives the desired attenuation**

   The general form for attenuation:

$$20\log_{10}\frac{|\text{input to filter}|}{|\text{output from filter}|} = 20\log_{10}\frac{1}{|H_n(j\omega)|} = 20\log_{10}\sqrt{1+\left(\frac{\omega}{\omega_c}\right)^{2n}} = 10\log_{10}\left[1+\left(\frac{\omega}{\omega_c}\right)^{2n}\right].$$

   Solving for our specific goal:

$$10 \le 20\log_{10}\frac{|\text{input to filter at } 2\omega_c|}{|\text{output from filter at } 2\omega_c|} = 10\log_{10}\left[1+\left(\frac{2\omega_c}{\omega_c}\right)^{2n}\right] = 10\log_{10}(1+2^{2n}),$$

i.e. $10 \le 10\log_{10}(1+2^{2n})$, or $\log_{10}\left(1+2^{2n}\right) \ge 1 \Rightarrow 1+2^{2n} \ge 10 \Rightarrow 2^{2n} \ge 9 \Rightarrow n \ge \frac{1}{2}\log_2 9 = 1.584.$

2. **If the needed value of *n* is not an integer (it usually will not be an integer), round up to the next highest integer to pick the order of the filter.**

   In this case, we need $n \ge 1.584$, so we pick $n = 2$ as the order of the filter.

3. **Use the tables to find the normalized transfer function:** $H_2(s) = \dfrac{1}{s^2 + 1.4142s + 1}$.

4. **Substitute s/ω_c for s to get the desired cut-off frequency.**

   In this case, $\omega_c = 2.5 \times 10^3$ rad/s, yielding

$$H_{n,\omega_c}(s) = H_n\left(\frac{s}{\omega_c}\right) = \frac{1}{\left(\frac{s}{2500}\right)^2 + 1.4142\left(\frac{s}{2500}\right) + 1} = \boxed{\frac{1}{1.6\times10^{-7}s^2 + 5.6568\times10^{-4}s + 1}},$$

and

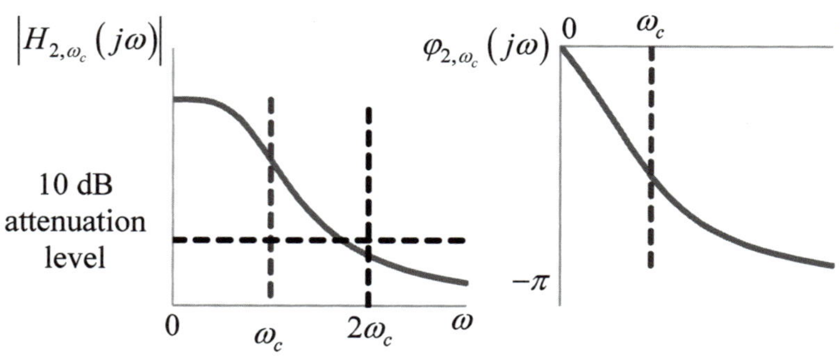

Fig. 7.2-6. System frequency response, left — magnitude, right — phase.

7.2.7 *Butterworth filter implementation*

We have two implementation options for this filter: using a passive circuit or an active one.

Option 1: Passive circuit (RLC):

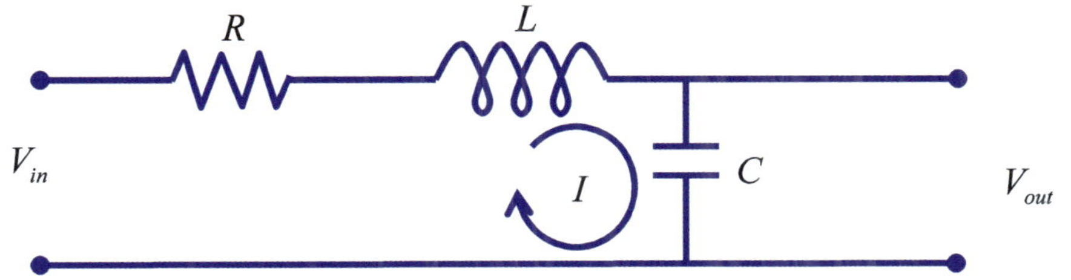

Fig. 7.2-7. Passive circuit (RLC) Butterworth filter implementation.

Using complex impedance:

$$\frac{V_{out}}{V_{in}} = \frac{1}{LCs^2 + RCs + 1}.$$

To obtain the transfer function

$$H(s) = \frac{1}{1.6\times10^{-7}s^2 + 5.6568\times10^{-4}s + 1}$$

we set LC = 1.6×10^{-7} and RC = 5.6568×10^{-4} and select any of the convenient R, L, C values.

Option 2: Active circuit: recall that we can use op-amps to add, multiply, and integrate voltages.

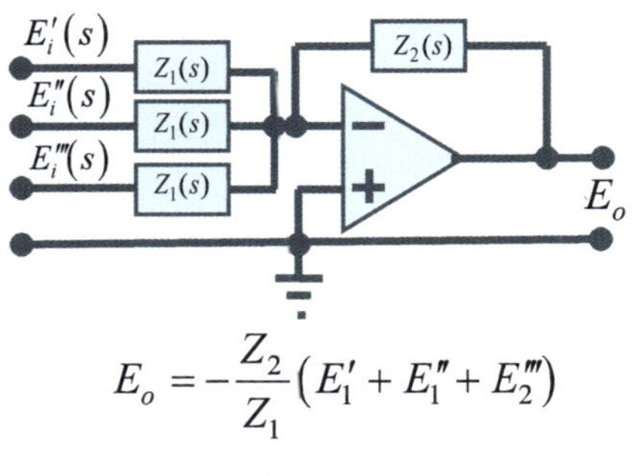

Fig. 7.2-8. Inverting summer.

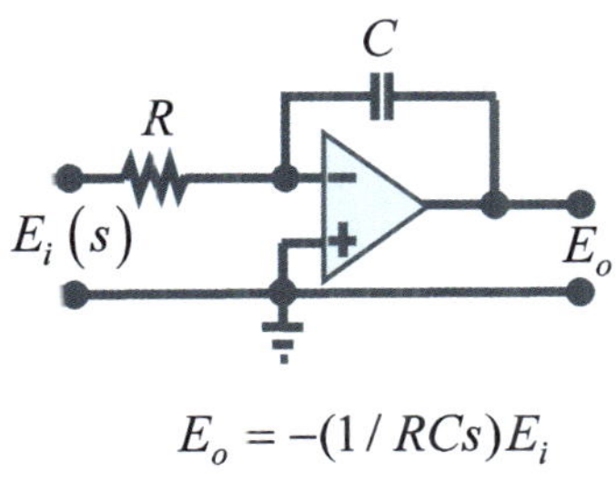

Fig. 7.2-9. Inverting integrator.

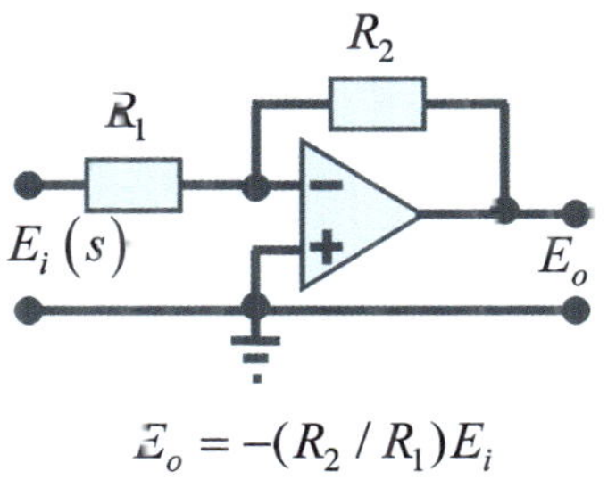

Fig. 7.2-10. Inverting amplifier.

Using these components, we can recreate the transfer function in an electrical circuit. First, rewrite the transfer function as an ODE:

$$\frac{V_{out}}{V_{in}} = \frac{1}{1.6\times10^{-7}s^2 + 5.6568\times10^{-4}s + 1} \Rightarrow \left(1.6\times10^{-7}s^2 + 5.6568\times10^{-4}s + 1\right)V_{out} = V_{in}$$

$$\Rightarrow 1.6\times10^{-7}\ddot{v}_{out} + 5.6568\times10^{-4}\dot{v}_{out} + v_{out} = v_{in}.$$

Second, create a circuit that is modeled by this ODE:

$$1.6\times10^{-7}\ddot{v}_{out} = -5.6568\times10^{-4}\dot{v}_{out} - v_{out} + v_{in}.$$

This circuit block diagram is demonstrated below:

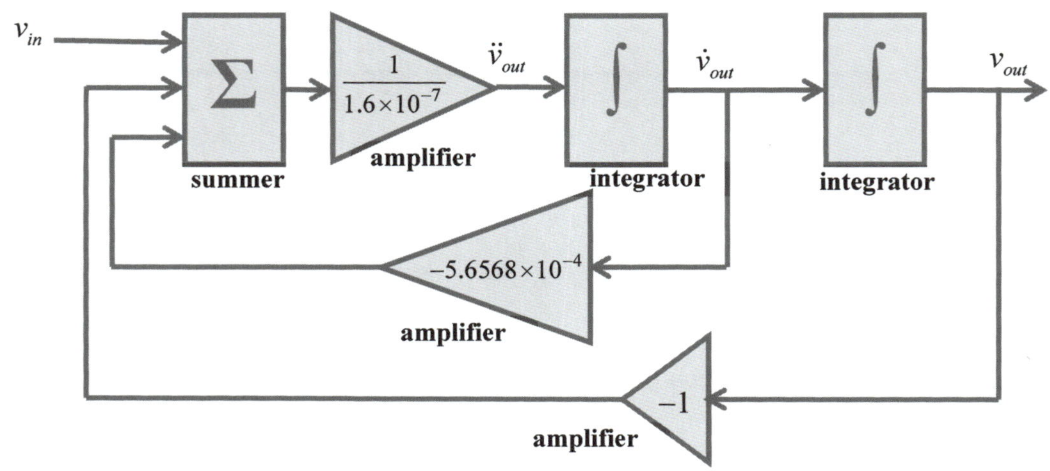

Fig. 7.2-11. Circuit block diagram.

This method can be used to implement any filter, as shown below in Subsection 7.3.4.

7.3 CT: High-Pass, Band-Pass, and Band-Stop Filters. General Filter Implementation. Classical Low Pass Filters: 5th Order Example

Below we list the main features of the rest of the basic filters: high-pass, band-pass, and band-stop.

7.3.1 *High-pass filters*

Single-pole filter: magnitude response is shown in Figure 7.3-1:

Blocking zero at zero, no blocking zeros at ∞

$$\text{Transfer function: } H(s) = \frac{H_0 s}{s+\omega_c}, \quad (7.3\text{-}1)$$

$$\text{Phase: } \varphi(\omega) = \angle H_0 j\omega - \angle(j\omega + \omega_c) = \tan^{-1}(\omega/0) - \tan^{-1}(\omega/\omega_c) = \pi - \tan^{-1}(\omega/\omega_c). \quad (7.3\text{-}2)$$

Two-pole filter: magnitude response is shown in Figure 7.3-2:

Transfer function: $$H(s)=\frac{H_0 s^2}{s^2+\alpha\omega_c s+\omega_c^2}\,. \qquad (7.3\text{-}3)$$

Two blocking zero at zero, no blocking zeros at ∞

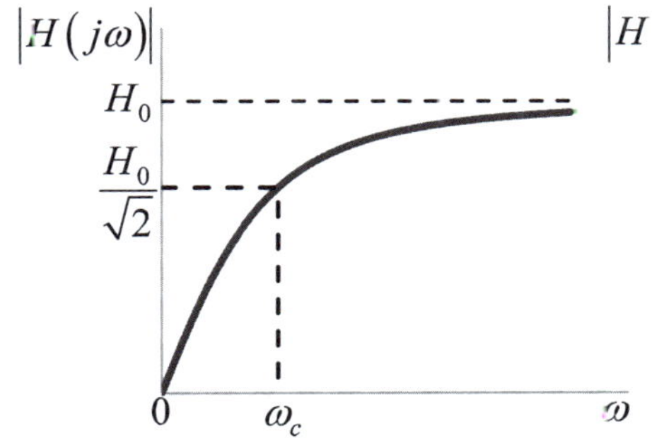

Fig. 7.3-1. Single-pole high-pass filter.

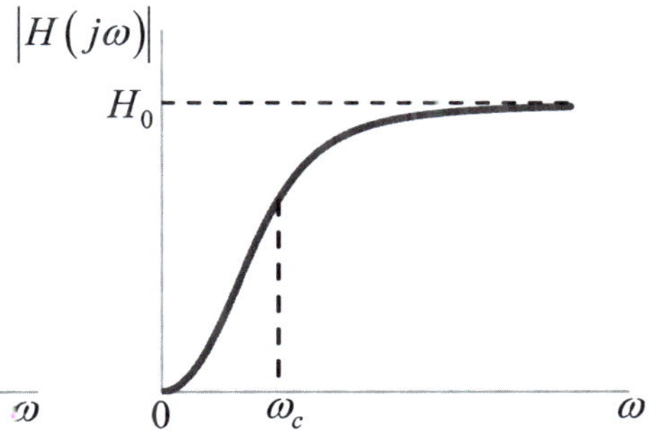

Fig. 7.3-2. Two-pole high-pass filter.

7.3.2 *Band-pass filter*

The simplest band-pass filter can be thought of as a high-pass filter cascaded with a low-pass filter:

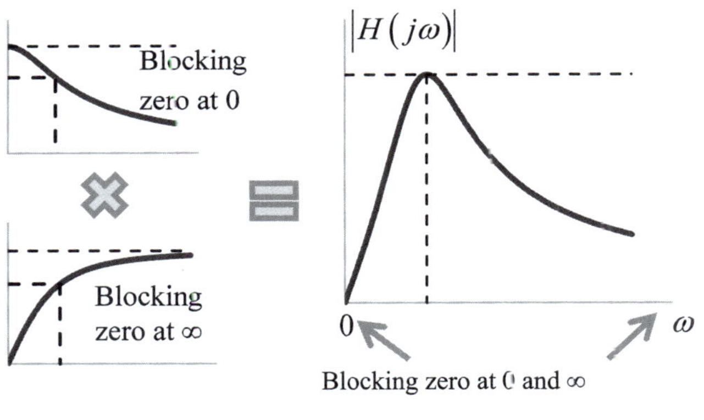

Fig. 7.3-3. Band-pass filter magnitude response.

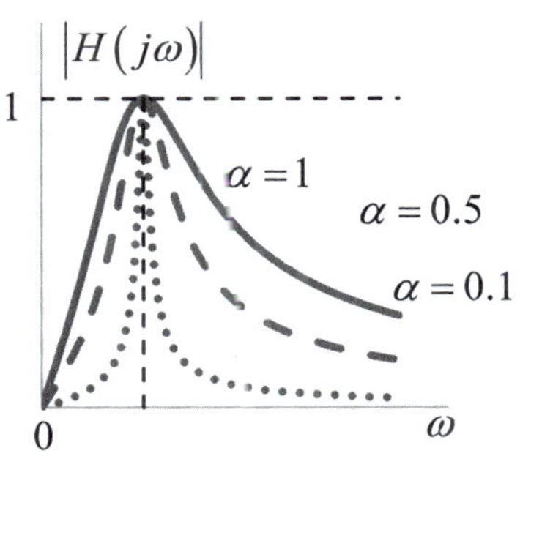

Fig. 7.3-4. Magnitude response with different α.

A more general form of the second order bandpass filter is given by the transfer function:

$$H(s)=\frac{H_0\alpha\omega_c s}{s^2+\alpha\omega_c s+\omega_c^2} \quad \text{where } \alpha=2\zeta\,.$$

Figure 7.3-4 indicates that the smaller the value of α is, the narrower the passband. Tradeoff: poles of $s^2+2\zeta\omega_c s+\omega_c^2=0 \Rightarrow s_{1,2}=-\zeta\omega_c \pm j\omega_c\sqrt{1-\zeta^2}$, $\zeta<1$, therefore as $\varsigma \to 0$ (or $\alpha \to 0$), the poles move closer to the imaginary axis, reducing stability margin.

7.3.3 *Band-stop filter*

The simplest way to obtain this filter is to move a blocking zero to the frequency we want to suppress as follows:

$$H(s) = \frac{H_0\left(s^2 + \omega_c^2\right)}{s^2 + \alpha\omega_c s + \omega_c^2}. \tag{7.3-5}$$

Blocking zero at $\omega = \omega_c$ then appears as follows: $(s^2 + \omega_c^2)_{s=j\omega} = j^2\omega^2 + \omega_c^2 = -\omega^2 + \omega_c^2 = 0$. *This simple mathematical relation will provide the backbone if the mold oscillator resonance suppression in Chapter 9.*
Note that (7.3-5) is essentially 1 minus a band-pass transfer function:

$$H(s) = H_0\Big(1 - \underbrace{\frac{\alpha\omega_c s}{s^2 + \alpha\omega_c s + \omega_c^2}}_{\textbf{Band-pass}}\Big). \tag{7.3-6}$$

Therefore, the effect of choice of α here is the same: a smaller α yields a thinner stopband, with poles closer to instability. This filter cuts out completely periodic signals with frequency ω_c, attenuating nearby harmonics. Due to the shape of its magnitude response it is called a **notch filter**.

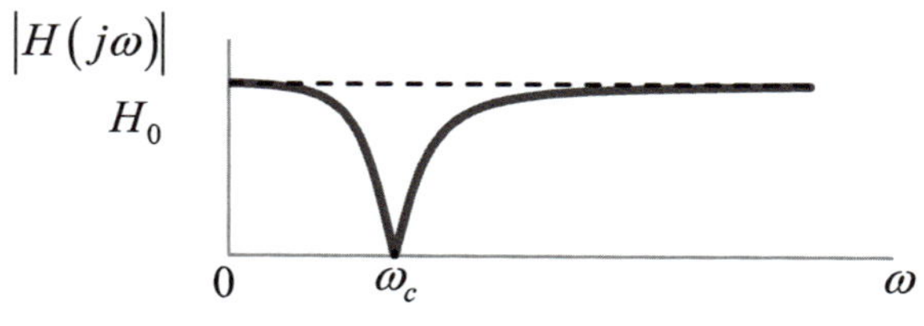

Fig. 7.3-5. Band-stop filter magnitude response.

7.3.4 *General active filter implementation*

Similarly to active Butterworth filter implementation, op amps provide a general method for implementing filters with arbitrary transfer functions. We will carry out filter implementation for a second-order transfer function, but it is possible to extend the method to any higher order.

The key idea here is a **state-variable realization**, or representation of filter transfer function.

A second-order transfer function with real coefficients has 2 complex conjugate poles:

$$\frac{V_o}{V_i}(s) = \frac{K}{s^2 + a_1 s + a_0} = \frac{K}{(s + a + jb)(s + a - jb)} = \frac{K}{s^2 + 2as + \left(a^2 + b^2\right)} = H(s).$$

Multiplying both sides by the denominators yields:

$$\left(s^2+2as+\left(a^2+b^2\right)\right)V_o=s^2V_o+2asV_o+\left(a^2+b^2\right)V_o=KV_i.$$

Now, taking the inverse Laplace transform yields:

$$\ddot{v}_o+2a\dot{v}_o+\left(a^2+b^2\right)v_o=Kv_i \;\Rightarrow\; \ddot{v}_o=(-2a)\dot{v}_o+\left[-\left(a^2+b^2\right)\right]v_o+Kv_i. \tag{7.3-7}$$

We can picture this ODE graphically in the state-variable form:

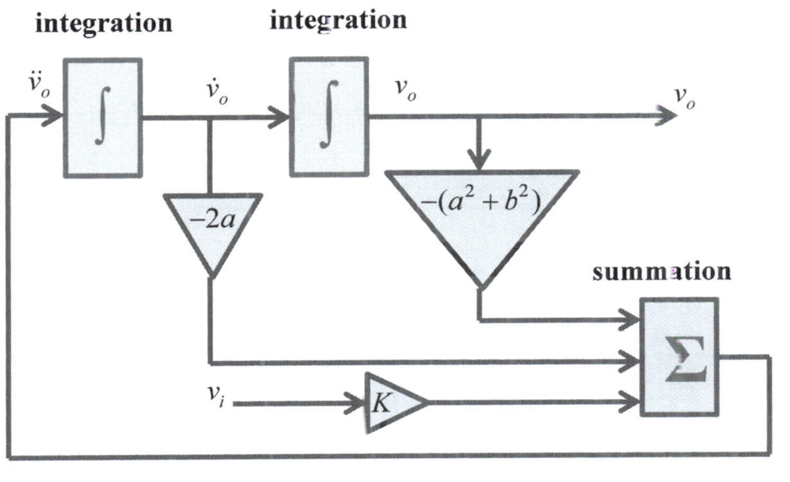

Fig. 7.3-6. Filter ODE diagram in state-variable form.

Redrawing this diagram in terms of op-amp components and accounting for inversion yields

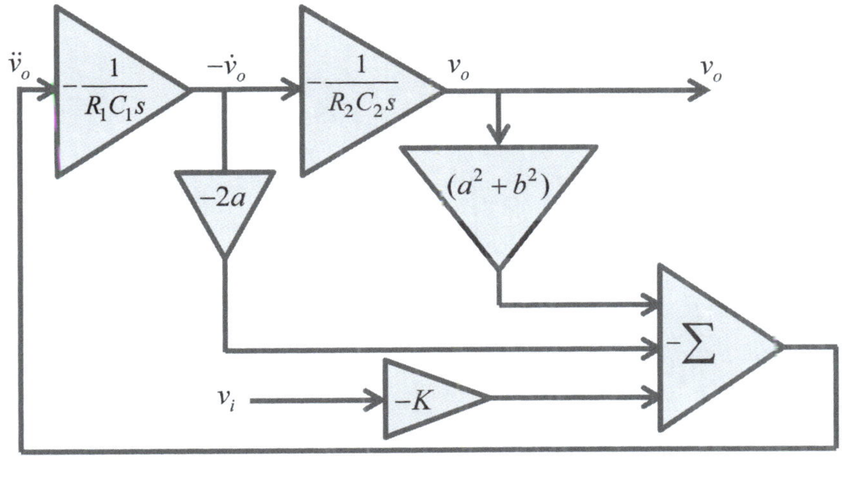

Fig. 7.3-7. The schematics re-drawn in terms of inverting op-amp components.

Replacing the schematics with the actual op-amps yields

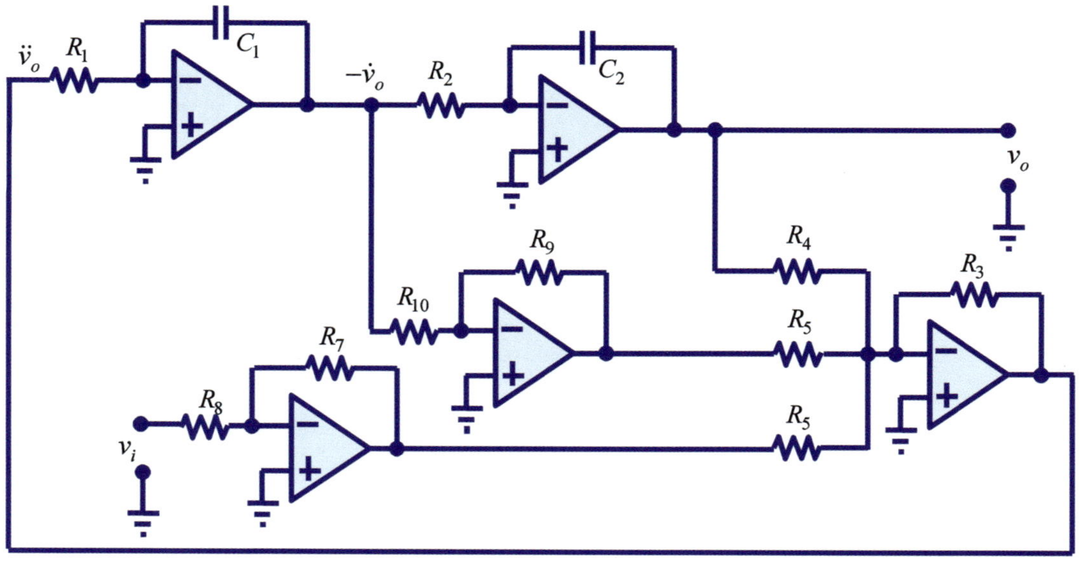

Fig. 7.3-8. The schematics with actual op-amps.

Choosing components to obtain correct gains yields values below, but others are possible.

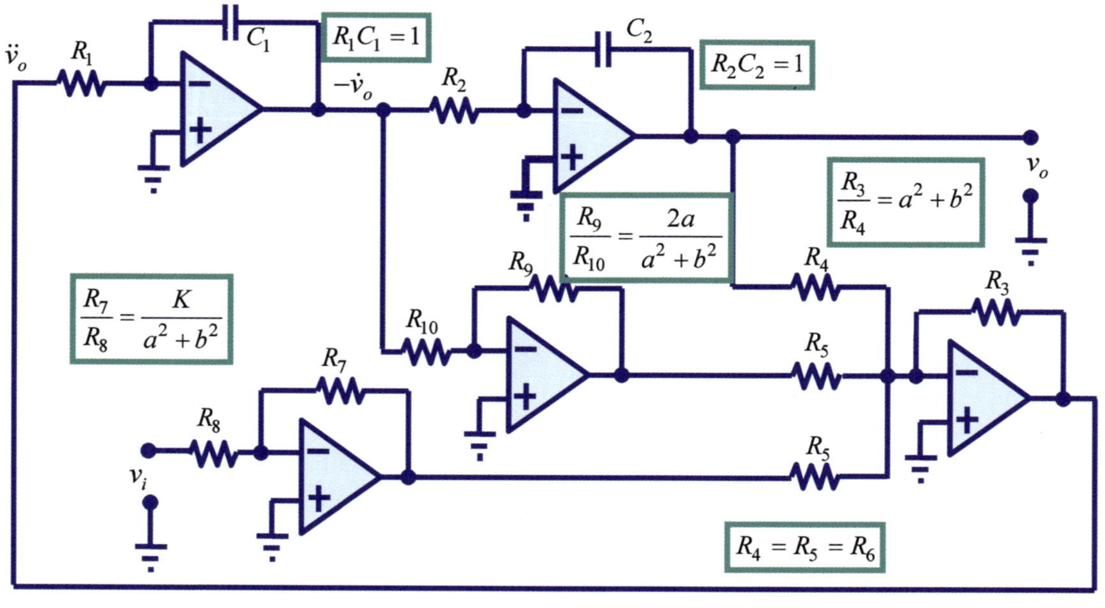

Fig. 7.3-9. Components with the correct gains.

7.3.5 *Classical low pass CT filter transfer functions and their comparison*

The Butterworth filter presented earlier is one the classical low pass filters, all of which approximate the ideal "brick wall" low pass filter according to different criteria. To more fully describe the phase response properties of these filters, we introduce several additional filter characterizations.

7.3.5.1 *Phase delay, group delay, and linear phase filter*

Recall the time lag t_1 and its relation to the phase:

$$\cos(\omega_1(t-t_1)) = \cos(\omega_1(t-\overbrace{(-\theta_1/\omega_1)}^{t_1})).$$

Select a group of i cosines with $\omega_i \neq \omega_j$, but with $t_i = t_j, \forall i, j$. Then $\theta_i/\omega_i = k, \forall i,\ k-const$ $\Rightarrow \theta_i = k\omega_i, \forall i$: linear variation of phase with frequency. Such relation is called ***linear phase***. It implies no relative phase distortion for this group of sinusoids, so that the shape of the signal formed by this cosine group with $t_i = 0$ is preserved when delayed.

For a filter phase response $\theta(\omega)$, the ***phase delay*** is given by $t_{PD} = -\theta(\omega)/\omega$, and the ***group delay*** is defined as $t_{GD} = -d\theta(\omega)/d\omega$. If the ***filter is linear phase***, i.e. $\theta(\omega) = k\omega,\ k-$constant, there is no relative phase distortion, and $t_{PD} = t_{GD} = k$

7.3.5.2 *Classical analog IIR low pass filters (LPFs): 5th order example*

The main classical LPFs include

i. Butterworth
ii. Bessel (Butterworth and Bessel have no finite zeros)
iii. Chebyshev (I,II)
iv. Elliptical (Cauer).

All of these are coded in Matlab. An example of the Butterworth filter calculation looks as follows:

Matlab Example:

```
>> [z,p,k] = butter(N,w,'s')
```

zeros → z; poles → p; constants → k; order → N; break frequency → w; continuous time → 's'

An example of the ideal LPF approximation with 1 rad/sec cutoff frequency by the 5th order transfer functions calculated according to these filter designs, as designated by subscripts, is given below:

Bessel filter: $$H_{Bess}(s) = \frac{1}{s^5 + 3.811s^4 + 6.777s^3 + 6.886s^2 + 3.936s + 1}$$

Butterworth filter: $$H_{Btwth}(s) = \frac{1}{s^5 + 3.236s^4 + 5.236s^3 + 5.236s^2 + 3.236s + 1}$$

Chebyshev Type I filter: $$H_{Ch-I}(s) = \frac{1}{s^5 + 1.172s^4 + 1.937s^3 + 1.31s^2 + 0.7525s + 0.1789}$$

Chebyshev Type II filter: $$H_{Ch-II}(s) = \frac{0.5025s^4 + 2.01s^2 + 1.608}{s^5 + 3.247s^4 + 5.145s^3 + 5.406s^2 + 3.305s + 1.608}$$

Elliptic (Cauer) filter: $$H_{Ellipt}(s) = \frac{0.3098s^4 + 0.8812s^2 + 0.5983}{s^5 + 1.153s^4 + 2.224s^3 + 1.767s^2 + 1.2s + 0.5983}$$

7.3.5.3 *Magnitude responses of the classical low pass filters: 5th order example*

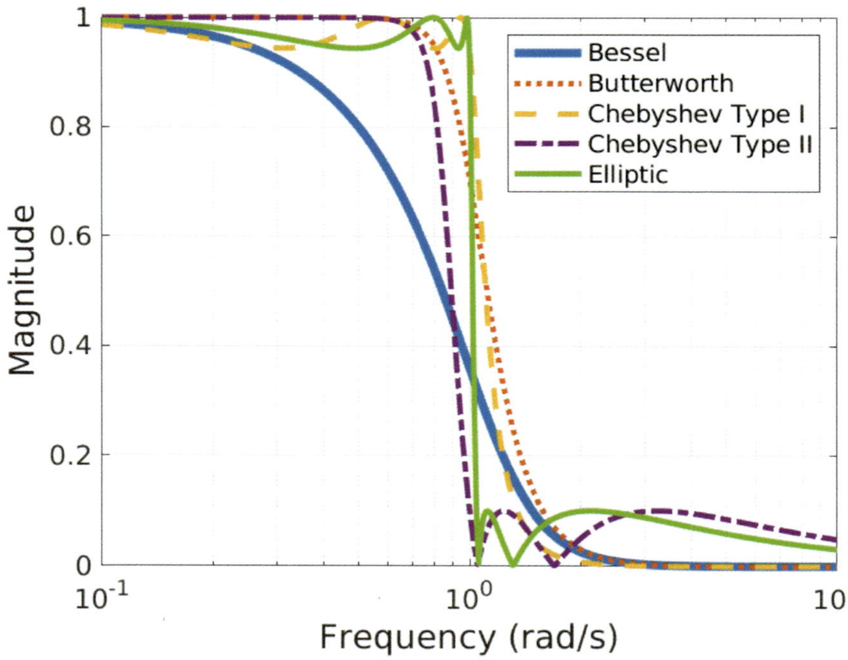

Fig. 7.4-1. Magnitude responses of the classical 5th order low pass filters.

Butterworth filters — nice rolloff, no ripples (maximally flat).
Bessel filters — rolloff not steep, need higher order for a good rolloff.
Chebyshev Type I and II filters minimize the absolute difference between the ideal and the actual frequency response over, respectively, the entire passband and stopband.
Elliptic (Cauer) filters are equiripple in both the passband and stopband. They generally meet filter requirements with the lowest order of any supported filter type.

7.3.5.4 *Phase responses of the classical low pass filters: 5th order example*

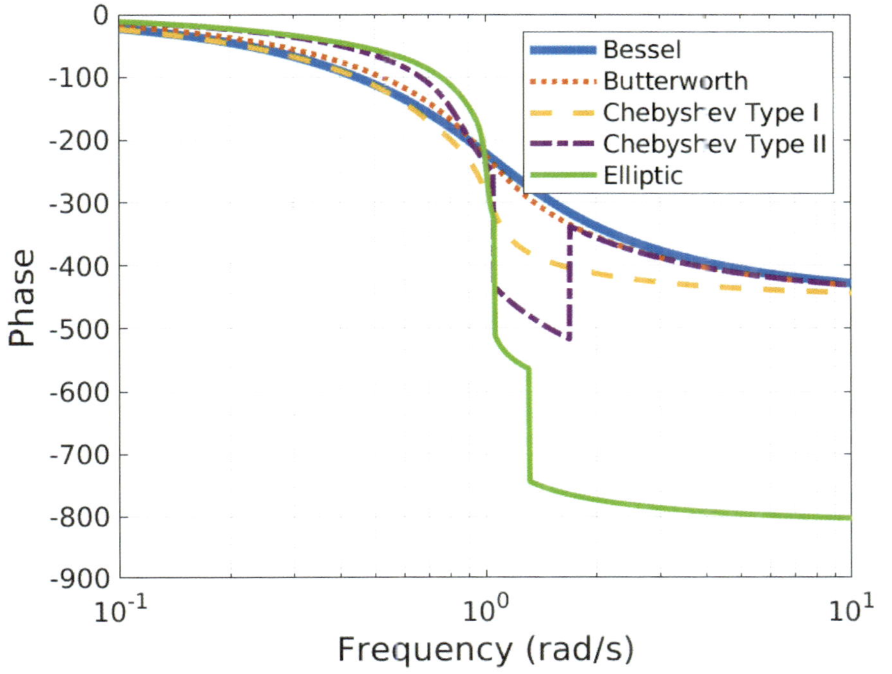

Fig. 7.4-2. Phase responses of the classical 5th order low pass filters.

Filter choice depends on application:
1) for audio, the phase distortion is more damaging than the magnitude distortion, so Bessel LPF would work — linear phase, so gives minimal relative phase distortion; but Bessel filter rolloff is not steep, so high oversampling, like in DSM, would help keep the filter order low to further minimize phase distortion.
2) For data acquisition in mechanical systems, like in mold oscillator, relative phase distortion is of no concern, so elliptic (Cauer) filter is often used. **Butterworth filters** — not the best rolloff/phase-lag tradeoff, but still widely used.
Bessel filters — phase response is linear phase — produces the same group delay.
Chebyshev Type I and II filters reasonable phase response.
Elliptic (Cauer) filters the best rolloff/phase lag tradeoff for many applications.

7.4 CT: Combined Active Filter Implementation. DT Filters: Z-Transform

In practice, a single optimized chip provides the desired filtering properties, as shown next.

7.4.1 *Practical filter circuit*

The practical circuit given in Figure 7.4-1 implements low-pass, band-pass, and high-pass filters, as seen from transfer functions (7.4-1)–(7.4-3) and Figures 7.4-2–7.4-6.

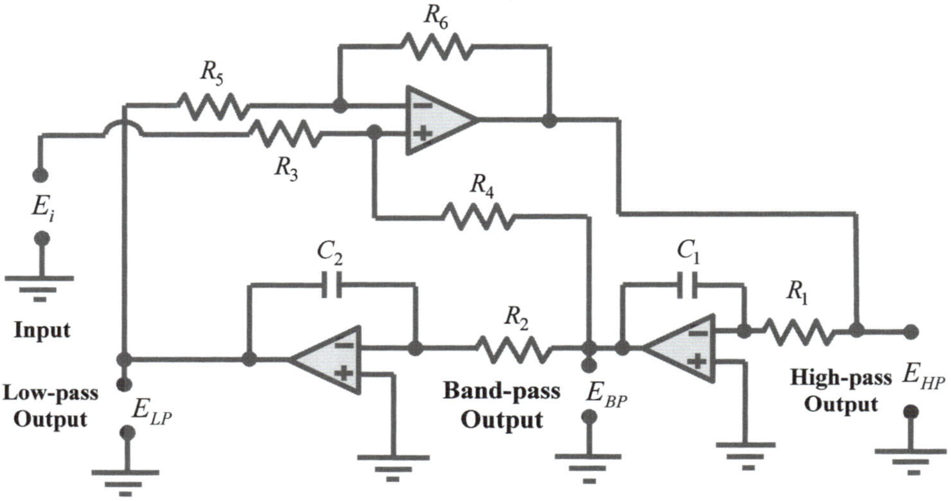

Fig. 7.4-1. Practical filter circuit.

For the low-pass filter transfer function, input/output relation has two zeros at infinity:

$$\frac{E_{LP}}{E_i} = \frac{[1/(R_1R_2C_1C_2)]\cdot[(1+R_6/R_5)/(1+R_3/R_4)]}{s^2+s[1/(R_1C_1)][(1+R_6/R_5)/(1+R_3/R_4)]+(R_6/R_5)\cdot[1/(R_1R_2C_1C_2)]}. \tag{7.4-1}$$

For the band-pass filter transfer function, input/output relation has one zero at zero, and another at infinity:

$$\frac{E_{BP}}{E_i} = \frac{[1/(R_1C_1)]\cdot[(1+R_6/R_5)/(1+R_3/R_4)]s}{s^2+s[1/(R_1C_1)][(1+R_6/R_5)/(1+R_3/R_4)]+(R_6/R_5)\cdot[1/(R_1R_2C_1C_2)]}. \tag{7.4-2}$$

For the high-pass filter transfer function, input/output relation has two zeros at zero.

$$\frac{E_{HP}}{E_i} = \frac{[(1+R_6/R_5)/(1+R_3/R_4)]s^2}{s^2+s[1/(R_1C_1)][(1+R_6/R_5)/(1+R_3/R_4)]+(R_6/R_5)\cdot[1/(R_1R_2C_1C_2)]}. \tag{7.4-3}$$

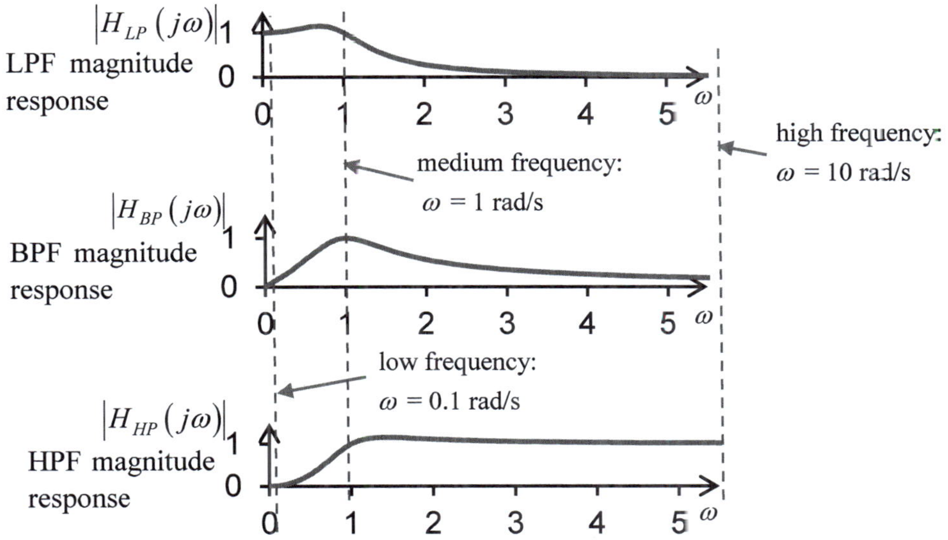

Fig. 7.4-2. Practical filter magnitude responses.

Simulation of the filter response for three filter types are given in Figures 7.4-3–7.4-6. Components in Figure 7.4-2 and simulation below are: $R_1 = R_2 = R_3 = R_4 = R_5 = R_6 = C_1 = C_2 = 1$.

7.4.1.1 *Practical filter circuit: Simulink diagram*

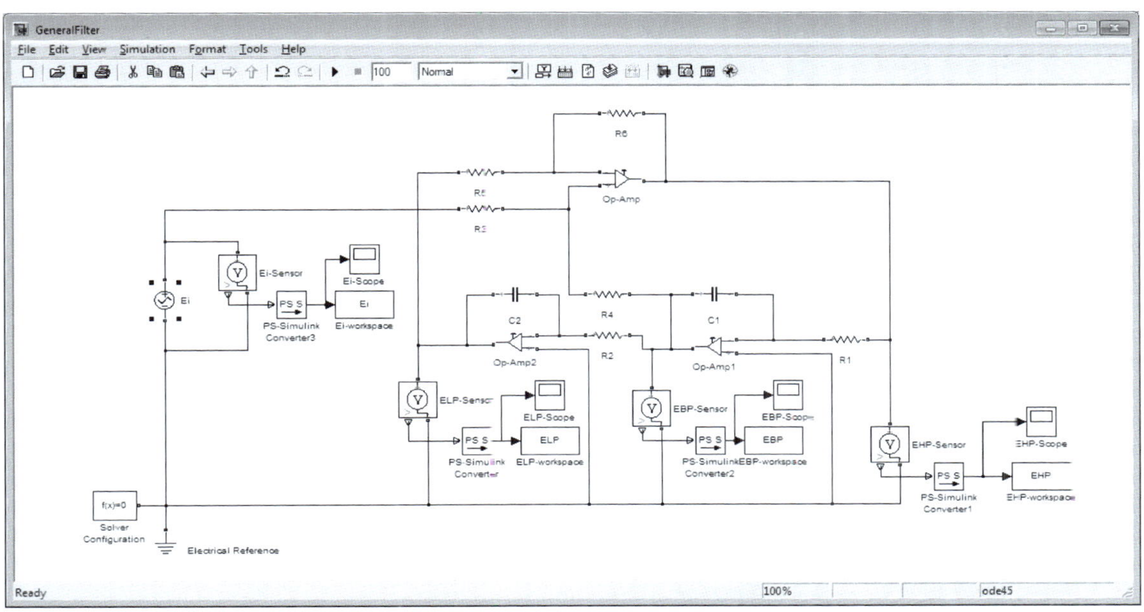

Fig. 7.4-3. Matlab Simulink diagram.

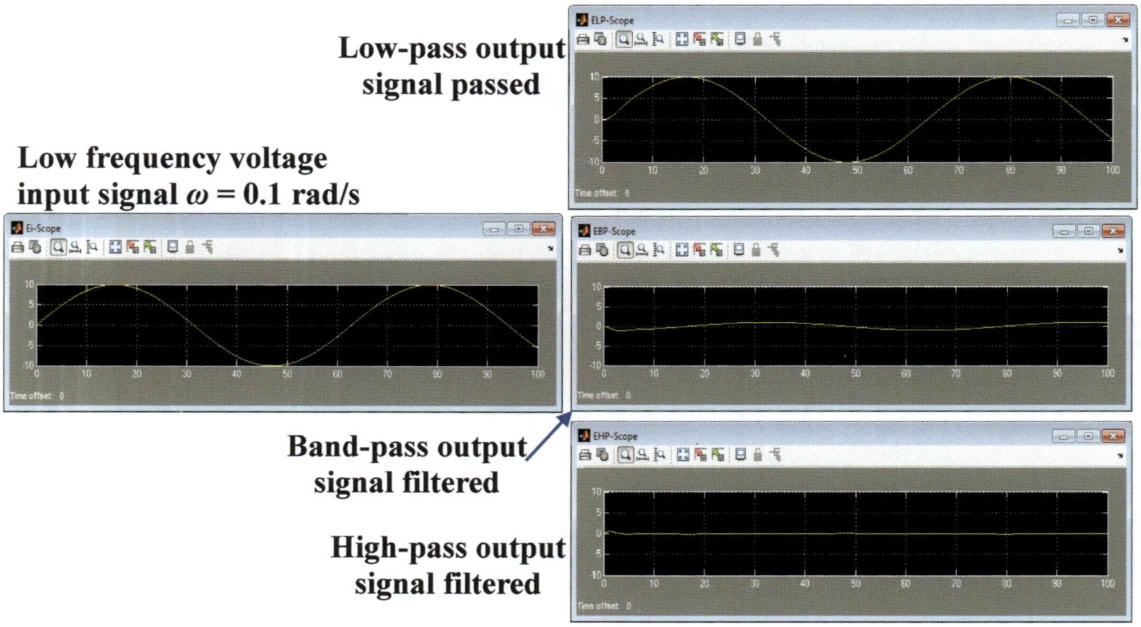

Fig. 7.4-4. Practical filter circuit: responses of filters to low frequency signal.

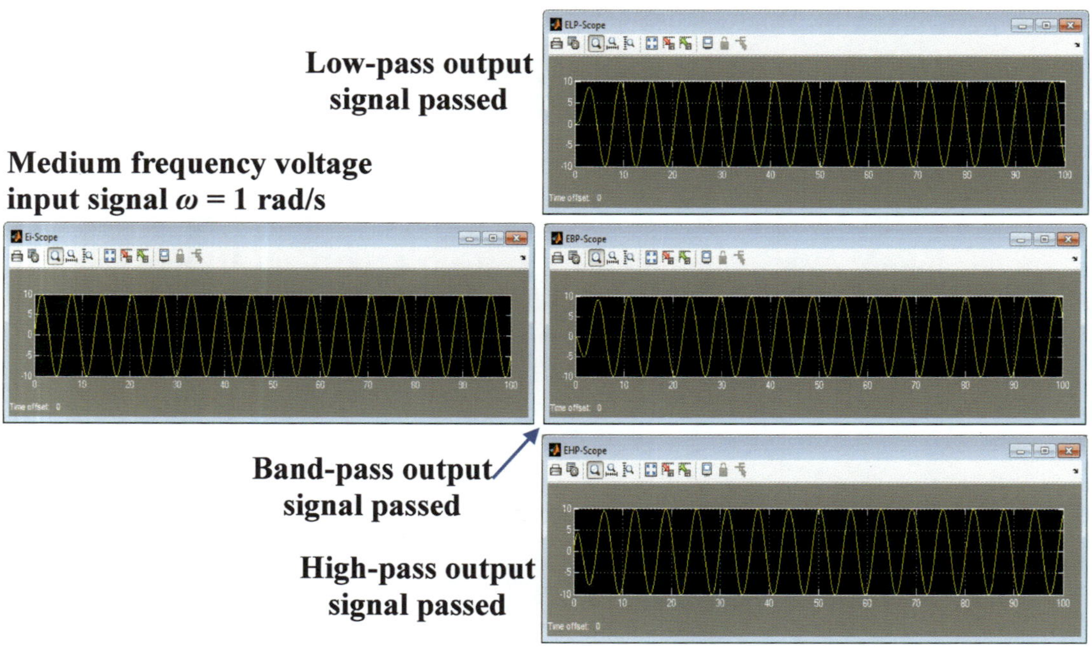

Fig. 7.4-5. Practical filter circuit: responses of filters to medium frequency signal.

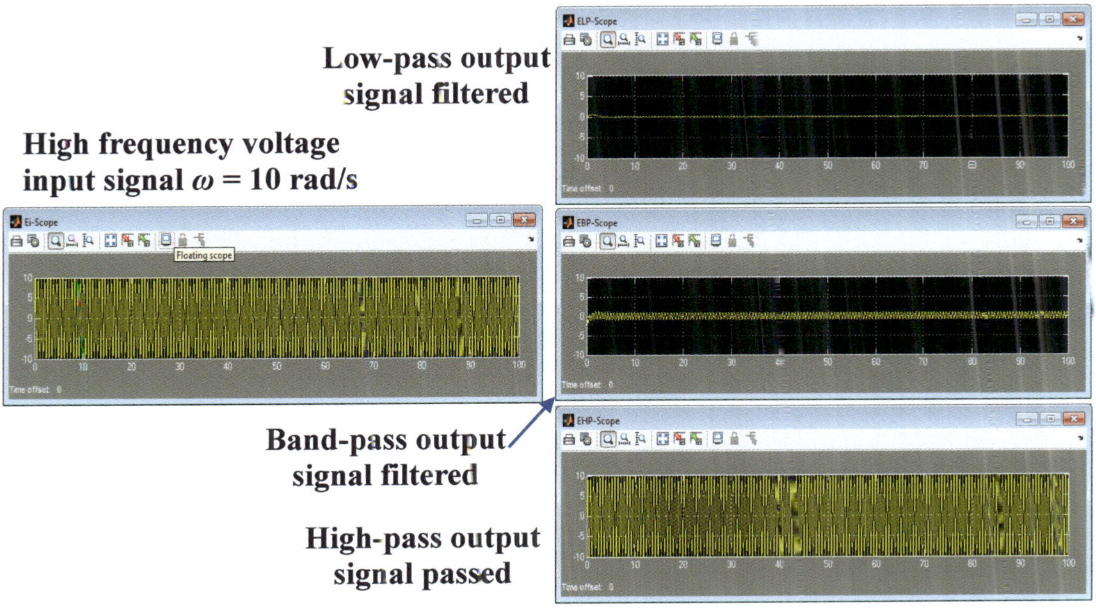

Fig. 7.4-6. Practical filter circuit: responses of filters to high frequency signal.

7.4.2 *DT systems and the Z-transform*

Before moving on to DT filters, we need to introduce a useful tool: the Z-transform. For causal CT systems and signals, we have the Laplace Transform and the Fourier Transform:

$$X(s)=\int_0^\infty x(t)e^{-st}dt; \quad X(\omega)=\int_0^\infty x(t)e^{-j\omega t}dt \tag{7.4-4}$$

$$X(j\omega)=X(s)\big|_{s=j\omega}$$

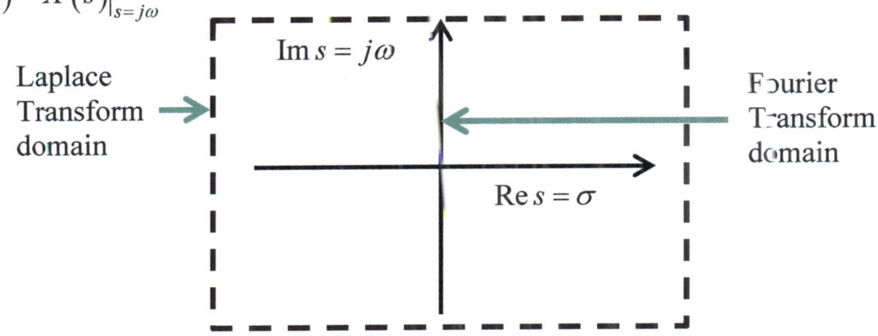

Fig. 7.4-7. Laplace and Fourier transform domains.

7.4.2.1 *Complex variable* e^{-st}

Looking at the causal Laplace transform, we see that it represents Fourier transform of the exponentially weighted CT signal:

$$\underbrace{X(s)}_{\text{Laplace Transform}}=\int_0^{\infty}x(t)e^{-st}dt=\int_0^{\infty}x(t)e^{-(\sigma+j2\pi f)t}df=\overbrace{\int_0^{\infty}\underbrace{x(t)e^{-\sigma t}}_{\text{Exponentially damped signal } x(t)}\underbrace{e^{-j2\pi ft}}_{\text{CT basis continuum}}df}^{\text{Weight (as a function of frequency) of projection of exponentially damped signal } x(t) \text{ onto CT basis continuum}}$$

$-s$ in e^{-st} yields the exponential damping factor $e^{-\sigma t}$: $\sigma>0 \Rightarrow e^{-\sigma t}$ decays as $t\to\infty$, $\sigma<0 \Rightarrow e^{-\sigma t}$ grows as $t\to\infty$.

7.4.2.2 *Complex variable* z^{-1}

Now, let's sample e^{-st}:

$$e^{-st}\xrightarrow{t\to nt_s}e^{-snt_s}=[(\underbrace{e^{st_s}}_{z})^{-1}]^n=[(\underbrace{e^{(\sigma+j2\pi f)t_s}}_{z})^{-1}]^n=[(\underbrace{e^{\sigma t_s}}_{r}e^{\frac{j2\pi F}{\Omega}})^{-1}]^n=$$

$[1/(\underbrace{re^{j\Omega}}_{z})]^n=z^{-n}=r^{-n}e^{-j\Omega n}$. This discretization produces an s-to-z map: $s\to z=e^{st_s}$.

Thus, we have obtained:

- DT basis continuum $e^{-j\Omega n}$
- DT exponential weighting r^{-n}

z is a complex number, with phase Ω (digital angular frequency) and magnitude r.
As n increases:

- $e^{-j\Omega n}$ rotates around the unit circle
- $r > 1$ (this holds if and only if $\sigma > 0$ since $t_s > 0$): r^n grows and r^{-n} decays
- $r < 1$ (this holds if and only if $\sigma < 0$ since $t_s > 0$): r^n decays and r^{-n} grows.

7.4.3 *Z-transform*

Recall the Discrete-Time Fourier transform for causal signals:

$$X_p(F)=\sum_{n=0}^{\infty}x[n]e^{-j2\pi Fn}=\sum_{n=0}^{\infty}x[n]e^{-j\Omega n}=\sum_{n=0}^{\infty}x[n]\left(e^{j\Omega}\right)^{-n}. \quad (7.4\text{-}5)$$

Just like in CT Fourier transform, let's now introduce into DTFT the damping factor r^{-n} to admit its convergence for exponentially growing $x[n]$:

$$\sum_{n=0}^{\infty} x[n] r^{-n} e^{-j\Omega n} = X(z) = \sum_{n=0}^{\infty} x[n] z^{-n}. \tag{7.4-6}$$

This is called the (single-sided) **Z-transform**. For causal systems, the DTFT and the Z-transform are related similarly to the way the Fourier transform relates to the Laplace transform, as shown in Figure 7.4-8.

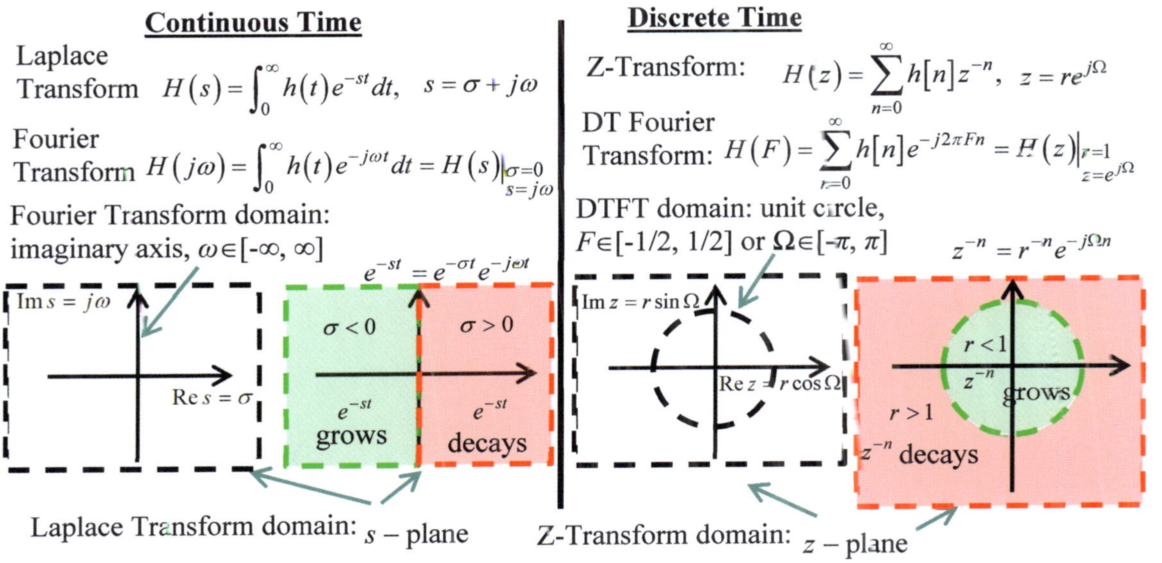

Fig. 7.4-8. Domains of Laplace and Z-transforms and the corresponding CT and DT Fourier transforms.

7.4.3.1 *Z-transform of DT signals*

The Z-transform for DT signals is much easier to calculate than the Laplace transform for CT signals:

$$X(z) = \sum_{n=-\infty}^{\infty} x[n] z^{-n} \text{ or for causal signal } X(z) = \sum_{n=0}^{\infty} x[n] z^{-n}. \tag{7.4-7}$$

Example 7.4-1:

$$x[n] = \left\{1, 3, 0, \overset{\downarrow}{4}, 5, 2\right\} \quad \Rightarrow \quad X(z) = z^3 + 3z^2 + 4 + 5z^{-1} + 2z^{-2}.$$

7.4.3.2 *Z-Transform of shifted signals*

An important property of a signal Z-transform is the change of the latter when the signal is shifted:

$$x[n-1] \;\Rightarrow\; \sum_{n=0}^{\infty} x[n-1]z^{-n} = \sum_{n=0}^{\infty} x[n-1]z^{-(n-1)}z^{-1} = z^{-1}\sum_{n=0}^{\infty} x[n]z^{-n} = z^{-1}X(z).$$

Repeating this k times yields:

$$\boxed{x[n-k] \;\Rightarrow\; z^{-k}X(z).}$$

7.4.3.3 *Z-transform of LCCDE*

We describe linear, time-invariant DT systems as linear, constant-coefficient *difference equations* (LCCDE):

$$y[n]+A_1y[n-1]+\ldots+A_Ny[n-N]=B_0x[n]+B_1x[n-1]+\ldots+B_Nx[n-M],\; M\le N.$$

Taking the Z-Transform using the previous result yields:

$$\left(1+A_1z^{-1}+\ldots+A_Nz^{-N}\right)Y(z)=\left(B_0+B_1z^{-1}+\ldots+B_Mz^{-M}\right)X(z)\Rightarrow H(z)=\frac{Y(z)}{X(z)}=$$

$$\frac{B_0+B_1z^{-1}+\ldots+B_Mz^{-M}}{1+A_1z^{-1}+\ldots+A_Nz^{-N}}=\frac{B_0z^N+B_1z^{N-1}+\ldots+B_Mz^{N-M}}{z^N+A_1z^{N-1}+\ldots+A_N}=\frac{B_0z^M+B_1z^{M-1}+\ldots+B_M}{z^N+A_1z^{N-1}+\ldots+A_N}z^{N-M}.$$

$H(z)$ is a DT transfer function. Like a CT transfer function, it has **poles** (roots of the denominator) and **zeros** (roots of the numerator), but now in z-plane:

zeros

$$H(z)=\frac{Y(z)}{X(z)}=K\frac{(z-z_1)(z-z_2)\cdots(z-z_M)z^{N-M}}{(z-p_1)(z-p_2)\cdots(z-p_N)},\; N-M: \text{ pole excess.} \tag{7.4-8}$$

poles

Pole-zero map in the z-plane for system with two complex-conjugate poles, one nonzero zero and *one zero at 0 from pole excess*

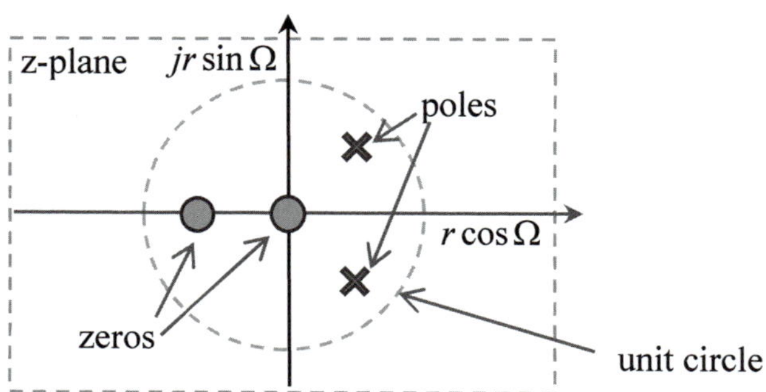

Fig. 7.4-9. An example of the pole-zero map of a 2nd order DT system.

Example 7.4-2: system — LCCDE: $y[n]=\alpha y[n-1]+x[n]$.

Transfer function: $Y(z)=\alpha z^{-1}Y(z)+X(z)\Rightarrow(1-\alpha z^{-1})Y(z)=X(z)$

$$\therefore H(z)=\frac{Y(z)}{X(z)}=\frac{1}{1-\alpha z^{-1}}\cdot\frac{z}{z}=\frac{z}{z-\alpha}.$$

Zero $z_1=0$
Pole $p_1=\alpha$

Inverse z-transform. To find the function of DT corresponding to $H(z)$, i.e. the inverse z-transform of $H(z)$, we simply note that for $x[n]=\delta[n]$

$$X(z)=\sum_{n=0}^{\infty}\delta[n]z^{-n}=1\cdot z^{0}+0\cdot z^{-1}+....=1.$$

Therefore, by convolution theorem $Y(z)=H(z)X(z)=H(z)\cdot 1=H(z)\overset{Z}{\rightleftharpoons}h[n]$, i.e. $H(z)$ is the z-transform of the system impulse response $h[n]$. Hence, $h[n]$ is the inverse z-transform of $H(z)$ and can simply be computed by setting $x[n]=\delta[n]$ and recursively calculating, for example for the system above, $h[n]=\alpha y[n-1]+\delta[n]$, $y[-1]=0$, $n=0,1,2,...$, yielding α^n, $n\geq 0$ shown in Figure 7.4-9. In general, the z-transform inversion is performed through partial fractions expansion and tables.

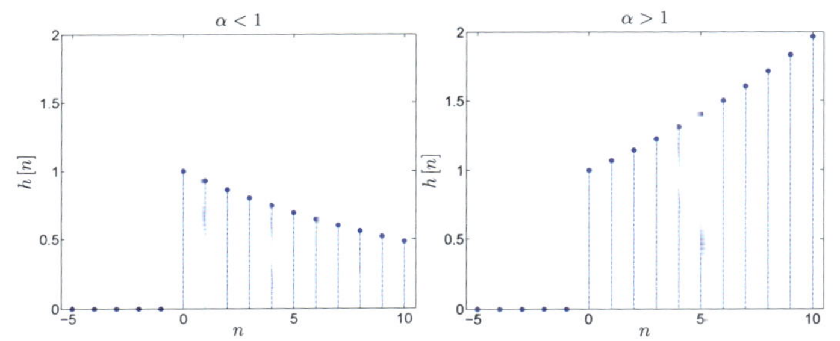

Fig. 7.4-9. DT 1st order system impulse response.

7.4.3.4 *Z-transform: stability*

The latter suggests the stability criterion for a DT system based on its Z-transform transfer function: **Stability Criterion**: all poles must lie inside the unit circle (magnitude less than 1).

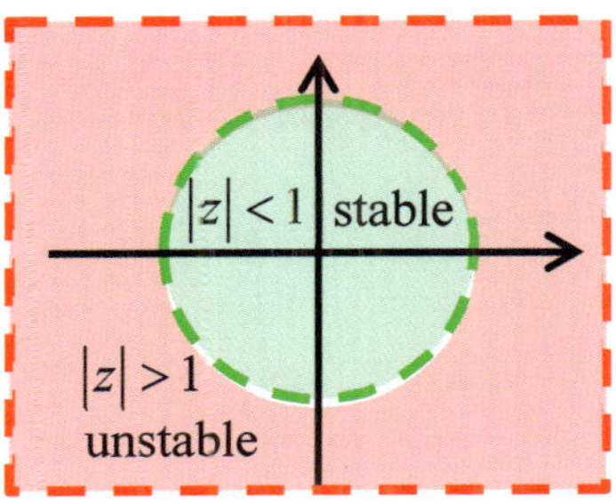

Fig. 7.4-10. Stability criterion for DT systems.

7.5 Digital Filter Design and Implementation

7.5.1 *Digital filter design*

So far, we've talked about designing analog filters. What about digital filters?

Recall:

Real analog filters are *physical systems* described by *differential equations, e.g.*:

$$\ddot{y}(t)+a_1\dot{y}(t)+a_0y(t)=x(t). \tag{7.5-1}$$

Real digital filters are *computations* described by *difference equations, e.g.*:

$$y[n]+a_1y[n-1]+a_2y[n-2]=x[n]. \tag{7.5-2}$$

Continuing the analogy, for CT systems we found the transfer function through the Laplace transform:

LCCODE: $$\ddot{y}(t)+a_1\dot{y}(t)+a_0y(t)=x(t), \tag{7.5-3}$$

Laplace Transform: $$s^2Y(s)+a_1sY(s)+a_0Y(s)=X(s), \tag{7.5-4}$$

Transfer Function: $$\frac{Y(s)}{X(s)}=\frac{1}{s^2+a_1s+a_0}. \tag{7.5-5}$$

For DT systems, we found the transfer function via the Z-Transform:

LCCDE: $$y[n]+a_1y[n-1]+a_2y[n-2]=x[n], \tag{7.5-6}$$

Z Transform: $$Y(z)+a_1z^{-1}Y(z)+a_2z^{-2}Y(z)=X(z), \tag{7.5-7}$$

Transfer Function: $$\frac{Y(z)}{X(z)}=\frac{1}{1+a_1z^{-1}+a_2z^{-2}}. \tag{7.5-8}$$

We will perform DT filter design through ***emulation*** idea — assuming we can convert CT transfer functions to DT transfer functions, we can design a CT filter and then just find the DT filter equivalent. Then, as outlined in Figure 7-4-11, DT filter design will have three steps: i) finding the CT filter satisfying the desired performance specifications (specs), ii) emulation — mapping the resulting CT filter into its DT counterpart, iii) implementing the latter either as IIR algorithm through the recursive LCCDE calculation, or finding the filter impulse response, truncating it, and implementing through convolution sum in FIR form. The simplest way to do emulation is to apply Euler's finite difference approximation to the LCCODE; however, we are going to use another intuitively clear technique — trapezoidal integration, which preserves CT filter stability in DT.

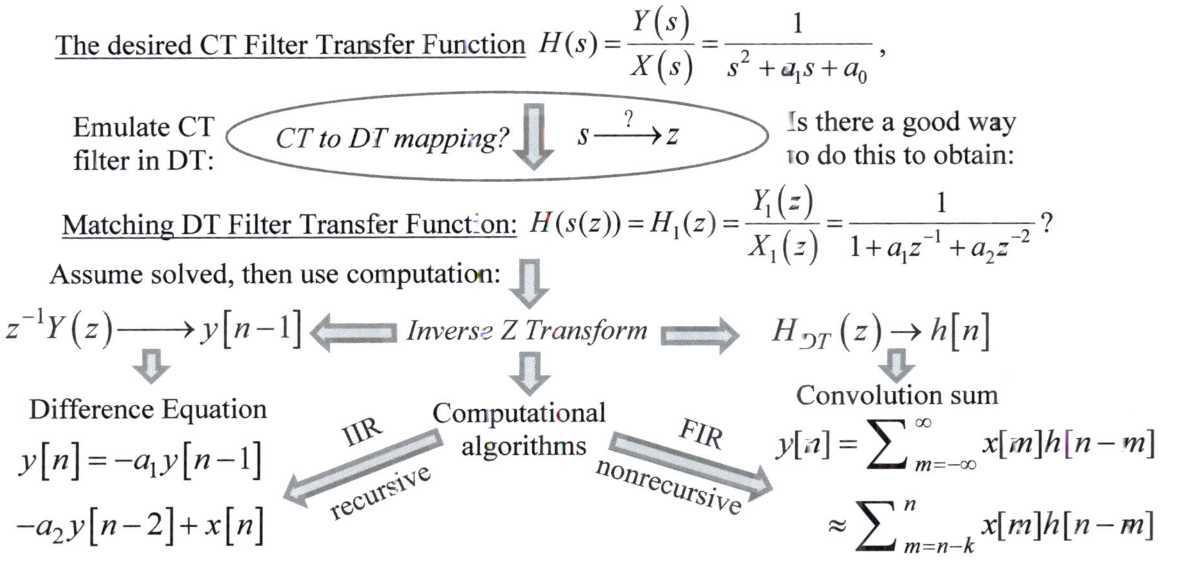

Fig. 7.4-11. Digital filter emulation design problem and computational implementation of its solution.

7.5.2 *Bilinear transformation*

If we use the exact s-to-z map $s \rightarrow z = e^{st_s}$, then

$$s = (1/t_s)\ln z \text{ and } H(s) = H[(1/t_s)\ln z] \doteq H_{EDT}(z),$$

where the subscript EDT stands for "exact discrete time," and $H_{EDT}(z)$ is an irrational function of z (not a ratio of two polynomials). Can we simplify the exact map to obtain rational approximation of $H_{EDT}(z)$?

Let us approximate the exact s-to-z map as follows:

$$z = e^{st_s} = e^{st_s/2}e^{st_s/2} = \frac{e^{st_s/2}}{e^{-st_s/2}} = \frac{1+\frac{1}{2}st_s+\frac{1}{2}\left(\frac{1}{2}st_s\right)^2+\frac{1}{3!}\left(\frac{1}{2}st_s\right)^3+\ldots}{1-\frac{1}{2}st_s+\frac{1}{2}\left(\frac{1}{2}st_s\right)^2-\frac{1}{3!}\left(\frac{1}{2}st_s\right)^3+\ldots} \approx \frac{1+0.5st_s}{1-0.5st_s},$$

i.e., highly nonlinear exponential relation between z and s, $s \rightarrow z = e^{st_s}$, is approximated by a "less" nonlinear (referred to as bilinear) relation: $z = e^{st_s} \approx \frac{1+0.5st_s}{1-0.5st_s} = \frac{2+st_s}{2-st_s}$. For high sampling rate (small t_s), relative contribution of neglected high order terms becomes smaller, both sides tend to 1, and approximation error goes to zero.

Then, inverting this approximation and substituting into $H(s)$:

$$z \approx \frac{1+0.5st_s}{1-0.5st_s} \Rightarrow z-0.5zst_s \approx 1+0.5st_s \Rightarrow z-1 \approx 0.5zst_s+0.5st_s \Rightarrow z-1 \approx 0.5(z+1)st_s$$

$$\Rightarrow s \approx \frac{2}{t_s}\frac{z-1}{z+1} \Rightarrow H_{EDT}(z) \approx H(s)\Big|_{s=\frac{2}{t_s}\frac{z-1}{z+1}} = H\left(\frac{2}{t_s}\frac{z-1}{z+1}\right) \equiv H_{DT}(z),$$

i.e., we obtain rational approximation $H_{DT}(z)$ of $H_{EDT}(z)$ given rational $H(s)$.

7.5.3 *Tustin's rule*

There is actually a number of other reasonable ways to map s to z, each with their own trade-offs. Let's consider a map called **Tustin's rule** (also called the **trapezoidal rule**). It arises from general idea: given signal $x(t)$ and its exact integral $y(t)$, for both of which by definition:

$$y(t+t_s) = y(t) + \int_t^{t_s} x(\tau)d\tau,$$

with their Laplace transforms related as $Y(s) = \frac{1}{s}X(s)$, approximate integration by the trapezoidal rule described next:

Let's plot $x(t)$ and consider $y[n]$ to be its integral $y(t)$ sampled using sampling interval t_s. Then:

$y((n-1)t_s)$: area under $x(t)$ for t from 0 to $(n-1)t_s$

$y(nt_s)$: area under $x(t)$ for t from 0 to nt_s

⇨ Shaded area is approximately $y(nt_s) - y((n-1)t_s) = y[n] - y[n-1]$

Calculating the shaded area in terms of $x(t)$ yields: $0.5t_s\left(x[n]+x[n-1]\right)$.

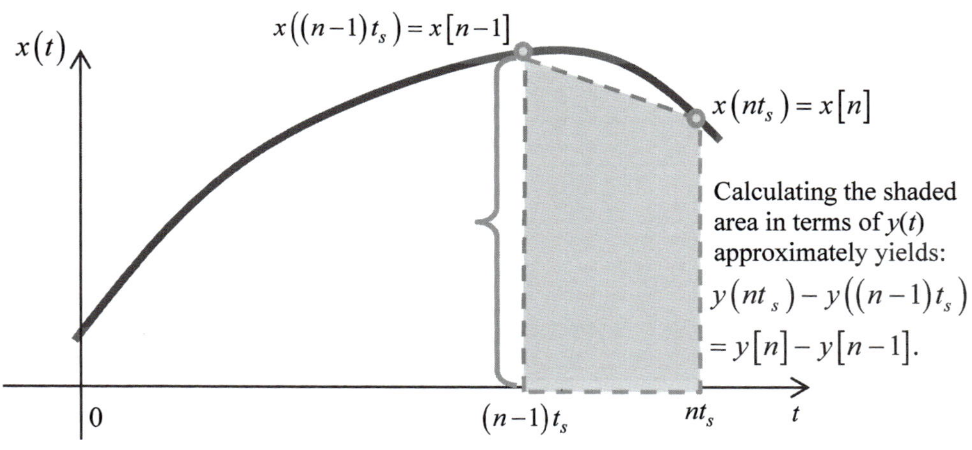

Fig. 7.5-1. Integral of $x(t)$.

Thus, $y[n]-y[n-1]\approx 0.5t_s\left(x[n]+x[n-1]\right)$. Taking the Z-transform and neglecting an approximation error yields:

$$\left(1-z^{-1}\right)Y(z)\approx\frac{t_s}{2}\left(1+z^{-1}\right)X(z)\Leftrightarrow\frac{Y(z)}{X(z)}=\frac{t_s}{2}\cdot\frac{1+z^{-1}}{1-z^{-1}}=\frac{t_s}{2}\cdot\frac{z+1}{z-1}.$$

Comparing the original CT integrator $\frac{Y(s)}{X(s)}=\frac{1}{s}$ with its DT approximation:

$\frac{Y(z)}{X(z)}=\frac{t_s}{2}\cdot\frac{z+1}{z-1}$ yields an s-to-z map: $\frac{1}{s}\rightarrow\approx\frac{t_s}{2}\cdot\frac{z+1}{z-1}$ or, neglecting the approximation error:

$$\textbf{Tustin's Rule (Trapezoidal Rule)}\quad s\rightarrow\frac{2}{t_s}\cdot\frac{z-1}{z+1}. \tag{7.5-9}$$

Thus, Tustin's rule is exactly the bilinear approximation of a nonlinear map $z=e^{st_s}$!

7.5.3.1 *Inverting Tustin's rule*

Thus, we have Tustin's rule to go from CT transfer function to DT transfer function

$$s\rightarrow\frac{2}{t_s}\cdot\frac{z-1}{z+1}\Rightarrow H(s)\rightarrow H_{DT}(z)=H\left(\frac{2}{t_s}\cdot\frac{z-1}{z+1}\right)$$

and from DT to CT one, replacing the arrow henceforth by the equality sign to mean direct substitution:

$$z=\frac{2+t_s s}{2-t_s s}\Rightarrow H(s)=H_{DT}\left(\frac{2+t_s s}{2-t_s s}\right). \tag{7.5-10}$$

There is also a number of other reasonable ways to map s to z, each with their own trade-offs.

7.5.3.2 *Tustin's rule — stability*

A useful feature of Tustin's rule is that it is *stability preserving*. Suppose the CT transfer function has a pole at $s_p=\sigma_p+j\omega_p$ in the complex plane.

Then, using Tustin's rule, (7.5-10) yields:

$$z_p=\frac{2+t_s s_p}{2-t_s s_p}=\frac{2+t_s\sigma_p+jt_s\omega_p}{2-t_s\sigma_p-jt_s\omega_p}.$$

The magnitude of this pole is $\left|z_p\right|=\frac{\sqrt{\left(2+\sigma_p t_s\right)^2+\left(\omega_p t_s\right)^2}}{\sqrt{\left(2-\sigma_p t_s\right)^2+\left(\omega_p t_s\right)^2}}$, yielding a map:

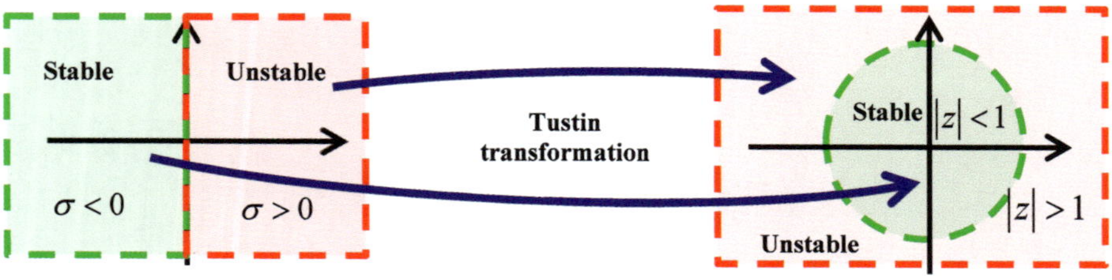

Fig. 7.5-2. Tustin rule: CT to DT stability criterion map.

By inspection:

$$\text{If } \sigma_p > 0, |z_p| > 1, \text{ If } \sigma_p < 0, |z_p| < 1.$$

That is, the open left half-plane in the s plane is mapped to the open unit ball in the z plane. Recall, the stability conditions for CT and DT systems:

- If all poles $s_p = \sigma_p + j\omega_p$ in a CT transfer function satisfy $\sigma_p < 0$, the CT system is stable.
- If all poles z_p in a DT transfer function satisfy $|z_p| < 1$, the DT system is stable.

So, under Tustin's rule, the DT system is stable if and only if the CT system was stable. Same holds for the original bilinear approximation.

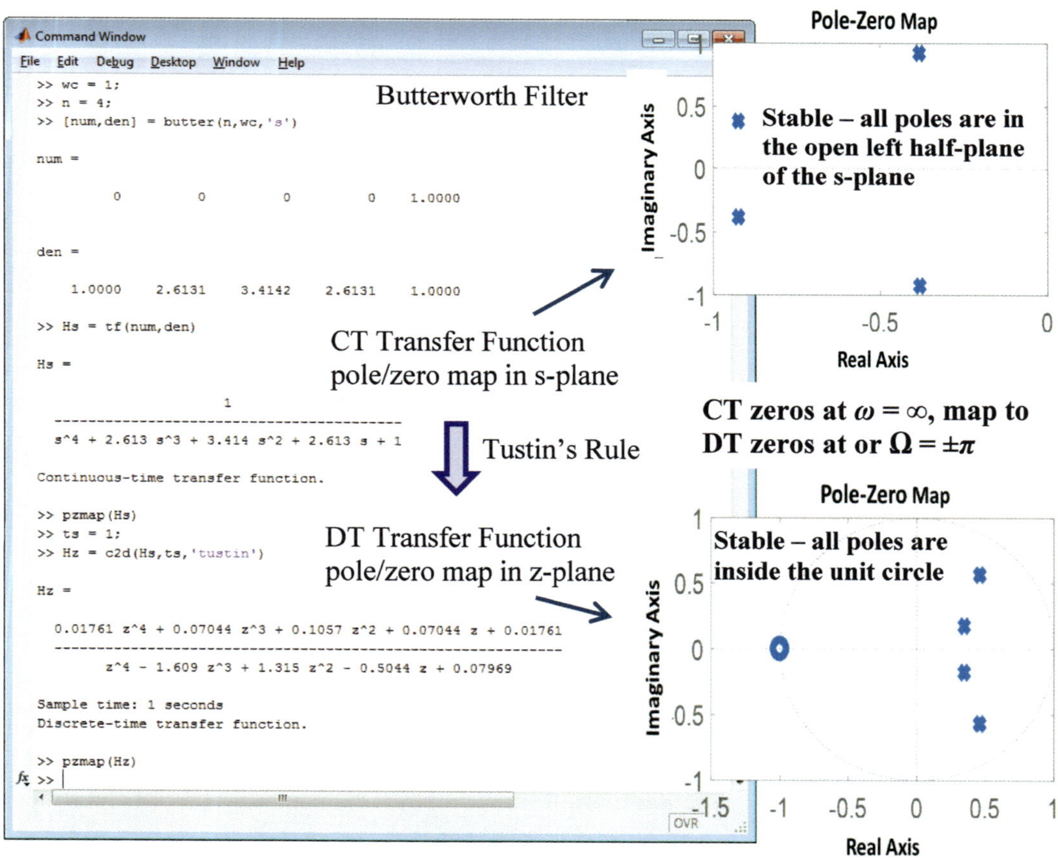

Fig. 7.5-3. Use of Tustin's rule in Matlab: stable CT system poles are mapped into stable DT system ones.

7.5.3.3 *Tustin's rule — CT to DT Matlab conversion*

Replacing $2/t_s$ by a general variable C we obtain a general **bilinear transformation**

$$s = C\frac{z-1}{z+1} \iff z = \frac{C+s}{C-s}.$$

For $s = j\omega$ or $r = 1$ in the exact map between z and s we had

$z = e^{j\omega t_s} = e^{j2\pi \frac{f}{S_f}} = e^{j2\pi F} = e^{j\Omega} \Rightarrow \Omega = \omega t_s$ — a linear relation. Now we have

$$z = \frac{C+j\omega}{C-j\omega} = \frac{\sqrt{C^2+\omega^2}\,e^{j\tan^{-1}\left(\frac{\omega}{C}\right)}}{\sqrt{C^2+\omega^2}\,e^{-j\tan^{-1}\left(\frac{\omega}{C}\right)}} = e^{j2\tan^{-1}\left(\frac{\omega}{C}\right)} = e^{j\Omega} \Rightarrow \Omega = 2\tan^{-1}(\omega/C). \qquad (7.5\text{-}11)$$

This shows that the bilinear transformation distorted (*warped*) the CT angular frequency: instead of $\Omega = \omega t_s$ we have $\Omega = 2\tan^{-1}(\omega/C)$ — a nonlinear relation.

Because of the inverse tangent relationship, $\Omega \to \pi$ as $\omega \to \infty$. The two are equal at $\omega = 0$ and one additional point that depends on the choice of C.This *warping* effect can be partially accounted for by *prewarping* the filter w.r.t. ω using the inverse of $\Omega = 2\tan^{-1}(\omega/C)$:

$$\Omega = 2\tan^{-1}(\omega/C) \Rightarrow \omega = C\tan(\Omega/2). \qquad (7.5\text{-}12)$$

For high sampling rates the product ωt_s is very small in a broad frequency range of ω. In this case, for example, for Tustin's map, where $C = 2/t_s$, small ωt_s moves Ω into the near-linear range of $\tan^{-1}$:

$$\Omega = 2\tan^{-1}(\omega t_s/2) \approx 2(\omega t_s/2) = \omega t_s, \qquad (7.5\text{-}13)$$

removing the need for prewarping.

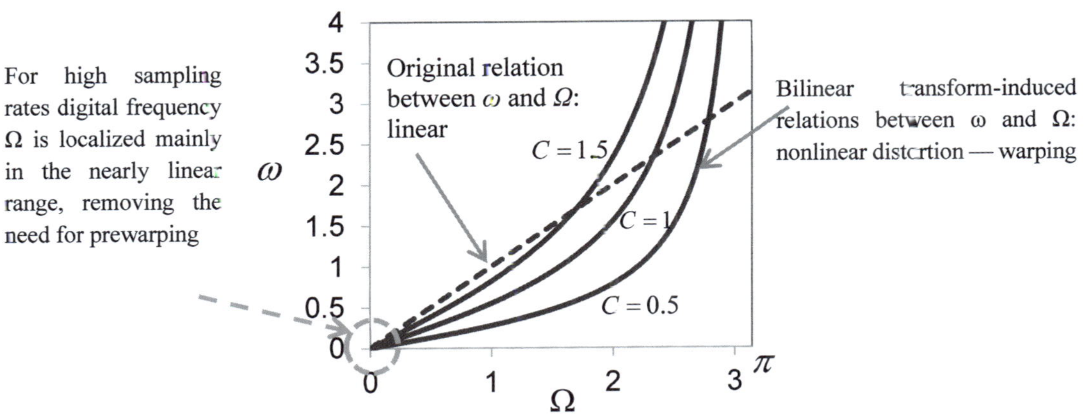

Fig. 7.5-4. Distortion of the original analog angular frequency by the bilinear transformation.

7.5.3.4 *Prewarping*

Suppose we want to divide noise removal between analog and digital sides of our data acquisition system. Assume that we oversample the output of analog filter with respect to its cutoff frequency to avoid noticeable aliasing around cutoff frequency: $S_F >> f_c$, or $2\pi S_F >> 2\omega$. To further reduce the noise that was passed beyond cutoff frequency into the sampled signal, we pass the latter signal through a DT filter with the characteristics that recreate similar characteristics of the antialiasing analog filter. For this purpose, we obtain the DT counterpart cutoff frequency Ω_c of the CT cutoff frequency ω_c of the analog antialiasing filter simply as $\Omega_c = \omega_c / S_F$, but if we want to use bilinear transformation to map CT filter to DT, we need to apply pre-warping for low sampling rates.

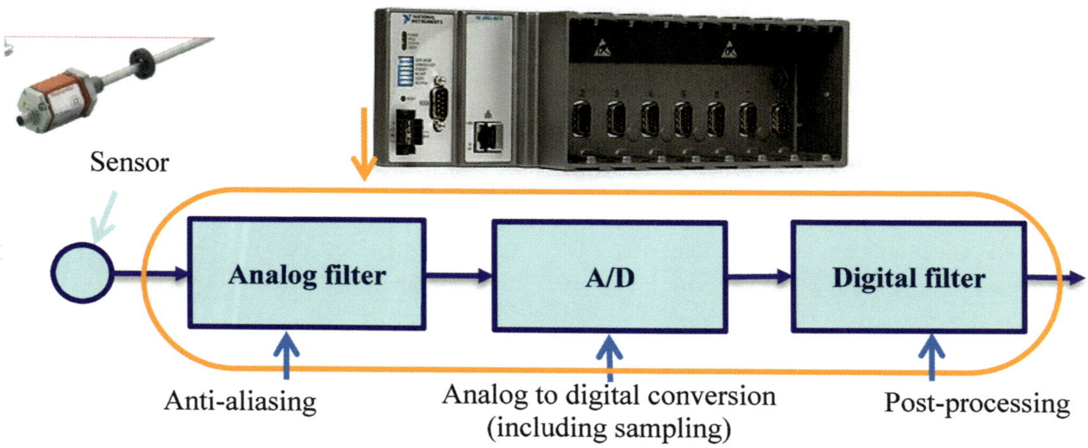

Fig. 7.5-5. Prewarping process.

Example: CT to DT, Tustin's rule

Assume that the CT antialiasing filter is the first-order Butterworth filter with the cut-off frequency $\omega_c = 1 = 2\pi f_c$ rad/sec, so that $f_c = 1/2\pi = 0.16$ Hz and $2 f_c = 0.32$ Hz.

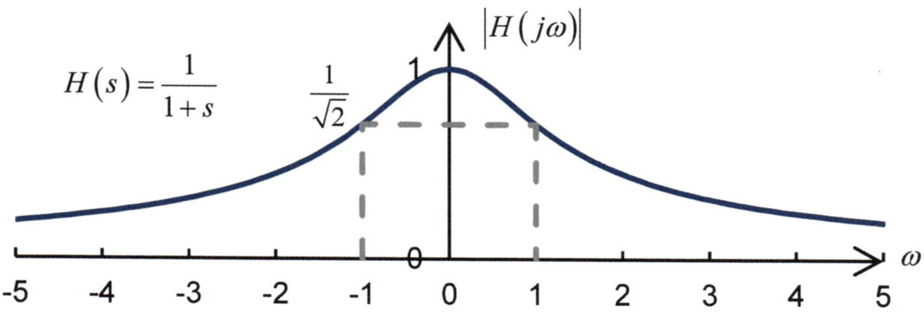

Fig. 7.5-6. CT Butterworth filter magnitude response.

Suppose we sample analog signal with $S_F = 0.5$ Hz, i.e. with $S_F/2\,f_c = 0.5/0.32 = 1.57$ or ≈ 1.5 times oversampling with respect to ω_c. Per earlier discussion, for a DT filter based on this CT filter, we choose the cut-off frequency to be the same as the cut-off frequency of the CT filter, sampled, that is: $\Omega_c = \omega_c / S_F = 1/0.5 = 2$ rad/sec, and just apply Tustin's rule:

$$S_F = 0.5 \Rightarrow t_s = 2 \Rightarrow s = \frac{2}{t_s}\frac{z-1}{z+1} = \frac{2}{2}\frac{z-1}{z+1} = \frac{z-1}{z+1} \Rightarrow H_1(z) = \frac{1}{1+\dfrac{z-1}{z+1}} = \frac{z-1}{2z},$$

so that, $H_{DT}(z) = \dfrac{z+1}{2z} = \dfrac{1}{2} + \dfrac{1}{2z}$. What is the cut-off frequency of this filter? Evaluating H_{DT} at $z = e^{j\Omega}$ yields filter frequency response:

$$H_{DT}(\Omega) \equiv H_{DT}\left(z = e^{j\Omega}\right) = \frac{1}{2} + \frac{1}{2}e^{-j\Omega}$$

which yields magnitude response shown in Figure 7.5-7:

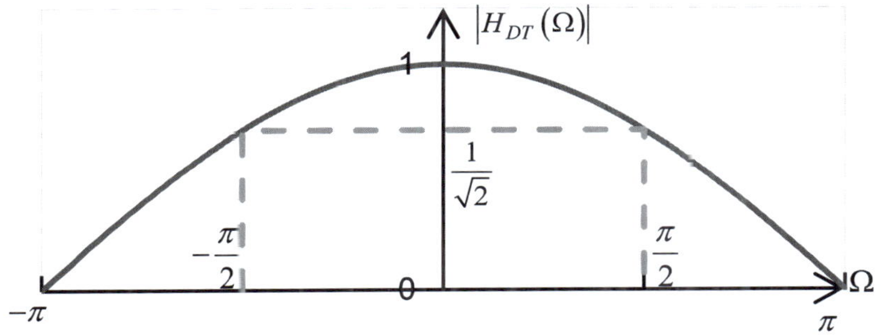

Fig. 7.5-7. DT Butterworth filter magnitude response after s-to-z Tustin rule mapping.

To find the cutoff frequency Ω_c: $|H_{DT}(\Omega_c)| = \sqrt{1/2}$, we square the latter relation and use spectral factorization: $|H_{DT}(\Omega_c)|^2 = 1/2 \Rightarrow H_{DT}(\Omega_c)H^*_{DT}(\Omega_c) = 1/2$. Then, substituting expression for $H_{DT}(\Omega_c)$ and recalling that for any two complex numbers z_1 and z_2 we have $(z_1 + z_2)^* = z_1^* + z_2^*$, yields

$$H_{DT}(\Omega_c)H^*_{DT}(\Omega_c) = \left(\frac{1}{2}+\frac{1}{2}e^{-j\Omega_c}\right)\left(\frac{1}{2}+\frac{1}{2}e^{-j\Omega_c}\right)^* = \left(\frac{1}{2}+\frac{1}{2}e^{-j\Omega_c}\right)\left(\frac{1}{2}+\frac{1}{2}e^{j\Omega_c}\right) = \frac{1}{2}+\frac{1}{4}(e^{-j\Omega_c}+e^{j\Omega_c})$$

$$= \frac{1}{2} \Rightarrow e^{-j\Omega_c} + e^{j\Omega_c} = 0 \Rightarrow e^{-j\Omega_c} = \underbrace{e^{\pm j\pi}}_{-1} e^{j\Omega_c} \Rightarrow e^{j(-\Omega_c)} = e^{j(\pm\pi+\Omega_c)} \Rightarrow -\Omega_c = \pm\pi + \Omega_c \Rightarrow \Omega_c = \pm\frac{\pi}{2}, \tag{7.5-14}$$

so that $\Omega_c = \pi/2$ for $\Omega_c \in [0, \pi]$. The cut-off frequency, however, is no longer the one we chose when we designed the filter:

$$\Omega_c = \frac{\pi}{2} \approx 1.57 \neq \frac{\omega_c}{S_F} = \frac{1}{0.5} = 2.$$

Example: CT to DT, prewarping What if we wanted to keep the cut-off frequency the same? That is,

$$\Omega_c = \frac{\omega_c}{S_F} = \frac{1}{0.5} = 2.$$

We can use prewarping to do this. The bilinear transform converts CT frequencies to DT frequencies using the relationship

$$\omega = C\tan\left(\frac{\Omega}{2}\right).$$

So, we just need to set $\Omega_c = \omega_c/S_F$, and solve for C:

$$\omega_c = C\tan\left(\frac{\omega_c / S_F}{2}\right) \Rightarrow 1 = C\tan\left(\frac{1/0.5}{2}\right) \Rightarrow C = \cot(1) \approx 0.642.$$

Then: $s = 0.642\dfrac{z-1}{z+1} \Rightarrow H_{DT}(z) = \dfrac{1}{1+0.642\dfrac{z-1}{z+1}} = \dfrac{z+1}{1.642z+0.358}.$

Prewarping yields: $s = 0.642\dfrac{z-1}{z+1} \Rightarrow H_{DT}(z) = \dfrac{z+1}{1.642z+0.358}.$

What is the cut-off frequency of this filter? Calculation of the DTFT:

$$H_{DT}(\Omega) \equiv H_{DT}\left(z = e^{j\Omega}\right) = \frac{e^{j\Omega}+1}{1.642e^{j\Omega}+0.358}$$

yields magnitude response shown in Figure 7.5-8:

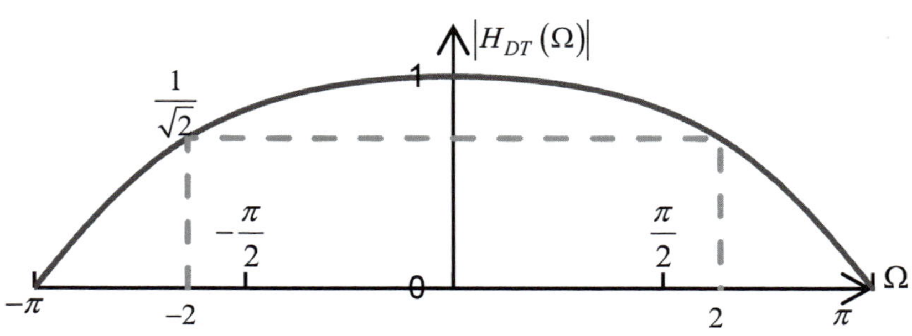

Fig. 7.5-8. Prewarped DT Butterworth filter magnitude response.

This time, the cut-off frequency exactly maps to that of the CT filter: $\Omega_c = 2 = \omega_c / S_F = 1/0.5$.

Example: CT to DT conversion: The shape of the prewarped DT filter differs from that of the CT filter. The tradeoff for a good match at the cut-off frequency is warping at the other frequencies:

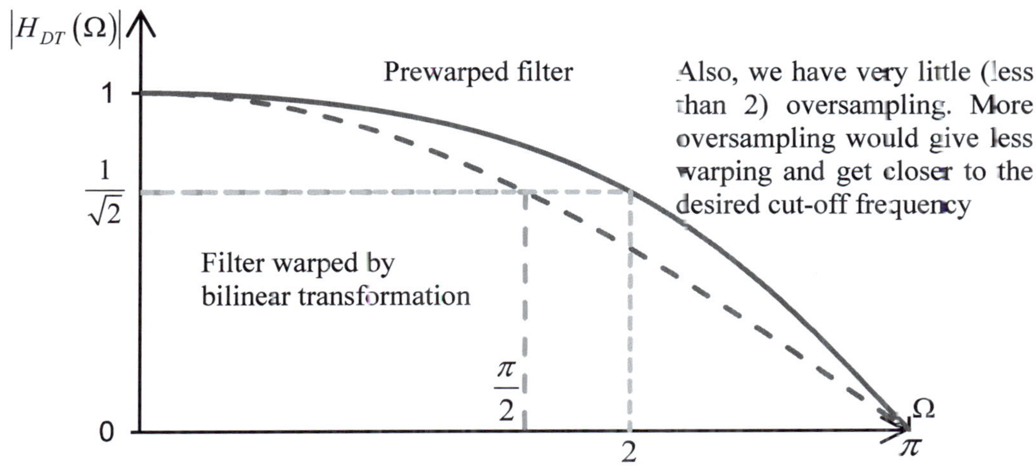

Fig. 7.5-9. DT Butterworth filter magnitude response: prewarping distortion.

Example: Tustin's rule with oversampling. Let's now assume that we oversampled well.

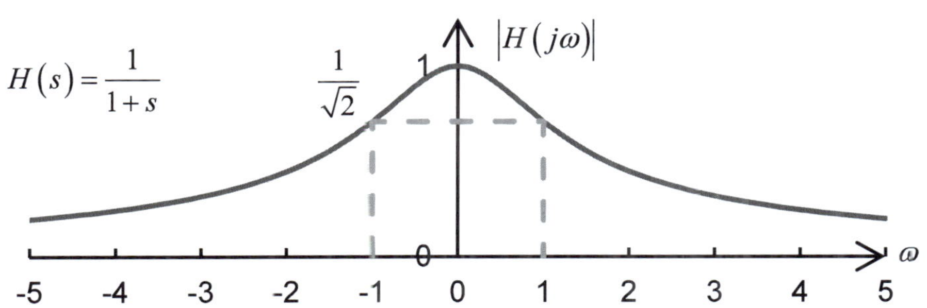

Fig. 7.5-10. CT Butterworth filter magnitude response.

As before, we would like to match the cut-off frequency of the DT filter with the sampled cut-off frequency, i.e. $\Omega_c = \omega_c / S_F$.

However, this time we sample with $S_F = 20$ Hz, or with 62.8 times oversampling compared to the CT filter cut-off frequency, and apply Tustin's rule again:

$$s = \frac{2}{t_s}\frac{z-1}{z+1} = \frac{2}{0.05}\frac{z-1}{z+1} = 40\frac{z-1}{z+1} \Rightarrow H_{DT}(z) = \frac{1}{1+40\frac{z-1}{z+1}} = \frac{z+1}{41z-39}.$$

What is the cut-off frequency of this filter? Calculating the filter DTFT, setting again

$$H_{DT}(\Omega)=H_{DT}\left(z=e^{j\Omega}\right)=\frac{e^{j\Omega}+1}{41e^{j\Omega}-39},\quad \left|H_{DT}(\Omega_c)\right|^2=\frac{1}{2}\Rightarrow H_{DT}(\Omega_c)H^*_{DT}(\Omega_c)=\frac{1}{2},$$

carrying out calculations as in (7.5-14), and using Euler's formula $e^{j\Omega}=\cos\Omega+j\sin\Omega$ yields

$$H_{DT}(\Omega_c)H^*_{DT}(\Omega_c)=\frac{(e^{j\Omega_c}+1)(e^{j\Omega_c}+1)^*}{(41e^{j\Omega_c}-39)(41e^{j\Omega_c}-39)^*}=\frac{1+e^{-j\Omega_c}+e^{j\Omega_c}+1}{1681-1599e^{-j\Omega_c}-1599e^{j\Omega_c}+1521}$$

$$=\frac{2+e^{-j\Omega_c}+e^{j\Omega_c}}{3202-1599e^{-j\Omega_c}-1599e^{j\Omega_c}}=\frac{1}{2}\Rightarrow(2+e^{-j\Omega_c}+e^{j\Omega_c})2=3202-1599e^{-j\Omega_c}-1599e^{j\Omega_c}$$

$$\Rightarrow 3198=1601(e^{-j\Omega_c}+e^{j\Omega_c})\Rightarrow 3198=1601(2\cos\Omega_c)\Rightarrow\cos\Omega_c=3198/3202\Rightarrow$$

$$\Omega_c=\cos^{-1}(3198/3202)\approx 0.04998959\approx\frac{\omega_c}{S_F}=\frac{1}{20}=0.05.$$

The resulting filter magnitude response is given in Figure 7.5-11 and is now seen to be very close to that of the original analog filter:

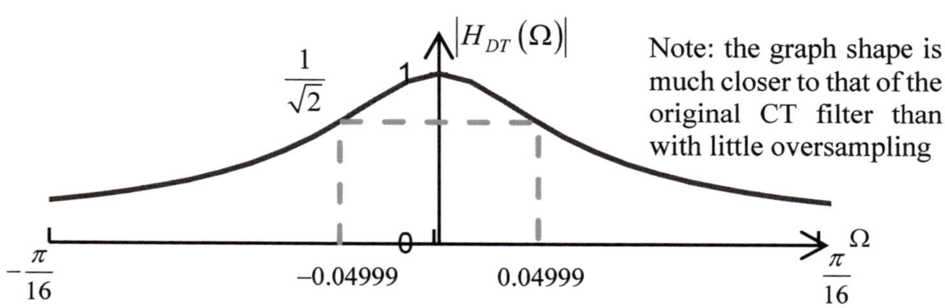

Fig. 7.5-11. Highly oversampled DT Butterworth filter magnitude response.

Thus, more oversampling means less distortion when approximating CT filters (or systems) by DT versions, often allowing us to ignore the DT nature of implementation.

7.5.4 *DT filter implementation*

Once the filter is designed, how do we implement it? Equivalently, how do we model the output of a sampled system? We have two methods to do that:

Method 1: Using recursive formula given by a difference equation (DE) corresponding to the Infinite Impulse Response (IIR) filter.

The DE is obtained from the z-transform transfer function, so this technique requires finding the latter. Also, it can discard old data sooner than Method 2.

Method 2: Using convolution of filter input signal with the impulse response of the Finite Impulse Response (FIR) filter.

For a physical system, this representation can be directly obtained from measured impulse response. For most real filters, the impulse response is infinite, so it must be cut-off for calculation purposes. This technique requires keeping a longer data set than Method 1.

7.5.4.1 *DT filter implementation: recursive form*

Due to the shift property of z-transform, we can think of z^{-1} as the "backward shift operator," i.e.:

$$z^{-1}x[n]=x[n-1].$$

So, considering the filter from this viewpoint:

Filter output ↘ (points to $y[n]$)

Transfer function →

$$H_{DT}(z)=\frac{y[n]}{x[n]}=\frac{z+1}{1.642z+0.358}=\frac{1+z^{-1}}{1.642+0.358z^{-1}}, \tag{7.5-15}$$

Filter input ↗ (points to $x[n]$)

we solve for $y[n]$ in terms of its past values and current and past input values:

$$\begin{aligned}
&(1.642+0.358z^{-1})y[n]=(1+z^{-1})x[n]\\
&\Rightarrow \underbrace{1.642y[n]+0.358y[n-1]}_{\text{Autoregressive (AR) part}}=\underbrace{x[n]+x[n-1]}_{\text{Moving average (MA) part}}\\
&\Rightarrow y[n]=-0.218y[n-1]+0.609x[n]+0.609x[n-1].
\end{aligned} \tag{7.5-16}$$

This is a recursive description of the filter: $y[n]$ depends on the past values of y: $y[n-1]$, etc. It is very quick and easy to calculate, with the example given in Figure 7.5-12:

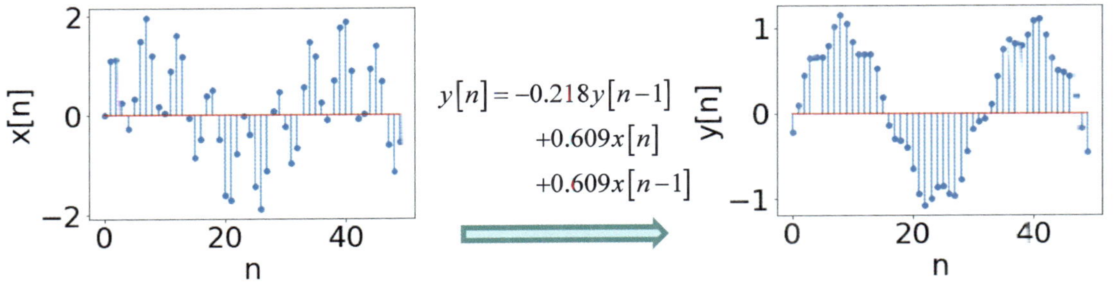

Fig. 7.5-12. DT filter implementation: recursive (ARMA — autoregressive moving average) form.

7.5.4.2 *DT filter implementation: finite impulse response*

Recall that

Filter output ↘ Filter input ↙

$$y[n]=x[n]*h[n]=\sum_{m=-\infty}^{\infty}x[m]h[n-m].$$

Impulse response ↗

The impulse response $h[n]$ can be found:

- experimentally, measured from feeding the system an impulse
- analytically, using the recursive input-out relationship.

Using equation (7.1-15) to obtain truncated impulse response of filter (7.1-14) yields

$$x[n]=\delta[n] \quad\Rightarrow\quad h[n]=-0.218h[n-1]+0.609\delta[n]+0.609\delta[n-1],$$

which is plotted as:

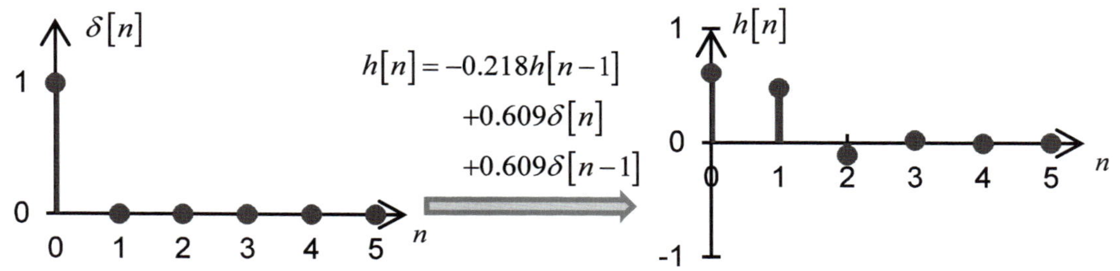

Fig. 7.5-13. DT filter implementation: finite impulse response.

The impulse response is essentially 0 for $n \geq 4$.

- Theoretically, we need the entire, infinitely long impulse response to get the actual system behavior.
- In practice, we can cut off the impulse response when the values are small enough.

How many data points are needed depends on the decay rate of the system. Consider the following DT filter

$$H_1(z)=\frac{z+1}{5z-3}=\frac{1+z^{-1}}{5-3z^{-1}}\Rightarrow 5Y(z)-3Y(z)z^{-1}=X(z)+X(z)z^{-1}\Rightarrow \\ y[n]=0.6y[n-1]+0.2x[n]+0.2x[n-1],$$

with impulse response plotted as

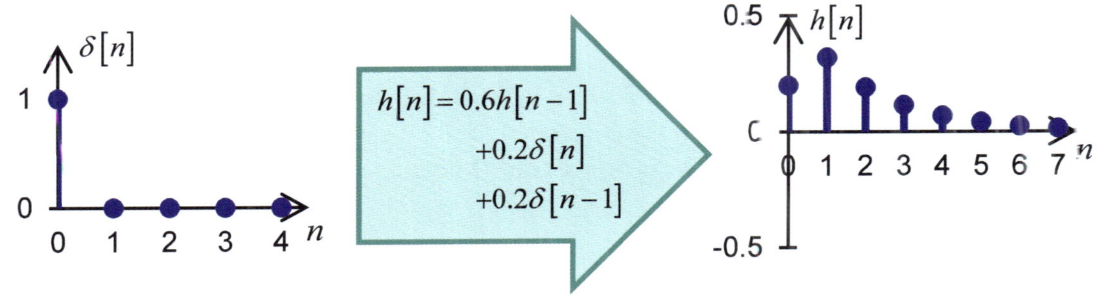

Fig. 7.5-14. DT filter implementation: finite impulse response.

This filter has a slower decay rate than that in Figure 7.5-13, so more data points are needed for its FIR approximation.

7.5.4.3 *DT filter implementation: non-recursive form; Savitzky-Golay smoothing tradeoff*

Let's consider now DT FIR filter with impulse response given in Figure 7.5-15:

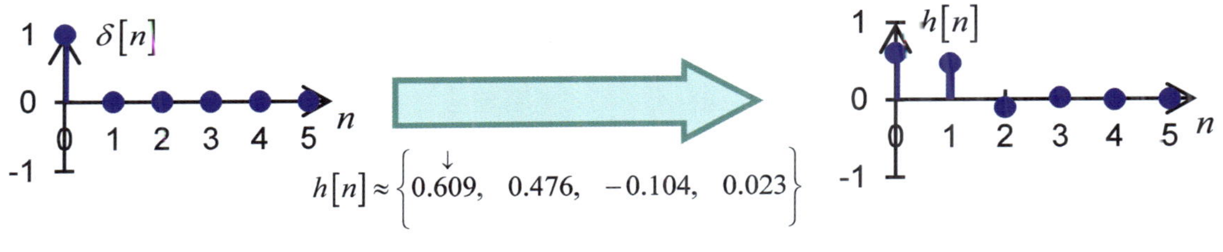

Fig. 7.5-15. DT filter implementation: non-recursive form.

The FIR filter implementation is given by:

$$y[n]=x[n]*h[n]=\sum_{m=-\infty}^{\infty}x[m]h[n-m]\approx\sum_{m=n-3}^{n}x[m]h[n-m]=x[n-3]h[n-(n-3)]+\ldots=$$
$$x[n-3]h[3]+\ldots=x[n-3]0.023+\ldots=0.609x[n]+0.476x[n-1]-0.104x[n-2]+0.023x[n-3].$$

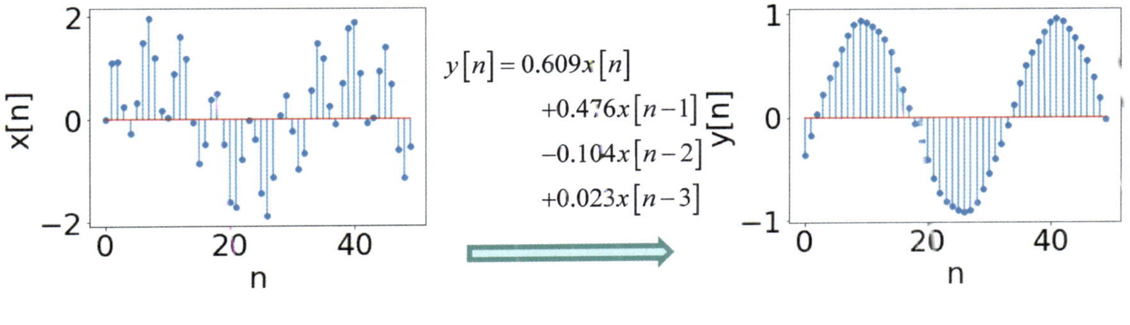

Fig. 7.5-16. DT filter implementation: non-recursive (MA — moving average) form.

As seen from this form, FIR filter can be viewed as a moving average (MA) approximation of an IIR filter in ARMA form. Since the signal bandwidth can vary, it can exceed that of FIR LPF, leading to the informative sharp signal peaks loss. A tradeoff towards retaining this high frequency signal content is removing less noise. This is accomplished by generalizing the FIR averager to the popular least-squires smoothing ***Savitzky-Golay filter*** (Matlab: sgolayfilt(x, order, framelen)).

Homework 9: Solved Examples and Solution Guide

Problem 1.

1) Derive the 1st, the 2nd, and the 3rd order Butterworth denominator polynomials for $\omega = \omega_c = 5\pi$ rad/s. Find the poles of the polynomials for each case, draw s-plane map with pole location for each case, comment on the distribution of poles in the s-plane. Plot the magnitude responses of the corresponding three low pass filters on the same graph and comment on their relative filtering properties. Plot the phase responses of these three filters on the same graph and comment on their relative phase distortion.

Solution. First order:

$$T(s) = \frac{1}{as+1} \Rightarrow |T(j\omega)|^2 = \left|\frac{1}{ja\omega+1}\right|^2 = \frac{1}{1+(a\omega)^2} = \frac{1}{1+a^2\omega^2},$$

$$|T(j\omega)|_{\omega=5\pi} = \frac{1}{\sqrt{2}} \Rightarrow \frac{1}{\sqrt{1+(5\pi a)^2}} = \frac{1}{\sqrt{2}} \Rightarrow 5\pi a = 1 \Rightarrow a = \frac{1}{5\pi},$$

$$T(s) = \frac{1}{as+1} = \frac{1}{s/(5\pi)+1} = \frac{5\pi}{s+5\pi}.$$

Pole: $s = -5\pi$ — lies on a circle of $R = 5\pi$ in the s-plane.

Second order:

$$T(s) = \frac{1}{as^2+bs+1} \Rightarrow |T(j\omega)|^2 = \left|\frac{1}{-a\omega^2+jb\omega+1}\right|^2$$

$$= \frac{1}{(1-a\omega^2)^2+b^2\omega^2} = \frac{1}{a^2\omega^4+(b^2-2a)\omega+1}$$

thus, $b^2 - 2a = 0$, and

$$|T(j\omega)|_{\omega=5\pi} = \frac{1}{\sqrt{2}} \Rightarrow \frac{1}{\sqrt{1+(5\pi)^4 a^2}} = \frac{1}{\sqrt{2}} \Rightarrow (5\pi)^4 a^2 = 1$$

$$\Rightarrow a = \frac{1}{(5\pi)^2}; \quad b^2 = 2a \Rightarrow b = \frac{\sqrt{2}}{5\pi}.$$

So,

$$T(s) = \frac{1}{as^2+bs+1} = \frac{1}{s^2/(5\pi)^2+s\sqrt{2}/(5\pi)+1} = \frac{(5\pi)^2}{s^2+5\pi\sqrt{2}s+(5\pi)^2},$$

poles: $s = 5\pi(-\frac{\sqrt{2}}{2} \pm \frac{\sqrt{2}}{2} j)$. The poles lie on a circle of $R = 5\pi$ in the LHS of s-plane.

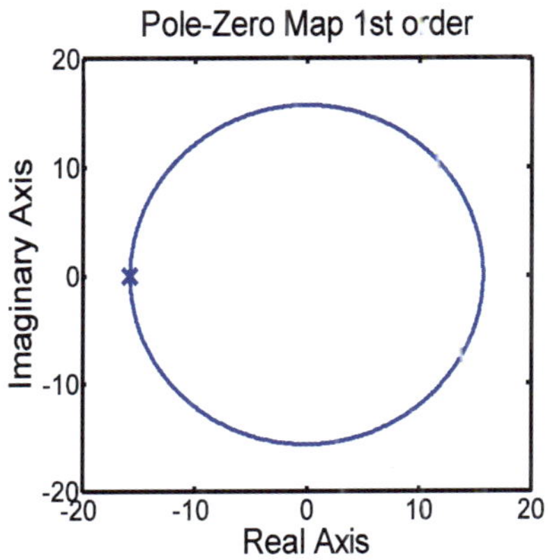

Fig. Hw9-1. First order s-plane map.

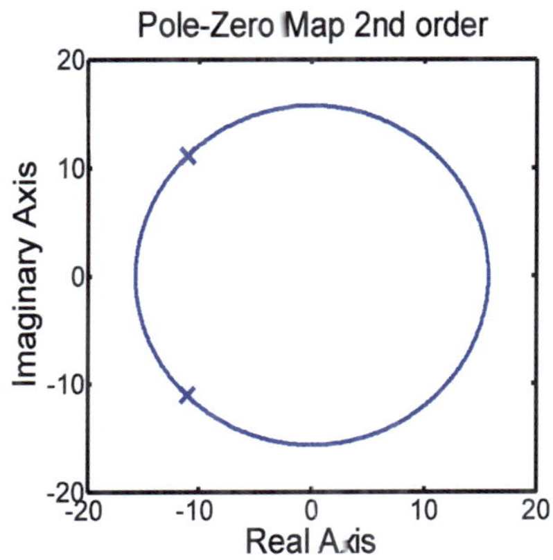

Fig. Hw9-2. Second order s-plane map.

Third order: $T(s) = \frac{1}{as^3 + bs^2 + cs + 1}$

$$\Rightarrow |T(j\omega)|^2 = \left| \frac{1}{-ja\omega^3 - b\omega^2 + jc\omega + 1} \right|^2 = \frac{1}{(1-b\omega^2)^2 + (c\omega - a\omega^3)^2} = \frac{1}{a^2\omega^6 + (b^2 - 2ac)\omega^4 + (c^2 - 2b)\omega^2 + 1}.$$

Thus, $c^2 - 2b = 0$, $b^2 - 2ac = 0$, and

$$|T(j\omega)|_{\omega=5\pi} = \frac{1}{\sqrt{2}} \Rightarrow \frac{1}{\sqrt{1 + (5\pi)^6 a^2}} = \frac{1}{\sqrt{2}} \Rightarrow (5\pi)^6 a^2 = 1 \Rightarrow a = \frac{1}{(5\pi)^3},$$

$c^2 = 2b \quad \Rightarrow c^4 = 4b^2 = 8ac \Rightarrow c^3 = 8a \Rightarrow c = \frac{2}{5\pi}, \quad b = \frac{c^2}{2} = \frac{2}{(5\pi)^2}$. Therefore

$$T(s) = \frac{1}{as^3 + bs^2 + cs + 1} = \frac{1}{s^3/(5\pi)^3 + 2s^2/(5\pi)^2 + 2s/(5\pi) + 1} = \frac{(5\pi)^3}{s^3 + 10\pi s^2 + 2*(5\pi)^2 s + (5\pi)^3}.$$

Poles: $s = -5\pi,\ 5\pi * (-\frac{1}{2} \pm \frac{\sqrt{3}}{2} j)$. The poles lie on a circle of $R = 5\pi$ in the LHS of s-plane:

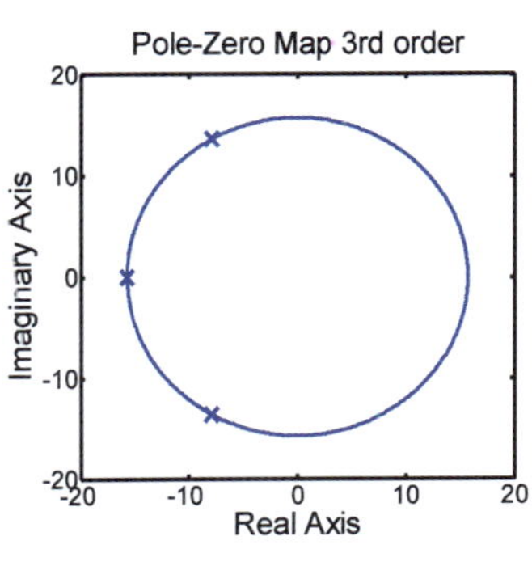

Fig. Hw9-3. Third order s-plane map.

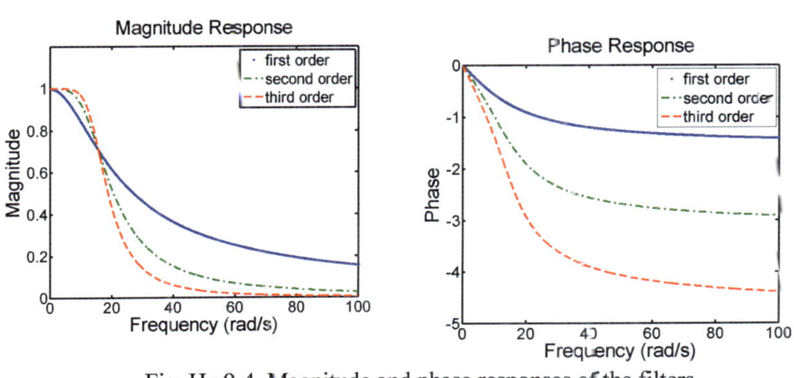

Fig. Hw9-4. Magnitude and phase responses of the filters.

From the above plot, the higher the order of the filter is, the flatter the passband is and the steeper the transition from the passband to the stopband is.

As $\omega \to \infty$, the phase will go to $-n90^o$ where n is the order of the filter.

2) Using MATLAB function "butter" in Signal Processing Toolbox, design the 1st, the 2nd, and the 3rd order Butterworth low pass filters for $\omega = \omega_c = 5\pi$ rad/s and check if the result reproduces the filters calculated in 1).

MATLAB commands for this task are:

```
>>[a1,b1]=butter(1, 5*pi, 's');
>>[a2,b2]=butter(2, 5*pi, 's');
>>[a2,b2]=butter(3, 5*pi, 's');
>>sys1=tf(a1,b1)
Transfer function:
15.71
--------------
s + 15.71
>>sys2=tf(a2,b2)
Transfer function:
246.7
-----------------------------
s^2 + 22.21 s + 246.7
>>sys3=tf(a3,b3)
Transfer function:
3876
-----------------------------------------------
s^3 + 31.42 s^2 + 493.5 s + 3876
```

Matlab produces the same results as those obtained in 1).

3) Find the transfer function of a Butterworth filter with attenuation of at least 14 dB at twice the cutoff frequency $\omega_c = 5\pi$ rad/s. Plot its magnitude and phase responses, each in its own figure.

The L dB attenuation for a Butterworth filter of order n is given by:

$$-10\log_{10}|T_n(j\omega)|^2 = 10\log_{10}(1+\omega^{2n}) \geq L\ dB.$$

Note: Remember that a quantity x can be expressed as a ratio with respect to a reference x_{ref} by using the decibel (dB):

$$x_{dB} = 10\log_{10}\left(\frac{x^2}{x_{ref}^2}\right) = 20\log_{10}\left(\frac{x}{x_{ref}}\right).$$

Let's start by finding the normalized Butterworth filter at $\omega = 2\omega_c = 2$ rad/s ($\omega_c = 1$, normalized), $|T_n(j\omega)|_{\omega=2} = \frac{1}{1+2^{2n}}$.

Then: $-10\log_{10}|T_n(j2)|^2 = 10\log_{10}(1+2^{2n}) \geq 14$, from which antilog $_{10}(1.4) = 1+2^{2n}$. The latter yields: $1+2^{2n} \geq 10^{1.4} \Rightarrow 2^{2n} \geq 10^{1.4} - 1 = 24.12$. The order of the filter must be then $n = (\ln_2 24.12)/2 = 2.2961$.

Hence, the third order filter calculated in 1) satisfies this requirement. For a cutoff frequency $\omega_c = 5\pi$ rad/s, the transfer function found was:

$$T(s) = \frac{(5\pi)^3}{s^3 + 10\pi s^2 + 2(5\pi)^2 s + (5\pi)^3}.$$

The magnitude and the phase responses are:

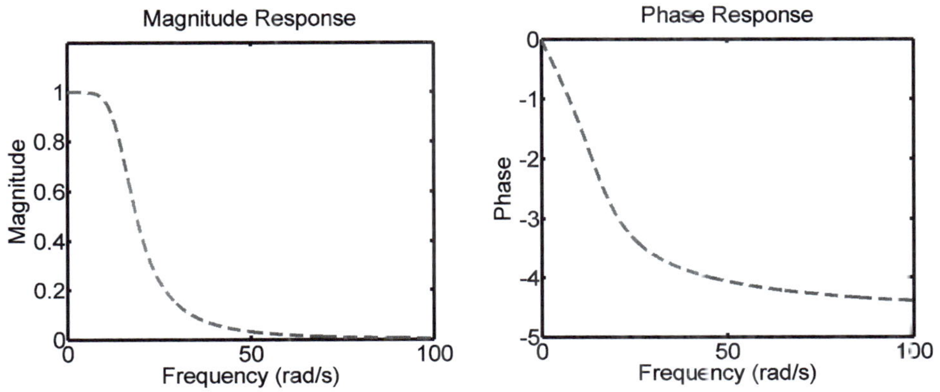

Fig. Hw9-6. Third order Butterworth filter magnitude and phase responses.

4) Implement all three filters in state-variable form in SIMULINK. Pass through each filter the sawtooth waveform (generated by the sawtooth function in the Signal Processing Toolbox) with the period $T = 0.6$ given by *t*, $0 \leq t \leq 0.6$, and plot the resulting output waveforms in the time domain.

Using SIMULINK in MATLAB, we obtain:

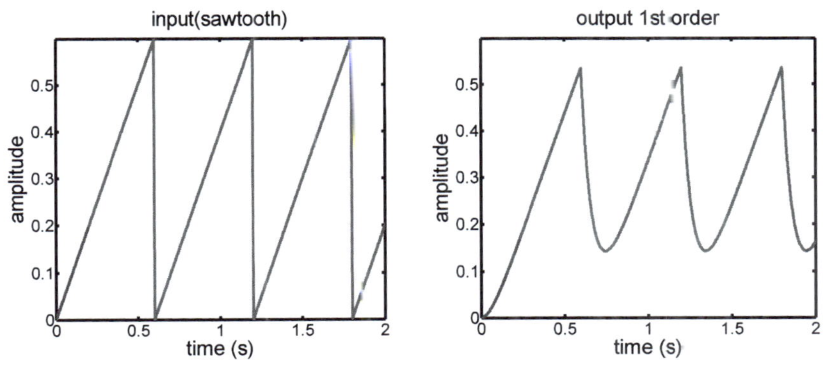

Fig. Hw9-7 (a). Passing a sawtooth waveform through the 1st order Butterworth filter.

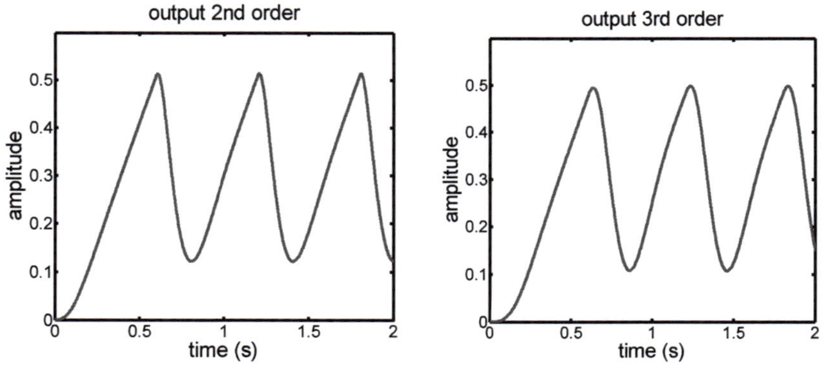

Fig. Hw9-7 (b). Passing a sawtooth waveform through the 2nd and the 3rd order Butterworth filters.

5) For each output waveform, calculate analytically the single-sided magnitudes of the two harmonics adjacent to the filter cut-off frequency — one to the left and the other to the right, and plot the single-sided magnitudes of these harmonics on the same graph as the function of the filter order. Indicate if the changes in the magnitudes of these harmonics as a function of filter order in the frequency domain are visible on the time domain graphs of the filter output waveforms.

The input sawtooth wave given by $t, 0 \le t \le 0.6$ is a periodic signal with period $T_1 = 0.6$. Thus, $f_1 = \frac{1}{T_1} = \frac{10}{6} = \frac{5}{3} \approx 1.67$ Hz. The waveform trigonometric Fourier series expansion has only sine terms (odd hidden symmetry) — all cosine terms, i.e. a_k, are zero except $k = 0$. The coefficients are $a_o = \frac{1}{T}\int_0^T t dt = \frac{10}{6}\int_0^{0.6} t dt = 0.3$,

$$b_k = \frac{2}{T}\int_0^T t\sin(2\pi k f_o t)dt = \frac{20}{6}\int_0^{0.6} t\sin(\tfrac{20}{6}\pi k t)dt = \tfrac{20}{6}[\tfrac{-6t}{20\pi k}\cos(\tfrac{20}{6}\pi k t) + \tfrac{36}{400(\pi k)^2}\sin(\tfrac{20}{6}\pi k t)]_0^{0.6} = \tfrac{-0.6}{\pi k}.$$

The cutoff frequency is: $\omega_c = 5\pi$ rad/s or 2.5 Hz, and

$$f_1 k = 1.67k < 2.5 \Rightarrow k = 1,\ 1.67k > 2.5 \Rightarrow k \ge 2.$$

So the 1st and the 2nd harmonics are adjacent to the filter cutoff frequency. The corresponding frequencies are:

$\omega_1 = 2\pi f_1 = 3.33\pi$, and $\omega_2 = 2(2\pi f_1) = 6.67\pi$. For the first order filter,

$|T(j\omega)|_{\omega=3.33\pi} = \frac{1}{|j\omega/5\pi+1|} = \frac{1}{\sqrt{(3.33/5)^2+1}} = 0.8321 \Rightarrow$ the magnitude of the 1st harmonic of the output is $0.8321 \cdot \frac{0.6}{\pi} = 0.1589$.

$|T(j\omega)|_{\omega=6.67\pi} = \frac{1}{|j\omega/5\pi+1|} = \frac{1}{\sqrt{(6.67/5)^2+1}} = 0.6 \Rightarrow$ the magnitude of the 2nd harmonic of the output is: $0.6 \cdot \frac{0.6}{2\pi} = 0.0573$. For the second order filter,

$$|T(j\omega)|_{\omega=3.33\pi} = \frac{1}{|-\omega^2/5\pi^2 - j\omega\sqrt{2}/5\pi+1|} = \frac{1}{|1-(3.33/5)^2+j3.33\sqrt{2}/5|} = \frac{1}{\sqrt{(1-(3.33/5)^2)^2+(3.33\sqrt{2}/5)^2}}$$

$$= 0.9138$$

$\Rightarrow$ the magnitude of the 1st harmonic of the output is: $0.9138 \cdot \frac{0.6}{\pi} = 0.1745$.

$|T(j\omega)|_{\omega=6.67\pi} = \frac{1}{\sqrt{(1-(6.67/5)^2)^2+(6.67\sqrt{2}/5)^2}} = 0.4903$ $\Rightarrow$ the magnitude of the 2nd harmonic of the output is: $0.4903 \cdot \frac{0.6}{2\pi} = 0.0468$. For the third order filter,

$$|T(j\omega)|_{\omega=3.33\pi} = \frac{1}{\left\|-j\omega^3/(5\pi)^3 - 2\omega^2/(5\pi)^2 - 2j\omega/(5\pi)+1\right\|} = \frac{1}{\sqrt{(1-2(3.33/5)^2)^2+(2 \cdot 3.33/5-3.33^3/5^3)^2}} = 0.9588$$

$\Rightarrow$ the magnitude of the 1st harmonic of the output is: $0.9588 \cdot \frac{0.6}{2\pi} = 0.1831$.

$|T(j\omega)|_{\omega=6.67\pi} = \frac{1}{\sqrt{(1-2(6.67/5)^2)^2+(2 \cdot 6.67/5-6.67^3/5^3)^2}} = 0.3887$ $\Rightarrow$ the magnitude of the 2nd harmonic of the output is: $0.3887 \cdot \frac{0.6}{2\pi} = 0.0371$. The filtered harmonic magnitudes are compared in Figure HW9-6

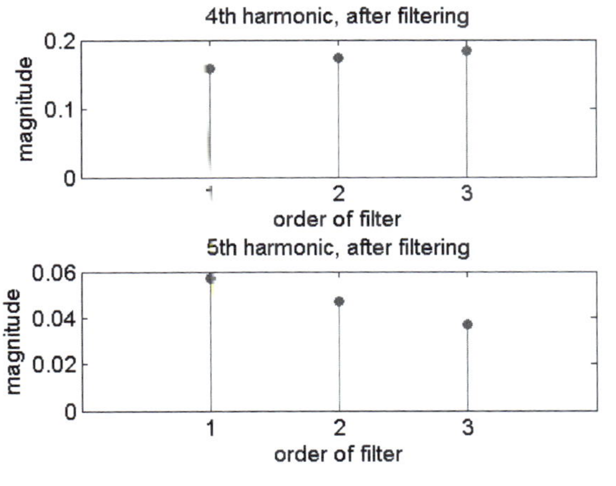

Fig. Hw9-8. Magnitude spectrum.

The changes of the magnitudes in above two plots are visible on the time domain graphs of the filter outputs: shape of the output of the 3rd filter is visibly more distorted than shape of the output of the first filter since higher harmonics are attenuated more.

6) Implement the first two filters in the state-variable form using inverting op amps.

The first order filter is given by $\frac{V_{out}}{V_{in}} = T(s) = \frac{5\pi}{s+5\pi} = \frac{15.708}{s+15.708} \Longrightarrow V_{out} = -15.708V_{out} + 15.708V_{in}$. The block diagram satisfying above differential equation is:

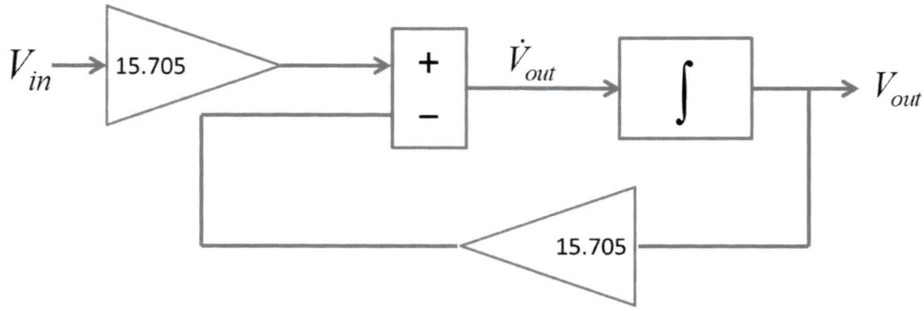

Fig. Hw9-9. The first order filter block diagram.

Replace the integrators and gain blocks with the equivalent circuits:

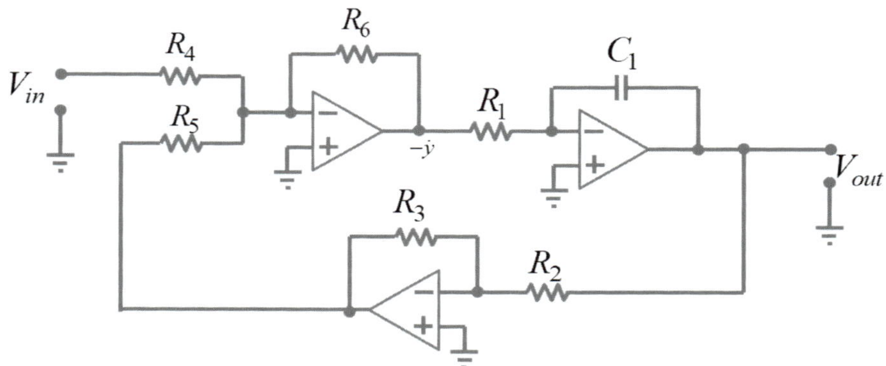

Fig. Hw9-10. The first order filter circuit diagram.

Pick the R and C values to satisfy the following relations:

$$R_1C_1 = 1;\ \frac{R_3}{R_2} = 1;\ \frac{R_6}{R_4} = 15.708;\ \frac{R_6}{R_5} = 15.708.$$

The second order filter is given by

$$\frac{V_{out}}{V_{in}} = T(s) = \frac{(5\pi)^2}{s^2 + 5\pi\sqrt{2}s + (5\pi)^2} = \frac{246.74}{s^2 + 22.214s + 246.74}$$

$\Rightarrow \ddot{V}_{out} = -22.214\ \dot{V}_{out} - 246.74V_{out} + 246.74V_{in}$. The block diagram satisfying above differential equation is:

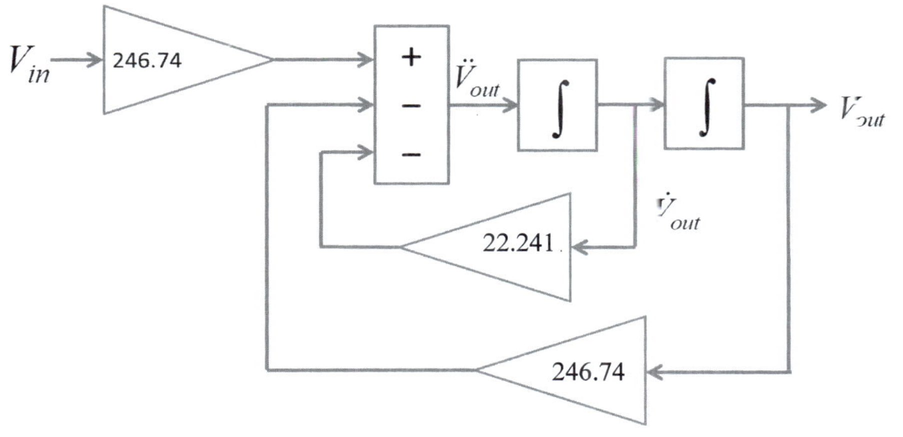

Fig. Hw9-11. The second order filter block diagram.

Replace the integrators and gain blocks with the equivalent circuits:

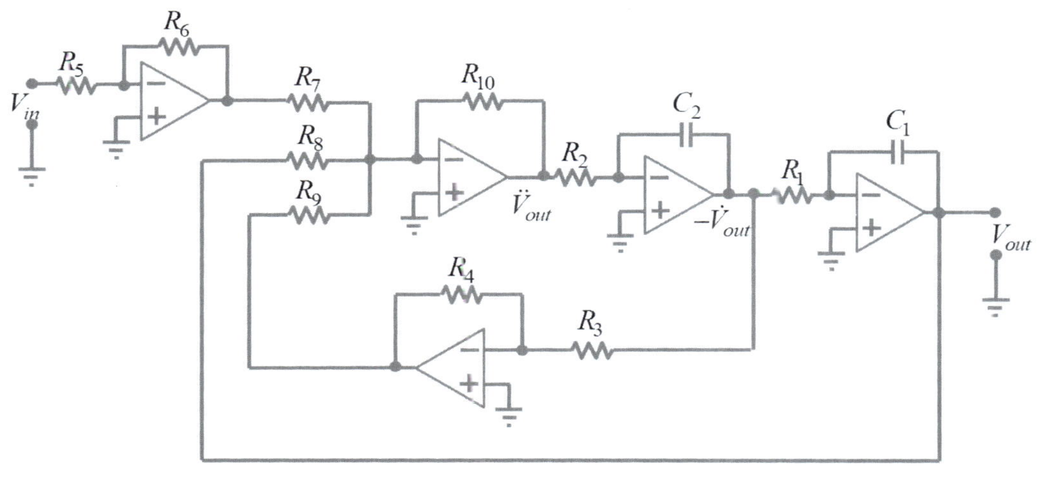

Fig. Hw9-12. The second order filter circuit diagram.

Pick the R and C values to satisfy the following relations:

$$\ddot{V}_{out} = -22.214\ \dot{V}_{out} - 246.74V_{out} + 246.74V_{in} \Rightarrow$$

$$R_1C_1 = R_2C_2 = 1;\ \frac{R_4}{R_3} = 22.241;\ \frac{R_6}{R_5} = 246.74;\ \frac{R_{10}}{R_8} = 246.74;\ \frac{R_{10}}{R_7} = \frac{R_{10}}{R_9} = 1.$$

7) Oversampling the cut-off frequency of each CT filter by 10 times and choosing C in each case so that the digital filter cut-off frequency corresponds to the resulting oversampled CT filter cut-off frequency, map the first two filters into digital filters with pre-warping using

general bilinear transform and give the transfer functions and the corresponding recursion relation (discrete equation) for each filter.

The general bilinear transformation from the Laplace transform to the z-transform is:

$$s = C\frac{z-1}{z+1}.$$

As described in Figure 7.5-2, this transformation is stability preserving and prevents aliasing by warping the CT frequency range $-\infty < \omega < \infty$ into the DT frequency range $-\pi < \Omega = 2\pi F < \pi$ according to the relationship $\Omega = 2\tan^{-1}(\omega / C)$. We would like to choose C so that the cut-off frequency corresponds to the same sampled frequency, i.e.

$$\Omega_c = \omega_c / S_F = 5\pi / S_F.$$

If we oversample the cut-off frequency by 10 times,

$$S_F = 10 \times 2 \times f_b = 20 \times \frac{5\pi}{2\pi} = 50,$$

then, the target cut-off frequency for the digital filter is $\Omega_c = \omega_c / S_F = 5\pi / 50 = \pi / 10$. Now, we need to solve

$$\Omega_c = 2\tan^{-1}(\omega / C) \Rightarrow \tan\left(\frac{1}{2}\Omega_c\right) = \frac{\omega_c}{C} \Rightarrow C = \frac{\omega_c}{\tan(\Omega_c / 2)} = \frac{5\pi}{\tan(\pi / 20)}.$$

From here on, all we need to do is to substitute

$$s = \left(\frac{5\pi}{\tan(\pi / 20)}\right)\frac{z-1}{z+1} = 5\pi\cot(\pi / 20)\frac{z-1}{z+1}$$

into the corresponding transfer functions. This is carried out next.

First order:

$$T(s) = \frac{5\pi}{s+5\pi} \Rightarrow T_{DT}(z) = \frac{5\pi}{5\pi\cot(\pi / 20)\frac{z-1}{z+1} + 5\pi} = \frac{z+1}{\cot(\pi / 20)(z-1) + (z+1)}$$

$$= \frac{z+1}{(1+\cot(\pi / 20))z + (1-\cot(\pi / 20))}.$$

Rewriting the latter in terms of the backward shift operator $z^{-1}x[n] = x[n-1]$ yields

$$T_{DT}\left(z^{-1}\right) = \frac{1+z^{-1}}{(1+\cot(\pi / 20)) + (1-\cot(\pi / 20))z^{-1}}.$$

This transfer function yields the recursion equation as follows:

$$\frac{y[n]}{x[n]}=\frac{1+z^{-1}}{(1+\cot(\pi/20))+(1-\cot(\pi/20))z^{-1}}$$

$$\Rightarrow\left((1+\cot(\pi/20))+(1-\cot(\pi/20))z^{-1}\right)y[n]=\left(1+z^{-1}\right)x[n]$$

$$\Rightarrow(1+\cot(\pi/20))y[n]+(1-\cot(\pi/20))y[n-1]=x[n]+x[n-1]$$

$$\Rightarrow(\cot(\pi/20)+1)y[n]=(\cot(\pi/20)-1)y[n-1]+x[n]+x[n-1]$$

$$\Rightarrow y[n]=\frac{\cot(\pi/20)-1}{\cot(\pi/20)+1}y[n-1]+\frac{1}{\cot(\pi/20)+1}x[n]+\frac{1}{\cot(\pi/20)+1}x[n-1].$$

Second order:

$$T(s)=\frac{(5\pi)^2}{s^2+5\pi\sqrt{2}s+(5\pi)^2}\Rightarrow$$

$$T_{DT}(z)=\frac{(5\pi)^2}{(5\pi\cot(\pi/20))^2\left(\frac{z-1}{z+1}\right)^2+5\pi\sqrt{2}(5\pi\cot(\pi/20))\left(\frac{z-1}{z+1}\right)+(5\pi)^2}$$

$$=\frac{(z+1)^2}{\cot^2(\pi/20)(z-1)^2+\sqrt{2}\cot(\pi/20)(z+1)(z-1)+(z+1)^2}$$

$$=\frac{z^2+2z+1}{\cot^2(\pi/20)\left(z^2-2z+1\right)+\sqrt{2}\cot(\pi/20)\left(z^2-1\right)+\left(z^2+2z+1\right)}$$

$$=\frac{z^2+2z+1}{\left(1+\sqrt{2}\cot\left(\frac{\pi}{20}\right)+\cot^2\left(\frac{\pi}{20}\right)\right)z^2+2\left(1-\cot^2\left(\frac{\pi}{20}\right)\right)z+\left(1-\sqrt{2}\cot\left(\frac{\pi}{20}\right)+\cot^2\left(\frac{\pi}{20}\right)\right)}$$

$$\approx\frac{z^2+2z+1}{49.7925z^2-77.7269z+31.9345}.$$

Rewriting this in terms of the backward shift operator $z^{-1}x[n]=x[n-1]$ yields

$$T_{DT}\left(z^{-1}\right)=\frac{1+2z^{-1}+z^{-2}}{49.7925-77.7269z^{-1}+31.9345z^{-2}}.$$

This transfer function yields the recursion equation as follows:

$$\frac{y[n]}{x[n]} = \frac{1+2z^{-1}+z^{-2}}{49.7925-77.7269z^{-1}+31.9345z^{-2}}$$
$$\Rightarrow \left(49.7925-77.7269z^{-1}+31.9345z^{-2}\right)y[n] = \left(1+2z^{-1}+z^{-2}\right)x[n]$$
$$\Rightarrow 49.7925y[n]-77.7269y[n-1]+31.9345y[n-2] = x[n]+2x[n-1]+x[n-2]$$
$$\Rightarrow 49.7925y[n] = 77.7269y[n-1]-31.9345y[n-2]+x[n]+2x[n-1]+x[n-2]$$
$$\Rightarrow y[n] = 1.5610y[n-1]-0.6414y[n-2]+0.0201x[n]+0.0402x[n-1]+0.0201x[n-2].$$

8) Give the FIR 10-point length sequence for each filter and write a convolution sum for each of them.

By definition, the FIR (finite impulse response) is the output of the filter when the input is the DT impulse:

$$\delta[n] = \begin{cases} 1, & 0, \\ 0, & else. \end{cases}$$

Putting this function as an input into the recursion relationship above yields

First order:

$$h[n] = \frac{\cot(\pi/20)-1}{\cot(\pi/20)+1}h[n-1] + \frac{1}{1+\cot(\pi/20)}\delta[n] + \frac{1}{1+\cot(\pi/20)}\delta[n-1]$$
$$\approx 0.7265h[n-1]+0.1367\delta[n]+0.1367\delta[n-1].$$

The latter yields the impulse response:

$$h[n] = 0, \qquad n<0$$

$$h[0] = 0.7265h[-1]+0.1367\delta[0]+0.1367\delta[-1]$$
$$= 0.7265\cdot(0)+0.1367\cdot(1)+0.1367\cdot(0) = 0.1367,$$

$$h[1] = 0.7265h[0]+0.1367\delta[1]+0.1367\delta[0]$$
$$= 0.7265\cdot(0.1367)+0.1367\cdot(0)+0.1367\cdot(1) = 0.2361,$$

$$h[2] = 0.7265h[1]+0.1367\delta[2]+0.1367\delta[1]$$
$$= 0.7265\cdot(0.2361)+0.1367\cdot(0)+0.1367\cdot(0) = 0.1715,$$

$$h[3] = 0.7265h[2]+0.1367\delta[3]+0.1367\delta[2]$$
$$= 0.7265\cdot(0.1715)+0.1367\cdot(0)+0.1367\cdot(0) = 0.1246,$$

$$h[4] = 0.7265h[3]+0.1367\delta[4]+0.1367\delta[3]$$
$$= 0.7265\cdot(0.1246)+0.1367\cdot(0)+0.1367\cdot(0) = 0.0905,$$

$$h[5] = 0.7265h[4] + 0.1367\delta[5] + 0.1367\delta[4]$$
$$= 0.7265\cdot(0.0905) + 0.1367\cdot(0) + 0.1367\cdot(0) = 0.0658,$$

$$h[6] = 0.7265h[5] + 0.1367\delta[6] + 0.1367\delta[5]$$
$$= 0.7265\cdot(0.0658) + 0.1367\cdot(0) + 0.1367\cdot(0) = 0.0478,$$

$$h[7] = 0.7265h[6] + 0.1367\delta[7] + 0.1367\delta[6]$$
$$= 0.7265\cdot(0.0478) + 0.1367\cdot(0) + 0.1367\cdot(0) = 0.0347,$$

$$h[8] = 0.7265h[7] + 0.1367\delta[8] + 0.1367\delta[7]$$
$$= 0.7265\cdot(0.0347) + 0.1367\cdot(0) + 0.1367\cdot(0) = 0.0252,$$

$$h[9] = 0.7265h[8] + 0.1367\delta[9] + 0.1367\delta[8]$$
$$= 0.7265\cdot(0.0252) + 0.1367\cdot(0) + 0.1367\cdot(0) = 0.0183,$$

Cutting off here yields a 10-point FIR to be used further in the convolution sum

$$\{h[n]\}_{n=0}^{9} = \left\{\overset{\downarrow}{0.1367},\ 0.2361,\ 0.1715,\ 0.1246,\ 0.0905,\ 0.0658,\ 0.0478,\ 0.0347,\ 0.0252,\ 0.0183\right\}.$$

Then we can approximate the recursive ARMA filter by convolution with the finite impulse response:

$$y[n] = x[n] * h[n] = \sum_{m=-\infty}^{\infty} x[m]h[n-m] \approx \sum_{m=n-9}^{n} x[m]h[n-m] =$$
$$0.1367x[n] + 0.2361x[n-1] + 0.1715x[n-2] + 0.1246x[n-3] + 0.0905x[n-4]$$
$$+0.0658x[n-5] + 0.0478x[n-6] + 0.0347x[n-7] + 0.0252x[n-8] + 0.0183x[n-9].$$

The latter filter differs conceptually from the recursive one, having only the MA part, but in has the same input/output behavior, ignoring the error due to truncating the actual, infinitely long impulse response.

<u>Second order:</u>

$$y[n] = 1.5610y[n-1] - 0.6414y[n-2] + 0.0201x[n] + 0.0402x[n-1] + 0.0201x[n-2].$$

Then, setting $x(n) = \delta[n]$, we can calculate the impulse response values as follows:

$$h[n] = 0,\ n < 0,$$

$$h[0] = 1.5610h[-1] - 0.6414h[-2] + 0.0201\delta[0] + 0.0402\delta[-1] + 0.0201\delta[-2]$$
$$= 1.5610\cdot(0) - 0.6414\cdot(0) + 0.0201\cdot(1) + 0.0402\cdot(0) + 0.0201\cdot(0) = 0.0201,$$

$$h[1] = 1.5610h[0] - 0.6414h[-1] + 0.0201\delta[1] + 0.0402\delta[0] + 0.0201\delta[-1]$$
$$= 1.5610\cdot(0.0201) - 0.6414\cdot(0) + 0.0201\cdot(0) + 0.0402\cdot(1) + 0.0201\cdot(0) = 0.0715,$$

$$h[2] = 1.5610h[1] - 0.6414h[0] + 0.0201\delta[2] + 0.0402\delta[1] + 0.0201\delta[0]$$
$$= 1.5610\cdot(0.0715) - 0.6414\cdot(0.0201) + 0.0201\cdot(0) + 0.0402\cdot(0) + 0.0201\cdot(1) = 0.1188,$$

$$h[3] = 1.5610h[2] - 0.6414h[1] + 0.0201\delta[3] + 0.0402\delta[2] + 0.0201\delta[1]$$
$$= 1.5610\cdot(0.1188) - 0.6414\cdot(0.0715) + 0.0201\cdot(0) + 0.0402\cdot(0) + 0.0201\cdot(0) = 0.1396,$$

$$h[4] = 1.5610h[3] - 0.6414h[2] + 0.0201\delta[4] + 0.0402\delta[3] + 0.0201\delta[2]$$
$$= 1.5610\cdot(0.1396) - 0.6414\cdot(0.1188) + 0.0201\cdot(0) + 0.0402\cdot(0) + 0.0201\cdot(0) = 0.1418,$$

$$h[5] = 1.5610h[4] - 0.6414h[3] + 0.0201\delta[5] + 0.0402\delta[4] + 0.0201\delta[3]$$
$$= 1.5610\cdot(0.1418) - 0.6414\cdot(0.1396) + 0.0201\cdot(0) + 0.0402\cdot(0) + 0.0201\cdot(0) = 0.1317,$$

$$h[6] = 1.5610h[5] - 0.6414h[4] + 0.0201\delta[6] + 0.0402\delta[5] + 0.0201\delta[4]$$
$$= 1.5610\cdot(0.1317) - 0.6414\cdot(0.1418) + 0.0201\cdot(0) + 0.0402\cdot(0) + 0.0201\cdot(0) = 0.1147,$$

$$h[7] = 1.5610h[6] - 0.6414h[5] + 0.0201\delta[7] + 0.0402\delta[6] + 0.0201\delta[5]$$
$$= 1.5610\cdot(0.1147) - 0.6414\cdot(0.1317) + 0.0201\cdot(0) + 0.0402\cdot(0) + 0.0201\cdot(0) = 0.0946,$$

$$h[8] = 1.5610h[7] - 0.6414h[6] + 0.0201\delta[8] + 0.0402\delta[7] + 0.0201\delta[6]$$
$$= 1.5610\cdot(0.0946) - 0.6414\cdot(0.1147) + 0.0201\cdot(0) + 0.0402\cdot(0) + 0.0201\cdot(0) = 0.0741,$$

$$h[9] = 1.5610h[8] - 0.6414h[7] + 0.0201\delta[9] + 0.0402\delta[8] + 0.0201\delta[7]$$
$$= 1.5610\cdot(0.0741) - 0.6414\cdot(0.0946) + 0.0201\cdot(0) + 0.0402\cdot(0) + 0.0201\cdot(0) = 0.0550.$$

Stopping at this point yields a 10-point FIR:

$$\{h[n]\}_{n=0}^{9} = \underset{\downarrow}{\{0.0201}, 0.0715, 0.1188, 0.1396, 0.1418, 0.1317, 0.1147, 0.0946, 0.0741, 0.0550\}.$$

Then the FIR filter described by convolution is:

$$y[n] = x[n] * h[n] = \sum_{m=-\infty}^{\infty} x[m]h[n-m] \approx \sum_{m=n-9}^{n} x[m]h[n-m] =$$
$$0.0201x[n] + 0.0715x[n-1] + 0.1188x[n-2] + 0.1396x[n-3] + 0.1418x[n-4] +$$
$$0.1317x[n-5] + 0.1147x[n-6] + 0.0946x[n-7] + 0.0741x[n-8] + 0.0550x[n-9].$$

Chapter 8

Introduction to Discrete-Continuous Spectral Analysis

8.1 Working with Real Signals: Introduction to Discrete-Continuous (Mixed) Spectral analysis. Correlation and Spectral Density for Energy Signals

8.1.1 *Motivation: a) capturing and tracking periodic signals hidden in noise, b) system model building using random signals*

So far we have introduced

— filtering out noise in signal ***outside*** of the frequency range of interest,

— modeling of systems through analysis of response to ***impulsive*** or ***periodic*** inputs.

But what if

— noise in the signal is ***inside*** the frequency range of interest, say burying small periodic components, or

— a system ***doesn't admit*** impulsive impact or swept-sine test as a precaution against, respectively, potential breakage or damaging resonance excitation?

Examples of the latter two settings are:

— a noisy eddy-current sensor signal of the fluctuating steel level in the mold where base level and periodic components with slowly varying frequencies need to be tracked for steel level control; the latter is done by the submerged entry nozzle (shown in Figure 1.2-2) steel flow manipulation,

— a tiny periodic precursor to an incipient fault in a worn out bearing,

— a mold oscillation system, where beam assembly resonance frequency and magnitude need to be determined in each production run prior to turning on mold oscillation resonance controller.

To proceed with problem solving in the latter settings we need to know how to generate, collect, and process random data, such as that shown in Figure 8.1-1, to extract the meaningful information.

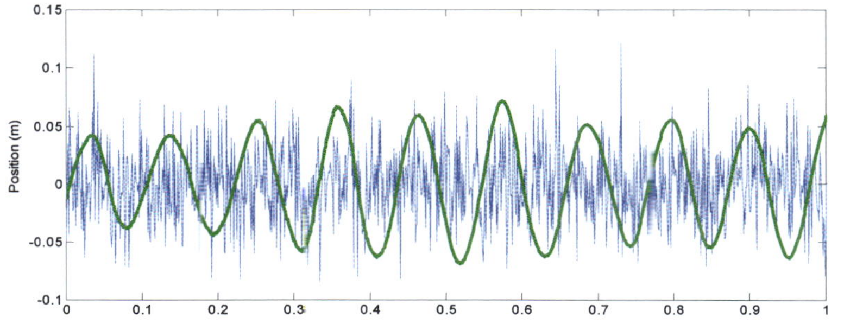

Fig. 8.1-1. Beam model system response (green) to random signal generated by piston (blue).

8.1.2 *Real signals*

In reality, only a **finite data record** of any given signal is available, i.e. **only energy signals**. Moreover, the signal of interest is often corrupted by **persistent noise,** often modeled by random signals (cf. Appendix C), and **transients** decaying during part of the data acquisition time interval. *How do we analyze real signals using the spectral methods that we have developed in this book?*

First, we will set up **ideal signal models** that admit the error-free use of spectral methods. Then, we will see how to make the error small enough in applying these spectral methods to **real data records** to permit extracting the necessary information.

We will consider real signals to be reasonably described by the following ideal signal models shown in Figure 8.2-1:

1. Energy signals: $x_e(t)$
   - Finite energy — which usually means they decay to 0
   - Usually causal
2. Power signals: $x_p(t)$
   - Periodic
3. White noise: $\sigma\varepsilon(t)$
   - Nonperiodic, but non-decaying
   - Contains no distinct periodic components, but crosses horizontal axis extremely large number of times during any arbitrarily small time interval
   - Zero mean, with variance σ^2
   - The noise is assumed to be i) wide sense stationary, i.e. the mean is constant in time and the covariance depends only on the separation in time between the two samples, and ii) second moment ergodic, i.e. statistical properties, such as mean and variance can be found from a single data record
   - Power signal? We'll discuss this later
4. Ideal mixed signals:
   - Combine one or more of the first three types — best for modeling real signals:

$$x(t)=x_e(t)+x_p(t)+\sigma\varepsilon(t) \tag{8.1-1}$$

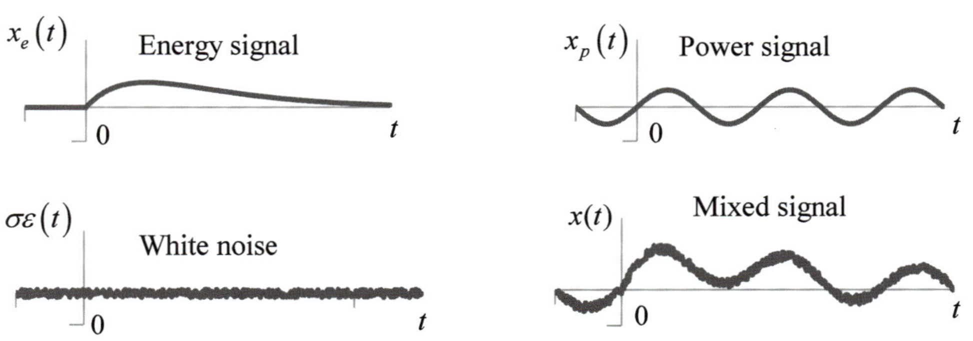

Fig. 8.1-2. Ideal signal models.

8.1.2.1 *Power spectrum*

Let's recall that power signals include periodic signals, which have discrete power spectra that can be found from signal Fourier series:

$$x_p(t) \xrightleftharpoons{\text{Fourier Series}} X_s[k] \qquad P_s[k] = |X_s[k]|^2$$

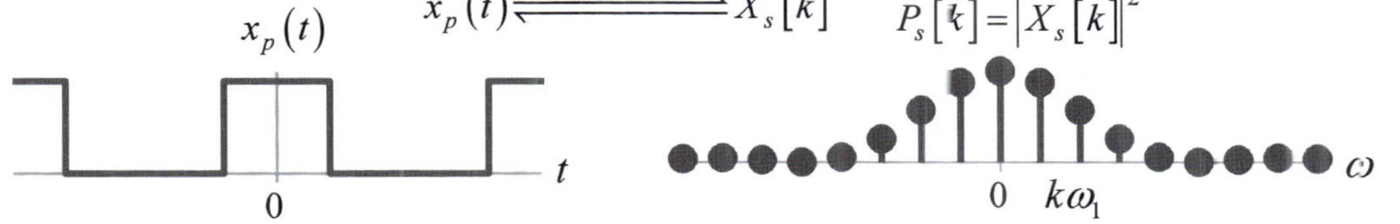

Fig. 8.1-3. Periodic signal in time domain and its power spectrum.

Signal power can then be computed both from signal time domain description and, using Parseval's Theorem, from signal power spectrum as:

$$P_x = \frac{1}{T_1}\int_{T_1} |x_p(t)|^2\, dt = \sum_{k=-\infty}^{\infty} \overbrace{|X_s[k]|^2}^{P_s[k]} \qquad (8.1\text{-}2)$$

Energy signals differ from power signals in that they must either have finite support, or must be decaying if they have infinite support, *both obviously precluding single frequency concentration*. Therefore, ***energy signals can't have discrete power or energy spectra***.

8.1.2.2 *Energy spectral density*

We have already introduced through an example in Chapter 6 the key frequency-based characteristic the energy signals can have — ***energy spectral density***. Let's now revisit the latter by deriving it as a limiting case of power spectrum to further clarify the relationship between two ways of computing the energy of the signal: i) from its time domain description and ii) from its Fourier transform:

$$E_x \rightleftharpoons x_e(t) \xrightleftharpoons{\text{Fourier transform}} X_e(j\omega) \xrightleftharpoons{?} E_x$$

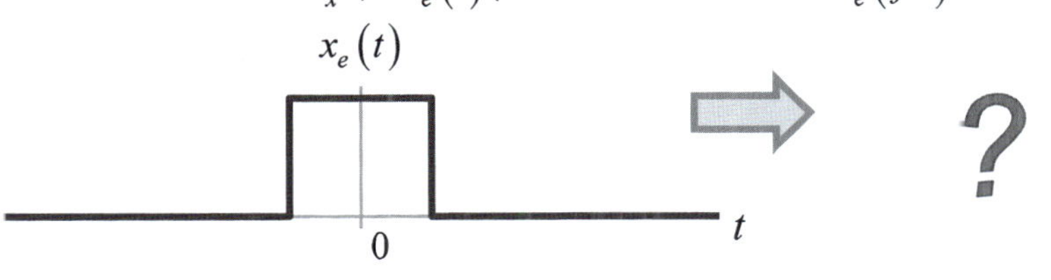

Fig. 8.1-4. Computing signal energy using signal Fourier transform.

To simplify presentation, we will again resort to notation

$$X(\omega) \equiv X(j\omega), \quad X(f) \equiv X(j2\pi f).$$

Let's start with the relationship between Fourier series and Fourier transform. If $x_p(t)$ is the periodic extension of $x_e(t)$, i.e.

$$x_p(t)=x_e(t),\ t\in[0,T_1]\Rightarrow x_p(t)=\sum_{m=-\infty}^{\infty}x_e(t-mT_1)$$

Fig. 8.1-5. Periodic extension of a pulse.

then

$$x_p(t)=\sum_{m=-\infty}^{\infty}x_e(t-mT_1)\ \Leftrightarrow\ X_s[k]=\frac{1}{T_1}X_e(\omega)\big|_{\omega=k\omega_1}.$$

Therefore, using Parseval's theorem:

$$P_x=\frac{1}{T_1}\int_{T_1}|x_p(t)|^2\,dt=\sum_{k=-\infty}^{\infty}|X_s[k]|^2=\frac{1}{T_1^2}\sum_{k=-\infty}^{\infty}|X_e(k\omega_1)|^2.$$

Multiplying both sides by T_1 yields

$$\int_{-\frac{T_1}{2}}^{\frac{T_1}{2}}|x_p(t)|^2\,dt=\frac{1}{T_1}\sum_{k=-\infty}^{\infty}|X_e(k\omega_1)|^2=\frac{\omega_1}{2\pi}\sum_{k=-\infty}^{\infty}|X_e(k\omega_1)|^2=\frac{1}{2\pi}\sum_{k=-\infty}^{\infty}|X_e(k\omega_1)|^2\,\omega_1.$$

Thus

$$\int_{-\frac{T_1}{2}}^{\frac{T_1}{2}}|x_p(t)|^2\,dt=\frac{1}{2\pi}\sum_{k=-\infty}^{\infty}|X_e(k\omega_1)|^2\,\omega_1.$$

Now, letting $T_1\rightarrow\infty$:

$$x_p(t)\rightarrow x_e(t)\ \text{ and}\begin{cases}\omega_1=\dfrac{2\pi}{T_1}\rightarrow d\omega,\\ k\omega_1\rightarrow\omega,\\ \displaystyle\sum_{k=-\infty}^{\infty}\omega_1\rightarrow\int_{-\infty}^{\infty}d\omega.\end{cases}$$

That is, power signal $x_p(t)$ becomes energy signal $x_e(t)$ and the summation on the right-hand side of the equation becomes an integral.

Thus, letting $T_1 \to \infty$, we obtain

$$\underbrace{\int_{-\infty}^{\infty} |x_e(t)|^2 \, dt}_{\substack{\text{Signal energy} \\ \text{(energy units)}}} = \frac{1}{2\pi}\int_{-\infty}^{\infty} |X_e(\omega)|^2 \, d\underbrace{\omega}_{2\pi f} = \overbrace{\int_{-\infty}^{\infty} \underbrace{|X_e(2\pi f)|^2}_{\substack{\text{Energy units/Hz -} \\ \text{energy spread over frequency:} \\ \text{ENERGY SPECTRAL DENSITY}}} \underbrace{df}_{\text{Hz}}}^{\text{Signal energy}} . \tag{8.1-3}$$

This is Parseval's relationship for energy signals. It allows us to compute signal energy via the signal Fourier transform. Thus,

$$\int_{-\infty}^{\infty} |x_e(t)|^2 \, dt = \int_{-\infty}^{\infty} |X_e(f)|^2 \, df, \tag{8.1-4}$$

so that given $x_e(t) \xrightleftharpoons{\text{Fourier Transform}} X_e(f)$, we can just square the magnitude spectral density, $|X_e(f)|$, versus f and plot $|X_e(2\pi f)|^2 \equiv |X_e(f)|^2$ versus f to show the **energy spectral density** or **energy density spectrum:**

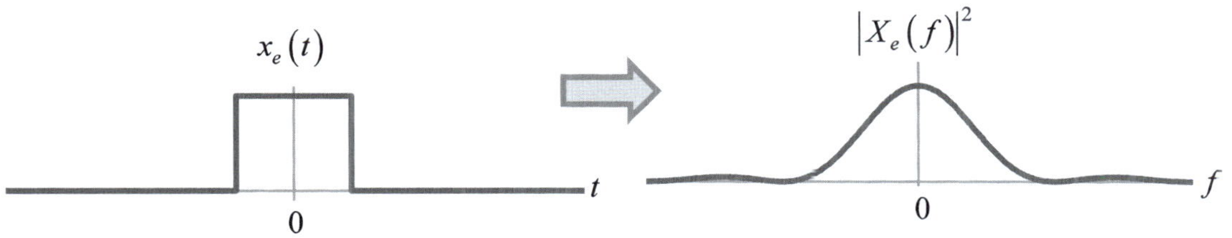

Fig. 8.1-6. Energy spectral density plot

Example 8.1-1: Low-pass RC filter

$$H(\omega) = \frac{1}{1 + j\omega\tau} \Rightarrow |H(f)|^2 = \frac{1}{1 + (\omega/\omega_c)^2} = \frac{1}{1 + (\underbrace{2\pi f}_{\omega} \underbrace{\tau}_{1/\omega_c})^2}.$$

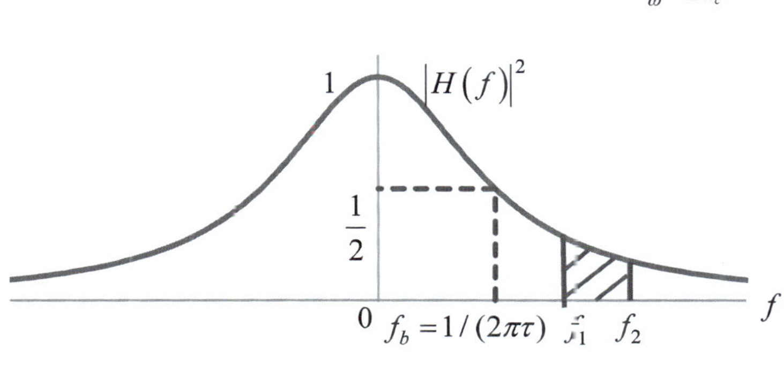

Fig. 8.1-7. Low-pass filter energy spectral density: shaded area — energy of the low-pass filter impulse response $h(t)$ in the $[f_1, f_2]$ frequency range.

As indicated in Chapter 6, the filter bandwidth is defined by the break frequency, f_b, where the energy spectral density takes the value that corresponds to passing through half of the input signal power. The area under the energy spectral density in any $[f_1, f_2]$ frequency range is the energy of $x_e(t)$ in this frequency range.

8.1.3 *Correlation for energy signals: CT*

Correlation analysis provides a) a conceptual basis for processing noisy data and b) a direct way to compute signal spectra, and c) the least intrusive system modeling tool using random excitation.

For an energy signal, $x(t)$, the **autocorrelation** and its Fourier Transform are defined as

$$r_{xx} = x**x \triangleq r_{xx}(t) = \int_{-\infty}^{\infty} x(\tau)x(\tau - t)d\tau,\ -\infty \le t \le \infty \xrightleftharpoons{FT} R_{xx}(f),\ -\infty \le f \le \infty. \quad (8.1\text{-}5)$$

For two energy signals, $x(t)$ and $y(t)$, the **crosscorrelation** and its Fourier Transform are defined as

$$r_{xy} = x**y \triangleq r_{xy}(t) = \int_{-\infty}^{\infty} x(\tau)y(\tau - t)d\tau,\ -\infty \le t \le \infty \xrightleftharpoons{FT} R_{xy}(f),\ -\infty \le f \le \infty. \quad (8.1\text{-}6)$$

These correlations are referred to as ***regular*** or ***linear***, as opposed to periodic ones considered further. **Unless otherwise stated, any use of correlation will further refer to linear correlation.**

Example 8.1-2: Compute autocorrelation of $x(t) = e^{-t}u(t)$.
Solution:

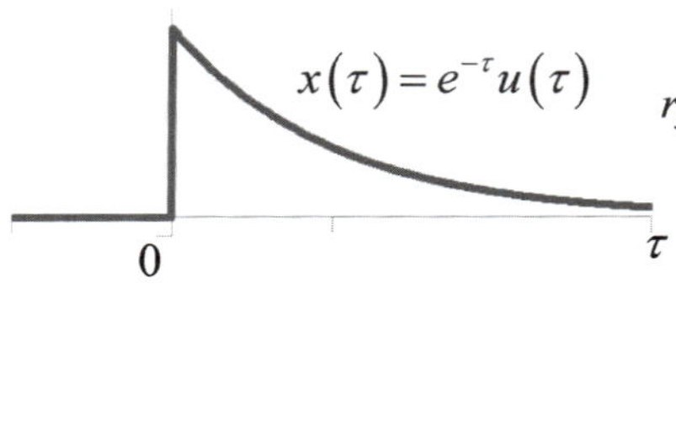

$$r_{xx}(t) = \int_{-\infty}^{\infty} x(\tau)x(\tau - t)d\tau = \int_{-\infty}^{\infty} e^{-\tau}u(\tau)e^{-(\tau-t)}u(\tau - t)d\tau$$

$$= \begin{cases} t \ge 0,\ \int_t^{\infty} e^{-\tau}e^{-(\tau-t)}d\tau \\ = e^t\int_t^{\infty} e^{-2\tau}d\tau = e^t\left[\dfrac{e^{-2\tau}}{-2}\right]_t^{\infty} = e^t\left(\dfrac{1}{2}e^{-2t}\right) = \dfrac{1}{2}e^{-t} \\ t < 0,\ \int_0^{\infty} e^{-\tau}e^{-(\tau-t)}d\tau \\ = e^t\int_0^{\infty} e^{-2\tau}d\tau = e^t\left[\dfrac{e^{-2\tau}}{-2}\right]_0^{\infty} = e^t\left[0 + \dfrac{1}{2}\right] = \dfrac{1}{2}e^t \end{cases}$$

$$= \frac{1}{2}e^{-|t|}$$

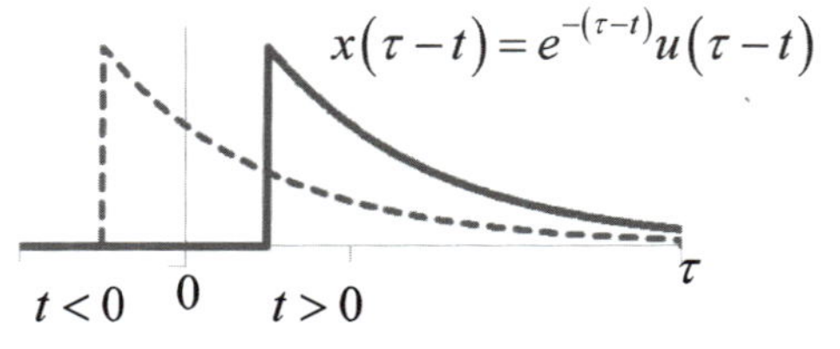

Note that $r_{xx}(t)$ is even: $r_{xx}(t) = r_{xx}(-t)$

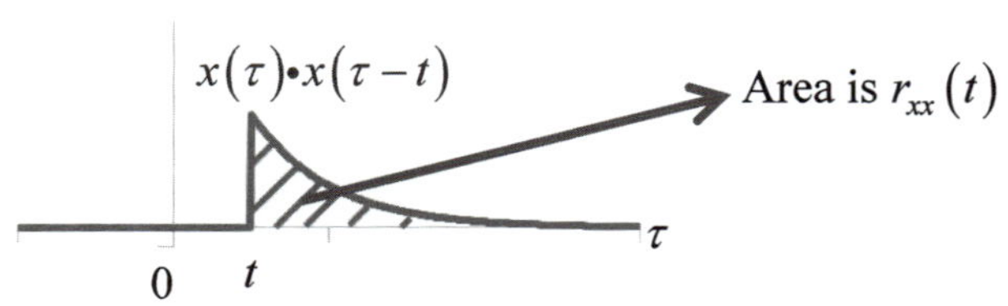

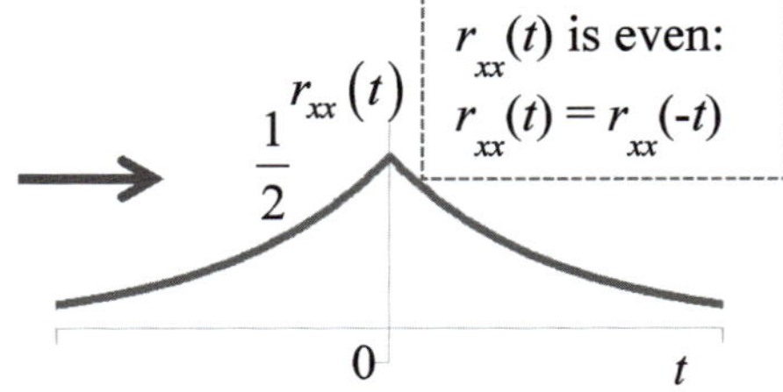

Example 8.1-3: Compute crosscorrelation of $x(t) = e^{-t}u(t)$ and $y(t) = Ke^{-t}u(t)$.
Solution:

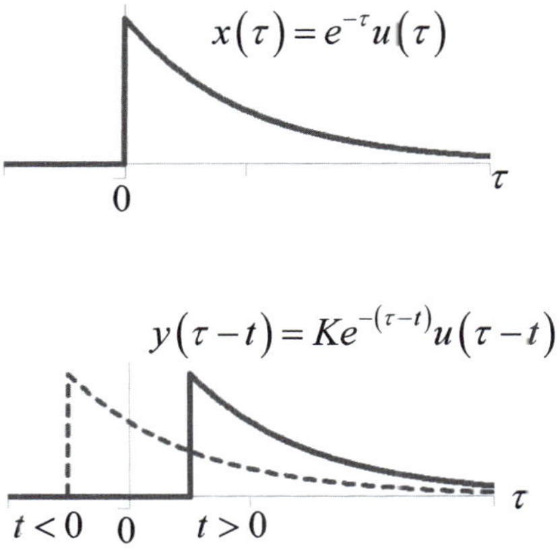

$$
\begin{aligned}
r_{xy}(t) &= \int_{-\infty}^{\infty} x(\tau)y(\tau - t)d\tau \\
&= \int_{-\infty}^{\infty} e^{-\tau}u(\tau)Ke^{-(\tau-t)}u(\tau - t)d\tau \\
&= K\int_{-\infty}^{\infty} e^{-\tau}u(\tau)e^{-(\tau-t)}u(\tau - t)d\tau \\
&= Kr_{xx}(t) \\
&= \frac{K}{2}e^{-|t|}
\end{aligned}
$$

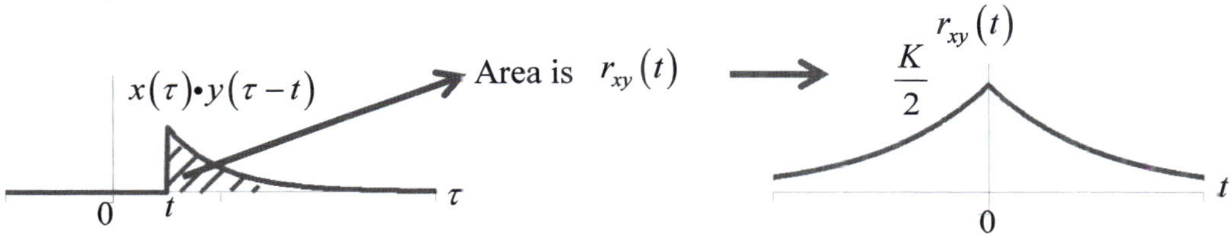

8.1.3.1 *Correlation for energy signals: DT*

For an energy signal, $x[n]$, the **autocorrelation** and its DTFT are defined as

$$r_{xx}[n] = \sum_{k=-\infty}^{\infty} x[k]x[k-n], \ -\infty \le n \le \infty \xrightleftharpoons{DTFT} R_{xx}(F), \ -1/2 \le F \le 1/2. \qquad (8.1\text{-}5)$$

Shorthand notation: $r_{xx} = x**x$.

For two energy signals, $x[n]$ and $y[n]$, the **crosscorrelation** and its DTFT are defined as

$$r_{xy}[n] = \sum_{k=-\infty}^{\infty} x[k]y[k-n], \ -\infty \le n \le \infty \xrightleftharpoons{DTFT} R_{xy}(F), \ -1/2 \le F \le 1/2. \qquad (8.1\text{-}6)$$

Shorthand notation: $r_{xy} = x**y$.

Example 8.1-4: Compute autocorrelation of $x[n] = \alpha^n u[n], \ (|\alpha| < 1)$.

Solution:

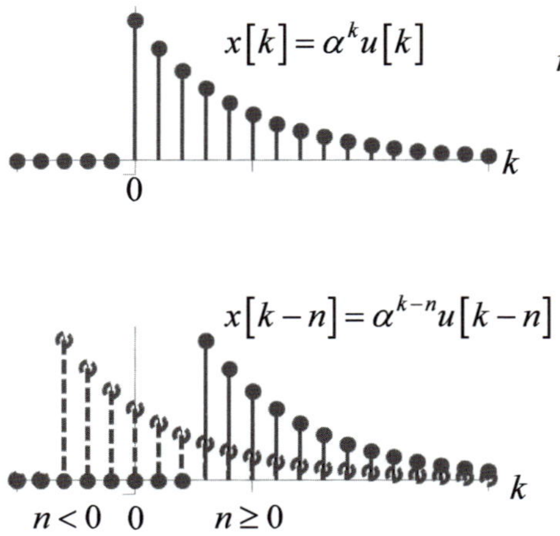

$$r_{xx}[n]=\sum_{k=-\infty}^{\infty}x[k]x[k-n]=\sum_{k=-\infty}^{\infty}\alpha^k u[k]\alpha^{k-n}u[k-n]$$

$$=\begin{cases} n\geq 0, \sum_{k=n}^{\infty}\alpha^k\alpha^{k-n}=\sum_{k=n}^{\infty}\alpha^{2k-n}=\alpha^n\sum_{k=n}^{\infty}\alpha^{2k-2n} \\ \qquad =\alpha^n\sum_{k-n=0}^{\infty}\left(\alpha^2\right)^{k-n}=\dfrac{\alpha^n}{1-\alpha^2} \\ n<0, \sum_{k=0}^{\infty}\alpha^k\alpha^{k-n}=\alpha^{-n}\sum_{k=0}^{\infty}\alpha^{2k}=\dfrac{\alpha^{-n}}{1-\alpha^2} \end{cases}$$

$$=\frac{\alpha^{|n|}}{1-\alpha^2}$$

Note that $r_{xx}[n]$ is even: $r_{xx}[n]=r_{xx}[-n]$

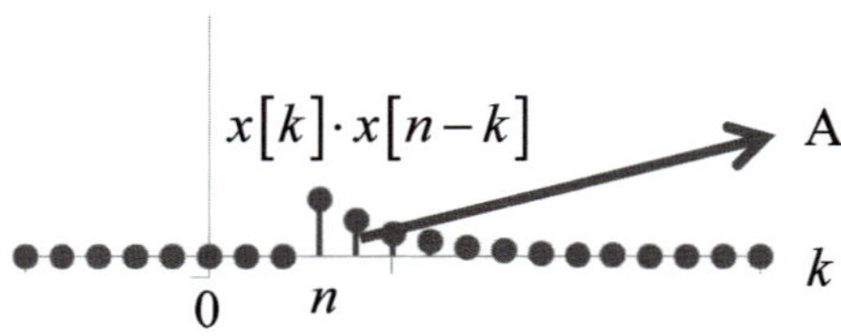

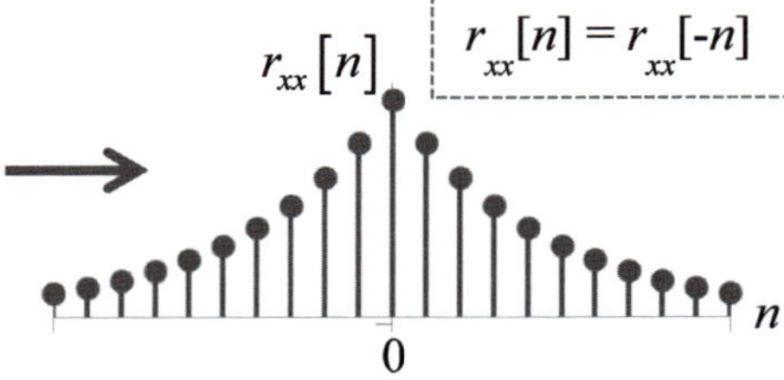

Example 8.1-5: Compute crosscorrelation of $x[n]=\alpha^n u[n]$ and $y[n]=k\alpha^n u[n]$, $(|\alpha|<1)$.

Solution:

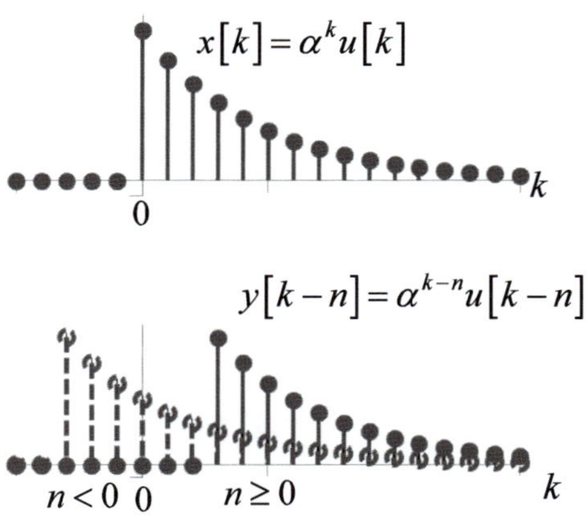

$$r_{xy}[n]=\sum_{k=-\infty}^{\infty}x[k]y[k-n]?$$

$$=\sum_{k=-\infty}^{\infty}\alpha^k u[k]K\alpha^{k-n}u[k-n]$$

$$=K\sum_{k=-\infty}^{\infty}\alpha^k u[k]\alpha^{k-n}u[k-n]$$

$$=Kr_{xx}[n]$$

$$=\frac{K\alpha^{|n|}}{1-\alpha^2}$$

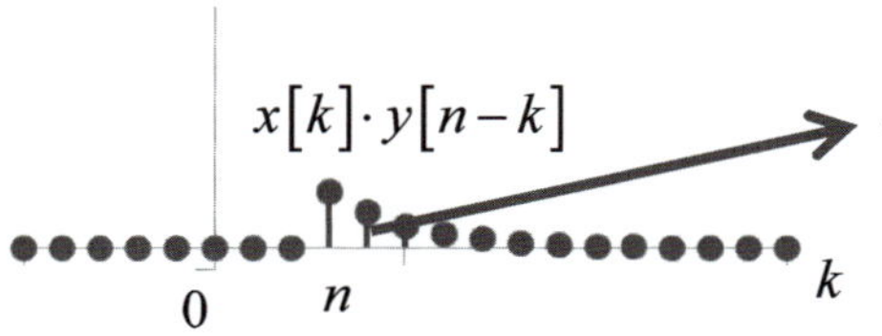

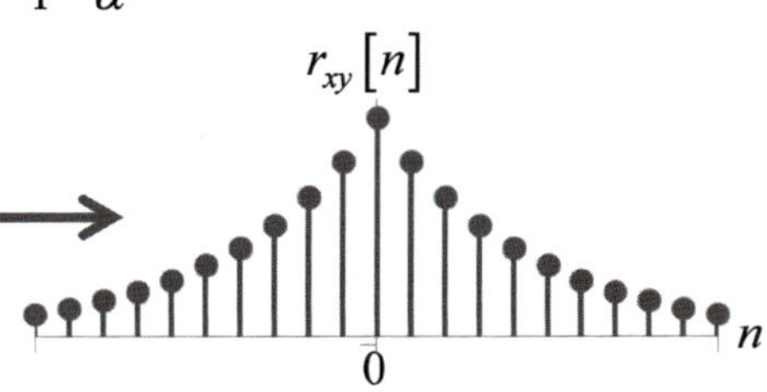

8.1.3.2 *Relationship between correlation and convolution*

It is important not to confuse *correlation* and *convolution* (which is easy to do, because the ideas, calculations, and notations are very similar).

CT:

- Correlation: $x**y = r_{xy}(t) = \int_{-\infty}^{\infty} x(\tau)y(\tau - t)d\tau$
- Convolution: $x*y = \int_{-\infty}^{\infty} x(\tau)y(t-\tau)d\tau$

$$x(t)**y(t) = x(t)*y(-t)$$

DT:

- Correlation: $x**y = r_{xy}[n] = \sum_{k=-\infty}^{\infty} x[k]y[k-n]$
- Convolution: $x*y = \sum_{k=-\infty}^{\infty} x[k]y[n-k]$

$$x[n]**y[n] = x[n]*y[-n]$$

8.1.3.3 *Correlation and energy*

We have

$$r_{xy}(t) = x(t)**y(t) = x(t)*y(-t). \tag{8.1-9}$$

Since this is true in the time domain, in the frequency domain we must have:

$$r_{xy}(t) \xleftrightarrow{\text{Fourier Transform}} R_{xy}(f) = X(\omega)Y(-\omega) = X(\omega)Y^*(\omega), \text{ or } X(f)Y^*(f). \tag{8.1-10}$$

In particular, for the autocorrelation:

$$r_{xx}(t) \xleftrightarrow{\text{Fourier Transform}} R_{xx}(f) = X(f)X^*(f) = |X(f)|^2. \tag{8.1-11}$$

Therefore, **the Fourier Transform of the autocorrelation of a signal is its energy spectral density (ESD),** and

$$r_{xx}(0) = \int_{-\infty}^{\infty} x(\tau)x(\tau)d\tau = \int_{-\infty}^{\infty} |x(\tau)|^2 d\tau = \text{Signal energy}. \tag{8.1-12}$$

8.2 Correlation and Spectral Density for Power Signals

8.2.1 *From energy to power*

Let us recall that *energy signals* (which are *non-periodic*, because otherwise they would have infinite energy) have a continuous spectrum.

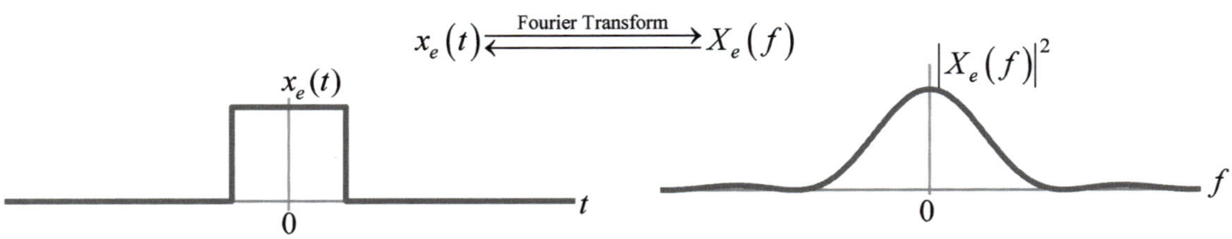

Fig. 8.2-1. Energy spectral density.

The *energy* of a signal and the total signal energy computed through its energy spectral density are equal to each other:

$$E = \int_{-\infty}^{\infty} \left| x_e(t) \right|^2 dt = \int_{-\infty}^{\infty} \left| X_e(f) \right|^2 df. \tag{8.2-1}$$

Let us also recall that *periodic signals*, which are noncausal *power signals*, because they have nonzero finite energy over a period, have a discrete spectrum:

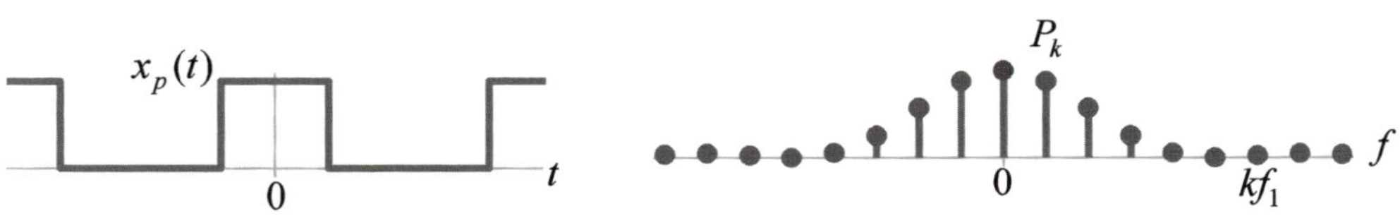

Fig. 8.2-2. Power spectrum of periodic signals.

The *power* of the signal, and the sum of the power spectrum components are equal to each other:

$$P = \frac{1}{T_1} \int_{-T_1/2}^{T_1/2} \left| x_p(t) \right|^2 dt = \sum_{k=-\infty}^{\infty} P_k. \tag{8.2-2}$$

What about noncausal *non-periodic power signals*?

They have

- No discrete spectrum
- Finite power, but since there is no period, power is calculated using the entire time from $-\infty$ to ∞:

$$P = \lim_{T_1 \to \infty} \frac{1}{T_1} \int_{-T_1/2}^{T_1/2} \left| x(t) \right|^2 dt. \tag{8.2-3}$$

Is there still a relationship between the frequency domain and the power of the signal?

$$x(t) \overset{?}{\rightleftharpoons} ?$$

$$P = \lim_{T_1 \to \infty} \frac{1}{T_1} \int_{-T_1/2}^{T_1/2} |x(t)|^2 \, dt \rightleftharpoons ?$$

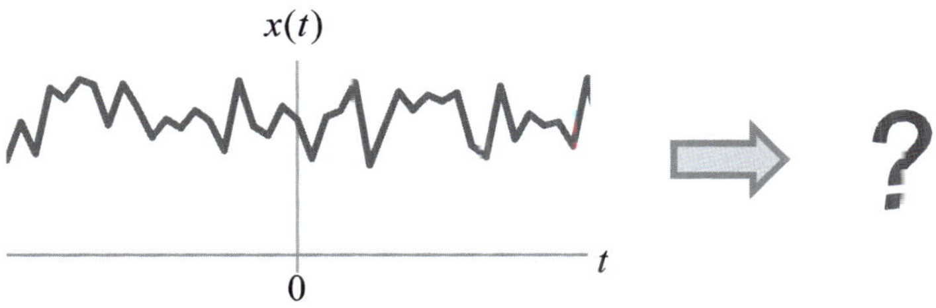

Fig. 8.2-3. Frequency domain representation of nonperiodic power signals: what to expect?

What is this relationship and how to calculate it?

Suppose we truncate the power signal $x(t)$ outside of the time interval $[-t_{tr}/2, t_{tr}/2]$:

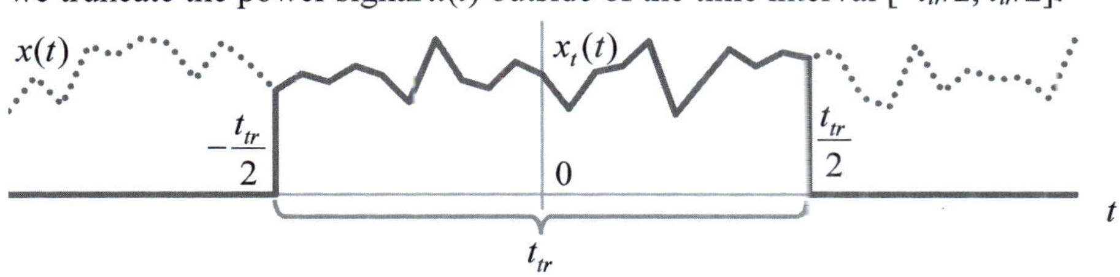

Fig. 8.2-4. Signal $x(t)$ and its truncation — signal $x_t(t)$.

This truncated signal, $x_t(t)$, is a noncausal *energy signal*, so we can calculate its energy through *energy spectral density*:

$$x_t(t) \xrightleftharpoons{\text{Fourier Transform}} X_t(f)$$

$$\int_{-\infty}^{\infty} x_t^2(t)\,dt = \int_{-\infty}^{\infty} |X_t(f)|^2 \, df = \int_{-t_{tr}/2}^{t_{tr}/2} x_t^2(t)\,dt = \int_{-t_{tr}/2}^{t_{tr}/2} x^2(t)\,dt.$$

To go from energy to power, let us increase truncation time interval $[-t_{tr}/2, t_{tr}/2]$ to recover more energy of the original signal. Divide this energy by the size of truncation time interval t_{tr} both for the time and the frequency domains energy expressions. For $t_{tr} \to \infty$ recover the original signal power through both time and frequency:

$$P_x = \lim_{t_{tr} \to \infty} \frac{1}{t_{tr}} \int_{-t_{tr}/2}^{t_{tr}/2} x_t^2(t)\,dt = \lim_{t_{tr} \to \infty} \frac{1}{t_{tr}} \int_{-\infty}^{\infty} |X_t(f)|^2 \, df = \int_{-\infty}^{\infty} \underbrace{\lim_{t_{tr} \to \infty} \frac{1}{t_{tr}} |X_t(f)|^2}_{S(f)} \, df. \quad (8.2\text{-}4)$$

8.2.2 *Power spectral density*

Thus, the *total power* of the signal is

$$P_x = \int_{-\infty}^{\infty} S(f)df \text{ where } S(f) = \lim_{t_{tr}\to\infty} S_t(f) = \lim_{t_{tr}\to\infty} \frac{1}{t_{tr}}\left|X_t(f)\right|^2,\ S_t(f) \triangleq \frac{1}{t_{tr}}\left|X_t(f)\right|^2. \quad (8.2\text{-}5)$$

Therefore, $S(f)$ represents spread of power over frequency, or the **power spectral density (PSD)**. This PSD is given in terms of the truncated signal energy. How to actually compute $S(f)$ from the data record?

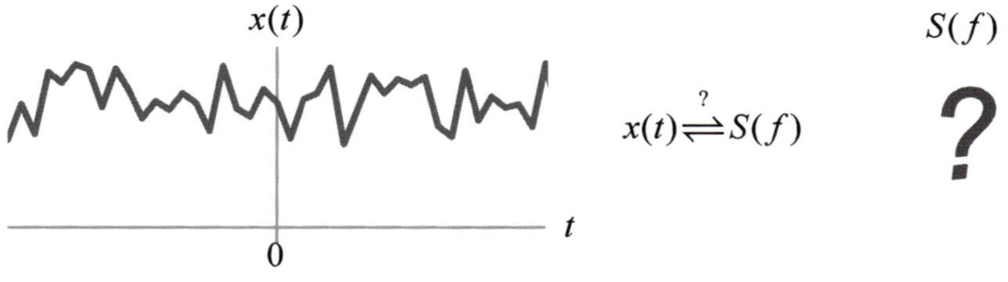

Fig. 8.2-5. Calculating power spectral density $S(f)$ from the data record $x(t)$.

8.2.3 *Calculating the PSD using pre-limit approximation*

Let's try to use data to directly compute

$$S(f) = \lim_{t_{tr}\to\infty} \frac{1}{t_{tr}}\left|X_t(f)\right|^2. \quad (8.2\text{-}6)$$

From (8.1-11) it follows that

$$r_{x_t x_t}(t) \xrightleftharpoons{\text{Fourier Transform}} R_{x_t x_t}(f) = \left|X_t(f)\right|^2. \quad (8.2\text{-}7)$$

Therefore, by linearity of the Fourier Transform:

$$\frac{1}{t_{tr}} r_{x_t x_t}(t) \xrightleftharpoons{\text{Fourier Transform}} S_t(f) = \frac{1}{t_{tr}} R_{x_t x_t}(f) = \frac{1}{t_{tr}}\left|X_t(f)\right|^2. \quad (8.2\text{-}8)$$

So, in the limit,

$$\lim_{t_{tr}\to\infty} \frac{1}{t_{tr}} r_{x_t x_t}(t) \xrightleftharpoons{\text{Fourier Transform}} \lim_{t_{tr}\to\infty} \frac{1}{t_{tr}} R_{x_t x_t}(f) = \lim_{t_{tr}\to\infty} \frac{1}{t_{tr}}\left|X_t(f)\right|^2 = \lim_{t_{tr}\to\infty} S_t(f) = S(f). \quad (8.2\text{-}9)$$

Thus, for sufficiently large t_{tr}:

$$\frac{1}{t_{tr}} r_{x_t x_t}(t) \xleftrightarrow{\text{Fourier Transform}} \frac{1}{t_{tr}} R_{x_t x_t}(f) = \frac{1}{t_{tr}} \left| X_t(f) \right|^2 \equiv S_t(f) \approx S(f). \qquad (8.2\text{-}10)$$

To approximate $S_t(f)$, we can sample $x_t(t)$ and then

1. Compute DT autocorrelation $r_{x_t x_t}[n]$ using sampled $x_t(t)$, and then compute FFT of $r_{x_t x_t}[n]$ to obtain $R_{x_t x_t}[k]$ and relate the result to $S_t(f)$, or
2. Directly compute FFT of sampled $x_t(t)$ to obtain $X_{DFT}[k]$ and relate the result to $S_t(f)$.

We will, however, develop an alternative to the first procedure through the use of periodic autocorrelation, the FFT of which will directly yield discrete approximation of (8.2-8). For this purpose, we will first take a look at computing the linear correlation of a finite data record.

Example 8.2-1: Computing CT crosscorrelation.
Let's consider $x(t) = \text{rect}(t + 0.5)$ and $h(t) = (t - 0.5)\text{rect}(t - 1)$, shown in Figure 8.2-6:

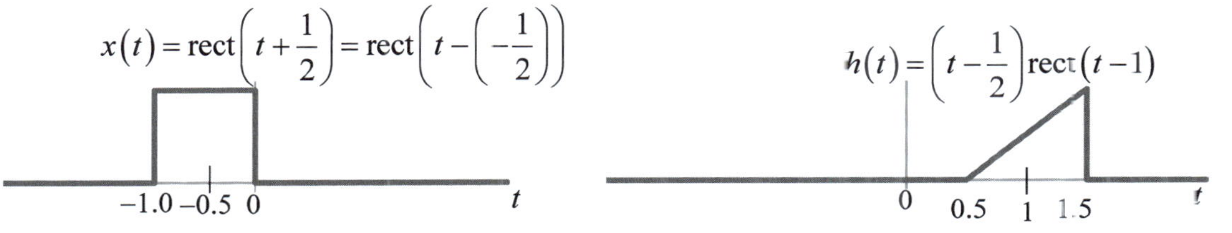

Fig. 8.2-6. Signals $x(t)$ and $h(t)$.

The crosscorrelation integral of these two functions,

$$r_{xh}(t) = \int_{-\infty}^{\infty} x(\tau) h(\tau - t)\, d\tau,$$

will then have to be calculated differently for different time intervals:
The first interval — no overlap — front of $h(\tau\text{-}t)$ behind the tail of $x(\tau)$:

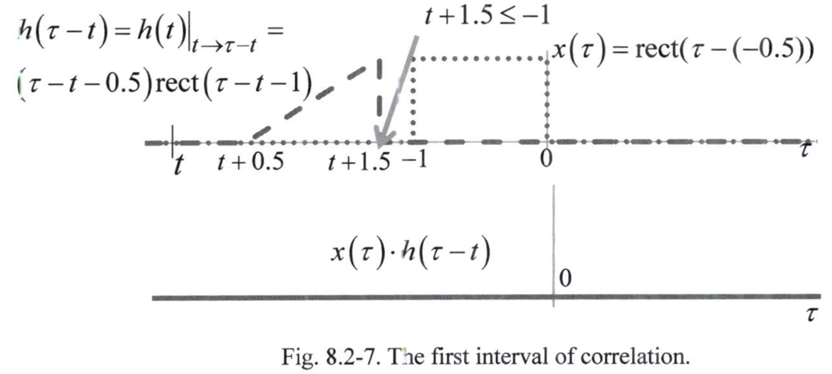

Fig. 8.2-7. The first interval of correlation.

$$t + 1.5 \le -1 \Rightarrow -\infty < t \le -2.5 \Rightarrow r_{xh}(t) = 0.$$

The second interval — overlap.
The front of $h(\tau\text{-}t)$ is in front of the tail of $x(\tau)$, but behind the front of $x(\tau)$: $-1 \le t + 1.5 \le 0$.
The tail of $h(\tau\text{-}t)$ is behind the tail of $x(\tau)$: $t + 0.5 \le -1$.

$$\left.\begin{array}{r} -1 \le t+1.5 \le 0 \\ t+0.5 \le -1 \end{array}\right\} \Rightarrow \boxed{-2.5 \le t \le -1.5,}$$

$$r_{xh}(t) = \int_{-1}^{t+1.5} (\tau - t - 0.5)\,d\tau = \int_{-1}^{t+1.5} \tau\,d\tau - (t+0.5)\int_{-1}^{t+1.5} d\tau = \boxed{\frac{(t+1.5)^2}{2} - \frac{1}{2} - (t+0.5)(t+2.5).}$$

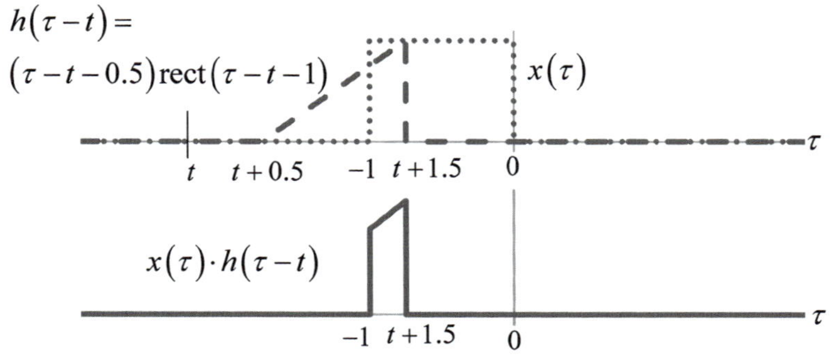

Fig. 8.2-8. The second interval of correlation.

The third interval: $0 \le t + 1.5$, $-1 \le t + 0.5 \le 0$.

$$\left.\begin{array}{r} 0 \le t+1.5 \\ -1 \le t+0.5 \le 0 \end{array}\right\} \Rightarrow \boxed{-1.5 \le t \le -0.5}$$

$$r_{xh}(t) = \int_{t+0.5}^{0} (\tau - t - 0.5)\,d\tau = \int_{t+0.5}^{0} \tau\,d\tau - (t+0.5)\int_{t+0.5}^{0} d\tau = -\frac{(t+0.5)^2}{2} - (t+0.5)[-(t+0.5)]$$

$$= \boxed{\frac{1}{2}(t+0.5)^2.}$$

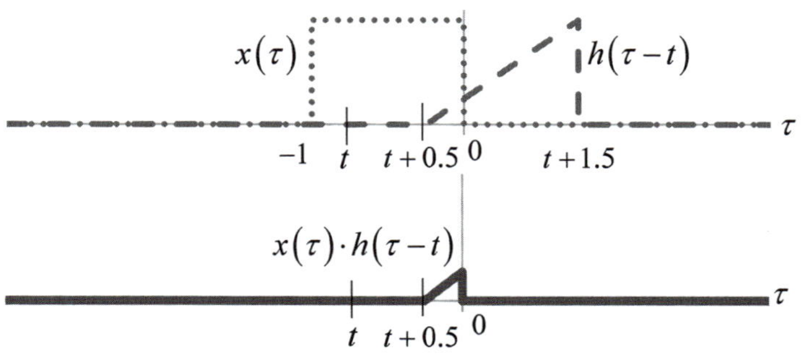

Fig. 8.2-9. The third interval of correlation.

The last interval: $0 \le t + 1.5,\ 0 \le t + 0.5$

$$0 \le t + 1.5 \Rightarrow -0.5 \le t < \infty$$
$$r_{xh}(t) = 0$$

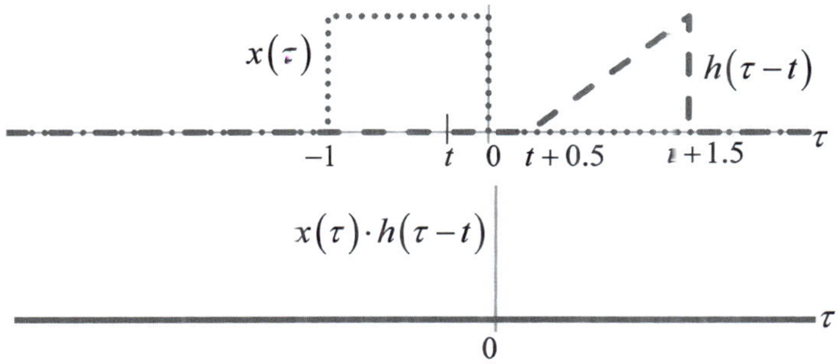

Fig. 8.2-10. The last interval of correlation.

Combining all the intervals yields signal given below and plotted in Figure 8.2-11:

$-\infty < t \le -2.5$ $\quad r_{xh}(t) = 0$,

$-2.5 \le t \le -1.5$ $\quad r_{xh}(t) = 0.5(t + 1.5)^2 - 0.5 - (t + 0.5)(t + 2.5)$,

$-1.5 \le t \le -0.5$ $\quad r_{xh}(t) = 0.5(t + 0.5)^2$,

$-0.5 \le t < \infty$ $\quad r_{xh}(t) = 0$,

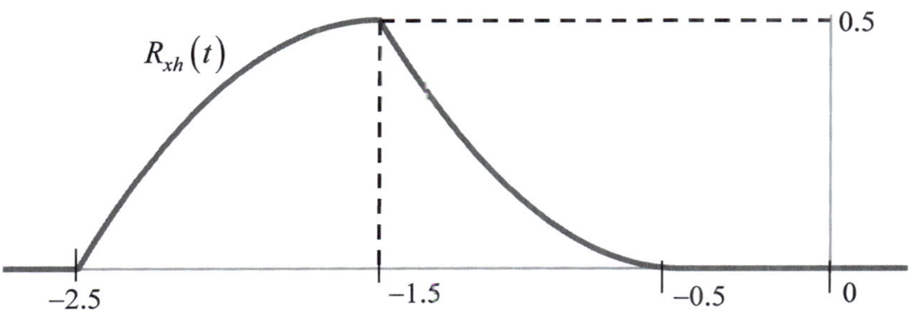

Fig. 8.2-11. Correlation of signal $x(t)$ and $h(t)$.

8.2.4 *Energy spectral density and correlation are not affected by the time shift*

The lack of phase information in the signal ESD makes the latter invariant to the signal time shift. To show this, let's first derive the change in Fourier Transform $X(f)$ of a signal $x(t)$ due to delay t_1.

$$x(t) \rightleftharpoons X(f) \Rightarrow x_1(t) \triangleq x(t - t_1) \rightleftharpoons X_1(f) = \int_{-\infty}^{\infty} x(t - t_1) e^{-j2\pi ft} dt \underset{\tau = t - t_1 \Rightarrow t = \tau + t_1}{=} \tag{8.2-11}$$
$$\int_{-\infty}^{\infty} x(\tau) e^{-j2\pi f(\tau + t_1)} d\tau = X(f) e^{-j2\pi f t_1}.$$

Then

$$\begin{aligned} r_{x_1x_1}(t) &\rightleftharpoons R_{x_1x_1}(f) = |X_1(f)|^2 = X_1(f)X_1^*(f) = X(f)e^{-j2\pi ft_0}[X(f)e^{-j2\pi ft_0}]^* \\ &= X(f)e^{-j2\pi ft_0}X(f)^* e^{j2\pi ft_0} = |X(f)|^2 = R_{xx}(f) \rightleftharpoons r_{xx}(t). \end{aligned} \tag{8.2-12}$$

Thus, given an energy signal $x(t)$ and its delayed version $x_1(t) = x(t\text{-}t_1)$, the autocorrelations and the ESDs of these two signals are, respectively equal to each other: $r_{xx}(t) = r_{x_1x_1}(t)$ and $R_{xx}(f) = R_{x_1x_1}(f)$. This fact permits processing physical data records, which are causal signals, through the noncausal frequency domain techniques of Fourier, or spectral, analysis as will be shown next.

8.2.5 *Calculating PSD using periodic extension and periodic correlation*

8.2.5.1 *Correlation for power signals — periodic correlation*

As seen from (8.2-9) and (8.2-10), for sufficiently large t_{tr}, linear autocorrelation $r_{x_tx_t}(t)$ of $x_t(t)$ yields $x(t)$ PSD approximation

$$\frac{1}{t_{tr}} r_{x_tx_t}(t) \xrightleftharpoons{\text{Fourier Transform}} \frac{1}{t_{tr}}|X_t(f)|^2 \equiv S_t(f) \approx S(f). \tag{8.2-13}$$

The approximation on the left of (8.2-13) is based on linear correlation with the length twice that of the data record and is rather awkward to use in the analysis of PSD of signals with periodic components.

We also note that from Subsection 8.2.4 it follows that both $\lim_{t_{tr}\to\infty}(1/t_{tr})r_{x_tx_t}(t)$ and $S(f)$ are invariant to the time shift in $x(t)$ due to the wide sense stationarity of the $x(t)$ white noise component.

Periodic correlations — definition. Using our familiar tool — periodic extension, let's now develop a much more convenient theoretical and computational tool for approximating PSD of power signals and detecting hidden periodic components in noise.

For this purpose, let us assume causal data record, rename $T_1 \equiv t_{tr}$, and assume that $x_t(t)$ is T_1-periodically extended to $x_{tp}(t)$:

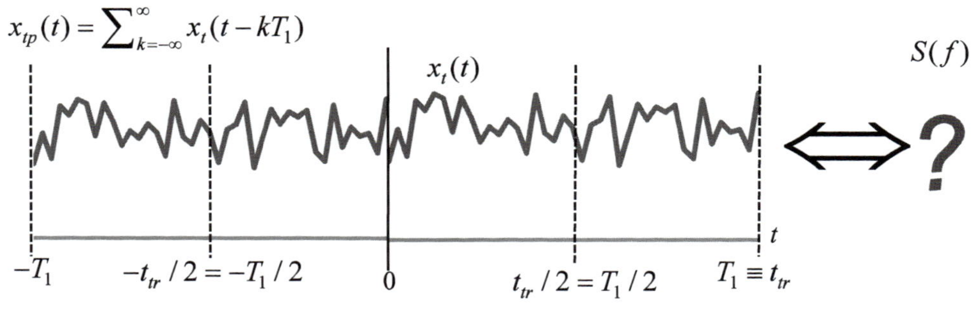

Fig. 8.2-12. Calculating PSD from the periodic extension of the data record.

Then, using symbol $\circledast$ (found, for example, in Cambria font), we can define ***periodic autocorrelation*** for signal in Figure 8.2-12 as

$$r^{\circledast}_{x_{tp}x_{tp}}(t) \equiv x_{tp}(t)\circledast\circledast x_{tp}(t) \triangleq \frac{1}{T_1}\int_0^{T_1} x_{tp}(\tau)x_{tp}(\tau-t)\,d\tau = \frac{1}{T_1}\int_{-T_1/2}^{T_1/2} x_{tp}(\tau)x_{tp}(\tau-t)\,d\tau, \quad (8.2\text{-}14)$$

and ***periodic crosscorrelation*** as

$$r^{\circledast}_{x_{tp}h_{tp}}(t) \equiv x_{tp}(t)\circledast\circledast h_{tp}(t) \triangleq \frac{1}{T_1}\int_0^{T_1} x_{tp}(\tau)h_{tp}(\tau-t)\,d\tau = \frac{1}{T_1}\int_{-T_1/2}^{T_1/2} x_{tp}(\tau)h_{tp}(\tau-t)\,d\tau. \quad (8.2\text{-}15)$$

Periodic autocorrelation and crosscorrelation are both seen to be periodic functions with period T_1 and the latter of the two relates to periodic convolution introduced in Chapter 6 as

$$x_{tp}(t)\circledast\circledast h_{tp}(t) \triangleq \frac{1}{T_1}\int_0^{T_1} x_{tp}(\tau)h_{tp}(\tau-t)\,d\tau = \frac{1}{T_1}\int_0^{T_1} x_{tp}(\tau)h_{tp}(-(t-\tau))\,d\tau = x_{tp}(t)\circledast h_{tp}(-t). \quad (8.2\text{-}16)$$

Periodic correlations are also seen to be invariant to the identical time shifts in signals, such as for example shift of $x_t(t)$ to the left by $T_1/2$ to match noncausal signal $x_t(t)$ of Figure 8.2-4.

Periodic autocorrelation and power spectral density. To relate periodic autocorrelation of causal finite duration signal $x_t(t)$ depicted in Figure 8.2-12 to power spectral density defined in (8.2-9) for a noncausal infinite duration signal $x(t)$ shown in Figure 8.2-4, we shift the causal signal $x_t(t)$ to the left by $T_1/2$. Then, for the resulting signal, as $T_1 \to \infty$ we pick up the entire noncausal signal $x(t)$ of Figure 8.2-4: $\lim_{T_1\to\infty} x_{tp}(t) = \lim_{T_1\to\infty} x_t(t) = x(t)$. Then, the periodic autocorrelation definition (8.2-14) extends to the nonperiodic noncausal infinite duration signals as follows:

$$\begin{aligned} r^{\circledast}_{x_{tp}x_{tp}}(t) \underset{T_1\to\infty}{\to} r^{\circledast}_{xx}(t) &\triangleq \lim_{T_1\to\infty}\frac{1}{T_1}\int_{-T_1/2}^{T_1/2} x_{tp}(\tau)x_{tp}(\tau-t)\,d\tau = \lim_{T_1\to\infty}\frac{1}{T_1}\int_{-T_1/2}^{T_1/2} x_t(\tau)x_t(\tau-t)\,d\tau \\ &\equiv \lim_{T_1\to\infty}\frac{1}{T_1} r_{x_t x_t}(t) = \lim_{T_1\to\infty}\frac{1}{T_1}\int_{-T_1/2}^{T_1/2} x(\tau)x(\tau-t)\,d\tau \equiv \lim_{T_1\to\infty}\frac{1}{T_1} r_{xx}(t). \end{aligned} \quad (8.2\text{-}17)$$

Therefore, denoting Fourier transforms of $r^{\circledast}_{x_{tp}x_{tp}}(t)$ and $r^{\circledast}_{xx}(t)$ as, respectively, $R^{\circledast}_{x_{tp}x_{tp}}(f)$ and $R^{\circledast}_{xx}(f)$, from (8.2-9) and (8.2-17) we obtain

$$r^{\circledast}_{x_{tp}x_{tp}}(t) \underset{T_1\to\infty}{\to} r^{\circledast}_{xx}(t) = \lim_{T_1\to\infty}\frac{1}{T_1} r_{x_t x_t}(t) \xrightleftharpoons{\text{Fourier Transform}} R^{\circledast}_{x_{tp}x_{tp}}(f) \underset{T_1\to\infty}{\to} R^{\circledast}_{xx}(f) = \lim_{T_1\to\infty}\frac{1}{T_1}\left|X_t(f)\right|^2 = S(f), \quad (8.2\text{-}18)$$

i.e. ***Fourier transform*** $R^{\circledast}_{xx}(f)$ ***of periodic autocorrelation*** $r^{\circledast}_{xx}(t)$ ***extended to a general nonperiodic power signal x(t) is the power spectral density*** $S(f)$ ***of this signal.***

If $x(t)$ is a T_1-periodic signal, $x_p(t)$, its periodic autocorrelation $r^{\circledast}_{xx}(t)$ computed over $-\infty \le t \le \infty$ is obviously equal to that computed over a single period:

$$r_{xx}^{\circledast}(t) \equiv \lim_{T_1\to\infty}\frac{1}{T_1}r_{x_tx_t}(t) = \frac{1}{T_1}r_{x_px_p}(t) \equiv r_{x_px_p}^{\circledast}(t).$$

Then, denoting Fourier transform of $r_{x_px_p}^{\circledast}(t)$ as $R_{x_px_p}^{\circledast}(f)$, due to (8.2-18) this immediately implies that $R_{x_px_p}^{\circledast}(f) = R_{xx}^{\circledast}(f) = S(f)$, i.e. ***Fourier transform*** $R_{x_px_p}^{\circledast}(f)$ ***of periodic autocorrelation*** $r_{x_px_p}^{\circledast}(t)$, ***computed over period*** T_1 ***of a*** T_1***-periodic signal*** $x_p(t)$ ***is the power spectral density*** $S(f)$ ***of this signal.***

We will also demonstrate below that the discrete version of $S_t(f)$ is directly given by the DFT of the discrete version of $r_{x_{tp}x_{tp}}^{\circledast}(t)$, making DT periodic correlation convenient for PSD estimation.

One must exercise care in using the correct definitions, noting that the latter depend on:

- the type of signal considered: power (periodic or non-periodic) or energy,
- the kind of information sought: PSD or ESD.

The discrete time counterparts of the above definitions are straightforward and are illustrated by the example below.

Example 8.2-2. Compute periodic autocorrelation

$$r_{x_px_p}^{\circledast}[n] \triangleq \frac{1}{N_1}\sum_{0}^{N_1-1} x_p[k]x_p[k-n],$$

of the following periodic signal $x_p[k]$:

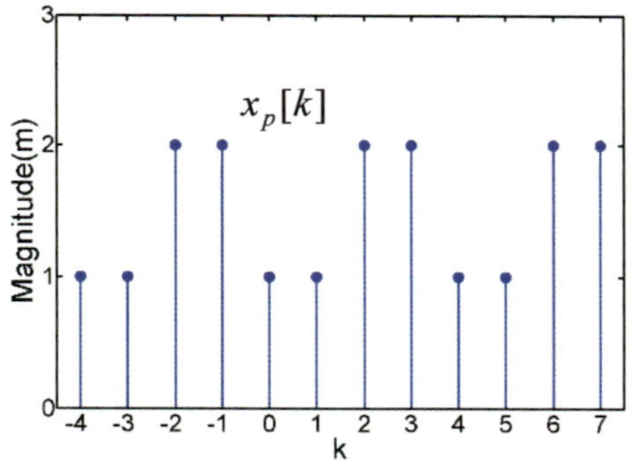

Fig. 8.1-13. Periodic signal $x_p[k]$.

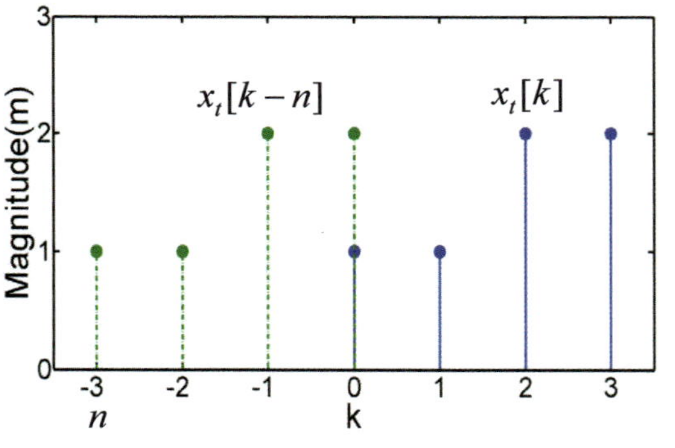

Fig. 8.1-14. Signals $x_t[k]$ and $x_t[k-n]$ for $n = -3$.

First, let us truncate this signal to only one period denoted as $x_t[k]$ and add signal $x_t[k-n]$ to the left of $x_t[k]$ as shown in Figure 8.1-14. Then, let's slide the left signal to the right in the direction of increasing n, while keeping the right signal stationary, and compute nonperiodic autocorrelation

$$r_{x_tx_t}[n] = \sum_{k=-\infty}^{\infty} x_t[k]x_t[k-n]:$$

when $n \le -4$, $x_t[k]$ and $x_t[k\text{-}n]$ have no overlap, so $r_{x_t x_t}[n] = 0$,
when $-3 \le n < 4$, $x_t[k]$ and $x_t[k\text{-}n]$ have overlap, as in the above figure for $n = -3$,
when $4 \le n$, $x_t[k]$ and $x_t[k\text{-}n]$ have no overlap, so $r_{x_t x_t}[n] = 0$.
Shifting, multiplication, and summation then yield $r_{x_t x_t}[n]$ of the form:

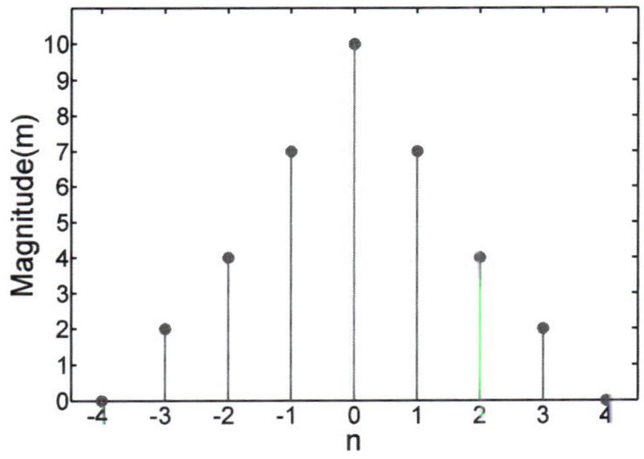

Fig. 8.1-15. Graph of $r_{x_t x_t}[n]$.

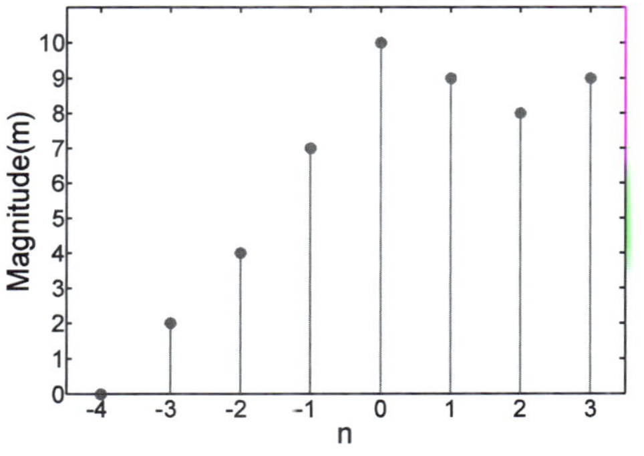

Fig. P8.1-16. Graph of $r_{x_t x_t}[n]$ with the cylic correction for n from 0 to 3.

Let's now focus on the values of this graph for n from 0 to 3. The trailing edge of the function $x_t[k - n]$ passes through these values on the way out of overlap. If we now assume that $x_t[k - n]$ is followed by $x_t[k - n - N_1]$, $N_1 = 4$, then the latter signal will also have overlap with $x_t[k]$ for the values of n from 0 to 3. We should then add the resulting values from the latter overlap to the graph of $r_{x_t x_t}[n]$. This procedure, called **cyclic calculation**, yields the graph shown in Figure P8.1-16.

The above plot is not complete, but we only need to look at the part from $n = 0$ to 3. The actual periodic autocorrelation $r^{\circledast}_{x_p x_p}[n]$ of $x_p[k]$ will then be *the periodic extension of* $r_{xx}[n]$ with cyclic correction from $n = 0$ to 3, *divided by* $N_1 = 4$. The result is symmetric with respect to the origin.

We also note that we can obtain the values of the graph in Figure P8.1-16 for $n = 0$ to 3 by simply stacking the left part from $n = -4$ to $n = -1$ of the graph in Figure 8.1-15 on top of its right part, respectively, from $n = 0$ to $n = 3$. The latter procedure is called **wraparound**.

8.3 Correlation and Spectral Density of Ideal Mixed Signals

8.3.1 *Building a power spectral density graph*

Once the periodic autocorrelation is computed, taking its DFT yields approximation of signal PSD, as will be shown further. Now, it would be very helpful to know what to expect — what does a typical PSD look like and what can we see from it.

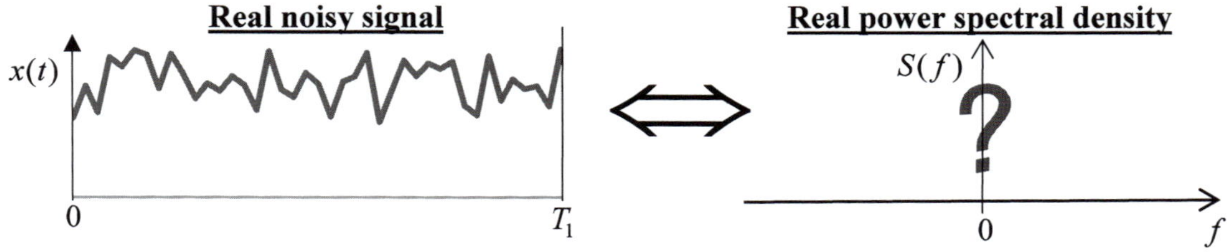

Fig. 8.3-1. How does noisy signal power spectral density look like?

Going back to the ideal model (8.1-1) of a real signal (*ideal signal*, for brevity):

$$x(t) = x_e(t) + x_p(t) + \sigma\varepsilon(t),$$

Energy signal — decaying transients **Periodic signal — information of interest** **White noise**

what are the PSD's of these different components? To calculate the PSD we start with the autocorrelation in the power signal sense, i.e. periodic autocorrelation.

8.3.1.1 *Autocorrelation of ideal signals*

- ***Energy signals***

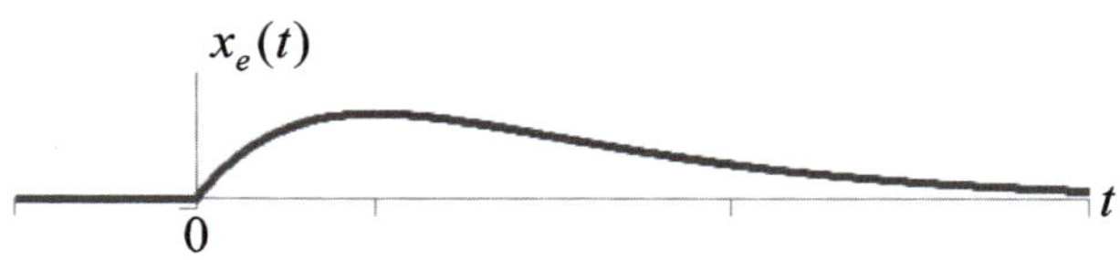

Fig. 8.3-2. Energy signal.

Since the signal is an energy signal, its linear autocorrelation (used for energy signals) is finite:

$$r_{x_e x_e}(t) = \int_{-\infty}^{\infty} x_e(\tau) x_e(\tau - t) d\tau = \lim_{t_{tr}\to\infty} \int_{-t_{tr}/2}^{t_{tr}/2} x_e(\tau) x_e(\tau - t) d\tau.$$

However, since we apply periodic autocorrelation to the entire ideal signal, we have to use it here:

$$r^{\circledast}_{x_{tp}x_{tp}}(t) \underset{T_1\to\infty}{\to} \lim_{T_1\to\infty} \frac{1}{T_1}\int_{-T_1/2}^{T_1/2} x_{tp}(\tau) x_{tp}(\tau - t) d\tau = \lim_{T_1\to\infty} \frac{1}{T_1}\int_{-T_1/2}^{T_1/2} x_e(\tau) x_e(\tau - t) d\tau = \lim_{T_1\to\infty} \frac{1}{T_1} r_{x_e x_e}(t) = 0.$$

This shows that the PSD of an energy signal is 0, which makes sense: energy signal has no power.

Thus, for the ideal signals, spectral information related to the energy signals is not seen on the PSD graph. In practice, *the longer the data set (the larger the time T_1 in the above calculation with limits dropped), the smaller the effect energy signals have on the real PSD.*

- ***Periodic signals***

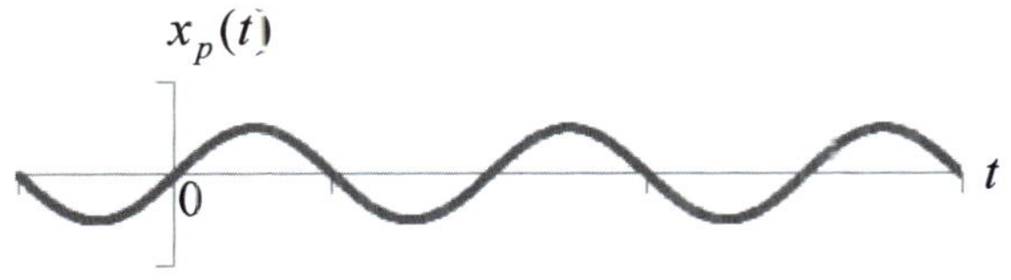

Fig. 8.3-3. Periodic signal.

Representing a periodic signal in terms of its Fourier series (FS):

$$x_p(t)=\sum_{k=-\infty}^{\infty} X_s[k]e^{j2k\pi f_1 t}\ ,\ f_1=\frac{1}{T_1},$$

noting that $x_p(t)=x_p^*(t)$, since $x_p(t)$ is real, and computing its periodic autocorrelation yields:

$$\begin{aligned}
r_{x_p x_p}^{\circledast} &= \frac{1}{T_1}\int_0^{T_1} x_p(\tau)x_p(\tau-t)d\tau=\frac{1}{T_1}\int_0^{T_1} x_p(\tau)x_p^*(\tau-t)d\tau \\
&=\frac{1}{T_1}\int_0^{T_1}\underbrace{\left(\sum_{k=-\infty}^{\infty} X_s[k]e^{j2k\pi f_1\tau}\right)}_{x_p(\tau)}\underbrace{\left(\sum_{m=-\infty}^{\infty} X_m^*[k]e^{-j2m\pi f_1(\tau-t)}\right)}_{x_p^*(\tau-t)}d\tau \\
&=\sum_{k=-\infty}^{\infty}\sum_{m=-\infty}^{\infty} X_s[k]X_s^*[m]e^{j2m\pi f_1 t}\frac{1}{T_1}\int_0^{T_1} e^{j2(k-m)\pi f_1\tau}d\tau.
\end{aligned}$$

Recalling that

$$\frac{1}{T_1}\int_0^{T_1} e^{j2(k-m)\pi f_1\tau}d\tau=\begin{cases}1, & m=k,\\ 0, & \text{otherwise,}\end{cases}$$

yields:

$$r_{x_p x_p}^{\circledast}(t)=\sum_{k=-\infty}^{\infty} X_s[k]X_s^*[k]e^{j2k\pi f_1 t}=\boxed{\sum_{k=-\infty}^{\infty}\left|X_s[k]\right|^2 e^{j2k\pi f_1 t}.}$$

Thus, ***autocorrelation of a periodic signal has FS expansion with coefficients given by signal power spectra***. The latter could also be directly obtained from the periodic convolution property:

$$x_p(t)\circledast h_p(t)=\sum_{k=-\infty}^{\infty} X_s[k]H_s[k]e^{jk\omega_1 t}\ ,$$

by noting that $r_{x_p x_p}^{\circledast}(t)\equiv x_p(t)\circledast\circledast x_p(t)=x_p(t)\circledast x_p(-t)$ and $x_p(-t)\Leftrightarrow\{X_s^*[k]\}_{k=-\infty}^{\infty}$.

- **White noise**

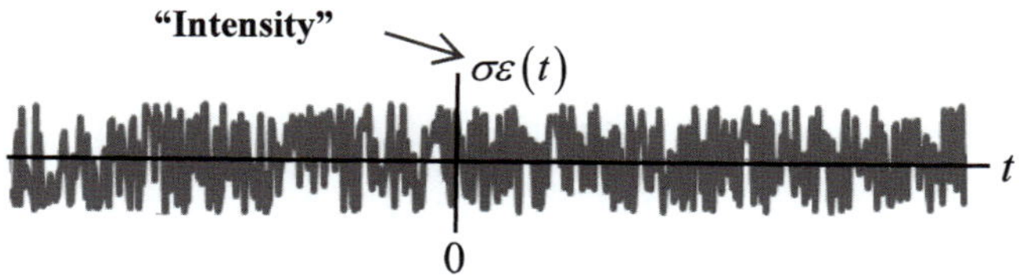

Fig. 8.3-4. White noise.

"White noise" is an infinitely fast signal nonrepeatable over any time interval. It crosses horizontal axis infinitely many times within any finite time interval. Its randomness, i.e. lack of periodic components coupled with its agility implies that it can correlate only with itself, i.e. at zero time shift. But when it correlates, the value of autocorrelation is ∞, i.e

$$r^{\circledast}_{\sigma\varepsilon\sigma\varepsilon}(t)=\sigma^2\delta(t).$$

8.3.1.2 *Summary: autocorrelations of components of ideal mixed signal*

Thus, for the components of the mixed ideal signal $x(t)$:

$$x(t)=x_e(t)+x_p(t)+\sigma\varepsilon(t)$$

Energy signal — decaying transients ($x_e(t)$)

Periodic signal — information of interest ($x_p(t)$)

White noise ($\sigma\varepsilon(t)$)

we have autocorrelations given as follows:

energy signal: $$r^{\circledast}_{x_ex_e}(t)=\lim_{t_{tr}\to\infty}\frac{1}{t_{tr}}r_{x_ex_e}(t)\equiv 0, \tag{8.3-1}$$

periodic signal: $$r^{\circledast}_{x_px_p}(t)=\sum_{k=-\infty}^{\infty}\left|X_s[k]\right|^2 e^{j2k\pi f_1 t},\ x_p(t)=\sum_{k=-\infty}^{\infty}X_s[k]e^{j2k\pi f_1 t}, \tag{8.3-2}$$

and

white noise: $$r^{\circledast}_{\sigma\varepsilon\sigma\varepsilon}(t)=\sigma^2\delta(t). \tag{8.3-3}$$

What about the entire signal? Calculating the total autocorrelation includes the autocorrelation of each component plus cross-correlations:

$$r^{\circledast}_{xx}(t)=r^{\circledast}_{x_ex_e}(t)+r^{\circledast}_{x_px_p}(t)+\sigma^2r^{\circledast}_{\varepsilon\varepsilon}(t)+r^{\circledast}_{x_px_e}(t)+\sigma r^{\circledast}_{\varepsilon x_e}(t)+r^{\circledast}_{x_ex_p}(t)+\sigma r^{\circledast}_{\varepsilon x_p}(t)+\sigma r^{\circledast}_{x_e\varepsilon}(t)+\sigma r^{\circledast}_{x_p\varepsilon}(t).$$

By definition, white noise is uncorrelated with any signal other than itself, yielding:

$$\sigma r^{\circledast}_{\varepsilon x_e}(t)=\sigma r^{\circledast}_{\varepsilon x_p}(t)=\sigma r^{\circledast}_{x_e\varepsilon}(t)=\sigma r^{\circledast}_{x_p\varepsilon}(t)\equiv 0.$$

Power signal crosscorrelations with an energy signal $x_e(t)$ (that has zero power) yield:

$$r^{\circledast}_{x_e x_p}(t) = \lim_{T_1\to\infty}\frac{1}{T_1}\int_{-T_1/2}^{T_1/2} x_e(\tau)x_p(\tau - t)d\tau = 0 = r^{\circledast}_{x_p x_e}(t).$$

This is quite easy to see from computing periodic convolution of a causal transient $h(t) = e^{-t}u(t)$ with an even periodic function $x_p(t) = \cos(t)$:

$$h(t)\circledast\circledast x_p(t) = h(t)\circledast x_p(-t) \underset{x_p(-t)=x_p(t)}{=} h(t)\circledast x_p(t),$$

which is exactly the same as applying linear (regular nonperiodic) convolution to a causal system to compute response to a sinusoid and then applying the limit $\lim_{T\to\infty} 1/T$ to the result:

$$y(t) = \underbrace{h(t)}_{e^{-t}u(t)} * \underbrace{x_p(t)}_{\cos(t)} = x_p(t) * h(t) = \int_{-\infty}^{\infty}\cos(\tau)e^{-(t-\tau)}u(t-\tau)d\tau = e^{-t}\int_{-\infty}^{t}\cos(\tau)e^{\tau}d\tau =$$

$$e^{-t}\left[\frac{1}{2}e^{\tau}(\sin(\tau)+\cos(\tau)\right]_{-\infty}^{t} = e^{-t}\left[\frac{1}{2}e^{t}(\sin(t)+\cos(t)\right] = \frac{1}{2}[(\sin(t)+\cos(t)] \Rightarrow$$

$$\cos(t)\circledast e^{-t}u(t) = \lim_{T\to\infty}\frac{1}{T}\int_{-T/2}^{T/2}\cos(\tau)e^{-(t-\tau)}u(t-\tau)d\tau = \lim_{T\to\infty}\frac{1}{T}\frac{1}{2}[(\sin(t)+\cos(t)] = 0.$$

Collecting the resulting components and dropping the zero-valued terms yields:

$$r^{\circledast}_{xx}(t) = r^{\circledast}_{x_p x_p}(t) + \sigma^2 r^{\circledast}_{\varepsilon\varepsilon}(t). \tag{8.3-4}$$

The latter permits us now to proceed to the calculation of the mixed ideal signal PSD.

8.3.2 *PSD of a mixed ideal signal*

8.3.2.1 *PSD of an energy signal*

Since periodic autocorrelation of an energy signal is

$$r^{\circledast}_{x_e x_e}(t) = \lim_{t_{tr}\to\infty}\frac{1}{t_{tr}} r_{x_e x_e}(t) = 0,$$

its PSD is also 0:

$$r^{\circledast}_{x_e x_e}(t) \equiv 0 \xrightleftharpoons{\text{Fourier Transform}} S_{x_e}(f) \equiv S_e(f) \equiv 0:$$

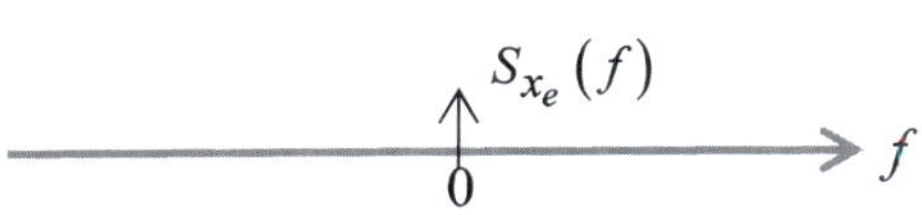

Fig. 8.3-5. PSD of energy signal.

8.3.2.2 *PSD of a periodic signal*

Let's start by recalling the definition of the Fourier Transform:

$$X(\omega)=\int_{-\infty}^{\infty} x(t)e^{-j\omega t}dt, \text{ or } X(f)=\int_{-\infty}^{\infty} x(t)e^{-j2\pi ft}dt.$$

Using this definition, what is the Fourier transform of a periodic signal?

Keeping in mind that any periodic signal can be written as a combination of (possibly infinitely many) complex exponential terms (Fourier series), let's just apply the Fourier transform definition above to a single complex exponential:

$$x(t)=e^{j\omega_1 t}=e^{j2\pi f_1 t} \Rightarrow$$

$$X(f)=\int_{-\infty}^{\infty} e^{j2\pi f_1 t}e^{-j2\pi ft}dt = \lim_{T_1\to\infty}\int_{-T_1/2}^{T_1/2} e^{j2\pi(f_1-f)t}dt = ?.$$

This integral does not converge! We need another way to compute this transform.

Inverse Fourier Transform

Starting with the Fourier series signal representation and extending fundamental period to ∞ yields

$$x_p(t)=\sum_{k=-\infty}^{\infty} X_s[k]e^{jk2\pi f_1 t} = \sum_{k=-\infty}^{\infty} T_1\underbrace{\frac{1}{T_1}}_{f_1}X_s[k]e^{jk2\pi f_1 t} = \sum_{k=-\infty}^{\infty} T_1 X_s[k]e^{jk2\pi f_1 t}f_1 \Rightarrow$$

$$T_1\to\infty \Rightarrow \begin{cases} \frac{1}{T_1}=f_1\to 0 \Rightarrow x_p(t)\to x(t) \text{ nonperiodic}, \\ kf_1\to f, \\ T_1X_s[k]\to X(kf_1)\to X(f), \\ \sum_{k=-\infty}^{\infty} f_1 \to \int_{-\infty}^{\infty} df, \end{cases}$$

$$\therefore x(t)=\int_{-\infty}^{\infty} X(f)e^{j2\pi ft}df = \frac{1}{2\pi}\int_{-\infty}^{\infty} X(\omega)e^{j\omega t}d\omega, \tag{8.3-5}$$

which is the **Inverse Fourier transform** representation of the resulting nonperiodic signal.

Fourier transform of periodic signals

Let us now take a look at the resulting integral:

$$x(t)=\int_{-\infty}^{\infty} X(f)e^{j2\pi ft}df$$

as follows. Noting that $e^{j2\pi ft}$ is a smooth function of f, the sifting property of delta-function in frequency domain $\delta(f-f_1)$ can be used to take $e^{j2\pi ft}$ out of the integral at $f=f_1$:

$$x(t)=\int_{-\infty}^{\infty}\delta(f-f_1)e^{j2\pi ft}df=e^{j2\pi f_1 t}.$$

This means that Fourier transform of $e^{j2\pi ft}$ is an impulse at f :

$$x(t)=e^{j2\pi f_1 t}\Leftrightarrow X(f)=\delta(f-f_1).$$

Therefore, the Fourier transform of an arbitrary periodic signal is an impulse train:

$$x_p(t)=\sum_{k=-\infty}^{\infty}X_s[k]e^{jk2\pi f_1 t}\xleftrightarrow{\text{Fourier Transform}}X_{x_p}(f)=\sum_{k=-\infty}^{\infty}X_s[k]\delta(f-kf_1).$$

PSD of a single sinusoid — a common sense look

Let's now move to a single sinusoid. What is the PSD of a single sinusoid $x_p(t)=A_1\cos(2\pi f_1 t)$? Clearly, sinusoid is a power signal with all its power concentrated at a single frequency:

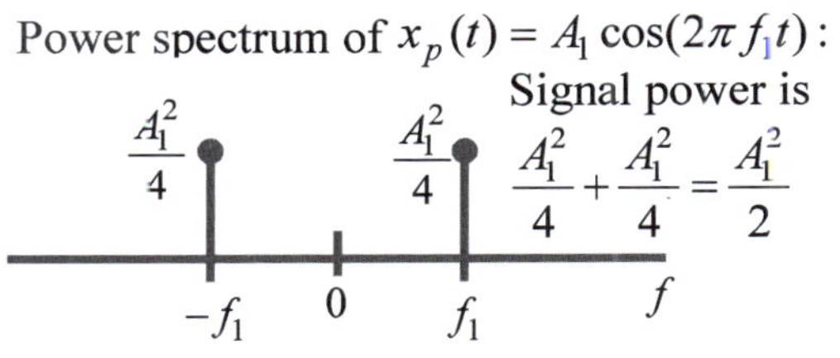

Fig. 8.3-6. Power spectrum of a single sinusoid.

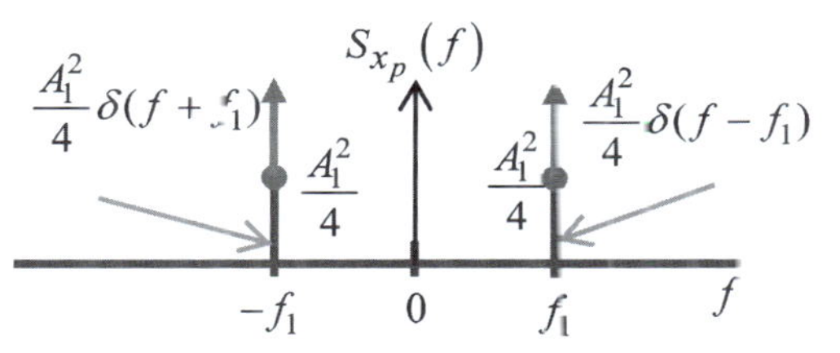

Fig. 8.3-7. Power spectral density of $A_1\cos(2\pi f_1 t)$.

Now, it becomes obvious that to obtain the PSD graph of this signal from its power spectrum graph we simply replace the spectral lines with delta-functions with the weight given by signal power:

Thus PSD of $A_1\cos(2\pi f_1 t)$ is

$$S(f)=\frac{A_1^2}{4}\delta(f+f_1)+\frac{A_1^2}{4}\delta(f-f_1),$$

and the power computed from this PSD is

$$\int_{-\infty}^{\infty}S(f)df=\int_{-\infty}^{\infty}\left[\frac{A_1^2}{4}\delta(f+f_1)+\frac{A_1^2}{4}\delta(f-f_1)\right]df=\frac{A_1^2}{2}.$$

Discrete PSD: a formal look at periodic signal PSD

Taking now a formal look at representing periodic signal in terms of its Fourier transform, we have: for the signal itself:

$$x_p(t)=\sum_{k=-\infty}^{\infty} X_s[k]e^{jk2\pi f_1 t} \xleftrightarrow{\text{Fourier Transform}} X_{x_p}(f)=\sum_{k=-\infty}^{\infty} X_s[k]\delta(f-kf_1):$$

Fig. 8.3-8. Magnitude spectrum and magnitude spectrum density of periodic signal.

and for the signal power:

$$r^{\circledast}_{x_p x_p}(t)=\sum_{k=-\infty}^{\infty} |X_s[k]|^2 e^{j2k\pi f_1 t} \xleftrightarrow{\text{Fourier Transform}} S_{x_p}(f)=\sum_{k=-\infty}^{\infty} |X_s[k]|^2 \delta(f-kf_1):$$

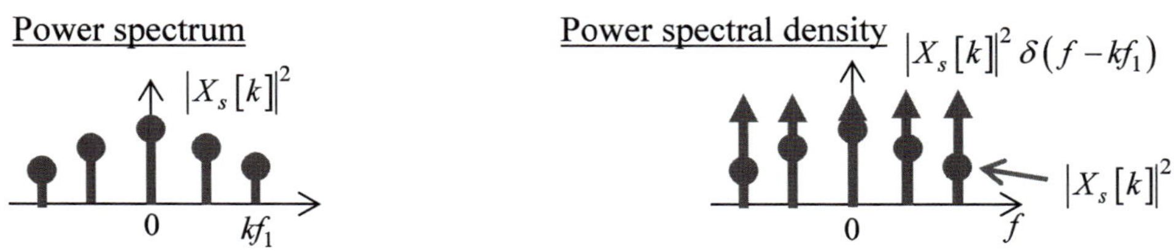

Fig. 8.3-9. Discrete power spectrum and power spectrum density of periodic signal.

The phase doesn't map into "phase density" and is the same for both.

Summary: Thus, for $x_p(t)$, periodic autocorrelation and Fourier transform yield the following:

$$r^{\circledast}_{xx}(t)=\sum_{k=-\infty}^{\infty} |X_s[k]|^2 e^{j2k\pi f_1 t} \text{ and } e^{j2k\pi f_1 t} \xleftrightarrow{\text{Fourier Transform}} \delta(f-kf_1) \Rightarrow$$

$$r^{\circledast}_{x_p x_p}(t)=\sum_{k=-\infty}^{\infty} |X_s[k]|^2 e^{j2k\pi f_1 t} \xleftrightarrow{\text{Fourier Transform}} S_{x_p}(f) = \boxed{\sum_{k=-\infty}^{\infty} |X_s[k]|^2 \delta(f-kf_1):}$$

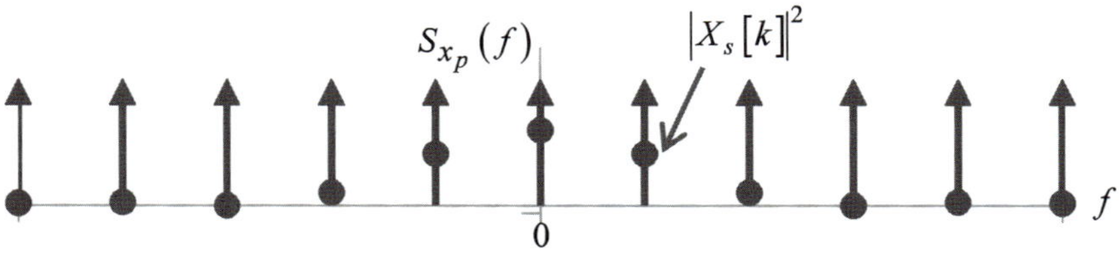

Fig. 8.3-10. PSD of a periodic signal with powers indicated as impulse weights.

8.3.2.3 *PSD of white noise*

With autocorrelation:

$$r_{\sigma\varepsilon\sigma\varepsilon}^{\circledast}(t) = \sigma^2\delta(t)$$

"white noise," depicted in Figure 8.3-4, is nonperiodic and has continuous spectrum, meaning it has no distinct periodic components in its PSD. That is,

$$r_{\sigma\varepsilon\sigma\varepsilon}^{\circledast}(t) \xrightleftharpoons{\text{Fourier Transform}} R_{\sigma\varepsilon\sigma\varepsilon}^{\circledast}(f) = S_{\sigma\varepsilon}(f) \equiv \sigma^2 \ \forall(\text{for all}) f \in (-\infty, \infty).$$

This relation is depicted in Figure 8.3.11:

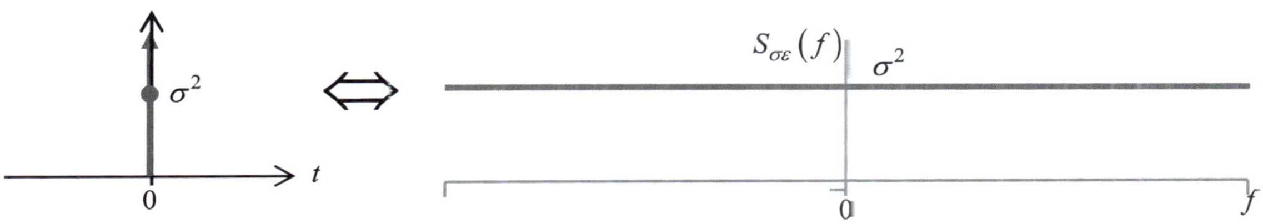

Fig. 8.3-11. Autocorrelation and PSD of white noise signal.

Thus, representation of white noise is quite logical and simple: frequency domain constant (white — all frequencies present), the inverse Fourier Transform (autocorrelation) of which is an impulse.

Computing signal power from its PSD or autocorrelation yields

$$r_{\sigma\varepsilon\sigma\varepsilon}^{\circledast}(t) = \sigma^2\delta(t) \Rightarrow P_{\sigma\varepsilon} = r_{\sigma\varepsilon\sigma\varepsilon}^{\circledast}(0) = \sigma^2\delta(0) = \infty$$

$$\Updownarrow$$

$$S_{\sigma\varepsilon}(f) \equiv \sigma^2 \ \forall f \Rightarrow P_{\sigma\varepsilon} = \int_{-\infty}^{\infty} S_{\sigma\varepsilon}(f)\,df = \int_{-\infty}^{\infty} \sigma^2 df = \infty.$$

This makes white noise not a power signal.

In spite of that, however, white noise is seen to have a nice (that's what ideal means) representation in frequency domain in terms of a clearly defined power spectral density given just by a straight line. The infinite area under the latter is what's rendering the signal nonpower.

8.3.2.4 *Putting the components together: PSD plot of a mixed ideal signal*

Getting back to the entire ideal signal

$$x(t) = x_e(t) + x_p(t) + \sigma\varepsilon(t)$$

White noise ($\sigma\varepsilon(t)$)

Periodic signal: information of interest ($x_p(t)$)

Energy signal: decaying transients ($x_e(t)$)

we have:

$S_{x_e}(f) \equiv 0$: periodic autocorrelation is 0,

$S_{x_p}(f) = \sum_{k=-\infty}^{\infty} |X_s[k]|^2 \delta(f - kf_1)$: $X_s[k]$ and f_1 come from the Fourier Series of $x_p(t)$,

$S_{\sigma\varepsilon}(f) = \sigma^2 S_\varepsilon(f) \equiv \sigma^2.$

Thus, since the autocorrelation of the mixed ideal signal is given by (8.3-4) as

$$r_{xx}^{\circledast}(t) = r_{x_p x_p}^{\circledast}(t) + \sigma^2 r_{\varepsilon\varepsilon}^{\circledast}(t), \tag{8.3-6}$$

due to the linearity of the Fourier Transform,

$$r_{xx}^{\circledast}(t) = r_{x_p x_p}^{\circledast}(t) + \sigma^2 r_{\varepsilon\varepsilon}^{\circledast}(t) \xrightleftharpoons{\text{Fourier Transform}} S(f) = S_{x_p}(f) + \sigma^2 S_\varepsilon(f) \tag{8.3-7}$$

$$\therefore S(f) = \sigma^2 + \sum_{k=-\infty}^{\infty} |X_s[k]|^2 \delta(f - kf_1):$$

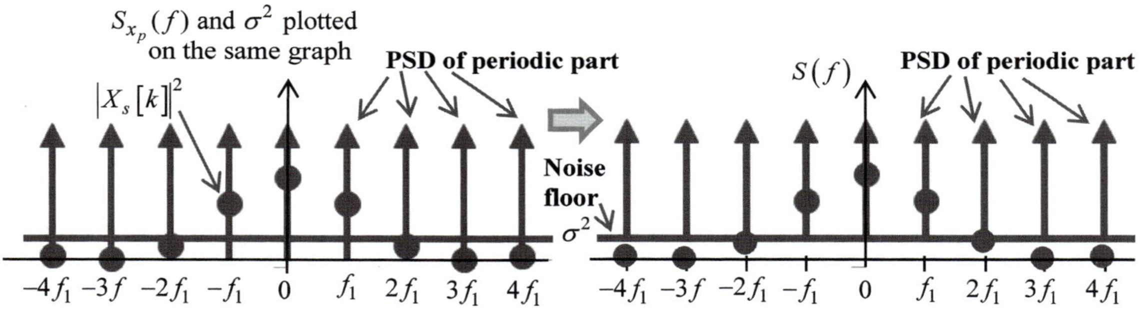

Fig. 8.3-12. Mixed ideal signal PSD: while the noise floor blocks all the information under it, PSDs of all periodic components spike to ∞ no matter how small their respective weights are.

Thus, we draw: i) ***discrete*** power spectral densities for periodic signals, ii) ***continuous*** power (and energy) spectral densities for non-periodic signal, and iii) ***mixed discrete-continuous*** power spectral densities for mixed signals. ***This shows that magnitude and power spectral densities, <u>in theory</u>, are great diagnostic tools: they <u>convert</u> the <u>multiscale</u> magnitude and power spectra with components not visible on the same graph into the respective <u>single-scale</u> spectral densities with components going to ∞!***

Let's now take a look at how to reduce this theory to practice.

8.4 Discrete-Continuous (Mixed) Spectral Analysis of a Sampled Signal. Spectral Analysis Applications: Diagnostics, System Identification

8.4.1 *Spectral analysis of sampled data*

8.4.1.1 *Approximating PSD with DFT in analog frequency*

In reality, we have a DT signal (sampled from the actual CT system) as the data record.

The PSD is approximated by taking the DFT (usually using FFT) of either $r^{\circledast}_{xx}[n]$, or $x[n]$:

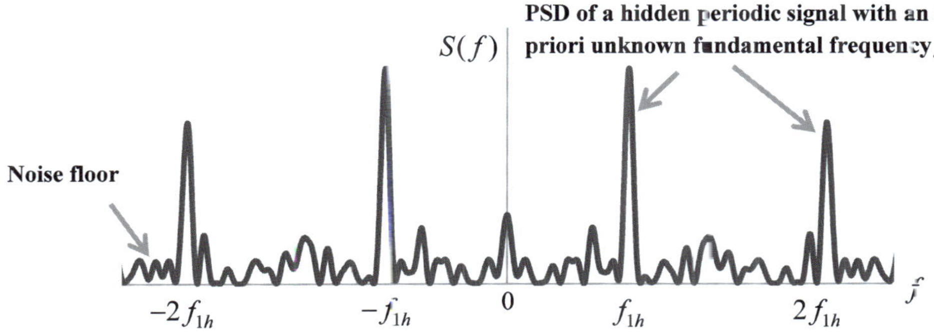

Fig. 8.4-1. PSD calculated from the actual data.

Let's recall expression (5.6-2) showing that the Discrete Fourier Transform (multiplied by t_s) is an approximation of the actual Fourier Transform:

$$t_s X_{DFT}[k] \approx X_t(f)\big|_{f=\frac{k}{T_1}} \approx X(f)\big|_{f=\frac{k}{T_1}}. \tag{8.4-1}$$

The DFT uses a finite causal data record, so it actually transforms the sampled version of truncated signal $x_t(t)$. As $t_s \to 0$, the summation used to calculate the DFT multiplied by t_s better approximates the integral used to calculate $X_t(f)$.

DFT also "samples" the FT in the analog frequency domain. As $t_{tr} = T_1 \to \infty$, and, the truncated signal becomes closer to the actual signal, and the discrete samples of analog frequency move closer together. ***Since DFT is always applied to truncated sequences, it will be used without subscript t.***

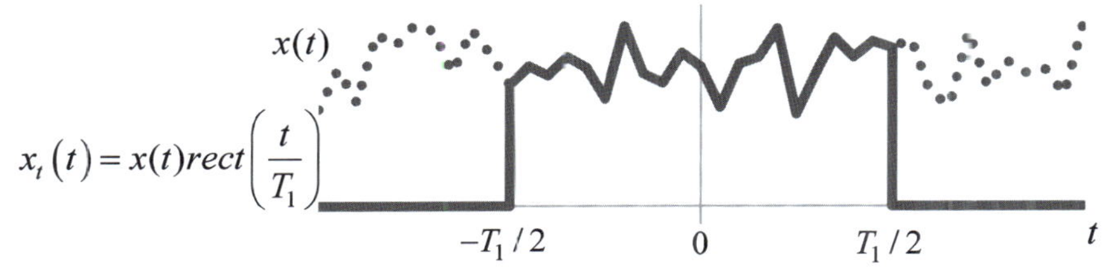

Fig. 8.4-2. Data record truncated through multiplication by not bandlimited rect function.

Thus, we have

$$t_s X_{DFT}[k] \approx X_t(f)\big|_{f=\frac{k}{T_1}}, \quad S_t(f) = \frac{1}{T_1}|X_t(f)|^2, \quad \lim_{T_1\to\infty} S_t(f) = S(f).$$

The accuracy of the PSD estimate based on the sampled data record and (8.4-1) is limited for two main reasons:

- the presence of noise,
- compact support of CT data: *multiplication* of bandlimited signal by rect *in the time domain* corresponds to *convolution* of Fourier transform of this signal with not bandlimited sinc function *in the frequency domain.*

Each of these factors makes finite CT data record not bandlimited and leads to aliasing during sampling. Let's now address these problems.

Noise problem. As an ideal signal, white noise has infinite bandwidth. Therefore making t_s small doesn't help to overcome aliasing error coming from it: ***white noise can't be sampled without aliasing.*** But making t_s small does help with reducing aliasing error coming from ***wide band noise***, whose PSD tapers off as frequency increases:

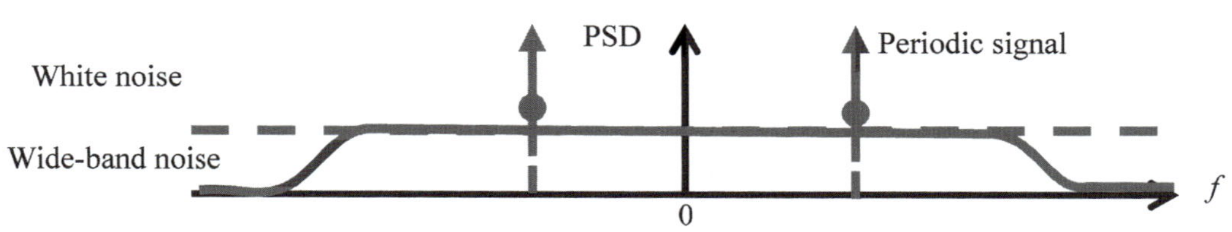

Fig. 8.4-3. Approximating PSD of CT signal with wide band noise.

Finite duration problem. Increasing the data record size reduces frequency domain distortion induced by sinc function.

Therefore, we can reduce aliasing in approximating the PSD of CT signal with wide band noise and compact support using measured sampled data by taking:

- small enough t_s
- large enough CT data record length T_1 and hence number of DT samples N_1.

Then, setting $T_1 = t_s N_1$, we can approximate the continuous time signal PSD in frequency f by using measured sampled data as:

$$S(f)\big|_{f=\frac{k}{T_1}} \underbrace{\approx}_{\substack{\text{Truncation}\\ \text{of CT data}\\ \text{record}}} S_t\left(\frac{k}{T_1}\right) = \frac{1}{T_1}|X_t(f)|^2\Big|_{f=\frac{k}{T_1}} \underbrace{\approx}_{\substack{\text{Sampling of}\\ \text{not bandlimited}\\ \text{CT data}}} \tag{8.4-2}$$

$$S_{ta}\left(\frac{k}{T_1}\right) \triangleq \frac{1}{N_1 t_s}|t_s X_{DFT}[k]|^2 = \frac{t_s}{N_1}|X_{DFT}[k]|^2$$

where subscript a in S_{ta} stands for aliasing, or

$$S(f)\Big|_{f=\frac{k}{T_1}} \approx S_{ta}\left(\frac{k}{T_1}\right) = \frac{t_s}{N}\left|X_{DFT}[k]\right|^2. \tag{8.4-3}$$

This can be calculated

i) from the magnitude of the data record DFT or

ii) by taking the DFT of the periodic autocorrelation introduced in Example 8.3-1.

From (8.4-3) it is seen that this PSD approximation can be associated with plotting $S_{ta}(f)$ as a piecewise constant function piecewise continuous in *continuous* frequency f:

$$S(f) \approx S_{ta}\left(\frac{k}{T_1}\right) = \frac{t_s}{N_1}\left|X_{DFT}[k]\right|^2, \; kf_1 \le f < (k+1)f_1, \; 0 \le k \le N_1 - 1, \tag{8.4-4}$$

i.e., as a bar graph of *"power density bins"*, each having width $\Delta f = 1/N_1 t_s$, and with the k'th bin located at the interval $kf_1 \le f_1 < (k+1)f_1$, $0 \le k \le N_1$.

The power of the k'th bin, denoted by $P[k]$, equals

$$\boxed{P[k] = \Delta f \, S_{ta}\left(\frac{k}{T_1}\right) = \Delta f \frac{t_s}{N_1}\left|X_{DFT}[k]\right|^2 = \frac{1}{N_1 t_s}\frac{t_s}{N_1}\left|X_{DFT}[k]\right|^2 = \frac{1}{N_1^2}\left|X_{DFT}[k]\right|^2.} \tag{8.4-5}$$

8.4.1.2 *DT PSD in analog frequency F*

Define "digital" PSD, or DT PSD, denoted as $S_{DFT}[k]$, as a piecewise constant function that is also piecewise continuous in continuous digital frequency F and has the same power of the k'th bin as $S_{ta}(k/T_1)$, namely

$$P[k] = \frac{1}{N_1^2}\left|X_{DFT}[k]\right|^2.$$

To ensure this, divide the above $P[k]$ by $\Delta F = 1/N_1 = F_1$. This yields ***periodogram***, which can be reasonably considered to be digital counterpart of CT PSD:

$$\begin{aligned} P[k]/\Delta F &= P[k]/(1/N_1) = P[k]N_1 = \frac{1}{N_1^2}\left|X_{DFT}[k]\right|^2 N_1 \\ &= \frac{1}{N_1}\left|X_{DFT}[k]\right|^2 \equiv S_{DFT}[k], \; kF_1 \le F < (k+1)F_1, \; 0 \le k \le N_1 - 1. \end{aligned} \tag{8.4-6}$$

Relation between approximate CT PSD in analog frequency *f* and DT PSD in digital frequency *F*. As it is seen, relation between $S_{DFT}[k]$ plotted in f and $S_{DFT}[k]$ plotted in F is given by

$$S_{ta}(k/T_1) = t_s S_{DFT}[k], \; 0 \le k \le N_1 - 1, \tag{8.4-7}$$

which yields the following corresponding graphs:

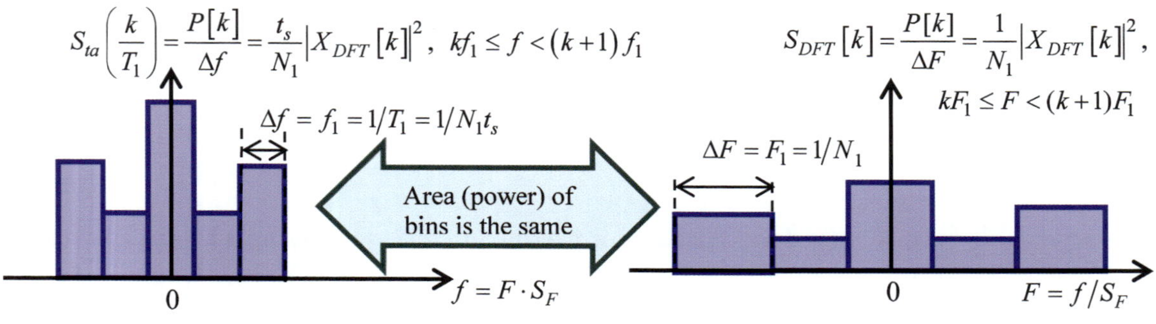

Fig. 8.4-4. Correspondence of the DT PSD in digital frequency F to the approximate CT PSD in analog frequency f.

8.4.2 *Detecting CT sinusoids in noise*

8.4.2.1 *Computing DT PSD in digital frequency F using periodic autocorrelation*

Consider N_1-length sampled data record $x_t[n]$ and its DFT $X_{DFT}[k]$: $x_t[n] \overset{DFT}{\rightleftharpoons} X_{DFT}[k]$. Let $x_{tp}[n]$ be N_1-periodic of extension of $x_t[n]$. Then, using **DT periodic autocorrelation**

$$r^{\circledast}_{x_{tp}x_{tp}}[n] \triangleq \frac{1}{N_1}\sum_{m=0}^{N_1-1} x_{tp}[m]x_{tp}[m-n] = \frac{1}{N_1}\sum_{m=0}^{N_1-1} x_{tp}[m]x_{tp}[-(n-m)] = \frac{1}{N_1}(x_{tp}[n] \circledast x_{tp}[-n]), \qquad (8.4\text{-}8)$$

and **DT periodic convolution** relation (6.3-26), the ***periodogram***, or DT PSD of $x_t[n]$ is given by:

$$r^{\circledast}_{x_{tp}x_{tp}}[n] = \frac{1}{N_1}(x_{tp}[n] \circledast x_{tp}[-n]) \overset{DFT}{\rightleftharpoons} \frac{1}{N_1}X_{DFT}[k]X_{DFT}[-k] =$$
$$\frac{1}{N_1}X_{DFT}[k]X^*_{DFT}[k] = \frac{1}{N_1}|X_{DFT}[k]|^2 \equiv S_{DFT}[k],\ kF_1 \le F < (k+1)F_1,\ 0 \le k \le N_1 - 1. \qquad (8.4\text{-}9)$$

Detecting hidden sinusoids in CT noisy data using DFT (computed through FFT)

We have already shown analytically in (5.7-1) and (5.7-2) that *when we sample a CT sinusoid, if we decrease t_s and/or increase N_1, then DFT magnitude at the digital frequency corresponding to this sinusoid goes to infinity.* How does this result help in detecting hidden (hence subscript h) in noise periodic signals with frequencies $mf_{1h} = mN_{1h}f_1 = m_h f_1$?

Here f_{1h} is unknown fundamental frequency $f_{1h} = F_{1h}S_F = N_{1h}F_1S_F = N_{1h}f_1$, where $F_1 = 1/N_1$ and $f_1 = 1/T_1$ are known, but integer N_{1h} is unknown. The existence of the latter implies small leakage, since f_{1h} is assumed to be the N_{1h}'th harmonic of the data record periodic extension frequency f_1. This holds if for a sufficiently small t_s, data record length $T_1 = t_sN_1$ is large enough.

DFT provides the key sinusoid detection capability under noise as well, but ***in manipulating the sampling interval t_s and the data record length $T_1 = t_sN_1$ we will distinguish three cases***:

i) **Pure digital**: *DT PSD* computation based on sampled data record *plotted along <u>digital</u> F axis;*

ii) **Pure analog**: *CT PSD* approximation using i) *plotted along <u>analog</u> f axis*

iii) **Mixed digital/analog**: *DT PSD* taken *unchanged from* i), *but plotted along <u>analog</u> $f = FS_F$ axis.*

The latter is our favorite! Because in this case, (8.4-4) shows that we are not plotting CT PSD approximation $S_{ta}(f)$, but rather ***CT PSD approximation $S_{ta}(f)$ divided by the sampling interval t_s***:

$$\frac{S(f)}{t_s} \approx \frac{S_{ta}(k/T_1)}{t_s} = \frac{1}{N_1}\left|X_{DFT}[k]\right|^2,\ kf_1 \le f < (k+1)f_1,\ 0 \le k \le N_1 - 1. \qquad (8.4\text{-}10)$$

- First, we recall that for two-sided spectra, power of periodic signal is square of its magnitude
- Therefore, the analysis carried out for periodic signal DFT computation in (5.7-1) and (5.7-2) applies to DT PSD computation, with the only difference that *the nonperiodic component will give rise to nonzero values between periodic component frequencies.*

This is considered next in three cases indicated above, with two practical scenarios for the first two cases.

8.4.2.2 *Discrete time power spectra density (DT PSD), F axis*

Case i) DT1: *CT data collection time interval T_1 fixed, sampling rate t_s decreases, length of DT data record $N_1 = T_1/t_s$ increases:*

- the power bins in $S_{DFT}[k]$ (8.4-6) become finer
- powers of periodic components concentrate around the corresponding frequency points, leaving little power between these points
- the magnitude squares of DFT of these components go to infinity, while the DFT magnitude values between these components grow very slowly:

$$T_1 = t_s N_1 \text{ fixed} \Rightarrow t_s \to 0 \Rightarrow N_1 \to \infty:\ S_{DFT}[k] = \frac{1}{N_1}\left|X_{DFT}[k]\right|^2 = \frac{1}{N_1}\left|N_1 X_{DFS}[k]\right|^2 \xrightarrow[t_s \to 0]{}$$

$$\underbrace{\left[N_1\left|X_{DFS}[k=m_h]\right|^2 \xrightarrow[N_1\to\infty]{} \infty\right]}_{\text{fast growing hidden components with frequencies } m_h F_1} + \underbrace{N_1\left|X_{DFS}[k \ne mN_{1h}]\right|^2 \xrightarrow[N_1\to\infty]{} \infty}_{\text{slow growing DT PSD noise floor with no hidden harmonics}}. \qquad (8.4\text{-}11)$$

Case i) DT2: *CT data collection time interval $T_1 = kT_k = t_s kN_k$ increases, t_s fixed, $N_1 = kN_k$ increases:*

- if t_s is small enough and N_k is big enough, then the power of the k'th sinusoid is already concentrated enough at its frequency,
- DFT magnitude square at this frequency adds up and grows faster than noise as k, the number of cycles in the data, grows
- this will also decrease leakage if we missed on taking the exact integer number of cycles:

$$t_s \text{ fixed, } T_k = t_s N_k \text{ fixed} \Rightarrow N_k \text{ fixed; } T_1 = kT_k \xrightarrow[k\to\infty]{} \infty \Rightarrow N_1 = kN_k \to \infty:\quad S_{DFT}[k] =$$

$$\frac{1}{N_1}\left|X_{DFT}[k]\right|^2 = \underbrace{\left[kN_k\left|X_{DFS}[k=m_h]\right|^2 \xrightarrow[k\to\infty]{} \infty\right]}_{\text{fast growing hidden components with frequencies } m_h F_1} + \underbrace{kN_k\left|X_{DFS}[k \ne m_h]\right|^2 \xrightarrow[k\to\infty]{} \infty}_{\text{slow growing DT PSD noise floor with no hidden harmonics}}. \qquad (8.4\text{-}12)$$

Case ii) CT1: *Data collection time interval $T_1 = t_s N_1$ fixed, t_s decreases, N_1 increases:*

- the power bins in (8.4-4) approximation become finer
- powers of periodic components concentrate around the corresponding frequency points, leaving little power between these points
- the DFT-based approximation magnitudes of these components ***remain about the same due to the factor t_s*** (don't like that!) while decreasing between these components (good):

$$T_1 \text{ fixed, } t_s \to 0,\ N_1 \to \infty:\ S(f)\big|_{f=\frac{k}{T_1}} \approx \frac{1}{T_1}\left|X_t(f)\right|^2\Big|_{f=\frac{k}{T_1}} \approx \frac{1}{N_1 t_s}\left|t_s X_{DFT}[k]\right|^2 = \frac{t_s}{N_1}\left|X_{DFT}[k]\right|^2$$

$$\xrightarrow[t_s\to 0]{} \frac{t_s}{N_1}\left|N_1 X_{DFS}[k]\right|^2 = \underbrace{\overbrace{t_s N_1}^{Fixed=T_1}\left|X_{DFS}[k=m_h]\right|^2}_{\substack{\text{not growing components } T_1 X_s[m_h] \\ \text{with frequencies } m_h f_1}} + \underbrace{t_s N_1\left|X_{DFS}[k\neq m_h]\right|^2}_{\substack{\text{decaying DT PSD noise floor} \\ \text{with no hidden harmonics}}}. \tag{8.4-13}$$

Not great, since densities of small magnitude harmonics don't' rise up as we decrease the sampling rate. How to fix this? By plotting Case iii): $S_{ta}(k/T_1)/t_s = N_1\left|X_{DFS}[k = mN_{1h}]\right|^2$ + slow growing DT PSD noise floor.

Case ii) CT2: *Data collection time interval $T_1 = kT_k = t_s kN_k$ increases, t_s fixed, $N_1 = kN_k$ increases:*

- if t_s is small enough and N_k is big enough, then leakage is small and the power of k'th sinusoid is already concentrated enough at its frequency,
- DFT magnitude square at this frequency adds up and grows faster than noise as *k*, the number of cycles in the data, grows.
- this also further decreases leakage if we missed on the exact integer number of cycles:

$$t_s, T_k = t_s N_k, N_k \text{ all fixed; } T_1 = kT_k = t_s kN_k \xrightarrow[k\to\infty]{} \infty:\ S(f)\big|_{f=\frac{k}{T_1}} \approx \frac{1}{T_1}\left|X_t(f)\right|^2\Big|_{f=\frac{k}{T_1}} \approx$$

$$\frac{1}{kN_k t_s}\left|t_s X_{DFT}[k]\right|^2 = \frac{t_s}{kN_k}\left|X_{DFT}[k]\right|^2 = \frac{t_s}{kN_k}\left|kN_k X_{DFS}[k]\right|^2 \tag{8.4-14}$$

$$= \underbrace{t_s kN_k\left|X_{DFS}[k=m_h]\right|^2 \xrightarrow[k\to\infty]{} \infty}_{\substack{\text{fast growing hidden components} \\ t_s kN_k X_s[m_h] \text{ with frequencies } m_h f_1}} + \underbrace{t_s kN_k\left|X_{DFS}[k\neq m_h]\right|^2 \xrightarrow[k\to\infty]{} \infty}_{\substack{\text{slow growing DT PSD noise floor} \\ \text{with no hidden harmonics}}}.$$

Summary: Thus, both data collection enhancements — decrease of t_s and increase of T_1 — are seen to be important, and the same analysis applies to DT PSD and approximate CT PSD.

The only difference between them is the $t_s = 1/S_F$ factor relating DT PSD plotted in *F* to approximate CT PSD plotted in *f*:

$$\text{i) } S_{DFT}[k] = \frac{1}{N_1}\left|X_{DFT}[k]\right|^2,\ \text{ii) } S_{ta}(f)\big|_{f=\frac{k}{T_1}} = \frac{t_s}{N_1}\left|X_{DFT}[k]\right|^2 = t_s S_{DFT}[k] \Rightarrow \text{iii) } S_{DFT}[k] = [S_{ta}(f)/t_s]\big|_{f=\frac{k}{T_1}}.$$

This factor makes approximate CT PSD based on DT PSD and plotted in *f* behave differently than DT PSD plotted in *F* for the same values of power bins, as just shown, because the support of each power bin in digital frequency $F = f/S_F$ of DT PSD shrinks as sampling rate S_F increases. To have these estimates look identical we will plot $S_{DFT}[k]$ with respect to both digital and analog frequency axes, the latter yielding

Case iii): $S_{ta}(k/T_1)/t_s = S_{DFT}[k]$ versus f. As indicated above in (8.4-10), to see spikes for fixed T_1 and decreasing t_s we can enhance ii) CT1 by mixing it with i) DT1 — plotting DT PSD on f axis.

Example 8.4-1: Approximating CT PSD with DFT, actual CT data.
Figures 8.4-6 and 8.4-7 below show noisy data with hidden periodic components generated by

$$x(t) = 1 + 3\cos(2\pi \cdot 2 \cdot t) + 2\cos(2\pi \cdot 5 \cdot t) - noise$$

in Excel, with uniformly distributed noise. We expect the CT PSD of a function of this type to have impulse spikes rising above the noise floor:

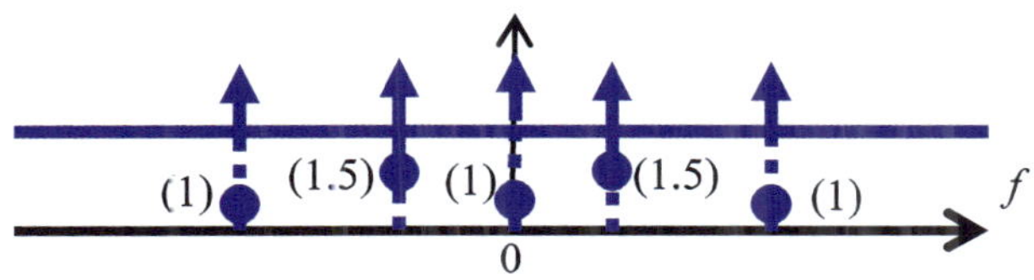

Fig. 8.4-5. CT PSD.

Applying Case II) CT1: decreasing t_s, fixed T_1, increasing N_1, yields:

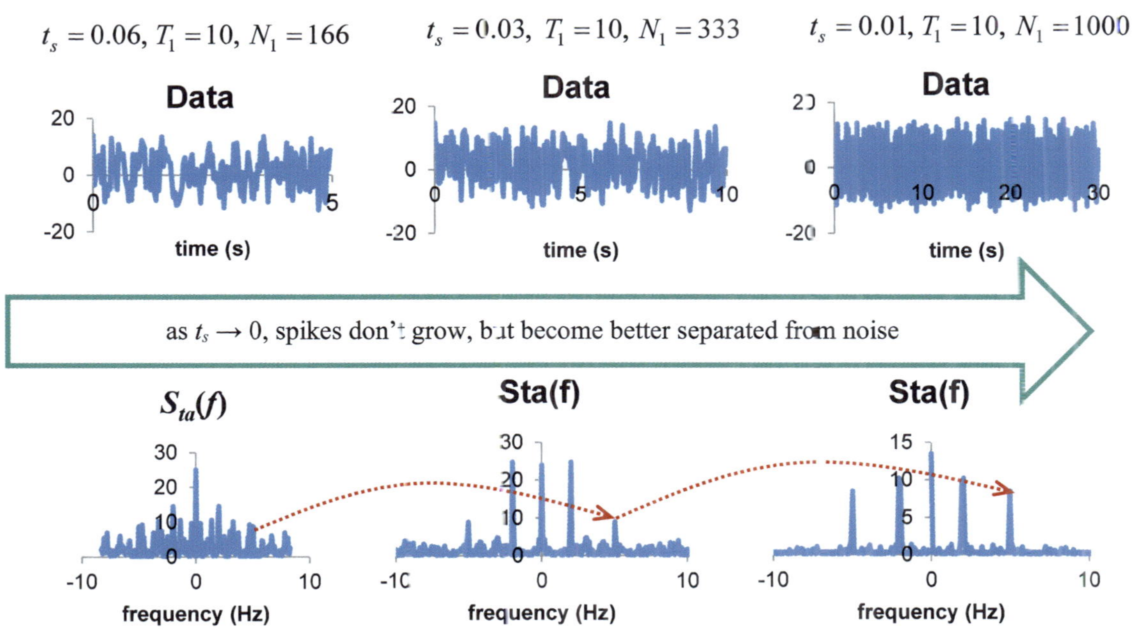

Fig. 8.4-6. Approximating CT PSD with DFT: example for Case ii) CT1: decreasing t_s, fixed T_1, increasing N_1.

Applying Case II) CT2: fixed t_s, increasing T_1 and N_1, yields:

$t_s = 0.03,\ T_1 = 5,\ N_1 = 166$ $\qquad t_s = 0.03,\ T_1 = 10,\ N_1 = 333$ $\qquad t_s = 0.03,\ T_1 = 30,\ N_1 = 1000$

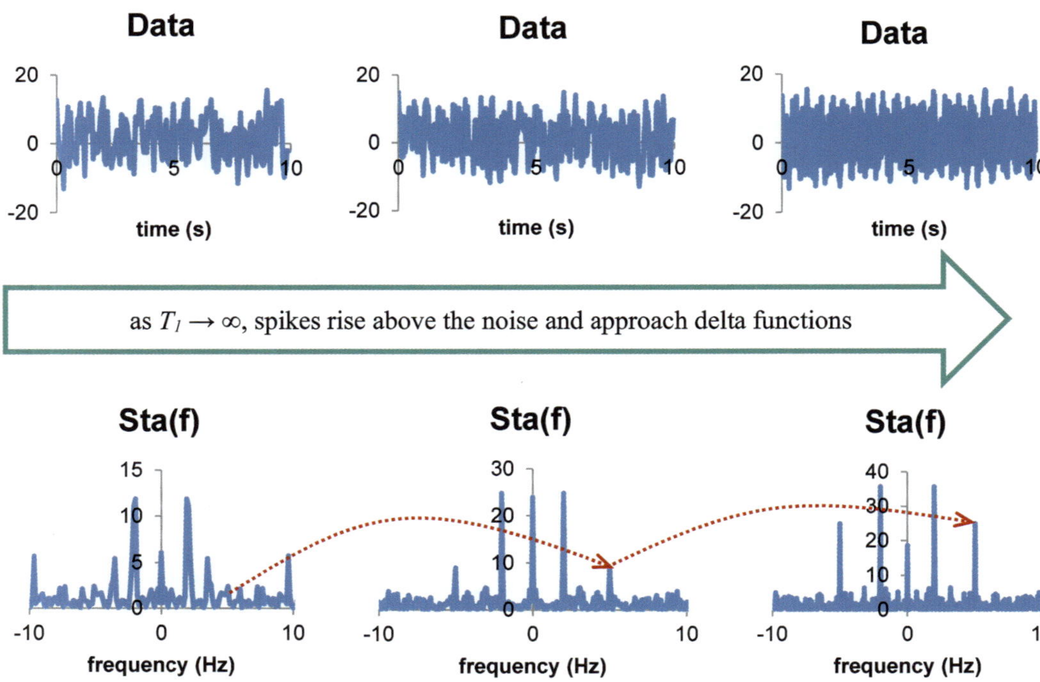

Fig. 8.4-7. Approximating CT PSD with DFT: example for Case ii) CT2: fixed t_s, increasing T_l and N_l.

Why sinusoids buried in noise stick out of the noise floor on the DT PSD graph? Let's now discuss why, when computing DT PSD through taking the data record DFT, we can make the hidden sinusoids show well by multiplication the DFS by N_1. Won't the noise floor be amplified simultaneously?

To answer this question, we first note that we have a theoretical picture to guide us in separating a sinusoid from noise: PSD plot with delta functions sticking out of the noise floor for the ideal signal power spectral density. This picture then answers this question as follows.

On one hand, the delta function at a single frequency means that the whole signal power is concentrated at a single point of the frequency axis, and that's possible only for the periodic signal located at that frequency — that's what single frequency means.

On the other hand, and just as important, the noise power at a single frequency is zero (!) even for a very big, but finite noise floor. This is because there is no area under any single point of the PSD graph of the noise. Indeed, noise can't produce power at a single frequency, since it has no periodic components.

For applications, assuming we can reduce t_s and increase T_1, this means that if we have a good data taking of the noisy signal (small t_s and large T_1), then there would be no power concentrated at a single frequency between the hidden harmonics, and hence nothing that would make the PSD between the hidden harmonics rise up at any single point to infinity — it will just be reaching the finite noise floor value. So the noise will not be amplified simultaneously as we reduce t_s and increase N_1, since the power bins between the hidden sinusoids will drop in magnitude as T_1 increases and t_s decreases.

Why we consider all five cases? Now, let's discuss why we need the five cases mentioned in pp. 500–04, and specifically why would the X_{DFS} associated with the noise floor become smaller if we decrease t_s while keeping T_1 fixed?

To answer this, imagine now that we would like to rescale our theoretical PSD graph to make the magnitudes of the periodic signal spikes on a PSD-type graph finite. In this case we would need to divide all the values on the graph by infinity, or alternatively, multiply by 0 (or by infinitesimally small t_s) — this would then make the power spikes at the periodic signal frequencies become finite, but would completely wipe out the noise floor — rescale it to zero. This is exactly what happens when we keep T_1 fixed, reduce t_s to very small value, get then N_1 to go to infinity, and plot the result along f axis — as in the Case ii) CT1 on p. 502 and Example 8.4-1 on p. 503.

This case shows that to see how to manipulate the tools at our disposal, how to plot the result, and what would it mean, we need to consider all the possibilities that arise. These are exactly all the cases considered: a) reduction of t_s, b) increase of T_1 — the real-time window of the data taking, c) DFT computation, and d) plotting of the resulting graph, The cases show how to extract the most information under the practical constraints (e.g., can we still get the result for not small enough t_s? Yes, by spending more time on data taking — as in Case ii) CT 2), and also how to best present the resulting PSD — through plotting not CT PSD vs f Hz, but rather (CT PSD)/t_s vs f Hz axis, which is Case iii): DT PSD plotted along f Hz.

8.5 Applications of PSD and FFT

8.5.1 *Diagnostics*

Finding small magnitude periodic components buried in noise and monitoring change of their magnitudes is important. It is often exactly these components, such as, for example, a tiny periodic precursor to an incipient fault in a worn out bearing or small magnitude periodic harmonics caused by actuator nonlinearity that are the key to problem understanding and solution.

Below, diagnostics through using tools introduced so far is demonstrated on several examples.

Example 8.5-1: Maintenance of fighter aircraft on an aircraft carrier.

<u>Problem:</u> balancing act of too-early maintenance (cost increase, lack of availability for mission) versus too late (risk of major damage to aircraft and carrier itself).

<u>Goal:</u> early detection of incipient faults.

<u>Tool:</u> spectral analysis of sensor data — detection of emergence of high frequency small magnitude components in aircraft rotating machinery motion.

Example 8.5-2: Characterization of industrial rotating or vibrating machinery operation.
Problem: showing clearly the change of principal modes with conditions, and bringing out especially interesting or problematic operating points.
Tool. Time-frequency, or waterfall plots — 3-D plots where the x-axis is ***frequency***, z-axis is FFT representing ***displacement, velocity, or acceleration magnitude spectrum or PSD***, and y-axis is some ***quazistationary operating condition of interest*** (e.g. ***rotation speed or vibration frequency***).

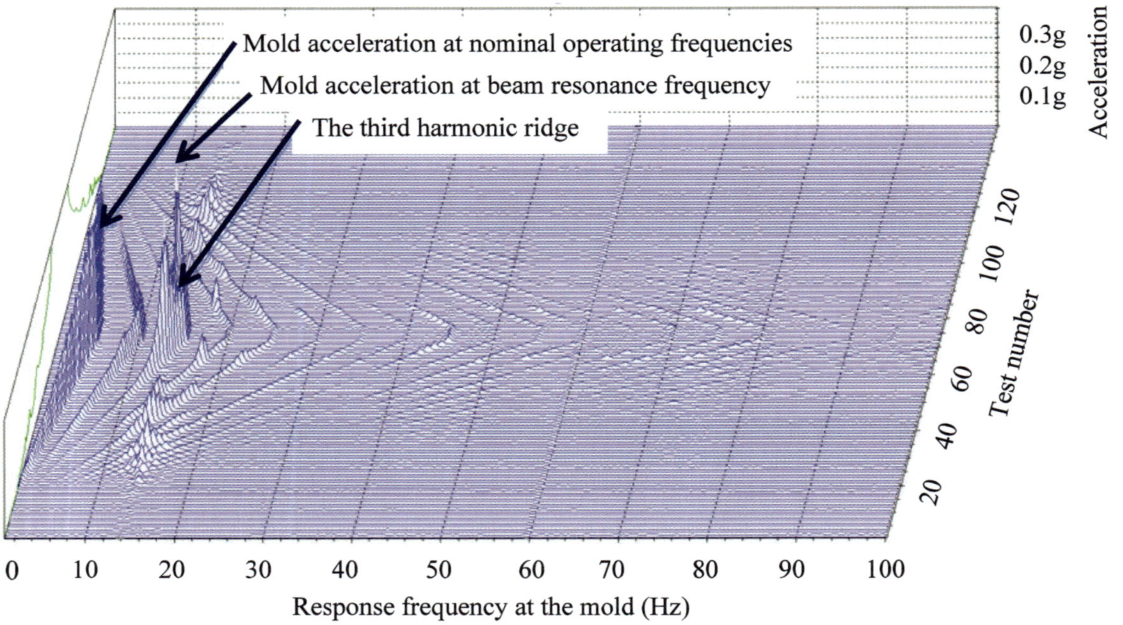

Fig. 8.5-1. Resonance problem: frequency domain waterfall plot based on mold acceleration data shows large force at the primary beam resonance for operating frequency equal to one third of resonance frequency.

A waterfall plot given in Figure 8.5-1 is based on the data from accelerometers on the mold and clearly shows that distortion is caused by the primary beam resonance. It represents a sequence of the steady state frequency response plots obtained by incrementing the mold oscillator operating frequency — the source of the ridge on the far left of the plot — from 0 Hz to around 8 Hz, and then decreasing it back to 0 Hz. When the nominal operating frequency enters an interval between 4 and 6 Hz, the third harmonic gets close to the resonance frequency of the beam, about 13.2 Hz. At the primary beam resonance frequency $f = 13.2$ Hz, or $\omega = 2\pi 13.2 = 82.37$ rad/sec mold position $a\sin(\omega t)$ has magnitude a, mold velocity $a\omega\cos(\omega t)$ has magnitude $a\omega = 82.37a$, and mold acceleration $a\omega^2\sin(\omega t)$ has magnitude $a\omega^2 = 6784.49a$ producing large forces.

Example 8.5-3: Mold oscillator distortion zoom-in.
Problem. Previous attempts at eliminating velocity distortions focused on the beam because sensor data showed the driving piston displacement to be a pure sinusoid, whereas the mold waterfall plot, given above, showed strong mold motion distortion at the driving frequency multiples.

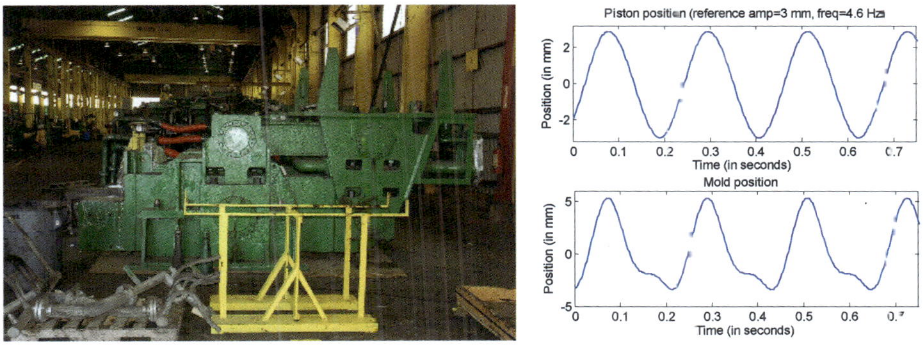

Fig. 8.5-2. Mold oscillator problem: clean sine driving input — distorted mold motion.

So, the distortion source was thought to be somewhere between the piston and the mold. However, the distortion origin was not found. <u>Solution:</u> apply magnitude spectrum as magnifying glass:

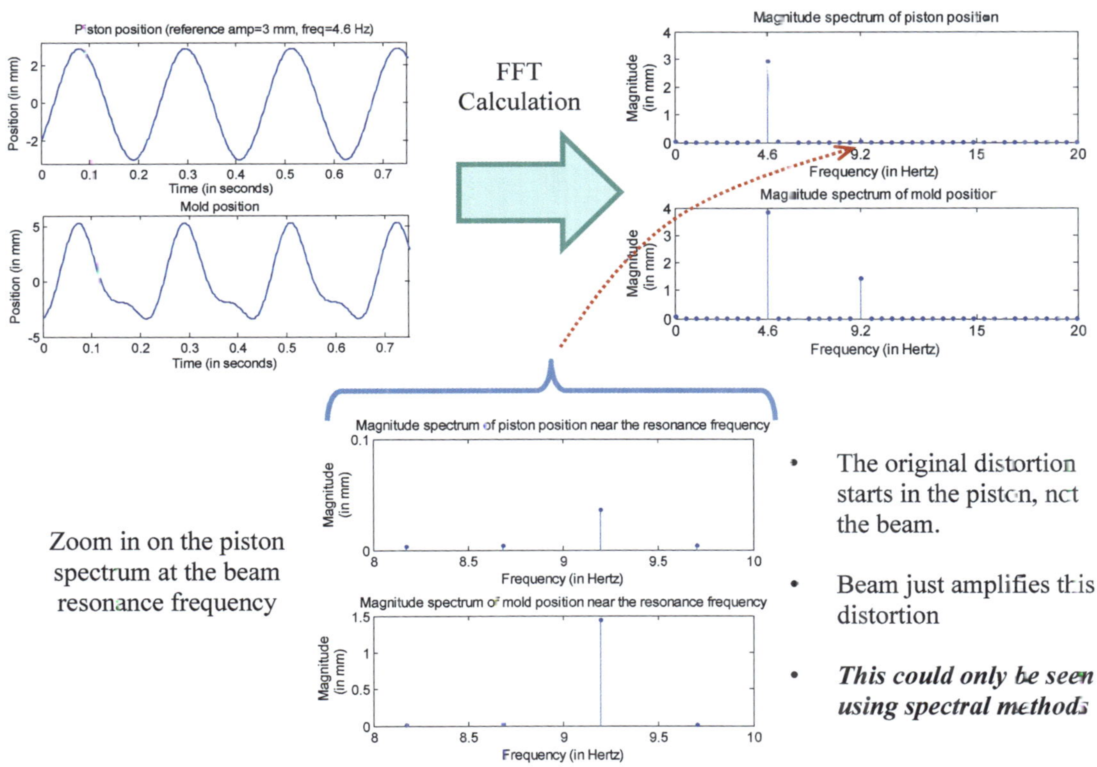

Fig. 8.5-3. Resonance problem: through zooming in on the piston magnitude spectrum at the beam resonance frequency, the distortion is seen to originate in the piston, not the beam.

Example 8.5-4: Mold oscillator hydraulics spectral analysis.

<u>Problem</u>: Zooming in on the piston magnitude spectrum at the beam resonance frequency looks

odd — why zoom there? Because of the hypothesis that the distortion due to beam resonance originates at the beam input, and the piston might generate harmonics at those frequencies!

This hypothesis, however, looked questionable, and the opinion was that the distortions were caused mainly by an imperfect pivot performance.

Tool and solution: No frequency location hypothesis needed, just analyzing piston signal: catching the culprit with a naked eye — using case iii) DT power spectral density plotted along f axis.

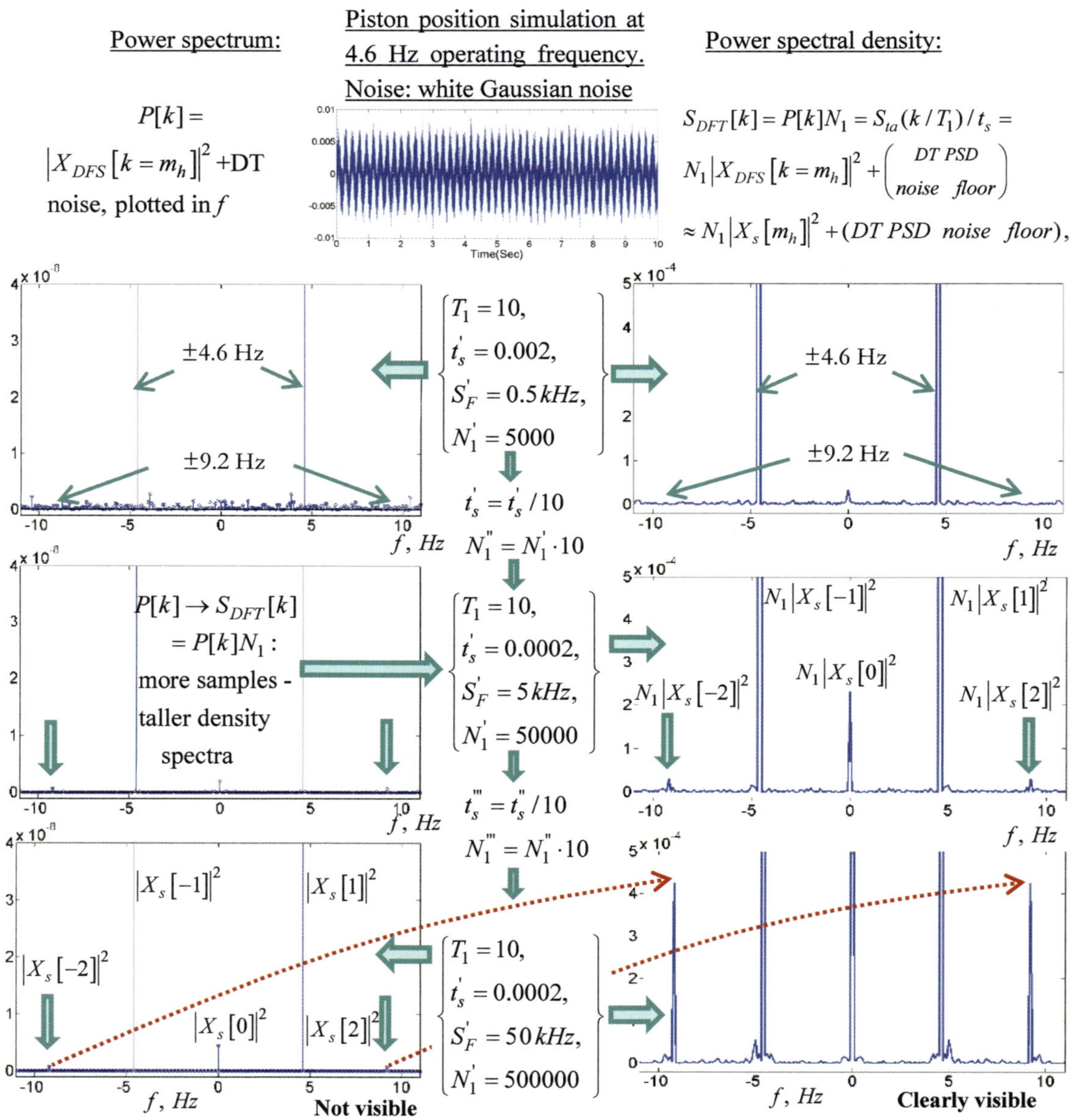

Fig. 8.5-4. Comparison of detection of small periodic components hidden in noise through power spectrum and power spectral density: fixed CT data record and increasing sampling rate. Plots on the right approximate graph $S_{ta}(f)/t_s$ vs f.

8.5.2 *FFT-based detection of hidden CT sinusoids — DFT ⇒ DFS ⇒ FS: summary*

Plotting DT MSD in analog frequency

Making aliasing/leakage as small as possible, take FFT of noisy N_1 samples data record $x_t[n]$, $1 \le n \le N_1$, (take small enough t_s and large enough N_1)

- plot magnitude of the resulting FFT in analog frequency
- this yields DT magnitude spectral density plotted in analog frequency, which for hidden in the noise sinusoids with fundamental frequency $f_{1h} = F_{1h}S_F = N_{1h}F_1S_F = N_{1h}f_1$ is:

$$\underbrace{|X_{DFT}[k]|}_{\text{DFT values}} = N_1|\underbrace{X_{DFS}[k = \underbrace{mN_{1h}}_{m_h}]}_{\text{Hidden DFS coefficients}}| + (\underbrace{DT\ MSD}_{\text{noise floor}}) \approx N_1|\underbrace{X_s[m_h]}_{\text{Hidden CT Fourier Series coefficients!}}| + (\underbrace{CT\ MSD}_{\text{noise floor}}),\ kf_1 \le f < (k+1)f_1$$

- if the noise is small, the small magnitudes $|X_s[m_h]|$ of the actual analog hidden sinusoids of the physical signal will show well due to multiplication by N_1.

Plotting DT PSD in analog frequency

- if the noise is relatively big, we can use the previous FFT or take FFT of DT periodic autocorrelation of the $x_{tp}[n]$ (periodically extended $x_t[n]$) to compute periodogram (DT PSD)

$$FFT\{r^{\circledast}_{x_{tp}x_{tp}}[n]\} = |X_{DFT}[k]|^2 / N_1 = S_{DFT}[k] = P[k]N_1 = N_1|X_{DFS}[k]|^2,\ kF_1 \le F < (k+1)F_1,$$

and plot the latter in analog frequency where it becomes

$$\underbrace{|X_{DFT}[k]|^2}_{\text{DFT values}} / N_1 = S_{DFT}[k] = S_{ta}(k/T_1)/t_s = P[k]N_1 = N_1|\underbrace{X_{DFS}[k = mN_{1h}]}_{\text{Hidden DFS coefficients}}|^2 +$$

$$(DT\,PSD\ \ noise\ \ floor) \approx N_1|\underbrace{X_s[m_h]}_{\text{Hidden CT Fourier Series coefficients!}}|^2 + (DT\,PSD\ \ noise\ \ floor),\ \ kf_1 \le f < (k+1)f_1$$

- for $|X_s[m_h]| << 1$, $|X_s[m_h]|^2$ might be too small, requiring much larger N_1
- the latter could be attained by ***strongly reducing t_s, e.g. by two orders of magnitude*** as in the mold oscillations example, and/or strongly increasing analog record length T_1.

Plotting DT MSD square in analog frequency

We can also try direct squaring of FFT magnitude to obtain another multiplication of power spectra $P[k]$ by N_1:

$$|X_{DFT}[k]|^2 = N_1S_{DFT}[k] = N_1S_{ta}(k/T_1)/t_s = P[k]N_1^2 =$$

$$N_1^2|X_{DFS}[k = mN_{1h}]|^2 \approx N_1^2|X_s[m]|^2 + N_1(DT\,PSD\ \ noise\ \ floor).$$

Here, the spikes are seen to grow faster, but with visibility no better than in the previous case.

There are techniques more refined than DT PSD (periodogram) for estimating CT PSD from a *single data record* for *large noise*, however, ***the DT PSD estimate of CT PSD can be improved if periodograms of several data records from the process of interest are averaged.***

What cannot be diagnosed through CT PSD? Nonperiodic fault precursors arising in processes like mixed two-phase flows. In these cases more advanced methods are needed (e.g. wavelets).

8.5.3 *System identification using PSD: from signal processing to control*

System identification is an important technique for generation of the data-based models that can be used for system design, simulation, diagnostics and health monitoring, and the model-based control, a glimpse of which is given in Chapter 9. System identification techniques discussed so far include:

1. Time-domain methods
   - Fitting curves or matching step or impulse response characteristics of the unknown system to those of the known one
2. Sine-sweep
   - Applying a sinusoid at system input, slowly varying the frequency, and measuring the amplitude and phase-shift of the output
     — Time consuming
     — Dangerous — may not know system gain at resonance frequency and break the system if input magnitude at resonance frequency is too big:

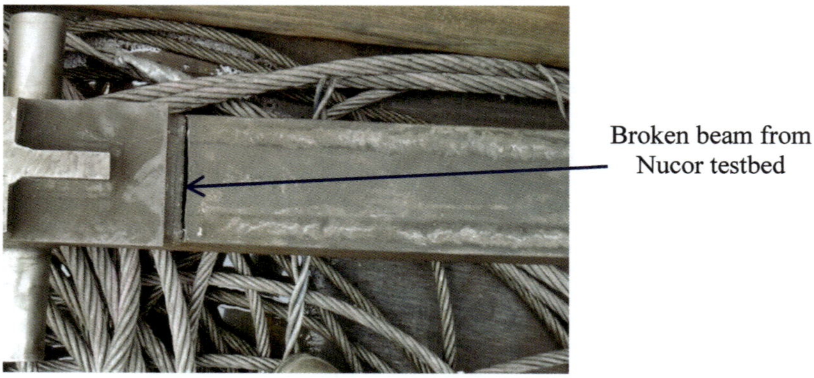

Fig. 8.5-5. Broken beam in the Nucor hardware testbed.

3. Fourier Transform of impulse response
   - Actual impulsive inputs are difficult to generate and apply in practice — they are approximated by a pulse, and not all systems easily admit pulse input.

Thus, it is of interest to find frequency response of a poorly damped system without exciting its resonance — an unavoidable swept-sine test feature. The mixed spectrum analysis suggests another possibility for finding a system's spectrum — through noise excitation.

8.5.4 *Filtering action of physical system*

Putting together the noise-based system identification technique relies on the fact established in applying periodic input to a physical LTI system: the *system output magnitude spectrum* is the *system input magnitude spectrum* multiplied by the *system magnitude spectral density*:

for periodic power signals $x_p(t)$ and $y_p(t)$:

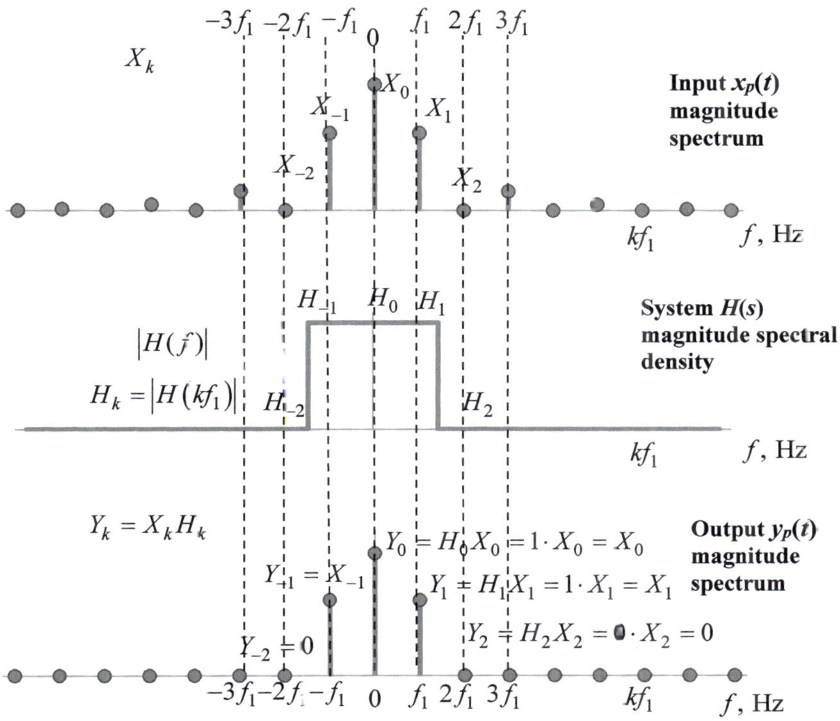

Fig. 8.5-6. Filtering action of a dynamic system for a power signal.

and for energy signals $x(t)$ and $y(t)$:

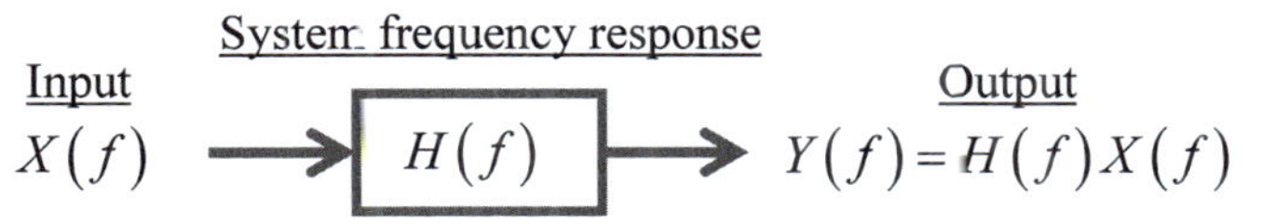

Fig. 8.5-7. Filtering action of an LTI dynamic system for an energy signal.

This means that $|Y(f)|^2 = |H(f)|^2 |X(f)|^2$.

Assume now $x(t)$ and $y(t)$ are nonperiodic power signals. Then, recalling our definitions for PSD:

$$\overbrace{R_{yy}^{\circledast}(f)}^{S_y(f)} = \lim_{t_{tr}\to\infty} \frac{|Y_t(f)|^2}{t_{tr}} = \lim_{t_{tr}\to\infty} \frac{|H(f)|^2 |X_t(f)|^2}{t_{tr}} = |H(f)|^2 \lim_{t_{tr}\to\infty} \frac{|X_t(f)|^2}{t_{tr}} = |H(f)|^2 \overbrace{R_{xx}^{\circledast}(f)}^{S_x(f)}.$$

This actually doesn't follow directly from the previous statements, but it is true for well-posed systems.

Thus, for a nonperiodic power signal input $x(t)$ to a physical system $H(s)$ ***the power spectral density $S_y(f)$ of the output $y(t)$ is the product of the power spectral density $S_x(f)$ of the input $x(t)$ and the energy spectral density $|H(f)|^2$ of the system $H(s)$:***

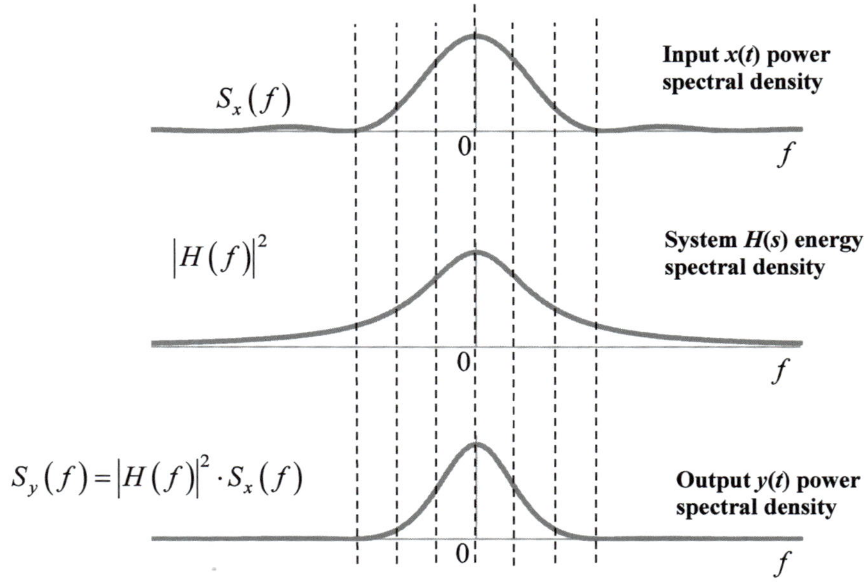

Fig. 8.5-8. Power spectral density propagation through LTI system.

What would happen if the input power spectral density is 1? The output *power spectral density* of the output becomes numerically equal to the *energy spectral density* of the system!

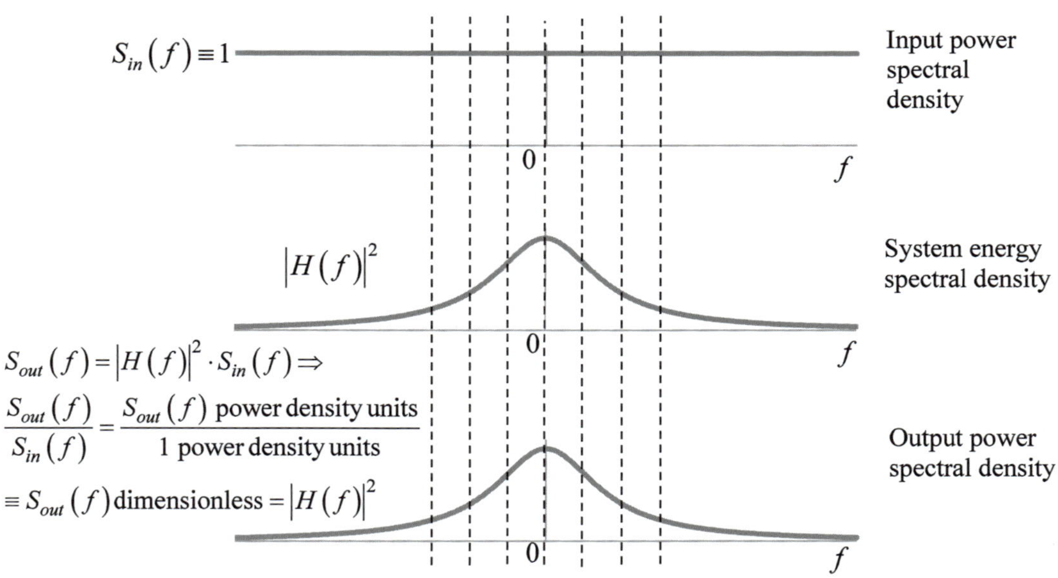

Fig. 8.5-9. System energy spectral density recreated as the output power spectral density.

Noise Excitation

The relation established above suggests a possible nonparametric method for identifying systems:

— excite a system with white noise (Gaussian, or normally-distributed random noise)

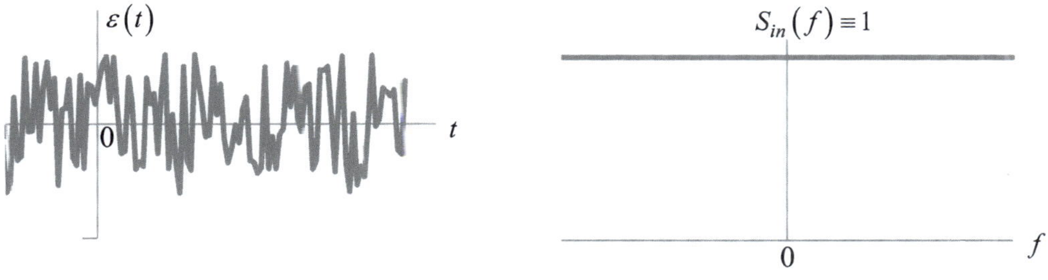

Fig. 8.5-10. White noise and its PSD.

— take FFT of the sampled output and compute CT PSD approximation, or

— compute periodic autocorrelation of the output data record and take its FFT.

At the outset it is hard to believe that the technique "input the noise, get the system model" actually works. To ascertain that it does, it can be tried out in the hardware lab or on the software testbed, as shown next.

Example 8.5-5: System identification using noise excitation.

Let's apply the techniques proposed to the state-space beam model of mold oscillator testbed:

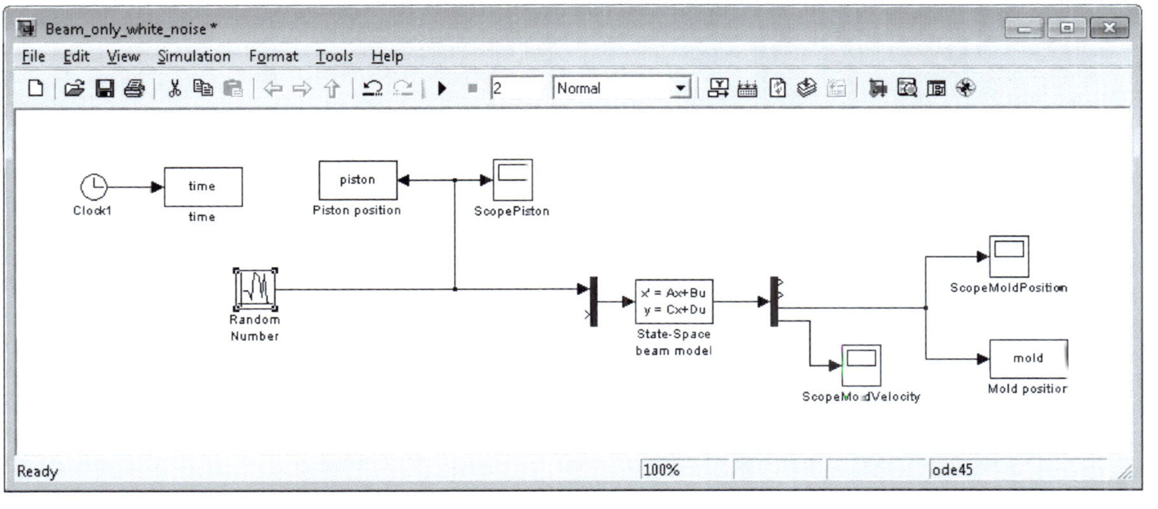

Fig. 8.5-11. Simulink state-space beam model with input noise excitation.

The resulting input and output are given by

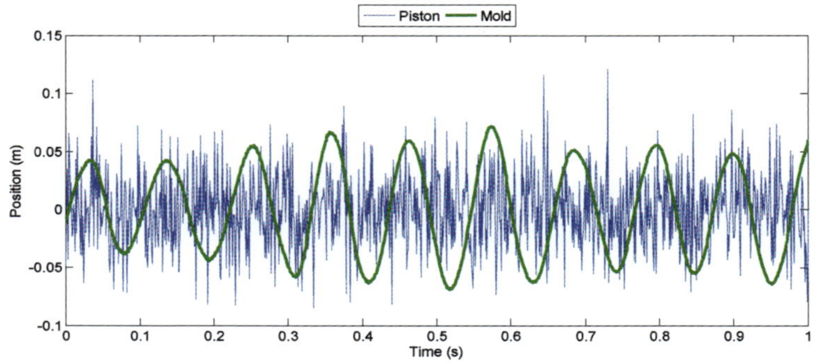

Fig. 8.5-12. Beam model system response to random signal generated by piston.

Calculated PSD averaged over 5 simulations has the form

Power spectral density of the output ≡ Energy spectral density of the system

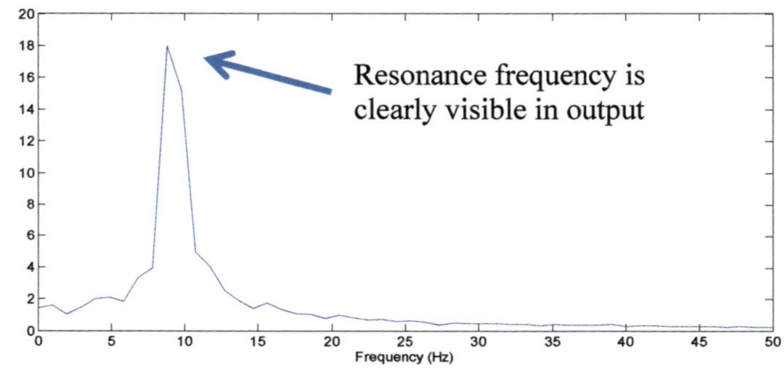

Fig. 8.5-13. PSD of Beam model system response.

Clearly showing the resonance peak and frequency without causing strong resonance excitations.

System identification using crosscorrelation

More generally, let $x(t)$ be an energy signal input to the system, $y(t)$ the output, and $h(t)$ the system impulse response. Then $y(t)=h(t)*x(t)$.

By linearity of the convolution

$$r_{yx}(t)=\underbrace{y(t)*x(-t)}_{\text{Alternative way of writing correlation}}=\underbrace{(h(t)*x(t))}_{\text{Solution of LTI system using impulse response}}*x(-t)=\underbrace{h(t)*(x(t)*x(-t))}_{\text{This is true for convolutions}}=h(t)*r_{xx}(t),$$

Therefore: $r_{yx}(t)=h(t)*r_{xx}(t)\overset{F}{\rightleftarrows}R_{yx}(f)=H(f)R_{xx}(f)\Rightarrow H(f)=\dfrac{R_{yx}(f)}{R_{xx}(f)}.$

This expression represents a non-parametric identification showing that a large class of LTI transfer functions can be estimated using an arbitrary input.

Welch's PSD calculation method. Averaging over 5 simulations to obtain the graph in Figure 8.5-13 is a simplified version of the Welch's PSD calculation method implemented in Matlab. This method uses averaging over short, modified periodograms.

Per Matlab description, **pxx = pwelch(x)** returns the power spectral density (PSD) estimate, pxx, of the input signal, x, found using Welch's overlapped segment averaging estimator.

By default, x is divided into the longest possible segments to obtain as close to, but not exceed, 8 segments with 50% overlap. Each segment is windowed with a Hamming window. The modified periodograms are averaged to obtain the PSD estimate. If the length of x cannot be divided exactly into an integer number of segments with 50% overlap, x is truncated accordingly.

The functions of MATLAB tools include:

- taking high-sampling-frequency data and running FFTs to uncover amplitude spikes that show the oscillating frequencies
- using windowing techniques and low-pass filtering to reduce noise components in very noisy signals
- computing the PSD of a signal using Welch's method, and using those power levels to indicate the presence of oscillatory phenomena.

Project 2: Solved Examples and Solution Guide

Project 2 includes analytical — Problems 1-6 — and computational — Examples P2-1 and P2-2 — parts.

<u>**Analytical part.**</u>

Problem 1: Let $x(t)=2\text{rect}(t+1)$ **and** $h(t)=2t\,\text{rect}(t-1)$**; a) Find the autocorrelation** $r_{xx}(t)$**; b) Find the autocorrelation** $r_{hh}(t)$**; c) Find the cross-correlation** $r_{hx}(t)$**; d) Find the cross-correlation** $r_{xh}(t)$**; (e) How are the results of parts (c) and (d) related?**

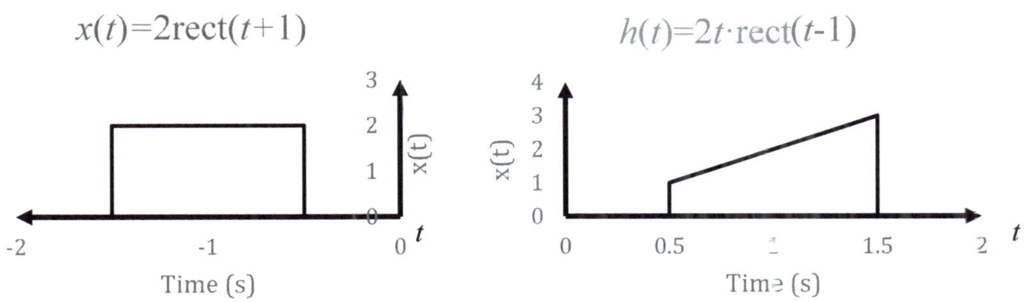

Fig. P2-1. Plots of $x(t)$ and $h(t)$.

a) Find the autocorrelation $r_{xx}(t)$. The definition of the autocorrelation is:

$$r_{xx}(t)=\int_{-\infty}^{\infty}x(\tau)x(\tau-t)d\tau,$$

which (just by substituting values into the formulas above) yields

$$x(\tau)=2\text{rect}(\tau+1) \text{ and } x(\tau-t)=2\text{rect}(\tau+1-t).$$

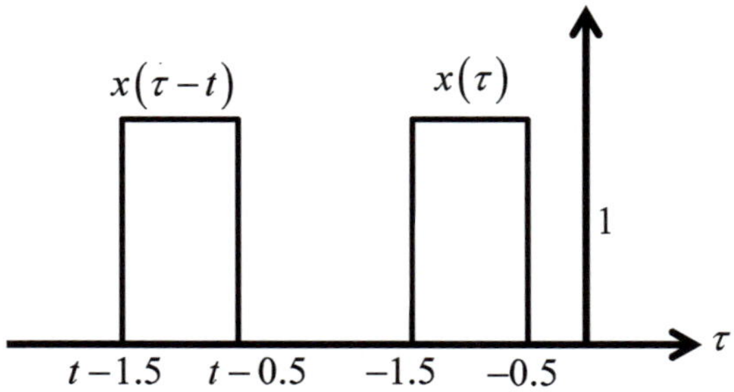

Fig. P2-2. $x(\tau)$ and $x(\tau-t)$.

From Figure P2-2, it is seen that as t increases, $x(\tau-t)$ moves to the right. Therefore, by observation, we can break the calculation into four different regions based on the value of t and use the corresponding limits of integration in each interval:

Region 1: prior to the overlap of two functions ($x(\tau-t)$ is to the left of $x(\tau)$):

$$t-0.5\leq-1.5\Rightarrow t\leq-1.$$

Since there is no overlap, $r_{xx}(t)=0$.

Region 2: $x(\tau-t)$ is moving to overlap $x(\tau)$ from the left:

$$-1.5\leq t-0.5\leq-0.5\Rightarrow-1\leq t\leq0,$$

$$\therefore r_{xx}(t)=\int_{-1.5}^{t-0.5}x(\tau)x(\tau-t)d\tau=\int_{-1.5}^{t-0.5}2\cdot2d\tau=4\tau\Big|_{-1.5}^{t-0.5}=4(1+t).$$

Region 3: $x(\tau-t)$ is moving to the right of $x(\tau)$, but there is still an overlap:

$$-1.5\leq t-1.5\leq-0.5\Rightarrow0\leq t\leq1,$$

$$\therefore r_{xx}(t)=\int_{t-1.5}^{-0.5}x(\tau)x(\tau-t)d\tau=\int_{t-1.5}^{-1.5}2\cdot2d\tau=4\tau\Big|_{t-1.5}^{-0.5}=4(1-t).$$

Region 4: $x(\tau-t)$ has moved completely past $x(\tau)$:

$$-0.5\leq t-1.5\Rightarrow1\leq t.$$

Since there is no overlap, $r_{xx}(t)=0$.

Assembling all the intervals together yields

$$r_{xx}(t)=\begin{cases}0, & t\le -1,\\ 4(1+t), & -1\le t\le 0,\\ 4(1-t), & 0\le t\le 1,\\ 0, & 1\le t,\end{cases}$$

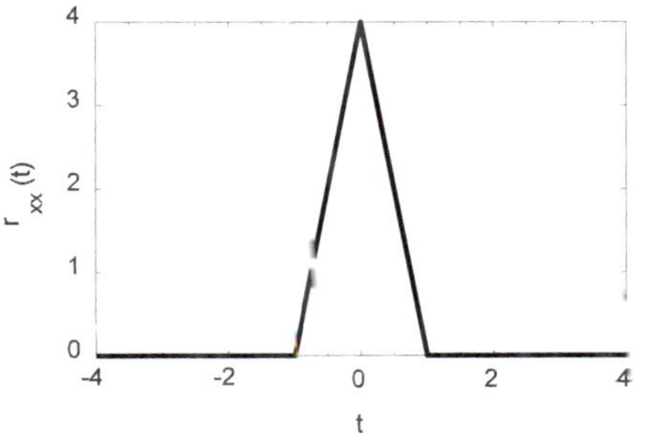

Fig. P2-3. Autocorrelation of $x(t)$.

b) Find the autocorrelation $r_{hh}(t)$.

The autocorrelation is

$$r_{hh}(t)=\int_{-\infty}^{\infty}h(\tau)h(\tau-t)d\tau$$

where

$$h(\tau)=2\tau\,\mathrm{rect}(\tau-1)\ \text{and}\ h(\tau-t)=2(\tau-t)\mathrm{rect}(\tau-t-1).$$

<u>Region 1</u>: prior to overlap

$$t+1.5\le 0.5\Rightarrow t\le -1,\ \text{so}\ r_{hh}(t)=0.$$

<u>Region 2</u>: $h(\tau-t)$ is moving to overlap $h(\tau)$ from the left:

$$0.5\le t+1.5\le 1.5\Rightarrow -1\le t\le 0,$$

$$\therefore r_{hh}(t)=\int_{0.5}^{t+1.5}h(\tau)h(\tau-t)d\tau=\int_{0.5}^{t+1.5}2\tau\cdot 2(\tau-t)d\tau=4\int_{0.5}^{t+1.5}\left(\tau^2-t\tau\right)d\tau=4\left(\frac{1}{3}\tau^3-\frac{1}{2}t\tau^2\right)\Bigg|_{0.5}^{t+1.5}$$

$$=\frac{4}{3}(t+1.5)^3-2t(t+1.5)^2-\frac{4}{3}(0.5)^3+2t(0.5)^2=\frac{4}{3}t^3+6t^2+9t+\frac{9}{2}-2t^3-6t^2-\frac{9}{2}t-\frac{1}{6}+\frac{1}{2}t$$

$$=-\frac{2}{3}t^3+5t+\frac{13}{3}.$$

<u>Region 3</u>: $h(\tau-t)$ is moving to the right of $h(\tau)$, but there is still an overlap:

$$0.5\le t+0.5\le 1.5\Rightarrow 0\le t\le 1,$$

$$\therefore r_{hh}(t) = \int_{t+0.5}^{1.5} h(\tau)h(\tau - t)\,d\tau = 4\left(\frac{1}{3}\tau^3 - \frac{1}{2}t\tau^2\right)\Bigg|_{t+0.5}^{1.5}$$

$$= \frac{4}{3}(1.5)^3 - 2t(1.5)^2 - \frac{4}{3}(t+0.5)^3 + 2t(t+0.5)^2 = \frac{9}{2} - \frac{9}{2}t - \frac{4}{3}t^3 - 2t^2 - t - \frac{1}{6} + 2t^3 + 2t^2 + \frac{1}{2}t$$

$$= \frac{2}{3}t^3 - 5t + \frac{13}{3}.$$

Region 4: $h(\tau - t)$ has moved completely past $h(\tau)$: $1.5 \le t + 0.5 \Rightarrow 1 \le t$.

There is no overlap, so $r_{hh}(t) = 0$. Assembling all the intervals together yields

$$r_{hh}(t) = \begin{cases} 0, & t \le -1, \\ -\frac{2}{3}t^3 + 5t + \frac{13}{3}, & -1 \le t \le 0, \\ \frac{2}{3}t^3 - 5t + \frac{13}{3}, & 0 \le t \le 1, \\ 0, & 1 \le t, \end{cases}$$

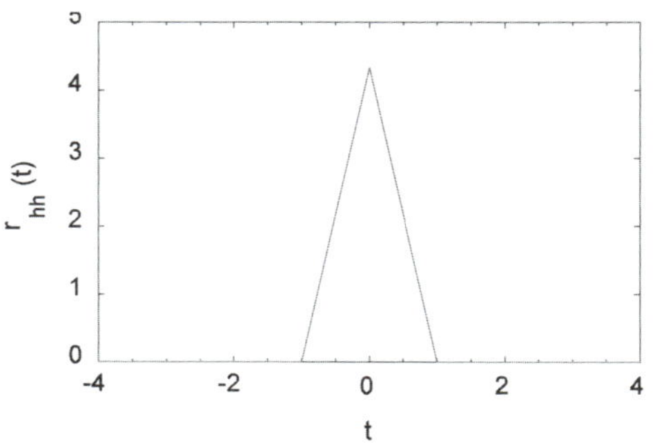

Fig. P2-4. Autocorrelation of h(t).

c) Find the crosscorrelation $r_{hx}(t)$. The cross-correlation formula is

$$r_{hx}(t) = \int_{-\infty}^{\infty} h(\tau)x(\tau - t)\,d\tau$$

where

$$h(\tau) = 2\tau\,\text{rect}(\tau - 1) \text{ and } x(\tau - t) = 2\,\text{rect}(\tau + 1 - t).$$

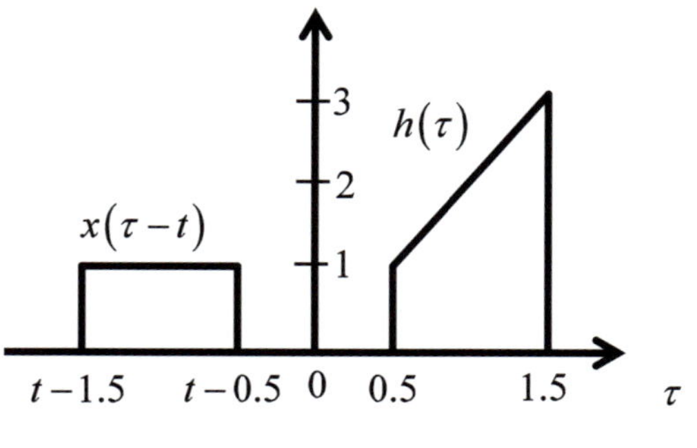

Fig. P2-5: $h(\tau)$ and $x(\tau - t)$.

From Figure P2-5 we can see that as t increases, $x(\tau - t)$ moves to the right. Therefore, by observation, we can break the calculation into four different regions based on the value of t and use the corresponding limits of integration in each interval:

Region 1: prior overlap

$$t - 0.5 \le 0.5 \Rightarrow t \le 1,$$

$$\therefore r_{hx}(t) = 0.$$

Region 2: $x(\tau - t)$ is overlapping $h(\tau)$ on the left:

$$0.5 \le t - 0.5 \le 1.5 \Rightarrow 1 \le t \le 2,$$

$$\begin{aligned}\therefore r_{hx}(t) &= \int_{0.5}^{t-0.5} h(\tau)x(\tau - t)d\tau = \int_{0.5}^{t-0.5} 2\tau \cdot 2d\tau = 4\int_{0.5}^{t-0.5} \tau d\tau = 2\tau^2\Big|_{0.5}^{t-0.5} \\ &= 2(t-0.5)^2 - 2(0.5)^2 = 2t^2 - 2t.\end{aligned}$$

Region 3: $x(\tau - t)$ is moving out of $h(\tau)$, still overlapping:

$$0.5 \le t - 1.5 \le 1.5 \Rightarrow 2 \le t \le 3,$$

$$\begin{aligned}\therefore r_{hx}(t) &= \int_{t-1.5}^{1.5} h(\tau)x(\tau - t)d\tau = \int_{t-1.5}^{1.5} 2\tau \cdot 2d\tau = 2\tau^2\Big|_{t-1.5}^{1.5} \\ &= 2(1.5)^2 - 2(t-1.5)^2 = -2t^2 + 6t.\end{aligned}$$

Region 4: $x(\tau - t)$ has moved completely past $h(\tau)$:

$$1.5 \le t - 1.5 \Rightarrow 3 \le t.$$

There is no overlap, so $r_{hx}(t) = 0$.

Assembling all the intervals together yields

$$r_{hx}(t) = \begin{cases} 0, & t \le 1, \\ 2t^2 - 2t, & 1 \le t \le 2, \\ -2t^2 + 6t, & 2 \le t \le 3, \\ 0, & 3 \le t, \end{cases}$$

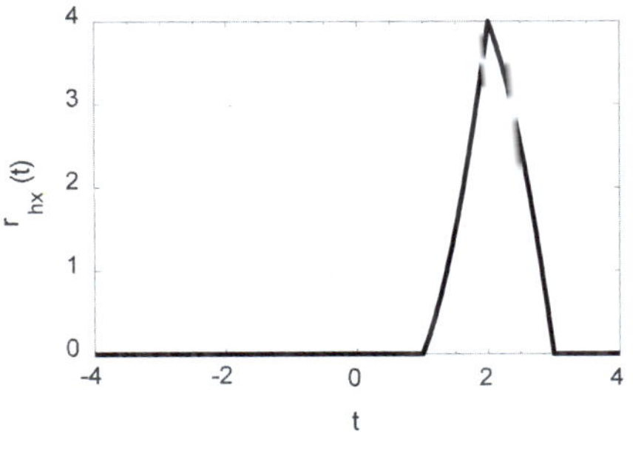

Fig. P2-6: Crosscorrelation of $r_{hx}(t)$.

d) Find the crosscorrelation $r_{xh}(t)$. The crosscorrelation formula is

$$r_{xh}(t)=\int_{-\infty}^{\infty}x(\tau)h(\tau-t)d\tau$$

where

$$x(\tau)=2\operatorname{rect}(\tau+1) \text{ and } h(\tau-t)=2(\tau-t)\operatorname{rect}(\tau-1-t).$$

Region 1: before overlap:

$$t+1.5\leq -1.5\Rightarrow t\leq -3.$$

There is no overlap, so $r_{xh}(t)=0$.

Region 2: $h(\tau-t)$ is overlapping $x(\tau)$ from the left:

$$-1.5\leq t+1.5\leq -0.5\Rightarrow -3\leq t\leq -2,$$

$$\begin{aligned}\therefore r_{xh}(t)&=\int_{-1.5}^{t+1.5}x(\tau)h(\tau-t)d\tau=\int_{-1.5}^{t+1.5}2\cdot 2(\tau-t)d\tau=4\int_{-1.5}^{t+1.5}(\tau-t)d\tau=2t^2-4t\tau\Big|_{-1.5}^{t+1.5}\\&=2(t+1.5)^2-4t(t+1.5)-2(-1.5)^2+4t(-1.5)=2t^2+6t+4.5-4t^2-6t-4.5-6t\\&=-2t^2-6t.\end{aligned}$$

Region 3: $h(\tau-t)$ is moving out of $x(\tau)$, still overlapping:

$$-1.5\leq t+0.5\leq -0.5\Rightarrow -2\leq t\leq -1,$$

$$\begin{aligned}\therefore r_{xh}(t)&=\int_{t+0.5}^{-0.5}x(\tau)h(\tau-t)d\tau=\int_{t+0.5}^{-0.5}2\cdot 2(\tau-t)d\tau=2t^2-4t\tau\Big|_{t+0.5}^{-0.5}\\&=2(-0.5)^2-4t(-0.5)-2(t+0.5)^2+4t(t+0.5)=0.5+2t-2t^2-2t-0.5+4t^2+2t\\&=2t^2+2t.\end{aligned}$$

Region 4: $h(\tau-t)$ has moved completely past $x(\tau)$:

$$-0.5\leq t+0.5\Rightarrow -1\leq t,$$

there is no overlap, so $r_{xh}(t)=0$.

Assembling all the intervals together yields

$$r_{hx}(t)=\begin{cases}0, & t\leq -3,\\ -2t^2-6t, & -3\leq t\leq -2,\\ 2t^2+2t, & -2\leq t\leq -1,\\ 0, & -1\leq t,\end{cases}$$

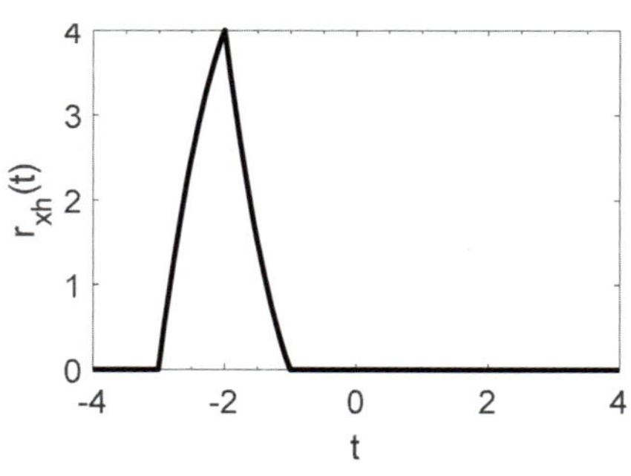

Fig. P2-7: Crosscorrelation of $r_{hx}(t)$.

(e) How are the results of parts (c) and (d) related?

The crosscorrelations are symmetric: $r_{xh}(t) = r_{hx}(-t)$.

Problem 2: Find the Fourier Transform $X(f)$ of each periodic signal $x(t)$ and sketch the magnitude spectrum $|X(f)|$ and phase spectrum $\angle X(f)$:
(a) $x(t) = 5 - 2\sin(10\pi t) + 3\sin(15\pi t)$; **(b)** $x(t) = 6\sum \delta(t - 3n)$; **(c)** $x(t) = |\cos(\pi t)|$.

(a) $x(t) = 5 - 2\sin(10\pi t) + 3\sin(15\pi t)$. The first step is to convert this function to an exponential Fourier series:

$$x(t) = 5 - 2\sin(10\pi t) + 3\sin(15\pi t) = 5 - 2\frac{1}{2j}\left(e^{j10\pi t} - e^{-j10\pi t}\right) + 3\frac{1}{2j}\left(e^{j15\pi t} - e^{-j15\pi t}\right)$$

$$= 5 + je^{j10\pi t} + (-j)e^{-j10\pi t} + \left(-\frac{3}{2}j\right)e^{j15\pi t} + \frac{3}{2}je^{-j15\pi t}$$

$$= 5e^{j0}e^{j2\pi 0t} + e^{j\frac{\pi}{2}}e^{j2\pi 5t} + e^{j\left(-\frac{\pi}{2}\right)}e^{-j2\pi 5t} + \frac{3}{2}e^{j\left(-\frac{\pi}{2}\right)}e^{j2\pi 7.5t} + \frac{3}{2}e^{j\frac{\pi}{2}}e^{-j2\pi 7.5t}.$$

Then we can take the Fourier transform of the latter function using the Fourier transform pair

$$e^{j2\pi kf_1 t} \xrightarrow{F} \delta(f - kf_1).$$

The fundamental frequency by inspection is $f_1 = 2.5$. The Fourier Transform and the spectra are

$$X(f) = 5e^{j0}\delta(f) + e^{j\frac{\pi}{2}}\delta(f-5) + e^{j\left(-\frac{\pi}{2}\right)}\delta(f+5) + \frac{3}{2}e^{j\left(-\frac{\pi}{2}\right)}\delta(f-7.5) + \frac{3}{2}e^{j\frac{\pi}{2}}\delta(f+7.5),$$

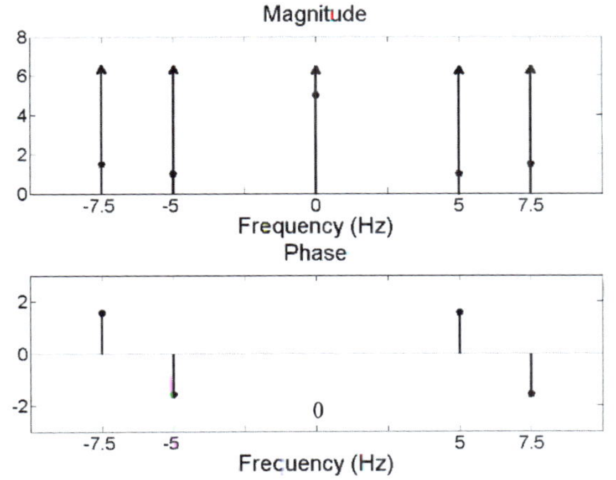

Fig. P2-8. Magnitude and phase spectrum of the signal given above.

(b) $x(t)=6\sum\delta(t-3n)$. The general Fourier Transform for this type of signal is

$$\sum\delta(t-Tn)\xrightarrow{F}\frac{1}{T}\sum\delta\left(t-\frac{k}{T}\right).$$

In this case $T=3$, so

$$X(f)=\frac{6}{3}\sum\delta\left(t-\frac{k}{3}\right)=2\sum\delta\left(t-\frac{k}{3}\right):$$

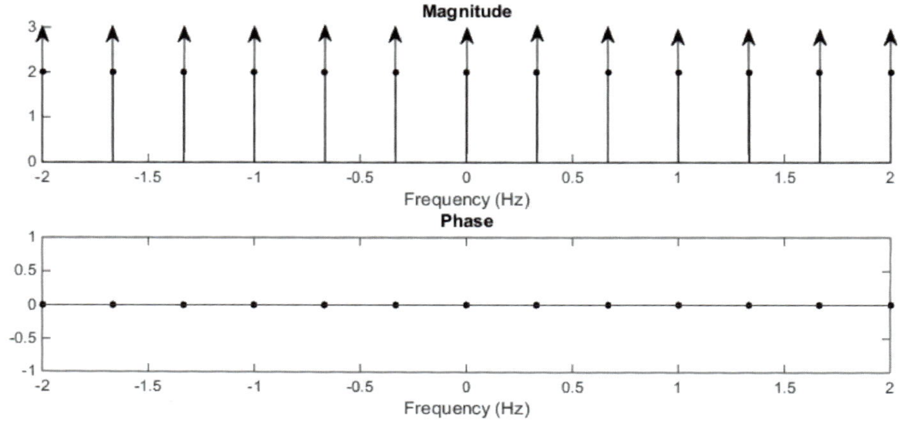

Fig. P2-9. Magnitude and phase spectrum of the signal given above.

(c) $x(t)=|\cos(\pi t)|$.

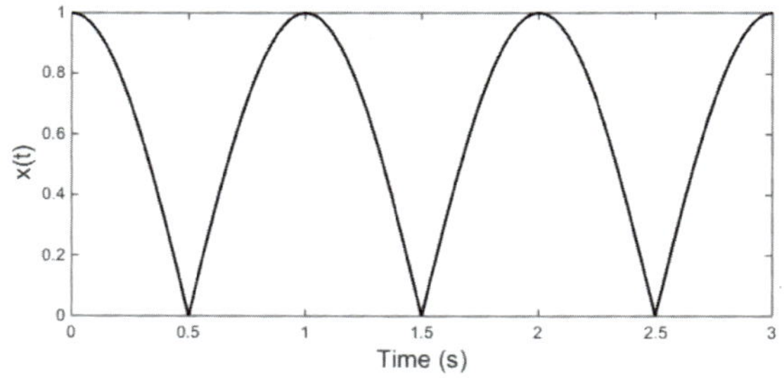

Fig. P2-10. Plot of function $x(t)=|\cos(\pi t)|$.

As in part (a), the easiest way to find the Fourier Transform is by finding the exponential Fourier Series. In this case, the fundamental frequency is 1, due to frequency doubling caused by the $|\bullet|$. Using the formula for the Fourier Series coefficients:

$$X_s[k]=\frac{1}{T_1}\int_0^{T_1} x(t)e^{-j2\pi k f_1 t}dt=\int_{-0.5}^{0.5}\cos(\pi t)e^{-j2\pi k t}dt=\int_{-0.5}^{0.5}\frac{1}{2}\left(e^{j\pi t}+e^{-j\pi t}\right)e^{-jk2\pi t}dt=$$

$$\frac{1}{2}\int_{-0.5}^{0.5}\left(e^{-j\pi(2k-1)t}+e^{-j\pi(2k+1)t}\right)dt=\left.\frac{1}{-j2\pi(2k-1)}e^{-j\pi(2k-1)t}+\frac{1}{-j2\pi(2k+1)}e^{-j\pi(2k+1)t}\right|_{-0.5}^{0.5}$$

$$=\left(\frac{1}{-j2\pi(2k-1)}e^{-j\pi k}e^{j0.5\pi}+\frac{1}{-j2\pi(2k+1)}e^{-j\pi k}e^{-j0.5\pi}\right)-\left(\frac{1}{-j2\pi(2k-1)}e^{j\pi k}e^{-j0.5\pi}+\right.$$

$$\left.\frac{1}{-j2\pi(2k+1)}e^{j\pi k}e^{j0.5\pi}\right)=\left(\frac{1}{-j2\pi(2k-1)}\cos(k\pi)j+\frac{1}{-j2\pi(2k+1)}\cos(k\pi)(-j)\right)-$$

$$\left(\frac{1}{-j2\pi(2k-1)}\cos(k\pi)(-j)+\frac{1}{-j2\pi(2k+1)}\cos(k\pi)j\right)=\left(\frac{1}{-2\pi(2k-1)}(-1)^k+\right.$$

$$\left.\frac{1}{2\pi(2k+1)}(-1)^k\right)-\left(\frac{1}{2\pi(2k-1)}(-1)^k+\frac{1}{-2\pi(2k+1)}(-1)^k\right)=2\left(-\frac{(-1)^k}{2\pi(2k-1)}+\right.$$

$$\left.\frac{(-1)^k}{2\pi(2k+1)}\right)=2\left(\frac{-(2k+1)}{2\pi(2k-1)(2k+1)}(-1)^k+\frac{(2k-1)}{2\pi(2k+1)(2k-1)}(-1)^k\right)$$

$$=2\left(\frac{(-1)^k(-2)}{2\pi(2k-1)(2k+1)}\right)=\frac{2}{\pi\left(1-4k^2\right)}(-1)^k.$$

Therefore,

$$x(t)=\sum_{k=-\infty}^{\infty}X_s[k]e^{j2\pi k f_1 t}=\frac{2}{\pi}\sum_{k=-\infty}^{\infty}\frac{1}{1-4k^2}(-1)^k e^{j2\pi k t}\ \Rightarrow\ X(f)=\frac{2}{\pi}\sum_{k=-\infty}^{\infty}\frac{(-1)^k}{1-4k^2}\delta(f-k).$$

Since the coefficients are negative (except for $k=0$) the phase could be either π or $-\pi$. The usual choice is to have $-\pi$ for $k>0$ and π for $k<0$.

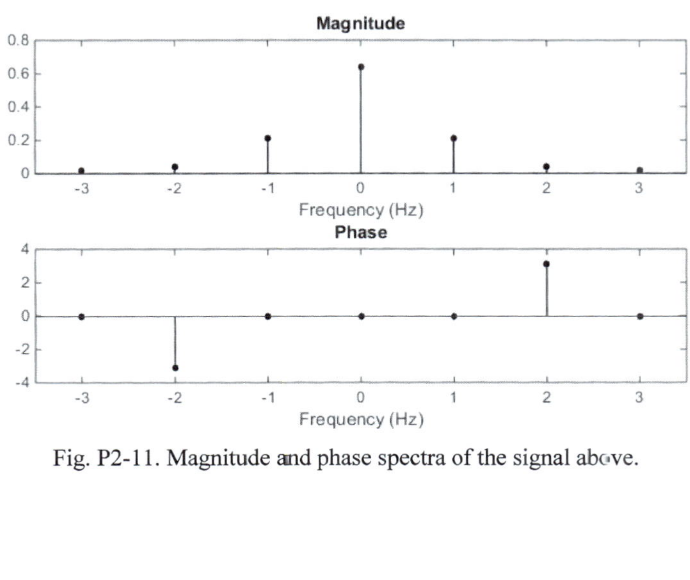

Fig. P2-11. Magnitude and phase spectra of the signal above.

Problem 3: Sketch the power spectral density (PSD) of:

$$x(t)=6-7\cos\left(6\pi t+\frac{\pi}{4}\right)+5\cos(10\pi t).$$

First, we write this as an exponential Fourier Series:

$$x(t)=6-7\cos\left(6\pi t+\frac{\pi}{4}\right)+5\cos(10\pi t)=6-\frac{7}{2}\left(e^{j\left(6\pi t+\frac{\pi}{4}\right)}+e^{-j\left(6\pi t+\frac{\pi}{4}\right)}\right)+\frac{5}{2}\left(e^{j10\pi t}+e^{-j10\pi t}\right)$$

$$=6-\frac{7}{2}e^{j\frac{\pi}{4}}e^{j2\pi 3t}-\frac{7}{2}e^{j\left(-\frac{\pi}{4}\right)}e^{j2\pi(-3)t}+\frac{5}{2}e^{j2\pi 5t}+\frac{5}{2}e^{j2\pi(-5)t}$$

$$=6e^{j2\pi 0t}+\frac{7}{2}e^{j\left(-\frac{3\pi}{4}\right)}e^{j2\pi 3t}+\frac{7}{2}e^{j\left(\frac{3\pi}{4}\right)}e^{j2\pi(-3)t}+\frac{5}{2}e^{j2\pi 5t}+\frac{5}{2}e^{j2\pi(-5)t}.$$

For a periodic function, the autocorrelation (periodic) is defined as

$$r^{\circledast}_{x_p x_p}(t)=\frac{1}{T_1}\int_0^{T_1}x_p(\tau)x_p(\tau-t)d\tau=\sum_{k=-\infty}^{\infty}\left|X_s[k]\right|^2 e^{j2k\pi f_1 t}.$$

In this case,

$$r^{\circledast}_{x_p x_p}(t)=36e^{j2\pi 0t}+\frac{49}{4}e^{j2\pi 3t}+\frac{49}{4}e^{j2\pi(-3)t}+\frac{25}{4}e^{j2\pi 5t}+\frac{25}{4}e^{j2\pi(-5)t}.$$

Finally, the PSD is the Fourier Transform of the autocorrelation, yielding

$$S(f)=36\delta(f)+\frac{49}{4}\delta(f-3)+\frac{49}{4}\delta(f+3)+\frac{25}{4}\delta(f-5)+\frac{25}{4}\delta(f+5).$$

This is an impulse train, so it is drawn as arrows pointing up to indicate an impulse, with markers to indicate the weight, as shown:

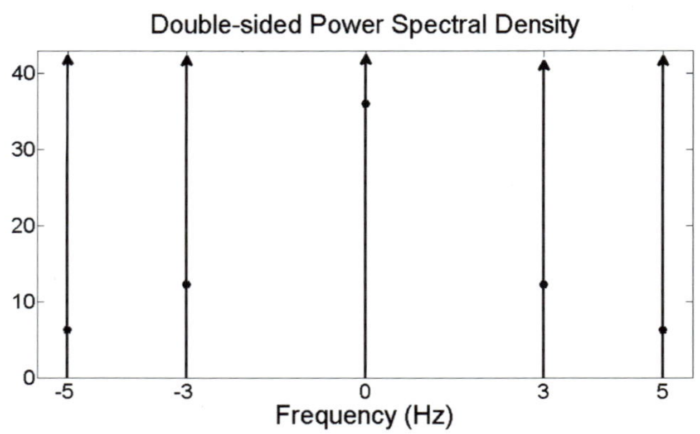

Fig. P2-12. Magnitude and phase spectrum.

Problem 4: Derive the autocorrelation function for the pulse train with $T_1 = 12T_p$ and draw it as a function of time including only the DC component and the first two harmonics. Derive and draw the power spectral density (PSD) for this pulse train.

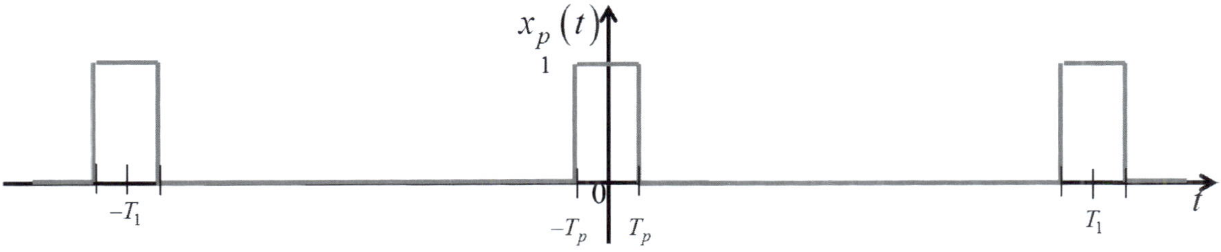

Fig. P2-13. Specified pulse train $x_p(t)$.

The exponential Fourier series pulse train representation derived in Chapter 4 has the form

$$x_p(t)=\sum_{k=-\infty}^{\infty}\frac{2T_p}{T_1}\text{sinc}\left(2k\frac{T_p}{T_1}\right)e^{j2\pi kf_1t}.\ \text{Setting } T_1=12T_p,\text{ yields } x_p(t)=\sum_{k=-\infty}^{\infty}\frac{1}{6}\text{sinc}\left(\frac{k}{6}\right)e^{j2\pi kf_1t}.$$

Including the DC component and the first 2 harmonics:

$$x_p(t)=\frac{1}{6}\text{sinc}\left(-\frac{2}{6}\right)e^{j2\pi(-2)f_1t}+\frac{1}{6}\text{sinc}\left(-\frac{1}{6}\right)e^{j2\pi(-1)f_1t}+\frac{1}{6}\text{sinc}\left(\frac{0}{6}\right)e^{j2\pi 0f_1t}$$
$$+\frac{1}{6}\text{sinc}\left(\frac{1}{6}\right)e^{j2\pi 1f_1t}+\frac{1}{6}\text{sinc}\left(\frac{2}{6}\right)e^{j2\pi 2f_1t}=\frac{1}{6}\text{sinc}\left(\frac{2}{6}\right)e^{j2\pi(-2)f_1t}+\frac{1}{6}\text{sinc}\left(\frac{1}{6}\right)e^{j2\pi(-1)f_1t}$$
$$+\frac{1}{6}e^{j2\pi 0f_1t}+\frac{1}{6}\text{sinc}\left(\frac{1}{6}\right)e^{j2\pi 1f_1t}+\frac{1}{6}\text{sinc}\left(\frac{2}{6}\right)e^{j2\pi 2f_1t}.$$

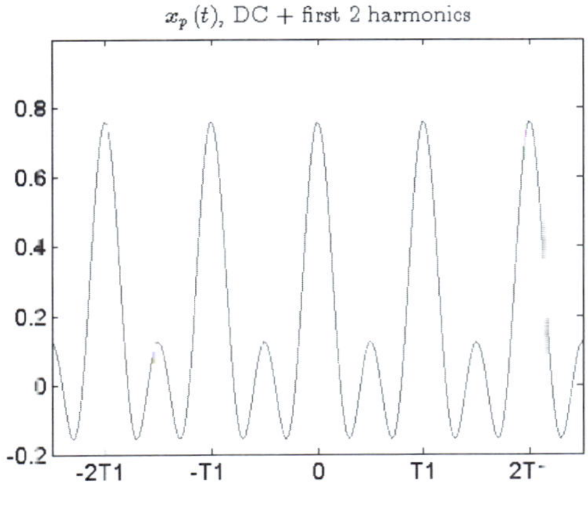

Fig. P2-14. Reconstructed $x_p(t)$ with the DC component and the first 2 harmonics.

As in Problem 3, we can immediately find the autocorrelation for an exponential Fourier Series.

$$x_p(t)=\sum_{k=-\infty}^{\infty} X_s[k]e^{j2\pi k f_1 t} \Rightarrow r^{\circledast}_{x_p x_p}(t)=\sum_{k=-\infty}^{\infty} \left|X_s[k]\right|^2 e^{j2k\pi f_1 t}.$$

For this problem,

$$r^{\circledast}_{x_p x_p}(t)=\frac{1}{36}\operatorname{sinc}^2\left(\frac{2}{6}\right)e^{j2\pi(-2)f_1 t}+\frac{1}{36}\operatorname{sinc}^2\left(\frac{1}{6}\right)e^{j2\pi(-1)f_1 t}+\frac{1}{36}e^{j2\pi 0 f_1 t}$$
$$+\frac{1}{36}\operatorname{sinc}^2\left(\frac{1}{6}\right)e^{j2\pi 1 f_1 t}+\frac{1}{36}\operatorname{sinc}^2\left(\frac{2}{6}\right)e^{j2\pi 2 f_1 t}:$$

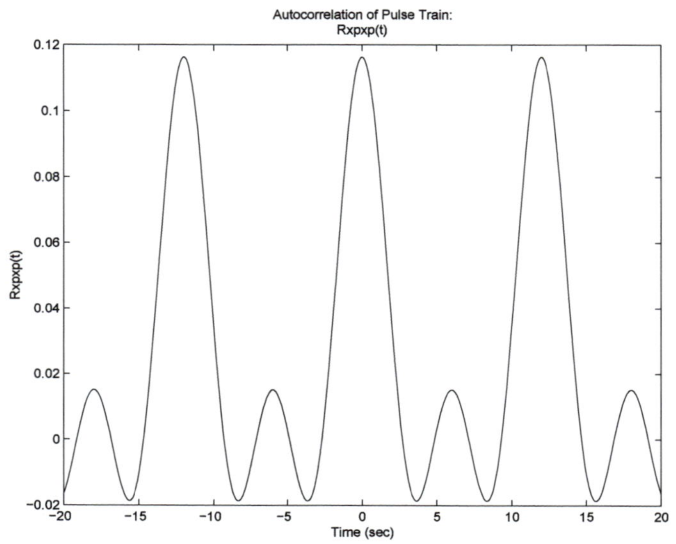

Fig. P2-15. Periodic autocorrelation of the signal in Figure P2-14 for $T_1 = 12$ sec.

The exact periodic autocorrelation of the pulse train in Figure P2-13 is found through cyclic computation or wraparound introduced in Example 8.3-1. However, due to large gaps between pulses, regular correlation over one period divided by the period value is sufficient. The resulting triangle pulse-train is shown in Figure P2-16, with the Figure P2-15 graph being its approximation.

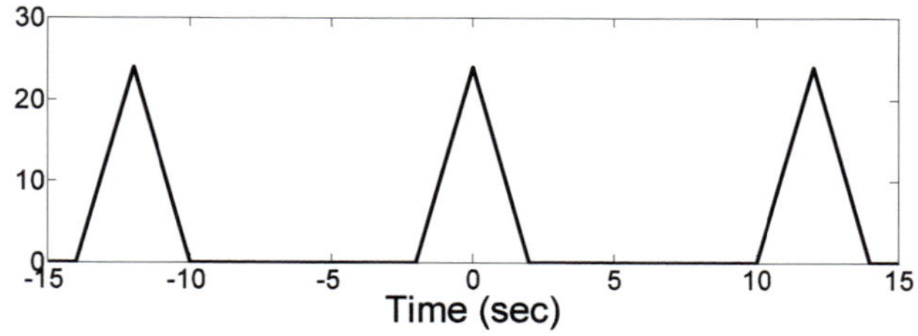

Fig. P2-16. Exact periodic autocorrelation of pulse train.

Finally, the PSD is the Fourier transform of the autocorrelation. Since the autocorrelation as written above is already in the form of an exponential Fourier series, the PSD takes the form:

$$r^{\circledast}_{x_px_p}(t)=\sum_{k=-\infty}^{\infty}\left|X_s[k]\right|^2 e^{j2k\pi f_1 t}=\frac{1}{36}\sum_{k=-\infty}^{\infty}\operatorname{sinc}^2\left(\frac{k}{6}\right)e^{j2\pi k f_1 t}\Rightarrow S(f)=\frac{1}{36}\sum_{k=-\infty}^{\infty}\operatorname{sinc}^2\left(\frac{k}{6}\right)\delta(f-kf_1):$$

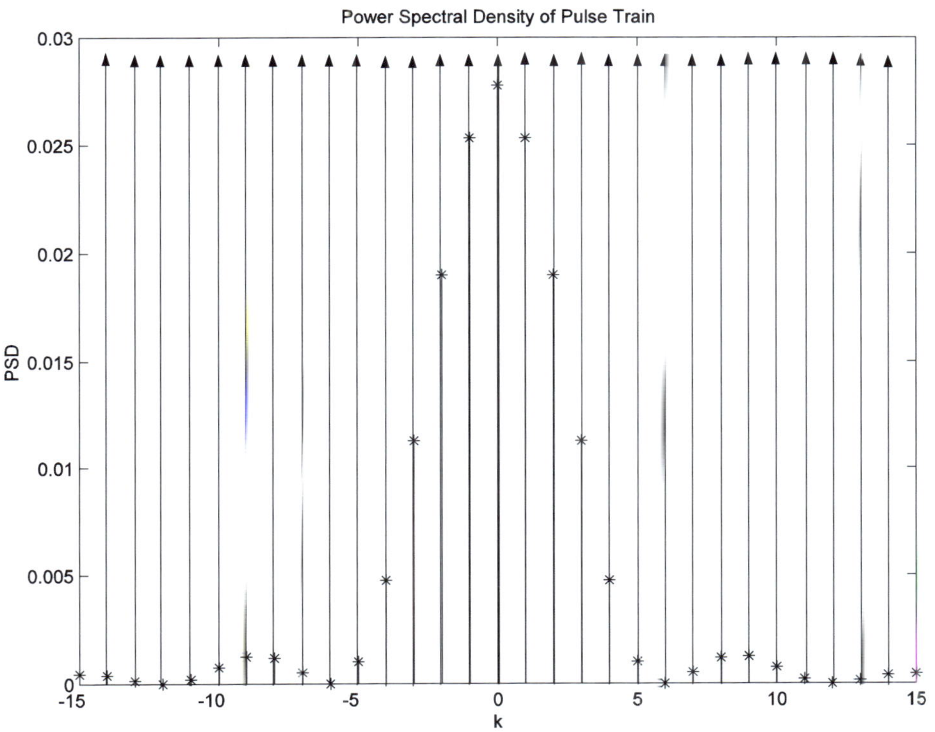

Fig. P2-17. PSD of the pulse train given in Figure P2-14.

Problem 5. Now, suppose the pulse train specified in Figure P2-14 is corrupted by white noise of intensity 2. Repeat the previous two tasks for the noisy pulse train.

Let's call the noisy signal $y(t)$:

$$y(t)=x_p(t)+2\varepsilon(t)$$

where $\varepsilon(t)$ is white noise of unit intensity. Since the noise has intensity 2, and the height of the pulse train is only 1, the graph of the original signal would actually be covered up by noise, as shown in Figure P2-18.

Taking the periodic autocorrelation:

$$r^{\circledast}_{yy}(t)=\left(x_p+2\varepsilon\right)\circledast\circledast\left(x_p+2\varepsilon\right)=r^{\circledast}_{x_px_p}(t)+2r^{\circledast}_{x_p\varepsilon}(t)+2r^{\circledast}_{\varepsilon x_p}(t)+2^2 r^{\circledast}_{\varepsilon\varepsilon}(t).$$

But, by definition, white noise is uncorrelated with any signal other than itself. Therefore $r^{\circledast}_{x_p\varepsilon}(t) = r^{\circledast}_{\varepsilon x_p}(t) = 0$ and $r^{\circledast}_{\varepsilon\varepsilon}(t) = \delta(t)$. What we're left with is

$$r^{\circledast}_{yy}(t) = r^{\circledast}_{x_p x_p}(t) + 2^2 r^{\circledast}_{\varepsilon\varepsilon}(t) = \frac{1}{36}\sum_{k=-\infty}^{\infty} \text{sinc}\left(\frac{k}{6}\right) e^{j2\pi k f_1 t} + 4\delta(t),$$

where $r^{\circledast}_{x_p x_p}(t)$ was calculated in Problem 4.

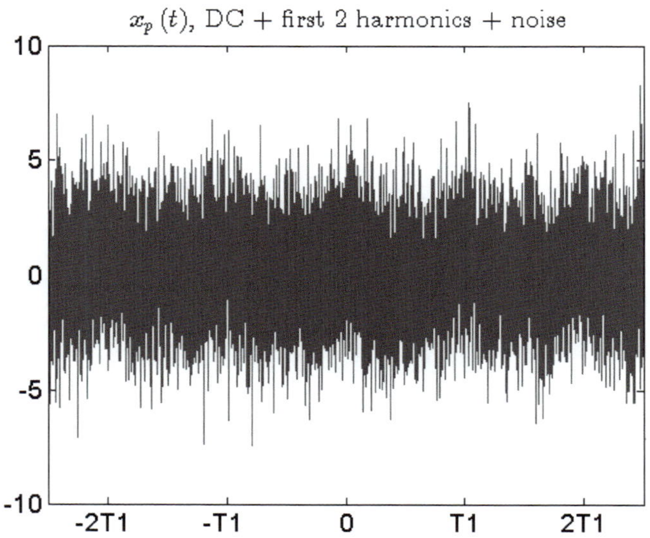

Fig. P2-18. Pulse train approximation corrupted by noise.

Now, we can find the PSD by taking the Fourier Transform:

$$S_y(f) = 4 + \frac{1}{36}\sum_{k=-\infty}^{\infty} \text{sinc}^2\left(\frac{k}{6}\right)\delta(f - kf_1).$$

The graph of this expression would have a noise floor at 4 with an impulse train. The weights of the impulses would all lie below the noise floor, but the spikes of the impulses would rise above the floor, allowing the periodic components to be seen in the PSD plot, as shown in Figure P2-19.

Problem 6. Using the concepts of truncation, bandlimiting, sampling, and periodic extension, explain computationally, in discrete time, how to detect periodic signal hidden in noise and how would the noisy signal look on the screen of the spectrum analyzer.

One way to find this experimentally is to sample the signal with the sampling rate t_s over a finite length of time, say, T_1 to get a finite, DT signal $x_t[n]$ with N_1 samples. Then, take the periodic extension of the data, $x_{tp}[n]$, compute the periodic autocorrelation $r^{\circledast}_{x_{tp}x_{tp}}[n]$ of the periodic signal using, for example, wraparound, and take the FFT of the result: $FFT\{r^{\circledast}_{x_{tp}x_{tp}}[n]\} = |X_{DFT}[k]|^2 / N_1$. Then plot the resulting periodogram wrt f to see where the spikes are in analog frequency:

$$|X_{DFT}[k]|^2 / N_1 = S_{ta}(k/T_1)/t_s \approx S(k/T_1)/t_s \approx N_1 |X_s[m_h]|^2 + (DT\ PSD\ \ noise\ \ floor), kf_1 \le f < (k+1)f_1 .$$

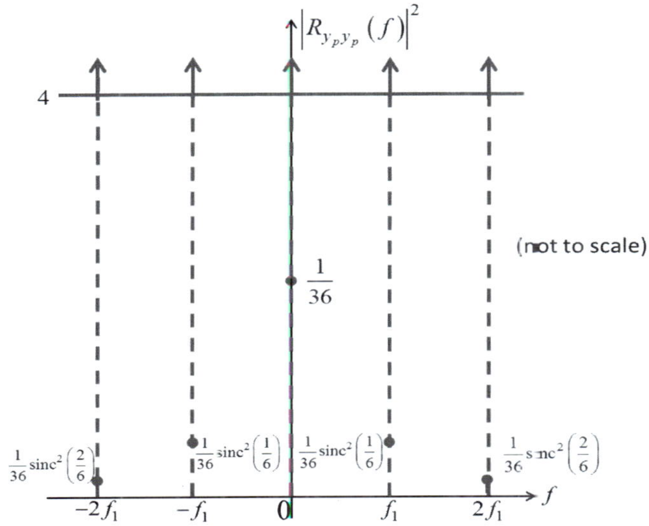

Fig. P2-19. PSD of $y(t)$ in Problem 5.

On a spectrum analyzer screen we would see a graph with spikes at each harmonic, where the height is approximately proportional to the power of the signal at that harmonic. If the signal is sampled at a sufficiently high rate, there will be spikes visible above the noise floor at the harmonic frequencies of the periodic function even for very small harmonic magnitudes.

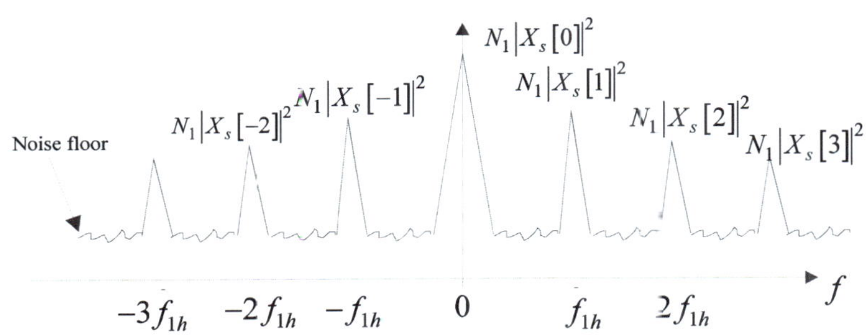

Fig. P2-20. DFT of periodic autocorrelation of noisy signal with periodic component.

<u>**Computational part**</u>

Example 1. Diagnostics of Harmonic Distortion in Hydraulic Actuator

a) Simulate software testbed under load, with 4.6 Hz 4 mm sinusoidal reference input. Obtain sampled data record, choosing sampling rate and duration to clearly display the distortion, if any.

First of all, when we simulate the testbed under load we need to set the 'Gravity constant' to be '1'. Then we can change the sampling time setting in the workspace of 'time' and 'piston'. Running for different sampling time values yields:

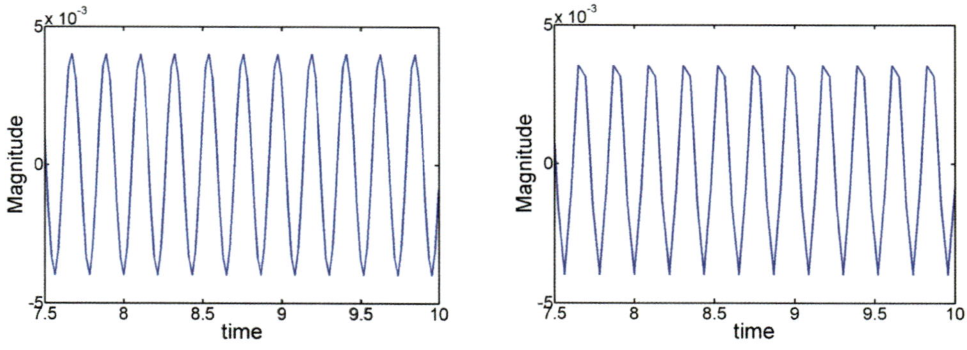

Fig. P2-21. Piston position for sampling times 1/46 (left) and 1/23 (right).

However, the distortion seen in Figure P2-21 in the right plot is not caused by the harmonic multiples arising due to piston nonlinearity, but rather by the low resolution due to small sampling frequency. Due to the wide magnitude scale separation between the driving frequency and its harmonic multiples, it is impossible to visually detect any distortion arising due to piston nonlinearity by plotting piston displacement vs time and visually examining the plots.

b) Compute periodic autocorrelation of the sampled data record (use Matlab xcorr). Taking FFT of this autocorrelation, compute DT power spectral density of the sampled data record and plot it on *f*, Hz, axis (the mixed CT1/DT1 case).

The sampling time used for this problem is 1/46. T1 (sec.) in the code below is a large enough analog data record length, chosen to minimize the leakage, as in Example 5.7-2 on p. 309.

Matlab code is as follows:

```
T1 = 10;
ts = 1/46;
N1 = T1/ts;
x = xcorr(piston); %Output of xcorr, give rxx[n] from n = −N1 to N1
```

$$r^{\circledast}_{x_p x_p}[n] \triangleq \frac{1}{N_1}\sum_{k=0}^{N_1-1} x_p[k]x_p[k-n]$$

```
%DT data is from 0 to N1.
%The following 3 lines are used to convert autocorrelation of non-periodic signal
%to periodic signal, details are shown in the above examples.
```

```
x2 = x((N1 + 1):(2*N1));
x1 = x(1:N1);
xnew = (x2 + x1)/N1; %cyclic correction divided by N1, see Example 8.2-2
%xnew is the 1 period of autocorrelation of xp(n)
%Treat xnew as our new data record, use fft command to get SDFT
SDFT = fft(xnew); %SDFT[k], is FFT of new data record
```

$$r^{\circledast}_{x_p x_p}[n] = \frac{1}{N_1}\sum_{k=0}^{N_1-1} x_p[k]x_p[k-n] \overset{DFT}{\rightleftharpoons} \frac{1}{N_1}X_{DFT}[k]X^*_{DFT}[k]$$

$$= \frac{1}{N_1}\left|X_{DFT}[k]\right|^2 \equiv S_{DFT}[k],\ kF_1 \le F < (k+1)F_1$$

```
%The following 3 lines are used to transfer from the single-sided to the two-sided graph
SDFT_positive = SDFT(2:N1);
SDFT_negative = flipud(SDFT_positive);
SDFT_ts = [SDFT_negative; SDFT];
%The following lines are used for plotting
k = -(N1-1):(N1-1);
f = k/T1; %f = k/T1
%plot(f,abs(STA),'LineWidth',4)
plot(f,abs(SDFT_ts),'LineWidth',4);
axis([-10,10,0,0.00001]);
ylabel('Magnitude','FontSize',50);
xlabel('frequency (Hz)','FontSize',50);
set(gca,'FontSize',40,'LineWidth',4);
```

The plot is shown in Figure P2-22. The DT power spectral density can be explicitly indicated on the vertical axis in this figure by replacing the word 'Magnitude' in the ylabel('Magnitude', 'FontSize', 50) Matlab display setting by any common abbreviation, such as 'DT PSD'.

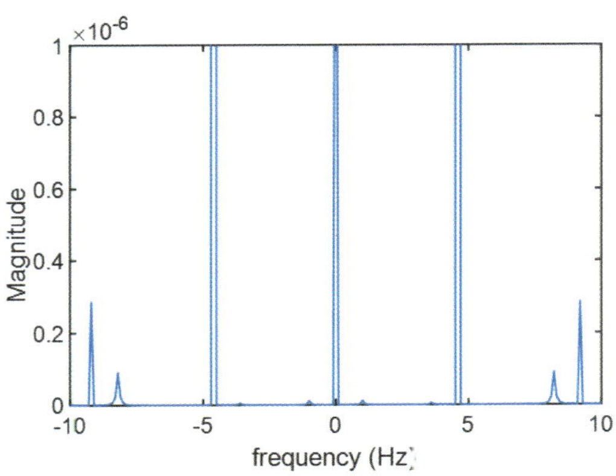

Fig. P2-22. DT power spectral density computed from periodic autocorrelation and FFT.

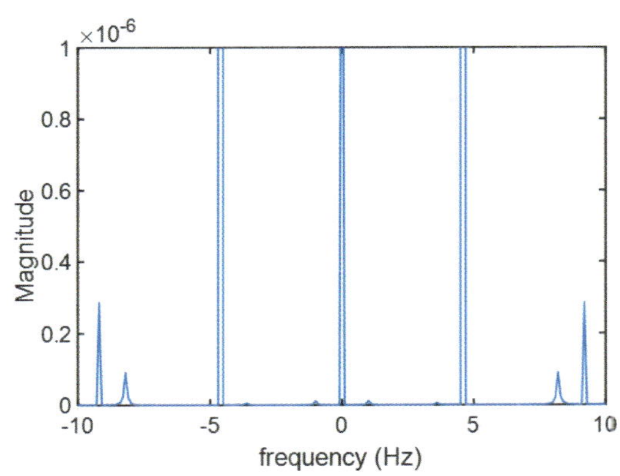

Fig. P2-23. DT power spectral density computed from the FFT of the data.

c) Compute FFT of the sampled data record directly. Using it compute DT power spectral density of the sampled data record and plot it on *f*, Hz, axis (the mixed CT1/DT1 case).

Matlab code is as follows:

```
T1 = 10;
ts = 1/46;
N1 = T1/ts;
piston1 = piston(1:N1); % take N1 data record
XDFT = fft(piston1); % Directly transfer from data record to DFT
SDFT = 1/N1*XDFT.*conj(XDFT);% From DFT to SDFT
%In the above line, we need to calculate the square of magnitude of the XDFT, since XDFT is a vector and '.*' is the command used in MATLAB, means that each number in the vector XDFT is multiplied to the number with the same index in the other vector and stored in SDFT.
%The following 3 lines are used to transfer from the single-sided to the two-sided graph
SDFT_positive = SDFT(2:N1);
SDFT_negative = flipud(SDFT_positive);
SDFT_ts = [SDFT_negative; SDFT];
%The following lines are used for plotting
k = -(N1-1):(N1-1);
f= k/T1; %f = k/T1
plot(f,(abs(SDFT_ts)),'LineWidth',4)
axis([-10,10,0,0.000001]);
ylabel('Magnitude','FontSize',50);
xlabel('frequency (Hz)','FontSize',50);
set(gca,'FontSize',40,'LineWidth',4);
```

The plot is shown in Figure P2-23 above and is seen to be identical to that in Figure P2-22.

Example 2. Identification of Nucor Mold Oscillator[1]

Problem definition

Consider the problem of continuous casting of steel where the mold executes a sinusoidal vertical motion of specific frequency and amplitude, imposed by a mold oscillation system as described in Chapter 1: the mold oscillation system consists of a subsystem of beams that supports a heavy mold (~25,000 Kg) at one end and is subject to sinusoidal motion by an electro-hydraulic servo actuator with piston attached to this subsystem at the other end, to drive the mold.

To carry out experiments, a testbed of the mold oscillation system was built at Nucor Steel, Decatur. The testbed has a hinged hollow beam that supports a heavy mass (~2500 Kgs), resembling a mold, on one end and has a servo actuator located at the other end.

The frequency response data of the beam is provided in Table 1. The corresponding frequency response plot is shown in Figure P2-24.

---

[1]Natarajan, Vivek; Bentsman, Joseph; Proceedings of American Control Conference (ACC) 2011, pp. 3289–3294, June 29–July 1, 2011.

Table 1.

Frequency (Hz)	Gain (output/input)	Phase (rad)
5.6	1.5331	−0.0201
5.8	1.5972	−0.0149
6	1.6701	−0.0132
6.2	1.7422	0.0161
6.4	1.841	0.0125
6.6	1.9547	0.0169
6.8	2.0843	0.0225
7	2.2456	0.0281
7.2	2.4334	0.034
7.4	2.6576	0.0418
7.6	2.9459	0.0461
7.8	3.3189	0.0449
8	3.8161	0.0436
8.2	4.5397	0.0418
8.4	5.6946	0.028
8.7	8.3371	−0.0639
9	24.6164	−1.3805
9.3	22.5136	−2.5025
9.6	10.1904	−2.885
9.9	5.86	−2.9632
10.2	3.9779	−2.989
10.5	2.9434	−2.9988
10.8	2.2853	−2.9996
11.1	1.8357	−2.9992
11.4	1.5123	−2.9799
11.7	1.2616	−2.9536
12	1.0654	−2.9026
12.3	0.958	−2.8354
12.6	0.8953	−2.7629
12.9	0.855	−2.6867

System Identification

Background.

A system is a process or device that takes an input, operates on such input, and produces an output. Systems can be dynamic or static. We are interested in finding a model of a dynamical system based on input-output data.

Mathematically, a dynamical system can be described, or modeled, as a function that maps an input to an output. Our goal now is to find such mathematical model. It will allow us to further study the properties of the system, usually with the purpose of understanding how to select the inputs such that the system produces the desired output, i.e. for its control.

In general, there are two ways to build this model:

- By means of the first principles, e.g. Newton's second law, in order to construct a set of differential equations.
- By means of the actual measured data, together with statistical methods. This approach is called System Identification and is extremely useful when the system's internal structure is unknown or too complex to describe with the first principles.

So, in a dynamical system, there exists a relationship between inputs and outputs and we will approximate this relationship through a model using actual measured data.

Identified Transfer Function

Using MATLAB's System Identification Toolbox, the following transfer function was found as the "best" match to the data above:

$$\frac{Y(s)}{U(s)} = \frac{5.307s + 2950}{s^2 + 1.484s + 3268}.$$

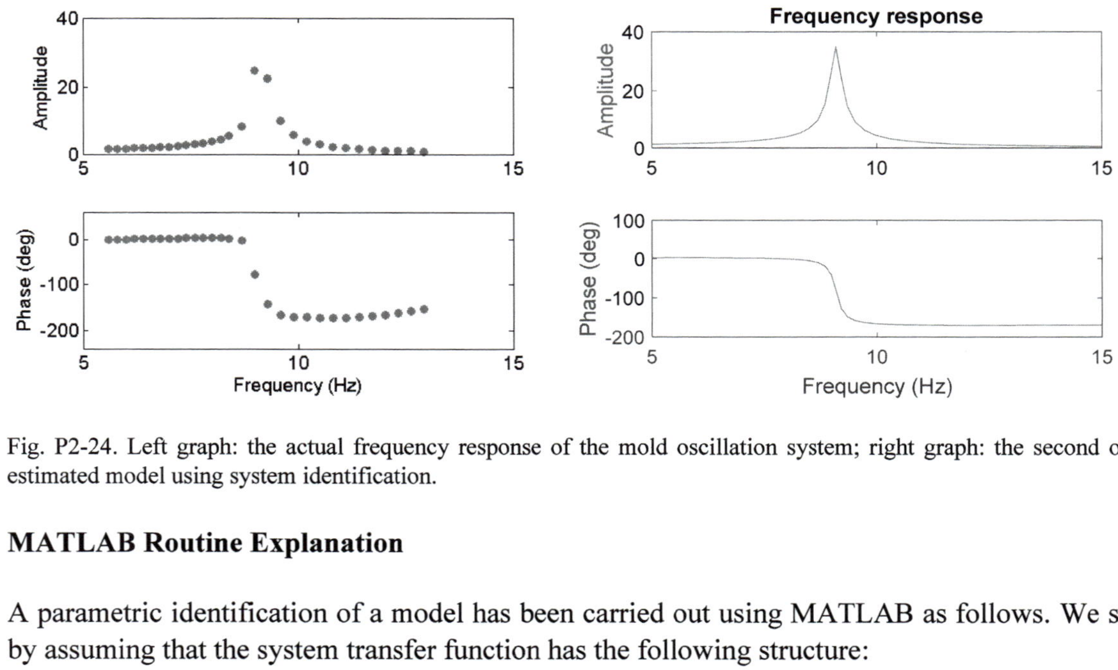

Fig. P2-24. Left graph: the actual frequency response of the mold oscillation system; right graph: the second order estimated model using system identification.

MATLAB Routine Explanation

A parametric identification of a model has been carried out using MATLAB as follows. We start by assuming that the system transfer function has the following structure:

$$H(s) = \frac{\alpha_m s^m + \alpha_{m-1} s^{m-1} + \ldots + \alpha_1 s + \alpha_0}{s^n + \beta_{n-1} s^{n-1} + \ldots + \beta_1 s + \beta_0}$$

where n is the number of poles and m is the number of zeros, both chosen by the user based on the system behavior, e.g., choosing $n = 1$ for a system with oscillatory step response is not a good idea — the system must then be non-oscillatory in its step response, i.e. rather very well damped. We would like to determine the coefficients β_i and α_i. This task represents a parameter identification problem:

finding the finite number of parameters that best characterize the system. One way to do carry this task out is through least-squares (LS) minimization of the sum of the squares of the error (computed, for example, using Levenberg-Marquardt algorithm introduced in Chapter 10):

$$S = \sum_{i=1}^{n} (f(x_i, \theta) - y_i)^2 \text{ by setting the gradient to zero: } \frac{\partial S}{\partial \theta_j} = 0 .$$

We will now perform data collection using the software testbed and identify the beam model.

System Identification Assignment

1. **Collect enough frequency response measurements (please, see the procedure description below) from the Simulink model of the testbed beam to identify a transfer function model.**
2. **Use the Matlab System Identification Toolbox to fit the 2nd order transfer function to the measured frequency response data. (Please, see the detailed instructions below.)**
3. **Compare your transfer function to the transfer function for the physical system given above, graphically by comparing the magnitude and phase responses, and analytically by comparing the poles and zeros. Try to explain any discrepancies.**

Data Collection Instructions

The frequency response will be found through using a set of oscillatory inputs with fixed frequencies, phases, and magnitudes, rather that the swept-sine and waiting until the magnitude and the phase in the response for each input no longer change, i.e. until the transients die out.

Find the frequency response of the Simulink model of the beam for the discrete set of frequencies: 5, 8, 8.4, 8.8, 9, 9.2, 9.4, 9.6, 10, 10.4, and 15 Hz.

It would be helpful to modify the Simulink model first to make finding the amplitude and the phase shift easier.

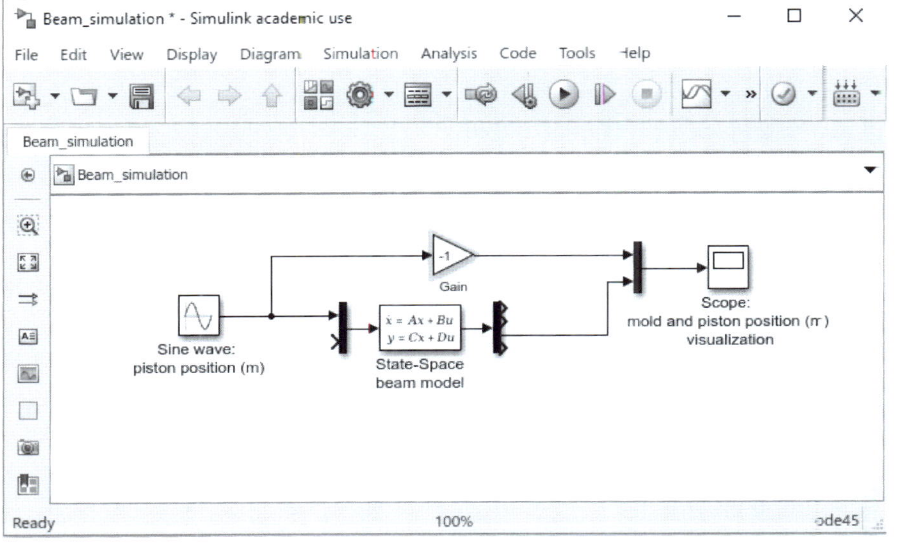

Fig. P2-25. Modified Beam simulation Simulink model.

- The gain block is included because when the piston end of the beam is up, the mold end is down and vice-versa. Adding the gain of -1 means that both ends move exactly in phase, rather than with the opposite phase, when the phase shift is 0.
- The "Mux" block (in the "Signal Routing" library) superimposes the two signals, so that they can be viewed on the same scope.

Once this is done, zooming in at the output on the scope and making sure to take the measurement after the initial transient has died down should yield:

- The amplitude of the input
- The amplitude of the output
- The peak time of the input
- The peak time of the output.

Using the measurements find:

$$\text{gain} = \frac{\text{amplitude of output}}{\text{amplitude of input}},$$

$$\text{phase (rad)} = \left(\text{peak time of input (s)} - \text{peak time of output (s)}\right) \cdot \left(\text{frequency (rad/s)}\right).$$

System Identification

Using the data points above, identify the testbed of the mold system by using MATLAB's System Identification Toolbox (**MATLAB R2021a**).

1. Input test data and generate plots
   a. Put the gain, the phase, and the frequency data into Matlab variables
   b. Convert the phase data to degrees
   c. Plot the data for later comparison. Remember to
      i. Label axes and units
      ii. Pick a font size that is readable
      iii. Use markers with no lines to indicate measurements
2. Start Matlab System Identification Toolbox
   a. Go to APPS → System Identification (Alternately, type "ident" in the command prompt.)
3. Import data to toolbox
   a. Select *Freq. Domain Data* under the *Import Data* pull-down menu.
   b. Select *FRD Amplitude and Phase* (in MATLAB R2021a) or *Freq. Func. (Amp/Phase)* (in the previous versions).
   c. Type the variable names you gave to the amplitude, phase (degrees), and frequency data in the respective entries. Remember to convert frequency data to radians if not already converted.
   d. Type 0.0001* for *Sample time*
      *Note: The sample time is not arbitrary and should be chosen to match the sampling rate of your data. This will differ between simulations, possibly even for different runs of the

same simulations. Make sure you calculate this value from the time vector generated by your simulation
  e. Click import
  f. Close window
4. Estimate a 2nd order linear parametric model
  a. Select *Polynomial Models...* under the *Estimate* pull-down menu.
  b. Change structure to *ARX*
  c. Type [2 2 1] under *Orders*
  d. Change the *Focus* to *Stability* using the pull-down menu
  e. Click estimate
  f. Close window
5. Plot estimated frequency response
  a. Select *Frequency resp* under *Model views*
  b. Select *Frequency (Hz)*, *Linear Frequency Scale*, *Linear Amplitude Scale* under *Style*.
  c. Select *Set Axes Limits* under *Options* to pick reasonable axes
  d. Click apply
  e. Choose *File* → *Copy figure* to copy the figure into a new window
  f. Adjust the size and font to be readable
6. Import model to workspace and obtain transfer function
  a. Drag the arx221 icon and release it over the *To workspace* icon.
  b. Return to workspace

Type: tf(d2c(arx221)) to see the fitted model in the transfer function format.

Exam 2: Solved Examples and Solution Guide

Problem 1: Consider a mass-spring-damper system (*m* = *1*, *b* = *0.1*, *k* = *1*) with force input.
a) Sketch the impulse response of the mass-spring-damper system. Consider mass displacement as output. Derive and sketch the magnitude response of this system and classify it as a filter.

System equation of motion: $m\ddot{x} + b\dot{x} + kx = F$.
Take the Laplace Transform of the system equation:
$ms^2X(s) + bsX(s) + kX(s) = F(s)$ where F is the input and X is the output. Then, substituting the parameter values, the system transfer function, the natural frequency, and the damping ratio are, respectively, given by

$$H(s) = \frac{X(s)}{F(s)} = \frac{1}{ms^2 + bs + k} = \left(\frac{1}{m}\right)\frac{1}{s^2 + \left(\frac{b}{m}\right)s + \left(\frac{k}{m}\right)} = \frac{1}{s^2 + 0.1s + 1}$$

$$\frac{1}{s^2 + 0.1s + 1} = \frac{\omega_n^2}{s^2 + 2\zeta\omega_n s + \omega_n^2} \Rightarrow \omega_n = 1;\ \zeta = 0.05$$

The system poles location in the open left half-plane of the s-plain at $s = -0.05 \pm j$ corresponds to stable, decaying, oscillatory response shown in Figure E2-1.

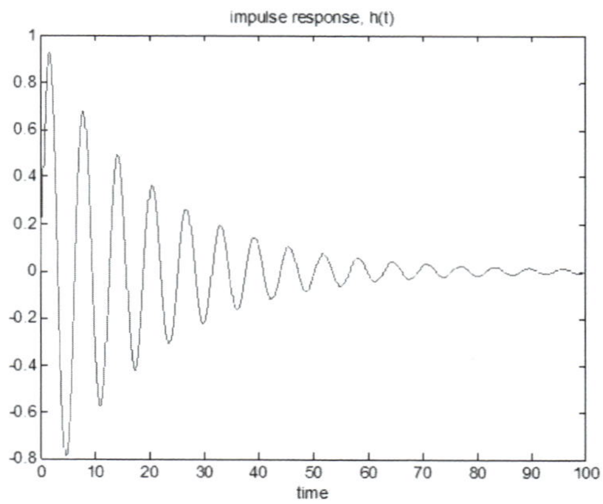

Fig. E2-1. System impulse response.

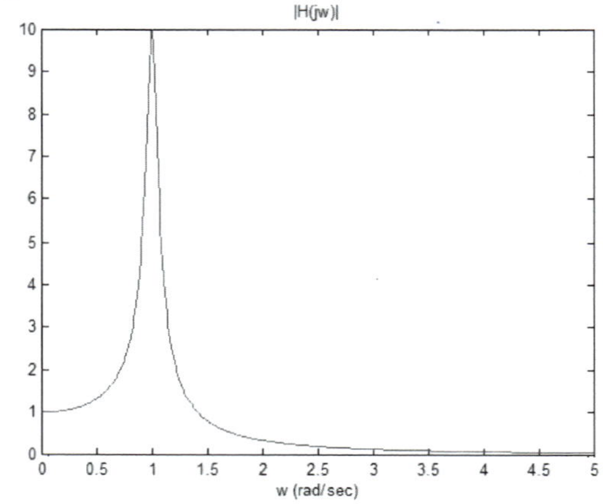

Fig. E2-2. System magnitude response.

System magnitude response is computed as follows:

$$H(s) = \frac{1}{s^2 + 0.1s + 1}\bigg|_{s=j\omega} = H(j\omega) = \frac{1}{j^2\omega^2 + 0.1j\omega + 1} = \frac{1}{(1-\omega^2) + j0.1\omega} \Rightarrow$$

$$\text{Magnitude: } |H(j\omega)| = \frac{1}{\sqrt{(1-\omega^2)^2 + (0.1\omega)^2}},$$

and is shown in Figure E2-2. The filter is seen to act as a bandpass filter.

b) Implement the equivalent filter using op-amps.

The op amp circuit drawn to implement

$$\ddot{x} = -0.1\dot{x} - x + F \text{ (from } \ddot{x} + 0.1\dot{x} + x = F\text{)}, \text{ input:} F, \text{ output: } x$$

is shown in Figure E2-3.

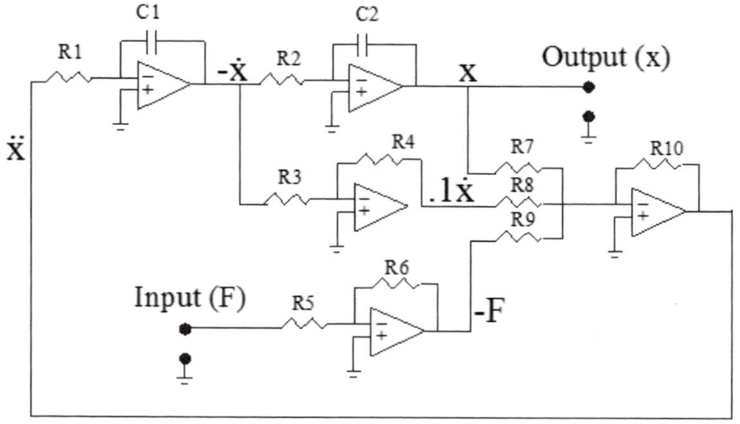

Fig. E2-3. Op amp circuits for MSP system.

Constraints on the passive components:

$$\bullet \ \frac{1}{R_1C_1} = \frac{1}{R_2C_2} = \frac{R_6}{R_5} = \frac{R_{10}}{R_7} = \frac{R_{10}}{R_8} = \frac{R_{10}}{R_9} = 1, \quad \bullet \ \frac{R_4}{R_3} = 0.1$$

c) Suppose that the impulse response of a mechanical system is an oscillatory decaying function that looks like $h(t) = exp(-5t)\, cos(4\pi t)$ and we can sample it using high quality data acquisition system until the response values decay to the noise level. Explain how to evaluate transfer function $H(j\omega)$ of this system in terms of DFT. Use the concept of the truncation $h_t(t)$ of $h(t)$, the periodic extension $h_{tp}(t)$ of the truncated signal and its bandlimiting, and the Fourier series and the Discrete Fourier series of the periodic extension.

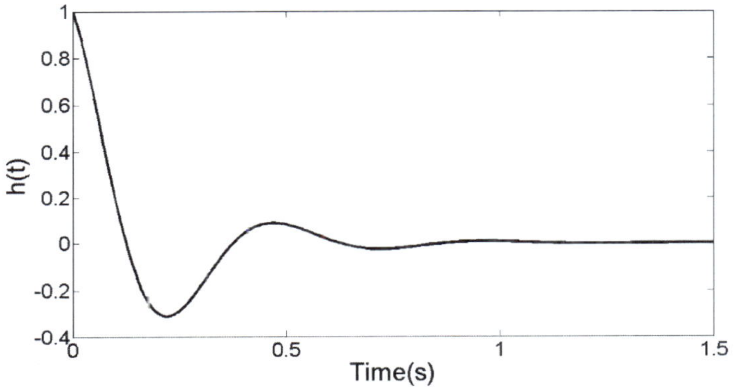

Fig. E2-4. The specified impulse response $h(t)$ of a MSD system.

First, truncate the original signal in order to obtain a finite data record. Call this signal $h_t(t)$. Then create a periodic extension of the truncated signal with some finite period T_1. Call this signal $h_{tp}(t)$. Then, this periodic signal can be represented by the Fourier series coefficients:

$$h_{tp}(t) \Leftrightarrow \underbrace{\{H_s[k]\}_{k=-m}^{m}}_{\text{Fourier series coefficients}} .$$

Assume now that the magnitudes of $H_s[k]$ decay fast at high frequencies, i.e. that the signal $h_{tp}(t)$ is close to being bandlimited.

If $h_{tp}(t)$ is sampled with negligible aliasing, then $h_{tp}(t) \Leftrightarrow \{H_s[k] \approx H_{DFS}[k]\}$, i.e., F.S. coefficients of $h_{tp}(t)$ and DFS coefficients of $h_{tp}[n]$ are almost equal.

Then, since $T_1 H_s[k] \approx H(j\omega)\Big|_{\omega = k\omega_1 = k\frac{2\pi}{T_1}}$, for $T_1 = t_s N_1$ and negligible aliasing,

$T_1 H_s[k] \approx \underbrace{t_s N}_{T_1} H_{DFS}[k] = t_s \underbrace{N_1 H_{DFS}[k]}_{H_{DFT}[k]}$, so that $H(j\omega)\big|_{\omega=k\omega_o} \approx t_s \cdot H_{DFT}[k]$ where $H_{DFT}[k]$ is computable from the data record.

Problem 2: Consider $x(t) = 2rect\left[\frac{(t-2)}{4}\right]$ and $h(t) = (t+6)rect\left[\frac{(t-1)}{2}\right]$.

(a) Draw these signals. Clearly label both axes and indicate signal values on the graph

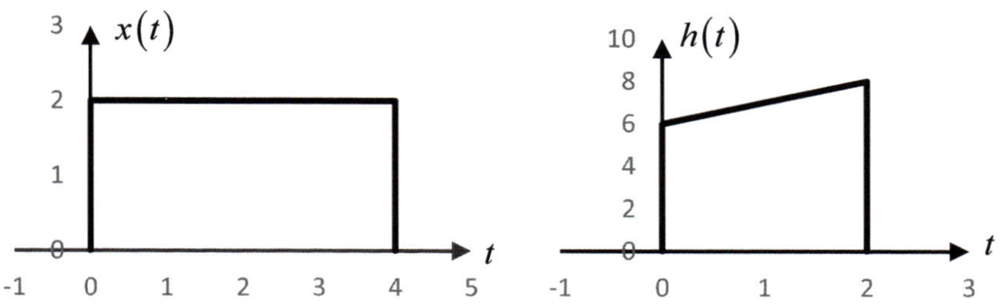

Fig. E2-5. Signals $h(t)$ and $x(t)$.

(b) Derive the analytical expression for the autocorrelation function $r_{xx}(t)$ and draw its graph.

Set

$$x(t) = 2rect\left[\frac{(t-2)}{4}\right], \quad x(\tau - t) = 2rect\left[\frac{(\tau - t - 2)}{4}\right].$$

According to the definition of autocorrelation, $r_{xx}(t) = \int_{-\infty}^{\infty} x(\tau)x(\tau - t)d\tau$, we break the computation the evaluation of the integral into four regions with the different expressions of the limits of integration:

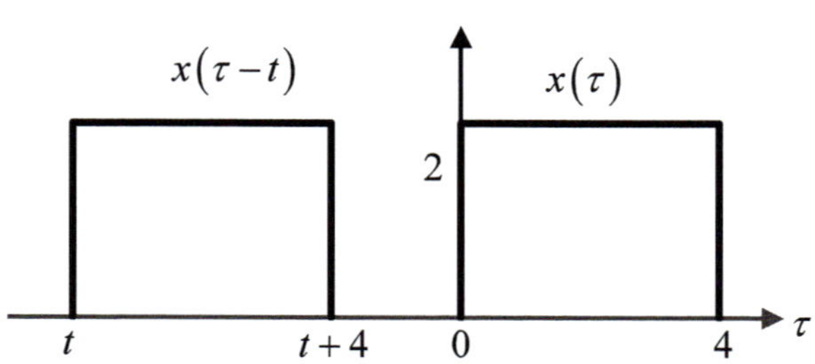

Fig. E2-6. Signal $x(\tau)$ and $x(\tau - t)$.

Region 1: $t+4<0 \Rightarrow t<-4$

$\therefore$ No overlap: $r_{xx}(t)=0$,

Region 2: $0 \le t+4 \le 4 \Rightarrow -4 \le t \le 0$

$\therefore$ $r_{xx}(t)=\int_0^{t+4} x(\lambda)x(\lambda-t)d\lambda = \int_0^{t+4} 2\cdot 2d\lambda = 4\lambda\Big|_0^{t+4} = 4t+16$,

Region 3: $0 \le t \le 4$

$\therefore$ $r_{xx}(t)=\int_t^{4} 2\cdot 2d\lambda = 4\lambda\Big|_t^{4} = 16-4t$,

Region 4: $t>4$

$\therefore$ No overlap: $r_{xx}(t)=0$. The full expression for $r_{hx}(t)$ and the corresponding graph are given below.

$$r_{xx}(t)=\begin{cases}0,\ t<-4,\\ 4t+16,\ -4\le t\le 0,\\ 16-4t,\ 0\le t\le 4,\\ 0,\ t>4,\end{cases}$$

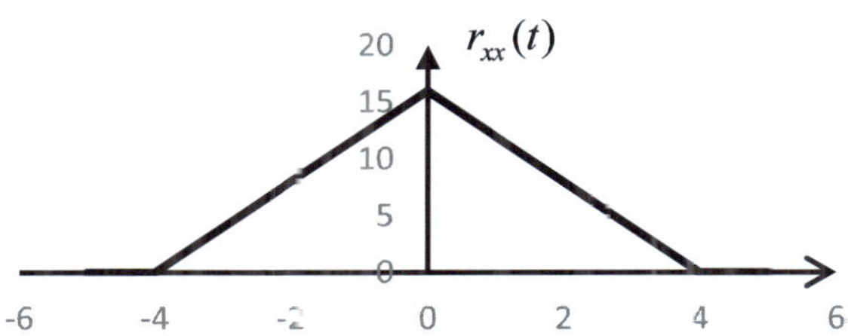

Fig. E2-7. Autocorrelation $r_{xx}(t)$.

(c) Derive the analytical expression for the cross-correlation function $r_{hx}(t)$ and draw its graph.

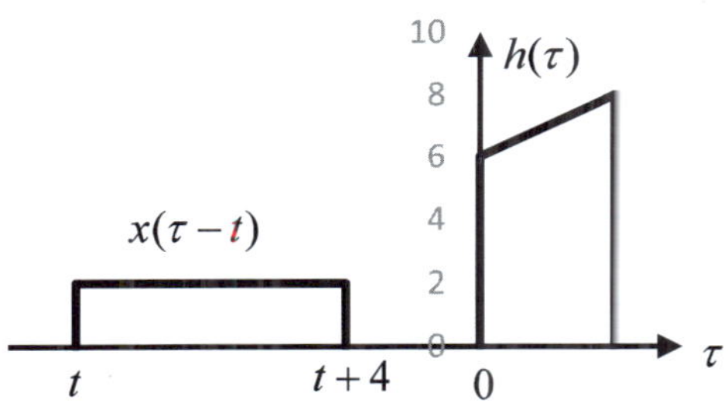

Fig. E2-8. Signals $h(\tau)$ and $x(\tau-t)$.

There are 5 regions over which we can evaluate the cross correlation function:

Region 1: $t+4<0 \Rightarrow t<-4$

$\therefore$ No overlap: $r_{hx}(t)=0$,

Region 2: $0 \le t+4 \le 2 \Rightarrow -4 \le t \le -2$

$\therefore$ $r_{hx}(t)=\int_0^{t+4} (\tau+6)\cdot 2d\tau = (\tau^2+12\tau)\Big|_0^{t+4} = t^2+20t+64$,

Region 3: $-2 \le t \le 0$

$\therefore$ $r_{hx}(t)=\int_0^{2} (\tau+6)\cdot 2d\tau = (\tau^2+12\tau)\Big|_0^{2} = 28$,

Region 4: $0 \le t \le 2$

$\therefore$ $r_{hx}(t)=\int_t^{2} (\lambda+6)\cdot 2d\lambda = (\lambda^2+12\lambda)\Big|_t^{2} = -t^2-12t+28$,

Region 5: $t > 2$

$\therefore$ No overlap: $r_{hx}(t) = 0$. The full expression for $r_{hx}(t)$ and the corresponding graph are given below.

$$r_{hx}(t) = \begin{cases} 0, t < -4, \\ t^2 + 20t + 64, \ -4 \le t \le -2, \\ 28, -2 \le t \le 0, \\ -t^2 - 12t + 28, \ 0 \le t \le 2, \\ 0, \ t > 2, \end{cases}$$

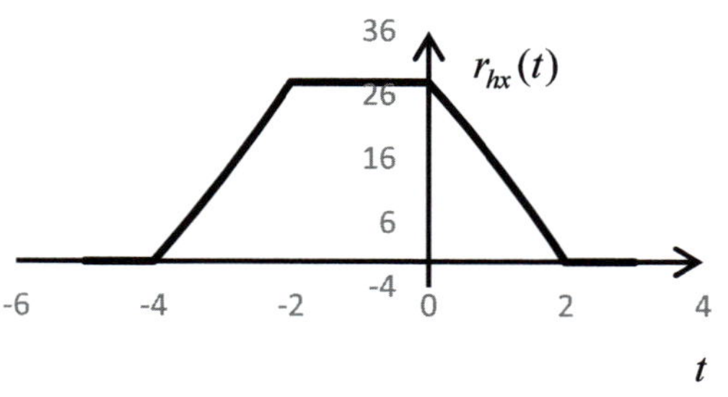

Fig. E2-9. Crosscorrelation $r_{hx}(t)$.

Problem 3: Consider a periodic signal: $x(t) = 4\cos(30\pi t) + 6\cos(60\pi t)$.

(a) Derive its Fourier transform

Expressing the signal in terms of the exponential Fourier series yields:

$$x(t) = 4\cos(30\pi t) + 6\cos(60\pi t) = 4\cos(2\pi(1)f_1 t) + 6\cos(2\pi(2)f_1 t) \Rightarrow f_1 = 15 \Rightarrow$$
$$= 2e^{j2\pi(-1)15t} + 2e^{j2\pi(1)15t} + 3e^{j2\pi(-2)15t} + 3e^{j2\pi(2)15t}.$$

Now, using Fourier transform pairs for complex harmonics, we obtain

$$\mathrm{e}^{-j2\pi\alpha t} \Leftrightarrow \delta(f+\alpha), \ \mathrm{e}^{j2\pi\alpha t} \Leftrightarrow \delta(f-\alpha), \ 1 \Leftrightarrow \delta(f)$$
$$\therefore X(f) = 2[\delta(f+15) + \delta(f-15)] + 3[\delta(f+30) + \delta(f-30)].$$

(b) Derive its periodic autocorrelation function $r_{xx}^{\circledast}(t)$.

$$r_{xx}^{\circledast}(t) = \sum_{k=-\infty}^{\infty} |X_s[k]|^2 e^{j2k\pi f_1 t} = (2)^2 e^{j2\pi(-1)f_1 t} + (2)^2 e^{j2\pi(1)f_1 t} + (3)^2 e^{j2\pi(-2)f_1 t} + (3)^2 e^{j2\pi(2)f_1 t}$$
$$= 4 \cdot 2\cos(2\pi 15t) + 9 \cdot 2\cos(2\pi 30t) = 8\cos(2\pi 15t) + 18\cos(2\pi 30t).$$

(c) Write the analytical expression for and draw the power density spectrum *S(f)* of this signal. Indicate the relation between the $r_{xx}^{\circledast}(t)$ **and *S(f)*.**

$S(f)$ is the Fourier Transform of the periodic autocorrelation $r_{xx}^{\circledast}(t)$ of $x(t)$:

$$r_{xx}^{\circledast}(t) \overset{Fourier\ Transform}{\Leftrightarrow} S(f) \ \Rightarrow \ S(f) = (2)^2[\delta(f+15) + \delta(f-15)] + (3)^2[\delta(f+30) + \delta(f-30)]:$$

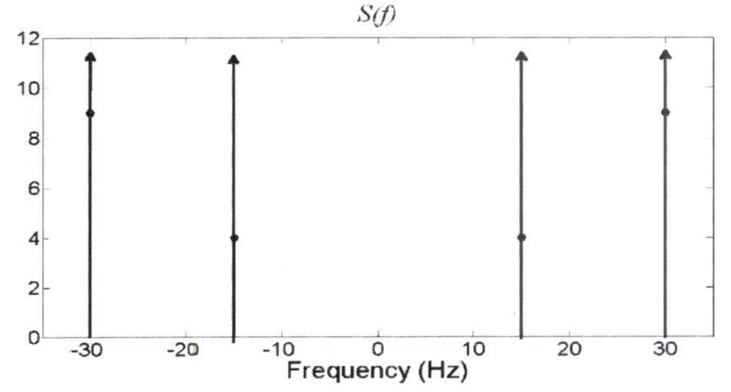

Fig. E2-10. PSD of signal $x(t)$.

(d) Consider now signal $x(t)$ corrupted by white noise: $y(t) = x(t) + 2\varepsilon(t)$. Give analytical expressions for the periodic autocorrelation function and the power spectral density of $y(t)$. Draw the periodic autocorrelation function and the power spectral density of $y(t)$.

$$r_{yy}^{\circledast}(t) = r_{xx}^{\circledast}(t) + r_{\sigma\varepsilon\sigma\varepsilon}^{\circledast}(t) = r_{xx}^{\circledast}(t) + (2)^2\delta(t) = 8\cos(30\pi t) + 18\cos(60\pi t) + 4\delta(t): \text{Figure E2-11.}$$

$$S(f) = 4 + 4[\delta(f+15) + \delta(f-15)] + 9[\delta(f+30) + \delta(f-30)]: \text{Figure E2-12.}$$

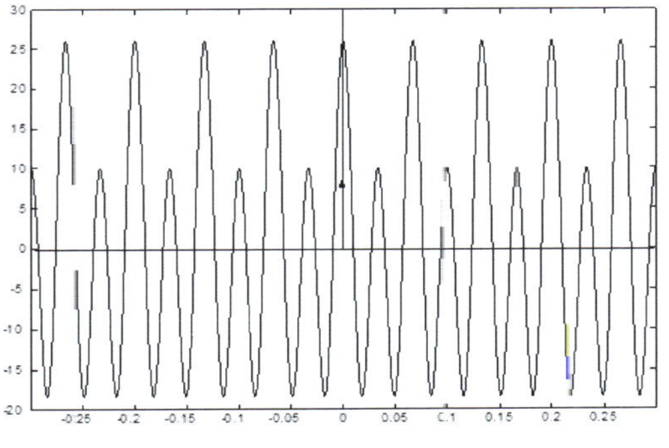

Fig. E2-11. Periodic autocorrelation of signal $y(t)$.

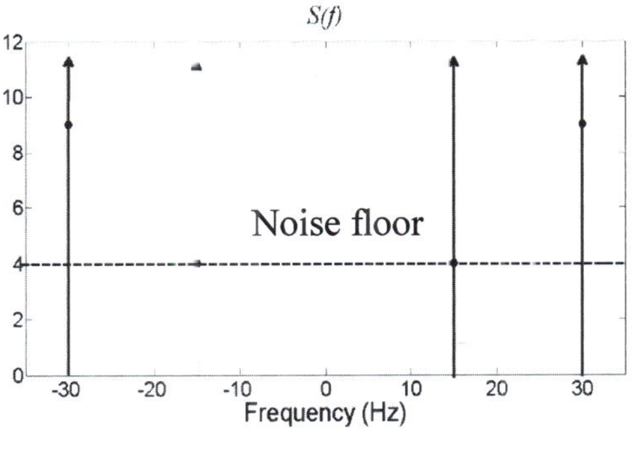

Fig. E2-12. PSD of signal $y(t)$.

(e) Using the concept of truncation, explain mathematically, in continuous time, how to detect periodic signal buried in noise. Give a sketch of what the corresponding power spectral density estimate would look like in practice, and indicate on the graph how to find periodic components hidden in noise.

Analytical explanation. In theory, we have a noncausal nonperiodic signal, $y(t)$. Truncate this signal around zero, $-T_1/2$ to the left and $T_1/2$ to the right, and call the truncated signal $y_t(t)$. Now, we can proceed two ways:

1) Using autocorrelation $r_{y_t y_t}(t)$ of $y_t(t)$. Compute $r_{y_t y_t}(t)$ and take its Fourier transform. This yields energy spectral density $E_{y_t}(t)$ of $y_t(t)$. Now, multiply $r_{y_t y_t}(t)$ and $E_{y_t}(t)$ by $\frac{1}{T_1}$ to obtain

$$\frac{1}{T_1} r_{y_t y_t}(t) \xleftrightarrow{\text{Fourier Transform}} \frac{1}{T_1} E_{y_t}(t) \equiv S_t(f)$$

where $S_t(f)$ is the approximate power spectral density of $y(t)$.

2) Using magnitude spectral density $Y_t(f)$ of $y_t(t)$. Take Fourier transform of $y_t(t)$ to obtain $Y_t(f)$. Then, the energy spectral density $E_{y_t}(t)$ of $y_t(t)$ is given by $|Y_t(f)|^2$: $E_{y_t}(t) = |Y_t(f)|^2$, and the approximate power spectral density — by $S_t(f) = (1/T_1)E_{y_t}(t) = (1/T_1)|Y_t(f)|^2$.

Then, for both cases, for sufficiently large T_1 the $S_t(f)$ approximates well the actual PSD $S(f)$ of $y(t)$:

$$\frac{1}{T_1} r_{y_t y_t}(t) \approx r_{yy}^{\circledast}(t) = \lim_{T_1 \to \infty} \frac{1}{T_1} r_{yy}(t) \quad \overset{\text{Fourier transform}}{\Leftrightarrow}$$

$$S(f) = \lim_{T_1 \to \infty} \frac{1}{T_1} E_{y_t \to y}(t) \approx S_t(f) = \frac{1}{T_1} E_{y_t}(t) = \frac{1}{T_1} |Y_t(f)|^2 .$$

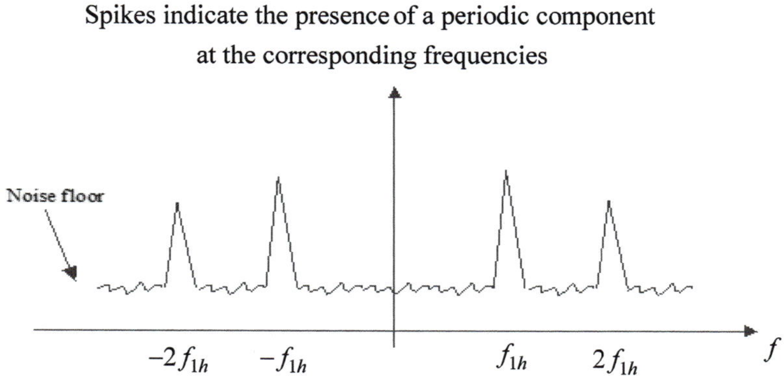

Fig. E2-13. A rough estimate of the PSD of a noisy signal with hidden periodic components revealed by the spikes.

Chapter 9

Introduction to Control Systems: Basic Control Actions and Basic Controller Design

9.1 Control Objectives, System Layout, Open-Loop and Closed-Loop Systems, Block Diagrams, System Performance Assessment: Tracking, Disturbance Rejection, Robustness, Position and Velocity Servomechanisms

Flyball governor as a feedback stabilizer of angular velocity in the steam-driven rotating machinery ushered the first industrial revolution. Since then the control field has taken off with no bounds in sight. This chapter provides the basic introduction to this area.

9.1.1 *Control system examples*

1. The first example is the centerpiece of this book — mold oscillator in a modern production unit. In this chapter we will design controller for solving a mold oscillation distortion problem.

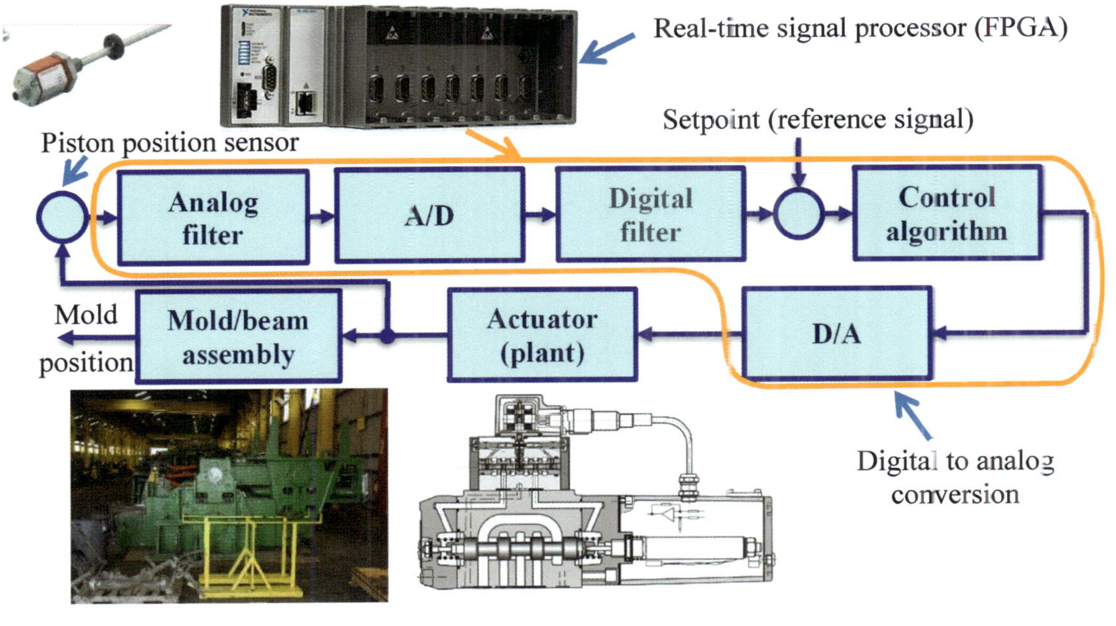

Fig. 9.1-1. Thin slab continuous steel casting mold oscillator at Nucor Steel.

2. The second example is about lasers.

A. In a lecture presentation, instructor *acquires the target* (say, finds a formula) on the screen with his/her *target acquisition system* (the eyes) and gives *command* (generates a *reference signal*) to position a laser-pointer-generated spot on this formula to the *closed-loop system*. The latter consists of the *process* (a lit laser pointer), the *sensor* (the eyes again, but in a different role), the *controller* (the brain), and the *actuator* (the hand/body motion delivery mechanism). The command is accomplished by the controller-coordinated closed-loop activity of driving the error between the spot and the formula positions to zero. Once the spot, after some *transient*, reaches the target, it is held fixed at the formula long enough to convey formula meaning.

B. Now the "star wars." The conceptually insignificant difference between the scenario above and that depicted in the Figure 9.1-2 examples is that the laser-generated spot, when held fixed at the point of interest long enough, means to burn a hole through the illuminated surface.

Fig. 9.1-2. Laser-based weapon systems; image — US Navy.

C. More star wars: taking care of modern rust belt (http://phys.org/news/2011-03-nasa-laser-space-junk.html).

Fig. 9.1-3. Modern rust belt — space junk; image — NASA.

We've got a big problem: space junk is multiplying through a "cascade of collisions" and will start knocking out satellites if no measures are taken, even if nothing else is launched. What to do? Concept: swarm of small agile satellites flying in formation, firing solar-powered lasers at a space junk or an asteroid. Current state: a funded project to zap orbital debris with lasers from Earth, likely to work in 10 years.

D. Now backyard star wars: laser mosquito zapper (https://spectrum.ieee.org/backyard-star-wars): by Intellectual Ventures Laboratory (https://photonicsentry.com/) to fight malaria and zika:

- *Mosquito detector* — 1.3-megapixel video camera — 30 frames per second, light source.
- *Connections:* FireWire or gigabit Ethernet.
- *Computer:* high data rate and low latency — *small delay between pixels capture and* processing.

- *Servo actuator (to steer a laser beam), sensor, and controller:* galvo — a DC motor with a mirror mounted on the shaft where motor produces a torque proportional to the current through the galvo; controlled motion is achieved through position encoder; individual galvos are often mounted in pairs to provide x-y motion (https://www.cambridgetechnology.com/).
- *Servo sensor*: a built-in encoder — *feeds back* the mirror position for *closed-loop control*.
- *Laser:* red (can pick up used one from Nine Inch Nails live concert).

Fig. 9.1-4. Intellectual Ventures Laboratory laser mosquito zapper: early prototype.

3. Third example is about obtaining the reference (command) signal: example — command generation in missile defense is accomplished through target acquisition.

The latter is performed by a radar, such as for example Russia's Voronezh-M missile-attack warning radar capable of detecting ballistic and long-range cruise missiles at blastoff and in-flight. Tracking range: 6,500 kilometers (about 4000 miles). Operation: pulsed radio waves are generated by the radar, reflected by the moving object, and reflection is processed to determine location and object speed based on frequency change of reflected radio wave due to the Doppler Effect.

Fig. 9.1-5. The Voronezh-M radar system.

4. Now, it's time for a short break. Problem: partying in 3D. Concept: personal quadro embedded into a 3D social quadro network. Current state: just came to my mind, ripe for crowdfunding, technologically feasible: networked control has been under development for some time, with path generation and collision avoidance being worked out (http://www.intelinair.com/). So, we will now get to making this happen.

Fig. 9.1-6. Personal quadro.

9.1.2 *Plant, and closed-loop and open-loop control systems layouts and objectives*

We will define the basic control concepts in detail in Figures 9.1-7–9.1-10 using laser beam steering and mold oscillation setups. The pictorial schematics will be referred to as ***layout***.

Example 9.1-1: Laser beam steering:

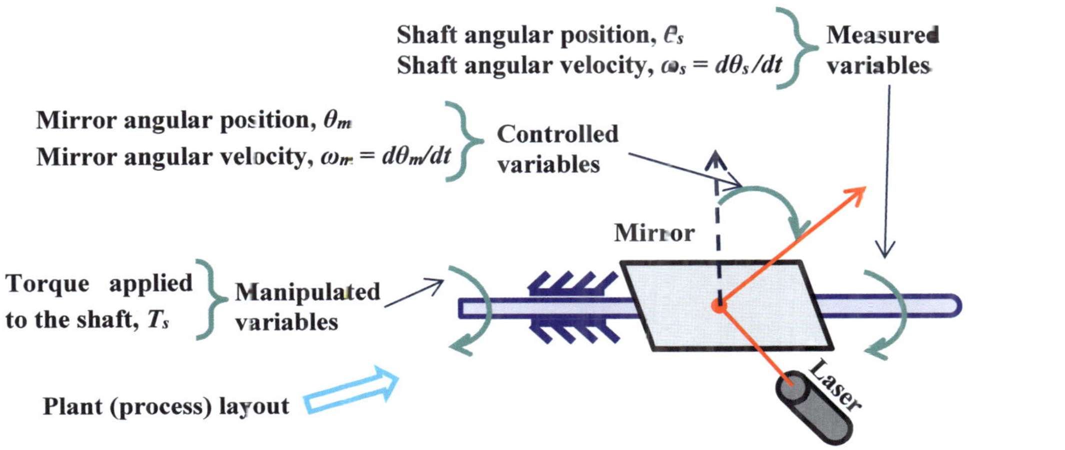

Fig. 9.1-7. Laser beam steering plant (process) layout.

Position servomechanism *(angular position servo)* is characterized by
Position Tracking Objective: actual angular *position*, θ_m = desired angular *position*, θ_d.
Velocity servomechanism *(angular velocity servo)* is characterized by
Velocity Tracking Objective: actual angular *velocity*, ω_m = desired angular *velocity*, ω_d.

Example 9.1-2: Mold oscillator in continuous casting:

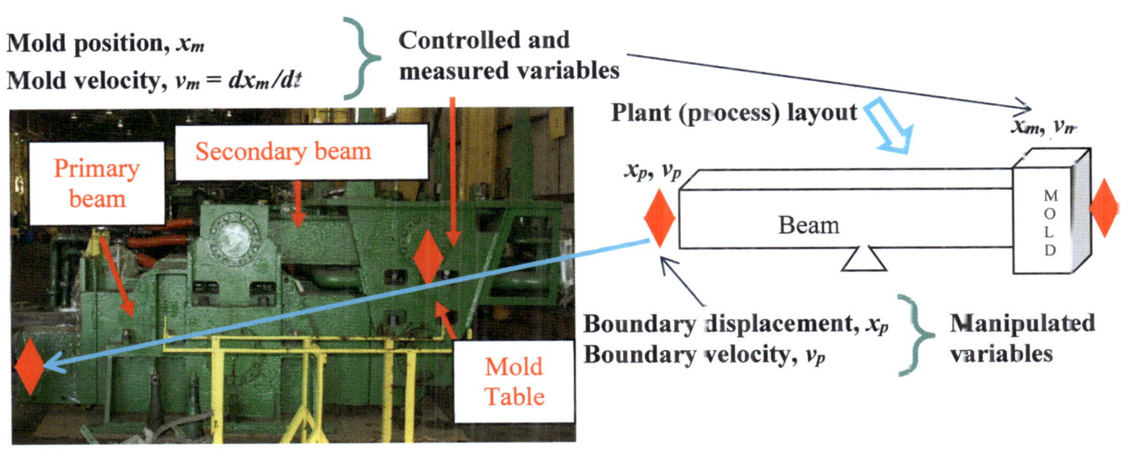

Fig. 9.1-8. Continuous casting mold oscillation plant (process) layout.

For mold oscillation, *velocity servomechanism* (*velocity servo*) is of interest to provide sufficient duration of time interval when mold velocity exceeds shell velocity in the same direction (negative strip time) during each oscillation cycle, posing

Velocity Tracking Objective: the actual mold *velocity*, v_m = the desired mold *velocity*, v_i. For practical implementation, we use instead *position servomechanism (position servo), posing*

Position Tracking Objective: the actual mold *position*, x_m = the desired mold *position*, x_i.

Open-loop system versus Closed-loop system layout

Example 9.1-3: Laser beam steering.

Position Tracking Objective: actual angular *position*, θ_m = desired angular *position*, θ_i.

Velocity Tracking Objective: actual angular *velocity*, ω_m = desired angular *velocity*, ω_i.

- Open-loop system and its input and output

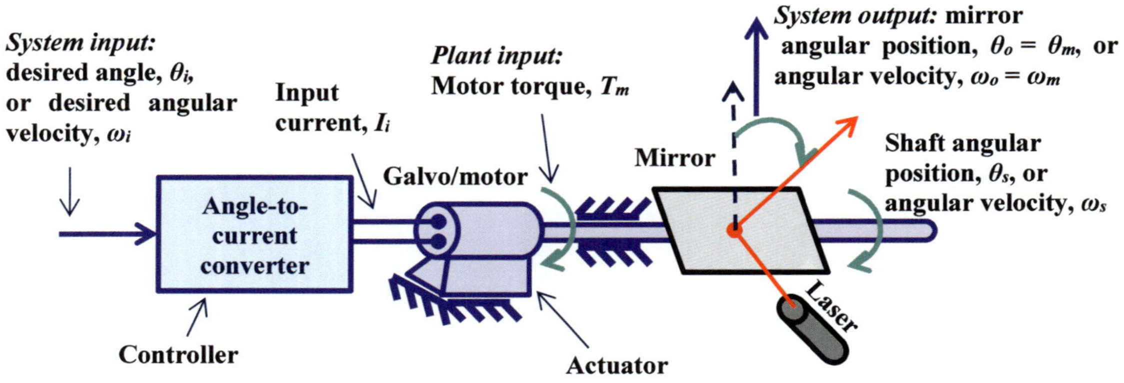

Fig. 9.1-9. Open-loop laser beam steering system layout.

- Close-loop system and its input and output

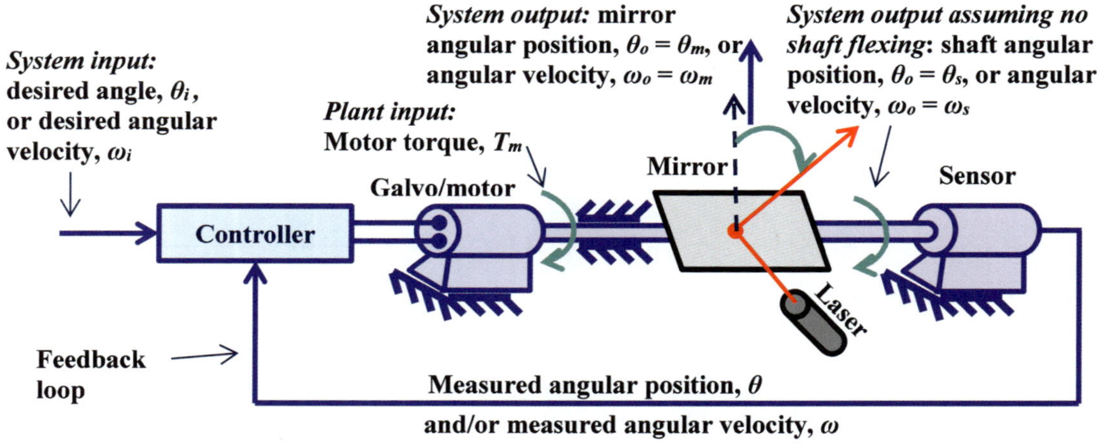

Fig. 9.1-10. Closed-loop laser beam steering system layout.

Plant input, parameters, and disturbances

Example 9.1-4: laser beam steering

1. **Plant input**:
   - Torque from the motor
2. **Plant parameters**:
   - Rotational inertia of the system
   - Damping due to the moving parts
3. **Disturbance** (any input to the system that is outside of our control)
   - Ambient air fluctuations affecting mirror

The closed-loop system layout with input disturbance is given in Figure 9.1-11.

9.1.3 *Block diagrams*

9.1.3.1 *Block assignment*

Example 9.1-5: Laser beam steering assuming no shaft flexing.

Terminology: we typically break this up into three sub-systems:

1. **Plant:** the dynamics of the output we want to control in response to the various inputs. We'll denote this as $G_p(s)$
2. **Controller:** determines how to set the motor torque based on desired angle (or velocity) and measured angle (or velocity. We'll denote this as $G_c(s)$
3. **Sensor:** how the controller measures the output. We'll denote this as $H(s)$

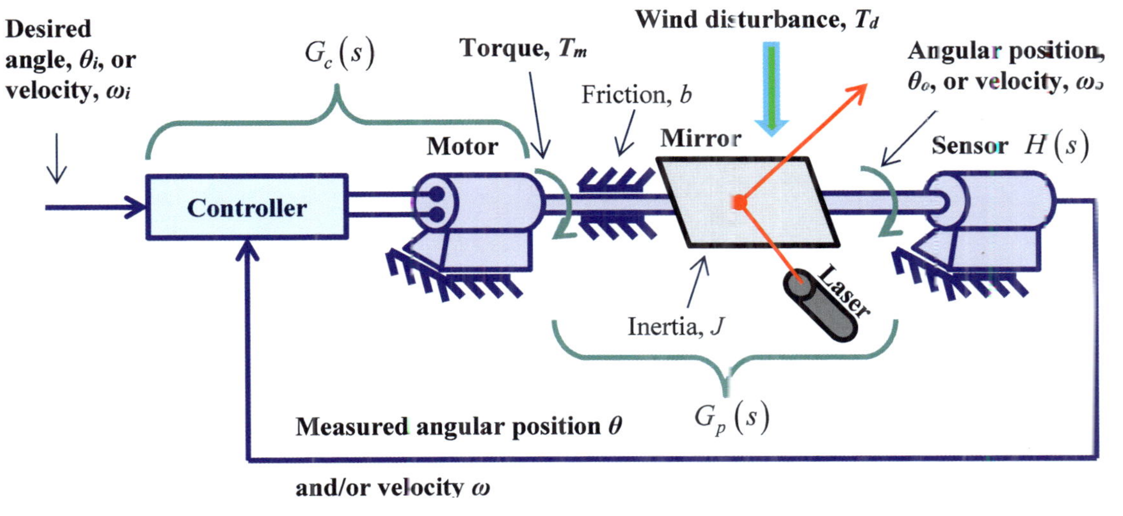

Fig. 9.1-11. Laser beam steering system layout with disturbance and assigned transfer function blocks.

9.1.3.2 *Closed-loop block diagram*

Example 9.1-6: Laser beam steering.
These transfer functions permit us to greatly simplify the picture:

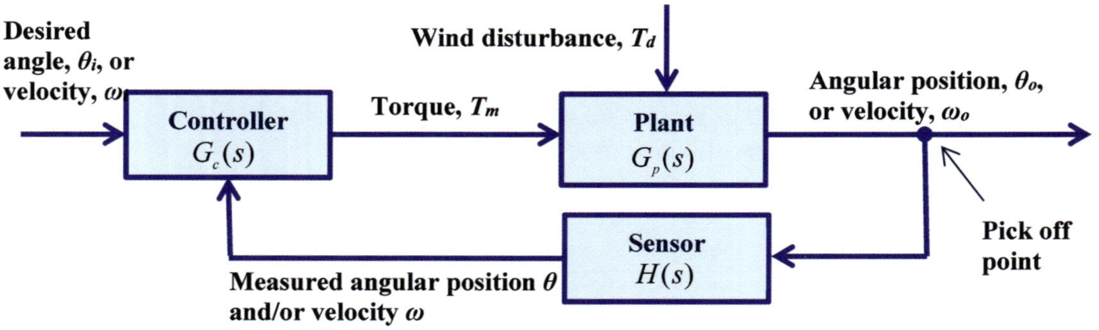

Fig. 9.1-12. Laser beam steering system closed-loop block diagram.

9.1.3.3 *Open-loop block diagram*

Objective-based dynamic plant model: position tracking.
Example 9.1-7: Laser beam steering.
In this case, the disturbance-to-output transfer function is the same as the control-to-output transfer function, but this is not always true.

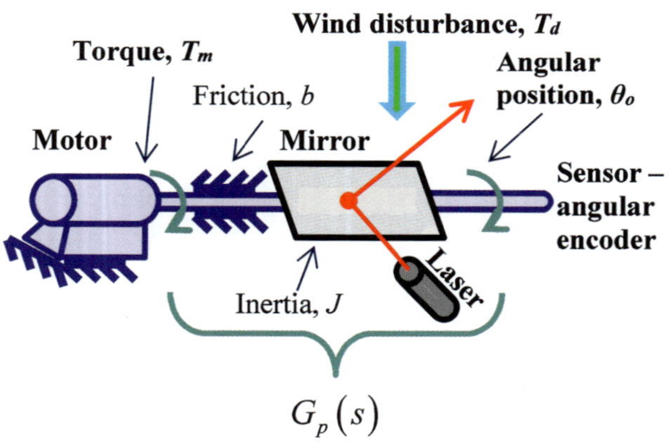

$$J\ddot{\theta}_o(t) + b\dot{\theta}_o(t) = T_m(t) + T_d(t)$$

$$Js^2\theta_o(s) + bs\theta_o(s) = T_m(s) + T_d(s)$$

$$\left(Js^2 + bs\right)\theta_o(s) = T_m(s) + T_d(s)$$

$$\theta_o(s) = \underbrace{\frac{1}{Js^2 + bs}}_{G_p(s)} T_m(s) + \underbrace{\frac{1}{Js^2 + bs}}_{G_p(s)} T_d(s)$$

Control to output *Disturbance to output*

Fig. 9.1-13. Laser beam steering system open-loop position tracking layout.

Objective-based dynamic plant model: velocity tracking.
Example 9.1-8: Laser beam steering.
Here the disturbance-to-output transfer function is also the same as the control-to-output transfer function.

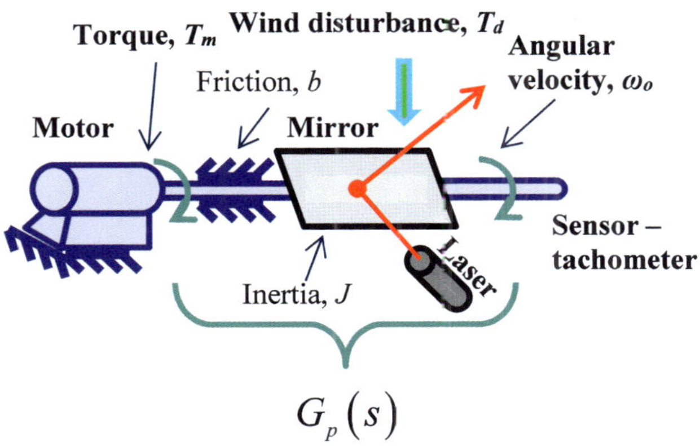

$$J\dot{\omega}_o(t) + b\omega_o(t) = T_m(t) + T_d(t)$$

$$Js\omega_o(s) + b\omega_o(s) = T_m(s) + T_d(s)$$
$$(Js + b)\omega_o(s) = T_m(s) + T_d(s)$$

$$\omega_o(s) = \underbrace{\frac{1}{Js+b}}_{G_p(s)} T_m(s) + \underbrace{\frac{1}{Js+b}}_{G_p(s)} T_d(s)$$

Control to output   Disturbance to output

Fig. 9.1-14. Laser beam steering system open-loop velocity tracking layout.

Open-loop control: open-loop block diagram and transfer function
A controller that does not use any feedback from the system is called ***open-loop.***

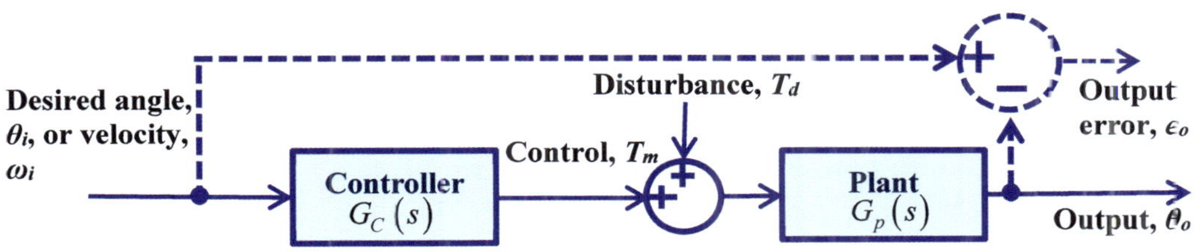

Fig. 9.1-15. Laser beam steering system open-loop block diagram.

Open-loop topology is seen to yield transfer function

$$\theta_o = G_p \underbrace{G_c\theta_i}_{T_m} + G_p T_d \;\; or \;\; \omega_o = G_p \underbrace{G_c\omega_i}_{T_m} + G_p T_d.$$

How well does this system perform in practice? First we, need to decide how we judge performance. The size of the difference between the desired and the actual angle or velocity, e_o, called ***output error***, looks like a reasonable ***performance measure***.

9.1.3.4 *Open-loop control: position tracking, disturbance rejection, and robustness*

__Position Tracking:__ How well does the output (θ_o) follow the desired value (θ_i)? Let's ignore disturbance for now. For **perfect position tracking,** $\theta_i = \theta \Rightarrow e_o = \theta_o - \theta_i = 0$. Therefore for

$$\theta_o = \underbrace{G_p G_c}_{W_{tr,ol}(s)} \theta_i + G_p T_d$$

perfect position tracking is achieved when $W_{tr,ol}(s) = G_p(s)G_c(s) = 1$. This may be possible in practice for the frequency range of interest only if the plant is stable.

Disturbance rejection: How well does the output (θ_o) ignore the disturbance (T_d)? Let's ignore tracking for now. Then, for

$$\theta_o = G_p G_c \theta_i + \underbrace{G_p}_{W_{d,ol}(s)} T_d$$

perfect disturbance rejection is achieved when $W_{d,ol} = 0$. This is impossible, since we have no effect on $W_{d,ol}$ from the controller.

Robustness: How well does the output (θ_o) ignore the differences in the plant dynamics (G_p)? Considering both tracking and disturbance rejection

$$\theta_o = \underbrace{G_p G_c}_{W_{tr,ol}(s)} \theta_i + \underbrace{G_p}_{W_{d,ol}(s)} T_d,$$

requires choosing the controller $G_c(s)$ for tracking so that $W_{tr,ol}(s) = G_p(s)G_c(s) = 1$. If the plant dynamics $G_p(s)$ change, or were modeled incorrectly or inexactly, tracking is lost.

9.2 Closed-Loop System Topology, Negative and Positive Feedback. Block Diagram Reduction. Closed-Loop System Tracking. Disturbance Rejection, Robustness, Response Shaping (Stabilization), and Output Error

A controller that does use feedback from measurements of the plant is called **closed-loop**. The closed-loop system topology, along with the ***forward path*** and the ***feedback path***, is shown in Figure 9.2-1. Similarly to the open-loop case, we would like to analyze the tracking, disturbance rejection, and robustness. For this purpose, we need to determine $W_{tr,cl}(s)$ and $W_{d,cl}(s)$ in

$$\theta_o = W_{tr,cl}\theta_i + W_{d,cl}T_d.$$

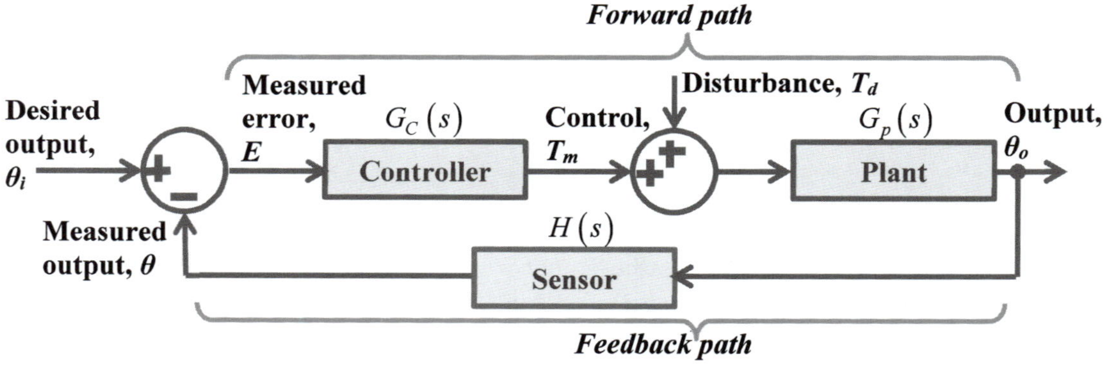

Fig. 9.2-1: Block diagram of the closed loop.

9.2.1 *General closed-loop system*

9.2.1.1 *Closed-loop transfer function for tracking assessment: input/output, negative and positive feedback*

We find $W_{tr,cl}(s)$ by setting $T_d = 0$:

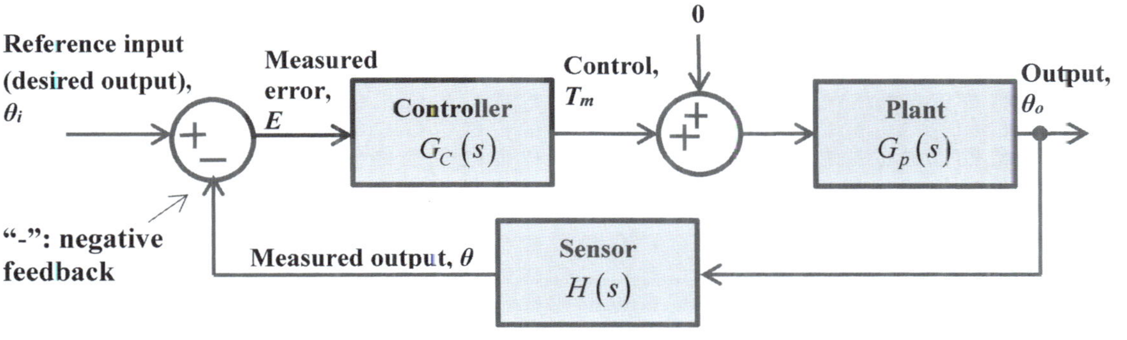

Fig. 9.2-2. Negative feedback system block diagram for tracking assessment (disturbance set to zero).

This diagram yields:

$$\theta_o = G_p G_c E$$

$$E = \theta_i - \theta = \theta_i - H\theta_o = \theta_i - HG_pG_cE$$

$$\theta_i = E + HG_pG_cE = \left(1 + HG_pG_c\right)E$$

$$\frac{E}{\theta_i} = \frac{1}{1 + HG_pG_c}$$

$$\boxed{\frac{\theta_o}{\theta_i} = \frac{G_pG_c}{1 + HG_pG_c}}$$

In the positive feedback case we find $W_{tr,cl}(s)$ also by setting $T_d = 0$:

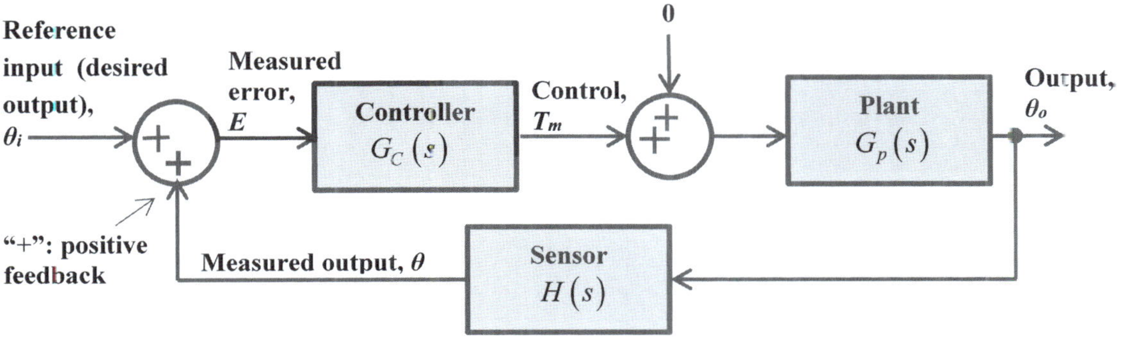

Fig. 9.2-3. Positive feedback system block diagram for tracking.

This diagram yields:

$$\theta_o = G_p G_c E$$

$$\Downarrow$$

$$E = \theta_i + \theta = \theta_i + H\theta_o = \theta_i + HG_p G_c E$$

$$\Downarrow$$

$$\theta_i = E - HG_p G_c E = \left(1 - HG_p G_c\right) E$$

$$\Rightarrow \quad \frac{E}{\theta_i} = \frac{1}{1 - HG_p G_c}$$

$$\Downarrow$$

$$\boxed{\frac{\theta_o}{\theta_i} = \frac{G_p G_c}{1 - HG_p G_c}}$$

9.2.1.2 *Forward and feedback gains*

Let's define *forward, feedback, and open-loop, often simply referred to as loop, gains* as follows:

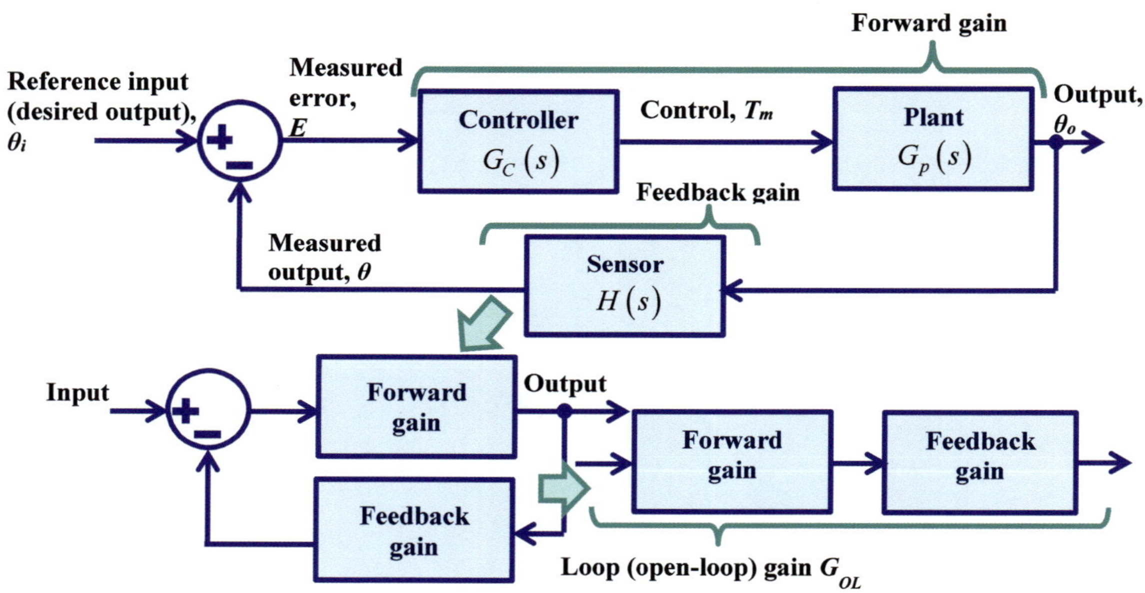

Fig. 9.2-4. General closed-loop block diagram.

General closed-loop input-output transfer function in terms of forward and feedback gains

Then a general block diagram reduction shortcut looks as follows:

$$\left.\begin{array}{r}\text{"}-\text{": negative}\\ \text{feedback}\end{array}\right\} \frac{\theta_o}{\theta_i} = \frac{G_p G_c}{1 + HG_p G_c} = \frac{(\text{forward gain})}{1 + (\text{forward gain}) \cdot (\text{feedback gain})} \tag{9.2-1}$$

$$\left.\begin{array}{r}\text{"}+\text{": positive}\\ \text{feedback}\end{array}\right\} \frac{\theta_o}{\theta_i} = \frac{G_p G_c}{\underbrace{1 - HG_p G_c}_{\text{SPECIFIC}}} = \frac{(\text{forward gain})}{\underbrace{1 - (\text{forward gain}) \cdot (\text{feedback gain})}_{\text{GENERAL}}} \tag{9.2-2}$$

9.2.1.3 *Closed-loop transfer function for disturbance rejection assessment: disturbance-to-output*

We find $W_{d,cl}(s)$ by setting $\theta_i = 0$:

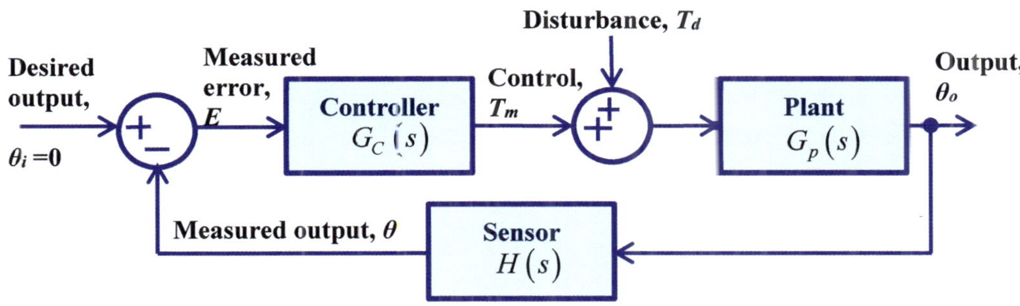

Fig. 9.2-5. Feedback system block diagram.

Instead of carrying out the detailed calculation of transfer function again, let's find it by using the feedback relationship. First, we rearrange the system so it has the right topology.

We "roll" the loop counterclockwise to have the disturbance input appear in place of the reference input. Then we drop the reference input summation, and since the topology is linear, we move the subtraction to the upper left side.

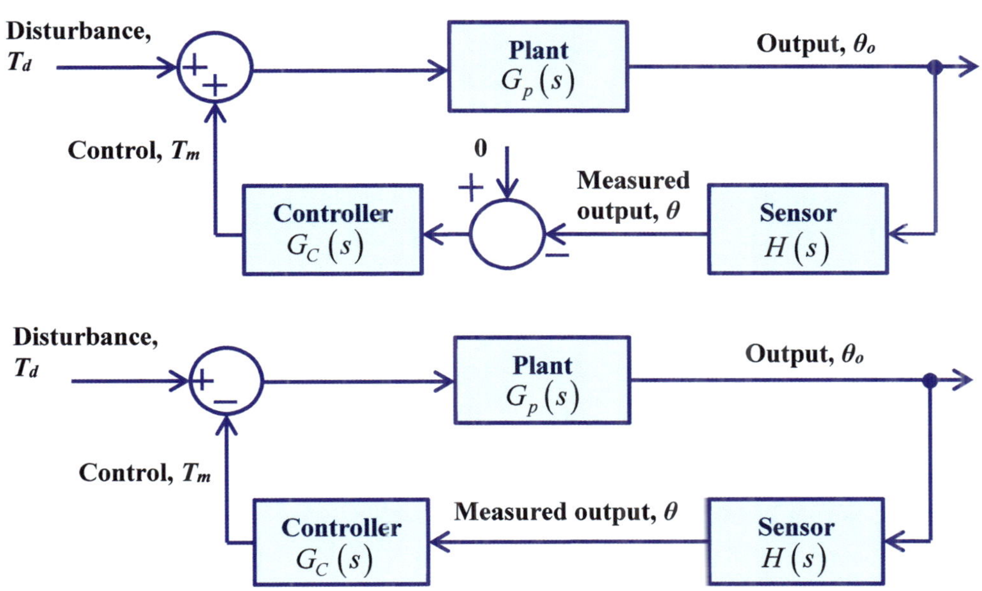

Fig. 9.2-6. Closed-loop disturbance-to-output block diagram.

Now we can use the reduction rule:

$$\frac{\text{output}}{\text{disturbance}} = \frac{(\text{forward gain})}{1+(\text{forward gain})\cdot(\text{feedback gain})} = \boxed{\frac{\theta_o}{T_d} = \frac{G_p}{1+HG_pG_c}}. \quad (9.2\text{-}3)$$

9.2.1.4 *Closed-loop transfer function for tracking assessment: input-to-(measured-error)*

For tracking assessment we use the measured error as system output:

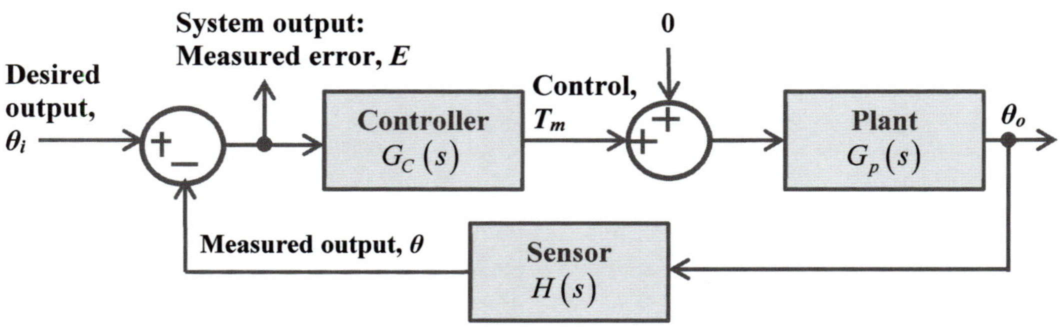

Fig. 9.2-7. Block diagram with measured error as the output of interest.

Converted to the latter topology to the explicit feedback form yields:

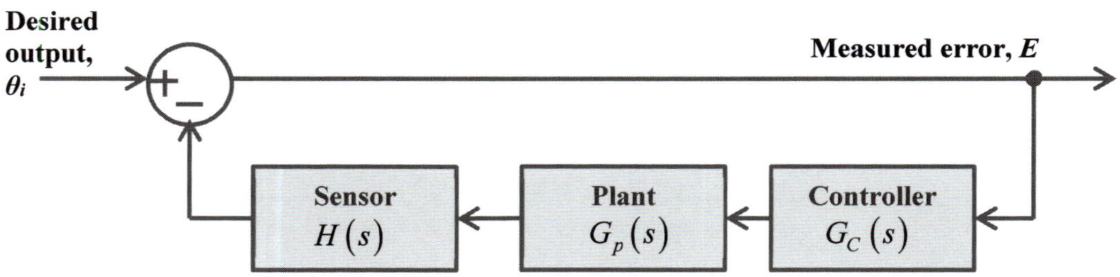

Fig. 9.2-8. Block diagram: input-to-(measured-error).

Applying feedback rule yields:

$$\frac{\text{measured error}}{\text{input}} = \frac{(\text{forward gain})}{1+(\text{forward gain})\cdot(\text{feedback gain})} = \boxed{\frac{E}{\theta_i} = \frac{1}{1+HG_pG_c}}. \quad (9.2\text{-}4)$$

9.2.2 *Closed-loop transfer function: unity feedback (perfect sensing)*

The closed-loop unity feedback (perfect sensing) topology implies $H(s) = 1$:

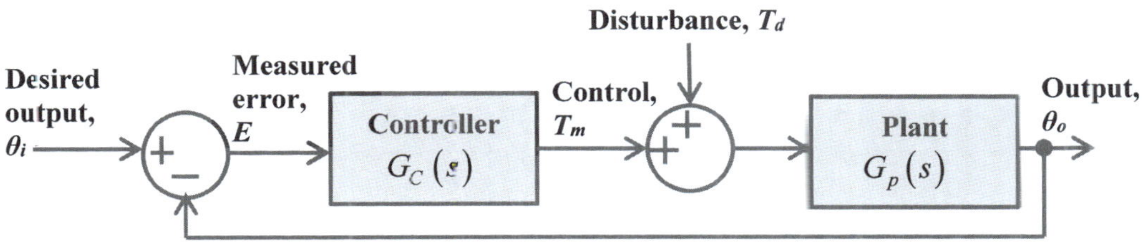

Fig. 9.2-9. Block diagram with unity feedback.

The two-input/one-output transfer function is then given by

$$\theta_o = \underbrace{\frac{G_pG_c}{1+G_pG_c}}_{W_{tr,cl}(s)}\theta_i + \underbrace{\frac{G_p}{1+G_pG_c}}_{W_{d,cl}(s)}T_d. \tag{9.2-5}$$

9.2.2.1 *Unity feedback topology performance characterizations*

Closed-loop control: tracking ($T_d = 0$ in Figure 9.2-9):

$$\theta_o = \underbrace{\frac{G_pG_c}{1+G_pG_c}}_{W_{tr,cl}(s)}\theta_i. \tag{9.2-6}$$

Similarly to the open-loop case, **perfect tracking** is achieved when $W_{tr,cl}(s) = 1$. Looking at (9.2-6), this requires $1+ G_pG_c \to G_pG_c$. Recalling (4.4-1), the latter can happen when controller gain $|G_c(s)| \to \infty$. In practice, this means $|G_c|$ becomes very large (***high gain***) compared to $|G_p|$.

Closed-loop control: disturbance rejection ($\theta_i = 0$ in Figure 9.2-9):

$$\theta_o = \underbrace{\frac{G_p}{1+G_pG_c}}_{W_{d,cl}(s)}T_d. \tag{9.2-7}$$

For **perfect disturbance rejection**, $W_{d,cl} = 0$. Remember that the perfect tracking controller required that $|G_c| = \infty$. In that case, we also have $W_{d,cl} = 0$. The perfect tracking controller also gives perfect disturbance rejection!

Closed-loop control: robustness (changing G_p in Figure 9.2-9):

$$\theta_o = \underbrace{\frac{G_p G_c}{1+G_p G_c}}_{W_{tr,cl}(s)}\theta_i + \underbrace{\frac{G_p}{1+G_p G_c}}_{W_{d,cl}(s)}T_d. \tag{9.2-8}$$

If $|G_c|$ is much larger than $|G_p|$, then it will still be larger if G_p is somewhat different than expected. So, for $|G_c| = \infty$ and finite modeling errors in G_p we will still have $W_{tr,cl}(s) = 1$ (perfect tracking) and $W_{d,cl} = 0$ (perfect disturbance rejection). This property is called **robustness (performance robustness, robust performance)** — performance preservation in spite of modeling errors or drift in plant dynamics, such as that shown in Figure 9.2-10.

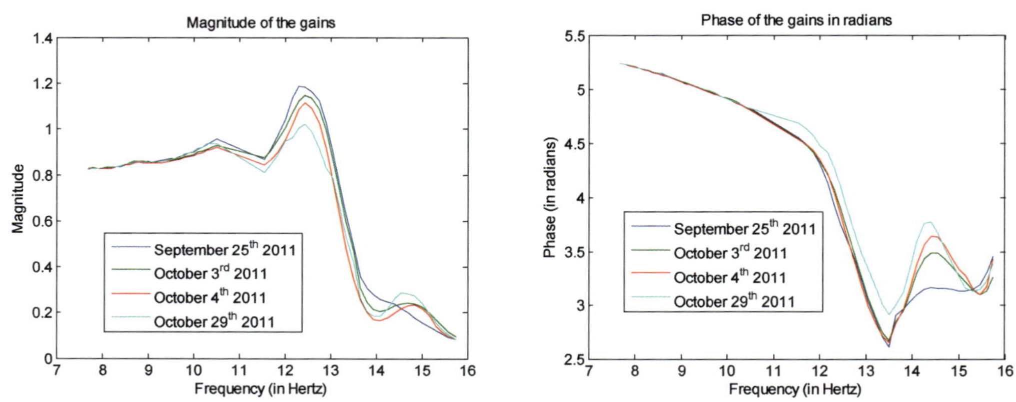

Fig. 9.2-10. Plant dynamics uncertainty represented by changes is magnitude (left) and phase (right) responses.

Nucor mold oscillation production unit plant gain $|G_p|$ changes (by less than 100% over a 1 month period). Plant gain is typically collected every 2 weeks. Although resonance peak changes in magnitude, resonance frequency is seen to stay almost constant. Therefore, distortion rejection controller to be introduced later requires infrequent retuning.

Closed-loop control: summary:

For a closed-loop unity feedback control system described by transfer function

$$\theta_o = \underbrace{\frac{G_p G_c}{1+G_p G_c}}_{W_{tr,cl}(s)}\theta_i + \underbrace{\frac{G_p}{1+G_p G_c}}_{W_{d,cl}(s)}T_d$$

the following holds under the closed-loop stability assumption:

Closed-loop performance as $|G_c| \to \infty$:

- Tracking performance
  - $W_{tr,cl} \to 1$
- Disturbance rejection performance
  - $W_{d,cl} \to 0$
- Performance robustness:
  - if $G_{p,true} = G_{p,model} + \Delta$ (the actual plant differs from the expected model), then we still have $W_{tr,cl} \approx 1$ and $W_{d,cl} \approx 0$

Advantages

Closed-loop control has major advantages over open-loop control:

- It can achieve tracking that is perfect in the limit and very good in actual application in spite of modeling errors and/or variations in the plant dynamics.
- It can reject unanticipated disturbances.

9.2.2.2 *Closed-loop control: plant stabilization*

Another advantage is that closed-loop control allows for **response shaping**. The most simple and common example of this is **stabilization of unstable plants:**

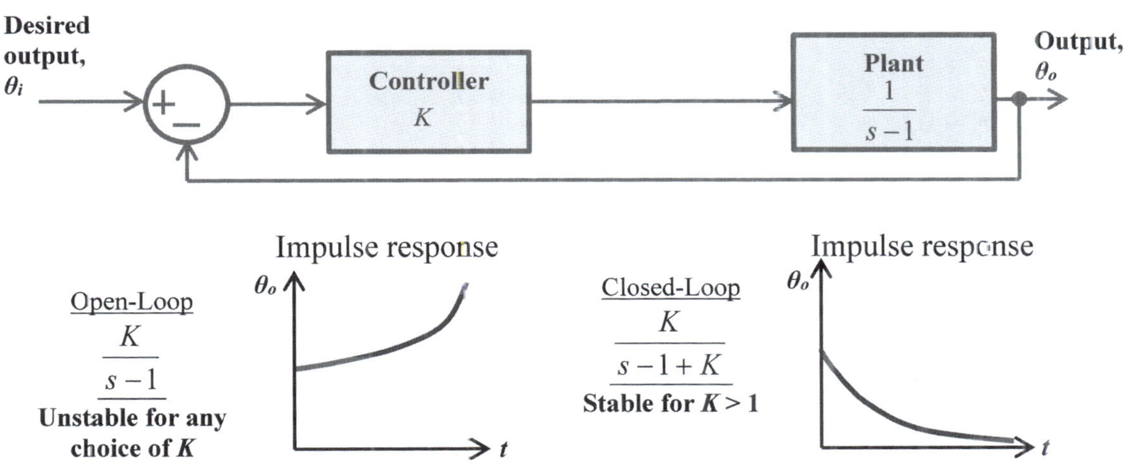

Fig. 9.2-11. Stabilization of an unstable plant by a closed-loop control.

In addition, by choosing K, the decay rate can be made as fast or as slow as we like. This decay rate tuning capability is what we mean by **response shaping**.

9.2.2.3 *Input-to-(output-error) transfer function*

In Figure 9.2-2 we used the input-to-(measured-error) transfer function

$$\frac{E}{\theta_i} = \frac{1}{1 + HG_p G_c}$$

as an intermediate step to finding the closed-loop transfer function.

In general, though, the (actual) **output error** is of more concern.

Actual output error — definition:

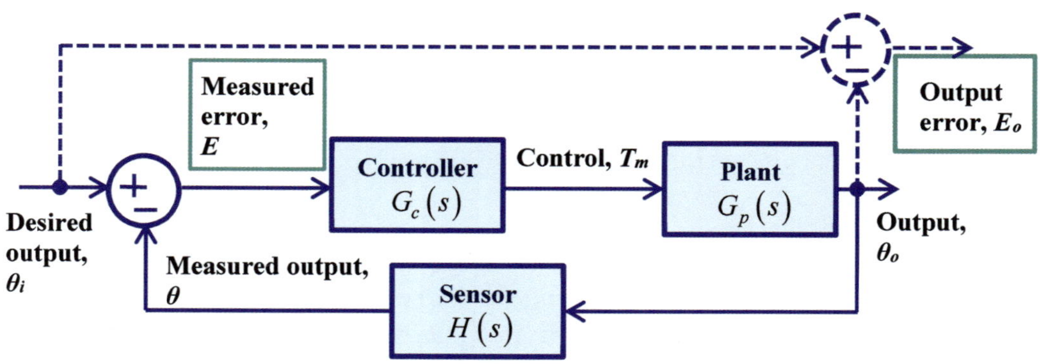

Fig. 9.2-12. System block diagram: output error.

More generally,

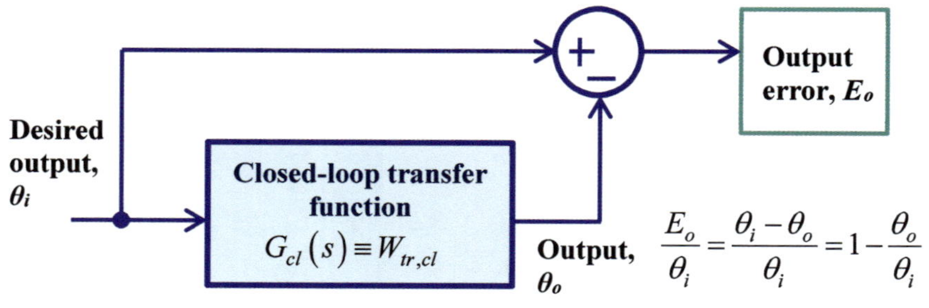

Fig. 9.2-13. System block diagram: output error general case.

From Figure 9.2-2 we had (ignoring disturbances)

$$W_{tr,cl} = \frac{\theta_o}{\theta_i} = \frac{G_p G_c}{1 + HG_p G_c}.$$

Therefore,

$$G_{oe,cl} \triangleq \frac{E_o}{\theta_i} = 1 - \frac{G_p G_c}{1 + HG_p G_c} = \boxed{\frac{1 + (H-1)G_p G_c}{1 + HG_p G_c}}. \tag{9.2-9}$$

In the special case of unity feedback ($H(s) = 1$):

$$\frac{E_o}{\theta_i} = \frac{1 + (1-1)G_p G_c}{1 + 1 \cdot G_p G_c} = \boxed{\frac{1}{1 + G_p G_c}}. \tag{9.2-10}$$

This expression is identical to the input-to-(measured-error) transfer function with $H = 1$. This observation indicates that unity feedback means perfect sensing.

Steady-state output error calculation through the Final Value Theorem. *The steady-state output error, $e_{o,ss}$, is the constant the output error converges to in response to the reference input after the transients have died out:* $e_{o,ss} = \lim_{t\to\infty} e_o(t)$. It can be found through the Final Value Theorem: if $f(t)$ and $\dot{f}(t) \equiv df(t)/dt$ have Laplace transforms, and if $\lim_{t\to\infty} f(t)$ exists, then the value of the limit is

$$\lim_{t\to\infty} f(t) = \lim_{s\to 0} s\mathcal{L}\left[f(t)\right]. \tag{9.2-11}$$

Example 9.2-1: steady-state closed-loop output error for unit step reference input. Applying the definition of $E_o(s)$ above and (9.2-11) to the input-to-(output-error) transfer function (9.2-9) and recalling that the Laplace transform of the unit step is $1/s$ shows that $G_{oe,cl}(s)$ zeros shape this error:

$$e_{o,ss} = \lim_{t\to\infty} e_o(t) = \lim_{s\to 0} sE_o(s) = \lim_{s\to 0} sG_{oe,cl}(s)R(s) = \lim_{s\to 0} sG_{oe,cl}(s)\frac{1}{s} = \lim_{s\to 0} G_{oe,cl}(s). \tag{9.2-12}$$

Note on the usage of the Final Value Theorem: an example of a function for which the limit doesn't exist is $\cos(\omega t)$, since it doesn't settle at a constant. This fact, however, does not imply that this theorem can't be used to check the steady-state error in tracking of the sinusoidal reference input, since in this case the final value theorem is applied *not to the sinusoidal output itself, but to the difference between the input and the output, both sinusoidal, i.e. to the tracking error.*

9.2.2.4 *Shaping zeros through loop closure: open-loop poles become zeros of the closed-loop input-to-(output-error) transfer function*

Now we come to one of the main uses of feedback — placement of blocking zeros. For the general open-loop transfer function $G_{OL} = Z(s)/P(s)$

$$G_{oe,cl}(s) = \frac{E_o}{\theta_i} = \frac{1}{1 + G_{OL}} = \frac{1}{1 + \frac{Z(s)}{P(s)}} = \frac{P(s)}{P(s) + Z(s)}. \tag{9.2-13}$$

- Poles (infinite gains) of the open-loop transfer function become zeros (zero gains) of the closed-loop input-to-(output-error) transfer function!
- Therefore, to create a blocking zero at a particular frequency in the input-to-(output error) closed loop transfer function, we set this zero to be a pole at that frequency in the loop gain.

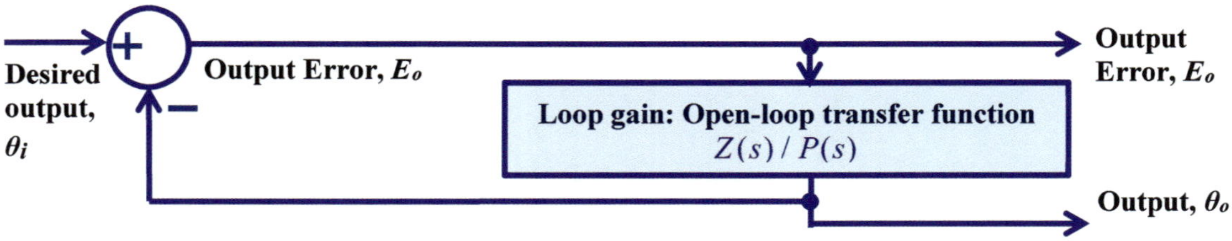

Fig. 9.2-14. System block diagram indicating open-loop poles and zeros.

Example 9.2-2: the open-loop integrator creates the closed-loop input-to-(output-error) differentiator.

This effect of closing the loop is illustrated by Figures 9.2-15 and 9.2-16:

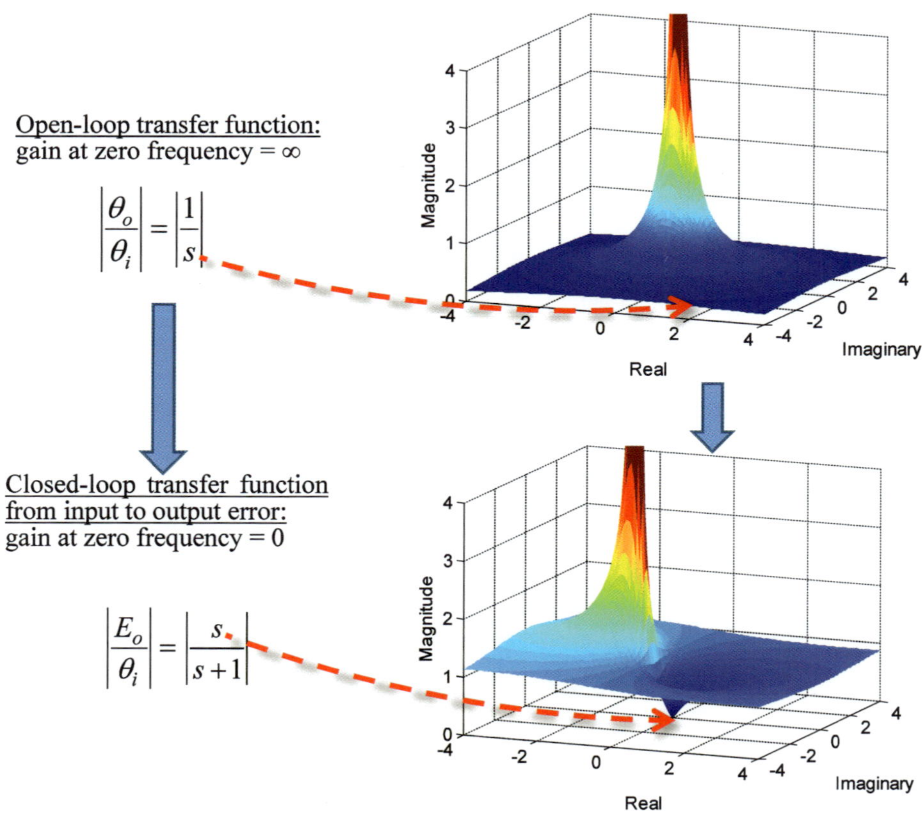

Fig. 9.2-15. 3D pole-zero maps of the open-loop and the closed-loop transfer functions — the pole of the open-loop transfer function becomes the zero of the closed-loop input-to-(output-error) transfer function.

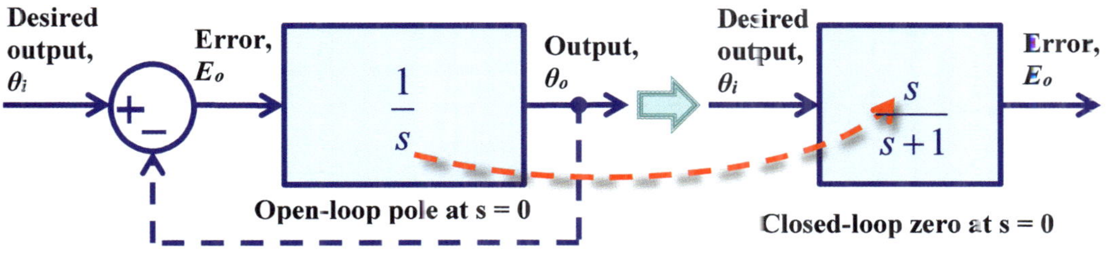

Fig. 9.2-16. The effect of the loop closure on the poles and the zeros — the pole of the open-loop transfer function becomes the zero of the closed-loop input-to-(output-error) transfer function.

From (9.2-12) and Figure 9.2-16 it is seen that the open-loop pole at 0 sets the closed-loop steady state output error in response to the unit step input to zero through creating the blocking zero at 0.

9.3 Basic Control Actions

9.3.1 *Basic control actions*

Let's now focus on how analog controller in Figure 9.3-1 acts on the error to produce control signal.

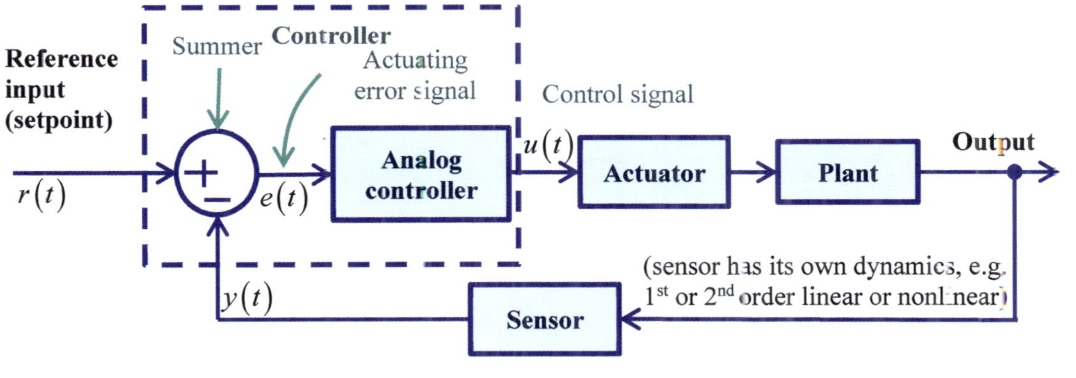

Fig. 9.3-1. Controller block: controller inputs: setpoint and sensor output, controller output — control signal.

Example 9.3-1: hardware testbed.
System layout and block diagram of the latter are shown in Figure 9.3-2. Control is performed by a ***proportional controller*** (gain = 2) through measuring the piston position x_p and using the error between the reference r and the piston position x_p.

In the short stroke mode, mold position x_m is measured for monitoring purposes but not used for the closed-loop control.

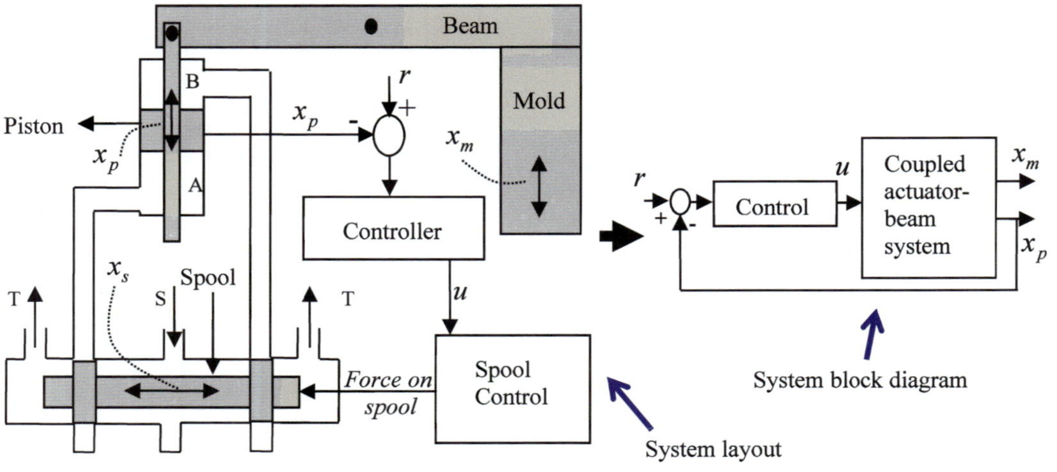

Fig. 9.3-2. Testbed system layout and block diagram.

Self-operated: Sensor and actuator are one mechanical device.
Example 9.3-2: Fly-ball (Watt) governor in steam engines. As rotational velocity ω increases, the rotating arms swing up, pulling up the valve and cutting off the amount of steam to the engine:

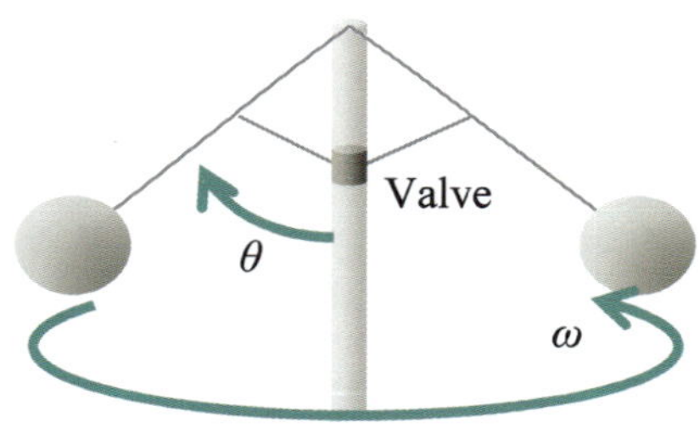

Fig. 9.3-3. Flyball.

Example 9.3-3: Bimetallic strips in thermostats.

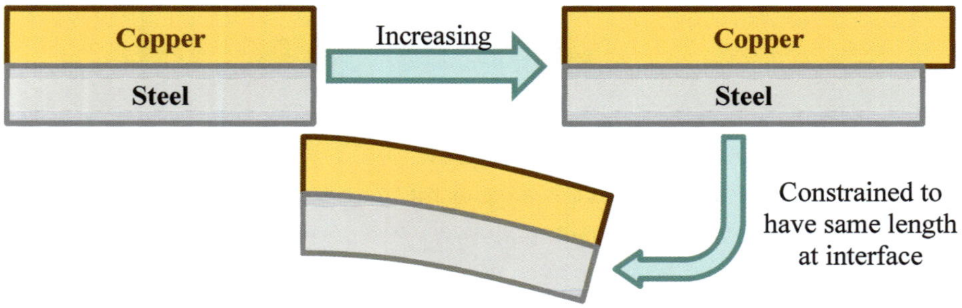

Fig. 9.3-4. Bimetallic strips in thermostats.

Two strips of metals with different coefficients of thermal expansion, attached together. When temperature changes, one of the metals expands/shrinks more than the other, causing the strip to bend. This can be used to open or close an electric switch.

Two-position (on-off): Control takes only one of two values.

Example 9.3-4: Thermostat:

$$u(t)=\begin{cases} U_1, e(t)>0, \\ U_2, e(t)<0, \end{cases}$$

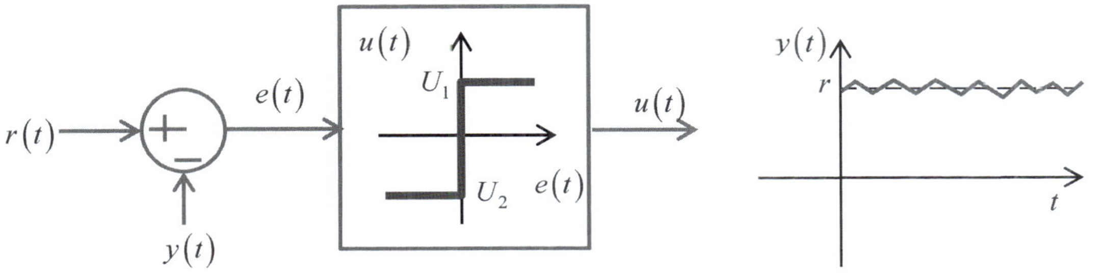

Fig. 9.3-5. Thermostat operation.

One problem with this type of control is that it tends to lead to a lot of chattering around the reference. Chattering can be reduced by introducing "hysteresis" where the control depends on the previous error, not just the current error. The controller turns on or off shortly *after* reaching 0.

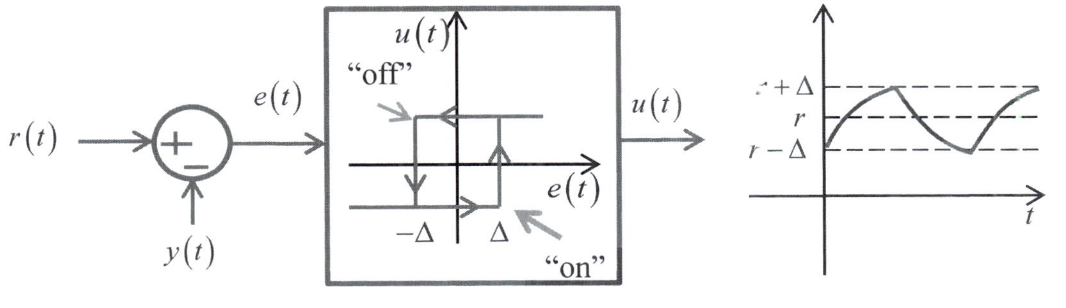

Fig. 9.3-6. Thermostat with "hysteresis".

Proportional, or P: Control is proportional to the error. K_p is called the **proportional gain.**

$$u(t)=K_p e(t)\Rightarrow\frac{U(s)}{E(s)}=K_p. \tag{9.3-1}$$

Integral, or I: Control is proportional to the time-integral of the error. K_i is called the **integral gain.**

$$u(t) = K_i \int_0^t e(s)ds + u(0) \Rightarrow \frac{U(s)}{E(s)} = \frac{K_i}{s} \text{ for } u(0) = 0. \tag{9.3-2}$$

$$e(t) = 0 \Rightarrow u(t) = u(0) = const$$

Pole at zero

Derivative, or D: Control is proportional to the time-derivative of the error. K_d is called the **derivative gain.**

$$u(t) = K_d \frac{de(t)}{dt} \Rightarrow \frac{U(s)}{E(s)} = K_d s. \tag{9.3-3}$$

PI (Proportional + Integral): Another convention used for the I gain: $K_i = K_p/T_i$ where T_i is called the **integral time**, and $1/(T_i\ 60)$ is called reset rate (repeats per minute — how many times per minute K_p is added to the control signal under constant error), so that

$$u(t) = K_p e(t) + \frac{K_p}{T_i} \int_0^t e(s)ds \Rightarrow \frac{U(s)}{E(s)} = K_p \left(1 + \frac{1}{T_i s}\right) = \frac{K_p s + K_i}{s}. \tag{9.3-4}$$

Integral controller pole at zero is retained

Suppose $e(t)$ is the unit step, $1(t)$.

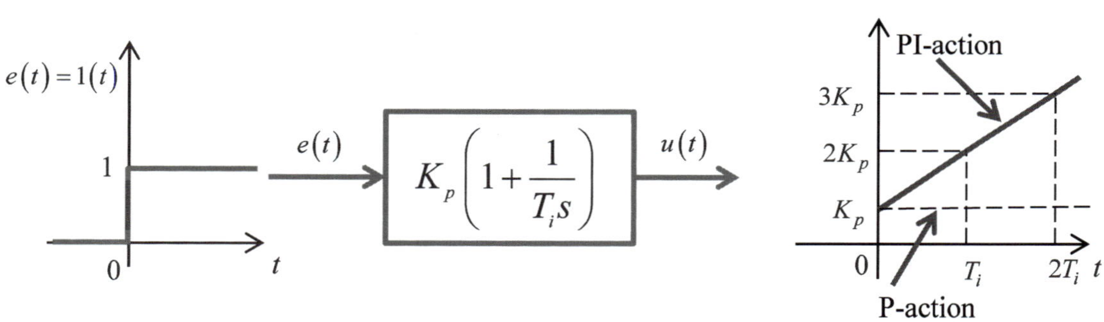

Fig. 9.3-7. PI controller action.

Every T_i second, the value of K_p is seen to be added to the control signal.

PD (Proportional + Derivative):

$$u(t) = K_p e(t) + K_p T_d \frac{de(t)}{dt} \Rightarrow \frac{U(s)}{E(s)} = K_p (1 + T_d s).$$

T_d is called the **derivative time**. Suppose $e(t)$ is the unit ramp, then

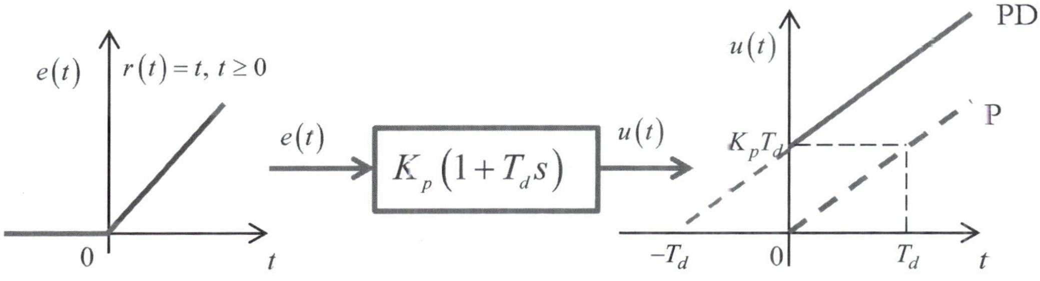

Fig. 9.3-8. PD controller action.

$$e(t)=t \Rightarrow \begin{cases} u_P(t)=K_p t, \\ u_{PD}(t)=K_p e(t)+K_pT_d\dfrac{de(t)}{dt}=K_p t+K_pT_d=K_p(t+T_d)=K_p[t-(-T_d)], \end{cases}$$

i.e. for ramp input D-part advances action of P-controller by T_d. PD controller is used to damp transients, but can cause actuator saturation and amplify noise. Thus, for ramp input PD control is the P control advanced by T_d seconds. For this reason, it is often referred to (incorrectly) as *anticipatory*. However, it is still a causal controller, so it can't anticipate signal change.

PID (Proportional + Integral + Derivative):

$$u(t)=K_p e(t)+\frac{K_i}{T_i}\int_0^t e(s)ds+K_pT_d\frac{de(t)}{dt}\Rightarrow\frac{U(s)}{E(s)}=K_p\left(1+\frac{1}{T_i s}+T_d s\right)=\frac{K_pT_d s^2+K_p s+K_i}{s}$$

Integral controller pole at zero is retained

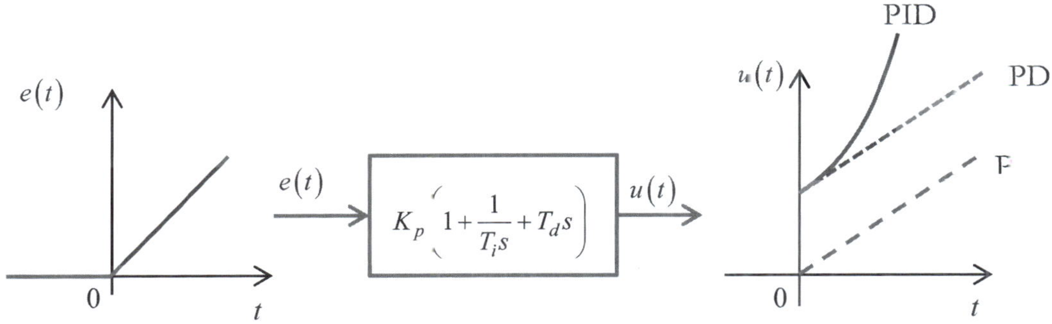

Fig. 9.3-9. PID controller action.

9.3.2 *Second-order systems*

For many systems, the first or the second-order models are adequate. Why?

- For higher-order systems, only one or two poles could be dominant.

- There are well-known model reduction techniques that can reduce the order of a model.
- We can treat the second-order system as the nominal (the design) model and the un-modeled poles as uncertainty or noise which can be dealt with by the controller.

Examples:

- The beam, which is a flexible structure described by a PDE, can be approximated by a second-order underdamped system (Project 2).
- Motor can be modeled by the first-order system under PI controller in closed-loop.

Consider the following general model of a closed-loop system given by a second-order transfer function with unity DC gain, natural frequency ω_n, and amping coefficient ζ:

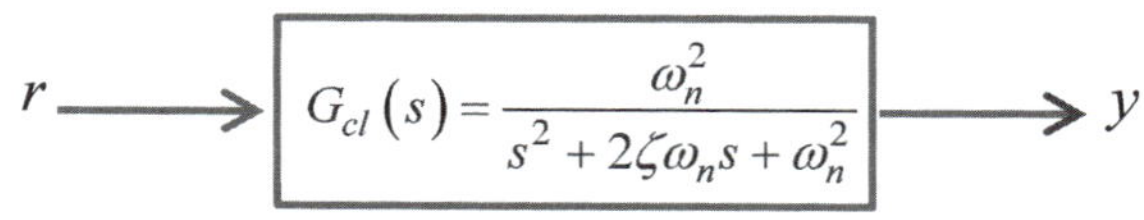

9.3.2.1 *Underdamped second order system*

Unit step response: Assume that the transfer function of the system,

$$G_{cl}(s) = \frac{Y(s)}{R(s)} = \frac{\omega_n^2}{s^2 + 2\zeta\omega_n s + \omega_n^2}, \tag{9.3-7}$$

has $0 < \zeta < 1$, i.e. the system is *underdamped*. Then its step response is given by

$$y(t) = 1 - e^{-\zeta\omega_n t}\left(\cos(\omega_d t) + \frac{\zeta}{\sqrt{1-\zeta^2}}\sin(\omega_d t)\right), \quad t > 0, \; \omega_d = \omega_n\sqrt{1-\zeta^2}. \tag{9.3-8}$$

This function represents the system response to a setpoint step change, so we can use it to identify some important performance characteristics and set the performance goals.

Poles: using quadratic formula yields

$$s = -\zeta\omega_n \pm j\omega_n\sqrt{1-\zeta^2} = -\zeta\omega_n \pm j\omega_d. \tag{9.3-10}$$

The real part determines the decay rate of the envelope and the imaginary part determines the frequency of damped oscillations:

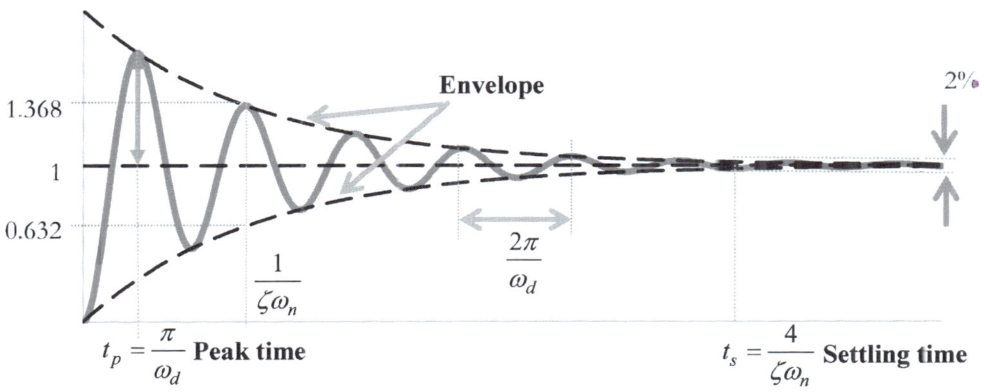

Fig. 9.3-10. Unit step response of an underdamped second order system.

Maximum overshoot: We find peak time first. At t_p, y attains a maximum, so

$$y(t) = 1 - e^{-\zeta\omega_n t}\left(\cos(\omega_d t) + \frac{\zeta}{\sqrt{1-\zeta^2}}\sin(\omega_d t)\right), t > 0, \Rightarrow$$

$$\frac{dy}{dt} = \frac{\omega_n}{\sqrt{1-\zeta^2}} e^{-\zeta\omega_n t_p} \sin(\omega_d t_p) = 0 \Rightarrow t_p = \frac{\pi}{\omega_d}.$$

Then, $y(t_p)$, defines the maximum percent overshoot:

$$y(t_p) = 1 - e^{-\zeta\omega_n t_p} = 1 - e^{-\zeta\omega_n \frac{\pi}{\omega_d}} = 1 - e^{-\frac{\zeta\pi}{\sqrt{1-\zeta^2}}} \Rightarrow M_p = 1 - y(t_p) = e^{-\frac{\zeta\pi}{\sqrt{1-\zeta^2}}}. \quad (9.3\text{-}11)$$

Settling time: Settling time is the time required for the step response curve to reach 2% of the final value and stay within this range. It can be computed by analyzing the envelope:

$$1 - y(t) = e^{-\zeta\omega_n t}\left(\cos(\omega_d t) + \frac{\zeta}{\sqrt{1-\zeta^2}}\sin(\omega_d t)\right).$$

We see that this will happen when $e^{-\zeta\omega_n t} < 0.02$, so that

$$t > -\frac{\ln(0.02)}{\zeta\omega_n} \approx \frac{4}{\zeta\omega_n}. \quad (9.3\text{-}12)$$

Closed-loop tracking of constants — steady-state response, y_{ss}, to the unit step input: We can find it three ways: 1) in time domain by setting $t = \infty$ in (9.3-8), 2) by setting $s = 0$ in the transfer function (9.3-7) to find the DC gain and multiplying the step value after discontinuity by this gain, and 3) using the Final Value Theorem. Applying this theorem to (9.3-7) for the unit step input yields perfect steady state tracking, with the closed-loop DC gain shaping the steady-state response to the unit step input:

$$y_{ss} = \lim_{t\to\infty} y(t) = \lim_{s\to 0} sY(s) = \lim_{s\to 0} sG_{cl}(s)R(s) = \lim_{s\to 0} sG_{cl}(s)\frac{1}{s} = \lim_{s\to 0} G_{cl}(s) = \lim_{s\to 0} (\underbrace{\frac{\omega_n^2}{s^2 + 2\zeta\omega_n s + \omega_n^2}}_{G_{cl}(s)}) = 1.$$

9.3.2.2 *Underdamped second-order system: resonance frequency and peak*

As presented in Chapter 3, for the second order system we have following calculation:

$$H(j\omega) = \frac{\omega_n^2}{s^2 + 2\xi\omega_n s + \omega_n^2}\bigg|_{s=j\omega} = \frac{1}{\frac{1}{\omega_n^2}s^2 + \frac{2\xi}{\omega_n}s + 1}\bigg|_{s=j\omega}$$

$$= \frac{1}{\frac{(j\omega)^2}{\omega_n^2} + \frac{2\xi}{\omega_n}(j\omega) + 1} = \frac{1}{-\left(\frac{\omega}{\omega_n}\right)^2 + 2\xi j\frac{\omega}{\omega_n} + 1} = \frac{1}{\left(1 - \left(\frac{\omega}{\omega_n}\right)^2\right) + j\left(2\xi\frac{\omega}{\omega_n}\right)} \Rightarrow$$

$\omega_r = \omega_n\sqrt{1-2\xi^2}$ - resonance frequency, $M_r = |H(j\omega)|_{\max} = \frac{1}{2\xi\sqrt{1-\xi^2}} = -$ resonance peak.

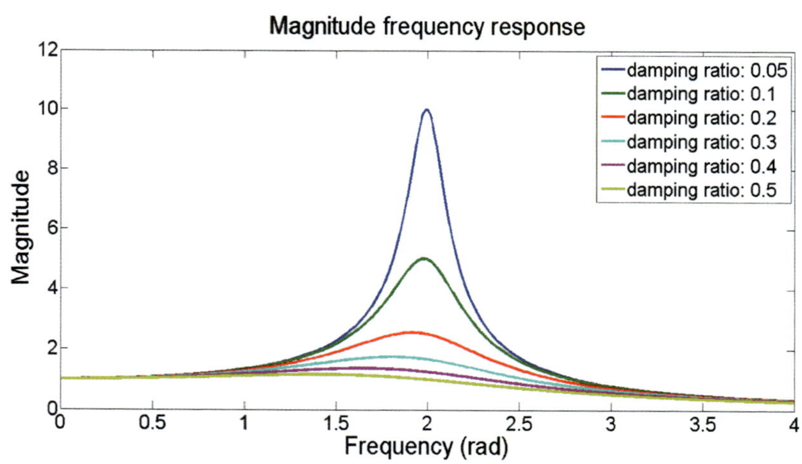

Fig. 9.3-11. Frequency response of second order system with different damping ratio.

Summary:
For a closed-loop system

$$G_{cl}(s) = \frac{\omega_n^2}{s^2 + 2\zeta\omega_n s + \omega_n^2},$$

the basic characteristics, often used as ***closed-loop controller design specifications*** (***specs***) are:

ξ — damping ratio, ω_n — natural frequency,

$\omega_d = \omega_n\sqrt{1-\xi^2}$ — damped natural frequency,

$\omega_r = \omega_n\sqrt{1-2\xi^2}$ — resonance frequency,

$M_r = \left|H(j\omega)\right|_{\max} = \dfrac{1}{2\xi\sqrt{1-\xi^2}}$ — resonance peak,

$M_p = e^{-\frac{\zeta\pi}{\sqrt{1-\zeta^2}}}$ — maximum percent overshoot,

$t_s = \dfrac{4}{\zeta\omega_n}$ — settling time,

DC gain=1, $y_{ss}=1$ — steady state response to a unit step.

Homework 10: Solved Examples and Solution Guide

Problem 1. The mold testbed at Nucor Steel Decatur is a scaled-down and highly simplified physical (meaning — hardware) model of the mold oscillation system. It consists of a large mass, which approximates the mold in the actual system, attached to a beam, with the other end of the beam driven by a hydraulic piston. Let us model the piston system by the second-order equation

$$m_p\ddot{x}_p + b_p\dot{x}_p = u$$

where x_p is the position of the piston, m_p is the piston mass = 0.001, b_p is damping due to friction = 1, and u is the force on the piston, which we control.

The disturbances caused by nonlinearities in the hydraulics can be modelled as a position disturbance x_d added to the piston position,

$$x_{p,d} = x_p + x_d.$$

The force applied by the piston to the beam at one end drives the beam/mold dynamics that can be modelled by treating the first mode of the beam as an underdamped mass-spring-damper system, so that the relation between the piston dynamics and the mold position attached to the other end of the beam takes the form

$$m_m\ddot{x}_m + b_b\dot{x}_m + k_b x_m = b_b\dot{x}_{p,d} + b_b x_{p,d},$$

where x_m is the position of the mold, m_m is the mass of the mold = 1, b_b is the damping of the beam = 100, and k_b is the stiffness of the beam = 3600. The goal is to have the mold move

according to the desired pattern (reference signal) $r(t)$. The engineers at the mill have chosen a proportional controller that sets the piston force relative to the error in mold position,

$$u = k_p(r - y)$$

where y is the measurement of the mold position x_m. This measurement is corrupted by sensor noise n:

$$y = x_m + n.$$

The proportional gain $k_p = 2$ provides good nominal performance at low frequencies.

(a) Draw this system as a block diagram

This drawing is given in the figure below:

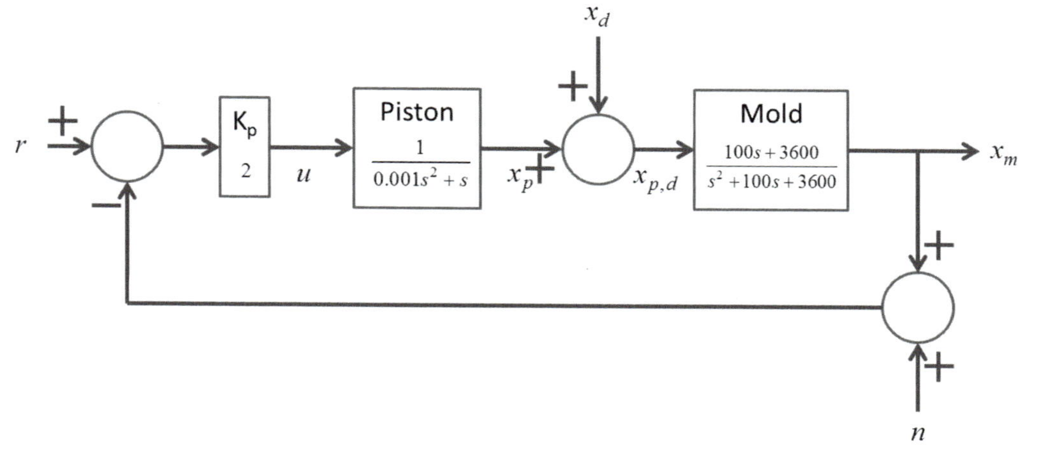

Fig. Hw10-1. System block diagram with reference input, disturbance, and measurement noise.

(b) Find the transfer functions from r, x_d, and n to the mold position x_m. Check stability of the closed-loop.

We can carry this out for each input one by one by setting all the other inputs equal to 0. Starting with the reference r:

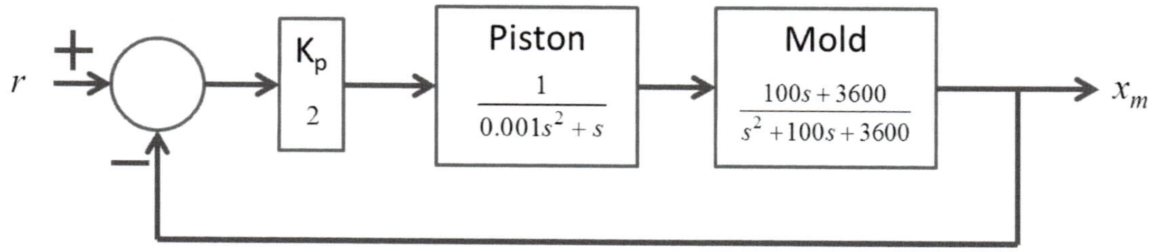

Fig. Hw10-2. System block diagram with the reference input.

This is just a standard closed-loop transfer function calculation:

$$\frac{X_m}{R}=\frac{2\cdot\frac{1}{0.001s^2+1s}\cdot\frac{100s+3600}{s^2+100s+3600}}{1+2\cdot\frac{1}{0.001s^2+1s}\cdot\frac{100s+3600}{s^2+100s+3600}}.$$

Simplifying the numerator and denominator gives

$$\begin{aligned}\frac{X_m}{R}&=\frac{2(100s+3600)}{(0.001s^2+s)(s^2+100s+3600)+2(100s+3600)}\\&=\frac{200s+7200}{0.001s^4+1.1s^3+103.6s^2+3800s+7200}.\end{aligned}$$

Moving on to the disturbance x_d, the block diagram for this input will look like

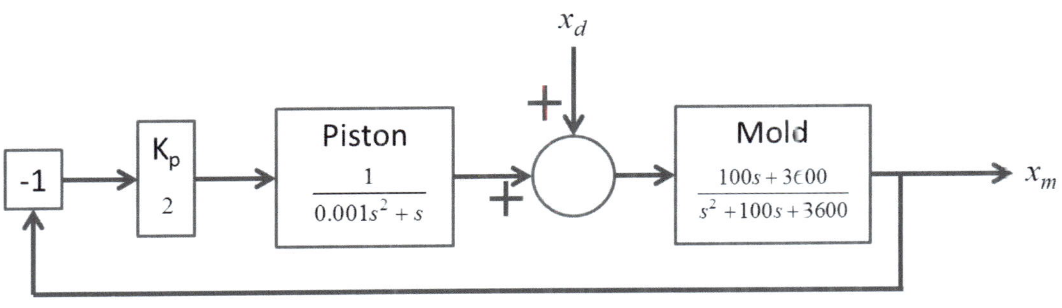

Fig. Hw10-3. System block diagram with disturbance input.

Since the output was subtracted from the reference input which is not 0, we have to multiply the output by -1 to retain the same loop behavior. Now, we rearrange this diagram as a closed-loop with x_d as the input and x_m as the output:

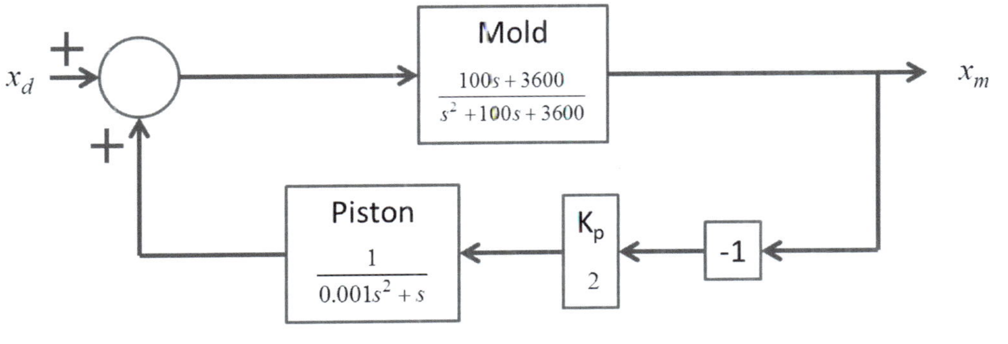

Fig. Hw10-4. Rearranged block diagram for Figure HW10-3.

This closed loop has positive instead of negative feedback, but there is a negative gain block in the feedback path which yields

$$\frac{X_m}{X_D}=\frac{\dfrac{100s+3600}{s^2+100s+3640}}{1-(-1)\cdot 2\cdot\dfrac{1}{0.001s^2+s}\cdot\dfrac{100s+3600}{s^2+100s+3600}}.$$

Simplifying the numerator and the denominator gives

$$\frac{X_m}{X_d}=\frac{(0.001s^2+s)(100s+3600)}{(0.001s^2+s)(s^2+100s+3600)+2(100s+3600)}$$

$$=\frac{0.1s^3+103.6s^2+3600s}{0.001s^4+1.1s^3+103.6s^2+3800s+7200}.$$

Finally, for the noise n the block diagram is

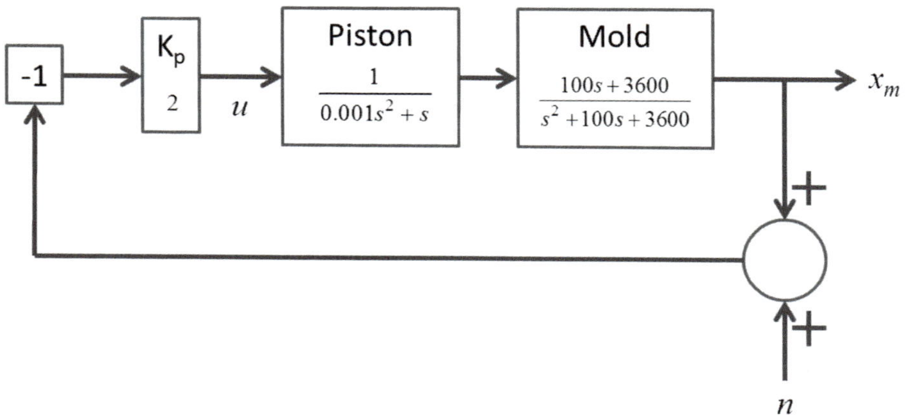

Fig. Hw10-5. System block diagram with noise input.

Rearranging it into a feedback loop:

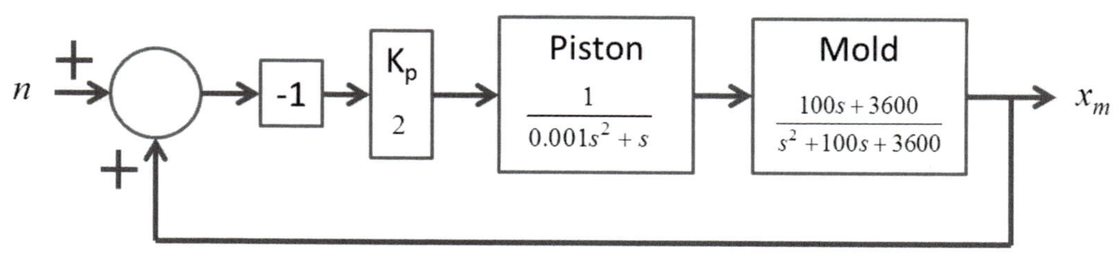

Fig. Hw10-6. Rearranged system block diagram for Figure HW10-5.

and denoting Laplace transform of n as N yields

$$\frac{X_m}{N}=\frac{-1\cdot 2\cdot\dfrac{1}{0.001s^2+s}\cdot\dfrac{100s+3600}{s^2+100s+3600}}{1-(-1)\cdot 2\cdot\dfrac{1}{0.001s^2+s}\cdot\dfrac{100s+3600}{s^2+100s+3600}}.$$

Simplifying the numerator and denominator gives

$$\begin{aligned}\frac{X_m}{N}&=\frac{-2(100s+3600)}{(0.001s^2+s)(s^2+100s+3600)+2(100s+3600)}\\&=\frac{-200s-7200}{0.001s^4+1.1s^3+103.6s^2+3800s+7200}.\end{aligned}$$

All of these transfer functions have the same denominator (as it should be, unless there are cancelled pole-zero pairs). The roots of the denominator are (using Matlab)

$$p=\{-1000,-48.9+j34.7,-48.9-j34.7,-2.00\},$$

all of which have negative real parts. Therefore the closed-loop is stable.

(c) Determine the steady-state output error $e_{o,ss}=r-x_{m,ss}$ for a unit step applied to each input using the Final Value Theorem. Based on the result, what can you say about this controller's response to low-frequency references and disturbances?

In part (b), we found the transfer functions for several inputs, with x_m as the output. However, to obtain clear performance results through the use of the Final Value Theorem, we need to first find the transfer functions with $e_o=r-x_m$ as the output. Once again, we can find each transfer function one by one for each input by setting the other inputs equal to 0.
Therefore, for the reference r,

$$\frac{E_o}{R}=\frac{R-X_m}{R}=1-\frac{X_m}{R}.$$

Substituting the transfer function from part (b),

$$\frac{E_o}{R} = 1 - \frac{X_m}{R} = 1 - \frac{200s + 7200}{0.001s^4 + 1.1s^3 + 103.6s^2 + 3800s + 7200}$$
$$= \frac{\left(0.001s^4 + 1.1s^3 + 103.6s^2 + 3800s + 7200\right) - \left(200s + 7200\right)}{0.001s^4 + 1.1s^3 + 103.6s^2 + 3800s + 7200}$$
$$= \frac{0.001s^4 + 1.1s^3 + 103.6s^2 + 3600s}{0.001s^4 + 1.1s^3 + 103.6s^2 + 3800s + 7200}.$$

If $R = 1/s$ for a unit step,

$$E_o = \frac{E_o}{R} \cdot R = \frac{0.001s^4 + 1.1s^3 + 103.6s^2 + 3600s}{0.001s^4 + 1.1s^3 + 103.6s^2 + 3800s + 7200} \cdot \frac{1}{s}$$
$$= \frac{0.001s^3 + 1.1s^2 + 103.6s + 3600}{0.001s^4 + 1.1s^3 + 103.6s^2 + 3800s + 7200}.$$

This transfer function has all negative-real-part poles, so the Final Value Theorem applies. Using this theorem yields

$$e_{o,ss} = \lim_{t\to\infty} e_o(t) = \lim_{s\to 0} sE_o(s) = \lim_{s\to 0} s\frac{0.001s^3 + 1.1s^2 + 103.6s + 3600}{0.001s^4 + 1.1s^3 + 103.6s^2 + 3800s + 7200} = 0.$$

For the disturbance x_d, we set the reference r to 0, so that

$$\frac{E_o}{X_d} = \frac{R - X_m}{X_d} = \frac{0 - X_m}{X_d} = -\frac{X_m}{X_d}.$$

Substituting the transfer function from part (b),

$$\frac{E_o}{X_d} = -\frac{0.1s^3 + 103.6s^2 + 3600s}{0.001s^4 + 1.1s^3 + 103.6s^2 + 3800s + 7200},$$

and taking $X_d = 1/s$, i.e. as a unit step, yields

$$E_o = \frac{E_o}{X_d} \cdot X_d = -\frac{0.1s^3 + 103.6s^2 + 3600s}{0.001s^4 + 1.1s^3 + 103.6s^2 + 3800s + 7200} \cdot \frac{1}{s}$$
$$= -\frac{0.1s^2 + 103.6s + 3600}{0.001s^4 + 1.1s^3 + 103.6s^2 + 3800s + 7200}.$$

Using the Final Value Theorem yields

$$e_{o,ss} = \lim_{t\to\infty} e_o(t) = \lim_{s\to 0} sE_o(s) = \lim_{s\to 0} s\left(-\frac{0.1s^2 + 103.6s + 3600}{0.001s^4 + 1.1s^3 + 103.6s^2 + 3800s + 7200}\right) = 0.$$

Similarly, for the measurement noise n,

$$\frac{E_o}{N} = \frac{R - X_m}{N} = \frac{0 - X_m}{N} = -\frac{X_m}{N} \Rightarrow$$

$$\frac{E_o}{N} = \frac{200s + 7200}{0.001s^4 - 1.1s^3 + 103.6s^2 + 3800s + 7200} \Rightarrow$$

$$E_o = \frac{E_o}{N} \cdot N = \frac{200s + 7200}{0.001s^4 + 1.1s^3 + 103.6s^2 + 3800s + 7200} \cdot \frac{1}{s} \Rightarrow$$

$$e_{o,ss} = \lim_{t \to \infty} e_o(t) = \lim_{s \to 0} sE_o(s) = \lim_{s \to 0} s\left(\frac{200s + 7200}{0.001s^4 + 1.1s^3 + 103.6s^2 + 3800s + 7200} \cdot \frac{1}{s}\right) = \frac{7200}{7200} = 1.$$

From these results, constant (0-frequency) reference signals or disturbances will not cause any error in the output. However, constant measurement noise signals will lead to 100% error. Very low-frequency measurement noise signals will behave similarly, since they are close to 0 frequency.

(d) Verify these results by simulating the response of this system to a unit step reference input with no disturbances.

This can be easily done in Simulink, as illustrated below.

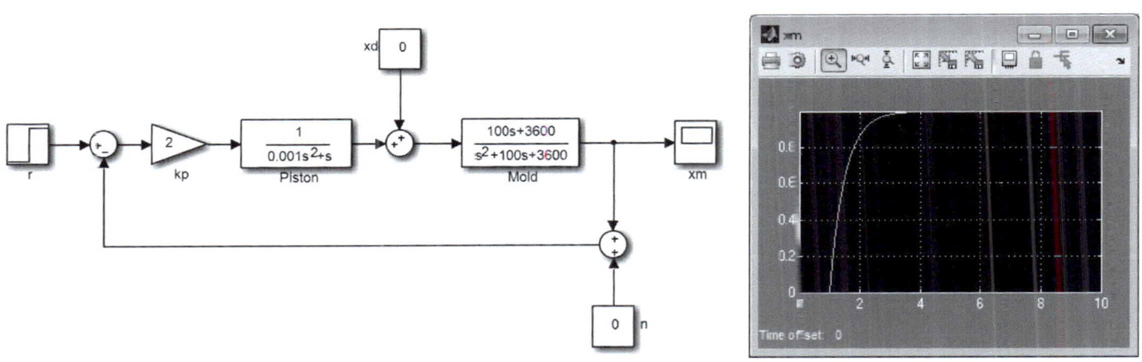

Fig. Hw10-7. Simulink model of system block diagram and output result.

Note that x_m converges to 1 in the steady-state, verifying the results of part (c).

Problem 2.

When the oscillation frequency is increased, there is a possibility of disturbances appearing at the resonance frequency of the beam (approximately 46 rad/s). The engineers plan to change the control loop to eliminate this disturbance. This new controller will have the following block diagram.

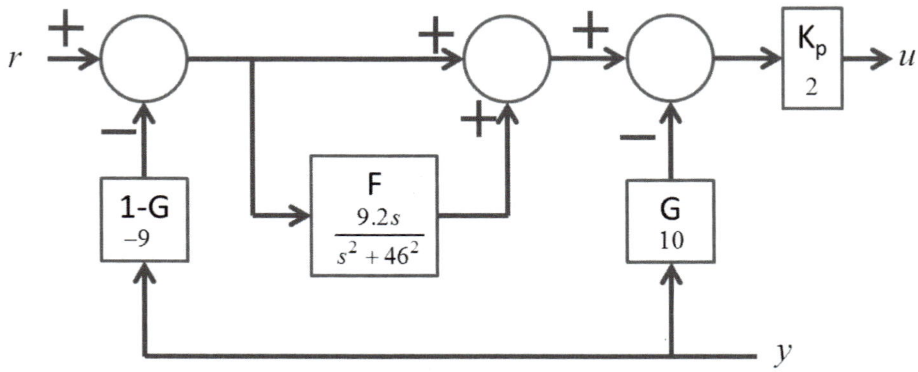

Fig. Hw10-8. New controller for Problem 2.

(a) Find the new transfer function from *r* to x_m. Check closed-loop stability.

The new block diagram takes the form (ignoring the disturbances)

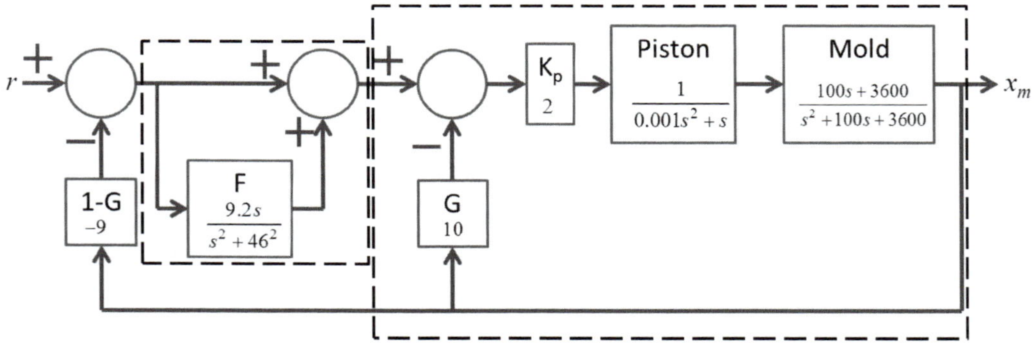

Fig. Hw10-9. System block diagram with the new controller.

To find the overall transfer function, we start by looking for ways to simplify the block diagram by combining blocks. The dashed box on the left can be simplified easily into a single transfer function by noting that an upper straight line is actually a block with gain = 1 and adding two gains together, since they are connected in parallel:

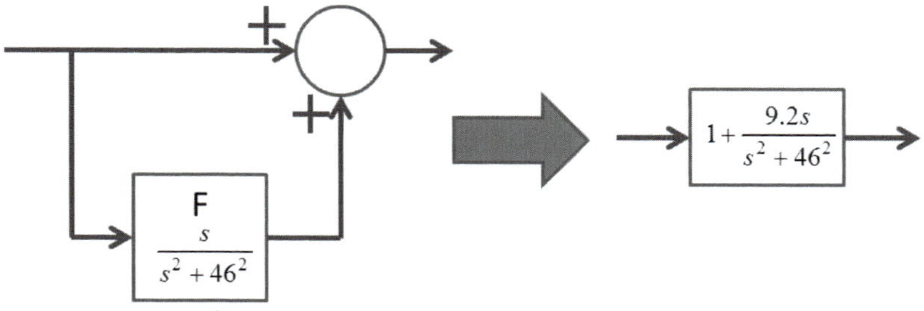

Fig. Hw10-10. The first part of the feedback loop of block diagram in Figure HW10-9.

This block can be further simplified to

$$1+\frac{9.2s}{s^2+46^2}=\frac{s^2+9.2s+2116}{s^2+2116}.$$

The dashed box on the right is a closed-loop, so we can calculate its transfer function:

$$\frac{2\cdot\dfrac{1}{0.001s^2+s}\cdot\dfrac{100s+3600}{s^2+100s+3600}}{1+10\cdot 2\cdot\dfrac{1}{0.001s^2+s}\cdot\dfrac{100s+3600}{s^2+100s+3600}}=\frac{2(100s+3600)}{(0.001s^2+s)(s^2+100s+3600)+10\cdot 2\cdot(100s+3600)}$$

$$=\frac{200s+7200}{0.001s^4+1.1s^3+103.6s^2+5600s+72000}.$$

Making these two substitutions, the block diagram is now:

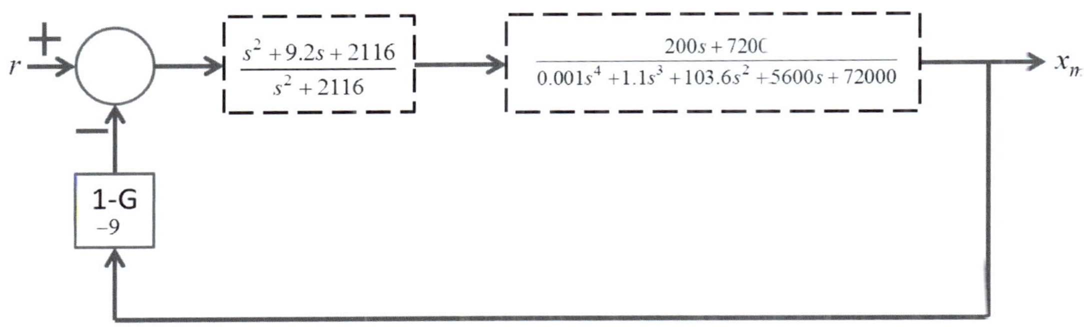

Fig. Hw10-11. Simplified system block diagram of Figure HW10-9.

This leaves one more feedback loop to calculate:

$$\frac{X_m}{R}=\frac{\dfrac{s^2+9.2s+2116}{s^2+2116}\cdot\dfrac{200s+7200}{0.001s^4+1.1s^3+103.6s^2+5600s+72000}}{1+(-9)\cdot\dfrac{s^2+9.2s+2116}{s^2+2116}\cdot\dfrac{200s+7200}{0.001s^4+1.1s^3+103.6s^2+5600s-72000}}$$

$$=\frac{(s^2+9.2s+2116)(200s+7200)}{(s^2+2116)(0.001s^4+1.1s^3+103.6s^2+5600s+72000)-9(s^2+9.2s+2116)(200s+7200)}$$

$$=\frac{200s^3+9040s^2+489440s+15235200}{0.001s^6+1.1s^5+105.716s^4+6127.6s^3-209857.6s^2+7444640s+15235200}.$$

The roots of the denominator are, using Matlab,

$$p=\{-1000,-48.8+j36.6,-48.8-j36.6,-2.17,-0.723+j43.8,-0.723-j43.8\}.$$

All the roots have negative real parts, so the closed-loop system is stable.

(b) Determine the steady-state output error $e_{o,ss} = r - x_{m,ss}$ for a unit step from r using the Final Value Theorem.

As in Problem 1(c), for the reference r,

$$\frac{E_o}{R} = \frac{R - X_m}{R} = 1 - \frac{X_m}{R}.$$

Substituting the transfer function X_m/R from part (b) yields

$$\frac{E_o}{R} = 1 - \frac{X_m}{R} = 1 - \frac{200s^3 + 9040s^2 + 489440s + 15235200}{0.001s^6 + 1.1s^5 + 105.716s^4 + 6127.6s^3 + 209857.6s^2 + 7444640s + 15235200}$$
$$= \frac{0.001s^6 + 1.1s^5 + 105.716s^4 + 5927.6s^3 + 200817.6s^2 + 6955200s}{0.001s^6 + 1.1s^5 + 105.716s^4 + 6127.6s^3 + 209857.6s^2 + 7444640s + 15235200}.$$

If $R = 1/s$ — a unit step,

$$E_o = \frac{E_o}{R} \cdot R = \frac{0.001s^6 + 1.1s^5 + 105.716s^4 + 5927.6s^3 + 200817.6s^2 + 6955200s}{0.001s^6 + 1.1s^5 + 105.716s^4 + 6127.6s^3 + 209857.6s^2 + 7444640s + 15235200} \cdot \frac{1}{s}$$
$$= \frac{0.001s^5 + 1.1s^4 + 105.716s^3 + 5927.6s^2 + 200817.6s + 6955200}{0.001s^6 + 1.1s^5 + 105.716s^4 + 6127.6s^3 + 209857.6s^2 + 7444640s + 15235200}.$$

Using the Final Value Theorem yields

$$e_{o,ss} = \lim_{t\to\infty} e_o(t) = \lim_{s\to 0} sE_o(s)$$
$$= \lim_{s\to 0} s \frac{0.001s^5 + 1.1s^4 + 105.716s^3 + 5927.6s^2 + 200817.6s + 6955200}{0.001s^6 + 1.1s^5 + 105.716s^4 + 6127.6s^3 + 209857.6s^2 + 7444640s + 15235200} = 0.$$

9.4 Basic Controller Design: Shaping the Second Order System Step Response with P and PD Controllers. Steady State Error for Unit Step and Unit Ramp Inputs. Closed-Loop Stability and Closed-Loop Poles

9.4.1 *Closed-loop position and velocity servos with P-controller under unity feedback*

Angular mirror position control system using P controller has the form

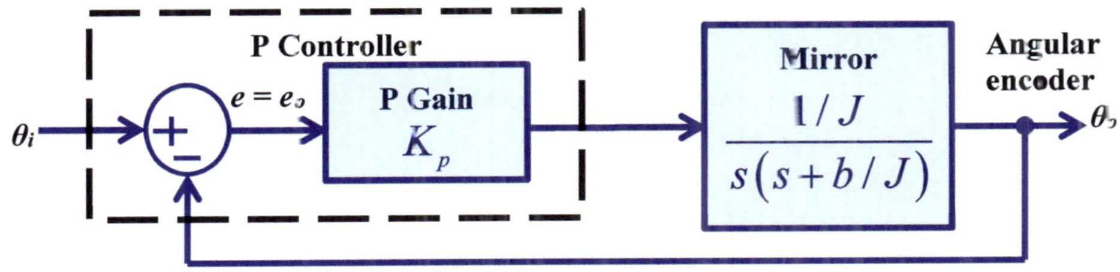

Fig. 9.4-1. Angular mirror position control system block diagram under unity feedback.

Angular mirror velocity control system using P-controller has the form

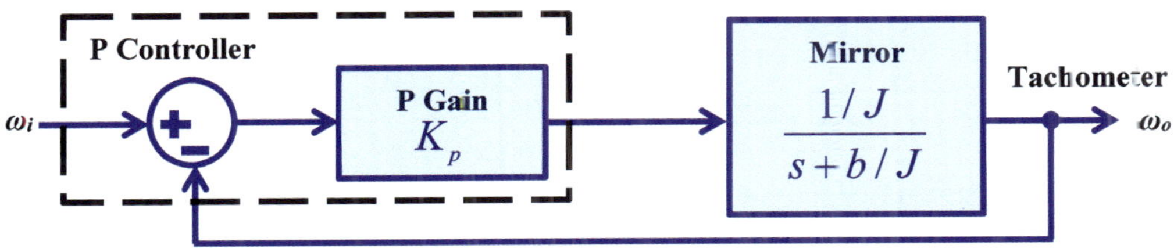

Fig. 9.4-2. Angular mirror velocity control system block diagram under unity feedback.

Steady state error

The transfer function from input to measured error with perfect sensing, i.e. to output error, as shown in Figure 9.4-3 yields:

$$\frac{E_o}{R} = \frac{1}{1+G_pG_c} \Rightarrow E_o = R\frac{1}{1+G_pG_c}. \tag{9.4-1}$$

Therefore, the steady state error for a constant input (i.e. for a unit step) using the Final Value Theorem is:

$$e_{o,ss} = \lim_{s\to 0} sR\frac{1}{1+G_pG_c} = \lim_{s\to 0} s\frac{1}{s}\frac{1}{1+G_pG_c} = \lim_{s\to 0}\frac{1}{1+G_pG_c}. \tag{9.4-2}$$

As seen in the diagram, for tracking objective "reference input $r(t)$ = desired output", the output error should be zero except for a brief transient:

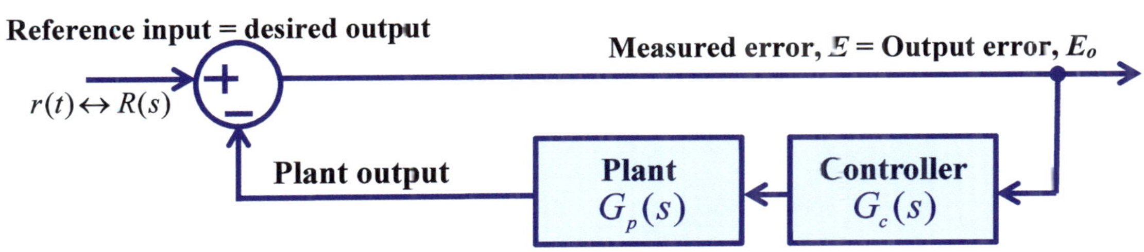

Fig. 9.4-3. Rearranged system block diagram.

Response shaping using P controller: design objective and specifications (specs)

Example 9.4-1: position control of a mirror using P controller (Figure 9.4-1).

Goal: Design a P controller for a laser steering system with $J = 1$, $b = 2$, so that

- Maximum overshoot < 5 %
- Settling time < 5 sec.

Find the steady state output error, $e_{o,ss}$, for constant and ramp inputs.

Solution:

i) Closed-loop transfer function: From Figure 9.4-1 we have

$$\frac{\theta_o}{\theta_d} = \frac{K_p \dfrac{1/J}{s(s+b/J)}}{1+K_p \dfrac{1/J}{s(s+b/J)}} = \frac{K_p/J}{s^2+(b/J)s+K_p/J} = \frac{K_p}{s^2+2s+K_p}.$$

ii) P-controller gain calculation by matching closed-loop coefficients: Matching the closed-loop transfer function coefficients to the coefficients of the 2nd order specs-generating transfer function (9.3-7) yields:

$$\frac{\theta_o}{\theta_d} = \frac{K_p}{s^2+2s+K_p} = \frac{\omega_n^2}{s^2+2\zeta\omega_n s+\omega_n^2} \Rightarrow \begin{cases} \omega_n^2 = K_p \Rightarrow \omega_n = \sqrt{K_p}; \\ 2\zeta\omega_n = 2 \Rightarrow \zeta = \dfrac{1}{\omega_n} = \dfrac{1}{\sqrt{K_p}}. \end{cases}$$

iii) P-controller gain calculation using maximum overshoot: Maximum percent overshoot < 5%:

$$M_p = e^{-\frac{\zeta\pi}{\sqrt{1-\zeta^2}}} = 0.05 \Rightarrow \zeta = 0.69 \Rightarrow K_p = \frac{1}{\zeta^2} = \frac{1}{0.69^2} \approx \boxed{2.1.}$$

iv) Settling time provided by P controller: Settling time < 5:

$$t_s = \frac{4}{\zeta\omega_n} = 4 < 5.$$

Settling time spec is satisfied.

v) P-controller response shaping limitations: Thus, we can meet specs by choosing $K_p = 2.1$:

$$\frac{\theta_o}{\theta_d} = \frac{K_p}{s^2+2s+K_p} = \frac{\omega_n^2}{s^2+2\zeta\omega_n s+\omega_n^2} = \frac{2.1}{s^2+2s+2.1}.$$

Specs are satisfied only because they are not tight. We can't actually change both M_p and t_s with just a P controller.

vi) Output error under P controller: Now, for the output error we have:

$$\frac{E_o}{\theta_d} = \frac{1}{1+G_pG_c} = \frac{1}{1+2.1\frac{1}{s(s+2)}} = \frac{s^2+2s}{s^2+2s+\boxed{2.1}} \leftarrow \textbf{P gain}$$

vii) Steady-state error for unit *step* input under P controller: For step input $\theta_d(s)=1/s$ the steady-state error is given by

$$E_o = \frac{1}{s}\frac{s^2+2s}{s^2+2s+2.1} \Rightarrow e_{o,ss} = \lim_{s\to 0} sE_o = \lim_{s\to 0} s\frac{1}{s}\frac{s^2+2s}{s^2+2s+2.1} = 0.$$

Thus, closed loop has perfect steady-state tracking for unit step input. Why? Zero steady-state error is obtained due to the presence of the integral action provided by the lack of the restoring force as indicated in Figure 9.4-4 below:

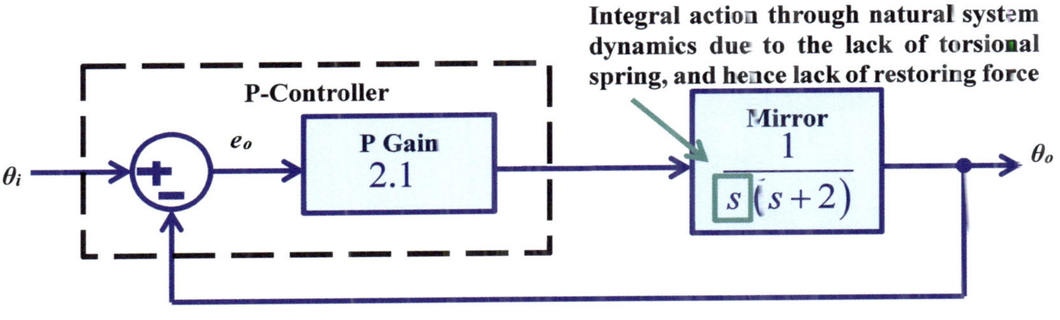

Fig. 9.4-4. Angular Position Control-Block Diagram (J = 1, b = 2, P GAIN = 2.1).

This configuration differs from velocity control with I controller of the 1st order systems, such as motor or mirror (Figure 9.4-2) in velocity servos. Why?

Because the resulting closed-loop systems with I controller, although mathematically identical to the system in Figure 9.4-4, perform speed control, not position control. The system in Figure 9.4-4 performs the latter, and employs for this purpose also an integrator, but in the *plant* transfer function, instead of the *controller* transfer function, giving 0 steady state error with just P control action in the controller block.

viii) Steady-state error for unit *ramp* input under P-controller: For ramp input $\theta_d(s)=1/s^2$ the steady-state error is given by:

$$E_o = \frac{1}{s^2}\frac{s^2+2s}{s^2+2s+2.1} \Rightarrow e_{o,ss} = \lim_{s\to 0} sE_o = \lim_{s\to 0} s\frac{1}{s^2}\frac{s^2+2s}{s^2+2s+2.1} = \lim_{s\to 0}\frac{s+2}{s^2+2s+2.1} = \frac{2}{2.1} \approx 1.$$

Thus, closed loop does not provide perfect steady-state tracking for unit ramp input, as shown in Figure 9.4-5:

Ramp input $\theta_d(s) = \frac{1}{s^2} \Rightarrow e_{ss} = \frac{2}{\text{P gain} = 2.1} \approx 1 \Rightarrow$

$\theta_d(t) - \theta_0(t) \approx 1 \Rightarrow \theta_0(t) \approx \theta_d(t) - 1.$

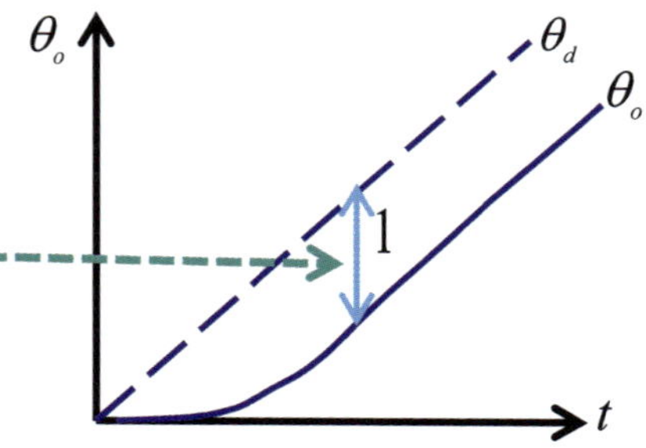

Fig. 9.4-5. System output tracking error with ramp input.

ix) Closed-loop stability under P controller: Closed-loop transfer function is

$$\frac{\theta_o}{\theta_d} = \frac{K_p}{s^2 + 2s + K_p}.$$

The closed-loop poles are the roots of the 2nd order polynomial with positive coefficients. This guarantees closed-loop stability (poles with negative real part) for any value of K_p, no matter how big.

9.4.2 *Response shaping using PD controller*

Example 9.4-2: Mirror position control using unity feedback PD-controller given in Figure 9.4-6:

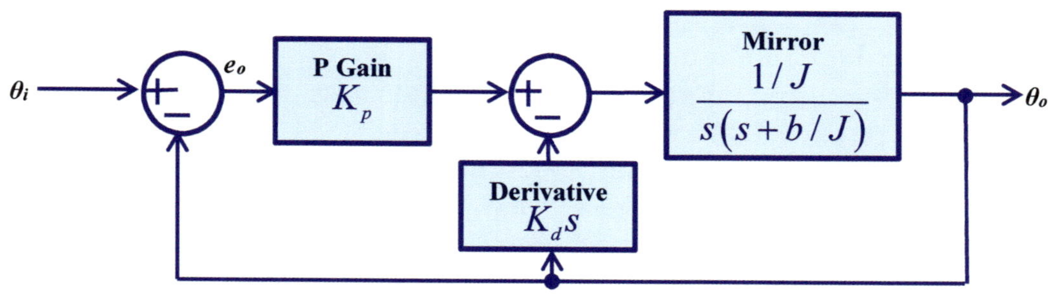

Fig. 9.4-6. Block diagram for Example 9.4-2. This D controller placement reduces control signal spike due to step input.

Goal: Design a PD controller for a mirror with $J = 1$, $b = 2$, so that

- Maximum overshoot < 5%
- Settling time < 2 sec.

Find the steady state error with step and ramp inputs.

Solution: There are actually two topologies, shown in Figure 9.4-7, of implementing PD controller given above in practice — through *measuring and differentiating the displacement to obtain velocity*, or through *measuring the full state* (both displacement and velocity) by adding a velocity sensor.

Closed-loop topologies under PD controller: Both topologies have the same transfer function.

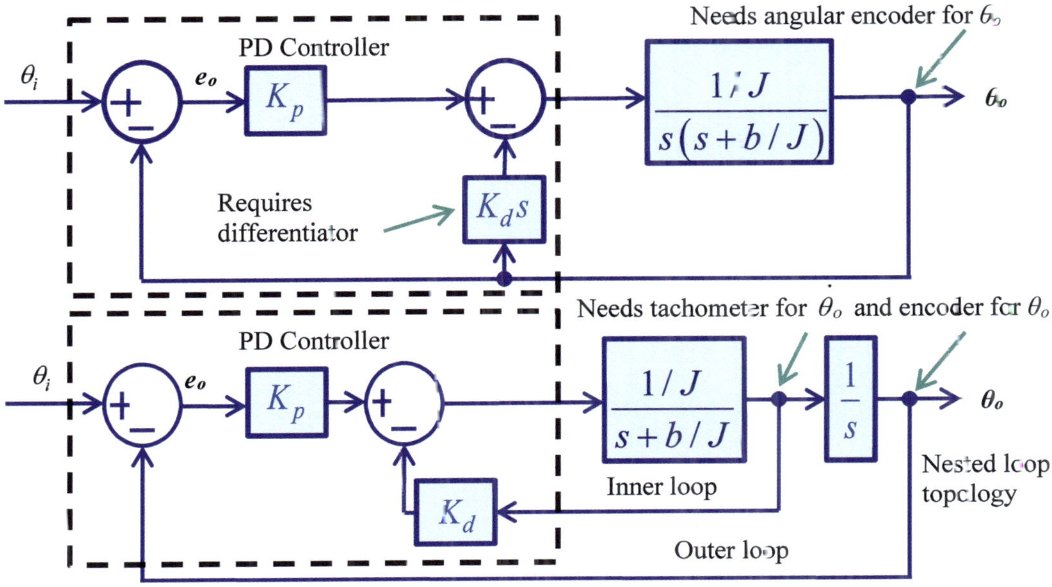

Fig. 9.4-7. Two PD controller implementation topologies.

i) Closed-loop transfer function: The latter can be found through calculating closed-loop transfer functions of first the inner loop and then the outer loop in Figure 9.4-7, or by decomposing PD controller topology in Figure 9.4-6 into feedback path controller and prefilter as

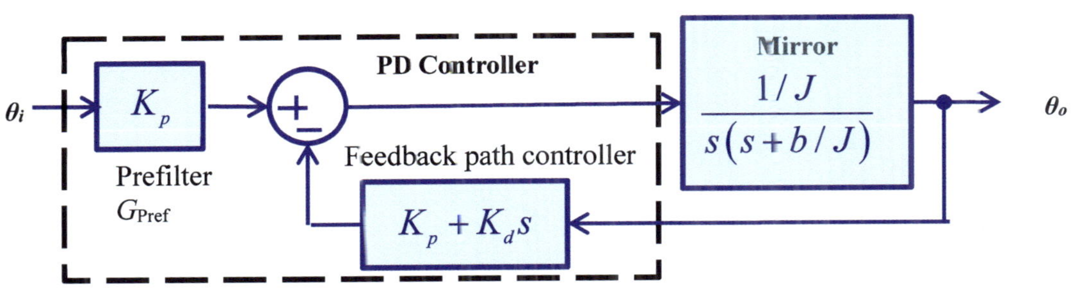

Fig. 9.4-8. Rearranged system block diagram for Figure 9.4-6.

yielding

DC gain = 1, i.e. in steady state $\vartheta_o = \vartheta_d$

$$\frac{\theta_o}{\theta_d} = G_{\text{Pref}} \frac{\begin{pmatrix}\text{Forward}\\ \text{Gain}\end{pmatrix}}{1+\begin{pmatrix}\text{Feedback}\\ \text{Gain}\end{pmatrix}\begin{pmatrix}\text{Forward}\\ \text{Gain}\end{pmatrix}} = K_p \frac{\dfrac{(1/J)}{s(s+b/J)}}{1+\left(K_p + K_d s\right)\dfrac{(1/J)}{s(s+b/J)}} = \frac{\dfrac{K_p}{J}}{s^2 + \dfrac{K_d + b}{J}s + \dfrac{K_p}{J}} = \frac{K_p}{s^2 + \left(K_d + 2\right)s - K_p}.$$

ii) PD-controller gain calculation by matching closed-loop coefficients:

$$\frac{\theta_o}{\theta_d} = \frac{K_p}{s^2 + (2 + K_d)s + K_p} = \frac{\omega_n^2}{s^2 + 2\zeta\omega_n s + \omega_n^2} \Rightarrow \begin{cases} \omega_n^2 = K_p \Rightarrow \omega_n = \sqrt{K_p}; \\ 2\zeta\omega_n = 2 + K_d \Rightarrow \zeta = \dfrac{2+K_d}{2\omega_n} = \dfrac{2+K_d}{2\sqrt{K_p}}. \end{cases}$$

So, we can use the relationships found earlier to calculate maximum percent overshoot and settling time:

iii) D gain calculation through settling time: Unlike the P controller case, here D gain can be used to tune settling time through changing damping ζ, or, conversely, given settling time (2 sec per specs) we can find D gain:

$$t_s = \frac{4}{\zeta\omega_n} = 2 \Rightarrow 2 = \zeta\omega_n = \frac{2+K_d}{2\sqrt{K_p}}\sqrt{K_p} = \frac{2+K_d}{2} \Rightarrow K_d = 2.$$

iv) P gain calculation through maximum overshoot: Coefficients matched above yield P gain as follows:

$$\zeta = \frac{2+K_d}{2\sqrt{K_p}} \Rightarrow 2\zeta\sqrt{K_p} = 2 + K_d \Rightarrow K_p = \left(\frac{2+K_d}{2\zeta}\right)^2.$$

Calculating damping ζ from maximum percent overshoot < 5% yields the value of P gain:

$$M_p = e^{-\frac{\zeta\pi}{\sqrt{1-\zeta^2}}} = 0.05 \Rightarrow \zeta = 0.69 \Rightarrow K_p = \left(\frac{2+K_d}{2\zeta}\right)^2 = \left(\frac{2}{0.69}\right)^2 \approx 8.4.$$

Substituting the values of control gains into block diagram in Figure 9.4-6 yields the desired closed-loop system:

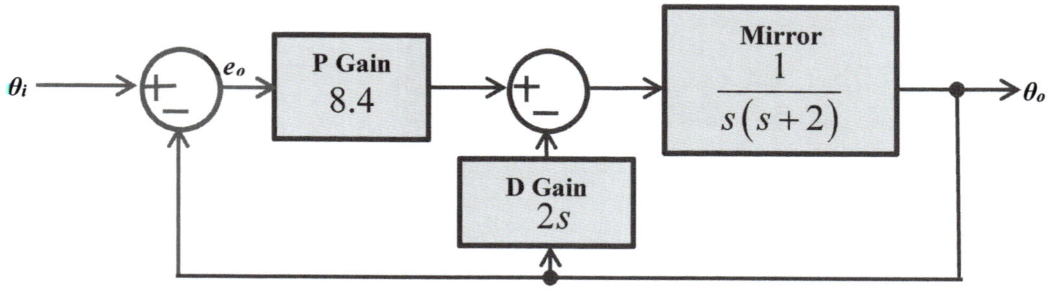

Fig. 9.4-9. System block diagram meets design goal.

This topology doesn't need prefilter to track constant inputs for a given plant.

v) PD-controller step response shaping capabilities: Focusing on the denominator coefficients of the closed-loop systems with PD controller, it is seen that natural frequency and damping ratio can be independently set to arbitrary values by tuning P and D controller gains:

$$\frac{\theta_o}{\theta_d}=\frac{K_p}{s^2+\left(2+K_d\right)s+K_p}=\frac{\omega_n^2}{s^2+2\zeta\omega_n\,s+\omega_n^2}.$$

vi) Output error under PD-controller: For the output error we have:

$$\frac{\theta_o}{\theta_d}=\frac{8.4}{s^2+4s+8.4}\Rightarrow\frac{E_o}{\theta_d}=\frac{\theta_d-\theta_0}{\theta_d}=1-\frac{\theta_o}{\theta_d}=1-\frac{8.4}{s^2+4s+8.4}=\frac{s^2+4s}{s^2+4s+8.4}.$$

This transfer function has the same form as that for the closed-loop error with P controller, yielding the steady-state errors:

$$\text{Step input: } \theta_d\left(s\right)=\frac{1}{s}\quad\Rightarrow\quad e_{o,ss}=\lim_{s\to 0}sE_o(s)=\lim_{s\to 0}s\frac{s^2+4s}{s^2+4s+8.4}\frac{1}{s}=0,$$

$$\text{Ramp input: } \theta_d\left(s\right)=\frac{1}{s^2}\quad\Rightarrow\quad e_{o,ss}=\lim_{s\to 0}sE_o(s)=\lim_{s\to 0}s\frac{1}{s^2}\frac{s^2+4s}{s^2+4s+8.4}=\frac{4}{8.4}\approx 0.5.$$

vii) Closed-loop stability under PD controller: Closed loop is stable for any positive values of P and D controller gains.

The previous example had only inertia and friction. What if there is a restoring force acting on the system?

Example 9.4-3: Design PD controller for mirror positioning system with mirror rotation subject to restoring force, as shown in Figure 9.4-10:

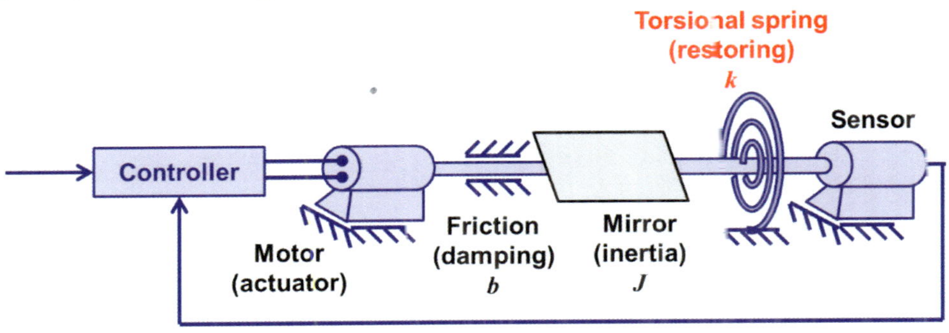

Fig. 9.4-10. Layout of the mirror positioning system with restoring force.

System parameters are: $J=1$, $b=2$, $k=4$. Closed-loop system performance specs are:

- Settling time < 2
- Maximum overshoot < 5%
- Perfect tracking of step (constant) inputs, i.e. transfer function at $s=0$ is equal to 1.

i) Closed-loop transfer function: For this plant we will use PD controller in prefilter form, shown in Figure 9.4-11. The latter is chosen to address the loss of the closed-loop unity DC gain, as seen next.

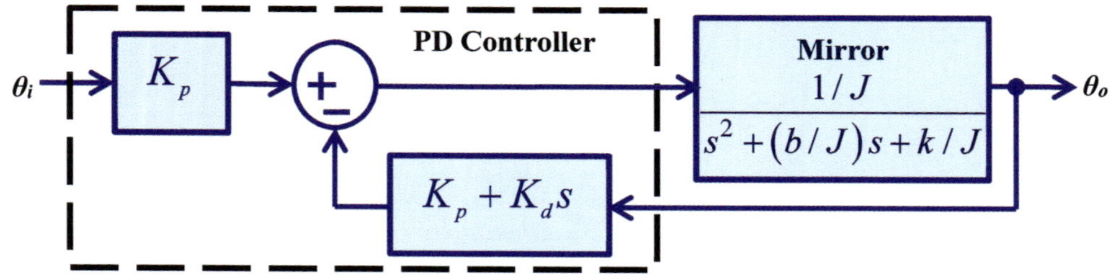

Fig. 9.4-11. Block diagram of the mirror positioning system with restoring force.

Substituting $J = 1$, $b = 2$, $k = 4$, the closed-loop transfer function then is given by

$$\frac{\theta_o}{\theta_d} = K_p \frac{\dfrac{1}{s^2+2s+4}}{1+\left(K_p+K_d s\right)\cdot\dfrac{1}{s^2+2s+4}} = \frac{K_p}{s^2+\left(2+K_d\right)s+\left(4+K_p\right)}.$$

ii) Setting closed-loop transfer function DC gain to one: The prefilter form is chosen to clearer show the required change in prefilter gain. Denoting DC gain as K_{DC} and checking its value yields

$$K_{DC} = \left.\frac{K_p}{s^2+\left(2+K_d\right)s+\left(4+K_p\right)}\right|_{s=0} = \frac{K_p}{4+K_p} \neq 1.$$

This shows that the DC gain is not 1, making this PD controller unable to track constant inputs!

To make closed-loop DC gain $K_{DC} = 1$ we simply multiply the previous closed-loop transfer function by the inverse of the resulting K_{DC}, i.e multiply it by $4 + K_p$ and divide by K_p:

$$\frac{\theta_o}{\theta_d} = \frac{4+K_p}{K_p}\frac{K_p}{s^2+\left(2+K_d\right)s+\left(4+K_p\right)} = \frac{4+K_p}{s^2+\left(2+K_d\right)s+\left(4+K_p\right)}.$$

This resets the prefilter gain from K_p to $4 + K_p$ as given in the block diagram in Figure 9.4-12.

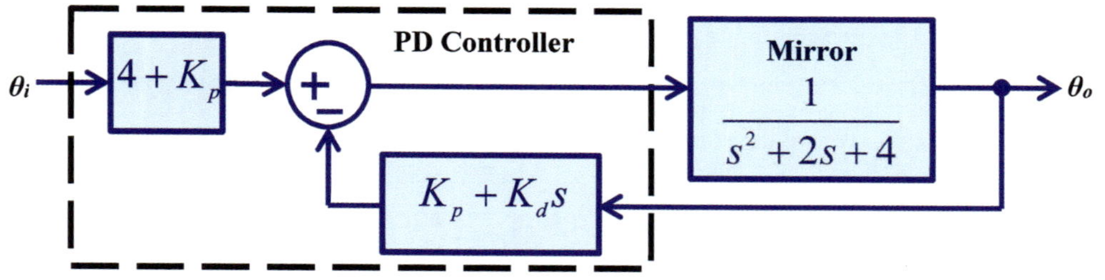

Fig. 9.4-12. Block diagram of the mirror positioning system with restoring force and DC gain = 1.

iii) PD-controller gain calculation by matching closed-loop coefficients:

$$\frac{\theta_o}{\theta_d}=\frac{4+K_p}{s^2+(2+K_d)s+\left(4+K_p\right)}=\frac{\omega_n^2}{s^2+2\zeta\omega_n s+\omega_n^2} \Rightarrow \begin{cases} 2+K_d=2\zeta\omega_n \Rightarrow \zeta\omega_n=\dfrac{2+K_d}{2}; \\ 4+K_p=\omega_n^2 \Rightarrow \omega_n=\sqrt{4+K_p}. \end{cases}$$

iv) D gain calculation through settling time: Settling time < 2:

$$t_s=\frac{4}{\zeta\omega_n}=2 \Rightarrow 2=\zeta\omega_n=\frac{2+K_d}{2} \Rightarrow K_d=2.$$

v) P gain calculation through maximum overshoot: Coefficients matched above yield P gain as follows:

$$\zeta=\frac{\zeta\omega_n}{\omega_n}=\frac{2+K_d}{2\sqrt{4+K_p}}=\frac{4}{2\sqrt{4+K_p}}=\frac{2}{\sqrt{4+K_p}} \Rightarrow K_p=\left(\frac{2}{\zeta}\right)^2-4.$$

Calculating damping ζ from maximum percent overshoot < 5% yields the value of P gain:

$$M_p=e^{-\frac{\zeta\pi}{\sqrt{1-\zeta^2}}}=0.05 \Rightarrow \zeta=0.69 \Rightarrow K_p=\left(\frac{2}{0.69}\right)^2-4\approx 4.4.$$

Substituting the values of control gains into block diagram in Figure 9.4-12 yields the desired closed-loop system.

9.5 Basic Controller Design: Disturbance Rejection Using Internal Model Principle, Measurement Noise Rejection through Loop Frequency Response Shaping

In this section we will connect infinite controller gain to the closed-loop blocking zeros and recast the feedback capability of turning the open-loop poles into the closed-loop zeros into a general controller design methodology. The latter will then be used it Section 9.6 for solving the mold velocity distortion problem.

9.5.1 *Controller gain over the entire frequency range*

Infinite controller gain over the entire frequency range: filter interpretation.
Let's recall the values of closed-loop tracking and disturbance rejection transfer functions and the corresponding conditions on controller gain for perfect close-loop performance

$$\theta_o=\underbrace{\frac{G_pG_c}{1+G_pG_c}}_{W_{tr,cl}(s)}\theta_i+\underbrace{\frac{G_p}{1+G_pG_c}}_{W_{d,cl}(s)}T_d \tag{9.5-1}$$

Closed-loop performance as $|G_c| \rightarrow \infty$

- Tracking performance
  - $W_{tr,cl} \rightarrow 1$
- Disturbance rejection performance
  - $W_{d,cl} \rightarrow 0$
- Performance robustness:
  - if $G_{p,true} = G_{p,model} + \Delta$ (actual plant is different from expected model)

  then we still have $W_{tr,cl} \approx 1$ and $W_{d,cl} \approx 0$

Let's now look at what $|G_c| = \infty$ means in the frequency domain, viewing the closed-loop transfer functions as *filters for the inputs*. The output is then seen to be sum of *the filtered versions of the inputs*:

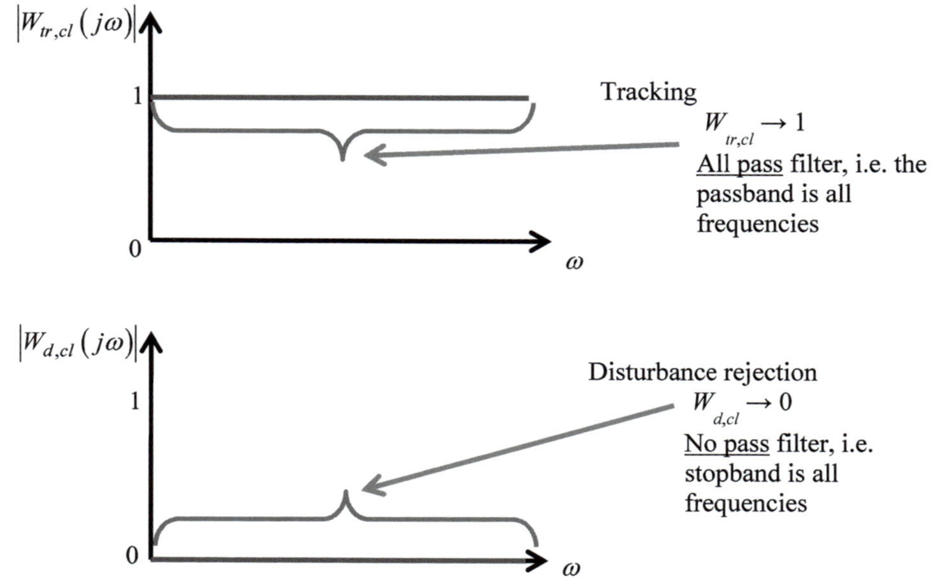

Fig. 9.5-1. Closed-loop magnitude response for perfect tracking and disturbance rejection.

In reality, we can't achieve $|G_c| = \infty$ over the entire frequency range, so what can we do?

Key idea — infinite controller gain over the limited frequency range.
If the *frequency ranges of the reference and the disturbance inputs are limited and we know them*, we can make the design goal less restrictive: tune the passband of the tracking transfer function and the stopband of the disturbance transfer functions using G_c as a design tool as follows:

- Tracking performance: $W_{tr,cl} \rightarrow$ *pass signals only in the tracking frequency range.*
- Disturbance rejection performance: $W_{d,cl} \rightarrow$ *reject signals only in the disturbance frequency range.*

With the latter in mind, we can pick G_c so that $|G_c| \to \infty$ *only for frequencies in the range of interest* as long as there is no severe conflict between these two performance objectives.

Example:

- for constant signals ($\omega = 0$), integral control gain, $|G_c| = |1/s|_{s=j\omega} = |1/j\omega|$, is infinite
- for quickly changing signals ($\omega = \infty$), derivative control, $|G_c| = |s|_{s=j\omega} = |j\omega|$, is infinite.

Infinite controller gain over the limited frequency range: performance trade-off.

We can try to pick G_c so that $|G_c| \to \infty$ *only for frequencies in the range of interest*

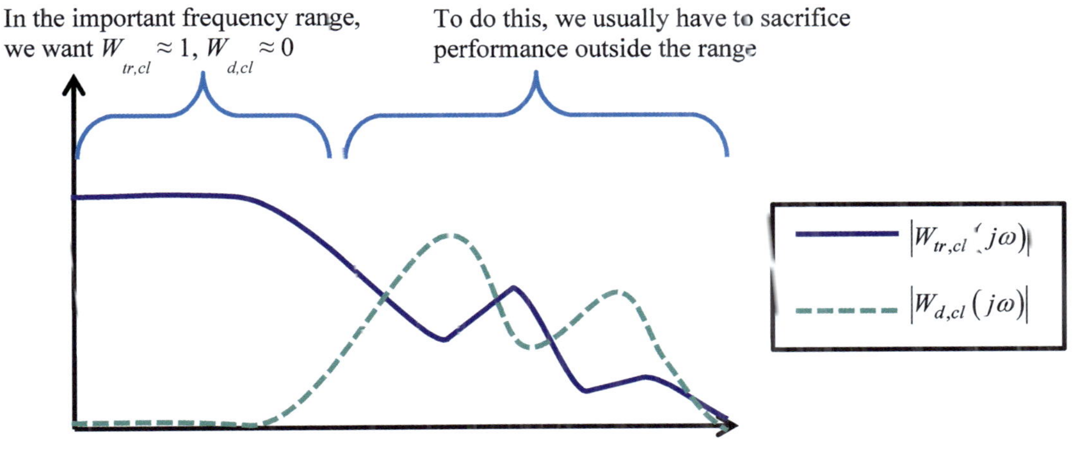

Fig. 9-5.2. Output result with $G_c \to \infty$.

Closed-loop transfer functions from reference input and disturbance to output error.

Let's go back to the basic closed-loop system with unity feedback, Recall the position servo transfer functions:

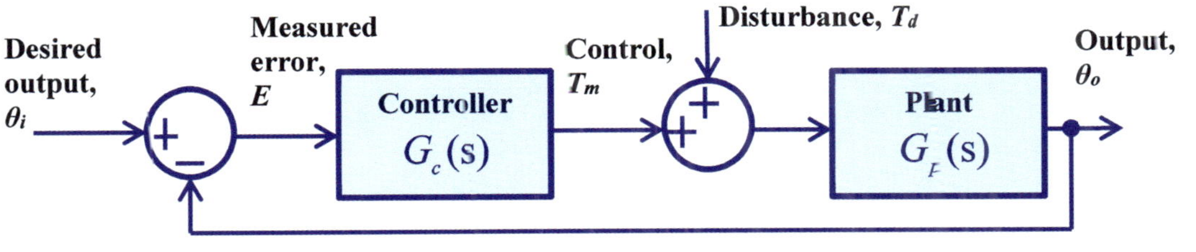

Fig. 9.5-3. Basic control loop system with unity feedback.

$$\theta_o = \frac{G_p G_c}{1 + G_p G_c}\theta_i + \frac{G_p}{1 + G_p G_c}T_d.$$

Since unity feedback means perfect sensing, the output error and the measured error will be the same. Instead of output, let's consider the transfer functions for output error:

$$E_o = \theta_i - \theta_o = \theta_i - \left(\frac{G_p G_c}{1+G_p G_c}\theta_i + \frac{G_p}{1+G_p G_c}T_d \right) = \frac{1}{1+G_p G_c}\theta_i - \frac{G_p}{1+G_p G_c}T_d. \quad (9.5\text{-}2)$$

9.5.2 *Closed-loop transfer function for infinite controller gain at zero frequency: I-controller*

Let's now see how the idea of restricting infinite gains only to frequencies of interest works.

Example 9.5-1: Assume that the plant is given by the first-order system. Examples of the first-order systems are:

- Cooling/heating dynamics of a thermal mass.
- Angular velocity ("speed") dynamics of a velocity servo with a mirror or a motor.

Let's specify the controller: to be the I-controller, which has infinite gain at $\omega = 0$. This yields

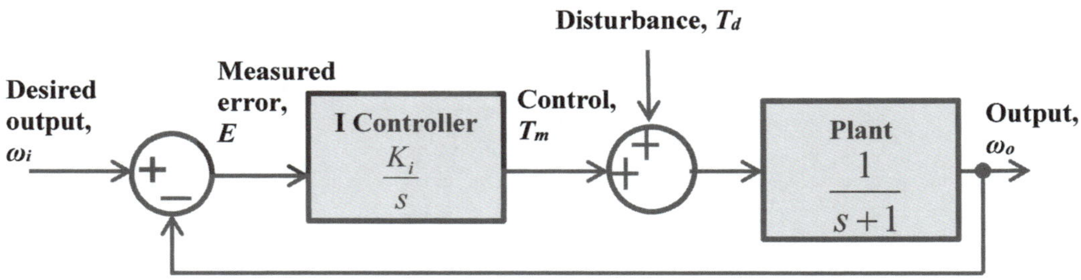

Fig. 9.5-4. The first-order velocity servo closed-loop system.

The closed-loop output error is then given by

$$E_o = \frac{1}{1+G_p G_c}\omega_i - \frac{G_p}{1+G_p G_c}T_d = \frac{1}{1+\frac{K_i}{s}\frac{1}{s+1}}\omega_i - \frac{\frac{1}{s+1}}{1+\frac{K_i}{s}\frac{1}{s+1}}T_d$$

$$= \frac{s(s+1)}{s^2+s+K_i}\omega_i - \frac{s}{s^2+s+K_i}T_d = \frac{(s+1)}{s^2+s+K_i}s\omega_i - \frac{1}{s^2+s+K_i}sT_d.$$

Integral control action gives rise to blocking zeros at zero in the output error equation

Output error response to the unit step input (for tracking performance check) $\omega_i(s) = 1/s$, $T_d(s) = 0$, is

$$E_o = \frac{s(s+1)}{s^2+s+K_i}\frac{1}{s} - \frac{s}{s^2+s+K_i}0 = \frac{s+1}{s^2+s+K_i} \Rightarrow$$

$$e_{o,ss} = \lim_{s\to 0} sE_o = \lim_{s\to 0} s\frac{s+1}{s^2+s+K_i} = \lim_{s\to 0}\frac{s^2+s}{s^2+s+K_i} = \frac{0}{K_i} = 0,$$

yielding perfect steady-state tracking.

Output error response to the unit disturbance input (for disturbance rejection performance check) $T_d(s)=1/s,\ \omega_i(s)=0$, is

$$E_o=\frac{s(s+1)}{s^2+s+K_i}0+\frac{s}{s^2+s+K_i}\frac{1}{s}=\frac{1}{s^2+s+K_i}\Rightarrow$$

$$e_{o,ss}=\lim_{s\to 0}sE_o=\lim_{s\to 0}s\frac{1}{s^2+s+K_i}=\lim_{s\to 0}\frac{s}{s^2+s+K_i}=\frac{0}{K_i}=0,$$

yielding perfect steady-state disturbance rejection.

9.5.3 *Internal Model Principle (IMP) controller*

Based on the example of Figure 9.5-4 it is seen that in the steady state, an integral controller:

- Perfectly tracks a unit step (constant) reference.
- Perfectly rejects a unit step (constant) disturbance.

Why?

The I controller represents a model, 1/s, of both exogenous signals — constant input and constant disturbance — and puts this model inside the closed loop.

In this example, I controller is a realization of one of the key control concepts, referred to as **the Internal Model Principle**: to reject an exogenous input, its model must be inside the loop.

For this reason, I controller here represents an **internal model principle (IMP)** controller.

9.5.3.1 *Application of the internal model principle to disturbance rejection*

Summarizing the results of Example 9.5-1 leads to conclusion that to reject a disturbance **i) the disturbance model must be in the loop** and **ii) the closed loop must be stable**. Let's now consider a sinusoidal disturbance.

Example 9.5-2: Assume that a velocity servo in Figure 9.5-4 has a sinusoidal disturbance T_d and frequency ω_{dist}. Then the controller/(internal model) candidates are

$$T_d(t)=\begin{cases}\text{i) } \cos(\omega_{dist}t)\cdot 1(t)\overset{\mathcal{L}}{\rightleftharpoons}\dfrac{s}{s^2+\omega_{dist}^2}, \\ \text{or} \\ \text{ii) } \sin(\omega_{dist}t)\cdot 1(t)\overset{\mathcal{L}}{\rightleftharpoons}\dfrac{\omega_{dist}}{s^2+\omega_{dist}^2}.\end{cases}$$

Taking the first disturbance model i) yields the system

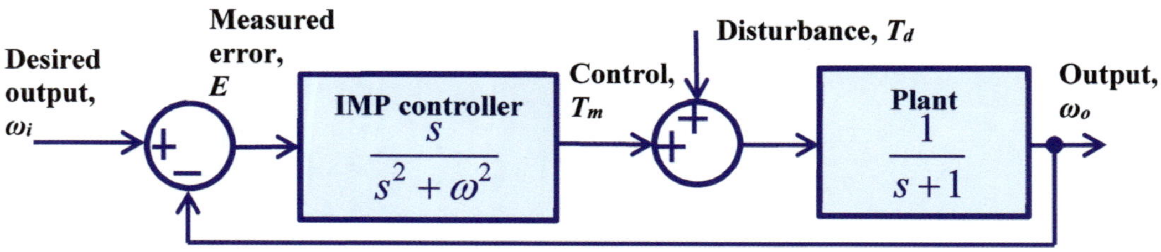

Fig. 9.5-5. System in Figure 9.5-4 with IMP controller.

and the disturbance-to-(output-error) transfer function:

$$\frac{E_o(s)}{T_d(s)} = \frac{-G_p}{1+G_pG_c} = \frac{-\frac{1}{s+1}}{1+\frac{1}{s+1}\frac{s}{s^2+\omega_{dist}^2}} = \frac{-(s^2+\omega_{dist}^2)}{(s+1)\left(s^2+\omega_{dist}^2\right)+s} = \frac{-(s^2+\omega_{dist}^2)}{\underbrace{1}_{a_3}s^3+\underbrace{1}_{a_2}s^2+\underbrace{\left(1+\omega_{dist}^2\right)}_{a_1}s+\underbrace{\omega_{dist}^2}_{a_0}}.$$

This transfer function is stable if each coefficient is positive and $a_2a_1 > a_3a_0$, which holds for any choice of ω: $1\cdot\left(1+\omega_{dist}^2\right) > 1\cdot\omega_{dist}^2$. Had we chosen $G_c(s) = \frac{\omega_{dist}}{s^2+\omega_{dist}^2}$ for the controller, the closed loop would not have been unconditionally stable. Finally, the steady-state output error is:

$$e_{o,ss} = \lim_{s\to 0} sE_o(s) = \lim_{s\to 0} s\left(\frac{E_o(s)}{T_d(s)}\right)T_d(s) = -\lim_{s\to 0} s\left(\frac{s^2+\omega_{dist}^2}{s^3+s^2+\left(1+\omega_{dist}^2\right)s+\omega_{dist}^2}\right)\frac{s}{s^2+\omega_{dist}^2}$$

$$= -\lim_{s\to 0}\frac{s^2}{s^3+s^2+\left(1+\omega_{dist}^2\right)s+\omega_{dist}^2} = -\frac{0}{\omega_{dist}^2} = 0.$$

Thus, the disturbance is rejected in the steady-state! The magnitude response of the transfer function $E_o(j\omega)/T_d(j\omega)$ given in Figure 9.5-6 shows a drop to zero at the disturbance frequency.

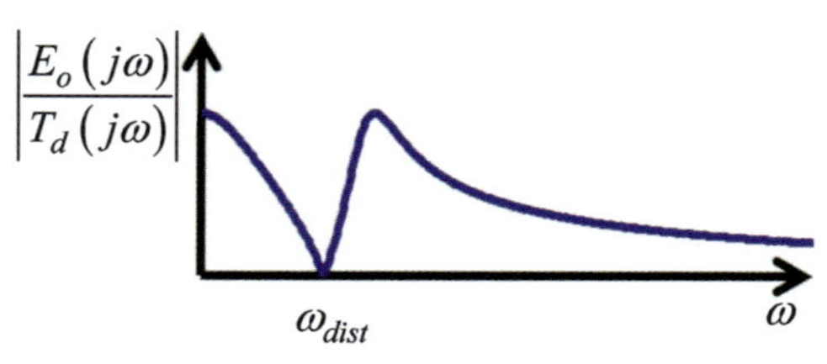

Fig. 9.5-6. The disturbance-to-output-error frequency response.

9.5.3.2 *Internal model controller selection for closed-loop stability of position servo*

Example 9.5-3: Consider a more complicated case — a 2nd order plant, e.g. position servo, given by

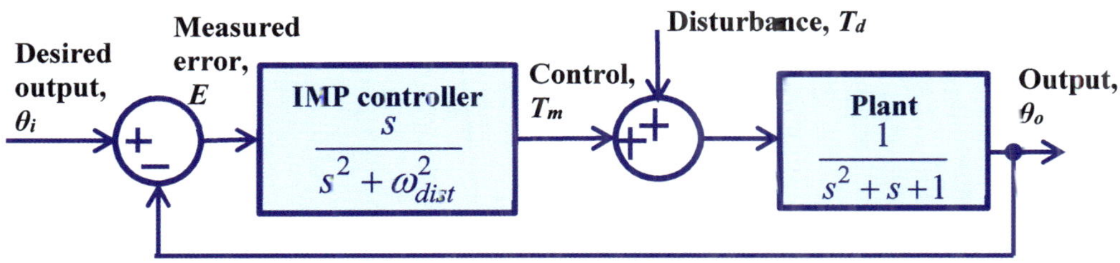

Fig. 9.5-7. A second order system with IMP controller.

The transfer function $E_o(s)/T_d(s)$ from disturbance to output error is then given by

$$\frac{E_o(s)}{T_d(s)} = \frac{-G_p}{1+G_pG_c} = \frac{-\dfrac{1}{s^2+s+1}}{1+\dfrac{1}{s^2+s+1}\dfrac{s}{s^2+\omega_{dist}^2}} = \frac{-(s^2+\omega_{dist}^2)}{\underbrace{1}_{a_4}s^4+\underbrace{1}_{a_3}s^3+\underbrace{\left(1+\omega_{dist}^2\right)}_{a_2}s^2+\underbrace{\left(1+\omega_{dist}^2\right)}_{a_1}s+\underbrace{\omega_{dist}^2}_{a_0}}.$$

Stability criterion for the fourth-order characteristic polynomial (red — violation):

$a_n > 0, \quad n = 0,\ldots,4$ ✓

$a_3a_2 > a_4a_1$: $1\cdot\left(1+\omega_{dist}^2\right) > 1\cdot\left(1+\omega_{dist}^2\right)$ ✗

$a_3a_2a_1 > a_4a_1^2 + a_3^2a_0$: $1\cdot\left(1+\omega_{dist}^2\right)\cdot\left(1+\omega_{dist}^2\right) > 1\cdot\left(1+\omega_{dist}^2\right)^2 + (1)^2\cdot\omega_{dist}^2$ ✗

yields unstable closed loop! We can check that internal model of sine will also give an unstable loop.

Let's now check out the following idea: just like in a PI controller, let's ***add a proportional gain*** to IMP controller to obtain the PI controller generalization — the "P-IMP" controller.

This yields the system of the form

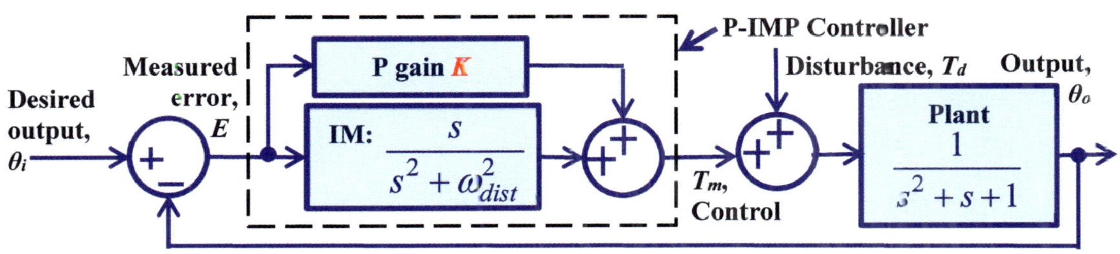

Fig. 9.5-8. System with an IMP controller and a P controller connected in parallel to form a P-IMP controller.

with the transfer function $E_o(s)/T_d(s)$ from disturbance to output error given by

$$\frac{E_o(s)}{D(s)}=\frac{-G_p}{1+G_pG_c}=\frac{-\dfrac{1}{s^2+s+1}}{1+\dfrac{1}{s^2+s+1}\left(K+\dfrac{s}{s^2+\omega_{dist}^2}\right)}=\frac{-(s^2+\omega_{dist}^2)}{\underbrace{1}_{a_4}s^4+\underbrace{1}_{a_3}s^3+\underbrace{\left(1+\omega_{dist}^2+K\right)}_{a_2}s^2+\underbrace{\left(1+\omega_{dist}^2\right)}_{a_1}s+\underbrace{\omega^2+K\omega_{dist}^2}_{a_0}}.$$

Checking stability for a fourth-order polynomial yields:

$a_n>0,\ n=0,\dots,4$ ✓

$a_3a_2>a_4a_1 \Downarrow$

$1\cdot\left(1+\omega_{dist}^2+K\right)>1\cdot\left(1+\omega_{dist}^2\right)$ ✓

$a_3a_2a_1>a_4a_1^2+a_3^2a_0 \Downarrow$

$1\cdot\left(1+\omega_{dist}^2+K\right)\cdot\left(1+\omega_{dist}^2\right)>1\cdot\left(1+\omega_{dist}^2\right)^2+(1)^2\cdot\left(\omega_{dist}^2+K\omega_{dist}^2\right)\Downarrow$

$K\left(1+\omega_{dist}^2\right)>\left(\omega_{dist}^2+K\omega_{dist}^2\right)$

✓ if $K>\omega_{dist}^2$, ✗ if $K\le\omega_{dist}^2$

Consequently, the closed-loop system can be made stable for any ω_{dist} by choosing K large enough.

Thus, although IMP is an effective controller design tool, we still need to ascertain system stability and assess the resulting system behavior. Now, we are ready to solve the Nucor problem.

9.5.3.3 *Mold oscillator velocity distortion removal in the testbeds and the production unit*

The problem is solved through introducing a P-IMP controller similar to that in Figure 9.5-8:

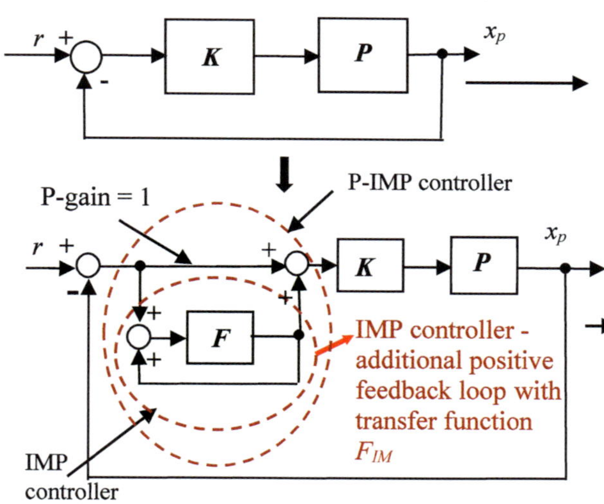

Problem: x_p tracks sinusoidal signal r at frequency f_{ref} well, but has small undesirable sinusoidal components at frequencies kf_{ref}, $k=2, 3$, which strongly distort mold velocity.

Solution: inserting into the forward path a P-IMP controller, where the IMP part is implemented as a positive feedback loop containing filter F, ensures that x_p continues tracking r well and has no undesirable sinusoids at frequencies kf_{ref}, $k=2, 3$.

Fig. 9.5-9. Mold velocity distortion suppression through introduction of the P-IMP controller.

$F(s)$ of (H-9, Appendix H) yields the single resonance frequency IMP controller, $F_{IM}(s)$, as follows:

$$F=\frac{2e\omega_{res}s}{s^2+2e\omega_{res}s+\omega_{res}^2}\Rightarrow F_{IM}=\frac{F}{1-F}=\frac{\dfrac{2e\omega_{res}s}{s^2+2e\omega_{res}s+\omega_{res}^2}}{1-\dfrac{2e\omega_{res}s}{s^2+2e\omega_{res}s+\omega_{res}^2}}=\frac{2e\omega_{res}s}{s^2+2e\omega_{res}s+\omega_{res}^2-2e\omega_{res}s}=e\frac{2\omega_{res}s}{s^2+\omega_{res}^2},$$

where F_{IM} is seen to be realized through positive feedback loop around F and e serves as the IMP controller tuning knob to reduce the effect of F_{IM} on closed loop at frequencies other than ω_{res}.

To simultaneously cancel the unwanted sinusoids at resonance frequencies $\omega_{1res} = 2\pi 2f_{ref}$ and $\omega_{2res} = 2\pi 3f_{ref}$, a similar multi-frequency controller F_{IM} with F given in Ref. 3, Section 4.4:

$$F = \frac{\prod_{i=1}^{2}\left(s^2 + 2\zeta\omega_{ires}s + \omega_{ires}^2\right) - \prod_{i=1}^{2}\left(s^2 + \omega_{ires}^2\right)}{\prod_{i=1}^{2}\left(s^2 + 2\zeta\omega_{ires}s + \omega_{ires}^2\right)} \Rightarrow F_{IM} = \frac{F}{1-F} = \frac{\prod_{i=1}^{2}\left(s^2 + 2\zeta\omega_{ires}s + \omega_{ires}^2\right) - \prod_{i=1}^{2}\left(s^2 + \omega_{ires}^2\right)}{\prod_{i=1}^{2}\left(s^2 + \omega_{ires}^2\right)}$$

is implemented at Nucor with 0.001 sec sampling rate. The topology of Figure 1.6-3, Chapter 1, used in Problem 2 of Homework 10, offers a wider range of controller K gains for stable operation when IMP controller is turned on and off. Its production variant is described in Ref. 3, Chapters 5 and 7.

Figure 9.5-10 shows that for the reference frequency $f_{ref} = 4.4$ Hz, the mold oscillator resonance peaks at $f_{1res} = 2f_{ref} = 8.8$ Hz and $f_{2res} = 3f_{ref} = 13.2$ Hz are significantly reduced. This reduction produced practically distortion-free mold velocity profiles during casting, solving the problem.

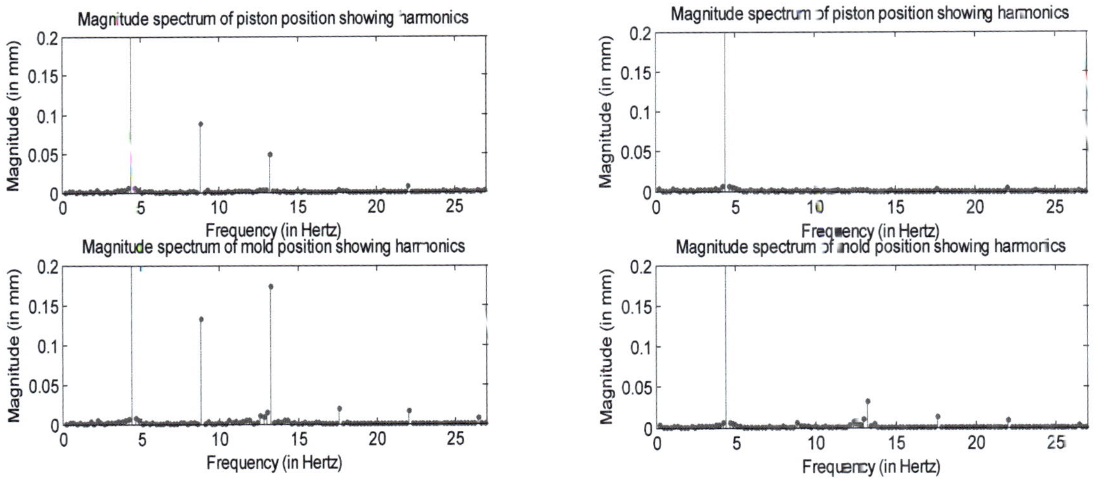

Fig. 9.5-10. Magnitude spectrum of piston and mold position: left — without IMP controller, right — with IMP controller.

The singe resonance frequency testbed distortion problem solution is detailed in Section 9.6.

9.5.4 *Measurement noise rejection. 2DOF controller topology. RST structure*

Measurement noise, like disturbances, is an input to the system that is outside of our control. However, unlike disturbance, it acts on the *measurement*, not the *plant* itself, which makes it harder to deal with. Why? Let's look at the transfer function.

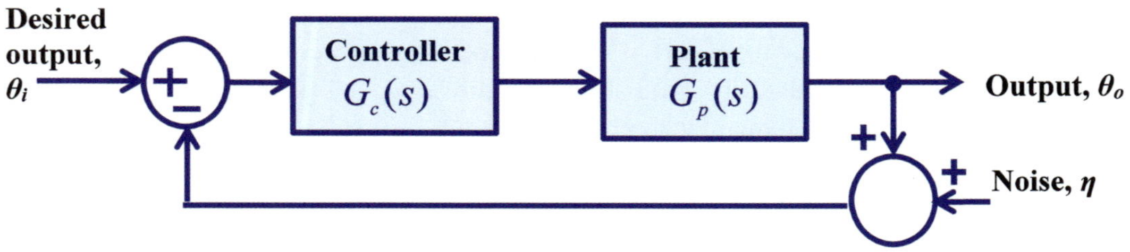

Fig. 9.5-11. Basic control loop system with unity feedback and noise.

Ignoring the reference for now, let's redraw the loop with noise being the input:

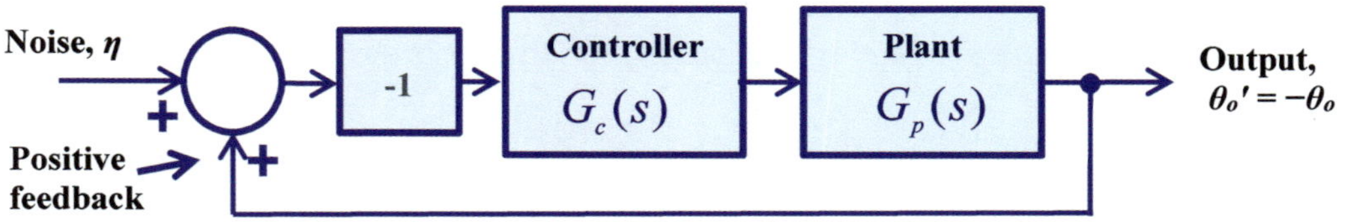

Fig. 9.5-12. Block diagram transformation by setting noise as the input for Figure 9.5-10.

This yields noise-to-output transfer function

$$G_{cl,noise} = \frac{\text{output}}{\text{noise}} = \frac{(\text{forward gain} = \text{-}G_cG_p)}{1-(\text{forward gain} = \text{-}G_cG_p)\cdot(\text{feedback gain} = 1)} = \boxed{\frac{\theta_o}{\eta} = -\frac{G_cG_p}{1+G_cG_p}.} \quad (9.5\text{-}4)$$

Let's now look at the resulting closed-loop transfer function $G_{cl,noise} = \dfrac{\theta_o'}{\eta} = \dfrac{G_cG_p}{1+G_cG_p}$.

This transfer function, except for a "-" sign difference between θ_o' and θ_o, is identical to the input/output one, which means that perfect tracking (for which transfer function = 1, acting as all pass filter) also implies *no noise rejection*! The only option in this case is *loop shaping*: making the closed-loop gain close to 1 and to 0 in, respectively, the reference and the noise frequency range.

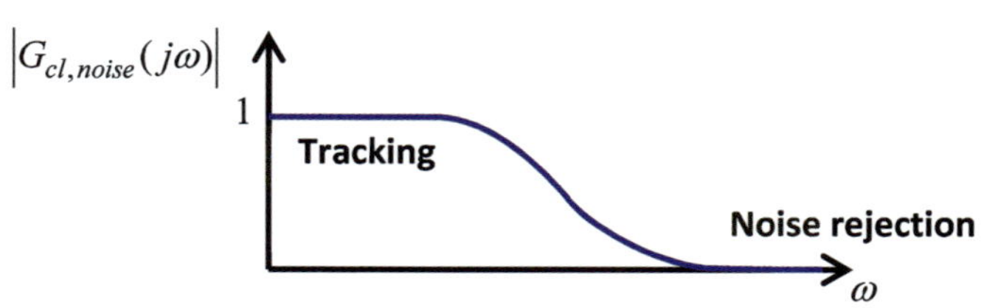

Fig. 9.5-13. Frequency domain loop shaping.

How? Put a transfer function into the feedback path, as shown in Figure 9.5-14. This yields the closed-loop transfer function (9.5-5), where $|G_{c2}|$ can be made small in the noise frequency range, e.g. can be a low-pass filter to attenuate propagation of the high frequency noise to the output.

$$\frac{\theta_o}{\theta_i} = \frac{G_{c1}G_p}{1+G_{c1}G_{c2}G_p}; \quad \frac{\theta_o}{\eta} = -\frac{G_{c1}G_{c2}G_p}{1+G_{c1}G_{c2}G_p}, \tag{9.5-5}$$

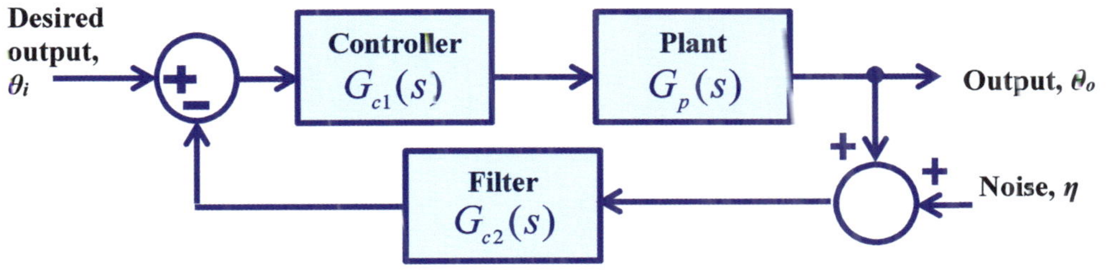

Fig. 9.5-14. Block diagram with the filter in the feedback path.

Adding a prefilter, G_{pref}, to the topology in Figure 9.5-14 yields a ***"2DOF" control system***, also referred to as the ***RST structure***, where R, S, and T stand respectively, for G_{c2}, $1/G_{c1}$, and G_{pref}.

9.6 Feedback-Based Mold Oscillator Problem Solution on the Hardware Testbed. PID Gain Tuning, Integrator Anti-Windup

9.6.1 *Mold oscillator problem*

Let's now consider in detail removal of the mold oscillation distortion depicted in Figure 9.6-1.

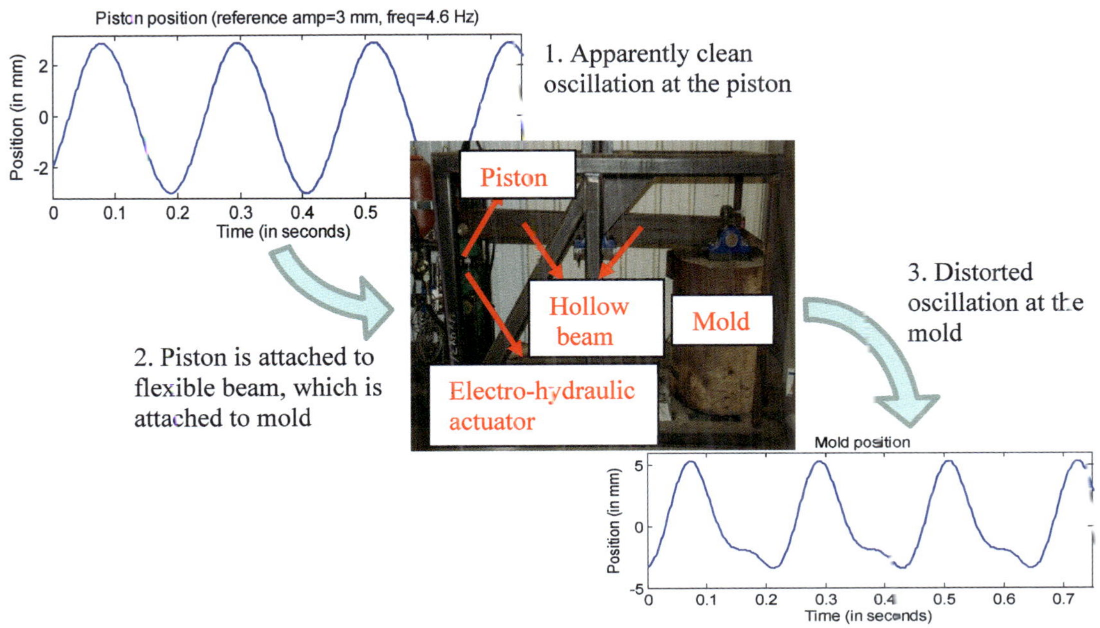

Fig. 9.6-1. Mold oscillator problem illustration by hardware testbed: distortion of the mold position sinusoidal profile.

Comparing the motion of piston on the left in Figure 9.6-2 with that of the mold on the right, Nucor engineers were concluding that it was the beam assembly that distorted mold oscillation.

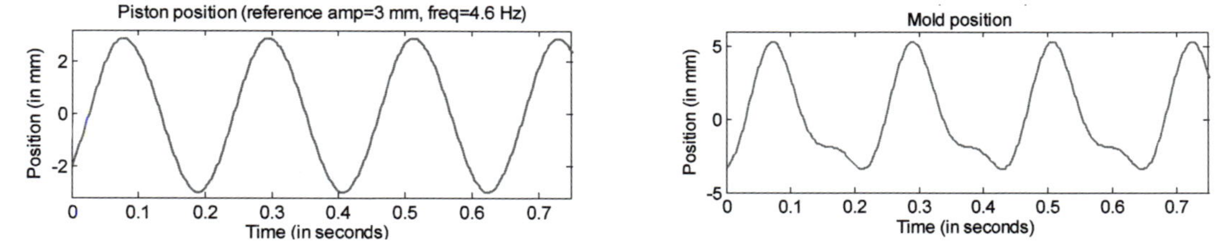

Fig. 9.6-2. Time domain data: piston and mold position.

Upon closer examination, "distortions" in the mold oscillation were found to be actually sinusoidal and take place at exactly integer multiples of the driving oscillation frequency, as shown in the right plot in Figure 9.6-3.

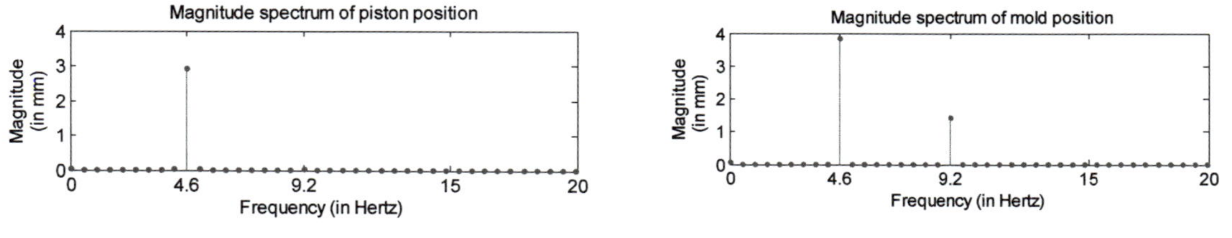

Fig. 9.6-3. Magnitude spectra of piston and mold position.

Further signal analysis revealed that the mold oscillation signal was actually multiscale — it was found to contain the same integer multiples of the driving oscillation frequency, but with the magnitudes so small that they were not seen on either time domain, or frequency domain graphs, requiring zooming, as in Figure 9.6-4, or using PSD and FFT, to see them.

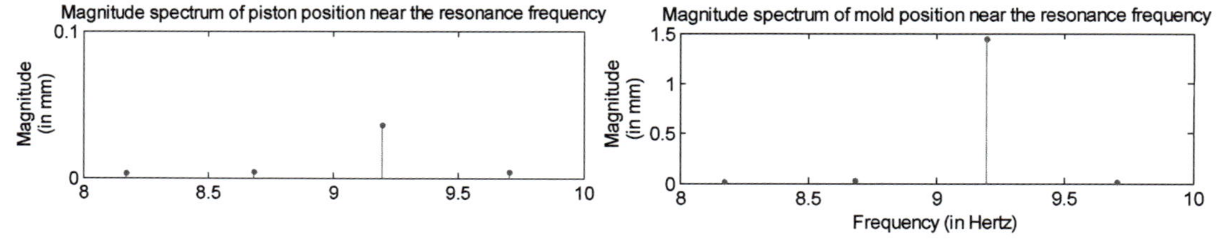

Fig. 9.6-4. Zoomed in spectrum: piston and mold position.

Further on, by collecting data, it was found that the beam acted as a strongly underdamped second-order system (similarly to a mass-spring-damper system), with the transfer function given by (9.6-1) and frequency response shown in Figures 9.6-5 and 9.6-6.

$$\frac{Y(s)}{U(s)} = \frac{4.9161s + 2918}{s^2 + 1.482s + 3272} \tag{9.6-1}$$

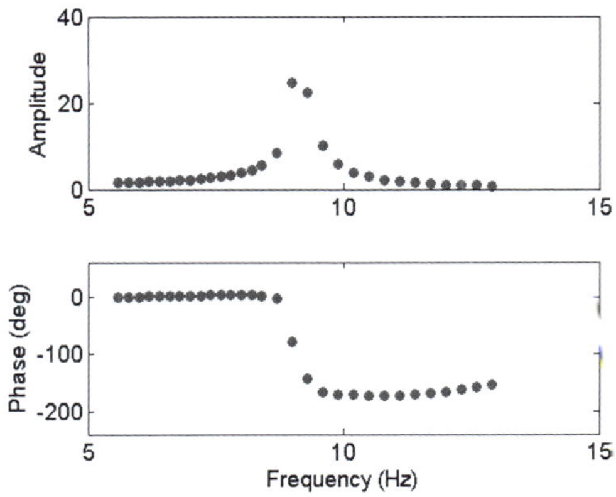

Fig. 9.6-5. Data collected during sine-sweep.

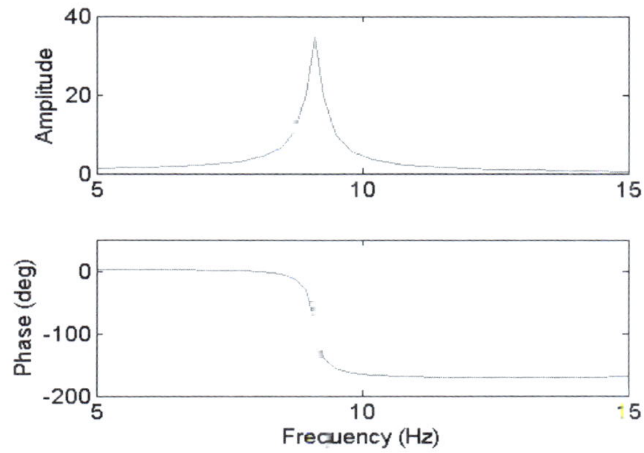

Fig. 9.6-6. Frequency response of the identified model.

Thus, it has become clear that the beam was not creating the distortions, but rather amplifying at its resonance frequency the distortions originating in the piston.

9.6.2 *Solution of the mold velocity distortion problem*

Thus, by *eliminating distortions in the piston* motion we can *eliminate distortions in the mold displacement* and, hence, mold velocity profiles. These ***distortions*** are *generated by hydraulics nonlinearity*, ***not removable*** through actuator ***resizing*** or redesign.

Approximating nonlinearity by a linear system and an external disturbance.
The *main part of the actuator response* to the reference signal — the actuator response fundamental harmonic — is, however, easily checked to be *linear with respect to the reference input* (through displaying the same magnitude scaling).

Therefore, the effect of the hydraulics nonlinearity is seen to be *limited only to the appearance of harmonic multiples* in the piston motion. This allows us to *treat harmonic multiples as external disturbance* (exogenous input) and approximate a ***single nonlinear actuator model*** with ***two*** distinct models:

i) the linear actuator model, and

ii) the exogenous periodic signal *added to linear actuator model at its output and contaiting only the harmonic multiples of interest.*

This makes the entire actuator closed-loop system linear, as shown in Figure 9.6-7. The disturbance here is added to the plant output, rather than plant input, as in Figure 9.5-8.

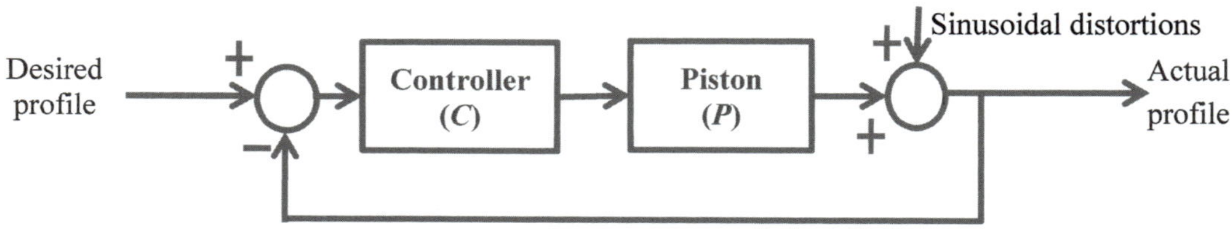

Fig. 9.6-7. Linear hydraulic actuator representation through adding external disturbance at the actuator output.

How do we remove the disturbances of specific frequencies (e.g. the second multiple of the fundamental frequency) from the system? We are going to use for this purpose the Internal Model Principle.

Solution of the mold velocity distortion problem: internal model principle
Intuitive idea: we want controller C to have infinite gain at the frequency of the distortions, so that

$$\left|\frac{\text{Actual Profile Error}}{\text{Distortions}}\right| = \left|\frac{1}{1+CP}\right|_{|C|=\infty} = 0. \tag{9.6-2}$$

Another wrinkle: the current controller runs well for other frequencies, and we don't want to lose that. So, we will add the internal model principle controller as an augmentation filter instead:

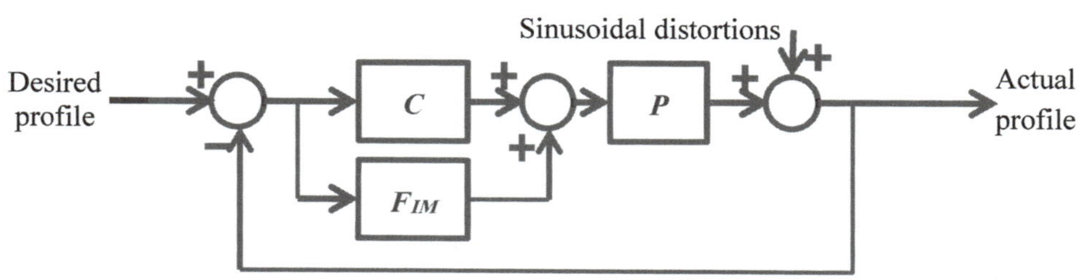

Fig. 9.6-8. Solution of the mold velocity distortion problem: controller topology.

where C is the nominal controller and F_{IM} is the internal model principle augmentation filter.

Example 9.6-1: Design P-IMP controller for the hardware testbed to provide tracking of the reference sinusoid at 4.6 Hz.
Solution: We will use the approach of Figure 9.5-9, presented in a simplified form in Figures 9.6-7 and 9.6-8.
i) Plant and controller structure choice: Let's take P to be the linearized piston model, with mass and friction

$$P(s) = \frac{1}{2s^2 + 1000s}.$$

Let's take C to be constant (proportional) gain, chosen to give good tracking of the reference sinusoid at 4.6 Hz, as

$$C(s)=10^5.$$

Let F_{IM} be internal model of the 9.2 Hz sinusoid given by

$$F_{IM}(s)=\frac{s}{s^2+(2\pi 9.2)^2}\approx\frac{s}{s^2+3342}.$$

ii) Closed-loop block diagram and transfer function: The block diagram of the system then takes the form

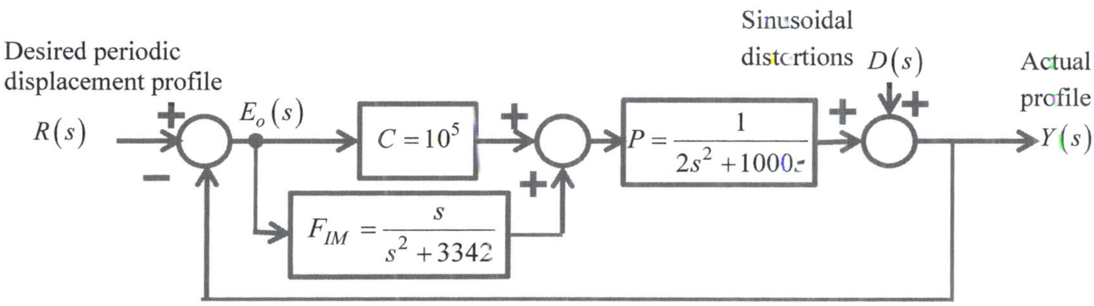

Fig. 9.6-9. Solution of the mold velocity distortion problem: controller values.

Without internal model (i.e. with $F = 0$), the closed-loop transfer functions are:

Disturbance-to-(output-error): $$\frac{E_o(s)}{D(s)}=-\frac{1}{1+CP}=-\frac{2s^2+1000s}{2s^2+1000s+10^5},$$

Reference-to-output: $$\frac{Y(s)}{R(s)}=\frac{CP}{1+CP}=\frac{10^5}{2s^2+1000s+10^5}.$$

With internal model:

Disturbance-to-(output-error): $$\frac{E_o(s)}{D(s)}=-\frac{1}{1+(C+F_{IM})P}=-\frac{(2s^2+1000s)(s^2+3342)}{(2s^2+1000s)(s^2+3342)+10^5s^2+s+3342\cdot 10^5},$$

Reference-to-output: $$\frac{Y(s)}{R(s)}=\frac{(C+F_{IM})P}{1+(C+F_{IM})P}=\frac{10^5s^2+s+3342\cdot 10^5}{(2s^2+1000s)(s^2+3342)+10^5s^2+s+3342\cdot 10^5}.$$

iii) Disturbance rejection: We will assess the efficacy of the IMP controller by checking the disturbance rejection with the latter and without it. Setting $R(s) = 0$ in Figure 9.6-9 and taking the disturbance as

$$D(s)=\frac{s}{s^2+(2\pi 9.2)^2}=\frac{s}{s^2+3342},$$

yields the disturbance-induced output errors as follows.

Without internal model: $$E_o(s) = \frac{E_o(s)}{D(s)}D(s) = -\frac{2s^2+1000s}{2s^2+1000s+10^5}\cdot\frac{s}{s^2+3342}.$$

This signal transform has two oscillatory poles, i.e. the corresponding signal has no fixed value steady-state, *so we calculate the closed-loop response magnitude scaling and phase shift for the oscillatory steady state*:

$$\frac{E_o}{D}(j2\pi 9.2) = -\frac{2(j2\pi 9.2)^2+1000(j2\pi 9.2)}{2(j2\pi 9.2)^2+1000(j2\pi 9.2)+10^5} \approx 0.530e^{-j2.01}.$$

The magnitude scaling shows that the output magnitude due to the disturbance is quite large — half of the disturbance magnitude is passed!

With internal model: $$E_o(s) = \frac{E_o(s)}{D(s)}D(s) = -\frac{(2s^2+1000s)(s^2+3342)}{(2s^2+1000s)(s^2+3342)+10^5s^2+s+3342\cdot 10^5}\cdot\frac{s}{s^2+3342}$$

$$= -\frac{2s^3+1000s^2}{(2s^2+1000s)(s^2+3342)+10^5s^2+s+3342\cdot 10^5}.$$

Oscillatory poles are cancelled, so we can use the Final Value Theorem:

$$e_{o,ss} = \lim_{s\to 0} sE_o(s) = \frac{0}{3342\cdot 10^5} = 0.$$

Disturbance is rejected!

iv) Reference tracking: We still need to make sure that the output tracks the reference sinusoid well. Block diagram of the system with the disturbance set to zero and the desired reference is

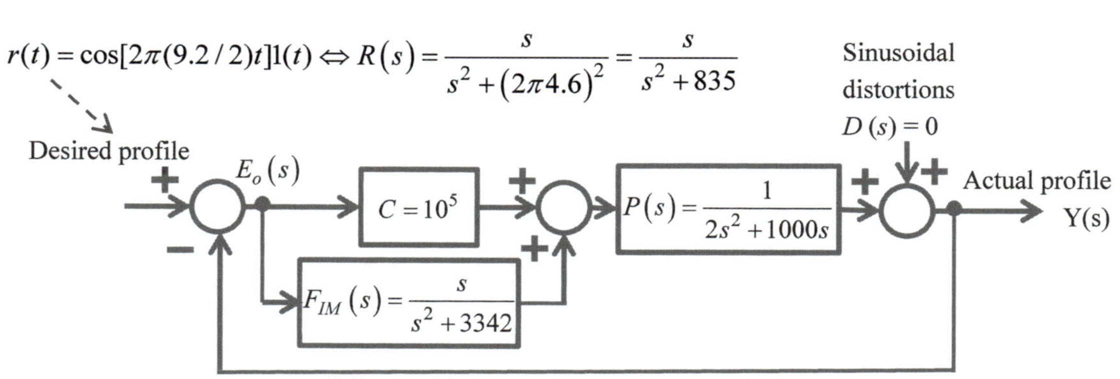

Fig. 9.6-10. Closed-loop system with the desired reference signal for assessing the steady-state tracking performance.

This yields two output signal transforms:

Without internal model: $Y(s)=\frac{Y(s)}{R(s)}R(s)=\frac{10^5}{2s^2+1000s+10^5}\cdot\frac{s}{s^2+835}$,

With internal model: $Y(s)=\frac{Y(s)}{R(s)}R(s)=\frac{10^5s^2+s+3342\cdot10^5}{(2s^2+1000s)(s^2+3342)+10^5s^2+s+3342\cdot10^5}\cdot\frac{s}{s^2+835}$.

Each transform is seen to have two oscillatory poles, so as in the disturbance rejection case above, we calculate the closed-loop steady-state oscillatory response magnitude scaling. This gives $|Y(j2\pi4.6)|\approx|0.976e^{-j0.286}|=0.976$ for both, i.e. reference is passed identically in both systems!

The resulting tracking error magnitude is $1-|Y(j2\pi4.6)|\approx1-0.976=0.024$. Further on, we can prefilter closed loop with gain 1/0.976 to eliminate tracking error. Thus, using superposition, this controller can simultaneously track reference at 4.6 Hz *and* reject disturbance at 9.2 Hz(!):

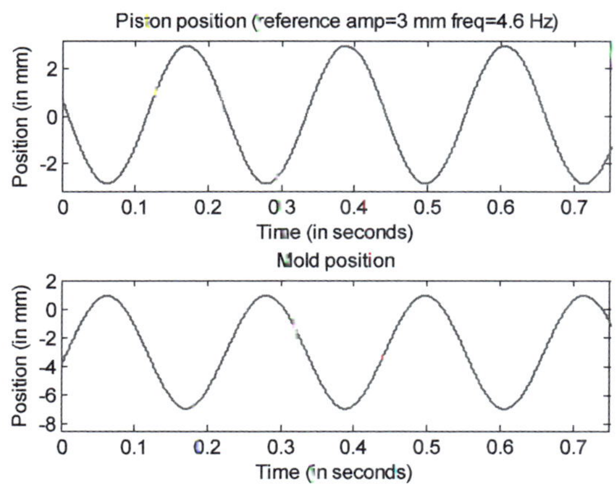

Fig. 9.6-11. Almost distortion-free mold position shows simultaneous reference tracking and disturbance rejection.

v) Stability and robustness: What happens if the plant model has some uncertainty? Say, it has an unknown gain, *K*, so that the block diagram in Figure 9.6-9 has new plant transfer function

$$P=\frac{K}{2s^2+1000s}.$$

Then, the denominator for any closed-loop transfer function takes the form

$$\begin{aligned}&(2s^2+1000s)(s^2+3342)+10^5Ks^2+Ks+3342\cdot10^5K\\&=2s^4+1000s^3+(6684+10^5K)s^2+(3342\cdot10^3+K)s+3342\cdot10^5K.\end{aligned}$$

Applying stability condition of Example 9.5-3, all closed-loop poles retain negative real parts for *K* larger than ~ 0.067. Therefore, this design has significant gain robustness margin.

9.6.3 *PID gain tuning*

Let's look at some basic rules for *controller tuning* — a major problem in many applications.

9.6.3.1 *P gain tuning for mold oscillator with mold position sensing*

The controller used at Nucor in low reference signal frequency range is P controller. Let's look at its *response shaping* capability in more detail. Assume for simplicity that the oscillator model is

$$P(s)=\frac{3272}{s^2+1.482s+3272}=\frac{\omega_n^2}{s^2+2\zeta\omega_n s+\omega_n^2}, \tag{9.6-3}$$

and consider the original controller configuration with the position sensor at the mold:

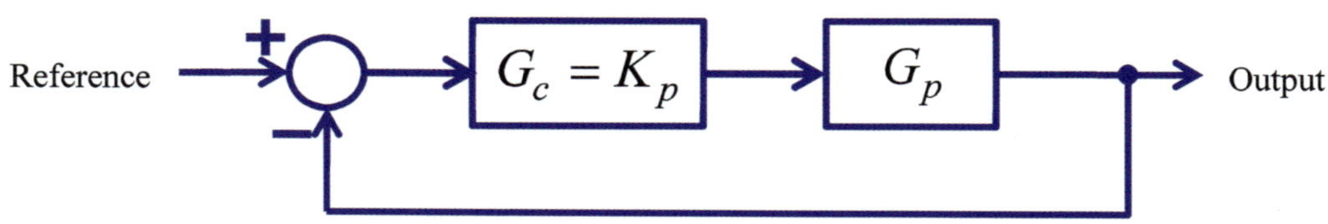

Fig. 9.6-12. Block diagram of mold oscillator with mold position sensor and P controller.

Then, the closed-loop transfer function is given by

$$G_{cl}(s)=\frac{\text{Output}}{\text{Reference}}=\frac{G_cG_p}{1+G_cG_p}=\frac{K_p 3272}{s^2+1.482s+\left(1+K_p\right)3272},\ G_{cl}(0)\equiv K_{ss}=\frac{K_p}{1+K_p}.$$

The effect of increasing k_p on the closed-loop step response is shown in the graphs below:

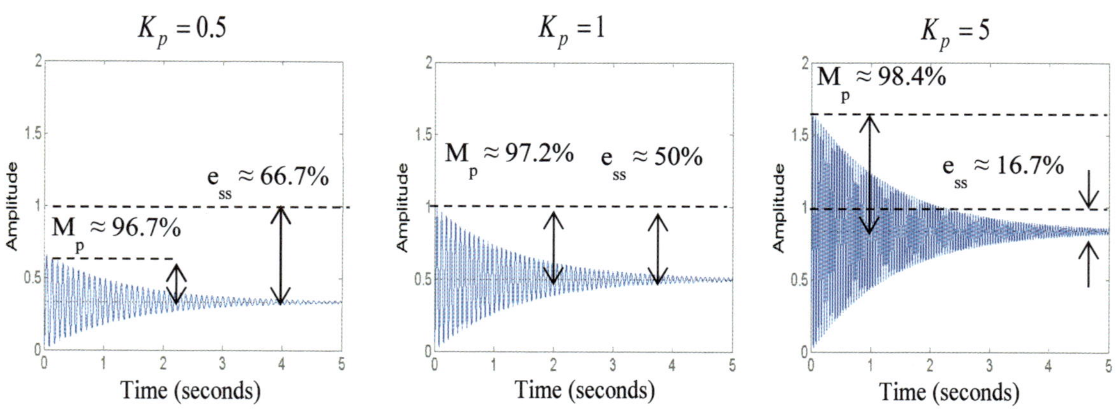

Fig. 9.6-13. P-gain tuning example for system in Figure 9.6-12.

The maximum percent overshoot (M_p) is barely changing when the steady state error ($e_{o,ss}$) decreases. What is happening? Let's take a closer look at the closed-loop transfer function:

$$G_{cl}(s)=\frac{\overbrace{k_p 3272}}{s^2+1.482s+\underbrace{(1+k_p)3272}}=\frac{\overbrace{k_{ss}\omega_n^2}}{s^2+2\zeta\omega_n s+\underbrace{\omega_n^2}}.$$

The formula (9.3-11) for maximum percent overshoot for the underdamped case, $M_p = e^{-\frac{\zeta\pi}{\sqrt{1-\zeta^2}}}$, shows that the latter depends on the damping — the damping margins give the following maximum overshoot values:

$\zeta \to 0$ (no damping), $M_p \to 100\%$,
$\zeta \to 1$ (critical damping), $M_p \to 0\%$.

But the proportional gain has only a secondary effect on the latter, as seen in the closed-loop transfer function matching above. The poles of the closed-loop system for the underdamped case are:

$$p_{1,2}=-\zeta\omega_n \pm j\underbrace{\omega_n\sqrt{1-\zeta^2}}_{\omega_d}. \tag{9.6-5}$$

This means that we can rewrite the formula for maximum overshoot as follows:

$$M_p=e^{-\pi\frac{\zeta\omega_n}{\omega_n\sqrt{1-\zeta^2}}}=e^{-\pi\frac{\operatorname{Re}p_i}{\operatorname{Im}p_i}},\ i=1,2, \tag{9.6-6}$$

so that fixing M_p specifies the s-plane "rays" of location of root pares with fixed (Re p)/(Im p) ratio:

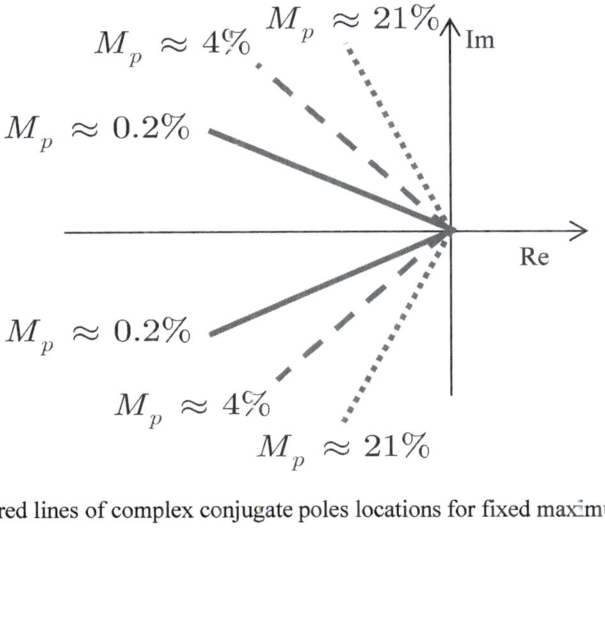

Fig. 9.6-14. Pared lines of complex conjugate poles locations for fixed maximum overshoot.

Thus, we can achieve our response shaping goal of maximum percent overshoot only through the appropriate placing of poles in the complex plane. Can we do it through P-gain tuning, i.e through changing k_p? Let's examine Figure 9.6-13 step responses from the root location viewpoint.

The effect of increasing k_p in the transfer function

$$G_{cl}(s) = \frac{k_p 3272}{s^2 + 1.482s + (1 + k_p)3272}$$

on the step response and on the corresponding root location is shown below in Figure 9.6-15:

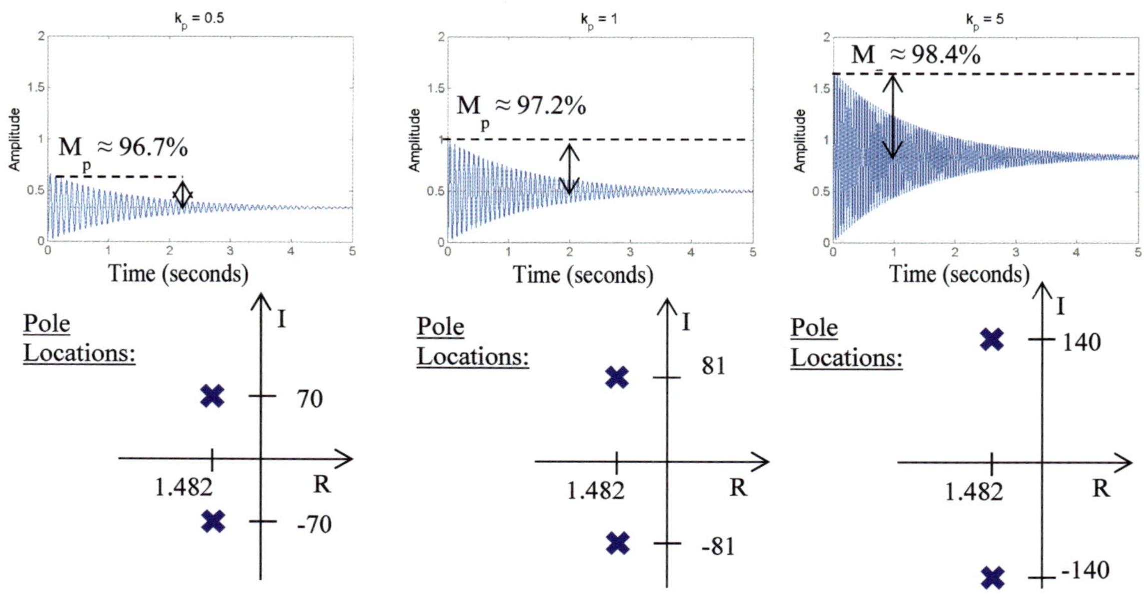

Fig. 9.6-15. Pole locations for different values of maximum overshoot.

As k_p increases, the poles only move up/down in the imaginary direction. With only proportional control, the smallest maximum percent overshoot that can be achieved is 96%. Figure 9.6-17 collects the poles into a "root locus" plot — one of the key tools in ***classical controller design***.

9.6.3.2 *PD gain tuning*

Let's now add a derivative controller:

$$G_{cl}(s) = \frac{k_p 3272}{s^2 + (1.482 + 3272k_d s) + (1 + k_p)3272},$$

to obtain a familiar closed-loop topology:

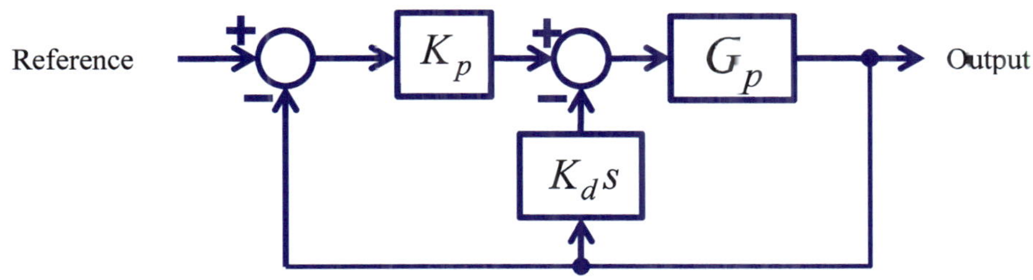

Fig. 9.6-16. Block diagram with PD controller.

Now, we can move the poles in the real direction as well. This, then, permits us to place poles into arbitrary locations through changing K_p and K_d, and thereby set an arbitrary maximum percent overshoot:

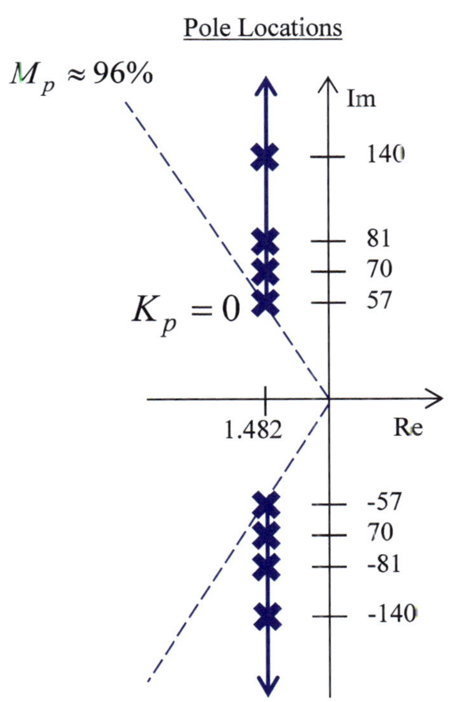

Fig. 9.6-17. Poles location under changing P gain.

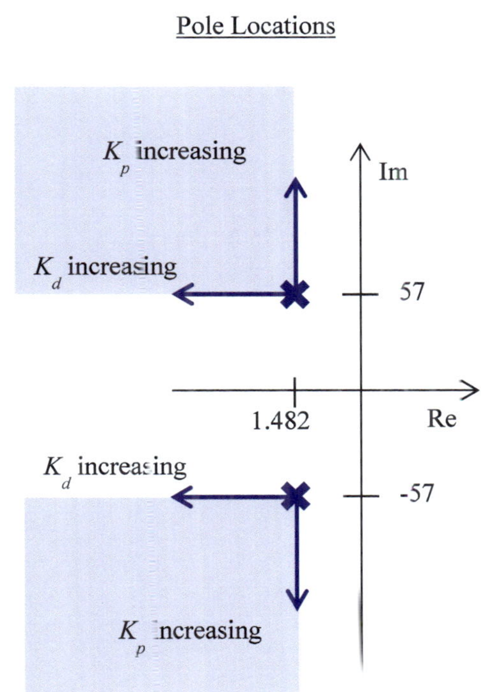

Fig. 9.6-18. PD gain tuning: arbitrary pole placement.

Let us now go back to the nested loop topology of Figure 9.4-7. Combining the full system state measurement with the presence of both gains, this topology implements what is called a *full state feedback*. This type of feedback has been shown to permit, for system of any order, arbitrary *pole placement* — one of the key tools in ***modern controller design***.

9.6.3.3 *PID tuning: heuristic rules*

The Zeigler-Nichols tuning rules are a set of rules for automatically setting PID gains based on the system characteristics. First the "ultimate" proportional gain, k_u, is found, at which the system just oscillates continuously without decaying. The period of this oscillation is denoted as T_u.

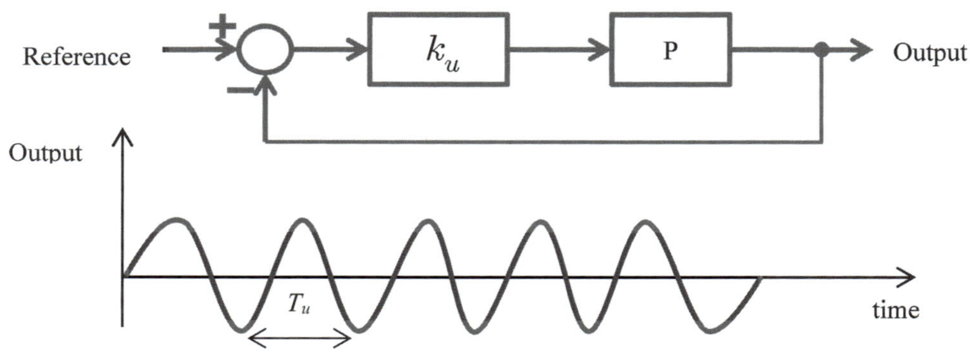

Fig. 9.6-19. System with continuous oscillation without decay.

Ziegler-Nichols PID tuning rules
Then the Ziegler-Nichols PID gains are:

Control Type	k_p	k_i	k_d
P	$0.5k_u$		
PI	$0.45k_u$	$1.2k_p / T_u$	
PID	$0.6k_u$	$2k_p / T_u$	$k_p T_u / 8$

These tend to have an aggressive response, not suitable for all applications.

Similar heuristic tuning rules, i.e the ones also depending on k_u and T_u, exist with different characteristics.

Ziegler-Nichols: open-loop tuning
Another method uses step response data to find the PID gains for a process characterized by the following step response:

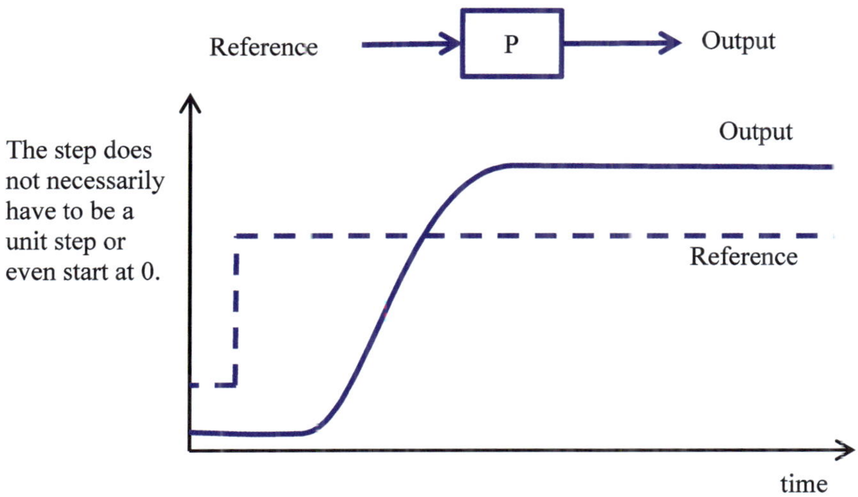

Fig. 9.6-20. Use step response data to find PID gains.

First, we need to calculate some characteristics of the step response as follows:

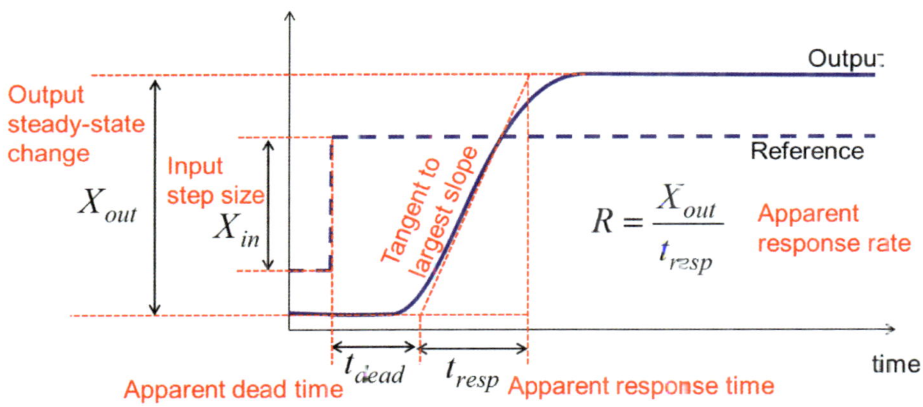

Fig. 9.6-21. Step response characteristics.

Then, calculating $k_0 = (X_{in} / X_{out}) \cdot (t_{resp} / t_{dead})$, the Ziegler-Nichols PID gains are:

Control Type	k_p	k_i	k_d
P	k_0		
PI	$0.9k_0$	$3.3t_{dead}$	
PID	$1.2k_0$	$2t_{dead}$	$0.5t_{dead}$

9.6.4 *Anti-windup*

9.6.4.1 *Actuator saturation*

Every actuator is characterized by a ***nominal output range*** beyond which the desired actuator response to the input signal is not guaranteed and typically becomes highly distorted.

This change in behavior outside of nominal range is called ***actuator saturation*** and essentially *breaks the loop*.

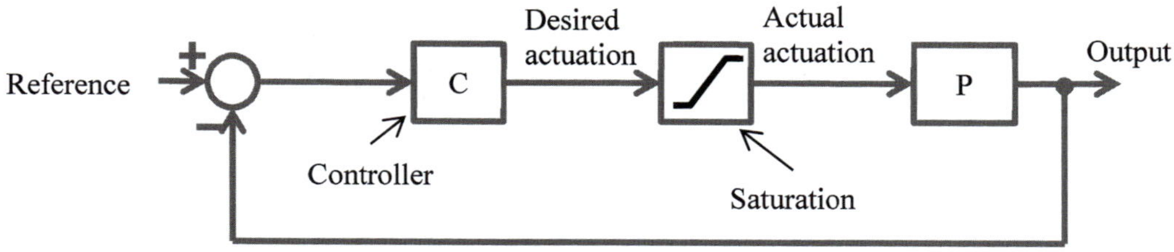

Fig. 9.6-22. Close loop with actuation.

This phenomenon, if not properly taken into account, is extremely dangerous: in the case of an aircraft it results in the loss of hundreds of lives, but in the case of a nuclear power plant (Chernobyl) — loss of thousands of lives.

Chernobyl engineers *underestimated actuator limitations* during routine plant maintenance testing, took reactor slightly beyond allowable pressure/temperature margin, and couldn't stop reaction runaway.

9.6.4.2 *Integrator windup*

A major problem with integral control is that it adds overshoot. This is especially true when there is a strict limit on the amount of actuation (saturation) and the integral controller action dictates going over this saturation level.

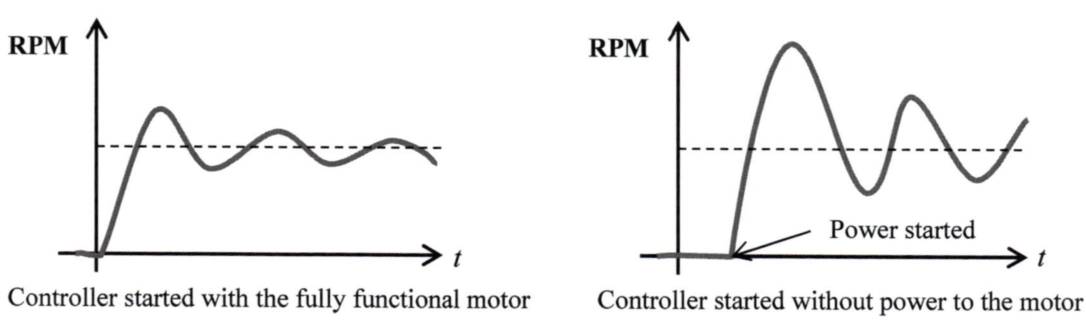

Fig. 9.6-23. RPM for motor with and without power.

If in a motor velocity control application the controller is started while the power to the motor is interrupted, the closed-loop system will wildly overshoot when the power to the motor is turned on, as shown in Figure 9.6-23.

Let us now examine in more detail the behavior of each control action. The decomposition of the control signal into two components, one generated by P controller, and the other — by I controller, is shown in Figure 9.6-24 under normal startup, with the power to the motor supplied at the outset.

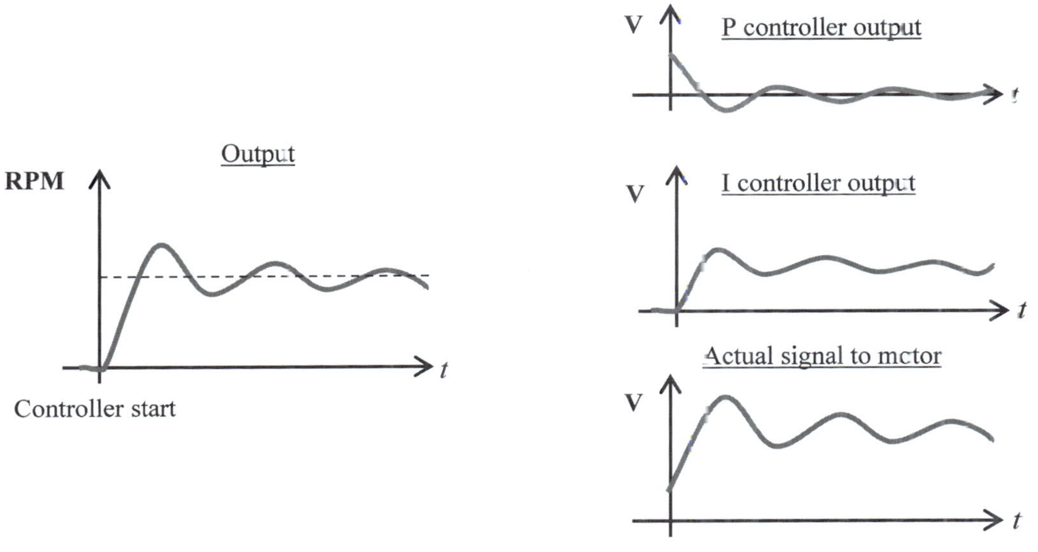

Fig. 9.6-24. Decomposition of the controller action into the P-controller and the I-controller outputs under normal startup conditions.

If the controller is started when the power to the motor is absent, the loop is essentially broken. In this case there is no plant output error reduction due to controller activity. Therefore, a large nondecreasing error signal gets integrated by the I-controller, producing growing signal at the I controller output.

This excessive accumulation of error in the form of a large I controller output signal is called the ***integrator windup***. It occurs any time the loop with the I controller gets broken, usually due the actuator saturation.

The interruption of power supply to the motor recreates the conditions of the actuator saturation by breaking the loop. The corresponding windup of the integrator in this case is shown in Figure 9.6-25. Once the loop gets restored, the already accumulated I controller output produces a jump in the actual control signal applied to the motor, as shown in the lower right graph. It is this jump due to the integrator windup that leads to the large closed-loop system overshoot when the power to the motor is turned on, as shown in the left graph.

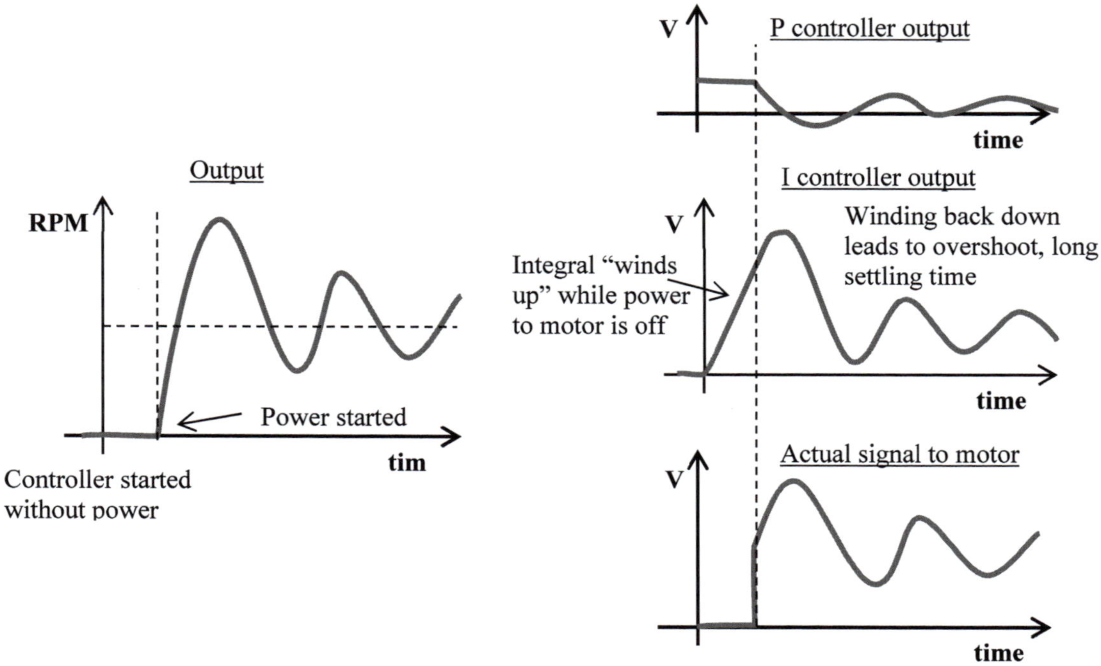

Fig. 9.6-25. Integrator windup and the resulting motor velocity overshoot.

9.6.4.3 *Anti-windup*

Control engineers have come up with a class of heuristic techniques to prevent windup, called "anti-windup" methods. The general idea is to prevent the excessive error integration by, for example, reducing the error going to the integrator though using the difference between the desired and actual (after saturation) actuation, as shown in Figure 9.6-26. The actual actuator output measured in this schematics is not always accessible, e.g. fluid flow rate, so actuator models are often used instead.

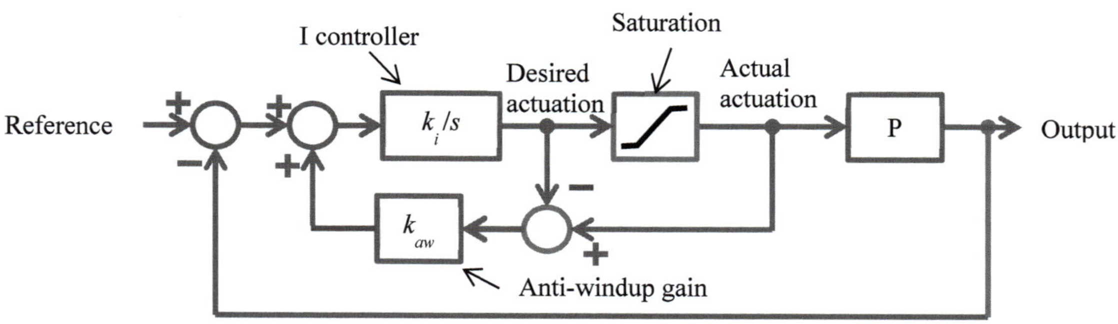

Fig. 9.6-26. System block diagram with anti-windup.

This results in the integrator action shown in the middle graph on the right in Figure 9.6-27. The jump in the actual control signal applied to the motor is seen to be significantly reduced in comparison to that without anti-windup shown by the dotted curve, so that once the actuation is restored, the overshoot is mostly contained as shown in the left graph.

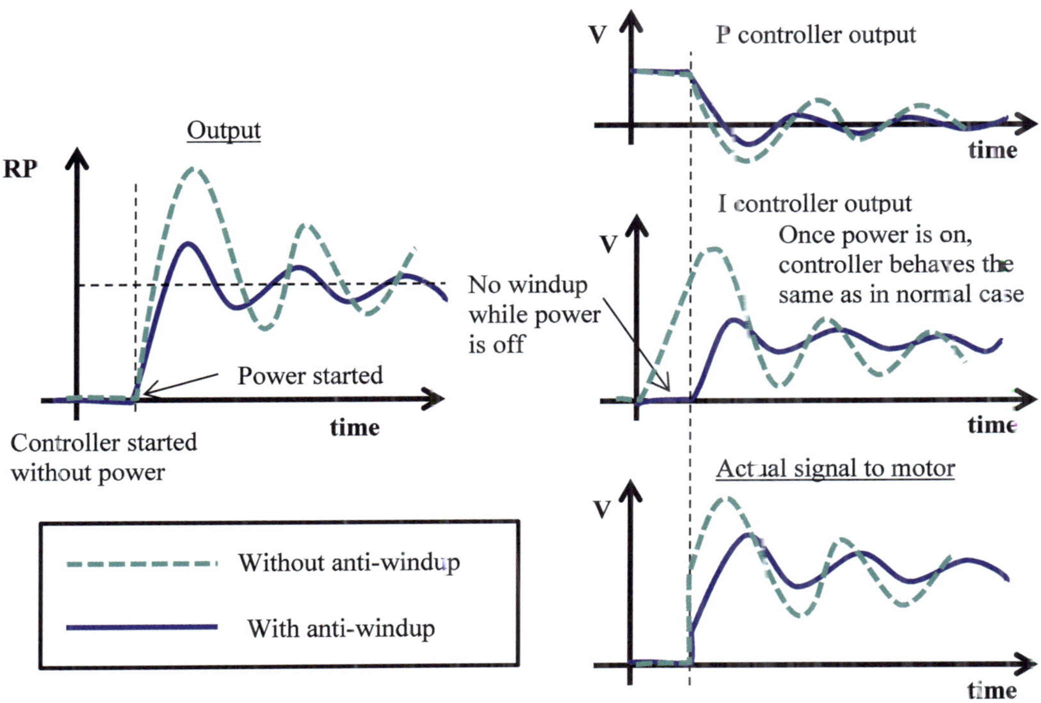

Fig. 9.6-27. Effect of anti-windup.

Since most of the actuators have saturation limits, the anti-windup safeguard is a standard feature in any industrial controller implementation.

Project 3: Solved Examples and Solution Guide

Problem 1:

a) Compute analytically and plot the output of the P, I, PI, PD, and PID controllers for input given by i) unit step, and ii) unit ramp. Take proportional gain = 3, integral time = 4 s, derivative time = 0.8 s.

b) Implement each of the controllers with parameters given in a) using op amps.

Solution:

a.i) Compute analytically and plot the output of the P, I, PI, PD, and PID controllers for input given by unit step. Take proportional gain = 3, integral time = 4 s, derivative time = 0.8 s.

A unit step error means $e(t) = 1(t)$, and $E(s) = \frac{1}{s}$. We are using here $1(t)$, rather than $u(t)$, to denote the unit step function, since the latter is usually used in control literature to denote the controller output signal (control action).

P controller: The transfer function for a P controller is $\frac{U}{E} = k_p = 3$.

Therefore, $U(s) = k_p E(s) = \frac{3}{s}$, and, taking the inverse Laplace Transform, $u(t) = 3 \cdot 1(t)$.

I controller: $\frac{U}{E} = k_i \cdot \frac{1}{s} = \frac{k_p}{T_i s} = \frac{3}{4s}$.

Therefore, $U(s) = \frac{k_p}{T_i s} E(s) = \frac{3}{4s} \cdot \frac{1}{s} = \frac{3}{4} \cdot \frac{1}{s^2}$, and $u(t) = 0.75t \cdot 1(t)$.

PI controller: $\frac{U}{E} = k_p + k_i \frac{1}{s} = k_p \left(1 + \frac{1}{T_i s}\right) = 3\left(1 + \frac{1}{4s}\right)$.

Therefore, $U(s) = k_p \left(1 + \frac{1}{T_i s}\right) E(s) = 3\left(1 + \frac{1}{4s}\right)\frac{1}{s} = 3 \cdot \frac{1}{s} + \frac{3}{4} \cdot \frac{1}{s^2}$, and $u(t) = (3 + 0.75t) \cdot 1(t)$.

Note that we could have just added the results for the P controller and the I controller together to get the answer.

PD controller: $\frac{U}{E} = k_p + k_d s = k_p (1 + T_d s) = 3(1 + 0.8s)$.

Therefore, $U(s) = k_p (1 + T_d s) E(s) = 3(1 + 0.8s)\frac{1}{s} = 3 \cdot \frac{1}{s} + 2.4$, and $u(t) = 3 \cdot 1(t) + 2.4\delta(t)$.

PID controller: $\frac{U}{E} = k_p + k_i \cdot \frac{1}{s} + k_d s = k_p \left(1 + \frac{1}{T_i s} + T_d s\right) = 3\left(1 + \frac{1}{4s} + 0.8s\right)$.

Therefore, $U(s) = k_p \left(1 + \frac{1}{T_i s} + T_d s\right) E(s) = 3\left(1 + \frac{1}{4s} + 0.8s\right)\frac{1}{s} = 3 \cdot \frac{1}{s} + 0.75 \cdot \frac{1}{s^2} + 2.4$, and $u(t) = 3 \cdot 1(t) + 0.75t \cdot 1(t) + 2.4\delta(t)$.

The outputs for the unit step input for all the above controllers are, respectively, shown in Figures P3-1–P3-5:

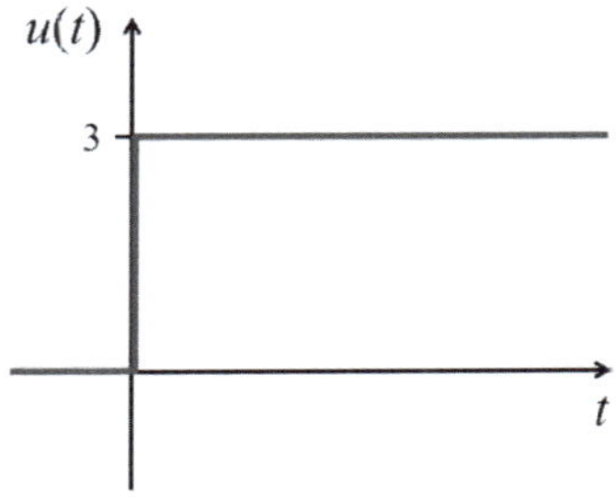

Fig. P3-1. Output with P controller.

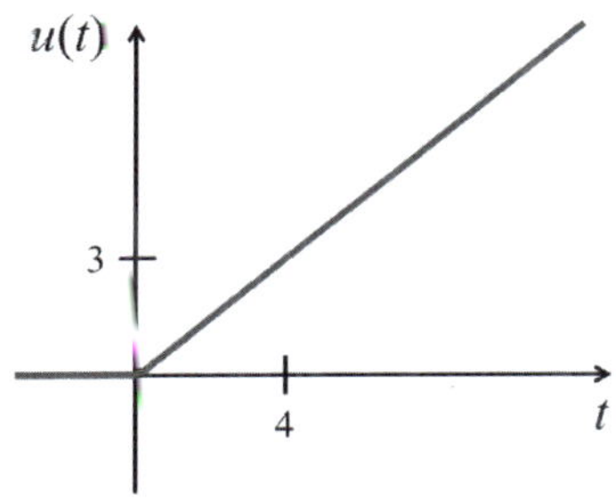

Fig. P3-2. Output with I controller.

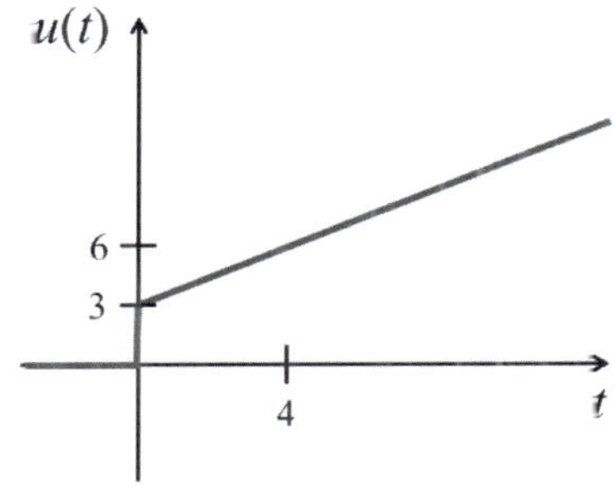

Fig. P3-3. Output with PI controller.

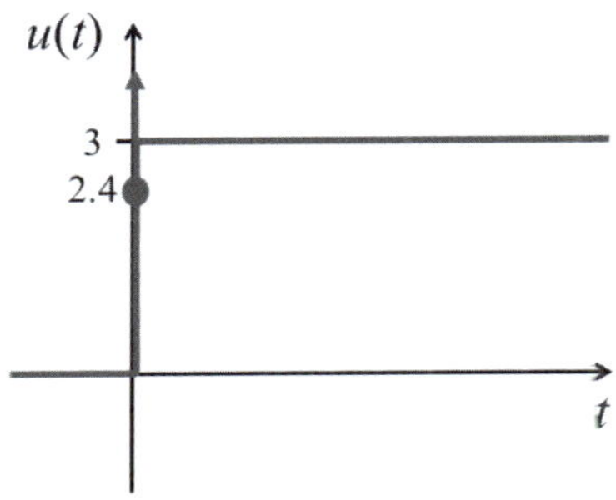

Fig. P3-4. Output with PD controller.

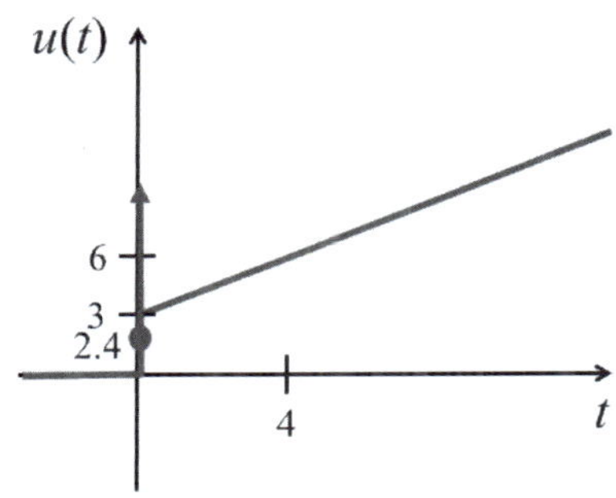

Fig. P3-5. Output with PID controller.

a.ii) Compute analytically and plot the output of the P, I, PI, PD, and PID controllers for input given by unit ramp. Take proportional gain = 3, integral time = 4 s, derivative time = 0.8 s.

A unit ramp error means $e(t)=r(t)=t\cdot 1(t)$, and $E(s)=\frac{1}{s^2}$. The transfer functions are the same for each controller as they were for part (i).

P controller: $U(s)=k_p E(s)=\frac{3}{s^2}$, and $u(t)=3t\cdot 1(t)$.

I controller: $U(s)=\frac{k_p}{T_i s}E(s)=\frac{3}{4s}\cdot\frac{1}{s^2}=\frac{3}{4}\cdot\frac{1}{s^3}$, and $u(t)=0.75t^2\cdot 1(t)$.

PI controller: $U(s)=k_p\left(1+\frac{1}{T_i s}\right)E(s)=3\left(1+\frac{1}{4s}\right)\frac{1}{s^2}=3\cdot\frac{1}{s^2}+0.75\cdot\frac{1}{s^3}$ and

$u(t)=\left(3t+0.75t^2\right)\cdot 1(t)$.

PD controller: $U(s)=k_p(1+T_d s)E(s)=3(1+0.8s)\frac{1}{s^2}=3\cdot\frac{1}{s^2}+2.4\cdot\frac{1}{s}$ and $u(t)=(3t+2.4)\cdot 1(t)$.

PID controller: $U(s)=k_p\left(1+\frac{1}{T_i s}+T_d s\right)E(s)=3\left(1+\frac{1}{4s}+0.8s\right)\frac{1}{s^2}=3\cdot\frac{1}{s^2}+0.75\cdot\frac{1}{s^3}+2.4\cdot\frac{1}{s}$ and $u(t)=(3t+0.75t^2+2.4)\cdot 1(t)$.

The outputs for the unit ramp input are, respectively, shown in Figures P3-6–P3-10:

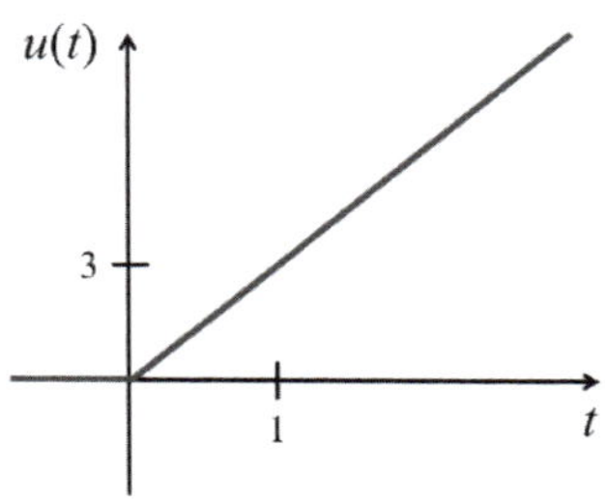

Fig. P3-6. Output with P controller.

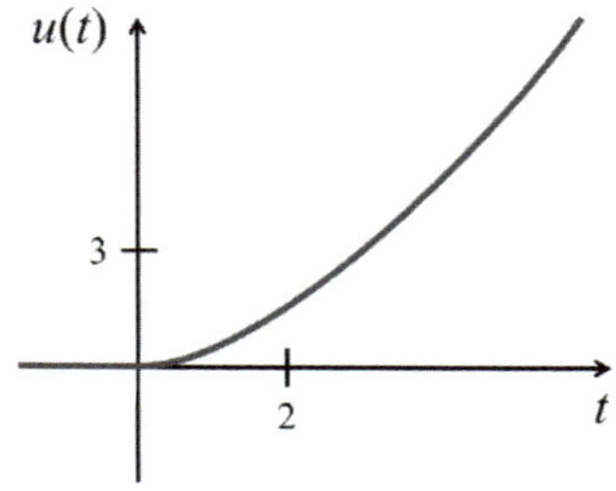

Fig. P3-7. Output with I controller.

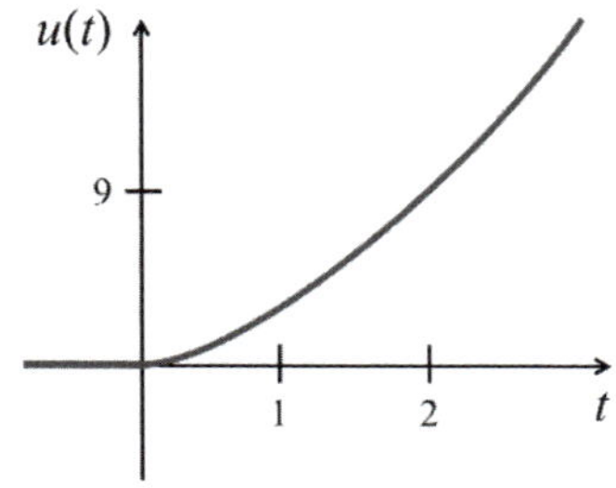

Fig. P3-8. Output with PI controller.

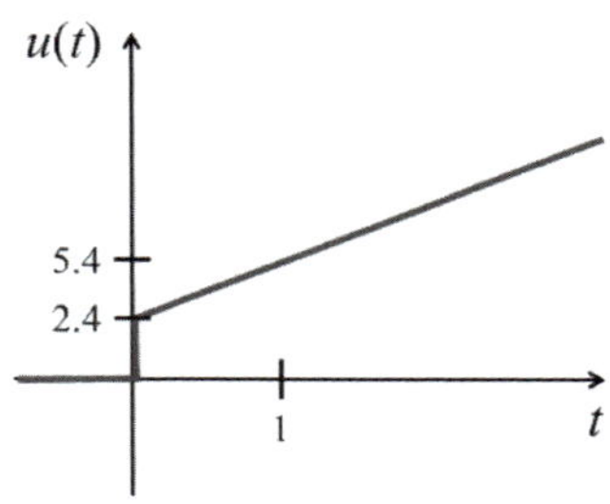

Fig. P3-9. Output with PD controller.

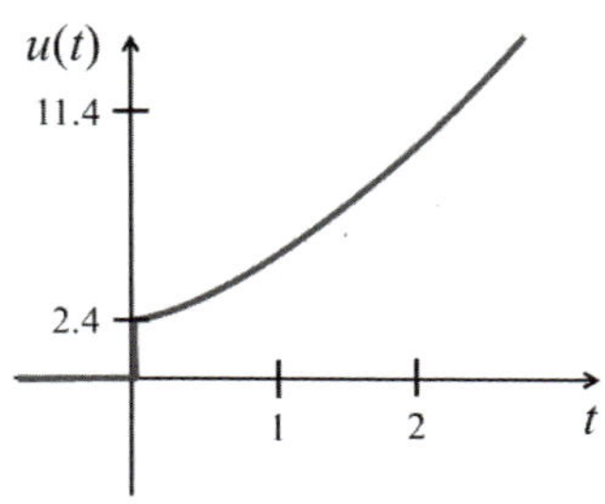

Fig. P3-10. Output with PID controller.

(b) Implement each controller using op-amps

For this one, we just need to implement the transfer functions from part (a) as op-amp circuits. This will require (inverting) amplifiers, integrators, differentiators, and summers.

P controller: $\frac{U}{E}=k_p=3$.

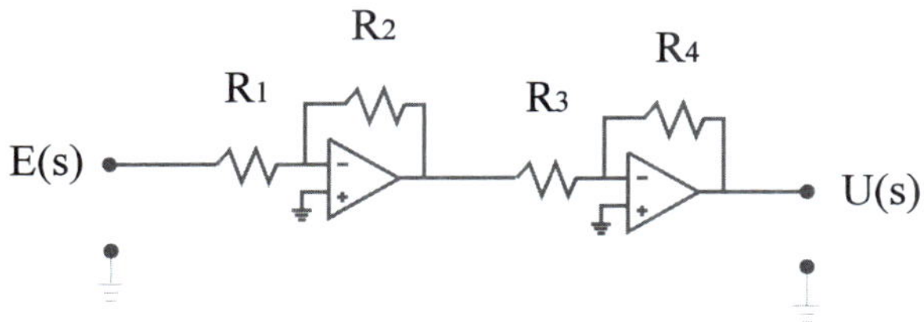

Fig. P3-11. Op-amp P controller implementation.

Constraints: $\frac{R_2}{R_1} \cdot \frac{R_4}{R_3} = k_p = 3.$

I controller: $\frac{U}{E} = k_p \frac{1}{T_i s} = 0.75\frac{1}{s}.$

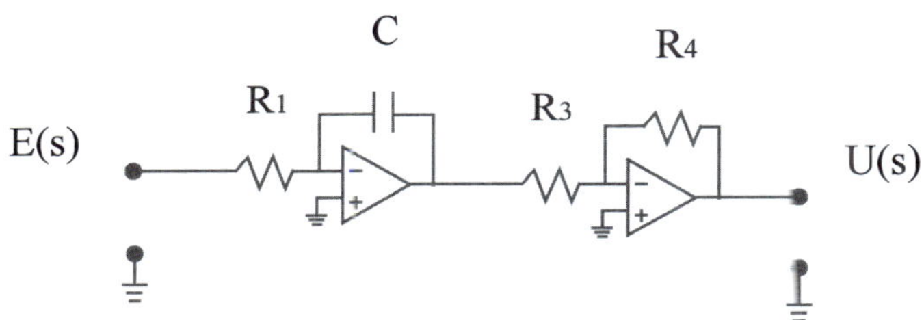

Fig. P3-12. Op-amp I controller implementation.

Constraints: $\frac{1}{R_1 C} \cdot \frac{R_4}{R_3} = k_i = \frac{k_p}{T_i} = 0.75.$

PI controller: $\frac{U}{E} = k_p\left(1 + \frac{1}{T_i s}\right) = 3 + 0.75\frac{1}{s}.$

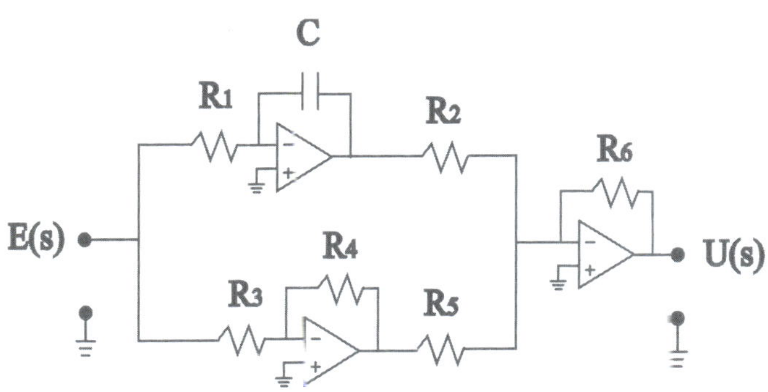

Fig. P3-13. Op-amp PI controller implementation.

Constraints: $\frac{R_4}{R_3}\cdot\frac{R_6}{R_5}=k_p=3;\ \frac{1}{R_1C}\cdot\frac{R_6}{R_2}=k_i=\frac{k_p}{T_i}=0.75.$

PD controller: $\frac{U}{E}=k_p\left(1+T_d s\right)=3+2.4s.$

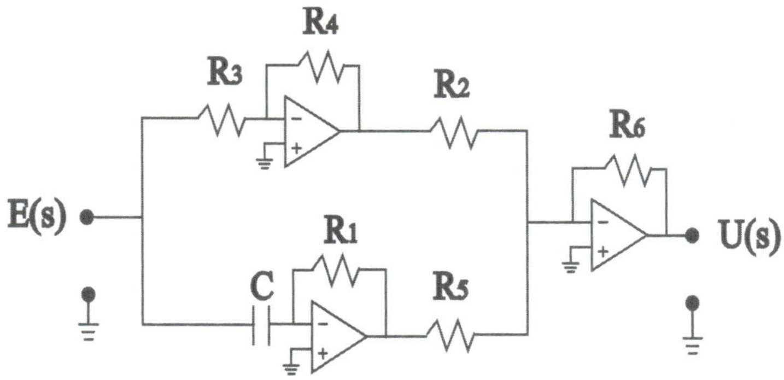

Fig. P3-14. Op-amp PD controller implementation.

Constraints: $\frac{R_4}{R_3}\cdot\frac{R_6}{R_2}=k_p=3;\ R_1C\cdot\frac{R_6}{R_5}=k_d=k_pT_d=2.4.$

PID controller: $\frac{U}{E}=k_p\left(1+\frac{1}{T_i s}+T_d s\right)=3+0.75\cdot\frac{1}{s}+2.4s.$

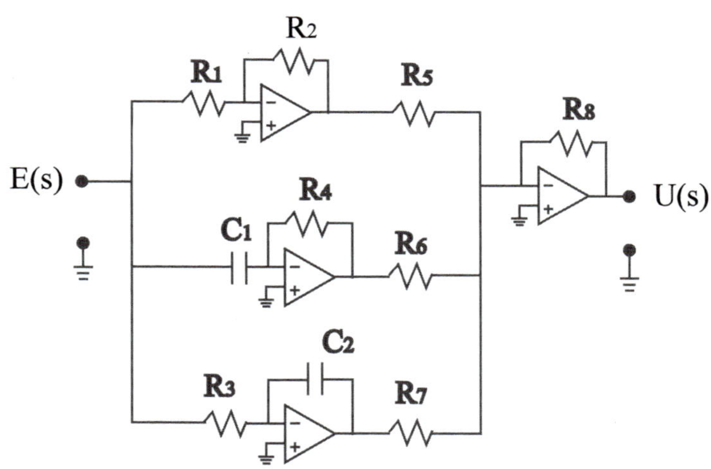

Fig. P3-15. Op-amp PID controller implementation.

Constraints: $\frac{R_2}{R_1}\cdot\frac{R_8}{R_5}=k_p=3;\quad \frac{1}{R_3C_2}\cdot\frac{R_8}{R_7}=k_i=\frac{k_p}{T_i}=0.75;\ R_4C_1\cdot\frac{R_8}{R_6}=k_d=k_pT_d=2.4.$

Problem 2: Consider a mold oscillator testbed that has been modeled as

$$G_p(s) = \frac{3000}{s^2 + 1.5s + 3300}.$$

Assume perfect actuation and mass displacement sensing.

a) Comment, based on this model, on what would be the worst frequency of the sinusoidal plant input disturbance $d(t) = \cos\omega_{dist}t$, $\omega_{dist} = 2\pi f_{dist}$ for this system. Determine $f_{dist} = f_{res}$ = frequency in Hz of the model resonance peak.
b) Find the open-loop and the closed-loop transfer functions from the input and the disturbance to the output for P and PD controllers (take PD controller implementation scheme given in Figure 9.4-11). For the closed-loop transfer functions, modify a prefilter, if necessary, to ensure the DC gain from reference to output is 1.
c) For overshoot less than 50%: check if it is possible to find analytically the P controller gain. If the answer is positive, determine the settling time of the system, calculate analytically and plot system response to a unit step input (using partial fraction expansion), find the steady state errors with respect to a step input and a ramp input. If the answer is negative, take P controller gain of your choice, determine the overshoot, and work out all the steps above.
d) Calculate analytically and draw magnitude response of the closed-loop system designed in b). Indicate how the bandwidth and the speed of the response are affected by the P controller gain and sketch the corresponding plots.
e) Find analytically the PD controller gains which reduce the settling time found in c) by half, calculate analytically and plot system response to a unit step input (using partial fraction expansion), find the steady state output errors with respect to the unit step input, and the unit ramp input.
f) Keeping only the proportional prefilter from b) and setting the PD block gain to 1 (i.e. omitting the PD block without breaking the loop), calculate the resulting output error magnitude and phase shift for a sinusoidal *reference* input $\cos\omega_{inp}t$, $\omega_{inp} = 2\pi f_{inp}$, $f_{inp} = 0.5 f_{res}$ Hz.
g) Still keeping only the proportional prefilter from b), insert into the forward path a controller block designed using the internal model principle to reject the sinusoidal plant input disturbance $d(t)$ found in a). Check closed-loop stability. Check if disturbance rejection takes place for the *disturbance* input $\cos\omega_{res}t$. For the new controller, calculate the output magnitude and the phase shift for a *reference* input $\cos\omega_{inp}t$, and compare these values to the result from f).

Solution:

a) The worst case frequency will be the resonance frequency, where the magnitude is maximum. From the problem statement,

$$G_p(s) = \frac{3000}{s^2 + 1.5s + 3300},$$

so,

$$\left|G_p(j\omega)\right| = \frac{3000}{\left|3300-\omega^2+j1.5\omega\right|} = \frac{3000}{\sqrt{\left(3300-\omega^2\right)^2+2.25\omega^2}} = \frac{3000}{\sqrt{\omega^4-6597.75\omega^2+10890000}}.$$

Maximizing this is the same as minimizing the denominator, so

$$\frac{d}{d\omega}\left(\omega^4+6597.75\omega^2+10890000\right) = 4\omega^3-13195.5\omega = 0 \Rightarrow \omega_{res} = \pm\sqrt{\frac{13195.5}{4}} \approx \pm 57.4 \text{ rad/s.}$$

Or, $f_{res} = \dfrac{\omega_{res}}{2\pi} \approx \dfrac{57.4}{2\pi} \approx 9.14$ Hz.

b) Find the open-loop and the closed-loop transfer functions from the input and the disturbance to the output for P and PD controllers. For the closed-loop transfer functions, modify a prefilter, if necessary, to ensure the DC gain from reference to output is 1.

Since the closed-loop has no integral action, we will choose the closed-loop block diagram for a PD controller of the form:

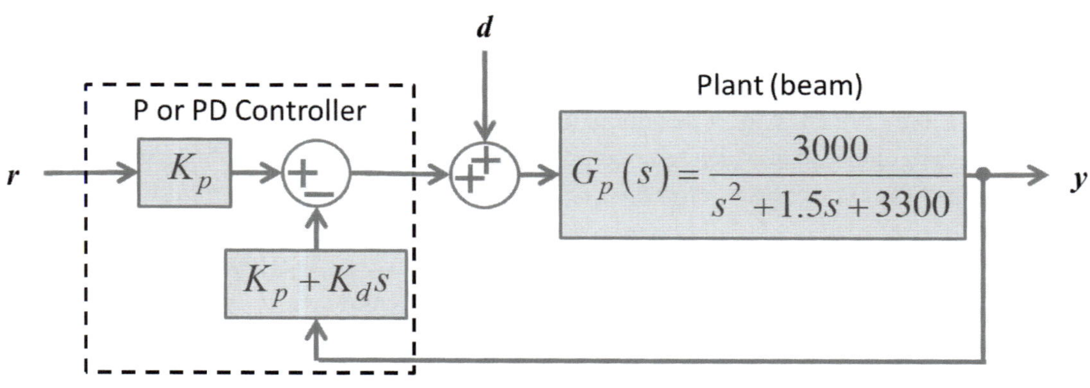

Fig. P3-16. Close loop block diagram with PD controller.

We call the K_p block a "prefilter" because it is outside the loop. We will eventually change it so that the closed-loop DC gain is 1, but first, we will calculate the open-loop transfer functions. For the open loop, we omit the feedback path, leaving only the prefilter gain:

$$G_{ol,r}(s) = k_p G_p(s) = \frac{3000K_p}{s^2+1.5s+3300}.$$

The disturbance-to-output transfer function will be

$$G_{ol,d}(s) = G_p(s) = \frac{3000}{s^2 + 1.5s + 3300}.$$

In the closed-loop diagram, we have $K_p + K_d s$ in the feedback path of the loop, $G_p(s)$ in the forward part of the loop, and K_p outside the loop. Therefore, the closed-loop transfer functions are given by

$$G_{cl,r}(s) = K_p \frac{G_p(s)}{1 + (K_p + K_d s) G_p(s)},$$

$$G_{cl,d}(s) = \frac{G_p(s)}{1 + (K_p + K_d s) G_p(s)}.$$

First, we adjust the prefilter so that the DC gain (at $s = 0$) be 1 for the reference-to-output transfer function. This will ensure that constant reference signals will be tracked exactly.

$$G_{cl,r}(s) = K_p \frac{G_p(s)}{1 + (K_p + K_d s) G_p(s)} = \frac{K_p \dfrac{3000}{s^2 - 1.5s + 3300}}{1 + (K_p + K_d s) \dfrac{3000}{s^2 + 1.5s + 3300}}$$

$$= \frac{3000K_p}{s^2 + (1.5 + 3000K_d)s + 3300 + 3000K_p} \Rightarrow$$

$$G_{cl,r}(0) = \frac{3000K_p}{3300 + 3000K_p} \neq 1.$$

We want the latter to be 1, so instead of using K_p for the prefilter, we will multiply the system DC gain by its inverse, so that

$$\frac{K_p}{\left(\dfrac{3000K_p}{3300 + 3000K_p} \right)} = \frac{3300 + 3000K_p}{3000} = 1.1 + K_p.$$

This yields the block diagram of the form

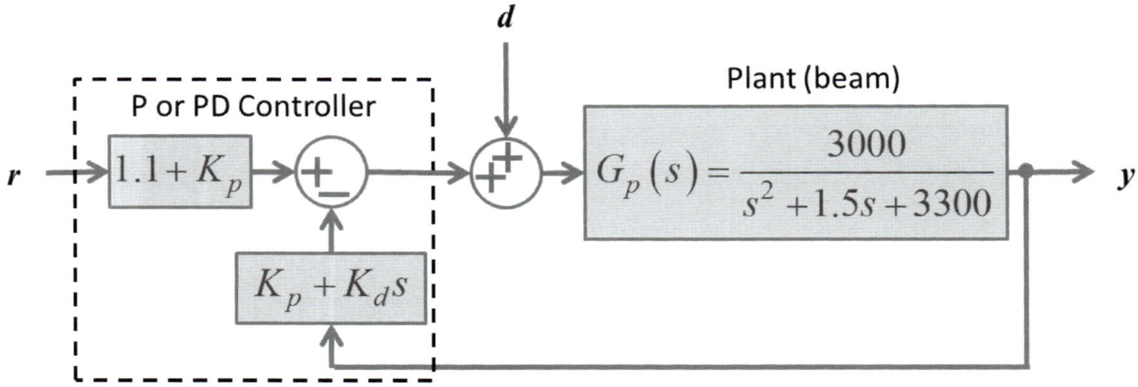

Fig. P3-17. New block diagram with prefilter.

P controller

For the P controller, the derivative gain (k_d) will be 0, yielding closed-loop transfer function

$$G_{cl,r}(s)=(1.1+K_p)\frac{G_p(s)}{1+K_pG_p(s)}=\frac{(1.1+K_p)\dfrac{3000}{s^2+1.5s+3300}}{1+K_p\dfrac{3000}{s^2+1.5s+3300}}=\frac{3300+3000K_p}{s^2+1.5s+3300+3000K_p}.$$

The closed-loop transfer function for the disturbance is

$$G_{cl,d}(s)=\frac{G_p(s)}{1+K_pG_p(s)}=\frac{\dfrac{3000}{s^2+1.5s+3300}}{1+K_p\dfrac{3000}{s^2+1.5s+3300}}=\frac{3000}{s^2+1.5s+3300+3000K_p}.$$

PD controller

For the PD controller, the closed-loop transfer function is

$$G_{cl,r}(s)=(1.1+K_p)\frac{G_p(s)}{1+(K_p+K_ds)G_p(s)}=\frac{(1.1+K_p)\dfrac{3000}{s^2+1.5s+3300}}{1+(K_p+K_ds)\dfrac{3000}{s^2+1.5s+3300}}$$

$$=\frac{3300+3000K_p}{s^2+(1.5+3000K_d)s+3300+3000K_p}.$$

The closed-loop transfer function for the disturbance is

$$G_{cl,d}(s)=\frac{G_p(s)}{1+\left(K_p+K_d s\right)G_p(s)}=\frac{\dfrac{3000}{s^2+1.5s+3300}}{1+\left(K_p+K_d s\right)\dfrac{3000}{s^2+1.5s+3300}}$$
$$=\frac{3000}{s^2+\left(1.5+3000K_d\right)s+3300+3000K_p}.$$

c) For overshoot less than 50%: check if it is possible to find analytically the P controller gain. If the answer is positive, determine the settling time of the system, calculate analytically (using partial fraction expansion) and plot system response to a unit step input, find the steady state errors with respect to a step input and a ramp input. If the answer is negative, take P controller gain of your choice, determine the overshoot, and work out all the steps above.

From b), the P controller closed-loop response is

$$G_{cl,r}(s)=\frac{3300+3000K_p}{s^2+1.5s+3300+3000K_p}.$$

This will be an underdamped second-order system, so the maximum percent overshoot will be

$$M_p=e^{-\frac{\zeta\pi}{\sqrt{1-\zeta^2}}}=0.5\Rightarrow\frac{\zeta^2\pi^2}{1-\zeta^2}=\ln^2 0.5\Rightarrow\zeta\approx 0.216$$
$$\Rightarrow\omega_n=\frac{1.5}{2\zeta}\approx 3.481$$
$$\Rightarrow 3300+3000k_p=3.481^2\Rightarrow k_p=-1.095.$$

This gain is negative, so this design problem is impossible. In this case, the problem asks us to use any gain, so let's use $K_p = 1$. This yields the closed-loop system

$$G_{cl,r}(s)=\frac{3300+3000\cdot 1}{s^2+1.5s+3300+3000\cdot 1}=\frac{6300}{s^2+1.5s+6300}.$$

Therefore, the final transfer function is

$$G_{cl,r}(s)=\frac{6300}{s^2+1.5s+6300}.$$

For this transfer function, $\omega_n=\sqrt{6300}\approx 79.4$, $\zeta=\dfrac{1.5}{2\omega_n}=\dfrac{1.5}{2\cdot 79.4}\approx 0.0094$, and

$$M_p=e^{-\frac{\zeta\pi}{\sqrt{1-\zeta^2}}}=e^{-\frac{0.0094\pi}{\sqrt{1-0.0094^2}}}\approx 0.971$$

So the maximum percent overshoot is about 97.1%.
We can still calculate the settling time, which will be

$$t_s = \frac{4}{\zeta\omega_n} = \frac{8}{2\zeta\omega_n} = \frac{8}{1.5} \approx 5.33 \text{ seconds.}$$

For a unit step input, the output will be

$$Y(s) = \frac{1}{s}\cdot G_{cl}(s) = \frac{1}{s}\cdot\frac{6300}{s^2+1.5s+6300} = \frac{1}{s}\cdot\frac{6300}{(s+0.75)^2+79.4^2} = \frac{A}{s} + \frac{B(s+0.75)}{(s+0.75)^2+79.4^2} +$$

$$\frac{C\cdot 79.4}{(s+0.75)^2+79.4^2} \Rightarrow 6300 = A\left(s^2+1.5s+6300\right) + B(s+0.75)s + C\cdot 81.2\cdot s =$$

$$A\left(s^2+1.5s+6300\right) + B\left(s^2+0.75s\right) + C(79.4s),$$

$$6300A = 6300 \Rightarrow A = 1,\quad As^2 + Bs^2 = 0 \Rightarrow B = -A = -1,$$

$$1.5As + 0.75Bs + 79.4Cs = 0 \Rightarrow C = -\frac{1.5A+0.75B}{79.4} \approx -0.0094$$

$$\Rightarrow Y(s) = \frac{1}{s} - \frac{s+0.75}{(s+0.75)^2+79.4^2} - 0.0094\frac{79.4}{(s+0.75)^2+79.4^2}$$

$$\Rightarrow y(t) = \left(1 - e^{-0.75t}\cos(79.4t) - 0.0094e^{-0.75t}\sin(79.4t)\right)u(t).$$

This step response is plotted in Figure P3-18.

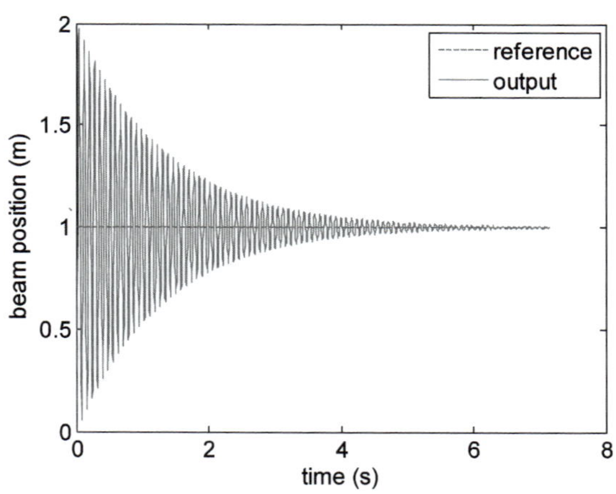

Fig. P3-18. Beam position output with unit step input.

The real parts of the poles of this system are both negative, so the final value theorem applies. To find the steady-state output error, we calculate

$$E_o(s) = R(s) - Y(s) = R(s) - G_{cl,r}(s)R(s) = \left(1 - G_{cl,r}(s)\right)R(s).$$

Therefore, the steady state output error to a unit step $\left(R(s)=1/s\right)$ is given by

$$E_o(s)=\left(1-G_{cl}(s)\right)\cdot\frac{1}{s}=\left(1-\frac{6300}{s^2+1.5s+6300}\right)\cdot\frac{1}{s}=\left(\frac{s^2-1.5s}{s^2+1.5s+6300}\right)\cdot\frac{1}{s}=\frac{s+1.5}{s^2+1.5s+6300}$$

$$\Rightarrow e_{o,ss}=\lim_{s\to 0} sE_o(s)=\lim_{s\to 0} s\cdot\frac{s+1.5}{s^2+1.5s+6300}=0\cdot\frac{1.5}{6300}=0.$$

The steady-state output error to a unit ramp $\left(R(s)=1/s^2\right)$ is:

$$E_o(s)=\left(1-G_{cl}(s)\right)\cdot\frac{1}{s^2}=\left(\frac{s^2+1.5s}{s^2+1.5s+6300}\right)\cdot\frac{1}{s^2}=\frac{s+1.5}{s^3+1.5s^2+6300s}\Rightarrow$$

$$e_{o,ss}=\lim_{s\to 0} sE_o(s)=\lim_{s\to 0} s\frac{s+1.5}{s^3+1.5s^2+6300s}=\frac{1.5}{6300}=\frac{1}{4200}.$$

d) Calculate analytically and draw magnitude response of the closed-loop system designed in b). Indicate how the bandwidth and the speed of the response are affected by the P controller gain and sketch the corresponding plots.

For the general K_p, from b) we have

$$G_{cl,r}(s)=\frac{3300+3000K_p}{s^2+1.5s+3300+3000K_p}\Rightarrow\left|G_{cl,r}(j\omega)\right|=\frac{3300+3000K_p}{\sqrt{\left(3300+3000K_p-\omega^2\right)^2+(1.5\omega)^2}}.$$

For the choice of $K_p=1$ the closed-loop magnitude response is

$$\left|G_{cl,r}(j\omega)\right|=6300/\sqrt{\left(6300-\omega^2\right)^2+(1.5\omega)^2},$$

with the response graph given by

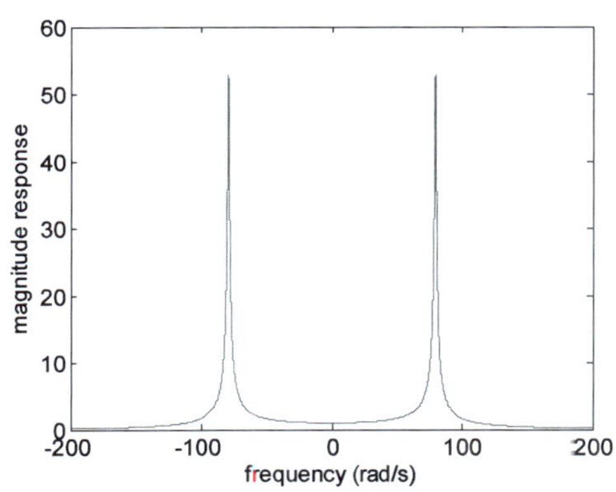

Fig. P3-19. Magnitude response of the closed-loop system.

Looking at the formula, as K_p increases, the natural frequency increases and the damping decreases. This means the bandwidth increases with K_p. As for "speed," the frequency of oscillations increases (and therefore the rise time decreases), but the settling time does not change.

Sketching this would look like:

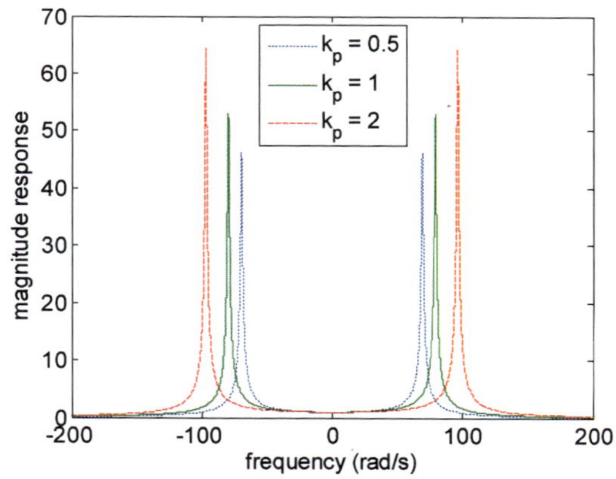

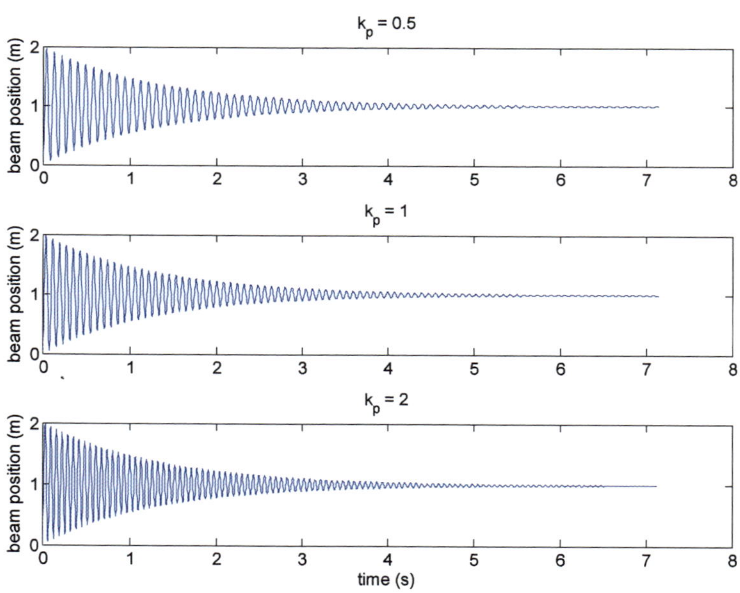

Fig. P3-20. Magnitude and step responses of closed-loop system with different K_p.

e) Find analytically the PD controller gains which reduce the settling time found in c) by half, calculate analytically (using partial fraction expansion) and plot system response to a unit step input, find the steady state output errors with respect to the unit step input and the unit ramp input.

Since the settling time is $t_s = \dfrac{4}{\zeta\omega_n}$ and the coefficient on the s term is $2\zeta\omega_n$, we just need to increase the coefficient by a factor of 2 to decrease the settling time in half. From a),

$$G_{cl,r}(s) = \frac{3300 + 3000K_p}{s^2 + (1.5 + 3000K_d)s + 3300 + 3000K_p}.$$

Therefore,

$$1.5 + 3000K_d = 2 \cdot 1.5 = 3 \Rightarrow K_d = 0.0005.$$

Now, we need to choose K_p that makes the overshoot remain the same. In this case

$$\zeta \approx 0.0094$$
$$\Rightarrow \omega_n = \frac{3}{2\zeta} \approx 158.7$$
$$\Rightarrow 3300 + 3000K_p = 158.7^2 \Rightarrow K_p \approx 7.3.$$

Substituting the controller gains into the closed-loop transfer function yields

$$G_{cl,r}(s) = \frac{25200}{s^2 + 3s + 25200}.$$

Now, taking the input to be a unit step, the time domain system response is calculated as follows:

$$Y(s) = \frac{1}{s} \cdot G_{cl}(s) = \frac{1}{s} \cdot \frac{25200}{s^2 + 3s + 25200} = \frac{1}{s} \cdot \frac{25200}{(s+1.5)^2 + 159^2}$$
$$= \frac{A}{s} + \frac{B(s+1.5)}{(s+1.5)^2 + 159^2} + \frac{C \cdot 159}{(s+1.5)^2 + 159^2} \Rightarrow$$
$$25200 = A(s^2 + 3s + 25200) + B(s+1.5)s + C \cdot 159 \cdot s$$
$$= A(s^2 + 3s + 25200) + B(s^2 + 1.5s) + C \cdot 159s \Rightarrow$$
$$25200A = 25200 \Rightarrow A = 1, \quad As^2 + Bs^2 = 0 \Rightarrow B = -A = -1,$$
$$3As + 1.5Bs + 159Cs = 0 \Rightarrow C = -\frac{3A + 1.5B}{159} \approx -0.0094 \Rightarrow$$
$$Y(s) = \frac{1}{s} - \frac{s+1.5}{(s+1.5)^2 + 159^2} - 0.0094\frac{159}{(s+1.5)^2 + 159^2} \Rightarrow$$
$$y(t) = \left(1 - e^{-1.5t}\cos(159t) - 0.0094e^{-1.5t}\sin(159t)\right)u(t).$$

The graph of the system step response is shown in Figure P3-21 below.

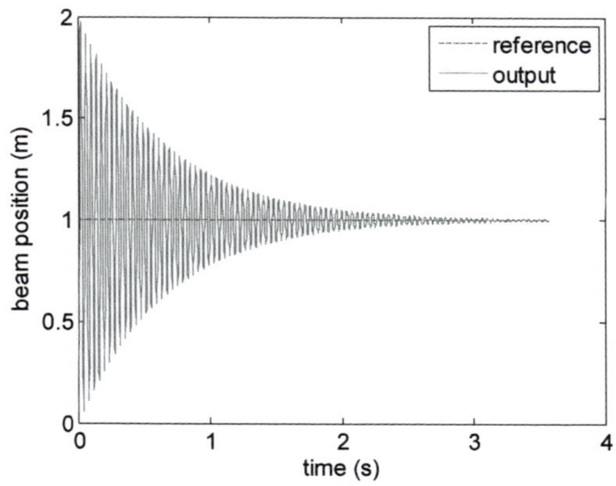

Fig. P3-21. Beam position output with unit step input.

To find the steady-state errors, we first compute the general error, which takes the form

$$E(s)=\left(1-G_{cl,r}(s)\right)\cdot R(s)=\left(1-\frac{25200}{s^2+3s+25200}\right)\cdot R(s)$$
$$=\frac{s^2+3s}{s^2+3s+25200}\cdot R(s).$$

This yields the following steady-state responses:

Unit step input:

$$e_{ss}=\lim_{s\to 0} sE(s)=\lim_{s\to 0} s\frac{s^2+3s}{s^2+3s+25200}\cdot\frac{1}{s}=0.$$

Unit ramp input:

$$e_{ss}=\lim_{s\to 0} sE(s)=\lim_{s\to 0} s\frac{s^2+3s}{s^2+3s+25200}\cdot\frac{1}{s^2}=\frac{3}{25200}=\frac{1}{8400}.$$

f) Keeping only the proportional prefilter from b) and omitting the PD block without breaking the loop, calculate the resulting output error magnitude and phase shift for a sinusoidal *reference* input $\cos\omega_{inp}t$, $\omega_{inp}=2\pi f_{inp}$, $f_{inp}=0.5\,f_{res}$ Hz.

The block diagram without the PD block takes the form shown in Figure P.3-22 below.

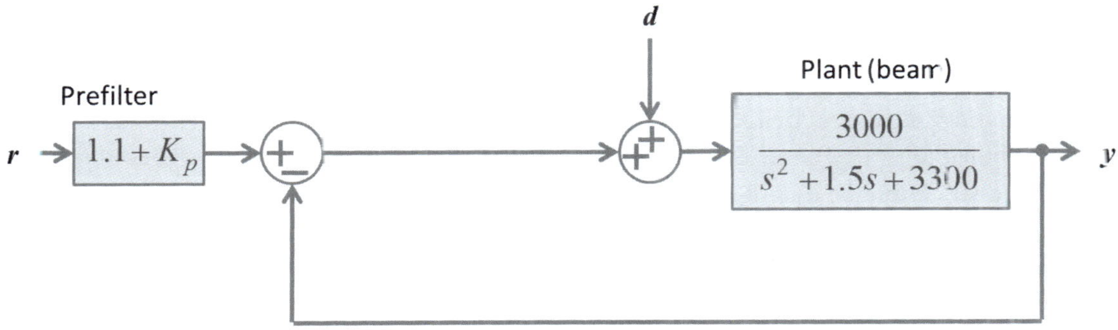

Fig. P3-22. Closed-loop system block diagram with PD block omitted and prefilter retained.

In b), we chose K_p = 1. For this controller setup, the general formula for the output error, as shown in part d), is

$$E_o(s)=R(s)-Y(s)=R(s)-G_{cl,r}\cdot R(s)=\left(1-G_{cl,r}(s)\right)\cdot R(s)$$
$$=\left(1-\frac{6300}{s^2+1.5s+6300}\right)R(s)$$
$$=\frac{s^2+1.5s}{s^2+1.5s+6300}R(s).$$

For a cosine input with frequency $\omega_{inp}=0.5\omega_{res}=28.7$ rad/s, the output error will therefore be

$$E_o(s)=\frac{s^2+1.5s}{s^2+1.5s+6300}R(s)=\frac{s^2+1.5s}{s^2+1.5s+6300}\cdot\frac{s}{s^2+(28.7)^2}.$$

This transform has two poles on the imaginary axis, which means that the final value theorem does not apply. Once the initial transient dies down, there will still be a sinusoid that persists, which will never reach a steady state. We could find the magnitude and the phase of the output sinusoid by partial fraction expansion, but it's easier to just calculate the frequency response:

$$\frac{E_o}{R}(j0.5\omega_{res})=E_o(j28.7)=\frac{(j28.7)^2+1.5(j28.7)}{(j28.7)^2+1.5(j28.7)+6300}$$
$$=\frac{-825+j43.1}{-825+j43.1+6300}=\frac{-825+j43.1}{5475+j43.1}$$
$$=\frac{\sqrt{(-825)^2+43.1^2}e^{j\tan^{-1}\left(\frac{43.1}{-825}\right)}}{\sqrt{(5475)^2+43.1^2}e^{j\tan^{-1}\left(\frac{43.1}{5475}\right)}}=\frac{826e^{-j0.0522}}{5476e^{j0.0079}}$$
$$=0.151e^{-j0.0601}.$$

Therefore, the output error will be a sinusoid with magnitude 0.151 and phase −0.0601, i.e. 15% of disturbance magnitude is propagated into the output at half the resonance frequency.

g) Still keeping only the proportional prefilter from b), insert into the forward path a controller block designed using the internal model principle to reject the sinusoidal plant input disturbance *d(t)* found in a). Check closed-loop stability. Check if disturbance rejection takes place for *disturbance* input $\cos\omega_{res}t$. For the new controller, calculate the output magnitude and phase shift for a *reference* input $\cos\omega_{inp}t$, and compare this to the result from f).

This means that we want to have a controller that looks like this:

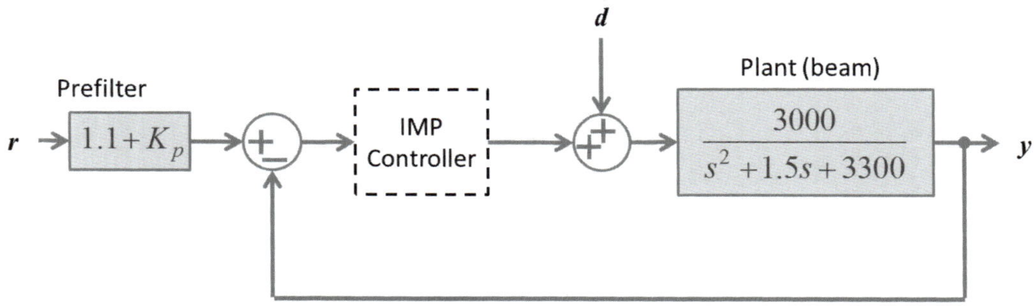

Fig. P3-23. System block diagram with Prefilter and IMP controller.

We need to figure out the form of the transfer function that should go into the block. Using the IMP, the obvious choice is the Laplace transform of a cosine with frequency at the resonance frequency of the system,

$$G_c(s) = \frac{s}{s^2+\omega_{res}^2} = \frac{s}{s^2+3299}.$$

Checking stability,

$$G_{cl,d}(s) = \frac{G_p}{1+G_cG_p} = \frac{\dfrac{3000}{s^2+1.5s+3300}}{1+\dfrac{s}{s^2+3299}\cdot\dfrac{3000}{s^2+1.5s+3300}}$$
$$= \frac{3000s^2+1.089\cdot 10^7}{s^4+1.5s^3+6599s^2+7948s+1.089\cdot 10^7}.$$

The poles of this transform, computed using Matlab, are

$$s = 2.18 \pm j60.0,\ -2.93 \pm j54.9.$$

Therefore the closed-loop system is unstable, so that no disturbance rejection can take place.

This instability is illustrated by the simple Simulink simulation:

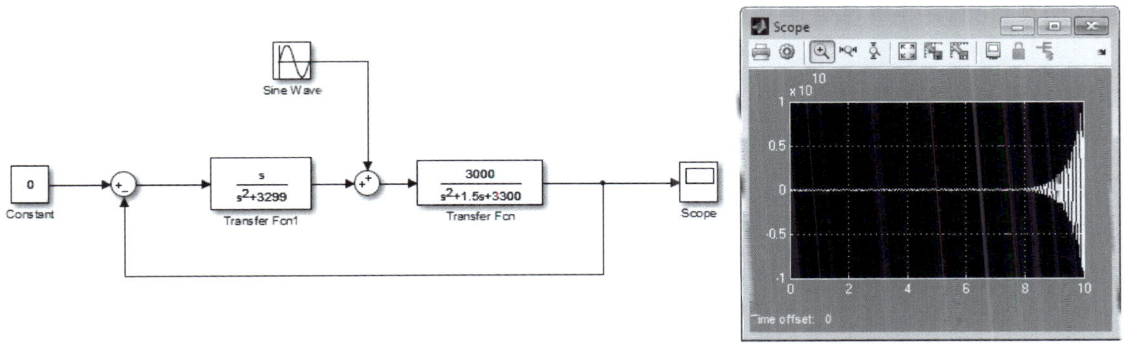

Fig. P3-24. Simulink model and system response.

Even though the closed-loop has a zero at the desired frequency, for any tiny nonzero initial conditions the system will go unstable. Therefore, we need a different controller.

There is a variety of ways we can address this instability. One possible solution is to introduce the counterpart of a PI controller by adding a P controller in parallel to the internal-model principle controller to obtain a P-IMP controller. Let's try just using 1 for the proportional part. This yields transfer function of the form

$$G_c(s) = 1 + \frac{s}{s^2 + \omega_{res}^2} = 1 + \frac{s}{s^2 + 3299} = \frac{s^2 + s + 3299}{s^2 + 3299},$$

and the closed-loop system shown in Figure P3-25 below.

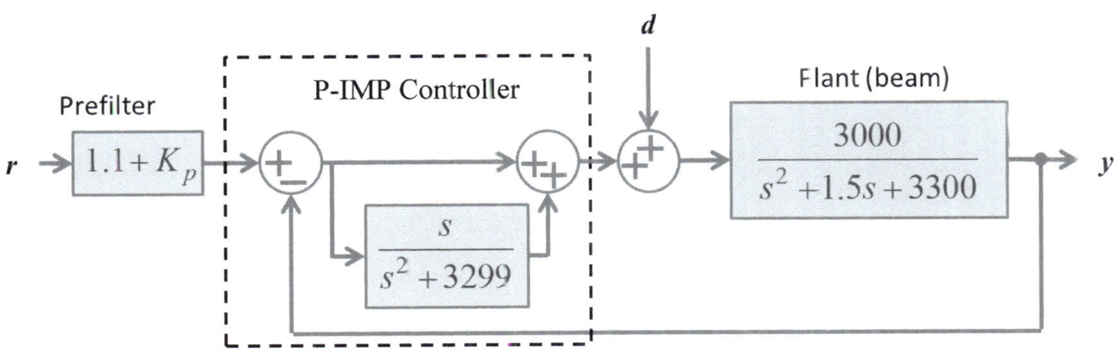

Fig. P3-25. System block diagram with new controller.

Calculating the closed-loop transfer function of this block diagram yields

$$G_{cl,d}(s) = \frac{G_p}{1+G_cG_p} = \frac{\dfrac{3000}{s^2+1.5s+3300}}{1+\left(\dfrac{s^2+s+3299}{s^2+3299}\right)\cdot\dfrac{3000}{s^2+1.5s+3300}}$$
$$= \frac{3000s^2+9.897\cdot 10^6}{s^4+1.5s^3+9599s^2+7948s+2.078\cdot 10^7}.$$

This transfer function has all poles with negative real parts:

$$s = -0.250 \pm j79.4,\ -0.500 \pm j57.4,$$

indicating that the closed-loop system is now stable.

Checking the disturbance rejection for an input sinusoid, which means setting $r(t)=0$ and $d(t)=\cos(\omega_{res}t)$ yields

$$E_o(s) = R(s)-Y(s) = -Y(s) = -G_{cl,d}(s)D(s)$$
$$= \frac{3000s^2+9.897\cdot 10^6}{s^4+1.5s^3+9599s^2+7948s+2.078\cdot 10^7}\cdot\frac{s}{s^2+(57.4)^2}$$
$$= \frac{3000(s^2+3299)}{s^4+1.5s^3+9599s^2+7948s+2.078\cdot 10^7}\cdot\frac{s}{s^2+3299}$$
$$= \frac{3000s}{s^4+1.5s^3+9599s^2+7948s+2.078\cdot 10^7}.$$

This time, the unstable poles are cancelled out by the internal model principle controller, and we can directly apply the final value theorem:

$$e_{o,ss} = \lim_{t\to\infty} e(t) = \lim_{s\to 0} sE(s)$$
$$= \lim_{s\to 0} s\cdot\frac{3000s}{s^4+1.5s^3+9599s^2+7948s+2.078\cdot 10^7} = 0.$$

Therefore, the desired disturbance frequency is completely rejected and a working P-IMP controller for this system is

$$G_c(s) = \frac{s^2+s+3299}{s^2+3299}.$$

Verifying this in Matlab yields the decaying step response shown in Figure P3-25 below.

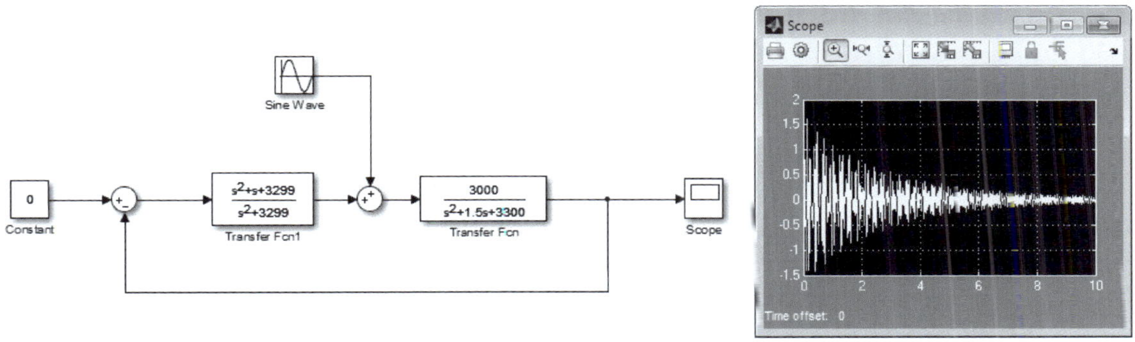

Fig. P3-25. Simulink model and system step response with P-IMP controller.

Finally, we need to check the reference tracking. The closed-loop transfer function from reference input $R(s)$ to output $Y(s)$ for this block diagram is

$$G_{cl,r}(s) = (1.1+K_p)\frac{G_cG_p}{1+G_cG_p}$$

$$= 2.1\frac{\left(\frac{s^2+s+3299}{s^2+3299}\right)\left(\frac{3000}{s^2+1.5s+3300}\right)}{1+\left(\frac{s^2+s+3299}{s^2+3299}\right)\left(\frac{3000}{s^2+1.5s+3300}\right)}$$

$$= \frac{6300s^2+6300s+2.078\cdot 10^7}{s^4+1.5s^3+9599s^2+7948s+2.078\cdot 10^7}.$$

Therefore, the transfer function from reference input $R(s)$ to the output error $E_o(s)$ is

$$\frac{E_o(s)}{R(s)} = \frac{R(s)-Y(s)}{R(s)} = 1-\frac{Y(s)}{R(s)} = 1-G_{cl,r}(s)$$

$$= 1-\frac{6300s^2+6300s+2.078\cdot 10^7}{s^4+1.5s^3+9599s^2+7948s+2.078\cdot 10^7}$$

$$= \frac{s^4+1.5s^3+3299s^2+1648s}{s^4+1.5s^3+9599s^2+7948s+2.078\cdot 10^7}.$$

Now, for an input sinusoid of frequency $\omega_{inp} = 0.5\omega_{res}$, we can find the output using the frequency response:

$$\frac{E_o}{R}(j0.5\omega_{res})=\frac{E_o}{R}(j28.7)=\frac{(j28.7)^4+1.5(j28.7)^3+3299(j28.7)^2+1648(j28.7)}{(j28.7)^4+1.5(j28.7)^3+9599(j28.7)^2+7948(j28.7)+2.078\cdot10^7}$$

$$=\frac{6.80\cdot10^5-j3.55\cdot10^4-2.73\cdot10^6+j4.73\cdot10^4}{6.80\cdot10^5-j3.55\cdot10^4-7.92\cdot10^6+j2.28\cdot10^5+2.078\cdot10^7}=\frac{-2.04\cdot10^6+j1.18\cdot10^4}{1.35\cdot10^7+j1.93\cdot10^5}$$

$$=\frac{\sqrt{(-2.04\cdot10^6)^2+(1.18\cdot10^4)^2}\,e^{j\tan^{-1}\left(\frac{1.18\cdot10^4}{-476\cdot10^6}\right)}}{\sqrt{(1.35\cdot10^7)^2+(1.93\cdot10^5)^2}\,e^{j\tan^{-1}\left(\frac{1.93\cdot10^5}{1.35\cdot10^7}\right)}}=\frac{2.04\cdot10^6e^{-j0.0058}}{1.35\cdot10^7e^{j0.0142}}$$

$$=0.151e^{-j0.0200}.$$

Therefore, the error will be a sinusoid with magnitude 0.151 and phase −0.0200. Note that the magnitude is the same as it was in part f). That is, we can conclude that the internal model principle controller designed to eliminate disturbances at frequency 57.4 rad/s has no significant effect on tracking a reference at half that frequency, 28.7 rad/s.

Exam 3: Solved Examples and Solution Guide

Problem 1: Consider a rotating mirror driven by a motor, where *b* is rotational damping of a mirror, *J* is the moment of inertia of the mirror, T_m and T_d are the motor and the disturbance torques, respectively, applied to the mirror, θ_i and θ_o are the input and the output angles, respectively, and $G_m(s)$ and $G_p(s)$ are the motor and the mirror transfer functions, respectively.

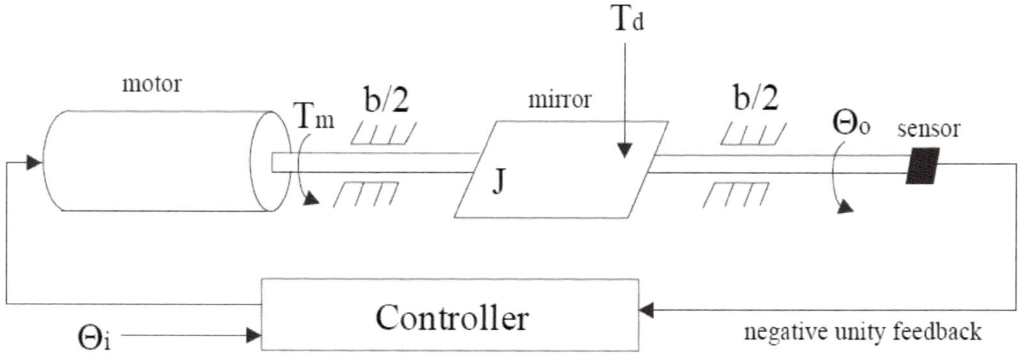

Fig. E3-1. Rotating mirror driven by a motor.

Derive the open-loop transfer function of the system. Explain what the transfer function of the open-loop should be to ensure perfect tracking and perfect disturbance rejection and give analytical expressions of such controllers. ($J = 2$, and $b = 0.4$).

Block diagram of rotating mirror system is shown as below, $G_c(s)$ is controller to be designed and $G_p(s)$ is the plant to be determined.

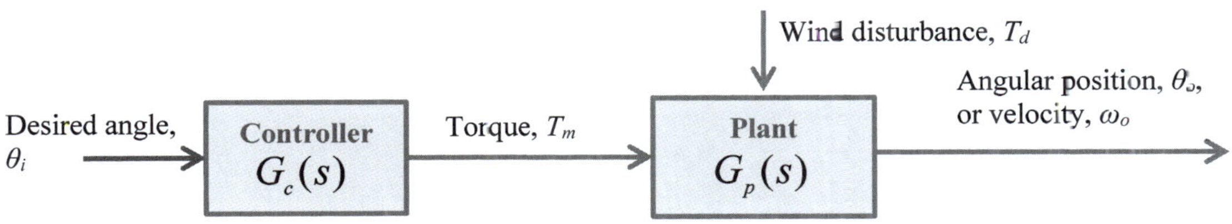

Fig. E3-2. Open-loop system block diagram.

For the system above we have: $J\ddot{\theta}_o(t) + b\dot{\theta}_o(t) = T_m(t) + T_d(t)$.

Taking Laplace transform yields:

$$Js^2\theta_o(s) + bs\theta_o(s) = T_m(s) + T_d(s) \Rightarrow \left(Js^2 + bs\right)\theta_o(s) = T_m(s) + T_d(s)$$

$$\theta_o(s) = \underbrace{\frac{1}{Js^2 - bs}}_{G_p(s)} T_m(s) + \underbrace{\frac{1}{Js^2 + bs}}_{G_p(s)} T_d(s).$$

Therefore we have: $G_p(s) = \dfrac{1}{Js^2 + bs} = \dfrac{1}{2s^2 + 0.4s}$.

For tracking: $\dfrac{\theta_o(s)}{\theta_i(s)} = G_c(s)G_p(s) = \dfrac{G_c(s)}{2s^2 + 0.4s}$.

For disturbance rejection: $\dfrac{\theta_o(s)}{T_d(s)} = G_p(s) = \dfrac{1}{2s^2 + 0.4s}$.

If we want to attain perfect tracking, we have to ensure that $\dfrac{\theta_o(s)}{\theta_i(s)} = 1$, which means

$$\frac{\theta_o(s)}{\theta_i(s)} = \frac{G_c(s)}{2s^2 + 4s} = 1 \Rightarrow G_c(s) = 2s^2 + 0.4s.$$

If we want to attain perfect disturbance rejection, we have to ensure $G_p(s)=\dfrac{1}{2s^2+0.4s}=0$. However, plant can't be 0, so perfect disturbance rejection is not attainable.

Problem 2: Now we want to use the rotating mirror system for the velocity tracking, so that the input of the system is ω_i instead of θ_i, and the output of the system is ω_o instead of θ_o.

a) Derive the open-loop transfer function of the system. Draw close loop block diagram with a P controller in the forward path and negative unity feedback, and express ω_o in terms of ω_i (reference input) for a closed loop.

For the system above we have: $J\dot{\omega}_o(t)+b\omega_o(t)=T_m(t)+T_d(t)$.

Taking Laplace transform yields:

$$Js\omega_o(s)+b\omega_o(s)=T_m(s)+T_d(s)\Rightarrow(Js+b)\omega_o(s)=T_m(s)+T_d(s).$$

Therefore: $\omega_o(s)=\underbrace{\dfrac{1}{Js+b}}_{G_p(s)}T_m(s)+\underbrace{\dfrac{1}{Js+b}}_{G_p(s)}T_d(s)$,

so that: $G_p(s)=\dfrac{1}{Js+b}=\dfrac{1}{2s+0.4}$.

The block diagram is

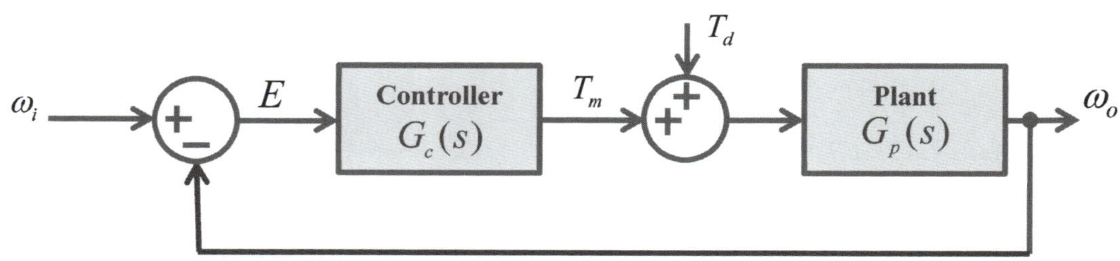

Fig. E3-3. The closed-loop system block diagram.

where $G_c=K_p$, $G_p=\dfrac{1}{2s+0.4}$ and the closed-loop transfer function is

$$\frac{\omega_o(s)}{\omega_i(s)}=\frac{G_cG_p}{1+G_cG_p}=\frac{K_p\cdot\dfrac{1}{2s+0.4}}{1+K_p\cdot\dfrac{1}{2s+0.4}}=\frac{K_p}{2s+0.4+K_p}.$$

Therefore, we have: $\omega_o = \dfrac{K_p}{2s+0.4+K_p}\omega_i$.

b) Suppose the mirror needs to execute a sawing motion with a constant velocity and for this purpose we specify a rectangular waveform $\omega_i(t)=\sum_{k=0}^{\infty} 4rect[(t-1-4k)/2]-2$, as the desired, or the reference, input. Draw this waveform. Find the value of the P-controller gain K_p that provides settling time equal to 1/10 of the input waveform period with respect to the step input $\omega_i(t)=4\cdot 1(t)$ based on 2% criterion. Give analytical expression of system response $\omega_o(t)$ to the unit step input $\omega_i(t)=4\cdot 1(t)$ for zero initial conditions and draw the response.

The plot of $\omega_i(t)=\sum_{k=0}^{\infty} 4rect[(t-1-4k)/2]-2$ is shown below:

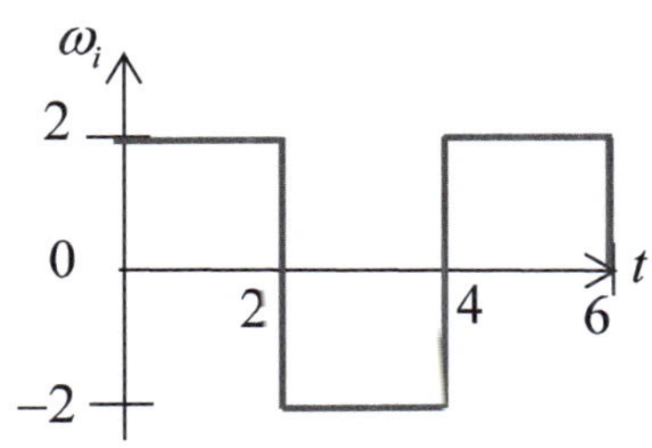

Fig. E3-4. Square wave reference signal.

Input waveform period: $T = 4s$.

Desired settling time: $t_s = \dfrac{1}{10}T = 0.4s$.

For step input $\omega_i(t)=4\cdot 1(t)$ taking Laplace transform, we have $\omega_i(s)=4\cdot\dfrac{1}{s}$

$$\Rightarrow \omega_o(s)=\frac{K_p}{2s+0.4+K_p}\omega_i(s)=\frac{4K_p}{s(2s+0.4+K_p)}=\frac{2K_p}{s\left(s+\dfrac{0.4+K_p}{2}\right)}.$$

Taking the inverse Laplace transform yields

$$\omega_o(t)=2K_p\cdot\frac{2}{0.4+K_p}\cdot(1-e^{-\left(\frac{K_p+0.4}{2}\right)t})u(t).$$

For settling time $t_s = 0.4s$

$$\omega_o(t) = \left(\frac{4K_p}{0.4+K_p}\right)\cdot(1-e^{-\left(\frac{K_p+0.4}{2}\right)t_s}) \Rightarrow \omega_o(t) = \left(\frac{4K_p}{0.4+K_p}\right)\cdot 0.98 \text{ (with 2\% criterion).}$$

Hence, $1-e^{-\left(\frac{K_p+0.4}{2}\right)(0.4)} = 0.98 \Rightarrow K_p = 19.16.$

Therefore, we obtain the analytical expression for $\omega_o(t)$ of the form

$$\omega_o(t) = 3.92\cdot(1-e^{-9.78t})1(t),$$

with the corresponding plot shown below:

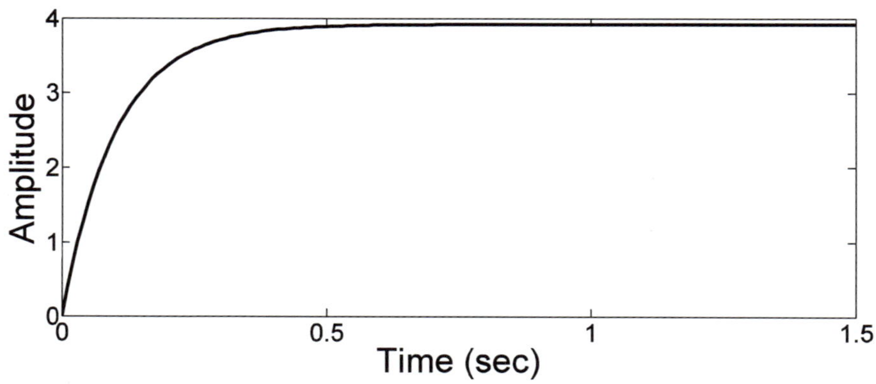

Fig. E3-5. System step response.

c) Find the closed-loop steady-state output error for the unit step and unit ramp inputs.

From above, we have $\dfrac{\omega_o(s)}{\omega_i(s)} = \dfrac{G_cG_p}{1+G_cG_p} = \dfrac{K_p}{2s+0.4+K_p} = \dfrac{19.16}{2s+19.56}.$

This gives the output error $\dfrac{E_o(s)}{\omega_i(s)} = \dfrac{\omega_i(s)-\omega_o(s)}{\omega_i(s)} = \dfrac{2s+0.4}{2s+19.56}.$

For the unit step input: $\omega_i(s) = \dfrac{1}{s}$, so that

$$e_{o,ss} = \lim_{s\to 0} sE_o(s) = \lim_{s\to 0}\frac{s(2s+0.4)}{2s+19.56}\cdot\frac{1}{s} = \frac{0.4}{19.56} = 0.02.$$

For unit ramp input: $\omega_i(s) = \frac{1}{s^2}$, yielding

$$e_{o,ss} = \lim_{s\to 0} sE_o(s) = \lim_{s\to 0} \frac{s(2s+0.4)}{2s+19.56} \cdot \frac{1}{s^2} = \infty.$$

d) Redo a) to express $\omega_o(s)$ in terms of $T_d(s)$.

The block diagram with $T_d(t)$ as an input takes the form

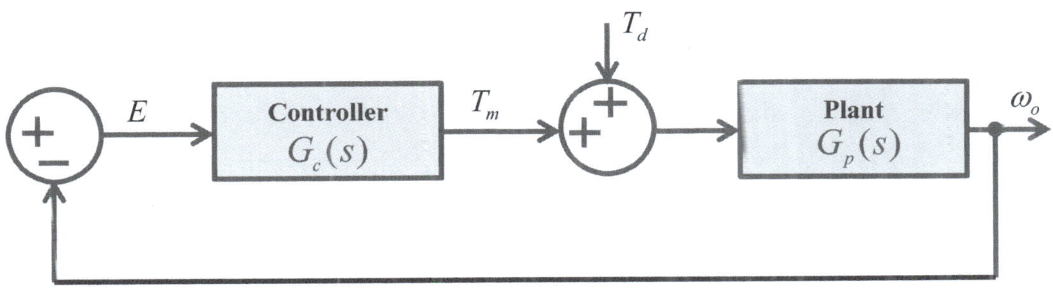

Fig. E3-6. Closed-loop system block diagram with the disturbance input.

where $G_c = K_p$ and $G_p = \frac{1}{2s+0.4}$, yielding disturbance-to-output transfer function

$$\frac{\omega_o(s)}{T_d(s)} = \frac{G_p}{1+G_cG_p} = \frac{\frac{1}{2s+0.4}}{1+K_p\cdot\frac{1}{2s+0.4}} = \frac{1}{2s+0.4+K_p}$$

and the closed-loop output $\omega_o(s) = \frac{1}{2s+0.4+K_p} T_d(s)$.

e) Assume that P gain is $K_p = 30$ and disturbance is periodic and given by $T_d(t) = \cos(2\pi 6t)$, and insert the additional controller into the loop per internal model principle to reject this disturbance. Check the close loop stability of the resulting system. Support your design by calculating the steady state error due to this periodic disturbance.

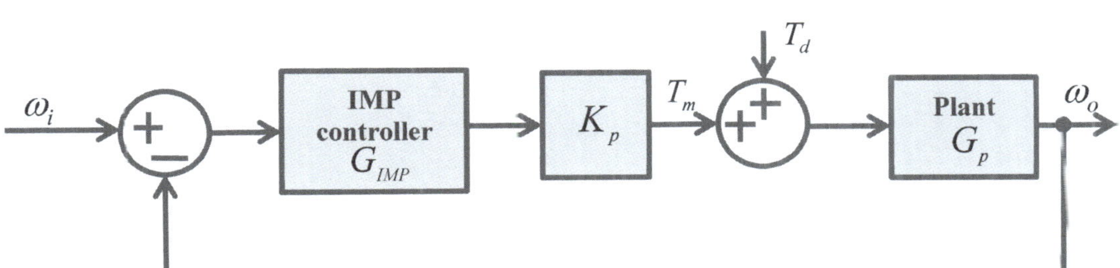

Fig. E3-7. Close loop system block diagram with IMP controller.

According to internal model principle (IMP), we should insert an IMP controller equal to the disturbance: $T_d(t) = \cos(2\pi 6t)$. Taking Laplace transform, we have $T_d(s) = \frac{s}{s^2 + (12\pi)^2}$.

Then the IM-controller $G_{CIM} = T_d(s) = \frac{s}{s^2 + (12\pi)^2}$ and $\frac{E_o(s)}{T_d(s)} =$

$$\frac{-\left(\frac{1}{2s+0.4}\right)}{1+30\left(\frac{s}{s^2+144\pi^2}\right)\left(\frac{1}{2s+0.4}\right)} = \frac{-(s^2+144\pi^2)}{(2s+0.4)(s^2+144\pi^2)+30s} = \frac{-(s^2+1421.22)}{2s^3+0.4s^2+2867.45s+568.5}.$$

The characters function of the system is $2s^3 + 0.4s^2 + 2867.45s + 568.5$. We now use stability inequality test for the 3rd degree polynomials: for characteristic polynomial $a_3s^3 + a_2s^2 + a_1s + a_0$ system is stable if and only if $a_2a_1 > a_3a_0$. Since

$$0.4 \cdot 2876.45 > 2 \cdot 568.5,$$

inequality holds, so the system is stable. Also

$$e_{o,ss} = \lim_{s\to 0} sE_o(s) = \lim_{s\to 0} s \cdot \frac{-(s^2+1421.22)}{2s^3+0.4s^2+2867.45s+568.5} \cdot \frac{s}{s^2+144\pi^2} = 0,$$

so that perfect disturbance rejection is seen to be attained.

Problem 3: For the rotating mirror system in Problem 1. Consider now a closed-loop system for this mechanism with a perfect output angle sensor, PD controller ($K_p + K_d s$) in the feedback path, and a prefilter (with gain K_p) for the reference input outside of the loop.

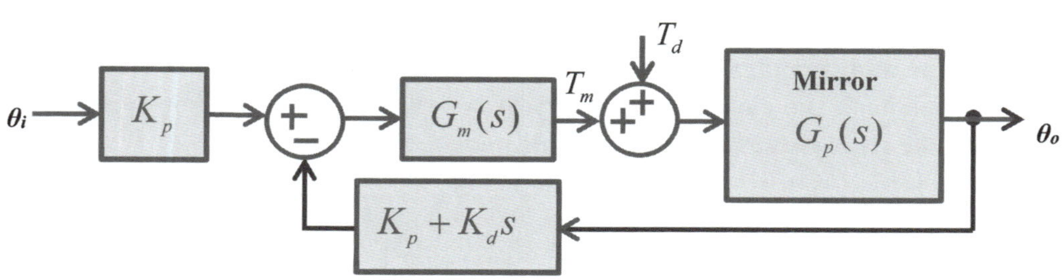

Fig. E3-8. System block diagram for rotating mirror system.

a) Find the closed-loop transfer function of this system from reference input to output.

First, we note that in this topology the controller transfer function, earlier denoted as G_c, is not K_p, but rather is split into the prefilter $G_{pref} = K_p$ and the feedback path controller $G_{FdbckPath} = K_p + K_p s$. Therefore, we need to use an input/output transfer function calculation of the form:

$$\frac{\theta_o(s)}{\theta_i(s)} = G_{Pref}\frac{G_{FwdPath}}{1+G_{FwdPath}G_{FdbckPath}} = K_p\frac{G_m/(Js^2+bs)}{1+[G_m/(Js^2+bs)](K_p+K_ds)} = \frac{G_mK_p}{Js^2+(b+G_mK_d)s+G_mK_p}.$$

b) For $G_m(s)$ = 1, J = 2, and b = 4 select the values of K_p and K_d so that the maximum overshoot for unit step input is less than 6% and settling time is less than 2 seconds.

$$\frac{\theta_o(s)}{\theta_i(s)} = \frac{0.5K_p}{s^2+(2+0.5K_d)s+0.5K_p} = \frac{\omega_n^2}{s^2+2\zeta\omega_n s+\omega_n^2} \Rightarrow$$

$$0.5K_p = \omega_n^2; \qquad (2+0.5K_d) = 2\zeta\omega_n$$

Specs: Max. overshoot: $M_p = \exp\left(-\frac{\zeta\pi}{\sqrt{1-\zeta^2}}\right) = 0.6 \Rightarrow \zeta = 0.6671.$

Settling time: $t_s = \frac{4}{\zeta\omega_n} = 2 \Rightarrow \omega_n = 2.9981 \;\therefore K_p = \frac{\omega_n^2}{0.5} \approx 18,\; K_d = \frac{(2\zeta\omega_n - 2)}{0.5} \approx 4.$

c) Find analytical expression of the output if the input is a unit ramp.

$$\theta_i(s) = \frac{1}{s^2} \Rightarrow \theta_o(s) = 9\left[\frac{1}{s^2(s^2+4s+9)}\right] = \frac{1}{\omega_n}\frac{\omega_n^3}{s^2(s^2+2\zeta\omega_n s+\omega_n^2)}.$$

Inverse Laplace transform yields:

$$\theta_o(t) = \frac{1}{3}\left[\omega_n t - 2\zeta + \frac{\omega_n}{\omega_d}\exp(-at)\sin(\omega_d t + 2\varphi)\right]1(t),$$

$$\omega_d = \omega_n\sqrt{1-\zeta^2} = 1.73, \qquad a = \zeta\omega_n = 2.04, \qquad \varphi = \cos^{-1}\zeta = 0.84\,rad$$

$$\therefore \theta_o(t) = \frac{1}{3}[3t - 1.33 + 1.73\exp(-1.73t)\sin(1.73t + 1.68)]1(t).$$

d) Find the steady state output tracking errors for unit step and unit ramp inputs.

To find the steady state tracking error, we first find the input-to-(output error) transfer function, which due to prefilter needs to be computed through direct input/output comparison:

$$H_{i\to o.e.}(s) = \frac{E_o}{\theta_i} = \frac{\theta_i - \theta_o}{\theta_i} = 1 - \frac{\theta_o}{\theta_i} = 1 - \frac{G_mK_p}{Js^2+(b+G_mK_d)s+G_mK_p} = \frac{Js^2+(b+G_mK_d)s}{Js^2-(b+G_mK_d)s+G_mK_p}.$$

Now, using the Final Value Theorem, we compute the steady-state output error for a given input:

$$e_{o,ss} = \lim_{s\to 0} s\cdot E_o(s) = \lim_{s\to 0} s\cdot H_{i\to o.e.}(s)\theta_i(s).$$

For the unit step input this yields

$$\theta_i = \frac{1}{s} \Rightarrow e_{o,ss} = \lim_{s\to 0} s \frac{s(Js + (b + G_m K_d))}{Js^2 + (b + G_m K_d)s + G_m K_p} \frac{1}{s} = 0,$$

i.e., for the unit step input, the steady-state output error is zero. For the unit ramp input

$$\theta_i = \frac{1}{s^2} \Rightarrow e_{o,ss} = \lim_{s\to 0} s \frac{s(Js + (b + G_m K_d))}{Js^2 + (b + G_m K_d)s + G_m K_p} \frac{1}{s^2} = \frac{b + G_m K_d}{G_m K_p} = \frac{4+4}{18} = 0.44,$$

i.e. for a ramp input, the steady state output error is nonzero and quite large.

e) Implement controller and feedback summer via op amps and draw the entire implementation schematics, including reference input signal, feedback summer, sensor, actuator, and plant.

The specified diagram is given in Figure E3-8.

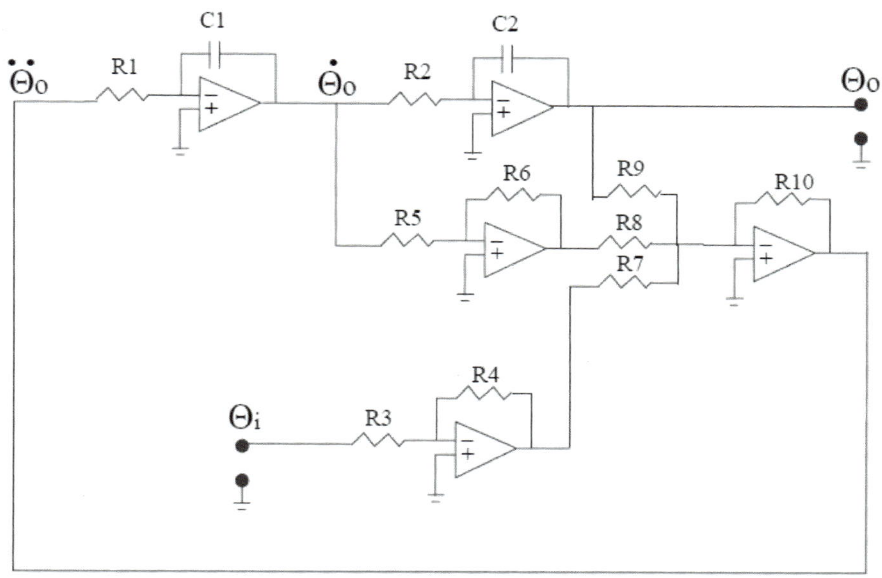

Fig. E3-8. Op amp implementation schematics for the system above.

The values of components are:

$$R_4 / R_3 = R_6 / R_5 = 1,\ R_{10} / R_9 = 9,\ R_{10} / R_8 = 4,\ R_{10} / R_7 = 9,\ 1/(R_1C_1) = 1/(R_2C_2) = 1.$$

Chapter 10

Introduction to Nonstationary Signal Analysis and Machine Learning

The advent of *machine learning* (ML) as one of the key technological tools is due to its runaway success in three areas — voice recognition and its use in automated two-way information exchange, where the *neural network* (NN) ML algorithm written over the weekend handily beat the IBM's algorithm created over a year's time, machine vision, where a *multiresolution convolutional* NN (CNN) permitted quick *image classification* and object recognition, and game winning, where *reinforcement learning* (RL) was used by a computer program (AlphaGo) to win the match with the world's top Go player. Notable breakthroughs have been also achieved in combination of machine learning and control — for example, *gradient-based learning* in combination with multiresolution process representation helped solve the chemical looping control problem. AI and ML are also widely used in transition to smart digital manufacturing, referred to as Industry 4.0. All these techniques will be briefly introduced in this chapter. To bring out the ML capabilities, we start with simple *sound equalization* in computer audio and *edge detection* in machine vision. Then, we present the single-resolution function approximation through generalized Fourier series, followed by the *time-frequency analysis* — *multiresolution* function approximation using *wavelets*, and nonstationary signals analysis through *Hilbert-Huang Transform* (HHT). These are followed by the basics of ML, including *perceptron*, *deep NN*, *convolutional NN*, *gradient descent* learning, and applications of the latter to nonlinear system identification and control, computer vision, voice recognition, and material phase change characterization. The rigorous proof that perceptron and wavelets are *universal approximators*, i.e. can approximate functions with arbitrary precision is left to references. The chapter ends with a set of solved homework problem examples.

10.1 Basic Elements of Computer Audio

10.1.1 *Basic audio file formats and storage*

Compression is a technique for producing a reduced size data set from the original one to facilitate the information storage and communication. Compression can be *lossy* or *lossless*. In lossy compression, the bits are eliminated according to a particular standard. In lossless compression, the data is partitioned into smaller segments, subsequently reconnected, to enable the full information retention. *Decompression* restores a compressed file to the initial form to the extent possible, depending on the compression used — lossy, enabling only the partial restoration or lossless, enabling the full restoration. A two-component computer program for compressing (encoding) and decompressing (decoding) media files is referred to as a *codec*. Programs, such as Windows Media Player, use codecs to play and create media (audio and video) files.

The most basic audio codec, and the file format associated with it, is MP3 invented in late 1980's. MP3 is an abbreviation for MPEG-1 Audio Layer III, where MPEG stands for Moving Pictures Experts Group, an international collaboration of engineers founded in 1988. It enables storing digitized audio information in a smaller file size than the PCM format (Subsection 3.3.5) used by CDs. MP3 codec performs a lossy compression of the original file, with the size depending on the bitrate, measured in *bps - bits per second*, and the sampling rate. A typical MP3 file is created with 44.1 kHz (CD) sampling rate and bitrate between 8 kbps and 320 kbps from a CD format file. The size of the resulting MP3 files is reduced up to tenfold, while still providing an acceptable audio reproduction, which for a 320 kbps file gets close to the CD sound.

Recently, the MP3 codec has been replaced by the codec referred to as the Advanced Audio Coding, or AAC, producing .m4a format — a current standard for music download, for example from Apple's iTunes Store. The AAC is still a lossy codec, but it generally achieves higher sound quality than MP3 for the same bit rate. Also, as the storage limitations receded, the so-called *container formats* — file formats allowing embedding of multiple data streams, such as audio and video, into a single file — have been introduced, with MP4 (MPEG-4 Part 14) being the most popular. The ACC m4a file is usually converted to MP4 file, with file extension .mp4. For streaming and storing high quality audio, an mp4a audio file format is employed as well (https://www.canto.com/blog/mp4a-audio-file/). MP4 also works with lossless formats, such as Apple Lossless Audio Codec (ALAC), offering MP4-ALAC — an audio download format also available from Apple's iTunes Store through "Kind" option, but also with file extension .m4a. The other popular lossless codec is Free Lossless Audio Codec (FLAC), which is widely used to download high resolution PCM audio files, discussed in Example 4.3-1 of Chapter 4.

In this section .mp4 format will be used for working with the audio files in Matlab. The Matlab command for reading and storing an audio file called, for example, Filename.mp4 is: [y_Filename, Fs] = audioread('Filename.mp4'); where Fs is the sampling rate. The audio can be played back using commands playerObj = audioplayer(y_Filename, Fs); play(playerObj).

10.1.2 *Signal equalization*

Human audio perception is tuned to the speech 1 kHz to 4 kHz frequency range, where it is the most sensitive. For example, a midrange sinusoid at 1.5 kHz will sound a lot louder than a low frequency sinusoid at 300 Hz with the same magnitude. Therefore, the adjustments are usually made to an audio signal to make the higher frequency, also referred to as the higher *pitch*, signals sound more pleasant to human ear. These adjustments are referred to as ***signal equalization***, and the analog electronic circuits performing a frequency-selective signal magnitude shaping or their digital counterparts are called ***signal equalizers***. The latter, therefore, are the filters that amplify ("boost"), attenuate ("cut"), or pass an audio signal frequency components, according to the user preference. In this subsection we briefly review the digital equalization, as more relevant to computer audio, considering both *batch* and *online* equalizers.

In batch equalization we assume that the entire signal record is available. We can then take FFT of the data record to obtain the DFT magnitude signal spectrum and perform a direct magnitude spectrum modification — magnitude windowing. In this case we selectively, and symmetrically to keep DFT inverse real, cut or boost only the magnitudes in the frequency ranges of interest, for example, boost high and low frequencies magnitudes, leaving the middle range intact, to produce a boom box effect. This modification acts as an ideal filter, not affecting the phase of the frequency components. The resulting magnitude-windowed, or equalized, spectrum, further referred to as Boom, and the audio time-domain signal corresponding to it can be reconstructed using the inverse DFT operation ifft(Boom) as follows: y_reconstructed = ifft(Boom); audiowrite('Boom.mp4', y_reconstructed, Fs);. The example of such equalization is given in Problem 1(a) of Hw11.

In the online equalization, we assume that we are acquiring data in real time as a bit stream, passing it through a bank of recursive IIR filters or convolution-based FIR calculations, which forms an equalizer, and getting the real-time filtered output signal. Other equalization scenarios are also possible, like application to the bit stream of a moving window FFT computation or short-term Fourier transform signal segmentation, described in Subsection 10.6.2, processing, and reconstitution, but they are easily derivable from these two basic configurations.

The online equalizers can be subdivided into three main types: *parametric*, *shelf*, and *graphic*. The parametric equalizers typically employ the second order band-pass or band-stop filters as the key elements and connect these elements, each centered at its specified frequency, in a cascade. Each filter has adjustable gain, and the shape of each filter is manipulated to set the desired value of the underdamping, usually expressed through quality factor Q (7.1-7) — roughly, the center frequency divided by the bandwidth. Tuning of Q determines the bandwidth around the center frequency, i.e. sets how wide or narrow the boost or cut would be around a particular frequency. Higher Q values yield a narrower bandwidth, while lower Q values affect a wider range of frequencies. The filter gain determines the signal magnitude ("level") cut or boost in a selected frequency range. The filter slope is measured in dB/octave. The higher the number, the steeper the roll-off around the corner frequency of the filter. Slopes of 6 db/octave to 24 db/octave are typical. The shelf equalizers are the highly simplified parametric equalizers, with only two filters, high-pass and low-pass, connected in cascade, each typically given by the first order digital filter with the user-tunable filter gain ("tone control"), but not bandwidth. An example of the tone control equalizer is given in Problem 10(b) of Hw11. The graphic equalizers lie between the above two types — they are formed by a bank of second order bandpass filters with tunable gains, but not bandwidth, connected in parallel, with each filter covering a separate part of the audible frequency range.

Another common computer audio procedure is ***pitch correction***. 90% of commercial songs go through pitch correction (https://www.musiciansonamission.com/pitch-correction/).

10.2 Basic Elements of Machine Vision

10.2.1 *Basic digital image units and formats*

A digital computer image is a grid (a 2D array), or a matrix, of discrete elements referred to as *pixels* (from "picture element"), the smallest image representation units. The terms *brightness* or *intensity* of a pixel are used to characterize the degree of pixel "greyness" — the position of a pixel along the grey scale range, as assignable by a byte — an 8-bit integer representing 256 brightness levels, with 0 assigned to black and 255 — to white, unless the opposite assignment brings out the image features of interest better. Thus, the pixel brightness is a term standing just for a shade of grey, only indirectly related to the physical concept of brightness as the luminance of an object. To represent a color image in RGB colorspace, a red, green, and blue intensity triplet (three bytes) is specified for each pixel. *Image resolution*, typically given by megapixels (MP — 1 million pixels), is a product of pixel grid columns and rows, giving a number of pixels in a digital image.

Just like audio files, digital image files can be compressed and have lossless and lossy compression formats. The most common format of lossy compression is JPEG, producing JPG files. The unrecoverable image detail loss takes place each time the file is saved in a lossy format (Subsection 10.2.4), so repeated saving of an image in JPEG significantly degrades the image quality. To avoid this, if the image is saved by a camera as a JPG file, it should be then saved for storage in one of the lossless formats: TIFF, PSD, PNG, GIF, RAW, BMP. Files containing audio and video are most often saved in mp4.

10.2.2 *Convolution and linear filters*

Section 7.5.4 of Chapter 7 presented FIR temporal filter implementation for discrete time signals using a convolution sum. A 2D spatial counterpart of this technique is implemented in most of modern computer vision systems through the convolution operators, ranging from convolution layers in neural networks to simple gradient filters. This subsection presents a number of basic linear (convolutional) operations and their applications in computer vision. Convolution is not straightforwardly applied in practice, requiring that the *padding* and the *stride* of the convolution be taken into account. We will apply these notions to the basic edge detection in images using gradient operators.

10.2.2.1 *Convolution on 2D images*

As shown in the non-recursive DT filter implementation in Subsubsection 7.5.4.3 of Chapter 7, convolution can be seen as a *sliding window operation* in which we slide a fixed-size window over an image, at each sliding step or several steps taking the weighted sum of each window as specified by a given *filter kernel,* a counterpart of a DT FIR filter impulse response. Let $I \in \mathbb{R}^{n \times m}$ be the input image, and $F \in \mathbb{R}^{p \times q}$ be the filter kernel, where n is the image height, m is the image width, $p \ll n$, and $q \ll m$. In the most basic case, we will shift by one pixel each time, given an output image size of $O \in \mathbb{R}^{k \times l}$, where $k = n - (p - 1)$ and $l = m - (q - 1)$. In this case, the elements of the output image are computed as follows:

$$O_{i,j} = \sum_{a=1}^{p} \sum_{b=1}^{q} \left(I[i : i + p - 1, j : j + q - 1] \right)_{a,b}, \tag{10.2-1}$$

where $I[i:i+p-1, j:j+q-1]$ denotes the matrix block (*slice*) with the top-left corner at $I_{i,j}$ and the bottom-right corner at $I_{i+p-1,j+q-1}$. We denote the convolution operation by $O = I * F$. An illustration of linear convolution is given in Figure 10.2-1.

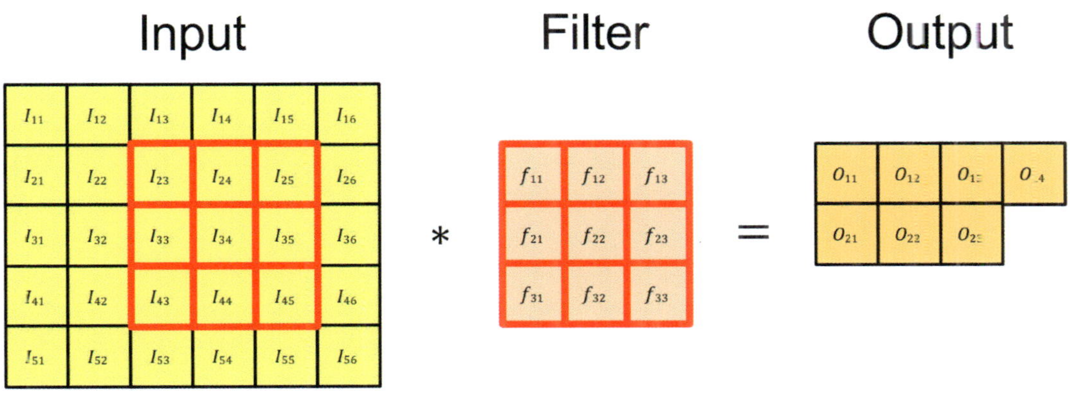

$$O_{23} = I_{23} \cdot f_{11} + I_{24} \cdot f_{12} + I_{25} \cdot f_{13} + \ldots + I_{45} \cdot f_{33}$$

Fig. 10.2-1. Linear convolution in action. The final output is of the size 3 by 4.

The output of a convolution can't have the same size as the input, unless the filter kernel is scalar. The output of the same size as the input can be obtained through *padding* the input image, that is, adding elements to the outer edges to obtain a padded input image size that produces the desired output size given the filter kernel. A number of padding operations is used in practice:

- Zero padding (adding zeros all around the edge)
- Wrap around (taking values of the opposing side)
- Edge copying (repeating the edge values)
- Reflection across edge (reflecting the inside values at the edge)

The choice of a particular padding method is immaterial when dealing with small padding sizes in practice, but when dealing with larger images, it is a common practice to instead *crop* the image (cut out image edges), and use the cropped edges as padding. In general, a filter kernel is chosen so that all elements add up to unity. This precludes a net change of the output image intensity.

To demonstrate the utility of linear filtering, we introduce three derivative filters: the Sobel, the Prewitt, and the Derivative of Gaussian (DoG) filters. The Sobel filter is simply a central differencing filter with a higher weight at the center. It uses the following two kernels, in the horizontal and vertical direction, respectively:

$$F_{x,\text{sobel}} = \begin{bmatrix} 1 & 0 & -1 \\ 2 & 0 & -2 \\ 1 & 0 & -1 \end{bmatrix}, \qquad (10.2\text{-}2)$$

$$F_{y,\text{sobel}} = \begin{bmatrix} 1 & 2 & 1 \\ 0 & 0 & 0 \\ -1 & -2 & -1 \end{bmatrix}. \tag{10.2-3}$$

The Prewitt filter is similar to the Sobel filter, but simpler — it does not assign a greater weight to the center:

$$F_{x,\text{prewitt}} = \begin{bmatrix} 1 & 0 & -1 \\ 1 & 0 & -1 \\ 1 & 0 & -1 \end{bmatrix}, \tag{10.2-4}$$

$$F_{y,\text{prewitt}} = \begin{bmatrix} 1 & 1 & 1 \\ 0 & 0 & 0 \\ -1 & -1 & -1 \end{bmatrix}. \tag{10.2-5}$$

Finally, the Derivative of Gaussian filter is based on a two-dimensional Gaussian's partial derivative. This idea can be extended to higher order derivatives (e.g., the Laplacian of Gaussian for the second derivatives). The shapes of these kernels are shown below:

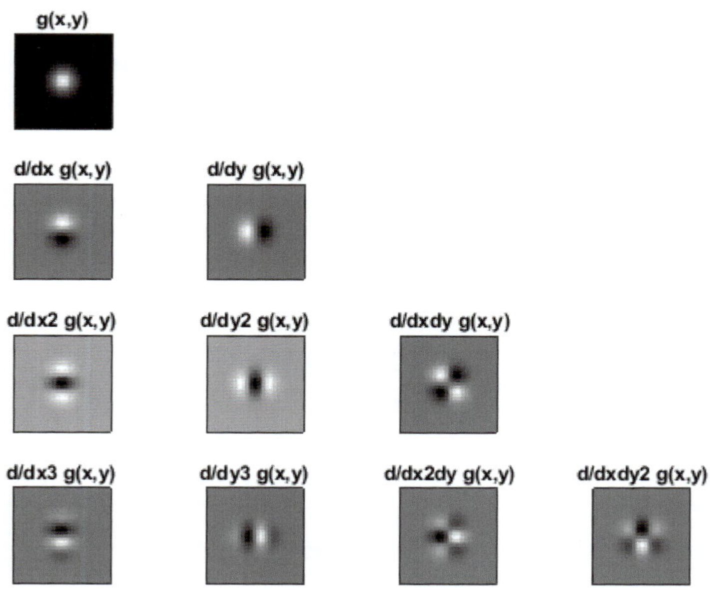

Fig. 10.2-2. Derivatives of a Gaussian filter.

The Gaussian filter is used to smoothen images with noise (also known as *Gaussian blurring*), making it potentially useful for dealing with noisy images, as shown next. We first apply all the presented derivative filters to a black and white image with clearly identifiable edges:

Fig. 10.2-3. Derivative filters applied to a clean image (https://www.publicdomainpictures.net/en/view-image.php?image=167212&picture=statue-of-liberty).

The Prewitt and the Sobel filters look almost identical, while the Derivative of Gaussian filter exhibits the aforementioned smoothing effect. Clearly, the Sobel and the Prewitt filters are desirable in the case of a virtually noiseless image. In the presence of the high frequency noise, these filters can produce image distortion in the form of the high intensity gradients. This is often unacceptable, making the Derivative of Gaussian filter more desirable in most computer vision algorithms. This effect is clearly seen in Figure 10.2-4, where high frequency noise is for the most part smoothed out by the Derivative of Gaussian filter, with a larger kernel mitigating more of the noise at the expense of a more blurred out picture.

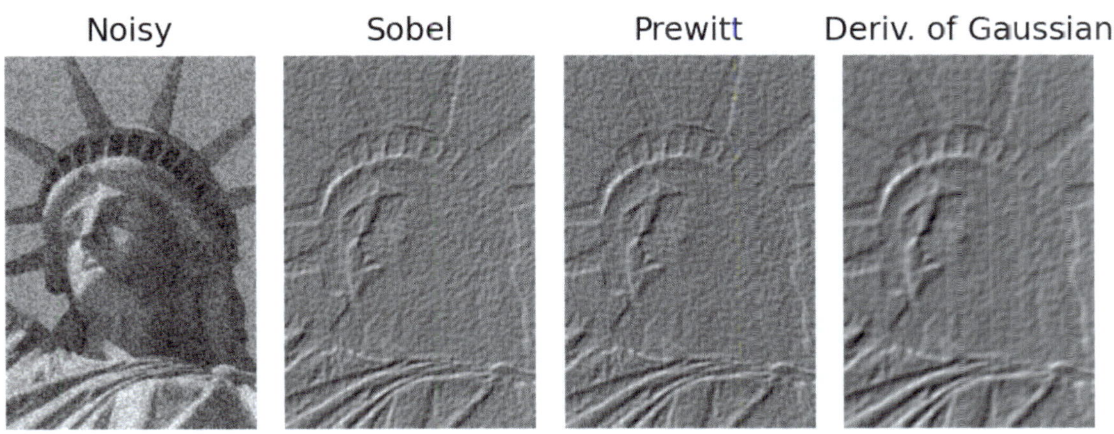

Fig. 10.2-4. Derivative filters applied to a noisy image.

Numerous other filters include blurring/smoothing (including the box filter), sharpening, shift (shifting by a predefined number of pixels), etc.

Just like in Chapter 7, many properties of the linear image filters can be characterized in the frequency domain, as presented next. The latter brings a concept of spatial frequencies, which greatly improve one's understanding of linear image filtering. The required two-dimensional

Fourier analysis is, however, rarely applied in practice, since algorithms in practical computer vision must be fast and easy to implement. Linear convolution filters require only elementary floating-point operations and demand modest storage requirements, whereas two-dimensional discrete Fourier transforms are demanding and hard to implement on real-time embedded hardware.

10.2.3 *Two-dimensional (2D) Fourier analysis in image processing*

To understand the relation of the frequency content to the features of the image, let's first discretize the CT variable t with very small sampling interval to obtain the DT variable n, and then reassign n to designate not the DT variable, but rather the *discrete spatial (DS)* variable.

Let's now consider the square wave in DS (*discrete space*) and assign the low intensity (black) $x[n] = 0$ to the bottom part of the square wave and high intensity (white) $x[n] = 255$ to its top part. Then, in the image of an $x[n]$ line in the 2D $x[m, n]$ grid, the sharp changes of the square wave value will be encoded into the color jumps from black to white and vice versa at the squire wave discontinuities. If we now decompose $x[n]$ into DT Fourier series, these intensity discontinuities will be built by many high frequency harmonics, as in Figure 5.3-3 in Chapter 5.

If we add another spatial dimension and equally spaced square waves in both directions, we obtain a 2D image $x[m, n]$ with the checked pattern shown on the bottom left in Figure 10.2-5, where for every fixed m and n pair, x takes the value of pixel intensity at this $[m, n]$ grid location. Any sharp intensity, or ***high gradient***, drop in this pattern forms an ***edge***. Thus, the latter is a high-frequency visual object.

To represent this image in the frequency domain, the DFT (5.6-3) and its corresponding DFS (5.6-4), both given in Chapter 5, are generalized to the 2-dimensional form as follows:

$$X_{DFS}[k] = \frac{1}{N_1}\sum_{n=0}^{N_1-1} x[n]e^{-j2\pi k\frac{1}{N_1}n} \Rightarrow X_{DFS}[k,l] = \frac{1}{M_1 N_1}\sum_{n=0}^{N_1-1}\sum_{m=0}^{M_1-1} x[m,n]e^{-j2\pi\left(\frac{k}{M_1}m+\frac{l}{N_1}n\right)}$$

and

$$X_{DFT}[k] = \sum_{n=0}^{N_1-1} x[n]e^{-j2\pi k\frac{1}{N_1}n} \Rightarrow X_{DFT}[k,l] = \sum_{n=0}^{N_1-1}\sum_{m=0}^{M_1-1} x[m,n]e^{-j2\pi\left(\frac{k}{M_1}m+\frac{l}{N_1}n\right)},$$

yielding the 2-dimensional Discrete Fourier Series (2D DFS) and Transform (2D DFT). Since $X_{DFT}[k, l]$ is a N_1 by M_1 grid of numbers, we can view them as pixels, assign 255 intensity to the biggest number and rescale the values of the entire grid to lie between 0 and 255.

Plotting the result using Matlab fft2( ) command and rescaling the magnitudes to lie in the (0, 255) pixel intensity range, with 255 assigned to the highest value, produces a greyscale frequency domain 2D representation of the original image shown on the lower right in Figure 10.2-5. To obtain a better visualization, this DFT magnitude spectrum image is plotted in the logarithmic magnitude scale. To shift the lowest frequencies to the center, the Matlab command fftshift( ) is used. The *whitest point* at the center represents the *highest magnitude* — that of the DC component, with the magnitudes of the low frequency components dominating in the vicinity of the center. The component magnitudes drop the most along the diagonals from the center to the corners, with the latter representing the magnitudes of the components with the highest frequencies. It is seen that many frequency components are required to build this sharp-edged checkered pattern.

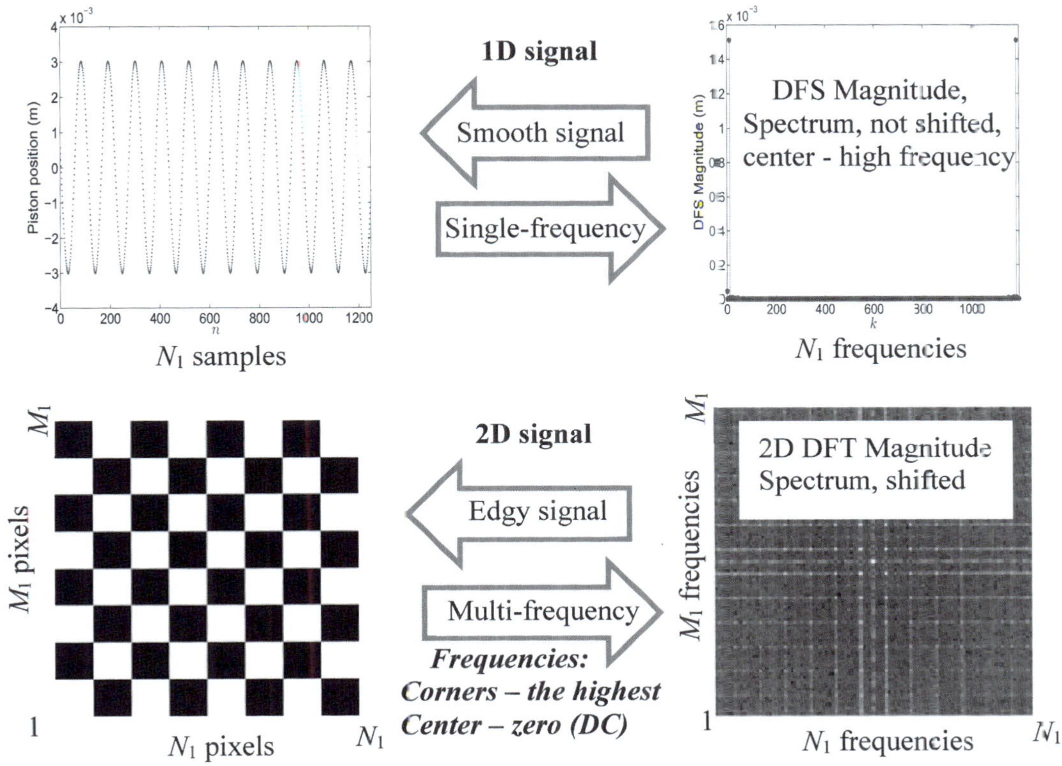

Fig. 10.2-5. Illustration of the 2D discrete Fourier transform. DT square wave is generalized to the checkered pattern. Logarithmic scale is used. The higher magnitudes are designated by the lighter colors.

The frequency unit of the digital frequency domain image is no longer given in cycle/sample, Associating the entire image with a single cycle, this unit becomes 1/pixel, or pixel^{-1}.

The fundamental features of the checkered pattern spectral characterization do not change if we scramble this pattern, as in Figure 10.2-6 (left), scattering the dark squares over the white color plane, but keeping the same gradients. Assigning now the *darker* colors to the *higher* magnitudes to make the latter better discernible in terms of contrast, the unshifted and the shifted spectra of this image are given by the center and the right-hand plots in Figure 10.2-6. It is seen that the lower frequencies, clustering around the origin in the frequency range from zero to about 20 pixel^{-1} in the unshifted plot, are the dominant ones. In the shifted plot on the right the lower frequencies cluster around the center, just like in Figure 10.2-5. However, the significant high frequency components presence, required to build the edges of the squares, is seen in both plots as the light shades away from the low frequency clusters.

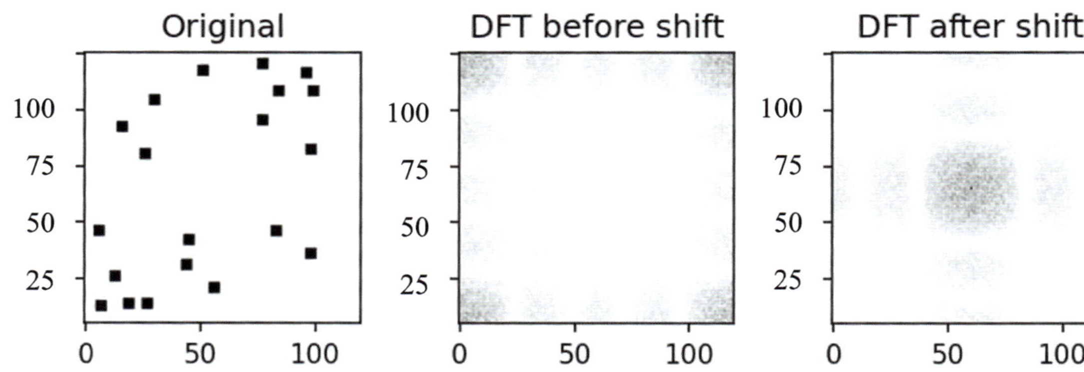

Fig. 10.2-6. Illustration of the effect of scrambling the checkered pattern, with the higher magnitudes designated by the darker colors. Before shifting, the low frequency content, seen to be dominant, clusters at the corners. After shifting, the low frequency content is concentrated at the center. The higher frequency components are seen as the lighter shades.

The right-hand spectral plots in Figures 10.2-5 and 10.2-6 bring up the idea that by eliminating the spectra outside of the center vicinity, we can smoothen the edges of both original images, and by eliminating the center harmonics we can isolate and display the edge information. These tasks are accomplished, respectively, by the low-pass and the high-pass image filtering described next.

10.2.4 *Low-pass image filtering. Low power harmonics filter for image compression*

To see how the 2D Fourier series approximates images, especially those with high gradients, let's consider the effect of a low-pass filter on the output image. In the image with squares in Figure 10.2-6 let's cut off all the frequency components, the powers of which fall below the power threshold, specified by the power percentile. Looking at the image on the right in Figure 10.2-6, it is seen that the very high power percentile filter will pass only the low frequency components belonging to the central cluster, but as the percentile decreases, the filter admits the components with the smaller powers from the higher frequency range, enhancing the image.

This approximative nature of the 2D DFS/DFT harmonic percentile filter is shown in Figure 10.2-7. At the 99.99th percentile, the harmonics passed are seen to build *a rotated 2D sinusoid* — the lowest frequency foundation of the squares image which roughly follows a diagonal pattern. The structure starts appearing in the 99th percentile filter output. With the black squares becoming distinctly visible, but smothered, this filter is seen to pass some higher frequency components.

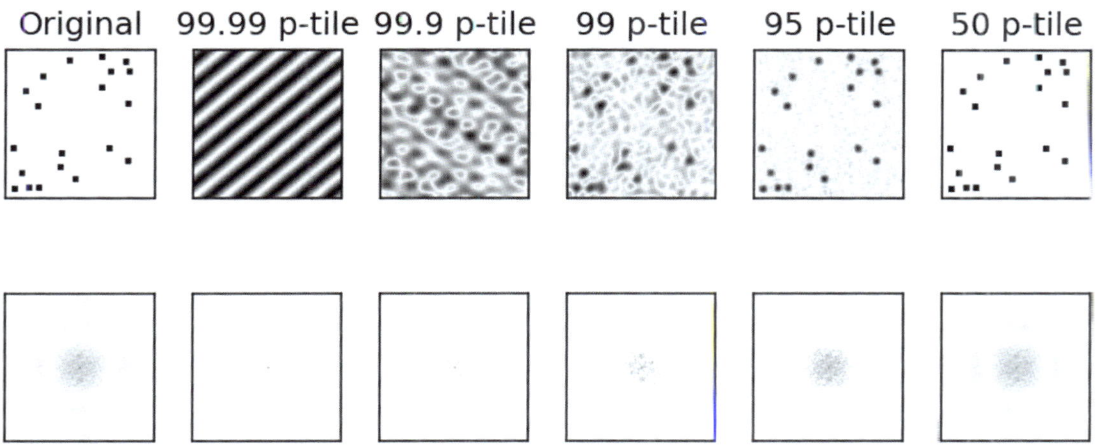

Fig. 10.2-7. Illustration of a 2D low-pass filter based on signal power and its effect on the resulting image. The darker the color, the higher the magnitudes in the DFT plots.

The remainder of the energy serves to construct the sharp edges around the squares, which gives a good idea about the image in the 95th percentile figure. Passing all the components in the 50th percentile shows that the lower power components can be completely discarded, no matter how much power of the signal they account for.

The technique described above is close to what drives the JPG image format (which uses direct cosine transforms instead of DFT). Lossy images often have off-white artifacts such as those seen in the 95th percentile image, which requires less memory to store since high frequencies are not accounted for, but comes at the cost of a potential distortion. It is also now clear that once we discard the lower 50th percentile components, we reduce the component set, and the next 50th percentile will be applied to the reduced components cohort, further discarding the components passed at the previous filtering stage. Thus, repeated application of lossy JPEG format iteratively reduces the number of components, severely degrading the image quality.

10.2.5 *High-pass image filtering and its use in image edge detection*

Since edges are built through the high frequency spectrum components, edge detection can be accomplished by passing an image through the high pass filter (HPF). To build an ideal high-pass filter we simply set the magnitudes of the lowest frequency components in the transform to zero, for example by *masking* them — blocking the magnitudes of all the frequency components that lie inside a circle of some fixed radius from the center. The frequency domain representation (DFT) of such high-pass masking filters is given by the center and the right-hand plots in the second row in Figure 10.2-7. In this row, *zero* values are designated by the *black* color, 1's — by the white color, and the values between zero and 1 — by the shades of grey. However, in the third row of the DFT magnitude plots, the black color stands for the higher magnitudes and the white color — for the lower ones to enhance the image.

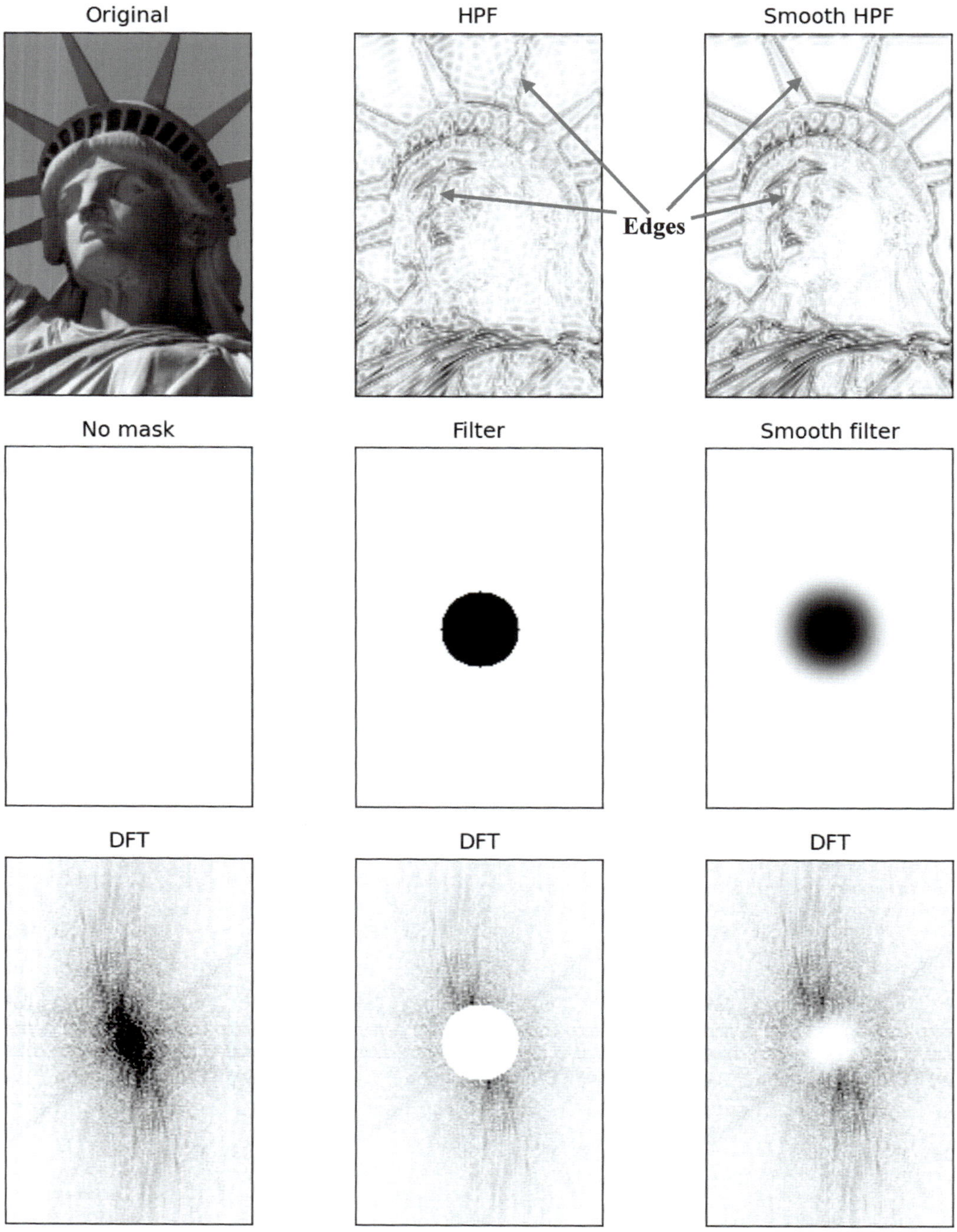

Fig. 10.2-7. Illustration of a 2D high-pass filter based on different filter designs. The black color in the second row masks stands for zero and the white color — for one. In the third row the black color designates the larger magnitudes, as opposed to what it designates in the middle row, with masks applied to two graphs on the right.

The image of a signal resulting from its filtered frequency domain representation can be reconstructed though the inverse DFT operation performed in Matlab by ifft2( ) function. The reconstructed images corresponding to the ideal HPF of the center mask and the smoothed (windowed) filter of the right-hand mask are given by the center and the right images in the first row in Figure 10.2-7 and *show the isolated edges of the original image*.

Zooming in on the parts of the reconstructed high-pass filtered images reveals echoes, as shown in the left image in Figure 10.2-8. Making the filter window smoother, using for example Gaussian filtering removes most of the echo effects, as seen in the right image in Figure 10.2-8.

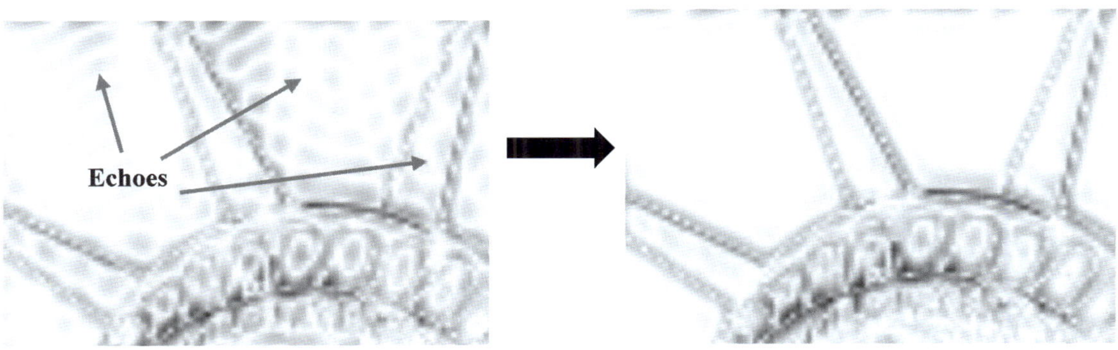

Fig. 10.2-8. Zoomed-in area showing artificial echoes (left) and artifacts removed with a mask smoothened using Gaussian filtering (right).

10.3 Beyond Power Spectral Density: Time-Frequency Analysis of Multiresolution and Nonstationary Signals

10.3.1 *Preliminaries: signal norms, Generalized Fourier Series as optimal periodic signal representation*

Let's first introduce the standard vector and signal norms used in signal processing. The *p*-norms for vector $x=[x_1,x_2,...,x_m]^T \in R^m$ are defined as:

$$1-norm \triangleq \left\{ x:\| x \|_1 = \sum_{i=1}^{m} | x_i | < \infty,\ p=1 \right\},$$

$$p-norm \triangleq \left\{ x:\| x \|_p = \left(\sum_{i=1}^{m} | x_i |^p \right)^{1/p},\ 1<p<\infty \right\},$$ for $p=2$ the subscript is often omitted,

$$\infty-norm \triangleq \left\{ x:\| x \|_\infty = \max_{1\le i\le m} | x_i |,\ p=\infty \right\}.$$

Then, the signal norms for $x(t)$, $t \in R$ are defined as follows:

$$L^1(R) = \left\{ x(t):\| x \|_1 = \int_{-\infty}^{\infty} | x(t) | dt < \infty,\ p=1 \right\},$$

$$L^p(R)=\left\{x:\|x\|_p=\left(\int_{-\infty}^{\infty}|x(t)|^p\,dt\right)^{1/p},\ 1<p<\infty\right\},$$

$$L^{\infty}(R)=\left\{x:\|x\|_{\infty}=\sup_{t\in R}|x(t)|,\ p=\infty\right\}.$$

To designate the space, subscripts are often used, as well, e.g. $L^2(R)\equiv L_2(R)\equiv L_2$.

$$L_m^p(R)=\left\{x(t):\|x\|_p=\left(\int_{-\infty}^{\infty}\sum_{i=1}^{m}|x_i(t)|^p dt\right)^{1/p}<\infty,\ 1\le p<\infty\right\},$$

$$L_m^{\infty}(R)=\left\{x(t):\|x\|_{\infty}=\max_{1\le i\le m}\sup_{t\in R}|x_i(t)|\right\}.$$

For $x[n]$, $n\in$ integers $\int_{-\infty}^{\infty}$ is replaced by $\sum_{-\infty}^{\infty}$ and L is often replaced by l.

Each norm defines its function space. A function is said to belong to a space defined by a norm, if the norm applied to this function is bounded. For example, $\cos(\omega t)$, $\omega=const$, $-\infty\le t\le\infty$, is bounded, but is neither absolutely integrable, nor square integrable, therefore

$$\cos(\omega t),\ \omega=const,\ -\infty\le t\le\infty,\ \in L^{\infty}(R),\ \notin L^1(R),\ \notin L^2(R).$$

Now, as a follow-up to Section 5.1 of Chapter 5, consider an orthogonal set of complex T_1-periodic functions $\{\psi_i(t)\}$, $-\infty\le t\le\infty$, with inner product

$$\int_{t_0}^{t_0+T_1}\psi_i(t)\psi_j^*(t)\,dt=\begin{cases}0, i\ne j,\\ \lambda_i, i=j,\end{cases} \tag{10.3-1}$$

where $\psi_j^*(t)$ is a complex conjugate of ψ_j. If $\lambda_i=1, \forall i\in[1,N]$, then our set $\{\psi_i\}$ is orthonormal. Thus, for a signal

$$y(t)=\sum_{i=1}^{N}A_i\psi(t) \tag{10.3-2}$$

a projection coefficient A_i of $y(t)$ on $\psi_i(t)$ is found by

$$A_i=\frac{1}{\lambda_i}\int_{t_o}^{t_o+T_1}y(t)\psi_i^*(t)dt=\frac{1}{\lambda_i}\int_{t_o}^{t_o+T_1}\left(\sum_{j=1}^{N}A_j\psi_j(t)\right)\psi_i^*(t)dt=\frac{1}{\lambda_i}A_i\lambda_i=A_i. \tag{10.3-3}$$

In the general case A_i's are the complex numbers.

Consider now an arbitrary function *x*(*t*) with *compact support* $t \in [t_o, t_o + T_1]$. Is it likely that this *x*(*t*) can be exactly represented in the form (10.3-2) on $t \in [t_c, t_o + T_1]$? Obviously not! At best, (10.3-2) can be a decent approximation of *x*(*t*):

$$x(t) \approx \sum_{i=1}^{N} A_i \psi_i(t),\ t \in [t_o, t_o + T_1],$$

for an appropriately chosen coefficients A_i. But how to compute these A_i's? It looks very tempting to substitute *x*(*t*) into (10.3-3) instead of *y*(*t*) and compute A_i's, in which case

$$x(t) \approx y(t) = \sum_{i=1}^{N} A_i \psi_i(t),\ t \in [t_o, t_o + T_1], \tag{10.3-4}$$

$$\text{with } A_i = \frac{1}{\lambda_i} \int_{t_0}^{t_o + T_1} x(t) \psi_i^*(t) dt. \tag{10.3-5}$$

Is it a good approximation or a bad one? It depends upon the criterion of "goodness" chosen. It looks reasonable to require that A_i be chosen so as to minimize the *energy difference* between *x*(*t*) and *y*(*t*). This energy difference can be represented by the so-called *integral square error*:

$$\varepsilon_N \triangleq \int_{t_o}^{t_o + T_1} \left| x(t) - y(t) \right|^2 dt,$$

where *N* stands for the number of terms in (10.3-4). Let us now show that the A_i's computed as in (10.3-5) minimize the integral square error ε_N. For simplicity denote

$$\int_{t_o}^{t_o + T_1} (\cdot) dt \text{ as } \int_{T_1} (\cdot) dt$$

and assume that $\lambda_i = 1$, then:

$$\varepsilon_N = \int_{T_1} |x(t) - y(t)|^2 dt = \int_{T_1} [x(t) - y(t)][x^*(t) - y^*(t)] dt$$

$$= \int_{T_1} \left[x(t) - \sum_{i=1}^{N} A_i \psi_i(t) \right] \left[x^*(t) - \sum_{i=1}^{N} A_i^* \psi_i^*(t) \right] dt$$

$$= \int_{T_1} x(t) x^*(t) dt - \int_{T_1} \left[\sum_{i=1}^{N} A_i \psi_i(t) \right] x^*(t) dt - \int_{T_1} x(t) \left[\sum_{i=1}^{N} A_i^* \psi_i^*(t) \right] dt$$

$$+ \int_{T_1} \left[\sum_{i=1}^{N} A_i \psi_i(t) \right] \left[\sum_{j=1}^{N} A_j^* \psi_j^*(t) \right] dt$$

$$= \int_{T_1} |x(t)|^2 dt - \sum_{i=1}^{N} A_i \int_{T_1} \psi_i(t) x^*(t) dt - \sum_{i=1}^{N} A_i^* \int_{T_1} x(t) \psi_i^*(t) dt + \sum_{i=1}^{N} A_i A_i^* +$$

$$+ \underbrace{\sum_{i=1}^{N} \left[\int_{T_1} x(t) \psi_i^*(t) dt \right] \left[\int_{T_1} x^*(t) \psi_i(t) dt \right]}_{add} - \underbrace{\sum_{i=1}^{N} \left| \int_{T_1} x(t) \psi_i^*(t) dt \right|^2}_{subtract}$$

$$= \int_{T_1} |x(t)|^2 dt - \sum_{i=1}^{N} \left| \int_{T_1} x(t) \psi_i^*(t) dt \right|^2 +$$

$$+ \underbrace{\sum_{i=1}^{N} \left[A_i A_i - A_i \int_{T_1} \psi_i(t) x^*(t) dt - A_i^* \int_{T_1} x(t) \psi_i^*(t) dt + \int_{T_1} x(t) \psi_i^*(t) dt \int_{T_1} x^*(t) \psi_i(t) dt \right]}_{square}$$

$$= \underbrace{\int_{T_1} |x(t)|^2 dt - \sum_{i=1}^{N} \left| \int_{T_1} x(t) \psi_i^*(t) dt \right|^2}_{\text{independent of } A_i\text{'s}} + \underbrace{\sum_{i=1}^{N} \left| A_i - \int_{T_1} x(t) \psi_i^*(t) dt \right|^2}_{\text{can only be positive or zero. If zero, then } \varepsilon_{N,\min}} .$$

Thus, $A_i - \int_{T_1} x(t) \psi_i^*(t) dt = 0 \Rightarrow$ coefficients

$$A_i = \int_{t_0}^{t_0+T_1} x(t) \psi_i^*(t) dt \text{ for } \lambda_i = 1 \text{ or } A_i = \frac{1}{\lambda_i} \int_{t_0}^{t_0+T_1} x(t) \psi_i^*(t) dt \text{ for } \lambda_i \neq 1 .$$

Now, minimizing energy difference between $x(t)$ and $y(t) = \sum_{i=1}^{N} A_i \psi_i(t)$ on $t \in [t_o, t_o + T_1]$ yields the minimum error

$$\varepsilon_{N,\min} = \int_{t_0}^{t_0+T_1} |x(t)|^2 dt - \sum_{i=1}^{N} \underbrace{\left| \int_{t_0}^{t_0+T_1} x(t) \psi_i^*(t) dt \right|}_{A_i}^2$$

$$
\begin{aligned}
&= \int_{t_0}^{t_0+T_1} |x(t)|^2 \, dt - \sum_{i=1}^{N} |A_i|^2 \text{ for orthonormal set and} \\
&= \int_{t_0}^{t_0+T_1} |x(t)|^2 \, dt - \sum_{i=1}^{N} \lambda_i |A_i|^2 \text{ for orthogonal set.}
\end{aligned}
\tag{10.3-6}
$$

Let us restrict the function $x(t)$ to be an *energy* function on $t \in [t_0, t_0 + T_1]$, i.e. $\int_{T_1} |x(t)|^2 < \infty$. Then the set $\{\psi_i(t)\}$ is called *complete* if $\lim_{N\to\infty} \varepsilon_{N,\min} = 0$.

In this case we can represent $x(t)$ on $t \in [t_o, t_o + T_1]$ in terms of the *generalized Fourier series* $x(t) = \sum_{i=1}^{\infty} A_i \psi_i(t)$, $t \in [t_o, t_o + T_1]$, with an understanding that if $x_p(t)$ and $y_p(t)$, $-\infty \le t \le \infty$, are the T_1-periodic extensions of $x(t)$ and $y(t)$, $t \in [t_o, t_o + T_1]$, this equality extends to $x_p(t)$ only in the power $(\lim_{N\to\infty} \varepsilon_{N,\min} = 0)$ sense! The actual function $x_p(t)$ may differ from $y_p(t) = \sum_{i=1}^{\infty} A_i \psi_i(t)$ on $-\infty \le t \le \infty$, but only at separate points; for example, the square wave

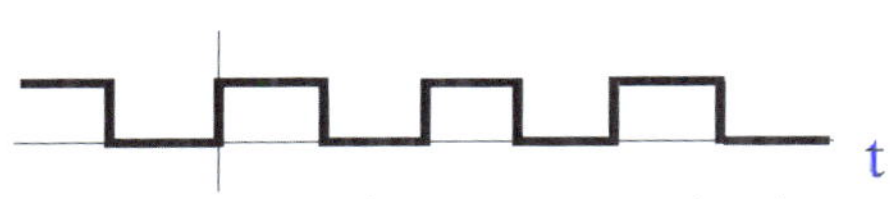

Fig. 10.3-1. The square wave signal.

has the corresponding pointwise Fourier series representation

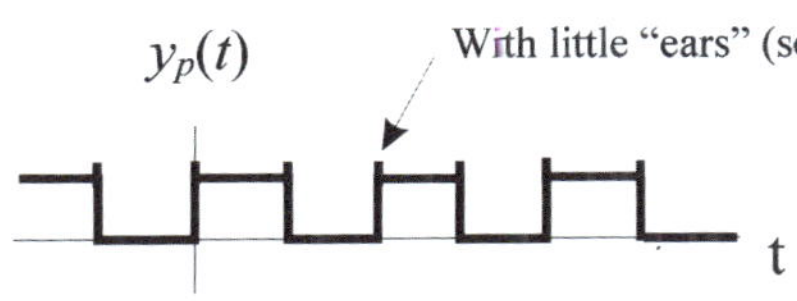

Fig. 10.3-2. Fourier series representation of the square wave signal in Figure 10.3-1.

If $\lim_{N\to\infty} \varepsilon_{N,\min} = 0$, then from (10.3-6)

$$\int_{T_1} |x(t)|^2 \, dt = \sum_{i=1}^{\infty} |A_i|^2 \text{ for an orthonormal set and}$$

$$\int_{T_1} |x(t)|^2 \, dt = \sum_{i=1}^{\infty} \lambda_i |A_i|^2 \text{ for an orthogonal set.}$$

i.e. the *energy* of $x(t)$ on $[t_0, t_0 + T_1]$ or the power of its T_1-periodic extension $x_p(t)$, $-\infty \le t \le \infty$, can be computed in terms of the *generalized Fourier coefficients* A_i. This is *Parseval's Theorem* generalized from the complex exponential functions of Chapter 5 to any orthogonal set of functions.

What is the best set of the basis functions $\psi_i(t)$ for a given $x(t)$, $t \in [t_o, t_o + T_1]$? This is a nontrivial problem. There are many orthogonal functions which can be chosen as the basis functions.

Example 10.3-1: Orthogonal functions

1. Walsh functions

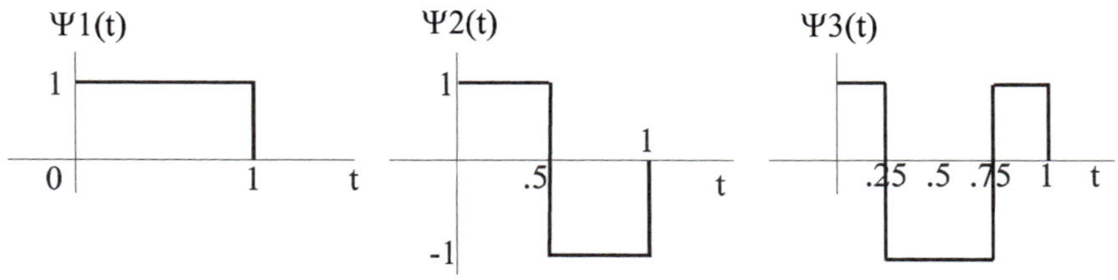

Fig. 10.3-3. Walsh functions.

Note that this set of functions is defined on [0, 1] time interval.

2. Exponential functions considered in Chapter 5:

$$\left\{e^{jn\omega_1 t}\right\}_{n=-\infty}^{\infty}.$$

3. Haar system

Let's now slightly modify the set of Walsh functions shown above by defining each function as follows:

$$\psi_{k2^{-n}}(t) = \begin{cases} 0 \text{ if } t < (k-1)2^{-n}, \\ 2^{(n-1)/2} \text{ if } (k-1)2^{-n} \le t < k2^{-n}, \\ -2^{(n-1)/2} \text{ if } k2^{-n} \le t < (k+1)2^{-n}, \\ 0 \quad \text{if } (k+1)2^{-n} \le t < 1, \end{cases}$$

where k = odd < 2^n. This set of rectangular (discontinuous) functions is called *Haar system* and it forms an orthonormal basis on [0, 1] in energy sense discussed above, so that

$$\left\langle \psi_{i2^{-n}}(t), \psi_{j2^{-n}}(t) \right\rangle \equiv \int_0^1 \psi_{i2^{-n}}(t)\psi_{j2^{-n}}(t)dt = \delta_{ij} \text{ (Kronecker delta } =1,\ i=j, \text{ and } =0,\ i \ne j)$$

and any continuous function $f(t)$ on [0, 1] can be represented as

$$f(t) = \sum_{n=1}^{\infty} \sum_{k=odd \le 2^n} a_{k,n} \psi_{k2^{-n}}(t) + a_0 \psi_0,$$

$$\psi_0 \equiv 1, \quad a_{k,n} = \int_0^1 f(t) \psi_{k2^{-n}}(t) dt.$$

10.3.2 *Introduction to the multiresolution wavelet-based function approximation*

10.3.2.1 *Review of basic operations on signals*

Let us fix ω at $\omega_1 = 2\pi/T_1$ and shift $\cos(\omega_1 t)$ by t_1:

$$\cos\left(\omega_1 (t-t_1)\right) = \cos\left(\frac{2\pi}{T_1}(t-t_1)\right) = \cos\left(\frac{t-t_1}{T_1/2\pi}\right) = \cos\Big(\frac{t}{\underbrace{T_1/2\pi}_{\sigma}=1/\omega_1} - \frac{t_1}{\underbrace{T_1/2\pi}_{\sigma}=1/\omega_1}\Big) = \cos\left(\frac{t}{\sigma} - \frac{t_1}{\sigma}\right).$$

Time and frequency are seen to be localized through the time shift scaled by σ and period T_1 scaled by 2π. Time shift t_1 is not changed by changing σ.

However, we can't play music with such a signal, since it has no starting point in time and also doesn't decay. But to produce music, the notes should start at some specified time moment and after some time decay. So we need a time-frequency localization to play the different height notes at the different times.

10.3.2.2 *Wavelets, nested subspaces, and successive approximation property*

As we have seen, the generalized Fourier series represents a periodic signal in terms of the orthogonal set of basis functions with fixed projection coefficients. Therefore, it is incapable of properly analyzing signals with frequencies varying over time. One proposed method to overcome the above limitation is the *Short-Time Fourier transform (STFT)* (Ref. [1]). When applying STFT, the long duration signals that change over time are segmented into the short time intervals, and then Fourier transform is applied to each of these intervals. As a result, it is possible to determine what frequencies exist in each time interval. That is, the shorter the width of the time window segmenting the signal, the better the time localization, which can be interpreted as the time resolution, and, conversely, the larger the width of the time window, the better the frequency resolution.

However, the short-time Fourier transform cannot improve the frequency resolution and the time resolution at the same time because of their fundamental trade-off interdependence. To overcome the limitations of the short-time Fourier transform, the *wavelet transform* uses the *mother wavelet* as the basis function while the Fourier transform uses the infinite duration periodic functions. Mother wavelets are typically signals that have compact support, and they are generated in a special manner so that they can not only form an orthogonal basis, but also exhibit specific properties such as the *nesting property*. Similarly to $\cos(\omega_1(t-t_1)) = \cos((t/\sigma)-(t_1/\sigma))$, considered above, but with the function acting only on compact support, the frequency domain can be accounted for by defining frequency at a particular time:

$$\Psi\left(\frac{x-\tau}{\sigma}\right)=\Psi\left(\frac{x}{\sigma}-\frac{\tau}{\sigma}\right),\quad \begin{matrix}\sigma-\text{scale parameter (frequency localization)}\\ \tau-\text{translation parameter (time localization)}\end{matrix}.$$

Let us take a closer look at an example of the Gaussian derivative as the mother wavelet as shown in Figure 10.3-4.

$$w_{gd}(x)=-Axe^{-\frac{1}{2}x^2}$$

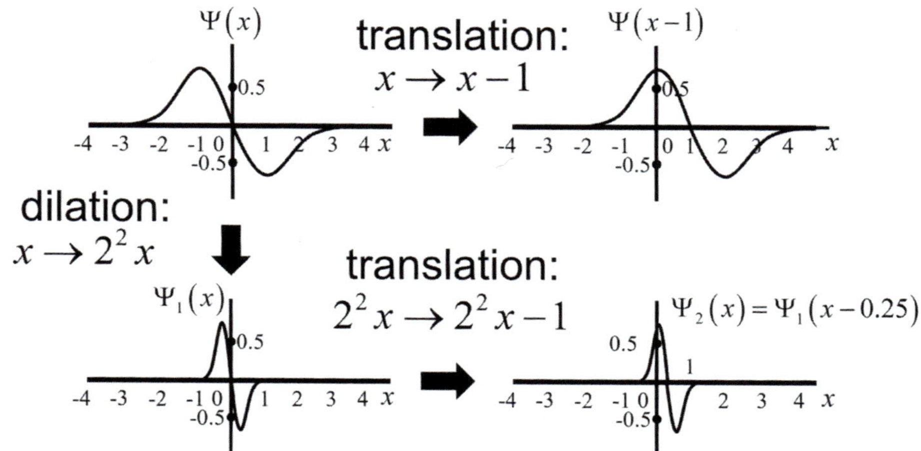

$$\Psi_1(x)=\Psi(2^2x)=\Psi\left(\frac{x}{0.25}\right)\rightarrow\Psi_2(x)=\Psi(2^2x-1)=\Psi\left(2^2(x-\frac{1}{2^2})\right)=\Psi\left(\frac{x-0.25\cdot 1}{0.25}\right)=\Psi_1(x-0.25)$$

Fig. 10.3-4. Example of dilation and translation using Gaussian derivative as the mother wavelet.

As shown in Figure 10.3-4, the larger is $1/\sigma$ (the stronger is the shrinking $\sigma\rightarrow 0$), the finer is the detail resolution, and the larger is the translation $\tau\rightarrow\infty$, the more the wavelet covers the entire axis.

The discretized wavelet basis can be defined as:

$$\psi_{j,k}(x)=A2^{j/2}\psi(2^jx-k),\ j,\ k\in\mathbb{Z}. \tag{10.3-7}$$

As j increases in (10.3-7), the support and the shift k of Ψ shrink with respect to the x-axis. Consider a scaling function $\xi\in L_2$. Then

$$\xi_{j,k}(x)=2^{j/2}\xi(2^jx-k),\ j,\ k\in\mathbb{Z}. \tag{10.3-8}$$

If j is fixed in (10.3-8), then the spread of ξ is fixed, and the dilated function is translated by the scaled integer, $k/2^j$, along x axis. By running k from $-\infty$ to ∞, the entire axis can be covered. For fixed j, by introducing θ to change the magnitude, the following space of functions is defined as:

$$V_j=\left\{f\in L_2\mid f(x)=\sum_{k=-\infty}^{\infty}\theta_{j,k}\xi_{j,k}(x)\right\}.$$

Considering intervals of different lengths,

$$\frac{1}{2}:\left[\frac{n}{2},\ \frac{n}{2}+\frac{1}{2}\right], 1:[n,\ n+1],\ 2:[2n,\ 2n+2],\ n\in\mathbb{Z},$$

define the corresponding functions a) $x_1(t)$, b) $x_0(t)$, and c) $x_{-1}(t)$ in L_2 where the functions are constant in the intervals:

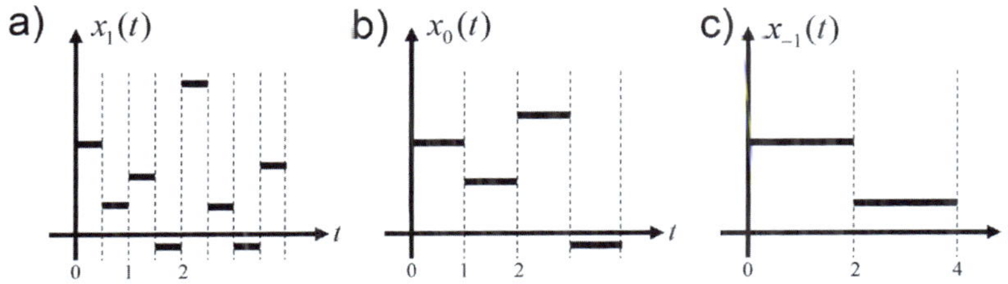

Fig. 10.3-5. Three typical functions that are constant on intervals of length 1/2, 1, and 2.

Defining the corresponding spaces a) $x_0(t)\in\mathcal{V}_0$, b) $x_1(t)\in\mathcal{V}_1$, and c) $x_{-1}(t)\in\mathcal{V}_{-1}$, a linear combination of two functions in any one space is a function of the same type. Then, $\mathcal{V}_{-1}\subset\mathcal{V}_0\subset\mathcal{V}_1$, since all the functions that are constant on the larger intervals are also constant on the smaller intervals, but not vice versa. By doubling the approximation resolution in going from $\mathcal{V}_0$ to $\mathcal{V}_{+1}$:

$$x(t)\in\mathcal{V}_0\Rightarrow x(2t)\in\mathcal{V}_{+1}.$$

This relation is called the *multi-resolution property*.

Thus, define subspaces $\mathcal{V}_j$ that are constant on the intervals of length $(1/2)^j$ of the form $[(1/2)^j n,\ (1/2)^j(n+1)]$, which are nested subspaces. For spaces $\mathcal{V}_j\subset L_2(\mathbb{R}),\ \forall j$, the relation $\mathcal{V}_j\subset\mathcal{V}_{j+1}$ will be referred to as *the nesting property*. Now, orthogonal projection in the function spaces is defined as follows.

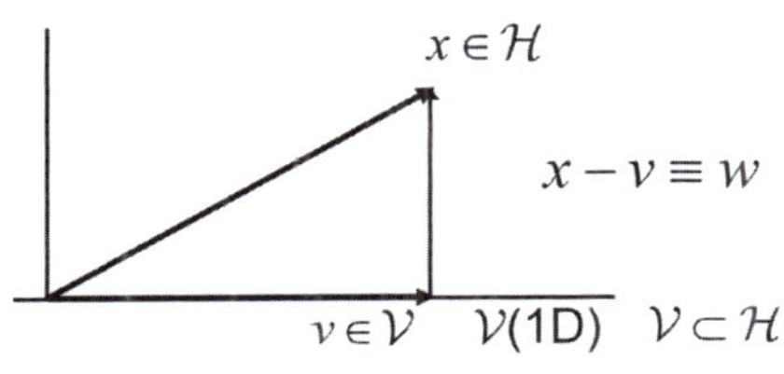

Fig. 10.3-6. Orthogonal projections in the function spaces

As shown in Figure 10.3-6, in vector space $\mathbb{R}^2\equiv\mathcal{H}$(2D), that is the inner product (Hilbert) space, the orthogonal projection v of x onto $\mathcal{V}$ which minimizes $|x-v|$ uniquely exists. Then, w spans the space $\mathcal{V}^{\perp}$ (orthogonal complement) of $\mathcal{V}$ with respect to $\mathcal{H}$, so that $\mathcal{H}$ is direct sum of $\mathcal{V}$ and $\mathcal{V}^{\perp}$:

$$\mathcal{H}=\mathcal{V}\oplus\mathcal{V}^{\perp}.$$

Similarly to the function space, define a nested set of closed subspaces $\mathcal{V}_j$ satisfying the completeness properties,

$$\bigcup_{j\in\mathbb{Z}} \mathcal{V}_j = L_2(\mathbb{R}), \ \bigcap_{j\in\mathbb{Z}} \mathcal{V}_j = \{0\}. \tag{10.3-9}$$

Expression (10.3-9) is said to have the *successive approximation property*. For such a set of nested subspaces we also have

$$\lim_{j\to\infty} \mathcal{V}_j = L_2(\mathbb{R}), \ \lim_{j\to-\infty} \mathcal{V}_j = \{0\}. \tag{10.3-10}$$

As j approaches positive infinity in (10.3-10), the support becomes infinitely small, which can express any function. To the contrary, as j approaches negative infinity, the support becomes infinitely large, losing ability to express anything. Then, any square integrable function in L_2 space can be approximated by an orthogonal projection onto any $\mathcal{V}_j$, and the nested sequence of $\mathcal{V}_j$ will look like:

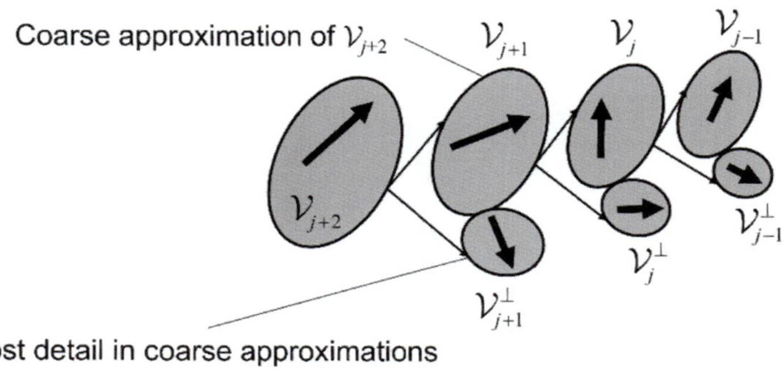

Fig. 10.3-7. Approximation of a vector using nested subspaces and their orthogonal complements.

The nested set of successive approximation subspaces $\mathcal{V}_j$ with multi-resolution property is said to form a *multi-resolution analysis* set, often referred to as just *MRA*.

10.3.2.3 *Multiresolution analysis (MRA)*

We can define orthonormality on compact support as follows:

$$\langle \xi_{0,0}, \xi_{0,k} \rangle = \int \xi(x)\xi(x-k)dx = \delta_{0,k},$$

$$\delta_{n,m} = \begin{cases} 1, & \text{if } n = m, \\ 0, & \text{else.} \end{cases}$$

Then, for a function $g \in L_2$, which we wish to approximate by a function $f \in \mathcal{V}_j$ with minimal error $\|f - g\|$, the optimal values of the coefficients $\theta_{j,k}$ in $\mathcal{V}_j = \{f \in L_2 \mid f(x) = \sum_{k=-\infty}^{\infty} \theta_{j,k}\xi_{j,k}(x)\}$ are given by the Fourier coefficients

$$\theta_{j,k} = \int g(x)\xi_{j,k}(x)dx. \tag{10.3-11}$$

Then, the j-th and the i-th coefficients in (10.3-11) are independent.

Due to MRA definition, $\{\xi_{j,n}\}_{n\in\mathbb{Z}}$ is orthonormal basis for $\mathcal{V}_j$, that is, translations are enough to fully decompose projection f of $g \in L_2$ onto $\mathcal{V}_j$, and by moving to finer projections to approximate $f \in L_2$ with any $\varepsilon > 0$ accuracy. However, we cannot use mixed resolution with ξ's, since if we change dilation from j to k in the scaling function $\xi_{j,n}$ with shift n, then $\xi_{j,n}$ is not necessarily orthogonal to $\xi_{k,n}, k \neq j$, that is, dilations and translations of ξ do not provide orthonormal basis for L_2: $\mathcal{V}_j \subset \mathcal{V}_{j+1} \not\Rightarrow \mathcal{V}_j \oplus \mathcal{V}_{j+1}$.

Then, let us bring to the rescue the orthogonal complement of $\mathcal{V}_j$ in $\mathcal{V}_{j+1}$ and call it $\mathcal{W}_j$:

$$\mathcal{V}_{j+1} = \mathcal{V}_j \oplus \mathcal{W}_j,$$

and introduce the wavelet family, that is, the function ψ (mother wavelet) generated by the scaling function ξ (father wavelet) such that the translates of ψ are mutually orthonormal, and orthogonal to $\xi_{0,n}$.

$$\int \psi(x-k)\psi(x-m)dx = \delta_{k,m}, \quad \forall k,m \in \mathbb{Z},$$

$$\int \psi(x-k)\xi(x-n)dx = 0, \quad \forall k,n \in \mathbb{Z}.$$

Then, we can form the basis for $\mathcal{W}_0$. The wavelets of resolution j are then defined as

$$\psi_{j,k}(w) = 2^{j/2}\psi(2^j w - k), \quad j,k \in \mathbb{Z},$$

and the set $\{\psi_{j,k}\}_{k\in\mathbb{Z}}$ forms the orthonormal basis for $\mathcal{W}_j$.

According to Daubechies, for MRA, there exists ψ constructed from ξ. Then,

$$L_2(R) = \cdots \oplus \mathcal{W}_{-1} \oplus \mathcal{W}_0 \oplus \mathcal{W}_1 \oplus \cdots = \bigoplus_{j=-\infty}^{\infty} \mathcal{W}_j,$$

which is constructed of all orthogonal complements of $\mathcal{V}_j$'s like $\mathbb{R}^1$ is orthogonal complement of $\mathbb{R}^1$ to form $\mathbb{R}^2$, of $\mathbb{R}^2$ to form $\mathbb{R}^3$, of $\mathbb{R}^3 = \mathbb{R}^1 \oplus \mathbb{R}^1 \oplus \mathbb{R}^1$ to form $\mathbb{R}^4$. Therefore, in energy sense,

$$f \in L_2(R) \Rightarrow f(w) = \sum_{j=-\infty}^{\infty} \sum_{k=-\infty}^{\infty} \langle f, \psi_{j,k} \rangle \, \psi_{j,k}(w).$$

That is, any L_2 function can be written as the unique linear combination of orthogonal wavelets of different resolutions:

$$f(w) = \cdots + g_{-1}(w) + g_0(w) + g_1(w) + \cdots, \; g_j \in w_j, \; \langle g_i, g_j \rangle \propto \delta_{i,j}.$$

As a result, *only wavelets have both multi-resolution and orthogonal decomposition properties at the same time.*

Now, substituting $\mathcal{V}_j = \mathcal{V}_{j-1} \oplus \mathcal{W}_{j-1}$ into $\mathcal{V}_{j+1} = \mathcal{V}_j \oplus \mathcal{W}_j$ and using similar sequential substitutions yields

$$\mathcal{V}_{j+1} = \mathcal{V}_{j-1} \oplus \mathcal{W}_{j-1} \oplus \mathcal{W}_j = \mathcal{V}_{j-2} \oplus \mathcal{W}_{j-2} \oplus \mathcal{W}_{j-1} \oplus \mathcal{W}_j \Rightarrow$$

$$\mathcal{V}_{j_0} = \mathcal{V}_{j_1} \oplus \mathcal{W}_{j_1} \oplus \mathcal{W}_{j_1+1} \oplus \cdots \oplus \mathcal{W}_{j_0-1},\ j_1 < j_0 - 1 \Rightarrow$$

$$\lim_{j\to\infty} \mathcal{V}_j = L_2 = \mathcal{V}_{j_1} \oplus \mathcal{W}_{j_1} \oplus \mathcal{W}_{j_1+1} \oplus \to \infty = \mathcal{V}_{j_1} \oplus \bigoplus_{j=j_1}^{\infty} \mathcal{W}_j .$$

Therefore, for function approximation, we can use both scaling functions and wavelets. Continuing the process to the left to -∞ while increasing resolution *j* we have

$$L_2(\mathbb{R}) = \cdots \oplus \mathcal{W}_{-1} \oplus \mathcal{W}_0 \oplus \mathcal{W}_1 \oplus \cdots = \bigoplus_{j=-\infty}^{\infty} \mathcal{W}_j$$

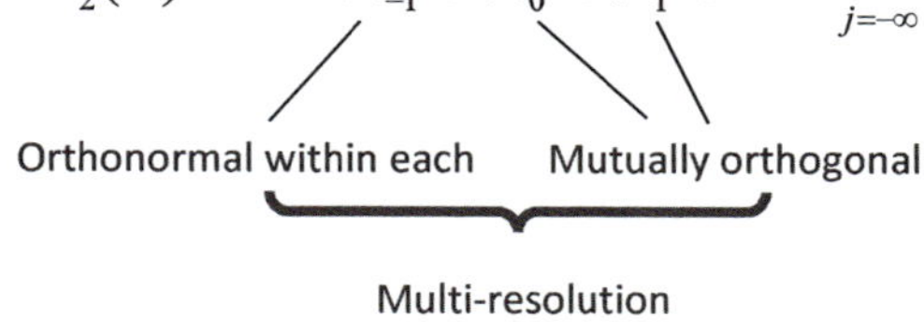

Only mother-wavelet based wavelets have both multi-resolution and orthogonal decomposition properties. Let ξ be scaling function, ψ — compactly supported orthogonal wavelet, and $\xi_{j,k}(w)$ span $\mathcal{V}_j$. Then, for any ε, $j_0 \in \mathbb{Z}$ and $\hat{f}(w;\mu) = \sum_{k=-\infty}^{\infty} \mu_k \xi_{j_0,k}(w)$ exist such that $\left\| f(w) - \hat{f}(w;\mu) \right\| < \varepsilon$. If *f* has compact support, then there would be only finite number of translates, so that

$$\hat{f}(w;\mu) = \sum_{k=L}^{U} \mu_k \xi_{j_0,k} \text{ for some } L, U \in \mathbb{Z}. \tag{10.3-12}$$

If ε is small, j_0 in (10.3-12) is large so that $\hat{f}$ is summation of translates at high resolution, and we have truncation in space $\mathcal{V}_{j_0}$ because of finite summation. Then, from

$$\mathcal{V}_{j_0} = \mathcal{V}_{j_1} \oplus \mathcal{W}_{j_1} \oplus \mathcal{W}_{j_1+1} \oplus \cdots \oplus \mathcal{W}_{j_0-1}, \text{ for } j_1 \ll j_0$$

we have

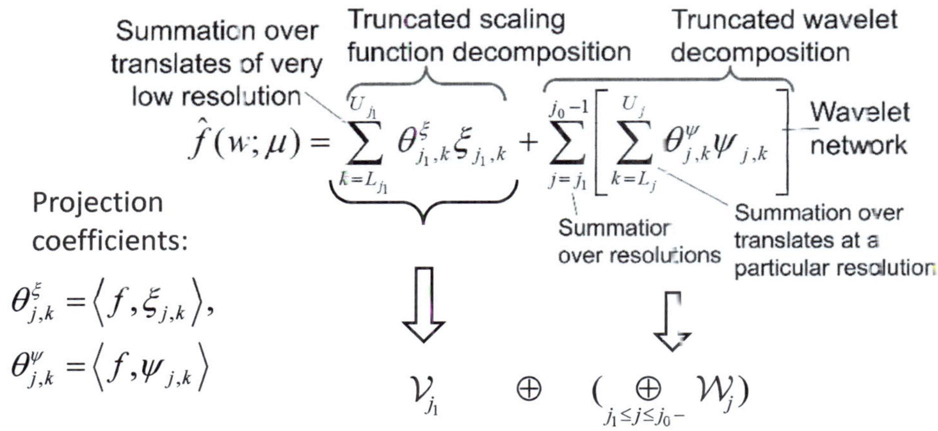

$$\hat{f}(w;\mu) = \sum_{k=L_{j_1}}^{U_{j_1}} \theta^{\xi}_{j_1,k}\xi_{j_1,k} + \sum_{j=j_1}^{j_0-1}\left[\sum_{k=L_j}^{U_j}\theta^{\psi}_{j,k}\psi_{j,k}\right]$$

$$\theta^{\xi}_{j,k} = \langle f, \xi_{j,k}\rangle,\quad \theta^{\psi}_{j,k} = \langle f, \psi_{j,k}\rangle$$

$$\mathcal{V}_{j_1} \oplus \left(\underset{j_1\le j\le j_0-}{\oplus} \mathcal{W}_j\right)$$

Example 10.3-2: Wavelet construction using the sinc function

The nesting direction, or the resolution increase, is set through the definition of $\psi_{j,k}(x)$ as follows:

$$j\uparrow \Rightarrow 2^{j/2}\psi(2^j x-k),$$
$$j\downarrow \Rightarrow 2^{-j/2}\psi(2^{-j} x-k).$$

Consider space $\mathcal{V}_0$ of functions bandlimited to $[-\pi, \pi]$ (includes $\cos(\pi t)$ at the boundary, but not $\sin(\pi t)$ orthogonal to it) with resolution increase direction $j\downarrow$.

This direction implies that $\mathcal{V}_{-1}$ — the space of functions bandlimited to $[-2\pi, 2\pi]$ — has smaller support or larger frequency. Now, denote by $\mathcal{W}_0$ the space of functions bandlimited to $[-2\pi, -\pi)\cup(\pi, 2\pi]$. This space includes $\sin(\pi t)$, but not $\cos(\pi t)$. Therefore $\mathcal{V}_{-1} = \mathcal{V}_0 \oplus \mathcal{W}_0 \Rightarrow f^{(-1)} \in \mathcal{V}_{-1}$ projected onto $\mathcal{V}_0$ is a low pass approximation $f^{(0)}$ of $f^{(-1)}$, while the difference $d^{(0)} = (f^{(-1)} - f^{(0)}) \in W_0$ is the ripple that is wavelet. Also

$$V_{-1} = \overset{\infty}{\underset{m=0}{\oplus}} W_m .$$

Denoting $\varphi(t) = \mathrm{sinc}_T(t) = \dfrac{\sin(\pi t/T)}{\pi t/T}$, we have

$$\int_{-\infty}^{\infty} \underbrace{\mathrm{sinc}_1(t-i)}_{\frac{e^{j\pi(t-i)}-e^{-j\pi(t-i)}}{2j\pi t}}\mathrm{sinc}_1(t-j)dt = 0,\ i\neq j,$$

where $\{\mathrm{sinc}_1(t-n)\}$ forms the orthogonal basis for V_0, that is,

$$f^{(0)} = \varphi_0(t) \equiv \varphi(t) = \mathrm{sinc}_1(t) = \frac{\sin \pi t}{\pi t} \overset{\mathcal{F}}{\longleftrightarrow} \mathrm{rect}\left(\frac{\omega}{2\pi}\right):$$

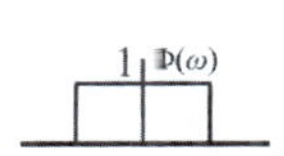

is a scaling function for the space V_0 of functions bandlimited to $[-\pi, \pi]$. After computing orthogonal complement in the frequency domain, to generate a wavelet we simply compute the inverse Fourier Transform of the orthogonal complement and find the function of time, or by linearity of Fourier Transform we can directly subtract the corresponding functions of time:

$$\Phi_{-1}(\omega) = \text{rect}\left(\frac{\omega}{4\pi}\right) \Leftrightarrow \varphi_{-1}(t) = 2\,\text{sinc}(2t) \in V_{-1},$$

$$\Phi_0(\omega) = \text{rect}\left(\frac{\omega}{2\pi}\right) \Leftrightarrow \varphi_0(t) = \text{sinc}(t) \equiv \text{sinc}_1(t) \in V_0,$$

$$\Psi_0(\omega) = \text{rect}\left(\frac{\omega}{4\pi}\right) - \text{rect}\left(\frac{\omega}{2\pi}\right) \Leftrightarrow d^{(0)}(t) = \psi_0(t) \in W_0,$$

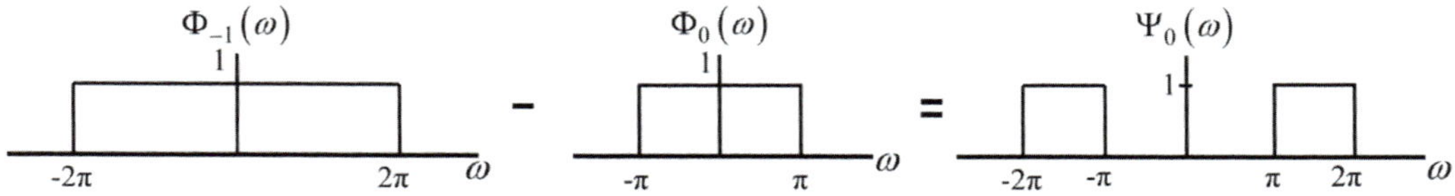

Fig. 10.3-8. Orthogonal complement computation of sinc function in the frequency domain.

Integrating directly the graph on the right in Figure 10.3-8 yields

$$\psi_0(t) = \frac{1}{2\pi}\int_{-2\pi}^{-\pi} e^{j\omega t}d\omega + \frac{1}{2\pi}\int_{\pi}^{2\pi} e^{j\omega t}d\omega = 2\,\text{sinc}(2t) - \text{sinc}(t)$$
$$= 2\frac{\sin(2\pi t)}{2\pi t} - \frac{\sin(\pi t)}{\pi t} = \frac{\sin(\pi t/2)}{\pi t/2}\cos(3\pi t/2).$$

From Fourier analysis, the corresponding wavelet that decomposes $\mathcal{W}_0$ is also given by

$$\psi_0(t) \equiv \psi(t) = \sqrt{2}\sum_n (-1)^{-n+1} g_0[n]\varphi(2t+n-1) = \frac{\sin(\pi t/2)}{\pi t/2}\cos(3\pi t/2),$$

$$g_0[n] = \frac{1}{\sqrt{2}}\frac{\sin(\pi n/2)}{\pi n/2}.$$

It can be also shown that $\langle \psi(t),\ \psi(t-n)\rangle = \delta[n]$, that is, translates of ψ form an orthogonal basis for $\mathcal{W}_0$. Shifting by 1/2 to exclude $\cos(\pi t)$ yields

$$\left\{\psi\left(t-n-\frac{1}{2}\right)\right\},\ n \in \mathbb{Z}$$

as orthogonal basis for $\mathcal{W}_0$. Therefore, wavelet basis for $L_2(\mathbb{R}) = \bigoplus_{m=-\infty}^{\infty} W_m$ is

$$\{\psi_{m,n}(t)\} = \left\{2^{-m/2}\psi(2^{-m}t-n-1/2)\right\},\ m,n \in \mathbb{Z},$$

which is the basis for functions with Fourier transform supported in the frequency domain on

$$\left[-2^{-m+1}\pi,-2^{-m}\pi\right]\cup\left[2^{-m}\pi,2^{-m+1}\pi\right]:$$

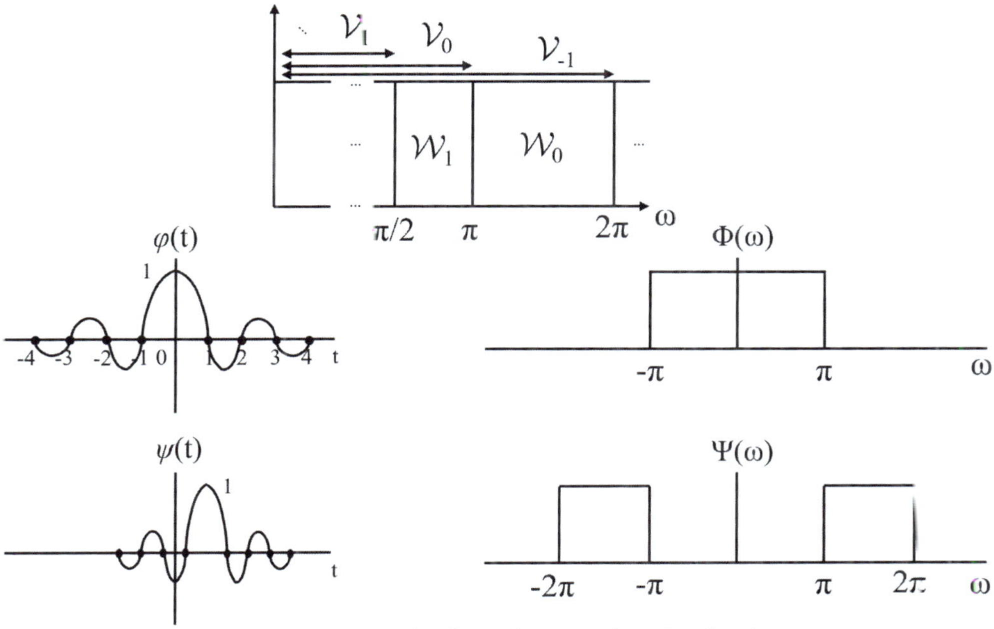

Fig. 10.3-9. Example of wavelet generation: sinc function.

Example 10.3-3: Multidimensional scaling and wavelet function: Mexican hat (radial Marr)

A drawback of sinc function is its noncompact support. Therefore, we usually employ other functions, like Radial Marr (Mexican hat). The scaling and the wavelet functions for the latter function set are, respectively, as follows:

$$\begin{aligned} f(z) &= \varphi^{[n]}(x) = e^{-0.5\|x\|^2},\ x\in\mathbb{R}^n,\ \|x\|^2 = x^T x = \sum\nolimits_{i=1}^{n} x_i^2, \\ g(z) &= \psi^{[n]}(x) = (\dim(x) - \|x\|^2)e^{-0.5\|x\|^2}. \end{aligned}$$

This function is used in Matlab Wavenet. An example of wavelet time-frequency decomposition of an audio file using another function is given in Problem 3 of Homework 11.

10.3.3 *Hilbert-Huang Transform and its application to nonstationary signals*

Hilbert-Huang transform (HHT) is an empirical technique for analyzing nonlinear and nonstationary signals. It comprises the *empirical mode decomposition* and *Hilbert spectral analysis*. It aims at decomposing a signal into the so-called *intrinsic mode functions (IMF)*, leading to the time-frequency-energy description — instantaneous frequency and amplitude — of the input signal, rather than the orthogonal basis decomposition as in Fourier Transform.

10.3.3.1 *Hilbert Spectral Analysis*

The two main components of the Hilbert-Huang Transform are the Empirical Mode Decomposition (EMD) and the Hilbert Transform applied to the results of EMD. An overview of Hilbert Transform is given next. Hilbert amplitude spectrum is obtained by first applying Hilbert Transform to all IMF components, thus obtaining the complex conjugate $y(t)$ of the real valued signal $x(t)$. The Hilbert Transform is the convolution of a real function $x(t)$ with $1/(\pi t)$, producing another function of a real variable $H[x(t)]$. 1/t is not integrable across t = 0, therefore the Hilbert Transform is defined using a Cauchy Principal Value, defined as PV. The Hilbert Transform is given by:

$$H[x(t)] = \frac{1}{\pi} PV \int_{-\infty}^{+\infty} \frac{x(\tau)}{t-\tau} d\tau . \tag{10.3-13}$$

The signal can then be defined as:

$$z(t) = x(t) + iy(t) = a(t)e^{i\theta(t)}, \tag{10.3-14}$$

where the instantaneous amplitude is

$$a(t) = \sqrt{x^2 + y^2}, \tag{10.3-15}$$

and the phase is given by

$$\theta(t) = \tan^{-1}\left(\frac{y}{x}\right), \tag{10.3-16}$$

where the instantaneous frequency is the derivative of the phase:

$$w = \frac{d\theta}{dt}. \tag{10.3-17}$$

Then, the original signal can be expressed as the real part *Re* given by:

$$x(t) = \mathrm{Re} \sum_{j=1}^{k} a_j(t) e^{i\int w_j(j)dt}, \tag{10.3-18}$$

with $a_j(t)$ and $w_j(t)$ representing, respectively, the amplitude and the frequency of the j component.

The energy spectrum is obtained by integrating $H^2(w, t)$ over frequency:

$$I_E = \int_w H^2(w,t)\, dw, \tag{10.3-19}$$

and the Hilbert marginal spectrum, representing the cumulative amplitude that each frequency contributes, is given by

$$h(w) = \int_0^T H(w,t) dt. \tag{10.3-20}$$

10.3.3.2 *Empirical mode decomposition (EMD)*

EMD decomposes the data into IMF components. IMF — Intrinsic Mode Function — represents a simple oscillatory mode but can have variable amplitude and frequency unlike a simple harmonic. An IMF is a function that must satisfy the requirements:

1. The number of extrema and Zero-crossings must be equal or differ by one
2. The mean value of the envelope is zero at any point

The process for extracting the IMF components is iterative and stops when no more IMF can be extracted (the function becomes monotonic). Additionally, each IMF extraction is an iterative process that involves "sifting" to extract an IMF that satisfies the requirement above.

To extract an IMF, we begin by identifying all local extrema (maximum and minimum) of the input data. All local maximum extrema are then connected by a cubic spline to define the upper envelope. Correspondingly, all local minimum extrema are connected by a cubic spline to define the lower envelope. There are algorithms alternative to the cubic spline approach, but cubic spline is common. A cubic spline has the form:

$$S_j(x) = a_j + b_j(x - x_j) + c_j(x - x_j)^2 + d_j(x - x_j)^3 .$$

The mean of this envelope is defined as m_1. This process is shown in the Figure 10.3-10. The first IMF is the difference between the input data and the mean and is defined as:

$$h_1 = X(t) - m_1 . \qquad (10.3\text{-}21)$$

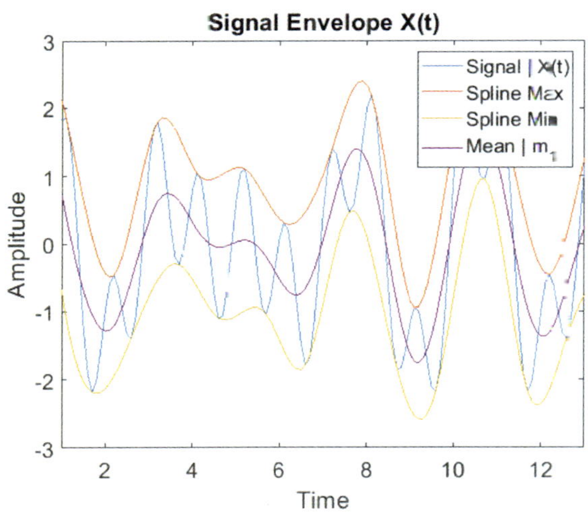

Fig. 10.3-10. First proto-IMF computation.

This should satisfy the requirement for IMF, set above, however, often it does not for the first iteration, and is then treated as a "proto-IMF". This is the first step in the sifting process, and the iterations proceed until a satisfactory IMF is obtained, denoted as $h_{1(k)}$.

The steps are then repeated, but the input is now replaced with h_1 as shown in Figure 10.3-11. An envelope is defined by the maximum and the local extrema, and $m_{1(2)}$ is defined as the mean of the new envelope. $h_{1(2)}$ is now defined as the difference between the input data and the mean:

$$h_{1(2)} = h_1 - m_{1(2)}. \tag{10.3-22}$$

This is repeated k times until the IMF is obtained.

The resulting IMF is then denoted as:

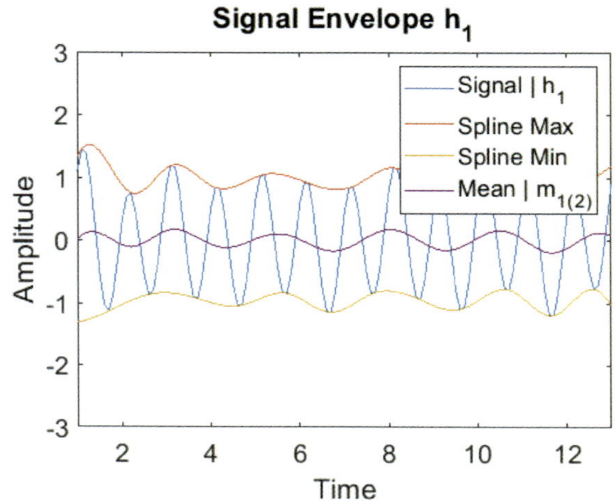

Fig. 10.3-11. Second proto-IMF computation.

$$c_1 = h_{1(k)}. \tag{10.3-23}$$

Some stopping criteria must now be defined to know when an IMF that satisfies the requirement is achieved. The original stopping criterion is a convergence test which stops the sifting process when the criterion values becomes smaller than a pre-defined one — the standard deviation of the difference between the two adjacent IMFs in DT over the number of samples T of interest:

$$SD_k = \frac{\sum_{t=0}^{T} \left| h_{k-1}(t) - h_k(t) \right|^2}{\sum_{t=0}^{T} h_{k-1}^2(t)}, \text{ index } t \text{ stands for the DT variable.} \tag{10.3-24}$$

The latter relies on some arbitrary selection of a stopping value. Another method relies on the number of zero-crossings and extrema. The stopping criterion is met if the number of zero-crossings is equal to the number of extrema or they differ by one and have stayed the same for a predefined number of iterations.

We can now obtain c_2 by repeating the previous sifting process, but the input now becomes

$$r_1 = X(t) - c_1, \tag{10.3-25}$$

with iterations

$$r_n = r_{n-1} - c_n. \tag{10.3-26}$$

This procedure is repeated until r_n becomes a monotonic function (number of extrema less than 2), meaning no more IMF can be extracted. Thus, a decomposition of $X(t)$ into modes has been achieved and we have:

$$X(t) = \sum_{j=1}^{n} c_j + r_n. \tag{10.3-27}$$

Then, equations (10.3-13)–(10.3-20) yield the Hilbert transform on the obtained IMFs.

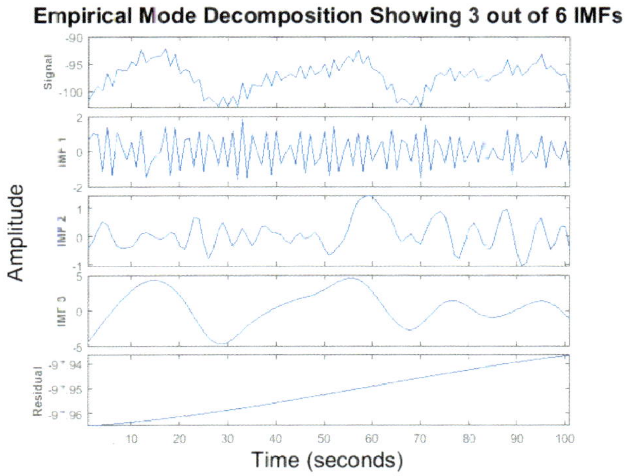

Fig. 10.3-12. IMFs of signal.

10.3.3.3 *Implementation*

Example 10.3-4: MATLAB implementation of HHT for generating 3D plots:

```
imf = emd(signal);
[hs,f,t] = hht(imf,'FrequencyLimits',[0 max]);
mesh(seconds(t),f,hs);
xlabel('Time(s)');
ylabel('Frequency(Hz)');
zzlabel('Instantaneous Energy');
```

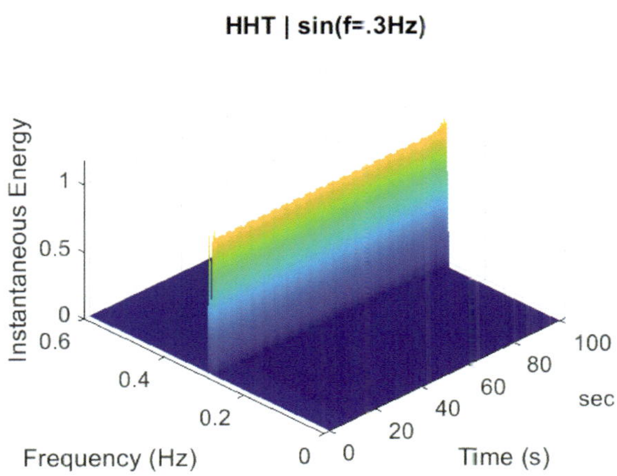

Fig. 10.3-13. HHT of a sinusoidal signal.

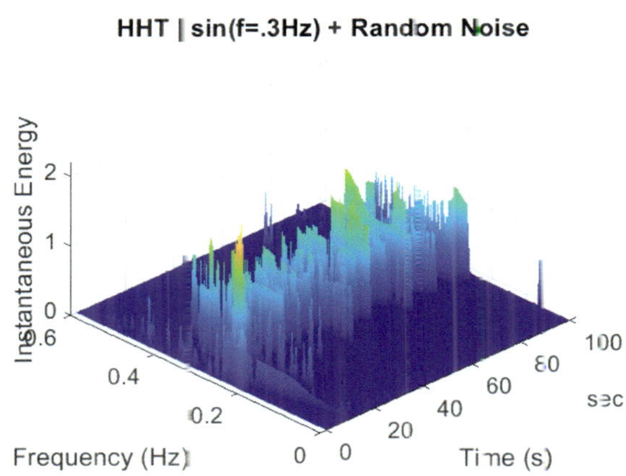

Fig. 10.3-14. HHT of a corrupted sinusoidal signal.

Generating a pure sine wave signal with frequency of .3HZ produces the result in Figure 10.3-13. As expected, a clear point at 0.3 Hz is seen and remains constant through time. To see how HHT picks up the noisy behavior, a stationary noise is added to the previous pure sine signal. This results in the plot shown in Figure 10.3-14, with components seen to be scattered along the path of the pure sine frequency of the previous figure, but still representing a stationary dataset. But for a dataset where it is necessary to detect the transient behavior, noise-corrupted or not, HHT is desirable as opposed to FFT which would not provide a visible representation of the transients.

To demonstrate the latter, consider a nonperiodic signal, where not the magnitude is transient, but the frequency. Applying FFT to such a frequency-varying signal will typically not give sufficiently meaningful results. However, HHT works well with such signals. Indeed, the signal seen in Figure 10.3-15 has constant magnitude, but time-varying frequency. The HHT of this signal is given in Figure 10.3-16. As seen, the HHT tracks the frequency of this signal in time very accurately.

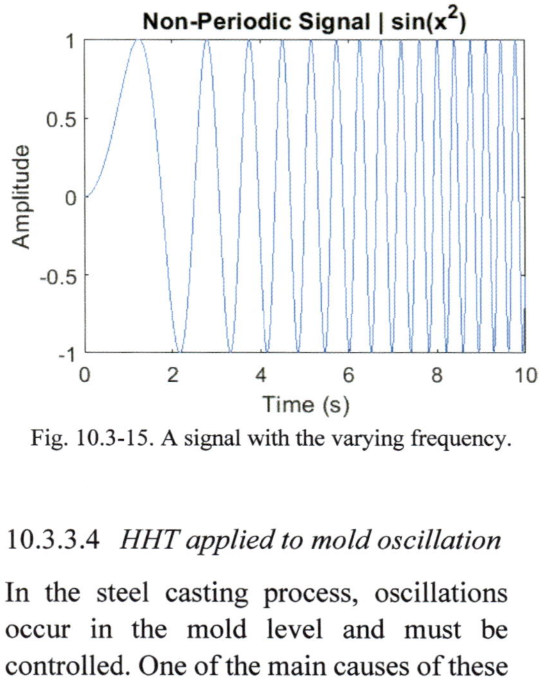

Fig. 10.3-15. A signal with the varying frequency.

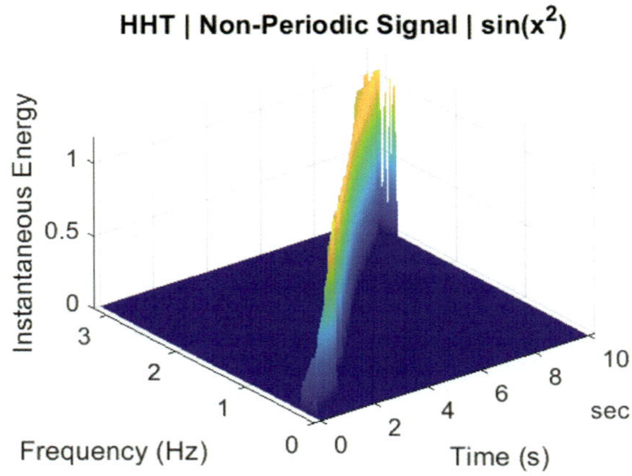

Fig. 10.3-16. HHT instantaneous tracking of the varying frequency.

10.3.3.4 *HHT applied to mold oscillation*

In the steel casting process, oscillations occur in the mold level and must be controlled. One of the main causes of these oscillations is the steel shell bulging. The shell bulges out between the containment rolls due to the ferrostatic pressure, when partially solidifying, and the bulges formed are forced back in when passing by the roll. This reduces the volume enclosed by the shell, forcing the molten steel up and causing a change in the mold level.

Mold level signal for the entire data set is seen in Figure 10.3-17. "Mold level w/ control" is the measured plant signal of a

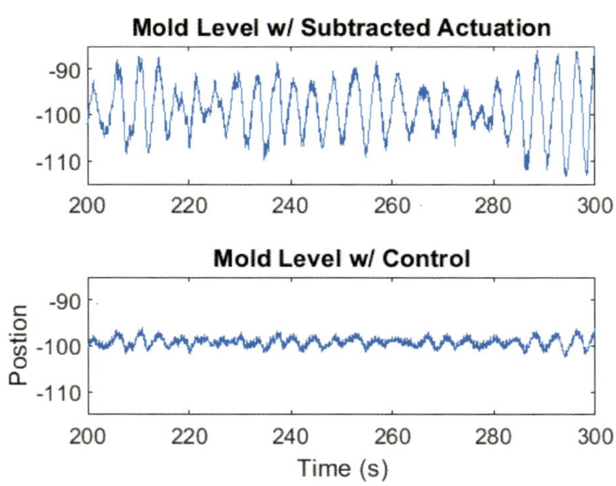

Fig. 10.3-17. Mold level oscillation signals.

controlled mold level. "Mold level w/ subtracted actuation" refers to the mold level signal estimate without control — after the subtraction of the controller rod induced effect. HHT application to these signals is depicted in Figures 10.3-18 and 10.3-19, respectively. The vertical scales of the magnitude spectra in these two figures are adjusted to permit visualization of the Hilbert Spectrum for mold level with control.

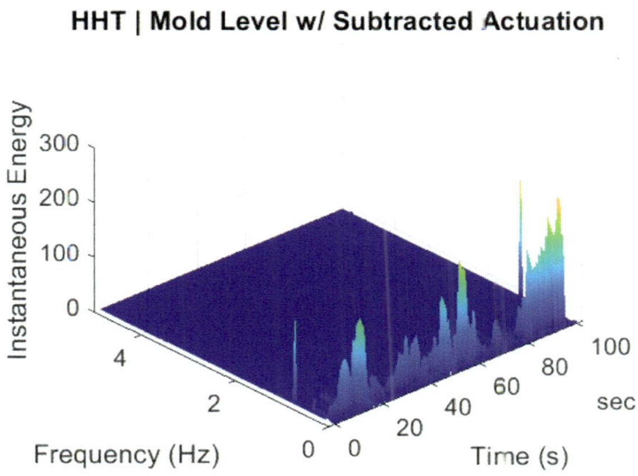

Fig. 10.3-18. HHT of the mold level with subtracted actuation.

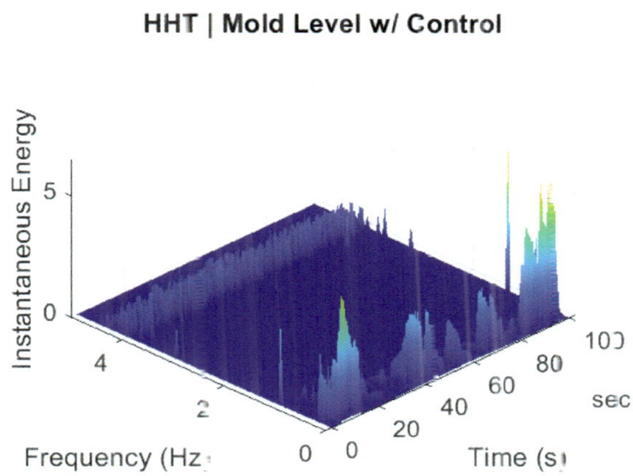

Fig. 10.3-19. HHT of the controlled mold level signal.

Mold level without control has energy that is 2 orders of magnitude greater than mold level with control. This shows that the mold level oscillations are greatly reduced in magnitude by control. Both signals are close in frequency (~0.3 Hz), but significantly differ in magnitudes.

HHT can also capture the ~4–4.5 Hz mold level perturbation induced by the mold oscillator! This is seen on the HHT w/ Control graph, but not on the mold level w/out control one. This is because the scale of the energy from the latter signal at 0.3 HZ is too large, making the energy of the 4.4 Hz signal not visible on the same plot. However, focusing on the 2–5 HZ range and zooming in on the magnitude pulls out the details of the mold-oscillator-induced perturbations that could not be seen before, as shown in Figure 10.3-20.

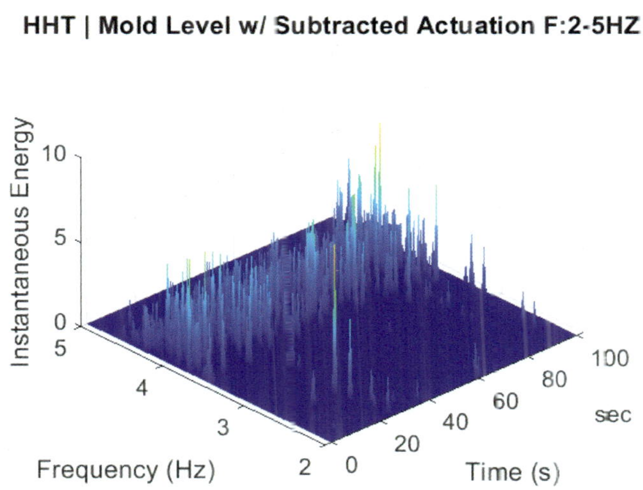

Fig. 10.3-20. HHT of the mold level with subtracted actuation, with the magnitude zoomed in the 2–5 Hz frequency range.

HHT Applied to mold level with subtracted Actuation. The uncontrolled mold level oscillation signal over a long period of time can be seen in Figure 10.3-21. The signal significantly changes at approximately 3000 s because of the bulging being quenched. This change in the system signal can be clearly seen in the HHT of the signal over the entire dataset in Figure 10.3-22. At 3000 s the instantaneous energy significantly decreases in magnitude.

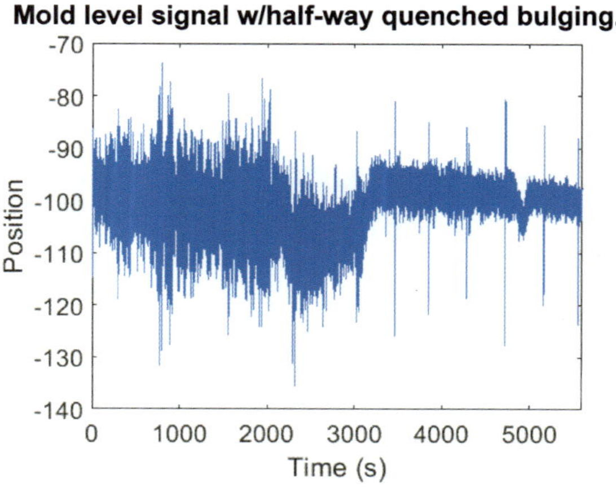

Fig. 10.3-21. Mold level signal of the entire dataset with bulging (courtesy Nucor Steel Decatur).

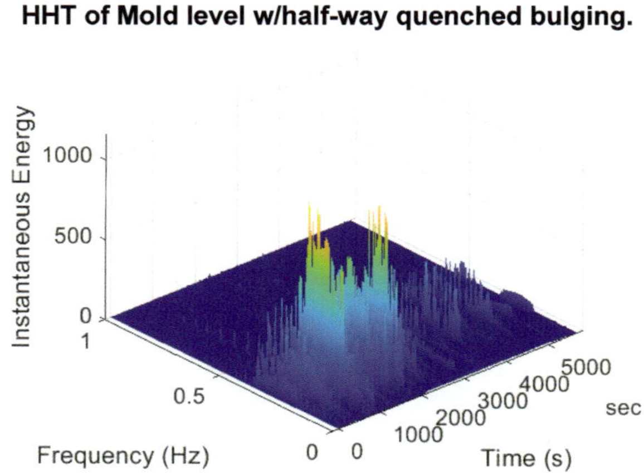

Fig. 10.3-22. HHT of the entire dataset with bulging.

Window size selection. Choice of the window size must be considered carefully. The window size and the number of points used alter the results and the resolution of the transformation. The HHT of the entire mold level dataset with the half-way quenched bulging in the 2–5 Hz range is seen in Figure 10.3-23. The graph is noisy and the mold-oscillator-induced disturbance at 4–4.5 Hz is not clearly seen. Narrowing the window *in time, but not in frequency*, and *looking at just 100 seconds of data* cleans this up and clearly shows in Figure 10.3-24 the expected results.

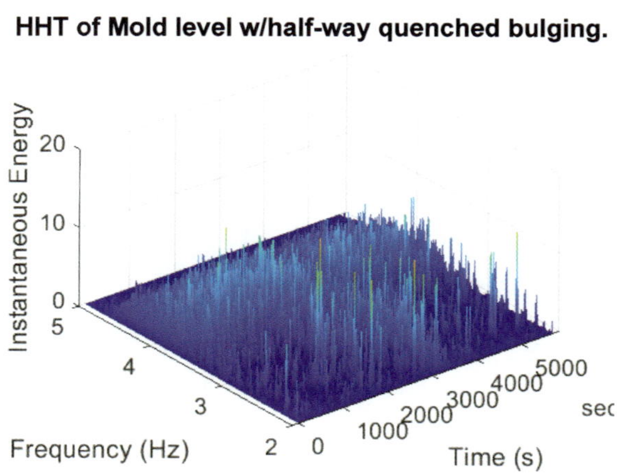

Fig. 10.3-23. HHT of the entire dataset with bulging, plotted in the 2–5 Hz frequency range.

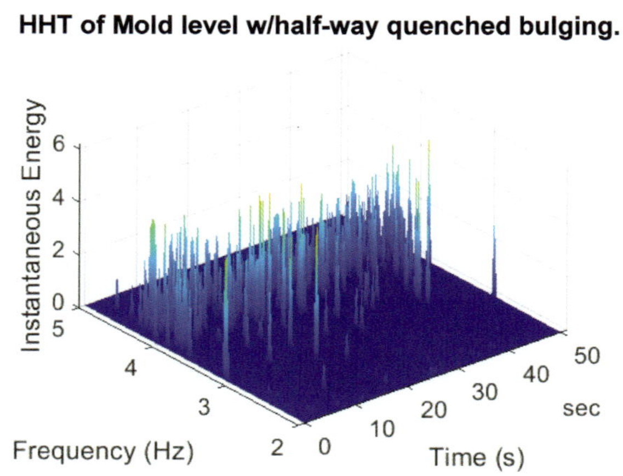

Fig. 10.3-24. HHT of the smaller window from the dataset with bulging, plotted in the 2–5 Hz frequency range.

Comparison to FFT in identifying bulging zones. Bulging is a significant problem in steel casting and as a result it is desirable to identify the supporting rolls zone locations at which it occurs. Bulging between rolls varies depending on the zone. As the piece of steel moves down, the shell thickens and the bulging severity changes, therefore identifying the location of these zones is desirable. We can begin by approximating the frequency of the zones as the casting speed divided by the roll pitch (the distance between the two adjacent rolls in a zone, typically fixed in a particular zone, but varying from zone to zone) and looking for these frequencies in the processed data.

The important question is which signal processing method can extract the desired information. If there is non-periodic information, then we can hypothesize that HHT will be able to pick up the information that FFT cannot. A direct comparison of the results from using HHT and FFT is shown in Figures 10.3-25 and 10.3-26, respectively, with the frequencies of the zones shown in Figure 10.3-25. FFT clearly points to zones 1-8, and HHT shows excitation at these frequency ranges as well. The question is if there is more information that HHT can extract that FFT cannot, and from these comparisons that is not clear.

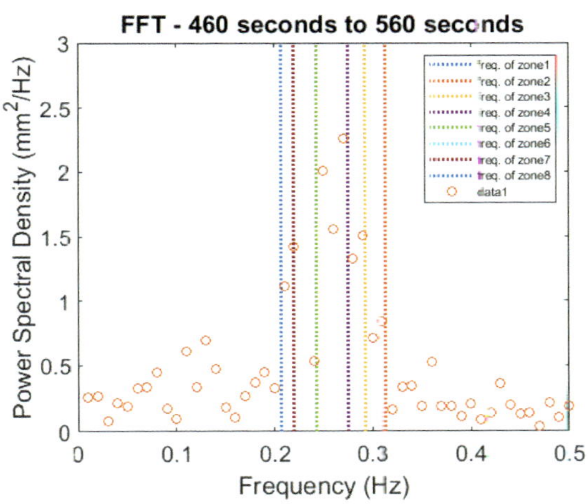

Fig. 10.3-25. FFT identifying the bulging zones, with the moving window size 100 s.

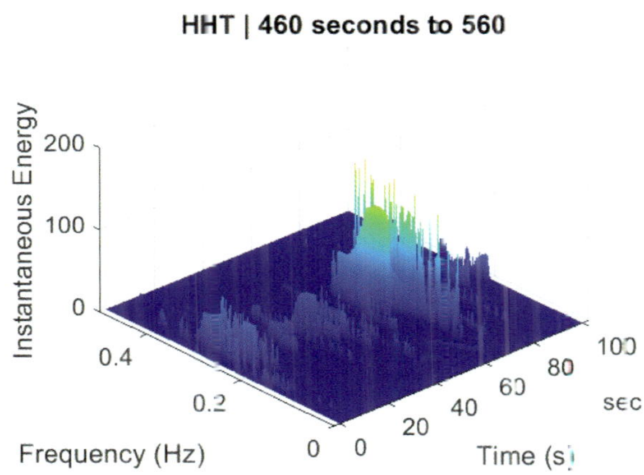

Fig. 10.3-26. HHT showing excitation at the frequencies of the zones.

Because it is unclear if there is an additional information to be extracted, a useful excersice is to construct our own known signal approximating the mold level oscillations. Using the first 8 zones, a mock signal can be defined as:

$$Signal = a \cdot \sin(freq_{zone1}) + b \cdot \sin(freq_{zone2}) + \cdots + h \cdot \sin(freq_{zone8}). \quad (10.3\text{-}28)$$

Application of FFT and HHT to this signal is shown in Figures 10.3-27 and 10.3-28, respectively. The window size for FFT is the same as that for HHT, 200 seconds. It is seen that FFT clearly identifies each zone. This is because this signal is purely periodic (composed of the sine waves).

However, in the case of the production mold level data, it is unknown if the bulging contributions to the signal are periodic. If they are not, HHT will be useful in identifying any additional non-periodic or transient behavior that can help in better pointing out the zones where bulging is occurring.

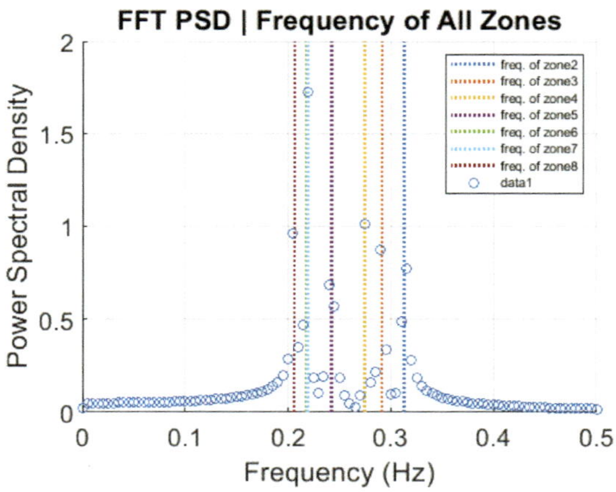

Fig. 10.3-27. FFT showing the mock signal modes of the mold level oscillation, window size 200 s.

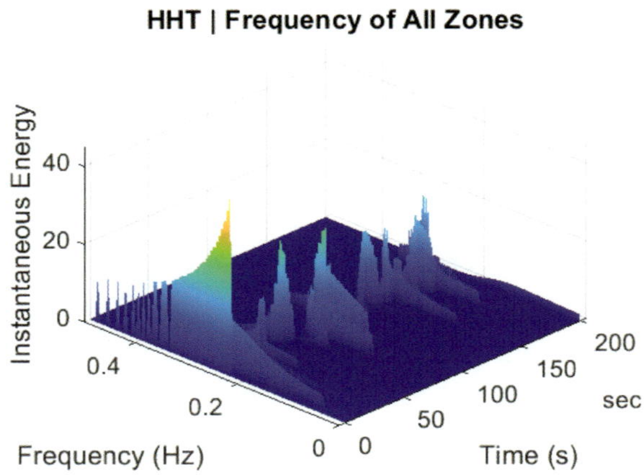

Fig. 10.3-28. HHT showing the mock signal modes of the mold level oscillation.

10.4 Basic Approaches to Machine Learning

10.4.1 *Neural Networks (NN) and Deep Learning (DL)*

This section introduces the basic concepts in machine learning, taking also a glimpse at *deep* and *shallow* neural networks. As a running example, we consider a problem of classifying the input data into one of known categories. Classically, a classification pipeline has the structure shown in Figure 10.4-1(a). Here, a predesigned feature extraction algorithm processes the input data, while the classifier processes only the feature extractor output. In the modern pipeline shown in Figure 10.4-1(b) the feature extractor is given by a trainable neural network.

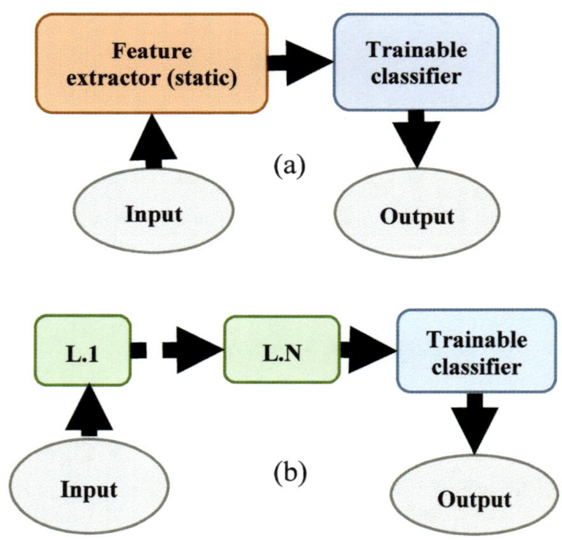

Fig. 10.4-1. Classical (a) and Neural Network (b) classification pipelines.

The feature extractor yields output properties advantageous over the raw input data, such as *order reduction.*

The main goal is to train a (linear) *regressor*, which maps a (high-dimensional) input to a vector that matches the number of categories. In general, for classification, negative entries of this vector indicate that this category is not identified, while positive values tell the opposite. If one wishes to find a confidence in each potential category, one can simply normalize all positive entries of the vector. If only one category is sought, the entry with the largest positive value is chosen. Training a regressor is often based on a least-squares like algorithm, and is often affine, i.e. based on a linear regressor with a bias vector, both of which are trained. Thus a single linear classifier has the form $\mathbf{x} \mapsto \mathbf{w} \cdot \mathbf{x} + \mathbf{b}$, where $\mathbf{w}$ is a vector of weights and $\mathbf{b}$ is a bias vector. In the case of multiple classes, we simply stack each linear regressor to end up with a matrix of weights and biases.

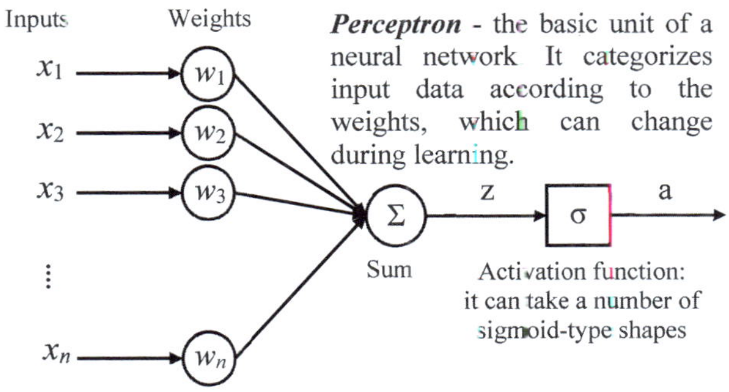

Fig. 10.4-2. The data from an input coming into perceptron gets multiplied by a weight assigned to this particular input, summed up, and passed through the activation function to get an output.

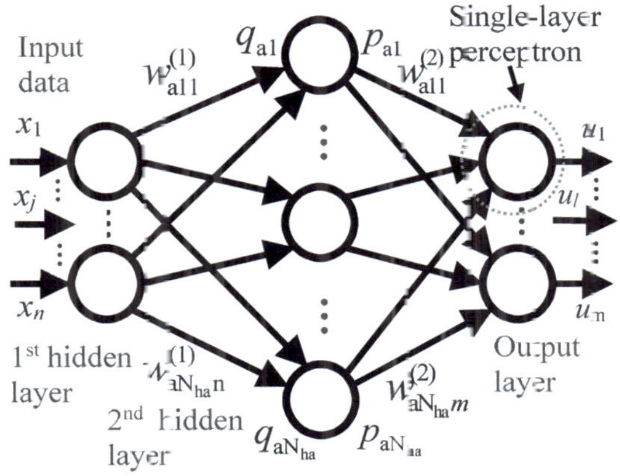

Fig. 10.4-3. Multi-layer perceptron — a shallow feedforward neural network with two hidden layers.

A regressor layer is often referred to as a *fully connected* (FC) layer in the machine learning literature (also known as a *single-layer perceptron,* shown in Figure 10.4-2). Considering a static feature extractor, the main task of the classification pipeline designer is then to come up with a feature extractor that transforms the input data through a series of *nonlinear* operations into some vector whose vector space is amenable to partitioning into *linear* subspaces, each of which (approximately) belongs to one of the categories to be classified.

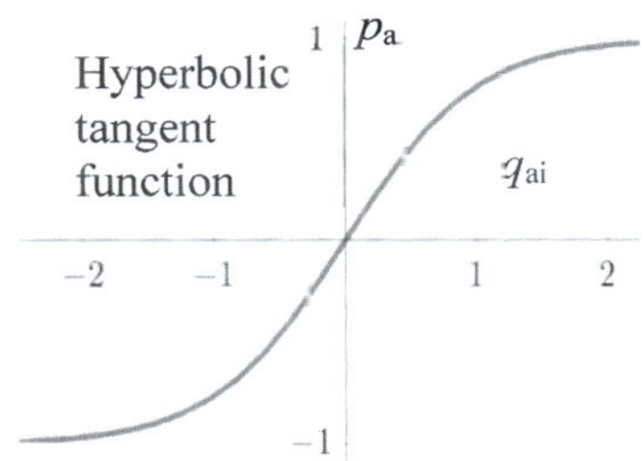

Fig. 10.4-4. Activation function.

The core idea in neural networks is to *delegate* not just the responsibility of training the classifier, but also the *design of the feature extractor itself to the training algorithm.* All that is provided is a model structure, in the form of *hidden layers,* which expose additional trainable parameters to the training algorithm. A *deep neural network* is a *neural network with more than two hidden layers.* The elements of a NN, such as that shown in Figure 10.4-3 and ***referred to as feedforward***, are typically given by

$$q_{ai}(t) = \sum_{j=1}^{n} w_{aij}^{(1)}(t) x_j(t) \Rightarrow p_{ai}(t) = \tanh q_{ai}(t) = \frac{e^{q_{ai}(t)} - e^{-q_{ai}(t)}}{e^{q_{ai}(t)} + e^{-q_{ai}(t)}} \Rightarrow u_l(t) = \sum_{i=1}^{N_{ha}} w_{ali}^{(2)}(t) p_{ai}(t),\ l = 1, \ldots, m,$$

$i = 1, \ldots N_{ha}$, N_{ha} — number of hidden nodes, $q_{ai}(t)$ and $p_{ai}(t)$ are hidden nodes inputs and outputs. q_{ai} and p_{ai} are related through the activation function given above and shown in Figure 10.4-4.

10.4.1.1 *Training a neural network*

Training a neural network is similar to training just the classifier — it is based on some form of *gradient descent*. Consider minimizing the *empirical loss*, $l(\mathbf{w}, \mathbf{x}_i, y_i)$, where l is a loss function, and $(\mathbf{x}_i, y_i)$ is an input-output pair. The common first choice is the *hinge loss*, defined as $l(\mathbf{w}, \mathbf{x}_i, y_i) = \max(0, -y_i \mathbf{w}^T \mathbf{x}_i)$. Commonly used loss functions are shown in Figure 10.4-5.

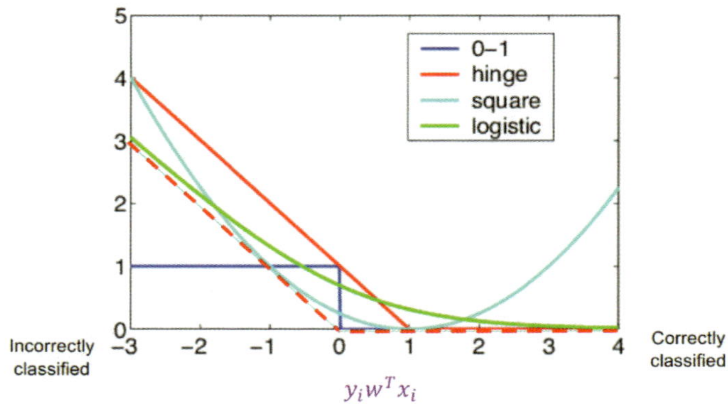

Fig. 10.4-5. Some common loss functions.

Consider the case with no hidden layers, and directly inject the data into a trainable classifier. Having chosen a loss function, we aim to find $\mathbf{w}$ to minimize $\hat{L} = (1/n)\sum_{i=1}^{n}(\mathbf{w}, \mathbf{x}_i, y_i)$. If we have multiple hidden layers, then we just compute the loss function given that $\mathbf{x}$ is transformed by the operators that depend on some set of parameters $\mathbf{W}$.

The gradient of the empirical loss can be computed using the chain rule, as will be shown in Subsection 10.5.2; this gradient may then be used in a *gradient descent algorithm* to train the parameters. This setup is employed by most machine learning algorithms in use today. The application of this approach to system identification and control is presented in Section 10.5 and to image processing, natural language recognition, and material phase change characterization — in Section 10.6.

10.4.2 *Reinforcement Learning (RL)*

Neural networks require, by design, access to the labeled input-output pairs in order to facilitate training. This may not be a problem for data that is available in large volumes, with labels already available as metadata, but in other applications this may mean that NNs are not applicable. One such problem is the case of training a robot to move around in a complex space. There are no "correct" input-output pairs, and outputs do not allow for meaningful labeling in general. This is precisely where *reinforcement learning* enters the stage.

In reinforcement learning, an *agent* is told through a *value function* what *reward* it gets for reaching a certain state by the current *action* from the state attained through a prior action. In effect, the agent tries to learn, based on its current state, what admissible action will yield the greatest reward. Over time, it will become apparent to the agent what actions are most applicable to the current situation (state), and suboptimal actions will be discarded in favor of (close to) optimal ones. This decision algorithm, which takes the current state and presents the action of choice, is known as a *policy*. To train a policy, *the value function assigns a reward to each state*. Once a trained policy is obtained, it could in principle be applied to guide an agent in the real world. We now present a formal problem structure, in which a policy may be sought.

10.4.2.1 *Markov decision processes (MDPs)*

Markov processes are introduced in Appendix C. Let S be the set of agent states, and A be the set of agent actions. Let $\#S$ and $\#A$ denote the number of states and actions in sets S and A, respectively. We assume that all actions are admissible at all times. Let $P_a(s,s') = \Pr(s_{t+1} = s' : s_t = s, a_t = a)$ be the probability of transition from state s to another state s' under action a. Let $R_a(s,s')$ be the immediate reward produced by the value function $R_a(\cdot,\cdot)$ after transition $s \to s'$ under action a. Our goal is to learn a policy $\pi : S \to A$ that maximizes the expected cumulative reward.

10.4.2.2 *Q-learning*

One of the main methods of obtaining a policy to solve an MDP is Q-learning. Q-learning is a model-free training algorithm applicable to problems with stochastic transitions and rewards. We first define a discount factor $\gamma \in [0,1]$ to favor rewards obtained early on in a run by exponentiating it by the current time — the bigger the value of time variable, the smaller the discount value becomes. Q-learning then trains a function that rates the quality of a state-action combination using the relation referred to as the ***Bellman's equation***:

$$Q^{new}(s_t,a_t) \leftarrow \underbrace{Q(s_t,a_t)}_{\text{old value}} + \underbrace{\alpha}_{\text{learning rate}} \cdot \overbrace{\Big(\underbrace{\underbrace{r_t}_{\text{reward}} + \underbrace{\gamma}_{\text{discount factor}} \cdot \underbrace{\max_a Q(s_{t+1},a)}_{\text{estimate of optimal future value}}}_{\text{new value (temporal difference target)}} - \underbrace{Q(s_t,a_t)}_{\text{old value}} \Big)}^{\text{temporal difference}} .$$

From the above update equation, we find that $Q \in \mathbb{R}^{\#S \times \#A}$. To implement the trained policy, for the current state s_t we take the action $a^*_{s_t} = \operatorname{argmax}_{a \in A} Q(s_t,a)$. A great property of Q-learning is that the algorithm can be trained using random system runs, also allowing training while the system is running in real-life. A Q-learning example is given next.

Example 10.4-1: Q-learning

```
MDP Example.
Format: (action, reward, transition prob.)

     (a0,1,0.6)
+---------------------------------+
|                                 |
v    (a0,1,0.5); (a1,2,0.2)       |  (a1,5,0.4)
s0 ----------------------------> s1 ----------------------> sf
^ \                              ^
| |                              | (a0,-1,0.8)
| |  (a0,1,0.5); (a1,2,0.8)      |
| +----------------------------> s2
|                                |
+--------------------------------+
     (a1,1,0.2)
```

Fig. 10.4-6. An MDP example system.

Given the MDP in Figure 10.4-6, we take [gamma = 0.9], and learning rate [alpha = 0.5]. We present 1000 random system runs that truncate after 10 steps to the Q-learning algorithm, resulting in the following Q matrix:

$$Q = \begin{bmatrix} 3.02 & 3.17 \\ 1.82 & 4.22 \\ 1.40 & 1.29 \\ 0.00 & 0.00 \end{bmatrix}.$$

To show the performance of this matrix, note the following ten policy-based system runs in the (states, reward) format:

```
([0, 1, 1, 1, 1, 1, 1, 1, 3], 7)
([0, 2, 2, 2, 1, 3], 8)
([0, 2, 1, 1, 3], 8)
([0, 1, 1, 1, 1, 3], 7)
([0, 2, 2, 2, 2, 1, 1, 1, 1, 1, 1, 1, 3], 8)
([0, 2, 2, 1, 3], 8)
([0, 2, 1, 3], 8)
([0, 2, 2, 1, 3], 8)
([0, 2, 2, 2, 2, 2, 1, 1, 3], 8)
([0, 2, 2, 2, 1, 3], 8)
```

Fig. 10.4-7. Policy implementation on the MDP of Figure 10.4-6.

It is also possible to set up a problem so that the value function itself could be subject to modification. In this case, the RL will be performing the policy/value iteration.

10.5 Application of Machine Learning to System Identification and Control

10.5.1 *The Gradient Descent (GD) function minimization*

The simplest ML technique, which is also widely used, is the *gradient descent (GD)*. As presented below on a simple example, this technique finds a local extremum of a sufficiently smooth dynamical system by setting up a related system where the smooth rhs (right-hand side) of the original system is replaced by the rhs gradient with the appropriate sign. The motion of the latter system then converges to the extremum, as shown below.

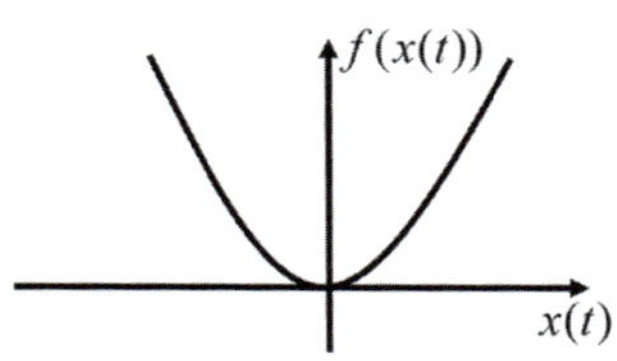

Fig. 10.5-1. The system considered for applying the gradient descent.

$$\frac{dx(t)}{dt} = f(x(t)) = x^2(t), \quad \frac{dx}{dt} = -\frac{\partial f(x)}{\partial x}(\text{function gradient}) = -2x$$

$$\Rightarrow \left\{ \begin{matrix} x > 0 \Rightarrow \dfrac{dx}{dt} < 0 \Rightarrow x \downarrow \Rightarrow x \to 0 \\ x < 0 \Rightarrow \dfrac{dx}{dt} > 0 \Rightarrow x \uparrow \Rightarrow x \to 0 \end{matrix} \right\} \text{always goes to } 0.$$

10.5.2 *Gradient descent based nonlinear system identification*

To demonstrate a GD application to nonlinear system identification we select the system model structure in the form of the so-called *nonlinear autoregressive exogenous model* (NARX), and embed into this structure a single-layer dynamic multiresolution neural network, referred to in Matlab System Identification Toolbox as *wavelet network*.

NARX relates the current value of a time series to the past values of the same series, and the current and the past values of the exogenous input:

$$y(t+1) = f(y(t), \cdots, y(t-(n-1)), u(t), \cdots, u(t-(n-1))$$

where u and y are system input and output, respectively, and $n_u = n_y = n$. NARX model can be simplified to the sum of the nonlinear autoregressive part and the exogenous input part using Narendra's *NARX-L1 approximation*, where *input enters linearly* as a weighted sum:

$$\text{NARX-L1}: y(t+1) = f_0(y(t), \cdots) y(t-(n-1)) + \sum_{i=0}^{N-1} g_i(y(t), \cdots, y(t-(n-1)) u(t-i)) .$$

Linearly parametrized nonlinear (MRA) regression functions are used to estimate nonlinear function f from data:

Modeling accuracy — Parameter vector (estimate) trained online

$$f = \sum_{i=1}^{m} \theta_i g_i, \quad \hat{\theta} \equiv \theta = \{\theta_i\}$$

$$g_i \in \{\varphi_{j,k}, \psi_{j,k}\}$$

n-dim scaling and wavelet functions of past input and outputs

If we identify f, then using past data, we can estimate the output from the model. And then, we can compare it to the actual $y(t)$ that we measure to see the error. A model can be expressed in terms of scaling function $\varphi_{J,n}$ and wavelet function $\psi_{j,n}$. Then, the *identified model* will produce the output estimate:

$$\hat{y}(t) = f(y(t-1), \cdots, y(t-n_y), u(t-1), \cdots, u(t-n_u)) = \sum_n \underbrace{< f, \varphi_{J,n} >}_{\theta^{\varphi}_{J,n}} \varphi_{J,n} + \sum_{j=J}^{N} \sum_n \underbrace{< f, \psi_{j,n} >}_{\theta^{\psi}_{j,n}} \psi_{j,n}$$

where J and j are the resolutions that we choose, and the finite integer n stands for as many translates as we need if f has compact support. To find the coefficients, we can excite the system with a pseudorandom binary sequence (PRBS, white noise).

Fig. 10.5-2. Black box system identification using PRBS.

This is carried out by using Marr φ and ψ, and the regressor, for example, $[y(t-1), y(t-2), u(t-1), \ldots, u(t-8)]$, where the dimension $n = 10$. We can use the initial coarse layer and a finer resolution, for example, $J = 3$ and $N = 6$. And then, for wavelet identifier (SISO), we can form the output error as follows:

Output error — Linear in plant model parameter estimates θ

$$e(n) = y(n) - \hat{y}(n)$$

Quadratic in θ

$$J_1(n) = \frac{1}{2}e^2(n)$$

Now, let us consider the system below to organize gradient descent. In discrete time domain, we will take partial derivative of $J_1(n) = \frac{1}{2}e^2(n)$ with respect to θ at discrete time n:

$$\theta(n+1) - \theta(n) = -\gamma_\theta \frac{\partial J_1}{\partial \theta}(n) \equiv \gamma_\theta \nabla_\theta J_1(n) \Rightarrow \theta(n+1) = \theta(n) - \gamma_\theta e(n)\frac{\partial e}{\partial \theta}(n),$$

where γ_θ is the learning rate parameter, $\nabla_\theta J_1(n)$ is the gradient of J_1 with respect to θ at discrete time n, and $\nabla_\theta e(n) = \frac{\partial e}{\partial \theta}(n)$ is the sensitivity derivative at time n that shows how the error is affected by θ. Then, substituting e by $y(n) - \hat{y}(n)$ yields

$$\frac{\partial e}{\partial \theta} = \frac{\partial(y(n) - \hat{y}(n))}{\partial \theta} = -\frac{\partial \hat{y}}{\partial \theta} = -\frac{\partial f}{\partial \theta} \equiv -g \Rightarrow \theta(n+1) = \theta(n) + \gamma_\theta e(n) g(n).$$

The equation on the right is the *parameter vector update equation with the GD machine learning.*

Example 10.5-1: Construct the NARX wavelet-based model to match the behavior of the strongly nonlinear Goodwin equation [2] (continued in Problem 6 of Homework 11)

$$\ddot{y}(t) + a\frac{y^2(t) - 1}{y^2(t) + 1}\dot{y}(t) + by(t) + cy^3(t) = u(t), \quad a = 0.1, b = -0.5, c = 0.5.$$

1) This equation can be implemented in Simulink as follows:

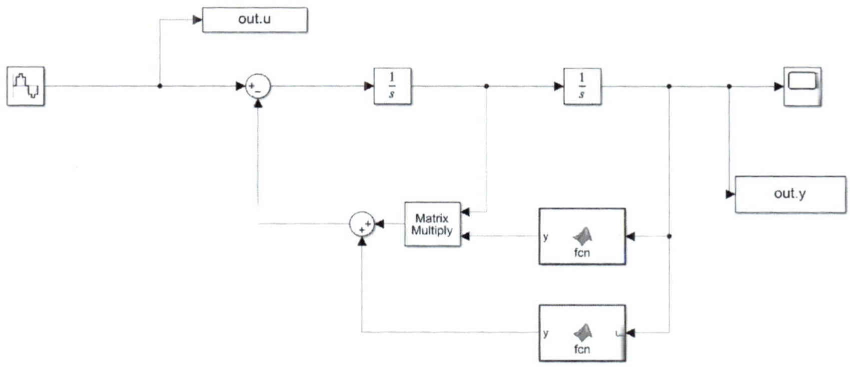

Fig. 10.5-3. Implementation of the Goodwin equation in Simulink.

2) Running the Simulink diagram above with initial conditions $\dot{y}(0) = y(0) = 0$, input $u(t)$=Acos(t) with A = 37, sampling the output $y(t)$ (considered as the measured output) with t_s = 0.02 sec., and plotting 3000 resulting samples $y[n]$ yields the curve in blue (partially visible) in Figure 10.5-4(a).

3) Matlab commands wavenet and nlarx permit identifying the nonlinear autoregressive models with exogenous input from the input/output (i/o) data. Now, we use the output generated in 2) above along with the input $u(t)$ sampled with t_s = 0.02 sec. and the Wavenet command to create an NARX input/output model. For this purpose we set the model structure as follows:

$$F(x) = (x-r)\cdot P\cdot L + \sum_{k=1}^{n_s} a_{s,k} f(b_{s,k}\cdot((x-r)\cdot Q - c_{s,k})) + \sum_{k=1}^{n_w} a_{w,k} g(b_{w,k}\cdot((x-r)\cdot Q - c_{w,k})) + d$$

where P and Q are linear and nonlinear subspaces, respectively,

L is linear coefficient, r is regressor mean, d is output offset,

$a_{s,k}$, $b_{s,k}$ and $c_{s,k}$ are scaling coefficient, dilation, and translation,

$a_{w,k}$, $b_{w,k}$ and $c_{w,k}$ are wavelet coefficient, dilation, and translation,

f is the scaling function and g is the wavelet function.

The model is trained using only the first 1000 samples of the i/o data set, and then evaluated using the full data set. Wavelet-based nonlinear system identification is performed by the following commands:

Matlab code:
```
ze = iddata(out.y(1:1000), out.u(1:1000), 1);
z = iddata(out.y, out.u, 1);
w = wavenet;
m = nlarx(ze, [2 1 1], w);
yp = predict(m, z, 1);
```

4) Now we use the sampled input of step 2) for this model to obtain the output, and draw the following graphs to see if the model works: (a) an overlap of the output of the NARX model with the measured output, (b) the discrepancy between the output of the NARX model and the measured output, and (c) the expanded interval of graph (a) for a more detailed visualization.

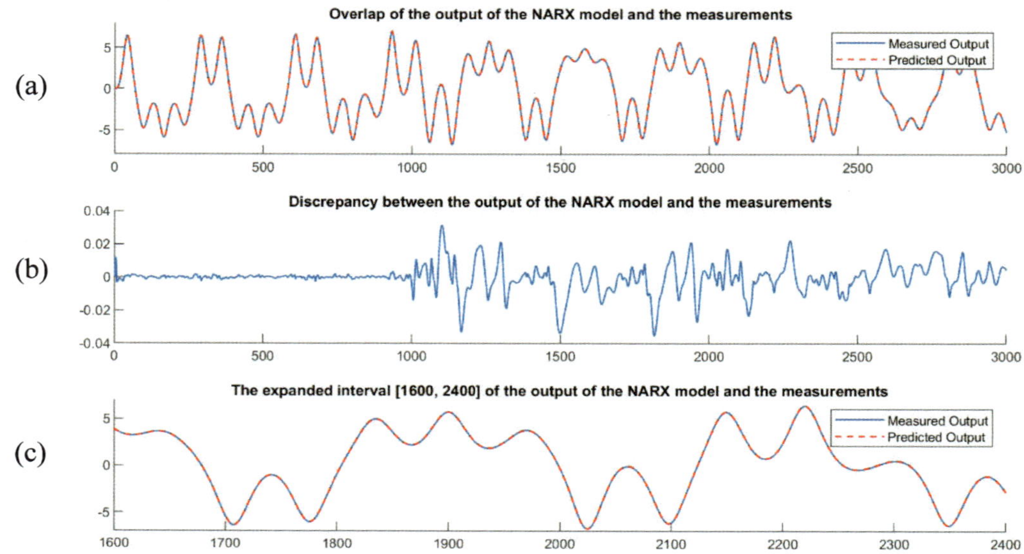

Fig. 10.5-4. (a) Overlap of the output of the NARX model and the measurements $y[n]$, (b) The discrepancy between the output of the NARX model and the measurements, (c) The expanded interval [1600, 2400] of the output of the NARX model and the measurements.

10.5.3 *Machine Learning application to control*

Cross-fertilization of ML and control is vast and growing quickly. The most basic application of ML to control is tuning of the PID controller gains, with the techniques ranging from local GD based tuning [3] to global one [4] based on PSO (particle swarm optimization), genetic algorithms (GA), and simulated annealing (SA), all available in Matlab Global Optimization Toolbox.

10.5.3.1 *Strongly nonlinear wavelet-based adaptive predictive control with gradient descent machine learning*

ML learning is often combined with adaptive control — an advanced control methodology that accommodates uncertainly of a physical object being controlled. For example, a modeling technique described in Subsection 10.5.2 has been used in the GD based adaptive configuration in [5]. In this effort, control of the multi-phase (gas-liquid-solid) flow was achieved using a multiresolution nonlinear adaptive control of the gas-particle flow in a chemical looping testbed, schematically shown in Figure 10.5-5. The chemical looping is a zero emission technology for extraction of energy from coal using solid oxidizer.

Wavelet MRA-based GPC scheme. Next we demonstrate the embedding of the NARX model structure with the parameter adaptation law introduced in Subsection 10.5.2 into the quadratic *GPC (generalized predictive control)-type* performance criterion to obtain the self-tuning MRA (multiresolution analysis)-based control law.

The basic methodology of GPC is to calculate the current control actions online at each sampling instant in order to solve a *finite horizon, open-loop, quadratic optimal control problem* where the first control in the optimal control sequence is applied to the plant. In this subsection, we present the online wavelet MRA/GPC-based predictive control strategy for the stable tracking problem of a single loop CL system.

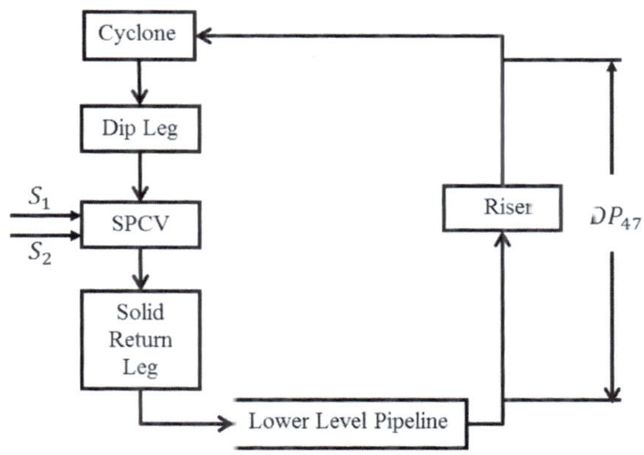

Fig. 10.5-5. Block diagram of a single-loop cold flow GE/Alstom Chemical Looping (CL) test rig.

Referring to Figure 10.5-5, let DP_{47}, the pressure drop across the riser, be the actual system output y and S_1, the fluidizing air flow rate, be the control input u, with the second fluidizing air flow rate S_2 set to a constant value. Let $\hat{y}$ denote the approximated system output in the identified wavelet MRA based model

$$\hat{y}(t) = f(y(t-1),\cdots,y(t-n_y),\ u(t-1),\cdots,u(t-n_u))$$

of Subsection 10.5.2. Suppose the future set-point signals $y_m(n+k), k=1,2,\cdots$ are available. Let's now embed the latter signals and the system outputs predicted through the identified model into the GPC-type loss function as follows:

$$J_2 = (1/2)\left\{\sum_{k=N_1}^{N_2} (y_m(n+k) - \hat{y}(n+k))^2 + \sum_{k=1}^{N_u} \rho_k \Delta u(n+k-1)^2\right\}, \tag{10.5-1}$$

where N_1 and N_2 are the minimum and the maximum output prediction horizons, respectively, N_u is the control horizon, Δ is the difference operator: $\Delta u(n) = u(n) - u(n-1)$, and ρ_k is the k-th control weighting factor. Assuming $N_1 = 1, N_2 = L = N_p$, and the identical control weighing factor $\rho_k = \rho$, Equation (10.5-1) can be rewritten in the vector form as:

$$J_2 = \frac{1}{2}\left\{\| Y_m(n+1) - \hat{Y}(n+1) \|^2 + \rho \| \Delta U(n) \|^2\right\}, \tag{10.5-2}$$

$$Y_m(n+1) = [y_m(n+1), y_m(n+2),\cdots, y_m(n+L)]^T,\quad \hat{Y}(n+1) = [\hat{y}(n-1), \hat{y}(n+2),\cdots,\hat{y}(n+L)]^T,$$
$$U(n) = [u(n), u(n+1),\cdots,u(n+N_u-1)]^T,\quad \Delta U(n) = [\Delta u(n), \Delta u(n+1),\cdots,\Delta u(n+N_u-1)]^T.$$

The loss function J_2 is now minimized to drive the system output y to the reference signal y_m given that the wavelet MRA identifier accurately approximates the real process dynamics online. At each sampling instant, an optimal control sequence is calculated using the future predicted output values of the identified model, but only the first one is applied to the system. To minimize J_2, the GD method is implemented again to recursively calculate the N_u -dimensional control increment sequence ΔU as follows:

$$\Delta U(n) = -\gamma_u \nabla_U J_2(n), \qquad (10.5\text{-}3)$$

where γ_u is the adaptation gain for the control input vector U . Noting that for any vector $y(x)$, $\nabla_x \| y \|^2 = 2(\nabla_x y) y$, from Equations (10.5-2) and (10.5-3), the gradient of the loss function J_2 with respect to U can be obtained as:

$$\begin{aligned}\nabla_U J_2(n) &= \frac{1}{2}\nabla_U \left\{ \| Y_m(n+1) - \hat{Y}(n+1) \|^2 + \rho \| \Delta U(n) \|^2 \right\} \\ &= \nabla_U \left\{ Y_m(n+1) - \hat{Y}(n+1) \right\} \left\{ Y_m(n+1) - \hat{Y}(n+1) \right\} + \rho \nabla_U \left\{ \Delta U(n) \right\} \left\{ \Delta U(n) \right\},\end{aligned}$$

yielding the control law of the form $\Delta U(n) = (I + \gamma_u \rho H)^{-1} \gamma_u G \left(Y_m(n+1) - \hat{Y}(n+1) \right)$, where I is the $N_u \times N_u$ identity matrix. G can be computed from the chosen wavelet MRA model structure. The full details of the derivation, the matrix elements, and *the input constraints* are given in [5].

The proposed control schematic is shown in Figure 10.5-6. As a result, the *highly nonlinear multi-phase tracking problem for a single loop cold flow system has been solved by the wavelet MRA-based adaptive GPC control strategy*. The resulting tracking performance, not attainable otherwise, is shown in Figure 10.5-7. It was acknowledged by GE/Alstom as a breakthrough.

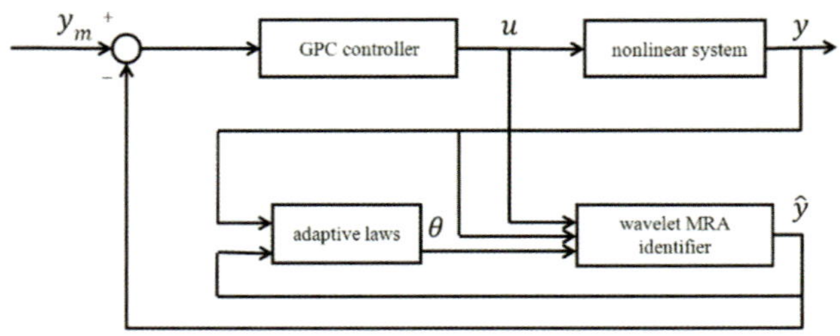

Fig. 10.5-6. Adaptive control system block diagram.

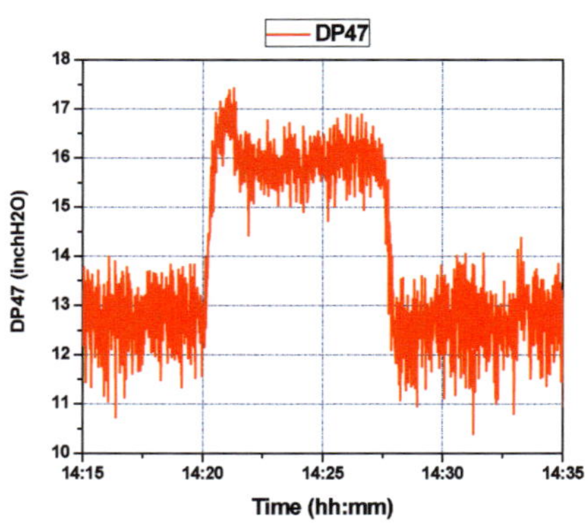

Fig. 10.5-7. Riser (DP47) pressure difference response to step setpoint changes in the CL cold flow test rig. Courtesy GE.

Recent merging of robust adaptive control with reinforcement learning (RL) [6] permitted the performance conservatism reduction of the simulation-trained RL when moving from simulated environment to the real one under dynamic variations.

An extensive support for integrating deep learning and reinforcement learning into various control applications is now available through Matlab as well as several noncommercial packages.

10.6 Applications of Machine Learning to Image Processing, Natural Language Recognition, and Material Phase Change Characterization

10.6.1 *Applications of ML to image processing using Convolutional Neural Networks (CNN)*

We have already discussed fully connected layers in Section 10.4.1, where the entirety of the input data is exposed to a single layer. When dealing with pictures, however, many of the fine details are prone to be lost when looking at the entire picture at once. This gives rise to the idea that somehow limiting the field of perception would be beneficial to a neural network's performance. Even more so, one could extend this idea to a *sliding* motion, much like one would read a text letter-by-letter, or word-by-word. In Section 10.2.2 we saw precisely such a sliding action when applying linear convolution. The latter forms the backbone of the *convolution neural networks* (CNNs), which differ from the fully connected networks (the multi-layer perceptrons) as shown in the figure below:

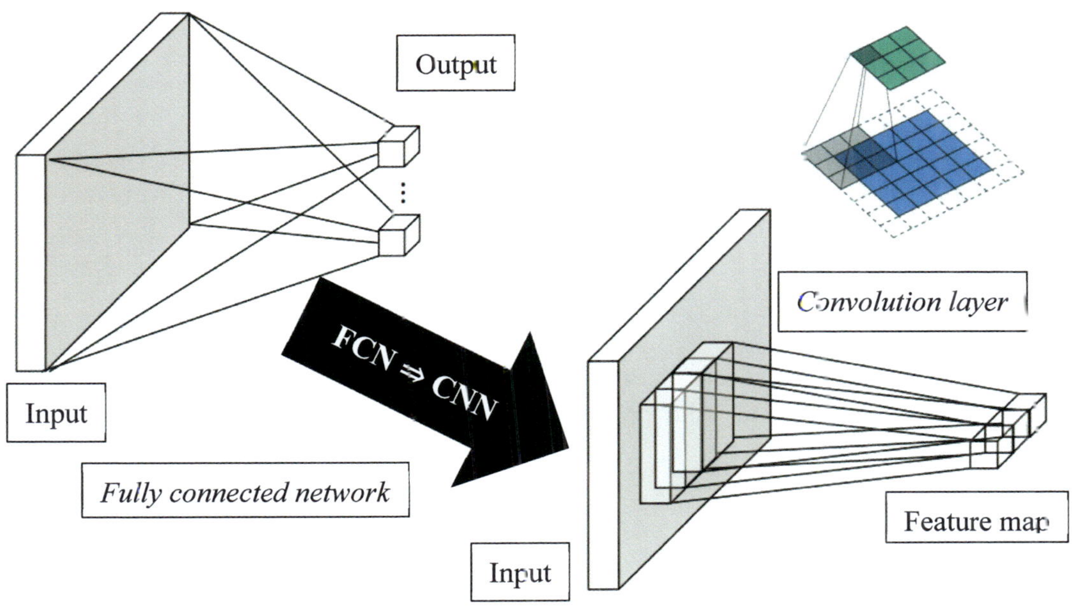

Fig. 10.6-1. Relationship between the fully connected network and the convolutional neural network.

Instead of using only a single filter kernel per layer, it is conventional to use multiple filters. After each convolution, a nonlinear operation is applied (much like those shown in Figure 10.4-5 but mirrored). This permits efficient manipulation through convolution, while still allowing for nonlinear manipulation of the data. The classification layer is still taken to be a fully connected layer, as discussed in Section 10.4.1.

CNNs have successfully been applied to many problems, including image classification, recognition, segmentation, and beyond. Today, pretrained CNNs can even be embedded into real-time hardware, with many vendors producing single-chip solutions to CNN driven low-cost hardware. One such application secured CNNs a premier spot in machine learning in the early 2010s and effectively dethroned classical image processing techniques, namely image classification, as detailed below.

A premier challenge for image classification is the ImageNet Large Scale Visual Recognition Challenge (ILSVRC), in which 1000 image categories are included, ranging from artificial objects to animals and vegetables. In 2012, the CNN-based AlexNet architecture, consisting of 60 million parameters, reported a classification error of 16.4%, compared to the 5% error by a trained human expert. A mere three years later, the ResNet architecture for the first time outperformed the 5% human error, producing a mere 3.8% error. Given these successes, the ImageNet challenge was discontinued in 2017, giving a decisive signal that image classification was "completely solved" by CNNs.

The addition of the wavelets to CNN structure produces the ***wavelet CNNs*** — the state-of-the-art image processing tools.

10.6.2 *Deep Learning (DL) in natural language recognition*

Natural language recognition is central to communicating with virtual assistants (VAs), such as Apple's Siri. Natural language VA has been proposed by the author's close friend Alex Kurganov (https://patents.google.com/patent/US10629206B1/en?inventor=Kurganov&oq=Kurganov) in Chicago in late 1990's as a tool to directly interact with the Internet search engines. As Alex's patent, granted in 2000, states: "embodiment of the present invention is a system and method for allowing users to browse information sources, such as web sites, by using naturally spoken, conversational voice commands spoken into a voice enabled device. Users are not required to learn a special language or command set in order to communicate with the voice browsing system of the present invention. Common and ordinary commands and phrases are all that is required for a user to operate the voice browsing system. The voice browsing system recognizes naturally spoken voice commands and is speaker-independent; it does not have to be trained to recognize the voice patterns of each individual user." Alex's patent was upheld in 2021 in the claim against all major users of VAs for the information retrieval — the main VA use (https://www.prnewswire.com/news-releases/ptab-affirms-patentability-of-parus-claims-301373111.html).

The key task in VA functioning is automatic speech recognition (ASR). It is nontrivial, as evidenced by the 2022 $19.7 billion acquisition by Microsoft of Nuance, a speech recognition company, whose software is used in Apple's Siri, even though Microsoft already has a strong speech products team, which developed Cortana. For a long period of time, speech recognition was carried out using hidden Markov models and phonemes — the smallest units of sound in speech, like voiced letters. Recently, however, this task has moved to DL through conversion of audio signals into images, further processed by various DL architectures, especially CNNs.

To enable this process, an audio record is transformed into a signal portrait called the ***spectrogram***, which aggregates the key features of an audio signal — variable magnitudes and frequencies — into a specially constructed image well-suited for DL. Both Matlab and Python have advanced audio processing tools for this purpose. For example, spectrograms with different features are simply constructed through commands in the Time-Frequency Analysis section of Matlab Signal Processing Toolbox. Matlab has also Audio Toolbox, where many audio processing procedures are automated. Below, we present the basic spectrogram construction.

A typical audio record is represented by a rather irregular signal with fast time-varying magnitude, such as that shown in Figure 10.6-2. It is this irregularity that carries the information.

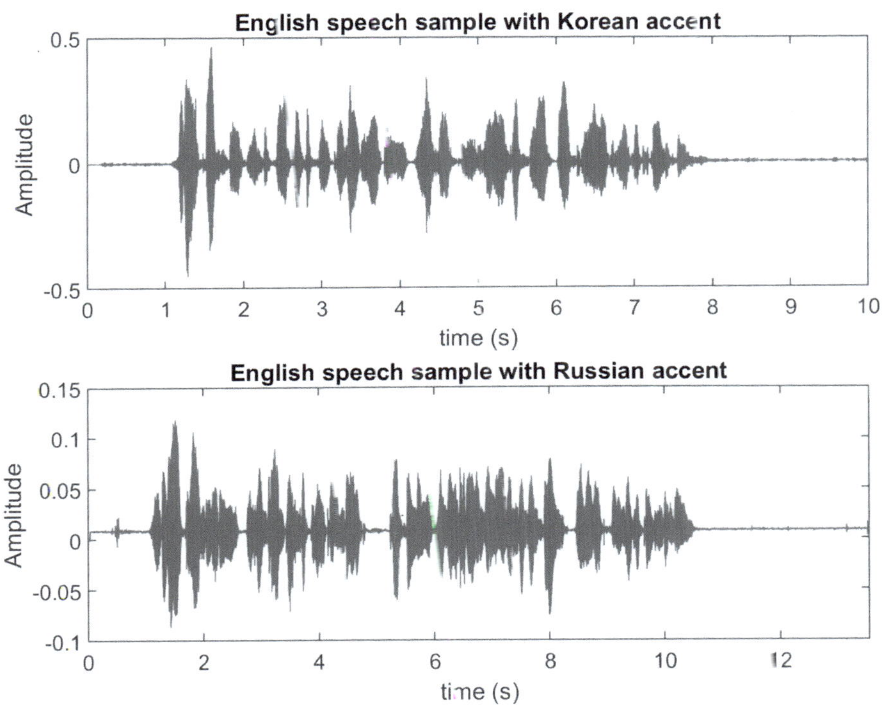

Fig. 10.6-2. Data records of a speech sample "Signal Processing is a broad discipline concerned with the manipulation and analysis of both analog and digital signals" with Korean and Russian accents.

This signal is composed of multifrequency sounds with frequencies varying over the relatively long time interval, but staying relatively constant over the very short time intervals. This is because our vocal chords need to vibrate during long enough periods of time to encode our thoughts into sounds perceptible by others. To capture this variability, the data record is segmented into the so-called short-time overlapping sections and FFT, referred to as short-time FFT, is computed for each section, as shown in Figure 10.6-3:

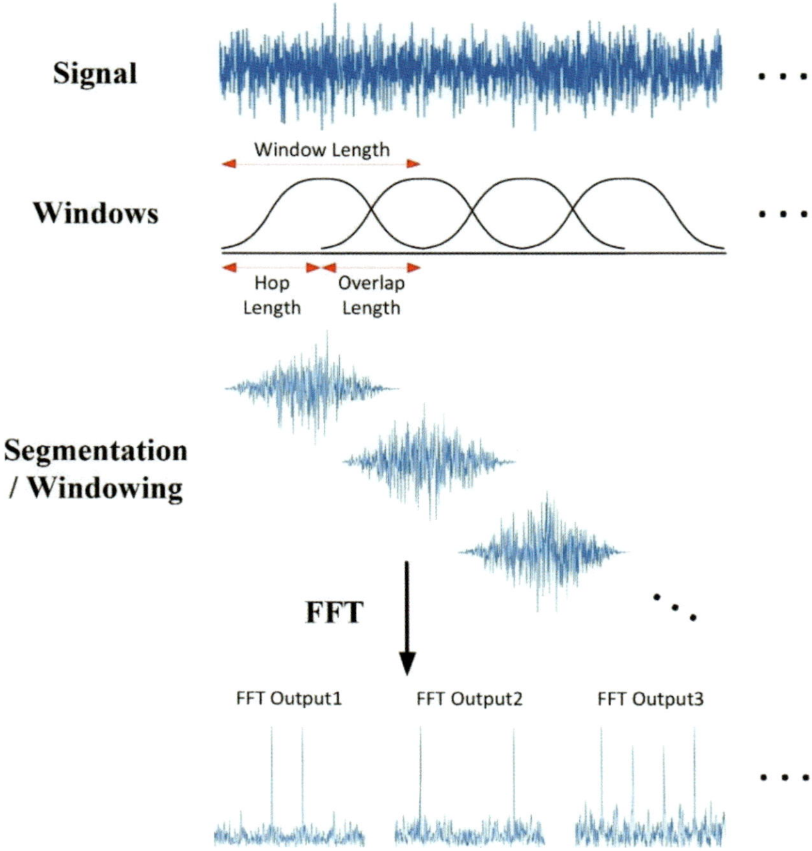

Fig. 10.6-3. Short-time Fourier transform (STFT) overview (courtesy Ref. [1]).

The next step is to bring the frequency and the magnitude signal representation into a form compatible with the human sound perception. Everyone is familiar with the aircraft jet engine noise, the level of which is measured in dB, and is about 120 dB. The log scale is used because loudness is perceived by a human ear logarithmically — we are sensitive to the magnitude difference of the very low magnitude sounds, but our sensitivity to this difference decreases with the increase in the sound magnitude. In my personal experience, a concert of Yes at UIUC State Farm Center Arena was played at the relatively low sound level, allowing to enjoy instrumental virtuosity and clear vocals of the band, but at the much louder Metallica concert at the same venue instruments and voices were mushed together, coming through as just an indiscernible wall of noise, no better than that from a jet engine.

A very similar situation is with the perception of pitch (sound frequency) difference — high sensitivity to this difference at low frequencies, such as around 200 Hz, and much lower sensitivity at high frequencies. In this case, the logarithmic frequency scale is expressed in mels (from melody) and is given by a few established formulas, such as

$$m = 2595 \log_{10}\left(1 + \frac{f}{700}\right)$$

plotted in Figure 10.6-4.

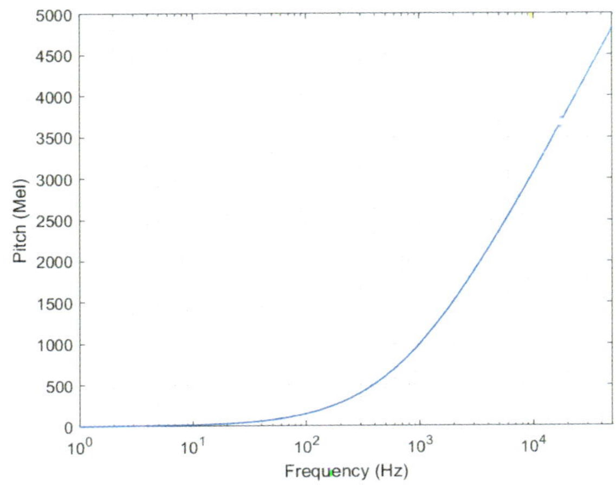

Fig. 10.6-4. Mel Scale measures human perception of pitch.

Now, using the mel scale instead of frequency on the vertical axis, the time on the horizontal axis, and the dB scale instead of magnitude to indicate colors (the lighter the color, the higher the magnitude) yields a mel spectrogram, such as that shown in Figure 10.6-5.

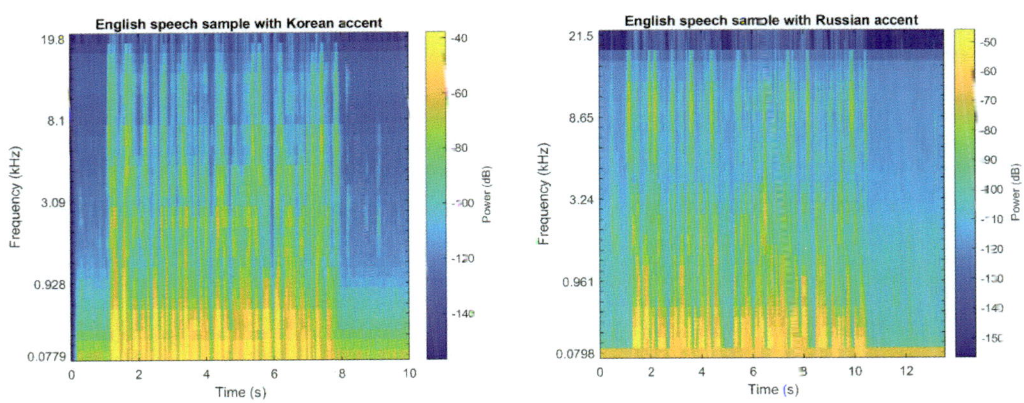

Fig. 10.6-5. Mel spectrograms.

A mel spectrogram is then processed through a DL-based sequence of steps, like the one shown in Figure 10.6-6.

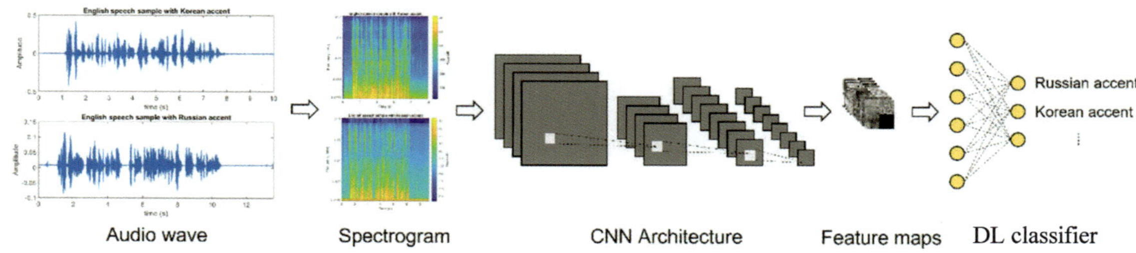

Fig. 10.6-6. Audio classification application.

This concludes a brief introduction to this topics. For further details on automatic speech recognition presented in an easily accessible form, the reader is referred to https://towardsdatascience.com/audio-deep-learning-made-simple-automatic-speech-recognition-asr-how-it-works-716cfce4c706 and the corresponding series of articles, on which this presentation is largely based.

In the inverse procedure — ***sound synthesis*** from text (as well as artificial image synthesis), a different type of NNs — Generative Adversarial Networks (GANs) are typically used.

Music synthesis is another area that has been at the forefront of the signal processing applications, offering computer-aided synthesis of orchestral performances, with the results as good as those of the live orchestras: https://www.vsl.co.at/en/Products.

10.6.3 *Deep Learning (DL) in material characterization during the phase change*

Continuous casting plays a dominant role in modern steel making, supplying 95% of the world's steel. Due to many casting variables and cost, the experimental capability to further learn the physics behind the steel continuous casting is rather limited. Therefore, a multiphysics model of steel solidification could be very helpful with managing the steel casting process. In [7] DL is shown to serve as a useful technique for estimating steel characteristics associated with the material phase change taking place during steel solidification. This approach is briefly reviewed next.

10.6.3.1 *Thermo-mechanical model of steel solidification*

The following thermo-mechanical model with a solidifying slice in the Lagrangian frame is used to describe the physics behind the solidification:

$$\rho\left(\frac{\partial \mathrm{H}}{\partial t}\right) = \nabla\cdot\left(k\nabla T\right),$$

where k is the thermal conductivity, ρ is the density, T is the temperature and H is the specific enthalpy, including the latent heat during phase transformation. Since inertial effects do not play an important role in the solidification process, the quasi-static mechanical equilibrium equation

$$\nabla\cdot\sigma(x) + b = 0$$

can be used, where $\boldsymbol{\sigma}$ is the Cauchy stress tensor and $\boldsymbol{b}$ is the body force density vector. The two equations given above are coupled. Together with visco-plastic models, which depend on casting settings, these coupled equations can be solved using the finite element methods, yielding the temperature and the stress histories of the solidification slice. In [7], the resulting database is used for training of a temporal convolutional network (TCN) for capturing the solidification behavior of steel during continuous casting.

10.6.3.2 *Deep Learning based solidification behavior predictive model development*

The temporal convolutional network (TCN) is a one-dimensional version of CNN introduced in Section 10.6.1, with the kernel sliding only along the time axis. It has the following structure:

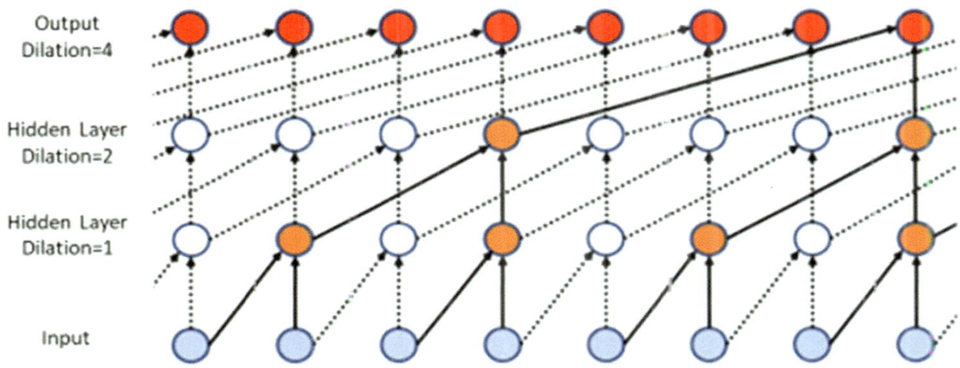

Fig. 10.6-7. Temporal convolutional neural network.

This structure, as seen in Figure 10.6-7, is characterized by a sequentially dilated convolution, with the dilation factor (enlargement of the receptive field by forming the gaps in the entries of the subsequent convolution layers) increasing exponentially between two layers.

After the training procedure, the TCN deep learning model has shown good results: it successfully predicted the highly complex and coupled steel solidification thermo-mechanical phenomena, generating the tensile and the compressive residual stresses on the chilled steel surface and in its through-thickness vicinity. These stresses and strains are good indicators of initiation of cracks and other defects in the continuously cast steel. The resulting DL TCN model can provide an online diagnostic tool for steel condition monitoring and cracks prevention.

Homework 11: Solved Examples and Solution Guide

Problem 1. Computer audio — signal equalization

(a) Batch equalization using FFT.

(a1) Use the Matlab *fft( )* function to compute the DFT, $X[k]$, of an mp4 music file of your choice. Perform the boom box filtering using the latter DFT as described in Subsection 10.1.2. Give the code for taking the inverse FFT to obtain the filtered data record and listen to it.

(a2) Plot the single-sided unfiltered and filtered magnitude spectra with a logarithmic vertical axis, using *semilogy*( ). Scale the horizontal axis in terms of continuous frequency units, Hz, using the sampling rate for converting the digital index values into the corresponding Hz values.

Solution:

(a1) As described in Subsection 10.1.2, the boom box filtering boosts the high and the low frequencies magnitudes, leaving the middle range intact. For this purpose, the low and the high cutoff frequencies are arbitrarily selected as 1200 Hz and 16000 Hz, respectively, and the magnitudes of the components with frequencies lower than 1200 Hz or higher than 16000 Hz are boosted three times. For this problem, any short mp4 file can be taken. The example mp4 file with a 44.1 kHz sampling rate used in this solution is downloadable from the link: https://uofi.box.com/s/tprw77d78rw4lyhqyfjhx735njdvel0o. The boom box filtering is performed by the following commands:

Matlab code:

```
[y_electronic, Fs] = audioread('Electronic.mp4');
playerObj = audioplayer(y_electronic, Fs); % build audio
play(playerObj); % play the audio
T = 1/Fs;
L = length(y_electronic);
t = (0:L-1)*T;
f = Fs*(0:(L/2))/L;
DFT_X = fft(y_electronic);
lowf = 1200;
highf = 16000;
Boom = DFT_X;
for i = 2:L
    if i <= lowf/(Fs/L) || (highf/(Fs/L) <= i && i <= L-highf/(Fs/L)+2) || i
>= L-lowf/(Fs/L)+2
        Boom(i) = 3*DFT_X(i);
    end
end
P2 = abs(Boom/L);
P1 = P2(1:L/2+1);
```

```
P1(2:end-1) = 2*P1(2:end-1);

y_reconstructed = ifft(Boom);
audiowrite('Boom.mp4', y_reconstructed, Fs);
[y_Boom, Fs_Boom] = audioread('Boom.mp4');
playerObj = audioplayer(y_Boom, Fs_Boom);
play(playerObj);
```

(a2)

```
figure;
semilogy(f,P1)
title('Single-sided magnitude spectrum of the boom box filtered signal')
xlabel('f (Hz)')
ylabel('Magnitude')
```

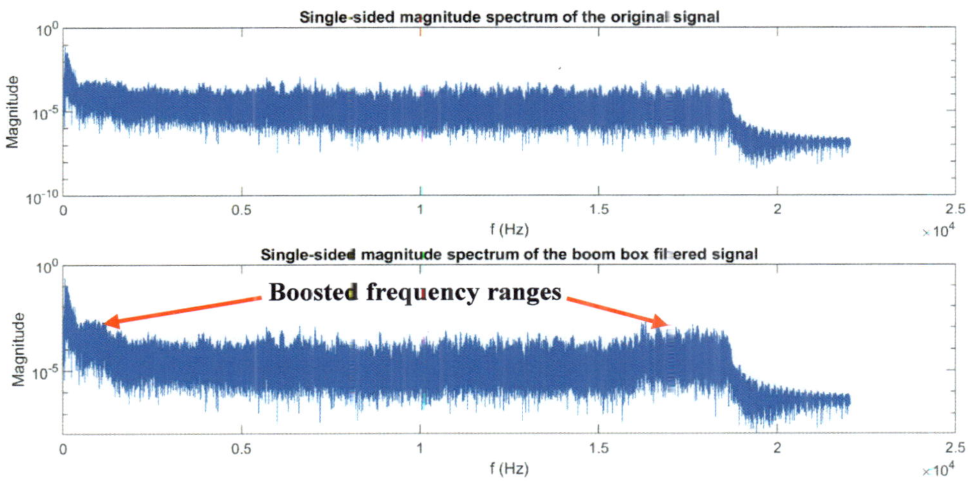

Fig. HW11-1. The single-sided magnitude spectra of the input and the output of the boom box filter.

(b) Online equalization using recursive digital filters represented by z-transform.
Following Subsubsection 7.5.4.1 of Chapter 7 and Subsection 18.5.2 of (Ambardar, 1999), consider the use of the first-order low pass and high pass filters $H_{LP}(z)$ and $H_{HP}(z)$, $H_{LP}(z) + H_{HP}(z) = 1$, given, respectively, by the transfer functions

$$H_{LP}(z) = \left(\frac{1-\alpha}{2}\right)\frac{z+1}{z-\alpha}, H_{HP}(z) = \left(\frac{1+\alpha}{2}\right)\frac{z-1}{z-\alpha}.$$

Setting α to α_L and α_H, where the subscripts L and H stand for the low pass and the high pass stages, respectively yields two sets of filter pairs:

$$H_{LPL}(z) = \left(\frac{1-\alpha_L}{2}\right)\frac{z+1}{z-\alpha_L}, H_{HPL}(z) = \left(\frac{1+\alpha_L}{2}\right)\frac{z-1}{z-\alpha_L},$$

$$H_{LPH}(z)=\left(\frac{1-\alpha_H}{2}\right)\frac{z+1}{z-\alpha_H}, H_{HPH}(z)=\left(\frac{1+\alpha_H}{2}\right)\frac{z-1}{z-\alpha_H}.$$

Then, assigning the adjustable real gain G to the low pass and the high pass filters in the first and the second filter pairs, respectively, yields the low pass and the high pass shelving filters

$$H_{SL}=GH_{LPL}(z)+H_{HPL}(z)=1+(G-1)H_{LPL}(z),$$

$$H_{SH}=GH_{HPH}(z)+H_{LPH}(z)=1+(G-1)H_{HPH}(z).$$

(b1) Connect in series the above low pass and high pass shelving filters to obtain the transfer function of the entire tone control filter. Give the detailed transfer function of the filter.
(b2) Plot its magnitude and phase responses to clearly see the filtering action (choose the axes scaling as needed — log, semilog).
(b3) Select and plot a three-frequencies (low, medium, high) DT signal as input to the filter, then pass this signal through the filter and plot the output for three gains — attenuated lows and highs, undistorted (to the degree possible) lows and highs, and boosted lows and highs.

Solution: (b1) Connecting LP and HP shelving filters in series yields the tone filter of the form:

$$\begin{aligned}H(z)&=H_{SL}(z)H_{SH}(z)=\left(1+(G-1)H_{LPL}(z)\right)\left(1+(G-1)H_{HPH}(z)\right)\\&=1+(G-1)H_{LPL}(z)+(G-1)H_{HPH}(z)+(G-1)^2H_{LPL}(z)H_{HPH}(z),\\&=1+(G-1)\left(\frac{1-\alpha_L}{2}\right)\frac{z+1}{z-\alpha_L}+(G-1)\left(\frac{1+\alpha_H}{2}\right)\frac{z-1}{z-\alpha_H}\\&\quad+(G-1)^2\left(\frac{1-\alpha_L}{2}\right)\left(\frac{1+\alpha_H}{2}\right)\frac{z+1}{z-\alpha_L}\frac{z-1}{z-\alpha_H}=\frac{b_2z^2+b_1z+b_0}{a_2z^2+a_1z+a_0},\end{aligned}$$

where $a_0=4\alpha_L\alpha_H$, $a_2=4$, $a_1=-4(\alpha_L+\alpha_H)$, $b_1=-4(\alpha_L+\alpha_H)-4(G-1)(\alpha_L+\alpha_H)$,

$$b_0=4\alpha_L\alpha_H-2(G-1)\alpha_H(1-\alpha_L)+2(G-1)\alpha_L(1+\alpha_H)-(G-1)^2\left(1-\alpha_L\right)\left(1+\alpha_H\right),$$

$$b_2=4+4(G-1)-2(G-1)(\alpha_L-\alpha_H)+(G-1)^2(1-\alpha_L)(1+\alpha_H).$$

(b2) Using the transfer function of the entire filter system, the magnitude and the phase responses of the filter are plotted in Figure HW11-2. The filtering action is seen to vary with varying the gain G values. The gain lower than 1 attenuates the low and the high frequency components, while the gain higher than 1 boosts them. The gain 1 passes lows and highs undistorted.

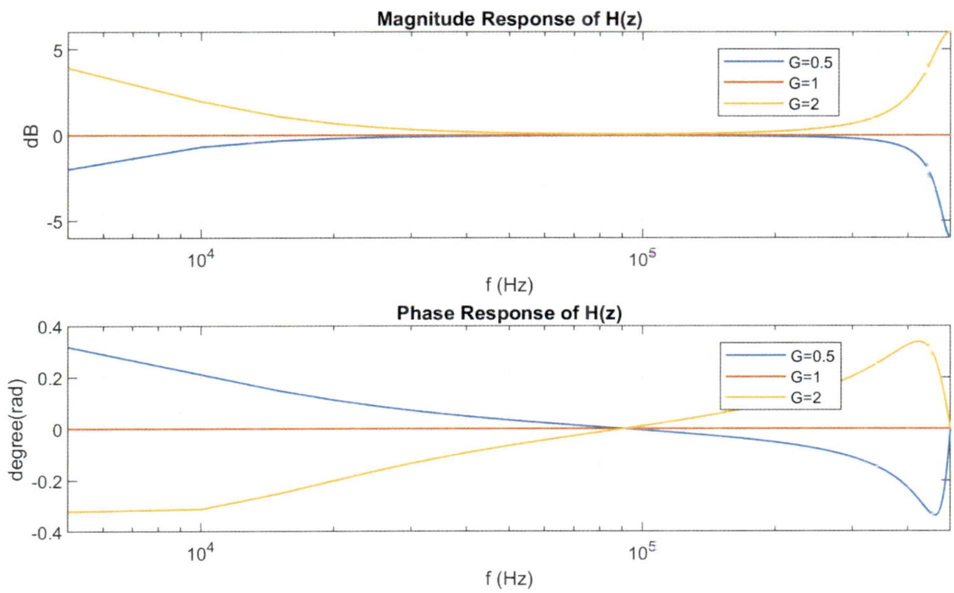

Fig. HW11-2. The magnitude and the phase responses of the "tone control" shelving filter formed by the low pass and the high pass shelving filters connected in series. In this example solution, the sampling rate is 1 MHz, and each alpha value is 0.97 for low pass shelving filter and −0.7 for high pass shelving filter.

(b3) A three-frequencies (low, medium, high) DT input signal to the filter, shown in Figure HW11-3, is obtained by sampling with 1 MHz sampling rate the CT signal

$$x(t) = \cos(2\pi 1000t + \pi) + 2\cos(2\pi 100000t + \pi/2) + 3\cos(2\pi 400000t + \pi/3)\,.$$

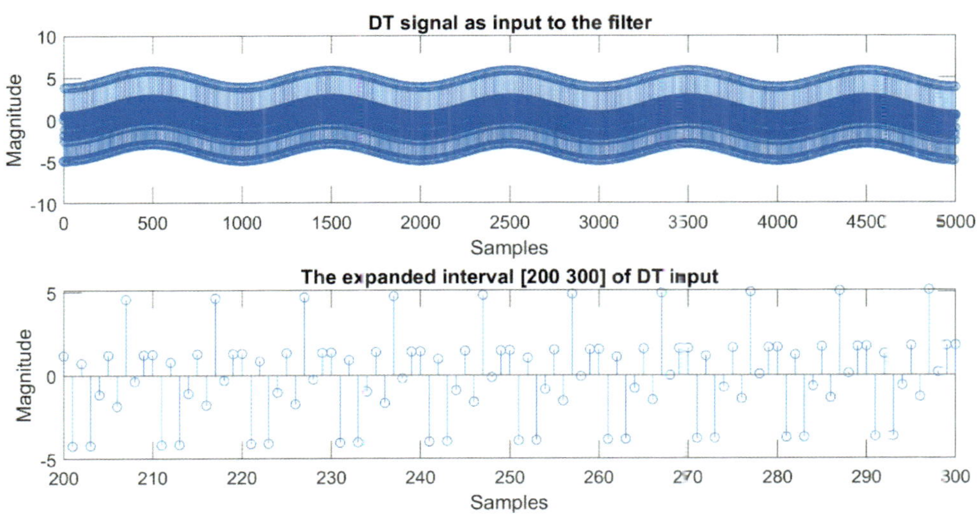

Fig. HW11-3. Three-frequencies (low-1000 Hz, medium-100000 Hz, high-400000 Hz) DT signal as an input.

Using `filter` command, we can easily observe how the entire filter affects the input signal from the output for three gains shown in Figures HW11-4, 5, and 6.

The output for gain 0.5:

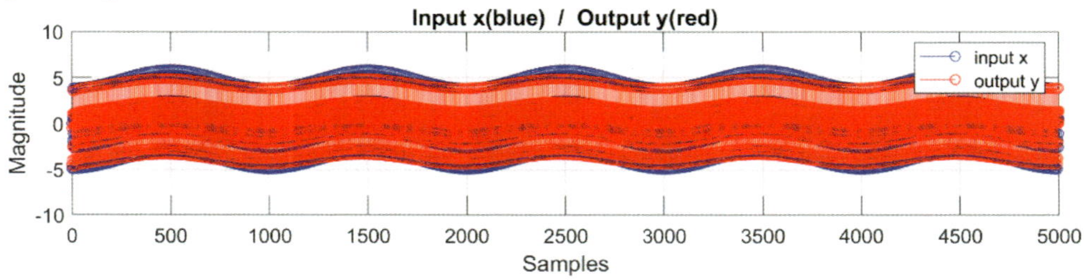

Fig. HW11-4. The output signal from the filter for gain 0.5. Lows and highs are attenuated.

The output for gain 1:

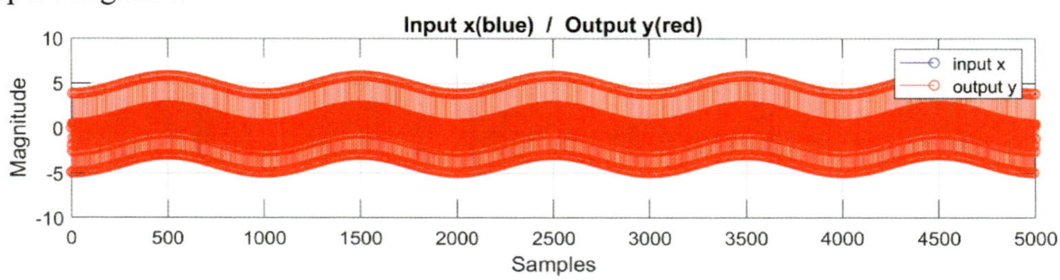

Fig. HW11-5. The output signal from the filter for gain 1. The output overlaps with the input.

The output for gain 2:

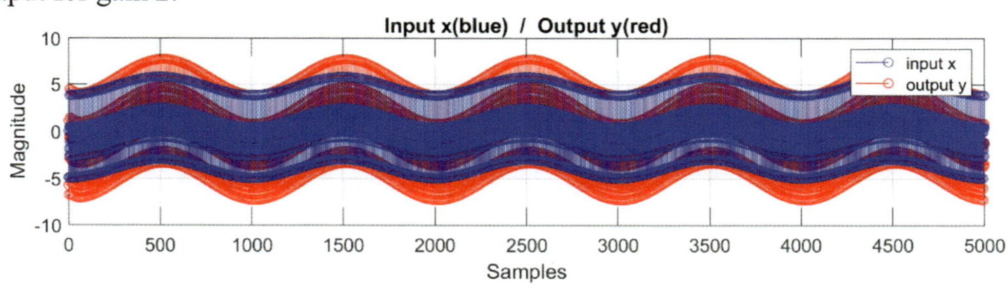

Fig. HW11-6. The output signal from the filter for gain 2. Lows and highs are boosted.

Matlab code:

```
fs=1000000;
f=[0:5000:500000];
Fn=fs/2;
w=2*pi*f/fs;
z=exp(w*j);
aL = 0.97;
```

```
aH = -0.7;
fcL = acos(2*aL/(1+aL^2))/(2*pi)*fs;
fcH = acos(2*aH/(1+aH^2))/(2*pi)*fs;
G = 0.5;
HLPL = (1-aL)/2*(z+1)./(z-aL);
HHPL = (1+aL)/2*(z-1)./(z-aL);
HSL = G*HLPL+HHPL;
HLPH = (1-aH)/2*(z+1)./(z-aH);
HHPH = (1+aH)/2*(z-1)./(z-aH);
HSH = G*HHPH+HLPH;
Hz = HSL.*HSH;
figure(1);
subplot(2,1,1);
Hz(Hz==0)=10^(-8);
semilogx(f,20*log10(abs(Hz)),'LineWidth', 1);
grid on
ylabel('dB')
xlabel('f (Hz)')
title('Magnitude Response of H(z)')
xlim([0 500000]);
subplot(2,1,2), semilogx(f,angle(Hz),'LineWidth', 1)
grid on
title('Phase Response of H(z)')
ylabel('degree(rad)')
xlabel('f (Hz)')
xlim([0 500000]);
fL=1000;
fM=100000;
fH=400000;
T=1/200;
twopi_fL_t=(1:fs*T)*2*pi*fL/fs;
twopi_fM_t=(1:fs*T)*2*pi*fM/fs;
twopi_fH_t=(1:fs*T)*2*pi*fH/fs;
A1L=1;
A1M=2;
A1H=3;
phiL=pi;
phiM=pi/2;
phiH=pi/3;
x = A1L*cos(twopi_fL_t + phiL) + A1M*cos(twopi_fM_t + phiM) +
A1H*cos(twopi_fH_t + phiH);
figure(2);
plot(x)
xlabel('Samples')
ylabel('Magnitude')
a=[4 -4*(aL+aH) 4*aL*aH];
b=[(4+4*(G-1)-2*(G-1)*(aL-aH)+(G-1)^2*(1-aL)*(1+aH)) (-4*(aL+aH)-4*(G-
1)*(aL+aH)) 4*aL*aH-2*(G-1)*aH*(1-aL)+2*(G-1)*aL*(1+aH)-(G-1)^2*(1-
aL)*(1+aH)];
y = filter(b,a,x);
```

```
figure(3);
plot(x,'b-'); hold on;
plot(y,'r-');
grid on
title('input x(blue)  /  output y(red)')
xlabel('Samples')
ylabel('Magnitude')
legend('input x', 'output y')
hold off;
```

Problem 2. Basic image processing — edge detection, 2D filtering
Carry out the edge detection on the coins image in Matlab.

Solution:

1. Load the coins.png image in Matlab library:
   `im = imread('coins.png');`
2. Use the Sobel filter to compute the gradient:
   `[im_grad, im_grad_dir] = imgradient(im);`
3. Show both images side-by-side:
   `figure; imshowpair(im, im_grad, 'montage');`
4. Threshold the gradient field to obtain an edge map (vary the threshold (0.5) to improve results):
   `b = imbinarize(rescale(im_grad), 0.5);`
5. Plot the gradient map and the threshold map side-by-side:
   `figure; imshowpair(im_grad, b, 'montage');`
6. Improve results by introducing a smoothing filter to the gradient map, or by applying morphological binary operations on the binary image (such as `imdilate()`)

Fig. HW11-7. Sample results of edge detection after thresholding and binary dilation.

Problem 3. Wavelet signal time-frequency decomposition

Use the mp4 audio record from Problem 1(b). Plot the signal in the time domain, and then give its wavelet decomposition graph.

Solution:
A 3-level wavelet decomposition is performed using the order 2 Daubechies wavelet through `wavedec` command in Matlab:

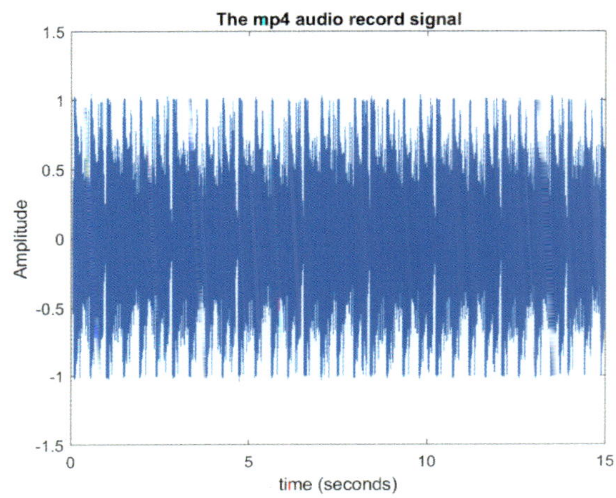

Fig. HW11-8. The time domain signal of the mp4 audio record from Problem 1.

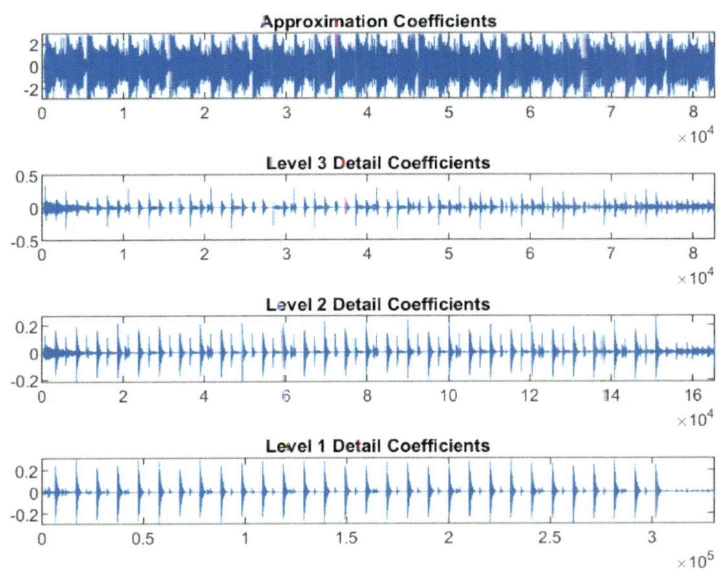

Fig. HW11-9. The wavelet decomposition of the mp4 audio record from Problem 1 using Daubechies wavelet.

Instead of Daubechies wavelet, other wavelets such as Haar wavelet can be used for wavelet decomposition depending on the characteristics of the original signal.

Matlab code:

```
[y_Boom, Fs] = audioread('Boom.mp4');
figure;
t = (1:length(y_Boom))/Fs;
plot(t, y_Boom)
xlabel('time (seconds)')
ylabel('Amplitude')
title('The mp4 audio record signal')
xlim([0 15])

[c,l] = wavedec(y_Boom,3,'db2');
approx = appcoef(c,l,'db2');
[cd1,cd2,cd3] = detcoef(c,l,[1 2 3]);

figure;
subplot(4,1,1)
plot(approx)
title('Approximation Coefficients')
xlim([0 length(approx)])

subplot(4,1,2)
plot(cd3)
title('Level 3 Detail Coefficients')
xlim([0 length(cd3)])

subplot(4,1,3)
plot(cd2)
title('Level 2 Detail Coefficients')
xlim([0 length(cd2)])

subplot(4,1,4)
plot(cd1)
title('Level 1 Detail Coefficients')
xlim([0 length(cd1)])
```

Problem 4. Basic NN problem

Predict the temperature and the humidity at a weather station in Massachusetts based on the dew temperature and air pressure. In particular, determine the dew point on Dec. 24, 2021, given the data of two days prior.

Solution:

For this purpose, we will train a feedforward network with 10 hidden layers. We use two days' worth of data, sampled once every minute

Setup:

data = thingSpeakRead(12397,'Fields',[2 3 4 6],'DateRange',[datetime('December 22, 2021'), datetime('December 23, 2021')], 'outputFormat','table');
tempC = (5/9)*(data.TemperatureF-32); % convert to degC
b = 17.62; % water vapor
c = 243.5; % barometric pressure
gamma = log(data.Humidity/100) + b*tempC ./ (c+tempC); % ideal gas ratio of specific heats
dewPointC = c*gamma ./ (b-gamma);
inputs = [dewPointC'; data.PressureHg'];
targets = [tempC'; data.Humidity'];

Now, define a feedforward net using the command net = feedforwardnet(10);
Training is straightforwardly done using [net,tr] = train(net,inputs,targets);
Now, predict the temperature and the humidity given that the forecast says 0 degrees Celsius and 29.8 inHg for Dec. 24, 2021. This can be done using net([0; 29.8]);
See if you can improve the performance of your network by introducing additional hidden layers, as well as more past data. Which of the two methods produces a lower validation error (perhaps both combined)? Report your validation and regression performance as shown below:

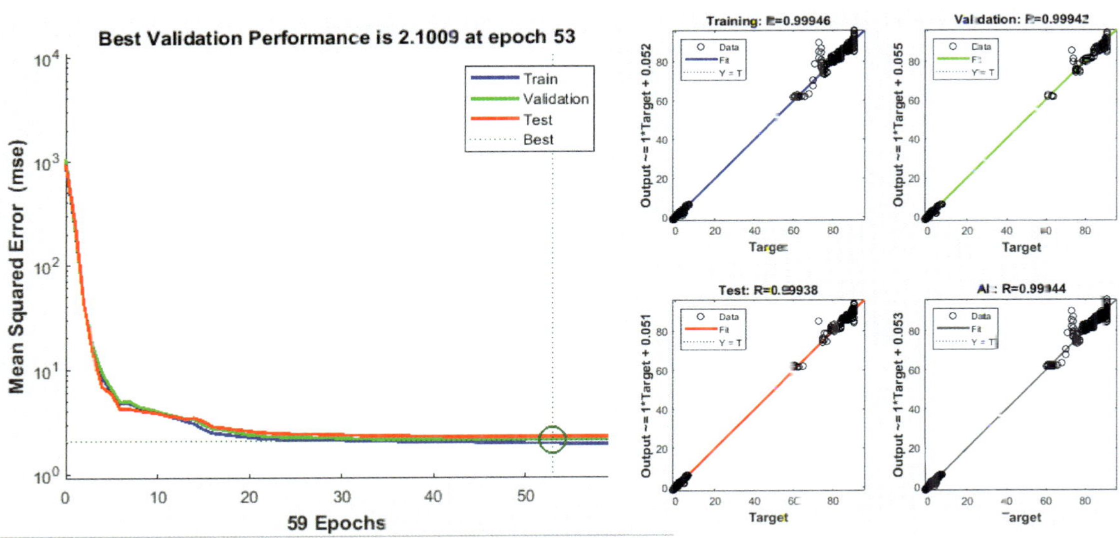

Fig. HW11-10. Sample training validation results.

Problem 5. Basic RL problem

We consider an agent in a grid world with the following rules:

1. The grid world is 5-by-5 and bounded by borders, with four possible actions (North = 1, South = 2, East = 3, West = 4).
2. The agent begins from cell [2,1] (second row, first column).
3. The agent receives a reward +10 if it reaches the terminal state at cell [5,5] (blue).
4. The environment contains a special jump from cell [2,4] to cell [4,4] with a reward of +5.
5. The agent is blocked by obstacles (black cells).
6. All other actions result in –1 reward.

The grid world is shown in the figure below:

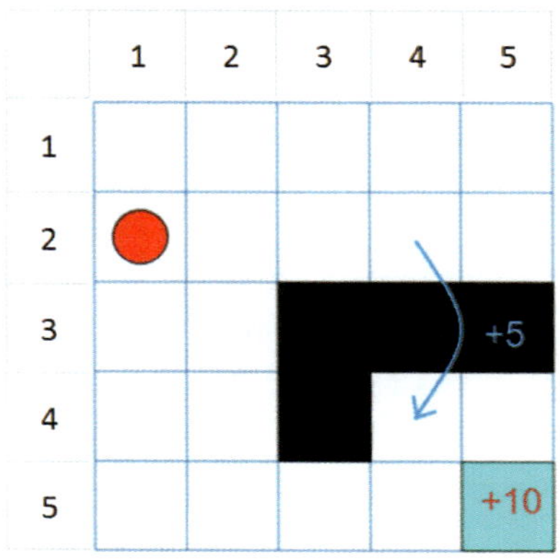

Fig. HW11-11. Grid world example configuration.

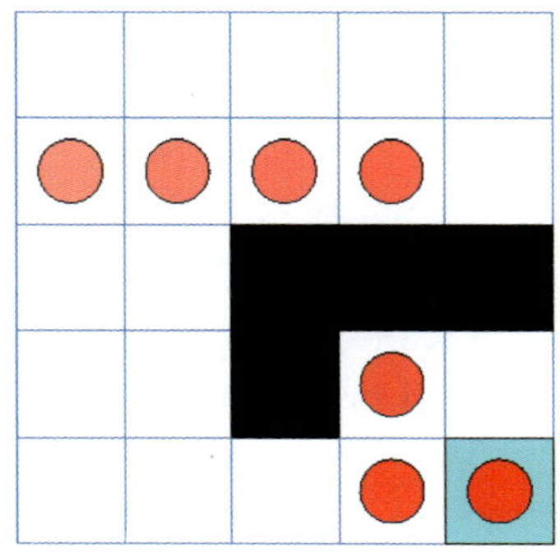

Fig. HW11-12. Example system run.

We load the environment as follows:

```
env = rlPredefinedEnv("BasicGridWorld");
env.ResetFcn = @() 2;
```

Now, specify the Q-learning settings as follows:

```
% Q-learning settings
qTable = rlTable(getObservationInfo(env),getActionInfo(env));
qRepresentation =
rlQValueRepresentation(qTable,getObservationInfo(env),getActionInfo(env));
qRepresentation.Options.LearnRate = 1;
```

Assignment a) Experiment with different learning rates and report any changes in performance.

We initialize the agent settings:

```
% Agent settings
agentOpts = rlQAgentOptions;
agentOpts.EpsilonGreedyExploration.Epsilon = .04;
qAgent = rlQAgent(qRepresentation,agentOpts);
```

We set the training options and commence training as follows:

```
% Training
trainOpts = rlTrainingOptions;
trainOpts.MaxStepsPerEpisode = 50;
trainOpts.MaxEpisodes= 200;
trainOpts.StopTrainingCriteria = "AverageReward";
trainOpts.StopTrainingValue = 11;
trainOpts.ScoreAveragingWindowLength = 30;

trainingStats = train(qAgent,env,trainOpts);
```

Assignment b) Vary the training parameters and report changes in overall performance (e.g., `MaxStepsPerEpisode`, and `StopTrainingValue`)

Validate your performance as follows, and report one of your system runs.

```
% Validation
plot(env)
env.Model.Viewer.ShowTrace = true;
env.Model.Viewer.clearTrace;

sim(qAgent,env)
```

Assignment c) Change your initial state (`env.ResetFcn = @() N; % N in [1, 25]`) and report if your system still runs as expected. Comment on whether adding random initial states would improve performance. In each of the above assignments, report your episode manager training progress (average episode reward).

Problem 6. Use of Wavenet for system identification

1) Repeat simulation of the Goodwin equation of Example 10.5-1, but with PRBS input with sampling interval 0.02 sec.
2) Repeat the tasks of Example 10.5-1 using the input/output data from task 1) above.
3) Linearize the NARX from Example 10.5-1 and from 2) above using Matlab.

Solution:

1) Instead of sine wave input, PRBS input is injected into the system.
Matlab code:

```
input = frest.PRBS('Order',12,'NumPeriods',1,'Amplitude',37,'Ts',0.02);
Input = generateTimeseries(input);
```

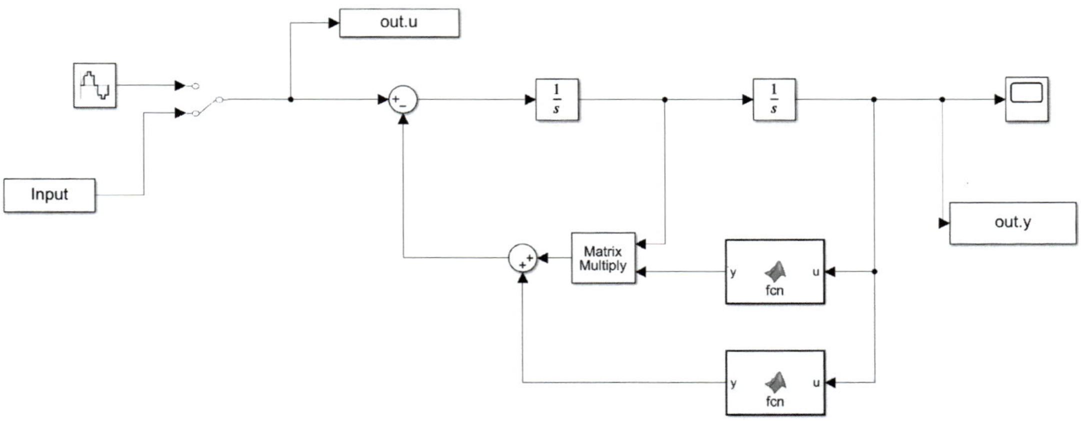

Fig. HW11-13. Implementation of the Goodwin equation with PRBS input in Simulink.

A different output signal is measured than in the case of the sine wave input:

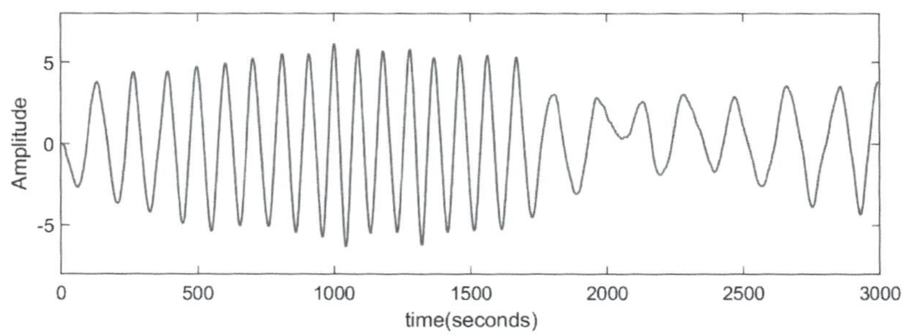

Fig. HW11-14. The measurements of the repeated simulation of the Goodwin equation with PRBS input.

2) In the same way as in Example 10.5-1,
Matlab code:

```
ze = iddata(out.y(1:1000), out.u(1:1000), 1);
z = iddata(out.y, out.u, 1);
w = wavenet;
m = nlarx(ze, [2 1 1], w);
yp = predict(m, z, 1);
```

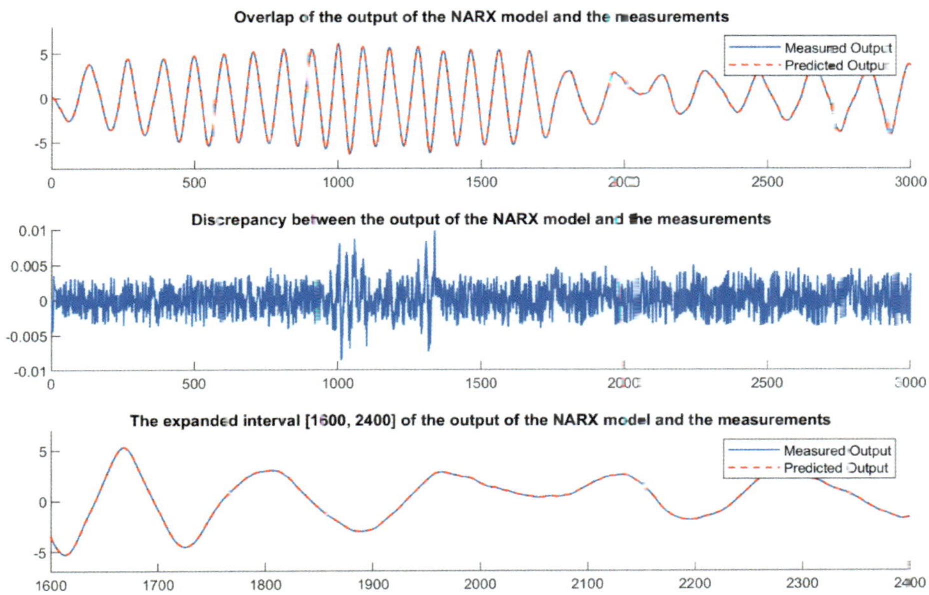

Fig. HW11-15. a) Overlap of the output of the NARX model and the measurements with PRBS input, b) Discrepancy between the output of the NARX model and the measurements with PRBS input, c) The expanded interval [1600, 2400] of the output of the NARX model and the measurements with PRBS input.

3) Matlab provides the `linearize` function for deriving a linear model from a NARX model.

```
lti = linearize(m, 0, [0 ,0 ,0]);
```

For a zero state and zero input, the following discrete time transfer functions are derived:

```
H = filt(lti.Numerator,lti.Denominator)
```

With a sine wave input,

$$\frac{Y(z)}{U(z)} = \frac{0.003664z^{-1}}{1-2.528z^{-1}+1.571z^{-2}} \Leftrightarrow \frac{B(q^{-1})}{A(q^{-1})} = \frac{0.003664q^{-1}}{1-2.528q^{-1}+1.571q^{-2}}$$
$$\Leftrightarrow (1-2.528q^{-1}+1.571q^{-2})y(t) = 0.003664u(t-1).$$

With a PRBS input,

$$\frac{Y(z)}{U(z)} = \frac{0.001094z^{-1}}{1-1.982z^{-1}+0.9889z^{-2}} \Leftrightarrow \frac{B(q^{-1})}{A(q^{-1})} = \frac{0.001094q^{-1}}{1-1.982q^{-1}+0.9889q^{-2}}$$
$$\Leftrightarrow (1-1.982q^{-1}+0.9889q^{-2})y(t) = 0.001094u(t-1).$$

Problem 7. CNN for handwriting recognition

This example deals with classifying the MNIST handwritten digit database using a CNN. Setup your environment by loading the dataset:

```
digitDatasetPath = fullfile(matlabroot,'toolbox','nnet','nndemos', ...
    'nndatasets','DigitDataset');

imds = imageDatastore(digitDatasetPath, ...
    'IncludeSubfolders',true, ...
    'LabelSource','foldernames');

numTrainFiles = 750;
[imdsTrain,imdsValidation] = splitEachLabel(imds,numTrainFiles,'randomize');

inputSize = [28 28 1];
numClasses = 10;
```

You may experiment with allotting a different number of training files. Too many training files will result in overfitting, whereas too few will result in underfitting. We proceed to setup the network layers. A rectified linear unit (ReLU, $x \mapsto \max(0, x)$) is used as the nonlinearity, preceded by a normalization layer and a convolution layer containing 20 filters of size 5 by 5. A fully connected layer followed by a soft maximum layer precedes the classification layer, the soft maximum layer being responsible for the final dimensionality reduction and nonlinearity prior to classification. The classification accuracy is printed at the end.

```
layers = [
    imageInputLayer(inputSize)
    convolution2dLayer(5,20)
    batchNormalizationLayer
    reluLayer
    fullyConnectedLayer(numClasses)
    softmaxLayer
    classificationLayer];

options = trainingOptions('sgdm', ...
    'MaxEpochs',4, ...
    'ValidationData',imdsValidation, ...
    'ValidationFrequency',30, ...
    'Verbose',false, ...
    'Plots','training-progress');
```

```
net = trainNetwork(imdsTrain,layers,options);

YPred = classify(net,imdsValidation);
YValidation = imdsValidation.Labels;
accuracy = mean(YPred == YValidation)
```

Show your prediction accuracy and your training report. Also report the effect of each nonlinearity (i.e., what happens if you remove it), and try to add a second convolution layer with ReLU (with and without normalization). Discuss if any of these changes results in improved classification performance.

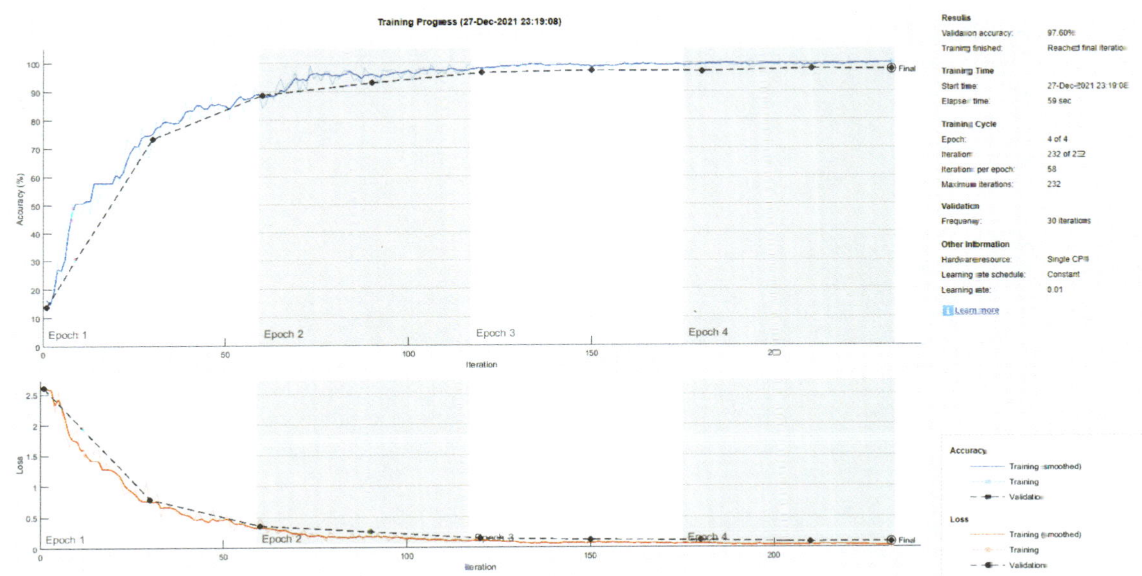

Fig. HW11-16. Sample training progress report.

References:

1. Jeon, H., Y. Jung, S. Lee, and Y. Jung (2020). "Area-Efficient Short-Time Fourier Transform Processor for Time–Frequency Analysis of Non-Stationary Signals," *Appl. Sci.*, 10, 7208 http://dx.doi.org/10.3390/app10207208.
2. Billings, S. A. and H.-L. Wei (2005), "A New Class of Wavelet Networks for Nonlinear System Identification," *IEEE Trans. Neur. Networks*, v. 16, no. 4, pp. 862–874.
3. Zhang, S., C. W. Taft, J. Bentsman, A. Hussey, and B. Petrus (2012). "Simultaneous Gains Tuning in Boiler/Turbine PID-Based Controller Clusters Using Iterative Feedback Tuning Methodology," *ISA Transactions*, 51:5, 609–621.

4. Zhang, S., D. Ye, J. Bentsman, C. W. Taft, and A. Hussey (2010). "Assessment of Global Optimizers: Particle Swarm Optimization, Simulated Annealing, and Genetic Algorithms in Local Simultaneous Multi-loop Tuning of PID Gains," *Proceeding of the 53rd Annual ISA POWID Symposium*, Summerlin, NV, June 7–10, 2010.
5. Zhang, S., J. Bentsman, X. Lou, C. Neuschaefer, Y. Lee, and H. El-Kebir (2020), "Multiresolution GPC-Structured Control of a Single-Loop Cold-Flow Chemical Looping Testbed," *Energies*, 13, 1759, 1–28, 2020. https://doi.org/10.3390/en13071759.
6. Cheng. Y., P. Zhao, M. Gandhi, B. Li, E. Theodorou, and N. Hovakimyan (2021). "Robustifying Reinforcement Learning Policies with L1 Adaptive Control," arXiv:2106.02249v4 [cs.LG] 9 Dec 2021. To appear in *Proceedings of 4th Learning for Dynamics and Control Conference*, 2022.
7. Koric, S. and D. W. Abueidda (2021). "Deep Learning Sequence Methods in Multiphysics Modeling of Steel Solidification," *Metals*, 11, 494. https://doi.org/10.3390/met11030494.

Appendix A

Useful Mathematical Formulas

A.1 Euler's Identity

$$e^{\pm j\theta}=\cos\theta\pm j\sin\theta;\ e^{\pm jk\pi}=\cos(k\pi);\ e^{\pm j\frac{\pi}{2}}=\pm j$$

$$\cos\theta=\frac{e^{j\theta}+e^{-j\theta}}{2};\ \sin\theta=\frac{e^{j\theta}-e^{-j\theta}}{2j};\ A\angle\pm\theta=Ae^{\pm j\theta}$$

A.2 Trigonometric Identities

$$\sin\alpha=\cos\left(\alpha-\frac{\pi}{2}\right);\ \cos\alpha=\sin\left(\alpha+\frac{\pi}{2}\right)$$

$$\sin 2\alpha=2\sin\alpha\cos\alpha;\ \cos 2\alpha=2\cos^2\alpha-1=1-2\sin^2\alpha=\sin^2\alpha-\cos^2\alpha$$

$$\sin(\alpha\pm\beta)=\sin\alpha\cos\beta\pm\cos\alpha\sin\beta;\ \cos(\alpha\pm\beta)=\cos\alpha\cos\beta\pm\sin\alpha\sin\beta$$

$$\sin\alpha\cos\beta=\frac{[\sin(\alpha+\beta)+\sin(\alpha-\beta)]}{2};\ \cos\alpha\sin\beta=\frac{[\sin(\alpha+\beta)-\sin(\alpha-\beta)]}{2}$$

$$\cos\alpha\cos\beta=\frac{[\cos(\alpha+\beta)+\cos(\alpha-\beta)]}{2};\ \sin\alpha\sin\beta=\frac{[\cos(\alpha+\beta)-\cos(\alpha-\beta)]}{2}$$

A.3 Indefinite Integrals

$$\int\sin(ax)\,dx=\frac{-\cos(ax)}{a};\ \int\cos(ax)\,dx=\frac{\sin(ax)}{a}$$

$$\int x^n dx=\frac{x^{n+1}}{n+1};\ \int\frac{1}{x}dx=\ln(x);\ \int\frac{1}{a^2+b^2x^2}dx=\frac{1}{ab}\tan^{-1}\left(\frac{bx}{a}\right)$$

$$\int x\sin(ax)\,dx=\frac{\sin(ax)-ax\cos(ax)}{a^2};\ \int x\cos(ax)\,dx=\frac{\cos(ax)+ax\sin(ax)}{a^2}$$

$$\int e^{-ax}dx = \frac{-e^{-ax}}{a};\ \int xe^{-ax}dx = \frac{-(1+ax)e^{-ax}}{a^2};\ \int x^2e^{-ax}dx = \frac{-\left(2+2ax+a^2x^2\right)e^{-ax}}{a^3}$$

$$\int \mathrm{e}^{-ax}\sin(bx)dx = -e^{-ax}\left[\frac{a\sin(bx)+b\cos(bx)}{a^2+b^2}\right];\ \int x\cos(ax)dx = \frac{\cos(ax)+ax\sin(ax)}{a^2}$$

A.4 Integration Properties

$$\int kf(x)dx = k\int f(x)dx, k \text{ is a constant.}$$

$$\int\left[f(x)\pm g(x)\right]dx = \int f(x)dx \pm \int f(x)dx$$

$$\int f(u)du = \int f\left[\varphi(x)\right]d\varphi(x) = \int f\left[\varphi(x)\right]\varphi'(x)dx,\ u=\varphi(x)$$

$$\int u(x)v'(x)dx = u(x)v(x) - \int v(x)u'(x)dx$$ **(Integration by parts)**

Example A-1: $\int e^x \cos x dx$

Solution: $\int e^x\cos x dx = \int e^x(\sin x)'dx = e^x\sin x - \int \sin x\left(e^x\right)'dx$

$$= e^x\sin x - \int \sin x e^x dx = e^x\sin x - \left(e^x(-\cos x) - \int(-\cos x)e^x dx\right)$$

$$= e^x\sin x + e^x\cos x - \int \cos x e^x dx$$

$$= e^x\sin x + e^x\cos x - \int e^x\cos x dx$$

Therefore, $\int e^x\cos x dx = \dfrac{e^x\sin x + e^x\cos x}{2} + C$

A.5 Definite Integrals Over $(0, \infty)$

$$\int_0^\infty \frac{\sin(ax)}{ax}dx = \frac{\pi}{2|a|};\ \int_0^\infty \frac{\sin^4(ax)}{(ax)^4}dx = \frac{\pi}{3|a|}$$

$$\int_0^\infty e^{-ax}dx = \frac{1}{a};\ \int_0^\infty xe^{-ax}dx = \frac{1}{a^2};\ \int_0^\infty x^n e^{-ax}dx = \frac{n!}{a^{n+1}}$$

$$\int_0^\infty e^{-ax}\cos(bx)dx = \frac{a}{a^2+b^2};\ \int_0^\infty\left[e^{-ax}\cos(bx)\right]^2dx = \frac{a}{4}\left(\frac{1}{a^2}+\frac{1}{a^2+b^2}\right)$$

$$\int_0^\infty e^{-ax}\sin(bx)dx = \frac{b}{a^2+b^2};\ \int_0^\infty \left[e^{-ax}\sin(bx)\right]^2 dx = \frac{a}{4}\left(\frac{1}{a^2}-\frac{1}{a^2+b^2}\right)$$

A.6 Differentiation Formulas

$$\frac{d}{dx}(x)=1;\ \frac{d}{dx}\left(a^x\right)=a^x\ln(a);\ \frac{d}{dx}\left(e^x\right)=e^x$$

$$\frac{d}{dx}(\ln(x))=\frac{1}{x},\ x>0;\ \frac{d}{dx}(\ln|x|)=\frac{1}{x},\ x>0;\ \frac{d}{dx}(\log_a(x))=\frac{1}{x\ln(a)}$$

$$\frac{d}{dx}(\sin x)=\cos x;\ \frac{d}{dx}(\cos x)=-\sin x;\ \frac{d}{dx}\left(\sin^{-1}x\right)=\frac{1}{\sqrt{1-x^2}};\ \frac{d}{dx}\left(\cos^{-1}x\right)=-\frac{1}{\sqrt{1-x^2}}$$

$$\frac{d}{dx}(\tan x)=\sec^2 x;\ \frac{d}{dx}(\cot x)=-\csc^2 x;\ \frac{d}{dx}\left(\tan^{-1}x\right)=\frac{1}{1+x^2}$$

$$\frac{d}{dx}(\sec x)=\sec x\cdot\tan x;\ \frac{d}{dx}(\csc x)=-\csc x\cdot\cot x$$

$$\frac{d}{dx}\int_a^x f(t)dt=f(x)$$

$(f\circ g)'(c)=f'(g(c))\cdot g'(c)$ **(Chain rule)**

Example A-2: $\left((2x+1)^3\right)'$

Solution: $\left((2x+1)^3\right)'=3(2x+1)^2\cdot(2x-1)'=6(2x+1)^2$

A.7 Finite Sums

$$\sum_{k=1}^{N}k=\frac{1}{2}N(N+1);\ \sum_{k=1}^{N}k^2=\frac{1}{6}N(N+1)(2N-1);\ \sum_{k=1}^{N}k^3=\frac{1}{4}N^2(N+1)^2;\ \sum_{k=0}^{N}\alpha^k=\begin{cases}\dfrac{1-\alpha^{N+1}}{1-\alpha}, & \alpha\neq 1\\ N+1, & \alpha=1\end{cases}$$

$$\sum_{k=1}^{N}\sin(kx)=\begin{cases}\dfrac{\sin(0.5Nx)}{\sin(0.5x)}\sin\left[\dfrac{1}{2}(N+1)x\right], & x\neq 2m\pi\ (m=0,1,2,\ldots)\\ 0, & x=2m\pi\ (m=0,1,2,\ldots)\end{cases}$$

$$\sum_{k=1}^{N}\cos(kx)=\begin{cases}\dfrac{\sin(0.5Nx)}{\sin(0.5x)}\cos\left[\dfrac{1}{2}(N+1)x\right], & x\neq 2m\pi\ (m=0,1,2,\ldots)\\ N, & x=2m\pi\ (m=0,1,2,\ldots)\end{cases}$$

$$\sum_{k=-N}^{N}e^{j2\pi k\alpha}=\frac{\sin[(2N+1)\pi\alpha]}{\sin(\pi\alpha)}=(2N+1)\frac{\operatorname{sin}c[(2N+1)\alpha]}{\operatorname{sin}c(\alpha)}$$

$$\sum_{k=1}^{N}k\alpha^k=\begin{cases}\dfrac{\alpha}{(1-\alpha)^2}\left[1-(N+1)\alpha^N+N\alpha^{N+1}\right], & \alpha\neq 1\\ \dfrac{1}{2}N(N+1), & \alpha=1\end{cases}$$

A.8 Sequences and Series

$$\sum_{k=1}^{\infty}\frac{1}{k^2}=\frac{\pi^2}{6};\ \sum_{k=1,odd}^{\infty}\frac{1}{k^2}=\frac{\pi^2}{8};\ \sum_{k=1}^{\infty}\frac{1}{k^4}=\frac{\pi^4}{90};\ \sum_{k=1,odd}^{\infty}\frac{1}{k^4}=\frac{\pi^4}{96};\ \sum_{k=-\infty}^{\infty}e^{-\alpha|k|}=\frac{1+e^{-\alpha}}{1-e^{-\alpha}},\ \alpha>0$$

$$\sum_{k=0}^{\infty}\alpha^k=\frac{1}{1-\alpha},\ |\alpha|<1;\ \sum_{k=1}^{\infty}\alpha^k=\frac{\alpha}{1-\alpha},\ |\alpha|<1;\ \sum_{k=1}^{\infty}k\alpha^k=\frac{\alpha}{(1-\alpha)^2},\ |\alpha|<1;\ \sum_{k=1}^{\infty}k^2\alpha^k=\frac{\alpha^2+\alpha}{(1-\alpha)^3},\ |\alpha|<1$$

$$\sum_{k=1}^{\infty}\alpha^k\sin(kx)=\frac{\alpha\sin(x)}{1-2\alpha\cos(x)+\alpha^2},\ |\alpha|<1;\ \sum_{k=1}^{\infty}\alpha^k\cos(kx)=\frac{\alpha\cos(x)-\alpha^2}{1-2\alpha\cos(x)+\alpha^2},\ |\alpha|<1$$

$$\sum_{k=1}^{\infty}\frac{\sin(kx)}{k}=\frac{1}{2}(\pi-x),\ 0<x<2\pi;\ \sum_{k=1}^{\infty}\frac{\sin^2(kx)}{k^2}=\frac{1}{2}x(\pi-x),\ 0<x<2\pi$$

A.9 Series Expansions

$$e^x=1+x+\frac{1}{2!}x^2+\frac{1}{3!}x^3+\cdots,\ |x|<1$$

$$\ln x=2\left(\frac{x-1}{x+1}\right)+\frac{2}{3}\left(\frac{x-1}{x+1}\right)^3+\frac{2}{5}\left(\frac{x-1}{x+1}\right)^5+\cdots,\ x>0$$

$$\ln(1+x)=x-\frac{1}{2}x^2+\frac{1}{3}x^3-\frac{1}{4}x^4+\cdots=-\sum_{n=1}^{\infty}\frac{(-x)^n}{n},|x|<1$$

$$\frac{\sin x}{x}=1-\frac{1}{3!}x^2+\frac{1}{5!}x^4-\frac{1}{7!}x^6+\cdots$$

$$(1+x)^{\pm n}=1\pm nx\pm\frac{1}{2!}n(n-1)x^2\pm\frac{1}{3!}n(n-1)(n-2)x^3\pm\cdots,\ |nx|<1$$

Appendix B

System Classification

B.1 Systems

"Naïve" Definition: A system is a combination of components acting together to achieve a specific objective. Usually a system is represented by a "black box" with input, output, and some internal states:

Fig. B-1. System expressed as black box.

or simply input → output. It depends on the designer how much to put into this box: electric motor, solar system, or the thinking and decision processes of a human being. Each will produce an output for some given inputs. To predict the behavior of a system, i.e., to assess beforehand what outputs will result from specific inputs; we must have its mathematical (analytical and/or software) model. Without external unknown disturbances, our resulting prediction will be as accurate as our model itself.

Notation:

- in control literature
  $u(t)$ — input,
  $y(t)$ — output,
  $x(t)$ — internal states;
- also could be
  $u(t)$ or $p(t)$ — input
  $x(t)$ — output.

Unless otherwise stated, we will use the first notation. We can think of a system in two ways:

- physical — as a "digestive machine" which consumes a specific input and generates a specific output;
- abstract — as a rule which, given a set (collection) of inputs and a set of outputs, matches a specific input with a specific output. This rule can be given by a table or by a mathematical operator.

Example B-1: Here our "rule" is the following procedure: $p(t) \to \int_0^t p(\tau)d\tau \to x(t)$:

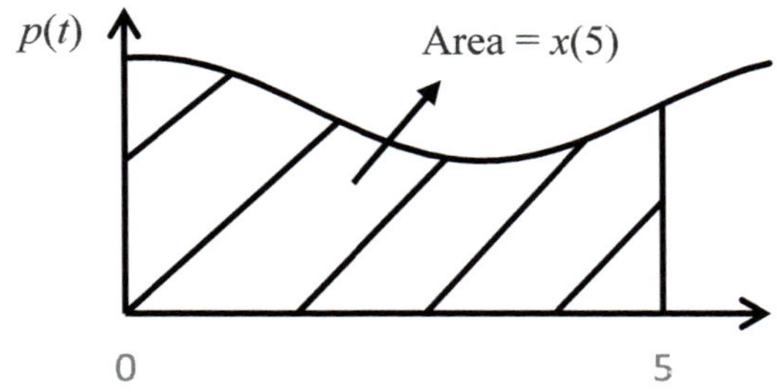

Fig. B-2. Signal $p(t)$ for Example B-1.

Solution: Fix t, say, $t = 5$. This means that the dummy variable τ takes values from 0 to 5, i.e., we must compute the area under $p(t)$, $0 \le t \le 5$, to find $x(5)$, Fix another t, say $t = 6$. Repeat computation of area under $p(t)$, $0 \le t \le 6$, and obtain $x(6)$; Repeat the procedure for as many values of t as desired to get a clear picture of $x(t)$.

Thus, we have constructed $x(t)$ for a given $p(t)$ using our rule (remember: this integral is a real number for every fixed t and does not depend on dummy variable τ).

B.2 System Classification

B.2.1 *Continuous time versus discrete time systems*

Continuous Time: independent variable, t, runs through all values on the real line between a and b:

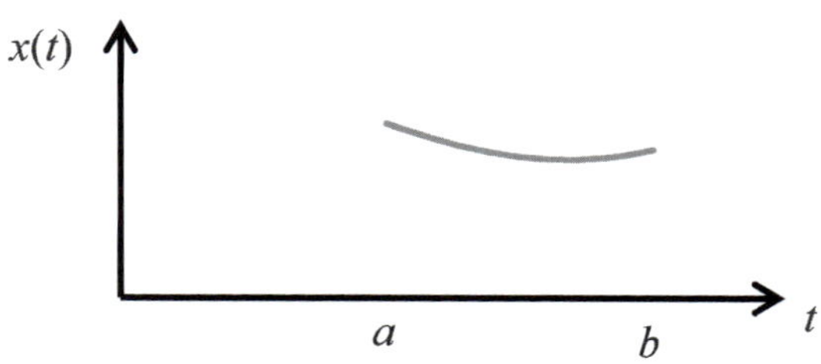

Fig. B-3. Continuous time signal $x(t)$.

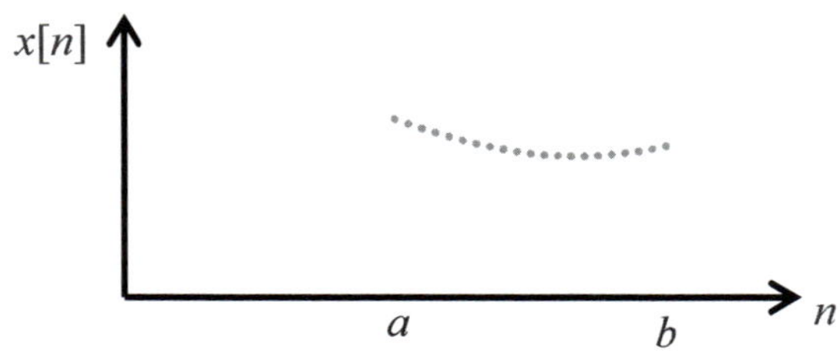

Fig. B-4. Discrete time signal $x[n]$.

Discrete Time: independent variable, t, runs only through certain separate values t_i ($i = 1, 2, 3, \ldots$) on the real line; usually these are integers, and t is replaced by n or k. Between integers $x[n]$ is not defined.

B.2.2 *Static versus dynamic system*

A system is said to be dynamic if and only if (iff) its output at some $t = t_0$ depends on input for $t < t_0$. This means that (i) if we choose some input, say, $p_1(t)$ and compute the corresponding output $x_1(t)$ at $t = t_0$, (ii) then choose another input $p_2(t)$, which at some time $t_1 < t_0$ is different from $p_1(t)$, and again compute the corresponding output $x_2(t)$ at $t = t_0$, and finally (iii) compare $x_1(t_0)$ and $x_2(t_0)$ and find them not equal to each other, then at time $t = t_0$ the system remembers the difference in inputs which occurred only once at $t_1(< t_0)$, i.e., system indeed depends upon past values of input and, hence, is dynamic (with memory).

A system which is not dynamic is called static (memoryless) Its output depends only upon the present (or the future) values of the input. Mathematically, "memory" can be represented by the integrals (or sums) of the input or by the derivatives of the output.

Example B-2: Determine if a system $x(t) = \int_0^t p(\tau)\,d\tau$ is dynamic or static.

Solution: Fix $t = 5$ and consider two inputs:

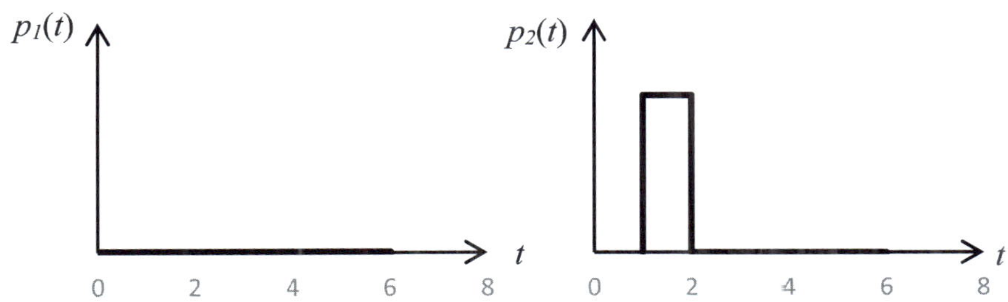

Fig. B-5. Signal $p_1(t)$ and $p_2(t)$ for Example B-2.

We have $x_1(5) = \int_0^5 p_1(\tau)d\tau = \int_0^5 (0)d\tau = 0$ and $x_2(5) = \int_0^5 p_2(\tau)d\tau = 1$. Obviously, $p_1(5) = p_2(5) = 0$, but $x_1(5) \neq x_2(5)$, i.e., at $t = 5$ the system remembers the difference in the inputs at $t \in [1,2]$.

Example B-3: $\dot{x}(t) = \dfrac{dx}{dt} = p(t)$ - dynamic, (same as $x(t) = x(t_0) + \int_{t_0}^{t} p(\tau)\,d\tau$).

$$\text{Indeed } \dot{x}(t) = \frac{dx(t)}{dt} = \frac{dx(t_0)}{dt} + \frac{d\left(\int_{t_0}^{t} p(\tau)d\tau\right)}{dt} = p(t).$$

Example B-4: Systems: $x(t) = k \cdot p(t)$, $k =$ constant, and $x(t) = \dot{p}(t) - \ddot{p}(t)$ are static.

System $x(n) = \sum_{k=-\infty}^{n} p(k)$ is dynamic (summation memorizes the function shape, just like integration).

If the equation contains the time derivatives of the output or the integrals of the input, the system is dynamic.

B.2.3 *Time varying versus time invariant systems*

A system is said to be time invariant (fixed, shift invariant, stationary, autonomous) iff a time shift in the input causes the output to be the time-shifted replica of the original output, i.e., iff $p(t) \rightarrow x(t) \Rightarrow p(t-t_0) \rightarrow x(t-t_0)$. Otherwise, the system is time varying.

Example B-5: Determine if system $x(t) = t \bullet p(t)$ is time invariant.

Solution: Consider input $p_1(t) = \begin{cases} 0, t \neq 2, \\ 1, t = 2, \end{cases}$

and its shifted by one replica $p_2(t) = p_1(t-1) \Rightarrow p_2(t) = \begin{cases} 0,\ t \neq 3, \\ 1,\ t = 3, \end{cases}$

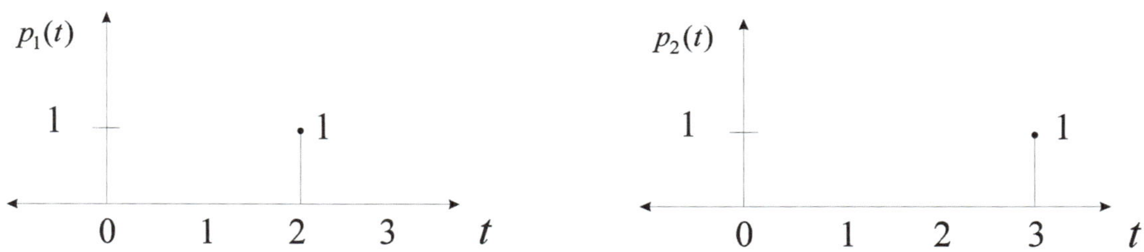

Fig. B-6. Signal $p_1(t)$ and shifted signal $p_2(t)$ for Example B-5.

The output for $p_1(t)$ is $x_1(t) = \begin{cases} 0,\ t \neq 2, \\ t \cdot p_1(t) = 2,\ t = 2. \end{cases}$

Thus, the property of time invariance demands that $x_2(t)$ for $p_2(t)$ be $x_2(t) = x_1(t-1)$. However, this is not the case since true $x_2(t)$ is

$$x_2(t) = \begin{cases} 0,\ t \neq 3, \\ t \cdot p_2(t) = 3,\ t = 3. \end{cases}$$

Hence, the system in time varying.

Example B-6: Is system $x(t) = t\int_0^t p(\tau)d\tau$ time invariant?

Solution: Consider an input

$$p_1(t) = \begin{cases} 0,\ t < 1,\ t > 2, \\ 1,\ 1 \leq t \leq 2, \end{cases}$$

and its shifted replica

$$p_2(t) = p_1(t-1) \Rightarrow p_2(t) = \begin{cases} 0, \ t < 2, \ t > 3, \\ 1, \ 2 \le t \le 3. \end{cases}$$

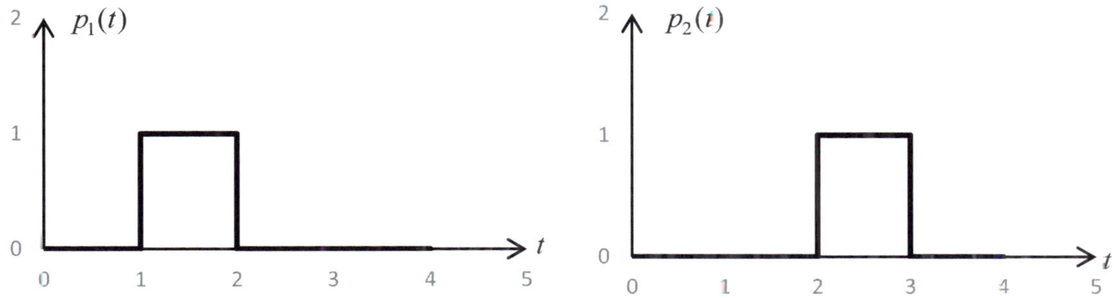

Fig. B-7. Signal $p_1(t)$ and shifted signal $p_2(t)$ for Example B-6.

The output $x_1(t)$ for the input $p_1(t)$ is:

$$x_1(t) = \begin{cases} t\int_0^t p_1(\tau)d\tau = t\int_1^t 1d\tau = t(\tau|_1^t) = t(t-1) = t^2 - t, \ t \in [1,2], \\ t, \ t > 2. \end{cases}$$

Note the lower limits of integration.
The output $x_2(t)$ for the input $p_2(t)$ is:

$$x_2(t) = \begin{cases} t\int_0^t p_2(\tau)d\tau = t\int_2^t 1d\tau = t^2 - 2t, \ t \in [2,3], \\ t, \ t > 3. \end{cases}$$

This system is time varying since $x_2(t) \ne x_1(t-1)$:

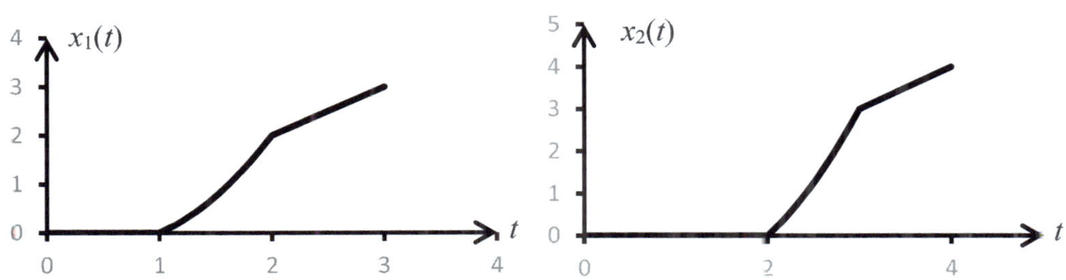

Fig. B-7. The outputs $x_1(t)$ and $x_2(t)$ for the input signals $p_1(t)$ and shifted signal $p_2(t)$ for Example B-6.

If the coefficients in front of $x(t)$, $p(t)$, and/or their integrals or derivatives are varying functions of time, system is time varying; if coefficients are constants, system is time invariant.

B.2.4 *Linear versus nonlinear systems*

Let us denote $x_1(t)$ as system response to $p_1(t)$ and $x_2(t)$ as system response to $p_2(t)$. A system is said to be linear iff $a_1x_1(t) + a_2x_2(t)$ is the system response to $a_1p_1(t) + a_2p_2(t)$, i.e.

$$\left.\begin{array}{l} p_1 \to x_1 \\ p_2 \to x_2 \end{array}\right\} \underset{linear}{\Rightarrow} a_1p_1 + a_2p_2 \to a_1x_1 + a_2x_2.$$

Linearity can also be defined in terms of system operator $H(\cdot)$ that maps input into output. Denoting a system as a transformation H of input $p(t)$ into output $x(t)$: $x(t) = H(p(t))$, so that $x_1(t) = H[p_1(t)]$ and $x_2(t) = H[p_2(t)]$, a system H is linear iff

$$H[a_1p_1(t) + a_2p_2(t)] = a_1\underbrace{H[p_1(t)]}_{x_1(t)} + a_2\underbrace{H[p_2(t)]}_{x_2(t)}.$$

The above property is called ***superposition***. This is an extremely important property, which immensely simplifies system analysis and design.

Example B-7: Determine whether the following systems are linear: (1) $x(t) = kp(t)$, k is constant; (2) $x(t) = \int_{-\infty}^{t} p(\tau)d\tau$; (3) $x(t) = ap(t) + b$; a,b is constant; (4) $x(t) = kp(t)^2$, k is constant.

Solution:

(1) Let $x_1 = kp_1$ (independent variable t will be often omitted) and $x_2 = kp_2$. Also define $p = a_1p_1 + a_2p_2$. Then, system (1) is linear, because

$$x = kp = k(a_1p_1 + a_2p_2) = ka_1p_1 + ka_2p_2 = a_1x_1 + a_2x_2$$

so that the linearity requirement is satisfied.

(2) Denoting $x(t) = H(p(t)) = \int_{-\infty}^{t} p(\tau)d\tau$, system (2) is also linear because

$$H[a_1p_1(t) + a_2p_2(t)] = \int_{-\infty}^{t} [a_1p_1(\tau) + a_2p_2(\tau)]d\tau = a_1\int_{-\infty}^{t} p_1(\tau)d\tau + a_2\int_{-\infty}^{t} p_2(\tau)d\tau$$
$$= a_1H[p_1(t)] + a_2H[p_2(t)].$$

(3) For the inputs p_1 and p_2 we have, respectively, the outputs $x_1 = ap_1 + b$ and $x_2 = ap_2 + b$. For the input $p = a_1p_1 + a_2p_2$ we have the output

$$x = a\left(a_1p_1 + a_2p_2\right) + b = a_1ap_1 + a_2ap_2 + b.$$

But for linearity we must have

$$x = a_1x_1 + a_2x_2 = a_1(ap_1 + b) + a_2(ap_2 + b) = a_1ap_1 + a_1b + a_2ap_2 + a_2b = a_1ap_1 + a_2ap_2 + (a_1 + a_2)b.$$

Clearly, for any a_1 and a_2 such that $a_1 + a_2 \neq 1$, system (3) is nonlinear! This happens due to offset b. Such systems are called ***incrementally linear*** or ***affine***, i.e., the difference between two responses for such systems is linear.

(4) System (4) is nonlinear, because

$$x = k\left(a_1 p_1 + a_2 p_2\right)^2 = k\left(a_1 p_1\right)^2 + k\left(a_2 p_2\right)^2 + 2ka_1 p_1 a_2 p_2.$$

Systems $(\dot{x})^2 = p$, $x(t) = \int_0^t p^2(\tau)d\tau$, $\dot{x}x = x\dot{p}$, and $x(n) = \sum_{k=-\infty}^{n} p^2(k)$ (k here is a dummy variable which runs from $-\infty$ to n) are all nonlinear. Systems $\ddot{x} + a_1(t)x = b_1(t)\dot{p} + bp$, $x(t) = a_1(t)\int_0^t p(\tau)d\tau$ are linear time varying.

If we have an offset or some powers of input and/or output or multiplications of input and output, a system is nonlinear.

B.2.5 *Causal versus noncausal systems*

A system is said to be causal (nonanticipatory) iff for any fixed $t = t_o$ the fact that inputs agree up to t_o implies that the outputs agree up to this same time t_o. Otherwise, system is noncausal (anticipatory).

Thus, given two inputs, p_1 and p_2 which satisfy $p_1 = p_2, t \in [0, t_0], p_1 \neq p_2, t > t_o$. Casual system would give two different outputs x_1 and x_2 which satisfy $x_1 = x_2$, $t \in [0, t_0]$, $x_1 \neq x_2, t > t_o$.

Noncausal system, however, at any time between *0* and t_o might produce different outputs for inputs p_1 and p_2, i.e., it will produce difference in the outputs before difference in the inputs will take place (just like dynamic system produces difference in the outputs after difference in the inputs takes place for a short time).

Example B-8: Is the system $x = H[p]$ given by $p(t) \rightarrow x(t) = [H(p)] = p(t+4)$, causal?
Solution: Consider two inputs

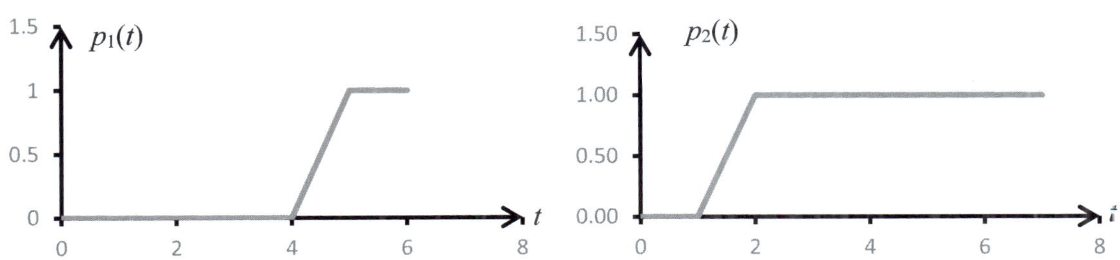

Fig. B-8. Input signal $p_1(t)$ and $p_2(t)$.

and the corresponding outputs

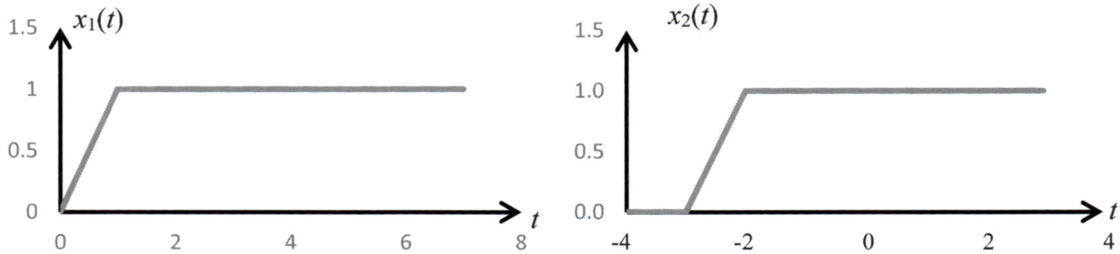

Fig. B-9. The corresponding output signals $x_1(t)$ and $x_2(t)$.

Now consider time interval $t \in [0,1]$. Obviously on this interval $p_1 = p_2$, i.e. inputs agree. Do outputs agree on $t \in [0,1]$? Clearly not, i.e. system anticipates (shows now) the difference which will happen in the future! Hence, the system is noncausal. This is clear also from "the common sense". Indeed, looking at $x(t) = p(t+4)$ we see that at time t (now) system shows the value of input which will happen 4 seconds later.

Example B-9: Is the system $x(n) = \sum_{k=-\infty}^{n} p(k+2)$ causal?

Solution: First we note that k is a dummy variable which runs from $-\infty$ to n, and that for any fixed n summation is a real number. Introduce substitution of dummy variable, $m = k+2$, so that $k = -\infty \Rightarrow m = -\infty$ and $k = n \Rightarrow m = n+2$. Thus, m runs from $-\infty$ to $n+2$ and we have $x(n) = \sum_{m=-\infty}^{n+2} p(m)$.

Now we clearly see that at time n the output $x(n)$ will contain the future values of the input $p(n+1)$ and $p(n+2)$. Hence, system is noncausal.

Example B-10: Check if system $x(t) = \int_{-\infty}^{t} p(\tau+2)d\tau$ is noncausal.

Solution: Since putting $\lambda = \tau + 2$, so that $\tau = -\infty$, $\Rightarrow \lambda = -\infty$, $\Rightarrow \tau = t \Rightarrow \lambda = t+2$ and $\tau = \lambda - 2 \Rightarrow d\tau = d(\lambda - 2) = d\lambda - d2 = d\lambda$. Now, $x(t) = \int_{-\infty}^{t+2} p(\lambda)d\lambda$, and again at time t the output $x(t)$ will have the values of the input from $p(t)$ to $p(t+2)$, i.e. system is noncausal.

A system is noncausal if it produces forward time shift or powers of independent variable of the input. It is causal if it produces backward time shift.

B.2.6 *Unstable versus stable systems*

There are many types of stability. One of the most important is a bounded-input-bounded-output (BIBO) stability.

A system is said to be BIBO stable iff for any bounded input it produces a bounded output. What is boundedness? A function $f(t), -\infty < t < \infty$, is bounded iff we can find such a fixed real number $M < \infty$ that $f(t) \le M, \forall t \in (-\infty, \infty)$ ($\forall t \in (-\infty, \infty)$ means for all t between $-\infty$ and ∞, not including $-\infty$ and ∞). For example, $f(t) = t, t \in (-\infty, \infty)$ is not a bounded function, since no matter how big M we choose, a larger *t* value may be found that is still less than infinity.

Thus, $p(t) < M = const < \infty \Rightarrow x(t) < N = const < \infty$ is a BIBO stable system. Otherwise, a system is BIBO unstable.

Example B-11: Is $x(t) = kp(t)$, $k =$ constant > 0, a stable system?

Solution: Let input be bounded, i.e. $p(t) < M, \forall t \in (-\infty, \infty)$. Multiply both sides of inequality $p(t) < M$ by *k*. Since $k > 0$, we have $kp(t) < kM$, i.e., "direction" of inequality is unchanged. But $kp(t)$ is, in fact, our output, *x*(*t*), therefore $x(t) = kp(t) < kM$. Noting that *k* and *M* are fixed real numbers, their product *kM* is also a fixed real number. Consequently x(*t*) is indeed bounded, and system $p(t) \to kp(t)$ is BIBO stable.

Example B-12: Is $p(t) \to \int_0^t p(\tau) d\tau$ a BIBO stable system?

Solution: First we note that to invalidate the property of BIBO stability it is enough to provide only one counter example, i.e., to find just one bounded input which produces an unbounded output. Take now $p(t) = a = const$. Then, $x(t) = \int_0^t p(\tau) d\tau = \int_0^t a d\tau = a\tau\Big|_0^t = at$. So system is not BIBO stable, since the output grows unbounded for any $t < \infty$.

Example B-13: Is system $x(n) = \sum_{k=0}^{n} p(k)$ stable?

Solution: The system is unstable, since setting $p(n) = a$, $n \ge 0$ yields $x(n) = \sum_{k=0}^{n} a = a \sum_{k=0}^{n} 1 = a(1+1+1+...+1) = a(n+1), n \in [0, \infty)$, which is unbounded.

B.2.7 *Lumped parameter versus distributed parameter systems*

Systems involving only time integrals or time derivatives are lumped parameter. They are usually described by ordinary differential equations (ODEs) or integral (time) equations. Systems involving spatial derivatives or integrals in addition to time derivatives are distributed parameter systems. They are described by partial differential equations (PDEs), integral equations, and functional equations. For example, one-dimensional heat equation $\frac{a^2 \cdot u(x,t)}{a \cdot x^2} = a^2 \frac{a \cdot u(x,t)}{a \cdot t}$ models distributed parameter process. This book uses Timoshenko beam model PDEs as the basis for the software testbed, as described in Chapter 1 and Appendices G and H.

B.2.8 *Stochastic versus deterministic systems*

Stochastic systems are those systems whose behavior is described not by the fixed values at every time moment but rather by random values with some probability distributions. Systems which are described by fixed numbers for every fixed time moment t are deterministic.

B.2.9 *Fuzzy versus exact systems*

Fuzzy systems try to model human perception. For example, "a tall man" is an imprecise notion. Fuzzy systems try to capture this imprecision by assigning "grades of membership" to physical objects.

Example B-14: Now that our classification is complete, let's classify a system.

$$x(n) = \sin(n) \sum_{k=-\infty}^{n} p^2(k+2).$$

Solution: This is:

- Discrete time system (obvious),
- Dynamic (summation involved, with upper limit at n and lower limit at time moment prior to n),
- Time varying, since $sin(n)$ is a time varying coefficient,
- Nonlinear, since square of input is involved,
- Noncausal, since by substitution we can show that at a time n system utilizes future values of input, $p(n+1)$ and $p(n+2)$,
- BIBO unstable, since taking $p(n) = 1$, the output grows without bound as n goes to ∞,
- Lumped parameter, since only summation of time is involved,
- Deterministic, for any input we can tell output at any moment n,
- Not fuzzy, no grades of membership shown.

B.2.10 *Linear operations in LTI systems*

It is convenient to introduce the differentiation operator $D = d / dt$, i.e.

$$Dp(t) \equiv \frac{d_p(t)}{dt}, D^n p(t) = \frac{d^n p(t)}{dt^n}$$

and integration operator $D^{-1} = \int_{-\infty}^{t} (\cdot) d\tau$, i.e.

$$D^{-1}p(t) = \int_{-\infty}^{t} p(\tau) d\tau.$$

Clearly $DD^{-1}p(t) = p(t)$, i.e. D^{-1} is the inverse of D. It is easy to check that operators D^{i}, $i = ..., -3, -2, -1, 0, 1, 2, 3, ... D^{-1}\{i\}$, $i = \ldots, -3, -2, -1, 0, 1, 2, 3, \ldots$ are linear, for example:

$$D(ap_1(t) + bp_2(t)) = aDp_1(t) + bDp_2(t).$$

For LTI system any linear operation (D, D^{-1}, multiplication by constant, summation, limit) on the input produces the same linear operation on the output, i.e.

$$p(t) \xrightarrow{LTI} x(t) \Rightarrow Dp(t) \xrightarrow{LTI} Dx(t) \Rightarrow D^{-1}p(t) \xrightarrow{LTI} D^{-1}x(t).$$

This happens because differentiation and integration are just the limiting cases of summation and multiplication, say

$$\frac{dp(t)}{dt} = \lim_{\Delta \to 0} \frac{p(t+\Delta) - p(t)}{\Delta}.$$

Let us use this rule in the following examples.

Example B-15: (the use of D^{-1} going through). Given the following circuit:

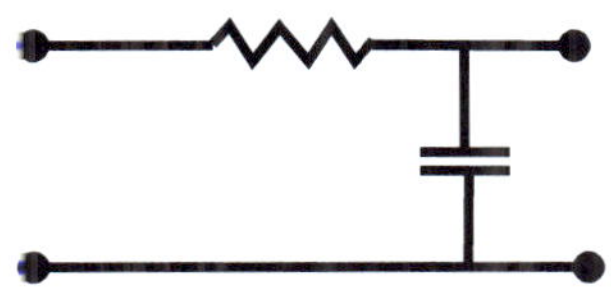

Fig. B-10. RC Circuits for Example B-15.

and the input signal $p(t) = r(t) - 2r(t-1) + r(t-2)$

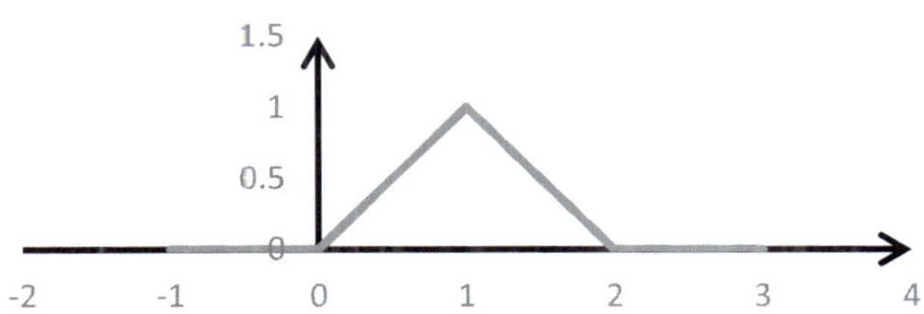

Fig. B-11. The input signal $p(t)$ in Example B-15.

find $x(t)$ using the response $y(t)$ to $1(t)$, i.e. using the step response.

Solution: First, we obtain the ramp response:

$$x_r(t) = \int_{-\infty}^{t} y(\tau)d\tau = \int_{-\infty}^{t} \left(1 - e^{-\tau/RC}\right)1(\tau)d\tau = \int_{0}^{t} \left(t - e^{-\tau/RC}\right)d\tau \cdot 1(t)$$
$$= \int_{0}^{t} d\tau \cdot 1(t) - \int_{0}^{t} e^{-\tau/TC} d\tau \cdot 1(t) = \underbrace{t \cdot 1(t)}_{r(t)} - RC \cdot \left(1 - e^{-t/RC}\right)1(t).$$

Now, using superposition yields

$$p_\Delta(t) = r(t) - 2r(t-1) + r(t-2) \ \underrightarrow{LTI} \ d_\Delta(t) = x_r(t) - 2x_r(t-1) + x_r(t-2).$$

B.2.11 *BIBO stability of LTI systems*

Recall from 2.6 a system is called BIBO if every bounded input produces a bounded output. Let's consider a continuous time bounded input signal: $|x(t)| < B$ for all t. Suppose we apply this input to an LTI system.

From the convolution theorem the output of LTI system is:

$$|y(t)| = \left|\int_{-\infty}^{\infty} h(\tau)x(t-\tau)d\tau\right| \leq \int_{-\infty}^{\infty} |h(\tau)||x(t-\tau)|d\tau \leq B\int_{-\infty}^{\infty} |h(\tau)|d\tau.$$

Therefore, the system is stable if the impulse response is absolutely integrable, i.e., if

$$\int_{-\infty}^{\infty} |h(\tau)|d\tau < \infty.$$

Appendix C

Basics of Random Signals

This appendix[1] reviews the basic probability concepts and their mathematical representation associated with random signals introduced in Chapters 2 and 8.

Practical systems often exhibit ***uncertainty***, which can be modeled by ***random, or stochastic process.*** A ***system*** incorporating such processes is called ***stochastic***, and some average measures of this process can be used to approximate the effect of uncertainty on the system. As the system evolves in time, for every time instant each of its uncertain dynamic features, such as states or parameters, is characterized by a ***random variable*** (denote it as x) — the totality of the randomly assumed feature values.

The single value which has taken place at a particular time moment is referred to as an ***event*** or an ***outcome***, and the collected data record of such outcomes is referred to as a ***random process realization***. Random process is then thought of as the multitude, or the ***ensemble***, of process realizations for a repeated stochastic system trials, such as daily flights of the same aircraft along the same route, say, from Champaign, Illinois, to "windy city" Chicago under varying weather conditions.

C.1 Probability basics

Referring to the flight example above, consider the ***event*** ζ being the crosswind velocity not exceeding an uncomfortable 30 mph over the entire route. The entire range of all maximum crosswind velocities over the entire route during past flights (assuming, say, 30,000 flights along this route over 70 years) can then be viewed as the ***ensemble space*** of all possible maximum crosswind velocities, and if we plan to fly over this route in a week, we would like to compute $P(\zeta)$, ***the probability of event*** ζ.

To compute the latter, we will subdivide the entire maximum wind velocity range into equal increments and compute the fraction of the number of flights falling into each velocity increment. This will produce a bar graph, such as that in Figure 8.4-4 of Chapter 8, called a ***histogram*** — it represents the flight density — the number of flights in each velocity range divided by the total number of flights, taken over the entire maximum velocity range. The sum of all such fractions will be clearly 1. We can now use this density to calculate the fraction of the flights conforming to ζ by just adding the fractions corresponding to the velocity increments starting from zero until we reach

[1]Appendix C partially follows Ordys, A. W., & Bentsman, J. (2004). Models of stochastic systems. In Ch. 18, Control of Stochastic Systems, *Vol. XI: Control Systems, Robotics and Automation, H. Unbehauen, Ed., UNESCO - Encyclopedia of Life Support Systems.*

30 mph. The result would then be an estimate of probability $P(\zeta)$ of event ζ — having maximum crosswind velocity under 30 mph over the flight path, and the bar graph will be an experimentally obtained ***probability density function,*** denoted as $\varphi(x)$.

C.1.1 *Probability distribution and density functions*

Based on the histogram — the probability density function $\varphi(x)$ approximated from data, ***probability distribution function*** is defined as:

$$P(x<\zeta)=\int_{-\infty}^{\zeta}\varphi(x)dx. \tag{C-1}$$

The ***uniform probability density function***, for example, flight density over each day of the week with respect to x — the maximum crosswind velocity over the entire flight, assuming the flights take place every day of the week, is roughly given by the expression (C-2) and shown in the left graph of Figure C-1, with the corresponding probability distribution shown in the right graph.

$$\varphi(x)=\begin{cases}\dfrac{1}{\beta-\alpha}, & \alpha\le x\le\beta,\\ 0, & \text{otherwise.}\end{cases} \tag{C-2}$$

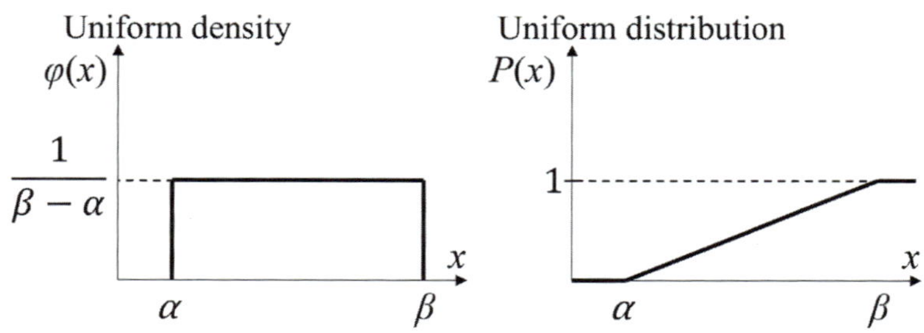

Fig. C-1. The uniform probability density (left) and distribution (right) functions.

C.1.2 *Expectation operator*

If $g(x)$ is a function of a random variable, with the latter described by the differentiable probability density $\varphi(x)$, the ***expectation operator*** of g is defined as:

$$\overline{g}=E\{g(x)\}=\int_{-\infty}^{\infty}g(x)\varphi(x)dx. \tag{C-3}$$

1. The ***mean value*** is defined as the expectation operator for the function $g(x)\equiv x$:

$$\overline{x}=\int_{-\infty}^{\infty}x\varphi(x)dx. \tag{C-4}$$

2. The ***covariance matrix*** P, or ***variance*** of x, will be defined as the expectation operator with the function $g(x)=(x-\overline{x})(x-\overline{x})^T$:

$$P = \int_{-\infty}^{\infty} (x - \bar{x})(x - \bar{x})^T \varphi(x) dx. \tag{C-5}$$

C.1.3 *The Gaussian probability density function*

The ***Gaussian*** probability density function of the random vector of dimension n is given by the expression (C-6), with the corresponding graphs shown in Figure C-2:

$$\varphi(x) = \frac{1}{\sqrt{(2\pi)^n \det(P)}} \exp\left[-\frac{1}{2}(x - \bar{x})^T P^{-1}(x - \bar{x}) \right]. \tag{C-6}$$

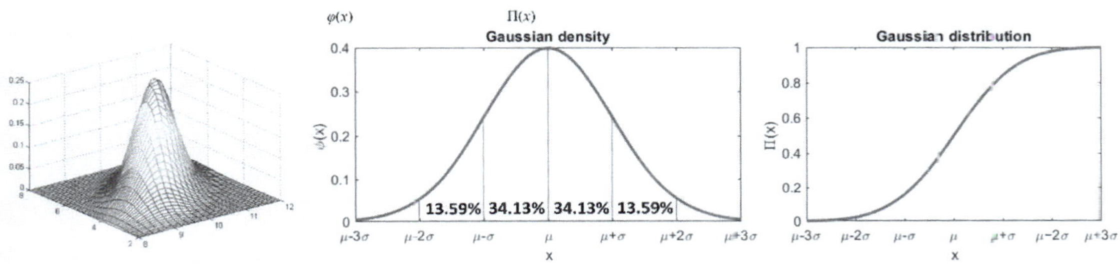

Fig. C-2. Two-dimensional (left) and one-dimensional (middle) Gaussian probability density functions and one-dimensional Gaussian probability distribution (right) function.

C.1.4 *Conditional probability*

Bayes rule describes the ***conditional probability*** of event A, based on the ***prior knowledge*** of the condition that B is true, as follows:

$$P(A \mid B) = \frac{P(B \mid A)P(A)}{P(B)}. \tag{C-7}$$

Applying the latter to the flight example, with condition B — flying in the fall, the maximum crosswind velocity range over 70 years for the fall flights could be narrower, providing a better idea of the maximum crosswind velocity to expect.

A ***joint*** probability density function, e.g. for the minimum flight altitude over 30,000 flights jointly with the maximum crosswind velocity over these flights, can be defined as:

$$\varphi_{joint}(x) = \varphi\left(\begin{bmatrix} x_1 \\ x_2 \\ x_3 \end{bmatrix} \right) = \varphi\left(\begin{bmatrix} x_{11} & \cdots & x_{1n} \\ x_{21} & \cdots & x_{2n} \\ x_{31} & \cdots & x_{3n} \end{bmatrix} \right). \tag{C-8}$$

Clearly, the altitude and the crosswind velocity are not completely independent. Then, the conditional probability density function determines the probability density function when some of the random variables are known. Assuming that the vector x_3 is known, from Bayes rule one obtains equation:

$$\varphi_{cond}(x_1, x_2 \mid x_3) = \frac{\varphi_{joint}(x)}{\varphi_3(x_3)}, \tag{C-9}$$

where $\varphi_3(x_3)$ denotes the probability density function of a vector random variable x_3 and can be calculated from:

$$\begin{aligned}\varphi_3(x_3) &= \int_{-\infty}^{\infty}\int_{-\infty}^{\infty} \varphi_{joint}(x_1, x_2, x_3) dx_1 dx_2 \\ &= \int_{-\infty}^{\infty}\cdots\int_{-\infty}^{\infty}\int_{-\infty}^{\infty}\cdots\int_{-\infty}^{\infty} \varphi_{joint}(x_1, x_2, x_3) dx_{11}\cdots dx_{1n} dx_{21}\cdots dx_{2m}.\end{aligned} \tag{C-10}$$

The ***conditional expectation*** of function $g(x_1, x_2)$ will be obtained if the probability density function $\varphi(x)$ is replaced by the conditional probability:

$$E\{g(x_1, x_2) \mid x_3\} = \int_{-\infty}^{\infty}\int_{-\infty}^{\infty} g(x_1, x_2)\varphi_{cond}(x_1, x_2 \mid x_3) dx_1 dx_2. \tag{C-11}$$

Two random vectors x_1 and x_2 are ***independent*** if

$$\varphi_{joint}(x) = \varphi\left(\begin{bmatrix} x_1 \\ x_2 \end{bmatrix}\right) = \varphi_1(x_1)\varphi_2(x_2), \tag{C-12}$$

resulting in

$$\varphi_{cond}(x_1 \mid x_2) = \varphi_1(x_1) \text{ and } \varphi_{cond}(x_2 \mid x_1) = \varphi_2(x_2). \tag{C-13}$$

For example the maximum crosswind velocity depends on the season, but is independent of the day of the flight.

C.2 Stochastic process

C.2.1 *Autocorrelation and cross-correlation*

For a stochastic process $x(t)$, the ***autocorrelation function*** (or just ***correlation***) is defined as the expectation operation on the variable $g(x, t_1, t_2) = x(t_1)x(t_2)^T$, i.e.:

$$r_{xx}(t_1, t_2) = \int_{-\infty}^{\infty}\int_{-\infty}^{\infty} x(t_1)x(t_2)^T \varphi_{joint}\left(x(t_1), x(t_2)\right) dx(t_1) dx(t_2). \tag{C-14}$$

Similar to the autocorrelation above, for two stochastic processes $x(t)$ and $y(t)$, a ***cross-correlation function*** is defined as:

$$r_{xy}(t_1, t_2) = \int_{-\infty}^{\infty}\int_{-\infty}^{\infty} x(t_1)y(t_2)^T \varphi_{joint}\left(x(t_1), y(t_2)\right) dx(t_1) dy(t_2). \tag{C-15}$$

C.2.2 *Spectra of continuous-time random processes*

For stationary stochastic processes, Fourier and Laplace transforms of the autocorrelation function

$$
\begin{aligned}
R_{xx}(\omega) &= \int_{-\infty}^{\infty} r_{xx}(\tau) e^{-j\omega\tau} d\tau, \\
R_{xx}(s) &= \int_{-\infty}^{\infty} r_{xx}(\tau) e^{-s\tau} d\tau,
\end{aligned}
\tag{C-16}
$$

can be used to analyze the propagation of random signals through the system. For a dynamic system with transfer function $G(s)$ and stochastic input $x(t)$, the output $y(t)$ is a stochastic process for which the cross-correlation function with the input process is determined by the transfer function, i.e.:

$$
\begin{aligned}
y(s) &= G(s)x(s), \\
R_{yx}(s) &= G(s)R_{xx}(s).
\end{aligned}
\tag{C-17}
$$

The autocorrelation of the output process can also be calculated as:

$$R_{yy}(s) = G(s)R_{xx}(s)G(-s). \tag{C-18}$$

C.2.3 *Spectra of discrete time (DT) random processes*

To accommodate discrete time signals, Fourier and Laplace transforms are, respectively, replaced by Discrete Time Fourier transform (DTFT) and z-transform. Then, for the discrete time stochastic process $x(n)$ with the autocorrelation function $r_{xx}(k)$ we have

$$R_{xx}(\Omega) = \sum_{k=-\infty}^{\infty} r_{xx}(k) e^{-j\Omega k} \tag{C-19}$$

and

$$R_{xx}(z) = \sum_{k=-\infty}^{\infty} r_{xx}(k) z^{-k}. \tag{C-20}$$

Therefore, for DT system with transfer function $G(z)$, input $x(k)$, and output $y(k)$, the power density spectra of the cross-correlation and the autocorrelation functions are:

$$
\begin{aligned}
R_{yx}(z) &= G(z)R_{xx}(z), \\
R_{yy}(z) &= G(z)R_{xx}(z)G(z^{-1}).
\end{aligned}
\tag{C-21}
$$

C.2.4 *White noise*

A stochastic process is called ***white noise*** if for each value of time t it has zero mean value and its realizations in time are independent, yielding the autocorrelation function:

$$
\begin{aligned}
r_{xx}(t,t+\tau) &= E\{x(t)x(t+\tau)\} \\
&= \int_{-\infty}^{\infty}\int_{-\infty}^{\infty} x(t)x(t+\tau)\varphi_{joint}\big(x(t),x(t+\tau)\big)dx(t)dx(t+\tau) \\
&= \int_{-\infty}^{\infty}\int_{-\infty}^{\infty} x(t)x(t+\tau)\varphi\big(x(t)\big)\varphi\big(x(t+\tau)\big)dx(t)dx(t+\tau) \\
&= \int_{-\infty}^{\infty} x(t)\varphi\big(x(t)\big)dx(t)\int_{-\infty}^{\infty} x(t+\tau)\varphi\big(x(t+\tau)\big)dx(t+\tau) \\
&= E\{x(t)\}\,E\{x(t+\tau)\} = \left\{\begin{matrix} 0,\ (\tau\neq 0) \\ E\{x(t)^2\} = \sigma_x(t)^2 \neq 0,\ (\tau = 0)\end{matrix}\right\} = \sigma_x(t)^2\delta(\tau),
\end{aligned}
\tag{C-22}
$$

where the value $\sigma(t)$ is called the ***variance*** of the process and $\delta(\tau)$ represents the Dirac function for continuous time processes or Kronecker function for discrete-time processes. White noise has a delta-function autocorrelation and a flat spectrum, as shown in Figure C-3 for DT white noise.

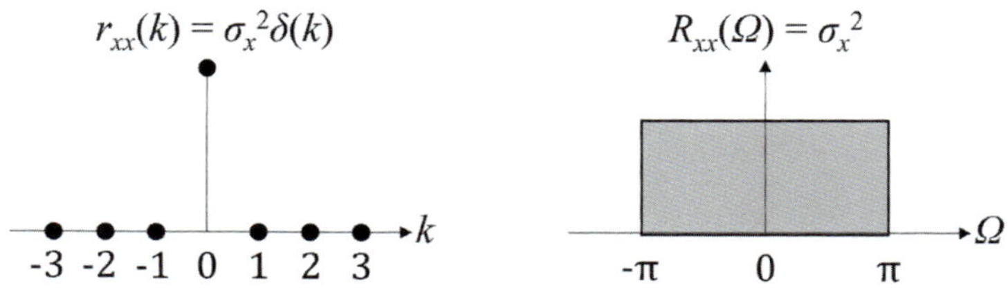

Fig. C-3. Autocorrelation (left) and power spectrum (right) of white noise.

C.2.5 *Wiener process*

A ***Wiener process***, or ***Brownian motion***, $w(t)$ is defined as the time integral of Gaussian white noise, i.e. if $x(t)$ is a Gaussian white noise then:

$$w(t) = \int_0^t x(\tau)d\tau. \tag{C-23}$$

This is a model for a ***random walk***, where successive increments of w, i.e. *dw*, are random and statistically independent with respect to one another. Brownian motion describes motion of small particles undergoing collisions with high frequency. Equation (C-23) permits viewing white noise as derivative of Wiener process, i.e. formally write

$$x(t) \cong \frac{dw(t)}{dt}. \tag{C-24}$$

The ***sample paths*** (realizations) of $w(t)$, however, although continuous, have unbounded variation (go up and down arbitrary many times over the arbitrarily short time interval) and are not differentiable in any sense.

C.2.6 *Series representation of noise*

Recall the definition of the orthogonal set of functions $\{\psi_i(t)\}$ though the inner product in terms of integration:

$$\int_{t_0}^{t_0-T_1} \psi_i(t)\overline{\psi}_j(t)dt = \begin{cases} 0, i \neq j, \\ \lambda_i, i = j, \end{cases} \tag{C-25}$$

along with ***Haar system***:

$$\psi_{k2^{-n}}(t) = \begin{cases} 0 \text{ if } t < (k-1)2^{-n}, \\ 2^{(n-1)/2} \text{ if } (k-1)2^{-n} \leq t < k2^{-n}, \\ -2^{(n-1)/2} \text{ if } k2^{-n} \leq t < (k+1)2^{-n}, \\ 0 \quad \text{if } (k+1)2^{-n} \leq t < 1, \end{cases} \tag{C-26}$$

where k = odd < 2^n,

$$\left\langle \psi_{i2^{-n}}(t), \psi_{j2^{-n}}(t) \right\rangle \equiv \int_0^1 \psi_{i2^{-n}}(t)\psi_{j2^{-n}}(t)dt = \delta_{ij} \text{ (Kronecker delta =1, } i = j\text{, and = 0, } i \neq j\text{)},$$

and any continuous function $f(t)$ can be represented as

$$f(t) = \sum_{n=1}^{\infty} \sum_{k=odd \leq 2^n} a_{k,n}\psi_{k2^{-n}}(t) + a_0\psi_0, \quad \psi_0 \equiv 1, \ a_{k,n} = \int_0^1 f(t)\psi_{k2^{-n}}(t)dt. \tag{C-27}$$

Now, we define

$$\xi_{k2^{-n}}(t) = \int_0^t \psi_{k2^{-n}}(s)ds.$$

Clearly, the newly defined functions are continuous and are formed by the connected triangles. We now construct noise process by setting

$$\varepsilon(t) = g_0\xi_0(t) + \sum_{n=1}^{\infty} \sum_{k=odd \leq 2^n} g_{k2^{-n}}\xi_{k2^{-n}}(t), \quad \psi_0 \equiv 1, \tag{C-28}$$

g_i - independent random variables with Gaussian distribution.

This process is easily seen to be continuous, and because it is constructed of triangles with the support of each triangle going to zero as n goes to infinity, it is also seen to be nondifferentiable anywhere on [0, 1].

This subsection is based on the noise representation given in: *Schuss, Z. (1980). Theory and applications of stochastic differential equations, Wiley.*

C.2.7 *Stationary processes and ergodicity — WSS SME process in signal processing*

A stochastic process is called (***strictly***) ***stationary*** if for any value of m and for any set of time instants $\{t_1, ..., t_m\}$ such that $\{t_1, ..., t_m\} \subset T$ and for any value of $\delta = const$ such that $\{t_1 + \delta, ..., t_m + \delta\} \subset T$ the joint probability density function remains constant when shifted by the value of δ:

$$\varphi_{joint}\left(x(t_1), x(t_2), ..., x(t_m)\right) = \varphi_{joint}\left(x(t_1+\delta), x(t_2+\delta), ...x(t_m+\delta)\right). \tag{C-29}$$

This definition means that all the characteristics of the process, such as the mean value and the variance, are constant in time. For a stationary process, the autocorrelation will not depend on the current time but only on the difference $\tau = t_2 - t_1$. This is denoted as $r_{xx}(\tau)$. Process with such $r_{xx}(\tau)$ and time-invariant mean, but not obeying (C-29), is called ***wide sense stationary (WSS).***

Time average of a stationary stochastic process is generally not equivalent to its mean value. If it is, the process is said to have the property of ***ergodicity***. That is, *if ensemble averages across the process by averaging over all realizations at one specific time instant and time averages along the process by averaging a single realization over time are equal, a stationary process is said to be ergodic. If this also holds for variance, the process is called* ***second moment ergodic (SME).***

C.2.8 *Statistical estimates for WSS SME random signals*

The main random variable characteristics — mean and variance — can be approximated by ***statistical estimates***, where the second average is usually taken over N-1 to yield ***sample variance***, with higher value (less certainty), rather than over N, which would yield ***population variance*** with lower value:

$$m_x = \frac{1}{N}\sum_{k=0}^{N-1} x_k, \quad \sigma_x^2 = \frac{1}{N-1}\sum_{k=0}^{N-1}(x_k - m_x)^2. \tag{C-30}$$

Given a length-N block of signal samples $x(n)$, $n = 0, 1, ..., N-1$, one can compute an ***estimate*** of the statistical quantity $r_{xx}(k)$ by the so-called ***sample autocorrelation*** obtained by *replacing the statistical average by the time average*:

$$\hat{r}_{xx}(k) = \frac{1}{N}\sum_{n=0}^{N-1-k} x(n+k)x(n) \tag{C-31}$$

for $k = 0, 1, ..., N-1$. Similarly, the more general sample cross correlation between two length-N signal blocks $x(n)$, $y(n)$, $n = 0, 1, ..., N-1$, is defined as:

$$\hat{r}_{xy}(k) = \frac{1}{N}\sum_{n=0}^{N-1-k} x(n+k)y(n) \tag{C-32}$$

for $k = 0, 1, ..., M$. The estimates are converging to statistical quantities as N increases in the mean-square sense.

The discrete-time Fourier transform of $\hat{r}_{xx}(k)$, the ***periodogram***, or ***periodogram spectrum***, $\hat{R}_{xx}(\Omega)$, where Ω rad/sample is digital frequency, can be thought of as an estimate of $R_{xx}(\Omega)$:

$$\hat{R}_{xx}(\Omega) = \sum_{k=-(N-1)}^{N-1} \hat{r}_{xx}(k)e^{-j\Omega k} = \frac{1}{N}|X_N(\Omega)|^2, \text{ where } X_N(\Omega) = \sum_{n=0}^{N-1} x(n)e^{-j\Omega n}. \quad \text{(C-33)}$$

The mean of the periodogram for the WSS SME random signals converges to the true power spectrum $R_{xx}(\Omega)$ for large N:

$$R_{xx}(\Omega) = \lim_{N\to\infty} E[\hat{R}_{xx}(\Omega)] = \lim_{N\to\infty} E\left[\frac{1}{N}|X_N(\Omega)|^2\right]. \quad \text{(C-34)}$$

However, the periodogram $\hat{R}_{xx}(\Omega)$ itself does not converge to $R_{xx}(\Omega)$ unlike its mean. To improve the periodogram, the averaging method can be applied by dividing the signal into contiguous blocks, computing the ordinary periodogram of each block, and then averaging the computed periodograms as shown in Figure C-4:

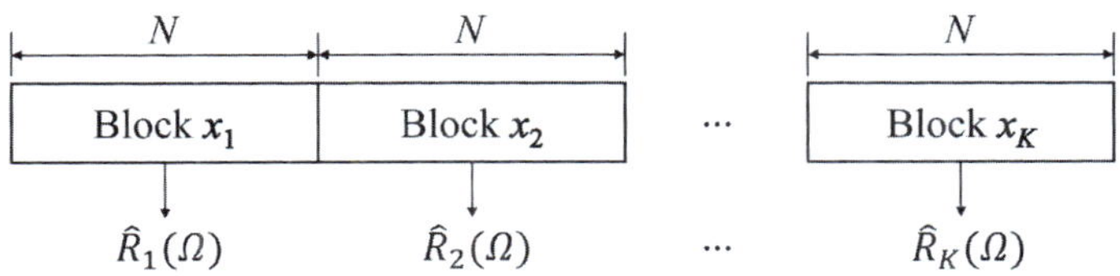

Fig. C-4. Periodogram averaging technique to improve power spectrum estimate.

where the total length of the stationary data record is divided into K blocks, each of length N. Denoting the i th block by $x_i(n)$, $n = 0, 1, ..., N-1$, the average of the K periodograms is:

$$\hat{R}(\Omega) = \frac{1}{K}[\hat{R}_1(\Omega) + \hat{R}_2(\Omega) + \cdots + \hat{R}_K(\Omega)] = \frac{1}{KN}\left[|X_1(\Omega)|^2 + |X_2(\Omega)|^2 + \cdots + |X_K(\Omega)|^2\right]. \quad \text{(C-35)}$$

$\hat{R}(\Omega)$ can be used as a good estimator of $R_{xx}(\Omega)$ for large K. The limitation of this method is that a large value of K requires a long total signal length and signal staying WSS SME over such long periods of time.

C.2.9 *The central limit theorem*

According to the ***central limit theorem***, *the sum or average of the independent noise distributions tends to a Gaussian distribution as the number N of distributions in the sum increases even if the original variables are not normally distributed*, as shown in Figure C-5.

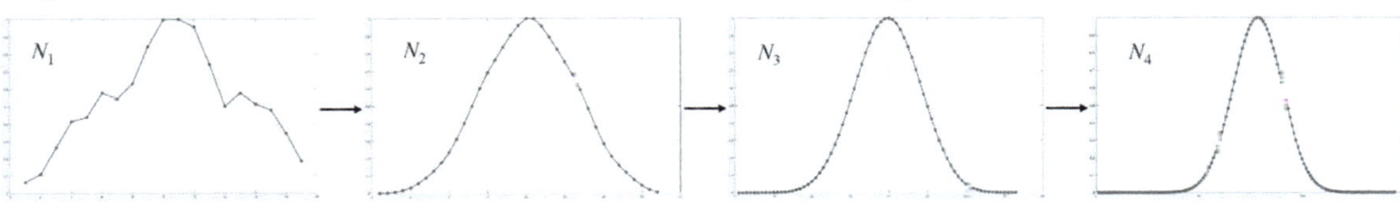

Fig. C-5. The distribution tends to a Gaussian distribution as the number N increases (N1 < N2 < N3 < N4).

C.2.10 *Filtering of random signals*

We need to consider the effect of filtering on the autocorrelation function and power spectrum for noise removing filter design. Suppose that a WSS SME signal $x(n)$ is entered into a stable filter $H(z)$, then the output signal $y(n)$ will also be a random signal:

$$
\begin{aligned}
y(n) &= \sum_m h(m)x(n-m), \\
R_{yy}(\Omega) &= |H(\Omega)|^2 R_{xx}(\Omega).
\end{aligned}
\tag{C-36}
$$

For example, in the case of a DT white noise input with a spectral density $R_{xx}(\Omega) = \sigma_x^2$,

$$
\begin{aligned}
R_{yy}(\Omega) &= |H(\Omega)|^2 \sigma_x^2, \\
R_{yy}(z) &= H(z)H(z^{-1})\sigma_x^2.
\end{aligned}
\tag{C-37}
$$

The autocorrelation function of the filtered output $y(n)$ is no longer a delta function, which means that the output $y(n)$ is no longer a white noise. Its power spectrum reflects the shape of the filter's spectrum. The variance of the output is:

$$
\sigma_y^2 = \int_{-\pi}^{\pi} R_{yy}(\Omega)\frac{d\Omega}{2\pi} = \sigma_x^2 \int_{-\pi}^{\pi} |H(\Omega)|^2 \frac{d\Omega}{2\pi}. \tag{C-38}
$$

From above, the ***noise reduction ratio (NRR)*** can be defined as:

$$
NRR = \frac{\sigma_y^2}{\sigma_x^2} = \int_{-\pi}^{\pi} |H(\Omega)|^2 \frac{d\Omega}{2\pi} = \sum_n h(n)^2, \tag{C-39}
$$

where $h(n)$ is the DT impulse response of the filter. If NRR is less than one, the input noise will be attenuated by the filter.

C.3 Finite dimensional approximations

It is impractical to write the joint probability density functions for infinitely many random variables. One way of overcoming this difficulty is through the autocorrelation functions, and another way is replacing such joint probability density functions with finite dimensional approximations. For the latter, the conditional probability density can be calculated using the Bayes theorem:

$$\varphi_{cond}\left(x(t_m) \mid x(t_{m-1}),\dots,x(t_1)\right) = \frac{\varphi_{joint}\left(x(t_1),x(t_2),\dots,x(t_m)\right)}{\varphi_{joint}\left(x(t_1),x(t_2),\dots,x(t_{m-1})\right)},$$

$$\varphi_{cond}\left(x(t_{m-1}) \mid x(t_{m-2}),\dots,x(t_1)\right) = \frac{\varphi_{joint}\left(x(t_1),x(t_2),\dots,x(t_{m-1})\right)}{\varphi_{joint}\left(x(t_1),x(t_2),\dots,x(t_{m-2})\right)}$$

$$\begin{aligned}&\Rightarrow \varphi_{joint}\left(x(t_1),x(t_2),\dots,x(t_m)\right) \\ &= \varphi_{cond}\left(x(t_m) \mid x(t_{m-1}),\dots,x(t_1)\right) \times \varphi_{cond}\left(x(t_{m-1}) \mid x(t_{m-2}),\dots,x(t_1)\right) \\ &\quad \times \cdots \times \varphi_{cond}\left(x(t_2) \mid x(t_1)\right) \times \varphi_{cond}\left(x(t_1)\right).\end{aligned} \tag{C-40}$$

That is, *the joint probability density can be calculated if all the conditional probabilities are known and the probability density at the time instant t_1 (beginning of the process) is known.*

C.3.1 *Markov process*

Markov process of order N is defined as the stochastic process in which the conditional probability density function $\varphi_{cond}(x(t_m)|x(t_{m-1}), \dots, x(t_1))$ depends on N past values of x:

$$\varphi_{cond}\left(x(t_m) \mid x(t_{m-1}),x(t_{m-2}),\dots,x(t_2),x(t_1)\right) = \varphi_{cond}\left(x(t_m) \mid x(t_{m-1}),\dots,x(t_{m-N})\right). \tag{C-41}$$

For example, the white noise is the Markov process of order 0 and the conditional probability does not depend on the previous values of the process, as shown below:

$$\begin{aligned}\varphi_{cond}\left(x(t_m) \mid x(t_{m-1}),\dots,x(t_1)\right) &= \frac{\varphi_{joint}\left(x(t_1),x(t_2),\dots,x(t_m)\right)}{\varphi_{joint}\left(x(t_1),x(t_2),\dots,x(t_{m-1})\right)} \\ &= \frac{\varphi\left(x(t_1)\right)\dots\varphi\left(x(t_m)\right)}{\varphi\left(x(t_1)\right)\dots\varphi\left(x(t_{m-1})\right)} = \varphi\left(x(t_m)\right).\end{aligned} \tag{C-42}$$

The Markov property of the process defines the propagation in time of the probability density functions.

For another example, assume that $\xi(t)$, $t = 0, \dots$ is a discrete time white noise and define the process $x(t)$ as:

$$\begin{aligned}x(1) &= \xi(0), \\ x(t+1) &= x(t) + \xi(t),\end{aligned} \tag{C-43}$$

where t and $t + 1$ represent the consecutive time instants, $t = 1, 2, \dots, \infty$. The process $x(t)$ is clearly the Markov process of order 1.

In many practical problems, however, the observation is a probabilistic function of a stochastic process, i.e. the process $x(t)$ is observable only through another stochastic process. For example, representation for a process with such observation can take the form:

$$\begin{aligned} x(1) &= \xi(0), \\ x(t+1) &= x(t) + \xi(t), \\ y(t) &= x(t) + \eta(t), \end{aligned} \quad \text{(C-44)}$$

where $\xi(t)$ and $\eta(t)$ are statistically independent white noise processes. The underlying stochastic process $x(t)$ can be referred to as ***hidden Markov process***, and the system is called ***hidden Markov model*** for $x(t)$. *Determining states $x(t)$ on the basis of observations $y(t)$ is referred to as the **state estimation problem**.*

Markov process is called ***homogeneous*** if its ***transition probability function,*** *i.e. the function describing the probability of the state y to take a certain value depending on the value of the current state,* depends only on the time difference, i.e. $T(s, x, t, y) = \Pi(\xi(t) < y \mid \xi(s) = x) = T(0, x, t-s, y) \equiv T(x, \tau, y)$, $\tau = t - s$. Although this process is nonstationary, its transition probability function converges to that of the stationary process as $t \to \infty$. This process, therefore, is typically used to represent paths of a stochastic system that converge to a steady state. Markov process is often called ***Markov chain***, and is typically used with order 1, especially for DT dynamical system, where the next state depends only on the current state, and is independent of the past states (and future states), i.e. is ***memoryless with respect to the past states***.

C.3.2 *Polynomial approximation*

Here, z^{-1}, interpreted as both the complex variable of z-transform and the backwards time shift operator, is often used as the polynomial argument, the probability densities are Gaussian, and a stochastic process is generated by a white noise propagating through a dynamic system.

1. MA model. The ***MA (Moving Average)*** model is defined as follows:

$$w^*(t) = \sum_{i=0}^{m} c_i \xi(t-i), \quad \text{(C-45)}$$

where $\xi(t)$ is the input signal, $w^*(t)$ is the output signal, and c_i, $i = 1, \ldots, m$ are constant coefficients.

2. AR model. ***The AR (Auto-Regressive)*** model is defined as:

$$\sum_{i=0}^{n} a_i w(t-i) = \xi^*(t), \quad \text{(C-46)}$$

where $\xi^*(t)$ is the input signal, $w(t)$ is the output signal, and a_i, $i = 1, \ldots, n$ are constant coefficients.

3. ARMA model. The ***ARMA (Auto-Regressive Moving Average)*** model is a series connection of the AR and MA models. Therefore $\xi^* \equiv w^*$, and using z-transform:

$$\begin{aligned} & w^*(z^{-1}) = C(z^{-1})\xi(z^{-1}), \\ & A(z^{-1})w(z^{-1}) = \xi^*(z^{-1}), \\ & \Rightarrow A(z^{-1})w(z^{-1}) = C(z^{-1})\xi(z^{-1}). \end{aligned} \quad \text{(C-47)}$$

C.4 Mixed stochastic-deterministic systems

C.4.1 *CARMA, CARIMA, and Box-Jenkins models*

Adding to ARMA model a deterministic model

$$r(z^{-1}) = \frac{B(z^{-1})}{A(z^{-1})} u(z^{-1}) \tag{C-48}$$

yields the ***Controlled Auto Regressive Moving Average (CARMA)***, or ***ARMAX***, model, where ***X*** stands for the ***exogenous input*** $u(t)$, given by (C-49) below and shown in Figure C-6:

$$y(z^{-1}) = \frac{B(z^{-1})}{A(z^{-1})} u(z^{-1}) + \frac{C(z^{-1})}{A(z^{-1})} \xi(z^{-1}) \Leftrightarrow A(z^{-1}) y(z^{-1}) = B(z^{-1}) u(z^{-1}) + C(z^{-1}) \xi(z^{-1}). \tag{C-49}$$

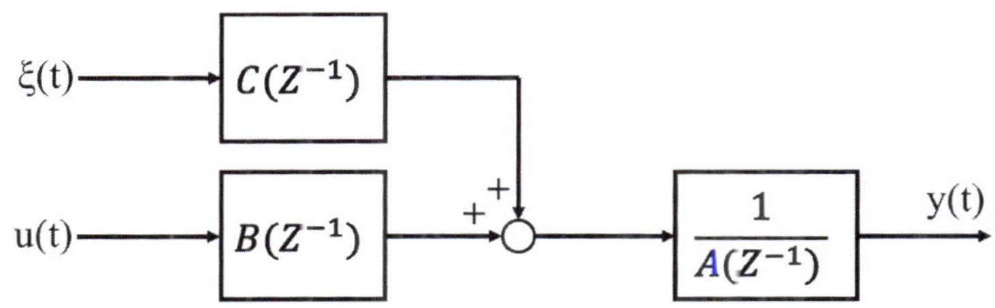

Fig. C-6. Controlled Auto Regressive Moving Average (CARMA) model.

Here A, B, C are polynomials given, respectively, by:

$$\begin{aligned} A(z^{-1}) &= 1 + a_1 z^{-1} + a_2 z^{-2} + \ldots + a_n z^{-n}, \\ B(z^{-1}) &= b_0 + b_1 z^{-1} + b_2 z^{-2} + \ldots + b_k z^{-k}, \\ C(z^{-1}) &= 1 + c_1 z^{-1} + c_2 z^{-2} + \ldots + c_m z^{-m}. \end{aligned} \tag{C-50}$$

The polynomial $C(z^{-1})$ is assumed to have no zeros on the unit circle $|z| = 1$.

Integrating C-polynomial: $C(z^{-1})/\Delta$, where $\Delta = 1 - z^{-1}$, yields Controlled Autoregressive Integrated Moving Average ***(CARIMA)***, or ***ARIMAX***, model:

$$A(z^{-1}) y(z^{-1}) = B(z^{-1}) u(z^{-1}) + \frac{C(z^{-1})}{\Delta(z^{-1})} \xi(z^{-1}) \Leftrightarrow A(z^{-1}) \Delta(z^{-1}) y(z^{-1}) = B(z^{-1}) \Delta(z^{-1}) u(z^{-1}) + C(z^{-1}) \xi(z^{-1}). \tag{C-51}$$

Comparison of the expression on the right in (C-51) with that of (C-49) shows that the system controlled input and the resulting total output are now given as increments with respect to their previous values.

A more general structure containing separate plant and noise denominators, referred to as ***Box-Jenkins model***, is given by (C-52) and shown in Figure C-7. In this case, the system is represented by two separate transfer functions, from the input sequence $u(t)$ and the disturbance sequence $\xi(t)$, to the output sequence $y(t)$:

$$y(t) = \frac{B(z^{-1})}{A(z^{-1})}u(t) + \frac{C(z^{-1})}{D(z^{-1})}\xi(t), \quad \text{(C-52)}$$

where $D(z^{-1}) = 1 + d_1 z^{-1} + d_2 z^{-2} + ... + d_j z^{-j}$.

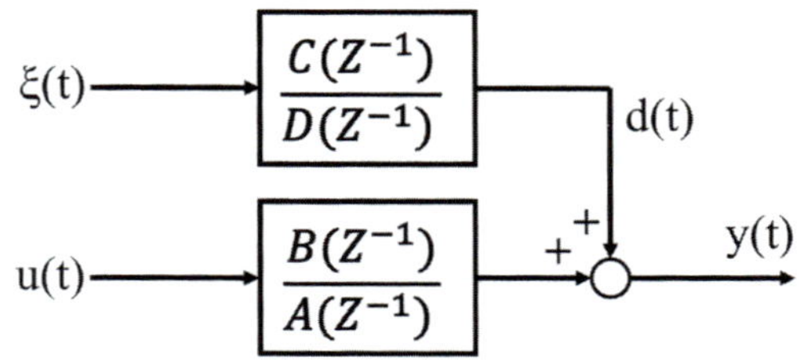

Fig. C-7. Box-Jenkins model.

C.4.2 *State-space representation*

The state-space description of a ***linear stochastic discrete time system*** has the form:

$$\begin{aligned} \chi(t+1) &= F\chi(t) + \beta u(t) + G\xi(t), \\ y(t) &= H\chi(t) + \xi(t), \end{aligned} \quad \text{(C-53)}$$

where

$\chi(t)$ is the state of the system (vector of size n),
$u(t)$ is the deterministic input (vector of size k),
$\xi(t)$ is the stochastic input, usually white noise (vector of size m),
$y(t)$ is the system output (vector of size j),
F is the state transition matrix ($n \times n$),
B, G, H are matrices of sizes ($n \times k$), ($n \times m$) and ($j \times n$).

*Notice that Eq. (C-53) is **a special case of the Markov process of order 1**. If the input process is Gaussian white noise, all the probability distributions in this system will be Gaussian as well.* From Eq. (C-53) it is easy to show that the output signal at the time instant $t + l$ (where $l = 1, 2, ..., N + 1$) can be expressed in the following way:

$$y(t+l) = HF^l \chi(t) + \sum_{j=1}^{l} HF^{l-j}\beta u(t+j-1) + \sum_{j=2}^{l} HF^{l-j}G\xi(t+j-1) + \xi(t+l) + HF^{l-1}G\xi(t).$$

(C-53)

Substituting $t = 0$, $\chi(0)$ — the initial state with Gaussian probability density function, $\xi(j), j = 0,1,...$ random signal with Gaussian probability density function, and $u(j), j = 0,...$ — deterministic signal, results in the probability density for the output signal being Gaussian.

Appendix D

Fourier Transform

This appendix presents the main properties of Fourier transform introduced in Chapter 5.

D.1 Fourier Transform Definition

The Fourier Transform (FT) provides the frequency domain description of the time domain signals and may be regarded as an extension of the Fourier series (FS) as the fundamental period of the latter goes to infinity. Recall that a periodic signal $x_p(t)$ with period T and its exponential Fourier series coefficients $X[k]$ are related by:

$$x_p(t)=\sum_{k=-\infty}^{\infty} X[k]e^{j2\pi kf_1 t};\ X[k]=\frac{1}{T_1}\int_{T_1} x(t)e^{-j2\pi kf_1 t}dt.$$

If the period T_1 of a periodic signal $x_p(t)$ is extended to infinity, the resulting signal no longer remains periodic, turning into a single pulse $x(t)$ with one period of $x_p(t)$. The frequency spacing $f_1=\frac{1}{T_1}$ approaches 0 and $kf_1 \to f$, therefore the Fourier series spectrum becomes continuous, and the signal transfers from power signal to energy signal.

The formula forming Fourier transform from to Fourier series is:

$$X(f)=\lim_{T_1\to\infty} T_1 X[k]=\int_{-\infty}^{\infty} x(t)e^{-j2\pi ft}dt, \text{ or}$$

$$X(\omega)=\int_{-\infty}^{\infty} x(t)e^{-j\omega t}dt.$$

The Fourier Transform provides a frequency domain representation of a time domain signal $x(t)$. It is a different way to represent a time domain signal.

D.2 Inverse Fourier Transform

A periodic signal $x_p(t)$ can be reconstructed through its Fourier series coefficients $X[k]$,

$$x_p(t)=\sum_{k=-\infty}^{\infty} X[k]e^{j2\pi kf_1 t}.$$

We can rewrite above formula as:

$$x_p(t)=\sum_{k=-\infty}^{\infty} T_1 X[k]e^{j2\pi kf_1 t}\frac{1}{T_1}=\sum_{k=-\infty}^{\infty} T_1 X[k]e^{j2\pi kf_1 t} f_1.$$

As $T_1 \to \infty$ and $kf_1 \to f$, the summation tends to become an integration over $(-\infty,\infty)$. With $f_1 \to df \to 0$, we obtain:

$$x(t)=\int_{-\infty}^{\infty} X[f]e^{j2\pi ft}df$$

$$\text{or } x(t)=\frac{1}{2\pi}\int_{-\infty}^{\infty} X[\omega]e^{j\omega t}d\omega .$$

This is the Inverse Fourier Transform, which allows us to obtain $x(t)$ from $X(f)$.

D.3 Properties of Fourier Transform

1. Superposition

If the signal $y(t)$ is the sum of the signals $x_1(t)$ and $x_2(t)$:

$$y(t)=c_1\cdot x_1(t)+c_2\cdot x_2(t)$$

then the same relationship is true for their transforms:

$$\begin{aligned}Y(f)&=\int_{-\infty}^{\infty}(c_1\cdot x_1(t)+c_2\cdot x_2(t))\cdot e^{-j2\pi ft}dt\\&=c_1\int_{-\infty}^{\infty}x_1(t)\cdot e^{-j2\pi ft}dt+c_2\int_{-\infty}^{\infty}x_2(t)\cdot e^{-j2\pi ft}dt\\&=c_1\cdot X_1(f)+c_2\cdot X_2(f).\end{aligned}$$

2. Convolution and multiplication

The convolution of signal $x(t)$ with the impulse signal $h(t)$,

$$y(t)=h(t)*x(t)=\int_{-\infty}^{\infty}h(\tau)x(t-\tau)d\tau,$$

corresponds to the multiplication of the signal transforms in the frequency domain:

$$\begin{aligned}
Y(f) &= \int_{-\infty}^{\infty} y(t)\cdot e^{-j2\pi ft}dt \\
&= \int_{-\infty}^{\infty}\left(\int_{-\infty}^{\infty} h(\tau)x(t-\tau)d\tau\right)e^{-j2\pi ft}dt \\
&= \int_{-\infty}^{\infty} h(\tau)\int_{-\infty}^{\infty} x(t-\tau)e^{-j2\pi ft}dtd\tau \\
&= \int_{-\infty}^{\infty} h(\tau)e^{-j2\pi f\tau}\int_{-\infty}^{\infty} x(t-\tau)e^{-j2\pi f(t-\tau)}d(t-\tau)d\tau \\
&= \int_{-\infty}^{\infty} h(\tau)e^{-j2\pi f\tau}d\tau\cdot X(f) \\
&= H(f)\cdot X(f).
\end{aligned}$$

Similarly, multiplication in the time domain:

$$y(t) = h(t)\cdot x(t)$$

corresponds to convolution in the frequency domain:

$$Y(f) = H(f) * X(f).$$

3. Parseval's theorem

The energy of the signal can be calculated in both the time and the frequency domains as

$$\int_{-\infty}^{\infty} x^2(t)dt = \int_{-\infty}^{\infty}|X(f)|^2 df = \frac{1}{2\pi}\int_{-\infty}^{\infty}|X(\omega)|^2 d\omega.$$

4. Time and frequency shift

The transform of a signal shifted in time by a constant, $y(t) = x(t-t_0)$, is

$$Y(f) = X(f)e^{-j2\pi ft_0}.$$

This is seen from

$$\begin{aligned}
\int_{-\infty}^{\infty} x(t-t_0)e^{-j2\pi ft}dt &\underset{\tau=t-t_0\Rightarrow t=\tau+t_0}{=} \int_{-\infty}^{\infty} x(\tau)e^{-j2\pi f(\tau+t_0)}d\tau \\
&= e^{-j2\pi ft_0}X(f).
\end{aligned}$$

Similarly, shift in the frequency domain, $Y(f) = Y(f-f_0)$, corresponds to time domain signal

$$y(t) = x(t)\cdot e^{j2\pi f_0 t}.$$

5. Time and Frequency Scaling

The FT of a signal in the time domain modified by scaling the time axis $y(t) = x(c\cdot t)$ is:

$$Y(f)=\frac{1}{|c|}X\left(\frac{f}{c}\right).$$

This is seen from

$$\begin{aligned}Y(f)&=\int_{-\infty}^{\infty}x(c\cdot t)e^{-j2\pi ft}dt=\int_{-\infty}^{\infty}x(\tau)e^{-j2\pi f\frac{\tau}{c}}d\frac{\tau}{c}\\&=\frac{1}{c}\int_{-\infty}^{\infty}x(\tau)e^{-j2\pi\frac{f}{c}\tau}d\tau\\&=\frac{1}{c}X\left(\frac{f}{c}\right).\end{aligned}$$

6. Derivative

The transform of the 1st derivative of $x(t)$ given by the signal $y(t)=x'(t)$ has the following property:

$$Y(f)=j2\pi f\cdot X(f).$$

7. Summarizing table

Property	$x(t)$	$X(f)$
Similarity	$x(t)$	$X(f)$
Time Scaling	$x(at)$	$\frac{1}{\lvert a\rvert}X\left(\frac{f}{a}\right)$
Folding	$x(-t)$	$X(-f)$
Time Shift	$x(t-a)$	$e^{-j2\pi fa}X(f)$
Frequency Shift	$e^{j2\pi at}x(t)$	$X(f-a)$
Convolution	$x(t)*h(t)$	$X(f)H(f)$
Multiplication	$x(t)h(t)$	$X(f)*H(f)$
Modulation	$x(t)\cos(2\pi at)$	$0.5[X(f+a)+X(f-a)]$
Derivative	$x'(t)$	$j2\pi fX(f)$
Times-t	$-j2\pi tx(t)$	$X'(f)$
Integration	$\int_{-\infty}^{t}x(t)dt$	$\frac{1}{j2\pi f}X(f)+0.5X(0)\delta(f)$
Conjugation	$x^*(t)$	$X^*(-f)$

Correlation	$x(t)**y(t)$	$X(f)Y^*(f)$
Periodic correlation	$r^{\circledast}_{x_{tp}h_{tp}}(t) \equiv x_{tp}(t)\circledast\circledast h_{tp}(t)$	$X_{tp}(f)Y_{tp}^*(f)$
Autocorrelation	$x(t)**x(t)$	$X(f)X^*(f)=\lvert X(f)\rvert^2$
Periodic autocorrelation	$r^{\circledast}_{x_{tp}x_{tp}}(t) \equiv x_{tp}(t)\circledast\circledast x_{tp}(t)$	$S_{tp}(f)=\lvert X_{tp}(f)\rvert^2$
PSD pairs	$\frac{1}{2t_{tr}} r_{x_t x_t}(t)$	$S_t(t)=\frac{1}{2t_{tr}}\lvert X_t(f)\rvert^2$
	$\lim\limits_{t_{tr}\to\infty}\frac{1}{2t_{tr}} r_{x_t x_t}(t)$	$S(f)=\lim\limits_{t_{tr}\to\infty}\frac{1}{2t_{tr}}\lvert X_t(f)\rvert^2$
	$x_p(t)=\sum\limits_{k=-\infty}^{\infty} X_s[k]e^{jk2\pi f_1 t}$	$X_{x_p}(f)=\sum\limits_{k=-\infty}^{\infty} X_s[k]\delta(f-kf_1)$:
	$r^{\circledast}_{x_p x_p}(t)=\sum\limits_{k=-\infty}^{\infty}\lvert X_s[k]\rvert^2 e^{j2k\pi f_1 t}$	$S_{x_p}(f)=\sum\limits_{k=-\infty}^{\infty}\lvert X_s[k]\rvert^2\delta(f-kf_1)$
	$r^{\circledast}_{\sigma\varepsilon\sigma\varepsilon}(t)=\sigma^2\delta(t)$	$S_{\sigma\varepsilon}(f)\equiv\sigma^2$
	$x_e(t)$	$S_{x_e}(f)\equiv 0$

Subscript "tp" in the time domain denotes truncation and subsequent periodic extension, and in the frequency domain just being a transform of truncated periodically extended time domain signal.

D.4 Useful Fourier Transform Pairs

$x(t)$	$X(f)$	$X(\omega)$
$\delta(t)$	1	1
$rect(t)$	$\mathrm{sinc}(f)$	$\mathrm{sinc}\left(\frac{\omega}{2\pi}\right)$
$tri(t)$	$\mathrm{sinc}^2(f)$	$\mathrm{sinc}^2\left(\frac{\omega}{2\pi}\right)$
$\mathrm{sinc}(t)$	$rect(f)$	$rect\left(\frac{\omega}{2\pi}\right)$
$\cos(2\pi f_0 t)$	$0.5[\delta(f+f_0)+\delta(f-f_0)]$	$\pi[\delta(\omega+2\pi f_0)+\delta(\omega-2\pi f_0)]$
$\sin(2\pi f_0 t)$	$j0.5[\delta(f+f_0)-\delta(f-f_0)]$	$\pi[\delta(\omega+2\pi f_0)-\delta(\omega-2\pi f_0)]$
$e^{-at}u(t)$	$\frac{1}{a+j2\pi f}$	$\frac{1}{a+j\omega}$

$te^{-at}u(t)$	$\frac{1}{(a+j2\pi f)^2}$	$\frac{1}{(a+j\omega)^2}$
$e^{-a\|t\|}$	$\frac{2a}{a^2+4\pi^2 f^2}$	$\frac{2a}{a^2+\omega^2}$
$e^{-\pi t^2}$	$e^{-\pi f^2}$	$e^{-\frac{\omega^2}{4\pi}}$
$\operatorname{sgn}(t)$	$\frac{1}{j\pi f}$	$\frac{2}{j\omega}$
$u(t)$	$0.5\delta(f)+\frac{1}{j2\pi f}$	$\pi\delta(\omega)+\frac{1}{j\omega}$
$e^{-at}\cos(2\pi\beta t)u(t)$	$\frac{a+j2\pi f}{(a+j2\pi f)^2+(2\pi\beta)^2}$	$\frac{a+j\omega}{(a+j\omega)^2+(2\pi\beta)^2}$
$e^{-at}\sin(2\pi\beta t)u(t)$	$\frac{2\pi\beta}{(a+j2\pi f)^2+(2\pi\beta)^2}$	$\frac{2\pi\beta}{(a+j\omega)^2+(2\pi\beta)^2}$
$\sum_{n=-\infty}^{\infty}\delta(t-nT)$	$\frac{1}{T}\sum_{k=-\infty}^{\infty}\delta\left(f-\frac{k}{T}\right)$	$\frac{2\pi}{T}\sum_{k=-\infty}^{\infty}\delta\left(\omega-\frac{2\pi k}{T}\right)$
$x_p(t)=\sum_{k=-\infty}^{\infty}X[k]e^{j2\pi kf_0t}$	$\sum_{k=-\infty}^{\infty}X[k]\delta(f-kf_0)$	$\sum_{k=-\infty}^{\infty}2\pi X[k]\delta(\omega-k\omega_0)$

Appendix E

Laplace Transform

E.1 Laplace Transform Definition

The application of Laplace transform for the solution of ODEs and other purposes is considered in Chapter 3 and its definition as an extension of Fourier Transform — in Subsection 5.5.1.2 in Chapter 5. Here we provide some facts about Laplace transform given by

$$X(s)=L\left[x(t)\right]=\int_{0-}^{\infty}x(t)e^{-(\sigma+j\omega)t}dt=\int_{0-}^{\infty}x(t)e^{-st}dt$$

where the complex variable $s=\sigma+j\omega,$ with ω measured in radians/second, and σ — in nepers/second. We say that "$X(s)$ is the Laplace transform of $x(t)$". The time variable t starts at 0−, meaning that it includes impulse at zero. This "–" is usually dropped.

E.2 Convergence

The Laplace transform is an integral operation. It exists if $x(t)e^{-\sigma t}$ is absolutely integrable. Obviously, this condition is satisfied for only certain choice of σ. The range of σ that ensures the Laplace transform existence defines its **region of convergence**. For example, the Laplace transform $X(s)$ of the signal $x(t)=e^{3t}u(t)$ exits only if $\sigma>3,$ which ensures that the product is a decaying exponential. Thus, $\sigma>3$ is the region of convergence of the transform $X(s)$. Since we deal only with the causal signals (whose region of convergence is $\sigma>a,\ a\geq 0$), we omit explicit mentioning of the region of convergence.

E.3 Properties of Laplace Transform

1. Linearity $\quad ax(t)+bx(t)\rightarrow aX(s)+bY(s).$
2. Time shifting $\quad x(t-t_1)u(t-t_1),\ t_1>0\rightarrow e^{-st_1}X(s).$
3. Frequency shifting, or s-shifting $\quad e^{s_1t}x(t)\rightarrow X(s-s_1).$
4. Real convolution $\quad g(t)*x(t)\rightarrow G(s)X(s).$
5. Time derivative $\quad \frac{d}{dt}x(t)\rightarrow sX(s)-x(0),$

$$\frac{d^2}{dt^2}x(t) \to s^2X(s) - sx(0) - x'(0),$$

$$\vdots$$

$$\frac{d^n}{dt^n}x(t) \to s^nX(s) - \sum_{k=1}^{n} s^{n-k}x^{(k-1)},$$

$$\frac{d^n}{dt^n}x(t) \to s^nX(s) \text{ (if initial value is 0).}$$

6. Time integral

$$\int_{-\infty}^{t} x(\tau)d\tau \to \frac{1}{s}X(s) + \frac{1}{s}\int_{-\infty}^{0} x(\tau)d\tau,$$

$$\iint x(t)(dt)^2 \to \frac{X(s)}{s^2} + \frac{\left[\int x(t)dt\right]_{t=0}}{s^2} + \frac{\left[\iint x(t)(dt)^2\right]_{t=0}}{s},$$

$$\vdots$$

$$\overbrace{\int\cdots\int}^{n} x(t)(dt)^n \to \frac{X(s)}{s^n} + \sum_{k=1}^{n} \frac{\left[\overbrace{\int\cdots\int}^{n} x(t)(dt)^n\right]_{t=0}}{s^{n-k+1}},$$

$$\overbrace{\int\cdots\int}^{n} x(t)(dt)^n \to \frac{X(s)}{s^n} \text{ (If initial value is 0).}$$

7. Complex derivative

$$tx(t) \to -\frac{d}{ds}X(s).$$

8. Complex convolution

$$x(t)y(t) \to \frac{1}{2\pi j}\int_{\delta_0-\infty}^{\delta_0+\infty} X(v)Y(s-v)dv.$$

9. Initial value theorem

$$x(0) \to \lim_{s\to\infty} sX(s).$$

10. Final value theorem

$$x(\infty) \to \lim_{s\to 0} sX(s).$$

E.4 Useful Laplace Transform Pairs

$x(t)$	$X(s)$	$x(t)$	$X(s)$
$\delta(t)$	1	$e^{-at}u(t)$	$\frac{1}{s+a}$
$u(t)$	$\frac{1}{s}$	$te^{-at}u(t)$	$\frac{1}{(s+a)^2}$
$tu(t)$	$\frac{1}{s^2}$	$t^n e^{-at}u(t)$	$\frac{n!}{(s+a)^{n+1}}$
$\frac{t^n}{n!}u(t)$	$\frac{1}{s^{n+1}}$	$(1-e^{-at})u(t)$	$\frac{a}{s(s+a)}$
$\sin\omega t\, u(t)$	$\frac{\omega}{s^2+\omega^2}$	$e^{-at}\sin\omega t\, u(t)$	$\frac{\omega}{(s+a)^2+\omega^2}$
$\cos\omega t\, u(t)$	$\frac{s}{s^2+\omega^2}$	$e^{-at}\cos\omega t\, u(t)$	$\frac{s+a}{(s+a)^2+\omega^2}$
$t\sin\omega t\, u(t)$	$\frac{2s\omega}{(s^2+\omega^2)^2}$	$\sin^2\omega t\, u(t)$	$\frac{2\omega^2}{s(s^2+4\omega^2)}$
$t\cos\omega t\, u(t)$	$\frac{s^2-\omega^2}{(s^2+\omega^2)^2}$	$\cos^2\omega t\, u(t)$	$\frac{s^2+2\omega^2}{s(s^2+4\omega^2)}$

E.5 Partial Fraction Expansion

In algebra, the partial fraction decomposition or partial fraction expansion (PFE) of a rational fraction (that is a fraction such that the numerator and the denominator are both polynomials) is the operation that consists in expressing the fraction as a sum of a polynomial (possibly zero) and one or several fractions with a simpler denominator. Below we give examples of the main PFE cases.

1. Single real roots.

If the denominator has n different single roots, irrespectively of whether they are real or complex, then

$$F(s)=\frac{A(s)}{B(s)}=\frac{A(s)}{(s-s_1)(s-s_2)\cdots(s-s_n)}=\sum_{i=1}^{n}\frac{K_i}{(s-s_i)}$$

where $K_i=(s-s_i)F(s)\big|_{s=s_i}$.

2. Multiple real roots.

The denominator has r same roots at $s = s_1$ and (n-r) different roots.

$$F(s)=\frac{A(s)}{B(s)}=\frac{A(s)}{(s-s_1)^r(s-s_{r+1})\cdots(s-s_n)}=\sum_{i=1}^{r}\frac{K_{1i}}{(s-s_1)}+\sum_{j=r+1}^{n}\frac{K_j}{(s-s_j)}$$

where $K_{1i}=\frac{1}{(r-i)!}\frac{d^{r-i}}{ds^{r-i}}\left[(s-s_1)^r F(s)\right]_{s=s_1}$, $K_j=(s-s_j)F(s)\big|_{s=s_j}$.

3. Conjugate complex roots.

For polynomials with real coefficients the complex roots appear in conjugate pairs: if the denominator has complex roots, $s_{1,2}=-\alpha\pm j\beta$, then

$$F(s)=\frac{A(s)}{(s+\alpha-j\beta)(s+\alpha+j\beta)}=\frac{K_1}{s+\alpha-j\beta}+\frac{K_2}{s+\alpha+j\beta}$$

where $K_1=K_2^*$.

Example E-1: Derive the inverse Laplace transform of the following $F(s)$:

$$F(s)=\frac{s+5}{(s+2)(s+3)}.$$

Solution:

$$F(s)=\frac{s+5}{(s+2)(s+3)}=\frac{K_1}{s+2}+\frac{K_2}{s+3}\Rightarrow$$

$$K_1=(s+2)\frac{s+5}{(s+2)(s+3)}\bigg|_{s=-2}=3,$$

$$K_1=(s+3)\frac{s+5}{(s+2)(s+3)}\bigg|_{s=-3}=-2.$$

Therefore, we have:

$$F(s)=\frac{3}{s+2}-\frac{2}{s+3}.$$

Using the corresponding Laplace transform pair from the E.4 table yields

$$f(t)=L^{-1}\{F(s)\}=\left(3e^{-2t}-2e^{-3t}\right)u(t).$$

Example E-2: Derive the inverse Laplace transform of the following $F(s)$:

$$F(s)=\frac{3s+5}{(s+1)^2(s+3)}.$$

Solution:

$$F(s)=\frac{3s+5}{(s-1)^2(s+3)}=\frac{K_{12}}{(s+1)^2}+\frac{K_{11}}{s+1}+\frac{K_3}{s+3}\Rightarrow$$

$$K_{12}=(s+1)^2\left.\frac{3s+5}{(s+1)^2(s+3)}\right|_{s=-1}=1,$$

$$K_{11}=\frac{d}{ds}\left[(s+1)^2\frac{3s+5}{(s+1)^2(s+3)}\right]\Bigg|_{s=-1}=1,$$

$$K_3=(s+3)\left.\frac{3s+5}{(s+1)^2(s+3)}\right|_{s=-3}=-1.$$

Therefore we have

$$F(s)=\frac{1}{(s+1)^2}+\frac{1}{s+1}-\frac{1}{s+3}.$$

Using the corresponding Laplace transform pairs from the E.4 table yields

$$f(t)=L^{-1}\{F(s)\}=\left(te^{-t}+e^{-t}-e^{-3t}\right)u(t).$$

Example E-3: Derive the inverse Laplace transform of the following $F(s)$:

$$F(s)=\frac{2s+8}{(s+2)^2+4}.$$

Solution: We can proceed two ways — use sine or cosine transform with s-shift theorem or use direct decomposition and Euler's formula. Here we use the latter:

$$F(s)=\frac{2s+8}{(s+2)^2+4}=\frac{2s+8}{(s+2-2j)(s+2+2j)}=\frac{K_1}{s+2-2j}+\frac{K_2}{s+2+2j}\Rightarrow$$

$$K_1=(s+2-2j)F(s)\big|_{s=-2+2j}=1-j=\sqrt{2}e^{-j\frac{\pi}{4}},$$

$$K_2=(s+2+2j)F(s)\big|_{s=-2-2j}=1+j=\sqrt{2}e^{j\frac{\pi}{4}}.$$

Therefore, we have

$$F(s)=\frac{\sqrt{2}e^{-j\frac{\pi}{4}}}{s+2-2j}+\frac{\sqrt{2}e^{j\frac{\pi}{4}}}{s+2+2j}.$$

The inverse Laplace transform

$$\begin{aligned}
f(t)=L^{-1}\{F(s)\}&=\left(\sqrt{2}e^{-j\frac{\pi}{4}}e^{(-2+2j)t}+\sqrt{2}e^{j\frac{\pi}{4}}e^{(-2-2j)t}\right)u(t)\\
&=\sqrt{2}e^{-2t}\left(e^{-j\frac{\pi}{4}}e^{2jt}+e^{j\frac{\pi}{4}}e^{(-2j)t}\right)u(t)=\sqrt{2}e^{-2t}\left(e^{j\left(2t-\frac{\pi}{4}\right)}+e^{-j\left(2t-\frac{\pi}{4}\right)}\right)u(t)\\
&=2\sqrt{2}e^{-2t}\cos\left(2t-\frac{\pi}{4}\right)u(t).
\end{aligned}$$

Appendix F

Basic Types of Sensors and Actuators

Rapidly expanding field of sensors and actuators provides the key elements of any data acquisition and control system. In this appendix we briefly survey the main sensor and actuator types and give references that contain the corresponding details and figures.

F.1 Sensors for Temperature

F.1.1 *Thermocouple*

Temperature gradients in conductive wires induce electromotive forces, not identical if the wire materials differ.

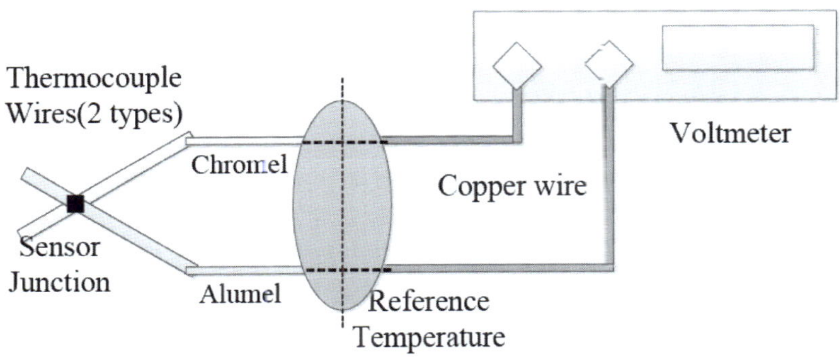

Fig. F-1. Example circuit using K type thermocouple as temperature sensor.

As a result, application of heat to the sensor junction that fuses wires with two different metallic materials, e.g. chromel and alumel in K-type thermocouples, as shown in Figure F-1, creates unequal voltage difference between the sensor junction and the reference temperature points. This mismatch in voltage differences induces a current flow. This phenomenon, called the Seebeck effect, is the basis for measuring the temperature at the junction. The details of the thermocouple circuitry are given in Texas Instruments guide https://www.ti.com/lit/an/sbaa274/sbaa274.pdf?ts=1632144279939&ref_url=https%253A%252F%252Fwww.google.com%252F. Thermocouples are classified based on the metal used and the temperature range. Base metal thermocouples such as K, J, T, and E type thermocouples are commonly used in most applications while noble metal thermocouples such as R, S, and B type are used for high temperature applications. The materials and the temperature ranges of each thermocouple type are summarized in Table 1.

Table F-1. Materials and temperature ranges of different thermocouple types.

Base metal thermocouples				
Type	K	J	T	E
Materials	Chromel/ Alumel	Iron/ Constantan	Copper/ Constantan	Chromel/ Constantan
Temperature range	−269 to 1260°C	−210 to 760°C	−270 to 370°C	−270 to 870°C

Noble metal thermocouples			
Type	R	S	B
Materials	Platinum 13% Rhodium/Platinum	Platinum 10% Rhodium/ Platinum (-)	Platinum 30% Rhodium/ Platinum 6% Rhodium
Temperature range	−50 to 1480°C	−50 to 1480°C	0 to 1700°C

The additional characteristics and information for several thermocouple materials are given in http://www.thermosensors.com/pdfs/Thermocouples_Application_and_Technical_Data.pdf and https://www.omega.com/en-us/.

F.1.2 *Resistance temperature detectors (RTDs)*

An RTD (resistance temperature detector) is a temperature sensor that uses metallic materials' electrical resistance changes as a function of temperature. RTD can be made of a variety of metals, such as nickel, copper, or tungsten, but platinum (Pt) is the most popular because of its high linearity and stability, e.g. Pt100 and Pt1000. Pt100 RTDs' nominal resistance at 0°C is 100 Ω while Pt1000 RTDs have nominal resistance of 1000 Ω at 0°C. RTDs are offered with 2, 3, or 4 lead configurations. 3-wire configuration is commonly used in most industrial applications, while 2-wire configuration is the least accurate, and 4-wire configuration is primarily used in the applications where high accuracy is required.

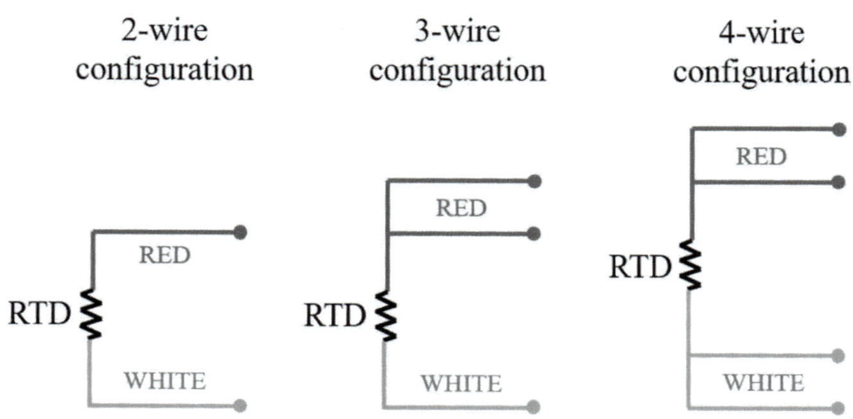

Fig. F-2. Lead configurations of 2, 3 and 4-wire RTD.

A typical circuit for 3-wire RTD is bridge configuration is given by

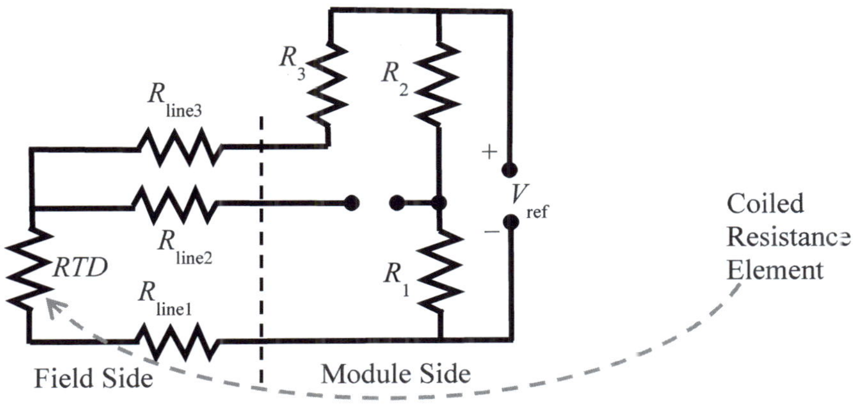

Fig. F-3. A typical circuit for the 3-wire RTD.

A reference on the RTDs provided by a manufacturer is: http://www.azom.com/article.aspx?ArticleID=5573.

The RTDs resistance versus temperature characteristics are given in https://www.dataforth.com/catalog/pdf/an105.pdf.

F.1.3 *Thermistors*

A thermistor (thermal resistor) is also a temperature sensor that uses materials' electrical resistance changes as a function of temperature like RTDs, but ceramic semiconductors in thermistors are composed of metal oxides, such as cobalt, manganese, or nickel oxide. Depending on materials used, thermistors are classified into two main types: NTC and PTC. Temperature coefficient of NTC thermistor is negative, that is, thermistor's resistance decreases as the temperature increases. Conversely, PTC thermistor has a positive temperature coefficient. Compared to thermocouples and RTDs, thermistors typically achieve a higher precision within a narrower temperature range. Simple voltage divider circuit or a Wheatstone bridge configuration is commonly used to read out the resistance relying on temperature change. The Wheatstone bridge provides higher precision and sensitivity through differential measurement, resulting in improving the signal to noise ratio.

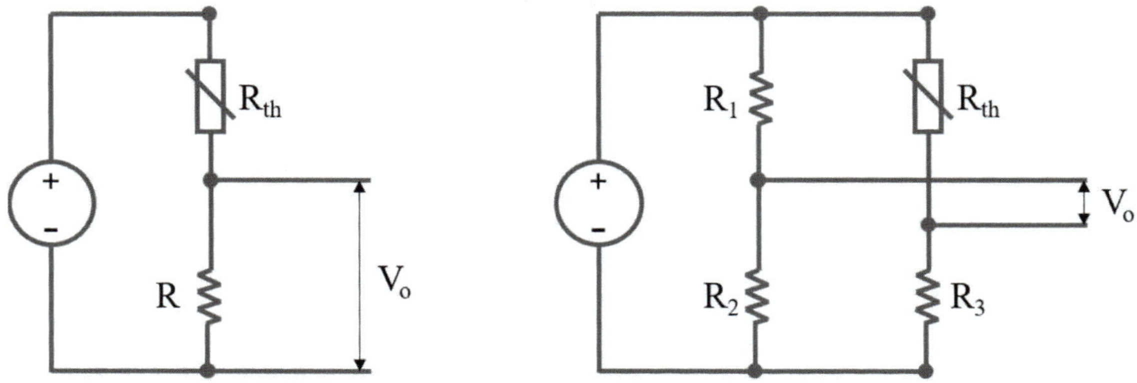

Fig. F-4. Simple voltage divider (left) and Wheatstone bridge circuit configuration (right) for the thermistor read-out.

While RTDs have the established standards, thermistors resistance versus temperature characteristics vary depending on the manufacturer. The additional characteristics and information for thermistors are given in https://www.te.com/usa-en/products/sensors/temperature-sensors/ntc-thermistors-sensors.html; https://www.amphenol-sensors.com/en/thermometrics/ntc-thermistors/epoxy/3447-type-nki; https://tewa-sensors.com/.

F.1.4 *Comparative characteristics*

Thermocouples, RTDs, and thermistors are all contact-type temperature sensors and have advantages and disadvantages. Therefore, it is necessary to consider them to decide which temperature sensor is the most appropriate for an application of interest. Thermocouples have the widest operating temperature range, the fastest response time, and low cost. Moreover, unlike RTDs and thermistors, thermocouples are self-powered, inducing electromotive forces, and require no external power source. Thus, the error known as self-heating does not occurs. For these reasons, thermocouples are commonly selected for typical or high temperature applications. However, the accuracy and noise susceptibility are worse compared to RTDs and thermistors, and an amplifier is required to amplify small electromotive signal. On the other hand, RTDs have the highest stability, excellent accuracy and reproducibility and a fairly linear relationship between temperature and electrical signals. However, due to high cost, slow response time and self-heating, RTDs are selected for specific applications. Lastly, thermistors have excellent accuracy, fast response time, low cost and good noise resistance while the temperature range is narrower than thermocouples and RTDs. Also, the linearity is worse than RTDs and self-heating error exists.

General characteristics of thermistors, RTDs, and thermocouples are summarized in Table F-2:

Table F-2. Comparison of general properties between thermocouples, RTDs, and thermistors.

	Thermocouple	RTD	Thermistor
Accuracy	0.5 to 5°C	0.1 to 1°C	0.05 to 1.5°C
Linearity	Non-linear	Fairly linear	Exponential
Temperature range	−200 to 1800°C	−260 to 850°C	−80 to 300°C
Response time	Fast (0.10 to 10s)	Slow (1 to 50s)	Fast (0.12 to 10s)
Noise susceptibility	Susceptible	Rarely susceptible	Rarely susceptible
Cost	Low	High	Low
Advantages	No Self-heating, Wide Temperature Range, Cost	Accuracy, Stability, Linearity	Sensitivity, Accuracy, Cost
Disadvantages	Accuracy, Stability, Non-linearity	Response time, Size, Cost	Non-linearity, Self-heating

F.1.5 *Thermographic infrared (IR) sensors*

A thermographic infrared sensor, or infrared thermometer, is a non-contacting sensor that remotely measures temperature using infrared radiation up to ~14 μm. Therefore, thermographic infrared sensors are preferred in situations where contact type sensors cannot be used. For example, these sensors can be selected in the case of the moving target object, or strong surrounding EM field. A basic thermographic infrared sensor consists of an optical lens to focus the infrared energy and a detector to convert the infrared energy into an electrical signal. There are various types of detectors depending on applications and wavelength range used such as micro-bolometer, photon-based detector, InGaAs, InSb, MCT (HgCdTe) and SLS (super lattice structure). By increasing the number of detector pixels, the device can create a thermal image of its field of view, functioning as an infrared thermal camera.

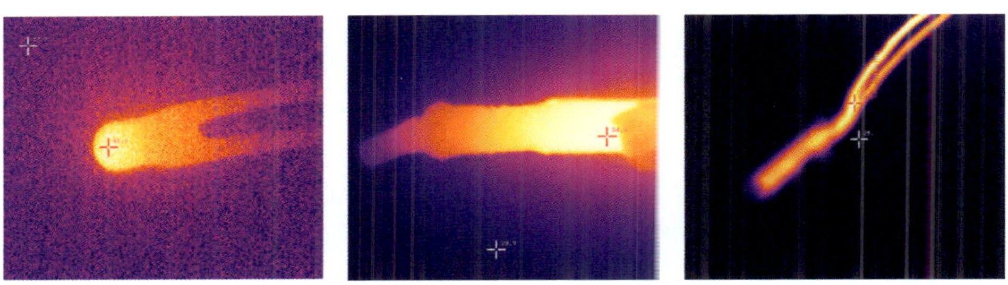

Fig. F-5. Examples of thermal images created by thermographic infrared camera.

One of the thermographic infrared cameras is shown in https://www.optris.com/optris-xi-400.

F.2 Sensors for displacement

F.2.1 *Linear variable differential transformer (LVDT)*

The LVDT is a displacement transducer that converts a position or linear displacement from a mechanical reference into a proportional electrical signal containing phase (for direction) and amplitude (for distance) information.

A detailed reference on LVDTs is: https://disensors.com/resources/principles-of-the-lvdt/.

F.2.2 *Proximity sensor*

A proximity sensor is a sensor able to detect the presence of nearby objects without any physical contact. Detailed references on various types of proximity sensors are given in http://www.fargocontrols.com/sensors.html.

F.2.2.1 *Acoustic sensor*

An acoustic proximity sensor uses a pulsed signal with a frequency above the range of human hearing that is generated by an oscillator and fed to a transducer that emits ultrasound pulses. These pulses are reflected by nearby objects and the reflections are captured by another transducer which converts the ultrasound back into high-frequency pulses. The delay between the transmitted and the received pulses is timed, indicating the distance to the obstruction.

F.2.2.2 *Inductive and capacitive sensors*

The operation of these sensors is based on a high frequency oscillator that creates a field in the close neighborhood of the sensing surface. The presence of a metallic object (inductive) or any material (capacitive) in the operating area causes a change of the oscillation amplitude. The rise or fall of oscillation magnitude is identified by a threshold circuit that changes the output state of the sensor.

F.2.2.3 *Photoelectric sensor*

These sensors use light sensitive elements to detect objects and consist of an emitter (light source) and a receiver. The photoelectric sensors can be subdivided into three configurations: i) direct reflection — emitter and receiver are housed together and use the light reflected directly off the object for detection, ii) reflection with reflector — emitter and receiver are housed together and a reflector is housed separately, iii) beam interruption — emitter and receiver are housed separately and detect an object when it interrupts the light beam between the emitter and the receiver.

F.2.2.4 *Magnetic sensor*

Magnetic sensors are actuated by the presence of a permanent magnet. They use two low reluctance ferromagnetic reeds enclosed in glass bulbs filled with inert gas. The reciprocal attraction of both reeds, caused by magnetic induction under magnetic field, establishes an electrical contact.

F.2.3 *Strain gauge*

The strain gage is one of the most widely used strain measurement sensors. Strain gages convert mechanical motion into an electrical signal. A change in capacitance, inductance, or resistance is proportional to the strain the sensor exhibits. For example, a wire held under tension gets longer and its cross-sectional area is reduced. This changes its resistance (R) in proportion to strain sensitivity (S) of the wire's resistance. When a strain is introduced, the strain sensitivity, which is also called the gage factor (GF), is given by:

$$GF = \frac{(\Delta R/R)}{(\Delta L/L)} = \frac{(\Delta R/R)}{\varepsilon}$$

where ε is the strain, L is the original length of the object, and ΔR is the change in resistance caused by strain. Strain gage transducers usually employ four strain gage elements electrically connected to form a Wheatstone bridge circuit. The latter is a divided bridge circuit used for the measurement of static or dynamic electrical resistance. The output voltage of the Wheatstone bridge is expressed in millivolts output per volt input:

$$V_O = V_{EX}\left[\frac{R_3}{R_3 + R_4} - \frac{R_2}{R_2 + R_2}\right].$$

The Wheatstone circuit is also well suited for temperature compensation.

The bridge configuration is given in Figure F-3. The bridge is considered balanced when $R_1/R_2 = R_4/R_3$ and, therefore, V_O equals zero. Any small change in the resistance of the sensing grid will throw the bridge out of balance, making it suitable for the detection of strain.

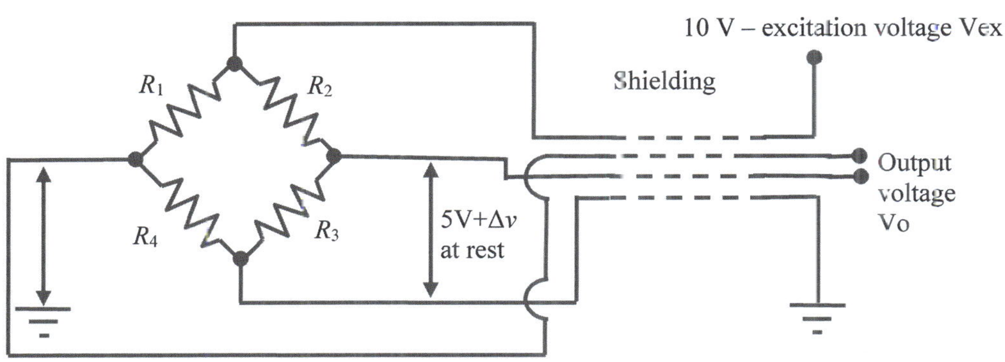

Fig. F-6. Wheatstone bridge circuit configuration for the strain gauge.

Omega Corporation is one of the main stain gauge manufacturers. Their tutorial is given at http://www.omega.com/literature/transactions/volume3/strain.html.
National Instruments tutorial is given in http://www.ni.com/white-paper/3642/en/.

F.2.4 *Extensometer*

Extensometers measure specimen elongation during tensile testing, characterizing strain. These instruments can perform contact and noncontact extensometry. Schematic diagram given in Figure F-3 shows the operating principles of various extensometers.

Clip-on extensometers are mounted directly onto the specimen, resulting in a high level of measurement accuracy. The range of a clip-on extensometer is limited to a few millimeters, and a load is applied directly to the specimen.

Sensor arm extensometers offer automatic operation and a large measurement range with high measurement accuracy. Precision designs with a very smooth and balanced mechanical operation apply minimum loading to the specimen

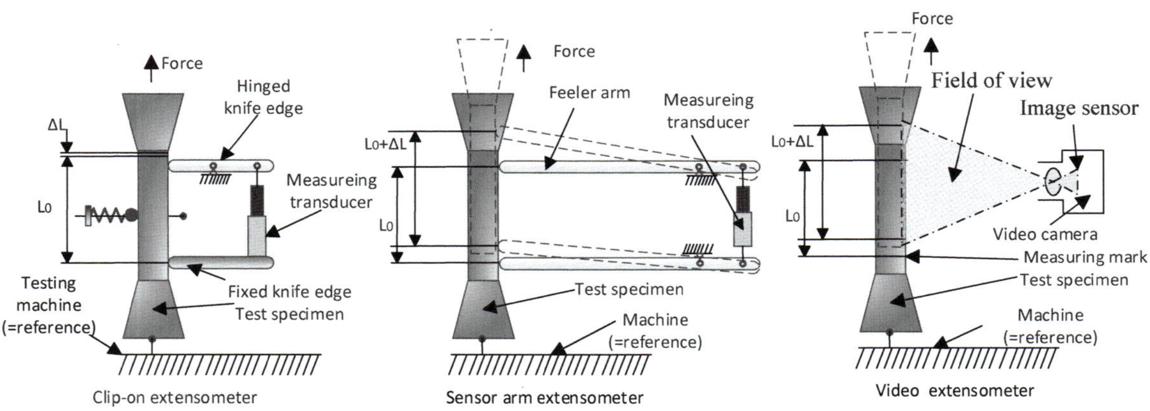

Fig. F-7. Extensometer operation schematics (adopted from the Zwick/Roell guide referenced below).

Noncontact video extensometers require measurement marks to be attached to the specimen, which are optically distinct from the surrounding area of the specimen, potentially introducing inaccuracies. The laser extensometer uses the unique structure of a specimen's surface as a "fingerprint" to generate a virtual measurement mark. Laser light directed on these measurement positions is reflected in various directions corresponding to the surface structure and creates a specific pattern of speckles.

Zwick/Roell provides detailed information on extensometers at
https://www.zwickroell.com/accessories/extensometers/.

F.2.5 *Encoder*

An encoder is an electromechanical device that measures motion or position. There are two types of mechanical configurations for optical encoders: rotary and linear. Most encoders use optical sensors to provide electrical signals in the form of pulse trains, which can, in turn, be translated into motion, direction, or position.

Rotary encoders are used to measure the rotational motion of a shaft. Figure 9 shows the fundamental components of a rotary encoder, which consists of a light-emitting diode (LED), a disk, and a light detector on the opposite side of the disk. The disk, which is mounted on the rotating shaft, has patterns of opaque and transparent sectors coded into the disk. As the disk rotates, the opaque segments block the light and, where the glass is clear, light is allowed to pass. This generates square-wave pulses, which can then be interpreted as position or motion.

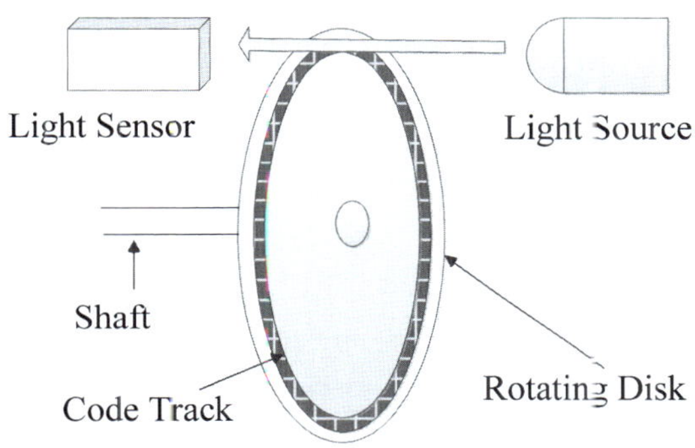

Fig. F-8. Rotary encoder for measuring the angular displacement and/or velocitty of a shaft.

National Instruments provides information on encooders at https://knowledge.ni.com/KnowledgeArticleDetails?id=kA03q000000x1riCAA&l=en-US.

F.2.6 *Magnetostrictive Sensor*

This sensor type has been introduced in Chapter 4. Extensive background on this sensor is given in the papers by Temposonics (formerly MTS Sensors) posted at https://www.temposonics.com/Products/IndustrialPositionSensors.

F.2.7 *Laser-based Sensor — LiDAR*

LiDAR (light detection and ranging) is a remote sensing instrument that uses a pulsed laser to recognize surrounding objects. LiDAR can determine the location of objects including distance, direction, and velocity by emitting a laser and measuring the time for the reflected light from objects. LiDAR works in much the same way as RADAR, but the only difference is that LiDAR uses a laser pulse instead of radio waves. For this reason, LiDAR provides higher resolution 3D images, with better accuracy and precision, than RADAR. On the other hand, RADAR is generally more robust against adverse environmental conditions than LiDAR, since radio waves have less absorption than the light waves when encountering objects. Moreover, the current high cost of LiDAR is another disadvantage. Due to their pros and cons, competition between LiDAR and RADAR is currently ongoing in autonomous driving sensor market.

F.3 Sensors for Velocity

F.3.1 *Variable reluctance sensor*

A variable reluctance sensor, depicted in Figure 6, converts mechanical motion into electric energy without direct contact.

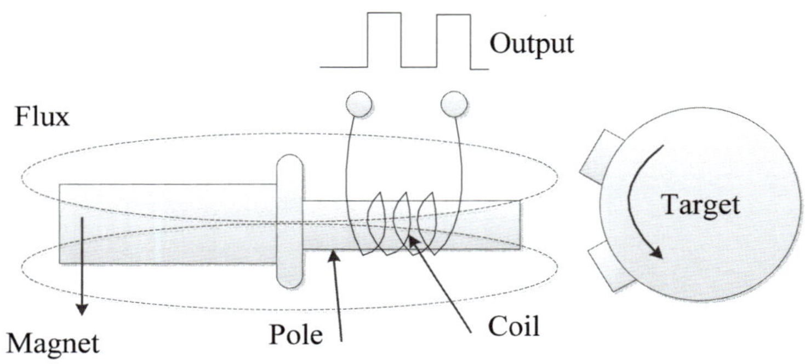

Fig. F-9. A variable reluctance sensor layout.

The sensor consists of a wire coil wound around a cylindrical typically ferrous magnetic material (pole piece). A magnet is attached behind the latter assembly, creating a magnetic field through the pole piece and the winding that protrudes beyond the latter assembly. When this magnetic field is interrupted by a moving object, the velocity of the latter can be calculated.

The details are given in https://www.ncbi.nlm.nih.gov/pmc/articles/PMC3264459/.

F.3.2 *Hall effect sensor*

The Hall voltage in a conductor or a semiconductor is generated when the latter two, while carrying current, are placed into a magnetic field. The induced voltage is perpendicular to both the current flow and the magnetic field and provides a very reliable reading for motion sensing. An excellent Honeywell tutorial on this type of sensors is given in https://sensing.honeywell.com/hallbook.pdf.

F.3.3 *Tachometer generator*

Tachometer generators are small AC or DC generators that output a voltage in proportion to the rotational speed of a shaft. They are capable of measuring speed and direction of rotation, but not

position. There are two basic types of tachometer generators: alternating current (AC) generators and direct current (DC) generators. Schematics of DC and AC tachometers are, respectively, given in
http://www.globalspec.com/learnmore/sensors_transducers_detectors/velocity_sensing/tachogenerators
and
http://electricaltopics.blogspot.jp/2015/02/construction-and-working-of-ac-tachometer-generator.html.

F.4 Force Sensors

F.4.1 *Accelerometer*

There are two gereal classes of accelerometers: with AC response, and with DC response. In an AC-response accelerometer the output is AC-coupled. An AC-coupled device cannot be used to measure static acceleration such as gravity and constant centrifugal acceleration. It is only suitable for measuring dynamic events. A DC-response accelerometer, on the other hand, is DC-coupled, and can respond down to zero Hertz. It therefore can be used to measure static, as well as dynamic acceleration. Layout of the latter accelerometer type, capacitive accelerometer, is given in Figure F-7.

An advanced accelerometer treatment is given in a Bruel and Kjaer 1987 book
http://www.bksv.com/doc/bb0694.pdf.

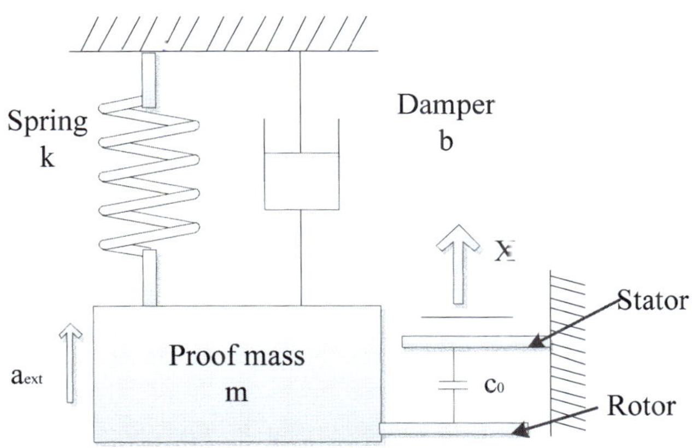

Fig. F-10. Layout of the capacitive accelerometer.

F.4.2 *Pressure sensor*

Pressure is defined as force per unit area that a fluid exerts on its surroundings. Pressure, P, is a function of force, F, and area, A:

$$P = \frac{F}{A}.$$

Pressure measurement is thoroughly covered in National Instruments tutorials http://www.ni.com/white-paper/3639/en/ and http://www.ni.com/white-paper/13034/en/.

F.4.3 *Load cell*

A load cell is a device that converts a force into an electrical signal. Load cells provide quick and precise measurements. Strain gauge load cell is the most common. Other load cells types are hydraulic (or hydrostatic), pneumatic, piezoelectric, capacitive, and piezo resistive. Details are in http://sensors-actuators-info.blogspot.com/2012/06/what-is-load-cell-and-how-load-cell.html.

F.5 PH sensor

The pH of a solution is a logarithmic measurement of the number of moles of hydrogen ions (H+) per liter of solution, which indicates how acidic or alkaline the latter is. A solution with a low pH value is called an "acid," while one with a high pH is called a "caustic." A good reference on pH sensing is http://www.sensorland.com/HowPage037.html.

F.6 Capacitive Touch Screen Sensor

A capacitive touch screen sensor is, at its core, a capacitor, which consists of two conductors, e.g. metal plates, separated by an insulator. The parameters which influence capacitance are:

$$C = \varepsilon \frac{A}{d}, \quad \varepsilon = \varepsilon_0 \cdot \varepsilon_r,$$

where C is the capacitance. ε_r is the relative permittivity, also called dielectric constant, of the insulating material between the plates. ε_0 is the permittivity of free space (8.854 × 10-12 F/m), A is the area of the plates, and d is the distance between the plates.Thus, the larger the area or the smaller the distance, the larger the capacitance is. The touch sensor electrode is one of the capacotor plates.

When this plate is touched (creating additional touch capacitance) or approached (inducing parasitic capacitance), the capacitance changes. Methods for measuring capacitance change include resonance frequency shift, frequency or amplitude modulation, and many others. Most methods require analog circuits that bring crosstalk, coupling, and noise sensitivity. A digital approach has smaller footprint and power consumption. An excellent reference on this sensor class is http://www.fujitsu.com/downloads/MICRO/fme/articles/fujitsu-whitepaper-capacitive-touch-sensors.pdf.

F.7 Gyroscopes

A gyroscope (gyro) is used for navigation and measurement of angular velocity. Gyroscopes measure rotational velocity in 1, 2, or 3 directions. 3-axis gyroscopes are often implemented with a 3-axis accelerometer to provide a full 6 degree-of-freedom motion tracking system.

Gyroscopes have evolved from mechanical-inertial spinning devices consisting of rotors, axles, and gimbals, like the one shown in Figure F-8, to various electronic and optical versions. Each exploits some physical property of the system for detecting rotational velocity about some axis.

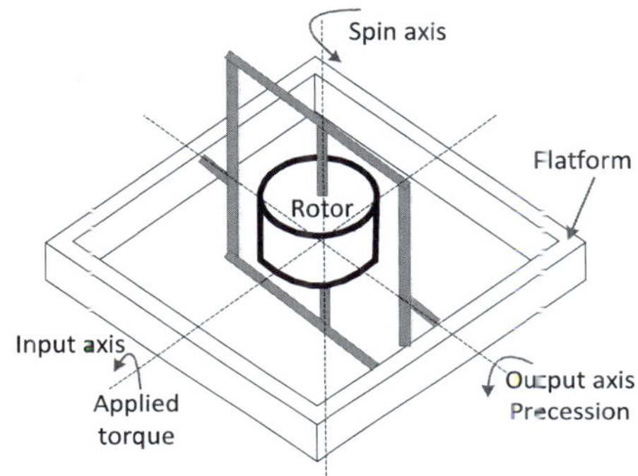

Fig. F-11. Layout of the classical gyroscope.

There are three basic types of gyroscope: rotary (classical), vibrating structure, and optical. At present, gyros are used eeverywhere, from drones to cellphones. Further information is given in https://en.wikipedia.org/wiki/Gyroscope.

F.8 Electromagnetic Actuator

F.8.1 *DC motor*

DC motor consists of stator and rotor. The stator includes the housing, permanent magnets, and brushes. The rotor consists of the output shaft, windings and commutator. There are numerous books on electric motors. A DC motor schematics and a brief description are given in http://hades.mech.northwestern.edu/index.php/Brushed_DC_Motor_Theory.

F.8.2 *Stepper motor*

A stepper motor is a "digitized" electric motor. The rotor moves in discrete steps, rather than rotating continuously. When stopped, but energized, a stepper (short for stepper motor) holds its load steady with a holding torque. The details are given in http://www.allaboutcircuits.com/vol_2/chpt_13/5.html.

F.8.3 *Brushless DC motor*

Brushless DC motors are similar to AC synchronous motors. The major difference is that synchronous motors develop a sinusoidal back EMF, as compared to a rectangular, or trapezoidal, back EMF for brushless DC motors. Both have stator created rotating magnetic fields producing torque in a magnetic rotor. Details are in http://www.allaboutcircuits.com/vol_2/chpt_13/6.html.

F.9 Fluid mechanical actuators

Hydraulics and pneumatics are two types of actuation that convert the fluid pressure into mechanical motion. Although their application is different, they are governed by the same principles and have similar configurations. We have introduced the description of the operation of the hydraulic actuator and its analytical model in Chapter 1. The numerical model is given in Appendices G and H.

Both hydraulics and pneumatics utilize the basic property of real fluids that "pressure exerted anywhere in a confined incompressible fluid is transmitted equally in all directions throughout the fluid." The details on various fluid actuator types are given in https://silo.tips/queue/actuation-systems-control-code-command-signal-microprocessor-or-microcontroller?&queue_id=-1&v=1632197973&u=MjYyMDowOmUwMDo0MDNiOjoxN2M=

F.10 Sensors/actuators for computer audio and machine vision

F.10.1 *Microphones and loudspeakers*

Modern smart computer audio interface uses a microphone, often abbreviated as mic, and a loudspeaker, or just speaker. A typical all-in-one (mic+DSP+speaker) card is RASPIAUDIO soundcard https://raspiaudio.com/audio-2-2. Mic is an audio transducer converting pressure fluctuations into electrical signals. Conversion is performed through a diaphragm attached to a wound coil or a ribbon moving in magnetic field (dynamic mic) or moving in a capacitive capsule (condenser mic). Speaker performs the opposite action, using the same two principles. References https://en.wikibooks.org/wiki/Engineering_Acoustics/Microphone_Design_and_Operation and https://en.wikipedia.org/wiki/Electrostatic_loudspeaker provide mic and speaker design basics.

F.10.2 *Digital image sensors*

Digital image sensors — convertors of visible light spectrum into an electrical signal — are key components in machine vision systems. Their design and classification details are presented in https://thinklucid.com/tech-briefs/understanding-digital-image-sensors/. A capable computer vision setup for autonomous operation is provided by Nvidia Jetson TX2 with the camera given in https://www.e-consystems.com/13mp-nvidia-jetson-tx2-camera-board.asp. Also, Raspberry Pi offers very inexpensive HQ Sony IMX477R 12.3 MP camera compatible with many DSP boards.

Appendix G

Euler-Bernoulli and Timoshenko Beam Models and Their Use in Software Testbed Development

G.1 Introduction

This appendix presents the theoretical background on the Euler-Bernoulli and Timoshenko beam models and the embedding of the latter into the software testbed. While the mold simulator software testbed development started with the former model, the latter one was selected due to its better numerical properties and tighter modeling fit to the actual hardware testbed behavior. To build the necessary background, the classical single beam results are first surveyed, based, in part, on [1]. However, since the testbed model, first introduced in [5], employs two coupled beams and has to be run sufficiently fast, its presentation is based on novel analytical and numerical results reported in [6]. The references are given at the end of this appendix to make it self-contained.

The beam theories address the transverse displacement of a uniform beam, where the bending effect is recognized as the most important factor [2].

The Euler-Bernoulli model [1], [4] includes the effects of strain energy due to bending and kinetic energy due to lateral displacement. This model is commonly used because of its simplicity and reasonable engineering approximations for many problems. However, the model tends to slightly overestimate the natural frequencies.

A more complete model [3], proposed by Stepan (Stephen) Timoshenko, commonly considered to be the father of modern Engineering Mechanics, adds the effect of shear and rotation to the Euler-Bernoulli beam. The model is a major improvement for non-slender beams and for high-frequency responses where shear or rotary effects are not negligible.

G.2 Derivation of Equations of Motion

G.2.1 *Equations of motion of the Euler-Bernoulli beam model*

The equations of motion are obtained by using Hamilton's variational principle. According to this principle, the "action" of a system over any time interval $[t_1, t_2]$, denoted as S, is defined as

$$S := \int_{t_1}^{t_2} L\big(q(t),t\big)\,dt.$$

In this definition, $q(t)$ is the "generalized coordinate" of the system, that is, the "state" of the system at time t. Depending on the complexity of the system, q could be a single number (e.g. temperature in a heat transfer problem), a vector (e.g. x, y, and z coordinates of a moving object), or a function (in our case, the displacement $v(x,t)$ of the beam at a point x). The Lagrangian L is defined as the difference between the kinetic and the potential energy ($KE - PE$) of the system.

Hamilton's principle focuses on the "variation", denoted as δS, of this action quantity. The "variation" can be thought of as the first-order (linear) approximation of the change in S for a small change in the generalized coordinate q:

$$\int_{t_1}^{t_2} L(q+\delta q,t)\,dt \approx S + \delta S = S + \frac{\delta S}{\delta q}\delta q.$$

Although the term $\delta S / \delta q$ looks like a derivative, it is not, since q may actually be a function. However, it has the same intuitive meaning, carrying the same general idea as the linear approximation of a function:

$$f(x+\Delta x) \approx f(x) + \Delta f = f(x) + f'(x)\Delta x.$$

Therefore, we can calculate the variation by looking at the difference

$$\delta S \approx \int_{t_1}^{t_2} L(q+\delta q,t)\,dt - \int_{t_1}^{t_2} L(q,t)\,dt \qquad \text{(G-1)}$$

and keeping only the terms that are linear in δq.

Hamilton's principle, also called the principle of least action, says that the action is minimized along trajectories of physical systems. Being similar to the first-order necessary conditions for a minimum of functions, Hamilton's principle says that the variation of the action must therefore be equal to zero for any variation of the generalized coordinate. This principle can also be generalized to systems with non-conservative forces (most commonly, gravity). In this case, the variation of the action of the system is equal to the work done by non-conservative forces, or in mathematical form

$$\delta S + \int_0^l \int_{t_0}^{t_1} f(q(x,t),t)\,\delta q\,dx\,dt = 0. \qquad \text{(G-2)}$$

G.2.1.1 *Potential energy due to the bending*

Suppose the beam at rest (no load) is divided into vertical cross-sections. These cross-sections can be assumed to remain flat and perpendicular to the neutral axis (dash-dotted line in Figure G-1) when a load is applied. This assumption is true only in the case of pure bending (constant moments and no transverse forces, which cause shear). But, in the case of longitudinal beams (length is much greater than the cross-section height and width), deformation due to the shear is very small, and therefore can be neglected.

In this appendix, x denotes the longitudinal direction (along the length of the beam) and y denotes the distance from the neutral axis. The presentation is confined to only two-dimensional bending. For the bent beam, $y(x,t)$ denotes the vertical displacement of the beam neutral axis at a point x and time t and $\psi(x,t)$ denotes the angle that the local cross-section of the beam rotates due to bending.

Figure G-1 shows an element, i.e. a very small portion, of the beam being bent under a load. The dashed line shows the original unloaded shape of the beam element. The figure shows how, under load, some areas of the beam are stretched, other are compressed, and only the areas on the neutral axis have the same length dx before and after bending. From this figure, the change in angle of the beam cross-section over the element, $d\psi$, and the length of the beam element, dx, are related by

$$r d\psi = dx, \tag{G-3}$$

where r is the beam curvature radius.

The strain (the relative longitudinal elongation or compression) of the beam in the x direction at a distance y_1 from the neutral axis is therefore equal to

$$\varepsilon = \frac{\Delta L}{L} = \frac{(r + y_1)d\psi - r d\psi}{r d\psi} = \frac{y_1}{r}.$$

The bending moment is given by

$$M = \int_{a_b} \sigma_x y_1 da_b,$$

where σ_x is the local normal stress in the x-direction, acting on the cross-section, and a_b is cross-section area of beam.

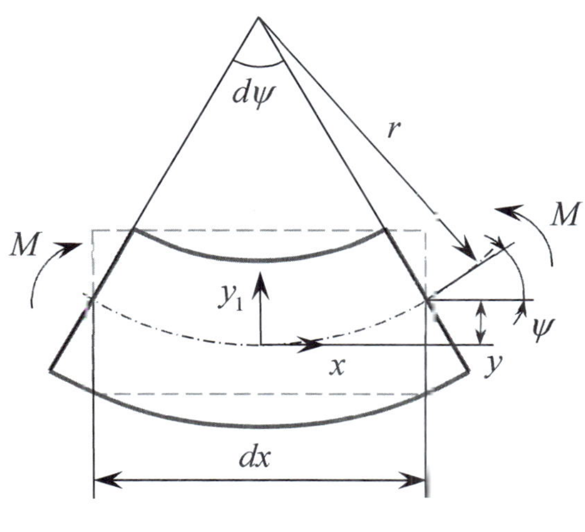

Fig. G-1. Element of a bent beam.

By using Hooke's law $\sigma = \varepsilon E$ and two previous equations we have

$$M = \int_A E \frac{y_1}{r} y_1 da_b = \frac{E}{r} \int_A y_1^2 da_b = \frac{EI}{r}, \tag{G-4}$$

where E is the modulus of elasticity, and I is the moment of inertia of the beam cross-section. Treating the beam in bending as a linear elastic spring, and combining equations (G-3) and (G-4), the change in energy needed to bend the beam element by an angle $d\psi$ using the moment M is

$$PE_{bending} = \frac{1}{2} M d\psi = \frac{1}{2} \frac{EI}{r} d\psi = \frac{1}{2} EI \left(\frac{d\psi}{dx} \right)^2 dx .$$

Now, adding up all the beam elements together, the total linear elastic potential energy caused by bending the beam is:

$$PE_{bending} = \frac{1}{2} \int_0^l M d\psi = \frac{1}{2} \int_0^l \frac{EI}{r} d\psi = \frac{1}{2} EI \int_0^l \left(\frac{d\psi}{dx} \right)^2 dx. \tag{G-5}$$

In practice, the deflection of the beam is usually small, so ψ can be approximated as

$$\psi \approx \tan \psi = \frac{\partial y}{\partial x}. \tag{G-6}$$

Finally, we have

$$PE_{bending} = \frac{1}{2} EI \int_0^l \left(\frac{\partial^2 y(x,t)}{\partial x^2} \right)^2 dx, \tag{G-7}$$

where $y(x,t)$ is the transverse deflection at the axial location x and time t, as seen in Figure G-1.

G.2.1.2 *Lateral kinetic energy*

The kinetic energy due to the lateral displacement in the dimensional notation can be found as

$$KE_{trans} = \frac{m(v(x,t))^2}{2} = \frac{1}{2} \int_0^l \rho a_b \left(\frac{\partial y(x,t)}{\partial t} \right)^2 dx, \tag{G-8}$$

where $v(x,t)$ is the velocity of the displacement y, m is the mass of the beam, ρ is the density of the beam, and a_b is the beam cross-sectional area.

G.2.1.3 *Equation of motion of the bending beam*

In order to find the equation of motion by using Hamilton's principle, we need to find Lagrangian first. From the equations (G-7) and (G-8), the Lagrangian is given by

$$L = \frac{1}{2}\int_0^l \left(\rho a_b \left(\frac{\partial y(x,t)}{\partial t} \right)^2 - EI \left(\frac{\partial^2 y(x,t)}{\partial x^2} \right)^2 \right) dx.$$

Let us consider the variation of the square of the function $q(x,t)$ from (G-1):

$$\delta q^2 = (q+\delta q)^2 - q^2 = 2q\delta q + (\delta q)^2.$$

Since the variation is the first order approximation we obtain

$$\delta q^2 = 2q\delta q. \tag{G-9}$$

Using (G-9) the variation of terms in Lagrangian is

$$\delta\left(\frac{\partial y}{\partial t}\right)^2 = 2\frac{\partial y}{\partial t}\delta\left(\frac{\partial y}{\partial t}\right), \delta\left(\frac{\partial^2 y}{\partial t^2}\right)^2 = 2\frac{\partial^2 y}{\partial t^2}\delta\left(\frac{\partial^2 y}{\partial t^2}\right).$$

Therefore, the variation of the action δS has the form

$$\delta S = \int_{t_0}^{t_1}\int_0^l \left(\rho a_b \frac{\partial y}{\partial t}\delta\left(\frac{\partial y}{\partial t}\right) - EI\frac{\partial^2 y}{\partial x^2}\delta\left(\frac{\partial^2 y}{\partial x^2}\right)\right) dxdt. \tag{G-10}$$

Integration of the first term gives

$$\int_{t_0}^{t_1}\int_0^l \rho a_b \frac{\partial y}{\partial t}\delta\left(\frac{\partial y}{\partial t}\right) dxdt = \int_0^l \left(\rho a_b \frac{\partial y}{\partial t}\delta y\bigg|_{t_0}^{t_1} - \int_{t_0}^{t_1} \rho a_b \frac{\partial^2 y}{\partial t^2}\delta y dt \right) dx. \tag{G-11}$$

The second integral has the form

$$\int_{t_0}^{t_1}\int_0^l EI\frac{\partial^2 y}{\partial x^2}\delta\left(\frac{\partial^2 y}{\partial x^2}\right) dxdt = \int_{t_0}^{t_1}\left(EI\frac{\partial^2 y}{\partial x^2}\delta\left(\frac{\partial y}{\partial x}\right)\bigg|_0^l - \int_0^l EI\frac{\partial^3 y}{\partial x^3}\delta\frac{\partial y}{\partial x} dx \right) dt$$

$$= \int_{t_0}^{t_1}\left(EI\frac{\partial^2 y}{\partial x^2}\delta\left(\frac{\partial y}{\partial x}\right)\bigg|_0^l - EI\frac{\partial^3 y}{\partial x^3}\delta y\bigg|_0^l + \int_0^l EI\frac{\partial^4 y}{\partial x^4}\delta y dx \right) dt.$$

It needs to be noted that the variation at the endpoints is equal to zero:

$$\delta y(x,t_0) = \delta y(x,t_1) = 0.$$

Therefore, the variation in (G-10) can be given as

$$\delta S = -\int_0^l \int_{t_0}^{t_1} \left(\rho a_b \frac{\partial^2 y}{\partial t^2} + EI \frac{\partial^4 y}{\partial x^4} \right) \delta y dt dx + \int_{t_0}^{t_1} \left(-EI \frac{\partial^2 y}{\partial x^2} \delta \left(\frac{\partial y}{\partial x} \right) + EI \frac{\partial^3 y}{\partial x^3} \delta y \right) \Bigg|_0^l dt.$$

Applying the extended Hamilton principle (G-2), the equation of motion for the bending beam takes the form

$$\rho a_b \frac{\partial^2 y(x,t)}{\partial t^2} + EI \frac{\partial^4 y(x,t)}{\partial x^4} = f(x,t), \tag{G-12}$$

with the boundary conditions determined by

$$\frac{\partial^2 y}{\partial x^2} \delta \left(\frac{\partial y}{\partial x} \right) \Bigg|_0^l = 0, \qquad \frac{\partial^3 y}{\partial x^3} \delta y \Bigg|_0^l = 0.$$

In these boundary conditions y is the displacement, the first derivative $\partial y/\partial x$ is the slope, the second derivative $\partial^2 y/\partial x^2$ is the moment, and the third derivative $\partial^3 y/\partial x^3$ is the shear. The condition $\delta y = 0$ means that the displacement is fixed. For practical cases, this can be assumed to mean "fixed at 0". There are four possible end conditions that satisfy the equation (G-12):

$$\begin{aligned} &(a)\ \frac{\partial^2 y}{\partial x^2} = 0,\ \ y = 0, \text{ for a hinged end;} && (b)\ \frac{\partial y}{\partial x} = 0,\ \ y = 0, \text{ for a clamped end;} \\ &(c)\ \frac{\partial y}{\partial x} = 0,\ \ \frac{\partial^3 y}{\partial x^3} = 0, \text{ for a sliding end;} && (d)\ \frac{\partial^2 y}{\partial x^2} = 0,\ \ \frac{\partial^3 y}{\partial x^3} = 0, \text{ for a free end.} \end{aligned} \tag{G-13}$$

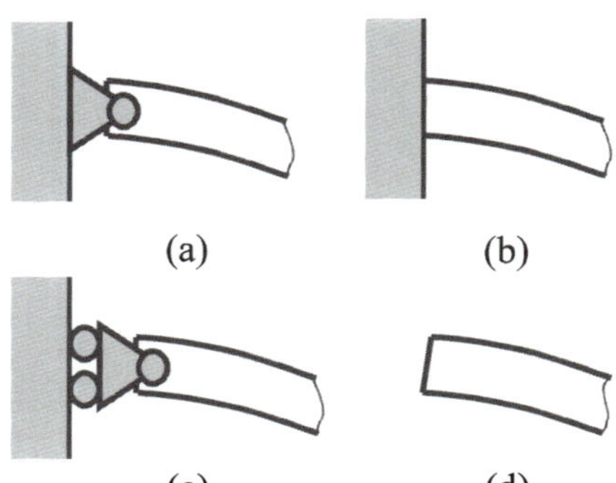

Fig. G-2. Beam end conditions: a — hinged, b — clamped, c — sliding, d — free.

G.2.2 *Equations of motion of the Timoshenko beam model*

The Timoshenko beam model adds kinetic energy due to the rotary inertia and deformation due to the shear distortion to the Euler-Bernoulli model. In this case, the total angle of rotation of the beam cross-section will be the sum of the angle of rotation due to bending, $\psi(x,t)$, and the angle of distortion due to shear, $\beta(x,t)$. The sum is approximately equal to the first derivative of the deflection $y(x,t)$ (Figure G-3). Therefore, instead of equation (G-6) for the Euler-Bernoulli beam model, the Timoshenko beam model has

$$\psi(x,t)+\beta(x,t)\approx\frac{\partial y(x,t)}{\partial x}.$$

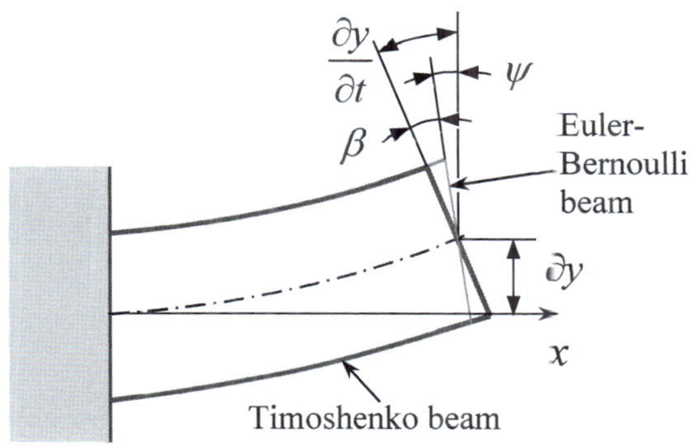

Fig. G-3. Deformation of a beam in different models.

Potential energy due to the bending

Using the equation above, the potential energy due to bending from equation (G-5) has the form

$$PE_{bending}=\frac{1}{2}\int_0^l EI\left(\frac{\partial\psi(x,t)}{\partial x}\right)^2 dx. \tag{G-15}$$

Potential energy due to the shear

The shear deformation of a beam element is shown in Figure G-4. The dot-dashed line is the shape of the beam element before the deformation caused by the shear force S. From the elementary bending theory and the Hooke's law the shear force is given by

$$S=k'Ga_b\beta,$$

where k' is the shape of the cross-section factor, G is the shear modulus, and β is the angle of rotation of the cross-section due to the shear.

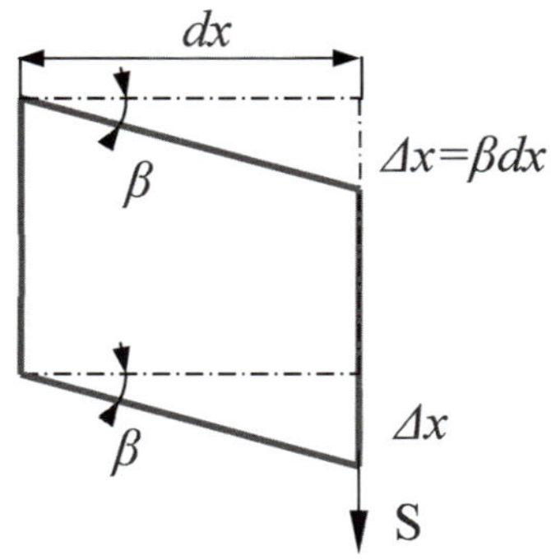

Fig. G-4. Shear.

Therefore, the potential energy due to the shear force is given by

$$PE_{shear} = \frac{1}{2}\int_0^l S\beta dx = \frac{1}{2}\int_0^l k'Ga_b\beta^2 dx.$$

Taking into account (G-14), the potential energy is given by

$$PE_{shear} = \frac{1}{2}\int_0^l k'Ga_b\left(\frac{\partial y(x,t)}{\partial x} - \psi(x,t)\right)^2 dx. \tag{G-16}$$

The kinetic energy of the rotation due to the bending

It is assumed here that the shear distortion does not add to the rotational kinetic energy. Therefore, rotational kinetic energy includes only the rotation due to bending and can be written as

$$KE_{rot} = \frac{m\left(\omega(x,t)r\right)^2}{2} = \frac{1}{2}\int_0^l \frac{\rho}{2}\left(\omega(x,t)\right)^2 \int_0^A r^2 dA dx = \frac{1}{2}\int_0^l \rho I\left(\frac{\partial \psi(x,t)}{\partial t}\right)^2 dx, \tag{G-17}$$

where $\omega(x,t) = \partial\psi/\partial t$ is the angular velocity of the beam rotation.

Equations of motion arising from Timoshenko model

Combining equations (G-15)–(G-17) with the equation of kinetic energy from the Euler-Bernoulli beam model (G-8), the Lagrangian is given by

$$L = \frac{1}{2}\int_0^l \left[\rho a_b\left(\frac{\partial y(x,t)}{\partial t}\right)^2 + \rho I\left(\frac{\partial \psi(x,t)}{\partial t}\right)^2 - EI\left(\frac{\partial \psi(x,t)}{\partial x}\right)^2 \right.$$
$$\left. -k'Ga_b\left(\frac{\partial y(x,t)}{\partial x} - \psi(x,t)\right)^2\right] dx.$$

The variations of terms in Lagrangian can be found from (G-9):

$$\delta\left(\frac{\partial y}{\partial t}\right)^2 = 2\frac{\partial y}{\partial t}\delta\left(\frac{\partial y}{\partial t}\right),\quad \delta\left(\frac{\partial \psi}{\partial t}\right)^2 = 2\frac{\partial \psi}{\partial t}\delta\left(\frac{\partial \psi}{\partial t}\right),\quad \delta\left(\frac{\partial \psi}{\partial x}\right)^2 = 2\frac{\partial \psi}{\partial x}\delta\left(\frac{\partial \psi}{\partial x}\right).$$

Let us consider the variation of the last term:

$$\delta\left(k'Ga_b\left(\frac{\partial y}{\partial x}-\psi\right)^2\right) = k'Ga_b\delta\left(\left(\frac{\partial y}{\partial x}\right)^2 - 2\frac{\partial y}{\partial x}\psi + \psi^2\right) = 2k'Ga_b\left(\frac{\partial y}{\partial x}\delta\left(\frac{\partial y}{\partial x}\right) - \delta\left(\frac{\partial y}{\partial x}\psi\right) + \psi\delta\psi\right).$$

The term in the middle can be presented as

$$\delta\left(\frac{\partial y}{\partial x}\psi\right) = \left(\frac{\partial y}{\partial x} + \delta\frac{\partial y}{\partial x}\right)(\psi + \delta\psi) - \frac{\partial y}{\partial x}\psi = \delta\frac{\partial y}{\partial x}\psi + \frac{\partial y}{\partial x}\delta\psi + \delta\frac{\partial y}{\partial x}\delta\psi.$$

Since the last term has the second order, but the variation is the first order approximation, we obtain

$$\delta\left(\frac{\partial y}{\partial x}\psi\right) = \delta\frac{\partial y}{\partial x}\psi + \frac{\partial y}{\partial x}\delta\psi.$$

Then, the variation of action δS after substituting formulas above has form

$$\begin{aligned}\delta S = \int_{t_0}^{t_1}\int_0^l \Bigg(&\rho a_b \frac{\partial y}{\partial t}\delta\left(\frac{\partial y}{\partial t}\right) + \rho I\frac{\partial \psi}{\partial t}\delta\left(\frac{\partial \psi}{\partial t}\right) - EI\frac{\partial \psi}{\partial x}\delta\left(\frac{\partial \psi}{\partial x}\right) \\ &- k'Ga_b\left(\frac{\partial y}{\partial x}\delta\left(\frac{\partial y}{\partial x}\right) - \delta\left(\frac{\partial y}{\partial x}\right)\psi - \frac{\partial y}{\partial x}\delta\psi + \psi\delta\psi\right)\Bigg)dxdt.\end{aligned} \quad \text{(G-18)}$$

The integrals of the terms in (G-18) could be found integrating by parts as in (G-11). For example

$$k'Ga_b\int_{t_0}^{t_1}\int_0^l \delta\frac{\partial y}{\partial x}\psi dxdt = k'Ga_b\int_{t_0}^{t_1}\left(\psi\delta y\Big|_0^l - \int_0^l \frac{\partial \psi}{\partial x}\delta y\right)dt.$$

Finally, the equation (G-18) has the form

$$\begin{aligned}\delta S = \int_0^l\int_{t_0}^{t_1}\Bigg(&\left(-\rho a_b\frac{\partial^2 y}{\partial t^2} + k'Ga_b\left(\frac{\partial^2 y}{\partial x^2} - \frac{\partial \psi}{\partial x}\right)\right)\delta y + \left(-\rho I\frac{\partial^2 \psi}{\partial t^2} + EI\frac{\partial^2 \psi}{\partial x^2} + k'Ga_b\left(\frac{\partial y}{\partial x} - \psi\right)\right)\delta\psi\Bigg)dtdx \\ &- \int_{t_0}^{t_1}\left(\frac{\partial \psi}{\partial x}\delta\psi + k'Ga_b\left(\frac{\partial y}{\partial x} + \psi\right)\delta y\Big|_0^l\right)dt.\end{aligned}$$

Application of the Hamilton principle (G-2) yields the equations of the beam motion in the form

$$\rho a_b \frac{\partial^2 y(x,t)}{\partial t^2} - k'Ga_b\left(\frac{\partial^2 y(x,t)}{\partial x^2} - \frac{\partial \psi(x,t)}{\partial x}\right) = f(x,t),$$
$$\rho I \frac{\partial^2 \psi(x,t)}{\partial t^2} - EI\frac{d^2\psi(x,t)}{\partial x^2} - k'Ga_b\left(\frac{\partial y(x,t)}{\partial x} - \psi(x,t)\right) = 0, \quad \text{(G-19)}$$

with the boundary conditions determined by

$$\frac{\partial \psi}{\partial x}\delta\psi\bigg|_0^l = 0, \qquad k'Ga_b\left(\frac{\partial y}{\partial x} - \psi\right)\delta y\bigg|_0^l = 0. \quad \text{(G-20)}$$

The physical quantities here are the same as in (G-12), but with different notation: y is the displacement, ψ is the angle of rotation due to the bending moment, $\partial\psi/\partial x$ is the moment, and $k'Ga_b\left(\partial y/\partial x - \psi(x,t)\right)$ is the shear. The four possible boundary conditions are:

$\frac{\partial \psi}{\partial x} = 0,\ \ y = 0$ for a hinged end; $\psi = 0,\ \ y = 0$ for a clamped end;

$\frac{\partial \psi}{\partial x} = 0,\ \ k'Ga_b\left(\frac{\partial y}{\partial x} - \psi\right) = 0$ for a sliding end; $\psi = 0,\ \ k'Ga_b\left(\frac{\partial y}{\partial x} - \psi\right) = 0$ for a free end.

G.3 Modal Analysis of Beam Models

G.3.1 *Natural frequencies for the Euler-Bernoulli beam model*

The equation of motion with both boundary and the initial conditions forms an initial-boundary-value problem, which can be solved by the separation of variables and eigenfunction expansion method.

In order to find natural frequencies and eigenfunctions, consider a homogeneous problem ($f(x, t) = 0$). The equation of motion can be separated into two ordinary differential equations by representing the lateral displacement $y(x, t)$ as the product of two functions $y(x, t) = y(x)\,T(t)$. This separation reduces solving the single PDE (G-12) to solving for y and T separately two ordinary differential equations:

$$\frac{d^2T(t)}{dt^2} + \omega^2 T(t) = 0, \qquad \frac{d^4 y(x)}{dx^4} - a^4 y(x) = 0,$$

where the term a is the ***wave number*** (***cycles per distance***), defined by the relationship

$$a^4 = \frac{\rho A\omega^2}{EI}. \quad \text{(G-21)}$$

The wave number shows the number of spatial cycles (wavelengths) in a 2π distance of the beam length. From the equation above the functions $T(t)$ and $y(x)$ have the form

$$\begin{aligned} T(t) &= d_1 \sin \omega t + d_2 \cos \omega t, \\ y(x) &= C_1 \sin ax + C_2 \cos ax + C_3 \sinh ax + C_4 \cosh ax, \end{aligned} \tag{G-22}$$

where d_i and C_i are constant coefficients.

The separated boundary conditions equations expressed in terms of $y(x)$ are given by

$$\frac{d^2y}{dx^2}\delta\left(\frac{dy}{dx}\right)\Bigg|_0^l = 0, \qquad \frac{d^3y}{dx^3}\delta y\Bigg|_0^l = 0.$$

The four possible end conditions in terms of $y(x)$ can be obtained as in equation (G-13).

In order to obtain the natural frequencies we need to apply the four sets of boundary conditions to the spatial solution. In Euler-Bernoulli model the frequency equations for the free-free and clamped-clamped cases are the same [1].

The actual frequency of vibration by using (G-21) after solving the frequency equations is

$$\omega = \sqrt{\frac{EI}{\rho A}}a^2,$$

where a is specific for different boundary conditions and depends on the length of the beam.

Example G-1:

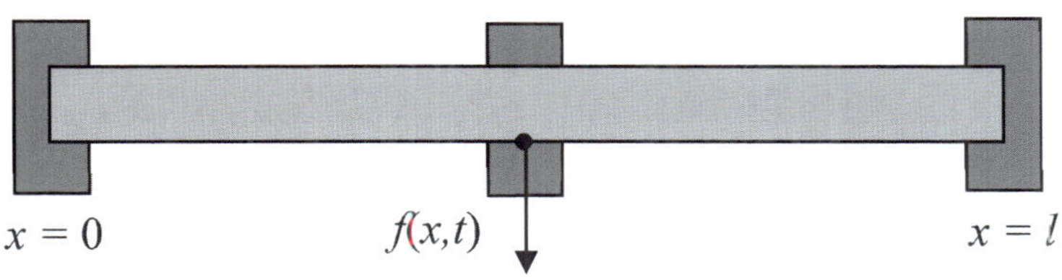

Fig. G-5. Beam-mass system.

The mechanical system with horizontal steel beam of rectangular cross-section clamped at both ends and operating on the midpoint by a large electromagnet is presented in Figure G-5 [4].

The governing differential equation from (G-12) and the boundary conditions are given by

$$m_b\frac{\partial^2 y}{\partial t^2} + EI\frac{\partial^4 y}{\partial x^4} = f(x,t),$$

$$y(x,t)\big|_0^l = 0, \qquad \frac{\partial y}{\partial x}\Bigg|_0^l = 0$$

In this equation, m_b is mass of beam per unit length, $m_b = \rho a_b$, and $f(x,t)$ is the distributed force per unit length.

Then the homogeneous problem solution can be given by

$$y(x,t) = \sum_{n=1}^{\infty} y_n(x)\left(d_{1n}e^{i\omega_n t} + d_{2n}e^{-i\omega_n t}\right)$$

where n is the number of the mode, d_{1n} and d_{2n} are the coefficients of the time dependent function, ω_n is the natural frequency of the beam, and the spatial function $y_n(x)$ is determined by (G-22). Application of the boundary conditions to the spatial solution leads to the eigenvalue equation

$$\cos a_n l \cosh a_n l = 1,$$

with an eigenvalue a_n given by

$$a_n \approx \frac{(2n+1)\pi}{2l}.$$

The first two mode shapes of the beam are plotted in Figure G-6 below.

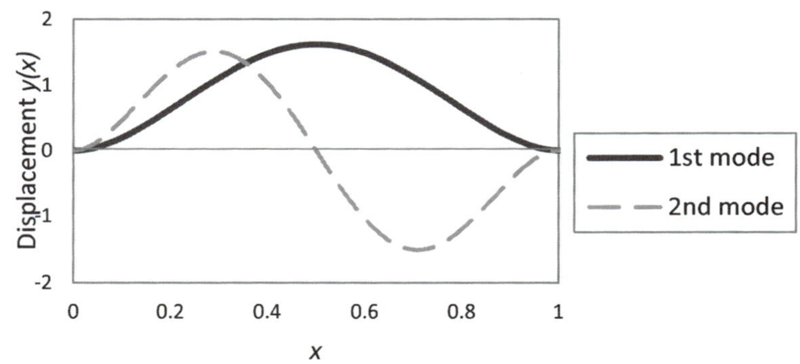

Fig. G-6. The first two mode shapes.

G.3.2 *Natural frequencies of the Timoshenko beam model*

As in the previous section, consider the homogeneous problem ($f(x,t) = 0$). Because there are two variables, the equation can be decoupled by, first, solving the first equation in (G-19) for $\partial\psi/\partial x$ in terms of y and its derivatives. Then, by taking the spatial derivative of second equation, and substituting the result from the first equation for $\partial\psi/\partial x$, an equation involving only y and its derivatives will be obtained. A similar process can be done for ψ. The resulting decoupled equations are

$$\frac{\partial^4 y}{\partial x^4} - \left(\frac{\rho}{E} + \frac{\rho}{k'G}\right)\frac{\partial^4 y}{\partial x^2 \partial t^2} + \frac{\rho a_b}{EI}\frac{\partial^2 y}{\partial t^2} + \frac{\rho^2}{k'GE}\frac{\partial^4 y}{\partial t^4} = 0,$$
$$\frac{\partial^4 \psi}{\partial x^4} - \left(\frac{\rho}{E} + \frac{\rho}{k'G}\right)\frac{\partial^4 \psi}{\partial x^2 \partial t^2} + \frac{\rho a_b}{EI}\frac{\partial^2 \psi}{\partial t^2} + \frac{\rho^2}{k'GE}\frac{\partial^4 \psi}{\partial t^4} = 0.$$

Both equations above have the same form and therefore, we can separate variables assuming that y and ψ are synchronized in time

$$\begin{bmatrix} y(x,t) \\ \psi(x,t) \end{bmatrix} = T(t)\begin{bmatrix} y(x) \\ \psi(x) \end{bmatrix}. \tag{G-23}$$

The governing differential equations (G-19), using this substitution are

$$\rho a_b y(x)\ddot{T}(t) - k'Ga_b\left(y''(x) - \psi'(x)\right)T(t) = 0,$$
$$\rho I \psi(x)\ddot{T}(t) - T(t)\left(Ea_b\psi''(x) + k'Ga_b\left(y'(x) - \psi(x)\right)\right) = 0,$$

where the prime notation is used to identify the derivative with respect to x and dot notations for the derivatives with respect to t. The separated ordinary differential equations are given by

$$k'Ga_b\left(y''(x) - \psi'(x)\right) + \omega^2 \rho a_b y(x) = 0,$$
$$\rho I \omega^2 \psi(x) + Ea_b\psi''(x) + k'Ga_b\left(y'(x) - \psi(x)\right) = 0, \tag{G-24}$$
$$\ddot{T}(t) + \omega^2 T(t) = 0.$$

If we decouple the functions $y(x)$ and $\psi(x)$ by using the procedure as before, we will see that the differential equations have the same form, i.e. the equation for y has the same coefficients as the equation for ψ:

$$y'''' + y''\left(\frac{\rho\omega^2}{k'G} + \frac{\rho I\omega^2}{Ea_b}\right) + y\left(\frac{\rho^2 I\omega^4}{k'GEa_b} - \frac{\rho\omega^2}{E}\right) = 0,$$
$$\psi'''' + \psi''\left(\frac{\rho\omega^2}{k'G} + \frac{\rho I\omega^2}{Ea_b}\right) + \psi\left(\frac{\rho^2 I\omega^4}{k'GEa_b} - \frac{\rho\omega^2}{E}\right) = 0.$$

Therefore we can assume that the solutions for $y(x)$ and $\psi(x)$ also have the same form with different constants

$$\begin{bmatrix} y(x) \\ \psi(x) \end{bmatrix} = due^{rx}, \text{ u — eigenvector, } r \text{ — eigenvalue, } d \text{ — const.} \tag{G-25}$$

By substituting this equation into the previous equation and putting the result into matrix form, we obtain

$$\begin{bmatrix} k'Ga_b r^2 + \rho a_b \omega^2 & -k'Ga_b r \\ k'Ga_b r & Ea_b r^2 + \rho I \omega^2 - k'Ga_b \end{bmatrix} \mathrm{u} = 0. \tag{G-26}$$

In order to find r and u we need to find a non-trivial solution of the matrix equation above. Therefore, the determinant of the matrix has to be zero, and we get the characteristic equation

$$r^4 + \left(\frac{\rho I}{Ea_b} + \frac{\rho}{k'G} \right) \omega^2 r^2 - \frac{\rho \omega^2}{E} + \frac{\rho^2 I}{k'GEa_b} \omega^4 = 0.$$

Let us introduce the notation

$$a = \sqrt{\left(\frac{I}{Ea_b} + \frac{1}{k'G} \right) \frac{\rho \omega^2}{2} + \sqrt{\left(\frac{I}{Ea_b} - \frac{1}{k'G} \right)^2 \frac{\rho^2 \omega^4}{4} + \frac{\rho \omega^2}{E}}},$$
$$b = \sqrt{-\left(\frac{I}{Ea_b} + \frac{1}{k'G} \right) \frac{\rho \omega^2}{2} + \sqrt{\left(\frac{I}{Ea_b} - \frac{1}{k'G} \right)^2 \frac{\rho^2 \omega^4}{4} + \frac{\rho \omega^2}{E}}}. \tag{G-27}$$

Then the roots of the characteristic equation can be represented as $r_{1,2} = \pm ja$, $r_{3,4} = \pm b$, where b is real if frequency ω is less than $\sqrt{k'Ga_b/\rho I}$ and imaginary if frequency ω is greater than $\sqrt{k'Ga_b/\rho I}$. This value is called the critical frequency ω_c. Therefore, when we obtain spatial solutions, two cases must be considered: $\omega < \omega_c$ and $\omega > \omega_c$.

In the first case, when $\omega < \omega_c$ the spatial solution is given by

$$\begin{bmatrix} y(x) \\ \psi(x) \end{bmatrix} = \sum_{i=1}^{4} d_i \mathrm{u}_i e^{r_i x} = d_1 \mathrm{u}_1 e^{bx} + d_2 \mathrm{u}_2 e^{-bx} + d_3 \mathrm{u}_3 e^{iax} + d_4 \mathrm{u}_4 e^{-iax}. \tag{G-28}$$

The values of u from the equation (G-26) are

$$\mathrm{u}_i = \begin{bmatrix} k'Ga_b r_i \\ k'Ga_b r_i^2 + \rho a_b \omega^2 \end{bmatrix}.$$

The spatial solution contains both sinusoidal and hyperbolic terms, and is given by

$$\begin{bmatrix} y(x) \\ \psi(x) \end{bmatrix} = \begin{bmatrix} N_1 \\ M_1 \end{bmatrix} \sin ax + \begin{bmatrix} N_2 \\ M_2 \end{bmatrix} \cos ax + \begin{bmatrix} N_3 \\ M_3 \end{bmatrix} \sinh bx + \begin{bmatrix} N_4 \\ M_4 \end{bmatrix} \cosh bx, \tag{G-29}$$

with the coefficients given by

$$\begin{bmatrix} N_1 \\ M_1 \end{bmatrix} = (d_3 \mathrm{u}_3 - d_4 \mathrm{u}_4) i, \qquad \begin{bmatrix} N_2 \\ M_2 \end{bmatrix} = d_3 \mathrm{u}_3 + d_4 \mathrm{u}_4,$$
$$\begin{bmatrix} N_3 \\ M_3 \end{bmatrix} = d_1 \mathrm{u}_1 - d_2 \mathrm{u}_2, \qquad \begin{bmatrix} N_4 \\ M_4 \end{bmatrix} = d_1 \mathrm{u}_1 + d_2 \mathrm{u}_2. \tag{G-30}$$

Because d_3 and d_4 are complex conjugates of each other, the relationships between coefficients can be given by

$$M_1 = -\frac{k'Ga^2 - \rho\omega^2}{k'Ga} N_2, \qquad M_2 = \frac{k'Ga^2 - \rho\omega^2}{k'Ga} N_1,$$
$$M_3 = \frac{k'Gb^2 + \rho\omega^2}{k'Gb} N_4, \qquad M_4 = \frac{k'Gb^2 + \rho\omega^2}{k'Gb} N_3. \tag{G-31}$$

In the case when $\omega > \omega_c$, the spatial solution is written in terms of sinusoidal terms, and is given by

$$\begin{bmatrix} y(x) \\ \psi(x) \end{bmatrix} = \begin{bmatrix} \tilde{N}_1 \\ \tilde{M}_1 \end{bmatrix} \sin ax + \begin{bmatrix} \tilde{N}_2 \\ \tilde{M}_2 \end{bmatrix} \cos ax + \begin{bmatrix} \tilde{N}_3 \\ \tilde{M}_3 \end{bmatrix} \sin \tilde{b}x + \begin{bmatrix} \tilde{N}_4 \\ \tilde{M}_4 \end{bmatrix} \cos \tilde{b}x,$$

where

$$a = \sqrt{\left(\frac{I}{Ea_b} + \frac{1}{k'G}\right)\frac{\rho\omega^2}{2} + \sqrt{\left(\frac{I}{Ea_b} - \frac{1}{k'G}\right)^2 \frac{\rho^2\omega^4}{4} + \frac{\rho\omega^2}{E}}},$$
$$\tilde{b} = \sqrt{\left(\frac{I}{Ea_b} + \frac{1}{k'G}\right)\frac{\rho\omega^2}{2} - \sqrt{\left(\frac{I}{Ea_b} - \frac{1}{k'G}\right)^2 \frac{\rho^2\omega^4}{4} + \frac{\rho\omega^2}{E}}},$$

with the relationships between coefficients $\tilde{C}_i$ and $\tilde{D}_i$ given by

$$\tilde{M}_1 = -\frac{k'Ga^2 - \rho\omega^2}{k'Ga} \tilde{N}_2, \qquad \tilde{M}_2 = \frac{k'Ga^2 - \rho\omega^2}{k'Ga} \tilde{N}_1,$$
$$\tilde{M}_3 = -\frac{k'G\tilde{b}^2 - \rho\omega^2}{k'G\tilde{b}} \tilde{N}_4, \qquad \tilde{M}_4 = -\frac{k'G\tilde{b}^2 - \rho\omega^2}{k'G\tilde{b}} \tilde{N}_3.$$

In these equations b and $\tilde{b}$ are related by

$$b = i\tilde{b}.$$

Let us consider the critical frequency, which can be written as

$$\omega_c = \sqrt{\frac{k'Ga_b}{\rho I}}.$$

Substituting the equation of the critical frequency into the wave numbers equations yields the critical wave numbers

$$a_c = \sqrt{\frac{k'G}{E} + \frac{a_b}{I}} = \sqrt{\frac{k'}{2(1+\nu)} + \frac{a_b}{I}}, \qquad b_c = \tilde{b}_c = 0.$$

Both shape factor k' and Poisson ratio ν do not vary much, so that the critical wave number essentially depends on the ratio A/I.

Because there are two wave numbers, in order to find the solution to the frequency equations, one of the wave numbers has to be expressed in terms of the other to make the frequency equation a function of one wave number only, for example a, as shown in the example below.

The next step is to determine the relationship between a and b from the boundary conditions by substituting equations (G-29) and (G-31), and finding the non-trivial solution for the matrix equation formed by the boundary conditions.

The natural frequency equation obtained from equations (G-27) by subtracting one from another after squaring is given by

$$\omega_n = \sqrt{\left(a_n^2 - b_n^2\right)\frac{k'GEa_b}{\rho\left(k'GI + Ea_b\right)}}.$$

Example G-2: ([5], [6])

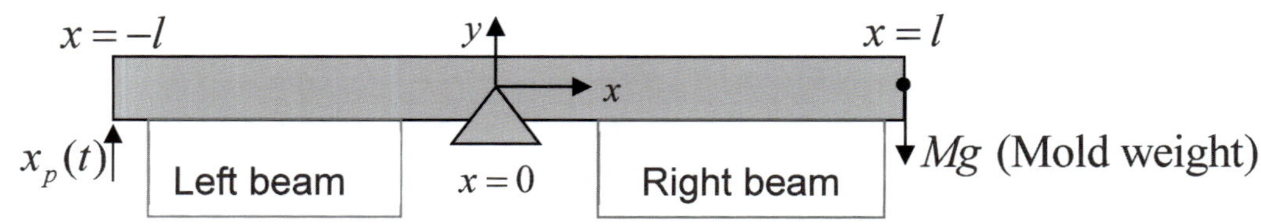

Fig. G-7: Mold oscillation system beam.

In this example the natural frequency of the mold oscillation software testbed beam described in Chapter 1 (Figure 1.5-1) is calculated. The beam in the mold oscillation testbed is loaded by the heavy mass at one end and is subject to the vertical sinusoidal motion by an electro-hydraulic servo actuator at the other end.

The single beam pivoted in the middle is represented by an equivalent description of two beams coupled together at the hinge via the boundary conditions. The governing equations are:

$$
\begin{aligned}
m_b \frac{\partial^2 y_L}{\partial t^2} + \gamma_y \frac{\partial y_L}{\partial t} &= \frac{\partial}{\partial x} k'Ga_b \left(\frac{\partial y_L}{\partial x} - \psi_L \right) - m_b g, \\
\frac{I}{a_b} m_b \frac{\partial^2 \psi_L}{\partial t^2} + \gamma_\psi \frac{\partial \psi_L}{\partial t} &= \frac{\partial}{\partial x} \left(EI \frac{\partial \psi_L}{\partial x} \right) + k'Ga_b \left(\frac{\partial y_L}{\partial x} - \psi_L \right), \\
m_b \frac{\partial^2 y_R}{\partial t^2} + \gamma_y \frac{\partial y_R}{\partial t} &= \frac{\partial}{\partial x} k'Ga_b \left(\frac{\partial y_R}{\partial x} - \psi_R \right) - m_b g, \\
\frac{I}{a_b} m_b \frac{\partial^2 \psi_R}{\partial t^2} + \gamma_\psi \frac{\partial \psi_R}{\partial t} &= \frac{\partial}{\partial x} \left(EI \frac{\partial \psi_R}{\partial x} \right) + k'Ga_b \left(\frac{\partial y_R}{\partial x} - \psi_R \right).
\end{aligned}
\tag{G-32}
$$

The boundary conditions for the each beam (for the case "hinged- free") are

$$
\begin{aligned}
&y_L(-l) = x_p(t),\ EI \frac{\partial \psi_L(-l)}{\partial x} = 0,\ y_L(0) = 0, y_R(0) = 0, \\
&\psi_L(0) = \psi_R(0),\ EI \frac{\partial \psi_L(0)}{\partial x} = EI \frac{\partial \psi_R(0)}{\partial x},\ EI \frac{\partial \psi_R(l)}{\partial x} = 0, \\
&k'Ga_b \left(\frac{\partial y_R(l)}{\partial x} - \psi_R(l) \right) + Mg + M \frac{\partial^2 y_R(l)}{\partial t^2} + \gamma_m \frac{\partial y_R(l)}{\partial t} = 0.
\end{aligned}
\tag{G-33}
$$

In these equations m_b, a_b, M stand for mass of beam per unit length, area of cross section of beam, and mold mass. In comparison to the system of equations (G-19) there are additional damping terms in equations: γ_y, γ_ψ, γ_m, which are beam transverse displacement, angular displacement, and mold damping, respectively.

In order to find the natural frequency of the beam the homogeneous problem, which presumes no external forces applied to the system, should be solved. In our case the external force is applied through the vertical displacement of the left end of the beam. Therefore, the boundary condition which includes the applied displacement produced by the servo actuator piston should be changed. Since the piston prevents vertical displacement of the beam at rest, the boundary condition at the piston side is set to zero, i.e.

$$
y_L(-l) = 0. \tag{G-34}
$$

The forces $m_b g$ and Mg in equations (G-32) and (G-33) stand for the static displacement and the angular distortion of the beam and do not affect the frequency response of the beam. Therefore, these forces are not considered in the modal analysis, the purpose of which is to determine the natural frequency of the beam.

The analysis below follows the steps done in (G-24)–(G-31). Assuming functions $y(x,t)$ and $\psi(x,t)$ represented by (G-23), the resulting equations take the form:

$$
\begin{aligned}
k'Ga_b\left(y_L''(x)-\psi_L'(x)\right)+C_1y_L(x)=0,\\
EI\psi_L''(x)+k'Ga_b\left(y_L'(x)-\psi_L(x)\right)+C_2\psi_L(x)=0,\\
k'Ga_b\left(y_R''(x)-\psi_R'(x)\right)+C_1y_R(x)=0,\\
EI\psi_R''(x)+k'Ga_b\left(y_R'(x)-\psi_R(x)\right)+C_2\psi_R(x)=0.
\end{aligned}
\tag{G-35}
$$

The separated boundary conditions are

$$
\begin{aligned}
y_L(-l)=0,\ EI\psi_L'(-l)=0,\ y_L(0)=0,\ y_R(0)=0,\\
\psi_L(0)=\psi_R(0),\ EI\psi_L'(0)=EI\psi_R'(0),\ EI\psi_R'(l)=0,\\
k'Ga_b\left(y_R'(l)-\psi_R'(l)\right)+\frac{4M^2\omega^2+\gamma_m^2}{4M}\psi_R(l)=0.
\end{aligned}
\tag{G-36}
$$

The equations for the time dependent functions of displacement and angular distortion for the both beams are identical, therefore they can be represented as

$$
\begin{aligned}
m_b\ddot{T}_1(t)+\gamma_y\dot{T}_1(t)+C_1T_1(t)=0,\\
\frac{m_bI}{a_b}\ddot{T}_2(t)+\gamma_\psi\dot{T}_2(t)+C_2T_2(t)=0,\\
M\ddot{T}_3(t)+\gamma_m\dot{T}_3(t)+\frac{4M^2\omega^2+\gamma_m^2}{4M}T_3(t)=0.
\end{aligned}
$$

In these equations coefficients C_1 and C_2 are given by

$$
C_1=\frac{4m^2\omega^2+\gamma_y^2}{4m},\quad C_2=\frac{4m^2I^2\omega^2+\gamma_\psi^2a_b^2}{4mIa_b}.
\tag{G-37}
$$

The separated equations can be represented in matrix form

$$
\begin{bmatrix} k'Ga_b & 0\\ 0 & EI\end{bmatrix}\begin{bmatrix} y''\\ \psi''\end{bmatrix}+\begin{bmatrix} 0 & -k'Ga_b\\ k'Ga_b & 0\end{bmatrix}\begin{bmatrix} y'\\ \psi'\end{bmatrix}+\begin{bmatrix} C_1 & 0\\ 0 & -k'Ga_b+C_2\end{bmatrix}\begin{bmatrix} y\\ \psi\end{bmatrix}=0.
$$

Assuming that functions $y(x)$ and $\psi(x)$ have the form (G-25), the resulting matrix equation is

$$
\begin{bmatrix} k'Ga_br^2+C_1 & -k'Ga_br\\ k'Ga_br & EIr^2-k'Ga_b+C_2\end{bmatrix}\mathrm{u}=0.
\tag{G-38}
$$

In order to obtain non-trivial solutions determinant of this matrix is taken to be zero, therefore the characteristic equation has form

$$r^4 + r^2\left(\frac{C_1}{k'Ga_b} + \frac{C_2}{EI}\right) - \frac{C_1}{EI} + \frac{C_1C_2}{k'Ga_bEI} = 0.$$

The wave numbers are given by:

$$a = \sqrt{\left(\frac{C_1}{k'Ga_b} + \frac{C_2}{EI}\right)\frac{1}{2} + \sqrt{\left(\frac{C_1}{k'Ga_b} - \frac{C_2}{EI}\right)^2 \frac{1}{4} + \frac{C_1}{EI}}}, \tag{G-39}$$

$$b = \sqrt{-\left(\frac{C_1}{k'Ga_b} + \frac{C_2}{EI}\right)\frac{1}{2} + \sqrt{\left(\frac{C_1}{k'Ga_b} - \frac{C_2}{EI}\right)^2 \frac{1}{4} + \frac{C_1}{EI}}}. \tag{G-40}$$

Then, the roots of the characteristic equation can be represented as $r_{1,2} = \pm ja$, $r_{3,4} = \pm b$. Here, a is always real, and b is real when the condition $C_2 \le k'Ga_b$ is satisfied.

From the equations (G-41) the coefficient C_2 can be represented in terms of C_1 as

$$m_b\omega^2 = C_1 - \frac{\gamma_y^2}{4m_b}, \quad C_2 = \frac{I}{a_b}\left(C_1 - \frac{\gamma_\psi^2}{4m_b}\right) + \frac{\gamma_\psi^2 a_b}{4m_b I}.$$

Substituting these equations into the equations (G-39) and (G-40), the relation

$$\begin{aligned} a^2b^2 &= s_1C_1^2 + s_2C_1, \\ a^2 - b^2 &= \alpha C_1 + \beta, \end{aligned} \tag{G-41}$$

is obtained, where

$$s_1 = -\frac{1}{Ek'Ga_b^2}, \quad s_2 = \frac{1}{EI} + \frac{1}{EIk'Ga_b}\left(\frac{\gamma_y^2 I}{4ma_b} - \frac{\gamma_\psi^2 a_b}{4mI}\right),$$

$$\alpha = \frac{1}{Ea_b} + \frac{1}{k'Ga_b}, \beta = -\frac{\gamma_y^2}{4MEa} + \frac{\gamma_\psi^2 a_b}{4mI^2E}.$$

From the equations (G-41) the coefficient C_1 in terms of a and b is

$$C_1 = \frac{a^2 - b^2 - \beta}{\alpha}.$$

Therefore, the equation resulting from (G-41) has the form

$$a^2b^2 = s_1\left(\frac{a^2 - b^2 - \beta}{\alpha}\right)^2 + s_2C_1.$$

After the transformation this equation can be given by

$$b^4 \frac{s_1}{\alpha^2} + b^2 \left(-a^2 - \frac{2(a^2-\beta)}{\alpha^2} s_1 - \frac{s_2}{\alpha} \right) + \left(\frac{(\alpha^2-\beta)^2}{\alpha^2} s_1 + s_2 \frac{(\alpha^2-\beta)}{\alpha} \right) = 0. \quad \text{(G-42)}$$

The solution of the system of equations (G-36) can be given as in (G-29), with the coefficients as in (G-36). The values u from the equation (G-26) are

$$\mathrm{u} = \begin{bmatrix} kGa_b r \\ k'Ga_b r^2 + C_1 \end{bmatrix}.$$

Therefore, the coefficients for equation in (G-30) have form

$$M_1 = -\frac{k'Ga_b a^2 - C_1}{k'Ga_b a} N_2 = -\eta_1 N_2, \quad M_2 = \frac{k'Ga_b a^2 - C_1}{k'Ga_b a} N_1 = \eta_1 N_1,$$
$$M_3 = \frac{k'Ga_b b^2 + C_2}{k'Ga_b b} N_4 = \eta_2 N_4, \quad M_4 = \frac{k'Ga_b b^2 + C_2}{k'Ga_b b} N_3 = \eta_2 N_3.$$

Application of the boundary conditions to the spatial solution (G-29) yields eight equations in terms of eight unknown coefficients and wave numbers:

$$\begin{aligned}
&-N_{1L}\sin al + N_{2L}\cos al - N_{3L}\sinh bl + N_{4L}\cosh bl = 0,\\
&\eta_1 a N_{1L}\sin al - \eta_1 a N_{2L}\cos al - \eta_2 b N_{3L}\sinh bl + \eta_2 b N_{4L}\cosh bl = 0,\\
&N_{2L} + N_{4L} = 0, \quad N_{2R} + N_{4R} = 0,\\
&\eta_1 N_{1L} + \eta_2 N_{3L} = \eta_1 N_{1R} + \eta_2 N_{3R},\\
&-\eta_1 a N_{2L} + \eta_2 b N_{3L} = -\eta_1 a N_{2R} + \eta_2 b N_{3R},\\
&-\eta_1 a N_{1R}\sin al - \eta_1 a N_{2R}\cos al - \eta_2 b N_{3R}\sinh bl - \eta_2 b N_{4R}\cosh bl = 0,\\
&k_1 N_{1R} + k_2 N_{2R} + k_3 N_{3R} + k_4 N_{4R} = 0,
\end{aligned} \quad \text{(G-43)}$$

where

$$k_1 = k'Ga_b(a-\eta_1)\cos al - \frac{\omega^2 M + \gamma_m}{4M}\sin al,$$
$$k_2 = k'Ga_b(-a+\eta_1)\sin al - \frac{\omega^2 M + \gamma_m}{4M}\cos al,$$
$$k_3 = k'Ga_b(b-\eta_2)\cosh bl - \frac{\omega^2 M + \gamma_m}{4M}\sin bl,$$
$$k_4 = k'Ga_b(b-\eta_2)\sinh bl - \frac{\omega^2 M + \gamma_m}{4M}\cosh bl,$$

and where the natural frequency from the equation (G-37) is given by

$$\omega = \sqrt{\frac{C_1}{m_b} - \frac{\gamma_y^2}{4m_b^2}}. \tag{G-44}$$

The equations (G-43) can be presented in matrix form by factoring out all the coefficients N. The matrix equation can be solved by equating the determinant to zero, which yields the equation in terms of a and b. Combining the resulting equation with the equation (G-43) gives the solution for the wave number a, from which the wave number b can be found. Then, the first natural frequency of the beam can be found from the equation (G-44).

G.4 Forced Beam Response

For the forced beam response the eigenfunction expansion is applied. According to this method, the applied force $f(x,t)$ and the displacement $y(x,t)$ can be presented as the summations of the products of spatial solutions for the homogeneous problem and time dependent functions

$$y(x,t) = \sum_{n=1}^{\infty} \eta_n(t) y_n(x), \quad f(x,t) = \sum_{n=1}^{\infty} \eta_n(t) y_n(x). \tag{G-45}$$

For Euler-Bernoulli model $y(x,t)$ denotes a displacement $y(x,t)$, for Timoshenko model $y(x,t)$ is a vector of displacements y and ψ functions.

In order to use eigenfunction expansion the orthogonality conditions should be established first. The spatial equations for the homogeneous problem can be written using the operator formalism

$$L(y_n) = \omega^2 M(y_n),$$

where for the Euler-Bernoulli model (G-12) the operators are

$$L(y_n) = \frac{\partial^4 y_n}{\partial x^4}, \quad M(y_n) = \rho a_b y_n,$$

and for the Timoshenko beam model (G-19) in the general case

$$L(y_n) = \begin{bmatrix} k'Ga_b \frac{\partial^2}{\partial x^2} & -k'Ga_b \frac{\partial}{\partial x} \\ k'Ga_b \frac{d}{dx} & Ea_b \frac{\partial^2}{\partial x^2} - k'Ga_b \end{bmatrix} \begin{bmatrix} y_n \\ \psi_n \end{bmatrix}, \quad M(y_n) = \begin{bmatrix} \rho a_b & 0 \\ 0 & \rho I \end{bmatrix} \begin{bmatrix} y_n \\ \psi_n \end{bmatrix}.$$

Operators L and M should be self-adjoint, which means that they should satisfy equations

$$\int_0^l \left[y_n^T L(y_m) - y_m^T L(y_n) \right] dx = 0,$$
$$\int_0^l \left[y_n^T M(y_m) - y_m^T M(y_n) \right] dx = 0.$$

The second equation satisfies all boundary conditions of both models. The first equation can be transformed to

$$\left(\omega_m^2 - \omega_n^2\right) \int_0^l y_n^T M(y_m) dx = 0.$$

In these equations, the squared natural frequencies are eigenvalues, which are unique for each eigenfunction. Therefore,

$$\int_0^l y_n^T M(y_m) dx = 0 \text{ for } m \neq n,$$

$$\int_0^l y_n^T M(y_m) dx = 1 \text{ for } m = n.$$

Combining these two equations yields

$$\int_0^l W_n^T M(W_m) dx = \delta_{mn}, \tag{G-46}$$

where δ_{mn} is the Kronecker delta.

Substituting the assumption (G-46) given above into the governing equations yields

$$\sum_{n=1}^{\infty} \rho a_b \frac{\partial^2 \eta_n(t)}{\partial t^2} y_n(x) + \sum_{n=1}^{\infty} \eta_n(t) L(y_n(x)) = \sum_{n=1}^{\infty} f_n(t) M(y_n(x)),$$
$$\sum_{n=1}^{\infty} \left(\frac{\partial^2 \eta_n(t)}{\partial t^2} + \eta_n(t) \right) M(y_n(x)) = \sum_{n=1}^{\infty} f_n(t) M(y_n(x)).$$

After multiplying by y_m^T and integrating over the length of the beam, using the orthogonality condition (G-45) results in the equation for the time dependent function

$$\rho a_b \frac{\partial^2 \eta_n(t)}{\partial t^2} + \eta_n(t) \omega^2 = f_n(t).$$

The solution of this equation is given by

$$\eta_n(t) = \frac{1}{\omega_n} \int_0^t F_n(t) \sin \omega_n (t - \tau) d\tau + \eta_n(0) \cos \omega_n t + \frac{1}{\omega_n} \frac{d\eta_n}{dt}\bigg|_{t=0} \sin \omega_n t.$$

Therefore, the resulting solution of the forced beam system can be found as

$$y(x,t)=\sum_{n=1}^{\infty}\eta_n(t)\,y_n(x). \tag{G-47}$$

G.5 Application of the Euler-Bernoulli and the Timoshenko Models to the Testbed Modeling: Analytical Description and Numerical Simulation Results

G.5.1 *Selection of the beam model: Euler-Bernoulli versus Timoshenko*

The Euler-Bernoulli model includes only translation and bending, whereas the Timoshenko model includes rotation and shear as well. It should be noted that deformation due to the shear effect is more significant than the rotary effect. In fact, the natural frequency predicted by the Euler-Bernoulli model approaches infinity as the slenderness ratio decreases whereas the natural frequencies predicted by the Timoshenko model level off at the same point. That is, for low slenderness ratios, the Euler-Bernoulli model is over-predicting the natural frequencies, which indicates that the model is inaccurate for the not very slender beams. Therefore, when the slenderness ratio is large: $\sqrt{I/A}>100$ [1], the Euler-Bernoulli model should be used, because the model will still be accurate and simpler to use. When the slenderness ratio is small, as in the case of the hardware testbed, the Timoshenko model is preferred because of its better accuracy. The latter model has been chosen for the software testbed. The rest of this appendix, therefore, derives the results for the software testbed using the latter model.

G.5.2 *Analytical model of Nucor Steel hardware testbed*

The analytical model of the beam in the Nucor Steel hardware testbed is represented by equations (G-32) and (G-33) with the system parameter values shown in Table G-1. The beam equations are coupled through the boundary conditions to the equation (H-4) of the hydraulic actuator given in Appendix H. For the actuator dynamics the corresponding equation for the spool motion is given in (H-5), the pressures in chambers are described in (H-6), and the flow rates equations are given in (H-7). In order to exclude the distortions in the piston position there are two controllers given by equations (H-6) and (H-8) were used in the testbed. They are described in detail in Appendix H.

Let us now find the wave numbers for the system in Table G-1. First, let find the coefficients in (G-41), which are

$$s_1=-\frac{1}{Ek'Ga_b^2}=-9.092\cdot10^{-18},$$

$$s_2=\frac{1}{EI}+\frac{1}{EIk'Ga_b}\left(\frac{\gamma_y^2 I}{4ma_b}-\frac{\gamma_\psi^2 a_b}{4mI}\right)=2.83\cdot10^{-7},$$

$$\alpha = \frac{1}{Ea_b} + \frac{1}{k'Ga_b} = 7.039 \cdot 10^{-9}, \ \beta = -\frac{\gamma_y^2}{4MEa} + \frac{\gamma_\psi^2 a_b}{4mI^2 E} = 5.088 \cdot 10^{-5}.$$

Then, by substituting these coefficients into (G-42), we get the wave number b in terms of a

$$b^2 = a^2\left(1+\frac{\alpha^2}{2s_1}\right)+\left(\frac{\alpha s_2}{2s_1}-\beta\right) \pm \sqrt{\left(a^2\left(1+\frac{\alpha^2}{2s_1}\right)+\left(\frac{\alpha s_2}{2s_1}-\beta\right)\right)^2 - \frac{\alpha^2}{s_1}\left(\frac{\left(\alpha^2-\beta\right)^2}{\alpha^2}s_1 + s_2\frac{\left(\alpha^2-\beta\right)}{\alpha}\right)}.$$

The boundary conditions equations (G-43) in the matrix form are given by

$$\begin{bmatrix} -\sin al & \cos al & -\sinh bl & \cosh bl & 0 & 0 & 0 & 0 \\ \eta_1 a \sin al & -\eta_1 a \cos al & -\eta_2 b \sinh bl & \eta_2 b \cosh bl & 0 & 0 & 0 & 0 \\ 0 & 1 & 0 & 1 & 0 & 0 & 0 & 0 \\ 0 & 0 & 0 & 0 & 0 & 1 & 0 & 1 \\ \eta_1 & 0 & \eta_2 & 0 & \eta_1 & 0 & \eta_2 & 0 \\ 0 & -\eta_1 a & 0 & \eta_2 b & 0 & -\eta_1 a & 0 & \eta_2 b \\ 0 & 0 & 0 & 0 & -\eta_1 a \sin al & -\eta_1 a \cos al & -\eta_2 b \sinh bl & -\eta_2 b \cosh bl \\ 0 & 0 & 0 & 0 & k_1 & k_2 & k_3 & k_4 \end{bmatrix} \begin{bmatrix} N_{1L} \\ N_{2L} \\ N_{3L} \\ N_{4L} \\ N_{1R} \\ N_{2R} \\ N_{3R} \\ N_{4R} \end{bmatrix} = 0.$$

This equation is solved by equating the determinant to zero with b from the equation above. Solving the resulting equation by Matlab gives the wave number $a = 0.3848$. Therefore, the wave number b is

$$b = \sqrt{-1.7245a^2 - 109.539 + \sqrt{\left(-1.7245a^2 - 109.539\right)^2 + 3.754 \cdot 10^{-4}}} = 0.3840.$$

The natural frequency is found by substituting a and b into (G-44), yielding

$$\omega_{nat} = \sqrt{\frac{C_1}{m} - \frac{\gamma_y^2}{4m^2}} = \sqrt{\frac{1}{m}\left(\frac{a^2 - b^2 - \beta}{\alpha}\right) - \frac{\gamma_y^2}{4m^2}} = 57.8053 \ rad/s.$$

Thus, the exact natural frequency of the mold oscillation system is $f = 9.2$ Hz. Applying the finite difference method to the analytical model — equations (G-32) and (G-33) — with parameters from Table G-1, a software testbed for Nucor Steel system is obtained that is described in Appendix H, with spatial and temporal discretization step sizes chosen as $n_x = 35$ and $n_t = 10^{-6}$ sec, respectively. Due to the numerical approximation error, the natural frequencies of the numerical model differ from the analytical ones and depend on the number of nodes. Thus, the natural frequency obtained numerically for $n_x = 35$ is 9.3305 Hz. The procedures for the choice of the beam parameters to match the beam desired natural frequency of as well as simulation guidelines that describe tradeoff between the simulation run time and the approximation error as determined by the discretization step size are presented, respectively, in Subsections G4.4 and G4.5 below.

Table G-1. Nominal values of the mold oscillation system parameters for thin slab caster beam model.

m_b	69.256	kg/m
a_b	0.0088	m^2
G	$7.7 \cdot 10^{10}$	Pa
E	$2 \cdot 10^{11}$	Pa
I	$2.2085 \cdot 10^{-5}$	m^4
γ_y	10	$kg/(m \cdot \sec)$
γ_ψ	10	$kg/(m \cdot \sec)$
γ_m	1	$kg/\sec$
k'	0.83	
M	2163.23	kg
l	0.88	m

The response of the beam under the load exerted by the piston and the mold can be found either analytically, or numerically through the testbed simulation. Analytical solution for the forced response of the beam in the testbed can be found through the representation of equations (G-32) and (G-33) as multiplication of the spatial solution and time dependent function as shown in (G-47) and then solving time dependent equations by means of the eigenfunction expansion as described before.

The testbed simulation results with the driving frequency 4.6 Hz are shown in Figure G-8. The figure clearly shows the distortions in the mold displacement caused by the small amplitude driving frequency harmonic multiple in the piston displacement exciting the beam at its resonance frequency.

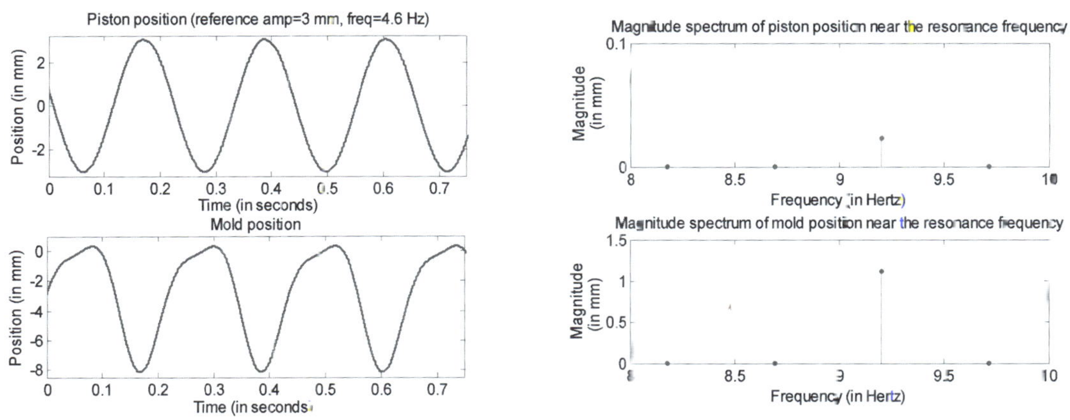

Fig. G-8. Piston and mold position; small magnitude of the piston position distortion harmonic at 9.2 Hz is significantly amplified in the mold position by beam resonance.

G.5.3 *Analytical model of AK Steel (currently Cleveland Cliffs) thick slab software testbed*

In this case the beam parameters in the testbed were chosen to make the natural frequency of the model $f = 5$ Hz. Therefore, applying modal analysis described earlier to the present problem, the length of the beam, cross-section area, and mass per unit length of the beam were changed to obtain the natural frequency of 5 Hz. The nominal values of the system parameters are given in the Table G-2.

Applying the finite difference method to the analytical model (G-32), with parameters from Table G-2, a software testbed for AK Steel system is obtained that is also described in Appendix H, where spatial and temporal discretization step sizes are $n_x = 153$ and $n_t = 10^{-6}$ sec respectively. The spatial discretization is much higher, which prevent change of the natural frequency of the beam after approximation. The distortions in the beam displacement at the excitation frequency 2.5 Hz caused by the resonance problem described before are shown in Figure G-9.

Table G-2. Nominal values of the mold oscillation system parameters for thick slab caster beam model.

m_b	104.23	kg/m
a_b	0.0132	m^2
G	$7.7 \cdot 10^{10}$	Pa
E	$2 \cdot 10^{11}$	Pa
I	$2.2085 \cdot 10^{-5}$	m^4
γ_y	10	$kg/(m \cdot \sec)$
γ_ψ	10	$kg/(m \cdot \sec)$
γ_m	1	$kg/\sec$
k'	0.83	
M	2250	kg
l	1.324	m

Piston position (reference magnitude = 3 mm, reference frequency = 2.5 Hz)

Magnitude spectrum of piston position near the resonance frequency

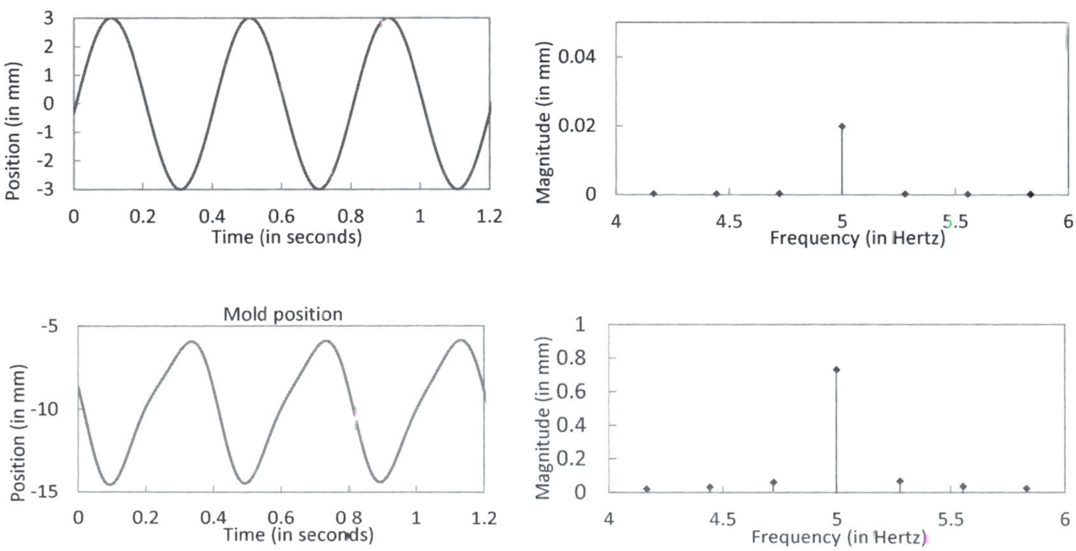

Fig. G-9. Piston and mold position: small magnitude of the piston position distortion harmonic at 5 Hz is significantly amplified in the mold position by beam resonance.

G.5.4 *Choice of the beam parameters to match the desired resonance frequency*

Matching by the model of the desired resonance frequency can be done by the variation of either the beam parameters itself or variation of the mold mass. Here the variation of the mold mass is considered. For this purpose, the equations of the mold oscillation system (G-32) are solved with the displacement and the angle of rotation functions represented as multiplication of the spatial and the complex exponential time-dependent functions

$$\begin{aligned} y(x,t) &= y(x)e^{j\omega t}, \\ \psi(x,t) &= \psi(x)e^{j\omega t}. \end{aligned} \tag{G-48}$$

The input function is considered to be complex exponential as well

$$x_p(t)=x_p e^{j\omega t}.$$

Substituting (G-48) into the equations (G-32) yields

$$-\omega^2 y + j\omega\frac{\gamma_y}{m_b}y = \frac{k'Ga_b}{m_b}\left(\frac{\partial^2 y}{\partial x^2} - \frac{\partial\psi}{\partial x}\right) - g,$$

$$-\omega^2\psi + j\omega\frac{\gamma_\psi a_b}{Im_b}\psi = \frac{Ea_b}{m_b}\frac{\partial^2\psi}{\partial x^2} + \frac{k'Ga_b^2}{Im_b}\left(\frac{\partial y}{\partial x} - \psi\right).$$

In the matrix form using (G-25):

$$\begin{bmatrix} \frac{k'Ga_b}{m_b}r^2 + \left(\omega^2 - j\omega\frac{\gamma_y}{m_b}\right) & -\frac{k'Ga_b}{m_b}r \\ \frac{k'Ga_b^2}{Im_b}r & \frac{Ea_b}{m_b}r^2 + \left(\omega^2 - j\omega\frac{\gamma_\psi a_b}{Im_b} - \frac{k'Ga_b^2}{Im_b}\right) \end{bmatrix}\begin{bmatrix} y(x) \\ \psi(x) \end{bmatrix} = 0. \quad \text{(G-49)}$$

This equation has non-trivial solution only if the determinant equals zero, which gives the equation

$$r^4 + r^2\left(\omega^2\frac{m_b}{a_b}\left(\frac{1}{E} + \frac{1}{k'G}\right) - j\omega\left(\frac{\gamma_\psi}{EI} + \frac{\gamma_y}{k'Ga_b}\right)\right) + \frac{m_b^2}{k'Ga_b^2}\left(\omega^4 - j\omega^3\left(\frac{\gamma_\psi a_b}{Im_b} + \frac{\gamma_y}{m_b}\right)\right.$$
$$\left. -\omega^2\left(\frac{\gamma_y\gamma_\psi a_b^2}{Im_b^2} + \frac{k'Ga_b^2}{Im_b}\right) + j\omega\frac{\gamma_y k'Ga_b^2}{Im_b^2}\right) = 0. \quad \text{(G-50)}$$

This characteristic equation has four solutions: $\pm r_1, \pm r_2$. The general solution can be represented as in (G-25) by

$$\begin{bmatrix} y(x) \\ \psi(x) \end{bmatrix} = C\begin{bmatrix} y \\ \psi \end{bmatrix}e^{rx} = C_1\begin{bmatrix} y_1 \\ \psi_1 \end{bmatrix}e^{r_1 x} + C_2\begin{bmatrix} y_2 \\ \psi_2 \end{bmatrix}e^{-r_1 x} + C_3\begin{bmatrix} y_3 \\ \psi_3 \end{bmatrix}e^{r_2 x} + C_4\begin{bmatrix} y_4 \\ \psi_4 \end{bmatrix}e^{-r_2 x},$$

where from (G-49) eigenvalues are

$$\begin{bmatrix} y_i \\ \psi_i \end{bmatrix} = \begin{bmatrix} k'Ga_b r_i / m_b \\ \left(k'Ga_b r_i^2 + \left(m_b\omega^2 - j\omega\gamma_y\right)\right)/m_b \end{bmatrix}.$$

Then, substituting the general solution into the boundary conditions (G-33) yields

$$
\begin{aligned}
&C_{1L}y_1e^{-r_1l} + C_{2L}y_2e^{r_1l} + C_{3L}y_3e^{-r_2l} + C_{4L}y_4e^{r_2l} = x_p,\\
&C_{1L}r_1\psi_1e^{-r_1l} - r_2C_{2L}\psi_2e^{r_1l} + r_2C_{3L}y_3e^{-r_2l} - r_2C_{4L}y_4e^{r_2l} = 0,\\
&C_{1L}\,y_1 + C_{2L}\,y_2 + C_{3L}\,y_3 + C_{4L}\,y_4 = 0,\\
&C_{1R}\,y_1 + C_{2R}\,y_2 + C_{3R}\,y_3 + C_{4R}\,y_4 = 0,\\
&C_{1L}\,y_1 + C_{2L}\,y_2 + C_{3L}\,y_3 + C_{4L}\,y_4 = C_{1L}\,y_1 + C_{2L}\,y_2 + C_{3L}\,y_3 + C_{4L}\,y_4, \qquad \text{(G-51)}\\
&C_{1L}\,r_1\psi_1 - C_{2L}\,r_1\psi_2 + C_{3L}\,r_2\psi_3 - C_{4L}\,r_2\psi_4 = C_{1R}\,r_1\psi_1 - C_{2R}\,r_2\psi_2 + C_{3R}\,r_2\psi_3 - C_{4R}\,r_2\psi_4,\\
&C_{1L}\,r_1\psi_1e^{r_1l} - C_{2L}\,r_2\psi_2e^{-r_1l} + C_{3L}\,r_2\psi_3e^{r_2l} - C_{4L}\,r_2\psi_4e^{-r_2l} = C_{1R}\,r_1\psi_1e^{r_1l} - C_{2R}\,r_2\psi_2e^{-r_1l}\\
&\qquad + C_{3R}\,r_2\psi_3e^{r_2l} - C_{4R}\,r_2\psi_4e^{-r_2l},\\
&C_{1R}e^{r_1l}D_1 + C_{2R}e^{-r_1l}D_2 + C_{3R}e^{r_2l}D_3 + C_{4R}e^{-r_2l}D_4 = 0,
\end{aligned}
$$

where

$$
\begin{aligned}
D_1 &= k'Ga_b\left(r_1y_1 - \psi_1\right) - y_1\left(M\omega^2 + j\omega\gamma_m\right),\\
D_2 &= -k'Ga_b\left(r_1y_2 - \psi_2\right) - y_2\left(M\omega^2 + j\omega\gamma_m\right),\\
D_3 &= k'Ga_b\left(r_2y_3 - \psi_3\right) - y_3\left(M\omega^2 + j\omega\gamma_m\right),\\
D_4 &= -k'Ga_b\left(r_2y_4 - \psi_4\right) - y_4\left(M\omega^2 + j\omega\gamma_m\right).
\end{aligned}
$$

Factoring out all the coefficients C_{iL} and C_{iR} in (G51), the matrix equation can be obtained. The latter is solved in Matlab by equating the determinant to zero with parameters from Tables G-1 and G-2. In order to find the parameters to match the desired resonance frequency, either the parameters of the beam (mass, cross-section area and moment of inertia) or mass of the mold should be changed. Therefore, by changing the beam mass in the specified range, one can find the mass of the mold required.

For the Nucor Steel caster the magnitude response indicating the first three modes is shown in Figure G-10. For mold mass M = 2163.23 kg the first natural frequency is exactly 9.2 Hz.

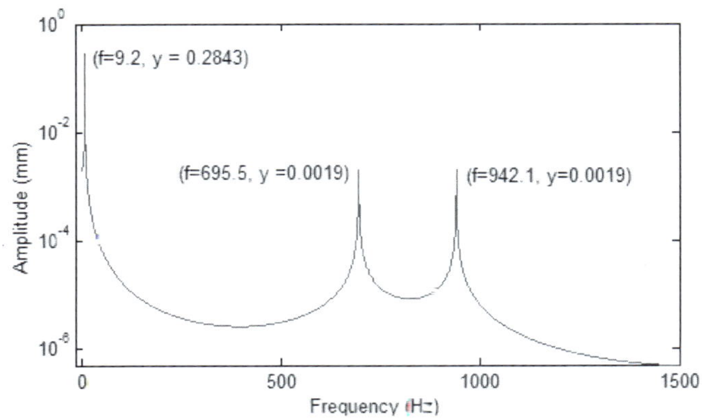

Fig. G-10. Magnitude response of the Nucor Steel beam.

G.5.5 *Simulation run-time/approximation error tradeoff*

In order to simulate the mold oscillation system, the beam is divided into several nodes by the length of the beam. Then, the quality of the beam simulation depends on the number of nodes causing the approximation error. Therefore, the bigger number of the nodes we take, the smaller error we have. After choosing the big enough number of nodes the natural and, hence, the (very close to it) resonance frequency of numerical simulation becomes equal to the actual one. However, at the same time the run-time of the simulation increases (Figure G-12). For example, the simulation run-time with 20 nodes takes 57 sec — nine times less than with 35 nodes, while the run-time for 10 seconds simulation with 100 nodes is 20 min.

Therefore, for smaller run-times, with the distortion phenomenon in the mold oscillation system retained, the natural frequency of the beam can be made to match the actual natural frequency by adjusting the beam parameters.

It needs to be noted that not any number of nodes could be taken for simulation, since the model which is given in state-space form (Appendix H) is stable only if the matrix A is negative definite. It means that all eigenvalues of the matrix A should be Hurwitz (lie in the left half-plane on the complex plane).

In the figure below the maxima of the real parts of eigenvalues for the numbers of nodes for which the matrix A is stable are plotted and the corresponding calculated natural frequencies are shown.

The natural frequencies are calculated from the step-input simulation data by the known formula

$$\omega_{num} = \frac{1}{2(t_{i+1} - t_i)}. \tag{G-52}$$

The points t_{i+1} and t_i are obtained by the linear approximation of intersections of the beam motion data with the horizontal axis.

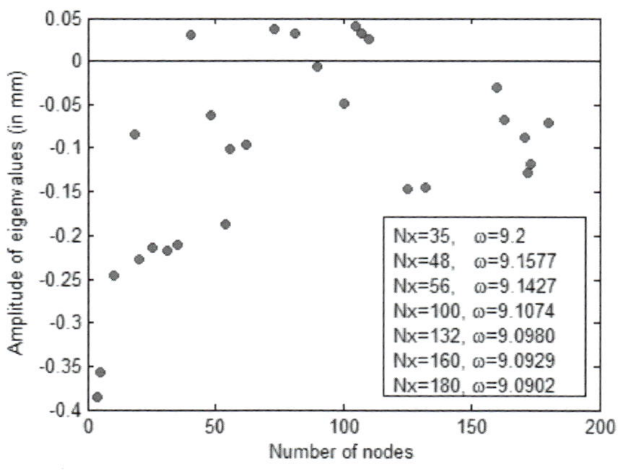

Fig. G-11. Eigenvalues of the Nucor Steel beam.

Experimentally, it was found that the dependence between the natural frequency and the mold mass for the Nucor Steel beam is linear and is equal approximately to the 0.02 Hz change in natural frequency if the mold mass is changed by 10 kg. The results of changing of the mold mass to match the natural frequency 9.2 Hz for different spatial discretizations are shown in Figure G-13 and Table G-3. As it is seen, one can obtain exactly 9.2 Hz.

Table G-3. Parameters for Nucor Steel model.

$\Delta x = l/n_x$	M, kg	ω_{num}, Hz
$l/10$	2552.2	9.2023
$l/20$	2296.9	9.2002
$l/35$	2225.2	9.2000
$l/48$	2204.7	9.2000
$l/56$	2198.4	9.2010
$l/62$	2192.7	9.2017
$l/100$	2179.6	9.2019

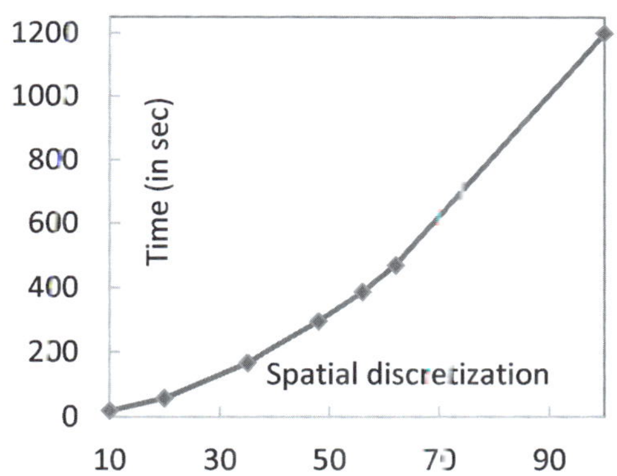

Fig. G-12. Spatial discretization vs. time of simulation.

However, changing of the mold mass could cause instability in the modeling system when the matrix A becomes positive definite at some choices of mold mass. In order to prevent that, the steps indicated in flow-chart in Figure G-14 should be taken, namely:

1. Set the starting values of the mold oscillation system parameters (Tables G-1 and G-2).

2. Select the number of nodes, which make the matrix A negative definite. It could be done by plotting the maximum of the real parts of matrix A eigenvalues for a given range of numbers of nodes.

Per Figure G-11, the numbers of nodes which make the maximum of the real parts of the eigenvalues lie below zero are required.

3. Specify the number of nodes based on the desired runtime and the precision of the modeling.

4. Adjust the mold mass to match the natural frequency of the mold oscillation system. The adjustment is done heuristically. First, the natural frequency of the approximated model needs to be determined from (G-52), and then the mass of mold needs to be changed and the natural frequency calculated again until sufficient accuracy of matching the resonance frequency of the mold oscillation system is attained.

5. Check the stability of the resulting system by simulation. If the mold oscillation system model is unstable, steps 2–5 need to be repeated and a different number of nodes should be chosen with the correspondingly adjusted mold mass.

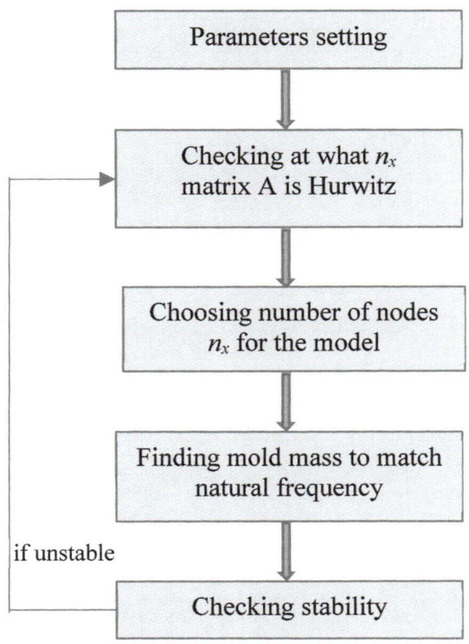

Fig. G-14. Parameters/nodes selection flow-chart.

G.5.6 *Convergence of the numerical solution to the analytical one*

In this section the correctness of the analytical solution is verified by the numerical simulation. As it can be seen on Figure G-16, increasing discretization makes the simulation results closer to the analytical solution. The time-histories of the mold displacement step responses proving that are shown below. The numerical solutions curves become closer as the discretization is increased.

The convergence of the numerical simulation to the analytical solution with the increase of the spatial discretization has an exponential behavior (Figure G-16), implying that the numerical results will be equal to the analytical result in 'infinity'.

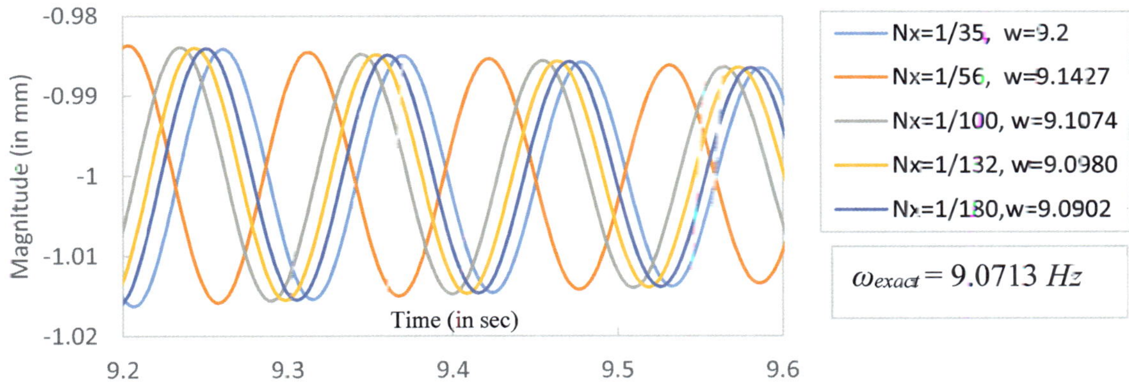

Fig. G-15. Beam simulation for different discretizations.

Table G-4. Approximation error convergence.

Δx	Damped Natural Frequency, Hz	Approximation error, Hz	Resonance Frequency, Hz	Approximation error, Hz
1/20	9.4794	0.2794	9.4790	0.2790
1/35	9.3305	0.1305	9.3301	0.1301
1/48	9.2875	0.0875	9.2871	0.0871
1/56	9.2725	0.0725	9.2721	0.0721
1/100	9.2368	0.0368	9.2364	0.0364
1/132	9.2270	0.0270	9.2265	0.0265
1/160	9.2219	0.0219	9.2215	0.0215
Analytical solution	9.2000	-	9.2000	-

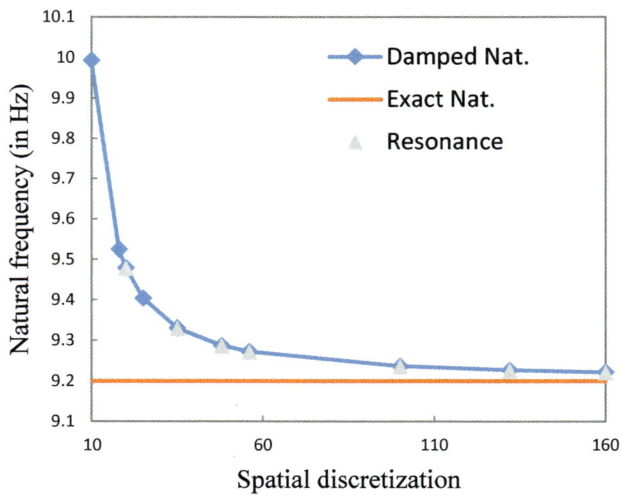

Fig. G-16. Numerical solution convergence to an analytical one.

References

1. Han, S. M., H. Benaroya, and T. Wei (1999). "Dynamics of the transversely vibrating beams using four engineering theories," *Journal of Sound and Vibration, Volume 225, Issue 5*, pp. 935–988, http://dx.doi.org/10.1006/jsvi.1999.2257. (http://www.sciencedirect.com/science/article/pii/S0022460X99922575)
2. Love, A. H. (1927). A Treatise on the mathematical theory of elasticity, New York: Dover.
3. Timoshenko, S. P. (1953). History of Strength of Materials, New York: Dover.
4. ME 360: Fundamentals of Signal Processing, Instrumentation and Control, Lab 3, (http://coecsl.ece.illinois.edu/me360/me360Exp3straingage.pdf).
5. Natarajan, V. and J. Bentsman (2011). "Robust Rejection of Sinusoids in Stable Nonlinearly Perturbed Unmodelled Linear Systems: Theory and Application to Servo," *Proc. of the 2011 Amer. Contr. Conf.*, pp. 3289–3294.
6. Angatkina, O., V. Natarajan, Z. Chen, and J. Bentsman (2015). "Capturing and suppressing resonance in steel casting mold oscillation systems using Timoshenko beam model," *Proc. of the 2015 Amer. Contr. Conf.*, pp. 559–564. Angatkina, O., V. Natarajan, Z. Chen, M. Ding, and J. Bentsman (2019). "Modeling and Control of Resonance Effects in Steel Casting Mold Oscillators", *Acta Mechanica*, 230, 2087–2104.

Appendix H

Software Testbed Matlab Programs

H.1 Introduction

This section presents the Matlab software describing a single-beam system, where the beam, pivoted in the middle, is driven at one end by the hydraulic actuator and has the mass attached to the other end, as described in Subsection 1.3.3 of Chapter 1. The hardware testbed built at Nucor Steel, captures the mold oscillation distortion of the thin slab casting production unit consisting of four beams. The software testbed models well the hardware testbed behavior. The model also admits parameter tuning for capturing distortions in the thick slab caster, such as those at AK Steel caster in Dearborn, Michigan, as reported in Ref. [3] of Chapter 1.

The software testbed is used in the book as a virtual laboratory for getting computational hands-on familiarity with basic linear system behavior, nonlinearity-induced distortion, analysis of the latter using the spectral methods to diagnose the distortion origin, system identification, and design of controller for distortion elimination.

The appendix is divided into four parts that respectively develop simulation of the beam dynamics, the piston hydraulics, the controller, and the entire system.

The full software testbed is given in the form of three programs — two simulation files, **Beam_simulation.slx** and **Testbed_simulation.slx**, and computational file **beam_state_space_form.m.** Simulink diagrams of the first two files and listing of the code for the computational file are provided in this appendix and are sufficient to run the software testbed. Detailed description of the use of these programs is given in the Homework 1: Assignment and Solution Guide. These files are also provided as the book supplement by contacting the author.

H.2 Beam Dynamics

H.2.1 *Beam equation*

The beam in the software testbed is described by equations (G-32) with the boundary conditions in (G-33). The equations are numerically approximated by the state-space LTI ODE system

$$\begin{aligned} x_t &= Ax + Bu, \\ y &= Cx + Du, \end{aligned} \tag{H-1}$$

where x_t is the time derivative of x.

The spatial derivatives in the equations are approximated using the finite difference method. To apply this method, beam is divided into n equal in length (Δx) sections and the spatial derivatives are represented as finite differences:

$$\frac{\partial^2 y}{\partial x^2}=\frac{y(x+\Delta x)-2y(x)+y(x-\Delta x)}{\Delta x^2},$$
$$\frac{\partial y}{\partial x}=\frac{y(x+\Delta x)-y(x-\Delta x)}{2\Delta x}.$$

The number of spatial discretization sections should be chosen so that all eigenvalues of matrix A are positive (for example, for the Nucor Steel model $n_x=35$, for the AK Steel $n_x=153$).

Using the finite differences defined above, the equations for the left beam take the form

$$\frac{\partial^2 y_L}{\partial t^2}=-\frac{\gamma_y}{m_b}\frac{\partial y_L}{\partial t}+\frac{k'Ga_b}{m_b}\frac{y_L(x+\Delta x)-2y_L(x)+y_L(x-\Delta x)}{\Delta x^2}-\frac{k'Ga_b}{m_b}\frac{\psi_L(x+\Delta x)-\psi_L(x-\Delta x)}{2\Delta x}-g,$$

$$\frac{\partial^2 \psi_L}{\partial t^2}=-\frac{\gamma_\psi a_b}{m_b I}\frac{\partial \psi_L}{\partial t}+\frac{Ea_b}{m_b}\frac{\psi_L(x+\Delta x)-\psi_L(x-\Delta x)}{2\Delta x}+\frac{k'Ga_b^2}{m_b I}\frac{y_L(x+\Delta x)-y_L(x-\Delta x)}{2\Delta x}-\frac{k'Ga_b^2}{m_b I}\psi_L(x), \text{ or}$$

$$\begin{aligned}\frac{\partial^2 y_L}{\partial t^2}&=C_1 y_L(x-\Delta x)+C_2\psi_L(x-\Delta x)-2C_1 y_L(x)-D_1\frac{\partial y_L}{\partial t}\\&\quad+C_1 y_L(x+\Delta x)-C_2\psi_L(x+\Delta x)-g,\\\frac{\partial^2 \psi_L}{\partial t^2}&=-D_2\frac{\partial \psi_L}{\partial t}+C_3\psi_L(x+\Delta x)-C_3\psi_L(x-\Delta x)+C_4 y_L(x+\Delta x)\\&\quad-C_4 y_L(x-\Delta x)-C_5\psi_L(x)\end{aligned} \tag{H-2}$$

where coefficients in the equations are

$$D_1=\frac{\gamma_y}{m_b},C_1=\frac{k'Ga_b}{m_b\Delta x^2},C_2=\frac{k'Ga_b}{2m_b\Delta x},D_2=\frac{\gamma_\psi a_b}{m_b I},C_3=\frac{Ea_b}{2m_b\Delta x},C_4=\frac{k'Ga_b^2}{2m_b I\Delta x},C_5=\frac{k'Ga_b^2}{m_b I}.$$

From the equations (H-2) the matrix of coefficients A_1 is given by:

$$A_1=\begin{bmatrix}0&0&0&0&0&1&0&0&0&0&0&0\\C_1&0&C_2&0&-2C_1&-D_1&0&0&C_1&0&-C_2&0\\0&0&0&0&0&0&0&1&0&0&0&0\\-C_4&0&C_3&0&0&0&(-2C_3-C_5)&-D_2&C_4&0&C_3&0\end{bmatrix}.$$

This matrix will form the multiplicative matrix of coefficients A for (H-1) later. It is seen that the first four columns of matrix A_1 correspond to the previous cross-section, the four columns in the middle correspond to the current cross-section, and the last four columns correspond to the next cross-section. Thus, matrix A is split into three matrices: A_{prev}, A_{curr}, and A_{next}. Equations for the left and the right beams have the same form and therefore matrices of the coefficients A_{left} and A_{right} formed for each beam have the same form

$$A_{left} = A_{right} = \begin{bmatrix} A_{curr} & A_{next} & 0 & \cdots & 0 \\ A_{prev} & A_{curr} & A_{next} & \cdots & 0 \\ 0 & 0 & \ddots & 0 & 0 \\ 0 & \cdots & A_{prev} & A_{curr} & A_{next} \\ 0 & \cdots & 0 & A_{prev} & A_{curr} \end{bmatrix}.$$

The matrix x in (H-1) for the entire beam is formed by vertical and angular displacements of the beam and their time derivatives:

$$x = \begin{bmatrix} y(-l+\Delta x,t) \\ y_t(-l+\Delta x,t) \\ \psi(-l+\Delta x,t) \\ \psi_t(-l+\Delta x,t) \\ \vdots \\ y(-\Delta x,t) \\ y_t(-\Delta x,t) \\ \psi(-\Delta x,t) \\ \psi_t(-\Delta x,t) \\ \vdots \\ y(\Delta x,t) \\ y_t(\Delta x,t) \\ \psi(\Delta x,t) \\ \psi_t(\Delta x,t) \\ \vdots \\ y(l-\Delta x,t) \\ y_t(l-\Delta x,t) \\ \psi(l-\Delta x,t) \\ \psi_t(l-\Delta x,t) \\ y(l,t) \\ y_t(l,t) \end{bmatrix}.$$

It is seen that the left end of the left beam, the cross-sections of both beams in the middle, and the angular displacement and its time derivative for right end of the right beam are not included. The vertical and angular displacements of these cross-sections will be specified by the boundary conditions below.

H.2.2 *Application of the boundary conditions*

The boundary conditions for the left end of the beam are

$$y_L(-l) = x_p(t),$$
$$EI\frac{\partial \psi_L(-l)}{\partial x} = 0. \tag{H-3}$$

After applying the finite difference method for the second equation we obtain

$$\frac{\psi_L(-l+\Delta x) - \psi_L(-l-\Delta x)}{2\Delta x} = 0,\ \psi_L(-l-\Delta x) \approx \psi_L(-l),$$
$$\psi_L(-l) \approx \psi_L(-l+\Delta x).$$

The displacement of the piston $x_p(t)$ is given as an input later. Then, for the cross-section $(-l + \Delta x)$ the unknown displacements at $(-l)$ in (H-2) can be formed by the matrix

$$A_{BCL} = \begin{bmatrix} 0 & 0 & 0 & 0 \\ 0 & 0 & C_2 & 0 \\ 0 & 0 & 0 & 0 \\ 0 & 0 & C_3 & 0 \end{bmatrix}.$$

For the right end of the beam the boundary conditions are

$$EI\frac{\partial \psi_R(l)}{\partial x} = 0,$$
$$k'Ga_b\left(\frac{\partial y_R(l)}{\partial x} - \psi_R(l)\right) + Mg + M\frac{\partial^2 y_R(l)}{\partial t^2} + \gamma_m\frac{\partial y_R(l)}{\partial t} = 0.$$

From the second equation by using the finite difference method we have

$$\frac{\partial^2 y_R(l)}{\partial t^2} = -\frac{\gamma_m}{M}\frac{\partial y_R(l)}{\partial t} - \frac{k'Ga_b}{M}\frac{y_R(l) - y_R(l-\Delta x)}{\Delta x} + \frac{k'Ga_b}{M}\psi_R(l-\Delta x) - g.$$

From this equation the matrix A_{mold} can be formed. The part of the program providing that is presented below. This matrix forms the coefficients for the $y_L(-l)$ and it's time derivative. Finally, the matrix A_{right} is expanded by including the matrix A_{mold}.

MATLAB Program 1: element of Beam_state_space_form

```
...
MoldDyn = [zeros(1,4*(n-2)) 0 1;
     zeros(1,4*(n-3)) kd*G*CSArea/dx/mold 0 kd*G*CSArea/mold 0
     -kd*G*CSArea/dx/mold -gamma3];

A_right = A_right + A_right_BCright;

A_right = vertcat(horzcat(A_right,[zeros(4*(n-3),2); 0 0; C1 0; 0 0; C4 0]),MoldDyn);
...
```

For the middle positions of both beams the boundary conditions are

$$
\begin{gathered}
y_L(0) = 0, y_R(0) = 0, \\
\psi_L(0) = \psi_R(0), \\
EI\frac{\partial \psi_L(0)}{\partial x} = EI\frac{\partial \psi_R(0)}{\partial x}.
\end{gathered}
$$

Using Taylor's series yields

$$
\begin{gathered}
\psi(\Delta x) \approx \psi(0) + \Delta x \psi'(0) + \frac{\Delta x^2}{2}\psi(0) + \ldots, \\
\psi(2\Delta x) \approx \psi(0) + 2\Delta x \psi'(0) + 2\Delta x^2 \psi(0) + \ldots .
\end{gathered}
$$

After multiplying the first equation by 4 and subtracting from the result the second equation, the boundary conditions for the right and the left beam become

$$
\begin{gathered}
4\psi_R(\Delta x) - \psi_R(2\Delta x) \approx 3\psi_R(0) + 2\Delta x \psi_R'(0) + \ldots \\
4\psi_L(-\Delta x) - \psi_L(-2\Delta x) \approx 3\psi_L(0) - 2\Delta x \psi_L'(0) + \ldots .
\end{gathered}
$$

Since $\psi_L(0) = \psi_R(0)$, after adding two equations, the angular displacement in the middle is

$$
\psi_L(0) \approx \psi_R(0) \approx -\frac{1}{6}\psi_L(-2\Delta x) + \frac{2}{3}\psi_L(-\Delta x) + \frac{2}{3}\psi_R(\Delta x) - \frac{1}{6}\psi_R(2\Delta x).
$$

The resulting matrix A for the entire beam with all the boundary conditions included is

$$A=\begin{bmatrix} A_{left} & 0 \\ 0 & A_{right} \end{bmatrix}+\begin{bmatrix} A_{BCL} & \cdots & 0 & 0 & \cdots & 0 \\ 0 & \ddots & 0 & 0 & \cdots & 0 \\ 0 & \cdots & A_{BCLM} & 0 & \cdots & 0 \\ \vdots & \cdots & 0 & A_{BCRM} & \cdots & \vdots \\ 0 & \cdots & 0 & 0 & \ddots & 0 \\ 0 & \cdots & 0 & 0 & \cdots & A_{BCR} \end{bmatrix}.$$

In order to account for the piston displacement in (H-3) and the gravity terms, which were not included yet, the matrix B with size $\left(2(n-2)+2\right)\times 2$ is created. The inputs u for the state-space form, respectively, are

$$u=\begin{bmatrix} x_p(t) \\ g \end{bmatrix}.$$

The matrix C is formed depending on the resulting displacements and their derivatives to be shown. Therefore, the matrix C contains coefficients for the mold displacement and its time derivative. The matrix D is assumed to be zero.

MATLAB Program 2: element of program Beam_state_space_form

```
...
B = [0 C1 0 -C4 zeros(1,4*(n-3) + 4*(n-2) + 2); [repmat([0 1 0 0],1,(n-2 + n-2)), 0, 1]*(-9.8)]';

C = [ 1 0 0 0 0 0 zeros(1,4*(n-2 + n-2)-4); 0 0 1 0 zeros(1,4*(n-2 + n-2)-2); zeros(1,4*(n-2 + n-2)),1,0;zeros(1,4*(n-2 + n-2)),0,1;];
...
```

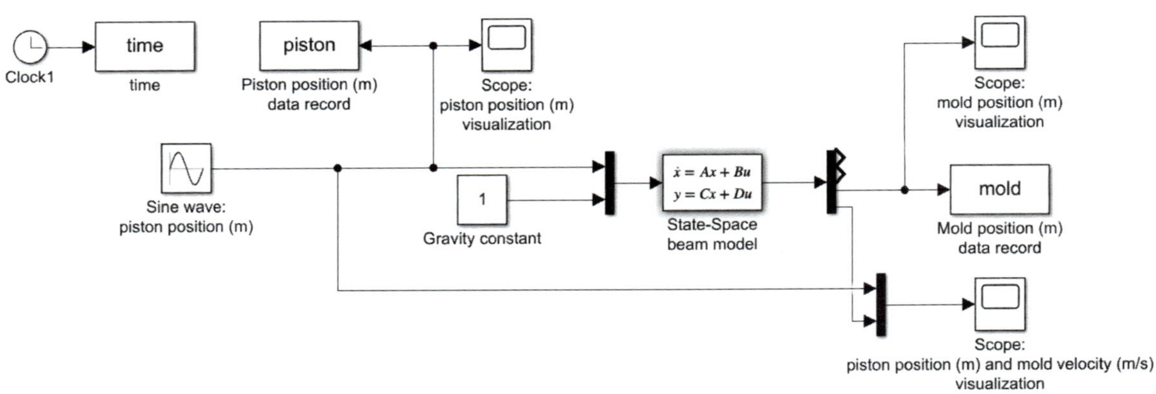

Fig. H-1. Beam Simulink model Beam_simulation.slx. The displacement units are in m.

Beam Simulink model **Beam_simulation.slx** is given in Figure H-1.

H.3 The Piston Hydraulics

This subsection presents the description of the dynamics of the servo providing the desired motion of the piston that is coupled with the beam.

The piston position x_p is governed by

$$m_p \ddot{x}_p + b x_p = (P_A - P_B) a_p - m_p g + F_B, \tag{H-4}$$

where m_p, b, a_p, P_A, P_B, g, F_B are mass of the piston, damping coefficient, area of the piston, pressures in chambers 'A' and 'B', gravity, and force from the beam. This part of software testbed is denoted as a 'Piston hydraulics' block in the Figure H-6.

In this equation the force exerted by the beam on the piston, which loads the piston, is given by

$$F_B = k' G a_b \left(\frac{\partial y_L(-l)}{\partial x} - \psi_L(-l) \right).$$

This force as the boundary condition for the displacement at the left end of the beam couples the nonlinear actuator and the beam. This block also includes the spool position dynamics x_S which is realized by

$$\ddot{x}_S + 2\zeta_S \omega_S \dot{x}_S + \omega_S^2 = \omega_S^2 u. \tag{H-5}$$

The transfer function, therefore, takes the form

$$\frac{x_s}{u} = \frac{\omega_s^2}{s^2 + 2\zeta \omega_s s + \omega_s^2},$$

where the input u is generated by the difference between the piston position $x_p(t)$ and the desired reference signal $r(t)$, multiplied by proportional gain k_p:

$$u = k_p (x_p(t) - r(t)). \tag{H-6}$$

Pressures in Chambers '*A*' and '*B*'

For the pressures in chambers '*A*' and '*B*' the governing equations are, respectively,

$$\begin{aligned} \dot{P}_A &= \beta (q_A - a_p \dot{x}_p) / (V_A + a_p (L + x_p)), \\ \dot{P}_B &= \beta (-q_B - a_p \dot{x}_p) / (V_B - a_p (L - x_p)) \end{aligned} \tag{H-7}$$

where β, q_A, q_B, V_A, V_B and L are bulk modulus of the actuator fluid, flow rates into chamber 'A' and out of chamber '*B*', volumes of tubes connected to chambers '*A*' and '*B*', and half of the stroke length of the piston, respectively. When the piston position x_p equals to zero, chambers '*A*' and '*B*' have the same volume.

The flow rates in the chambers A and B are given by

$$
q_A = \begin{cases} c(d-x_S)\sqrt{P_S-P_A}, & x_S<-d, \\ c(d-x_S)\sqrt{P_S-P_A}-c(d+x_S)\sqrt{P_A-P_t}, & -d<x_S<d, \\ -c(d+x_S)\sqrt{P_A-P_t}, & x_S>d, \end{cases}
$$
$$
q_B = \begin{cases} c(d-x_S)\sqrt{P_B-P_t}, & x_S<-d, \\ c(d-x_S)\sqrt{P_B-P_t}-c(d+x_S)\sqrt{P_S-P_B}, & -d<x_S<d, \\ -c(d+x_S)\sqrt{P_S-P_B}, & x_S>d, \end{cases} \tag{H-8}
$$

where $c=c_d w\sqrt{\dfrac{2}{\rho}}$.

In these equations c_d, w, ρ, d, P_S, P_t are the effective discharge coefficient, width of the port for fluid flow between chambers '*A/B*' and '*S/T*', density of the fluid, spool underlap length, supply pressure, and tank pressure, respectively.

The schematic description of the equations for the flow rate q_A and pressure in chamber '*A*' in Simulink is presented in the Figure H-2 below. The same was done for the chamber '*B*' by applying the equations for chamber '*B*' above.

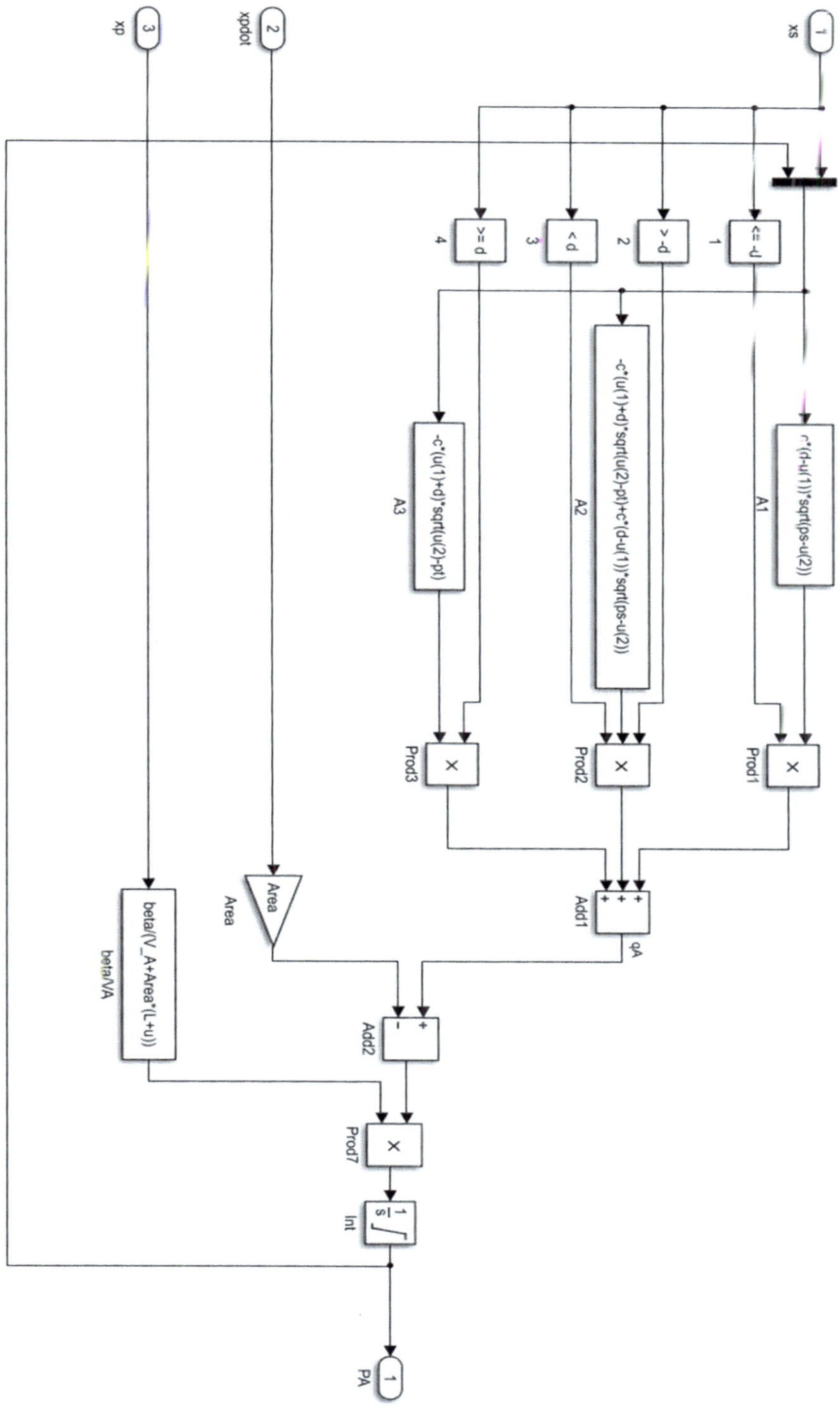

Fig. H-2. Simulink diagram of the piston PA part — a subsystem in the hydraulics servo.

Putting together schematics of both chambers yields the Simulink realization of the piston position dynamics, as shown in the diagram in Figure H-3.

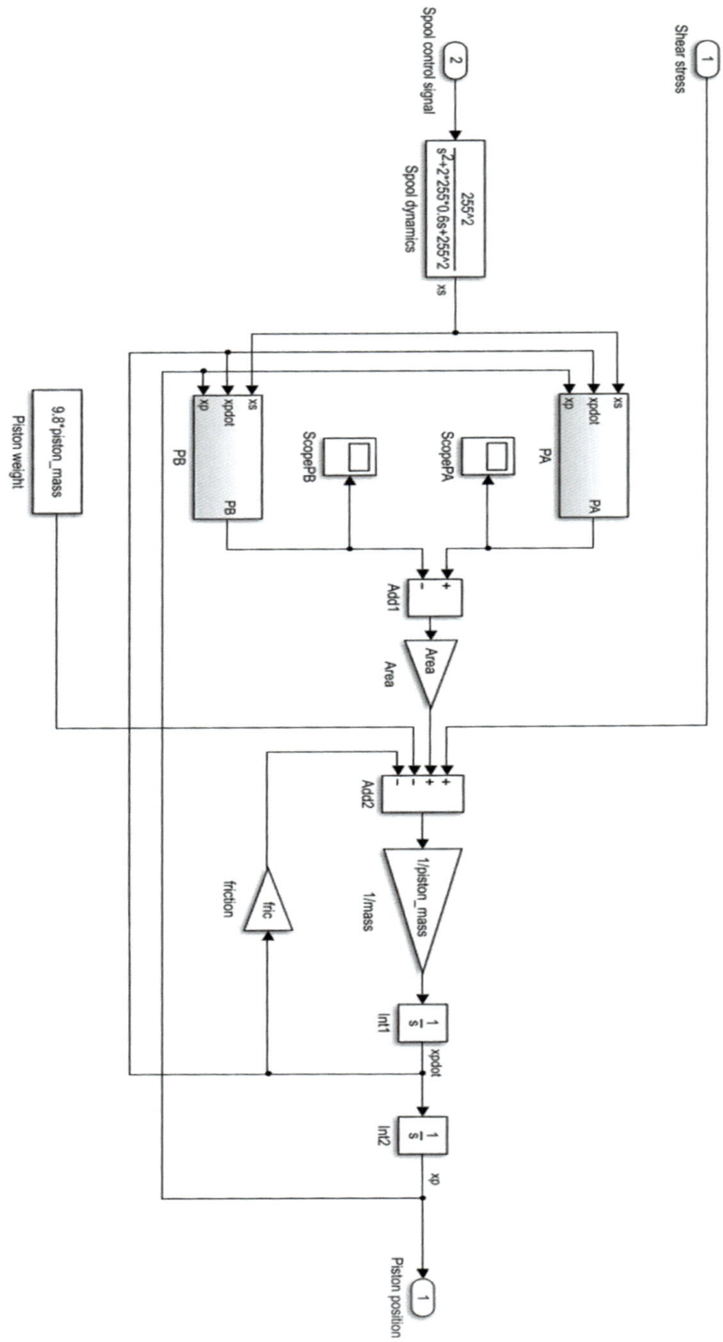

Fig. H-3. Detailed Simulink diagram of piston hydraulics — part of the Simulink program Testbed_simulation.slx.

H.4 The Controller Simulator

The controller configuration included into the testbed consists of two controllers. The first one is the P-controller, which just sets the original signal to be directly proportional to the error, multiplying the error by a constant number called "proportional gain." The second controller is a "tracking controller," introduced in Figure 1.6-2 of Chapter 1 as an additional positive feedback loop and realized, as described in Hw1 Problem 3 (Chapter 1), through a filter F given by

$$F(s)=\frac{2e\omega_{res}s}{s^2+2e\omega_{res}s+\omega_{res}^2}, \tag{H-9}$$

where ω_{res} is the first resonance frequency of the beam and $e = 0.1$.

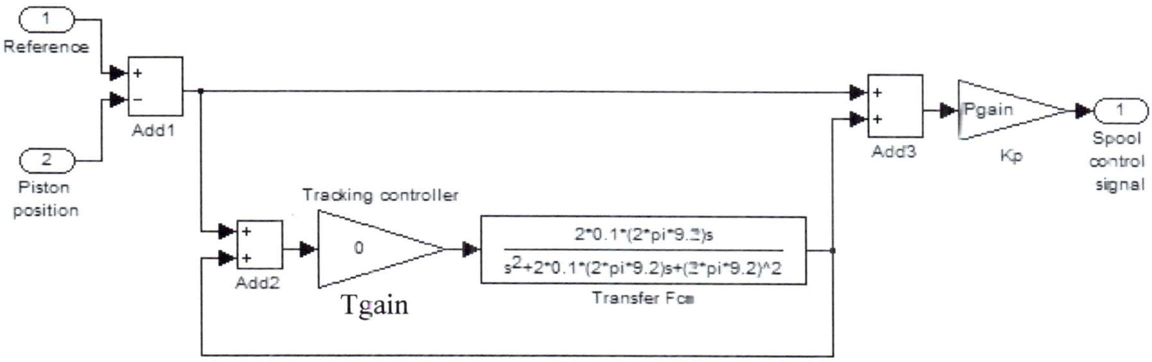

Fig. H-4. Detailed Simulink controller diagram — part of the Simulink program Testbed_simulation.slx.

The theory behind this control configuration is presented in detail in Chapter 9.

H.5 The Software Testbed Program

The testbed goal is to make the piston position follow the *reference* signal. The testbed schematic and Simulink diagram Testbed_simulation.slx are, respectively, given in Figures H-5 and H-6.

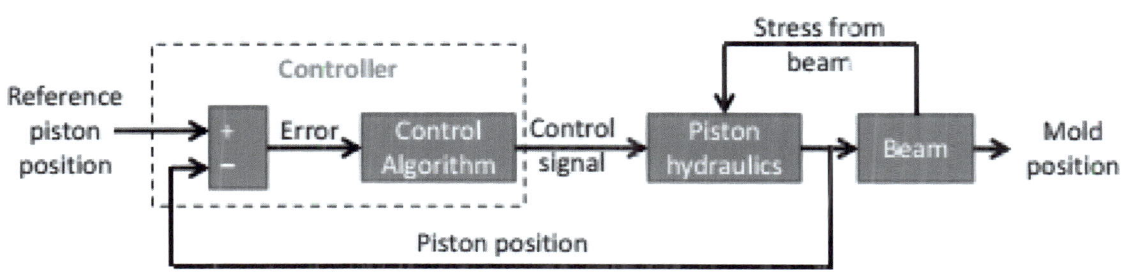

Fig. H-5. Schematic of the software testbed.

The *controller* calculates the *error* between the *actual piston position* and the *reference*, and uses a control algorithm to calculate a *control signal* to send to the *piston hydraulics*. Based on the control signal, the piston hydraulics change the piston position, which in turns moves the *beam* and the *mold* at the other end. However, the weight of the mold and the dynamics of the beam exert *loading* on the piston hydraulics that affects the movement of the piston.

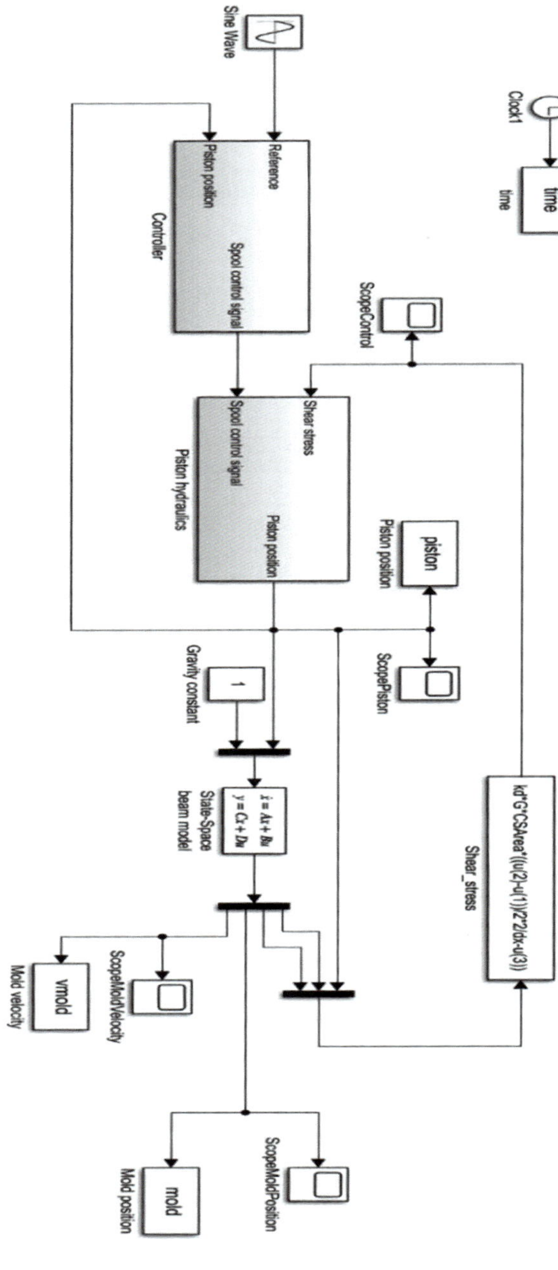

Fig. H-6. Testbed_simulation.slx — software testbed realization in terms of Simulink diagram.

Full testbed computational file **beam_state_space_form.m** is given below, with the parameters corresponding to the thin slab caster Nucor beam model.

To simulate AK Steel thick slab caster model, the parameters need to be changed according to the Table G-2 in Appendix G.

MATLAB Program 3 : Beam_state_space_form

```
Pgain = -0.6;
Tgain = 1; % make it zero if you want to stop tracking controller
d = 1.27e-6;
Area = 0.0046;
ps = 3000*6894.75;
pt = 30*6894.75;
c = 1.5e-4*2;
V_A = (2.375 + 0.5)*0.0254^3;
V_B = (3.8 + 0.5)*0.0254^3;
piston_mass = 2;
beta = 1500e6;
L = 0.015;
fric = 1000;
L_1 = 0.88;

Pa0 = 50*pt;
Pb0 = 50*pt;
x0 = 0;
xdot0 = 0;
t0 = 0;

dx = L_1/35;

%material properties and dimensions
G = 77e9;
rho = 7870;
E = 2e11;
kd = 0.83;
CSArea = 0.0088/3;
I = 2.2085e-5/1.25;
mass = rho*CSArea;
k = sqrt(I/CSArea);
gamma1 = 10;
gamma2 = 10;
gamma3 = 1;
```

```
n = L_1/dx + 1; %number of rod nodes


C1 = kd*G*CSArea/mass/dx^2;
C2 = kd*G*CSArea/mass/dx/2;
C3 = E*I/k^2/mass/dx^2;
C4 = kd*G*CSArea/k^2/mass/dx/2;
C5 = kd*G*CSArea/k^2/mass;
D1 = gamma1/mass;
D2 = gamma2/k^2/mass;
mold = 5000*0.45;


A1 = [0 0 0 0 0 1 0 0 0 0 0 0;
  C1 0 C2 0 -2*C1 -D1 0 0 C1 0 -C2 0;
  0 0 0 0 0 0 0 1 0 0 0 0;
  -C4 0 C3 0 0 0 -2*C3-C5 -D2 C4 0 C3 0];

A1_prev = A1(1:4,1:4);
A1_curr = A1(1:4,5:8);
A1_next = A1(1:4,9:12);

a = ones(n-2,1);
b = ones(n-3,1);

A_left_prev = A1_prev;
A_left_curr = A1_curr;
A_left_next = A1_next;

for i = 1:n-4
   A_left_prev = blkdiag(A_left_prev,A1_prev);
   A_left_curr = blkdiag(A_left_curr,A1_curr);
   A_left_next = blkdiag(A_left_next,A1_next);
end

A_left_curr = blkdiag(A_left_curr,A1_curr);

A_left = horzcat(vertcat(zeros(4,4*(n-3)),A_left_prev),
 zeros(4*(n-2),4)) + A_left_curr+vertcat(horzcat(zeros(4*(n-3),4),
 A_left_next),zeros(4,4*(n-2)));
```

```
A_right = A_left;

%left BC for left beam
A_left_BCleft = blkdiag([0 0 0 0; 0 0 C2 0; 0 0 0 0;0 0 C3 0;],
 zeros(4*(n-3)));

A_left = A_left + A_left_BCleft;

%right BC for right beam

A_right_BCright = blkdiag(zeros(4*(n-3)),[0 0 0 0; 0 0 -C2 0; 0 0 0 0;
 0 0 C3 0;]);

MoldDyn = [zeros(1,4*(n-2)) 0 1; zeros(1,4*(n-3)) kd*G*CSArea/dx/mold
 0 kd*G*CSArea/mold 0 -kd*G*CSArea/dx/mold -gamma3];

A_right = A_right + A_right_BCright;

A_right = vertcat(horzcat(A_right,[zeros(4*(n-3),2); 0 0; C1 0; 0 0;
 C4 0]),MoldDyn);

%middle BC
A = blkdiag(A_left,A_right);

A_left_middleBC = horzcat(zeros(4,4*(n-4)),
  [0 0 0 0; 0 0 -C2*(-1/6) 0; 0 0 0 0; 0 0 C3*(-1/6) 0],
  [0 0 0 0; 0 0 -C2*(2/3) 0; 0 0 0 0; 0 0 C3*(2/3) 0],
  [0 0 0 0; 0 0 -C2*(2/3) 0; 0 0 0 0; 0 0 C3*(2/3) 0],
  [0 0 0 0; 0 0 -C2*(-1/6) 0; 0 0 0 0; 0 0 C3*(-1/6) 0],
  zeros(4,4*(n-4) + 2));

A_right_middleBC = horzcat(zeros(4,4*(n-4)),
  [0 0 0 0; 0 0 C2*(-1/6) 0; 0 0 0 0; 0 0 C3*(-1/6) 0],
  [0 0 0 0; 0 0 C2*(2/3) 0; 0 0 0 0; 0 0 C3*(2/3) 0],
  [0 0 0 0; 0 0 C2*(2/3) 0; 0 0 0 0; 0 0 C3*(2/3) 0],
  [0 0 0 0; 0 0 C2*(-1/6) 0; 0 0 0 0; 0 0 C3*(-1/6) 0],
  zeros(4,4*(n-4) + 2));

A = A + vertcat( zeros(4*(n-3),4*(n-2 + n-2) + 2), A_left_middleBC,
 A_right_middleBC,zeros(4*(n-3) + 2,4*(n-2 + n-2) + 2));
```

```
B = [0 C1 0 -C4 zeros(1,4*(n-3) + 4*(n-2) + 2);
  [repmat([0 1 0 0],1,(n-2 + n-2)), 0, 1]*(-9.8)]';

C = [ 1 0 0 0 0 0 zeros(1,4*(n-2 + n-2)-4);
  0 0 1 0 zeros(1,4*(n-2 + n-2)-2);
  zeros(1,4*(n-2 + n-2)),1,0;
  zeros(1,4*(n-2 + n-2)),0,1;];

D = [0, 0; 0 0; 0 0;0 0; ];

a1 = fric/piston_mass;
a2 = 1/piston_mass;
a3 = Area/piston_mass;
a4 = 4*beta*Area/(V_A + V_B);
a5 = 4*beta*c/(V_A + V_B);
```

Bibliography

Introductory dynamical systems

Ogata, K. (2004). System dynamics, 4th ed., Prentice Hall.

Rowell, D. and D. N. Wormley (1997). System dynamics — an introduction, Prentice Hall.

Signals and systems

Bracewell, R. N. (1986). The Fourier transform and its applications, 2nd ed., McGraw-Hill.

McGillem, C. D. and G. R. Cooper (1991). Continuous and discrete signal and system analysis, 3rd ed., Oxford University Press.

Oppenheim, A. V., and A. S. Willsky, with H. Nawab (1997). Signals and systems, 2nd ed., Prentice Hall.

Palamides, A. and A. Veloni (2011). Signals and systems laboratory with MATLAB, Taylor & Francis.

Poularikas, A. D. and S. Seely (1994). Signals and systems, 2nd ed., Krieger.

Ziemer, R. E., W. H. Tranter and D. R. Fannin (1998). Signals and systems: continuous and discrete, 4th ed., Prentice Hall.

Instrumentation

Blackburn, J. A. (2001). Modern instrumentation for scientists and engineers. Springer-Verlag.

Nachtigal, C. L., ed. (1990). Instrumentation and control: fundamentals and applications. Wiley.

Electronics

Horowitz, P. and W. Hill (2015). The art of electronics, 3rd ed., Cambridge University Press.

Storey, N. (2009). Electronics — a systems approach, 4th ed., Prentice Hall.

Vibrations

Dimarogonas, A. and S. Haddad (1992). Vibration for engineers. Prentice Hall.

Inman, D. (2013). Engineering vibration, 4th ed., Prentice-Hall.

Meirovitch, L. (1997). Principles and techniques of vibrations. Prentice Hall.

Signal processing and digital filter design

Ambardar, A. (1999). Analog and digital signal processing, 2nd ed., Brooks/Cole.

Ambardar, A. (2006). Digital signal processing: a modern introduction, Cengage.

Mitra, S. K. (2011). Digital signal processing: a computer-based approach, 4th ed., McGraw-Hill.

Orfanidis, S. J. (1996), Introduction to Signal Processing, Prentice Hall; Copyright © (2010) by S. J. Orfanidis https://www.ece.rutgers.edu/~orfanidi/intro2sp/orfanidis-i2sp.pdf

Oppenheim, A. V. and R. W. Schaffer (2009). Discrete-time signal processing, 3rd ed., Prentice Hall.

Proakis, J. G. and D. K. Manolakis (2006). Digital signal processing: principles, algorithms, and applications, 4th ed., Prentice Hall.

Rabiner, L R. and B. Gold (1975). Theory and application of digital signal processing. Prentice Hall.

Window functions

Prabhu K. M. M. (2013). Window functions and their applications in signal processing, CRC Press.

Spectral estimation

Kay, S. M. (1988). Modern spectral estimation, Prentice Hall.

Analog filter design

Williams, A. (2013). Analog filter and circuit design handbook, McGraw-Hill.

Control systems

Åström, K. J. and R. M. Murray (2015). Feedback systems: an introduction for scientists and engineers, 2nd ed., Princeton University Press, http://www.cds.caltech.edu/~murray/amwiki/index.php/Second_Edition.

Franklin, G. F., J. D. Powell, and A. Emami-Naein (2014). Feedback control of dynamical systems, 7th ed., Prentice Hall.

Ogata, K. (2009). Modern control engineering, 5th ed. Prentice Hall.

Dynamical systems with system integration

Cochin, I. and W. Cadwallender (1997). Analysis and design of dynamical systems, 3rd ed., Addison-Wesley, 1997.

Kelly, S. G. (2007). Systems dynamics and response. Thomson.

Lobontiu, N. (2010). System dynamics for engineering students: concepts and applications, Academic Press.

Natarajan, V. (2012). Rejection of periodic disturbances in uncertain nonlinearly perturbed stable infinite dimensional systems, Ph.D. thesis, Department of Mechanical Science and Eng., Univ. of Illinois at Urbana-Champaign, 133p.

Nonstationary signal analysis

Najmi, A.-H. (2012). Wavelets: A Concise Guide, The Johns Hopkins University Press, Baltimore.

M. Vetterli and J. Kovacevic, Wavelets and Subband Coding, 1995 Prentice Hall, Englewood Cliffs, New Jersey. http://waveletsandsubbandcoding.org/Repository/VetterliKovacevic95_Manuscript.pdf

Daubechies, I. (1992). Ten Lectures on Wavelets, SIAM.

Machine Learning

Watt, J. R. Borhani, and A. Katsaggelos (2020). Machine Learning Refined: Foundations, Algorithms, and Applications, 2nd e., Cambridge University Press.

Sutton, R. S. and A. G. Barto (2018). Reinforcement Learning: An Introduction, 2nd es., The MIT Press.

Joshi, A. V. (2020). Machine Learning and Artificial Intelligence, Springer.

Russel, S. and P. Norvig (2021). Artificial Intelligence: A Modern Approach, 4th ed., Pearson.

Mathematical references

Abramovitz, M. and I. A. Stegun (1964). Handbook of mathematical functions with formulas, graphs, and mathematical tables, National Bureau of Standards, 10th printing: http://people.math.sfu.ca/~cbm/aands/.

Gradshtein, L. S., and I. M. Ryzhik (1980). Table of integrals, series, and products, Academic Press.

Korn, G. A. and T. M. Korn (1964). Mathematical handbook. McGraw-Hill.

Index

accelerometer, 763
activation function, 677
actuator saturation, 608
affine system, 721
aliasing, 207
almost periodic, 104
amplitude, 110
analog angular frequency, 103
analog frequency, 103
Analog-to-digital conversion (ADC), 182, 184, 213
anti-aliasing filter, 352
autocorrelation
  continuous time, 468
  discrete time, 469
bandlimit, 212
bandlimiting, 307
bandwidth, 372
basic control action
  PI (Proportional + Integral), 562
  self-operated, 560
  two-position (on-off), 561
basis vectors, 269
BIBO stable, 723
BIBO unstable, 723
bidirectional, 172
bilinear transformation, 441
black box identification, 241
blocking zero, 150
bode plot, 157
break frequency, 371
breakout, 4
brushless DC motor, 766
buffer, 173
Bellman's equation, 679
Bessel filter, 426, 427
Butterworth filter
  polynomial factorization, 408
  unity DC gain filter form, 408
  frequency response, 426, 427
Cauer (Elliptic) filter, 426, 427
casting speed, 3
causal system, 721
causality
  anti-causal, 48
  causal, 48
  non-causal, 48
central limit theorem, 735
Chain rule, 713
Chebyshev filter, 426, 427
classification, 676, 692
close-loop transfer function
  input-output, 550
codec, 642
commensurable, 104
Common mode gain, 187
Common mode rejection ratio (CMMR), 188
common mode voltage, 185, 188
comparator, 180
  latched, 182
complex conjugate, 216
compression, 56
Conditional probability, 729
continuous casting, 1
continuous time system, 716
control method of testbed, 11
controller, 545
conversion
  analog-to-digital, 213
convolution, 471, 742
  continuous time, 360
  discrete time, 361
  noncausal, 360
  2D, 644
Convolutional Neural Networks, 687

corner frequency, 160
correlation, 471
coupling, 172
  AC coupling, 176
cross-correlation
  continuous time, 468
  discrete time, 469
cross-coupling, 172
CT ideal filter
  magnitude response, 366
CT PSD approximation, 495
CT real filter, 369
CT sinusoidal signal, 102
cycle, 103
cyclic calculation, 481
DC motor, 765
decimation, 57
decoupling, 173
Deep Learning, 676
delta-function
  definition, 68
  sifting property, 68
  symmetry property, 70
  time-scaling property, 70
Delta-sigma modulation, 214, 352, 353
derivative gain, 562
derivative time, 562
detecting sinusoids in CT noisy data
  using DFT (FFT), 494
deterministic system, 724
digital audio file formats, 641
digital frequency, 205
digital frequency bounds, 210
digital image sensors, 766
digital period, 205
digital PSD, or DT PSD, 493
Digital-to-analog converter (DAC), 183, 214, 215
  Pulse Code Modulation (PCM), 353
  Pulse Density Modulation (PDM), 353
Discrete-continuous spectra, 490
discrete (exponential) Fourier series coefficients, 223
Discrete Fourier Series, 279, 343
Discrete Fourier Series, two-dimensional, 648
discrete Fourier series representation, 223
Discrete Fourier Transform, 306, 343
Discrete Fourier Transform, two-dimensional, 648
discrete frequency response spectra, 115
discrete PSD, 487
discrete real spectra, 112
discrete spectra, 220
discrete time Fourier transform (DTFT), 346
discrete time system, 716
discrete time system response, 362
disturbance, 139
disturbance rejection, 548
dot, 269
doublet, 147
DT periodic autocorrelation, 494
DT real filter, 375
dynamic range, 214
dynamic system, 717
edge detection, 651
eigenfunction, 218
eigenvalue, 218
electro-hydraulic actuator, 10, 14
Elliptic (Cauer) filter, 426, 427
encoder, 761
energy (spectral) density, 370
energy signal, 79
energy spectral density, 465
ergodicity assumption, 48
error
  actual output error, 556
Euler's Identity, 711
Euler-Bernoulli Beam Model, 767
equalization, 642, 694
exact system, 724
exponential form, 216
exponential Fourier series, 273
exponential Fourier series coefficients, 222
exponential Fourier series representation, 222
exponential signal, 292
extensometer, 760
Fast Fourier Transform (FFT), 313, 314
feedback path, 548
filters
  band-pass filter, 421
  band-stop filter, 422
  high-pass filter, 420
  ideal filter, 366
Final Value Theorem, 557, 565, 748
finite impulse response (FIR), 149
finite impulse response (FIR) filter, 299
flash ADC, 184
FIR approach, 342
folding, 211
folding frequency, 212

folding of signals, 60
forward path, 548
Fourier analysis, 239
Fourier reconstruction, 244
Fourier Series, 343
Fourier Series coefficients, 273, 289
Fourier Series for a periodic signal, 286
Fourier Transform, 343
Fourier Transform (FT), 741
frequency, 103
frequency response, 115, 339
function support
  compact support, 52
  noncompact support, 52
fundamental frequency, 110
fundamental period, 103, 110
fuzzy system, 724
Gibbs phenomenon, 82
Gradient Descent, 680
grey box identification, 241
gyroscope, 765
half-plane
  closed half-plane, 150
  open half-plane, 150
half-wave rectified sine, 291
Hall effect sensor, 762
Hamilton's principle, 768
hardware testbed, 9
harmonic frequency, 111
harmonic number, 111
harmonics, 111
Hilbert-Huang Transform, 667
Hooke's law, 773
hydraulics, 766
ideal sampling, 71
image processing filters
  Sobel, 645
  Prewitt, 646
  Derivative of Gaussian, 646
  low-pass, 650
  high-pass, 651
image separation, 250
impulse response, 70
impulse train, 289
infinite impulse response (IIR), 148
infinite impulse response (IIR) filter, 299
infrared (IR) sensor, 757
Initial value theorem, 748
inner product, 269
instrumentation amplifier, 186, 189
integrability
  absolutely integrable, 52
  square-integrable, 52
integral gain, 562
integral time, 562
Integration by parts, 712
Integrative LTI system portrait, 131
internal model principle, 589
internal model principle (IMP) controller, 589
interpolation
  Linear interpolation, 57
  zero interpolation, 57
  step interpolation, 57
inverse Fourier transform, 436
inverse Fourier Transform, 742
k'th real harmonic, 111
lack of commutativity, 169
Laplace transform, 747
leakage output voltage, 188
LIDAR, 761
linear system, 720
linear time invariant (LTI) system, 131
Linear Variable Differential Transformer (LVDT), 758
load cell, 764
loading, 170
low-pass filtering, 177
loss function, 678
magnetic sensor, 759
magnetostriction, 203
magnitude, 216
magnitude (spectral) density, 370, 371
magnitude and phase form, 274
Markov process, 737
Markov decision process (MDP), 679
mass-spring-two-dampers (MS2D), 133
maximum overshoot, 565
measurement noise, 139
mel scale, 691
mold oscillation system, 3
moving average (MA) filter, 149
multifrequency, 46
multiresolution, 21, 47. 659
multiscale. *See* multiresolution
negative strip time, 3
Neural Networks, 676

noncausal system, 721
nonlinear system, 720
norm, 270, 653
normal mode gain, 187
normal mode signal, 187
notch filter, 422
Nyquist rate, 212
Nyquist-Shannon-Kotelnikov sampling theorem, 212
operational amplifier, 178
orthogonal, 271
orthonormal, 272
orthonormal basis, 272
oscillation marks, 4
output error, 547
oversampling, 246
parametric system models
  linear constant coefficient ODE, 336
Parseval's relationship, 282
Parseval's Theorem, 282, 743
partial fraction expansion, 749
passive component, 166
perceptron, 677
performance measure, 547
period, 46, 103
periodic autocorrelation, 479
periodic convolution, 365
periodic correlation, 478
periodic crosscorrelation, 479
periodic extension, 81
periodicity
  nonperiodic, 46
  periodic, 46
pH sensor, 764
phase, 216
phase lag, 109
phase lead, 109
phase shift, 109
photoelectric sensor, 758
pitch
  shifting, 230
  sound, 690
  roll, 675
pixel, 643
plant, 545
pneumatics, 766
pointwise, 294
polar form, 216, 274
poles, 434
pole-zero map, 150
position tracking, 547
potentiometer (Pot), 184
power signal, 79
power spectral density (PSD), 474
pressure sensor, 764
prewarping, 442
principal range, 211
probability distribution function, 728
projection, 270
  scalar projection, 270
  vector projection, 271
proportional gain, 561
proximity sensor, 758
Psycho-acoustic revolution, 297
Pulse Code Modulation (PCM), 213. 214, 351
Pulse Density Modulation (PDM), 214, 352. 353
quadratic mean. *See* signal RMS
quantization, 213
quantization noise, 353
Q-learning, 679
randomness
  stochastic, 48
randomness
  deterministic signal, 47
real sampling, 72
Reinforcement Learning, 678
region of convergence, 747
regularity
  regular signals, 50
  singular signals, 50
resonance frequency and peak
  beam system., 163
  MS2D system, 162
  MSD system, 161
  second order system with 2 zeroes at the origin, 163
response shaping, 555
robust, 23
robustness, 548
RST structure, 595
RTD (resistance temperature detector), 754
sampling
  sampling frequency. *See* sampling rate
  sampling interval, 204
  sampling rate, 204
sampling time interval, 44
Savitzky-Golay filter, 450
sawtooth signal, 290
scalar product, 269

scale composition
multiscale (multiresolution), 47
single-scale (single-resolution), 47
scaling factor
direct scaling factor, 56
inverse scaling factor, 56
sensor, 545
settling time, 565
shifting of signals, 54
Short Time Fourier Transform (STFT), 690
signal energy, 78
signal equalization, 642
signal power, 78
signal processing, 43
signal root mean square (RMS), 78
signal speed, 293
sine-sweep, 341
sinusoid
magnitude, 110
phase, 109
sums of sinusoids, 104
Size preservation of CT signals, 63
Software testbed
Beam model, 13
Hydraulic servo model, 14
Numerical model, 16
spectral leakage, 309
spectral window, 296
spectrogram, 689
mel, 691
Stability Criterion, 435
state variable, 135
state-variable realization, 422
static system, 717
stationary
non-stationary signal, 48
stationary signal, 48
steady state error, 577
steel solidification, 692
stepper motor, 765
sticker, 4
stochastic system, 724
Stochastic process, 730
strain gauge, 185
strand, 3
stretching, 57
successive approximation property, 659
summability
absolutely summable, 52
square-summable, 52
superposition, 720
symmetry
odd signal, 45
even signal, 45
system
definition, 615
system identification
using cross-correlation, 508
using noise excitation, 507
using PSD, 504
tachometer generator, 763
techno-social revolution, 297
template, 208
temperature sensors, 753
temporal CNN, 693
thermographic infrared (IR) sensor, 757
time invariant system, 718
time varying system, 718
time-scaling of signals, 55
Timoshenko Beam Model, 773
Total Harmonic Distortion (THD), 190
touch screen sensor, 764
triangular pulse train, 290
trigonometric form, 275
Trigonometric Identities, 711
Tustin's rule, 439
two-sided Fourier Transform, 361
type of signal
analog signal, 43
continuous-time (CT) signal, 43
digital signal, 45
discrete time (DT) signal, 44
quantized signal, 44
sampled signal, 44
undersampling, 246
unidirectional, 172
unit circle, 217
variable reluctance sensor, 762
vector space, 269
Vertical sinusoidal oscillation, 3
voltage divider, 183
voltage ladder, 183, 184
warping effect, 441
waterfall plot, 7, 47, 500
Wavelets, 659
Wavenet, 683, 706

white noise, 484, 731
Wiener process, 732
windowing, 296
  spectral windowing, 297
  time-domain windowing, 297
wraparound, 481
zeros, 434
Ziegler-Nichols PID tuning rules, 606
Z-transform, 432